Fourth Conference of Computational Methods & Ocean Technology (COTech 2023)

IOP Conference Series: Materials Science and Engineering
Volume 1294

Stavanger, Norway
30 November - 1 December 2023

Editors:

Hirpa G. Lemu
Andres Garcia Granada
Dimitrios Pavlou
Gerhard Ersdal
Guang Yin
Idriss El-Thalji
Knut Erik Giljarhus
Mohsen Assadi
Emeritus Ove T. Gudmestad
Sudath C. Siriwardane
Tiago J.F. Marques Ferradosa
Tor H. Hemmingsen
Yanyan Sha
Yihan Xing
Yiliu Liu

ISBN: 978-1-7138-8731-7
ISSN: 1757-8981

Printed from e-media with permission by:

Curran Associates, Inc.
57 Morehouse Lane
Red Hook, NY 12571

Some format issues inherent in the e-media version may also appear in this print version.

Institute of Physics
Dirac House, Temple Back
Bristol BS1 6BE UK

Phone: 44 1 17 929 7481
Fax: 44 1 17 920 0979

techtracking@iop.org

Additional copies of this publication are available from:

Curran Associates, Inc.
57 Morehouse Lane
Red Hook, NY 12571 USA
Phone: 845-758-0400
Fax: 845-758-2633
Email: curran@proceedings.com
Web: www.proceedings.com

TABLE OF CONTENTS

IOP Conf. Series: Materials Science and Engineering 1294 (2023) 011001 doi:10.1088/1757-899X/1294/1/011001

Fourth Conference of Computational Methods & Ocean Technology (COTech 2023)

November 30 – December 1, 2023,
University of Stavanger, Stavanger, Norway

Preface

The COTech (Computational Methods & Ocean Technology) conference started as part of the research and dissemination activities of the Program Area for research "COTech - Computational methods in Offshore Technology" at Faculty of Science and Technology, University of Stavanger (UiS) with its first event in 2017. This Program Area for Research was founded in 2015 with seven professors, four associate professors, two adjunct professors and five research (PhD) students from the Department of Mechanical and Structural Engineering and Materials Science (IMBM), whose expertise and competence lies primarily within use of computational methods such as finite element methods, boundary and volume element methods, computational fluid dynamics and the like in marine and subsea technology, marine operations, design and analysis of mechanical systems, integrity and reliability of offshore structures and mechanical systems, renewable energy and wind engineering. In the ocean-related engineering area in particular, numerical computation approach is nowadays not only serving as a means to cultivate and realize innovative ideas, but also it is becoming the primary choice to solve complex engineering problems for the harsh and unfriendly environment in the Arctic.

Like the previous conference, this year's COTech conference is conducted as part of the dissemination activities of the Institute Strategic Program (ISP) entitled "Computational Methods and Ocean Technology" established under the research activities of the Department of Mechanical and Structural Engineering and Materials Science. This three-year program (2021 - 2023) intends to bring researchers of the department under four selected research areas:

1. *Ocean Energy Conversion* - installations & dynamic analysis of offshore wind turbines, structural health monitoring, corrosion assisted fatigue, ocean wave energy.
2. *Aquaculture Technology* - efficient farming techniques, advanced numerical modeling and computation, fluid-structure interaction, design effective maintenance programs for aquaculture.

3. *Marine and Subsea Technology* - sediment erosion (scour), scour prediction CFD models, structural integrity and fatigue life of offshore structures, adaptive control of ROVs in subsea application, predictive maintenance of subsea structures
4. *Innovative Solutions* - 3D printing based innovative solutions, inspection drones, augmented reality for maintenance training, ...

In addition to the thematic areas covered in the last three conferences, the 4th COTech conference accommodated some of the emerging technologies and applications such as machine learning techniques, additive manufacturing technologies, 3D printing in healthcare technology, smart energy storage and bio-inspired design.

In general, the conference is intended to provide a platform for academics and professionals working within diverse forms of the computational methods, Ocean/Offshore technology and Oil and Gas technologies to come together, present their recent works in the area, exchange ideas, and establish professional networks. It will serve as a forum for multidisciplinary research and bring together enable them to exchange their research experience and disseminate their results within the involved fields.

The conference is organized under 7 thematic areas, which will also serve as conference tracks.

1. *Wind Engineering and Industrial Aerodynamics*
2. *Advanced Computational Methods & Applications in Marine, Subsea and Offshore Technology*
3. *Computational Mechanics, Structural Integrity, and Life Extension of Structures*
4. *Cold Climate Region Technology*
5. *Smart Operations and Maintenance*
6. *Additive Manufacturing, Design Optimization and Composites*
7. *Smart Energy Storage, Integration and Utilization*

Among the submitted manuscripts, 63 full papers passed the review process. The review work was conducted according to the review policy of IOP conference series where each paper was reviewed by at least two reviewers involving both national and international reviewers. Six known experts in the area were invited to present keynote papers from different countries.

The conference committee and editors of this proceeding would like to thank all the reviewers and authors of the papers in this proceeding for their valuable contributions to 4th COTech conference. The financial support from Department of Mechanical and Structural Engineering and Materials Science at University of Stavanger through the ISP program is highly appreciated. Finally, the administrative support provided by Mrs Vanessa Grace Ochon Booc is highly appreciated.

Conference Organizing Committee

Prof. Hirpa G. Lemu
Prof. Dimitrios Pavlou
Prof. Muk Chen Ong
Prof. Jasna B. Jakobsen
Prof. Sudath C. Siriwardane
Prof. Yihan Xing
Assoc. Prof. Ove Mikkelsen

IOP Conf. Series: Materials Science and Engineering 1294 (2023) 011001 doi:10.1088/1757-899X/1294/1/011001

Volume Editors

Prof. Hirpa G. Lemu
Prof. Andrés García Granada
Prof. Dimitrios Pavlou
Adjunct Prof. Gerhard Ersdal
Dr. Guang Yin
Assoc. Prof. Idriss El-Thalji
Assoc. Prof. Knut Erik Giljarhus
Prof. Mohsen Assadi
Prof. Emeritus Ove T. Gudmestad
Prof. Sudath C. Siriwardane
Dr. Tiago J. F. Marques Ferradosa
Prof. Tor H. Hemmingsen
Assoc. Prof. Yanyan Sha
Prof. Yihan Xing
Prof. Yiliu Liu

Invited Keynote Speakers

Prof. Asfaw Beyene, San Diego State University, USA.
Prof. Mohammed Pourkashanian, University of Sheffield, United Kingdom
Prof. Nicholas Fantuzzi, University of Bologna, Italy
Prof. José A.F.O. Correia, University of Porto, Portugal
Prof. Jayantha Prasanna Liyanage, University of Stavanger, Norway
Prof. Odne S. Burheim, Norwegian University of Science and Technology (NTNU), Norway

Additional plenary session speakers

Dr. Dr. Xuanlie ZHAO, Harbin Engineering University, China
(Title of speech: Multi-Body Hydrodynamics and Hybrid Platforms: Insights from Recent Studies)

Dr. Stamatios Avlonitis, DEYATH Mesaria-Thira, Greece
(Title of speech: Reverse Osmosis Technology for Desalination)

Technical Committee Members and Reviewers

1.	Assoc. Prof. Nirosha Adasooriya	University of Stavanger, Norway
2.	Prof. Henning Agerskov	Technical University of Denmark (DTU), Denmark
3.	Ass. Prof. Paul Akangah	North Carolina A&T State University, USA
4.	Prof. Jafar Al-Zaili	City University, UK
5.	Prof. Jørgen Amdahl	Norwegian Univ. of Sci. and Tech (NTNU), Norway

6.	Prof. Mohsen Assadi	University of Stavanger, Norway
7.	Prof. Santanu Bandyopad	IIT Bombay, India
8.	Prof. Abbas Barabady	The Arctic University of Norway (UiT), Norway
9.	Prof. Jeroen van Beek	von Karman Institute for Fluid Dynamics, Belgium
10.	Assoc. Prof. Kostas Belibassakis	National Techn. Univ. of Athens, Greece
11.	Prof. André De Boer	University of Twente, The Netherlands
12.	Prof. Robert Brown	Memorial University (MU), Canada
13.	Prof. Hermes Carvalho	Federal University of Minas Gerais, Brazil
14.	Ass. Prof. José Miguel Castro	Faculty of Engineering, University of Porto, Portugal
15.	Dr Nianzhong Chen	Tianjing University, China
16.	Prof. Po Wen Cheng	University of Stuttgart, Germany
17.	Dr. Etienne Cheynet	University of Stavanger, Norway
18.	Dr. José António Correia	Faculty of Engineering, University of Porto, Portugal
19.	Prof. Tianjiao DAI	Huazhong Univers. of Science & Technology, China
20.	Prof. Hamidreza Gohari Darabkhani	Staffordshire Univ., UK
21.	Prof. Sreekanta Das	University of Windsor, Canada
22.	Assoc. Prof. Idris El- Thalji	University of Stavanger, Norway
23.	Prof. Dimitrios Emiris	Nazarbayev University, Kazakhstan
24.	Adj. Prof. Gerhard Ersdal	University of Stavanger, Norway
25.	Dr. Tiago Fazeres-Ferradosa	University of Porto, Portugal
26.	Prof. Lorenzo Ferrari	Pisa Univ., Italy
27.	Prof. Mario Luigi Ferrari	Genova Univ., Italy
28.	Dr. Arne Fjeldstad	DNV, Norway
29.	Adj. Prof. Kjell Tore Fosså	University of Stavanger, Norway
30.	Prof. Jan Frick	University of Stavanger, Norway
31.	Prof. David Fuhrman	Technical Univ. of Denmark (DTU), Denmark
32.	Dr. Aboma W. Gebisa	Yucogawa Techinvent AS, Norway
33.	Prof. Koh Chan Ghee	National Univ. of Singapore (NUS), Singapore
34.	Prof. Andrés-Amadir García-Granada	IQS School of Engineering, Spain
35.	Dr. Mahesh Gopal	Wallaga University, Ethiopia
36.	Prof. Ove T. Gudmestad	University of Stavanger, Norway
37.	Prof. Rui Miranda Guedes	University of Porto, Portugal
38.	Assoc. Prof Endalkachew M. Gutema	Wallaga University, Ethiopia
39.	Prof. Marya Haiden	Russian State Gubkin Univ. of Oil and Gas, Russia
40.	Prof. Tor Hemmingsen	University of Stavanger, Norway
41.	Prof. Lars Erik Holmedal	Norwegian Univ. of Sci. and Tech (NTNU), Norway

42.	Prof. Jasna B. Jakobsen	University of Stavanger, Norway
43.	Prof. Hrvoje Jasak	University of Zagreb, Croatia
44.	Assoc. Prof. Guomin Ji	Norwegian Univ. of Sci. and Tech (NTNU), Norway
45.	Prof. Zhiyu Jiang	University of Agder, Norway
46.	Dr. Audun Kristoffersen	Petroleum Safety Authority Norway, Norway
47.	Mr. Gen Li	Dalian Maritime University, China
48.	Assoc. Prof. Lin Li	University of Stavanger, Norway
49.	Prof. Jayantha P Liyanage	University of Stavanger, Norway
50.	Prof. Yiliu Liu	Norwegian Univ. of Sci. and Tech (NTNU), Norway
51.	Dr. Shengnan Liu	Subsea 7, Norway
52.	Dr. Inge Lotsberg	DNV GL Oslo, Norway
53.	Assoc. Prof. Stylianos Markolefas	Techn. Univ. of Central Greece, Greece
54.	Mr. Yucong Ma	University of Stavanger, Norway
55.	Dr. Gro Markset	OsloMet, Norway
56.	Assoc. Prof. Ove Mikkelsen	University of Stavanger, Norway
57.	Dr Per Olav Moslett	DNV, Norway
58.	Prof. Pantelis Nikolakopoulos	University of Patras, Greece
59.	Prof. Håkan Nilsson	Chalmers University of Technology, Sweden
60.	Dr. Piotr Omenzetter	University of Aberdeen, UK
61.	Prof. Muk Chen Ong	University of Stavanger, Norway
62.	Dr. Theodosios Papathanasiou	Brunel University, London, UK
63.	Prof. Dimitrios Pavlou	University of Stavanger, Norway
64.	Prof. Demosthenes Polyzos	University of Patras, Greece
65.	Dr. Jawad Raza	Moreld Apply AS, Norway
66.	Dr. Chao Ren	University of Stavanger, Norway
67.	Prof. Lorents Reinas	University of Stavanger and Equinor, Norway
68.	Dr. Zhengru Ren	Norwegian Univ. of Sci. and Tech (NTNU), Norway
69.	Prof. Dr. Paulo Rosa-Santos	Faculty of Engineering, University of Porto, Portugal
70.	Assoc. Prof. Amir Safari	University of South-Eastern Norway
71.	Dr. Mamdouh Salama	ConocoPhillips, USA
72.	Assoc. Prof. Yanyan Sha	University of Stavanger, Norway
73.	Dr. Srinivas Sriramula	University of Aberdeen, UK
74.	Dr. Lujie Shi	INSA Rouen Normandie, France
75.	Dr. Hailu G. Shimels	Addis Ababa Science And Techn. University, Ethiopia
76.	Ass. Prof. Hodjat Shiri	Memorial University (MU), Canada
77.	Prof. Jonas T. Snæbjörnsson	University of Stavanger, Norway

IOP Conf. Series: Materials Science and Engineering 1294 (2023) 011001 doi:10.1088/1757-899X/1294/1/011001

78. Prof. Sudath Siriwardane	University of Stavanger, Norway
79. Prof. Sudipta De	Jadavpor Univ., India
80. Dr. Jian Tan	The Delft University of Technology, The Netherlands
81. Prof. Dr. Francisco Taveira-Pinto	Faculty of Engineering, University of Porto, Portugal
82. Assoc. Prof. Mesay A. Tolcha	Jimms University, Ethiopia
83. Assoc. Prof. Yiannis Tsompanakis	Technical University of Crete, Greece
84. Assoc. Prof. Turkyilmaz	Nazarbayev University, Kazakhstan
85. Ass. Prof. Ling Wan	University of Newcastle, Singapore
86. Prof. Gary Wang	Simon Fraser University, Canada
87. Dr. Shuaishuai Wang	Norwegian Univ. of Sci. and Tech (NTNU), Norway
88. Dr. Xueliang Wen	University of Stavanger, Norway
89. Dr. Xiaopeng Wu	IKM Ocean Design, Norway
90. Dr. Guang Yin	University of Stavanger, Norway
91. Dr. Zhaolong Yu	Norwegian Univ. of Sci. and Tech (NTNU), Norway
92. Prof. Yihan Xing	University of Stavanger, Norway
93. Prof. Xiaoshen Xu	Jiangsu University of Science and Technology, China
94. Prof. Qian Xudong	National University of Singapore (NUS), Singapore
95. Prof. Stefanos Zaoutsos	University of Applied Sciences of Thessaly, Greece
96. Dr. Wuyang Zhang	Shanghai Electric, China
97. Dr. Xinying Zhu	Havfram, Norway
98. Prof. Sigmund Kyrre Ås,	Norwegian Univ. of Sci. and Tech (NTNU), Norway

IOP Conf. Series: Materials Science and Engineering 1294 (2023) 011002 doi:10.1088/1757-899X/1294/1/011002

Peer Review Statement

All papers published in this volume have been reviewed through processes administered by the Editors. Reviews were conducted by expert referees to the professional and scientific standards expected of a proceedings journal published by IOP Publishing.

- **Type of peer review:** Single Anonymous
- **Conference submission management system:** Indico
- **Number of submissions received:** 89
- **Number of submissions sent for review:** 72
- **Number of submissions accepted:** 62
- **Acceptance Rate (Submissions Accepted / Submissions Received × 100):** 69.7
- **Average number of reviews per paper:** 2
- **Total number of reviewers involved:** 44
- **Contact person for queries:**
  Name: Hirpa G. Lemu
  Email: hirpa.g.lemu@uis.no
  Affiliation: University of Stavanger, N-4036 Stavanger, Norway

IOP Conf. Series: Materials Science and Engineering 1294 (2023) 012001 doi:10.1088/1757-899X/1294/1/012001

Active damage mitigation of the blade leading edge erosion for a wind turbine during rainfall events

Á Ú Ripoll[1], Z Jiang[2], A Verma[3], and J Zhou[2]

1 Industrial Engineering, Polytechnic University of Catalonia, Barcelona, Spain
2 Department of Engineering Sciences, University of Agder, N-4898 Grimstad, Norway
3 Department of Mechanical Engineering, University of Maine, USA

E-mail: alvaro98ubeda@gmail.com, zhiyu.jiang@uia.no

Abstract. Leading edge erosion (LEE) of wind turbine blades is a major concern for the wind turbine industry. LEE leads to increased blade surface roughness and incurs significant repair and maintenance costs. To address this issue, the present paper proposes an active LEE mitigation method by operating wind turbines in an erosion-safe mode with reduced rotor speed during precipitation. The proposed method involves first numerically modelling the NREL 5-MW reference wind turbine in openFAST. Then, during a regular operation in the presence of rainfall, the erosion-safe mode is designed, implemented and initiated with a modified blade pitch controller in Matlab-Simulink. To evaluate the effectiveness of this mode, the long-term rainfall distribution of the De Kooy site in the Netherlands is examined. The analysis considers a comparison of power output, erosion lifetime, and levelized cost of energy (LCOE) with and without the proposed safe modes. The study shows that the erosion-safe mode can significantly extend the erosion lifetime at the cost of reduced power generation. The controller settings affect the wind turbine performance, and a reduced rotor speed of 11 revolutions per minute is recommended considering the LCOE (308.64€/MWh). While the proposed mitigation method shows promise, it requires further refinement for real-life implementation in wind turbines.

1. Introduction

The wind energy sector is currently facing a major challenge in the form of leading edge erosion (LEE) of wind turbine blades. This issue is hampering the performance and longevity of wind turbines. The damage is typically caused during rainfall when rain droplets hit the rotating blades, leading to an increase in the blade surface roughness and eventually material loss at the leading edge. As the smoothness of the blades is related to the aerodynamic performance [1], a turbine with LEE may operate with a lowered efficiency and lower energy generation [2]. Reports show that the repair and maintenance costs of LEE damages reach values over £56 million for the European offshore wind turbine sector [2, 3]. In addition, the current trends in wind turbine development indicate that the future will see turbines with larger rotor diameters. This offers multiple advantages such as lower operational and maintenance costs as well as increased energy extraction per turbine [5]. However, large wind turbines may be associated with higher tip speeds making LEE even more critical.

There have been several efforts considered to mitigate the LEE of wind turbine blades. These can be broadly classified into three categories of solutions [6] as shown in Figure 1. These categories include (a) erosion-resistant blade coatings (b) new methods to predict blade erosion

Figure 1: Image of a blade with severe LEE and mitigation solutions; the main topic of this paper marked in red [4].

and associated repair, and (c) erosion-safe controls for wind turbines. For instance, accelerated erosion tests have been performed extensively to compare the performance of several coating materials [7, 8, 9]. Computational models have also been developed to estimate fatigue life based on cyclic stresses over the service life of the blades [10, 11]. Other research has been directed at developing a control algorithm that automatically reduces the tip speed of the blade, thereby reducing the LEE damage [12]. The LEE-induced damage can also be characterised using cumulative damage models for accurate long-term blade damage predictions using site-specific wind and rain statistics [13]. More details on the review of mitigation methods can be found in [14].

The present paper focuses on the third aspect of the solution which is proposing an active LEE mitigation method by operating wind turbines in an erosion-safe mode with reduced rotor speed during precipitation. While there have been scientific work done in the past on developing similar erosion-safe control algorithm, the paper presents three novelties: (1) design and implementation of an active erosion-safe mode through numerical simulations (2) comprehensive analysis of the effect of the new control algorithm on the operational dynamics of wind turbines, and (3) systematic assessment of the performance of the control algorithms using a realistic rain and wind data from a coastal site in the Netherlands. The focus of this research article is primarily on rain-induced blade erosion, as rainfall serves as the major contributor to LEE on the blades [6]. The rest of the paper is organised into four sections. Section 2 discusses the underlying theoretical concepts that form the basis of the investigation. Following this, section 3 describes the design, implementation, and initiation of the erosion-safe control method during precipitation. A description of the numerical simulations and analysis using the long-term De Kooy site weather statistics is also presented. Following these analyses, the results and discussions are presented. Finally, the conclusions section summarises the key insights of the study.

2. Mitigation of LEE by erosion-safe control

The basis of the mitigation method considered in this work involves reducing the blade tip speed by reducing the rotor speed of the wind turbine during operation. This is done by introducing an erosion-safe mode of the controller which is activated during rainfall events. The NREL 5-MW reference wind turbine model is chosen in this project for case studies [15]. While the work can be extended to offshore conditions, this paper focused on an onshore wind turbine. In addition, selecting an open-source model is a practical approach, as it simplifies result verification, provides access to the wind turbine control, and facilitates the evaluation of system performance by comparing the findings with previous studies.

IOP Conf. Series: Materials Science and Engineering 1294 (2023) 012001 doi:10.1088/1757-899X/1294/1/012001

Figure 2 depicts the structure of a typical wind turbine control system consisting of two parts: the generator torque control and the blade pitch control. The generator torque control regulates the power output, and it relies on a look-up table that provides a torque for a generator speed. This system is more effective when the wind turbine is operating in below-rated conditions. The blade pitch angle control, on the other hand, controls the main shaft speed by collectively adjusting the pitch angles of the three blades. This control system is active when the rotor speed is above the rated one. By adjusting the pitch angles, the controller minimises the error between the measured and reference speeds, maintaining the rotor speed at rated values.

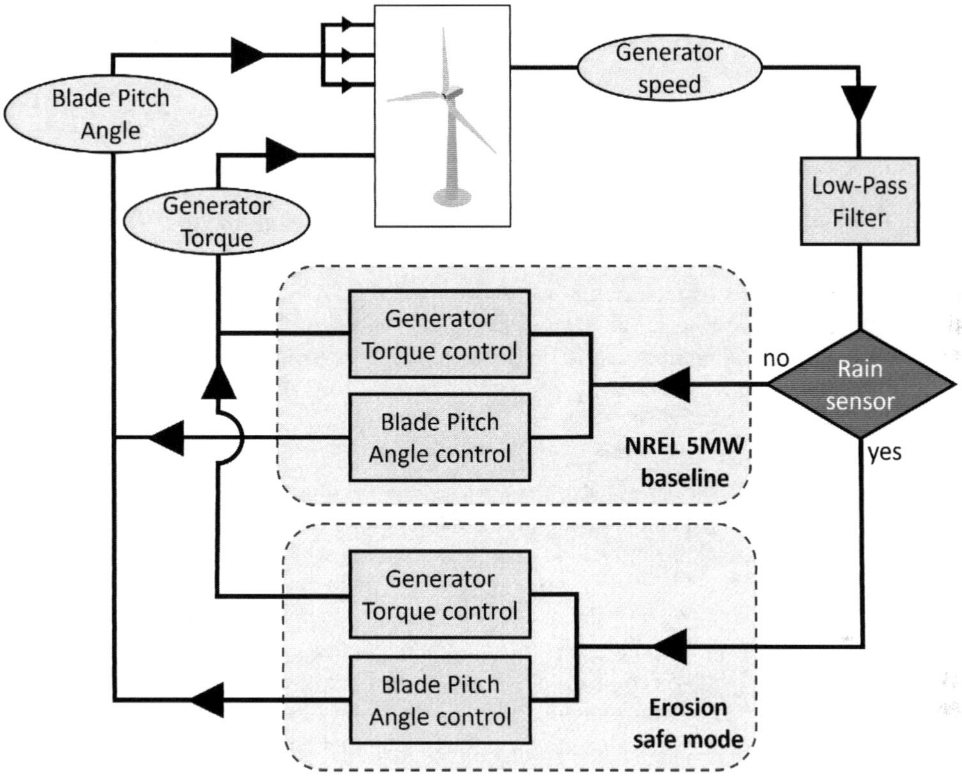

Figure 2: Schematic of the basic blocks of the proposed erosion safe mode

In this paper, a new active control algorithm is proposed to replace the original NREL controller upon the presence of erosion damages, as shown in Figure 2. The graphical representation illustrates the integration of the innovative control approach triggered by a fictitious rainfall sensor. In practice, sensors like disdrometers can be installed on a wind turbine. The new 'erosion safe mode' controller is a modified version of the 'DISCON' controller for the NREL 5-MW turbine with user-defined parameters [15]. These are the reference generator speed on the blade pitch control and the reference generator power output on the generator torque control. During rainfall events, the wind turbine's reference generator speed is adjusted to values below the rated ones. This causes the blades to pitch to a higher degree accordingly and reduce the rotational speed of the generator. Subsequently, the main shaft speed reduces and results in a reduction in the velocity at the blade tip. This is crucial in mitigating LEE as lower tip speeds mean reduced impact velocities with rain droplets. To maintain stable operation of the wind turbine, the generator torque control is also modified to have a lower target output values. Without this modification, the generator torque control would increase the generator torque to match the rated energy generation output, which is not desirable.

Reducing blade tip speeds significantly reduces the damage to the blades caused by LEE due to rainfall. Lower tip speeds result in reduced impact velocity between the surface of the blades and the rainfall droplets. The cyclic nature of the rain droplet impacts on the blade surface causes a fatigue damage phenomenon that can be characterised using the Palmgren-Miner's law for linear damage accumulation and the Springer model for damage rate [16]. This is a widely used damage accumulation model used in a few studies regarding LEE on wind turbines. [17, 18] This method considers a point of impact on the surface of the blade with which the droplets collide continuously during a rainfall. Every droplet contributes a specific amount of damage to the surface. Once the cumulative damage reaches a certain threshold, the material degradation on the wind turbine blade becomes significant and results in a noticeable decline in the aerodynamic efficiency.

$$\dot{D}_i = \frac{q \left| \vec{V}_{\text{imp}} \right| \beta_d}{\left(8.9/\phi_d^2\right) \left(S/p_{wh}\right)^{5.7}} \tag{1}$$

In the Springer model (Eq. 1), $\dot{D}_i$ is the fatigue damage rate [9] of the blade coating, $\vec{V}_{\text{imp}}$ is the impact velocity, q is the number of rain droplets per unit volume, ϕ_d is the rainfall droplet size, β_d is the impingement efficiency, S is the erosive strength of the coating material, and p_{wh} representing the water hammer pressure. Note that a reduction in $\vec{V}_{\text{imp}}$ will have a clear contribution to the reduction of fatigue damage from the droplets and therefore LEE. D_i represents the cumulative damage on the blade and a value greater than or equal to 1 represents the onset of material removal. This also represents the end of the incubation period.

3. Analysis procedure

The performance evaluation of the proposed erosion-safe mode controller was conducted in two stages as shown in Figure 3. In the first stage, the erosion-safe mode is designed and implemented through numerical simulations using the wind turbulence field conditions and wind turbine model. The resultant blade tip speeds were extracted and the damages caused to the blades during different rainfall conditions were analysed separately using the Springer erosion model. The results of the first stage led to the creation of a look-up table that was used in the second stage. The look-up table comprises data derived from numerical simulations conducted during the first stage. It presents information regarding the rate of damage accumulation and the power output of the wind turbine under various wind and rainfall conditions and controller settings. In the second stage, the controller's performance was tested using onsite data collected over 50 years in the Netherlands. The purpose is to compare different settings of the proposed erosion-safe control for a selection of parameters, such as reduced operational blade tip speeds.

3.1. Assumptions

The following are the main assumptions used in this study:

- Detection of rainfall is achieved by fictitious sensors and these sensors are assumed to be integrated into the wind turbine.
- There is no time delay between the start of a rainfall event and the triggering of the erosion-safe mode.
- The controller's performance is tested using a numerical framework. A real scale implementation is out of the scope of this paper.
- The LEE model used in this paper is based on the surface fatigue model.
- Only one representative rain droplet size is considered for a given rainfall intensity.
- Each of the examined control modes in the second stage imposes a specific rotational speed limit on the wind turbine during rainfall events, without adjustment to accommodate varying environmental conditions it may encounter.

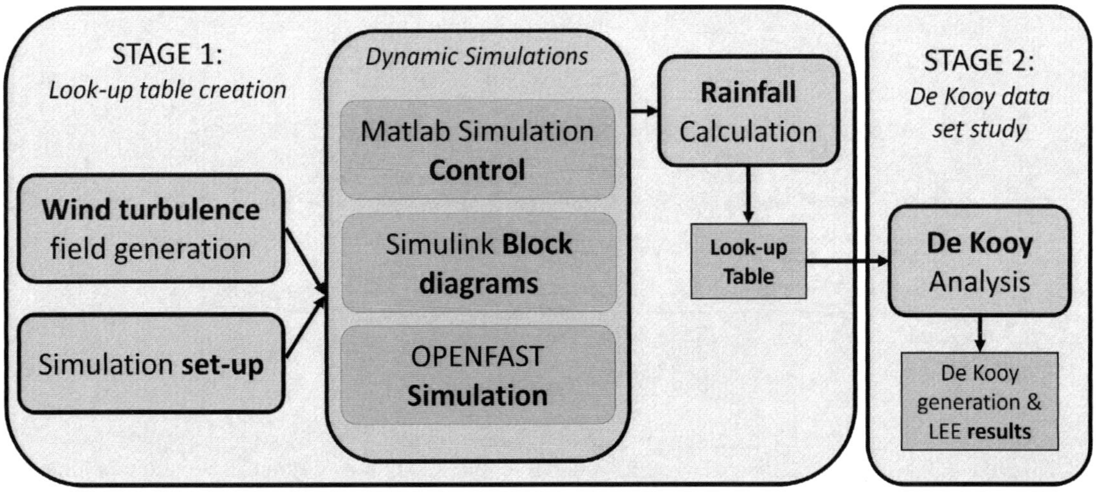

Figure 3: Overview of the workflow of the paper

3.2. Stage 1: Look-up table creation

Figure 3 shows an overview of the workflow. The first step involves generating a wind turbulence field for a class B wind, according to the Normal Turbulence Model (NTM). This determines the turbulence intensity for each wind speed considered in the study. In the second step, other aspects of the models are set up, such as the wind turbine structural models and blade properties. Finally, dynamic simulations are carried out using Matlab scripts to control the simulations, as well as Simulink block diagrams are developed to implement control algorithms together with openFAST scripts to run dynamic simulations.

Figure 4 shows the structure of the dynamic simulations. Each simulation lasts 10 minutes for a given wind condition, during which a rainfall event occurs between times t_s and t_i. The wind turbine operates using the system depicted in Figure 2. Initially, the NREL 5-MW baseline controller is used to achieve normal power production. At time t_s, the rainfall event occurs and the erosion-safe mode is activated immediately following detection by the rainfall sensor. The algorithm reduces the rotational speed of the wind turbine, which in turn minimises the LEE and reduces the energy output of the generator. This mode stays active until the rainfall event ends at time t_i. Once the rainfall sensor detects that the rain has ceased, the baseline NREL controller takes over again. Assuming the rainfall is detected by fictitious sensors, the Matlab scripts adjust the control algorithm settings according to the simulation time to activate and deactivate the baseline NREL and the erosion-safe mode controllers instantaneously at the times when the rainfall event starts, as indicated in Figure 4.

The timeline in Figure 4 demonstrates three phases that a wind turbine goes through while operating with the erosion safe mode. The first stage is the transition phase at time t_s. During this phase, the wind turbine switches from the baseline controller to the erosion-safe mode control. The second stage is the rainfall event, which is long enough for the system to stabilise with the new control mode. The third stage is the transition phase at time t_i, when the wind turbine switches back to the baseline controller. To verify the effectiveness of the second stage, it is necessary to include a reference case where only the baseline controller is active throughout the simulation. Therefore, for every environmental condition, there is also a reference case where only the baseline controller is active. The simulation process, as depicted in Figure 4, is consistent across all tests, but varies based on wind and control system conditions. We have simulated all possible combinations of the values listed in Table 1. After each simulation, we

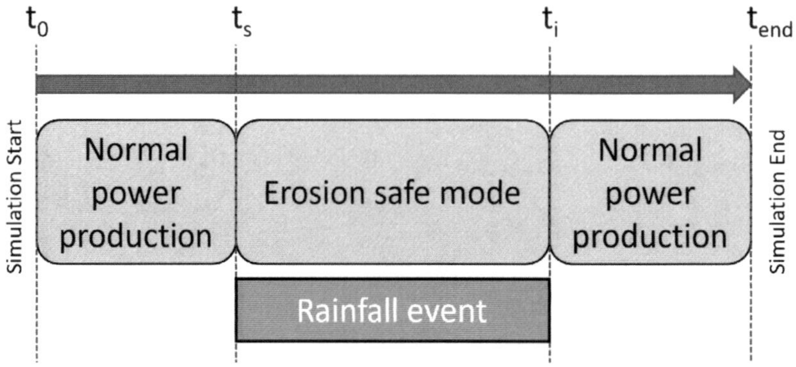

Figure 4: Overview of the basic events involved in the dynamic openFAST simulations

calculated the rainfall conditions for every possible combination of the parameters in the table by post-processing. For each combination of environmental and control parameters listed in Table 1, two variables are calculated: LEE damage rate (unit: 1/h) and generator electrical power (unit: kW). These variables are then compiled to obtain the final result of Stage 1, which is a look-up table that defines the LEE damage rate and power output value for each set of parameters.

Table 1: Wind, control and rainfall conditions parameters tested throughout all the simulations

Parameter	Interval	Values
Mean wind speed	3 m/s	6, 9,..., 24 m/s
Target reduced main-shaft speed	1 rpm	7, 8, ..., 11 rpm
Rainfall intensity	3 mm/h	2, 5,..., 50 mm/h

3.3. Stage 2: Case study of a Dutch site

In the case study, the De Kooy site in the Netherlands was chosen. The site's close proximity to the sea makes the erosion conditions more severe than those experienced by inland onshore wind turbines. Figure 5 presents the long-term wind and rainfall distribution for De Kooy. As shown, the wind speed is concentrated in the below-rated regions, with a mean wind speed of 5.79 m/s at the site. A noteworthy aspect of the site is that 78.26% of the time there is sufficient wind for the wind turbine to operate, out of which 7.21% of the time it is above the rated capacity. Figure 5 also shows relatively low rainfall intensities. However, rainfall occurred during 11.25% of the total hours over a period of almost 50 years. This information will be useful in evaluating the effectiveness of the new control mode. Few highly intensive rainfall events over 50 mm/hr are also observed, which is useful for studying the overall rainfall patterns.

The data collected at the De Kooy site comprises hourly records of rainfall intensity and mean wind speed. Using the mean wind speed, we can determine the turbulence intensity based on the turbulence class specified by the IEC standard. Additionally, we can use probabilistic rainfall models [14] to determine the mean droplet size for a given rainfall.

Each of the control parameters is tested separately, i.e., for every test of the 50 year period on De Kooy, the control parameters are fixed. Therefore, by matching the De Kooy environmental conditions to the ones found on the look-up table obtained on Stage 1, it is possible to assign a given value of power generation and LEE damage rate for every time interval of the dataset.

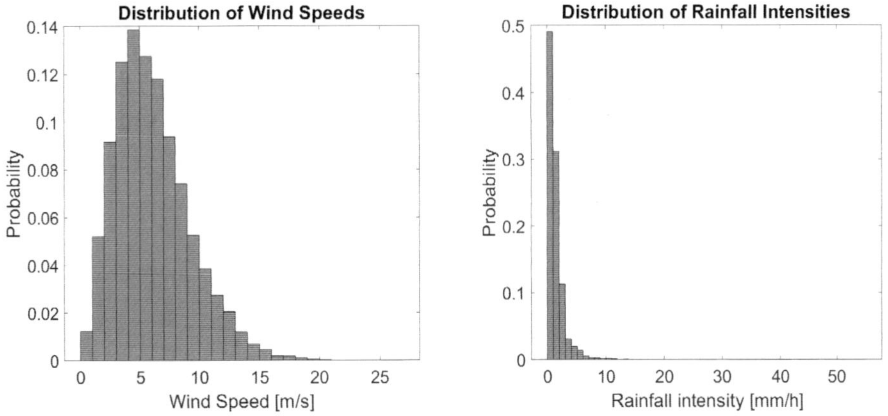

Figure 5: Distribution of wind speeds (left) and rainfall intensities (right) at the De Kooy from 1971-2020

In cases where the values of wind and rainfall conditions are not found in the look-up tables, approximate values from the look-up table close to the De Kooy conditions will be taken. By doing so, we are able to simulate the behaviour of the wind turbine during a time interval of 50 years. By studying the long-term results of these hourly behaviours, it is possible to compare the performance of the different tested control systems over an extended period of time. To facilitate a meaningful comparison we calculate the levelised cost of energy (LCOE). First, the CAPEX has been calculated to be 7180000 EUR according to [19] that estimates the CAPEX cost of a 2.6 MW wind turbine to be 1436 EUR/kW. Second, the only operational cost considered is the repair job of an LEE damaged wind turbine at 30000 EUR. [20]

4. Results and Discussion

This section is divided into two parts, each corresponding to the first and second stages shown in Figure 3. The first composed of an example of one of the simulated cases with one of the rainfall scenarios. Then, results for the De Kooy site are presented.

4.1. Time history of a numerical simulation

The parameters listed in Table 2 are considered in this numerical simulation. As the chosen wind speed is above rated, rated values of power generation are expected for simulation with the original controller. A high rainfall intensity of 50 mm/h is considered. Figures 6, 7 and 8 correspond to the timeline outlined in Figure 4. For the two periods at the start and end of the simulation, the wind turbine operates with the baseline controller. During the second stage with rainfall, the erosion safe mode is activated.

Table 2: Parameter selection for the time-series case analysis

Simulation parameter	Value
Mean wind speed	21 m/s
Turbulence intensity	0.1423
Target main-shaft speed	8 rpm
Rainfall intensity	50 mm/h
Median droplet size	1.4643 mm

Figures 6–7 show the influence on the rotational speed of the main shaft and the generator power when the proposed erosion safe mode is activated during a rainfall event. The rotational speed is reduced to 8 rpm as specified by the controller parameters in Table 2. To achieve a reduced rotor speed, the blade pitch angle must be adjusted to a higher degree. It can be seen in Figure 8 that the pitch angle increases significantly during the rainfall event, meanwhile in the reference case with the baseline controller, the fluctuation of the pitch angle remains relatively small in turbulent wind.

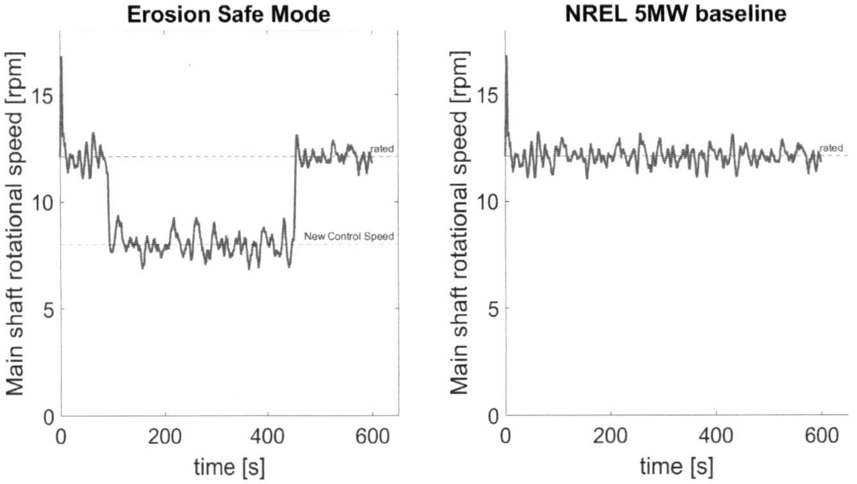

Figure 6: Rotational speed of the main shaft using the erosion safe mode (left) the baseline controller (right)

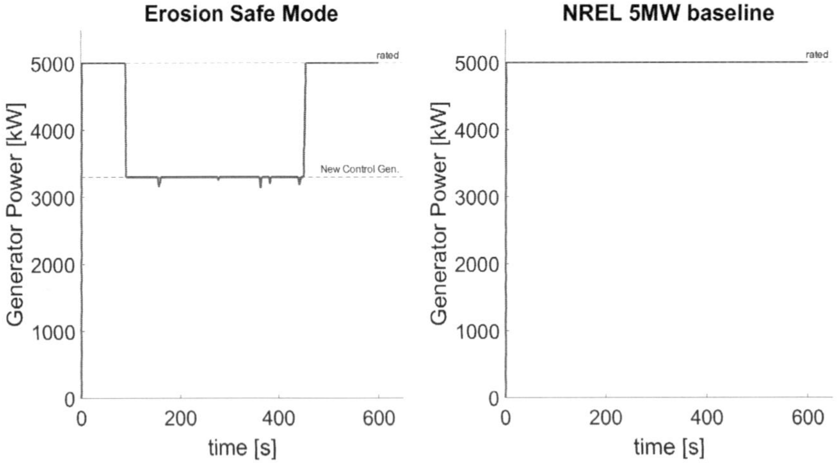

Figure 7: Generator torque during a simulation using erosion safe mode (left) and the baseline controller (right)

Figure 9 shows that the tip speed of the blades reduces substantially when the erosion control mode is activated. This reduction is due to the lowered rotor speed as seen in Figure 6. Comparing the two damage rates seen in Figure 10 it is clear that the proposed damage

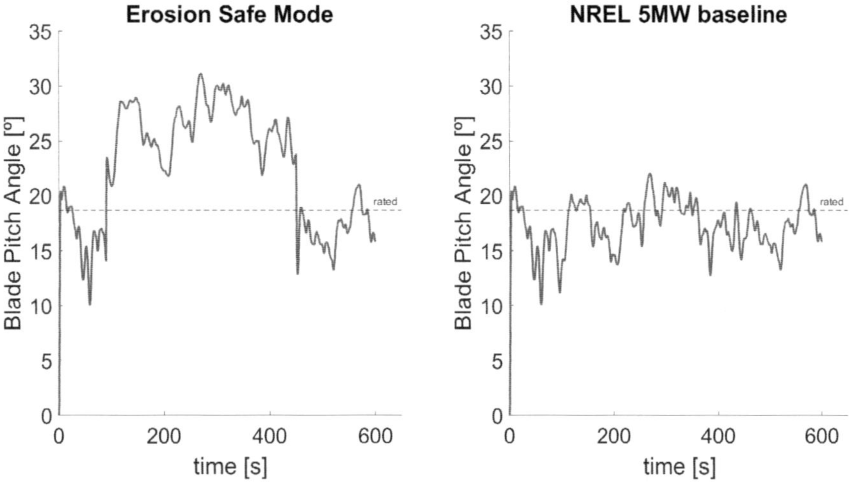

Figure 8: Collective blade pitch angle during a simulation using the erosion safe mode (left) and the baseline controller (right)

mitigation strategy by reducing the tip speeds is effective. As shown in Figure 11, the damage accumulation slopes of the baseline controller and of the erosion safe mode deviate significantly from each other, and the proposed erosion safe mode results in a significantly reduced damage rate.

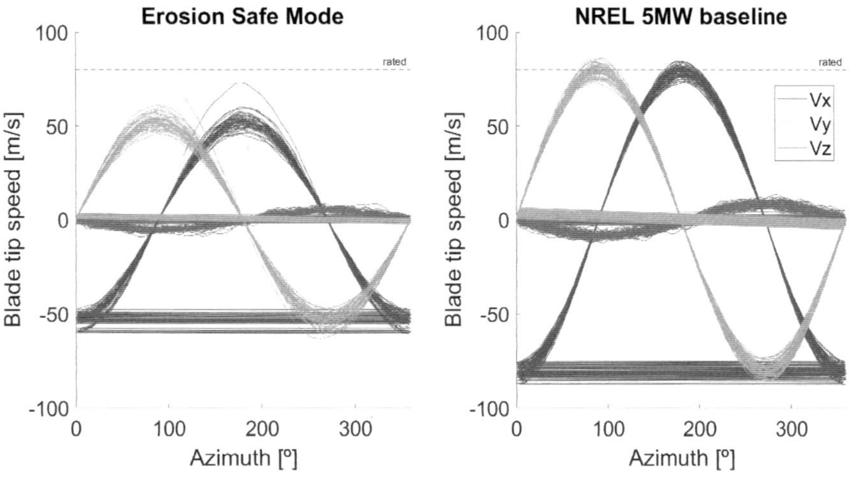

Figure 9: Blade tip speeds from a numerical simulation with the erosion safe mode (left) and baseline controller (right)

4.2. Case study results

After carrying out all the simulations with the parameters detailed in Table 1, we performed a case study simulation using the De Kooy dataset. The results show an expected difference in the energy generation using controllers with different target rotor speed, as can be seen in Figure 12. It is clear that all controllers start at the same value of energy generation (0 MWh)

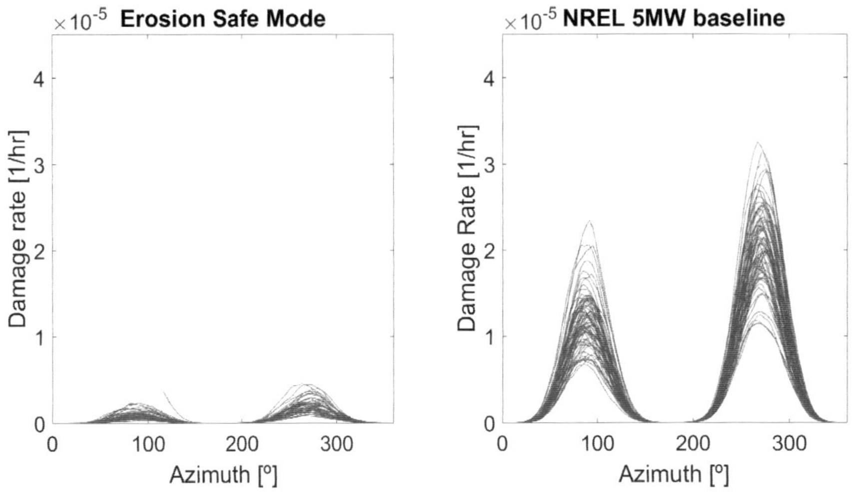

Figure 10: Resultant LEE damage rates calculated from the output data from the openFAST simulations

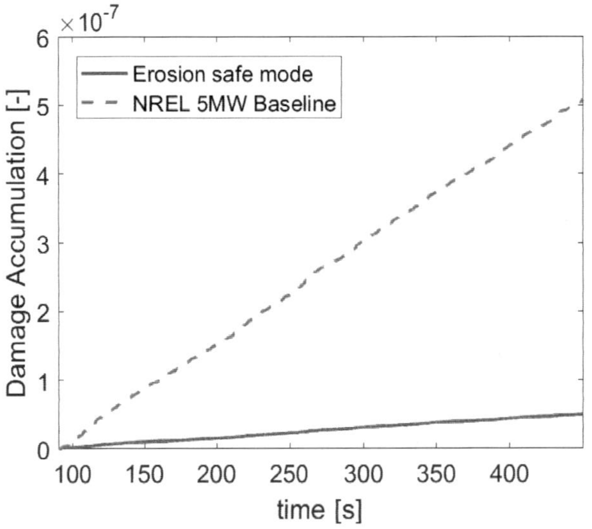

Figure 11: Damage accumulation time series through the rainfall event of the simulation for both erosion safe mode control and baseline control systems

and gradually increase the cumulative generated power over time. As expected, the baseline controller generates the most power from the beginning and keeps its power production rate during rainfall events.

Consequently, once the damage accumulation reaches a threshold, there is a need for repair and maintenance of the blades, and the energy generation reaches a plateau. This phenomena is shown in Figure 13. Here, the original baseline control results in a downtime of the wind turbine at day 773 for a duration of 3 days due to repair. In contrast, for erosion safe mode with target rotor speed of 10 and 11 rpm, the cumulative energy generation continues to increase during this period as the erosion damage has not reached a threshold yet. This example illustrates that operating the wind turbine with the erosion safe mode control can bring down costs and

increase the total power generation by avoiding premature failure and downtime.
While the 773-day (2.1 year) duration until failure may appear relatively brief, it aligns closely with findings reported in a previous study by J.I. Bech [12] which observed similar time frames for material degradation. It should also be noted that this case study is a site specific phenomenon that also appears in the paper from A.Verma [14] in which it appears as a location with highly severe rainfall intensities.

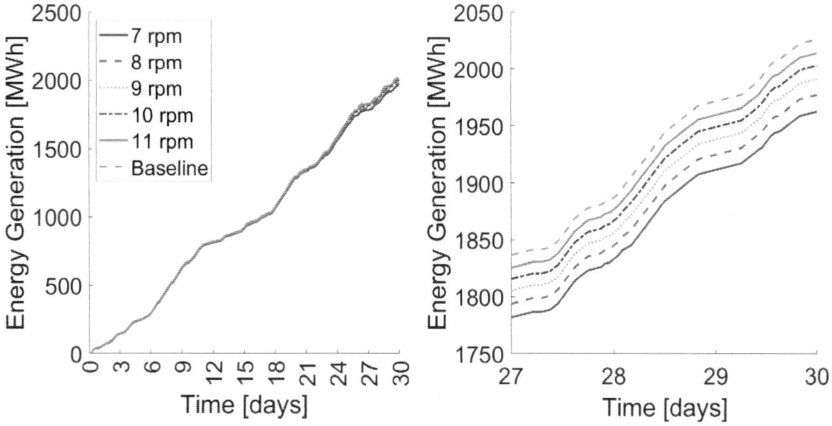

Figure 12: Cumulative energy generation for the first 30 days of the 50 year De Kooy data set (left) and for days 27 to 30 (right)

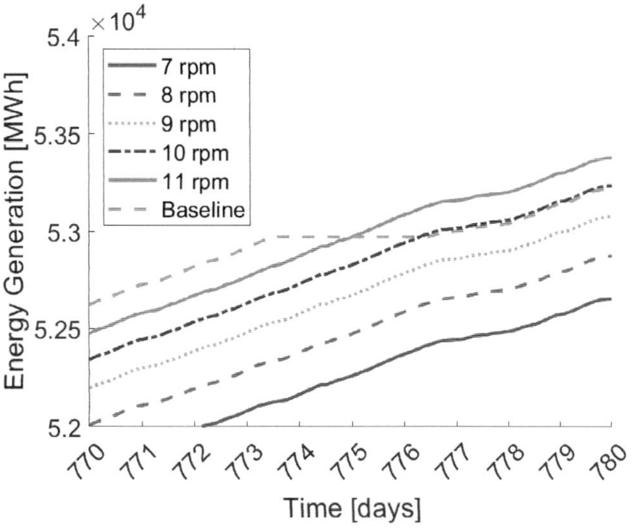

Figure 13: Cumulative energy generation for days 770-780 of the 50-year dataset of De Kooy in which the Baseline controller reaches it's first breaking point

Together with the total energy generation, the damage accumulation of the wind turbine's blades has also been analysed. The results can be seen in Figure 14. It is clear that the proposed erosion safe mode controllers are operating in a more damage-tolerant manner. Figure 15 shows that the expected lifetime of the blade's surface increases more significantly as the target rotor

speed is further reduced. Note that the lifetime expectancy depicted in Figure 15, is ideal as other failure modes of the wind turbine are not considered here.
Figure 15 shows the sensitivity of the levelized cost of energy (LCOE) to the target rotor speed during control in the erosion safe mode. It is interesting to notice that the lowest LCOE is achieved when the rotor speed is reduced to 11 rpm during a rainfall, and the NREL's baseline controller ranks next to this one. This interesting observation shows that although slowing down the rotor speeds during rainfall events may penalise the LCOE due to reduced power generation. The end result always depends on balancing the downtime and repair costs with the reduced generation in the erosion safe mode.

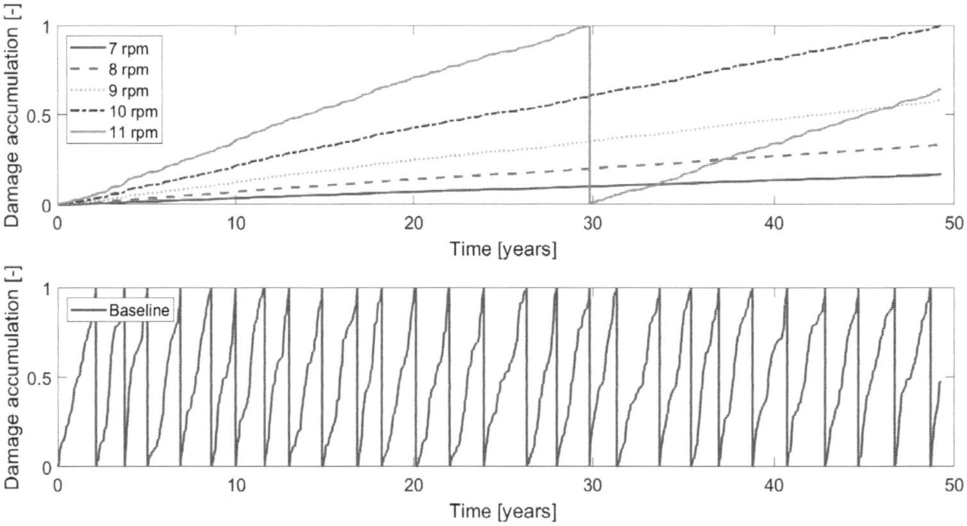

Figure 14: Damage accumulation for the erosion safe mode (top) and the baseline controller (bottom)

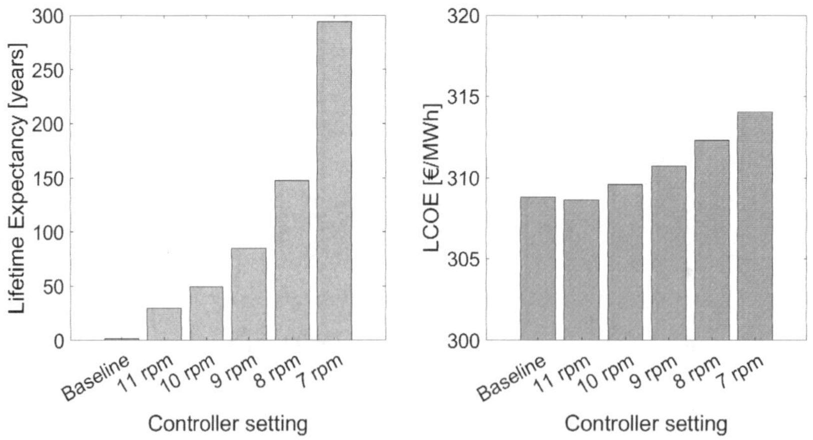

Figure 15: Lifetime expectancy(left) and LCOE (right) comparison for the different controller settings in the De Kooy 50 year data set scenario

5. Concluding remarks

The present paper conducts aero-hydro-servo-elastic simulations of a 5-MW wind turbine under various wind and rainfall conditions. To reduce the LEE damage encountered by the blades, the erosion safe mode has been proposed with several input parameters.
After a case study of the De Kooy site in the Netherlands, the main observations are as follows:

- The proposed erosion safe mode can work as an effective alternative to the original baseline controller during rainfall events. Through time-domain numerical simulations, it is demonstrated that the proposed control mode can be applied to operate the wind turbines in a stable and reliable way. The simulations also showcase that the transition from one controller to the other can be done in an efficient and safe way.
- The proposed control mode can increase significantly the lifetime expectancy of the blade's leading edge surface. This indicates that the erosion safe mode by slowing down the rotor speed is a valid method for LEE mitigation during a rainfall.
- The comparison between the several controllers on the De Kooy site is an efficient way to evaluate the performance of different controllers over an extended period of time. However, the results are highly site-specific and depend strongly on the set-up parameters.

6. Limitations and recommendations for future work

The present paper is hopefully setting the path for future efforts that may help contribute to the LEE problem on wind turbine blades. For instance, the following topics need further research:

- The design of an intelligent erosion safe mode that calculates and dynamically varies the optimal rotor speed according to the rainfall conditions.
- Consideration of the rainfall effects on the dynamic simulations. This would take into consideration any possible disturbances that the numerous rain droplet impacts could have on the behaviour of the turbine.
- Modelling the rainfall by considering the whole distribution of rain droplet sizes. In the present study, only the mean droplet size is considered for all impacts.
- Making use of the method used on this paper to compare performances other than erosion safe mode controllers. It would be possible to compare the performance of several coating materials for the leading edge of the blades or several aerodynamic designs of the blades.
- Instead of modelling the evolution of the wind turbine's leading edge state by determining the breakdown time, a gradual decrease in efficiency could be a better method of modelling the performance. It would be interesting to consider the possibility of extending the working state of a turbine even if the efficiency is lowered by some given amount due to LEE damages.

Acknowledgments

The first author acknowledges the ERASMUS program which supports the study exchange and the research work at the University of Agder, Norway.

References

[1] Sagol E, Reggio M, and Ilinca A. Issues concerning roughness on wind turbine blades. *Renewable and Sustainable Energy Rev.*, 23:514–525, 2013.
[2] Herring R., Dyer K., Martin F., and Ward C. The increasing importance of leading edge erosion and a review of existing protection solutions. *Renewable and Sustainable Energy Rev.*, 115:109382, 2019.
[3] Wiser R., Jenni K., Seen J., Baker E., Hand M., Lantz E., and Smith A. Forecasting wind energy costs and cost drivers: The views of the world's leading experts. *LBL Report*, 2016.
[4] Verma A. S., Jiang Z., Caboni M., Verhoef H., Meijer H. M., and Castro S. G.P.and Teuwen J. J.E. A probabilistic rainfall model to estimate the leading-edge lifetime of wind turbine blade coating system. *Renewable Energy*, 178:1435–1455, 2021.

[5] Hofmann M. and Sperstad I. B. Will 10 mw wind turbines bring down the operation and maintenance cost of offshore wind farms? *Energy Procedia*, 53:231–238, 2014. EERA DeepWind' 2014, 11th Deep Sea Offshore Wind R&D Conference.

[6] Hasager C., Mishnaevsky Jr. L., Bak C., Bech J.I., Fæster S., and Frost-Jensen Johansen N. How can we combat leading-edge erosion on wind turbine blades? *DTU Int. Energy Rep. 2021: Perspectives on Wind Energy*, pages 134–142, 2021.

[7] Slot H.M., IJzerman R.M., le Feber M., and Nord-Varhaug K.and van der Heide E. Rain erosion resistance of injection moulded and compression moulded polybutylene terephthalate pbt. *Wear*, 414-415:234–242, 2018.

[8] Keegan M. H. *Wind Turbine Blade Leading Edge Erosion: An Investigation of Rain Droplet and Hailstone Impact Induced Damage Mechanisms.* PhD thesis, University Of Strathclyde, 2014.

[9] Eisenberg D., Laustsen S., and Stege J. Wind turbine blade coating leading edge rain erosion model: Development and validation. *Wind Energy*, 21(10):942–951, 2018.

[10] Chen J., Wang J., and Ni A. A review on rain erosion protection of wind turbine blades. *J. of Coat. Technol. and Res.*, 16(1):15–24, 01 2019.

[11] Mishnaevsky L. Repair of wind turbine blades: Review of methods and related computational mechanics problems. *Renewable Energy*, 140:828–839, 2019.

[12] Bech J. I., Hasager C. B., and Bak C. Extending the life of wind turbine blade leading edges by reducing the tip speed during extreme precipitation events. *Wind Energy Sci.*, 3(2):729–748, 2018.

[13] Verma A. S., Castro S.G.P., Jiang Z., and Teuwen J.J.E. Numerical investigation of rain droplet impact on offshore wind turbine blades under different rainfall conditions: A parametric study. *Compos. Struct.*, 241:112096, 2020.

[14] Verma A.S., Jiang Z., Ren Z., et al. A probabilistic long-term framework for site-specific erosion analysis of wind turbine blades: A case study of 31 dutch sites. *Wind Energy*, 24(11):1315–1336, 2021.

[15] Jonkman J., Butterfield S., Musial W., and Scott G. Definition of a 5-mw reference wind turbine for offshore system development. *National Renewable Energy Laboratory, Tech. Rep., No. NREL/TP-500-38060*, 2009.

[16] Springer G. S. *Erosion by Liquid Impact.* John Wiley and Sons, New York, NY, USA, 1976.

[17] Verma A. S., Noi S. D., Ren Z., Jiang Z., and Teuwen J. J. E. Minimum leading edge protection application length to combat rain-induced erosion of wind turbine blades. *Energies*, 14(6), 2021.

[18] Nick Hoksbergen, Remko Akkerman, and Ismet Baran. The springer model for lifetime prediction of wind turbine blade leading edge protection systems: A review and sensitivity study. *Materials*, 15(3), 2022.

[19] Stehly T., Beiter P., Duffy P. 2019 cost of wind energy review. Technical Report NREL/TP-5000-78471, National Renewable Energy Laboratory (NREL), Golden, CO, 2020.

[20] Mishnaevsky L. and Thomsen K. Costs of repair of wind turbine blades: Influence of technology aspects. *Wind energy*, 23(12):2247–2255, 2020.

IOP Conf. Series: Materials Science and Engineering 1294 (2023) 012002 doi:10.1088/1757-899X/1294/1/012002

Flow over a step cylinder using partially-averaged Navier-Stokes equations with application towards dynamic subsea power cables

U Shaukat [0009-0000-9340-8555][1,*], **A Schnepf** [0000-0001-6164-1934][1,2] and **K E T Giljarhus** [0000-0002-4144-045][1]

[1] University of Stavanger, Department of Mechanical and Structural Engineering and Materials Science, Stavanger, Norway

[2] CoreMarine AS, Stavanger, Norway

E-mail: usman.shaukat@uis.no

Abstract. Understanding the flow characteristics of buoyancy sections in power cable and umbilical configurations is crucial for analyzing floating offshore wind turbine systems. Buoyancy sections decouple the motions from the floater and the destination point to ensure the integrity of the system. Using accurate drag coefficients in their design is essential, as they significantly impact the overall behavior of the configuration. Therefore, Computational Fluid Dynamic (CFD) analysis is performed in the present study on a single buoyancy module attached to a power cable represented by a step cylinder configuration. The Partially-Averaged Navier-Stokes (PANS) turbulence model is applied. Its accuracy is validated against experimental findings of a wall-mounted cantilever cylinder and an infinite cylinder. The numerical results reveal that the larger diameter cylinder (LDC) section reduces the drag experienced by the smaller diameter cylinder (SDC) section near the junction. The LDC has a sharp, chamfered, and filleted edge replicating the various shapes of buoyancy modules. The edge design of the LDC affects the drag forces and flow patterns. The SDC has an 8 % lower drag coefficient in the fillet edge case than the sharp edge case. The drag coefficient is 3.5 % lower for the LDC in the filleted edge case than the sharp edge case. The sharp edge causes a significant separation of the fluid upstream of the SDC. However, this separation is notably reduced when the LDC has a chamfered or filleted edge. Detailed drag coefficients for buoyancy section analysis of power cable configurations have been deduced from the presented results.

1. Introduction

Power cables transmit electrical energy generated by offshore wind turbines to consumers, whether on land or at sea, such as oil and gas platforms. In the case of floating offshore wind turbines, the power cable configurations have dynamic sections where the cable is designed to float freely in the water from its hang-off location to the seabed or the next floater. They are installed in specific configurations to maintain safe operation and prevent excessive loading. These configurations often require buoyancy modules attached to the power cable at certain sections to keep its shape and to decouple the motions from the floating offshore wind turbine to the next connection point.

IOP Conf. Series: Materials Science and Engineering 1294 (2023) 012002 doi:10.1088/1757-899X/1294/1/012002

Besides their use in offshore wind technologies, buoyancy modules have been applied in umbilical and riser configurations within the oil and gas industry. The recommended practice DNV-RP-C205 [1] specifically addresses risers equipped with buoyancy elements and provides guidance on their analysis. While the tangential drag coefficients are determined based on the experimental findings of Huse and Erling [2], DNV-RP-C205 recommends estimating the drag coefficients in transverse (normal) direction on an infinite cylinder. Drag coefficients for buoyancy modules are estimated by multiplying the infinite cylinder drag coefficient with a reduction factor according to the finite length. Selecting the appropriate transverse coefficient can have a substantial influence on the behavior of the system. The Morison equation is often applied in common software used for the hydrodynamic analysis of offshore systems. Therefore, when a current flows in the transverse direction towards the power cable, umbilical, or riser, it encounters a much larger resistance area than in the tangential direction. This can result in notable differences in simulation results when employing various values obtained from different methods to estimate the drag coefficient. For example, Schnepf et al. [3] observed a difference of about 8% in power cable tensions when using infinitely and finitely long cylinder approaches to determine the drag coefficient.

Due to the symmetric characteristic of the system, the focus can be on the specific region where a buoyancy module interfaces with the power cable. Therefore, for simplicity, this configuration can be replicated by a step cylinder. A step cylinder comprises a large diameter cylinder (LDC) and a small diameter cylinder (SDC) section merged together. Moreover, numerical and experimental studies on step cylinders were performed either at low Reynolds numbers with importance to flow structures [4, 5, 6] or on the high diameter ratio of the step cylinder [7].

The present study investigates transverse forces acting along smooth cylinders at higher Reynolds number flows in the subcritical regime. Different cylinder configurations are investigated, including infinite and finite cylinders and step cylinders with sharp, chamfered, and filleted edges.

2. Methodology and numerical setup

A smooth step cylinder is investigated in the present study, focusing on forces and changes in the flow pattern due to the diameter change between the LDC and SDC. In offshore systems, the Reynolds number can vary from subcritical to supercritical depending on the diameter of the cylinders and the current speed. For this study, the focus is specifically on Reynolds numbers in the subcritical flow regime because the power cable configurations are commonly designed below the critical regime. Therefore, the LDC and SDC sections have Reynolds numbers of 2.5×10^4 and 6250, respectively, which fall in the subcritical range. The flow is initially laminar at these Reynolds numbers, separating and transitioning to turbulence along the free shear layer around the cylinder [8]. The present study also investigates changes in the overall flow patterns and sectional drag forces due to chamfering and filleting of the edges of the LDC because practically the edges of buoyancy modules are usually not sharp; see [9].

Simulation results are presented using a scale-resolved turbulence model, in which large-scale eddies are resolved while small-scale eddies are modeled. These models have shown improved prediction in bluff body aerodynamics. This study applies the Partially-Averaged Navier-Stokes (PANS) equation model first developed by Girimaji [10]. The k ω SST PANS model is implemented in OpenFOAM v2206 and chosen for the study due to improved near-wall boundary layer modeling [11].

PANS switches between direct numerical simulation (DNS) and solving Unsteady Reynolds-Averaged Navier-Stokes (URANS) equations with the help of the blending factor f_k. This factor represents the resolved-to-unresolved turbulence ratio in the PANS equations [12]. The value of f_k ranges between 0 and 1, where 0 means DNS and 1 means URANS are applied.

Since PANS simulations are sensitive to the f_k parameter, many research methodologies have been established to dynamically adjust f_k value across the domain [12, 13, 14]. There is considerable uncertainty involved in the process of selecting an ideal dynamic formulation of f_k that can be universally applied to all cases. Moreover, Klapwijk et al. [15] argued that the current dynamic f_k approaches are not satisfactory with its sensitivity to grid resolution and suggested that the constant f_k approach is still a better way to apply PANS for industrial applications. Therefore, this study chooses a discrete constant f_k, and a validation study with two f_k values is initially performed for the benchmark case of a cantilever wall-mounted cylinder. The value that produces better flow predictions is chosen for the step cylinder study.

2.1. Numerical settings

SnappyHexMesh utility is used to generate mesh consisting of hexahedra, polyhedra, and prisms. A refinement level of four is applied at and near the cylinder surface, and a uniform boundary layer is generated close to the cylinder to resolve the viscous sublayer along the whole cylinder, as shown in Figure 2. Therefore, the average y^+, which is the nondimensional distance from the wall to the first grid point, approached a value of one near the wall of the cylinder. The spatial discretization relies on a second-order Linear-Upwind Stabilized Transport (LUST) scheme, which combines a 75 % linear weight to enhance accuracy and a 25 % linear upwind weight for stabilization. Concurrently, a second-order backward scheme is employed for temporal discretization. Throughout the simulation, temporal discretization is such that the maximum Courant Number for all time steps is smaller than one. Pressure and velocity equations are solved using the Pressure-Implicit with Splitting of Operators (PISO) algorithm.

2.1.1. Cantilever finite cylinder model validation setup In this study, the simulation setup reproduces the wind tunnel setup illustrated in Figure 1. The cylinder diameter of $D = 0.03\,\mathrm{m}$ and inlet velocity of $10\,\mathrm{m/s}$ generated a Reynolds number of approximately 2×10^4. The test section measures $24D \times 24D \times 20D$ ($L \times W \times H$) in three dimensions. The cylinder has a height of $6D$, resulting in a blockage ratio of 1.2 %. Slip boundary conditions are applied at the top and lateral patch. No slip conditions are applied to the cylinder and bottom patch. Uniform fixed inlet velocity is applied to the inlet patch. As a first step, a precursor RANS simulation is run up to 200 iterations, followed by an unsteady PANS simulation for 1.336 s for up to 445 vortex-shedding cycles. The fields were averaged after 112 vortex-shedding cycles.

2.1.2. Step and infinite cylinder models setup Considering the practical dimensions of an offshore subsea power cable buoyancy section, the ratio of the step cylinder is $D/d = 4$, while the aspect ratio is $L/D = 5$ in the present study. Here, D and d represent the diameter of the LDC and SDC, respectively, while L represents the length of the LDC. The diameter of the large cylinder D is taken as the reference length for analysis. Three different edge features, flat, fillet, and chamfered, are tested for the LDC, resembling different possible buoyancy element shapes. The chamfer angle is 45° with a distance of $\frac{13D}{32}$ away from the junction, while for the fillet, the radius is $\frac{D}{5}$. The diameter of the cylinder for the fillet and chamfer cases is raised to accommodate the loss in volume due to the machining process. The increase in diameter for the LDC is minimal, and the ratio of the diameters remains the same. A uniform inlet velocity condition is applied to the inlet patch.

Most of the numerical settings are similar to the previous validation setup in Section 2.1.1. The numerical tunnel is resized in the lateral and longitudinal direction to $16D$ and $10D$, respectively, resulting in a blockage ratio of 3.5 %. The boundary conditions remain identical to the previous case, with the difference at the bottom patch, which has a symmetry condition instead of a nonslip condition. The flow is simulated for 184 vortex-shedding cycles. Fields are averaged for the last 64 shedding cycles. Two different mesh refinements were considered, one

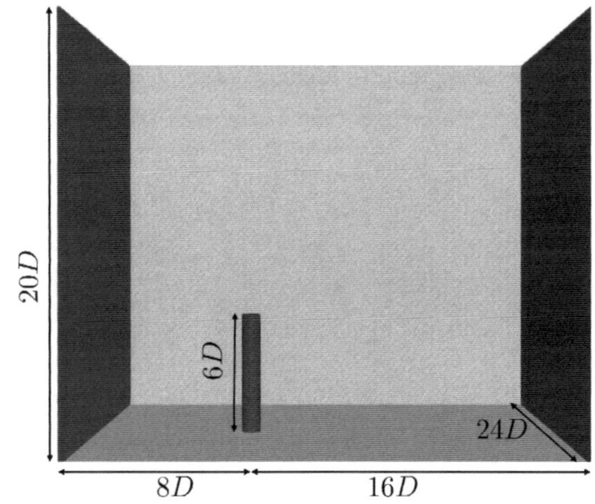

Figure 1: Computational domain for the validation case of the cantilever cylinder.

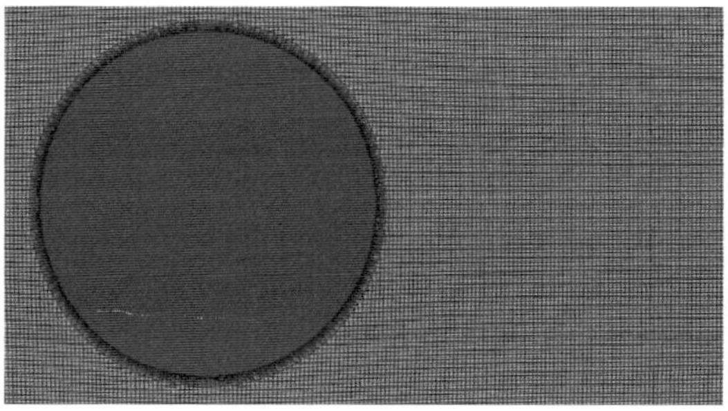

Figure 2: SnappyHexMesh is generated such that first layer is placed inside the viscous sublayer.

having a coarser mesh with an average y^+ value of 1.5 and the second with an average y^+ value less than 1. Meshes contain 4 million and 24 million cells, respectively.

To simulate the infinite cylinder, the LDC is removed and replaced by the symmetry boundary condition while the rest of the settings remain the same.

3. PANS model validation studies

To ensure the accuracy and reliability of the numerical model, validation studies are carried out for the cantilever and infinite cylinder models in this study.

3.1. Cantilever finite cylinder model

Firstly, the PANS model is validated against the wind tunnel experiment conducted by Park and Lee [16] on a cantilever wall-mounted cylinder. Previously, this setup had been validated with Large Eddy Simulation (LES) [17, 18] and PANS by Basara et al. [19]. The results of the surface pressure coefficient profile measured at different heights across the cylinder wall compared to the experimental findings are shown in Figures 3a to 3d. The flow across the cylinder is separated at an angle of $\theta = 76.9°$. Compared to the results from the simulations with $f_k = 0.5$, the PANS simulation with $f_k = 0.1$ resulted in more accurately predicting the surface pressure coefficient distribution across the cylinder at four different heights. The pressure reduction near the free end at $z/D = 0.75$ is due to the fluid movement towards the separation point at the top surface shown in Figure 4. The numerical results yield similar pressure profiles up to the midsection, with a small percentage of error in prediction occurring near the free end. This is the reason for the deviation in the pressure coefficient to the experimental setup. Overall, the results in present settings are comparable to the previous PANS study. A visualization of the mean flow pattern in Figure 4 successfully produced recirculation bubbles at the top surface and the leeward edge of the cylinder, consistent with the prior numerical prediction using four equations near wall PANS model [19]. The generation of vortex near the top edge emphasizes that the downwash effects are dominant in the upper half near the free end of the cylinder. The contours of the Q criterion, also called the second invariant of the velocity gradient tensor, show vortical features across the circular beam similar to the previous findings [17, 18, 19] shown in Figure 5. Frequency analysis of the total side force (lift force) of the entire cylinder resulted in Strouhal numbers of 0.150 and 0.148 for f_k values of 0.1 and 0.5, respectively. These values are close to the experimentally determined Strouhal number of 0.141 [16].

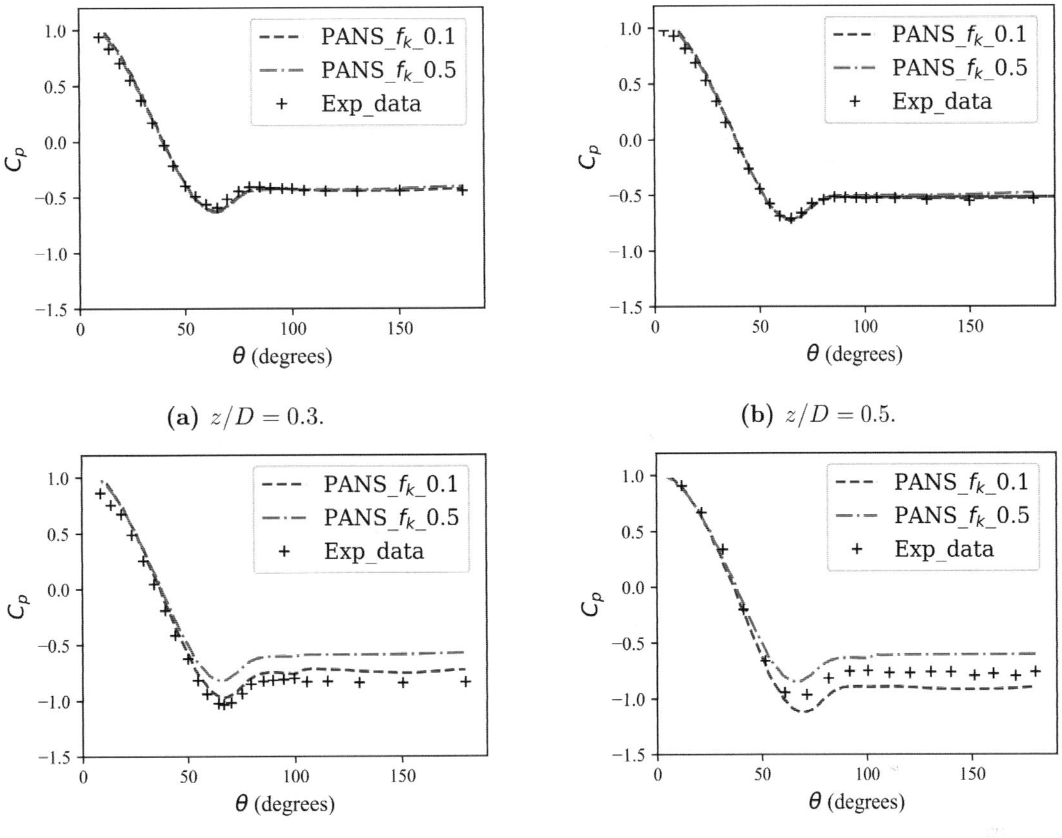

(a) $z/D = 0.3$. (b) $z/D = 0.5$.

(c) $z/D = 0.75$. (d) $z/D = 0.917$.

Figure 3: Pressure coefficients at different positions along the cylinder.

An isolated finite cylinder with a free end is also simulated at Reynolds number 2.5×10^4, and the results yield the drag coefficient of 0.805 consistent with the literature of free end cylinder with dimension $D/L = 5$ at this Reynolds number range [20].

3.2. Infinite cylinder model
For the infinite cylinder, a drag coefficient of 1.16 is obtained, which is close to the experimental value of 1.09 ± 0.5 at a Reynolds number of 6250 [21, 22]. Vortical features represented in Figure 6 are similar to the infinite circular cylinder studies.

4. Results and discussion
The focus of the study is to determine if simplified current practices for estimating forces in subsea power cables can be modified. Therefore, the study examines the flow interaction between step cylinders representing a subsea power cable with buoyancy modules, aiming to identify any changes in the drag force distribution and flow characteristics of the step cylinder. All simulation

IOP Conf. Series: Materials Science and Engineering 1294 (2023) 012002 doi:10.1088/1757-899X/1294/1/012002

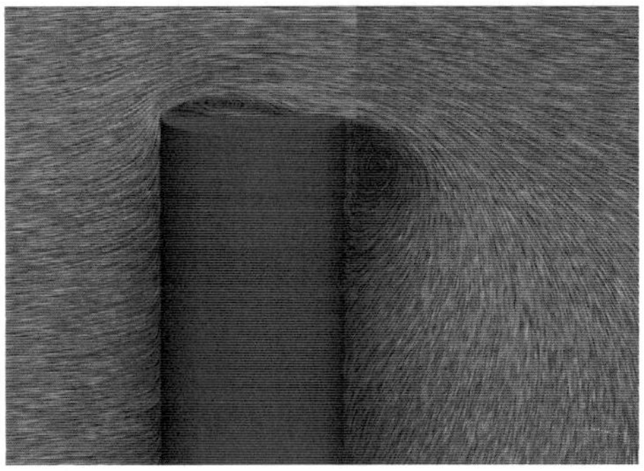

Figure 4: Line integral convolution of the mean velocity at the midsection of the y-plane, colored by mean velocity magnitude.

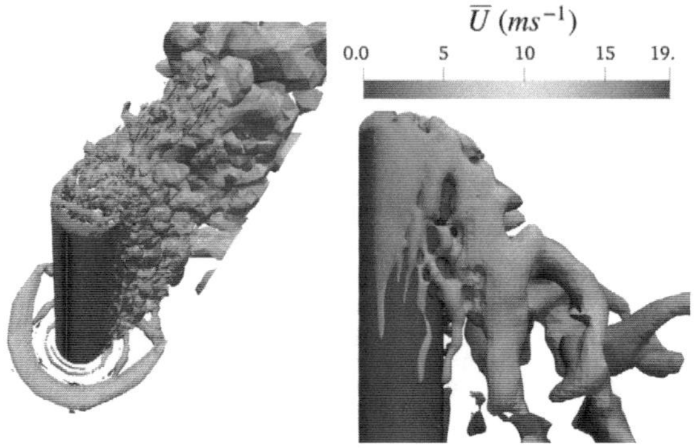

Figure 5: Contours of the Q-criterion (left) and pressure (right) at $Q = 2222\,\mathrm{s}^{-2}$ and $C_p = -0.4$, respectively; color by mean velocity magnitude.

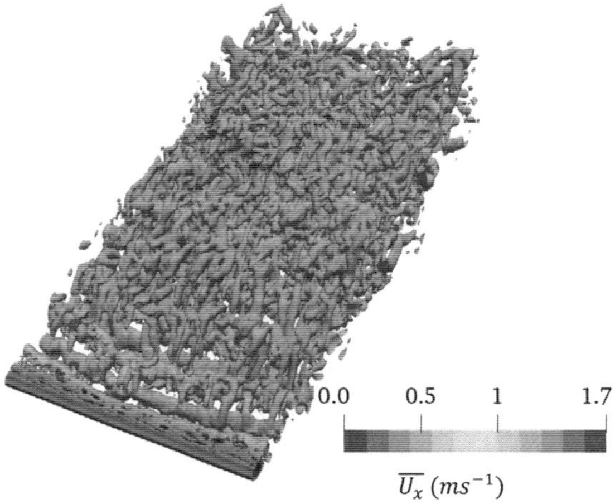

Figure 6: Contours of Q-criterion of the infinite SDC at $Q = 4\,\mathrm{s}^{-2}$; colored by mean velocity magnitude.

setups used a factor of $f_k = 0.1$ due to good results obtained for the cantilever cylinder validation case.

4.1. Force distribution

Figure 7 shows the averaged sectional drag coefficient distribution along the infinite cylinder, finite cylinder, and step cylinder with sharp, chamfered, and fillet edges. The obtained averaged drag coefficient for the step cylinder is comparable to a finite and an infinite cylinder with an open end. Overall, a reduction of the drag force on the SDC near the junction for all the step cylinder cases is observed, as shown in Figure 7a. The sectional drag along the SDC decreases at a distance of $2D$ from the interface of the two cylinders, where D represents the diameter of the LDC, as shown in Figure 7a. Moreover, looking along the SDC section for the chamfered case, the drag force is initially higher at a distance of $2D$ and then subsequently becomes lower up to the interface of the two cylinders. Absolute drag coefficient values are adjusted to align with the graph profile to provide a comparison of the change in the drag resulting from the modified edges of the LDC section due to uncertainty in simulations with temporal and spatial discretization for each case. The averaged drag values on the SDC section are 1.15, 1.45, and 1.10 for the sharp, chamfered, and filleted cases, respectively. Filleting causes the most significant overall drag force reduction by 8 % on the SDC section after a distance of $2D$ from the interface compared to the chamfered and sharp edges. However, the sharp edge causes a sudden increase in force experienced at the cylinder junction, consistent with previous experimental findings for step cylinders with the setting of $D/d = 2$ [7]. High oscillations occurring near the sharp edge shown in the iso-contour of the Q criterion in Figure 8 might cause the increase of force observed close to the interface for the sharp edge case.

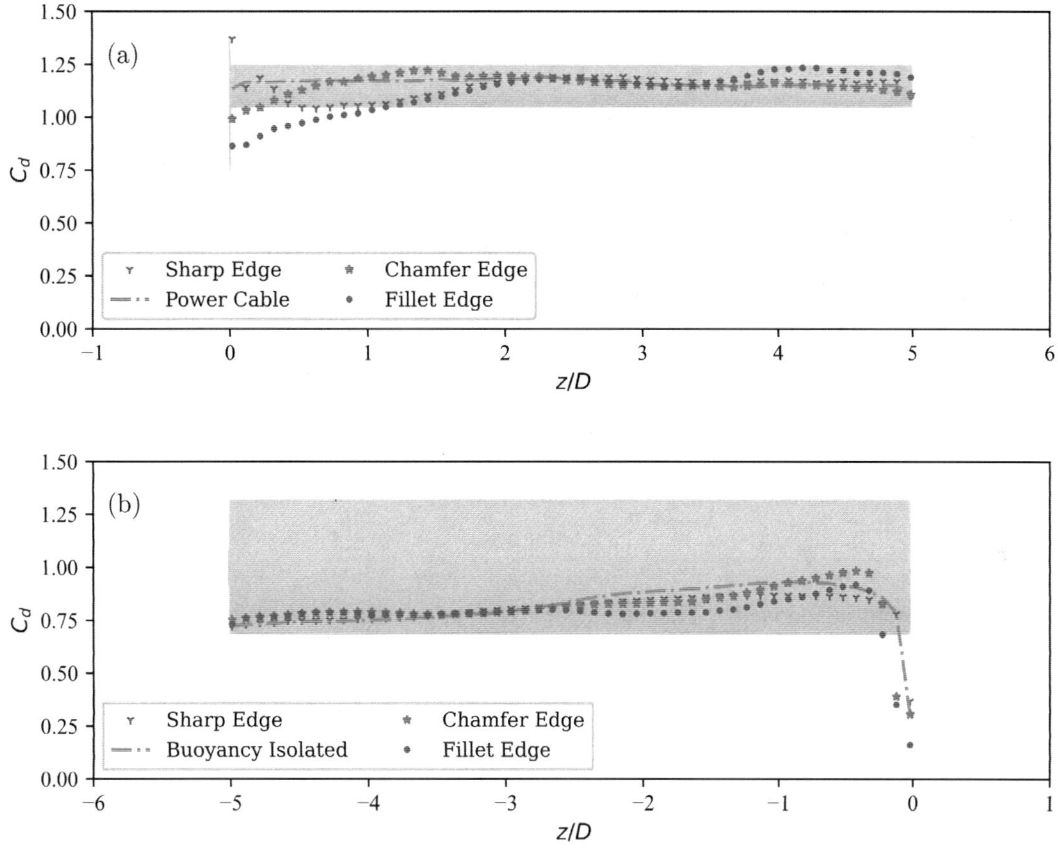

Figure 7: Sectional drag coefficient distribution along the SDC (upper) and LDC (lower).

The drag experienced by the LDC varies at a distance of $2.5D$ to the end of the cylinder junction, as visualized in Figure 7b. However, there is only a slight difference in drag values due to chamfering and filleting, with filleting resulting in the highest overall reduction in drag force by 3.34 %. The overall drag coefficients for sharp, chamfered, and fillet cylinders are 0.82, 0.82, and 0.79, respectively. There are two reasons for the sectional drag reduction observed in this case. Firstly, the chamfer and fillet shapes are more streamlined than sharp edges and guide the fluid towards the vortex at the top end of the LDC, which results in a reduction of pressure upstream near the top. Secondly, the chamfered and filleted edges have a reduced surface area, which decreases the drag coefficient.

IOP Conf. Series: Materials Science and Engineering 1294 (2023) 012002 doi:10.1088/1757-899X/1294/1/012002

4.2. Flow patterns

The flow observed is unsteady and three-dimensional. However, 2-D plane cuts are performed for better understanding. Velocity contours for the cross-section of the step cylinders with sharp, chamfered, and filleted edge configurations are depicted at three different distances in Figures 9 and 10. In Figure 9, flow separation occurs near the proximity of the sharp edge, leading to a pronounced thick region of recirculated flow upstream of the SDC at position $z = 0.01D$. However, higher positive streamwise velocity is observed upstream due to the absence of flow separation when the edges are chamfer and fillet. Only a small separation occurs at the end of the filleted and chamfered edges just upstream of the SDC due to their streamlined shape, which delays the flow separation. This flow separation phenomenon with the edges is similar to the other studies of bluff body aerodynamics [23].

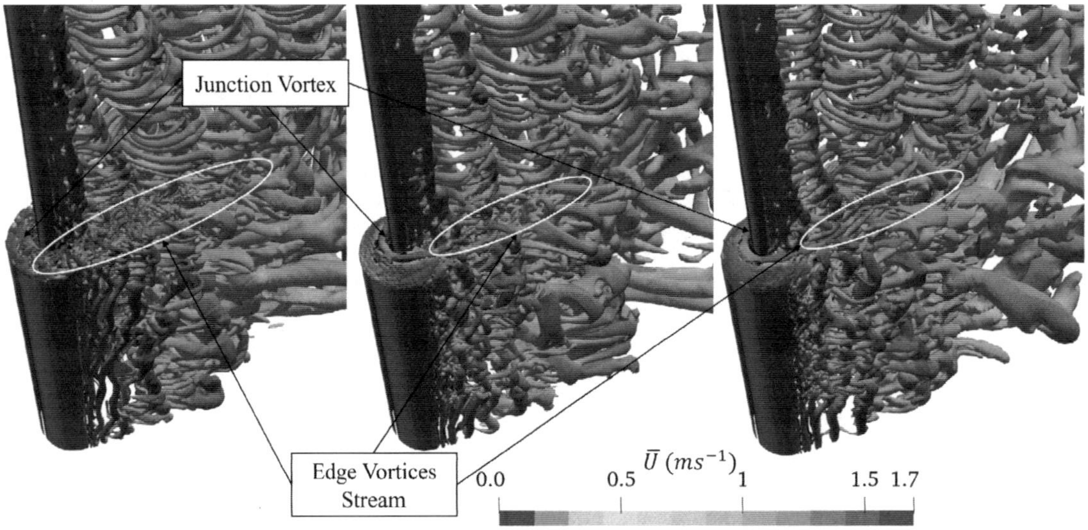

Figure 8: Isocontour of Q-criterion of the step cylinders with sharp (left), chamfered (middle), and fillet (right) edges at $Q = 4\,\mathrm{s}^{-2}$.

The comparison of velocity contours at the mid-section of the step cylinder and the single isolated cylinder cases is shown in Figures 11 and 12. The flow patterns in the wake and junction areas differ for the step cylinder shown in Figure 11 in comparison to the single isolated cylinder depicted in Figure 12. This disparity arises from the interactions between flow fields of SDC and LDC. The velocity contours downstream of the junction for the sharp edge case show accelerated flow and an absence of a substantial reverse flow regime downstream, as depicted in Figures 10 and 11. Conversely, reverse flow is evident for chamfered and filleted edges in these proximity areas. However, the sharp edge configuration leads to a downward flow and the formation of a sizable recirculation region downstream of the LDC shown in Figure 11. Reduction in the size of this recirculation region in the downwash region is seen in the results of chamfers and filleted edges.

The merging of two cylinders generates a distinct and complex coherent flow pattern, as shown in Figure 8. This pattern differs significantly from what was observed with an isolated cylinder case in Figure 6. The presence of a SDC as an obstruction induces the formation of vortices at the rear of the cylinder. The vortices generated from the SDC are pushed slightly downward due to a higher negative pressure region downstream. Notably, streams of edge vortices are visible for configurations with sharp edges similar to the tip vortices seen at the wingtips of an aircraft [24].

IOP Conf. Series: Materials Science and Engineering 1294 (2023) 012002 doi:10.1088/1757-899X/1294/1/012002

However, these vortex streams are less pronounced in cases where the edge of the LDC features a fillet or chamfer. This discrepancy emerges because of the pressure difference created between the top surface and sides of the cylinder. In the case of a sharp-edged cylinder, the thicker junction vortex upstream elucidates the pronounced reverse flow in the velocity contour plot depicted in Figure 9. On the contrary, in the fillet and chamfer cases, a thinner junction vortex

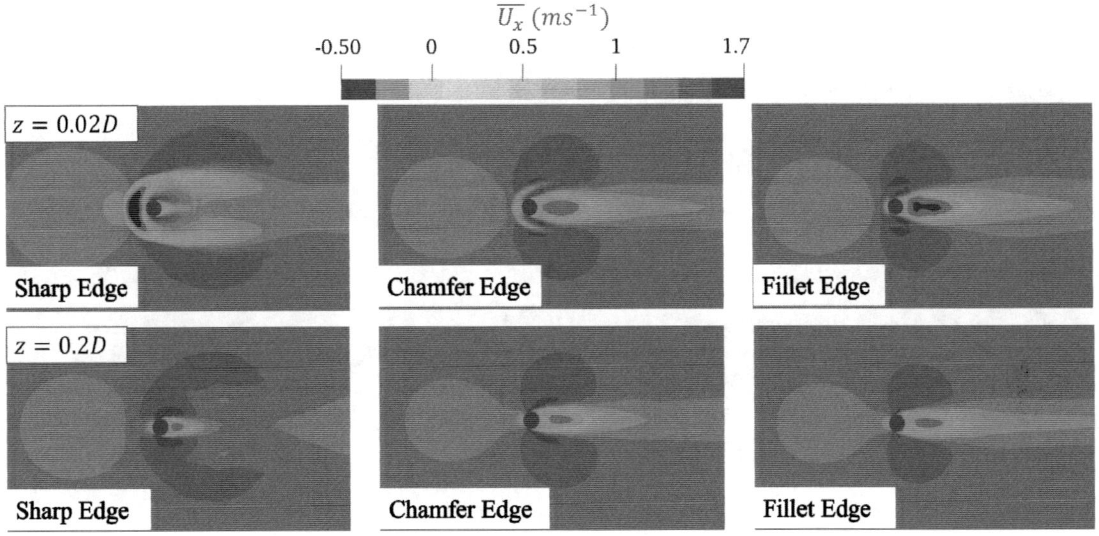

Figure 9: 2-D streamwise velocity contours for the three step cylinder cases at two different cross-sectional positions: $z = 0.02D$ and $z = 0.2D$ (arranged vertically) from the step cylinder junction along the SDC.

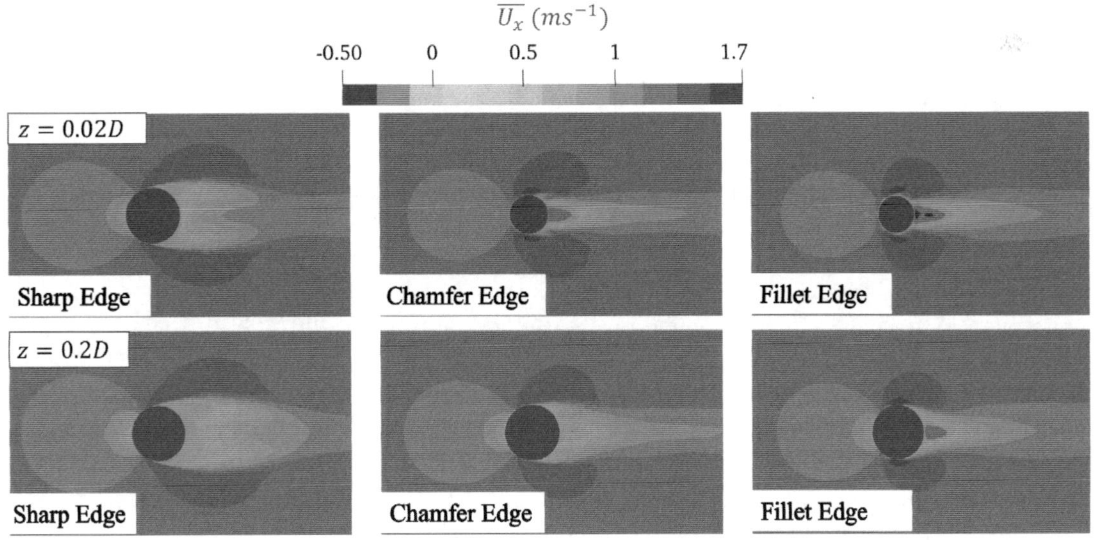

Figure 10: 2-D streamwise velocity contours for the three-step cylinder cases at two different cross-sectional positions: $z = 0.02D$, and $z = 0.2D$ from the junction of the step cylinder along the LDC.

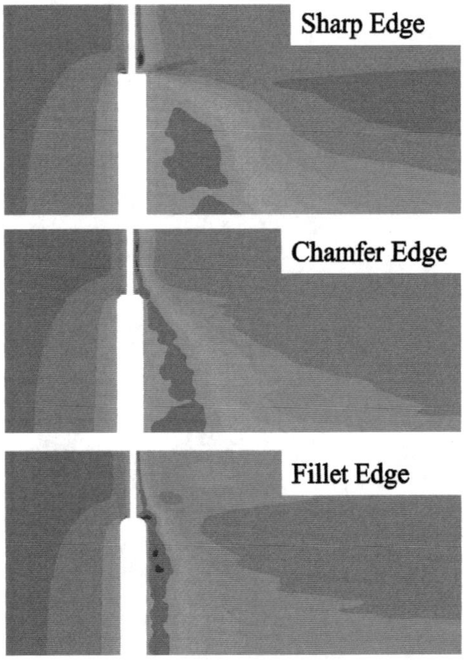

Figure 11: Streamwise velocity contours for the three different step cylinder cases at mid-section: sharp, chamfer, and fillet.

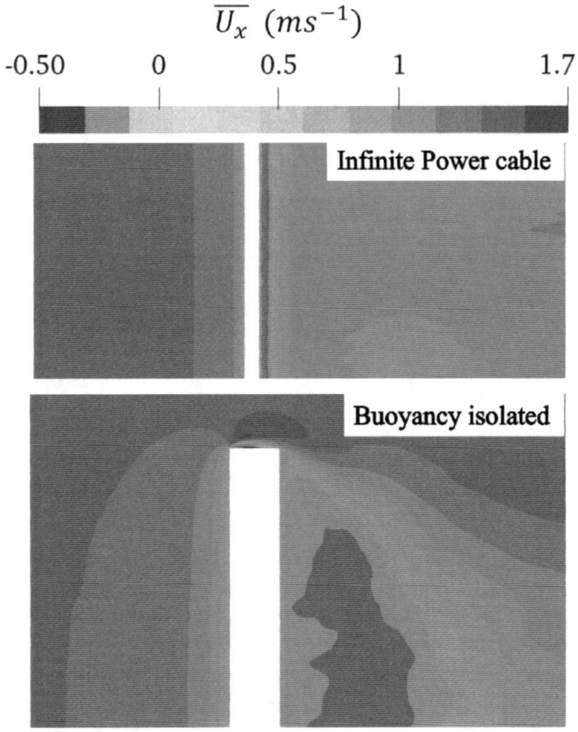

Figure 12: Streamwise velocity contours for the infinite power cable cylinder (top) and the sharp edged isolated buoyancy element finite cylinder (bottom) cases at the midsection.

in the form of a horseshoe is generated.

5. Conclusions

This study applied the PANS method to investigate flow structures and excitation forces on step cylinder configurations with a larger-to-smaller diameter ratio of $D/d = 4$. A comparative analysis to assess the influence on the flow of different edge shapes, namely sharp, fillet, and chamfer, on the LDC was carried out.

A validation study was performed on the cantilever wall-mounted cylinder, and pressure results obtained at the various heights of the cantilever wall-mounted cylinder closely matched experimental results. PANS simulations with the resolved-to-unresolved turbulence ratio $f_k = 0.1$ yielded more accurate results than when using $f_k = 0.5$. Hence, the step cylinder simulation was conducted using $f_k = 0.1$. The results are examined for sectional drag forces, velocity distribution contours, and vortical flow structures.

These are the findings for the step cylinder cases:

- The overall drag coefficient obtained on SDC and LDC is 1.15 and 0.82, respectively. Simulations show that the drag force on the SDC decreases at a distance $2D$ from the cylinder junction. Fillet edge case results in an overall sectional drag force reduction of 8% and 3.5% on the SDC and LDC, respectively, compared to the sharp edge case.
- On the top surface of the sharp cylinder, upstream of the interface, a wider reverse flow regime is observed as compared to the fillet and chamfer cylinders. Moreover, when the

sharp edge is converted to a chamfer and a fillet edge, it causes a reduction in the size of the recirculation region behind the LDC.

- For all step cylinder cases, a junction vortex is formed at the top surface of the LDC. Upstream of the SDC, the large recirculating junction vortex is formed at the sharp edge of the LDC, with a stream of edge vortices on the sides. However, chamfering and filleting the edge reduces the junction vortex size and minimizes the edge vortices intensity.

The DNV-RP-C205 practice estimates the transverse drag coefficient for riser sections with buoyancy elements by adjusting a two-dimensional drag coefficient for finite length effects. For the present study, the reduction factor is 0.68 for the buoyancy module represented by a LDC with a length-to-diameter ratio of 10, including the symmetry condition. With a drag coefficient of 1.2 for infinitely long smooth cylinders in the subcritical regime, the resulting buoyancy module drag coefficient is 0.82. This value is higher than those obtained for the LDC with a fillet edge. However, a drag coefficient of 1.2 is usually assumed for power cables, considering an infinitely long smooth cylinder in the subcritical regime. This value does not account for the decrease in the drag coefficient over a length of $2D$ from the buoyancy module observed in the present study.

In future research, it would be valuable to investigate the effects of varying diameter ratios of the different cylinders. If the distance between two buoyancy elements is less than $2D$, the drag coefficient will be different than in the present study since the step cylinder influences the drag on the power cable up to this length. Double-step cylinder simulations with a SDC in between with a length of $2D$ should be carried out, as the flow from the two LDCs will interact, influencing the drag. Due to the complex vortical interactions observed at the junction, a detailed analysis of the effects of vortex-induced vibrations can be conducted. Moreover, the impact of varying inlet current speed and angle between the cylinder axis and the incoming flow should be evaluated for further studies.

Acknowledgments

The second author would like to thank Markus Håheim from PartnerPlast AS for the insightful discussions on buoyancy modules.

References

[1] DNV AS 2021 Environmental conditions and environmental loads Recommended Practice DNV-RP-C205 Det Norske Veritas
[2] Huse E and Reitan O E 1989 Drag of Buoyancy Elements on Flexible Risers in Axial Flow Tech. Rep. 511171.00-1 MT51 89-0198 MARINTEK Trondheim, Norway
[3] Schnepf A, Lopez-Pavon C, Ong M C, Yin G and Johnsen Ø 2023 Feasibility study on suspended inter-array power cables between two spar-type offshore wind turbines *Ocean Eng.* **277** 114215 ISSN 00298018
[4] Chua L P, Liu C Y and Chan W K 1998 Measurements of a step cylinder *Int. Commun. Heat Mass Transf.* **25** 205–15
[5] Dunn W and Tavoularis S 2006 Experimental studies of vortices shed from cylinders with a step-change in diameter *J. Fluid Mech.* **555** 409–37
[6] Morton C, Yarusevych S and Carvajal-Mariscal I 2009 Study of flow over a step cylinder *Appl. Mech. Mater.* **15** 9–14
[7] Ko N W M and Chan A S K 1990 In the intermixing region behind circular cylinders with stepwise change of the diameter *Exp. Fluids* **9** 213–21
[8] Zdravkovich M and Bearman P 1998 *Flow around circular cylinders—Volume 1: Fundamentals* (Cambridge University Press)
[9] AS P 2023 Offshore & Subsea - Buoyancy systemes and cable protection solutions
[10] Girimaji S S 2005 Partially-Averaged Navier-Stokes Model for Turbulence: A Reynolds-Averaged Navier-Stokes to Direct Numerical Simulation Bridging Method *J. Appl. Mech.* **73** 413–21 ISSN 0021-8936
[11] Menter F R 1992 Improved two-equation k-omega turbulence models for aerodynamic flows Tech. rep. National Aeronautics and Space Administration Ames Research Center

[12] Girimaji S S and Abdol-Hamid K 2005 Partially-averaged navier stokes model for turbulence: Implementation and validation *43rd AIAA Aerospace Sciences Meeting and Exhibit* p 502

[13] Elmiligui A, Abdol-Hamid K, Massey S and Pao S 2004 Numerical study of flow past a circular cylinder using rans, hybrid rans/les and pans formulations *22nd Applied aerodynamics conference and exhibit* p 4959

[14] Davidson L and Friess C 2019 A new formulation of fk for the pans model *J. Turbul.* **20** 322–36

[15] Klapwijk M, Lloyd T and Vaz G 2019 On the accuracy of partially-averaged navier–stokes resolution estimates *Int. J. Heat Fluid Flow* **80** 108484

[16] Park C W and Lee S J 2002 Flow structure around a finite circular cylinder embedded in various atmospheric boundary layers *Fluid Dyn Res* **30** 197

[17] Afgan I, Moulinec C, Prosser R and Laurence D 2007 Large eddy simulation of turbulent flow for wall mounted cantilever cylinders of aspect ratio 6 and 10 *Int. J. Heat Fluid Flow.* **28** 561–74

[18] Krajnović S 2011 Flow around a tall finite cylinder explored by large eddy simulation *J. Fluid Mech.* **676** 294–317

[19] Basara B, Krajnovic S, Girimaji S S and Pavlovic Z 2011 Near-wall formulation of the partially-averaged navier stokes turbulence model *AIAA J.* **49** 2627–36

[20] Cengel Y and Cimbala J 2013 *Ebook: Fluid mechanics fundamentals and applications (si units)* (McGraw Hill)

[21] Clift R, Grace J R and Weber M E 2005 *Bubbles, drops, and particles* (Dover Publications)

[22] Wieselsberger C 1922 New data on the laws of fluid resistance Tech. rep. Aerodynamic Institute, Gottingen

[23] Yaghoubi M and Mahmoodi S 2004 Experimental study of turbulent separated and reattached flow over a finite blunt plate *Exp. Therm. Fluid Sci.* **29** 105–112

[24] Anderson J 2011 *EBOOK: Fundamentals of Aerodynamics (SI units)* (McGraw hill)

IOP Conf. Series: Materials Science and Engineering 1294 (2023) 012003 doi:10.1088/1757-899X/1294/1/012003

An image processing approach to reconstruct wind using long-range wind lidars

M Nafisifard[1,*], J B Jakobsen[1], J T Snæbjörnsson[1,2], H Ágústsson[3], M S Grønsleth[3], and O Undheim[3,4]

[1]Department of Mechanical and Structural Engineering and Material Science, University of Stavanger, Kjell Arholmsgate 41, Stavanger, Norway.
[2]Department of Engineering, Reykjavik University, Menntavegur 1 IS-101, Reykjavík, Iceland.
[3]Kjeller Vindteknikk, Norconsult Group, Norway
[4]Statkraft, Norway.

E-mail: *mohammad.nafisifard@uis.no

Abstract. Lidar-based wind sensing technology, originally used in the wind energy sector, is now being utilized in wind engineering to monitor wind action for designing fjord-crossing infrastructure like long-span bridges. Accurate estimation of design wind loads is crucial for the design process of such bridges. This paper examines wind data from two pairs of long-range lidars positioned along one side of the Sulafjorden, Norway, in a measurement campaign initiated by the Norwegian Public Roads Administration (NPRA). Two different scanning modes, the Plan Position Indicator (PPI) mode and the staring mode, with the fixed line-of-sight (LOS) orientation, are used to determine the line-of-sight wind speeds, the wind direction, and the related along wind speed. An image processing approach is used to compute along wind velocity information from all range gates based on the LOS wind speeds from two near-parallel lidar beams. The estimated along wind velocity is validated through wind data calculated at the intersection point of the two lidars, which provides velocity data with two horizontal components. The image-based reconstruction is found to produce reasonable wind speed estimates with a goodness-of-fit coefficient of $R^2 = 0.903$. The mean wind direction estimate is smoothened but still comparable to the actual wind direction. The image processing approach shows promising potential to provide wind speed and direction information for all range gates for the lidar setup used, which can supplement traditional single-point wind velocity characterization by dual lidars.

1. Introduction

The construction of long-span bridges across wide and deep fjords presents challenging design tasks. Accurate estimation of design wind loads is crucial in the design process, and this requires a thorough understanding of the local wind patterns. Fortunately, advancements in optical remote wind sensing instruments, such as Doppler wind lidars, have made it possible to gain valuable insights into wind patterns in hard-to-reach areas, supplementing conventional mast-based anemometry. Short-range Doppler lidars, capable of measuring winds within a few hundred meters, have been used to map turbulent flow around a suspension bridge [1, 2, 3]. Long-range wind lidars have also been used to access wind conditions relevant to fjord crossings. In 2014, Cheynet et al. [4] conducted a lidar measurement campaign, employing a long-range scanning pulsed coherent lidar positioned 1.75 km to the west of the Lysefjord Bridge. The outcomes of the analysis revealed that lidars have the potential to complement anemometers in the examination of wind effects on large, wind-sensitive structures. In 2017, Cheynet et al. [5] assessed the measurements of three synchronized long-range pulsed wind lidars at a planned bridge site in terms of wind spectrum for the along-wind component and lateral wind co-coherence. Using single pulsed Doppler wind lidar, Cheynet et al. [6] achieved coherence estimates in overall good agreements with IEC and Frøya coherence models [7, 8] from the PPI and the Range Height Indicator (RHI) scan for cross-flow separation from 20 m to 85 m.

In a recent investigation initiated by the NPRA and carried out by the Kjeller Vindteknikk from the Norconsult Group in Norway [9, 10], four long-range Doppler lidars, organized into two pairs, were used to map wind conditions in the Sulafjorden in Western Norway at a designated bridge location. The remote wind observations complement the key data acquired by the sonic anemometers on tall masts. The overall measurement campaign and the related numerical simulations are described in [11]. The resulting wind data plays a pivotal role in understanding the anticipated wind-related loads that will impact the proposed bridge crossing the fjord e.g. [12].

The proof-of-concept study presented in this paper utilizes data from [10]. An image processing approach is deployed to construct the along wind velocity and direction for any desired distance between the lidar and measurement point (also called the range gate centre) along a lidar beam with a fixed orientation, based on the relative comparison with nearly-parallel lidar beams. This approach is complementary and/or alternative approach to the traditional evaluation of 2D horizontal velocity components based on two crossing lidar beams. The approach adopted is based on an image processing technique [13] successfully applied in other fields of science [14].

The paper is organized into several sections. Section 2 provides an overview of the measurement site, including essential details about Sulafjorden and the instrumentation. Section 3 explains the setup of the lidar measurements and the image processing approach used to analyze wind data. Section 4 presents a comprehensive account of wind field characterization using PPI scanning mode within the fjord, along with the temporal evolution of wind direction and velocity fluctuations near the centre of the intended bridge location. Finally, Section 5 concludes the study and provides recommendations for future research.

2. Measurement Site and Monitoring System

Sulafjorden is situated around 10 km southwest of Ålesund. It is a fjord that divides the municipalities of Hareid on the Hareidlandet in the west and Sula in the east (see fig. 1-bottom right). The fjord runs along a south-southeast north-northwest axis and is approximately 12 km long, 3.1 km wide, and has a depth of 300 m. The mountain slopes above the fjord are steep and reach heights of 500 - 600 m above sea level at the bridge-shore connections.

The lidar measurement system consists of 4 scanning lidars, Windcube 400S [15], which are located on the west side of the fjord. Two of these lidars, Dolly and Minni, are situated at Kvitneset, the western end of the planned bridge. The other two lidars, Donald and Mikke, are located further south at Saudebøen, near Brandal. Since the lidars at Kvitneset are near the end of the planned bridge, one of the lidar beams can measure along the bridge span and form lidar pairs with both lidars at Saudebøen. This setup allows three lidars to measure the vector wind in two locations simultaneously, while the fourth lidar can perform independent descriptive scans. All lidars are positioned at an elevation similar to the height of the planned bridge deck at the fjord centre, 80 m above sea level, enabling the lidar beams to remain close to horizontal.

Two distinct scanning modes were employed: PPI (Plan Position Indicator) and LOS (sequential fixed Line-of-Sight measurement). The WINDCUBE 400S concurrently records wind velocities at multiple positions along the scanning beam. When considering an azimuth angle α, an elevation angle θ, and a radial distance r, the lidar is specifically designed to measure the wind velocity component along the beam's trajectory at specific ranges. In this paper, the azimuth angle is defined as the angle between the projection of the line of sight on the local

Figure 1. Long-range lidars at Kvitneset (top) at Saudebøen (bottom-left) the map from www.norgeskart.net (bottom-right), and lidar positions indicated by green dots on.

horizontal plane and the geometric North (varies between 0° and 360°), while the elevation angle is defined as the angle between the line of sight and the local horizontal plane (varies between -90° and 90°). The PPI and LOS scanning modes can be characterized solely by using the azimuth and elevation angles. The PPI scan involves scanning at a consistent elevation while varying azimuth angles, whereas the LOS scan maintains a fixed azimuth and elevation angle. For LOS modes of operation with the WindScanner software, two or more lidars have operated in a synchronized fashion, that is, the master software has ensured that each sampling is done synchronously with a time difference of 50 ms or less. In the LOS mode, the data are sampled at a rate of 1 Hz. In the PPI mode, each 1.5° sector is revisited once every 145 seconds. The key parameters of the scanning patterns are summarized in Table 1.

Table 1. Parameters used for each scanning scenario.

Parameters	PPI	LOS		
	Minni	Donald	Mikke	Dolly
Azimuth (°)	-39 to 170	34.95	36.08	76.67
Elevation (°)	0	0.44	0.28	0.76
Sampling frequency (Hz)	0.0069	1	1	1
Scan speed (°/s)	1.42	-	-	-
Accumulation time (s)	0.70	1	1	1
Approximate duration (min)	60	60	60	60
Range gate length (m)	75	75	75	75
Range gate step (m)	25	25	25	25

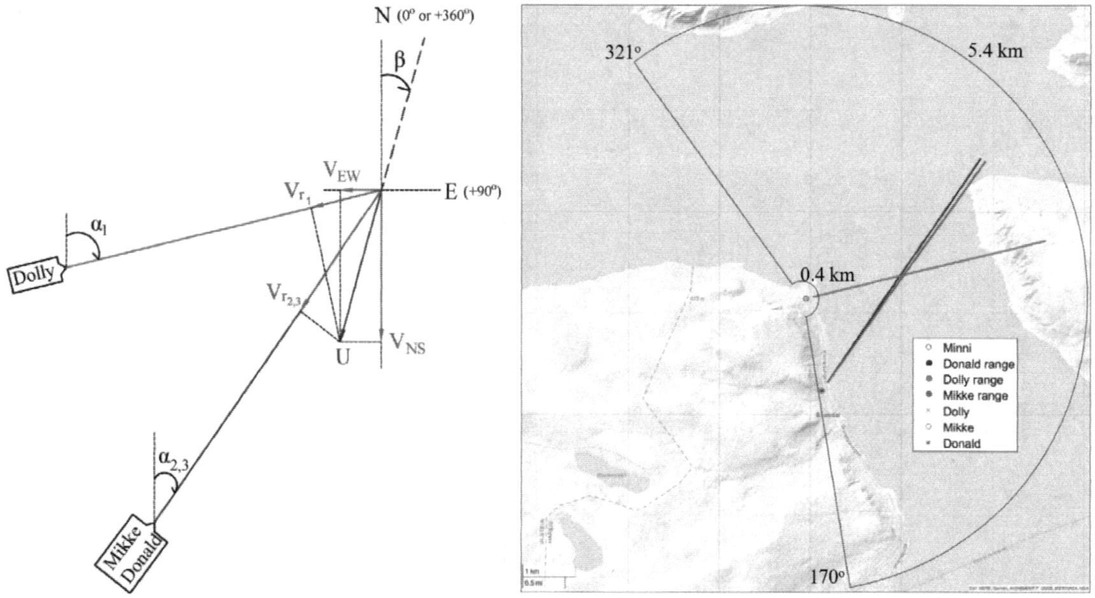

Figure 2. (left) Plan view of the measurement setup for LOS mode. (right) The measurement domain of the lidars.

3. Analysis Methods

3.1. Wind velocity direct formulation

Through the application of synchronized lidar-derived along-beam wind velocity data (v_{r_1} and v_{r_2}), in a horizontal plane (see fig. 2-left), one can ascertain the wind components in any direction in the horizontal plane with the provision that the scanning beams intersect at large enough angles. Through the following mathematical transformations, the components along the geographic North-South and East-West (v_{NS} and v_{EW}) are identified.

$$\begin{bmatrix} v_{EW} \\ v_{NS} \end{bmatrix} = \mathbf{M}^{-1} \begin{bmatrix} v_{r1} \\ v_{r2,3} \end{bmatrix} \tag{1}$$

where

$$\mathbf{M} = \begin{bmatrix} \sin(\alpha_1) & \cos(\alpha_1) \\ \sin(\alpha_{2,3}) & \cos(\alpha_{2,3}) \end{bmatrix} \tag{2}$$

The instantaneous horizontal wind data along the geographic North-South and East-West, can then be processed to retrieve the associated main wind direction and the along wind velocity, denoted by β and U, respectively. To compute β the following transformation is used.

$$\beta(t) = \arctan\left(\frac{v_{EW}(t)}{v_{NS}(t)}\right) \tag{3}$$

The wind velocity U is computed as

$$U(t) = \sqrt{v_{EW}(t)^2 + v_{NS}(t)^2} \tag{4}$$

This information is only available at the two points where the lidar beams from Donald and Mikke cross the lidar beam from Dolly. Therefore to evaluate the wind direction and the along wind velocity for other locations along the scanning path of Donald and Mikke, another approach is required. It is the objective of this paper to demonstrate the possibility of using parametric image alignment for that purpose.

3.2. Parametric Image Alignment

The objective of image alignment is to quantify the distortion between one or multiple input images with regard to a reference image, so-called template image, aiming to minimize a measure of dissimilarity between the template and the warped images. When prior knowledge about the type of difference occurring between the template and input images is available, this problem is commonly termed parametric image alignment [16]. Parametric image alignment can be used for various purposes, with one example being template tracking such as Enhanced Correlation Coefficient Maximization (ECC) [13]. In template tracking tasks, the number of variables involved is typically limited, making it possible to achieve satisfactory results using a general-purpose optimization approach.

The alignment challenge consists of identifying the set of coordinates in the warped image $I_w(\mathbf{y})$, at which the image has the greatest similarity with the template image $I_r(\mathbf{x})$. The coordinates in the region of interest in the template image $\mathbf{x}_k$, $k = 1, ..., K$ are mapped onto the coordinates in the warped image using a warping transformation $\psi(\mathbf{x}; \mathbf{p})$, $\mathbf{y}_k(\mathbf{p}) = \psi(\mathbf{x}_k; \mathbf{p})$, $k = 1, ..., K$. The coordinates $\mathbf{x}_k$ and $\mathbf{y}_k$ are partly shown in fig. 5 (bottom) as the red dots which form the basis of wind mapping images. In the ECC technique, the reference vector $\mathbf{i}_r$ and the associated warped vector $\mathbf{i}_w(\mathbf{p})$ are defined as

$$\mathbf{i}_r = [I_r(\mathbf{x}_1) \quad I_r(\mathbf{x}_2) \quad \cdots \quad I_r(\mathbf{x}_k)]^t \tag{5}$$

$$\mathbf{i}_w(\mathbf{p}) = [I_w(\mathbf{y}_1(\mathbf{p})) \quad I_w(\mathbf{y}_2(\mathbf{p})) \quad \cdots \quad I_w(\mathbf{y}_k(\mathbf{p}))]^t \tag{6}$$

and the following criterion to quantify the performance of the warping transformation is proposed

$$E_{ECC}(\mathbf{p}) = \left\| \frac{\bar{\mathbf{i}}_r}{\|\bar{\mathbf{i}}_r\|} - \frac{\bar{\mathbf{i}}_w(\mathbf{p})}{\|\bar{\mathbf{i}}_w(\mathbf{p})\|} \right\|^2 \tag{7}$$

where $\|\cdot\|$ denotes the Euclidean norm and $\bar{X}$ is the averaging operator. The problem is to find $\mathbf{p}$ where $\mathbf{p} = [p_1, ..., p_N]^t$ is a vector of unknown parameters in such a way that $E_{ECC}(\mathbf{p})$ be minimized. With certain assumptions in place (like $N < K$), this problem can be solved using an iterative procedure. More details of the problem-solving process can be found in [13, 16].

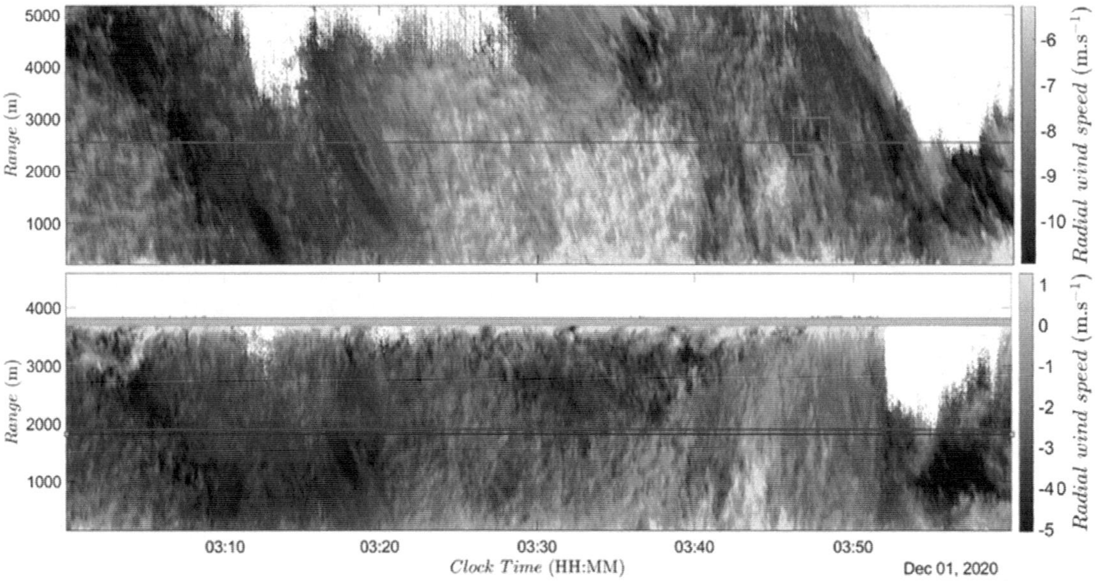

Figure 3. Spatiotemporal fluctuations of wind velocity recorded on 1 December 2020, 03:00 to 04:00 by (top) Donald and (bottom) Dolly. Intersection with other lidar(s) are indicated by colour horizontal lines.

3.3. Data availability

In order to measure the wind, a lidar depends on the existence of aerosols in the air that it probes. If there are no aerosols, there will be no measurements. If there are too many aerosols, for instance, heavy fog, the signal will be attenuated quickly, and the measurement range will be reduced. Therefore, the availability of lidar measurements varies with the atmospheric conditions.

The lidar internally calculates the Carrier-to-Noise Ratio (CNR) which is defined as the Signal-to-Noise Ratio (SNR) on each high-frequency sampling. This quantity depends on the aerosol concentration in the atmosphere. A high atmospheric backscatter coefficient leads to high CNR. CNR level depends on weather conditions. CNR can then be used in the filtering of the raw data. Good quality data typically has a CNR value in the range of -27 to 0 dB [17]. In this study, A constant threshold of -27 dB has been employed to define reliable data, which removes nearly all spurious data values. Subsequently, linear interpolation has been used to fill filtered data in the datasets recorded on December 2020, 03:00 to 04:00, at moderate wind speeds. The data is of good quality and used further in the image processing analysis.

4. Results

4.1. Wind velocities based on lidar data

The lidars deployed offer a suitable spatiotemporal resolution, allowing us to analyse the horizontal wind velocity field up to 5 kilometres from the instruments. An example of line-of-sight velocity captured by Donald and Dolly in a selected 1-hour interval can be seen in fig. 3.

Dolly records values between 1 m/s and -5 m/s along the lidar beam, while the Donald unit measures velocity components from -5 m/s to -11 m/s. Noticing the decreased wind speeds, Dolly observes a transformation in sign along the scanning line, while Donald's laser beam is more aligned with the main direction of the incoming wind. In the present campaign, the positive LOS velocity was adopted away from the lidar, opposite to a convention presented in fig. 2. The negative wind velocities in fig. 3 thus correspond to wind towards the lidar, from the north. Large gusts are observed in the entire monitored domain by both lidars during the first 12 minutes of the 1-hour interval examined.

Beyond 3800 meters, Dolly is seen to produce noisy or missing data (white colour) when the beam strikes the hard target (mountains). Another portion of noisy data is seen at the end of the 1-hour interval in both lidar signals, which may be associated with a reduced amount of aerosols in the air being measured. This is also seen at ranges above 3000 m between 3:10 and 3:30, however, that does not affect the extraction of the 2D horizontal wind components by direct formulation, at the intersection points between the lidar beams from Donald and Mikke with the lidar beam from Dolly.

The direct formulation explained in section 3.1 was used to construct the wind direction and along wind velocity, as reference information for the flow conditions in the middle of the fjord. The intersection point between Donald and Dolly gave a mean along wind velocity of 9.67 m/s with an average orientation of 4.59° from North, while the intersection point between Mikke and Dolly gave 9.66 m/s and 4.97°.

4.2. Mean wind direction based on PPI

Figure 4 shows the averages of the LOS wind velocity derived from PPI scans measured by Minni over 1.5° wide azimuth sectors in the interval studied for the December 2020 event. A wind speed reduction is illustrated at the boundaries of the mountain, especially the mountains in Sula island. Some disturbances are seen in the sectors between 30° and 60° at the 3.4 km and 5.4 km range. These are associated with large values of CNR and high atmospheric backscatter coefficients coming from a hard target located in the northeast. The sector positioned at the angle of 156°, recorded on 1 December 2020, 03:00 to 04:00, has the highest LOS wind velocity. The largest positive wind speeds are for the wind direction away from the lidar, i.e. for the wind direction from 336° (NW) in a common meteorological reference system. This indicates that the wind entering the fjord from NE is accelerated through the fjord "channel" defined by the steep hills and then it adopts the fjord orientation, veering to a NW direction.

4.3. Velocity based on parallel-positioned lidar beams

Having only a 1.13° deviation between the laser beams of Donald and Mikke makes the beams close to parallel and the LOS wind velocities mapped by them nearly identical in terms of wind speed levels. This allows us to compare the regions of interest in the flow mapped by Donald and Mikke to analyse the shift in time and space of the flow pattern between the two observation lines. An example of a shift in flow pattern tested is indicated as the region of interest (ROI) by a red square in fig. 3 (top) for Donald and fig. 5 (top) for Mikke. The contents of the red square in both figures are magnified in the bottom panels of fig. 5. Applying an ECC algorithm [18] on the two ROIs shows that a shift of 6.83 units in time and 2.50 units in space for the image created for Mikke gives the highest correlation with the template image provided by Donald. These shifts are illustrated with red lines in the image for Mikke's ROI. It is worth noting, that the resolution in time is 1 second and in space is 25 meters.

By knowing the amount of movement in space between the template and the shifted images, it is possible to compute the wind direction by comparing the ranges of the two lidars as illustrated in fig. 6, by the purple arrow assuming that the turbulence is frozen between the two beams. Using this analogy, the "instantaneous" wind direction is derived to be equal to 353.5°. Since the defined ROI includes the intersection between Donald and Mikke, the derived wind direction can be compared to the one calculated at the intersection point by the direct formulation explained in 3.1. For the duration of 185 s for which the ROI is defined, the mean wind direction at the intersecting point is calculated equal to 356.5° which is close to the value derived by the image processing approach i.e. ECC algorithm.

The lidar measures the along beam wind speed which has a relation with the main wind speed

IOP Conf. Series: Materials Science and Engineering 1294 (2023) 012003 doi:10.1088/1757-899X/1294/1/012003

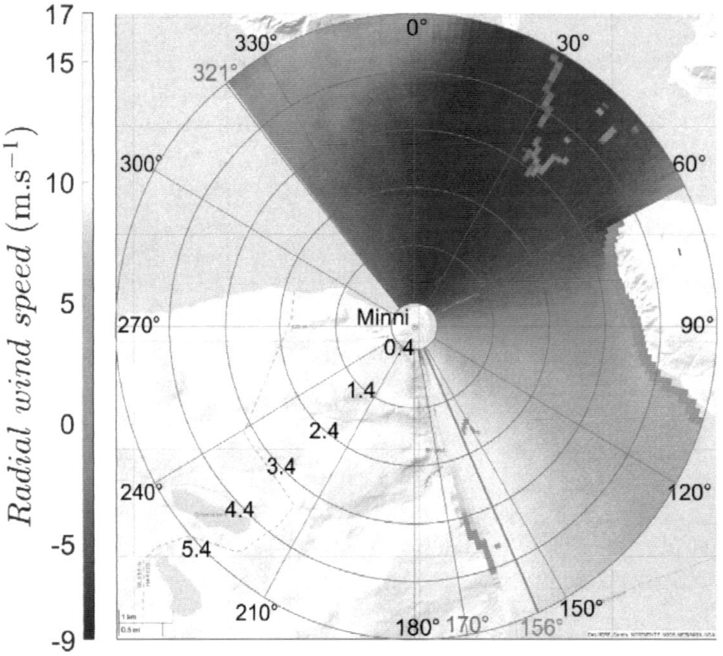

Figure 4. Mean of the along-beam wind velocity derived from PPI scans recorded on 1 December 2020, 03:00 to 04:00. Positive velocity is away from the lidar.

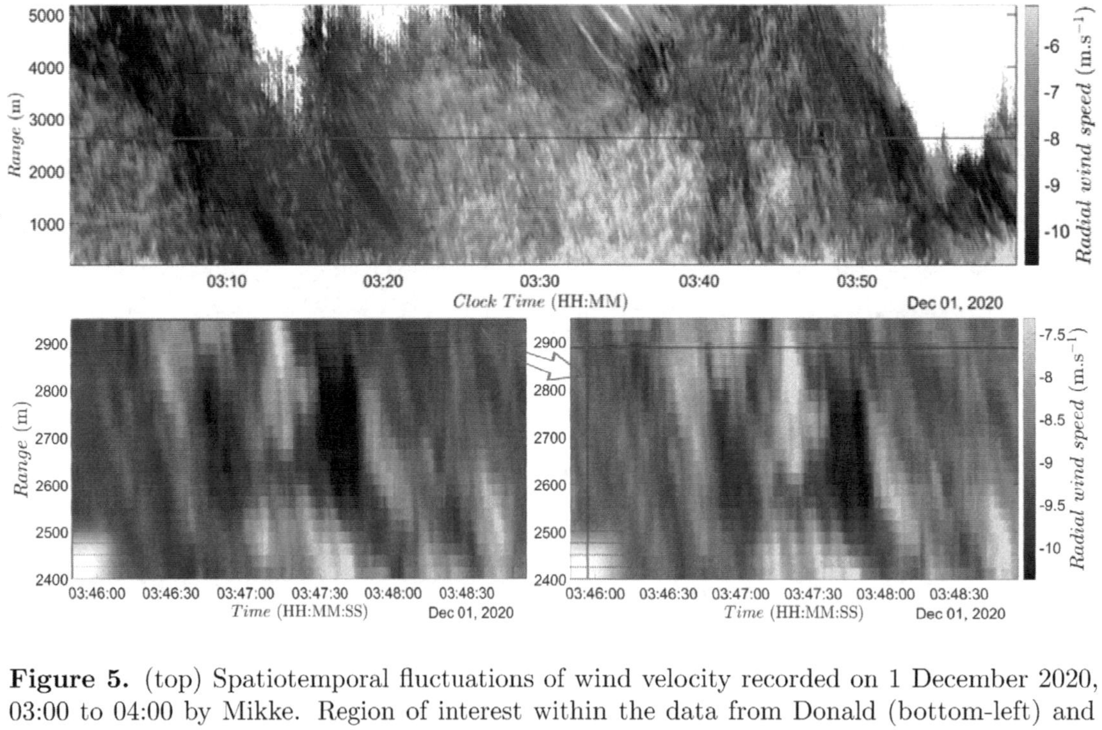

Figure 5. (top) Spatiotemporal fluctuations of wind velocity recorded on 1 December 2020, 03:00 to 04:00 by Mikke. Region of interest within the data from Donald (bottom-left) and Mikke (bottom-right). The red arrow indicates the time and distance shift in the flow pattern between the two lidar beams.

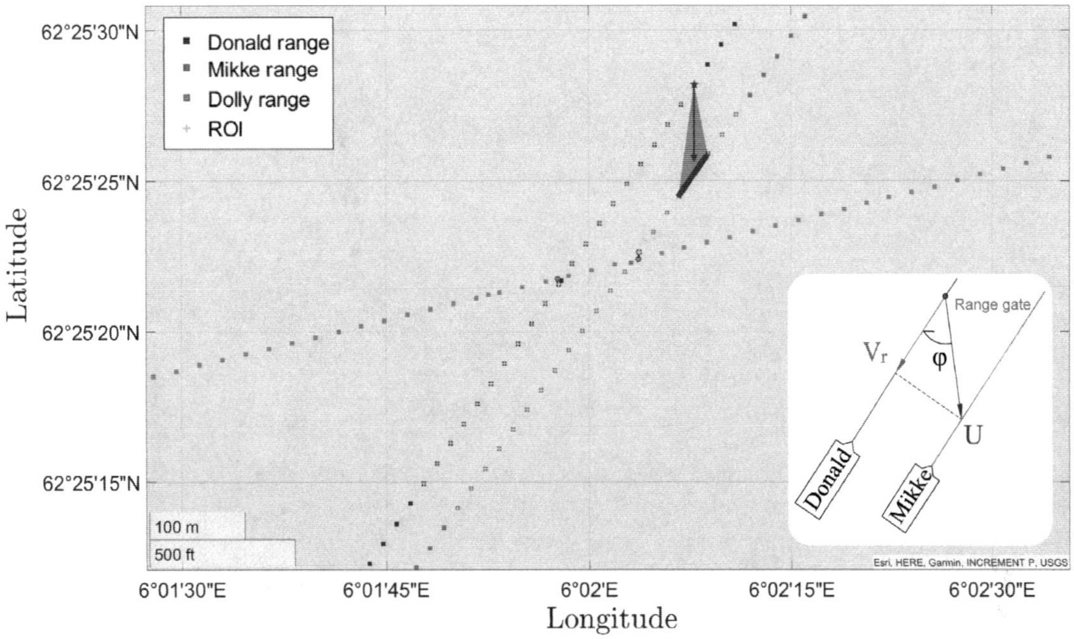

Figure 6. Plan view of the measurement domain of lidars for LOS mode in the middle of the fjord.

as

$$V_r(t) = U(t).\cos(\phi(t)) \tag{8}$$

where $U(t)$ is the along wind speed, and $\phi(t)$ is the angle between the along wind speed and beam orientation, computed by the ECC algorithm as illustrated in fig. 6. Once V_r and ϕ are known, the along wind speed can be easily obtained through Eq. (5). The algorithm has been applied to a strip with a length of 500 meters for the flow mapped by both Donald and Mikke, as indicated by green markers in fig. 6. The purple triangle in fig. 6 shows an envelope of computed $\phi(t)$ in the selected 1-hour interval.

Since the intersection point is embedded in the ROI, a comparison between the time series of wind direction and wind speed can be made between results from the image processing approach and the direct formulation. This is done in fig. 7, where the time series derived by the image processing approach is shown and compared with the time series based on the direct formulation for the intersection point of Donald and Dolly.

Due to the predefined resolution in space between observation points (25 m), the variation of the wind direction captured by the image processing approach is smoothened, compared to the time series at the intersection point, but is still a good estimate of the wind direction in the region being studied. The image processing approach is further seen to accurately reconstruct the wind velocity derived at the intersection point.

The correlation plot of the along wind velocity, U, computed by the two approaches is shown in fig. 8. Although some scattering is visible in the reconstructed wind speed, the regression line resembles $x = y$ quite well and yields a reasonable goodness-of-fit coefficient of $R^2 = 0.903$ for the period considered.

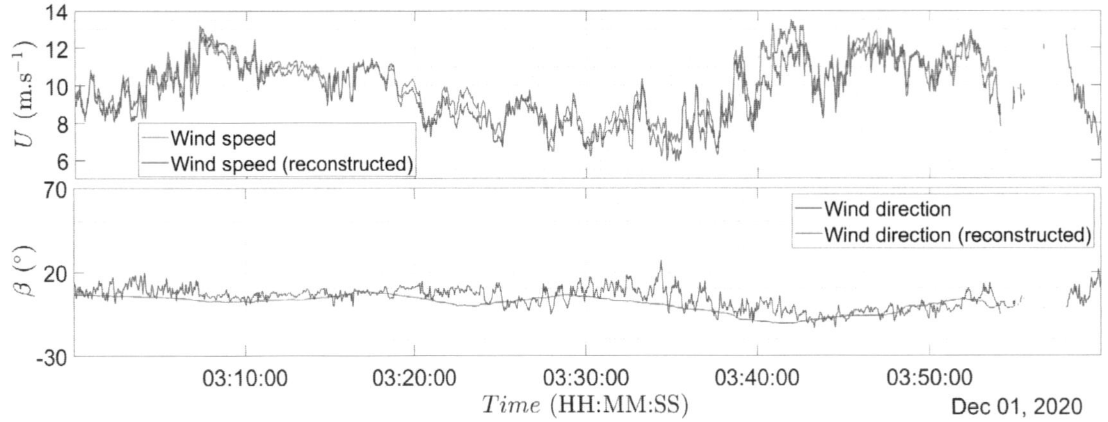

Figure 7. Time-series of the mean wind speed ($U(t)$) and wind direction (β) in the intersection point of Donald and Dolly on 1 December 2020, 03:00 to 04:00.

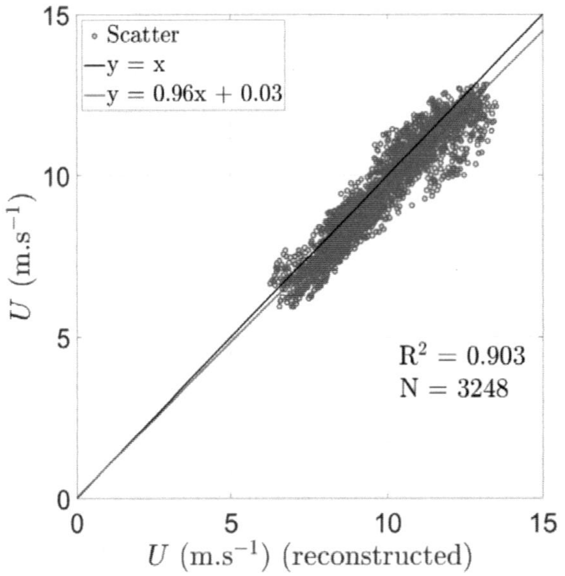

Figure 8. Correlation of the mean wind speed (U) recorded on 1 December 2020, 03:00 to 04:00.

5. Conclusions and Further Works

A system of 2 pairs of long-range lidars was deployed at Sulafjorden in Norway to monitor the horizontal wind field and record wind data to provide a design base for a planned fjord-crossing. One of the lidars was set to operate in PPI scanning mode while three others worked in LOS mode providing two intersection points in the middle of the fjord. The long-range lidar system was able to capture the wind in the fjord, and large gusts within the monitored domain were seen to be captured by all the lidars.

Measurements in PPI mode reveal that the wind entering the fjord from NE is accelerated through the fjord “channel” defined by the steep hills on either side, and that the flow then follows the fjord orientation, veering to an NW direction.

In order to resolve the wind speed and direction in an extended domain around the planned bridge span, the Enhanced Correlation Coefficient Maximization technique was used to evaluate the along wind speed based on image mapping of data from two nearly-parallel lidar beams. A resulting time series from a location close to an intersecting lidar beam was then compared to results provided by a direct approach based on the intersecting LOS velocities. The image-based analysis using the ECC technique showed reasonable performance with a goodness-of-fit coefficient of $R^2 = 0.903$ for the evaluation of the along wind speed. The wind direction estimates are also accurate in the mean, but due to the limited spatiotemporal resolution of the lidar LOS data, the natural fluctuations are somewhat smoothed.

The image processing approach introduced, provides an opportunity to evaluate the wind direction and wind velocity at any range gate observed for the two nearly parallel lidar beams. The approach may however be sensitive to the distance between the two parallel beams and more events need to be studied to evaluate the accuracy of the introduced image processing approach. An improved spatiotemporal resolution of the flow mapped by lidars would enable the estimation of more realistic fluctuations in the wind direction.

Acknowledgments

The authors acknowledge Kjeller Vindteknikk for supplying the data used from the lidar measurement campaign, which was financed by the Norwegian Public Roads Administration (NPRA). The presented data analysis is performed as part of the H2020-MSCA-ITN-2019 project funded by the European Union, grant number 858358.

References

[1] Nafisifard M, Jakobsen J B, Cheynet E, Snæbjörnsson J, Sjöholm M and Mikkelsen T 2021 *IOP Conf. Ser.: Mater. Sci. Eng.* vol 1201 (IOP Publishing) p 012008
[2] Nafisifard M, Jakobsen J B, Snæbjörnsson J, Sjöholm M and Mann J 2023 *J. Wind Eng. Ind. Aerodyn.* **240** 105491
[3] Nafisifard M, Jakobsen J B, Snæbjörnsson J, Sjöholm M and Mann J 2023 *J. Phys. Conf. Ser.* vol 2626 (IOP Publishing) p 012022
[4] Cheynet E, Jakobsen J B, Snæbjörnsson J, Reuder J, Kumer V and Svardal B 2017 *J. Wind Eng. Ind. Aerodyn.* **161** 17–26
[5] Cheynet E, Jakobsen J B, Snæbjörnsson J, Mann J, Courtney M, Lea G and Sverdal B 2017 *Remote Sens.* **9** 1–26
[6] Cheynet E, Jakobsen J B, Sverdal B, Reuder J and Kumer V 2016 *Energy Procedia* 462–477
[7] IEC61400-1 2005 *IEC 61400-1 Wind turbines, Part 1: Design requirements* (Norwegian electrotechnical publication)
[8] Andersen O J and Løvseth J 2006 *Mar. struct.* **19** 173–192
[9] Analysis: Lessons learned from wind lidar measurement campaigns for wind farms, bridges and airports during 2017-2021 https://www.vindteknikk.com/news/analysis-lessons-learned-from-wind-lidar-measurement-campaigns-for-wind-farms-bridges-and-airports-during-2017-2021/ kjeller Vindteknikk, 26 October 2021
[10] Ágústsson H 2022 E39, sulafjorden, møre & romsdal, analysis of lidar derived wind measurements, rap-kvt-lid-110-r1 / kvt/2021/r093/hÁ rev1 Tech. rep. Kjeller Vindteknikk, Norconsult (limited access)
[11] Furevik B R, Ágústsson H, Borg A L, Midjiyawa Z, Nyhammer F and Gausen M 2020 *Earth Syst. Sci. Data* **12** 3621–3640
[12] The Norwegian Public Roads Administration presented preliminary projects for possible bridge concepts for the sulafjorden https://www.vegvesen.no/vegprosjekter/europaveg/e39sulafjorden/nyhetsarkiv/statens-vegvesen-presenterte-forprosjekter-for-mulige-brukonsepter-for-sulafjorden/ statens Vegvesen, 10 June 2022
[13] Evangelidis G D and Psarakis E Z 2008 *IEEE Trans. Pattern Anal. Mach. Intell.* **30** 1858–1865
[14] Chizari A, Schaap M J, Knop T, Boink Y E, Seyger M M B and Steenbergen W 2021 *Sci. Rep.*
[15] Boquet M, Royer P, Cariou J P and Machta M 2016 *J. Atmos. Oceanic Technol.* **33** 977–987
[16] Alismail H 2016 *Direct Pose Estimation and Refinement (PhD thesis)* Ph.D. thesis Carnegie Mellon University Pittsburgh
[17] Kumar V M, Reuder J and Furevik B R 2014 *Energy Procedia* 214–220
[18] ECC image alignment algorithm (image registration) https://www.mathworks.com/matlabcentral/fileexchange/27253-ecc-image-alignment-algorithm-image-registration georgios Evangelidis, 4 October 2023

IOP Conf. Series: Materials Science and Engineering 1294 (2023) 012004 doi:10.1088/1757-899X/1294/1/012004

A novel design of a hybrid glulam-steel substructure for the IEA 15-MW floating wind turbine

H H Yousef, Y Ma*, K S Patel, and Y Xing

Faculty of Science and Technology, University of Stavanger, Norway

* Correspondence: yucong.ma@uis.no

Abstract. Wind energy has become increasingly recognised as a very promising type of renewable energy. In addition, floating offshore wind turbines have facilitated the development of electricity production in intermediate (45-150 m) and deep sea (>150 m) depths. Despite this, wind turbine manufacturing, installation, and operation may generate substantial greenhouse gas emissions. A novel hybrid glulam-steel floating substructure design is presented in this research, intended for the IEA 15 MW floating wind turbine. The objective is to contribute to advancing floating wind energy while minimising costs and carbon dioxide emissions. The objective of the novel design is to substitute steel with glued laminated lumber (glulam). It showcases an altered iteration of the UMaine VolturnUS-S semi-submersible platform originally created for the IEA 15 MW turbine. Before selecting one of three preliminary hybrid wood-steel models, the Ansys Workbench 2020 R1 is used to evaluate and appraise them per a set of criteria extracted from pertinent timber and steel standards. The chosen hybrid design conserves about 590 metric tonnes of steel mass compared to the UMaine VolturnUS-S semi-submersible platform. The selected model is then validated by executing a fully coupled aero-hydro-servo-elastic dynamic analysis with OpenFAST. Consideration is limited to the ultimate limit state design (ULS) for normal and severe operating situations. The utilisation factor of the glulam supporting structure for the IEA 15 MW turbine ranges from 74% to 94%, indicating that it is an effective load-bearing solution.

1. Introduction

The achievement of the net-zero emission objective by 2050 might be facilitated by the development of wind turbines with less CO_2 footprint [1]. The majority of offshore wind turbines are currently bottom-fixed monopile turbines. However, 80% of offshore wind potential is situated at depths beyond 60 m, where bottom fixed wind turbines are not cost-effective [2]. In contrast, floating offshore wind turbines (FOWTs) provide a significant benefit in such deep seas. As the International Electrotechnical Commission (IEC) [3] specified, floating wind turbines are designed to endure harsh offshore environmental conditions for at least two decades. Expanding offshore wind farms, particularly using floating wind turbines in connection with the global transition to renewable energy sources, requires more research and development in engineering, procurement, maintenance, and CO_2 emissions.

The findings of [4] indicate that the materials used in the construction and maintenance of a 6 MW spar buoy wind turbine and a 6 MW raft-buoy wind turbine have a substantial impact on the Carbon Intensity of Electricity (CIE), which varies between 26.1 and 78.7 CO_{2eq}/kWh. The aforementioned research underscores the need to develop novel technologies that reduce the overall steel mass to minimise the turbine's CO_2 footprint since steel production usually generates significant amounts of CO_2.

The strong, corrosion-resistant, stable, and environmentally friendly glued laminated wood (glulam) is superior to steel in many ways, including cost, strength-to-weight ratio, and prefabrication possibilities. With centuries of experience in the construction industry, we now have a greater understanding of efficient joint selection, constraints, and glulam construction procedures [5]. The glulam-only Mjøstårnet 18-story structure, which was finished in 2019, is the tallest wooden structure in the world [6]. In Bjorko, Sweden, Modvion constructed a 30-meter-tall timber tower in 2020 [7]. Stora Enso, a biomaterials and wood construction company, and Voodin Blade Technology GmbH, a German business specialising in the production of rotor blades for wind turbines, have just initiated a collaborative effort to substitute heavier, non-renewable wind turbine blades with wooden ones.

This study aims to propose a novel hybrid glulam-steel substructure design for the semi-submersible IEA 15 MW wind turbine (refer to Figure 1). The IEA 15 MW was chosen as the reference wind turbine because it corresponds to the currently largest state-of-the-art wind turbines in the market [8]. Three distinct hybrid models are initially constructed and evaluated using the ANSYS Workbench 2020 R1. Then, these three models were compared to identify the optimal one that maximises material mass while maintaining acceptable utilisation factors. A fully coupled aero-hydro-servo-elastic dynamic study is then conducted with OpenFAST to verify the identified optimal model. The design investigation is limited to the ultimate limit state (ULS) design of the turbine, which includes both severe and ordinary operating conditions.

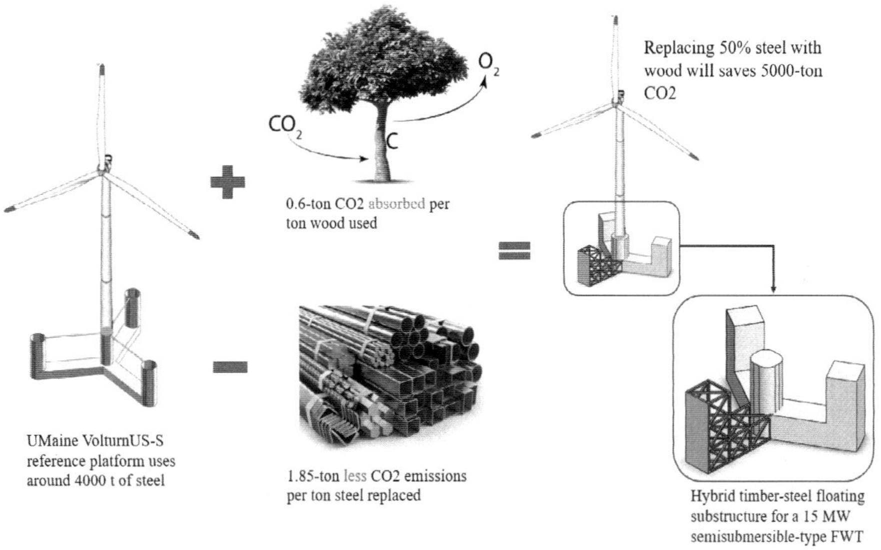

Figure 1. The environmental impact of using wood as a replacement for steel

2. System Description

This study employs the IEA 15-MW floating wind turbine (FWT) [9]. The FWT system will be described in the following sections in two parts. Following a description of the reference wind turbine, the characteristics of the first UMaine VolturnUS-S semi-submersible platform and the mooring system will be discussed.

2.1. IEA Wind 15-Megawatt reference FWT

The reference wind turbine (RWT) utilised in this study has a capacity of 15 MW. It was constructed following the International Electrotechnical Commission's (IEC) Class 1B wind regime and consists of three blades rotating clockwise upwind. Additionally, it is outfitted with a collective pitch and variable speed control system. Table 1 provides an overview of the IEA Wind 15-MW RWT.

Table 1. General parameters of IEA Wind 15-MW RWT [8,9]

Parameter	Value	Unit
Power rating	15	MW
Rotor orientation, configuration	Upwind, 3 blades	-
Control	Variable speed, collective pitch	-
Drivetrain	Low-speed, direct drive	
Cut-in, rated, cut-out wind speed	3, 10.59, 25	m/s
Rotor, hub diameter	240, 7.94	m
Hub height	150	m
Design tip-speed ratio	9	-
Minimum rotor speed	5	rpm
Maximum rotor speed	7.56	rpm
Maximum tip speed	95	m/s
Water depth	200	m
Total system mass	20,093	t
Platform mass	17,839	t
Rotor nacelle assembly mass	991	t
Tower mass	1,263	t
Tower base diameter	10	m

2.2. UMaine VolturnUS-S reference semi-submersible platform

The UMaine VolturnUS-S semi-submersible floating sub-structure design will serve as the basis for the novel hybrid design proposed in this study. The floater has three outer columns arranged in a circle, with a fourth column in the centre. A star-shaped pontoon with a triangular cross-section at its apex supports the columns. Three radial struts with a diameter of 0.9 metres are used to connect the columns at the uppermost point. The floater's location is maintained by the use of three catenary mooring lines (refer to Figure 2 (b)). Table 2 provides further information on the UMaine VolturnUS-S semi-floater, while Table 3 describes the mooring system in greater detail.

Table 2. Semi-submersible Platform Properties [9]

Parameter	Value	Unit
Hull displacement	20,206	m^3
Hull steel mass	3,914	t
Tower interface mass	100	t
Ballast mass (fixed/fluid)	2,540/11,300	t
Draft	20	m
Freeboard	15	m
Vertical center of gravity from SWL	-14.94	m
Vertical center of buoyancy from SWL	-13.63	m
Roll inertia about the center of gravity	1.251×10^{10}	$kg{\cdot}m^2$
Pitch inertia about the center of gravity	1.251×10^{10}	$kg{\cdot}m^2$
Yaw inertia about the center of gravity	2.367×10^{10}	$kg{\cdot}m^2$

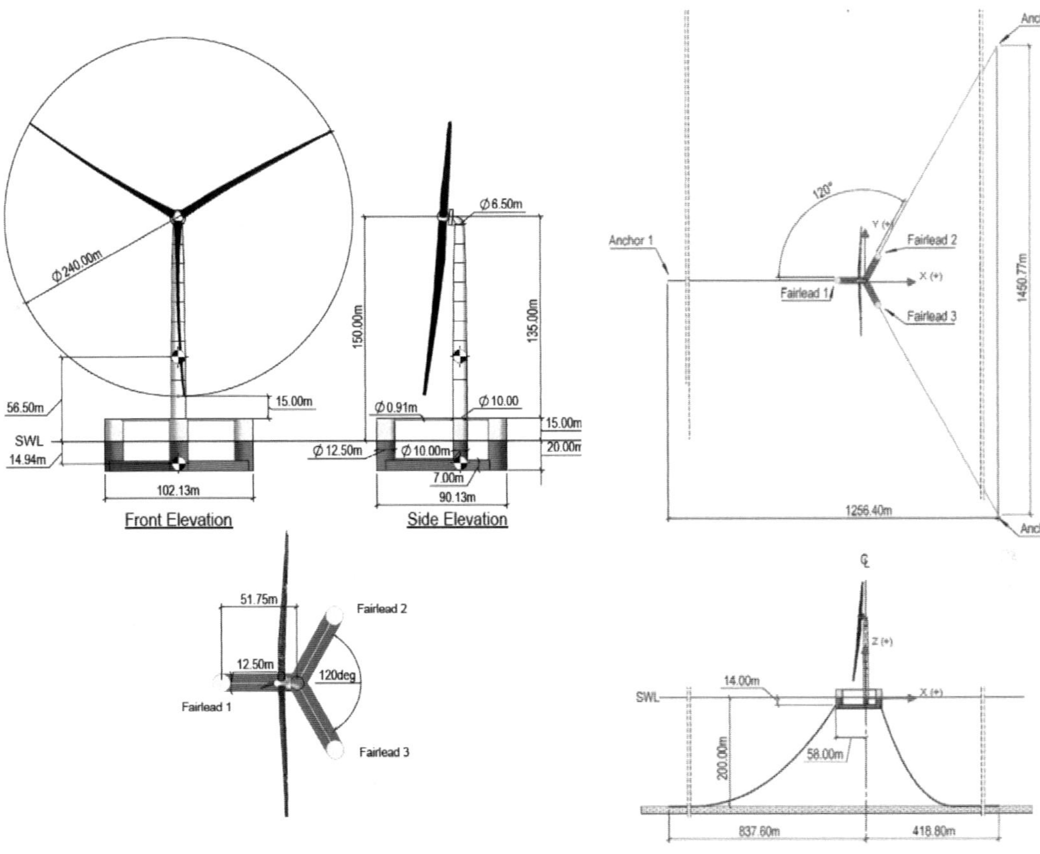

(a) (b)

Figure 2. (a) Main dimensions of the UMaine VolturnUS-S floater of the 15-MW wind turbine; (b) Sketch of the mooring system in the 15-MW FWT

Table 3. Mooring system's properties [9]

Parameter	Value	Unit
System type & number of lines	Chain Catenary & 3 Lines	-
Line type	Studless R3 Chain	-
Line breaking strength	22.286	kN
Fairlead depth	14	m
Dry line linear density	685	kg/m
Extensional stiffness	3270	MN
Line unstretched length	850	m
Fairlead pretension	2,437	kN
Fairlead angle from SWL	56.4	deg

3. Finite Element Analysis of Three Concept Configurations

As indicated in Figure 3, three configurations are recommended for this study. Figure 3 depicts the configuration of the glulam supporting system for a single pontoon and the middle region connecting all three pontoons. The outside dimensions of each configuration are similar to those of the UMaine platform. 0.05 m steel plates are used as the hull plates that are in contact with the water and are attached to the outer of the glulam structures. No glulam structures will be in contact with the water. Additional information about the design details may be found in Appendix A [10] of Yousef (2023).

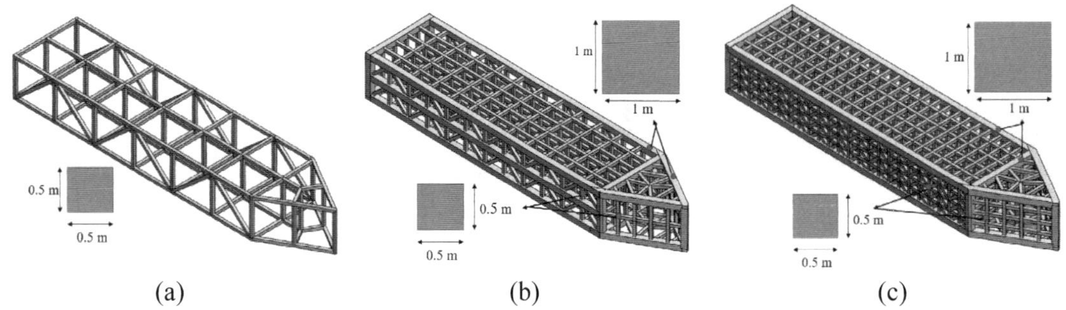

Figure 3. Glulam-based supporting structures for the three proposed configurations

3.1. Load and boundary conditions

To simplify the design procedure, only the maximum aerodynamic load is taken into account at the specified rated wind velocity, together with the corresponding hydrostatic pressure. Further details on loads and boundary conditions are available in [10].

3.2. Material properties

Table 4 presents the material properties used.

Table 4. Material properties of the plate's steel and glulam Gl30h assigned for the beams [11]

Property	Glulam G30h for beams	Steel for plates	Unit
Density	480	7850	kg/m^3
Young's Modulus X(L) direction	1.36×10^{10}		Pa
Young's Modulus Y(R) direction	3×10^8	2×10^{11}	Pa
Young's Modulus Z(T) direction	3×10^8		Pa
Poisson's Ratio XY	0.21		-
Poisson's Ratio YZ	0.21	0.3	-
Poisson's Ratio XZ	0.24		-
Shear Modulus XY	6.5×10^8		Pa
Shear Modulus YZ	6.5×10^8	7.692×10^{11}	Pa
Shear Modulus XZ	6.5×10^8		Pa
Tensile Yield Strength	2.4×10^7	2.5×10^8	Pa
Compressive Yield Strength	3×10^8	2.5×10^8	Pa

3.3. Mesh element type

Quadrilateral element types and more details on the element type are provided in Table 5.

Table 5. Element type details

Part	Element name IDs	Element shape
Plates	SHELL181	QUAD4
Beams	BEAM188	BEAM3

3.4. Design criteria

The following criteria were used for the design:

- Glulam under combined bending and axial tension criteria [12]

Criteria 1
$$\frac{\sigma_{t,0,d}}{f_{t,0,d}} + \frac{\sigma_{m,y,d}}{f_{m,y,d}} + k_m \cdot \frac{\sigma_{m,z,d}}{f_{m,z,d}} \leq 1 \qquad (1)$$

Criteria 2
$$\frac{\sigma_{t,0,d}}{f_{t,0,d}} + k_m \cdot \frac{\sigma_{m,y,d}}{f_{m,y,d}} + \frac{\sigma_{m,z,d}}{f_{m,z,d}} \leq 1 \qquad (2)$$

$\sigma_{t,0,d}$ is the design tensile stress along the grain, $\sigma_{m,y,d}$ and $\sigma_{m,z,d}$ are the design bending stresses around the y and z axes, $f_{t,0,d}$ is the design tensile stress along the grain, $f_{m,y,d}$, and $f_{m,z,d}$ are the corresponding design bending strengths, and k_m is a factor that makes allowance for the re-distribution of stresses and the effect of inhomogeneities of the material, and is 0.7 for glulam with a rectangular cross-section.

- Glulam under combined bending and axial compression criteria [12]

Criteria 3
$$\left(\frac{\sigma_{c,0,d}}{f_{c,0,d}}\right)^2 + \frac{\sigma_{m,y,d}}{f_{m,y,d}} + k_m \cdot \frac{\sigma_{m,z,d}}{f_{m,z,d}} \leq 1 \qquad (3)$$

Criteria 4
$$\left(\frac{\sigma_{c,0,d}}{f_{c,0,d}}\right)^2 + k_m \cdot \frac{\sigma_{m,y,d}}{f_{m,y,d}} + \frac{\sigma_{m,z,d}}{f_{m,z,d}} \leq 1 \qquad (4)$$

where $\sigma_{c,0,d}$ is the design compressive stress along the grain, $f_{c,0,d}$ is the design's compressive strength along the grain.

- Glulam beam stability criteria [12]

Criteria 5
$$\left(\frac{\sigma_{m,d}}{k_{crit} \cdot f_{m,d}}\right)^2 + \frac{\sigma_{c,0,d}}{k_{c,z} \cdot f_{c,0,d}} \leq 1 \qquad (5)$$

where $\sigma_{m,d}$ is the design bending stress, k_{crit} is a factor that considers the reduced bending strength due to lateral buckling, and $k_{c,z}$ is an instability factor. The values of k_{crit} and $k_{c,z}$ can be calculated based on equations given in EN 1995-1-1 standard [12].

- Steel plates yielding criteria [13]

Criteria 6
$$\gamma_m \frac{\sigma_{j,d}}{R_k} \leq 1 \qquad (6)$$

$\sigma_{j,d}$ is von Mises equivalent design stress, R_k is the characteristic strength and $\gamma_m = 1.15$ taken for plated structures.

The utilisation factor for configuration (b) as per the first criterion is shown in Figure 4. A comparison of the maximum utilisation factors for each configuration, as determined by the aforementioned criteria, is presented in Table 6. The results indicate that design (b) minimises glulam mass while maintaining an acceptable utilisation factor. Consequently, further investigations are only made on design (b).

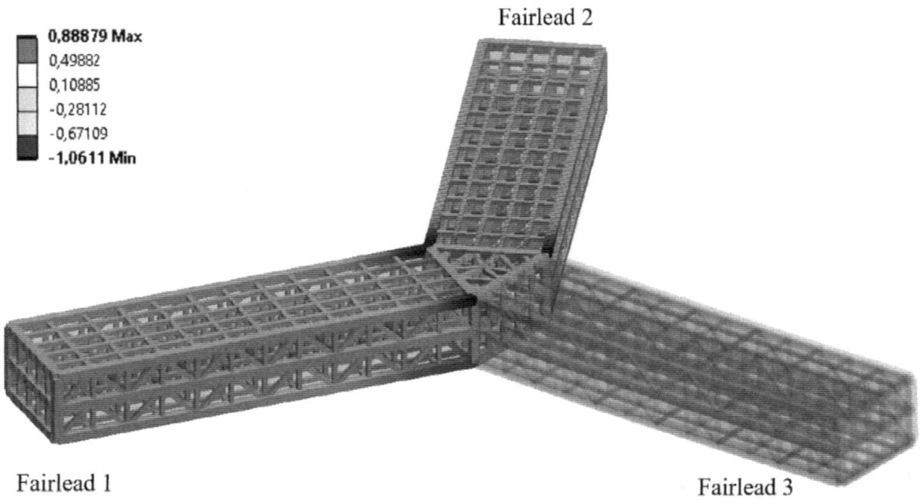

Figure 4. Utilisation factor corresponding to the 1st combined bending and tension criteria for glulam, configuration (b)

Table 6. Max utilisation factor for three hybrid configurations

Material	Design criteria	Config. (a)	Config. (b)	Config. (c)
Glulam	$UF_{max,\ 1}$	1.03	0.75	0.76
	$UF_{max,\ 2}$	1.04	0.80	0.82
	$UF_{max,\ 3}$	5.63	0.90	0.94
	$UF_{max,\ 4}$	5.36	0.90	0.94
	$UF_{max,\ 5}$	1.13	0.89	0.88
Steel	$UF_{max,\ 6}$	3.60	0.91	0.83

4. Hydro-Servo-Aero-Elastic analysis using OpenFAST

This study uses OpenFAST, an open-source modelling programme developed by the National Renewable Energy Laboratory (NREL), to analyse the 15-MW FWT in a fully coupled aero-hydro-elastic-servo dynamic manner. AeroDyn [14], HydroDyn [15], ServoDyn, ElastoDyn, TurbSim, InflowWind, and MoorDyn [16] are some of the software modules that the OpenFAST code integrates to simulate mooring system dynamics, control dynamics, hydrodynamic loads on rotor blades, and hydrodynamic loads on floaters.

Yousef [10] states that the hybrid floater exhibits the same CoB, CoG, mass, and moments of inertia as the UMaine VolturnUS-S platform, except for a substantial reduction in the mass of the steel and concrete components. Furthermore, the hybrid floater's outward dimensions and shape are also similiar to those of the UMaine VolturnUS-S platform. The aforementioned conditions provide automatic fulfilment of the stability and eigenfrequency criteria. Furthermore, the hydrodynamic characteristics pre-established for the UMaine VolturnUS-S platform [9] may be used immediately in the hydrodynamic analysis of the hybrid floater configuration (b).

Design load cases (DLC)

In order to assess the performance of the hybrid floater, a subset of IEC design load instances pertaining to the East Coast of the United States is used [17, 18]. See Table 7. All simulations were executed for 720 seconds under the assumption that the wind and waves are aligned at an angle of 0 degrees.

Table 7. IEC Design load cases

	DLC number	Wind condition	U_{hub} (m/s)	H_s (m)	T_p (sec)	Gamma shape factor
1.1	1	NTM	4.00	1.10	8.52	1.00
	2		24.00	4.52	9.45	1.89
1.3	3	ETM	4.00	1.10	8.52	1.00
	4		24.00	4.52	9.45	1.89
1.6	5	NTM	4.00	6.30	11.50	2.75
	6		24.00	9.80	14.10	2.75

where ETM is the Extreme turbulence model and NTM is the Normal turbulence model.

Using OpenFAST, all the total time-varying loads on the tower base and mooring line tension, such as TwrBsFxt, TwrBsFyt, TwrBsFzt, TwrBsMxt, TwrBsMyt, TwrBsMzt, FAIRTEN1, FAIRTEN2 are obtained.

5. Local Analysis

5.1. FEA Model

The loads obtained from OpenFAST are used to re-analyse model (b) (see Figure 5).

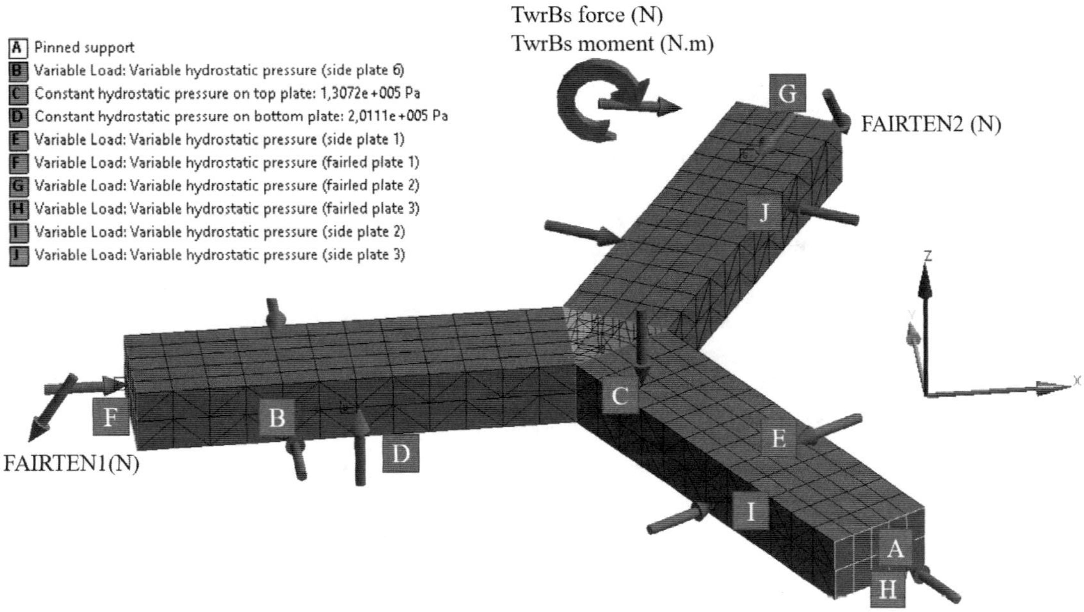

Figure 5. Actual loads acting on the hybrid design, configuration (b) (ANSYS)

5.2. *Response Surface Methodology (RSM)*

Utilising mathematical regression, response surface methodology (RSM) is a statistical technique that may be used to establish the relationships between several inputs and one or more outputs. Using a response surface minimises computing time by eliminating the need to execute thousands of simulations that account for various load combinations (obtained at each time step Δt over the simulation duration).

5.2.1. *Design of experiment (DoE)*

There are several methods for generating experimental design points in ANSYS Workbench, including the Box-Behnken design, central composite design, and optimum space-filling design, among others. In this study, the core composite design is used. 285 design points are generated in accordance with the higher and lower limits specified in Table 8.

Table 8. Upper and lower bounds assigned for input load variables

	TwrBsFxt	TwrBsFyt	TwrBsFzt	TwrBsMxt
Lower bound	-3×10^6	-2×10^6	-3×10^7	-8×10^7
Upper bound	9×10^6	2×10^6	3×10^7	2×10^8
	TwrBsMyt	TwrBsMzt	FAIRTEN1	FAIRTEN2
Lower bound	-4×10^8	-4×10^7	1×10^6	1×10^6
Upper bound	7×10^8	4×10^7	6×10^6	3×10^6

5.2.2. *Construction of response surface*

The response surface may be constructed using ANSYS Workbench in various methods, such as Kriging, traditional second-order regression, non-parametric regression, and more. This study generates the response surface by generic aggregation, which employs a genetic algorithm to solve several response surfaces concurrently while taking into account the response surface's correctness and stability at the design stage. The mathematical formulation of the generic aggregation strategy is given in Equation (7).

$$\hat{y}_{ens}(x) = \sum_{i=1}^{N_m} w_i \cdot \hat{y}_i(x) \tag{7}$$

where $\hat{y}_{ens}$ is the ensemble prediction, N_m is the total number of metamodels used and $\hat{y}_i$, w_i are the prediction and weight factor of the $i-th$ response surface.

The results derived from the ANSYS workbench's goodness of fit option indicate that the response surface produced satisfies the investigated model well (refer to Table 9).

Table 9. Goodness of fit results

	$UF_{max,\ 1}$	$UF_{max,\ 2}$	$UF_{max,\ 3}$	$UF_{max,\ 4}$	$UF_{max,\ 5}$	$UF_{max,\ 6}$
	Coefficient of Determination (best value=1)					
Learning Points	1	0.999	0.999	1	0.999	1
Cross-Validation on Learning Points	0.979	0.990	0.946	0.907	0.957	0.948
	Maximum Relative Residual (best value=0%)					
Learning Points	0	0.208	0.055	0	0.183	0
Verification Points	3.221	1.755	7.290	6.449	3.183	1.564

Cross-Validation on Learning Points	2.726	2.185	7.172	8.251	4.040	1.818
	Root Mean Square Error (best value=0)					
Learning Points	0	0	0	0	0	0
Verification Points	0.017	0.009	0.027	0.032	0.014	0.006
Cross-Validation on Learning Points	0.003	0.003	0.009	0.01	0.005	0.002
	Relative Root Mean Square Error (best value=0%)					
Learning Points	0	0.048	0.04	0	0.059	0
Verification Points	2.164	1.188	3.907	4.253	1.667	0.797
Cross-Validation on Learning Points	0.397	0.41	1.122	1.201	0.517	0.239

5.3. *Parameter correlation study*

To minimise the overall processing time, the correlation analysis seeks to identify the inputs that have the most significant impact on the output of interest. This allows for identifying worst-case load combinations with fewer inputs taken into account at their maximum and lowest levels.

The correlation coefficients between the load input factors and the output stress variables vary substantially following the direction of each individual load, according to the findings of the correlation research [10]. Nevertheless, the research emphasised that some inputs—TwrBsFxt, TwrBsFzt, TwrBsMyt, FAIRTEN1, FAIRTEN1, and FAIRTEN2—have a substantial impact on the resultant output and should be taken into account when determining the worst-case load combinations [10].

5.4. *Results*

Table 10 presents the highest utilisation factors for each design load situation, which are determined by the worst-case load combinations. Each combination of worst-case loads represents the most significant or most negligible value of one of the input load variables. The six utilisation factors for each of the specified load combinations are then retrieved using the response surface. The utilisation of glulam beams increases by as much as 78% when subjected to combined bending and compression and by as much as 85% when subjected to combined bending and tension. Nevertheless, the beams maintain 94% of their initial capability despite resisting buckling.

Table 10. The maximum utilisation factors expected for each of the design loads

DLC number	Wind condition	Glued laminated timber					Steel
		$UF_{max,\ 1}$	$UF_{max,\ 2}$	$UF_{max,\ 3}$	$UF_{max,\ 4}$	$UF_{max,\ 5}$	$UF_{max,\ 6}$
1	NTM	0.7679	0.8175	0.7417	0.7548	0.9085	0.8153
2		0.7908	0.8406	0.7793	0.7763	0.9339	0.8243
3	ETM	0.7776	0.8280	0.7581	0.7633	0.9196	0.8194
4		0.7914	0.8414	0.7805	0.7767	0.9349	0.8246
5	NTM	0.7834	0.8331	0.7667	0.7690	0.9253	0.8213
6		0.7943	0.8436	0.7834	0.7804	0.9372	0.8256

6. Conclusion

A novel hybrid timber-steel floating substructure for a 15 MW semi-submersible-type FWT is described in this article. An initial design study was carried out by simulating and comparing the three finite element models using ANSYS Workbench 2020 R1. The findings indicate that configuration (b) provides satisfactory utilisation factors while maintaining a minimum glulam mass (cost). In addition, the chosen hybrid configuration reduces steel mass by about 590 metric tonnes as compared to the UMaine VolturnUS-S semi-submersible platform. The hydrodynamic data supplied for the UMaine VolturnUS-S semi-submersible platform [9] were used in the hybrid model to conduct a completely nonlinear aero-hydro-servo-elastic simulation, taking into consideration the similarity in shape, mass properties, CoG, and CoB. A re-analysis was conducted on Model (b) using the loads generated by OpenFAST. Following this, the response surface approach is used to reduce the total computation time. In order to identify the input combinations with the highest utilisation factors (outputs) under the worst-case load conditions, parameter correlation analysis was undertaken. The results of the parameter correlation analysis indicated that the resultant utilisation factors exhibited a significant and variable association with TwrBsFxt, TwrBsFzt, TwrBsMyt, FAIRTEN1, FAIRTEN2, and FAIRTEN1. The conclusive findings indicate that glulam may serve as a viable substitute for structural steel in the context of IEA 15 MW. The utilisation factor for the various criteria fluctuates between 0.74 and 0.94, including both typical and severe operating situations. The lack of variation in the utilisation factor values among the several DLCs underscores the criticality of hydrostatic pressure as a load that drives the design.

Reference

[1] IEA 2020, *World energy outlook 2020.* (OECD Publishing).

[2] WindEurope 2018, *Floating offshore wind energy-a policy blueprint for europe floating offshore wind energy.*

[3] Igba J, Alemzadeh K, Durugbo C and Henningsen K 2015, "Performance assessment of wind turbine gearboxes using in-service data: Current approaches and future trends," *Renewable Sustainable Energy Rev.,* **50**, 144-159.

[4] Pulselli R, Maccanti M, Bruno M, Sabbetta A, Neri E, Patrizi N and Bastianoni S 2023, Benchmarking Marine Energy Technologies Through LCA: Offshore Floating Wind Farms in the Mediterranean. Front, *Perspectives for marine energy in the mediterranean area volume II,* **16648714**, p 18.

[5] Porteous J and Kermani A 2013, *Structural timber design to Eurocode 5* (John Wiley & Sons).

[6] Abrahamsen R, Mjøstårnet-Construction of an 81 m tall timber building. Available at: https://www.moelven.com/globalassets/moelven-limtre/mjostarnet/mjostarnet---construction-of-an-81-m-tall-timber-building_presentation.pdf (Accessed: 20.10.2023)

[7] Modvion 2020, The first wooden wind power tower has been erected in Sweden. Available at: https://modvion.com/news/the-first-wooden-wind-power-tower-has-been-erected-in-sweden/ (accessed: 20.10.2023)

[8] Gaertner E, Rinker J, Sethuraman L, Zahle F, Anderson B, Barter G E, Abbas N J, Meng F, Bortolotti P and Skrzypinski W 2020, *IEA wind TCP task 37: definition of the IEA 15-megawatt offshore reference wind turbine*, National Renewable Energy Lab.(NREL), Golden, CO (United States).

[9] Allen C, Viscelli A, Dagher H, Goupee A, Gaertner E, Abbas N, Hall M, and Barter G 2020, *Definition of the UMaine VolturnUS-S reference platform developed for the IEA Wind 15-megawatt offshore reference wind turbine*, National Renewable Energy Lab.(NREL), Golden, CO (United States).

[10] Yousef H H 2023, *Design and analysis of a hybrid timber-steel floating substructure for a 15 MW semi-submersible-type FWT*, (Master Thesis: University of Stavanger).

[11] European Committee for Standardization CEN 2013, *NS-EN 14080: 2013 NA: 2016 Timber structures-Glued laminated timber and glued solid timber Requirements.*

[12] European Committee for Standardization CEN 2006, *Eurocode 5: Design of timber structures - Part 1-1: General -Common rules and rules for buildings.*

[13] Det Norske Veritas (DNV) 2011, *DNV-OS-C101: Design of Offshore Steel Structures General (LRFD Method).*

[14] Moriarty P J and Hansen A C 2005, *AeroDyn Theory Manual,* NREL/TP-500-36881 (National

Renewable Energy Lab.: Golden).
[15] National Renewable Energy Laboratory (NREL) 2023. *OpenFAST Documentation*, 12.04.2023; https://openfast.readthedocs.io/en/dev/index.html.
[16] Hall M 2015, *MoorDyn user's guide*, (Department of Mechanical Engineering, University of Maine: Orono, ME, USA).
[17] Viselli A M, Goupee A J and Dagher H J 2015, Model test of a 1: 8-scale floating wind turbine offshore in the gulf of maine, *J. Offshore Mech. Arct. Eng.,* **137**, 2015.
[18] Stewart G M, Robertson A, Jonkman J, and Lackner M A 2016, The creation of a comprehensive metocean data set for offshore wind turbine simulations, *Wind Energy,* **19**, 1151-1159.

IOP Conf. Series: Materials Science and Engineering 1294 (2023) 012005 doi:10.1088/1757-899X/1294/1/012005

Numerical analysis of semiconductor-based energy conversion technologies for offshore applications

T Choi[†]**, J H Lee**[†] **and T Y Kim**[†*]

Department of Mechanical and Automotive Engineering, Seoul National University of Science and Technology, 232 Gongneung-ro, Nowon-gu, Seoul 01811, Republic of Korea

[†] These authors contributed equally to this work
[*] Correspondence: tykim@seoultech.ac.kr

Abstract. This study analyzes the physical principles and characteristics of thermoelectric energy conversion phenomena utilizing p-n semiconductor couples to propose a design methodology for thermoelectric generators (TEGs) capable of energy recovery in marine environments. To design TEGs with practical power generation and energy conversion efficiency, the development of a numerical analysis model that can predict and analyze the behavior characteristics of TEGs is required. We present the necessary conditions for a numerical analysis model that simulates the generation characteristics of TEGs at the system level, along with a comparison and analysis of experimental results. The electrical connection method of the thermoelectric modules (TEMs) significantly affects the power generation performance of TEGs; however, there is a lack of research on the electrical connections between TEMs in existing studies. This paper examines the reasons for performance changes due to electrical connections between TEMs and proposes an optimal electrical connection method. Additionally, it is demonstrated that improving the thermal flow distribution within the TEG can enhance system performance and maximize the effects of electrical connections between TEMs. We analyze the characteristics of technologies that can improve the level of thermal flow distribution within the TEG, such as finned and porous structures, and present optimal utilization strategies.

1. Introduction

With escalating concerns over environmental pollution, global warming, and the depletion of fossil fuels, there is a heightened focus across various industries and research sectors on improving fuel efficiency and overall energy efficacy [1]. Nations worldwide are strategizing to meet carbon neutrality and Net Zero targets in line with global accords, setting industry-specific carbon reduction goals. South Korea, aiming for carbon neutrality by 2050, has established carbon reduction objectives within the automotive, shipping, and electric generation sectors. Reducing carbon emissions is not about curtailing industrial activity or hindering technological progress but rather a long-term orientation towards sustainable growth. The approach must encompass the acceptance of mediating measures in politics, borders, and social realms against the backdrop of uninterrupted and continuous scientific and technological advancements. Energy recovery, as an area enabling sustained growth while reducing energy usage and improving efficiency, is gaining prominence. The levels of natural energy recovery technologies such as solar photovoltaic, solar thermal, wind, wave, and tidal power are rapidly advancing and permeating various aspects of life. Industrial heat recovery solutions are also receiving continual interest. Turbo compound technology, based on turbo machinery, employs turbines installed at the exhaust duct end to convert kinetic and thermal energy of exhaust gases into rotational turbine energy. Its advantage lies in the straightforward generation of mechanical energy by installing appropriately sized turbines, considering exhaust duct dimensions. However, disadvantages include vibrations and noise from turbine rotation and efficiency losses during the conversion to electrical energy.

To achieve higher energy recovery efficiency, the thermodynamic Organic Rankine Cycle (ORC) is utilized. ORC operates with organic refrigerants capable of evaporating at relatively low temperatures, recovering heat from low-temperature heat sources, and using high enthalpy refrigerant vapors to rotate a turbine, thus recovering mechanical energy [2]. Despite ORC's high energy conversion efficiency, its complexity and large system size, comprising numerous components like pumps, boilers, expanders, and condensers, remain drawbacks. To overcome these limitations, Thermoelectric Generators (TEGs) are gaining prominence across various industrial sectors [3]. TEGs, devoid of moving parts, eliminate operational noise and vibrations, directly converting thermal to electrical energy with outstanding operational stability as solid-state energy conversion technologies. Additionally, TEGs offer the flexibility of adjusting the number of Thermoelectric Modules (TEMs) for energy conversion based on the quantity and intensity of the heat source and are easily integrated with other engineering devices [4]. Thus, TEGs find applications in solar energy technologies, ORCs, domestic burners, body heat recovery, and energy recovery technologies for vehicle and ship engines. Among these, the application of TEGs in vehicle and ship engines is technically challenging, requiring consideration of both thermal transfer characteristics for high energy conversion and significant differential pressure features [5].

Despite significant advancements in the theoretical and numerical modeling for predicting the performance of Thermoelectric Generators (TEGs), there remains a notable deficiency in research outcomes aimed at improving the reliability and accuracy of these models at a system level. For instance, during the operation of TEGs, temperature variations across the Thermoelectric Modules (TEMs)—with a decrease at the high-temperature side and an increase at the low-temperature side—result in variations in power output and efficiency. This study proposes a numerical analysis model based on the physical phenomena occurring during thermoelectric generation, aiming to predict actual power generation by TEGs.

Furthermore, we explore strategies to enhance the flow and temperature distribution characteristics within TEGs using the derived numerical results. As the TEM utilizes temperature differences for power generation, a temperature gradient across TEMs can reduce the power output and efficiency at the system level, a phenomenon similarly observed in power source-based generation systems like photovoltaic devices [6]. This research also examines the design of internal structures within exhaust pipes to evenly distribute the high-inertia exhaust gas flow. The arrangement and electrical interconnection of TEMs are identified as critical factors affecting the energy recovery performance of TEGs. By analyzing the physical causes of how electrical interconnection methods influence TEG performance, this study seeks to derive a comprehensive system design approach for the optimization of TEG performance.

2. Numerical model of thermoelectric generator

2.1. Underlying physics of thermoelectric energy conversion

When a temperature difference is applied to a TEM, the potential proportional to the temperature difference can be represented by Equation (1) due to the Seebeck effect.

$$V_{oc} = n\alpha\Delta T \tag{1}$$

Where n_{cp}, α, ΔT represent the number of *p-n* couples, the Seebeck coefficient, and the temperature difference across the TEM, respectively. When the TEM is electrically connected to a load, a current flows, and it can directly convert thermal energy into electrical energy, thus constituting a Thermoelectric Generator (TEG) that can recover power. When a current flows through the TEM, not only is there heat transfer by conduction, but also heat pumping and Joule heating phenomena are induced by holes in the p-type semiconductor and electrons in the n-type semiconductor, respectively. The heat energy carried by the holes and electrons causes heat transfer from the hot side to the cold side of the TEM, while simultaneously generating heat due to internal electrical resistance. The flow of thermal energy due to conduction, heat transfer by holes and electrons, and Joule heating between the hot and cold sides of the TEM can be expressed as follows in Equations (2) and (3).

$$Q_h = n_{cp}\left[\alpha I T_h + K\Delta T - \frac{1}{2}I^2R\right] \tag{2}$$

$$Q_c = n_{cp}\left[\alpha I T_c + K\Delta T + \frac{1}{2}I^2R\right] \tag{3}$$

As shown in Equations (2) and (3), only the conduction term appears when no current flows, and complex electrical phenomena caused by holes and electrons occur when a current does flow. Therefore, developing a numerical analysis model that reflects the thermoelectric complex characteristics of TEG is essential to ensure the reliability of the model and accurately predict the power generation performance.

2.2. *Numerical model*

The numerical analysis model used in this study is designed with the same dimensions as the experimental apparatus and is shown in Figure 1. The dimensions of the exhaust gas channel are 183 mm (width) × 394 mm (length) × 50 mm (height), and the TEMs are arranged in a 3 × 5 array between the exhaust gas channel and the cooling water channel, as illustrated in Fig. 1 (d). An extended surface with a thickness of 2 mm is designed at equal intervals inside the exhaust gas channel, considering the position of the TEMs. The size of the cooling water channel is 183 mm (width) × 375 mm (length) × 10 mm (height), with the cooling water flowing in a counter-flow pattern. Considering the symmetry of the TEG, the analysis was conducted on the top plate where 15 TEMs are present, based on the center of the exhaust gas channel, utilizing symmetry conditions.

The exhaust gas inlet conditions for the numerical analysis model were set at 34.5 m/s, 551.3 K; 46.9 m/s, 558.7 K; and 51.0 m/s, 563 K, considering an Engine load and rotation speed of 0.8 BMEP and 1000, 1200, 1400 rpm, respectively. The inlet conditions for the cooling water were applied with boundary conditions considering a cooling water flow rate of 4 LPM and a temperature condition of 293 K. Taking into account strong air conditioning in the room and winter conditions, the outer walls of the exhaust gas channel were given a convection condition with a convective heat transfer coefficient of 50 W/m^2K and a temperature condition of 278 K.

For the numerical analysis technique, the realizable *k-ε* turbulence model was used along with the Scalable wall function. The SIMPLE algorithm was employed for pressure and velocity coupling. For the numerical analysis scheme, the Body-force weighted approach was used for pressure, and the Second-order upwind scheme was used for momentum, turbulent dissipation rate, turbulent kinetic energy, and energy.

2.3. *Underlying physics of thermoelectric energy conversion*

The thermal energy conversion phenomenon and the associated heat flow in the numerical analysis model were incorporated through a User-defined function (UDF). Heat source and Heat sink regions were assigned to each 1mm length at both ends of the total thickness of the TEM, and the analysis was conducted reflecting the first and third terms of Equations (2) and (3). To ensure the accuracy of the analysis, the thermal energy conversion phenomenon was reflected based on an Iterative method. The analysis procedure is as follows:

1) The Energy conservation equation is calculated, and the Open-circuit voltage is derived based on the Seebeck effect using the obtained temperatures.
2) The current flowing through each TEM is derived based on Kirchhoff's voltage and current law.
3) The current values flowing through each component are reflected in the Heat source and Heat sink boundary conditions through the UDF.

4) With the boundary conditions applied, the energy equation is solved again to derive a new temperature field.

By repeating steps 1 to 4, the numerically altered thermal flow field due to the thermal energy conversion phenomenon can be derived.

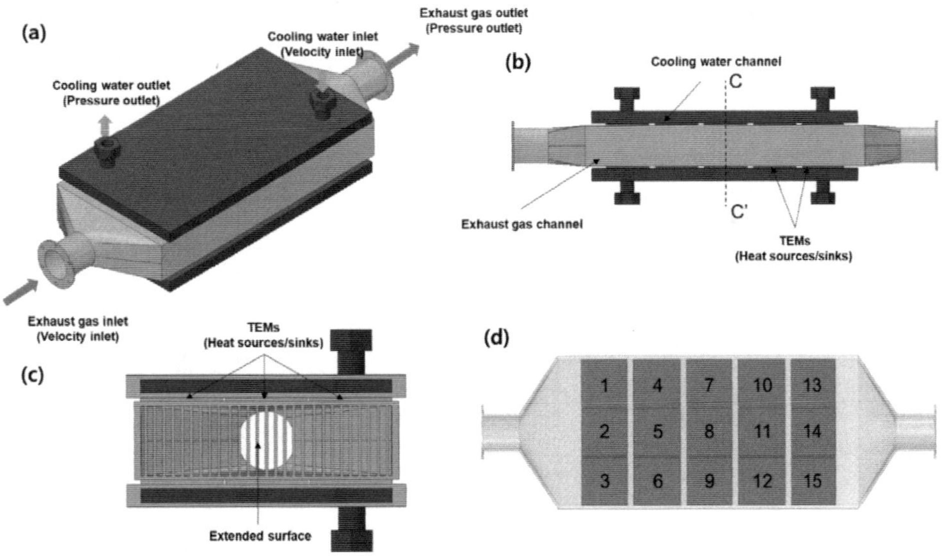

Figure 1. Schematic diagrams of different viewpoint of numerical model.

3. Effect of electrical array form of TEMs

3.1. Optimal current of TEMs

The output of the TEM can be represented as the value obtained by subtracting Equation (3) from Equation (2) and can be expressed as follows.

$$P_{out,TEM} = Q_h - Q_c = n_{cp}[\alpha I \Delta T - I^2 R] \quad (4)$$

The current that can derive the maximum output of the TEM is expressed as Equation (5), which is the result satisfying (dP/dI = 0) by differentiating Equation (4) with respect to current.

$$I_{opt} = \frac{n_{cp}\alpha\Delta T}{2R} \quad (5)$$

As shown in Equation (5), when a current proportional to the temperature difference in the TEM flows, the maximum output can be achieved. Therefore, as expressed in Step 2 of Section 2.3, the current flowing through the TEM is predominantly governed by the method of electrical connection, hence the output performance varies according to the electrical connection method.

Consequently, when the TEMs are connected using methods that do not consider the temperature difference, such as series or parallel connections, a difference arises between the sum of outputs derived from connecting the TEMs individually and the output derived from connecting all electrically. This is what can be called mismatch loss, and the electrical connection method that can minimize mismatch loss is the one that can allow a current proportional to the temperature difference to flow. The electrical connection method proposed in this study is to divide the layers based on the temperature difference, as shown in Figure 2. By reducing the number of parallel branches in the layer with TEMs that have a higher temperature difference, and increasing the number for layers with a lower temperature difference, a current proportional to the temperature difference can be allowed to flow. Therefore, by adjusting the number of TEMs in a layer and the number of parallel branches, an electrical connection method that minimizes mismatch loss can be derived.

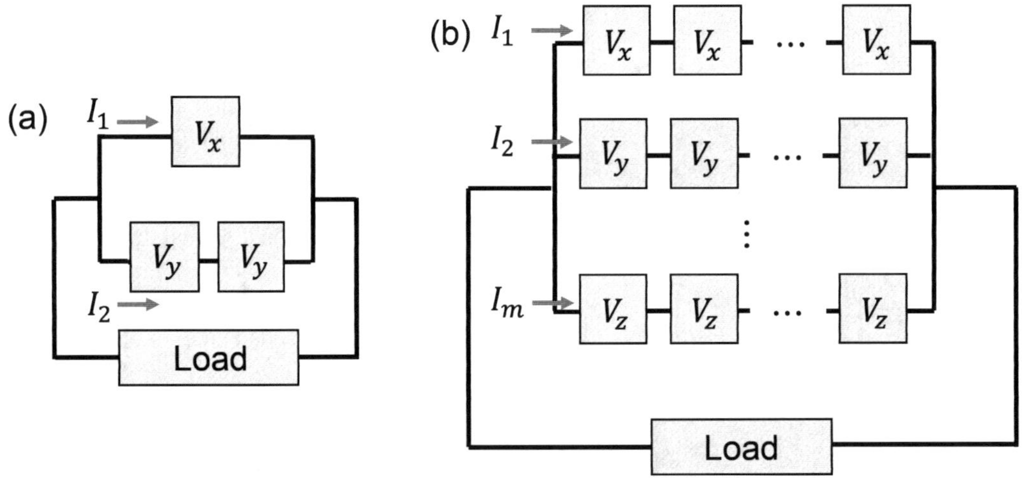

Figure 2. A simple schematic of layer-based electrical array configuration.

3.2. Concept of electrical array configuration for reducing mismatch loss

As discussed earlier, the current flowing in proportion to the temperature difference of each TEM minimizes the mismatch losses, yielding maximum output when the optimal current flows. This holds true when viewed from a system perspective as well. Figure 2(a) illustrates the simplest connection method capable of delivering the optimal current to TEMs with varying temperature differences. The V_x and V_y represent the open-circuit voltage of each thermoelectric module (TEM), with currents I_1 and I_2 passing through them, respectively. When the resistance of the TEM remains constant and the load resistance equals the electrical circuit's resistance, applying Kirchhoff's current and voltage laws reveals that the currents I_1 and I_2 vary based on the values of V_x and V_y, expressible by equations (6) and (7):

$$I_1 = \frac{2V_x - V_y}{3R} \tag{6}$$

$$I_2 = \frac{5V_y - V_x}{6R} \tag{7}$$

As derived from equation (5), to satisfy the optimal current for V_x and V_y, V_x must be twice the magnitude of V_y, resulting in zero mismatch losses. When $V_x = 2V_y$, as depicted in Figure 2 (a), an environment enabling the optimal current flow through each TEM is established. Analyzing based on the fundamental mismatch loss mechanisms in series and parallel connections, the following summary of connection methods can be deduced: In a series connection, due to a representative current flow, mismatch losses occur when the optimal current fails to flow through each TEM in the presence of temperature deviation. In a parallel connection, mismatch losses arise from potential differences between branches. Therefore, for the proposed mismatch loss mitigation in electrical connection methods in this study, arranging modules with similar or identical temperature differentials in series connections, as shown in Figure 2 (b), minimizes the mismatch losses caused by series connections. Additionally, arranging components to ensure the summation of their potentials in parallel connections minimizes mismatch losses caused by parallel connections.

4. Effect of flow and temperature distribution on TEG performance

4.1. Temperature and velocity fields

Numerical analysis results indicate the presence of non-uniform temperature and flow distributions within the Thermoelectric Generator (TEG). Figure 3 illustrates the temperature and velocity distributions across the Thermoelectric Modules (TEMs). The inlet geometry of the exhaust gas channel, transitioning from a narrow to a wider area, results in a flow that is concentrated towards the center due to the high inertia of the exhaust gases, leading to an uneven flow distribution. Additionally, flow recirculation occurs in the cone area as the exhaust gases collide with the foremost cooling fins, further contributing to the uneven flow. To address these issues, various structures have been proposed within the exhaust gas channel, such as incorporating metal foam. This study aims to examine the effect of implementing a perforated plate within the exhaust gas channel as a flow rectifier.

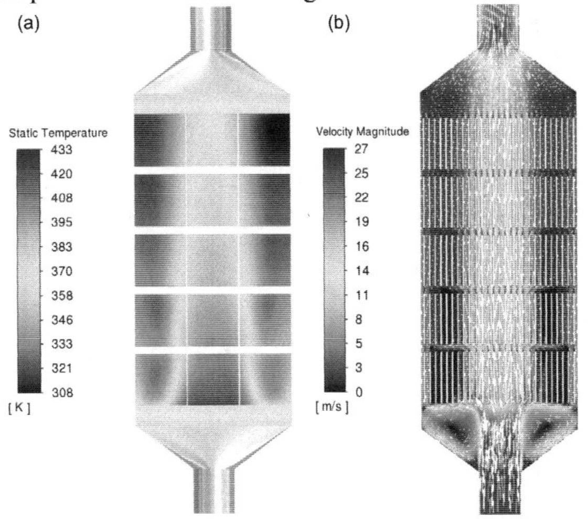

Figure 3. (a) Hot surface temperature profiles of TEMs and (b) Velocity profile of the exhaust gas flow

Figure 4 presents the numerical analysis model with the implementation of the perforated plate. The fundamental shape and dimensions are consistent with the numerical analysis model discussed in the previous chapter. To investigate the effects of the perforated plate's position on flow distribution and power generation characteristics, a 2.4 mm thick, porous media zone was modeled at the front end of each row of cooling fins. The positions of the perforated plate insertions have been designated as L1 to L6 from the exhaust gas inlet.

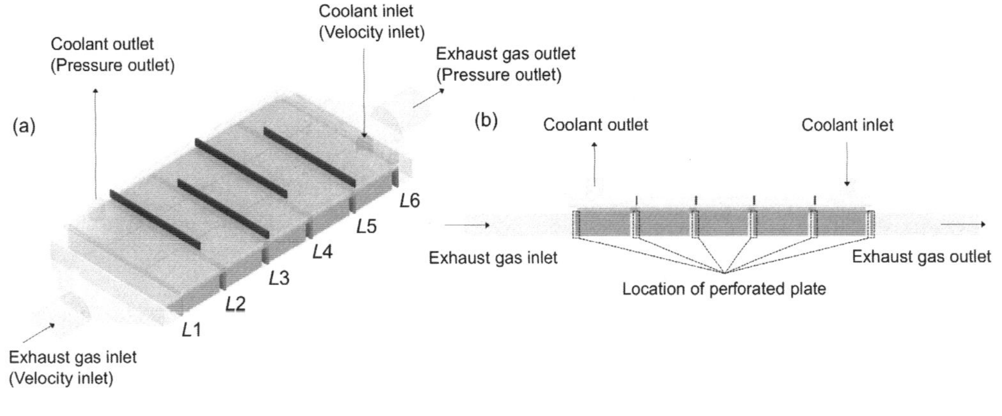

Figure 4. Schematic diagram of numerical model: (a) isometric view, (b) side view

4.2. Pressure drop characteristics of perforated plate

Figure 5 displays the pressure drop depending on the position of the perforated plate under the condition of an engine rotation speed of 1000 rpm. The pressure drop caused by the perforated plate was numerically simulated using the porous media modeling method. In this case, the pressure drop due to the perforated plate is reflected as an additional source term in the Navier-Stokes equations and is expressed as shown in Equation (8).

$$S = -\left(\frac{C_1}{t}U + \frac{C_2}{t}\frac{1}{2}\rho|U|U\right) \tag{8}$$

$$\Delta P = -\left(C_1 U + C_2 \frac{1}{2}\rho|U|U\right) \tag{9}$$

Where S, U, ρ, and t represent the source term, the fluid's superficial velocity, the fluid's density, and the thickness of the perforated plate, respectively, while C_1 and C_2 are constant terms. A numerical analysis method was used to derive these constants, which were then validated with a theoretical formula. Figure 5 illustrates the numerical analysis model and the results used to derive the constants for a perforated plate with a porosity of 0.335.

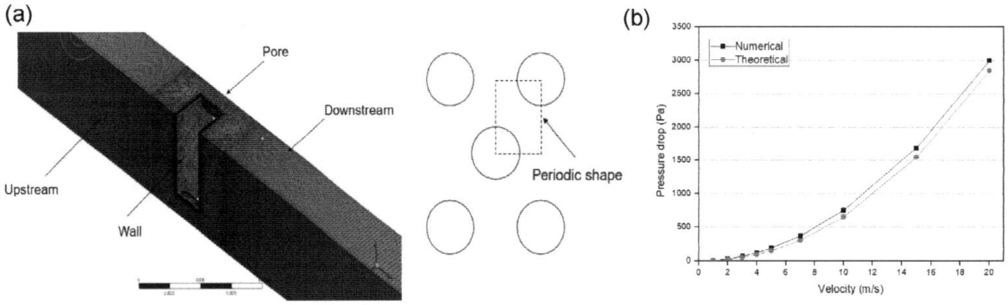

Figure 5. (a) Isometric view of a numerical model of unit perforated plate, and (b) numerical and theoretical pressure drop of the perforated plate ($\varepsilon = 0.335$) according to the superficial velocity.

The periodic shape of the perforated plate was modeled, and the differential pressure according to the superficial velocity was derived. The derived pressure drop is expressed as a second-order polynomial in superficial velocity as shown in equation (2), hence constants were derived after fitting it to a second-order polynomial. Furthermore, to verify the constants, the differential pressure was calculated using a theoretical formula for validation, and the numerical analysis results were demonstrated to be plausible.

4.3. Temperature and velocity characteristics

Figure 6 depicts the temperature field corresponding to the insertion location of the perforated plate under the condition of an engine rotation speed of 1000 rpm. The results without the insertion of the perforated plate are provided as the reference case. Upon examining the results without the perforated plate, it can be observed that the flow is concentrated towards the center, resulting in higher temperatures for the central devices while the devices placed at the sides exhibit relatively lower temperatures, with local differences reaching up to approximately 125K. Consequently, due to the lower high-temperature side of the devices placed at the sides, the overall heat absorption of the system decreases, leading to reduced power generation and efficiency.

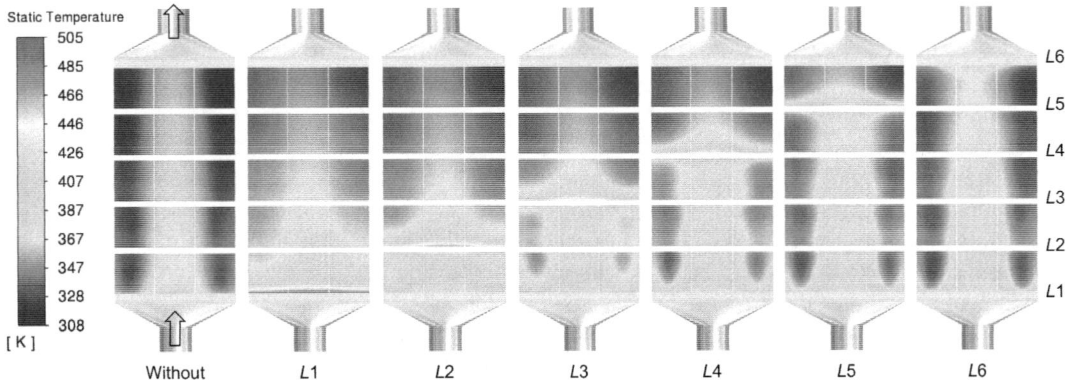

Figure 6. Contour map of the hot side temperature of TEMs according to the location of perforated plate ($\varepsilon = 0.335$)

With the insertion of the perforated plate, it can be seen that the temperature field becomes more uniform in the direction perpendicular to the flow. Additionally, even when the position of the perforated plate is at the center of the exhaust gas, the influence of the plate not only uniformizes the temperature after the flow has passed through but also before it reaches the plate, and the wave patterns of the temperature field created by the flow distribution can be observed.

Figure 7 illustrates the velocity distribution when the perforated plate is inserted at position *L*4. The blue arrows represent the velocity vectors, and the yellow line indicates the perforated plate inserted at the *L*4 position. As a result of inserting the perforated plate, the flow becomes stagnant before reaching the plate due to its effect, and after passing through, the flow mixes and distributes. Therefore, in the figure, it can be observed that the flow, which is concentrated in the middle between the heat dissipating fins due to stagnation just before the plate, is distributed.

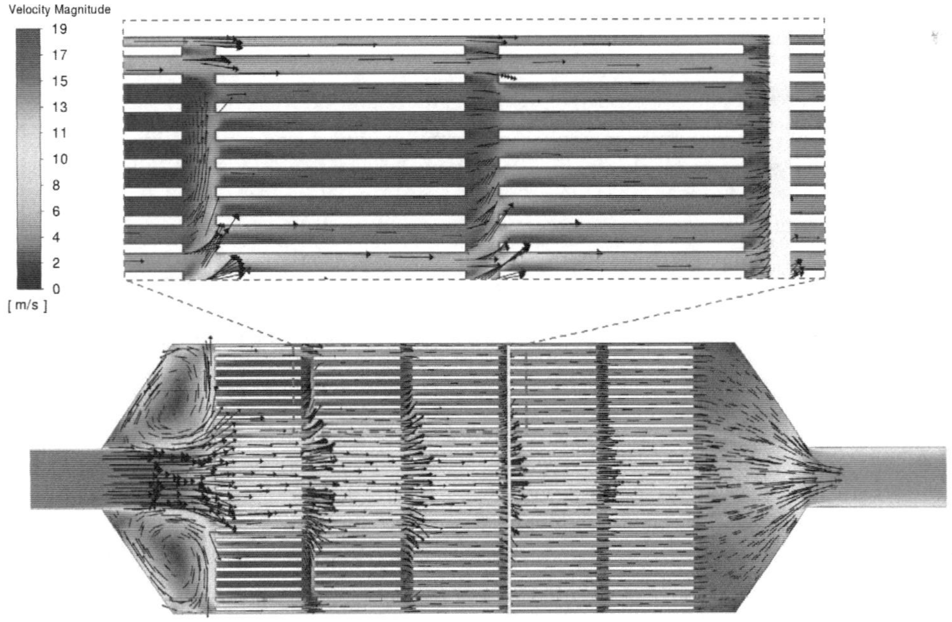

Figure 7. Contour map of the velocity magnitude of the exhaust gas for a case with a perforated plate inserted in the *L*4

4.4. Performance improvement of TEG
The degree of improvement in TEGs through enhanced thermal flow distribution characteristics varies depending on the porosity, position of the flow straightener with a porous structure, and the shape of the TEG exhaust duct. Generally, as the porosity of the structure decreases, the temperature and flow distribution characteristics improve, leading to an increase in power generation. However, a lower porosity significantly increases pressure drop, resulting in increased power loss. Therefore, when using a structure that can enhance temperature and flow distribution characteristics, it is necessary to consider both the gains in power generation and power losses, necessitating an analysis from the perspective of net output power. Net output power is the power generation of the TEG minus the power loss, which is the product of the exhaust gas volumetric flow and the pressure drop occurring in the TEG. Numerical studies suggest that while the increase in TEG power generation tends to have a linear relationship with decreasing porosity, the pressure drop shows nonlinear characteristics depending on porosity. Therefore, selecting a structure with a porosity outside the area where a sharp pressure drop characteristic appears is effective for increasing TEG power generation.

The position of the structure with flow distribution characteristics also greatly affects the performance of TEGs. Typically, as flow distribution characteristics appear before and after the structure, it is a good choice to insert the flow distribution structure in the middle position of the exhaust duct. This result, which greatly differs from conventional studies that install flow distributors at the front part of the exhaust duct, suggests that research to determine the optimal position of flow distributors will be greatly helpful in deriving the optimal design of heat transfer devices, including TEGs and heat exchangers.

5. Conclusions
In this study, the optimal design strategy for an energy recovery device utilizing the thermoelectric energy conversion effect was conducted from three perspectives. The first is to develop a numerical analysis model that is essential when designing a system with the maximum power generation and energy conversion efficiency in a given operational environment. Not only the existing flow and temperature analysis but also an analytical procedure to accurately simulate the thermoelectric energy conversion phenomenon was derived, and this method was integrated into commercial software as a user-defined function.

The second is to find an optimal electrical connection method for TEMs. The physical drawbacks of the existing series and parallel electrical connection methods were analyzed from the perspective of the optimal current required by the TEM. The mismatch loss of electrical array configuration of TEMs can be minimized by using the proposed concept of electrical array configurations.

The third is to examine flow and temperature distribution strategies. The flow and temperature concentration characteristics due to the high inertia of the exhaust gas were identified, and a flow distribution strategy using a porous structure was proposed. Considering the advantages of flow distribution with a porous structure and the disadvantage of increased pressure drop, an optimal utilization plan for the porous structure was presented from the perspective of net output power.

Acknowledgments
This work was supported by the "Development of a thermoelectric numerical model" funded by the National Research Foundation (NRF, 2021M3C1C3097580), Republic of Korea. Also, this work was supported by the "A nature-inspired high-efficiency solar energy collector mimicking energy metabolic mechanisms of plants" funded by the National Research Foundation (NRF, 2022R1C1C1005922), Republic of Korea.

References

[1] Luo D, Wang R, Yu W and Zhou W 2020 A numerical study on the performance of a converging thermoelectric generator system used for waste heat recovery, *Appl. Energy*, **270**, 115181, doi: 10.1016/j.apenergy.2020.115181.

[2] Hiranandani K, Aravind B, Kishore V R and Kumar S 2019 Development of a numerical model for performance prediction of an integrated microcombustor-thermoelectric power generator, *Energy*, **192**, 116624, doi: 10.1016/j.energy.2019.116624.

[3] Montecucco A, Siviter J and Knox A R 2014 The effect of temperature mismatch on thermoelectric generators electrically connected in series and parallel, *Appl. Energy*, **123**, 47–54, doi: 10.1016/j.apenergy.2014.02.030.

[4] Fang W, *et al.* 2017 Effect of topology structure on the output performance of an automobile exhaust thermoelectric generator, *IOP Conf. Ser.: Mater. Sci. Eng.*, **199**, 012056, doi: 10.1088/1757-899X/199/1/012056.

[5] Choi Y, Negash A and Kim T Y 2019 Waste heat recovery of diesel engine using porous medium-assisted thermoelectric generator equipped with customized thermoelectric modules," *Energy Convers. Manage.*, **197**, 111902, doi: 10.1016/j.enconman.2019.111902.

[6] Choi T and Kim T Y 2023 Current harmonized improvement method: A concept of the optimal electrical array configuration of thermoelectric modules for minimizing the mismatch loss of a thermoelectric generator, *Energy Convers. Manage.*, **286**, 117036, doi: 10.1016/j.enconman.2023.117036.

IOP Conf. Series: Materials Science and Engineering 1294 (2023) 012006 doi:10.1088/1757-899X/1294/1/012006

Life cycle cost analysis of a floating wind farm in the Norwegian Sea

F A Bjørni[1], S Lien[1], T Aa Midtgarden[1], L C Santos[2] and Z Jiang[1*]

[1] Department of Engineering Sciences, University of Agder, N-4898 Grimstad, Norway
[2]Department of Naval and Oceanic Engineering, Universidade da Coruña, Spain

* Correspondence: zhiyu.jiang@uia.no

Abstract. The offshore wind energy industry has witnessed rapid growth in the past decade. Still, there is a lack of commercial floating wind projects due to the relatively high development costs and other factors. To facilitate the holistic evaluation of floating wind farms in Norway, this article investigates the levelized cost of energy of a floating offshore wind farm and its economic feasibility. The Troll field west of Bergen, Norway, is assumed to be the target offshore site, and a farm size of 50 wind turbines with a lifespan of 25 years are considered. Each floating wind turbine has a 15-megawatt turbine mounted on a semi-submersible floater. Based on detailed analysis, the levelized cost of energy of the wind farm is estimated to be approximately 100.7 $/MWh. The capital expenditure is the most prominent cost and constitutes 63.1% of the total cost, and the operational expenditure constitutes the remaining 36.9%. Further, sensitivity analyses show the influence of the lifespan, capacity factor, and project discount rate on the levelized cost of energy. The present study contributes to techno-economic evaluation of floating wind projects at an early phase.

1. Introduction

The main advantage of offshore wind farms is their ability to bolster the effectiveness of renewable energy sources in alleviating the consequences of climate change [1]. With extensive maritime resources, industrial competence, and technical know-how from the petroleum sector, Norway may assume a prominent role in the production of floating offshore wind energy [2].

The levelized cost of energy (LCOE) is a concept employed to depict the average cost of generating electricity from an energy source over its entire lifespan [3]. In the case of floating wind farms (FWFs), the pricing of the electricity generated is influenced by numerous factors, such as the design of the support structure, grid connection, installation techniques [4], and operations and maintenance [5]. LCOE serves as a valuable metric for assessing the cost-effectiveness of different wind farm types.

The key obstacle to achieving commercial viability in floating wind power lies in the substantial initial capital investment, which significantly surpasses that of bottom-fixed turbines [6]. This underscores the challenge of addressing elevated costs, which is essential for facilitating the broad adoption of floating wind energy systems. Figure 1 illustrates a comparison of the estimated LCOE among different energy sources. Here, the global data are derived from [7-10] and the Norwegian data from [11]. As shown, the

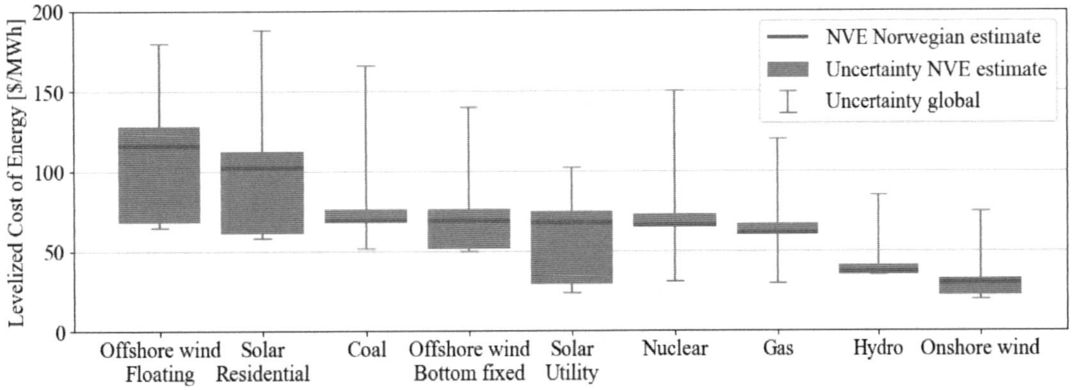

Figure 1. Comparison of the LCOE for different energy sources

costs linked to offshore floating wind energy are generally considered to be greater than those of other energy sources. Compared to the bottom-fixed offshore wind energy, the floating offshore wind energy has higher uncertainty in the LCOE estimated by either the Norwegian Water Resources and Energy Directorate (NVE) or by the global one. To reduce the uncertainties in the LCOE estimation, it is necessary to consider site-specific offshore conditions and turbine-specific technologies of the target offshore wind farm. To facilitate economic feasibility analysis of FWFs, we focus on a Norwegian offshore site and carry out a detailed LCOE analysis of an FWF consisting of 15-megawatt (MW) wind turbines for this site. The results of this paper contribute to an improved understanding of the cost elements for an offshore FWF and the key cost drivers of the LCOE.

The rest of the paper is structured as follows. Section 2 describes the approach for calculating the LCOE. Section 3 defines the floating wind turbine (FWT) system. Sections 4-5 specify the site conditions and the reference wind farm, and Section 6 present the results. Section 7 draws the concluding remarks.

2. Approach

2.1 Components of LCOE

The LCOE serves as a quantitative representation of the cost-efficiency of an energy project. LCOE can be computed by dividing the total project expenses by the total electricity production over the project's duration [3]. Equation (1) shows how the LCOE of an offshore wind farm is computed:

$$LCOE = \frac{(CapEx * FCR) + OpEx}{AEP_{net}} \tag{1}$$

where $CapEx$ is the investment expenditures, $OpEx$ is the operations and maintenance costs, FCR is the fixed charge rate, and AEP is the annual energy generation. Upon completing the LCOE calculation, the resulting figure is typically expressed in terms of dollars per MW hour [$/MWh].

2.2 Estimation of CapEx and OpEx

ORBIT [12] was applied to calculate the CapEx for all cost components of a wind farm except the wind turbine. ORBIT is an open-source Python tool designed to offer reasonably accurate estimates of component costs, sizes, and masses based on a limited set of user inputs. Given that ORBIT was initially introduced in 2017, subsequent developments have occurred. To ensure the precision of these estimates, it becomes imperative to source the wind turbine cost data from a current and up-to-date reference. Consequently, the wind turbine cost data is acquired from an analysis conducted by Rystad [13]. The costs provided by ORBIT are dollars from 2017, necessitating an adjustment for inflation up to 2023.

Two distinct inflation factors are employed to compute the adjusted CapEx. Specifically, the Producer Price Index (PPI) is used to modify general expenses, including production and installation costs, whereas steel prices are utilized to adjust costs associated with steel components, such as the material expenses for the substructure of an FWT. The calculation of these adjustment factors is presented in Table 1, and all the data values are sourced from the Organization for Economic Co-operation and Development [14].

Table 1. Calculation of CapEx inflation adjustment factors

Description	Reference value	2023 Value	Adjustment factor
PPI (general)	108.9 (2017)	147.8	1.35
PPI (wind turbine)	128.5 (2020)	147.8	1.15
Steel price index	1060.7 (2017)	1845.7	1.74

2.3 Calculation of the financial cost ratio

The financial cost ratio (FCR) represents the proportion of capital expenses required to cover capital-related costs [15]. The formula utilized to determine the FCR incorporates both the interest paid on debt and the return on equity. Consequently, when CapEx is multiplied by FCR, it yields a consistent annual annuity payment. The FCR is expressed in the following equation:

$$FCR = \frac{WACC}{1-(WACC+1)^{-n}} \tag{2}$$

where $WACC$ stands for the weighted average cost of capital and n the economic lifespan of the system. As $WACC$ is calculated to be 4.73% [16], and the economic lifespan of the system is set to 25 years, the FCR for yields is as follows:

$$FCR = \frac{4.73\ \%}{1-(4.73\ \%+1)^{-25}} = 6.9\% \tag{3}$$

3. Description of the floating wind turbine

The characteristics of the semi-submersible FWT are determined using models created by Jonkman et al. [17] and Gaertner et al. [18]. These reference wind turbine models were established to assist in conceptual studies of offshore wind technology.

The chosen semi-submersible substructure is designed based on the work of Offshore Code Comparison Collaboration Continuation (OC4) coordinated by the National Renewable Energy Laboratory (NREL) [17]. This semi-submersible floater features three side columns and one central column. with all these columns interconnected by a series of diagonal steel members and pontoons. Figure 2 displays a top and side view of the semi-submersible substructure and Figure 3 shows the full system with a mooring system. As details of mass and size for substructures of 15-MW semi-submersible are missing, we scale up the steel costs from the NREL 5-MW semi-submersible floater by considering the power rating (15 MW) of the target FWT. The wind turbine characteristics are selected based on the NREL 15-MW wind turbine model proposed by Gaertner et al. [18]. This is a conventional three-bladed turbine with active blade pitch control. Table 2 lists the main parameters of the wind turbine.

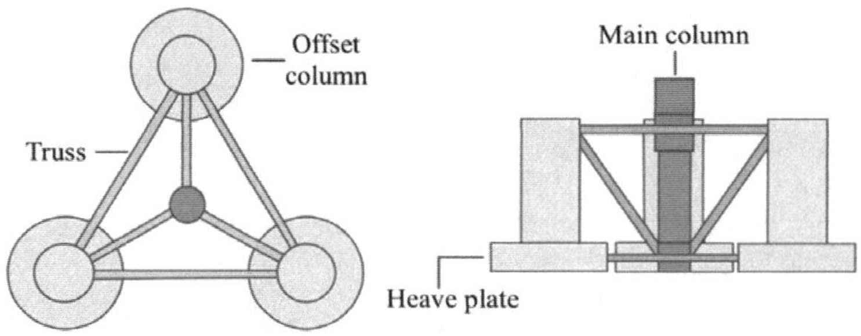

Figure 2. Top and side view of the semi-submersible floater.

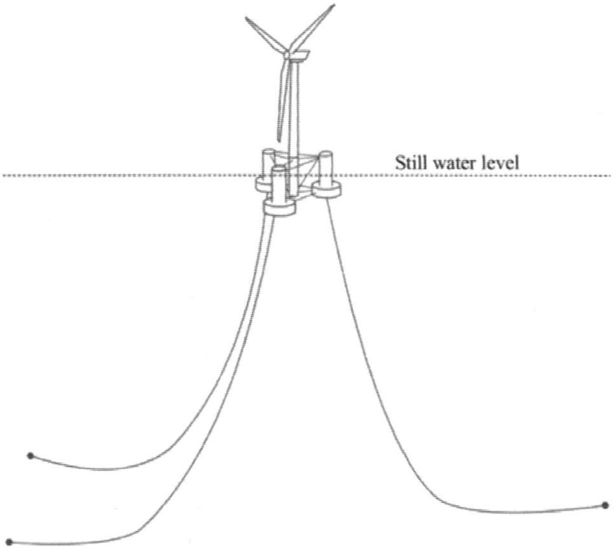

Figure 3. Illustration of the semi-submsersible FWT with moorings.

Table 2. Specifications of the wind turbine

Description		Value	Unit
General information	Power rating	15	MW
	Number of blades	3	-
	Rated wind speed	10.59	m/s
	Cut-in speed	3	m/s
	Cut-out speed	25	m/s
Tower	Length	150	m
	Mass	480	ton
Nacelle	Mass	797	ton
	Hub height	150	m
Blade	Length	120	m
	Mass	72	ton
	Rotor diameter	240	m

4. Description of site conditions

4.1 Geographical conditions

The target wind farm is located in the northern North Sea, along the Norwegian coast, where the water depth is approximately 325 m [19]. This particular area has not been previously explored for offshore wind projects. The only existing FWF in Norway, Hywind Tampen, is situated 128 km northwest of this site. The entire Troll field, which encompasses an area of 750 km^2, houses three operational oil platforms. As a result, the target wind farm has the potential to provide power to these existing oil field facilities. Figure 4 displays the geographical position of the Troll field.

IOP Conf. Series: Materials Science and Engineering 1294 (2023) 012006 doi:10.1088/1757-899X/1294/1/012006

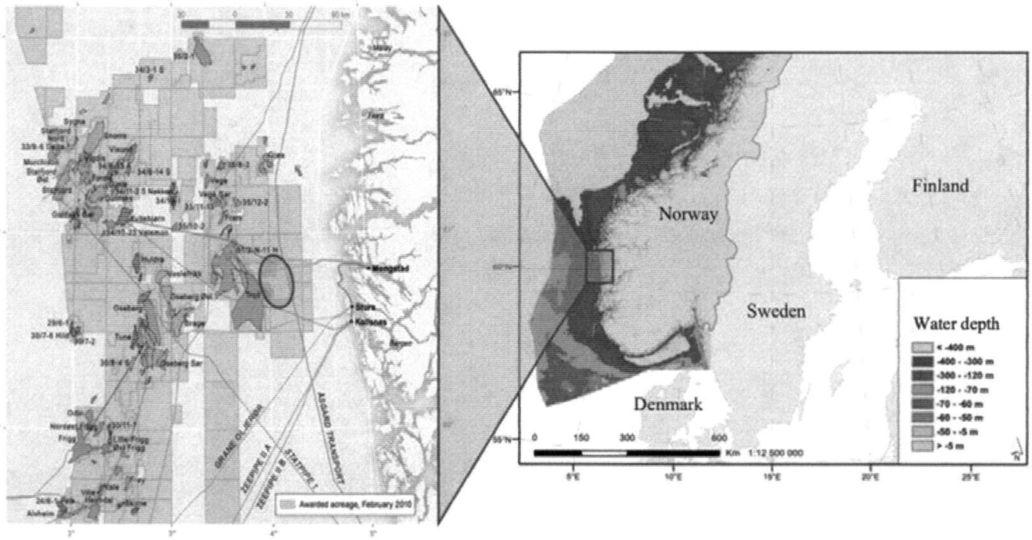

Figure 4. Geographical location of the Troll field.

4.2 Metocean conditions

To describe the long-term wind speed distribution at the Troll field, a two-parameter Weibull probability distribution model is used to fit the mean wind speed over 29 years. The fitted curve (blue) along with the scatter of mean wind speed is illustrated in Figure 5 where α and β are respectively the shape and scale parameters of the Weibull distribution. Figure 6 presents a wind rose depicting the dominant wind direction for different wind speeds. The wind rose data have been sourced from the Norwegian Centre for Climate Services [20]. As indicated, the prevailing wind direction at this site is from the southeast.

5. Case study of a reference wind farm

5.1 Wind farm layout

The reference wind farm is designed with an assumed operational lifespan of 25 years, comprising 50 turbines, each with the rated power of 15 MW. Consequently, the FWF has a theoretical total capacity of 750 MW. Figure 7 illustrates the layout of this reference wind farm.

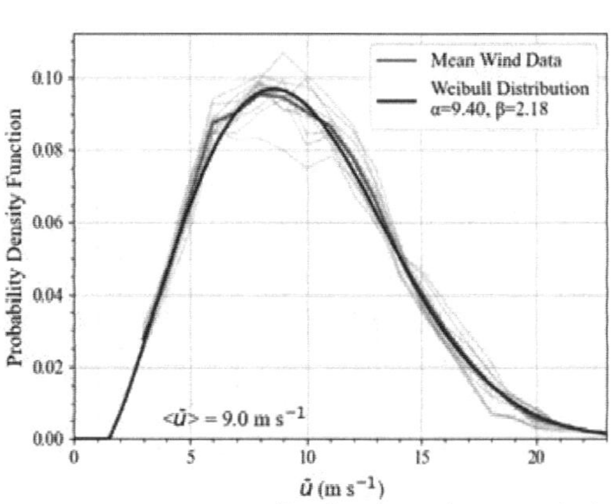

Figure 5. Long-term distribution of mean wind speed at Troll.

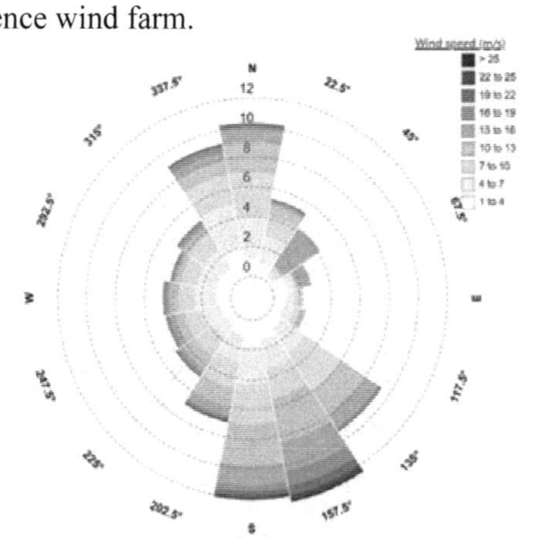

Figure 6. Wind rose at Troll.

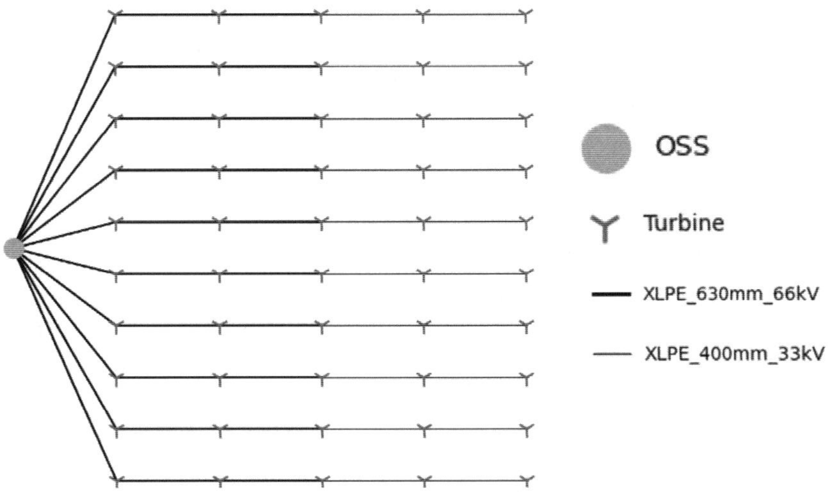

Figure 7. Layout of the studied FWF.

As depicted, the wind turbines are organized in a grid pattern with a layout of 5 turbines in a row and 10 rows. There are a total of 10 cables connected to the offshore substation (OSS), and these cables progressively increase in voltage and radius as they approach the OSS to sufficiently transmit electricity with minimal losses during transportation. To minimize the loss caused by aerodynamic wake effects, the spacing between the turbines is set to 8 times the rotor diameter in the dominant wind direction and 4 times the rotor diameter in the direction perpendicular to the dominant wind direction. As FWTs are in motion, these spacings are expected to be larger than those of bottom-fixed wind turbines. This layout is chosen for the sake of simplicity and no layout optimization is carried out in this work.

5.2 Capacity factor of the reference wind farm

To determine the capacity factor for the reference wind farm, statistical techniques are employed. The approach utilized is grounded in the Weibull distribution of wind speeds, combined with the specific power curve of the 15-MW wind turbines. Figure 8 illustrates the idealized power curve, sourced from [18], in conjunction with the probability distribution of the long-term wind speed.

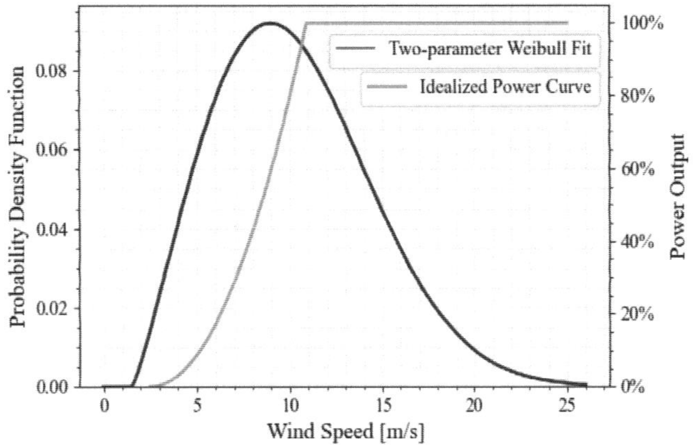

Figure 8. Power output curve and probability distribution of the mean wind speed.

The capacity factor for the 15-MW wind turbine at the Troll site is determined by integrating the Weibull distribution with the power curve as the upper limit. This process yields an approximate capacity factor of 57.75%. To consider grid or array losses and potential downtime, a 5% deduction has

IOP Conf. Series: Materials Science and Engineering 1294 (2023) 012006 doi:10.1088/1757-899X/1294/1/012006

been applied to the calculated capacity factor. Therefore, the capacity factor used for the reference wind farm is set at 52.75%.To account for grid or array losses, as well as potential downtime, 5% has been subtracted from the calculated capacity factor; thus, the capacity factor used for the reference wind farm is set to 52.75%. Larger fluctuations are expected in the wind speed and power output in turbulent wind, and FWF control will play an important role in the power quality.

5.3 Cost categories for the wind farm

To assess the cost breakdown, various subcategories have been established. These subcategories cover different aspects within both the design and installation phases, encompassing electrical components, turbines, and substructures. During the installation phase, the turbine and substructure are combined due to onshore assembly and the subsequent towing process. Similarly, mooring design and installation costs have been consolidated into one section since they represent a relatively small portion of the CapEx. Additionally, other sub-costs encompass all soft costs, project-related expenses, and OpEx. Table 3 provides an overview of the cost subcategories, outlining the components included in each sub-cost section.

Table 3. Cost elements included in the different cost categories

Cost category	**Cost elements**			
Electrical grid	Inter-array cables	Export cable	Offshore substation	Substation foundation
Turbine	Tower	Nacelle	Rotor blades	Hub
Substructure	Stiffened column	Truss	Heave plate	Secondary steel
Mooring line	Steel chain	Anchor		
Soft	Insurance	Financing	Commissioning	Decommissioning
Project	Site auction	Site assessment	Construction plan	Installation plan
OpEx	Operation	Maintenance		

6. Results and discussion

6.1 Initial calculation for the LCOE

According to the method outlined in Section 2, there are certain components that are not assessed in ORBIT but are essential for the calculation of the LCOE. These components are computed as follows:

$$OpEx_{annual} = \$\ 118\ 000/MW/yr * 15MW * 1yr * 50\ turbines = \$\ 88.5M \tag{4}$$

$$AEP = 15MW * 50\ turbines * 52.75\ \% * 8760hr/yr = 3\ 465\ 675\ MWh \tag{5}$$

6.2 Total LCOE estimation and cost description for the reference wind farm

Table 4 displays the breakdown of CapEx for the reference FWF. The cost categories and specific cost elements are described in Table 3. According to Table 4, the overall investment for the construction and installation of the reference wind farm, consisting of 50 wind turbines, amounts to \$3772572000. This results in a cost of \$75451440 for each installed turbine, encompassing all associated components.

Table 4. Cost breakdown of CapEx

Cost category	Cost	Percentage of CapEx
Electrical Grid	$ 384363000	10.2%
Electrical Grid Installation	$ 221400000	5.9%
Turbine	$ 759000000	20.1%
Substructure	$ 1275740000	33.8%
Substructure Installation	$ 56700000	1.5%
Mooring Line (Inc. Installation)	$ 218120000	5.8%
Soft	$ 653400000	17.3%
Project	$ 203850000	5.4%
Total	$ 3772572000	100%

Table 5 displays the key parameters applied during the LCOE calculations, along with the computed LCOE for the reference wind farm. In the table, FCR, OpEx and AEP are calculated using Eqs. (3)-(5), respectively, and the estimated annual cost per MW is from [5]. The table shows that the LCOE of the reference wind farm is estimated to be 100.69 $/MWh.

Table 5. Parameters for the LCOE calculation

Description	Reference Wind Farm
CapEx	$ 3772572 000
OpEx (annual)	$ 88500 000
AEP	3 465575 MWh
FCR	6.9%
Lifespan	25 years
LCOE	100.69 $/MWh

To evaluate the most influential components on LCOE, all costs related to energy production are displayed in Figure 9.

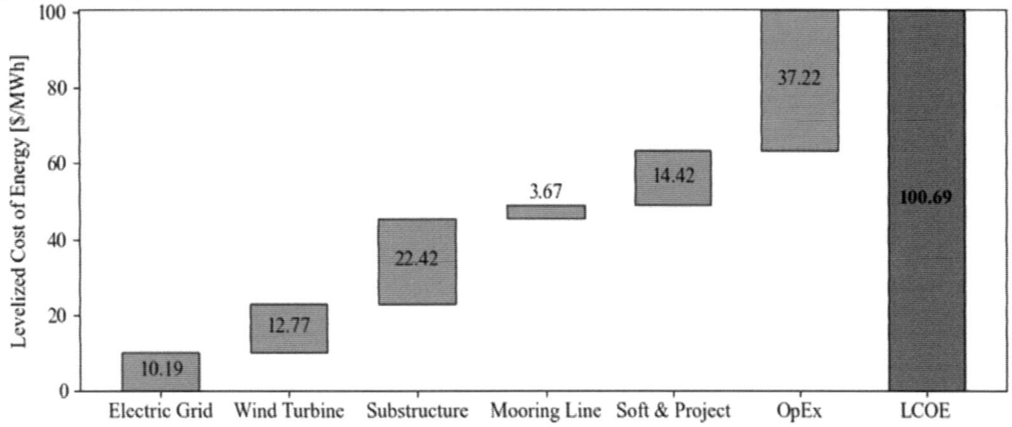

Figure 9. Total LCOE and its cost distribution.

Figure 9 integrates the installation costs within the total component costs. Among the components, the mooring line represents the smallest cost, totaling 3.67 $/MWh. On the other hand, OpEx is the most substantial contributor to the LCOE, amounting to 37.22 $/MWh. Notably, the costs associated with operation and maintenance make up approximately 37% of the LCOE throughout the 25-year production period. Despite the combined CapEx being higher in total, OpEx emerges as the most significant individual cost driver.

6.3 Sensitivity analysis of the reference wind farm

To analyze how LCOE varies with different input parameters, several sensitivity analyses have been performed. This will provide insights into factors that impact the LCOE for the reference FWF.

6.3.1 Key cost drivers of LCOE. Figure 10 illustrates the primary cost drivers affecting the LCOE. It reveals which factors exert the most significant influence on the LCOE and how they impact the LCOE when these cost drivers change. The baseline LCOE, set at 100.69 $/MWh, serves as the reference point. The diagram displays the extent to which the LCOE is affected by various cost drivers. Notably, it demonstrates that the lifespan, capacity factor, and project discount rate have the most substantial impact on the LCOE. These factors, when adjusted from the baseline, can significantly alter the LCOE.

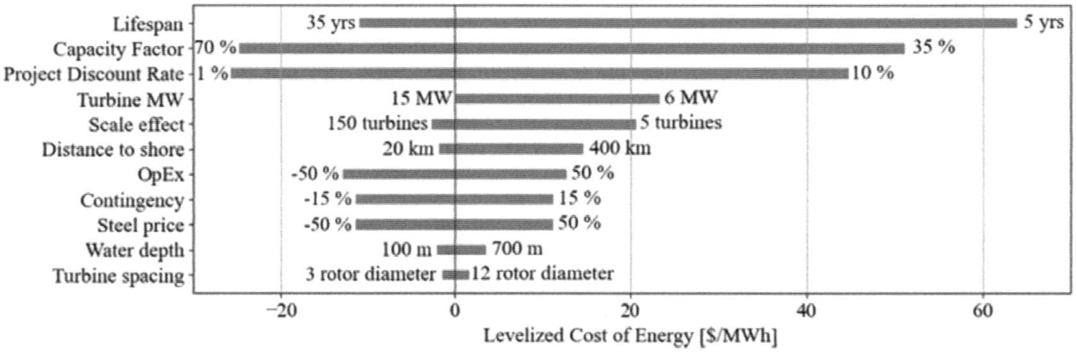

Figure 10. Key parameters affecting LCOE.

6.3.2 Effect of project lifespan and capacity factor. As the project lifespan and the capacity factor are two important factors in the cost of the LCOE, their influences are highlighted in Figure 11 and Figure 12.

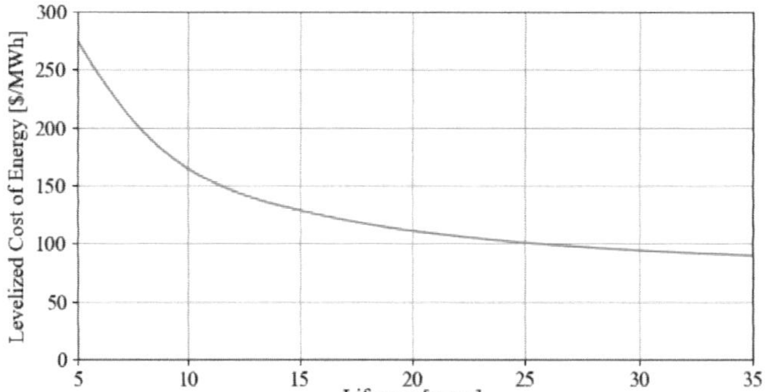

Figure 11. Variation of LCOE with project lifespan.

Figure 11 demonstrates that the LCOE decreases as the project's lifespan increases. For instance, at a five-year lifespan, the LCOE is 275.1 $/MWh, but it decreases significantly to 89.8 $/MWh for a 35-year lifespan, representing a reduction of 67.3%. Notably, this reduction in LCOE is more pronounced at the beginning of the project's lifespan, indicating that the benefits of longer project lifespans have a more substantial impact initially. It is important to acknowledge that the model does not account for adjustments in CapEx based on different lifespans. In practice, materials and installation for an FWF with an extended lifespan may be more time-consuming and costly, and the present model has not considered these factors.

Figure 12 illustrates that higher capacity factor leads to a reduced LCOE. Thus, it is of high importance to choose wind farm sites with the best wind resources to achieve a capacity factor. On the other hand, as most waves are wind-driven, such sites are often associated with large waves and higher material costs for mooring systems and substructures are expected.

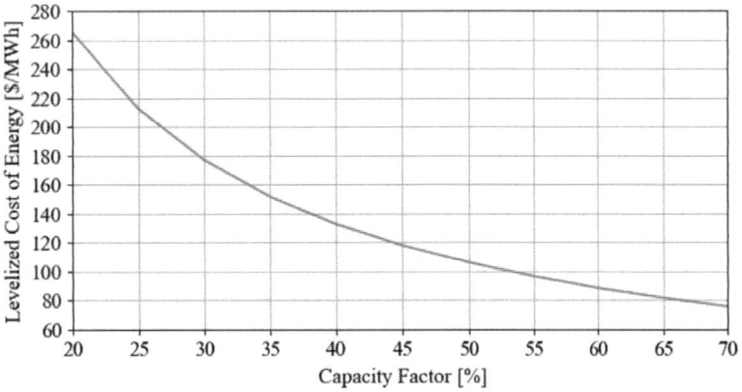

Figure 12. Variation of LCOE with capacity factor.

6.4 Financial modeling

To assess the economic viability of the wind farm, specific financial analyses have been performed, including the calculation of the Net Present Value (NPV), Internal Rate of Return (IRR), and Payback Period (PP). These calculations are derived from a discounted cash flow analysis, where the project discount rate is equivalent to the pre-tax WACC, set at 5.46%. Table 6 provides an overview of the cash flows for the initial five years and the final five years of the project, along with the resulting NPV, IRR, and PP values. These financial metrics serve as the basis for evaluating the project's economic feasibility and potential return on investment.

Table 6. Calculation of NPV, IRR, and PP

USDm	2024	2025	2026	2027	2028	2029	2046	2047	2048	2049	2050
Total Revenue	-	**333.2**	**333.2**	**333.2**	**333.2**	**333.2**	**333.2**	**333.2**	**333.2**	**333.2**	-
(-) CapEx	3 571.0	-	-	-	-	-	-	-	-	-	-
(-) OpEx	-	88.5	88.5	88.5	88.5	88.5	88.5	88.5	88.5	88.5	-
(-) Decommisioning	-	-	-	-	-	-	-	-	-	-	36.0
(+) Depreciation	-	142.8	142.8	142.8	142.8	142.8	142.8	142.8	142.8	142.8	
(-) Principal Payment	-	166.6	166.6	166.6	166.6	166.6	-	-	-	-	
(-) Interest Payment	-	118.0	110.1	102.3	94.4	86.5	-	-	-	-	-
(+) Interest Tax Deduction	-	26.0	24.2	22.5	20.8	19.0	-	-	-	-	-
Cash Flow	**- 3 571.0**	**128.9**	**135.0**	**141.2**	**147.3**	**153.4**	**387.6**	**387.6**	**387.6**	**387.6**	**- 36.0**
Cumulative Cash Flow	- 3 571.0	- 3 442.1	3 307.1	- 3 165.9	- 3 018.6	- 2 865.1	1 719.8	2 107.4	2 494.9	2 882.5	2 846.5
Disc. Rate 5.46 %		**NPV**	**-561.9**	***USDm***		**IRR**	**4.0 %**		**PP**	**18 years, 30 weeks**	

Given that the reference FWF exhibits a negative NPV exceeding $500000000, it suggests that the project is not financially viable without government subsidies. The high cost of manufacturing and installing an FWF leads to a substantial initial investment and a significant debt burden. An IRR of 4.0% is not inherently an unattractive return, but since this rate is lower than the WACC, it implies that the investment may warrant reassessment. An IRR lower than the WACC suggests that the cost of opportunity for this project may be higher than that of other alternative investments, making it potentially less profitable in comparison. Hence, it is crucial to explore alternative strategies or investment opportunities that might offer a more favorable return given the financial constraints and performance metrics of the reference FWF project.

7. Concluding remarks

This paper provides a comprehensive examination of an offshore floating wind farm situated in the Norwegian Sea. The research model involves the utilization of semi-submersible floaters, each supporting a 15-MW wind turbine, and the wind farm under investigation comprises 50 such turbines with a projected operational lifespan of 25 years. The calculated levelized cost of energy (LCOE) amounts to $100.69 per megawatt-hour, with capital expenditure accounting for 63.1% of the overall expenses. This highlights a significant challenge associated with floating offshore wind, which is the substantial upfront capital investment required for such projects.

A sensitivity analysis reveals that the most influential factors affecting the LCOE are the lifespan, capacity factor, and project discount rate. Furthermore, the financial analyses indicate that the reference wind farm is not economically viable due to a negative Net Present Value (NPV) and an Internal Rate of Return (IRR) lower than the Weighted Average Cost of Capital (WACC). This suggests that the project may not be financially feasible without government subsidies or alternative measures to address its financial challenges.

As this study is limited to a wind farm layout, idealized power curve for the wind turbines and simplified OPEX estimates, future work can address the motion characteristics of floating wind turbines, wind farm with multiple-rated wind turbines, and advanced operation and maintenance strategies. All these elements will influence the economic analysis.

Reference

[1] European Union 2020 An EU Strategy to harness the potential of offshore renewable energy for a climate neutral future.
[2] DNV, Energy Transition Norway, 2020.
[3] Emblemsvåg J 2020 On the Levelised Cost of Energy of Windfarms, *Int. J. Sustainable Energy.* **39**(7), 700 – 718.
[4] Jiang Z 2021 Installation of offshore wind turbines: A technical review. *Renewable and Sustainable Energy Rev.* **139**,110576.
[5] Ren Z, Verma AS, Li Y, Teuwen JJ and Jiang Z 2021 Offshore wind turbine operations and maintenance: A state-of-the-art review. *Renewable and Sustainable Energy Rev.* **144**,110886.
[6] Stehly T and Duffy D 2021 Cost of Wind Energy Review, National Renewable Energy Laboratory, 2022.
[7] Martinez A and Iglesias G 2022 Mapping of the levelised cost of energy for floating offshore wind in the European Atlantic, *Renewable and Sustainable Energy Rev.* **154**, 111889.
[8] Bilicic G and Scroggins S 2023 Lazard 2023 Levelized Cost Of Energy+.
[9] DNV 2023 Wind energy – going offshore.
[10] Trinomics 2020 Final Report Cost of Energy (LCOE) - Energy costs, taxes and the impact of government interventions on investments, Trinomics.
[11] NVE 2023 Kostnader for kraftproduksjon, Available: https://www.nve.no/energi/analyser-og-statistikk/kostnader-for-kraftproduksjon/.
[12] Nunemaker J, Shields M, Hammond R, and Duffy P 2020 ORBIT: Offshore renewables balance-of-system and installation tool, National Renewable Energy Laboratory, USA.
[13] Rystad Energy2023 Rystad Energy research and analysis Available: https://www.rystadenergy.com/.
[14] Organisation for Economic Co-operation and Development (OECD) 2023 available: https://stats.oecd.org/.
[15] Bosch J, Staffell I and Hawkes, A D 2019 Global levelised cost of electricity from offshore wind, *Energy*.
[16] Bjørni FA, Lien S, Midtgarden TA. Life cycle cost analysis of a floating wind farm located in the Norwegian sea, Master's thesis, University of Agder.
[17] Jonkman J, Robertson A, Masciola, A M, Song H, Goupee A, Coulling A and Luan C 2014 Definition of the Semisubmersible Floating System for Phase II of OC4, National Renewable Energy Laboratory, USA.
[18] Gaertner E, Rinker J, Sethuraman, L, Zahle, Anderson, F, Barter, B G, Abbas, N, Meng F, Bortolotti

P, Skrzypinksi W, Scott G, Feil R, Bredmose H, Dykes K, Shields M, Allen C and Viselli A 2020 Definition of the IEA 15-Megawatt Offshore Reference Wind Turbine National Renewable Energy Laboratory, USA.
[19] Equinor 2023 Equinor and partners consider 1 GW offshore wind farm off the coast of Western Norway, available: https://www.equinor.com/news/20220617-considering-1gw-offshore-wind-farm-off-western-norway.
[20] Norwegian Centre for Climate Services 2023 Observations and Weather Statistics, Oslo, Norway.

IOP Conf. Series: Materials Science and Engineering 1294 (2023) 012007 doi:10.1088/1757-899X/1294/1/012007

Two-body dynamic simulations of a combined semi-submersible floating offshore wind turbine and torus wave energy converter

C F Lee*, V De Santis Tavares, M C Ong

Faculty of Science and Technology, University of Stavanger, Norway

* Correspondence: chern.f.lee@uis.no

Abstract. Dynamic responses of a combined wind and wave energy system which consist of a 5MW semi-submersible floating offshore wind turbine (FOWT) and a torus-shaped wave energy converter (WEC) are investigated. In the investigated configuration, the torus WEC is constrained to move only in the heave direction with respect to the FOWT. This results in a two-body dynamic system that allows for the extraction of wave energy through the relative heave motion. A Modelica-based multi-body system (MBS) code is used to perform fully coupled dynamic analyses of the combined system. The wind turbine loads are calculated using the state-of-the-art Blade Element Momentum (BEM) method, while the wave power take-off (PTO) system is modelled as a linear spring and a viscous damper. A case study is presented where the mass of the torus is varied to investigate the impact on the two-body dynamics of the combined system. The system performance is assessed in terms of the maximum wave power absorption and the quality of wind power production.

1. Introduction

The concept of harnessing wave energy as a viable energy source has been actively pursued since its inception more than 200 years ago. Over the years, many types of wave energy converter (WEC) designs, each varying in working principle and power-take off (PTO) mechanism have been proposed. Nonetheless, progress in wave energy technology has been sluggish, with limited consensus on an optimal shape or operating principle having been reached. Of the thousands of proposed concepts, only a few designs have reached the stage of commercialized deployment, far lower in number as compared to other renewable sources, such as solar energy and wind energy [1]. The current poor state of commercial utilization of wave energy technology can be attributed mainly to the difficulty in achieving high energy conversion efficiency over a range of wave amplitudes, phases, and directions [2]. Additionally, the extreme environmental loadings on the WEC structures can be 100 times higher than the average loadings, which can lead to overdesign of load bearing structures and subsequently a significant increase in the cost of energy [3].

To improve the chance of commercialization, one of the solutions that researchers have been investigating is to integrate WECs with floating offshore wind turbine (FOWTs) in the form of combined energy systems. The advantages of using such combined systems lie in the synergies between two different disciplines of energy production. To begin with, as wind and waves are statistically correlated, sites with high wind energy resources usually come with significant wave energy density which makes the combined concept a sensible solution. In addition, the integration of multiple energy devices onto a single platform allows for higher energy production while sharing the same infrastructure, e.g., supporting structures, power cables, mooring systems, and substations. For a combined energy system, the motions of the floating platform can also be utilized to improve the performance of the WEC. Falnes [4] investigated the mechanism of wave energy conversion of a two-body WEC in heave mode and discovered that it is possible to achieve wave excitation forces on a two-body WEC that exceed those experienced by a single-body WEC. The conversion efficiency of a two-body WEC was also studied by Bijun Wu et al. [5]. It was found that the two-body motion characteristics are dependent on many parameters such as physical properties of the buoy and the incoming wave frequency. Liang and Zuo [6] described the dynamics of a two-body WEC by deriving the close-form solution of the power absorption as a function of mass and PTO parameters such as damping and stiffness. A notable commercial example of a two-body WEC is the Wavebob, developed by Wavebob Ltd. [7]. It uses the relative heave motion between the floating buoy and a second submerged body to extract wave energy. However, in 2013, the company was placed in liquidation being unable to attract sufficient financial investment [8].

To date, offshore wind technologies have reached a high technology readiness level, and a substantial reduction of 20% in the levelised cost of energy (LCOE) of offshore wind projects has been observed between 2010 and 2018 [9]. Taking advantage of the technological maturity of the wind industry, the combined exploitation of wave and offshore wind energy can potentially lead to a lower threshold of commercialization. This allows the wave energy technology development to accelerate with increased funding. The MARINA Platform project [10], which was funded by European Union (EU) shortlisted three floating wind and wave energy concepts for detailed investigation. The Spar-Torus-Combination (STC) [11], the Semi-submersible-Flap Combination (SFC) [12] and the combination of WindFloat semi-submersible wind turbine with an oscillating water column WEC array [13] make up the three characteristically different combined energy concepts. One of the common findings from the three concepts is that the introduction of WECs to a FOWT can modify the response characteristics of the floating platform. For the STC concept, Muliawan et al. [14] discovered that surge and pitch responses of the spar FOWT is reduced with the addition of a torus (donut-shaped) WEC, contributing to a stable pitch motion. Lee et al. [15] proposed a combined energy concept named STFC that integrates a semi-submersible FOWT with three flap type and one torus WECs. Their analysis showed that the presence of the WECs contributed to a reduced platform pitch motion and hence a reduction in wind turbine power fluctuation.

On the path towards complete commercialization of combined wind and wave energy systems, continued research is needed to improve conversion efficiency and to minimize the development costs. In this study, the two-body dynamics of a combined wind and wave energy system are investigated. A combined concept that integrates a column stabilized semi-submersible FOWT and a torus type WEC and is chosen as the case study, primarily due to its ease of integration and the straightforwardness of the wave power generation mechanism. To investigate the effect of torus buoy's physical properties on power absorption, the buoys with three different ballast mass are modelled. The performance of the combined system is compared in terms of wave power absorption and the resulting motion characteristic of the platform.

2. The combined wind and wave energy concept

The combined wind and wave energy concept in this study consists of a semi-submersible FOWT and a torus WEC as shown in Figure 1.

2.1. FOWT

The FOWT is of the 5-MW-CSC type that was originally proposed by Luan et al. [16] for a design water depth of 200 m. The braceless semi-submersible platform provides the motion stability and support required for the safe operation of a NREL 5-MW reference offshore wind turbine [17]. The supporting platform consists of four cylindrical columns of identical diameter, interconnected in a "Y" configuration via pontoons. The ballast water is distributed evenly across the pontoons and columns to attain an operating draft of 30 m. Station-keeping of the CSC under environmental loadings is achieved through a 3-legged wire-rope-catenary mooring system. The specifications of the 5-MW-CSC platform is shown in Table 1.

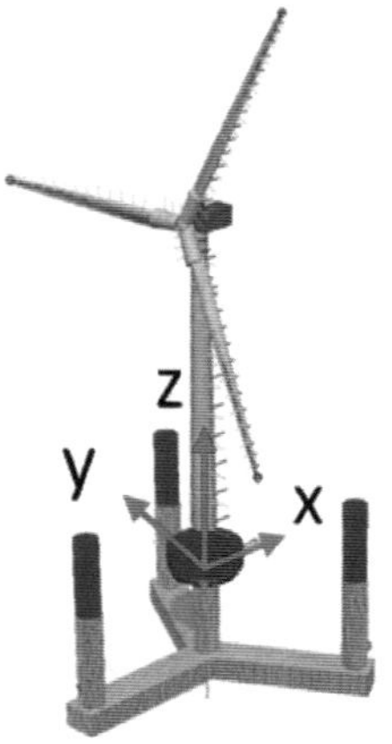

Figure 1. FOTW and Torus modelled in SIMA

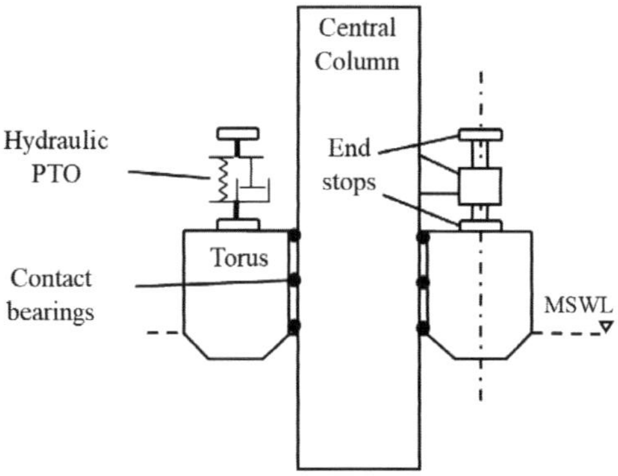

Figure 2. PTO mechanism of the torus WEC

Table 1. 5-MW-CSC platform specifications [16]

Draft (m)	30.0
Mass (kg)	9738000
Center of gravity (m)	-24.36
Center of buoyancy (m)	-22.48
Roll mass moment of inertia about COG ($kg \cdot m^2$)	4.72E+09
Pitch mass moment of inertia about COG ($kg \cdot m^2$)	4.72E+09
Yaw mass moment of inertia about COG ($kg \cdot m^2$)	8.24E+09

2.2. *Torus WEC*

The torus WEC comprises of two components; (1) a torus buoy installed through the central column of the FOWT and (2) the PTO system and mechanical connections. The PTO and mechanical connections facilitate the constrained relative motion between the buoy and the CSC. In its neutral position, the torus buoy is freely floating with a 2 m draft. The physical dimensions and properties of the torus WEC have been estimated according to published information of the STC [14,18]. Through contact bearings fitted around the inner diameter of the torus, it is constrained to move together with the CSC in surge, sway, roll, pitch and yaw. The PTO system, consisting of hydraulic cylinders and the hydraulic circuit converts the torus's heave motion into usable power. Mechanical end stoppers are used to limit the torus's heave motion to ±3 m in relation to the FOWT. The PTO mechanism of a torus WEC is illustrated in Figure 2. The power absorbed by the torus WEC can be derived from the relative heave velocity, v_{rel} between the CSC and the torus given by,

$$P_{\mathrm{Torus}} = B_{\mathrm{PTO}} \cdot v_{\mathrm{rel}}^2(t) \tag{1}$$

where B_{PTO} is the PTO damping coefficient. The PTO system in the present study utilizes a passive (constant damping and stiffness) control strategy which sees the force exerted by the PTO being represented by

$$F_{\text{Torus}}(t) = -B_{\text{PTO}} \cdot v_{\text{rel}}(t) - K_{\text{PTO}} \cdot x_{\text{rel}}(t) \tag{2}$$

where K_{PTO} is the PTO spring stiffness and x_{rel} is the relative heave position between the CSC and the torus. One of the primary limitations of passive control strategies is that they are pre-tuned to optimize energy absorption within a narrow frequency range. However, while passive control strategies are no longer considered state-of-the-art in modern WEC controllers, they remain a choice for developers willing to test a WEC at an industrial scale due to their ease of implementation and operational safety. Given the focus of this study on examining the impact of varying torus mass on WEC performance, a passive damping control strategy is employed.

Three tori with different ballast masses are modelled to investigate its effect on platform motions and power absorption. The difference in mass is achieved through ballasting of the tori to different draft levels. Slight modifications to the tori's submerged geometries as shown in Figure 3 are made in order to limit the change in hydrodynamic coefficients (added mass and potential damping) to a minimum. Table 2 provides a summary to the properties of the tori investigated in this study.

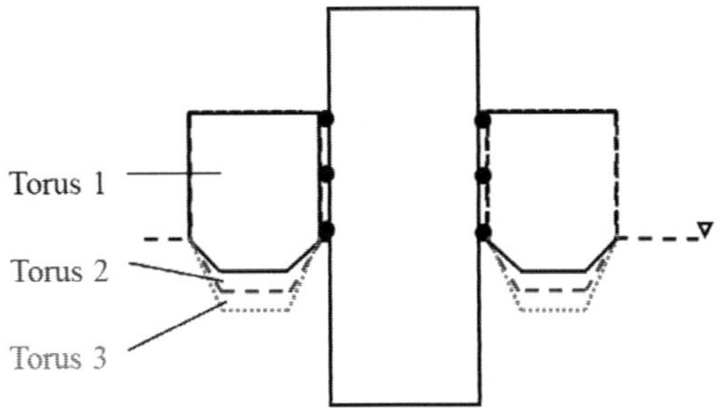

Figure 3. Comparison of submerged geometries for the three tori

Table 2. Torus specifications

	Torus #1	Torus #2	Torus #3
Draft (m)	2.0	2.5	3.0
Torus mass (kg)	423766	529710	635654
Center of gravity (m)	0.92	0.11	-0.52
Center of buoyancy (m)	-0.91	-1.13	-1.36
Roll mass moment of inertia about COG ($kg \cdot m^2$)	1.71E+07	2.04E+07	2.36E+07
Pitch mass moment of inertia about COG ($kg \cdot m^2$)	1.71E+07	2.04E+07	2.36E+07
Yaw mass moment of inertia about COG ($kg \cdot m^2$)	2.62E+07	3.21E+07	3.82E+07

3. Dynamic modelling and analysis of the combined concept

The combined concept is modelled using OpenModelica (OM), an open-source platform based on the Modelica language developed by the Open Source Modelica Consortium [19]. Using OM as the primary platform for development, the fully coupled aero-hydro-servo-elastic framework for the dynamic analysis of FOWT is described by El Beshbichi et al. [20].

3.1. Aerodynamics

The Blade Element Momentum (BEM) method as described by Glauert [21] is the most commonly used tool for calculating aerodynamic loads on wind turbine rotors due to it being computationally inexpensive while preserving reasonable accuracy. In the present study, the BEM calculations are implemented through AeroDyn v15, the wind turbine aerodynamic module within NREL's FAST v8.2

[22]. AeroDyn calculates wake influence on the wind turbine based on quasi-static BEM theory which means that the induction factors react instantaneously to the change in aerodynamic loads. In the BEM solution, the Glauert's empirical correction is applied for high axial induction factors while Prandtl tip-loss, Prandtl hub-loss, and Pitt and Peters skewed-wake are applied to account for the corrections due to 3-dimensional (3D) effects. Turbulent wind inputs are generated using TurbSim [23], a stochastic inflow turbulence tool by NREL.

3.2. *Hydrodynamics*
The hydrodynamic loads on the submerged body are calculated based on the linear wave theory. Frequency-domain hydrodynamic analysis is carried out using DNV's Wadam software [24] to compute the added mass, potential damping coefficients and first-order wave force transfer functions. In the present study, the added mass and potential damping couplings between CSC and the torus are assumed to be negligible. The calculations of first-order wave force transfer functions for the CSC take into account the presence of the torus, and similarly, the calculations of first-order wave force transfer functions for the torus consider the presence of CSC.

3.3. *Mooring line dynamics*
The station-keeping capability is achieved through the implementation of a quasi-static mooring line module for each mooring line. With the fairlead positions known at each time instance, the mooring line module solves for and output the tensions at fairleads, assuming that the mooring lines are in static equilibrium during each simulation time step.

3.4. *Connections between the FOWT and torus*
In OM, the contact between CSC and the torus is modelled as a cylindrical joint which allows for relative motion in the heave direction. In SIMA, however, the contact is realized using the "docking cone and pin" feature [25]. This feature keeps the torus and the center column of the CSC axially aligned by applying a restoring force that is a function of the radial offset. The PTO is modelled as a linear damper with damping coefficient of 8000 kNs/m and a linear spring with stiffness of 10 kN/m in parallel. The tuning of the PTO is according to the values suggested by Muliawan et al. [14].

3.5. *Rigid body and structural dynamics*
A multi-body approach is employed to model the floater dynamics and the wind turbine structural dynamics of the combined system. The floaters (CSC and torus) are modelled as rigid bodies with concentrated masses. The equations of motion of the rigid floaters can be given by,

$$[M + A_\infty]\ddot{q} + \int_0^t [K(t-\tau)]\,\dot{q}\,d\tau + [C]q = [F]_{\text{waves}} \tag{3}$$

where $[M]$ is the mass matrix, $[A_\infty]$ is the infinite frequency added mass matrix, $[K]$ is the wave-radiation-retardation kernel matrix, $[C]$ is the restoring coefficient matrix, $[F]_{\text{waves}}$ is the first-order wave force and q, $\dot{q}$, $\ddot{q}$ are the position, velocity and acceleration of the floaters, respectively. The wind turbine tower and blades are discretized as generalized beam elements (GBE) assuming small, linear, and elastic structural deformations. For each GBE, longitudinal, transverse, and torsional degrees-of-freedom (DOFs) are considered each with the corresponding stiffness values. More information on the modelling of structural dynamics can be referred to the work by El Beshbichi et al. [20].

4. Results and discussion
The numerical model established in OM is first validated against an identical twin created in SIMA, a dynamic analysis software developed by SINTEF Ocean [25,26]. Based on the validated model, the combined systems with different torus masses are simulated in regular wave conditions. One representative irregular waves and turbulent wind environmental condition (EC) is also simulated to look at the average hourly-power production of the WEC.

4.1. Code-to-code validation study

In the present study, regular waves with thirteen different wave periods as shown in Table 3 are simulated. Figures 4, 5, 6 and 7 show the resulting surge, heave, pitch and absorbed power response amplitude operators (RAOs), respectively, for the numerical models developed in OM and SIMA. Only the case with Torus no.1 is used in the validation study.

Table 3. Wave periods used for validation study and regular wave simulations in OM and SIMA

Wave periods (s)												
5.01	6.97	7.93	8.91	9.89	10.86	11.83	12.81	13.78	14.76	15.73	17.68	19.74

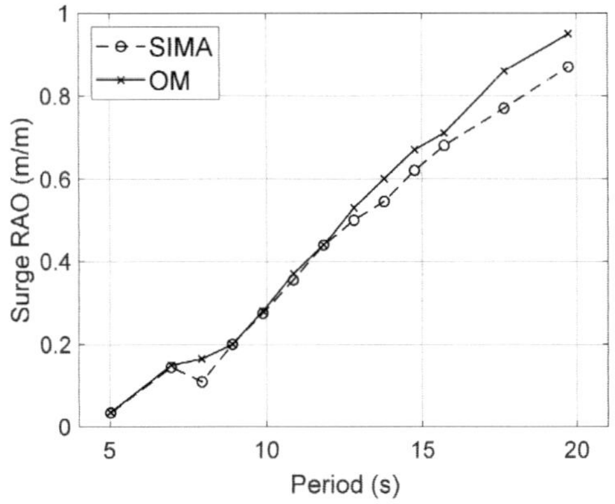

Figure 4. Surge RAO of the platform in SIMA and OM.

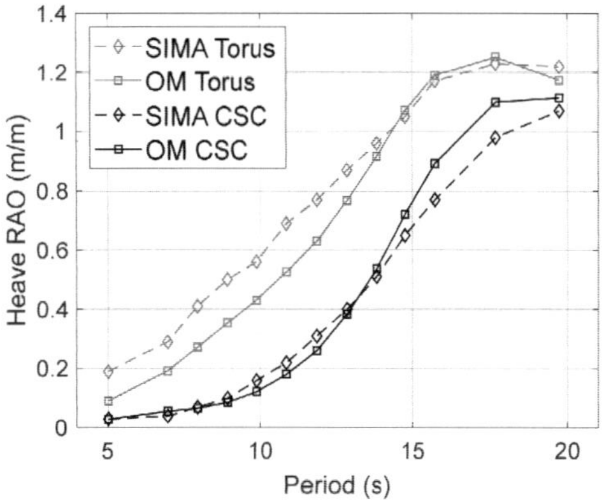

Figure 5. Heave RAO of the platform and torus in SIMA and OM.

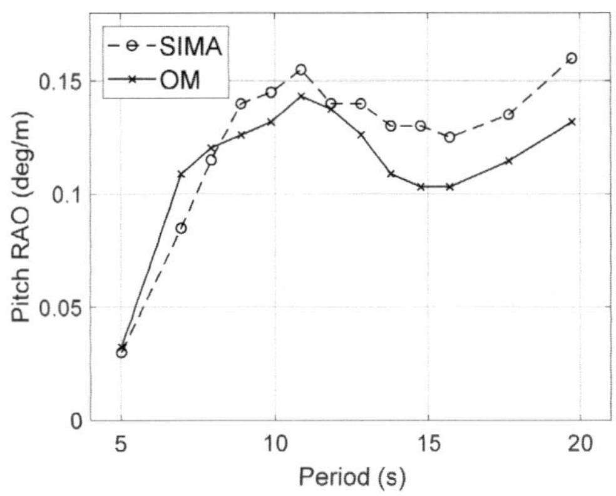

Figure 6. Pitch RAO of the platform in SIMA and OM.

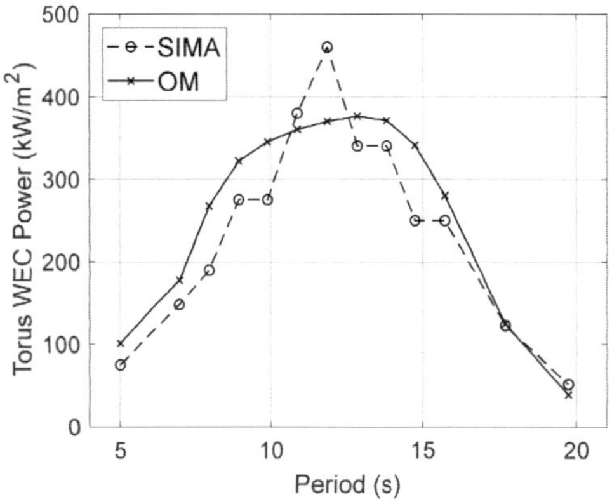

Figure 7. Absorbed power RAO of the WEC in SIMA and OM.

As shown in Figures 4 and 5, the surge and heave RAO values calculated by OM and SIMA are in close agreement. As compared to surge and heave RAOs, larger discrepancies are observed between the pitch RAO values estimated using different tools (Figure 6), and this is due to the difference in which the contact between the platform and torus is modelled. For the model in SIMA, the torus is allowed to have a slight relative pitch motion with respect to CSC while in OM, the torus follows CSC in pitch. In any case, within the investigated wave frequency range, pitch RAOs are in general less than 0.2° and thus the discrepancies are deemed insignificant for the purpose of this study, which revolves primarily around the heave DOF. As shown in Figure 7, the absorbed power RAO of the torus WEC peaks when the system is excited by regular waves with a period of approximately 12 s. Within the resonance frequency range, the absorbed power RAO calculated using SIMA are of higher values as compared to those calculated using OM. This can be attributed to a much-restricted pitch motion for the torus modelled in OM as shown in Figure 6.

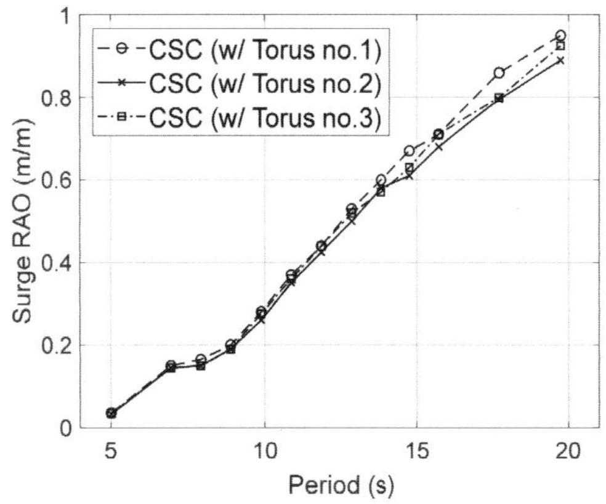

Figure 8. Surge RAO of the platform with three different torus masses

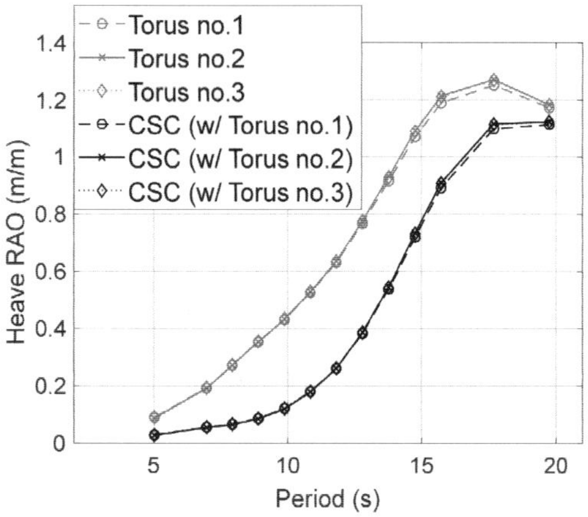

Figure 9. Heave RAO of the platform and the tori with three different masses

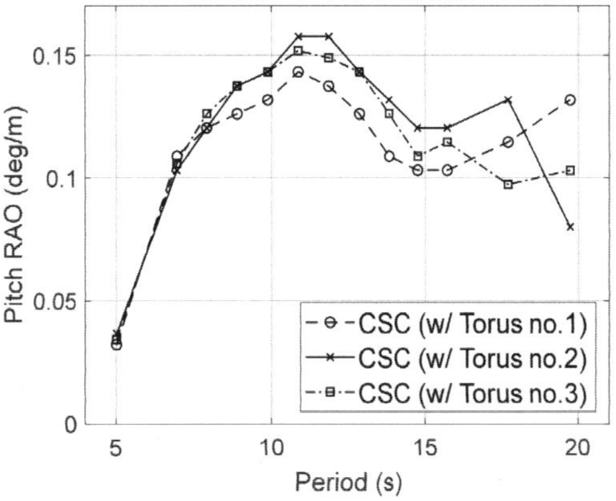

Figure 10. Pitch RAO the platform with three different torus masses

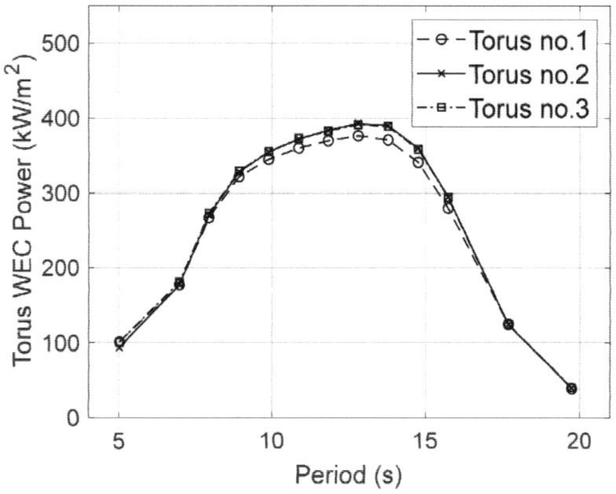

Figure 11. Absorbed power RAO of the WEC with three different torus masses

4.2. Regular wave simulations

In this section, regular wave simulations with the same wave periods as listed in Table 3 are carried out for the combined system by varying the torus properties (Torus no.1 is replaced with Torus no.2 and Torus no.3). The RAOs for the platform surge, platform heave, torus heave and platform pitch motions and the power absorbed by the PTO for the three investigated cases are compared in Figures 8 – 11, respectively.

As shown in Figures 8 and 9, the motions in the surge and heave DOFs are not significantly affected by the change in torus mass as the motion characteristics in these DOFs are influenced by the mooring system and the platform geometry of CSC. In pitch, the platform oscillates with higher RAO values with Torus 2 and Torus 3 as compared to Torus 1 as shown in Figure 10, with the pitch RAO of the platform with Torus 2 being the highest. This directly affects the absorbed power RAOs. As shown in Figure 11, with a wave period of approximately 13 s, a 6% increment in the absorbed power RAO is observed for the combined energy system with Torus no. 2.

4.3. Irregular waves and turbulent wind simulations

In this section, the combined energy system with three different torus masses is simulated in an irregular wave and turbulent wind condition. The selected sea state is based on the wind-wave joint probabilistic model proposed by Johannessen et al. [27] and has a significant wave height, H_s of 2.10 m and a peak period, T_p of 9.74 s. Three-dimensional turbulent wind fields of 5 m/s wind speed at wind turbine hub height are generated according to IEC Class C of the Kaimal turbulence model. Wind and waves are directionally aligned pointing in the positive x-direction as shown in Figure 1. Finally, to account for statistical variation, six random wind and wave realizations are simulated for the EC.

Figures 12, 13 and 14 show the average statistics of platform surge, platform heave, torus heave and platform pitch motions each presented in terms of mean offset and standard deviation of the corresponding motion. The platform's motion in surge is governed mainly by the mooring system. Hence, the mean offset and standard deviation for the platform surge motion remain largely unchanged when the torus mass increases as shown in Figure 12. Similar to the surge motion, the change in torus mass does not result in significant change in platform pitch motion as shown in Figure 13. This is due mainly to the unchanged water plane area (the water plane areas for all tori are the same). Figure 14 shows that the mean and standard deviation of the tori's heave motions are significantly higher than the platform. With a higher torus mass, an increase in torus heave standard deviation is observed. However, as the torus mass is increased further (from torus no.2 to torus no.3), smaller heave responses are detected for both CSC and the torus. This indicates that, with the current PTO parameter tuning, the mass ratio between CSC and torus no.2 reaches an optimal level.

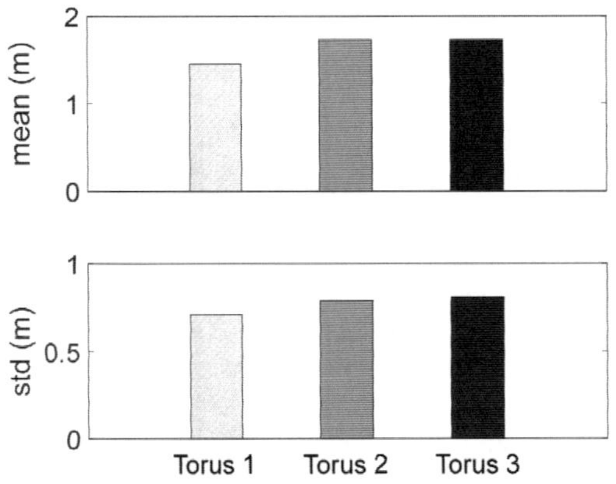

Figure 12. Platform surge motion statistics with three different torus masses

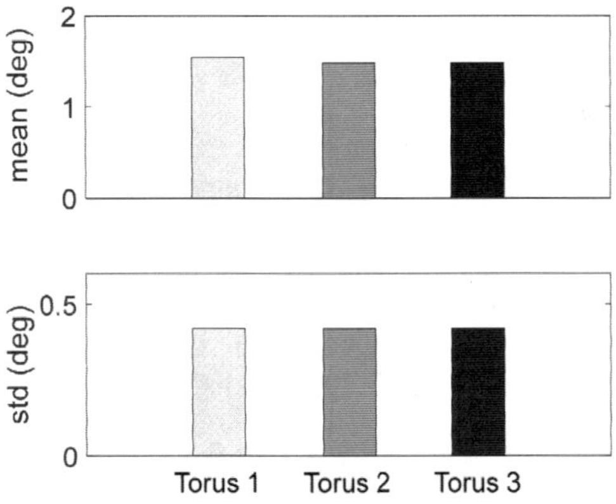

Figure 13. Platform pitch motion statistics with three different torus masses

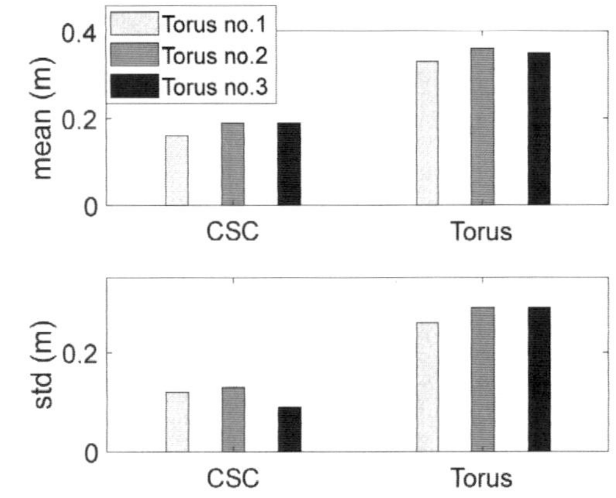

Figure 14. Platform and torus heave motion statistics with three different torus masses

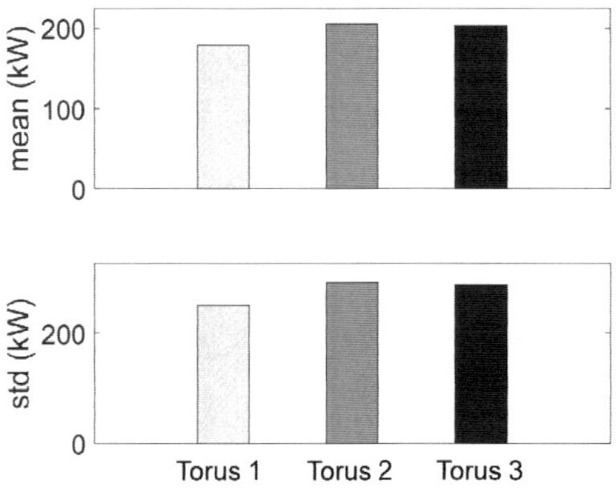

Figure 15. Absorbed power statistics for the WEC with three different torus masses

Figure 15 shows the average statistics of the WEC absorbed power. An increase in torus motion standard deviations as shown in Figure 14 directly results in an increase in mean absorbed power. For the investigated EC, the mean absorbed power of the combined system with Torus no. 2 and Torus no.3 increase by 15% and 14%, respectively. This increase in mean absorbed power is accompanied by an increase in the absorbed power standard deviation of approximately 15%. It can therefore be concluded that a smaller mass difference between the torus and platform will improve the performance of the combined wind and wave energy system in terms of wave power extraction.

5. Conclusion

A combined wind and wave energy system, consisting of a semi-submersible and a torus WEC, is proposed in the present study. The numerical model developed in OM is first validated using SIMA by comparing the results in regular wave simulations. Satisfactory agreement is found between the results calculated by OM and SIMA. To explore the impact of increasing torus mass on the platform's motion characteristics and the WEC's power absorption, two additional torus designs with different ballast masses have been introduced. RAO results show an increase in power absorption as the torus mass is increased while the platform motions remain insensitive to the increase in torus mass. One sea state is used to investigate the performance of the combined system in irregular waves and turbulent wind environmental conditions. With passive control system, it can be observed that an increase in torus mass positively impacts WEC power absorption, with a 15% rise in 1-hour mean power absorption under the examined environmental condition. Further improvements to the WEC system such as the implementation of a more advanced control system, further tuning of the control parameters (for each torus mass) and the optimization of torus shape are among the subjects for future studies.

References

[1] Aderinto T and Li H 2018, Ocean wave energy converters: status and challenges, *Energies*, **11**(5), 1250.

[2] Czech B and Bauer P 2012, Wave energy converter concepts: design challenges and classification", *IEEE Industrial Electronics Magazine*, 6(2), 4-16.

[3] Cruz J 2008, *Ocean Wave Energy: Current Satus and Future Perspectives*, Springer Berlin, Heidelberg.

[4] Falnes J 1999 Wave-energy conversion through relative motion between two single-mode oscillating bodies, *J. Offshore Mech. Arctic Eng.* **121**, 32-38.

[5] Wu B, Wang X, Diao X, Peng W and Zhang Y 2014, Response and conversion efficiency of two

degrees of freedom wave energy device, *Ocean Eng.* **76**, 10-20.

[6] Liang C, Zuo L, 2017 On the dynamics and design of a two-body wave energy converter. *Renewable Energy*, **101**, 265-274.

[7] Weber J, Mouwen F, Parish A and Robertson D 2009, Wavebob – research and development network and tools in the context of systems engineering, Proc. 8th European Wave and Tidal Energy Conference, Uppsala, Sweden.

[8] O'Halloran B, 2013, Ocean energy developer wavebob set to go under, The Irish Times, Ireland.

[9] Jiang Z, 2021 Installation of offshore wind turbines: A technical review, *Renewable and Sustainable Energy Rev.* **139**, 110576.

[10] Sojo M and Auer G 2014 Deliverable D1.12 Final Summary Report, Marine Renewable Integrated Application (MARINA) Platform.

[11] Muliawan M J, Karimirad M, Moan T, and Gao Z 2012 STC (Spar-Torus Combination): a combined spar-type floating wind turbine and large point absorber floating wave energy converter – promising and challenging", *Proc. of the ASME 31st Int. Conf. on Ocean, Offshore and Arctic Eng.*, Rio de Janeiro, Brazil.

[12] Luan C, Michailides C, Gao Z, and Moan T 2014 Modeling and analysis of a 5 MW semi-submersible wind turbine combined with three flap type wave energy converters, *Proc. of the ASME 33rd Int. Conf. on Ocean, Offshore and Arctic Eng.*, San Francisco, California, USA.

[13] O'Sullivan K P 2014 Feasibility of Combined Wind-wave Energy Platforms, Ph.D. thesis, National University of Ireland, Ireland.

[14] Muliawan M J, Karimirad M and Moan T 2013 Dynamic response and power performance of a combined Spar-type floating wind turbine and coaxial floating wave energy converter, *Renewable Energy*, **50**, 47-57.

[15] Lee C F, Tryfonidis C and Ong, M C 2022 Power performance and response analysis of a semi-submersible wind turbine combined with flap-type and torus wave energy converters, *J. Offshore Mech. Arctic Eng.*, **145**(4), 042001.

[16] Luan C, Gao Z, and Moan T 2016 Design and analysis of a braceless steel 5-mw semi-submersible wind turbine, *Proc. of the ASME 35th Int. Conf. Ocean, Offshore Arctic Eng.*, Busan, South Korea.

[17] Jonkman J, Butterfield S, Musial W and Scott G 2009 Definition of a 5-MW Reference Wind Turbine for Offshore System Development, NREL/TP-500-38060, NREL, CO, USA.

[18] Wan L, Gao Z and Moan T 2014 Dynamic response and power performance of a combined Spar-type floating wind turbine and coaxial floating wave energy converter, *Proc. ASME 33rd Int. Conf. Ocean, Offshore Arctic Eng.*, San Francisco, California, USA, 9A, V09AT09A010. DOI: 10.1115/OMAE2014-23213

[19] OSMC, 2021, OpenModelica User's Guide. Release v1.19.0.

[20] El Beshbichi O, Xing Y and M C Ong 2023, Modelica-AeroDyn: Development, benchmark, and application of a comprehensive object-oriented tool for dynamic analysis of non-conventional horizontal-axis floating wind turbines, *Wind Energy*, **26**(6), 538-572.

[21] Glauert H, 1963 Airplane propellers, W.F. Durand (Ed.), Aerodynamic theory, Dover Publications, New York.

[22] Moriarty P J and Hansen A C 2005 AeroDyn Theory Manual, NREL/TP-500-36881, NREL, CO, USA.

[23] Jonkman B J 2009 TurbSim User's Guide: Version 1.50, NREL/TP-500-46198, NREL, CO, USA.

[24] DNV, 2013, SESAM User Manual HydroD, Det Norske Veritas, Oslo, Norway.

[25] SINTEF Ocean, 2017, SIMO 4.10.1 User Guide, SINTEF Ocean: Trondheim, Norway.

[26] SINTEF Ocean, 2017, RIFLEX 4.10.1 User Guide, SINTEF Ocean: Trondheim, Norway.

[27] Johannessen K, Meling T S and Haver S 2002, Joint distribution for wind and waves in the northern North Sea, *Int. J. Offshore Polar Eng.*, **12**(2).

Selecting the inlet configuration for gas-liquid cylindrical cyclone separator

L V Sy[1*] and N T Lan[2]

[1] PetroVietnam University (PVU), Vietnam
[2] Ho Chi Minh University of Technology (HCMUT), Vietnam

[*] Correspondence: sylv@pvu.edu.vn

Abstract. Gas-Liquid Cylindrical Cyclone Separator (short for GLCC) is a compact separator that has been widely used in petroleum production. The inlet configuration has significant impact of the GLCC capacity separation. By utilization of both FEM simulation and practical experiments, three inlet configurations and twelve inclined angles are investigated to find the steadily operating inlet configuration for industrial applications. In this study, the flow pattern and kinematic behaviour of each configuration are thoroughly analysed. Besides, the distribution of three velocity components of the body flow in GLCC is compared with one of the practical experiments. The result shows that the current inclined-inlet angle of 27° is not the steadily operating angle for industrial applications. In steady, it is a range of inclined inlet angle from 0o to 35o for three inlet configurations. The dual-inlet configuration indicates better properties for separating efficiency.

1. Introduction

The Gas Liquid Cylindrical Cyclone (GLCC) separator is an attractive substitute for the traditional vessel-type separator which is extensively utilized in the petroleum industry and holds promise for various field applications (Figure 1). This separator has been extensively examined in numerous countries across the globe, with particular emphasis on the research conducted by scholars affiliated with Tulsa University in the United States (TUSTP). The predominant focus of researchers has been directed towards the examination of hydrodynamic flow behavior, the evaluation of separating efficiency, and the development of mechanistic models [1-7]. In a study conducted by Kouba [1], a series of experiments were conducted on three GLCC separators featuring a downward inclined inlet. These GLCCs were characterized by having a body and inlet with equal diameters. In the experiments, two GLCCs with diameters of 2.54 cm and 5.08 cm were utilized, while the TUSTP conducted a test on a GLCC separator with a larger diameter of 7.62 cm. Nissan [2] conducted the initial experimental investigation by introducing water into a GLCC through two tangential inlets. Subsequently, there is an increase in the area ratio between the GLCC body and inlet. Consequently, the axial velocity distribution undergoes a modification that alters the flow direction in the vicinity of the center of the GLCC separator. In a previous study, Wang [3] investigated the occurrence of liquid carry-over within GLCCs and proposed the utilization of a gradually reduced inlet nozzle configuration to enhance operational versatility. However, Wang did not address the flow characteristics and gas carry-under phenomena within these GLCCs. The decay of swirling flow within a GLCC body has been investigated by Iozia [4]. The researchers have conducted an examination of the distribution of tangential velocity and have

IOP Conf. Series: Materials Science and Engineering 1294 (2023) 012008 doi:10.1088/1757-899X/1294/1/012008

identified the presence of two distinct flow regions within the GLCC separator. These regions consist of a central area characterized by forced-vortex flow, as well as a surrounding region characterized by free-vortex flow. A number of investigators have successfully built mechanistic models of the GLCC separator. Movafaghian [5] collected experimental data for the GLCC separator at different operating pressures. The researcher then took measurements for the equilibrium liquid level and the operational envelope of liquid carryover. Furthermore, an initial mechanistic model has been created to precisely predict the general flow properties in a GLCC separator. The authors forecasted the hydrodynamic flow behavior in the GLCC using numerical codes and mathematical models. These models were then validated by comparing their results with experimental data.

Erdal [6] has conducted Computational Fluid Dynamics (CFD) simulations using a commercially available software package known as CFX. The simulations have provided comprehensive insights into the hydrodynamic flow characteristics within the GLCC for both single-phase and two-phase flow scenarios. A GLCC with an inclined inlet angle of 27° is the name of the simulation model that was used in their investigation as the experimental model. Axial and tangential velocities, as well as turbulent intensities, were measured at 24 different axial positions along the GLCC's diameter in a study by Erdal [7]. A Laser Doppler Velocimeter (LDV) was used to obtain these measurements. For a variety of inlet configurations, the experiment comprised injecting inlet water at flow rates of 10, 30, and 72 gpm with a wide range of Reynolds Numbers, from 5000 to 67,000. The acquired experimental data is employed to produce contour plots that illustrate the distribution of turbulent kinetic energy, tangential velocity, and axial velocity. Using CFD simulations, Hreiz [8] studied the swirling hydrodynamics of different GLCC separator inlet configurations. The author has conducted numerous turbulent models for Gas-Liquid Cylindrical Cyclone (GLCC) and has demonstrated that the high-Reynolds realizable k-epsilon model exhibits superior performance in predicting the local mean axial and tangential velocities.

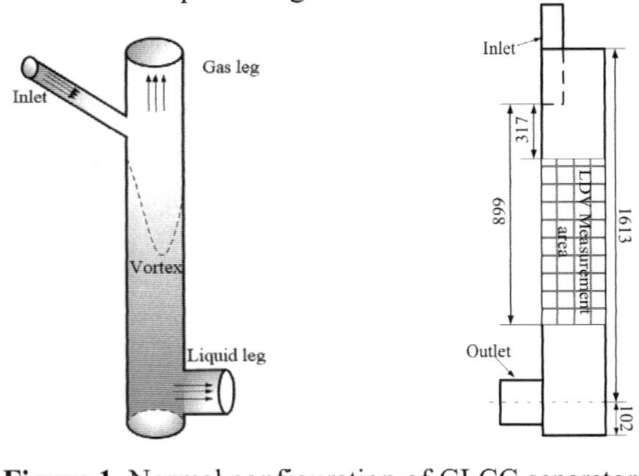

Figure 1. Normal configuration of GLCC separator.

The majority of the research conducted so far with the aim of providing a comprehensive understanding of the impact of various factors, such as inlet area, inlet dimensions, inlet section shape, number of inlets, and inlet position, on the flow pattern and performance of conventional hydro-cyclone separators [1-7]. Limited research has been conducted on the impact of inlet angle on performance, with conflicting findings reported [8-9]. There is a lack of mechanistic models and fundamental research available regarding the impact of various inlet configurations on the performance of compact separators for cylindrical cyclone separators [10-11]. The determination of a suitable range of inlet angle and inlet geometry for consistent operation in the petroleum sector's industrial applications, particularly for the GLCC where the inlet diameters are smaller than the body diameter, remains a topic of significant research interest among scholars. The corpus of experimental work on flow in cyclones that has already been done has mostly looked into how changes in geometry and operating parameters affect the efficiency of separation. Cyclone separation efficiency typically increases in proportion to increases in their inlet velocity. However, it is important to note that this enhancement in efficiency is accompanied by an elevation in the pressure drop experienced across the cyclone.

This study examines various configurations of GLCC with varying inclined inlet angles, number of inlets, and cross-sectional shapes to analyze the hydrodynamic characteristics of the inner flow within the GLCC. Numerical simulations are conducted for two cases: (1) a single inlet nozzle with a gradually reduced size, and (2) a dual inclined inlet nozzle configuration. The comparative analysis of simulation results utilizing various inlet configurations has been conducted in relation to experimental data pertaining to axial and tangential velocities. This study aims to explore the optimal approach for examining the consistent operational range of the inlet configuration of the GLCC in industrial settings. Eleven CFD models were used in the study to investigate how different inclined inlet angles affected GLCC separators with single and dual inlets. The distribution of radial, axial, and tangential velocity profiles, as well as their maximum magnitudes, were carefully taken into account in this study in relation to the variation of the inlet angle.

2. Experimental setup and CFD simulation

2.1. Experimental setup

Erdal [7] conducted localized measurements of axial and tangential velocities, as well as turbulence parameters, at TUSTP utilizing LDV device. Three distinct inlet configurations were used in the experimental setup: Figure 2a shows the single inclined inlet; Figure 2b shows the single reduced inlet; and Figure 2c shows the dual inclined inlet. The operating parameters for the single-phase flow are presented in Table 1. Measurements are taken at various positions along the diameter, ranging from 317 mm to 899 mm below the inlet. The angle of inclination of the inlet in all configurations is 27°. Via a comparison of the axial and tangential velocities acquired from real-world experiments, the accuracy of modeling models at this inclined angle was confirmed.

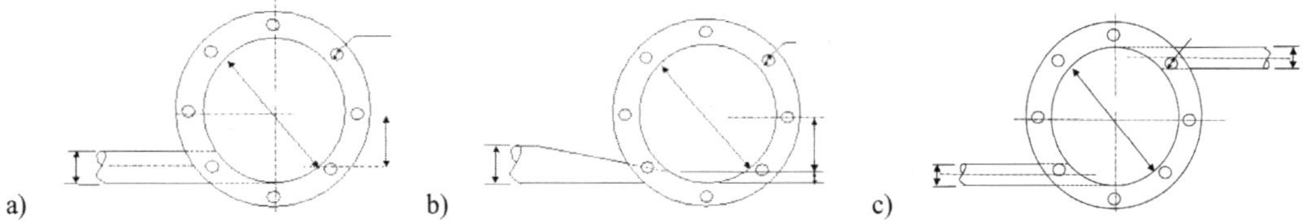

Figure 2. Configurations of inlet nozzle GLCC. a) Single inclined inlet; b) Single gradually reduced inlet; c) Dual inclined inlet

Table 1. Former experiment parameters.

Case	Working fluid	Flow rate (m^3/s)	Average velocity (m/s)	Reynold number	Viscosity (cP)
1	Water	0.00063	0.102	9285	1
1	Water	0.00454	0.731	66855	1
2	Water	0.00063	0.102	9285	1
2	Water	0.00454	0.731	66855	1
3	Water	0.00063	0.102	9285	1
3	Water	0.00454	0.731	66855	1

2.2. CFD simulation

The experimental inquiry aimed at gaining a comprehensive understanding of the intricate flow dynamics within the GLCC body is a method that incurs significant expenses and consumes a considerable amount of time. Conversely, the utilization of CFD simulations, incorporating an appropriate turbulence model and boundary condition, presents potential benefits such as the elimination of specialized measurements, reduced time requirements, and lower costs. CFD simulation is considered the most effective method for analyzing the operational performance of GLCC separator in a steady-state condition prior to their fabrication. The comparison of simulated results with experimental data holds significant importance as it demonstrates the efficacy of the CFD model in accurately predicting the intricate flow characteristics of the GLCC separator. This model can be employed to examine the impact of operational parameters on the performance of the GLCC, which have not been explored through empirical investigations.

The appropriate selection of a turbulence model for CFD simulation of swirl flow within a cyclone separator is a crucial undertaking that directly impacts the resultant outcomes. The flow pattern observed within a cyclone separator exhibits characteristics that can be described as nearly turbulent, displaying intricate and anisotropic behavior [3-10]. Sy [8] has demonstrated in their study on CFD simulation of GLCC separators that the Reynolds stress turbulence model is the most suitable option for accurately predicting the kinematic behavior of the flow within the GLCC body. This study involves the modeling of 22 different GLCC models with three distinct inlet configurations for the purpose of conducting CFD simulations. The mesh utilized for numerical analysis was created through the partitioning of the GLCC body into multiple geometric blocks, as depicted in Figure 3a, 3b, and 3c. The majority of these blocks are correctly identified for the purpose of generating a mesh with hexahedral elements, utilizing the intelligent tools provided by ANSYS Meshing 15.0.

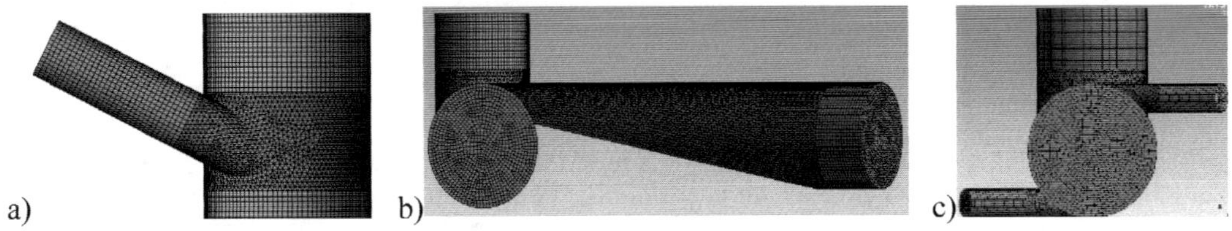

Figure 3. CFD meshing of three Inlet Nozzles of GLCC. a) Single inclined inlet; b) Single gradually reduced inlet; c) Dual inclined inlet

Three refinement meshes are generated to evaluate grid dependency, with the number of cells being 950.000, 1.500.000, and 1.650.000, respectively. All significant measures of mesh quality, such as mesh orthogonality, skewness, and aspect ratio, are maintained within the optimal range for high mesh quality [7-8]. A slight discrepancy is observed when comparing the outcomes of the model with 1.500.000 and 1.650.000 cells. In all simulations conducted in this study, the pressure-velocity coupling scheme employed the SIMPLEC algorithm (Semi-Implicit Method Pressure-Linked Equations Consistent). The evaluation of pressure interpolation schemes, including standard, linear, body force weighted, second order, and PRESTO (Pressure Staggered Option), has been conducted. The findings indicate that the utilization of a second-order interpolation scheme is most appropriate for pressure interpolation. In relation to the discretization of turbulent kinetic energy and turbulence dissipation rate, a comparison has been made between two schemes: QUICK (Quadratic Upstream Interpolation for Convective Kinetics) and first order upwind. The QUICK scheme demonstrated a high level of accuracy in its ability to predict the swirl flow of a GLCC separator. Following the completion of each CFD simulation, the flow velocity's axial and tangential components at each measuring position are extracted for the purpose of comparison with practical experiments.

3. Verification of simulated results

Since the axial velocity describes the flow of fluid in the direction of the outlet, it plays a crucial role in the flow within the GLCC body. The GLCC body's axial velocity profile demonstrates the existence of two separate flow streams: downward and upward flow. Whereas the flow near the wall shows a downward direction towards the bottom of the GLCC, more precisely at a narrow radial distance, the flow near the GLCC's cylindrical centerline is characterized by an upward direction towards the inlet. The study employs a set of 22 simulated models to extract the axial and tangential velocity components, utilizing three different inlet configurations. Figure 4 demonstrates a strong concurrence between the observed axial velocity contour obtained from Erdal's experimental data (on the left side) and the simulated contour (on the right side) for both inlet configurations of 2 and 3, with flow rates of 72 gpm and 10 gpm, respectively.

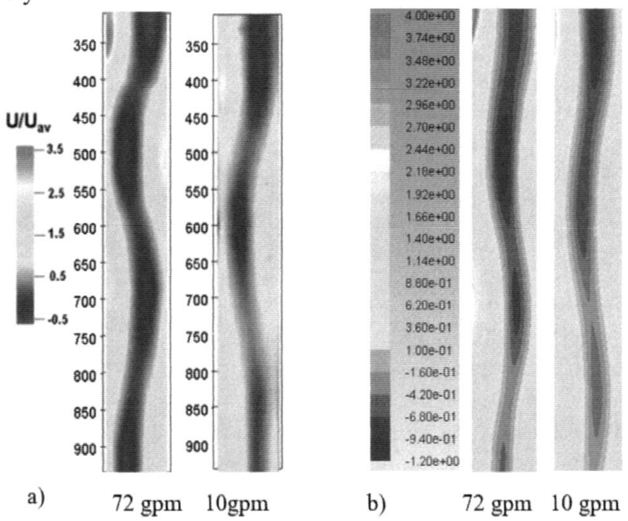

Figure 4. Comparison in contour of axial velocity between experiment with (a) the simulation and (b) Gradually reduced inlet

The axial velocity profiles are also influenced by the helical pitch (wavelength) of the vortex. The presence of high axial velocity profiles in the negative region, observed on either side of the centerline of the GLCC, can be attributed to the specific locations where measurements were taken and the influence of viscous flow conditions. These factors have an impact on the wavelength of the vortex occurring within the GLCC. Figure 5 illustrates how the axial velocity profiles change in form depending on the layout of the inlet. The single inlet configurations exhibit a similar shape to the axial velocity profiles, wherein the peak of the axial velocity is distributed on one side (see Figure 5a, b). Nevertheless, the distribution of axial velocity profiles is evenly spread across both sides, as depicted in Figure 5c. The simulated results of the axial velocity components exhibit a high level of agreement with the practical experiments conducted for three different inlet configurations.

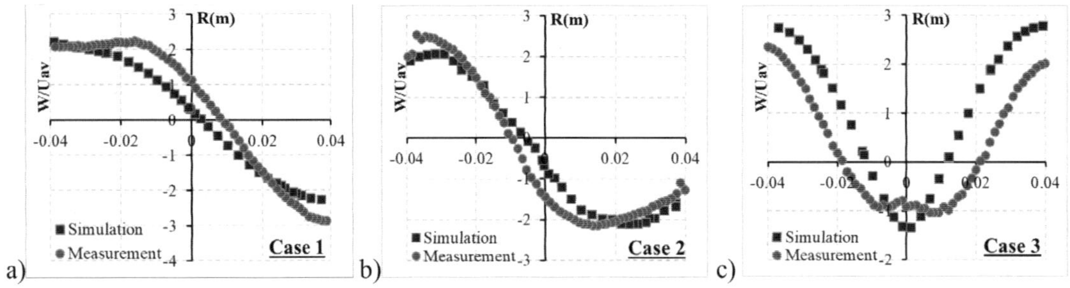

Figure 5. Comparison of axial velocity for three inlet configurations, (a) x = 467 mm, (b) x = 686 mm, (c) x = 317 mm

A GLCC separator's use of an inclined inlet helps to enhance multiphase flow stratification before it enters the GLCC body in a tangential fashion. The dominant role in generating the centrifugal force for phase separation within the GLCC is attributed to the tangential velocity component. The tangential inlet of the GLCC generates a significant centrifugal force that acts radially, directing the fluid flow towards the cylindrical wall. This results in an enhanced separation performance. The tangential velocity exhibits a positive value on the left side and a negative value on the right side, as depicted in Figure 6a and 6b. The observed phenomenon can be attributed to the rotational motion of the fluid flow. Negative velocities signify a flow that is directed toward the page, whereas positive velocities indicate a tangential velocity that is directed away from the page. In the vicinity of the wall region, the tangential velocity is greater, and it gets smaller as it gets closer to the center. However, because the two inlets are positioned asymmetrically within the GLCC body, the distribution of tangential velocity profiles in the dual inlet configuration stays equal (see Figure 6c). As the tangential velocity moves axially in the direction of the outlet or downwards in the axial direction, it decays. For a given incline inlet configuration, the location where the tangential velocity is zero or low has a helical (spiral) shape that is similar to the observed axial velocity contours shown in Figure 4..

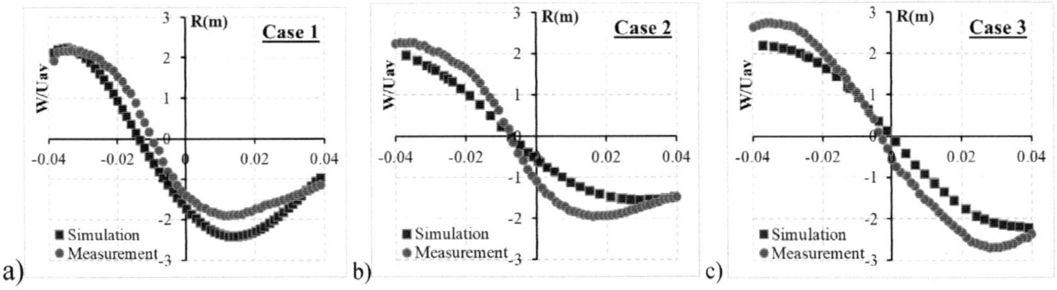

Figure 6. Comparison of axial velocity for three inlet configurations, (a) x = 899 mm, (b) x = 757 mm, (c) x = 317 mm.

3.1. *Single inclined inlet*

3.1.1. Axial velocity. In the case of a single inclined inlet configuration (Figure 7), the axial velocity profiles are examined at various inlet angles. These profiles are measured at a distance of 317 mm below the inlet for two different flow rates: 0.00063 m^3/s (Figure 7a) and 0.00454 m^3/s (Figure 7b). The axial velocity profiles in close proximity to the wall experience notable alterations due to variations in the inlet angle. A greater inlet angle results in an increased magnitude of axial velocity. However, it is important to note that this relationship is observed exclusively at the wall side. The phenomenon under consideration can be attributed to the internal circulation of swirling flow within the GLCC. The variation of axial velocities exhibits a relatively modest magnitude within the range of inlet angles spanning from 5° to 35°. Nevertheless, there is a substantial variation observed for inlet angles exceeding 35°. Within the range of inlet angles spanning from 5° to 35°, it is observed that the magnitude of axial velocities in the vicinity of the wall region experiences a decrease on one side and an increase on the other side. Conversely, in the region proximate to the central axis of the GLCC body, the axial velocities solely exhibit an increase (refer to Figure 7). This implies that the upward flow in the vicinity of the centerline of the GLCC consistently increases as the inlet inclined angle increases. The observed pattern of axial velocity remains consistent in the vicinity of the inlet section but exhibits a reversal about the centerline of the GLCC as indicated by the measured plans towards the lower region of the GLCC. The velocity of the downward flow in close proximity to the wall exhibits a significant positive magnitude, which gradually diminishes as the fluid flow progresses from the inlet towards the outlet. The decay of the vortex leads to an amplification of the wavelength in the axial direction towards the outlet, as depicted in Figure 7b. Therefore, the stretching of the vortex in this particular direction may contribute to the occurrence of the gas-carried under (GCU) phenomenon, wherein gas bubbles move downwards towards the liquid leg, thereby affecting the performance of the GLCC [10].

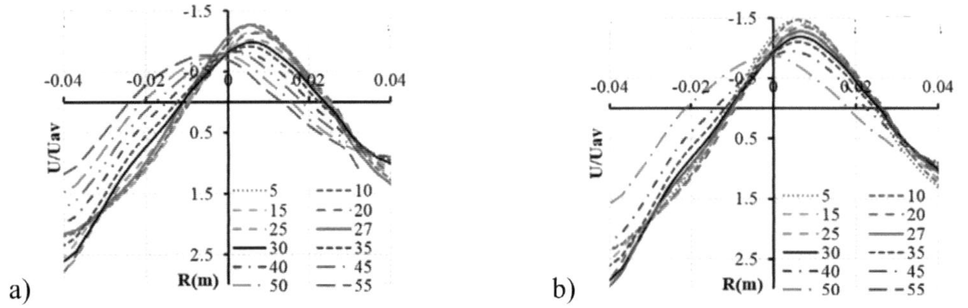

Figure 7. The distribution of axial velocity profiles at x = 317 mm with different inlet angles, and flow rates of (a) 0.00063 m^3/s, and (b) 0.00454 m^3/s

Figures 8a and 8b depict the variations in maximum axial velocities of upward and downward flow, respectively, in relation to different inlet angles across four measured planes. The graph presented provides a clear depiction of the minimal variation in maximum axial velocities within the range of 5°-35°. Within this range of conditions, there is evidence of a modest decline in the maximum axial velocity for both downward and upward flow. However, it should be noted that larger inlet angles demonstrate a contrasting trend in the behavior of the flows. Nevertheless, the upward trend in the maximum axial velocities of the downward flow directly beneath the inlet section is observed to be substantial as the inlet angle increases from 5° to 20°. However, a notable decline in these velocities is subsequently observed within the range of inlet angles between 45° and 50°. However, there is a significant increase observed at inlet angles exceeding 50°. The observed phenomenon can be attributed to the interaction between the incoming flow entering the GLCC body and the swirling flow in the vortex region. This interaction generates a negative pressure that directs the fluid flow towards the lower section of the GLCC. Therefore, the axial velocity of the downward flow experiences an increase in the vicinity of the inlet section. Specifically, the maximum axial velocity of downward flow exhibits significant fluctuations when the inlet angle is large, and the flow rate is high (see Figure 8b).

The dependence of the maximum axial velocity of upward flow on changes in inlet angles has been observed to be infrequent. There exists a minimal alteration in the maximum axial velocity at the inlet angle that exceeds 35°. It can be elucidated that the simulations with varying inlet angles maintain a consistent mass flow rate, characterized by an identical mean axial velocity.

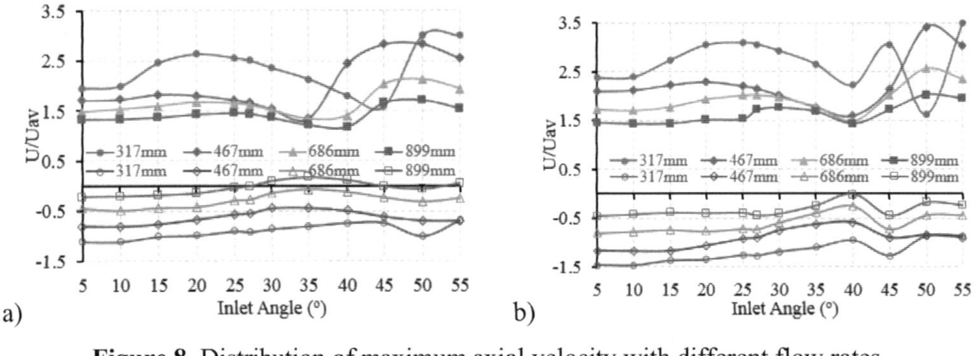

Figure 8. Distribution of maximum axial velocity with different flow rates.

3.1.2. Tangential velocity. Figure 9 illustrates the alterations in the tangential velocity profiles corresponding to different inlet angles, for both flow rates. The tangential velocity profiles within the GLCC body are segregated into positive and negative regions due to the presence of viscous flow rotation. Similar to the axial velocities, the tangential velocities experience a substantial decrease in magnitude as the GLCC flow progresses towards the outlet, as depicted in Figure 9. The flow distribution in the GLCC exhibits a helical shape due to the influence of centrifugal force. Consequently, the tangential velocities are relatively low in the radial region towards the center of the GLCC, while they are significantly higher in the vicinity of the wall. The impact of the inlet angle on the tangential velocity is found to be negligible for small angles (less than 35°), while it becomes more pronounced for inlet angles exceeding 35 degrees. Remarkably, the fluctuation of tangential velocity in relation to distinct inlet angles exhibits a significant disparity on one side, while remaining constant on the opposite side. The determination of the point at which zero tangential velocities occur for small inlet angles will govern the occurrence of significant variations in tangential velocity near the wall, as influenced by changes in the inlet angle. In Figure 9, it can be observed that when the location of zero-tangential velocity is situated on the left side of the y-axis, the high variation region of the tangential angle near the wall is found in the fourth quadrant. Conversely, when the location of zero-tangential velocity is on the right side of the y-axis, the high variation region of the tangential angle near the wall falls in the second quadrant.

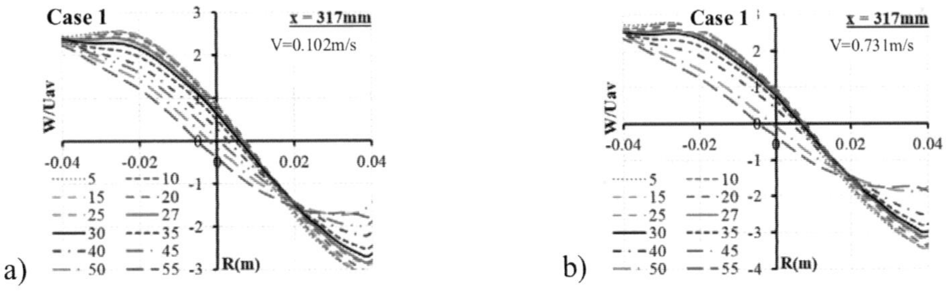

Figure 9. Tangential velocity profiles in different inlet angles.

The maximum tangential velocity is contingent upon the inlet angle, primarily occurring in the vicinity of the wall region (Figure 10). The aforementioned factor holds significant influence over the performance of the GLCC. In the context of the GLCC with a significant inlet angle, it can be observed that the tangential velocity experiences the most substantial variation in the vicinity of the wall region. Figure 9 illustrates the impact of the inlet angle on the maximum tangential velocity across four distinct measurement planes. The upper graph illustrates the variation in maximum positive and negative tangential velocities in relation to different inlet angles. The maximum tangential velocity is seldom influenced by variations in the inlet angle at the nearest and farthest points from the inlet plane, where the impact of swirl flow is minimal. The tangential velocity at the inlet angles larger than 35° experiences a slight increase due to the axial direction of the fluid flow, which interacts with the upward flow and consequently reduces the intensity of the tangential velocity. The GLCC's separating performance is adversely affected by the presence of high turbulence in the flow, particularly when accompanied by a large inlet angle.

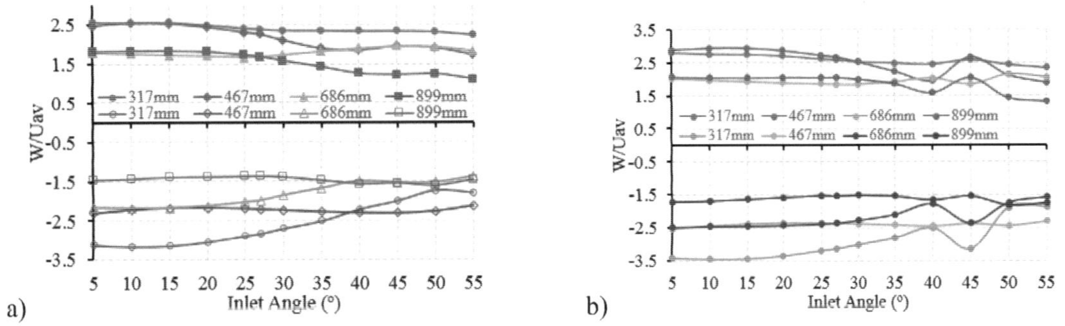

Figure 10. Distribution of maximum tangential velocity.

3.2. *Single gradually reduced inlet*

3.2.1. Axial velocity. Figure 11 displays the axial velocity profiles for various inlet angles, obtained from a flow rate of 0.0063 m^3/s, at a measurement distance of 317 mm. Similar to the single inclined inlet configuration, the velocity of the downward flow in close proximity to the wall exhibits a substantial positive magnitude and diminishes as the fluid flow progresses away from the inlet towards the outlet. The graph illustrates that the maximum axial velocities exhibit minimal variation within the range of 5° to 30°. However, the value experiences a significant increase and nearly reaches its maximum point when the inlet angle is set to 40°. The observed phenomenon can be attributed to the interaction between the incoming flow entering the GLCC body and the swirling flow in the vortex region. This interaction generates a negative pressure that directs the fluid flow towards the bottom of the GLCC.

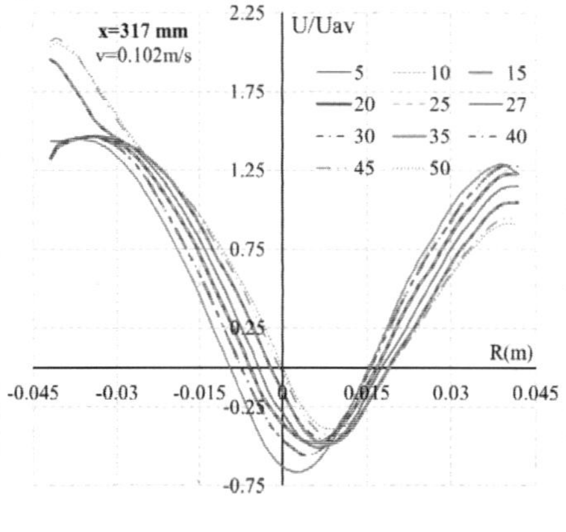

Figure 11. Axial velocity profiles with different inlet angles.

The alteration of axial velocities experiences a slight modification within the inlet angle range of 5° to 30°, as depicted in Figure 12. Nevertheless, there is a substantial variation observed for inlet angles exceeding 30°. The axial velocities near the wall region exhibit a decrease on one side and an increase on the other side, while in the region near the central axis of the GLCC body, they only experience an increase. This behavior is observed as the inclined inlet angle increases, as depicted in Figure 12. This implies that the upward flow in the vicinity of the center line of the GLCC consistently increases as the inlet inclined angle is escalated. The observed pattern of axial velocity remains consistent in the vicinity of the inlet section but exhibits a reversal about the centerline of the GLCC as indicated by the measured plans towards the lower region of the GLCC.

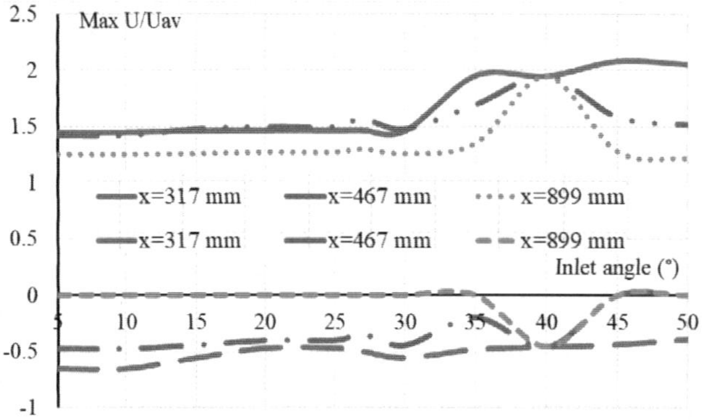

Figure 12. Distribution of maximum axial velocity in different inlet angles at U_{av} = 0.102 m/s.

3.2.2. Tangential velocity. Figure 13 displays the tangential velocity profiles' distribution for various inlet angles, with a flow rate of 0.0063 m^3/s, as measured at a distance of 317 mm. The maximum tangential velocity is contingent upon the inlet angle, predominantly occurring in the vicinity of the wall region. The above-mentioned factor holds significant influence over the performance of the GLCC. In the case of the GLCC featuring a high inlet angle, there is a discernible alteration in the tangential velocity in the vicinity of the wall region. The tangential velocity at the inlet angle greater than 35° experiences a substantial increase due to the axial propulsion of the fluid flow and its interaction with the upward flow, resulting in a reduction in the magnitude of the tangential velocity (Figure 14).

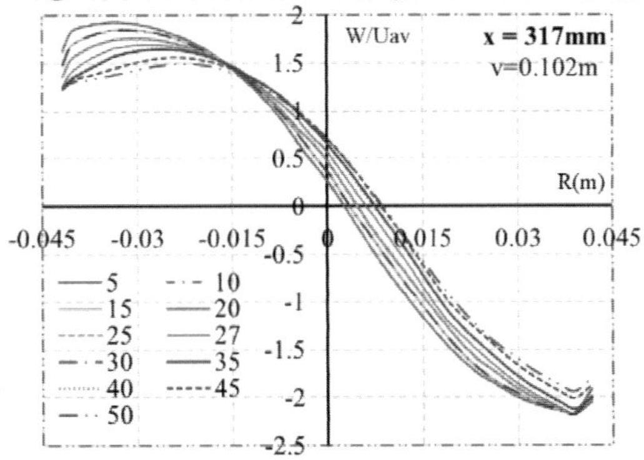

Figure 13. Tangential velocity profiles in different inlet angles.

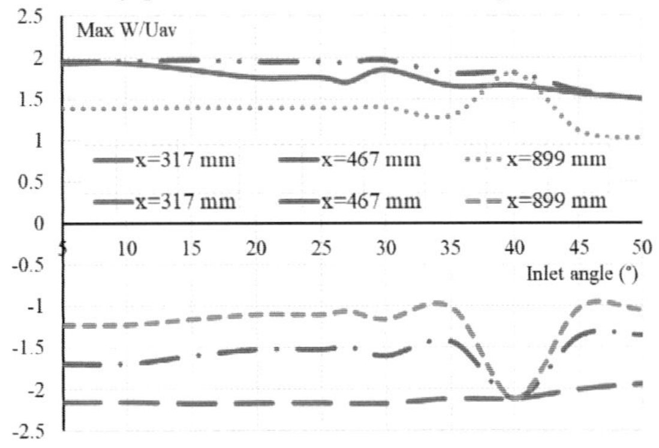

Figure 14. Distribution of maximum tangential velocity in different inlet angles at U_{av} = 0.102 m/s.

3.3. Dual inclined inlet

3.3.1. Axial velocity. Figure 15 illustrates the axial velocity profiles' distribution for various inlet angles in the case of a flow rate of 0.0063 m3/s, measured at a distance of 317 mm, in the context of a dual inclined inlet. In the case of two asymmetrical inlet configurations, it can be observed that the rate of swirl decay in the vicinity of the inlet is significantly minimal. According to Erdal [6-7], it has been observed that the decay of swirl is more rapid in the case of a single inlet compared to that of two inlets, which results in a more symmetrical flow. In the case of a single incline inlet configuration, the axial velocity at its minimum is predominantly concentrated towards the centre of the GLCC. A greater inlet angle results in an increased magnitude of axial velocity. The axial velocity exhibits a tendency to decrease as one moves from the wall towards the cylindrical centreline, followed by an increase towards the opposite wall in the context of a GLCC that possesses axisymmetric characteristics. In Figure 16, the alteration of axial velocity profiles exhibits a gradual transition when comparing the distribution of profiles with varying inlet angles, as opposed to the singular inclined inlet configuration.

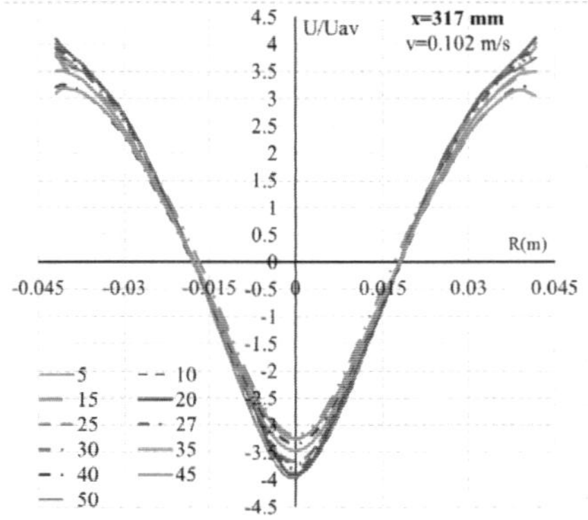

Figure 15. Axial velocity profiles in different inlet angles.

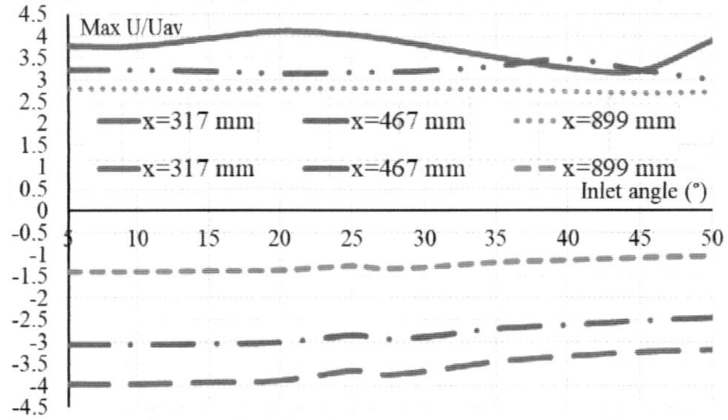

Figure 16. Distribution of maximum axial velocity in different inlet angles at U_{av} = 0.102 m/s.

3.3.2. Tangential velocity. Figure 17 displays the tangential velocity profiles' distribution for various inlet angles in the case of a flow rate of 0.0063 m^3/s. These measurements were taken at a distance of 317 mm for a dual inclined inlet configuration. In this scenario, the minimum tangential velocity is predominantly concentrated towards the centre of the GLCC. The variation in the magnitude of tangential velocity across different inlet angles in the GLCC is minimal. The tangential velocity exhibits a decreasing trend as one moves from the wall towards the cylindrical centreline in a GLCC. Subsequently, it increases in the opposite direction towards the other wall of the GLCC, while maintaining symmetry along the centreline. As one moves towards the lower portion, there is a progressive reduction in the tangential velocity. This implies that the turbulence of the flow decreases as it progresses towards the bottom of the GLCC. The distribution of the maximum tangential velocity component for the dual inclined inlet configuration (Figure 18) exhibits minimal variation. The angular range of 25° to 40° has resulted in a slight alteration in the distribution of maximum tangential velocity.

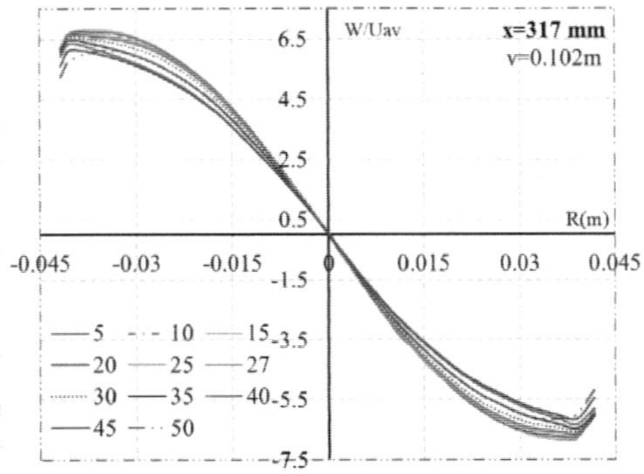

Figure 17. Tangential velocity profiles in different ii

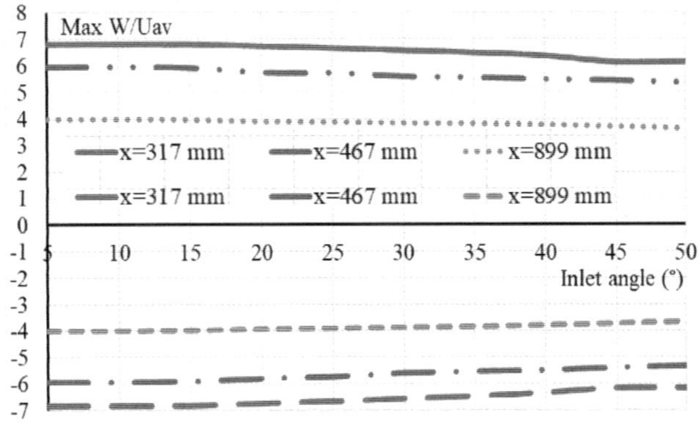

Figure 18. Distribution of maximum tangential velocity in different inlet angles at U_{av} = 0.102 m/s.

4. Conclusion

This study utilized computational fluid dynamics (CFD) simulations employing 22 gas-liquid cylindrical cyclone (GLCC) models in order to gain a comprehensive understanding of the intricate kinematic characteristics exhibited by complex flow within a GLCC separator. The utilization of three-dimensional models in computational fluid dynamics (CFD) simulations, coupled of with an appropriate turbulent model, has revealed alterations in the kinematic characteristics of gas-liquid concurrent flow in a vertical pipe when the inclination angle and inlet configuration are modified. The simulated results and practical experiments for three inlet configurations demonstrate a high level of agreement. The primary findings derived from this study indicate the consistent functionality of inlet angles ranging from 0° to 35° across three distinct inlet configurations. This outcome facilitates the manufacturers of GLCC separators in devising numerous design alternatives for inlet configurations within confined spaces on oil platforms. Additional conclusions can be derived from this study as following:

- The flow pattern of tangential and axial velocities and their maximum magnitude exhibit minimal changes within the range of inclined inlet angles from 5° to 35°. However, a notable increase is observed as the inclined inlet angle reaches 40°, resulting in a peak magnitude. The changes in velocity of a single inlet GLCC are more pronounced and easily discernible compared to those of the dual inlet GLCC.
- The axial velocities of the downward flow exhibit a decrease on one side and an increase on the other side. The velocities of downward flow diminish as the flow progresses from the inlet towards the outlet.
- The maximum tangential velocity is largely influenced by the inlet angle, particularly in the vicinity of the wall. The maximum tangential velocity is not significantly influenced by variations in the inlet angle at the nearest and farthest sections from the inlet. The minimum

tangential velocity primarily converges towards the centre of the GLCC. The single inlet GLCC exhibits a notable increase in maximum tangential velocity at an inlet angle of 35°, whereas the dual inlet GLCC does not display a significant change in this regard.

- The radial velocity profiles of both GLCC configurations demonstrate axial symmetry, with magnitudes considerably smaller than those of the axial and tangential velocities. Increasing the inclined inlet angle of a GLCC leads to an elevation in the maximum radial velocity. Significant variations in the maximum radial distribution are observed between the inlet region and the bottom region for various inclined angle configurations of the GLCC.

References

[1] Kouba G. E. A, O. Shoham 1996 Review of gas-liquid cylindrical cyclone technology. *Int. Conf. Prod. Sep. Syst.* (Aberdeen, UK: April, 1996).
[2] Nissan A.H. and Bresan V.P. 1961 Swirling flow in cylinders, *AIChE J.*, **7**(4), 543-547.
[3] Wang, S., Mohan, R.S., Shoham, O., Marrelli, J. D. and Kouba, G.E. 2000 Performance improvement of gas liquid cylindrical cyclone separators using integrated liquid level and pressure control systems *ASME Trans., J. Energy Res. Technol.* **122**(4): 185-192.
[4] Iozia D.L., D. Leith 1989 Effect of cyclone dimensions on gas flow pattern and collection efficiency, *Aerosol Sci. Technol.*, **10**(3), 491–500.
[5] S. Movafaghian, J.A. Jaua-Marturet, R. Mohan, O. Shoham, G. Kouba 2000 The effects of geometry, fluid properties and pressure on the hydrodynamics of gas–liquid cylindrical cyclone separators, *Int. J. Multiphase Flow* **26**(6), 999–1018.
[6] Erdal, F., Shirazi, S., Shoham, O., Kouba, G. 1997 CFD simulation of single-phase and two-phase flow in gas–liquid cylindrical cyclone separators. *SPE J.* **2**, 436–446.
[7] Erdal, F., Shirazi, S. 2011 Local velocity measurements and computational fluid dynamics (CFD) simulations of swirling flow in a gas–liquid cylindrical cyclone separator, *Eng. Technol. Conf. Energy*, (Houston, Texas: 2011, 23–30).
[8] Rainier H, Caroline G, Noël M. 2011 Numerical investigation of swirling flow in cylindrical cyclones. *Chem. Eng. Res. Design*, **89**, 2521–2539.
[9] Le Van Sy 2017 Influence of inlet angle on flow pattern and performance of gas-liquid cylindrical cyclone separator. *Part. Sci. Technol.* **35**(5), 555-564.
[10] Gomez L. E., Ram S. Mohan, O. Shoham, J. D. Marrelli, G. E. Kouba 1999 State-of-the-Art Simulator for Field Applications of Gas-Liquid Cylindrical Cyclone Separators. *SPE Annual Tech. Conf. Exhibition* (Houston, Texas, Oct-1999).
[11] Gupta, A., Kumar, R. 2007 Three-dimensional turbulent swirling flow in a cylinder: experiments and computations *Int. J. Heat Fluid Flow*, **28**, 249–261.

IOP Conf. Series: Materials Science and Engineering 1294 (2023) 012009 doi:10.1088/1757-899X/1294/1/012009

Assessment of a novel PTO system for swell energy convertors using digital twin modelling

C Ren[1], Y Xing[1] and L Moen[2,*]

[1] University of Stavanger, Norway
[2] Ocean Energies AS, Norway

[*] Correspondence: lyder.moen@oceanenergies.no

Abstract. There is no lack of ideas and prototypes for wave energy converters (WECs). However, existing WECs are designed to extract energy from waves at normal to high frequencies and typically focus on close-to-shore steep waves and short wavelengths to gain a frequent movement of interactive components. This limits the size of the buoyancy structure and results in the need to use many generating units to achieve a high-power output. We reverse this focus in small and coastal WECs mentioned above by harvesting swell energy through the Swell Energy Convertor (SWC). SWC uses an innovative power take-off (PTO) system, which Ocean Energies AS invented. This patented crankshaft mechanism transforms the linear motion from ocean waves into rotation for rotating a generator. When combined with a large buoy structure, this PTO system allows the SWC to focus on the swells (large wavelengths and low to high wave heights) further offshore and reduce/remove the drawbacks associated with the abovementioned coastal areas. The SWC (i) focuses on the high energy swells offshore, (ii) uses large buoy structures in combination with slight vertical movement, and (iii) can optimally extract energy in any direction. This results in SWC being a far less complex system with a much higher power output than existing WECs, which will significantly lower the cost of WECs. The SWC will target the area between Greenland and Scotland, with significantly high wave energy above 60 kW/m. This paper will present the design and analysis of the proposed PTO system in the SWC. First, the PTO model and working mechanism will be described. Second, a digital model of the PTO system will also be introduced. Third, three optimal design candidates are selected and analysed further to provide a deeper insight into the PTO design. Last, discussions and planned future works are presented.

1. Introduction

Wave energy is considered one of the most promising renewable energies with vast potential that remains largely unexploited. To date, many wave energy converters (WECs) have been developed and deployed in different countries. In addition, more than a hundred projects and more than one thousand patents have been developed worldwide, as summarised in [1]. A review of the existing WEC technologies can be found in some papers [2, 3]. The numerical benchmarking study of several existing WECs can also be noticed in [4]. However, existing wave energy plants are commonly small-scaled converters [5] and designed to extract energy from the wave at normal to high frequencies near the coast [6]. Most of them fail commercially due to low power generation compared to the installation cost and high maintenance requirements. A reason for the low power generation is the geographical location of the device. Near the coast, the wave patterns are generally focused in one direction and diminished in

IOP Conf. Series: Materials Science and Engineering 1294 (2023) 012009 doi:10.1088/1757-899X/1294/1/012009

amplitude in other directions. These waves also tend to break and put high loads on the structures. This means that the devices are often operating at suboptimal conditions.

To overcome the limitations of the small and coastal WECs mentioned above, this present work aims to harvest wave energy through the Swell Energy Converter (SWC). Swells often have a relatively long wavelength; long-wavelength swells carry more energy and dissipate slower than coastal waves. More importantly, the present SWC uses an innovative power take-off (PTO) system [7], as shown in Figure 1 and Figure 2, which Ocean Energies AS invented. In Figure 1, the "G," "FO," and "SF," respectively, represent the generator, floating object, and seafloor. The details of the PTO system are given in Figure 2. This patented crankshaft mechanism transforms the linear motion from ocean waves into rotation for rotating a generator. Combined with a large buoy structure, this PTO system allows the SWC to focus on the swells (large wavelength and heights) further offshore and reduce/remove the drawbacks associated with coastal areas.

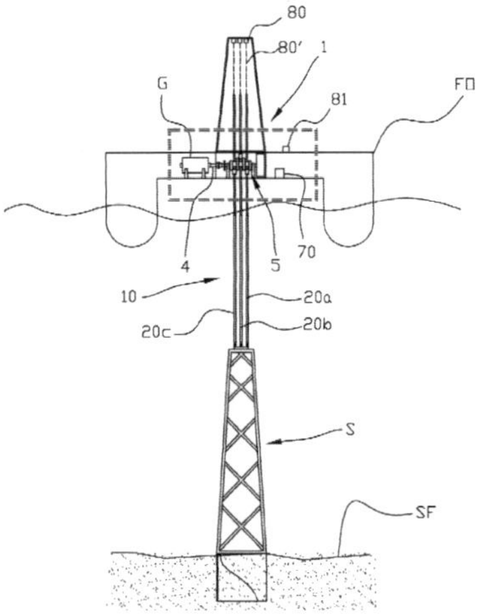

Figure 1. Swell energy converter [7]

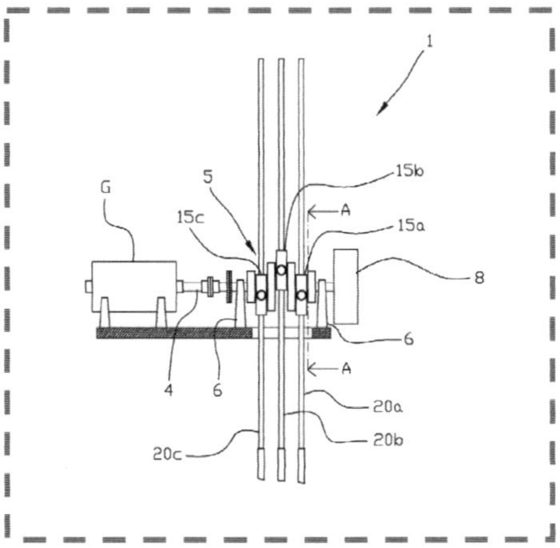

Figure 2 PTO system of SWC [7]

The present SWC will focus on the high energy swells offshore and use large buoy structures in combination with small vertical movement. This results in SWC being a far less complex system with a much higher power output than existing WECs which will significantly lower the cost of WECs. Also, the SWC will target the area between Greenland and Scotland, which has a significantly high wave energy of above 60 kW/m [8]. Scotland already has an established offshore electrical grid infrastructure that SWC can connect to.

2. PTO system working mechanism

As shown in Figure 2, the PTO mainly comprises a rotatable main shaft for driving a machinery, generator, and flywheel. Three tethers (20a, 20b, 20c) are connected to crank pins through three clutch apparatuses (15a, 15b, 15c) as depicted in Figure 3. The clutches allow gripping and releasing contact between tethers and crank pins, which can be operated by means of magnetism or hydraulics. A control system will also be developed to operate the clutch apparatuses. The operation of the clutches depends on the position of crank pins and the moving direction of the FO. Based on the movement of the FO compared to the sea floor, the operation of clutches in different positions can be classified into two cases.

2.1. An upward movement of the FO

When the floating object moves upward compared to the sea floor, the clutch apparatus located in the green portion, as depicted in Figure 4, will be set in clutch lock (CL) mode (gripping). Other clutches will be set into clutch unlock (CU) mode (releasing). The "gripping" term means that the relative movement between the tether and crankshaft is nearly zero. The "releasing" term means that the tether and crankshaft can move independently. Therefore, when in the gripping (CL) mode, any vertical movement energy of the floating object is transformed into the rotational energy of the crankshaft.

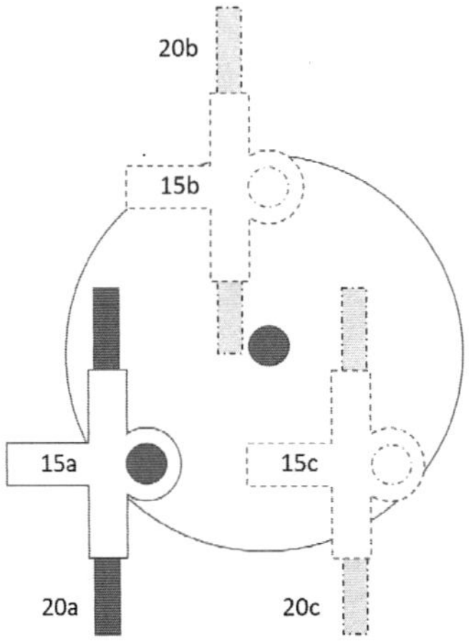

Figure 3. Three tethers and clutches in view of A-A of Figure 3

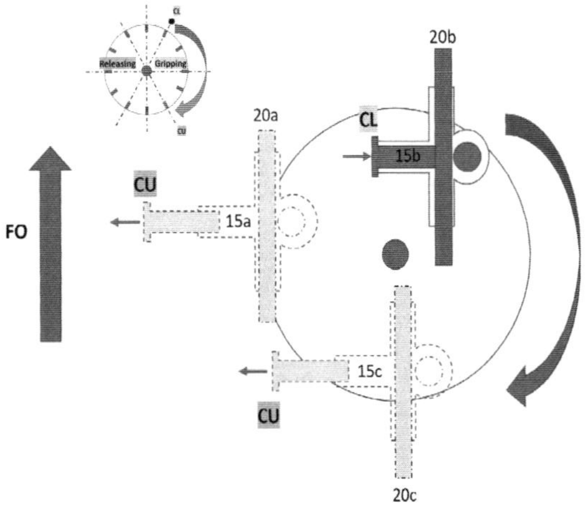

Figure 4. Upward movement of the FO

2.2. A downward movement of the FO

Similarly, when the floating object is moving downward compared to the sea floor, the clutch apparatus located in the green portion, as shown in Figure 5, will be set in CL mode. Other clutches will be set into CU mode. Depending on the up and down movement of the floating object, the switch of CL and CU mode ensures that the crankshaft can always rotate in the same direction.

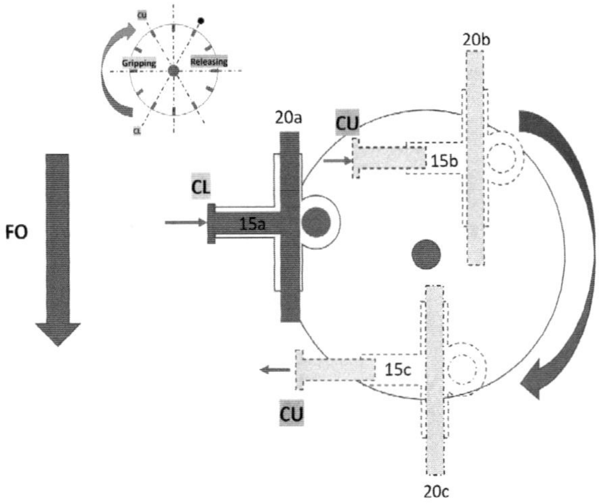

Figure 5. Downward movement of the FO

Due to this gripping and releasing configuration provided by the clutch apparatus, the crankshaft can be rotated several times during an upward movement and several times during a downward movement of the floating object. During such a rotation of the crankshaft caused by the upward or downward movement of the FO, a gripping contact will be at different positions along a longitudinal axis of the crankshaft. More importantly, this PTO system can be independent of wave amplitude and wavelength and can be installed in both shallow and deep seas. Additionally, more details of the PTO system working mechanism can be found in [7].

3. A digital twin model of the PTO system

Based on the working mechanism of the PTO system, a digital twin model of the PTO system is created in MATLAB toolbox. The PTO digital twin model mainly comprises two parts: Hydrodynamic forces and crankshaft and tether motion. The overview of the PTO digital twin system is given in Figure 6.

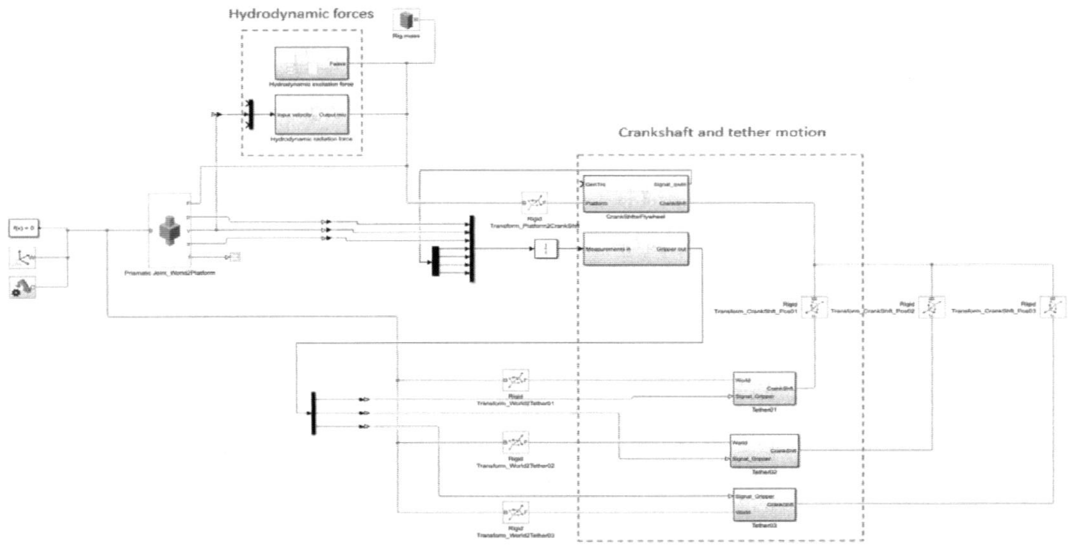

Figure 6. Overview of the PTO digital twin system

3.1. Hydrodynamic force subsystem

According to reference [9], the motion of the buoy can be formulated as follows based on Newton's second law:

$$\mathbf{M}\ddot{q} = \mathbf{F}_{ext} + \mathbf{F}_r + \mathbf{F}_b \tag{1}$$

where **M** is the inertia matrix of the oscillating body and $\boldsymbol{q}$ is the displacement from the equilibrium position. $\mathbf{F}_{ext}$, $\mathbf{F}_r$, and $\mathbf{F}_b$ are respectively excitation force, radiation force, and hydrostatic force. The radiation force can be represented as:

$$\mathbf{F}_r = -\mathbf{A}(\omega)\ddot{q} - \mathbf{B}(\omega)\dot{q} \tag{2}$$

where $\mathbf{A}(\omega)$ and $\mathbf{B}(\omega)$ are the frequency-dependent added mass matrix and damping matrix, respectively. Normally, it is assumed that the body is constrained to oscillate only in the heave direction. Hence, the hydrostatic force is calculated as:

$$\mathbf{F}_b = -\mathbf{G}\mathbf{q} = -\rho g S_w \mathbf{q} \tag{3}$$

Where **G** is the buoyancy stiffness. $\boldsymbol{\rho}$ and $\boldsymbol{g}$ are the water density and gravity acceleration, respectively. $\boldsymbol{S_w}$ is the equilibrium water plane area of the body. Then, Equation (1) can be rewritten as:

$$[\mathbf{M} + \mathbf{A}(\omega)]\ddot{q} + \mathbf{B}(\omega)\dot{q} + \mathbf{G}\mathbf{q} = \mathbf{F}_{ext} \tag{4}$$

And in the frequency domain:

$$\left[-\omega^2\left(\mathbf{M} + \mathbf{A}(\omega)\right) + i\omega\mathbf{B}(\omega) + \mathbf{G}\right]\mathbf{q} = \mathbf{F}_{ext} \tag{5}$$

In this present work, the matrices **M**, $\mathbf{A}(\omega)$, and $\mathbf{B}(\omega)$ are derived from the pre-simulation results in ANSYS AQWA. The $\mathbf{F}_{ext}$ is then simulated in MATLAB based on different wave theories. In this paper, only the regular waves are investigated, which gives:

$$\eta = A_\eta \cos(\omega t - kx) \tag{6}$$

where η is wave elevation, A_η is the wave amplitude, ω is the angular frequency, and k is the wave number.

3.2. Crankshaft and tether motion subsystem

Due to the excitation of the wave elevation, the floating object will go upward and downward. The movement of the floating object and the tether position will be measured in the subsystem. A control system is then developed to operate the clutch apparatuses for clutch lock and clutch unlock mode. The gripper forces provided by clutch apparatuses due to the motion of the floating object will drive the movement of the crankshaft. The gripper forces may come from different clutches due to the position of the tether and the movement of the floating object, as shown in Figure 7. In this work, it is assumed that the gripper force is constant for all the clutch apparatuses during the movement of the floating object.

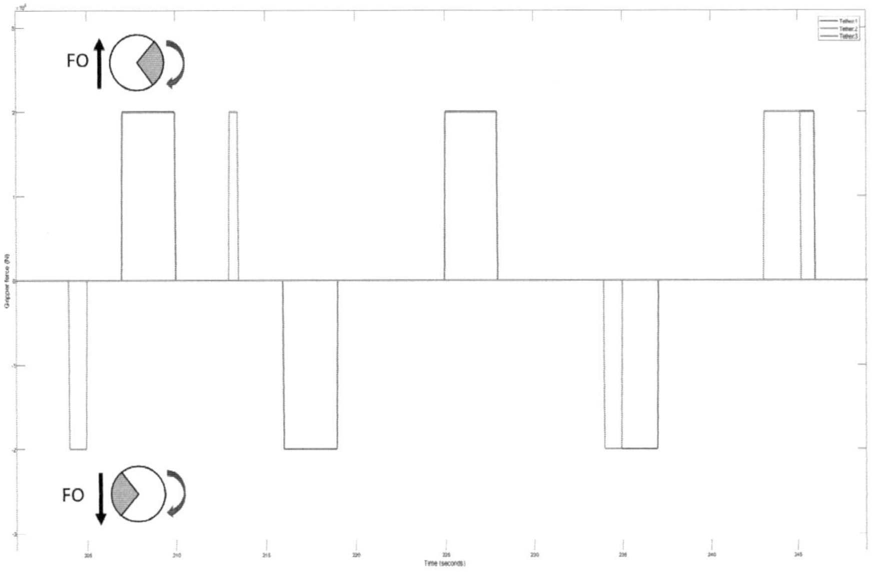

Figure 7. Gripper force from different clutches.

Additionally, to maximise the energy output, the floating object should be controlled to approach an optimum interaction between the SWC and wave. There are commonly three control strategies for wave energy converters: Reactive control, Latch control, and passive control. To simplify the modelling in this present work, a passive control system is adopted in this study.

4. Characteristic performance of the PTO system

In this part, the characteristic performances of the novel PTO system are investigated. At first, the influence of the floating object dimension is studied. The power matrix of the floating object is roughly calculated based on the Budal upper bound. The influence of the floater dimension on the power matrix is studied. Secondly, the sensitivity study of the PTO parameters is investigated. The influences of the key PTO parameters on averaged power generation are compared.

4.1. The influence of floater size on SWC power matrix calculation

In this subpart, the influence of the floater dimension is studied. The floater is assumed to be a cylinder and is semi-submerged in the seawater. The height of the cylinder floater is fixed at 10 meters, and the displacement of the floater is assumed at 5 m. Three different diameters (20 m, 60 m, 100 m) of cylinder floater are compared.

According to [10], for the heaving semi-submerged WECs, the WEC generated power is limited due to Budal upper bound (BUB), which is presented as follows:

$$P \leq P_A \quad \text{and} \quad P < P_B \tag{7}$$

Where P_A is known as left-hand upper bound, and P_B is the right upper bound. Figure 8 presents a schematic of Budal curves.

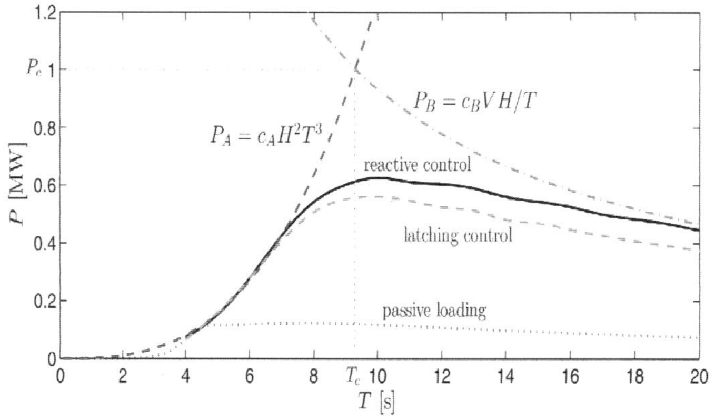

Figure 8. Budal upper bounds [10]

The equations of P_A and P_B are given as follows:

$$P_A = c_A H^2 T^3 \tag{8}$$

$$P_B = c_b V \frac{H}{T} \tag{9}$$

where H and T are, respectively, the wave height and wave period. c_A and c_B are respectively equal to:

$$c_A = \frac{\rho(\frac{g}{\pi})^3}{128} \tag{10}$$

$$c_B = \sigma \rho g \tag{11}$$

where ρ and g are seawater density and gravity constant. σ equals to $\pi/4$ in this study. As shown in Figure 8, the PTO control system greatly influences the average power generation. Typically, the ratio of the maximum absorbed power P_{max} to the intersection-point power P_c for each control strategy roughly follows:

$$\frac{P_{max}}{P_c}\begin{cases} 0.6 & for\ reactive\ control \\ 0.5 & for\ lathing\ control \\ 0.1 & for\ passive\ control \end{cases} \tag{12}$$

Based on the assumption of the passive control in this work, the power matric of different floater dimensions can be calculated, as shown in Figure 9 - Figure 11. With the increased floater dimension, (1) the WEC tends to have higher power density in the longer wave period. Also, (2) the left-hand upper (P_A) will affect the WEC power matrix calculation more. Additionally, because P_A is size independent, the left part (T_p:5-16s) of the power matrix of 60m and 100m diameter floater is nearly the same. (3) The highest power density will increase will the increase of floater size.

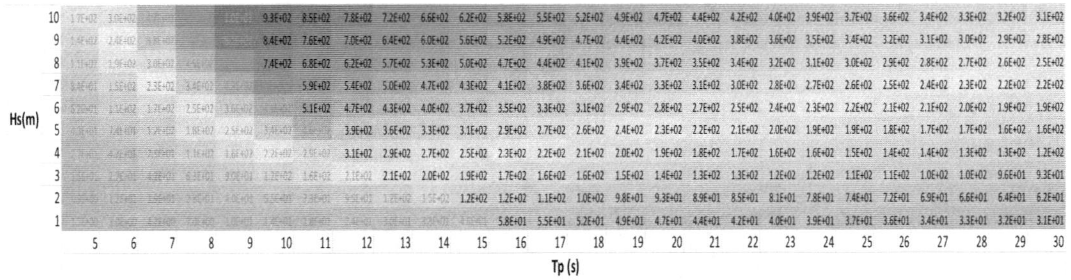

Figure 9. Floater power matric with a diameter equal to 20 m

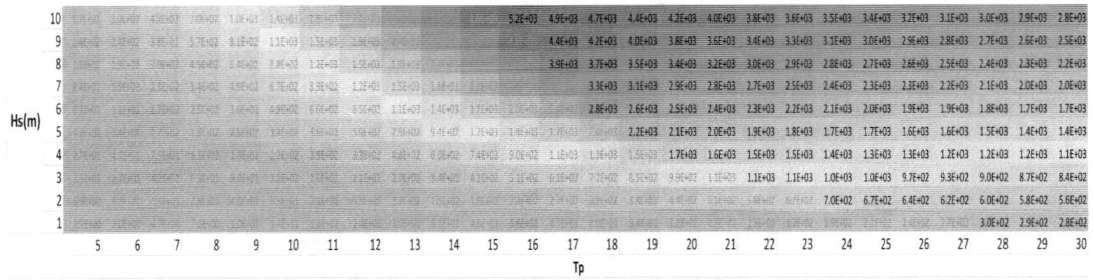

Figure 10 Floater power matric with a diameter equal to 60 m

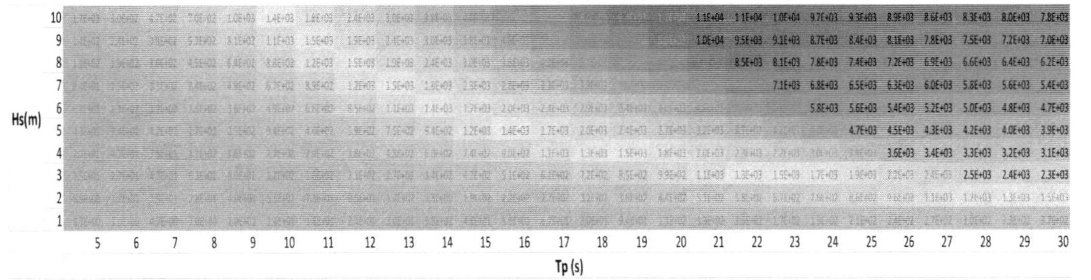

Figure 11. Floater power matric with a diameter equal to 100 m

However, it does not mean that the larger WEC is better. Considering the economic cost and ocean wave conditions, the WECs should be optimised based on different sites. In this work, two different sites, shown in Figure 12, demonstrate the influence of floater size and ocean wave conditions on WEC power generation. The wave statistics are given in Figure 13 and Figure 14 [4].

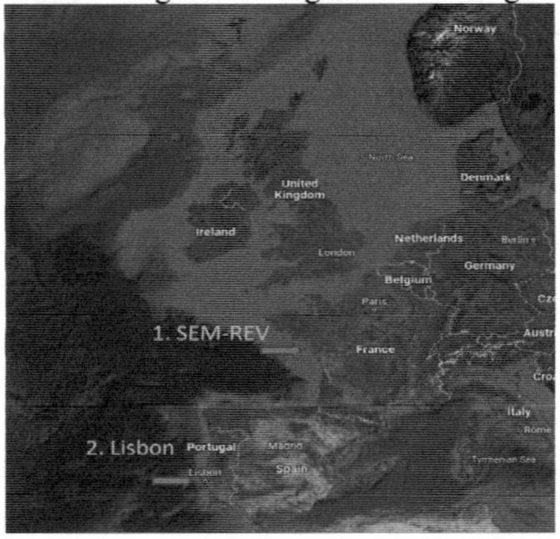

Figure 12. Sites where wave data statistics were taken.

IOP Conf. Series: Materials Science and Engineering 1294 (2023) 012009 doi:10.1088/1757-899X/1294/1/012009

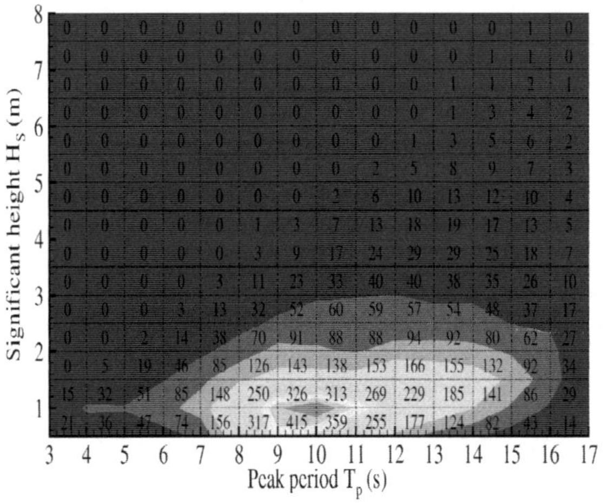

Figure 13. Scatter diagrams of wave statistics in SEM-REV [4]

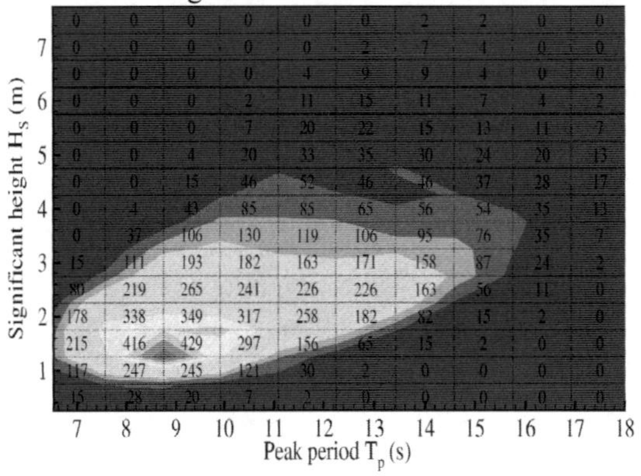

Figure 14. Scatter diagrams of wave statistics in Lisbon [4]

The average annual power generation can be roughly estimated based on the SWC power matrix and the wave data statistics. The results are given in Table 1 and Figure 15. As shown in Table 1 and Figure 15, in the same site, increasing the floater diameter within a certain size can effectively increase average power generation. However, when the floater is beyond a certain size, the floater diameter increase will not influence the power generation. More importantly, it is noticed that the different sites will have a great influence on power generation. As shown in Figure 15, site 2 can generate two times more power than site 1.

Table 1 Influence of floater size and ocean site

Site 1: SEM-REV	
Floater diameter (m)	Average power generation (kW)
20	70.74
60	102.13
100	102.13
Site 2: Lisbon	
Floater diameter	Average power generation (kW)
20	130.74
60	223.13
100	223.13

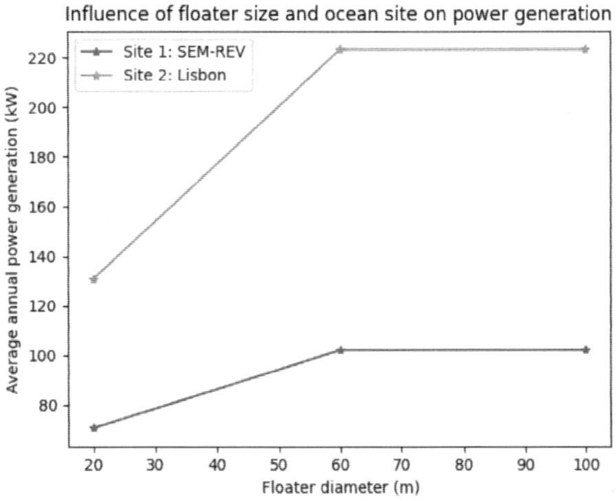

Figure 15. Influence of float size and ocean site on power generation

4.2. Performance study of the PTO system

The power generation of the WECs is not only influenced by the floater size and ocean wave condition but also by the PTO system. In the previous subpart, the influence of the floater size and ocean wave is studied. In this subpart, the influence of the PTO system parameters is considered. Four key parameters of the PTO system are investigated. They are, respectively diameter of the crankshaft (Dc), the inertia of the flywheel (I), the gripper force (F) and the generator damping coefficient (Cd). The results are summarised in Table 2. For the crankshaft diameter (Dc), it has an important influence on the averaged power generation. As depicted in Figure 16, the power generation will also increase with the increase of Dc. Contrary to the crankshaft diameter, the flywheel inertia seems to have a very limited influence on the power generation, as shown in Table 2 and Figure 17. With the change in flywheel inertia, the power generation is nearly the same. As for the gripper force, as shown in Figure 18, the increase of gripper force will significantly increase the power generation. It means that under the same wave excitation, the PTO system provides a higher gripper force that will generate more power. Additionally, with the increase of the generator damping coefficient, the SWC power generation will decrease a lot depicted in Figure 19. It is clear that Dc, F, and Cd have an important influence on power generation. To design the SWC, these three parameters need to be carefully chosen.

Table 2 Parametric study of the PTO system

Influence of Crankshaft diameter	
Dc (m)	Averaged power generation (kW)
2	0.97
4	1.80
6	6.63
Influence of flywheel inertia	
I (Kg*m^2)	Averaged power generation (kW)
2e6	0.97
6e6	0.97
1e7	0.97
Influence of Gipper force	
F (N)	Averaged power generation (kW)
2e4	0.13
6e4	0.97
1e5	5.43

Influence of Generator damping coefficient	
Cd (N*M/(rad*s))	Averaged power generation (kW)
2e5	0.97
6e5	0.38
1e6	0.11

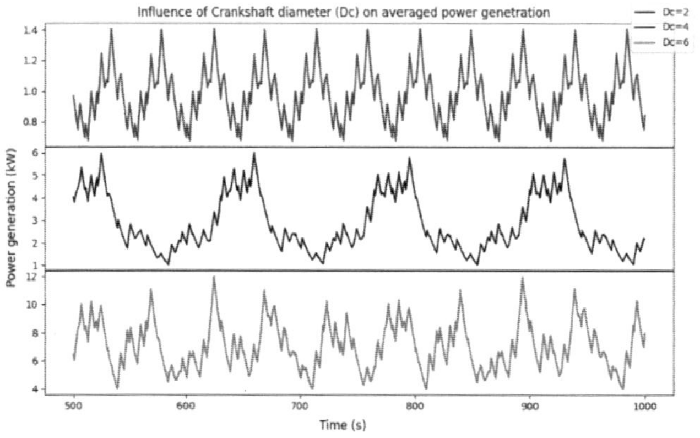

Figure 16. Influence of crankshaft diameter.

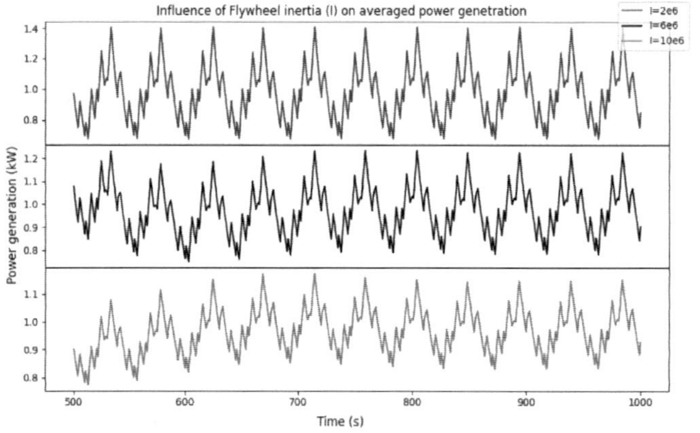

Figure 17 Influence of flywheel inertia.

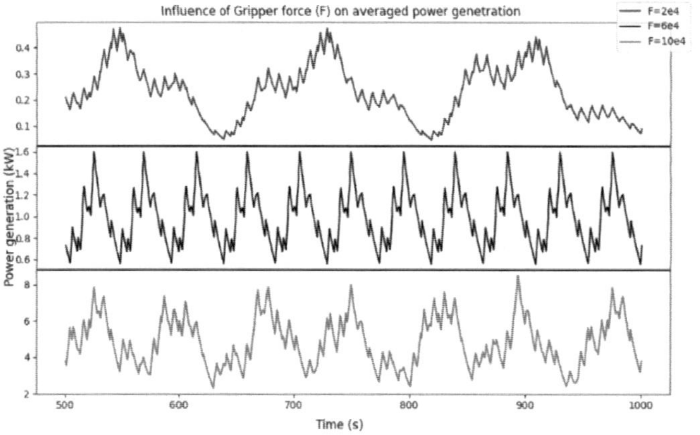

Figure 18. Influence of Gipper force.

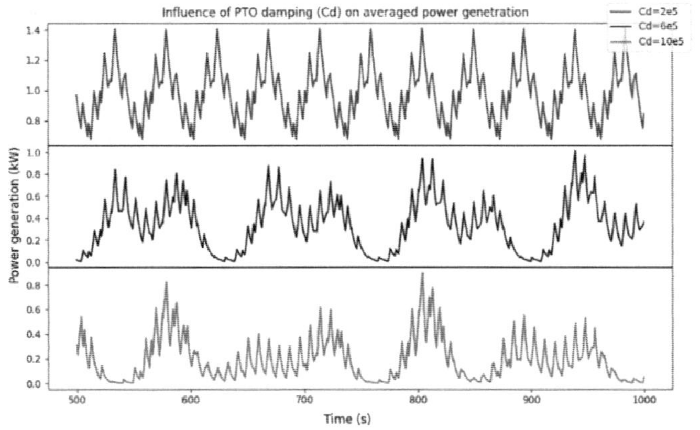

Figure 19. Influence of Generator damping coefficient.

5. Conclusion and future works

In this paper, a novel PTO system is presented to harvest energies from ocean waves (or swells). The novelty of the PTO system is to transform linear motion from ocean waves into rotation through a rotating shaft. The basic working mechanism of the novel PTO system is briefly introduced. A digital twin model of the PTO system is also created based on the MATLAB toolbox. The characteristic performance of the PTO system is also investigated. The results show that:

Firstly, from the study of floater size, it is noticed that the SWC power matrix tends to have power density in the long wave periods with the increase of the floater size. Also, with the increase of the floater size, the left-upper bound of the power limit will play a dominant role in the power matrix calculation. Additionally, based on the study of two different sites, the influences of ocean sites are important, and can even double the annual average power generation. Also, increasing floater size within a certain size in the same ocean site will effectively increase average power generation. However, when the floater is beyond a certain size, the floater diameter increase will not influence the average power generation. It is because, with the increase of the floater size, the left-upper bond of the power limit will play a dominant role in the power matrix, which is size-independent. The floater size needs to be optimised considering the economic cost and the specific ocean site.

Secondly, the parametric study of the PTO system is carried out based on the developed digital model. It is noticed that the crankshaft diameter, gripper force, and generation damping coefficient will play an important role in the SWC power generation. However, the flywheel inertia has little influence on the power generation, which can be ignored in the future optimisation process. More attention should be paid to the other three parameters for the parameter optimisation process.

Finally, the present work only focuses on the regular waves, and a passive control model is used in the PTO system. In the future, the influence of irregular waves will also be investigated. More importantly, based on the Scotland and Greenland ocean wave data, the optimal size of the floater will be investigated. The latch control and reactive control will also be applied to the novel PTO system, which will significantly improve the average power generation of the system.

References

Financial support from RFF Rogaland (Project: 336822 – RFFROGALAND Støtte fra andre enn Forskningsrådet Regionalt kvalifiseringsprosjekt Swell Energy Convertor – Proof of Principle Study for PTO system) is acknowledged.

References

[1] Day A H, Babarit A, Fontaine A, He Y P, Kraskowski M, Murai M, Penesis I, Salvatore F and Shin H K 2015 Hydrodynamic modelling of marine renewable energy devices: A state of the art review, *Ocean Eng.* **108**, 46-69.

IOP Conf. Series: Materials Science and Engineering 1294 (2023) 012009 doi:10.1088/1757-899X/1294/1/012009

[2] Ahamed, R, McKee K and Howard I 2020. Advancements of wave energy converters based on power take off (PTO) systems: A review, *Ocean Eng.* **204**, 107248.
[3] Sheng W 2019 Wave energy conversion and hydrodynamics modelling technologies: A review, *Renewable Sustainable Energy Rev.* **109**, 482-498.
[4] Babarit A, Hals J, Muliawan M J, Kurniawan A, Moan T and Krokstad J 2012 Numerical benchmarking study of a selection of wave energy converters, *Renewable Energy* **41**, 44-63.
[5] Foteinis, S 2022 Wave energy converters in low energy seas: Current state and opportunities, *Renewable Sustainable Energy Rev.* **162**, 112448.
[6] Ren, C., Tan, J., and Xing, Y 2023. ALK-PE: An efficient active learning Kriging approach for wave energy converter power matrix estimation. Ocean Engineering, 286, 115566.
[7] Moen L 2021 An Apparatus and a Method for Harvesting from Ocean Waves, PCT/NO2021/050077, *Norsk patentnr*, 345898.
[8] Straume I 2014 World wave energy resource map, https://no.wikibooks.org/wiki/Fil:World_wave_energy_resource_map.png
[9] Tristan P and Fossen T 2008 Time-vs. frequency-domain identification of parametric radiation force models for marine structures at zero speed, *Model. Identification Control* **29**(1), 1-19.
[10] Falnes J and Kurniawan A 2020 *Ocean waves and oscillating systems: linear interactions including wave-energy extraction* Vol. 8. (Cambridge University Press).

IOP Conf. Series: Materials Science and Engineering 1294 (2023) 012010 doi:10.1088/1757-899X/1294/1/012010

Research on dynamic responses of mooring line system of semi-submersible platforms in Vietnam sea conditions using random model

H H Pham[1,*], **T Mai**[1] **and Y Liu**[2]

[1] Hanoi University of Civil Engineering, Vietnam
[2] Norwegian University of Science and Technology, Norway

[*] Correspondence: hauph@huce.edu.vn

Abstract. Recently, due to the nearshore oil fields in Vietnam were nearly exhausted, oil and gas extraction proceeds to deeper water areas using the mooring floating offshore structures. Therefore, the durability of the mooring system in deep water should be studied through determining the dynamic tension of the mooring line system using random wave models. However, full stochastic dynamics analysis for mooring line system takes up a lot of time and requires large computing resources. This paper is going to study the hydrodynamic interaction of random waves onto a semi-submersible platform, thereby analysing the dynamic responses in time domain of the mooring system using the "time window" method to optimize computational time and computing resources. This method performs the dynamic simulations around the maximum quasi-dynamic tension values to determine the dynamic tension and the dynamic amplification factor (DAF) of a maximum load-bearing mooring line, thereby evaluating the dynamic responses for the whole mooring system. The method is applied to calculate for a semi-submersible platform in Dai Hung field in Vietnam Sea.

1. Introduction

The exploitation of oil and gas in Vietnam's offshore waters and deep continental shelf waters is in the development stage and has great potential due to the secondary exploitation of nearshore fields. The use of floating offshore structures including semi-submersible platforms to serve oil and gas exploitation in deep water areas is an inevitable need. The reason is that these structures have advantages such as flexibility in use, exploitation from shallow water to deep water, can be moved from one field to another and reusable.

A Semi-Submersible platform (also known as Semi- Floating Production Units – FPU or Semi-Floating Production System - FPS) consists of a rectangular deck structure supported by 4 to 8 surface-piercing vertical columns standing on submerged horizontal floaters. The vessels are moored by means of 8 to 12 catenary mooring lines consisting of chains or combinations of chain and wire (see Figure 1).

The study of the dynamic responses of mooring line system of semi-submersible platforms is one of the important problems for determining the durability safety of this structure. From the results of hydrodynamic force calculations on FPU, the dynamic response analysis of FPU's mooring system will be performed using quasi-dynamic and random dynamic simulations of mooring line tension according to the time domain analysis method. In this paper, the authors propose a sequence of practical computational steps for the problem of dynamic response analysis of mooring line systems to preliminarily quasi- dynamic calculate for the entire mooring system and for all sea states. From there, the most dangerous mooring line and sea state are selected to conduct detailed dynamic response through the dynamic amplification factor (DAF) of the mooring line. For each quasi-dynamic tension simulation during 3 h, the method of selecting a "time window" around the maximum quasi-dynamic tension value for line dynamic calculation is carried out, in order to optimize the time and resources for dynamic calculation. To calculate the design dynamic tension of the mooring line, in a simulation it is only necessary to take the maximum peak value. Therefore, selecting the "time window" around the max peak value of the quasi-dynamic simulation for dynamic calculation can ensure the accuracy of the results while optimizing the calculation time. The study has been applied to the mooring system of Dai Hung 01 semi-submersible platform (FPU DH01) at Dai Hung field, using the software of Bureau Veritas (HydroSTAR, ARIANE-3D and MCS-Cable) for dynamic analysis of mooring lines subjected to random waves.

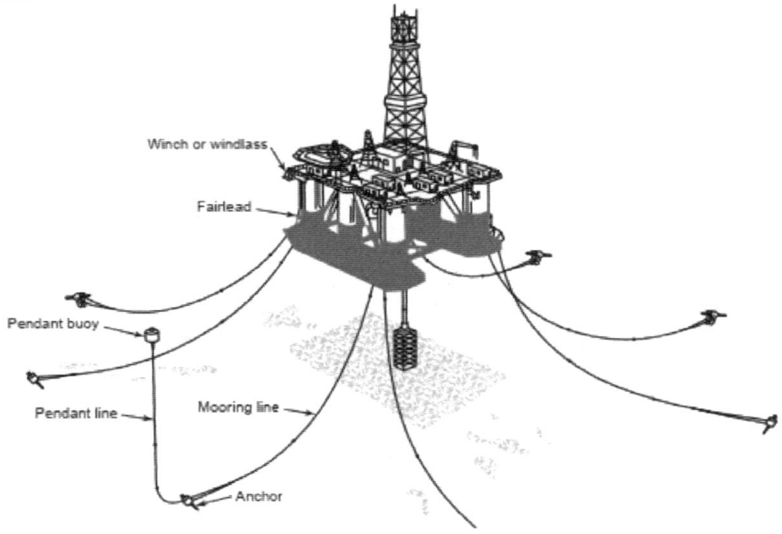

Figure 1. Typical semi-submersible platform [1]

2. Research steps

To implement the above research objectives, a sequence of steps as follows was performed:

- Research on the methodology of calculating the dynamic response of the mooring line of a FPU according to a random model: from the response amplitude operators (RAOs) transmission functions of the FPU's motions the hydrodynamic forces according to [7,8], using the analysis method in the time domain to solve the random dynamic problem in 2 steps: quasi-dynamic and dynamics analysis of mooring lines.
- Collection of the environmental data, the design parameters of FPU DH01 and its mooring system.
- Using HydroStar software to calculate the RAO of motions and hydrodynamic forces under the random waves acting on FPS-DH01 to calculate the mooring line system.
- The quasi-dynamic tension in mooring system is calculated according to the simulations during a 3-hour storm. In order to minimize time and resources for line dynamic calculation, select the "time window" around the maximum quasi-dynamic tension of the mooring line with the lowest safety factor in the most dangerous sea state to calculate dynamic simulation.
- Calculate the line dynamic tension in the above step, thereby determining the dynamic amplification factor (DAF) of the line tension.
- Evaluate the dynamic influence of the FPU's mooring system affected by random waves through the DAF.
- Determine the safety factor and the strength of the mooring lines from the design dynamic tension.

3. Methodology for calculating the dynamic response of the FPU's mooring system

The tension of the mooring line can be calculated under the action of extreme environmental forces according to two methods: Quasi-dynamic and dynamic analysis of the mooring lines.

In the quasi-dynamic analysis method, the problem can be solved in the frequency domain (also known as the spectral method) or in the time domain. However, the disadvantage of the spectral method is that this method can only solve the linear problem (quasi-dynamic analysis) and not the nonlinear problem (line dynamics analysis). Therefore, in the case of mooring line analysis according to both quasi-dynamic and dynamic methods, simulations of line tension should be analyzed in the time domain [3,9,10,11].

3.1. Quasi-dynamic analysis of the mooring lines in the time domain

According to [3], when analyzing the quasi-dynamic responses of the mooring lines in the time domain, the tension of the lines is determined from the quasi-static response of the lines due to the motions of FPU at fairlead (connection point between the line and FPU). The FPU is only affected by environmental forces acting on the floating structure, ignoring the environmental force acting on the mooring line.

Scope of application: Quasi-dynamic analysis method is suitable for calculating structures operating at shallow and medium water depths.

3.1.1. Quasi-dynamic simulation steps in the time domain. First, when there is no external force acting on the FPU, the FPU establishes the initial equilibrium position by the pre-tension in the mooring line. Then external forces are applied to the FPU, which are sub-divided as follows (see Figure 2):

– Steady component: the mean forces due to mean wind and current speeds and mean wave drift;
– Low frequency component: the slowly varying low frequency force (LF);
– High frequency component: the high frequency force oscillating with wave frequency (WF).

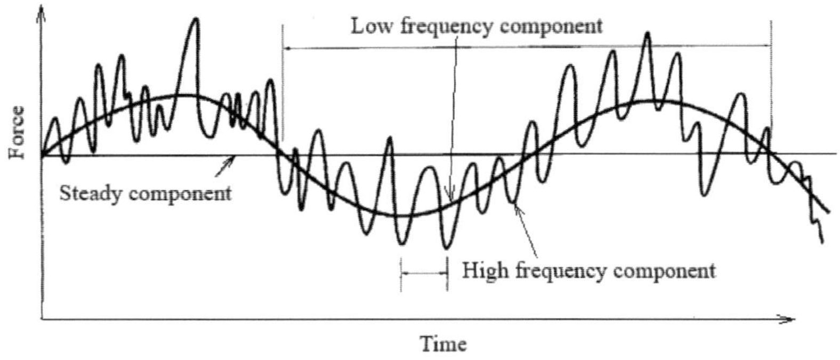

Figure 2. Components of external forces acting on semi-submersible platform (FPU) [3].

Under the action of external forces, the FPU is moved and create the mooring system response. The obtained signal of the tension is composed of three components:

- Mean response: static equilibrium by steady forces. The first component is due to the mean effect of wind, wave and current;
- Low frequency response: application of low-frequency forces at equilibrium position. The second component is due to the slow variation of the tension under the wave slow drift effect (second order wave drift load) and/or the wind gusts effect.
- Wave frequency response: addition of wave frequency responses on low-frequency response. The third signal is a variation of the tension at the wave frequency around the low frequency tension. This component is due to the wave-induced motion of the FPU (first-order wave motion).

(1) *STATIC ANALYSIS:* During the static analysis phase, under the action of the mean force of wind, current, and mean drift force of waves, the FPU will be changed to a new equilibrium position. The external forces are balanced by the tensions at fairlead of the mooring lines. Around this new

equilibrium position, the FPU performs the low-frequency and wave-frequency motions. After static analysis, the quasi-dynamic simulations in the time domain are performed.

(2) *LOW FREQUENCY:* At each time step t, the low-frequency displacement component is achieved by solving the following differential equation of motion [6,7,8,10]:

$$\sum_{j=1}^{6}\left\{\left[\boldsymbol{M}_{ij}+\boldsymbol{M}_{a_{ij}}(\omega)\right]\ddot{\boldsymbol{X}}_j+\boldsymbol{B}_{ij}(\omega)\dot{\boldsymbol{X}}_j+\boldsymbol{K}_{ij}\boldsymbol{X}_j\right\}=\boldsymbol{F}(t) \tag{1}$$

where $\boldsymbol{X}_j$, $\dot{\boldsymbol{X}}_j$, $\ddot{\boldsymbol{X}}_j$ are the motion, velocity, and acceleration vectors at the center of gravity G of the FPU, respectively; $\boldsymbol{M}_{ij}$ and $\boldsymbol{M}_{a_{ij}}$ are the mass matrix of floating structure and additional mass matrix, respectively; $\boldsymbol{K}_{ij}$ is the stiffness matrix of FPU; $\boldsymbol{B}_{ij}$ is the damping matrix; and $\boldsymbol{F}(t)$ is a vector with 3 components of the horizontal load applied at the center of the FPU at the time step *t*.

(3) *WAVE FREQUENCY:* At each time step after the numerical integration of the low frequency response, the 6 wave frequency motions of the FPU at centre of gravity are added to its low frequency position. The wave frequency motions derived from the Response Amplitude Operators (RAO) of the FPU.

3.1.2. Calculate the design quasi-dynamic tension of the mooring line. In order to avoid an undue numerical transitory response at the beginning of the simulation, it is recommended that the FPU not be suddenly submitted to the time varying components of environmental loads. The beginning of the response signal is therefore to be truncated. The time to start recording the simulation (t_l) is when the system is in equilibrium. According to the design experience of floating structures with catenary mooring lines (of FPU or FPSO/FSO), t_l is about from 2000s to 5000s, depending on the rigid of mooring system (conventional multi buoy mooring, yoke, soft-yoke, ...). In the ARIANE software, there is a function to find the initial equilibrium position in the static analysis. In this study for the application of FPU DH01, we use t_l is 2000s.

In order to calculate the design quasi-dynamic tension of the mooring line in the time domain, it is necessary to perform n quasi-dynamic simulations ($n \geq 5$) for each sea state of a short storm of 3 h (10,800 s). The simulations of the mooring line tension are calculated for each random wave seeding number (seed). Each simulation gives a maximum tension value, thereby calculating the mean and standard deviation of those n maximum tensions to calculate the design tension of the line in intact conditions, according to equations (2,3,4) [2,10]:

$$T_D = T_M + aT_S \tag{2}$$

$$T_M = \frac{1}{n}\sum_{k=1}^{n} T_k \tag{3}$$

$$T_S = \sqrt{\frac{1}{n-1}\sum_{k=1}^{n}\left(T_k - T_M\right)^2} \tag{4}$$

Where T_D is the design tension of the line; *n* is the number of simulation; T_k is the maximum tension at fairlead achieved in the *k*-th simulation; T_M, T_S are the average value and the standard deviation of T_k obtained by *n* simulations, respectively; *a* is the coefficient depending on *n* and the analysis method. If the quasi-dynamic method is applied, then the coefficient $a = 1.8$ and 0.5 corespoding to $n = 5$ and 20, respectively [2].

3.2. Line dynamic analysis

In the line dynamic analysis, the dynamic response of the FPU is still calculated as in the quasi-dynamic analysis, but the tension of the lines is calculated from a dynamic response of the lines. The mooring lines are meshed into elements along its length. The meshing of the mooring lines is automatically proposed but the user may also define the lengths of the elements. Meanwhile, with quasi-dynamic analysis, the line tension is only calculated at fairlead where the tension is maximum. In calculating the line dynamic response, it is necessary to take into account the environmental load acting on the mooring line, due to the current profile that varies with water depth both in velocity and direction. Meanwhile, in quasi-dynamic analysis, the environmental load acting on the mooring line is ignored. Thus, it can be seen that the line dynamic analysis is more accurate than the quasi-dynamic analysis. Therefore, dynamic analysis also requires smaller time step, in order to find the maximum tension Tk of the simulation more accurately. The time step of line dynamic analysis is 0.02 s, while the time step of quasi-dynamic analysis is 1 s. So, fully line dynamic analysis with realizations during 3 h for all sea states takes a lot of time and computer potential. Therefore, this study uses the practice method of dynamic calculation using the "time window" to minimize computational time and capacity.

3.3. Selection of "time window" to study the dynamic response of mooring lines

Limiting dynamic analysis to a "time window" (Figure 3) is a useful method to minimize computational time. But the problem is how to determine the time duration, location, number of time windows for a dynamic simulation?

These windows are selected around the maximum tension values of *dTqs/dt* in quasi-dynamic simulation during 3 hours. After practical studies [10,11], to mitigate this method, it is necessary to select a window around the maximum value of *dTqs/dt*.

The correlation between the peaks of the dynamic tension and the peaks of the quasi-dynamic tension dT_{qs}/dt allows to suggest a practical method for achieving the dynamic design tension of lines from the result of quasi-dynamic tension:

- Perform at least 5 quasi-dynamic simulations in 3 h of the most load-bearing line (n = 5);
- For each quasi-dynamic simulation, find out the maximum quasi-dynamic tension value;
- Around each of these maximum values, select a window (see Figure 3) whose interval is one low-frequency period to perform a dynamic simulation;
- For each dynamic simulation calculated in the time window, find the maximum value of dynamic tension (T_{dyn} max).

The line dynamic design tension is defined: T_D = average of T_{dyn} max + 1.8 standard deviations.

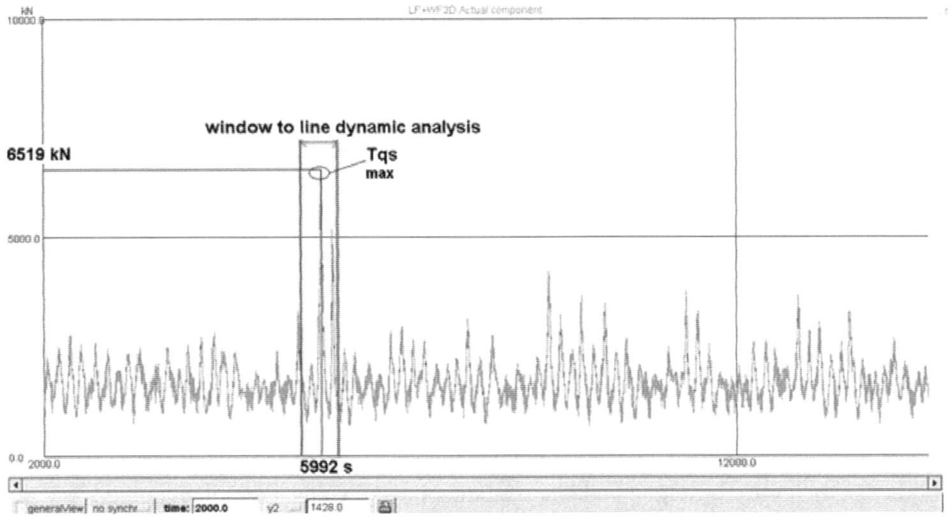

Figure 3. Selection of "time window" from quasi-dynamic simulation to line dynamic analysis.

3.4. Resistance safety factor of mooring line

After determining the design tension T_D of the mooring line, we calculate the resistance safety factor (SF) and evaluate the safety of the line compared to Bureau Veritas regulations [2]:

$$SF = \frac{T_{Br}}{T_D} \geq [SF] \tag{5}$$

Where T_{Br} is the minimum breaking load (MBL) of the line according to the manufacturer's catalog [14], depending on the material and diameter of the line; [*SF*] is the normative minimum safety factor. According to [2], with quasi-dynamic method [*SF*] = 1.75 and with dynamic method [*SF*] = 1.67 in intact condition. According to API [1], [*SF*] = 2.

3.5. Dynamic amplification factor of mooring line tension (DAF)

The purpose of determining the DAF (Dynamic Amplification Factor) is to evaluate the dynamic effect compared to the quasi-dynamic in the results of mooring line tension. The DAF is defined by the ratio between tension signals according to [3,10,11]:

$$DAF = \frac{T_{dyn} - T_{LF}}{T_{qs} - T_{LF}} \tag{6}$$

Where T_{LF} is a low frequency tension (T_{LF} = mean tension + low frequency tension variation); T_{qs} is a quasi-dynamic tension (T_{qs} = mean tension + low frequency tension variation + quasi-dynamic wave frequency tension); and T_{dyn} is a dynamic tension (T_{dyn} = mean tension + low frequency tension variation + dynamic wave frequency tension). Equation (6) is valid only if we take the T_{dyn} max, T_{qs} max, and T_{LF} at the same time, or at close times.

There is another formulation to calculate the DAF of mooring line tension according to [2,10]:

$$DAF = \frac{\sigma_{wfdyn}}{\sigma_{wfqs}} \tag{7}$$

Where σ_{WFdyn} is the standard deviation of time series simulation of line dynamic tension T_{WFdyn}; $T_{WFdyn} = T_{dyn} - T_{LF}$; and σ_{WFqs} is the standard deviation of time series simulation of line quasi-dynamic tension T_{WFqs}; $T_{WFqs} = T_{qs} - T_{LF}$.

Applications of DAF:

- Dynamic effect is evaluated through the DAF.
- Using the DAF to evaluate the resistance of mooring lines according to the dynamic method: first, we calculate the quasi-dynamic tensions for mooring lines, then select the most tension-bearing line to calculate the dynamic tension. We use the equation (5) to calculate the DAF for that line. Then we take the quasi-dynamic tensions of mooring line system multiplied by DAF to get the dynamic tensions for the entire mooring system.

4. Numerical application to evaluate the dynamic responses of mooring system of FPU DH-01

4.1. Simulation steps to calculate the dynamic tension of mooring line with ARIANE-3D

Applying the diagram in Figure 4, in the following application section, the authors perform the quasi-dynamic and dynamic response calculations for the mooring system of Dai Hung 01 semi-submersible platform (FPU DH01) at Dai Hung field according to [13], using the software of Bureau Veritas: HydroSTAR [4], ARIANE-3Dynamic and MCS-Cable [5].

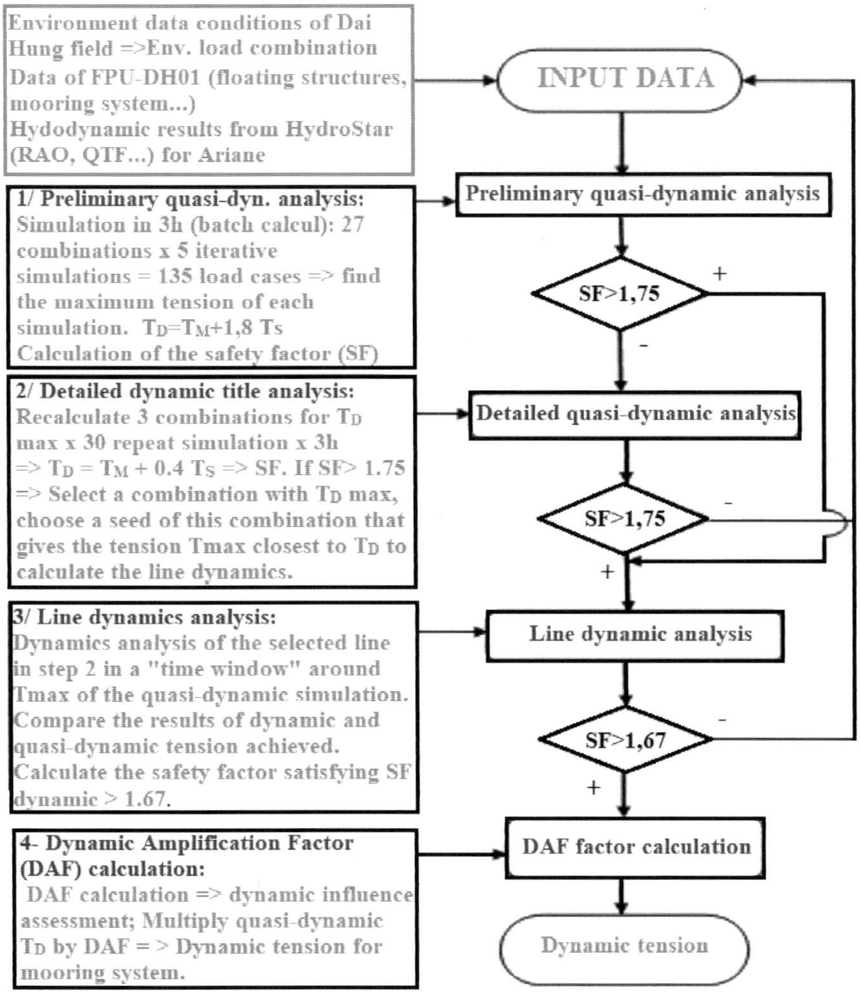

Figure 4. Practical calculation steps of dynamic response of mooring line with ARIANE-3D.

4.2. Data

Planning diagram of Dai Hung field with a water depth of 110 m is presented in Figure 5. The maximum random wave parameters with a 100-year return period taken in the Nord-Est direction (NE) which is the main direction of the impact wave at Dai Hung field. Environmental data in the NE direction according to [13] are as follows: Jonswap spectrum with spectral parameters: $\alpha = 0.0078$, $\gamma = 1.0504$, $\sigma_A = 0.1039$, $\sigma_B = 0.1079$. Significant wave height: $H_s = 10$ m; Period of peak spectral: $T_p = 15.96$ s; Average wind speed with return period of 100-year during 10': $V_w = 33.5$ m/s. Surface current velocity in SW direction (combined with NE wave direction): $V_c = 0.92$(m/s).

Data of FPU DH01 is taken according to [12]: Static weight is 10,866 tons ; FPU is structured according to AKER H-03 style, including: 02 pontons, 08 columns, cross braces, 8 oblique mooring lines: length of one mooring chain is 1,150 m, diameter 84 mm NVK4 RIG, the pre-tension is 70 tons, the minimum breaking load (MBL) of chains is 10,600 kN, linked to the seabed by 37-ton anchors of the Stevshark type.

IOP Conf. Series: Materials Science and Engineering 1294 (2023) 012010 doi:10.1088/1757-899X/1294/1/012010

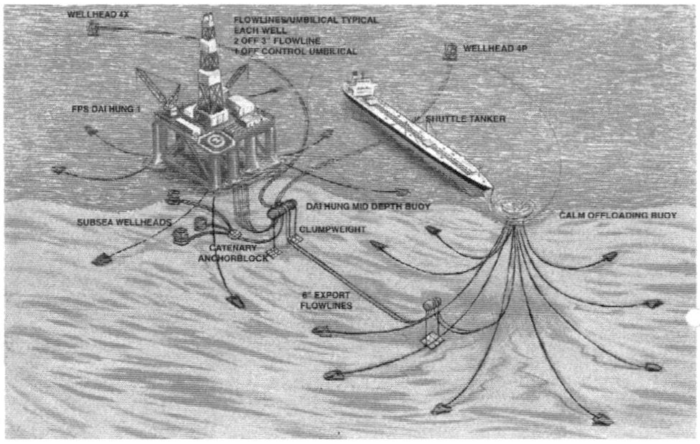

Figure 5. Dai Hung field schematic field lay-out [12].

4.3. Hydrodynamic analysis results using HydroStar for ARIANE calculation

Using HydroStar software and the theories in [4,6,7,8] to calculate the hydrodynamic force through the potential of incident, diffraction and radiation waves acting on semi-submersible platform. The outputs are RAO transfer functions of structural reactions (1st, 2nd order wave forces, motions of FPU...). These results are inputs to mooring line calculations using ARIANE software. The computational model image and some RAO results of the HydroStar software are showed in Figures 6 & 7.

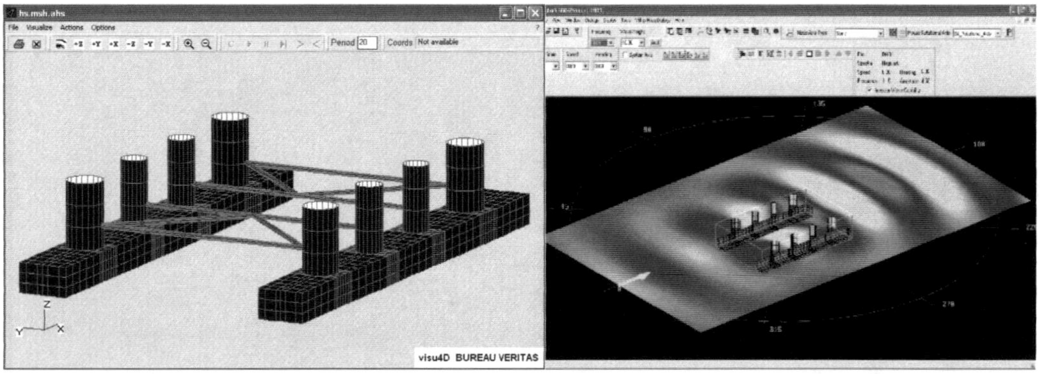

Figure 6. 3D meshing model (left) - Diffraction wave acting on the FPU (right) in HydroStar.

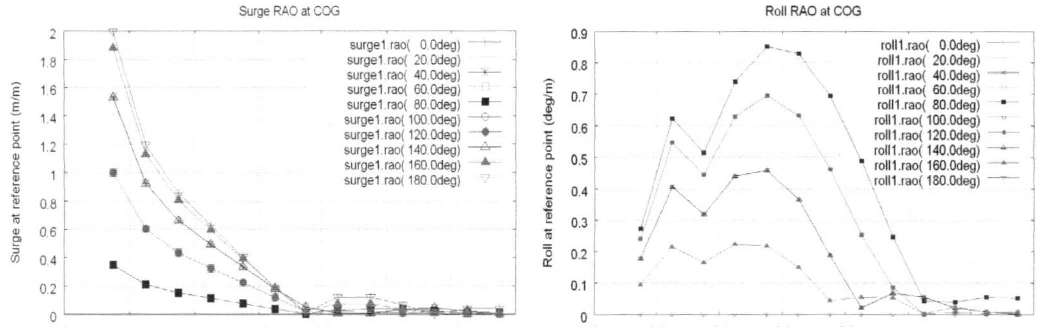

Figure 7. Surge RAO (left) - Roll RAO (right) of FPU DH01.

IOP Conf. Series: Materials Science and Engineering 1294 (2023) 012010 doi:10.1088/1757-899X/1294/1/012010

4.4. Dynamic calculation for the mooring system of FPU DH-01

4.4.1. Combination of extreme environmental parameters: Combinations of values: According to [2], there are two types of combinations: (1) Dominant wave; (2) Dominant current. In Vietnam's sea conditions, waves are dominant, so we combine the extreme wave parameters of 100 years with the wind and current speeds of 100 years multiplied by the reduction coefficient, which depends on the angle between wave and wind directions. Combinations of directions: As a rule [3], for a calculated direction (NE = 45°), it is necessary to account a variation of 22.5° around that direction for each type of impact load. So, we have the combinations: Wave = (45° ± 22.5°); Wind = (45° ± 22.5°); Current = (45°+22.5°). In total, we have 3x3x3=27 basic combinations (sea states). Each combination performs 5 iterative simulations in 3 hours, so there are 135 loading cases in total.

4.4.2. Quasi-dynamic tension calculation in time domain for 135 loading cases (batch): Calculate the design tensions (T_D) of the mooring system under the random wave loads for all loading cases (batch calculation) according to the quasi-dynamic method to find the most dangerous loading case for each mooring line. From there, calculate the safety factor (*SF*) and determine the line with the greatest tension in the mooring system to calculate in detail the quasi-dynamic simulation and dynamics for that line. Figure 8 shows a diagram and a 3D image of the mooring system of FPU DH01 modeling in ARIANE7.

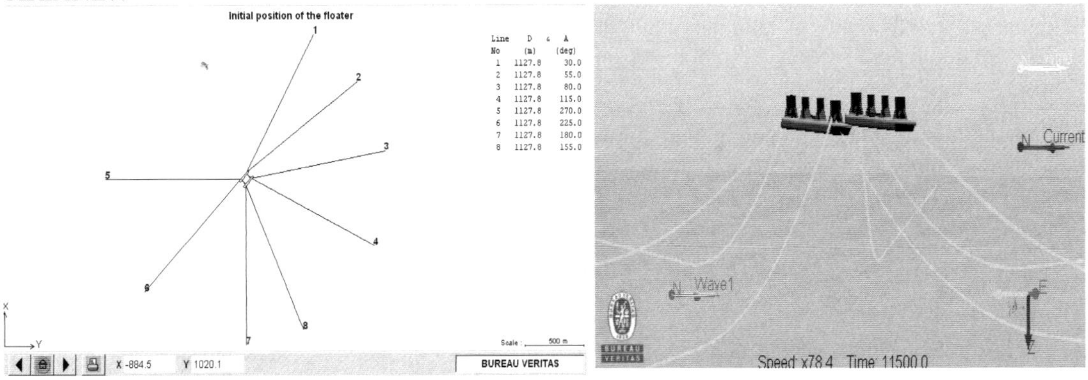

Figure 8. Modeling of FPU DH01's mooring system in ARIANE7.

The peaks T_k of the tension simulations output from ARIANE are processed and given T_D as shown in Table 1.

Table 1. Summarize the quasi-dynamic tension results of the mooring system in batch calculation.

Combination number	Line number	T_D max (kN)	Safety factor (SF)
6	5	1032.87	10.26
7	**8**	**2586.22**	**4.1**
9	6	564.19	18.79
20	7	1272.79	8.33
25	1	2148	4.93
	2	1064.39	9.66
27	3	568.99	18.63
	4	612.2	17.31

Comments: The lowest *SF* calculated by quasi-dynamic method for line 8 is *SF* min = 4.1 > [*SF*]=1.75. So, the mooring system is satisfying the durable condition. Line 8 has the largest design tension T_D, so it is necessary to calculate details according to single simulations for that mooring line.

4.4.3. Single simulation of line tension in time domain: Calculation steps as follow:

- *Step 1:* Quasi-dynamic simulation of 10,800 s for line 8 in combinations n° 7 with the seed 922 using ARIANE7 (Figure 9). The maximum longitudinal tension results $T_{x,max}$ = 2,527 kN at time t = 2,154 s.

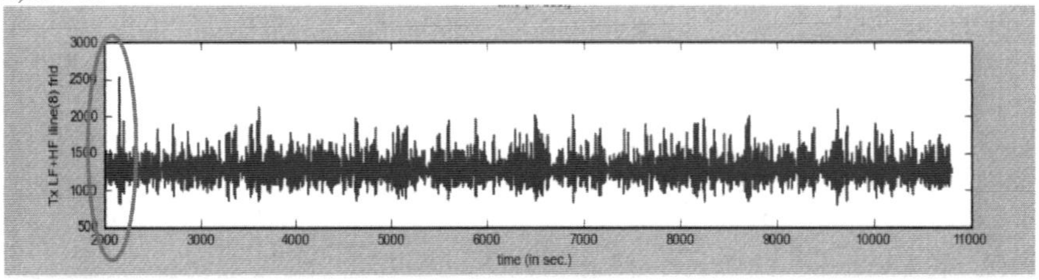

Figure 9. Quasi-dynamic tension simulation at fairlead of line 8 in 10,800 s using ARIANE7.

- *Step 2:* Select the "time window" of 200 s around $T_{x,max}$ of the simulation in step 1, recalculate the quasi-dynamic tension simulation in 200 s of line 8 with the same loading condition (Figure 10).

The results in the Figure 9&10 showed that the peak tension $T_{x,max}$ and the moment t of the peak were the same between the simulation in 10,800 s and the simulation in "time window" of 200 s around the peak. These results prove that the simulation method using time window is suitable.

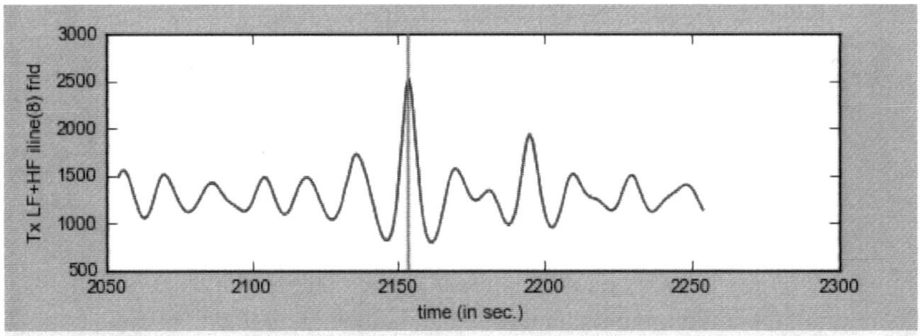

Figure 10. Quasi-dynamic tension simulation at fairlead of line 8 in "time window" of 200 s.

- Step 3: Single simulation of line tension in time domain using ARIANE6.3. The reason for using ARIANE6.3 to recalculate the quasi-dynamic tension is this version 6.3 integrates with the MCS Cable 3D module to dynamic analysis of the mooring lines. Note: Due to the different conventions of 2 versions of ARIANE, line 8 in the ARIANE7 is line 5 in ARIANE6.3. Therefore, we simulate the quasi-dynamics tension of line 5 in 10,800 s with ARIANE6.3 (see Figure 11).

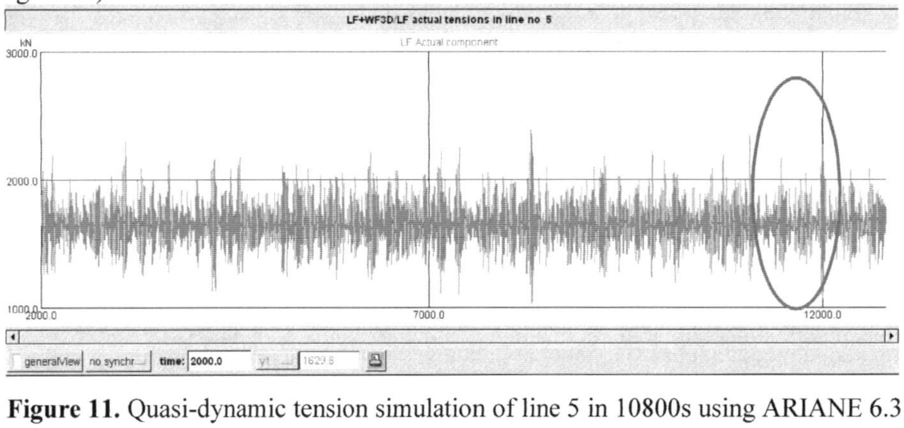

Figure 11. Quasi-dynamic tension simulation of line 5 in 10800s using ARIANE 6.3

4.4.4. Simulation of quasi-dynamic and dynamic tensions for line 5 in a " time window" of 200 s.
From the simulation in Figure 11, select a "time window" around the peak tension to recalculate the quasi-dynamic tension using ARIANE6.3 and dynamic tension using MCS Cable 3D. The results of these two software are shown in Figures 12. The results from Figure 12 show that T_{max} is of 2,475 kN at t = 11,992 s in quasi-dynamic and in dynamic simulation, T_{max} = 2,973 kN at t = 11,990 s.

T_{max} in quasi-dynamic < T_{max} in dynamic, so the dynamic influences need to be considered through the DAF. The time moment t of reaching T_{max} in two simulations is nearly identical, proving that the choice of "time window" around the T_{max} of quasi-dynamic tension for dynamic analysis is reasonable.

Then, we compare the quasi-dynamic and dynamic tension simulations of line 5 in a "time window" of 200 s in a same figure (see Figure 13). These two simulations in the Figure 13 are very similar, only differing in the magnitude of the tension. So, using DAF multiplied by quasi-dynamic tension to calculate the dynamic tension of the mooring system is suitable. Note that it's only necessary to determine the peak of simulation to calculate the design tension T_D from which to define the *SF*.

Safety factor of line 5 with dynamic tension: SF_{dyn} = 10,600/2,973 = 3.56 < $[SF]_{dyn}$ = 1.67, so the mooring system is satisfying the durable condition in dynamic analysis, according to [2].

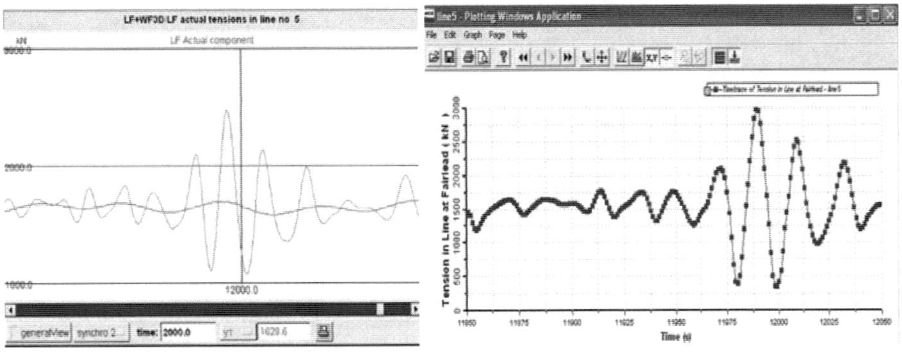

Figure 12. Quasi-dynamic tension (left) and dynamic tension simulations (right) of line 5 in 200 s.

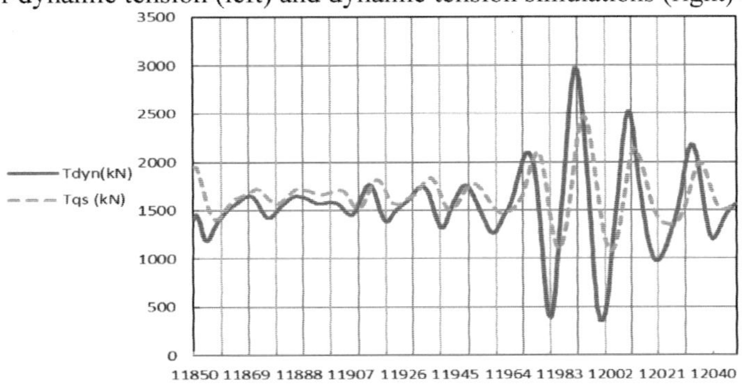

Figure 13. Comparison of quasi-dynamic simulation and dynamic simulation of line 5 in 200 s.

4.5. Dynamic amplification factor of mooring lines tension of FPU DH01 (DAF)
DAF is calculated by formula (6) or formula (7) in the section 3.5. In this application, we used formula (7) to calculate DAF. The DAF results for line 5 – the mooring line bearing the greatest tension in the most hazardous environment combination are shown in Table 2.

Table 2. Dynamic amplification factor of mooring lines tension of FPU DH01.

Line N_o – Load combination	"seed" N°	σ_{WFqs} (kN)	σ_{WFqs} (kN)	DAF
Line 5- Combination 7	922	232.5846	232.5846	1.792

Comments: The dynamic effect of mooring line tension of the FPU DH01 at Dai Hung field with 110m of water depth is evaluated through a DAF of 1.792. From there, we can use the quasi-dynamic method to calculate the mooring lines, then multiply by the DAF to determine the dynamic tension, serving to evaluate the durability of the mooring line system.

5. Conclusions

In this study, a simplified practical method has been suggested to optimize the calculation time for dynamic analysis. Accordingly, it is possible to select a time window around the maximum quasi-dynamic tension to simulate the dynamic tension the mooring lines. The maximum quasi-dynamic tension is smaller than the maximum dynamic tension, indicating that the needs to define DAF for dynamic calculation. The times (*t*) when the tension reaches the maximum values in 2 simulations are nearly identical, demonstrating that it makes sense to choose the time window around the maximum quasi-dynamic value.

From the current case study presented in the paper, we found that for Vietnamese waters, oil and gas exploitation projects have a common water depth from about 50m to 100m, the calculated DAF coefficient is relatively small. It is not necessary to analysis fully dynamically for the mooring line. So, the similar projects in Vietnam may refer to the practical calculation steps presented in the paper to minimize calculation time while ensuring the accuracy of calculation results. The safety factors (*SF*) of the mooring lines of the FPU DH01 are greater than the permissible safety factor, according to [2]. Therefore, the design tensions of the mooring system satisfy the resistance condition for extreme sea states of the environment (ULS).

The results of dynamic responses of the mooring lines can be applied to verify the resistance of lines according to the dynamic calculation method, giving more accurate results than quasi-dynamic method. The next research is using the dynamic tension graph of the lines in a time series (with the input of a 1-year statistical wave load) to analyze dynamic fatigue for the mooring lines.

References

[1] American Petroleum Institute 2005 Recommended Practice for Design and Analysis of Stationkeeping Systems for Floating Structures, API RP 2SK 3rd edition.
[2] Bureau Veritas 2004 Classification of Mooring Systems for Permanent Offshore Units, Guidance Note NI 493 DTM R00 E, Paris.
[3] Bureau Veritas 1998 Quasi-Dynamic Analysis of Mooring Systems using ARIANE software, Guidance Note NI 461 DTO R00 E, Paris.
[4] Bureau Veritas User guide Ariane 7 (361p).
[5] Bureau Veritas HydroStar For Experts – User Manual (176 p).
[6] Chen X-B 2004 Hydrodynamics in Offshore and Naval Applications - Part I, *Paper presented at the 6th Int. Conference on Hydrodynamics*, University of Western Australia, Perth..
[7] Chen X-B 1988 Étude des réponses du second-ordre d'une structure soumise à une houle aléatoire, *PhD. Thesis*, Université de Nantes (233p.).
[8] Molin Bernard 2002 Hydrodynamique des Structures Offshore, Guides pratiques sur les ouvrages en mer. Editions Technip, Paris, France.
[9] Pham H.H 2019 Methodology for total reliability evaluation of the mooring lines of floating offshore structures, *Proc. of the 1st VSOE2018*, ISBN:978-981-13-2305-8, 572-579.
[10] Pham H.H 2015 FPSO - Fiabilité des lignes d'ancrage avec prise en compte de fatigue, ISBN: 978-3-8381-7928-5, 336 p., Presses Académiques Francophones.
[11] Pham H.H 2009 Quasi-dynamic and line dynamic random analysis of mooring systems of FPSO

at White-tiger field using ARIANE-3D software, *Petro Vietnam J.*, **10**, 41-48.
[12] PVEP 2006 DH 01-FPU Structural and Mooring Systems Drawings.
[13] PVEP 2006 Final meteocean and environmental design criteria for Dai Hung fields, Report in VSP, Vietnam, 157p.
[14] Vicinay – Mooring Lines Catalogue. Available at: Vicinay Mooring Connectors - Products catalogue (vicinayconnectors.com) (Accessed: 22.10.2023)

IOP Conf. Series: Materials Science and Engineering 1294 (2023) 012011 doi:10.1088/1757-899X/1294/1/012011

Numerical study on the de-ballasting operation of a floating dock with a malfunctioning pump

J Zhang *, X Wen and M C Ong

Department of Mechanical and Structural Engineering and Materials Science
University of Stavanger, NO-4036, Stavanger, Norway

* Correspondence: jianan.zhang@uis.no

Abstract. In normal de-ballasting operations of floating docks, each ballast pump manages a specific group of ballast tanks, independently of the others. However, if one pump is malfunctioning, a connection valve between the group and its adjacent group is opened, allowing its adjacent pump to control two groups of tanks simultaneously. This study aims to investigate the de-ballasting dynamic process in the event of a ballast pump malfunction by utilizing a numerical model of a full-scale digital floating dock. In the present numerical model, the dock is represented as a rigid body with six degrees of freedom and is subjected to hydrostatic, hydrodynamic, and mooring forces. To maintain pitch and roll angles within predefined limits, a modified P-controller is employed to regulate the valve opening angles at all the ballast tanks. Hydraulic calculations for the ballast water system with a malfunctioning pump are conducted, and the numerical results demonstrate the effectiveness of the present modified P-controller when dealing with a malfunctioning Pump No. 1.

1. Introduction

Floating docks are important in ship repair and maintenance. During operations of ship repair and maintenance, effective ballast water adjustment is crucial for maintaining a floating dock's stability. Traditionally, this task is done manually by a skilled dock master, which is risky since it depends heavily on the expertise of the dock master. To enhance accuracy and efficiency, an automatic ballast control algorithm is needed to support the dock master in making informed decisions.

In recent years, researchers have proposed automatic ballast control algorithms for floating docks, based on normal ballast water systems. For example, Ohkawa et al. [1] developed a linearized algorithm to manage the dock's draught, heel, and trim based on grouped ballast tanks. They divided the dock's ballast tanks into groups and applied controllers to the valve opening angles to fill and empty these groups of ballast tanks. Guo et al. [2] detailed the hardware and software components required for the automatic ballast control, monitoring the draught at the dock's corners to calculate heel and trim angles. Control of valve opening angles and pump operations ensured the heel and trim angles remained within allowable ranges. Topalov et al. [3] employed physical sensor inputs to develop an algorithm that controlled heel and trim angles during submerging and emerging operations. Moreover, the dock's bending moment is also maintained below a defined threshold. Wen et al. [4], [5] proposed a modified P-controllers to stabilize the floating dock during ballasting and de-ballasting operations by controlling the valve opening angles at different ballast tanks. However, none of these studies involve the performance of their automatic ballast control algorithms under malfunctioning ballast water systems. Such malfunctions are a major risk in floating dock operations, leading to failure in controlling ballast water flow rates and thus resulting in uneven ballasting, tank flooding, excessive heeling or trimming, and structural damage. For instance, the power outage of a ballast pump in floating dock PD-50 [6] caused sinking and structural damage to a docked aircraft carrier. Additionally, a malfunctioning ballast valve in Dry Dock #3 resulted in tank flooding, causing the dock to tilt, and leading to the sinking of both the dock and the ship on top [7].

Given these risks, the research on automatic ballast control algorithms for floating docks is still insufficient, particularly concerning their performance under malfunctioning ballast water systems. Therefore, this study continues the work of Wen et al. [5] to investigate automatic ballast control for a floating dock, focusing on hydraulic calculations involving malfunctioning ballast pumps and conducting time-domain simulations of the de-ballasting operations for a full-scale floating dock. The numerical tools developed by Zhang et al. [8], [8] are utilized.

2. Numerical Model

2.1. Floating dock specifications

Figure 1(a) displays the investigated floating dock in full-scale, with dimensions of 168.48 m in length, 39.8 m in width, and 18.2 m in height, respectively. There are 18 ballast tanks installed inside the dock and their inner surfaces are shown in Figure 1(b). Six ballast pumps are located at the void spaces at the port side for de-ballasting operations. The mooring system consists of 12 mooring lines. The comprehensive details of the dock can be found in the Zhang et al. [8],[8].

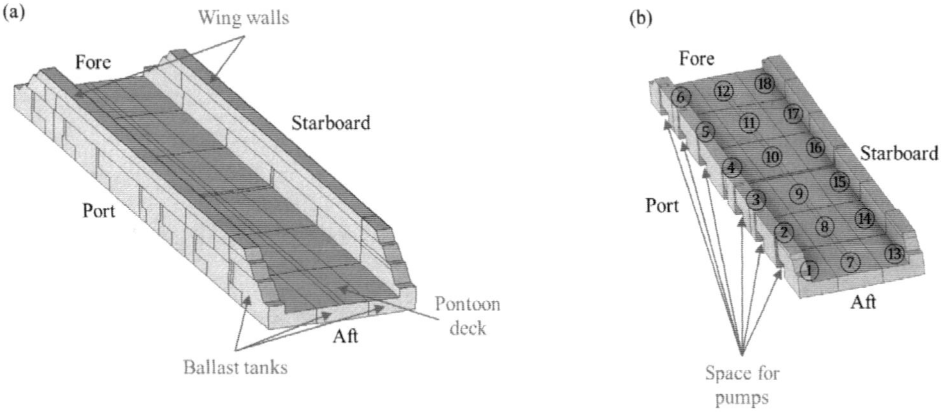

Figure 1. Outer surface of a full-scale floating dock and inner surfaces of 18 ballast tanks.

The dynamic analysis of the floating dock de-ballasting operation is performed using an in-house code, Digiflodock, developed by Zhang et al. [8]. The framework is outlined in Figure 2. This code manages various calculations, including hydrostatic, hydrodynamic and mooring loads of the dock. The hydraulic calculation of the ballast water system is performed, and automatic ballast control is enabled.

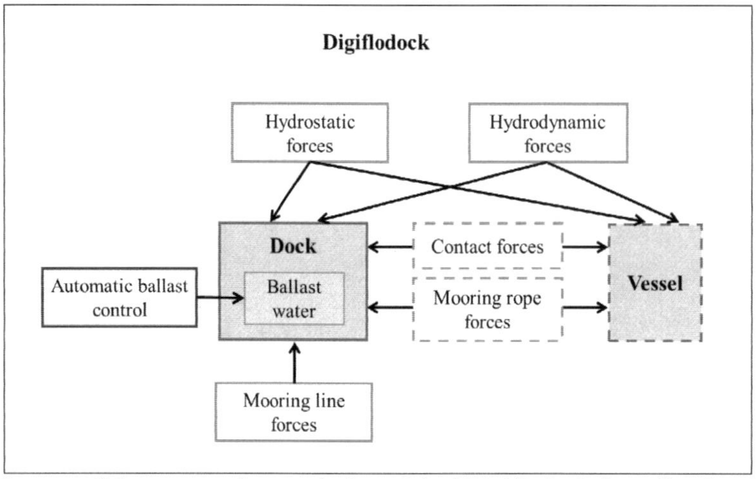

Figure 2. Diagram of the proposed numerical model of Digiflodock for a floating dock system.

2.2. 6-DOF model

The floating dock is modelled as a rigid body with six degrees of motions. The positions of their CoG in the global coordinate system are calculated based on Newton's second law [10]. The angular velocity vector in the body-fixed coordinate system can be determined using Euler's rotational equations of motion [10] for rigid body dynamics. The yaw φ, pitch ψ, and roll γ are determined using the kinematic equations which describes the relation between body angular velocities and Euler angles [10].

2.3. Hydrostatic and hydrodynamic force models

In the hydrostatic force model, the forces and moments resulting from the dock's displaced water and the ballast water are computed using a strip theory. The outer surfaces of the dock and the inner surfaces of the ballast tanks are sliced into number of sections. By integrating the hydrostatic loads acting on these sections along the longitudinal direction, the hydrostatic loads for the entire dock can be obtained [8].

In the hydrodynamic force model, the dock's added mass, added mass moment of inertia are approximated based on the corresponding results of 2-D plates and a 3-D correction [11]. The damping coefficients of the dock are calculated using the dock's critical damping and a damping ratio of 5% [8].

2.4. Mooring force model

The catenary equations [12], [13] are employed to determine the horizontal and vertical tensions at the fairleads for the twelve mooring lines without considering the hydrodynamic forces resulting from the current and wave. The Newton-Raphson method is adopted to solve the catenary equations [8].

2.5. Hydraulic model of the ballast water system

Figure 3 shows the ballast piping network of two adjacent pumps, each responsible for pumping water out of the port, centre and starboard tanks. The network includes main pipes (marked red), branch pipes connected to the tanks (marked green), and a connection pipe (marked black) connecting the pipe network of one pump to the other, as well as butterfly valves installed on these pipes.

During a normal de-ballasting operation, tank valves and the outlet valves are open, while the inlet and connection valves are closed. However, if Pump 1 is malfunctioning, the connection valve is open, allowing Pump 2 to manage the de-ballasting of six ballast tanks, as illustrated in Figure 3. Simultaneously, Outlet valve 1 is closed.

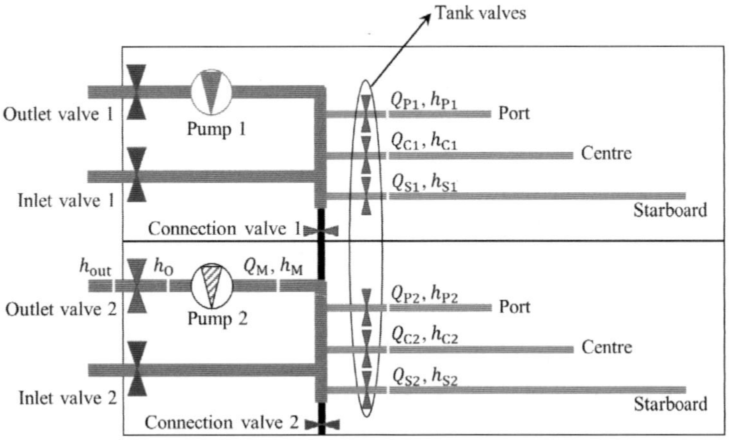

Figure 3. Ballast water system schematic in the de-ballasting operation when Pump 1 malfunctions.

The flow rates out of the main pipes and ballast tanks, Q_M, Q_{P1}, Q_{C1}, Q_{S1}, Q_{P2}, Q_{C2}, and Q_{S2}, are calculated using Equations (1) – (9), based on water head changes and the pump and valve characteristics. h_O is the pump's total water head with zero flow rate, and λ_{pump} stands for the pump coefficient. h_{out}, h_M, h_O, h_{P1}, h_{C1}, h_{S1}, h_{P2}, h_{C2}, and h_{S2} represent the water heads shown in Figure 3. λ_M, λ_{P1}, λ_{C1} λ_{S1}, λ_{P2}, λ_{C2} and λ_{S2} are the valve opening coefficients related to the valve diameter and the valve opening angle. They are calculated using Equation (10) [4], where K_V is the volume of water (unit: m3) that passes through a butterfly valve in one hour at a pressure drop of 1bar. After obtaining the flow rate of a ballast tank, the ballast water volume can be updated using Equation (11) [4], where V_j is the volume of the water in Tank No.j, and Q_j is the flow rate into/out of Tank No.j.

$$h_O - h_{out} = \lambda_M |Q_M| Q_M \tag{1}$$
$$h_{P1} - h_M = \lambda_{P1} |Q_{P1}| Q_{P1} \tag{2}$$
$$h_{C1} - h_M = \lambda_{C1} |Q_{C1}| Q_{C1} \tag{3}$$
$$h_{S1} - h_M = \lambda_{S1} |Q_{S1}| Q_{S1} \tag{4}$$
$$h_{P2} - h_M = \lambda_{P2} |Q_{P2}| Q_{P2} \tag{5}$$
$$h_{C2} - h_M = \lambda_{C2} |Q_{C2}| Q_{C2} \tag{6}$$
$$h_{S2} - h_M = \lambda_{S2} |Q_{S2}| Q_{S2} \tag{7}$$
$$h_O - h_M = h_0 - \lambda_{pump} |Q_M| Q_M \tag{8}$$
$$Q_M = Q_{P1} + Q_{C1} + Q_{S1} + Q_{P2} + Q_{C2} + Q_{S2} \tag{9}$$
$$\lambda = (K_V/36000)^{-2}/g \ \ [s^2/m^5] \tag{10}$$
$$dV_j/dt = Q_j \tag{11}$$

2.6. *Automatic ballast control using a P-controller*

The target of the ballast control system is to keep the dock's pitch ψ and roll γ within small ranges. The control objects are the tank valve opening angles which are divided into 6 groups, as shown in Table 1. The opening angles of the outlet valves for the normal pumps are set to be 90deg during the de-ballasting operation. The target values of valve opening angles are given in Equation (12) [5].

$$\theta^n_{target,i} = \theta_{max} \min\{1 + K^n_p c_{p,i} L\psi + K^n_r c_{r,i} B\gamma, 1\} \tag{12}$$

where

$$K_p^{n+1} = \begin{cases} K, & |\psi| > \psi_{upper} \\ 0, & |\psi| < 0.1\psi_{upper}, \\ K_p^n, & \text{otherwise} \end{cases} \quad K_r^{n+1} = \begin{cases} K, & |\gamma| > \gamma_{upper} \\ 0, & |\gamma| < 0.1\gamma_{upper} \\ K_r^n, & \text{otherwise} \end{cases} \tag{13}$$

Where ψ_{upper} and γ_{upper} satisfy $L\psi_{upper} = B\gamma_{upper} = 0.04$m, and L and B are the dock's length and width. θ_{max} represents the maximum valve opening angle. When the mean sea water level is below the pontoon deck, θ_{max} is set at 90°; otherwise, it is adjusted to 70°. To maintain pump efficiency and protect the pump, the minimum valve opening angle is considered as 35°. In this study, K represents the total control coefficient, fixed at 0.5. The pitch and roll control coefficients for different groups of valves during the de-ballasting operation, denoted as $c_{p,i}$ and $c_{r,i}$, are detailed in Table 1. Due to the valve discs' angular velocity limitation of $\omega_{\text{valve}} = 90\,°/\text{min}$, the actual valve opening angle $\theta_{\text{actual},i}$ is updated at an interval of $\Delta T = 8\Delta t$, using Equation (14) [5].

$$\theta_{\text{actual},i}^{n+1} = \begin{cases} \theta_{\text{actual},i}^n + \Delta t\omega_{\text{valve}}, \text{when } \theta_{\text{actual},i}^n < \theta_{\text{target},i}^n - \Delta t\omega_{\text{valve}} \\ \theta_{\text{actual},i}^n + \Delta t\omega_{\text{valve}}, \text{when } \theta_{\text{actual},i}^n > \theta_{\text{target},i}^n + \Delta t\omega_{\text{valve}} \\ \theta_{\text{actual},i}^n, \text{otherwise} \end{cases} \tag{14}$$

Table 1. Pitch and roll control coefficients for de-ballasting operation [5].

Control group of valves	No.1-3	No.4-6	No.7-9	No.10-12	No.13-15	No.16-18
Pitch control coefficient $c_{p,i}$	1	-1	1	-1	1	-1
Roll control coefficient $c_{r,i}$	1	1	0	0	-1	-1

3. Results and discussions

3.1. Case description

The dynamic process of the de-ballasting operation for the floating dock is examined in this section. As shown in Figure 4, Pump No.1 malfunctions, and Pump No.2 manages the de-ballasting of Tanks No. 1, 7, 13, 2, 8, and 14. Pumps No. 3-6 are responsible for other groups. The dock's initial draught is 12 m, and its initial heel and trim are both zero. The duration of simulation is 3500 s with a time step of 0.5 s.

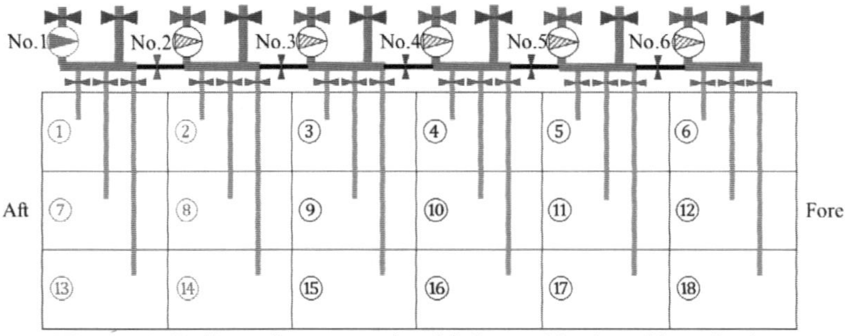

Figure 4. Ballast pipe network with Pump No.1 malfunctioning.

3.2. Dynamic process of the de-ballasting operation

Figure 5 illustrates the time histories of the dock's draught, roll, and pitch angles throughout the de-ballasting process, utilizing the P-controller described in Section 2.6. The dock initially submerges from a draught of 12 m and gradually reaches a level of 3.477 m at 3500 s.

Throughout the simulation, the roll and pitch angles show fluctuations, with maximum values of 0.0509deg and 0.0303°, respectively. These results prove the effectiveness of the present automatic ballast control algorithm which ensures the dock's stability with minimal roll and pitch angles, even with a malfunctioning pump.

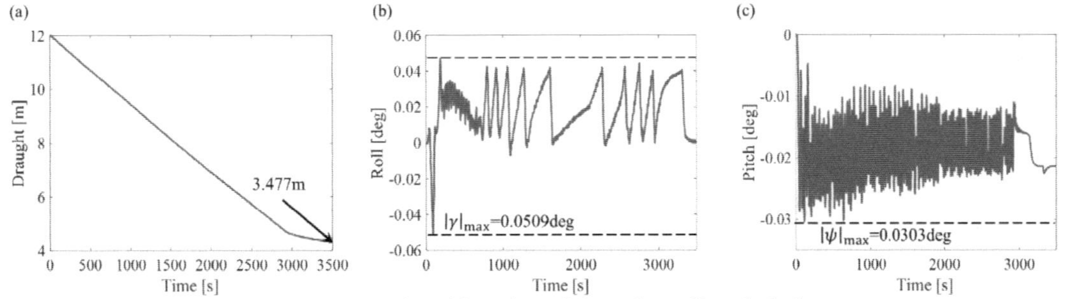

Figure 5. Time histories of draught, roll and pitch.

Figure 6 shows a comparison between the target and actual valve opening angles for Tank No.13. While the actual valve opening angle generally tracks the target value, a minor delay occurs due to the valve disc's angular speed.

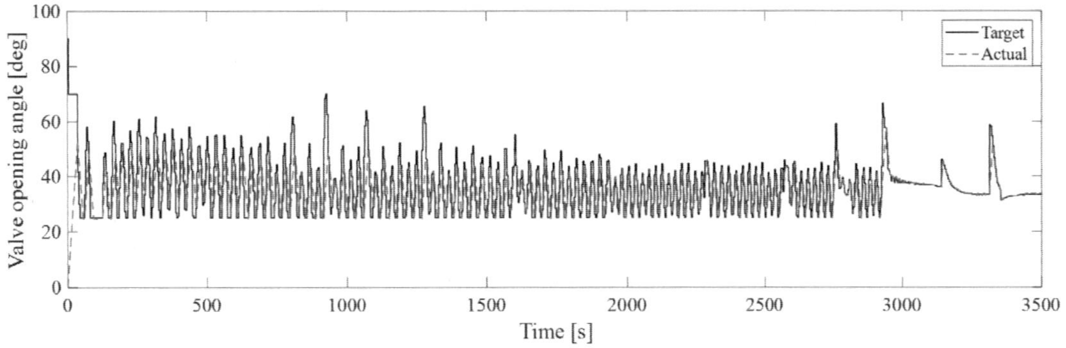

Figure 6. Comparison between target and real values of valve opening angles of tank No.13.

In Figure 7(a), the valve opening angles for Tanks No.13-15 rise to 70° and maintain stable when the sea water level remains below the pontoon deck. Once the sea water level reaches the pontoon deck, the opening angles increase to 90°. In contrast, Tanks No.16-18 in Figure 7(b) exhibit fluctuating valve opening angles, varying from 25° to 57.75°. This variation arises because the malfunctioning of Pump No.1 weakens the de-ballasting capacity on the aft side. Consequently, the valve openings on the fore side must be smaller than those on the aft side to maintain the dock's pitch angle within an acceptable range. This difference in valve opening angles between the aft and fore sides leads to different total flow rates through the pumps, illustrated in Figure 8 where the flow rates through Pumps No.2 and 3 are larger than those of other pumps. Moreover, the valve opening angles of the port and starboard sides are also different for maintaining allowable roll angles, but the difference is much smaller than that between aft and fore sides, as shown in Figure 9.

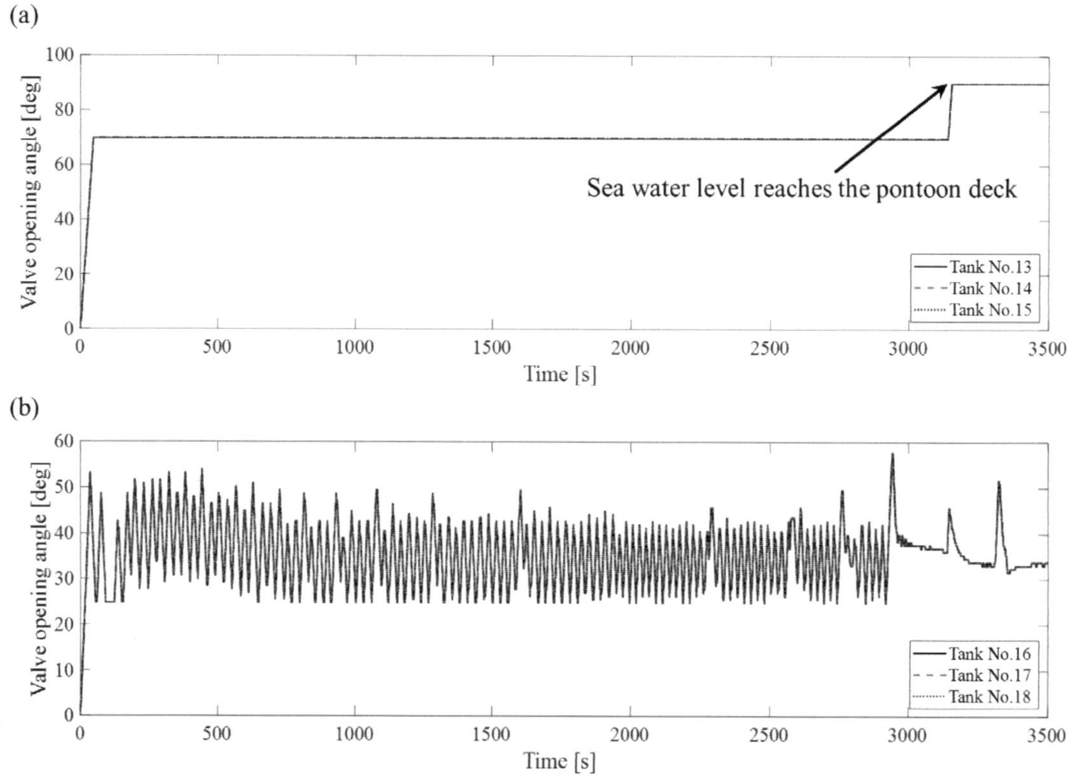

Figure 7. Real valve opening angles of tanks No.13-18.

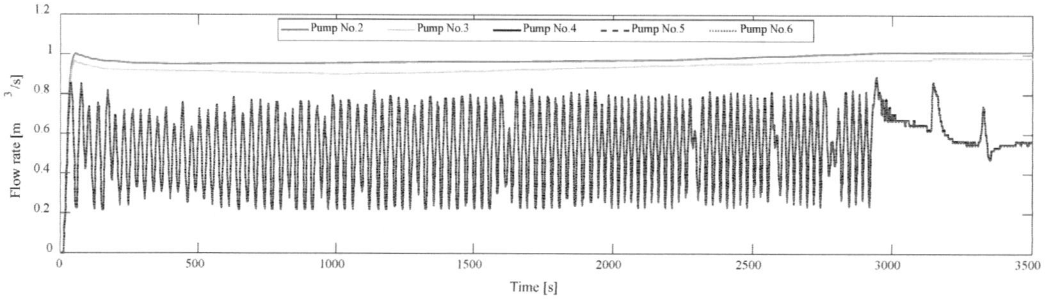

Figure 8. Flow rates through pumps No. 2-6.

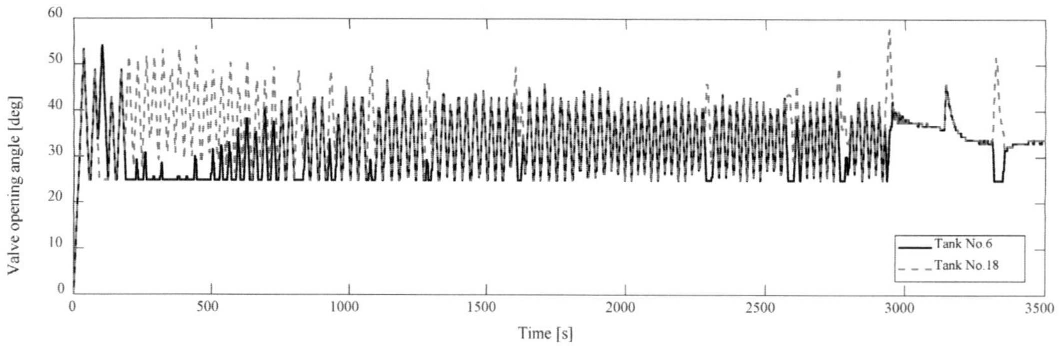

Figure 9. Real valve opening angles of tanks No. 6 and 18.

IOP Conf. Series: Materials Science and Engineering 1294 (2023) 012011 doi:10.1088/1757-899X/1294/1/012011

4. Conclusion

The effectiveness of the proposed automatic ballast control algorithm when dealing with a malfunctioning ballast pump is investigated by simulating the de-ballasting operation of a full-scale floating dock. The dock is modelled as a rigid body with six degrees of freedom, subjected to the hydrostatic, hydrodynamic, and mooring forces. The hydraulic calculation is conducted considering the malfunctioning pump located at the dock's aft. Time-domain results demonstrate that the roll and pitch angles maintain below 0.0509° and 0.0303°, respectively, by using the present Modified P-controller. Future research will extend the range of scenarios involving various pump malfunctions and diverse interconnections between the groups of ballast tanks.

References

[1] Ohkawa T, Itoh M, Matsunaga H, Horiguchi K and Sato K 1984 Automatic Ballast Control for Floating Dock *Proc. of 9th IFAC World Cong.* (Budapest, Hungary: July 2 - 6,1984). **17** (2): 2747-2752.

[2] Guo T, Zhang R, and Zhou S H 2014. Design of automatic control system of floating dock. *Appl. Mech. and Mater.* **527,** 237-241.

[3] Topalov A, Kozlov O, Gerasin O, Kondratenko G, and Kondratenko Y 2018. Stabilization and control of the floating dock's list and trim: Algorithmic solution. *Proc. of 14th Int. Conf. on Adv. Trends in Radio., Teleco. and Comput. Eng. (TCSET-2018)* (Lviv-Slavske, Ukraine: Feb 20-24).

[4] Wen X, García Conde A, Zhang J, and Ong M C 2023. Numerical modelling of a floating dock and its automatic control. *Proc. of the 42nd Int. Conf. on Offshore Mech. Arct. Eng.* (Melbourne, Australia: Jun 11 – 16, 2023).

[5] Wen X, García Conde A, Zhang J, and Ong M C 2023. Dynamic Analysis of a Floating Dock under Accidental Conditions. Appl. Ocean Res. (Under review).

[6] Rainsford S 2018. Russian aircraft carrier Admiral Kuznetsov damaged by crane. http://www.bbc.com/news/world-europe-46030113.

[7] National Transportation Safety Board 2012. Marine Accident Breif: Dockside Capsizing and Sinking of Towing Vessel Invader and Dry Dock #3. https://www.ntsb.gov/investigations/AccidentReports/Reports/MAB1316.pdf

[8] Zhang J, Wen X and Ong M C 2023. Development of a floating dock numerical model and the ballast water distribution strategy. *Proc. of the 42nd Int. Conf. on Offshore Mech. Arct. Eng.* (Melbourne, Australia: Jun 11 – 16, 2023).

[9] Zhang J, Wen X and Ong M C 2023. Development of numerical tools for a floating dock system. *Ocean Eng.* (Under review)

[10] Schaub H and Junkins J L 2003. *Analytical mechanics of space systems.* Aiaa. p 26, 77, 123.

[11] Pabst W 1930. *Theory of the landing impact of seaplanes*. Technical Memorandum No.580, (National Advisory Council for Aeronautics: Washington, USA).

[12] Faltinsen O M 1993. *Sea loads on ships and offshore Structures*, (Cambridge: Cambridge University Press), p 58-66.

[13] Masciola M 2018. *MAP++ Documentation Release 1*.15, p10-13.

Calculation and prediction of suspended span caused by erosion of deep water submarine pipelines

X Jia, K Ni*, Y Li and L Jia

Engineering Research & Design Dept. CNOOC Research Institute Ltd.

*Correspondence: nikk@cnooc.com.cn

Abstract. Submarine pipelines can be divided into three types based on their laying status: pipelines laid flat on the seabed, pipelines without buried, and pipelines buried in shallow trenches. In addition, there is also a suspended span state caused by waves and currents erosion. The four laying states of submarine pipelines indicate that the seabed, as the foundation supporting the pipeline, must provide sufficient support to ensure the stability of the pipeline. This article studies the erosion span data of existing pipelines, evaluates and predicts the pipeline span state through theoretical calculation and analysis, and then uses numerical simulation calculation methods to predict the pipeline span depth, and compares and verifies it with measured data.

1. Introduction

With the further development of marine oil and gas resources, the total length of submarine pipelines continues to rise. But at the same time, damage accidents are also gradually increasing. There have been hundreds of cases of oilfield shutdowns and ecological environmental pollution caused by submarine pipeline damage both domestically and internationally, resulting in huge economic losses and adverse social impacts.There were a total of 184 leakage accidents in the Gulf of Mexico from 1967 to 2012, and there were also 19 domestic incidents from 1998 to 2012[1].From the causes of these accidents, the pipeline span caused by submarine pipeline erosion is one of the main reasons for leakage accidents.This article will study the suspended span of the Liwan 3-1 pipeline.

Liwan 3-1 gas field is located in the Pearl River Mouth Basin of South China Sea, about 310 km southeast of Hong Kong. The project had built a subsea production system consisting of 8 sets of subsea wellhead, east and west manifolds, and central manifolds in the deep water area. Two of 22 inch multiphase pipelines, one of 6 inch MEG (Mono Ethylene Glycol) pipeline, a new central platform and a 30 inch of 260 km oil and gas transportation submarine pipeline are in the shallow water area. Figure 1. shows the schematic diagram of Liwan 3-1 development.

Figure 1. Schematic diagram of Liwan 3-1 development

The 30 inch pipeline of the Liwan 3-1 project has a long route length and the route direction is close to perpendicular to the bottom flow direction, making it highly likely to experience erosion. Under the influence of temperature and pressure, pipelines bear significant internal forces, which are also superimposed with earthquakes, external water pressure, and overlying soil pressure. Therefore, the combined forces acting on pipelines are extremely complex and significant. If external forces caused by pipeline suspension are added to this foundation, the safety of submarine pipelines will be a very serious problem. Therefore, regular in situ status surveys must be conducted on this pipeline, including the use of multi beam, side scan sonar, and other methods to investigate the burial status of the pipeline. Evaluate the situation of pipeline scouring suspended span through annual survey data and provide theoretical basis for suspended spans management.

The aim of this article is to study the erosion span data of existing pipelines, evaluates and predicts the pipeline span state through theoretical calculation and analysis, and then use numerical simulation calculation methods to predict the pipeline span depth, and compare and verify with measured data.

2. Evaluation method for pipeline suspension span

2.1. Theoretical analysis method for pipeline suspension span

The free span analysis of submarine pipelines is based on the DNV-RP-F105 standard [2], and the main analysis and calculation includes two parts: VIV (vortex induced vibration) calculation and ULS (ultimate limit strength) verification. There are two methods for VIV calculation: one is based on avoiding resonance frequency, with VIV Screening and Onset criteria; the other is VIV fatigue calculation, which is based on cumulative damage stress/strain fatigue calculation method. In theory, the results of Screening and Onset calculations are relatively conservative, while VIV fatigue calculations can quantitatively analyze the service life of offshore pipelines under working and environmental loads, which is closer to engineering practice.

According to the DNV-RP-F105 specification, when the length of the suspended span exceeds the allowable value of the Screening or Onset criteria, the VIV fatigue calculation method can be chosen for reanalysis. If the VIV fatigue calculation meets the requirements and the ULS strength verification also meets the requirements, the suspended span can be left untreated. Figure 2 shows the suspended span analysis process.

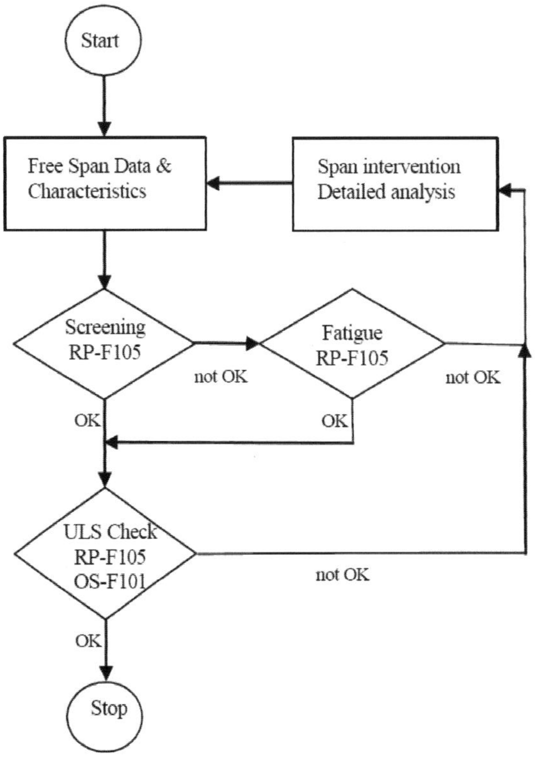

Figure 2. Suspended span analysis process

2.2. *VIV Suspension fatigue analysis*

The main calculation method for VIV suspension span fatigue is to verify the response model and mechanical model recommended in DNV-RP-F105 separately. The response model calculates the alternating stress of vortex induced vibration in the cross-flow direction and inline direction based on the curve of attenuation speed and vibration amplitude recommended by the experimental model. The mechanical model mainly calculates the alternating stress in the inline direction caused by wave drag force on a suspended pipeline based on empirical formulas.

After calculating the alternating stress, the fatigue life is calculated according to the standard S-N curve recommended in DNV-RP-C203 [3]. Firstly, based on the probability distribution of wave periods, the fatigue life of the entire distribution of ocean current velocity under a certain wave height, frequency, and wave direction is calculated. Based on the probability distribution of waves, frequency, and direction in the sea area, the overall fatigue life in both the cross flow and in line directions are calculated. After obtaining the fatigue life under the suspended span length, evaluation of whether the fatigue life meets the design life requirements is conducted.

3. Computational fluid dynamics numerical simulation prediction

Compared with analytical methods, numerical methods have significant advantages in simulating complex wave conditions and considering the characteristics of seabed soil. The numerical simulation of pipeline erosion span was carried out using the OpenFOAM program.

3.1. *Scour calculation model*

The Euler two-phase flow sediment model is composed of the flow phase and the sediment phase (Figure 3). Water flow and sediment are independent continuous media that satisfy the control equations of the fluid. At the same time, the model describes turbulent motion using momentum exchange, turbulent stress, and sediment particle stress based on the influence of the interaction between sediment particles and fluid sediment particles. Figure 3 shows the schematic diagram of submarine pipeline erosion, while the model diagram of submarine pipeline erosion is illustrated in Figure 4.

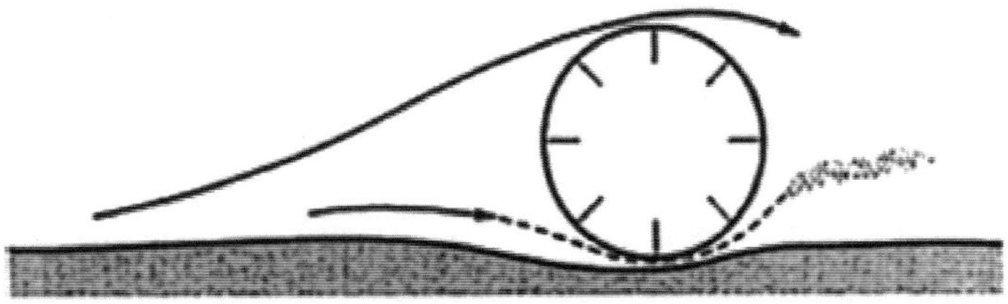

Figure 3. Schematic diagram of submarine pipeline erosion

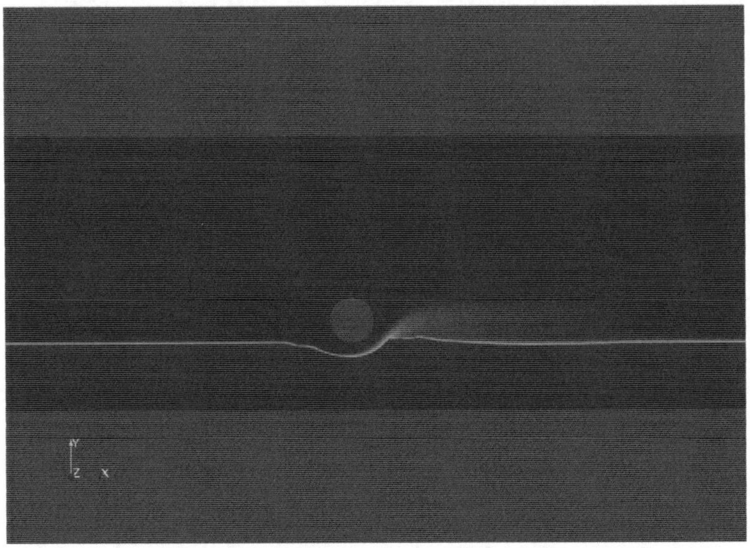

Figure 4. Model diagram of submarine pipeline erosion

3.2. Governing equation

Assuming no mass exchange occurs between the sediment phase and the water flow phase, the mass continuity equations for water flow and sediment can be obtained separately [3]:

$$\frac{\partial(1-c)}{\partial t}+\frac{\partial(1-c)u_i^f}{\partial x_i}=0 \tag{1}$$

$$\frac{\partial c}{\partial t}+\frac{\partial c u_i^s}{\partial x_i}=0 \tag{2}$$

Where c is volume fraction of sediment in water flow, u_i^f and u_i^s represent the velocity of water flow and sediment respectively, with i =1 and 2 representing the horizontal and vertical directions. The momentum continuity equation can be expressed as:

$$\frac{\partial\rho^f(1-c)u_i^f}{\partial t}+\frac{\partial\rho^f(1-c)u_i^f u_j^f}{\partial x_j}=(1-c)f_i-(1-c)\frac{\partial p^f}{\partial x_i}+\frac{\partial\tau_{ij}^f}{\partial x_j}+\rho^f(1-c)g\delta_{i3}+F_i^{fs} \tag{3}$$

$$\frac{\partial \rho^s c u_i^s}{\partial t}+\frac{\partial \rho^s c u_i^s u_j^s}{\partial x_j}=\phi f_i-\phi\frac{\partial p^f}{\partial x_i}-\frac{\partial p^s}{\partial x_i}+\frac{\partial \tau_{ij}^s}{\partial x_j}+\rho^s c g\delta_{i3}+F_i^{sf} \tag{4}$$

Where ρ^f and ρ^s denote the density of the water flow phase and the sediment phase, respectively. p^f and p^s are fluid pressure and sediment normal pressure, respectively. F_i^{sf} and F_i^{fs} are the interaction force between the fluid phase and the sediment phase, following the Newton's Third Law.

3.3. *Turbulent closure equation*

Based on the work of Hsu [5] and Cheng [6], an improved two-phase K-Epsilon turbulence model can be obtained from equation (5) and (6). The model coefficients are given in Table 1.:

$$\frac{\partial k^f}{\partial t}+u_j^f\frac{\partial k^f}{\partial x_i}=\frac{R_{ij}^{ft}}{\rho^f}\frac{\partial u_i^f}{\partial x_j}+\frac{\partial}{\partial x_j}\left[\left(\nu^f+\frac{\nu^{ft}}{\sigma_k}\right)\frac{\partial k^f}{\partial x_j}\right]-\epsilon^f-\frac{2\beta(1-\alpha)ck^f}{\rho^f(1-c)}-\frac{1}{(1-c)}\frac{\nu^{ft}}{\sigma_c}\frac{\partial c}{\partial x_j}(s-1)g_j \tag{5}$$

$$\frac{\partial \epsilon^f}{\partial t}+u_j^f\frac{\partial \epsilon^f}{\partial x_j}=C_{1\epsilon}\frac{\epsilon^f}{k^f}\frac{R_{ij}^{ft}}{\rho^f}\frac{\partial u_i^f}{\partial x_j}+\frac{\partial}{\partial x_j}\left[\left(\nu^f+\frac{\nu^{ft}}{\sigma_\epsilon}\right)\frac{\partial \epsilon^f}{\partial x_j}\right]-C_{2\epsilon}\epsilon^f\frac{\epsilon^f}{k^f}-C_{3\epsilon}\frac{\epsilon^f}{k^f}\frac{2\beta(1-\alpha)ck^f}{\rho^f(1-c)}$$
$$-C_{4\epsilon}\frac{\epsilon^f}{k^f}\frac{1}{(1-c)}\frac{\nu^{ft}}{\sigma_c}\frac{\partial c}{\partial x_j}(s-1)g_j \tag{6}$$

Table 1. Turbulence model coefficients

C_μ	$C_{1\epsilon}$	$C_{2\epsilon}$	$C_{3\epsilon}$	$C_{4\epsilon}$	σ_k	σ_ϵ	σ_c
0.09	1.44	1.92	1.2	0 or 1	1.0	1.3	1.0

Where C_μ, $C_{1\epsilon}$, $C_{2\epsilon}$, $C_{3\epsilon}$, are empirical coefficients. $C_{4\epsilon}$ is the coefficient associated with the buoyancy term,while it is set to 1 for unstably stratified condition and 0 for stably stratified condition. σ_c is the Schmidt number. ϵ^f is the rate of turbulent kinetic energy dissipation. k^f is the turbulent kinetic energy.

3.4. *Sediment stress closure equation*

In the model, the sediment volume fraction *c* is used as the basis for determining sediment stress [6]. When $c > c^*(c^* = 0.635)$, it was believed that sediment was stationary and sediment stress was ignored. When $c_* < c < c^*$ $(c_* = 0.57)$, the sediment was in the transition stage from solid to liquid, and the sediment stress was considered as continuous contact stress, calculated using the Johnson & Jackson friction stress model. When $0.001 < c < c_*$, the sediment phase was dominated by collision stress, and calculated using collision theory simulation. When $c < 0.001$, the sediment was considered as a diluted flow, ignoring the influence of interaction.

The calculation equation for the temperature of sediment particles θ is given as:

$$\rho^s\left(\frac{\partial c\theta}{\partial t}+\frac{\partial c\theta u_j^s}{\partial x_j}\right)=\tau_{ij}^{sc}\frac{\partial u_i^s}{\partial x_j}-\frac{\partial q_j}{\partial x_j}-\gamma+2\beta c(\alpha k-\theta) \tag{7}$$

Calculation equation for collision stress of sediment particles is done using:

$$\tau_{ij}^{sc}=-p^{sc}\delta_{ij}+\mu^{sc}\left(\frac{\partial u_i^s}{\partial x_j}+\frac{\partial u_j^s}{\partial x_i}\right)+\left(\lambda-\frac{2}{3}\mu^{sc}\right)\frac{\partial u_k^s}{\partial x_k}\delta_{ij} \tag{8}$$

while calculation equation for frictional force of sediment particles is done using:

$$\tau_{ij}^{sf}=-p^{sf}\delta_{ij}+\mu^{sf}\left(\frac{\partial u_i^s}{\partial x_j}+\frac{\partial u_j^s}{\partial x_i}\right)-\frac{2}{3}\mu^{sf}\frac{\partial u_k^s}{\partial x_k}\delta_{ij} \tag{9}$$

$$p^{sf} = \begin{cases} 0, & c < c_* \\ F\frac{(c-c_*)^m}{(c^*-c)^n}, & c \geq c^* \end{cases} \quad (10)$$

$$\mu^{sf} = \frac{p^{sf}\sin(\theta_f)}{\sqrt{2S_{ij}^s S_{ij}^s}} \quad (11)$$

Table 2. Sediment stress model coefficients

F	m	n
0.05	3	5

The empirical coefficients F, m and n were calibrated to ensure that a stable sediment bed can be established.

4. Case analysis calculation

4.1. Design basis

The typical cross-section of a 30 inch single layer pipeline with a cement weight layer is shown in Figure 5, where the specific design parameters of the pipeline are shown in Tables 3.

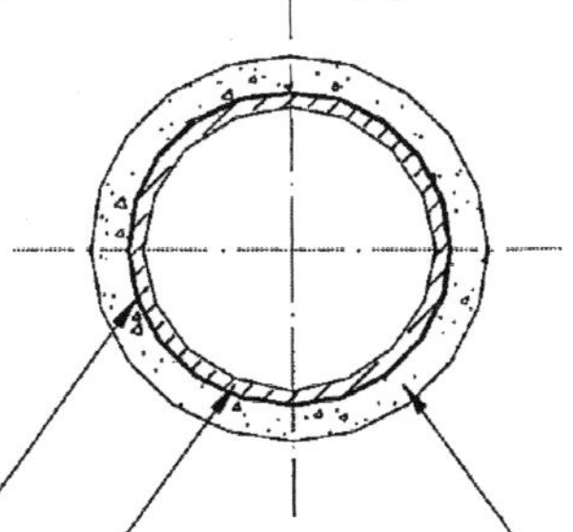

Figure 5. Cross section of submarine pipeline

Table 3. Pipeline properties and design parameters

KP	O.D. (mm)	Wall thickness (mm)	Corrosion allowance (mm)	Design pressure (MPa)	Design temperature (℃)	Material grade
0.0-0.50	762	30.2	2	23.6	60	API 5L X65 PSL2
0.50-192	762	28.6	2	23.6	60	API 5L X65 PSL2
192-260	762	30.2	2	23.6	60	API 5L X65 PSL2
260-261.026	762	31.8	2	23.6	60	API 5L X70 PSL2

By comparing and analyzing the survey data of pipeline in-service status over the years (Table 4), it can be concluded that the evolution trend of pipeline suspension spans is from short discontinuous suspension spans to long continuous suspension spans. The suspended span height generally shows an increasing trend.

Table 4. Comparison of 30 inch pipeline survey data

2014 Data				2016 Data				2018 Data			
KP	No.	Total length (m)	Max length (m)	KP	No.	Total length (m)	Max. length (m)	KP	No.	Total length (m)	Max length (m)
0-20	41	274.8	18.1	0-20	50	343.9	18	0-5	1	14.52	14.5
								15-20	13	209.9	12.8
20-30	5	25.2	9.0	20-30	54	457.7	24.2	20-30	49	1085.2	33.5
30-40	35	196.7	11.1	30-40	100	818.3	27.2	30-40	81	1521.6	27.8
40-50	40	267.3	17.9	40-50	149	1339.8	27.4	40-50	159	4206.1	42.2
50-60	35	292.0	17.8	50-60	58	495.9	19.1	50-60	73	1500.7	44.6
60-70	8	34.8	7.1	60-70	37	261.7	22.5	60-70	53	895.0	33.3
70-80	31	217.2	17.5	70-80	27	174.2	11.9	70-80	30	456.1	15.1
Total	195	1308.0	18.1	Total	475	3891.5	27.2	Total	459	9889.1	44.6

4.2. Theoretical analysis and calculation results

Based on the VIV fatigue theory analysis method mentioned earlier, fatigue analysis was conducted using the DNV professional software FATFREE for the 14 overhanging sections of the LW3-1 to Gaolan terminal subsea natural gas pipeline that exceeded the design value. The specific analysis results are shown in Table 5.

The analysis results show that the fatigue life of all suspended VIV spans is relatively large, far greater than the design life of 50 years, and there is no need for treatment.

Table 5. Fatigue analysis results of suspended span

No,	From KP	To KP	Span length (m)	Span height (m)	Allowable span length (m)	Fatigue life (year)
1	25.062	25.0956	33.54	0.36	25	2.98×10^4
2	25.1339	25.1603	26.41	0.66		2.97×10^4
3	25.5621	25.5878	25.71	0.52		2.97×10^4
4	30.6083	30.6361	27.83	0.63		2.92×10^4
5	40.0462	40.0818	35.69	0.41		2.97×10^4
6	42.7018	42.7288	26.96	0.55		2.8×10^4
7	47.3948	47.4288	33.95	0.48		2.96×10^4
8	48.3554	48.381	25.59	0.63		2.98×10^4
9	48.5254	48.5676	42.21	0.62		2.98×10^4
10	48.6179	48.6467	28.79	0.75		2.98×10^4
11	48.7408	48.7661	25.31	0.40		2.98×10^4
12	49.4085	49.4424	33.94	0.43		2.98×10^4
13	55.484	55.5286	44.56	0.47		2.98×10^4
14	61.3216	61.3549	33.32	0.41		2.98×10^4

4.3. Comparison of Numerical Simulation Prediction Results

In the Liwan 3-1 project, according to a survey conducted in 2018, it was found that there was a large scour hole between the Kp48.52-Kp48.57 sections of the 30 "pipeline. The results of the conducted suspension span prediction evaluation for this location is listed in Table 6.

Table 6. Pipeline properties and design parameters

Time (s)	Scour depth S (m)	S/D
0.00	0.00	-0.01
0.10	0.00	-0.03
0.30	-0.01	-0.10
0.50	-0.01	-0.19
0.80	-0.01	-0.27
1.00	-0.02	-0.32
2.00	-0.02	-0.44
5.00	-0.03	-0.56
10.00	-0.03	-0.62
20.00	-0.04	-0.73
30.00	-0.04	-0.76
40.00	-0.04	-0.78
50.00	-0.04	-0.80
60.00	-0.04	-0.83
70.00	-0.04	-0.89
80.00	-0.05	-0.91
90.00	-0.05	-0.94
100.00	-0.05	-0.94

The depth of erosion rapidly increases in the initial stage of erosion, and then the rate of change in the depth of the erosion pit decreases with time. The scouring depth enters a stable scouring state within 90 seconds of simulation, but by the 5th second of simulation, the maximum scouring depth can reach 60% of the stable scouring depth. Figure 6 shows the curve of the variation of the diameter ratio over time.

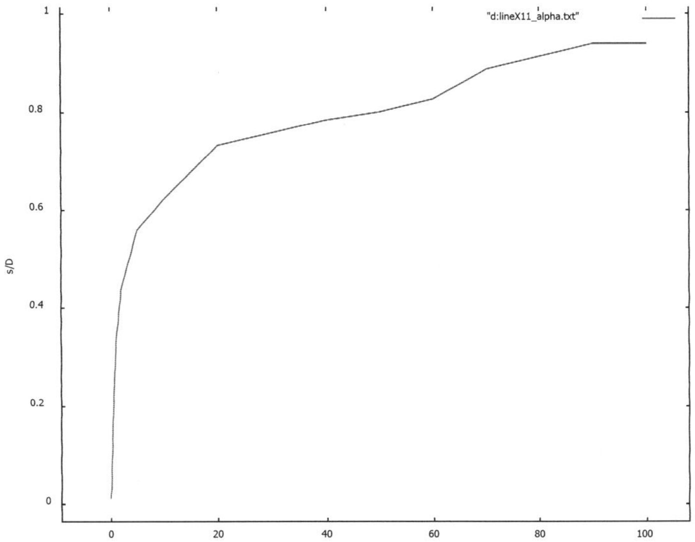

Times (s)

Figure 6. Curve of the variation of the diameter ratio over time

The maximum scouring depth of the pipeline is about 0.94D (D is the diameter of the pipeline), and it can be predicted that the maximum depth of scouring pits generated by the pipeline under this environmental condition is about 0.72 m. According to the statistical data of the 30 "pipeline in the key investigation section, the depth of the scouring pit at Kp48.52-Kp48.57 section is 0.62 m, which has not yet reached the predicted maximum value. The future pipeline erosion phenomenon will enter a slow development stage, gradually reaching the maximum erosion depth and then entering a stable state.

5. Conclusion

This article summarizes the development law of pipeline suspension span through the analysis of on-site survey data of pipelines over the years as follows:
(1) Developing from short discontinuous suspension spans to long continuous suspension spans;
(2) The suspended span height generally shows an increasing trend.

Based on the survey data, theoretical analysis and evaluation were conducted on the suspended span of the pipeline. The evaluation conclusion shows that the fatigue life of the pipeline at the suspended span meets the original design life requirements and can be temporarily left untreated. Review should be conducted at any time based on the latest evaluation conclusions.

The numerical simulation method is used to predict the height of pipeline erosion span, and the calculated results are basically consistent with the measured results. This method can be used for span prediction and provides a theoretical basis for pipeline span control.

References

[1] Wang H H, Liu G H 2017 Statistics and analysis of China National Offshore Oil Corporation submarine pipeline accidents. *China Offshore Oil and Gas*, **2017**(5).
[2] DNV-RP-F105 Free Spanning Pipeline. Available at: https://www.dnv.com/oilgas/download/dnv-rp-f105-free-spanning-pipelines.html
[3] DNV-RP-C203 Fatigue Strength Analysis of Offshore Steel Structures. Available at DNV-RP-C203 Fatigue design of offshore steel structures - DNV
[4] Xinwei H and Zhilin, S 2029 Two-phase flow simulation of submarine pipeline erosion under steady flow. *Sci. Technol. Bulletin* **2019**(2), 6-11 (in Chines).
[5] Hsu T J, Jenkins J T and Liu L F 2004 On two-phase sediment transport: sheet flow of massive particles. *Proc. Royal Soc. A Math. Phys. Eng. Sci.* **460**(2048), 2223-2250.
[6] Cheng Z, Hsu T J, Calantoni J 2017 SedFoam: A multi-dimensional Eulerian two-phase model for sediment transport and its application to momentary bed failure. *Coastal Eng.* **119,** 32-50.

IOP Conf. Series: Materials Science and Engineering 1294 (2023) 012013 doi:10.1088/1757-899X/1294/1/012013

Dynamic response of a single point mooring submersible fish cages in waves and current

X Wen*, H Cheng and M C Ong

Department of Mechanical and Structural Engineering and Materials Science
University of Stavanger, NO-4036, Stavanger, Norway

* Correspondence: xueliang.wen@uis.no

Abstract. Submersible fish cages are designed for installing in the open sea sites that are subjected to violent sea conditions to reduce the hydrodynamic forces acting on the fish cages. To assess the dynamic responses of a single point mooring submersible fish cage under the rough sea conditions, a numerical model including hydrodynamic and structural methods is proposed. The waves and current are modelled using the airy wave theory. For the floating collars and sinker tube, the Morison model is adopted to calculate the hydrodynamic force and the modal superposition method is proposed to calculate the structural responses. For the aquaculture net, the screen model is adopted to calculate the hydrodynamic force and the extended position-based dynamics (XPBD) method is proposed to obtain the structural deformations. The hydrodynamic dynamic forces and the deformations of the fish cages are compared between the fish cages on water surface and in deep water. Results show that when the fish cage is submerged 20m from the water surface, the average horizontal force is reduced by 30% and the variation of the horizontal force is reduced by 81%. Significant reductions in the deformation of the floating collars are observed. Therefore, using submersible fish cage can help reduce physical stress and avoid structural damage for the fish cage system. In addition, the coupling of XPBD and MS can provide a quick solution for the structural modelling of the fish farm system, which can be utilized at the initial stage of fish farm design.

1. Introduction

Aquaculture has been the world's fastest-growing food production method in the past 40 years [1]. Open sea aquaculture is a new development of the aquaculture industry because of the benefits of the fish welfare and the ecosystem through better water exchange and dispersal of waste [2]. However, it can pose a significant challenge to the design of open sea fish cages due to the strong waves and current in the open sea. Submersible fish cages were designed for installing in the open sea sites that are subjected to violent sea conditions to reduce the hydrodynamic forces acting on the fish cages. The cages are floated to the water surface when the sea is in calm condition. If the tidal currents and waves are strong when a furious storm or a typhoon is coming, the fish cage will descend to a specific position. This position is especially designed to minimize or avoid structural damage by reducing physical stress from the rough sea conditions. The descending is achieved by ballasting sea water into one of the floating collars and thus increasing the fish cage's weight. When the furious storm or the typhoon passes, the water inside the floating collar will be de-ballasted using a pump and high-pressure air. The fish cage ascends to the water surface due to the decreasing of the weight of the fish cage. To design such a submersible fish cage system, an effective numerical method should be proposed to assess the dynamic responses of the fish cage under rough sea conditions of the typhoon or the furious storm. Besides, the descending and ascending of the fish cage are risky under the rough sea conditions when the typhoon or the furious storm is approaching. It is also necessary to evaluate the structural responses during the dynamic processes of the descending and ascending of the fish cage.

The numerical methods to assess the dynamic response of the fish cage include the hydrodynamic models and the structural models. For the hydrodynamic models, there are three types of hydrodynamic models for aquaculture nets: Morison model and screen model and porous media model. The Morison models were proposed to calculate the hydrodynamic force acting on line-like structures of the aquaculture nets [3-4]. The aquaculture nets are usually very dense. The twine length can be smaller than 2cm. The mesh number of the net for the Morrison model becomes quite extensive if each twine is modelled. A mesh grouping method was proposed by Cheng et al. [5] to reduce the mesh number of the net. In screen models, the hydrodynamic forces are calculated based on a number of net panels. The twines and knots in the net panel are considered as an integrated structure with a solidity representing the ratio of the twines' projected area to the total area of the net panel. In some screen models, the hydrodynamic force acting on the net panel is decomposed into two components: a normal drag force and a tangential drag force [6], dependent on the orientation of the net panel. In other models [7-8], the hydrodynamic force acting on the net panel is decomposed into a drag force and a lift force according to the direction of the flow passing the panel. The porous media models were adopted in the modelling of the aquaculture net in the computational fluid dynamics methods [9-10]. For the structural methods, various methods were proposed to assess the structural responses of different components of the fish cage system. A mass-spring model [11] or a truss model [12] was used to calculate the motions and deformations of the aquaculture nets, mooring lines and bridles. A rigid body model [13] or beam model [12] was adopted to model the motions and deformations of the floating collars and the sinker tubes.

In the present study, the dynamic responses of a single point mooring fish cage under rough sea conditions are assessed. The hydrodynamic models include a screen model proposed by Aarsnes et al. [7] to obtain the hydrodynamic forces acting on the net panels and a Morrison model to obtain the hydrodynamic forces acting the floating collars and sinker tube. The structural models include an extended position-based dynamics (XPBD) method [14-15] to model the structures of the aquaculture nets, mooring lines and bridles and a mode superposition method to model the floating collars and the sinker tubes. The hydrodynamic dynamic forces and the deformations of the fish cages are compared between the fish cage on surface and in deep water.

2. Numerical model

2.1 Description of the single point mooring submersible fish cage

The investigated single point mooring submersible fish cage system is shown in Figure 1. The fish cage has a circumference of 160 m and a total height of 35 m. There are two floating collars on the top of the fish cage, a sinker tube at the depth of 30m, a sinker of 5000 N (weight in water) at the depth of 35m and the net pens between them. The mooring system includes a buoy, an anchor, one mooring line and three bridles. The three bridles connect the fish cage and the buoy. The buoy is a combination of a cylinder and an inverse cone, which have a diameter of 2m. The heights of the cylinder part and cone part are 1m and 2m, respectively. The floating collars of the fish cage is initially located at the mean sea water level. The origin of the present coordinate system is set at the centre of the floating collars. The x-axis is along the direction of the current. The z-axis points vertically upward and the y-axis is given by satisfying the right-hand rule. The conjunction point between the mooring line and the three bridles is initially located at a depth of −10 m. The anchor is installed at the sea bottom with a depth of -100 m. The density of the sea water is 1025 kg/m^3 and the gravitational acceleration is 9.81 m/s^2.

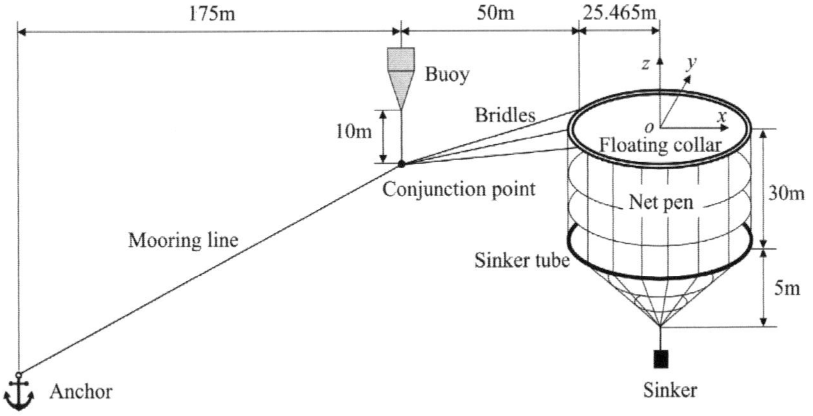

Figure 1. Schematic of a single point mooring fish cage.

Figure 2 shows the front, side and top views of the net pen as well as well as the description of the net twines and ropes. The frame of the net pen is made of net ropes and enclosed by the net ropes is net twines. The specifications of the single point mooring fish cage system are displayed in Table 1. The density and Young's modulus of different materials are listed in Table 2.

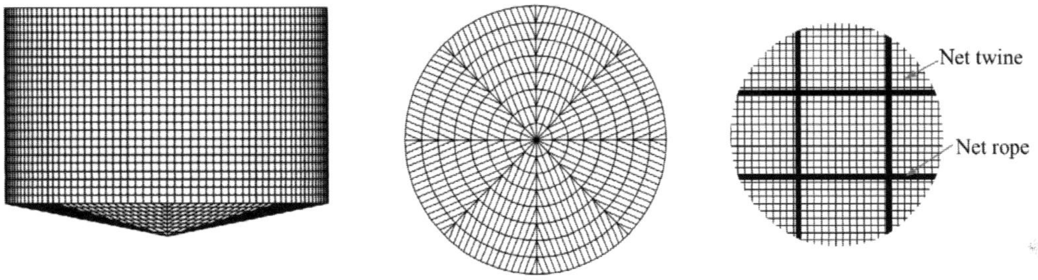

(a) Front and side views (b) Top view (c) Net twines and ropes

Figure 2. Front, side and top view of the net pen as well as the description of the net twines and ropes.

Table 1. Specifications of the single point mooring fish cage system.

Component	Parameter	Value
Net twine	Diameter	2.8 mm
	Mesh length	25 mm
	Solidity	0.2
	Material	Nylon
Net rope	Diameter	10mm
	Material	PE
Floating collar	Section diameter	355 mm
	Pipe thickness	21mm
	Material	HDPE
Sinker tube	Section diameter	250 mm
	Pipe thickness	21 mm
	Material	HDPE
Bridles	Diameter	100 mm
	Material	Nylon
Mooring line	Diameter	100 mm
	Material	Nylon

Table 2. Density and Young's modulus of different materials.

	Nylon	PE	HDPE
Density	1140 kg/m^3	960 kg/m^3	960 kg/m^3
Young's modulus	2.0 GPa	1.5 GPa	3.0 GPa

2.2 Hydrodynamic model

2.2.1 Current and wave conditions. The linear wave is assumed based on the Airy wave theory in the present study. The water level of the wave is given as

$$\eta = \eta_0 \sin(\omega t - n_x x - n_y y) \tag{1}$$

where η_0 is the amplitude, $\omega = gk\tanh(kH)$, and $k = 2\pi/\lambda$. λ is the wavelength. n_x and n_y are the x and y components of the wave direction. The velocities of the fluid are given as

$$u_x = \left\{\omega\eta_0 \sin(\omega t - n_x x - n_y y)\frac{\cosh[k(z+H)]}{\sinh(kH)} + U_0\right\} n_x \tag{2}$$

$$u_y = \left\{\omega\eta_0 \sin(\omega t - n_x x - n_y y)\frac{\cosh[k(z+H)]}{\sinh(kH)} + U_0\right\} n_y \tag{3}$$

$$u_z = \omega\eta_0 \cos(\omega t - n_x x - n_y y)\frac{\sinh[k(z+H)]}{\sinh(kH)} \tag{4}$$

2.2.2 Morison model. The line-type elements of the fish cage system include the floating collar, sinker tube, mooring line and bridles. The normal and tangential forces resulting from the viscous effects are given by Equations (5) and (6).

$$\boldsymbol{F}_n = \frac{1}{2} C_n \rho L d |\boldsymbol{U}_n| \boldsymbol{U}_n \tag{5}$$

$$\boldsymbol{F}_t = \frac{1}{2} C_t \rho L \pi d |\boldsymbol{U}_t| \boldsymbol{U}_t \tag{6}$$

where the directions of the $\boldsymbol{F}_n$ and $\boldsymbol{F}_\tau$ are shown in Figure 3. L and d are the length and diameter of the line element, respectively.

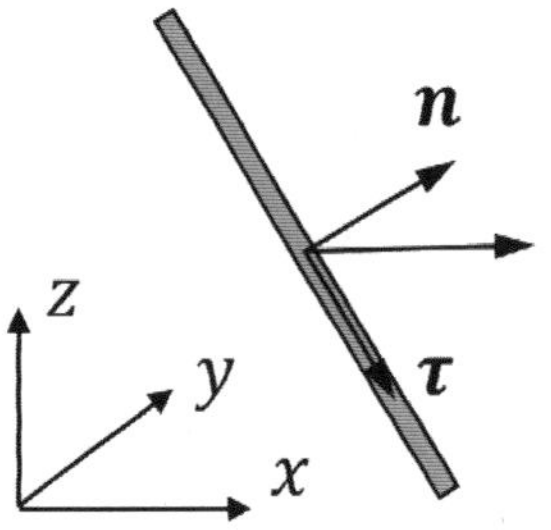

Figure 3. Schematic of the Morison model for the line-like elements.

The tangential and normal velocities are calculated as

$$\boldsymbol{U}_t = [(\boldsymbol{U} - \boldsymbol{v}) \cdot \boldsymbol{\tau}]\boldsymbol{\tau}, \qquad \boldsymbol{U}_n = (\boldsymbol{U} - \boldsymbol{v}) - \boldsymbol{U}_t \tag{7}$$

where $\boldsymbol{U}$ and $\boldsymbol{v}$ are the velocities of fluid and structure respectively, and $\boldsymbol{\tau}$ is the vector of the line element. $\boldsymbol{U}$ is given by Equations (2)-(4). C_n and C_t are normal and tangential drag coefficients, given by experimental results based on the Reynolds number of the line-type structures.

2.2.3 Screen model. The drag force $\boldsymbol{F}_D$ and lift force $\boldsymbol{F}_L$ are calculated as

$$\boldsymbol{F}_D = \frac{1}{2} C_D \rho A_t |\boldsymbol{U} - \boldsymbol{v}|^2 \boldsymbol{i}_D, \qquad \boldsymbol{F}_L = \frac{1}{2} C_L \rho A_t |\boldsymbol{U} - \boldsymbol{v}|^2 \boldsymbol{i}_L \tag{8}$$

where

$$\boldsymbol{i_D} = \frac{\boldsymbol{U} - \boldsymbol{v}}{|\boldsymbol{U} - \boldsymbol{v}|}, \qquad \boldsymbol{i_L} = \boldsymbol{i_D} \times \boldsymbol{n} \times \boldsymbol{i_D} \tag{9}$$

The values of C_D and C_L are given using the following the formulas [7].

$$C_D = 0.04 + (-0.04 + S_n - 1.24S_n^2 + 13.7S_n^3)\cos\theta \tag{10}$$

$$C_L = (0.57S_n - 3.54S_n^2 + 10.1S_n^3)\sin^2\theta \tag{11}$$

When the current passes one side of the net, the current velocity will decrease due to the friction on the net twines. The velocity reduction of the incoming flow is calculated using a reduction factor r [7-8,16],

$$r = 1 - 0.46C_{D(\theta=0)} \tag{12}$$

2.2.4 Strip model. The strip model is proposed to calculate the hydrodynamic force of the floating collars. The floating collars are divided into N circular sections and each section is a line-type element as discussed in Section 2.2.2. In each section, the water surface is assumed to be flat, as shown in Figure 4. Because the collar sections are not fully submerged, the hydrodynamic force in Equations (5) - (6) need to be corrected using a submerging factor κ_D.

$$\kappa_D = \begin{cases} 0, & \Delta z < -R \\ \frac{1}{2}\left(1 + \frac{\Delta z}{R}\right), & -R \le \Delta z \le R \\ 1, & \Delta z > R \end{cases} \tag{13}$$

where Δz is the distance from the undistributed water level to the centre of the section, R is the radius of the section and $\theta = \arccos\left(-\frac{\Delta z}{R}\right)$.

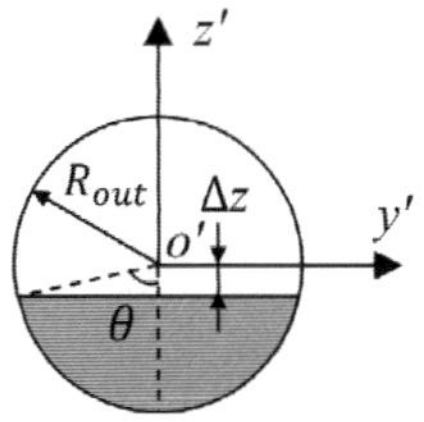

Figure 4. Illustrations of the strip mode for the floating collars.

2.3 Structural models

2.3.1 Extended position-based dynamics (XPBD). The structural responses of the net, mooring line and bridles are calculated using an XPBD method. In XPBD method, the positions of the nodes are updated based on the masses at the nodes and the stiffness k_j of the constraints between the nodes. After the nodes are updated using Newton's second law without considering the inner forces resulting from the constraints, the position update for the two nodes due to a single constraint C_j is given by evaluating the following expressions.

$$\begin{aligned} \boldsymbol{\Delta x_1} &= +\Delta\lambda_j w_{j,1}\boldsymbol{q} \\ \boldsymbol{\Delta x_2} &= -\Delta\lambda_j w_{j,2}\boldsymbol{q} \end{aligned} \tag{14}$$

Where $\boldsymbol{\lambda} = [\lambda_1, \lambda_1, \cdots, \lambda_M]^T$ is the Lagrange multiplier of the constraints $\boldsymbol{C} = [C_1, C_1, \cdots, C_M]^T$. $w_{j,1}$ and $w_{j,2}$ are the inverse mass of the nodes. $\boldsymbol{q}$ is the unit vector between the two nodes. $\boldsymbol{\Delta\lambda}$ is the change of $\boldsymbol{\lambda}$, and can be given by Equation (15), where $\tilde{\alpha}_j = \frac{1}{k_j \Delta t^2}$.

$$\Delta\lambda_j = \frac{-C_j}{w_{j,1} + w_{j,2} + \tilde{\alpha}_j} \tag{15}$$

Compared with the mass spring method, XPBD can provide a quick result for the net dynamics in an implicit way. However, XPBD has its inevitable error for calculating the tensions of the constraints. It is because the masses of the nodes are involved when evaluating the Lagrange multiplier change. It is against the physics intuition that the elastic force is only determined by the stiffness and the constraint. If $w_{j,1}$ and $w_{j,2}$ are removed from the denominator of Equation (15), the method will become unstable. To correct the results of XPBD and maintain the stable update process, a PBD correction is proposed by applying extra forces acting on the two nodes of the constraint.

$$\boldsymbol{F_1} = +\gamma_j \boldsymbol{q} \quad \boldsymbol{F_2} = -\gamma_j \boldsymbol{q} \tag{16}$$

where $\Delta\gamma_j$ is evaluated by Equation (17).

$$\gamma_j = \frac{-C_j(w_{j,1} + w_{j,2})}{(w_{j,1} + w_{j,2} + \tilde{\alpha}_j)\tilde{\alpha}_j \Delta t^2} \tag{17}$$

2.3.2 Mode superposition (MS). The structural responses of the floating collars and sinker tube are calculated using the mode superposition method. The structures of the floating collars and sinker tube are simplified as 3D beam models, where the sections are shown in Figure 5. The conjunctions between the two floating collars are neglected for simplicity, and the two floating collars are assumed to be a single circular beam. The section information of the floating collars and sinker tube are listed in Table 3. To address the inertia forces due to the added mass and additional mass of the water (see Figure 5b), the mass of each beam element includes the contributions the added mass of water with an added mass coefficient of $C_a = 1.$ while that of the sinker tube includes the contributions of the added mass and the additional mass of water.

Table 3. Section information of the floating collars and sinker tube.

Section	E (GPa)	G (GPa)	A (m^2)	J (m^4)	I_y (m^4)	I_z (m^4)
Floating collars	3	0.6	0.044	0.0068	6.2×10^{-4}	0.0062
Sinker tube	3	0.6	0.015	2×10^{-4}	1×10^{-4}	1×10^{-4}

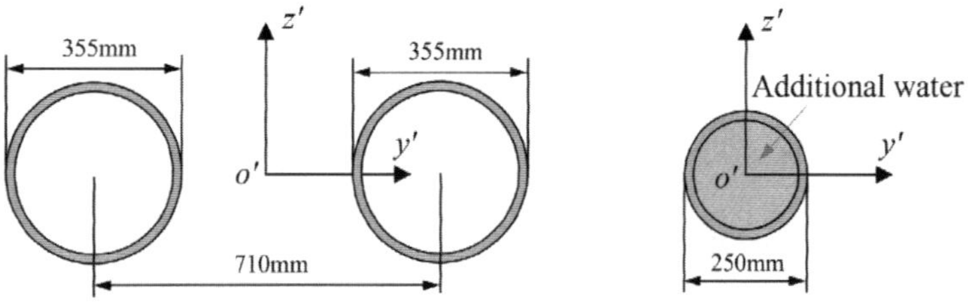

(a) Floating collars' section (b) Sinker tube's section

Figure 5. Descriptions of the floating collars' and sinker tube's sections.

The modal coordinate is updated based on the Newmark-beta method.

$$q_j^{(n+1)} = P_j / K_j \tag{18}$$

where

$$P_j = F_j^{(n)} + a_0 q_j^{(n)} + a_2 \dot{q}_j^{(n)} + a_3 \ddot{q}_j^{(n)} \tag{19}$$

$$K_j = a_0 + (2\pi f_j)^2 \tag{20}$$

The updates of the acceleration and speed are given as

$$\ddot{q}_j^{(n+1)} = a_0(q_j^{(n+1)} - q_j^{(n)}) - a_2 \dot{q}_j^{(n)} - a_3 \ddot{q}_j^{(n)} \tag{21}$$

$$\dot{q}_j^{(n+1)} = \dot{q}_j^{(n)} + a_6\ddot{q}_j^{(n)} + a_7\,\ddot{q}_j^{(n+1)} \tag{22}$$

The modal force is calculated as

$$F_j^{(n)} = \sum_{i=1}^{N} \boldsymbol{\phi}_{i,j} \cdot \boldsymbol{F}_i \tag{23}$$

where $\boldsymbol{F}_i$ is the external force vector acting on the ith node. $\boldsymbol{\phi}_{i,j}$ is the shape function vector at the ith node. The parameters of a_0, a_2, a_3, a_6 and a_7 are given as $a_0 = \frac{1}{\beta\Delta t^2}$, $a_2 = \frac{1}{\beta\Delta t}$, $a_0 = \frac{1}{2\beta} - 1$, $a_6 = \Delta t(1-\gamma)$, and $a_7 = \Delta t\gamma$, where $\beta = 1/4$ and $\gamma = 1/2$. The updates of the node positions and speeds are given by evaluating the following expressions.

$$\boldsymbol{x}_i^{(n+1)} = \boldsymbol{x}_{0,i} + \sum_{j=1}^{N} \boldsymbol{\phi}_{i,j}\, q_j^{(n+1)} \tag{24}$$

$$\boldsymbol{u}_i^{(n+1)} = \boldsymbol{u}_{0,i} + \sum_{j=1}^{N} \boldsymbol{\phi}_{i,j}\, q_j^{(n+1)} \tag{25}$$

2.3.3 Coupling between XPBD and MS. The coupling between XPBD and MS is conducted as follows.

(1) XPBD provides the tensions of the brindles and the net ropes for MS.
(2) MS provides the positions and speeds of the nodes located at the centre lines of the floating collars and sinker tube.

In the present study, the XPBD and MS share the same nodes at the centre lines of the floating collars and sinker tube. For XPBD, the inverse masses of these nodes are set to be zero, which means that their positions and speeds are not updated in XPBD. These motions are updated by the MS solver. For the MS solver, the tensions of the brindles and the net ropes are provided by XPBD.

3. Results and discussions

3.1 Case description

The rough sea condition is simplified as the waves with a heigh of 156 m [17] and a current with a speed of 1m/s. The water depth is 100 m. The wave period is 10 s. The free surface and the velocities of the incoming waves can be given by Equations (1) – (4) using the Airy wave theory. The initial locations of the fish cages on the water surface and in deep water in a current with a speed of 0.3m/s are shown in Figure 6. The total simulation time is 300 s, which includes 30 wave periods.

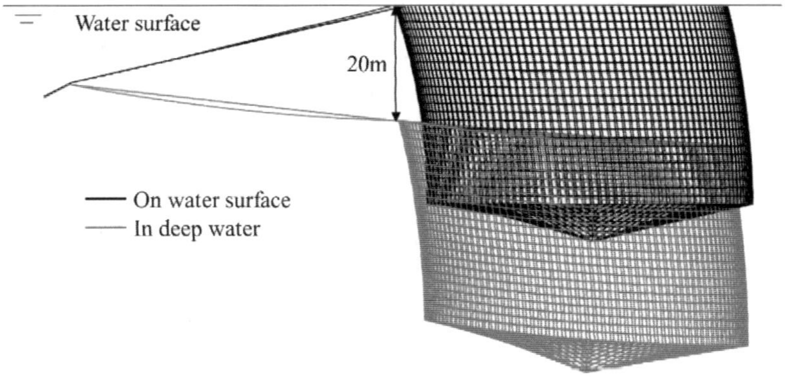

Figure 6. Fish cages on water surface and in deep water in a current with a speed of 0.3 m/s.

IOP Conf. Series: Materials Science and Engineering 1294 (2023) 012013 doi:10.1088/1757-899X/1294/1/012013

3.2 *Dynamic process*

Figure 7 shows the time-step sensitivity study of the horizontal force of the fish cage on the water surface. Three time-steps of 0.5 ms, 1 ms and 2 ms are tested. The results of 0.5 ms and 1 ms exhibit a good agreement, indicating that the convergent result is achieved when the time step is small as 1ms. Table 4 shows the computational time of the surface cases with the three time-steps. As the time-step decreases, the computational time has a rapid increase. Therefore, to achieve a balance between the accuracy and computational cost, the time-step of 1ms is selected for further calculations.

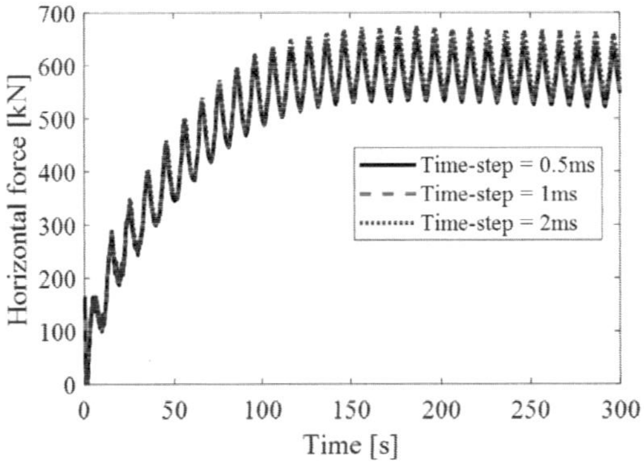

Figure 7. Time-step sensitivity study of the horizontal force of the fish cage on the water surface.

Table 4. Computational time of the surface case with different time-steps

Time-step (ms)	0.5	1.0	2.0
Computational time (s)	4185	1856	937

Figure 8 shows the comparisons of the horizontal force between the fish cages on the water surface and in deep water. The horizontal force has a significant reduction when the fish cage submerges to the deep water. The comparisons of the average results and the variation of the horizontal forces are shown in Table 4. When the fish cage is lowered by 20 m from the water surface, both the average results and the variation of the horizontal forces are significantly reduced.

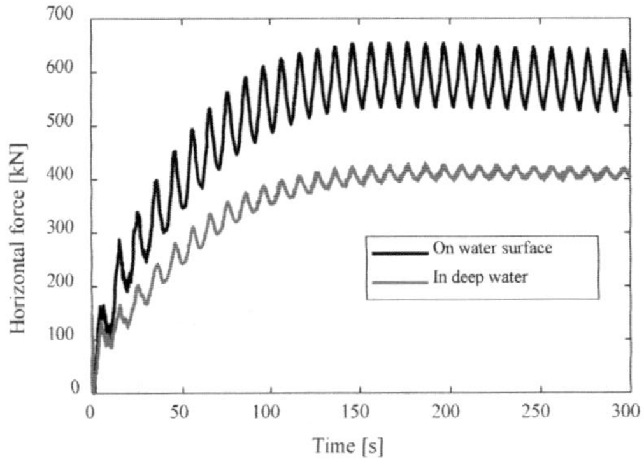

Figure 8. Comparisons of the horizontal force between the fish cages on the water surface and in deep water.

Table 5 Comparisons of the average results and the variation of the horizontal forces.

	On water surface	In deep water	Reduction
Average horizontal force (kN)	581	408	30%
Horizontal force variation (kN)	113	21.2	81%

Figure 9 shows the comparisons of the deformations between the fish cages on water surface and in deep water in a typical period from 290 s to 300 s. The deformation of the floating collars of the fish cage on water surface is much larger than that in deep water, which indicates that the floating collars of the fish cage on the water surface experience larger stress than that in deep water.

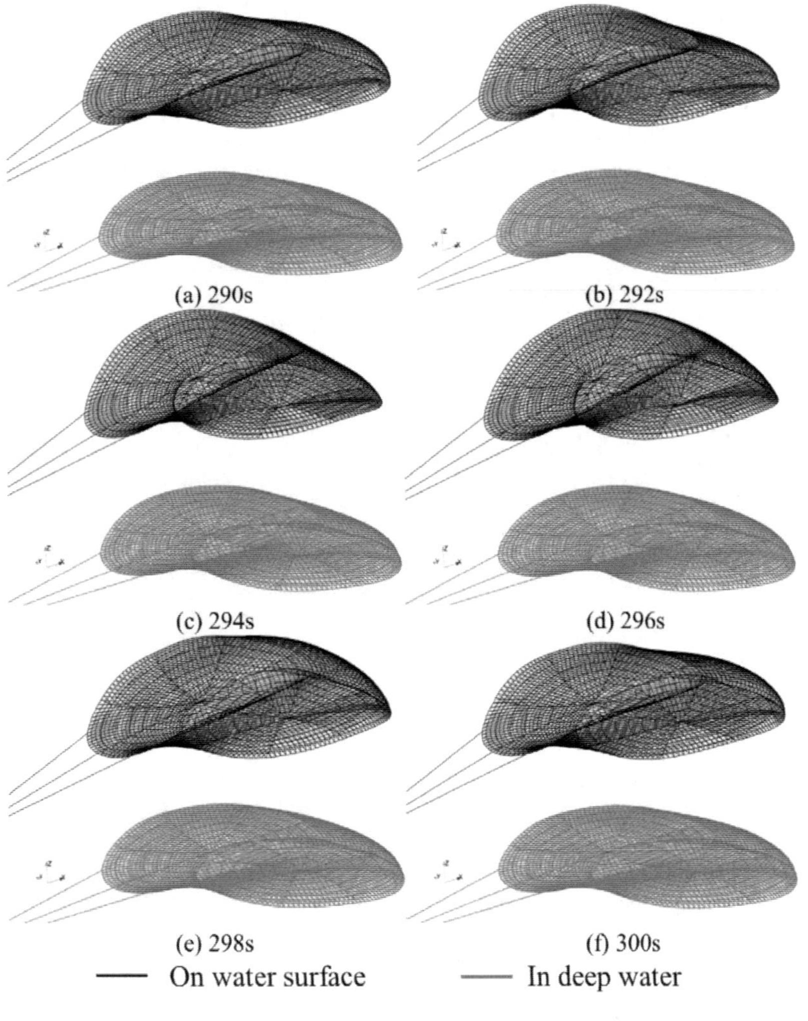

Figure 9. Comparisons of the deformations between the fish cages on the water surface and in deep water.

4. Conclusion

The dynamic responses of a single point mooring submersible fish cage in waves and current are assessed using numerical methods. The wave and current conditions are modelled using the airy wave theory. For the floating collars and sinker tube, the Morison model is adopted to calculate the hydrodynamic force and the modal superposition method is proposed for the structural response calculations. For the aquaculture nets, the screen model is adopted to compute the hydrodynamic force and the XPBD method is proposed to obtain the structural deformations. The hydrodynamic forces and the deformations of the fish cages are compared between the fish cage on water surface and in deep water. Results show that lowering the fish cage by 20m from the water surface leads to a 30% reduction in average horizontal force and an 81% decrease in the variation of horizontal force. Additionally, significant reductions in the deformation of the floating collars are observed. Consequently, using submersible fish cages can effectively mitigate physical stress and prevent structural damage for the fish cage system.

It should be mentioned that more cases should be performed, and more structural information should be provided to show the differences between the surface case and the deep-water case. The further study will also focus on the structural responses of the descending and ascending of the fish cage. In addition, the coupling of XPBD and MS can provide a quick solution for the structural modelling of the fish farm system, which can be utilized at the initial stage of fish farm design.

Acknowledgment: This study was supported by the Research Council of Norway through the project "Unleashing the sustainable value creation potential of offshore ocean aquaculture" (Project number: 328724).

References

[1] FAO, 2018. In: The State of World Fisheries and Aquaculture 2018 Meeting the sustainable development goals. Rome. Licence: CC BY-NC-SA 3.0 IGO.

[2] Cardia, F., Lovatelli, A., 2015. Aquaculture Operations in Floating HDPE Cages: A Field Handbook. FAO and Ministry of Agriculture of the Kingdom of Saudi Arabia.

[3] Li, L., Fu, S.X., Xu, Y.W., Wang, J.G., Yang, J.M., 2013. Dynamic responses of floating fish cage in waves and current. *Ocean. Eng.* **72**, 297–303.

[4] Moe-Føre, H., Lader, P., Lien, E., Hopperstad, O., 2016. Structural response of high solidity net cage models in uniform flow. *J. Fluids Struct.* **65**, 180–195.

[5] Cheng, H., Li, L., Aarsæther, K. G., Ong, M. C., 2020. Typical hydrodynamic models for aquaculture nets: A comparative study under pure current conditions. Aquacultural Engineering, 90, 102070.

[6] Fridman, A.Lv., 1973. Theory and Design of Commercial Fishing Gear (Trans. from Russ.). Jerusalem.

[7] Aarsnes, J.V., Løland, G., Rudi, H., 1990. Current Forces on Cage, Net Deflection. Eng. Offshore Fish Farm. p. 891–921.

[8] Løland, G., 1991. Current Forces on and Flow Through Fish Farms. *PhD Thesis,* Norwegian Institute of Technology.

[9] Zhao, Y. P., Bi, C. W., Dong, G. H., Gui, F. K., Cui, Y., Guan, C. T., Xu, T. J., 2013. Numerical simulation of the flow around fishing plane nets using the porous media model. *Ocean Eng. 62*, 25-37.

[10] Chen, H., Christensen, E. D., 2017. Development of a numerical model for fluid-structure interaction analysis of flow through and around an aquaculture net cage. *Ocean Eng.* **142,** 597-615.

[11] Cifuentes, C., Kim, M.H., 2017. Numerical simulation of fish nets in currents using a Morison force model. *Ocean. Syst. Eng.* **7**(2), 143–155.

[12] Berstad, A.J., Aarsnes, J.V., 2018. A case study for an offshore structure for aquaculture: comparison of analysis with model testing. *ASME 2018 37th Int. Conf. Ocean, Offshore Arctic Eng.: American Society of Mechanical Engineers* V11BT12A057-V11BT12A057.

[13] Xu, T. J., Zhao, Y. P., Dong, G. H., Gui, F. K. 2013. Analysis of hydrodynamic behavior of a submersible net cage and mooring system in waves and current. *Appl. Ocean Res.* **42**, 155-167.

[14] Müller, M., Heidelberger, B., Hennix, M., Ratcliff, J. 2007. Position based dynamics. *J. Visual Commun. Image Repr.* **18**(2), 109-118.

[15] Miles, M., Müller, M., Chentanez, N., 2016. XPBD: position-based simulation of compliant constrained dynamics. *Proc. 9th Int. Conf. Motion in Games.* P. 49 - 54

[16] Kristiansen, T., Faltinsen, O. M., 2012. Modelling of current loads on aquaculture net cages. *J. Fluids Struct.* **34**, 218–235.

[17] Lee, C. W., Lee, G. H., Choe, M. Y., Song, D. H., Hosseini, S. A., 2009. Dynamic behavior of a submersible fish cage. In: *Int. Conf. Offshore Mech. Arctic Eng.* **43444**, 201-206.

IOP Conf. Series: Materials Science and Engineering 1294 (2023) 012014 doi:10.1088/1757-899X/1294/1/012014

Applying a machine learning method for cumulative fatigue damage estimation of the IEA 15MW wind turbine with monopile support structures

C Ren* and Y Xing

Department of Mechanical and Structural Engineering and Materials Science, University of Stavanger, Norway

E-mail: chao.ren@uis.no

Abstract. Offshore support structures are critical for offshore bottom-fixed wind turbines, as they bear nearly all the mass and loading of wind turbine systems. In addition, the support structures are generally subjected to a harsh environment and require a design life of more than 20 years. However, the design validation of the support structure normally needs thousands of simulations, especially considering the fatigue limit state. Each simulation is quite time-consuming. This makes the design optimization of wind turbine support structures lengthy. Therefore, an effective approach for estimating the fatigue damage of wind turbine support structures is essential. This work uses a machine learning method named the AK-DA approach for cumulative fatigue damage of wind turbine support structures. An offshore site in the Atlantic Sea is studied, and the related joint probability distribution of wind-wave occurrences is adopted in this work. The IEA 15MW wind turbine with monopile support structure is investigated, and different wind-wave conditions are considered. The cumulative fatigue damage of the monopile support structure is estimated by the AK-DA approach. The numerical results showed that this machine learning approach can efficiently and accurately estimate the cumulative fatigue damage of the monopile support structure. The efficiency is increased more than 55 times with an error of around 1%. The AK-DA approach can highly enhance the design efficiency of offshore wind support structures.

1. Introduction

In pursuing sustainable energy solutions, wind power has emerged as a prominent source of renewable electricity generation. Many works [1, 2, 3, 4, 5, 6] have been carried out for wind turbine structures. Yang et al.[1] did a reliability-based design of wind turbine sub-structure optimization. Wang and Kolios [3] proposed a framework for system reliability assessment of offshore monopiles considering soil-solid interaction and harsh marine environments. Ren et al. [2, 5] compared different dynamic simulation approaches of wind turbine jacket foundations and carried out structural reliability analysis of jacket structures with machine learning approaches. Yu et al. [6] built a predictive model for the mooring line failure diagnosis and motion control. To increase wind power generation, the new trend is to install wind turbines offshore, as offshore wind has higher and uniform wind speeds. Offshore wind turbines are commonly classified into bottom-fixed and floating wind turbines (FWTs). Compared to the bottom-fixed wind turbines, the technologies of FWTs are not yet mature. Therefore, the installed offshore wind turbines are mostly bottom-fixed. Among the installed bottom-fixed wind turbines, more than 70% offshore wind turbines are with monopile support structures. Wind turbine support structures like monopiles are critical components that bear the weight of massive turbine components, endure the forces of dynamic wind loads, and ensure the safe and efficient operation of wind turbines throughout their operational lifespan. The design of these support structures is fundamental to the success of wind energy projects. According to the design codes [7] and [8], fatigue limit state and ultimate limit state are the two most essential criteria for offshore wind turbine structure design.

Among these two design criteria, the fatigue limit state is more complex as offshore wind turbine structures experience a large combination of wind and wave conditions during their lifetime. Further, the load responses are highly dynamic, fully coupled, and sensitive to wind and wave combinations. This means one needs to consider all wind-wave combinations individually when estimating fatigue. The fatigue estimation will, therefore, require the computation of a large number of load cases, requiring simulation times of a few thousand hours, for the accurate representation of fatigue damage. To tackle this problem, different solutions [9, 10] have been proposed for the fatigue damage assessment. They can be roughly classified into two types: (1). fatigue damage estimation in the frequency domain [9], and (2). reduced load cases for fatigue damage estimation [10]. In addition, due to the powerful capacity of machine-learning techniques, some researchers used machine-learning approaches for offshore wind turbine reliability assessment [11, 12] and fatigue damage estimation [13, 14]. More recently, Huchet et al. [15] and Ren and Xing [16] proposed efficient active learning approaches for cumulative fatigue damage estimation of offshore wind turbine structures. The proposed active learning approaches have demonstrated high efficiency and accuracy for the fatigue damage assessment of wind turbine towers.

Considering the importance of wind turbine support structures and the efficiency of active learning approaches, we apply an active learning approach named AK-DA proposed by [15] for the cumulative fatigue damage estimation of monopile support structures. The IEA 15MW wind turbine model with monopile support structure [17] is investigated in this work. In addition, the AK-DA approach assumes that the joint probability of wind-wave occurrence is already known. Thus, one offshore site in the Atlantic Sea and its joint probability distribution are considered in this work. The layout of this paper is organized as follows: Section 2 presents the IEA 15MW wind turbine models and monopile support structures. The classical fatigue damage estimation approach is also given in this work. Section 3 overviews the AK-DA approach and the related joint probability of wind-wave occurrences in the Atlantic sea. Section 4 gives the results of cumulative fatigue damage estimation. In the end, the conclusion and discussion are given.

2. Wind turbine model and support structures

2.1. The IEA 15MW wind turbine model with monopile support structures

The IEA 15MW wind turbine with a fixed-bottom monopile support structure [17] is shown in Figure 1. This wind turbine is a Class 1B direct-drive machine with a rotor diameter of 240 meters and a hub height of 150 meters, which is jointly designed by the National Renewable Energy Laboratory (NREL) and the Technical University of Denmark (DTU). The monopile structure is designed as an isotropic steel tube. More detailed information on the wind turbine model can be found in the original report [17]. In this work, the aero-hydro-servo-elastic simulation of the IEA 15MW wind turbine is conducted through OpenFAST codes [18].

2.2. Load-stress in the monopile support structure

For the wind turbine tower and monopile structures, the normal stress typically dominates the fatigue damage. The normal stress at the monopile structures can be calculated as follows:

Figure 1. The IEA 15MW wind turbine with monopile support structures [17]

$$\sigma(\alpha) = \frac{F_z}{A} - \frac{M_x}{I_x} R \sin\alpha + \frac{M_y}{I_y} R \cos\alpha \tag{1}$$

Where F_z is axial force; A is nominal cross section area; M_x and M_y are side-to-side bending moment and fore-after bending moment, respectively; I_x and I_y are sectional moments of area; R is the radius of the cross-section; α is the azimuth angle. The force (F_z) and moments (M_x and M_y) can be obtained from the OpenFast simulation. In this study, the fatigue damage caused by the normal stress at the mud line is assessed, and the dimension of the monopile structure at the mud line is given in Table 1.

Table 1. Mudline monopile parameters [17]

Location	Height (m)	Outer diameter (m)	Thickness (mm)
Mudline monpile	-30	10	55.341

2.3. Classicial cumulative fatigue damage estimation approach
The cumulative fatigue damage can be calculated as follows:

$$D = \sum_{i=1}^{n_c} \rho_i d_i \tag{2}$$

IOP Conf. Series: Materials Science and Engineering 1294 (2023) 012014 doi:10.1088/1757-899X/1294/1/012014

Where ρ_i is the probability of occurrence of ith wind-wave case, n_c is the total number of wind-wave cases. and d_i is the fatigue damage value of ith wind-wave case. The fatigue damage d_i value can be calculated based on the time series stress by Rainflow counting technique [19, 20] and the S-N curve of the material property, which gives:

$$d_i = \sum_s \frac{n_s}{N_s} \tag{3}$$

Where s is the stress range in MPa, n_s is the number of cycles obtained using the Rainflow counting method at the stress range s, and N_s is the number of cycles to failure at the stress range s. According to the design code [8], the N_s value can be calculated as follows:

$$\log_{10} N_s = \log_{10} \bar{a} - m \log_{10}\left(s\left(\frac{t}{t_{ref}}\right)^k\right) \tag{4}$$

Where m is the negative slope of the S-N curve and $\log_{10} \bar{a}$ is the intercept of the S-N curve. The parameters t and t_{ref} represent the thickness of the wind turbine structure and the reference thickness, respectively, depending on the connection type of steel structures. In the context of this study, t refers explicitly to the monopile thickness, while t_{ref} is a fixed value of 25mm. For the fatigue analysis of the present study, the S-N curve D in the seawater from the DNV code [8] is employed, with the exponent m set to 3, the logarithm of the reference stress amplitude $\log_{10} \bar{a}$ assigned to 11.764, and the constant k set to 0.20.

3. The AK-DA approach

3.1. The overview of the AK-DA approach

In the AK-DA approach [15], the Kriging model is employed for the prediction of simulated fatigue damage across various wind-wave cases. Subsequently, the expression for the estimated long-term cumulative fatigue damage ($\hat{D}$) is as follows:

$$\hat{D} = \sum_{i=1}^{n_c} \hat{d}_i \rho_i \tag{5}$$

Here, $\hat{d}_i$ represents the predicted damage value by the Kriging model for the ith wind-wave case, and ρ_i is the related probability of occurrences. The variance of the predicted cumulative fatigue damage can be expressed as follows:

$$\sigma_{\hat{D}}^2 = \mathbf{VAR}\Big(\sum_{i=1}^{N} p_i \hat{d}_i\Big) = \sum_{i=1}^{N}\sum_{j=1}^{N} p_i p_j \mathbf{COV}(\hat{d}_i, \hat{d}_j) \tag{6}$$

Where $\mathbf{COV}(\hat{d}_i, \hat{d}_j)$ denotes the covariance between $\hat{d}_i$ and $\hat{d}_j$. The coefficient of variation ($\delta_{\hat{D}}$) for the predicted cumulative fatigue damage can be computed as follows:

$$\delta_{\hat{D}} = \frac{\sigma_{\hat{D}}}{\hat{D}} \tag{7}$$

The AK-DA approach aims to reduce the coefficient of variation value ($\delta_{\hat{D}}$) by updating the Kriging model with the enriched samples. The enriched wind-wave cases are selected as follows:

$$i_* = \text{argmax} \quad p_i \sum_{j=1}^{N} p_j \mathbf{COV}(\hat{d}_i, \hat{d}_j) = \text{argmax} \quad C_i \tag{8}$$

Where i_* denotes the enriched wind-wave case number. The active learning process will stop when the convergence criterion ($\delta_{\hat{D}} \leq 0.01$) is satisfied. The general flowchart of the AK-DA approach can be summarized in Figure 2, where n_i is the total number of initial training samples.

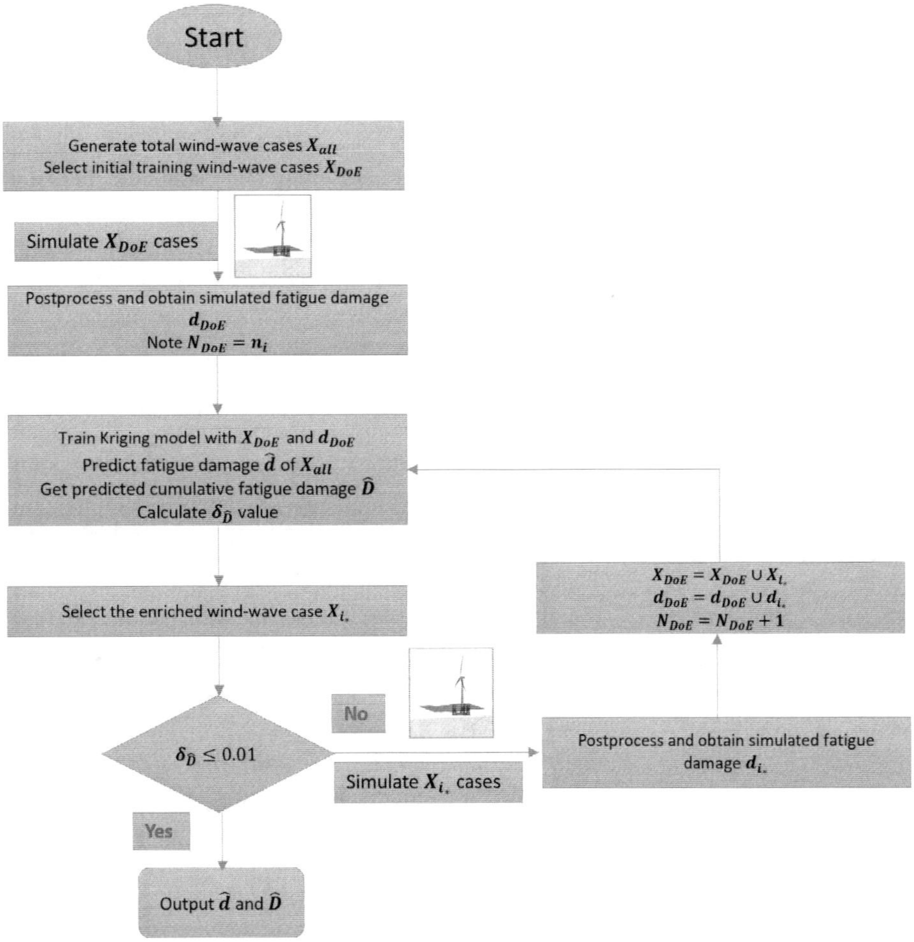

Figure 2. The general flowchart of AK-DA approach

3.2. Wind-wave probability of the occurrences

In the AK-DA approach, it is assumed that the probability of wind-wave occurrences has been pre-established. This paper adopts the joint wind and wave distribution presented in the work of Li et al. [21]. Utilizing the one-hour mean wind speed (U_w), significant wave height (H_s), and wave peak period (T_p), a long-term joint wave and wind distribution are constructed, yielding the following:

$$f_{U_{\mathrm{w}},H_{\mathrm{s}},T_{\mathrm{p}}}(u,h,t) \approx f_{U_{\mathrm{w}}}(u) \cdot f_{H_{\mathrm{s}}|U_{\mathrm{w}}}(h \mid u) \cdot f_{T_{\mathrm{p}}|H_{\mathrm{s}}}(t \mid h) \tag{9}$$

$f_{U_{\mathrm{w}}}()$ represents the marginal distribution of U_w, $f_{H_{\mathrm{s}}|U_{\mathrm{w}}}()$ denotes the conditional distribution of H_s given U_w, and $f_{T_{\mathrm{p}}|H_{\mathrm{s}}}()$ is the the conditional distribution of T_p given H_s. Li et al. [21] conducted a study on the joint wind-wave distribution of five different sites. In these sites, Site one, Sem Rev, with a water depth of 40 meters, is suitable for the IEA 15MW wind turbine with monopile support structures. Therefore, the simplified joint distribution derived from Site One is utilized in this investigation. Further information regarding the joint distribution can be found in the work by Li et al. [21]. It should be noted that this study assumes alignment between wind and wave direction, and the potential misalignment between wind and wave is not addressed.

4. Cumulative fatigue damage estimation of the monopile support structure

In this section, the cumulative fatigue damage of the mud line monopile support is estimated by the AK-DA approach. Different wind-wave cases are considered and a total of 4032 cases are investigated in this work. As listed in Table 2, the mean wind speed is from the cut-in wind speed (3 m/s) to the cut-out wind speed (25 m/s), with an interval of 2 m/s. The significant wave height is from 0.5m to 8m, with an interval of 0.5m. Also, the peak period time is from 5s to 15s, with an interval of 0.5s. The intervals of U_w, H_s and T_p are following the recommendations from the design code [7]. Also, a 4000-second simulation is conducted for each wind-wave case, but the first 400 seconds are not considered. Furthermore, the initial wind-wave cases for training the Kriging model in the AK-DA approach are also given in Table 2. The initial wind-wave cases can normally be selected by the grid sampling based on each parameter design space. Only 60 wind-wave cases are used to train Kriging model at the beginning of the AK-DA approach.

Table 2. Wind-wave cases considered for the cumulative fatigue estimation

Parameter	Domain of variation	Discretization	Number of cases	Initial grid samples
U_w(m/s)	[3, 25]	2	12	[3, 7, 11, 19, 25]
H_s(m)	[0.5, 8]	0.5	16	[1, 4, 7]
T_p(s)	[5, 15]	0.5	21	[5, 8, 11, 14]
	Total cases		$N = 4032$	$n_i = 60$

The global one-hour cumulative fatigue damage at the mud line of monopile support is calculated. The final results of the AK-DA approach and the related simulation results are given in Table 3. In Table 3, the simulation calls mean that a total of simulated wind-wave cases, D^{1H} denotes the global one-hour cumulative fatigue damage, $Diff.$ represents the absolute percentage difference between the simulated and predicted damage values, the increased efficiency is calculated based on the simulation calls between the simulation approach and the AK-DA approach. As shown in Table 3, only 72 wind-wave cases are required for the AK-DA approach to estimate the cumulative fatigue damage. The absolute difference between the simulated reference and the AK-DA approach is just around 1%. The efficiency is increased more than 55 times. The final prediction results of the AK-DA approach at different mean wind velocities are given in Figure 3, compared with the simulated values. As shown in Figure 3, the prediction results of AK-DA approaches are nearly the same compared to the simulated references.

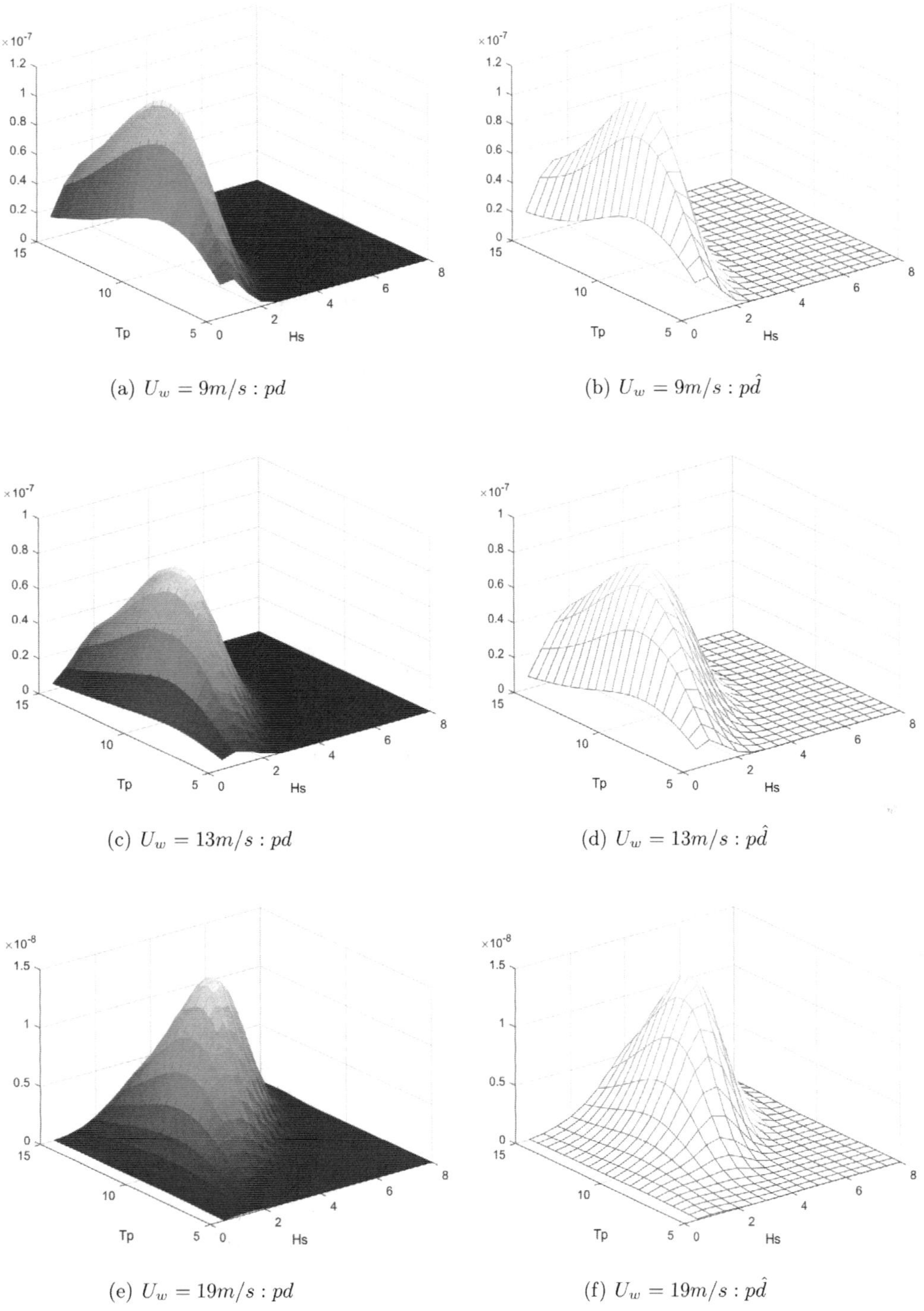

(a) $U_w = 9m/s : pd$

(b) $U_w = 9m/s : \hat{pd}$

(c) $U_w = 13m/s : pd$

(d) $U_w = 13m/s : \hat{pd}$

(e) $U_w = 19m/s : pd$

(f) $U_w = 19m/s : \hat{pd}$

Figure 3. Simulated values pd and Final predicted values $\hat{pd}$ at different mean wind velocities

Table 3. Results of AK-DA approach compared to the simulation reference

Approach	Simulation calls	D^{1H} (10^{-5})	Diff. (%)	Increased efficiency (times)
Simulated reference	4032	2.797	-	-
AK-DA	72	2.826	1.03	56

5. Conclusion

In this work, we apply a machine learning method, the AK-DA approach, to estimate the cumulative fatigue of monopile support structures in the IEA 15MW wind turbine. The joint wind-wave distribution at one offshore site in the Atlantic Sea is used in this paper. The global one-hour cumulative fatigue damage in the mud line of monopile structures is estimated by the AK-DA approach. The predicted results are compared to the simulated references. The results indicate:

The AK-DA approach can efficiently and accurately estimate the cumulative fatigue damage of the monopile support structures. The efficiency can be increased more than 55 times with an error of around 1%. This machine-learning method can significantly reduce the computational effort required for fatigue damage estimation in the design process of wind turbine support structures. The AK-DA approach can be an effective tool for wind turbine support designers, which can greatly reduce the fatigue damage estimation time and accelerate the iterative design process.

Of course, no work is perfect. This work only considers that the wind-wave direction is aligned, and no misalignment between wind-wave directions is considered. Also, only one location at the monopile is considered for fatigue damage estimation. In the future, the wind-wave direction misalignment should be included, and more locations in the monopile support structures will be considered.

References

[1] Yang H, Zhu Y, Lu Q and Zhang J 2015 *Renewable Energy* **78** 16–25
[2] Ren C, Aoues Y, Lemosse D and De Cursi E S 2020 Structural reliability assessment of offshore wind turbine jacket considering corrosion degradation *14th WCCM-ECCOMAS Congress 2020* vol 800
[3] Wang L and Kolios A 2017 A generic framework for reliability assessment of offshore wind turbine monopiles considering soil-solid interaction and harsh marine environments *Proceedings of the 5th International Conference on Marine Structures, MARSTRUCT, Lisbon, Portugal* pp 8–10
[4] Ren C, Aoues Y, Lemosse D and De Cursi E S 2021 Global sensitivity analysis of offshore wind turbine jacket *Proceedings of the 5th International Symposium on Uncertainty Quantification and Stochastic Modelling: Uncertainties 2020* (Springer) pp 35–48
[5] Ren C, Aoues Y, Lemosse D and De Cursi E S 2021 *Engineering Structures* **249** 113308
[6] Yu Y, Cheng S, Cui Y, Zhang X, Zhang P and Yu J 2023 *Ocean Engineering* **280** 114907
[7] IEC I 2009 *International Electrotechnical Commission, Geneva*
[8] DNV D N V *et al.* 2014 *Internet Requests for Comments, DET NORSKE VERITAS AS, OFFSHORE STANDARD DNV-OS-j101*
[9] Kvittem M I and Moan T 2015 *Journal of Offshore Mechanics and Arctic Engineering* **137**
[10] Stieng L E S and Muskulus M 2019 *Wind Energy* **22** 1472–1486
[11] Ren C 2022 *Reliability assessment of an offshore wind turbine jacket with active learning approaches* Ph.D. thesis Normandie Université
[12] Ren C, Aoues Y, Lemosse D and De Cursi E S 2023 *Ocean Engineering* **281** 114657
[13] Wilkie D and Galasso C 2021 *Structural Safety* **88** 102020
[14] Sun K, Xu Z, Li S, Jin J, Wang P, Yue M and Li C 2023 *Renewable Energy* **204** 733–749
[15] Huchet Q, Mattrand C, Beaurepaire P, Relun N and Gayton N 2019 *Wind Energy* **22** 638–652
[16] Ren C and Xing Y 2023 *Renewable Energy* 118977

[17] Gaertner E, Rinker J, Sethuraman L, Zahle F, Anderson B, Barter G, Abbas N, Meng F, Bortolotti P, Skrzypinski W, Scott G, Feil R, Bredmose H, Dykes K, Sheilds M, Allen C and Viselli A 2020 Definition of the IEA 15-megawatt offshore reference wind turbine Tech. rep. International Energy Agency URL `https://www.nrel.gov/docs/fy20osti/75698.pdf`
[18] Jonkman J M and Buhl M L 2005 *National Renewable Energy Laboratory, Golden, Colorado*
[19] Rychlik I 1987 *International journal of fatigue* **9** 119–121
[20] Amzallag C, Gerey J, Robert J L and Bahuaud J 1994 *International journal of fatigue* **16** 287–293
[21] Li L, Gao Z and Moan T 2015 *Journal of Offshore Mechanics and Arctic Engineering* **137**

IOP Conf. Series: Materials Science and Engineering 1294 (2023) 012015 doi:10.1088/1757-899X/1294/1/012015

Fixed type-oscillating water column front wall angle variation and impact on chamber performance: CFD numerical wave tank assessment

A H Samitha Weerakoon[1,*], W Thilan, H A De Silva[2] and M Assadi[1]

[1] Faculty of Science and Technology, University of Stavanger, Stavanger, Norway
[2] Faculty of Engineering, Ocean University of Sri Lanka

* Correspondence: abeysingha.h.weerakoon@uis.no

Abstract. This study investigates optimizing the front wall geometry of an Oscillating Water Column (OWC) chamber to enhance turbine performance and output efficiency when harnessing energy from progressive waves with consistent periods and wavelengths. Numerical Wave Tank (NWT) adopted simulations carried out using ANSYS-Fluent ® package, VOF method with multiphase flow (air-water) where 2-D wave motion of NWT implemented using a C+ computer code. Frontal wall angle of the air/water chamber changed from 0° to 80°, for 10 cases. OWC system was most effective in harnessing and converting wave energy at 0° angle, reaching the maximum power output of 605.08 W/m. In contrast, 80° angle exhibited lower efficiency, with the lowest power output of 26.55 W/m. The average power output over time reflects consistent energy conversion, with the 0° angle demonstrating the highest average power output of 123.72 W/m, while the 80° case exhibited less efficient pneumatic power potential with an average of 3.5 W/m. A uniform cross section provides higher pneumatic power when with PTO in OWC. The efficiency of power generation can be increased by approximately 10 to 20% by keeping 0 to 10° angle of front wall without any appendages fitted with into the wall.

1. Introduction

The utilization of ocean waves as a renewable energy resource has gained considerable attention due to its potential to contribute to global energy demands while reducing carbon emissions [1]. The increasing demand for clean and sustainable energy sources has fueled research and development in the field of wave energy conversion. Ocean waves, a vast and virtually untapped resource, possess significant potential to meet a portion of the world's energy needs [2]. Mainly WECs can be categorized into four main types: Oscillating Water Column (OWC), overtopping systems, attenuators, and point absorbers. Among these, the OWC stands out for its advantages, including low environmental impact and minimal maintenance requirements.

The OWC concept stands out as a promising technology due to its inherent simplicity, reliability, and environmental compatibility. Among the various technologies available for harnessing wave power, the OWC concept has emerged as a promising solution [3]. The most extensively studied and successful wave energy converter (WEC) concept appears to be the Oscillating Water Column (OWC) technology, which has progressed to deploying full-scale prototypes in open sea conditions [3-5]. Japan and Europe regions commercialized floating and fixed type OWC devices as early as 1965, and more recent efforts in Ireland involved testing a 1:4th-scale buoy converter [6]. Fixed OWC devices are typically deployed near the shore in shallow waters, which have lower wave energy compared to offshore areas due to energy dissipation from bottom friction and wave breaking, similar to small scale WEC's [7]. OWCs are celebrated for their elegant and straightforward design and operational principles. They have been implemented in various locations, including Tofteshallen, Norway (500 kW); Sakata, Japan (60 kW); Pico, Portugal (400 kW); Limpet, Scotland (500 kW); and Mutriku, Spain (300 kW), showcasing their versatility and potential for widespread use [8]. From the early days of Fixed type OWCs deployed near the shoreline to the recent emergence of floating OWCs in deeper waters and open seas. This shift towards floating OWCs is driven by the desire to tap into the immense wave energy available in these environments and to significantly scale up wave energy production. However, floating OWCs introduce intricate hydrodynamic interactions which are unbearable and extremely catastrophic under storms and high wave conditions. When waves encounter a floating OWC, they can be partially reflected, partially transmitted beneath the OWC, and partially captured by it.

Simultaneously, the incoming waves stimulate motion responses in the OWC, effectively transforming it into a wave generator that radiates waves in both seaward and leeward directions [9]. Several esteemed researchers, including Hong et al., Sphaier et al. (2010), Sykes et al., (2012) Toyota et al., (2015) and Sheng et al., (2016) have conducted extensive studies and simulations to delve into these complex hydrodynamic phenomena, striving to comprehend the behavior of floating OWCs and optimize their design [10]. Their work has led to valuable insights into factors such as heavy motion resonance, water column motion, and the damping coefficients that govern the performance of these innovative wave energy converters [9-11]. But the floating OWC structures are still considered to be not economically feasible and reliable with hash oceanic nature. However, selecting areas with locally concentrated wave energy due to topography can mitigate this issue [12]. Near-shore OWCs benefit from less exposure to harsh open-ocean wave conditions, such as seen with tidal energy conversion systems, making them more easily monitored and maintained, thus increasing their survivability [13]. Moreover, OWCs embedded in breakwaters or piers can serve dual functions, both generating energy and providing coastal protection [14], making them attractive for investment.

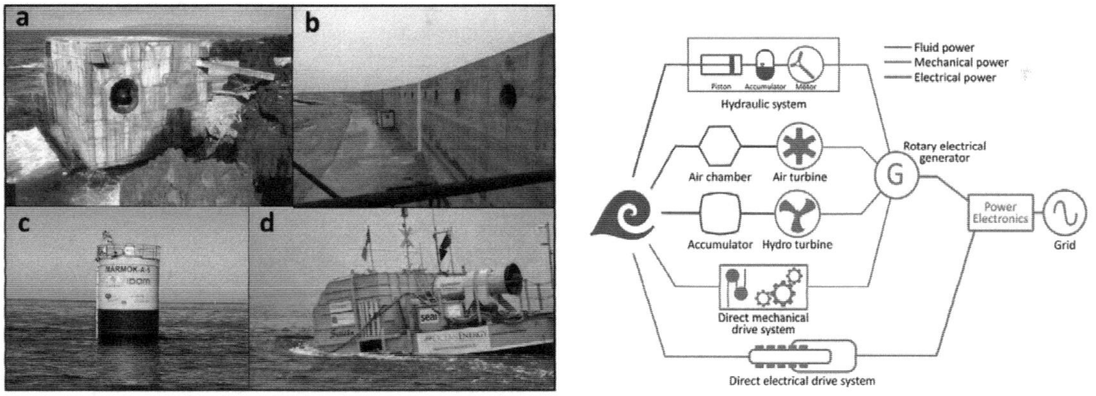

(i) (ii)

Figure 1. (i) OWC concepts into the sea: (a) shoreline (Pico Island, Portugal), (b) U-OWCs integrated into breakwater (Civitavecchia, Italy), (c) spar-buoy (Basque Country, Spain) (courtesy of IDOM), (d) backward-bent-duct-buoy BBDB (Galway Bay, Ireland) (courtesy of Kymaner) and (ii) Different paths for Power Take Off (PTO) from wave energy to electricity conversion [8-14]

An essential element in the makeup of wave energy converters (WECs), influencing the technical and economic effectiveness of WECs, is the power take-off (PTO) mechanism. This subsystem within WECs transforms the hydrodynamic forces acting on the WEC into valuable mechanical and usually electrical power. Diverse varieties of PTO systems are in existence, and the selection of a specific PTO for a given wave energy converter is frequently closely associated with the converter's type. The various primary routes for converting wave energy into electricity are depicted in Figure 1 (b). The classifications of PTO systems fall into five primary groups, each of which can be studied with available published materials Ref. [10-15]. And to perceive the aim and objectives of this present study detailed descriptions of PTO methods are omitted.

The fundamental principle of OWC wave energy converters revolves around the utilization of the periodic motion of ocean waves. The OWC device consists of a collector chamber, designed to extract power from the incoming waves. As waves impinge upon the front wall of the chamber, the air trapped inside undergoes oscillatory motion, resulting in pressure variations [15]. The simple form (Figure 2, provide a visual representation of how this technology harnesses wave energy to generate electricity) of OWC consists of a chamber with an "air turbine" (mainly the *wells turbine*) positioned on top of the structure. The lower part of the chamber is open to the sea, allowing waves to enter. When waves interact with the chamber, they cause the water inside to oscillate. This oscillation, in turn, pushes and pulls air through the connected air turbine and into the atmosphere. Remarkably, the turbine rotates in a consistent direction regardless of the airflow's direction. The OWC operates in two distinct steps: first, it transforms hydrodynamic energy into pneumatic energy within the chamber, and then it converts this pneumatic energy into electricity using the turbine. The schematic operation of an OWC is illustrated in Figure 2.

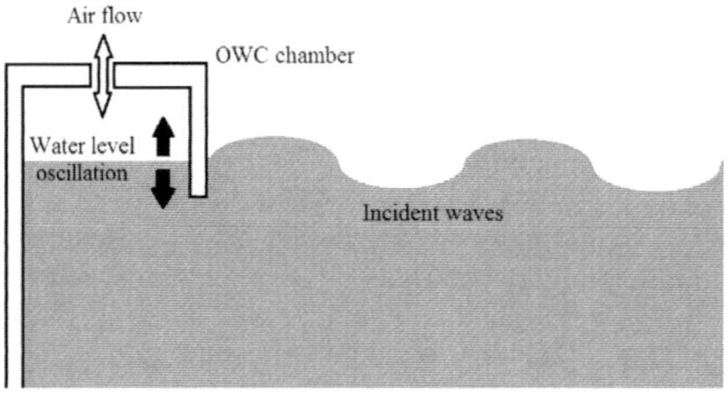

Figure 2. Fixed type OWC operation [16]

The schematic operation of an OWC is illustrated in Figure 1, providing a visual representation of how this technology harnesses wave energy to generate electricity. The pneumatic power is then converted into useful form of work through a Power Take Off (PTO) system [17]. The power train of an OWC system, which involves the conversion of wave energy into electrical energy. This process consists of several stages, first, the wave energy is transferred to the oscillating air column within the chamber and next, the energy embedded in the oscillating motion is transferred to the mechanical rotation of a turbine shaft and finally, the turbine shaft drives a generator that converts the rotation into electrical energy. The overall efficiency of an OWC system depends on the efficiencies of both the oscillating air column and the turbine itself [18-19]. Several researchers have extensively studied the impact of various bi-directional turbines on the overall operation and efficiency of OWC systems. These turbines were designed with symmetrical placement of blades, resulting in Uni-directional turbine shaft rotation. Factors such as blade arrangement, guide vanes, and non-return valves have also been subjects of research to optimize turbine efficiency [19]. Additionally, studies have explored energy dissipation and power output of turbines under different air flow conditions with a fixed chamber geometry [20].

1.1 Geometric optimization and output efficiency of OWC chamber

The geometric optimization of an OWC chamber plays a crucial role in maximizing wave energy capture and enhancing the efficiency of the system. This process involves designing the chamber shape, size, and turbine placement to optimize the oscillation of the air column and increase the overall power generation. One of the key parameters in the geometric optimization is the chamber length (l). The length of the chamber affects the wave propagation inside it and the resulting oscillation of the water column. Longer chambers allow for more wave interaction, increasing the energy capture potential.

However, there is an optimal length that balances wave energy extraction and excessive energy loss due to wave reflection. Determining this optimal chamber length involves considering factors such as wave characteristics, site-specific conditions, and the desired power output [8, 15]. The chamber width (b) is another important parameter in geometric optimization. Wider chambers provide a larger surface area for wave interaction, allowing for more efficient energy capture. However, excessively wide chambers can lead to increased wave reflection, resulting in energy losses. Thus, optimizing the chamber width involves finding a balance between maximizing energy capture and minimizing wave reflection [10, 17]. The height of the chamber (h) is also a crucial factor to consider. The height determines the volume of the air column and influences the magnitude of air compression and expansion during wave oscillation. A larger air column volume can result in higher air pressures and increased turbine performance. However, it is essential to ensure that the chamber height is within practical limits and feasible for construction and maintenance purposes [16-20].

The position of the turbine inside the chamber is another aspect of geometric optimization. Placing the turbine at the optimal location within the chamber is vital for efficient energy conversion. The turbine should be positioned to capture the maximum airflow generated by the OWC. Computational fluid dynamics (CFD) simulations and experimental studies are often employed to determine the optimal turbine placement. But in this study the CFD method is involved for the frontal wall shape optimization [7, 15-17]. In addition to the shape and size parameters, the overall chamber geometry needs to be optimized to enhance wave focusing and energy capture. Various chamber shapes, such as converging or concave designs and front wall optimization of the chamber have been studied to exploit the wave energy focusing effect. These shapes can help concentrate the wave energy towards the air column, increasing the power output. The optimization process considers factors like wave diffraction, wave reflection, and the desired energy conversion efficiency to determine the most suitable chamber shape [19-21].

1.2 Front wall angle variation and output efficiency of OWC chamber

The output efficiency of an OWC chamber is a vital performance parameter directly impacting the overall energy conversion efficiency [18]. It represents the ratio of electrical power output to the wave energy input. Several factors influence this efficiency:

- *Turbine Performance*: The efficiency of the turbine in converting air column kinetic energy into electrical power significantly affects output efficiency. Choosing an efficient turbine type and optimizing its design and operation can enhance the system's efficiency [19].
- *Geometric Design*: Optimizing the OWC chamber's dimensions, like length, width, and height, is crucial. Balancing wave energy capture and minimizing losses due to reflections and dissipation within the chamber is key. Geometric optimization improves energy extraction from incident waves [20].
- *Wave Characteristics*: Local wave conditions, including height, period, and direction, influence the energy content and the ability to induce efficient air column oscillation. Understanding and analyzing the wave climate through measurements and modeling is essential for optimizing output efficiency [21].
- *Generator Efficiency*: The electrical generator's efficiency in converting mechanical power from the turbine affects overall efficiency. Higher generator efficiency reduces energy loss during conversion [22].

Enhancing output efficiency requires a comprehensive approach, considering turbine, chamber design, wave characteristics, and generator efficiency. Optimizing these factors individually and ensuring their compatibility can lead to significant improvements. Real-world challenges such as varying wave conditions, operational limitations, and maintenance requirements must be considered.

Ongoing research in turbine design, chamber optimization, and system integration advances the state-of-the-art, improving practical OWC applications. Improving OWC system output efficiency is crucial for making wave energy economically viable and competitive as a renewable energy source. Maximizing energy conversion efficiency allows OWC technologies to better contribute to the global energy transition and harness the ocean's immense power potential [20-23]. Several researchers (Table 1) have conducted numerical and experimental investigations on the geometric optimization of OWC chambers.

Table 1. Key findings on chamber geometry optimization in OWC systems

Key finding on OWC chamber performance	Study (Ref.)
Numerical simulation of an oscillating water column device and investigating the effects of lip submergence on velocity and pressure.	[34]
Larger chamber diameter significantly enhances turbine performance in OWC systems.	[35]
Longer chamber length and higher turbine blade pitch angle lead to higher output efficiency in OWC systems.	[36]
Extending chamber length and reducing inlet height positively affect OWC system efficiency.	[37]
Genetic algorithm optimization of chamber geometry improves OWC system power output by 18%.	[38]
Increasing chamber diameter results in higher power output and efficiency in OWC systems.	[39]
Enlarging chamber diameter and adjusting draft tube diameter enhance power output and efficiency.	[40]
Multi-objective genetic algorithm optimizes chamber geometry based on wave conditions and desired power output.	[41]
Vortex shedding at the sharp edge of the OWC chamber enhances the spatial non-uniformity inside the OWC chamber through a resonant sloshing mechanism	[42]
Ideal gas adiabatic polytropic process and the real gas model were observed, a new value for the polytropic exponent was proposed, representing a non–adiabatic real gas behavior for the air–water vapor mixture of OWC chamber.	[43]
Rectangular chambers and Dual-Opening chamber with large aspect ratios yield higher power output in OWC systems.	[44]
Chamber height influences power output, and thick front wall and a large incidence angle of the wave can significantly narrow the hydrodynamic efficiency band and modify the resonant frequency in higher power generation.	[45]
A parabolic chamber shape with a specific height-to-length ratio maximizes power output, propose OWC numerical analysis for modified upper OWC chamber shape and four vertical plates inside the chamber and signify the improved peak efficiency.	[46]

While numerous OWC fixed type models and prototypes for harnessing ocean wave energy have been developed over many decades, achieving consensus on an economically competitive and reliable design remains a challenge [21-22]. The best method many researchers have focused with development is using Computational Fluid Dynamics (CFD) technology for optimization of OWC systems. Not only the CFD based Finite Volume Method (FVM) approaches developed, but also, studies concentration OWC optimization and design, several studies gained efforts related to the hydrodynamic properties and performance evaluation of OWC wave power devices, particularly within the context of numerical simulations and experiments [22-23]. Some key methods and approaches include: -

- *Linear Theory Analysis* by Hong et al. [6]: They presented numerical estimations of hydrodynamic properties for a floating OWC device using linear theory.
- *Optimal Moonpool Geometry* by Sphaier et al. [7]: Their experimental study aimed to find the optimal entrance geometry of a moonpool to minimize vertical water motion during waves, revealing resonant periods and their impact.
- *BEM Code Predictions* by Sykes et al. [8]: They employed a Boundary Element Method (BEM) code to predict the displacement and hydrodynamic properties of a floating undamped OWC, comparing results with experiments.
- *Time-Domain Numerical Method* by Toyota et al. [9]: They developed a time-domain numerical method to evaluate the performance of floating OWC-type wave energy converters, considering various factors simultaneously, and found good agreement with experimental results.
- *Hydrodynamic Studies and Damping Coefficients* by Sheng et al. [10,11]: Their studies involved experimental and numerical investigations of moored floating OWCs, including tuning damping coefficients and assessing hydrodynamic performance.
- *Numerical Wave Tanks* by Repalle et al. [12] and Horko [13]: They used commercial software to simulate wave run-up and OWC system behavior, respectively, in lieu of physical wave tank experiments.
- *CFD Simulations* by various authors: Multiple studies (Marjani et al. [14], Liu et al. [15], Finnegan and Goggins [16], Teixeira [1], Falcão [17]) utilized CFD-based numerical simulations to investigate wave elevation, pressure variation, air flow, and turbine optimization for floating OWCs.
- *Simplified Time-Domain Model* by Iturrioz [18]: They presented a time-domain model for a fixed detached OWC device, aiming to develop a foundation for modeling more complex floating multi-chamber OWCs.

In this research, CFD Fixed type OWC case study employs a 2D fully nonlinear Numerical Wave Tank (NWT) based on first principles and adopting several techniques in Table 01 and hydrodynamic theories developed for OWC structure analysis. This 2-D NWT is used to analyze 2nd order, linear wave system was implemented in OWCs. Ansys ® FLUENT v19.1 software employed to solve Navier-Stokes equations and equations of motion in a coupled manner, enabling the simulation of chamber motion in response to waves. This advanced tool serves as a digital laboratory for studying the intricate interactions between waves and chamber walls of OWCs. The use of dynamic mesh simulations enhances the accuracy of the analysis, offering valuable insights into the complex dynamics at play in these energy conversion systems. This paper presents a comprehensive investigation into the performance variation of an OWC wave energy converter by optimizing the front wall configuration and providing insights into the power development potential of the OWC structure.

2. Methodology

2.1 Case Study Theories: Fixed type OWC, NWT and Ocean Wave Mechanics

2.1.1 Formulating wave parameters and chamber design. The OWC chamber designed adopting case study by Falcão [24] the dimensions ratios and OWC structure is then matched for the present case study. As shown in the Figure 3, the OWC hypothetical case considered. A fixed-structure OWC considered where, the water depth is h. In calm water, the submerged potion taken as b. The thickness of the walls is assumed negligible.

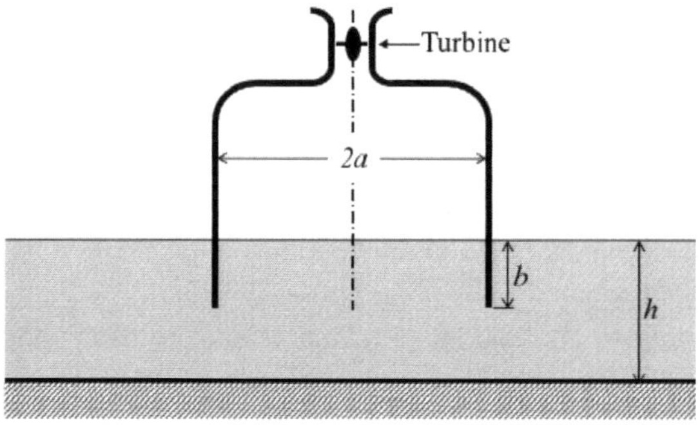

Figure 3. Fixed-OWC hypothetical structure [24]

The dimensionless hydrodynamic coefficients G* (radiation conductance) and H* (radiation susceptance) are defined as,

$$\{G^*, H^*\} = \frac{\rho_w g}{\omega \pi a^2}\{G, H\} \tag{1}$$

Where ρ_w is the water density, g is the acceleration of gravity and $\boldsymbol{\omega}$ is the wave radian frequency. For an axisymmetric OWC, the excitation flow rate coefficient $\boldsymbol{\tau}$ is related to G by (Ref. [25-26])

$$\tau = \left[\frac{2\rho_w g^2 D(kh) G}{\omega k}\right]^{1/2} \tag{2}$$

where k is the wavenumber and

$$D(kh) = \left[1 + \frac{2kh}{\sinh 2kh}\right]\tanh kh \tag{3}$$

which may be written in dimensionless form as

$$\tau^* = \left[\frac{2\pi D(kh) G^*}{a^* kh}\right] \tag{4}$$

where $\tau^* = g^{-1/2}a^{-3/2}\tau$ and $a^* = a/h$. In the equations above, kh need to be expressed as $kh = \omega^* c^* \sqrt{a^*}$ where $\omega^* = \omega\sqrt{a/g}$ is the dimensionless wave frequency, and dimensionless wave celerity $c^* = c/\sqrt{gh}$ is the solution of the dispersion relationship in finding the wave speed matching with the case study.

$$c^* = \frac{\sqrt{a^*}}{\omega^*} tanh \frac{\omega^*}{c^* a^*} \tag{5}$$

To use the wave celerity value from Eq. 05, specifically for the simulation purposes, we use the Stokes 2nd order wave theory. Where upstream wave amplitude and wavelength corresponding to the second-order Strokes theory, we therefore impose the velocity components provided as boundary conditions on the numerical wave wavemaker. The wave characteristics are chosen, and it is well represented by Stokes second order solution. Difference regions of validity for various theories are described by Mehaute et al., [25]. The wave considered in this study has a constant period T, wavelength λ and incoming wave height H (in Figure 4).

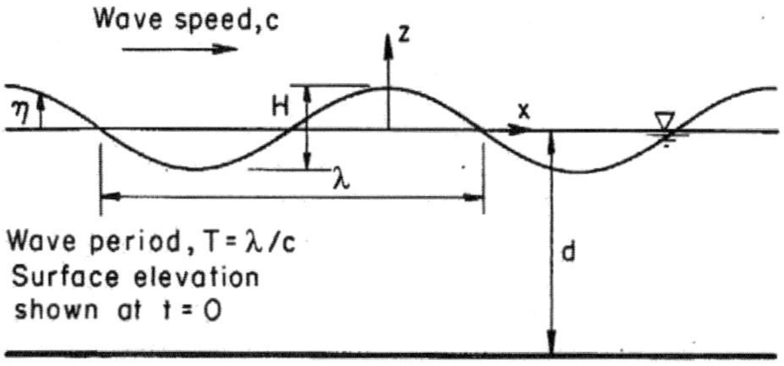

Figure 4. Defined properties of surface wave model (adopted from [27])

According to the second order theory, the equation of the free surface elevation by Dean et al., [25,26] is given as

$$\eta = \frac{H}{2}cos(kx - \omega t) + \frac{\pi H^2}{8\lambda}\frac{cosh(kh)}{sinh^3(kh)}[2 + cosh(2kh)\}cos\,(2kx - 2\omega t) \tag{6}$$

Where t is the time, h the water depth, k the wave number and ω is the angular frequency same as used before. The following relations give the horizontal component u and the vertical component w of the local fluid velocity:

$$\mathrm{u} = \frac{Hgk}{2\omega}\frac{cosh\,k(z+h)}{cosh(kh)}cos(kx - wt) + \frac{3H^2\omega k}{16}\frac{cosh\,2k(h+z)}{sinh^4(kh)}cos\,2(kx - \omega t) \tag{7}$$

$$w = \frac{Hgk}{2\omega}\frac{sinh\,k(z+h)}{cosh(kh)}sin(kx - wt) + \frac{3H^2\omega k}{16}\frac{sinh\,2k(h+z)}{sinh^4(kh)}cos\,2(kx - \omega t) \tag{8}$$

The dispersion relation is given by:

$$\lambda = \frac{gT^2}{2\pi}tanh\,(\frac{2\pi h}{\lambda}) \tag{9}$$

The corresponding wave incident power [27] per unit width is:

$$P_{in} = \frac{\rho g\,\lambda}{16T}H^2\left[1 + \frac{2kh}{sinh(2kh)}\right][1 + \frac{9}{64}\frac{H^2}{k^4h^6}] \tag{10}$$

Assuming that the water free surface in OWC behaves as a flat plate, the hydrodynamic power transferred to the air column inside the chamber can be determined as follows:

$$P_{hyd} = p_a b\frac{dZ_p}{dt} \tag{11}$$

Where Pa is the air column pressure, b is the size of the chamber and Zp is the free surface elevation inside the chamber, these parameters were defined for the ease of creating the geometry in the present work based on the Ref. [24] hypothetical case study. The OWC efficiency ε computed by the ratio of the hydrodynamic power to the incident power:

$$\varepsilon = \frac{P_{hyd}}{P_{in}} \tag{12}$$

In this case we focused on the OWC front wall chamber, in particular its immersion depth and orientation versus the flow direction. However, nine cases are to be distinguish as shown in figure. All cases are studied for a same progressive monochromatic wave that has constant height and wavelength. But for the simplicity of the waves, we use some assumptions. In this paper we used the C computer program source file to generate the required wave for the simulation. The code will be as; "X = 0.5 × 1.57 × cos (1.57 × time)" represents a cosine wave with an angular frequency of 1.57 and an amplitude of 0.5. The angular frequency determines how quickly the wave oscillates, while the amplitude determines its maximum displacement. The wavelength (λ) of a wave is the distance between two corresponding points on the wave, such as two consecutive peaks or troughs. In this case, the wavelength can be calculated using the formula: where ω represents the angular frequency.

$$\lambda = \frac{2\pi}{\omega} \tag{13}$$

In the code, angular frequency of 1.57, the calculated wavelength is approximately 4.012 units. The amplitude (A) of a wave is the maximum displacement of any point on the wave from its equilibrium position. In the code snippet, the coefficient 0.5 in front of the cosine function represents the amplitude. Therefore, the amplitude of the wave is 0.5.

Wave generation in NWT summary

- Wavelength (λ) ≈ 4.012 units: The distance between two corresponding points on the wave.
- Amplitude (A) = 0.5: The maximum displacement of any point on the wave from its equilibrium position.
- Dynamic pressure (Pd) is a critical parameter for calculating power in fluid systems. Representing the kinetic energy per unit volume of fluid and can be determined using the equation:

$$Pd = \frac{1}{2}\rho v^2 \tag{14}$$

where ρ denotes the density of the fluid and v represents the velocity of the fluid at a specific point. This formula establishes the foundation for estimating the dynamic pressure.

2.1.2 Flow rate measurement through OWC. Accurate measurement or calculation of the flow rate is essential for determining power. Flow rate quantifies the volume of fluid passing through a given point per unit time. This thesis discusses various methods of measuring flow rate, including the use of flow meters and calculations based on flow dimensions and velocity. The units of flow rate are typically in cubic meters per second (m3/s) or relevant units.

2.1.3 Power generation potential with OWC. Once the dynamic pressure (Pd) and flow rate (Q) are determined, the power (P) can be calculated using the equation.

$$P = Pd * Q \tag{15}$$

The resulting power value is expressed in units of energy per unit time, in W/m . It is important to ensure consistency in the units used for all variables in the calculation to obtain accurate results. The flow rate (Q) can be taken using following equation:

$$\mathrm{Q} = A * v \tag{16}$$

Where A represents the area and v represents the velocity of the fluid.

2.1.4 Considerations and limitations. While the presented methodology provides a useful estimation of power, it is crucial to acknowledge certain considerations and limitations. Factors such as fluid properties, flow characteristics, and system efficiency can influence the actual power output. This paper highlights the significance of accounting for these factors and provides guidance on addressing them effectively on CFD simulation basis. The theories developed according to suit frontal wall angle alteration (from 0° to 90°), adopted by present analysis. The OWC chamber performance and chamber behaviour by Vyzikas et al. [28], where validation of commercial CFD NWT model for OWCs against experimental results produced presented and adopted. Tests include irregular and regular wave conditions for performance, fixed-OWC chamber response. The OWCs testing method without PTO highlighted the potential uses where boundary condition of absorbing sea walls. Chen. et al. [29] presented correlation study of optimal chamber width with the relative front wall draught of onshore OWC structure identified and, in this study, two-phase wave flow model (VOF-Volume of Fluid) presented for NWT specially to be utilized in OWCs. The model helps to optimize the geometrical parameters of fixed OWC. One of the key features adopted from this study was the development of a fitting formula that describes the relationship between the optimal chamber width and the front wall draught. This formula provides a convenient and practical way to understand how the width of the OWC chamber should be adjusted in relation to the front wall draught and angle to optimize the chamber performance. Using this fitting formula significantly simplified the preliminary design process of onshore fixed OWC. This simplification saved time and resources in the design phase, making it more efficient and accessible [29].

2.2 OWC model and CFD setup (NWT-model)

The real-world wave tank models are expensive to operate and only limited access and locations available for testing with these facilities. Real world application setup combined with computationally modelled with the Numerical Wave Tank (NWT) concepts and several models were considered and referred [24-32]. The NWT was introduced with regards to the 0th angle of the wall initially, while for all other scenarios, the methodology remains the same as the 0th angle. In the initial phase, CAD modeling was performed using SolidWorks Software ®. In Figure 5, the specified dimensions of the NWT presented. The 2-D geometry was kept simple to remove uncertainties and solver mesh convergence errors with lowered mesh angle errors such as skewness angle error during solving. To facilitate the simulation process, a 2-D model was imported to ANSYS Fluent ® v19.1 and converted to ANSYS environment using Space Claim software, and from 0deg angle to 90deg angle with 10 deg, variants modeled using Space Claim package. This approach was adopted to simplify the analysis and streamline the calculations. The front wall of the chamber facing the wave geometric angle from O deg to 80 deg. Was changed and analyzed during numerical simulations to assess the pneumatic power potential variation inside the air column of OWC, illustrated in Figure 5.

Once the 2-D modelling for all 10 set angles created the NWT and OWC sections were meshed for CFD simulations. The meshing process involved generating a mesh on 2-D surface area. Mesh is mixed type tetrahedral wrapping with hexahedral dominant core. Due to highly turbulent behaviour of "Air" the hexa-core mesh helps to capture numeric properties accurately. The general sizing of the mesh set to maximum of 0.2 m, and minimum 0.005 m. These sizes can be changes with software control. However, for the near area of the column, refined the mesh with manual resized mesh 0.01m of the size of element. This finer mesh around the column allows for more accurate representation of the flow initiation and propagation behaviour in that region. And shown in Figure 6 (c) the mesh sizes made finer with sizing 0.005 m making coarse mesh. The total mesh count is 96,574 elements taken as approximately 0.1 million, which is adequate for this simulation task. Increasing the mesh would proportionately increase the computation power, capacity, solving time. To capture the wave propagation and air chamber fluid variations accurately. The dynamic mesh motion method was adopted in presenting surface waves. Figure 6 (b) presents the how the dynamic mesh in CFD simulation process contracts and retract accordingly.

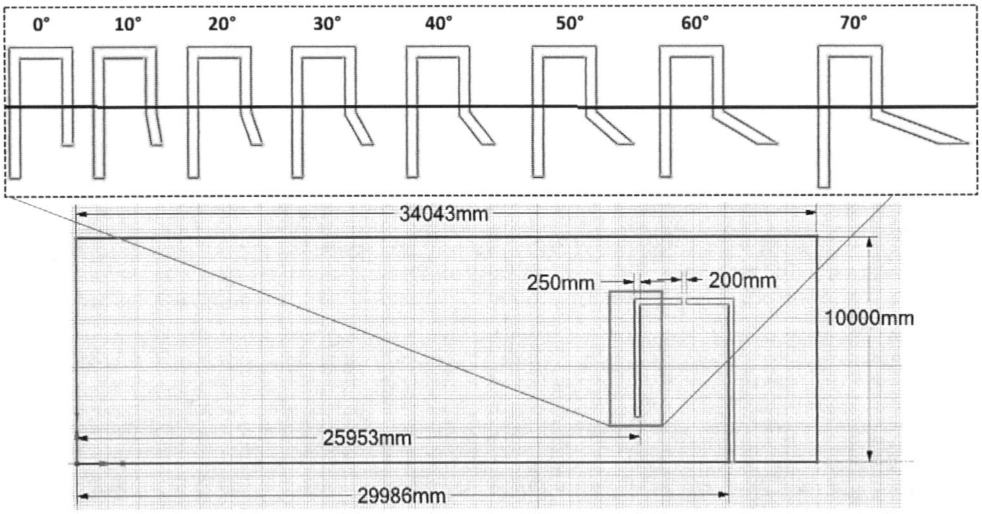

Figure 5. Geometry and dimensions of the NWT and studied OWC.

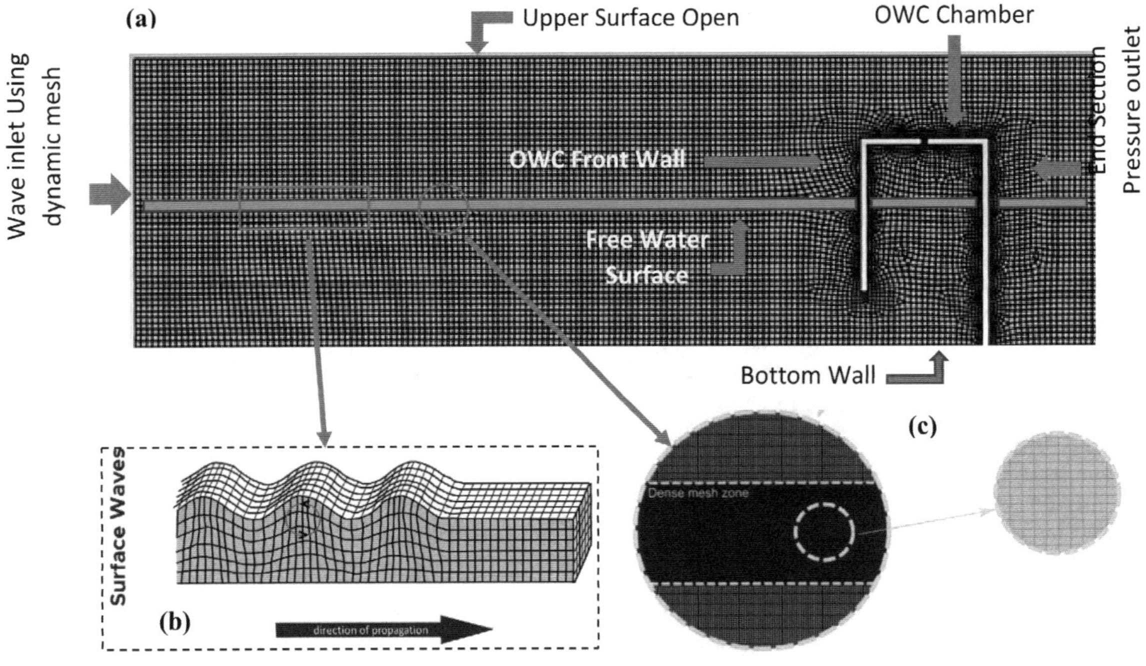

Figure 6. Computational NWT domain and the OWC chamber (a) Named sections and domain with NWT (b) Surface wave propagation model with mesh motion, and (c) The Air/Water surface denser mesh to capture accurate fluid properties.

After the mesh model created and refined with reasonable mesh quality, simulation setup using Ansys Fluent was done. The physics setup adopted the Multiphase VOF method. This model enables the simulation of flows that involve the presence of multiple fluid phases simultaneously, in this case Air being primary fluid and water secondary fluid with Volume fractions for each phase set as 0.5. The flow modelled using RANS (Reynolds-Average Navier Stokes) equations coupled modelling. Turbulent model set as k-Omega SST model, suitable for capturing complex flow phenomena. It combines the advantages of the k-Ω (k-Omega) and k-Ɛ (k-epsilon) models to accurately predict turbulence effects in various flow conditions by switching and mixing each model near the boundaries and away from the boundaries of flow attachment. By solving equations for turbulent kinetic energy (k) and specific dissipation rate (Ω-omega), the model determines turbulence intensity (in this case 5% and 10% for wave maker and air sections) and length scales.

Ensuring accurate predictions near walls and in free shear regions. Fluid properties defined for both air and the water-liquid mixture to ensure accurate simulation of their behaviour within the system. After that, boundary conditions set as shown in Fig. o5 (A). NWT top surface was set as a pressure outlet boundary condition, whereby the solver calculates the pressure at this boundary based on flow behaviour. The wave maker uses Dynamic mesh boundary condition with C program code input for wave maker action. The outlet or end section chosen as pressure outlet wall. With viscous wave damping constant ($\beta \sim 0.087$) added. The complete numerical simulation boundary conditions, settings, mesh statistics, solver controls, convergence targets, residuals and time step settings provided with Table 2. It is important to note the Dynamic Mesh control. In a computer code developed using C language source file integrated into ANSYS Fluent, providing the necessary code to generate sinusoidal motion for the moving wall. This motion induced a wave in the liquid phase of the simulation. The source file determined motion attributes, such as amplitude and frequency.

C Code for dynamic mesh motion at the wave maker boundary to create the desired waves as follows:

```
#include "udf.h"
DEFINE_CG_MOTION (movingwall,dt,cg_vel,cg_omega,time,dtime)
{
real kecepatan_X;
real kecepatan_Y;

kecepatan_X = 0.5*1.57*cos
    (1.57*time);
kecepatan_Y = 0.0;

cg_vel[0] = kecepatan_X;
cg_vel[1] = kecepatan_Y;
cg_vel[2] = 0.0;

cg_omega[0] = 0.0;
cg_omega[1] = 0.0;
cg_omega[2] = 0.0;
```

Dynamic mesh layering was employed to allow the mesh to deform and adapt to the changing geometry during the simulation, ensuring accurate representation of fluid-structure interaction. The meshing option "cell height" was set to 0.2 m, determining cell size and aiding in controlling simulation accuracy and detail. Once the CFD setup finalized the "Initialization" carried out. Hybrid initialization used, combining the benefits of pressure-based and density-based initialization methods to establish a robust and accurate starting point for the simulation. In Figure 7(a), the final computational setup fluid domains with boundary conditions illustrated for NWT. A VOF patch operation (in Figure 7 (b)) performed during initialization to make the Ansys FLUENT solver identify Air and Water separation mesh layer easily and to aid numerical calculations making them push faster for convergence. The Mesh region dominated by hexahedral-core area illustrated in Figure 7 (c). Also, the arrows point in boundary outlet mentioning pressure outlet with relative pressure set 0 Pa.

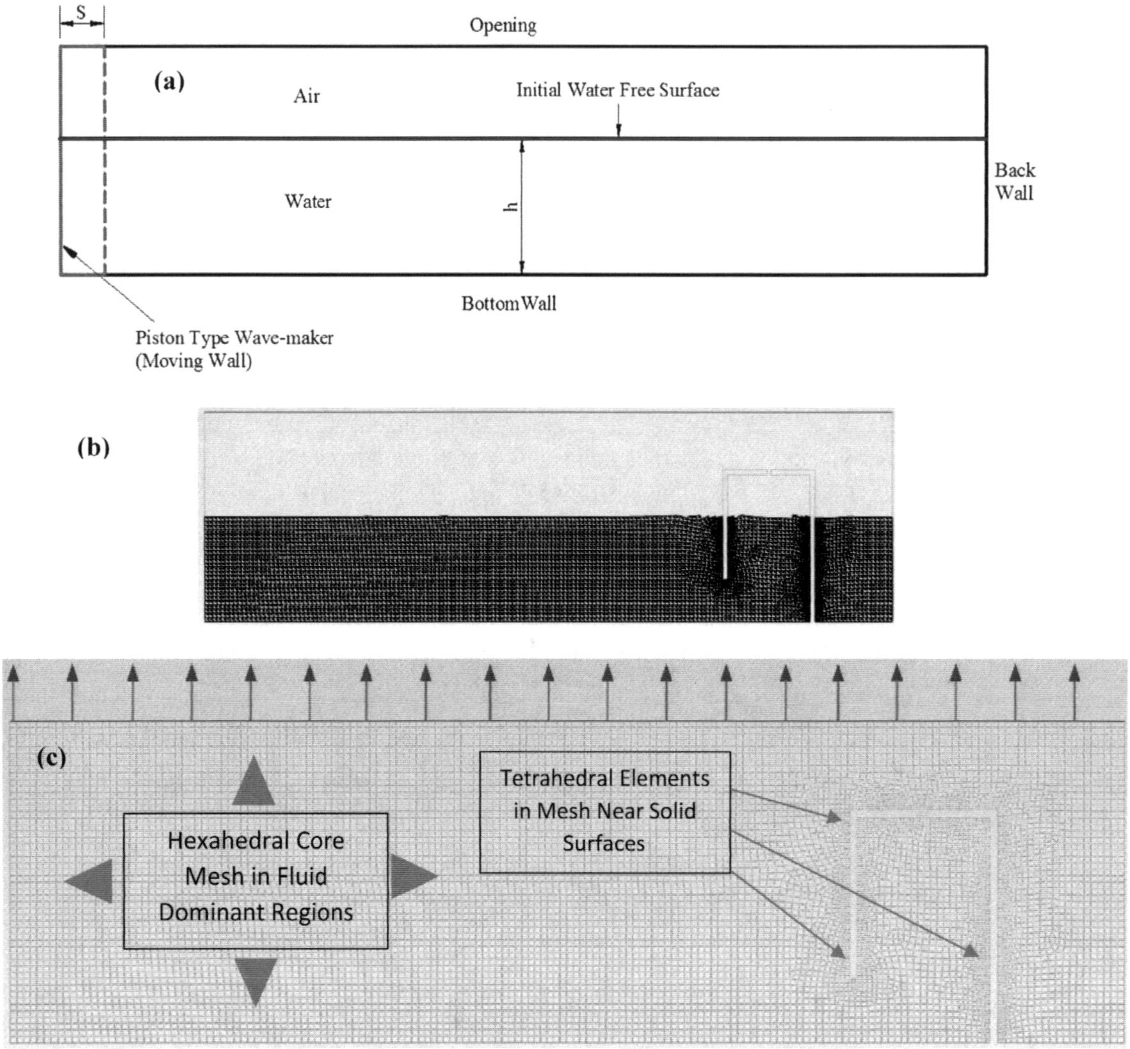

Figure 7 (a) The initialization VOF patch operation (b) Upper surface boundary condition: Pressure outlet setting and (c) the hexahedral-tetrahedral mixed but hexa-core dominant mesh model.

Table 2. Numerical simulation parameters in detail

(a) Solver	
Type	Pressure based
Velocity formulation	Absolute
Time	Transient
Simulation time	40s
Time step Size	0.05
Number of time steps	800
Iterrative loops	10
Maximum iterations	20
Convergane Target Residuals	0.0001
Simulations	1000
Iitialization	Yes

2D Space	Planar
Moving Wall	Rigid mody motion / C-Code Input
Fluid Strucuture Intercation (FSI)	ON
Gravity	Gravitational acceleration ON
Multiphase	Mixture model VOF
Fluids	Air and Water
(b) VOF Model	
VOF -Water	0.5
VOF -Air	0.5
Mixture Parameters	Slip velocity
Volume fraction parameters	Formulation - implicit
Interface modeling - type	Sharp/dispersed
Body force formulation	Implicit Body Force
Air-primary phase	Phase material - air
Water - secondary phase	Phase material - water
Phase Interaction (Air Water)	
Drag coefficient	Schiller-naumann
Slip velocity(m/s^2)	Mannninen-et-al
Surface Tention Coefficient(N/m)	Constant - 0.072
Global Options	
Surface tention force modeling	Model-continuum surface force
Viscous model	
Turbulance model	k-Omega - SST
(c) Boundary Conditions	
Top	Pressure outlet
Bottom	No-slip wall
Moving Wall	Wave maker input
Sides of NWT	Symmetry
Column	Wave maker column
Mesh type	2-D hexahedral domoinant core with Tetrahedral
Mesh Size	0.1 million (approximatly)
Type	Automatic mesh
Mesh Methods	Layering
Dynamic Mesh Zones	Moving wall - rigid body

VOF method, after initialization, produces the boundary between the air and the water surface (Air in blue and water in red). With simulation initialization 100% completed the volume fraction visually presented with Figure 8 (a). The whole idea of the NWT concept illustrated in Figure 8 (b) with the linear surface wave generated by the CFD code.

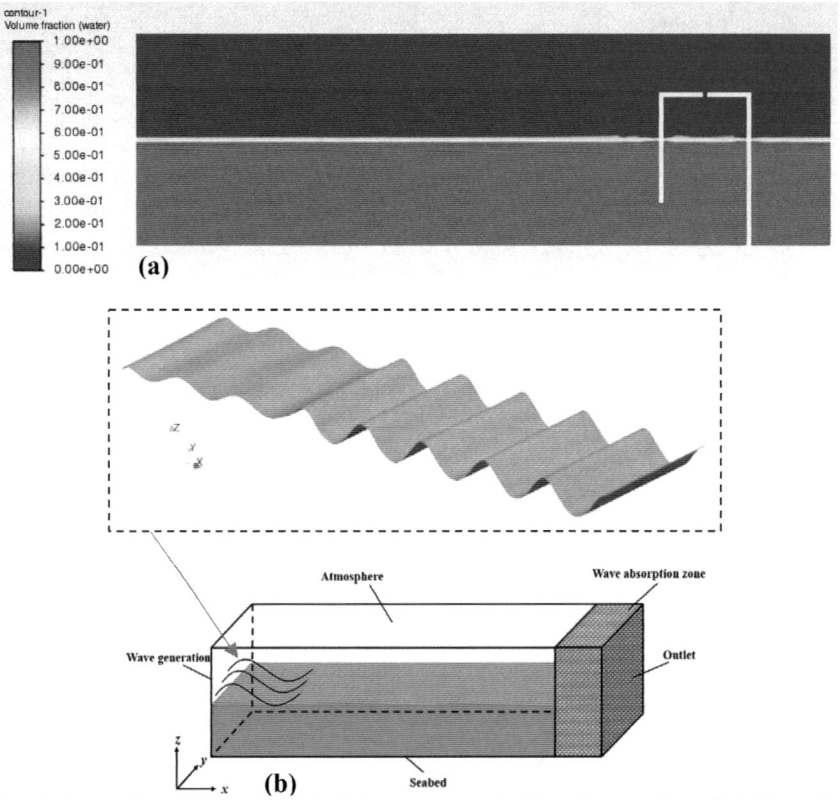

Figure 8. The Volume Fraction method defining primary fluid and secondary fluid based on Volume fraction (VF 0.5 each phase), (b) NWT operational concept with simulations.

Monitoring points implemented at a specific outlet location within the OWC chamber to track velocity and pressure values after solver finishes calculations. As shown in Figure.09 providing insights into flow behaviour at that specific point. The simulation run for 800-time steps, with each time step set at a size of 0.05. The time step size determined the temporal resolution of the simulation, influencing result accuracy and stability.

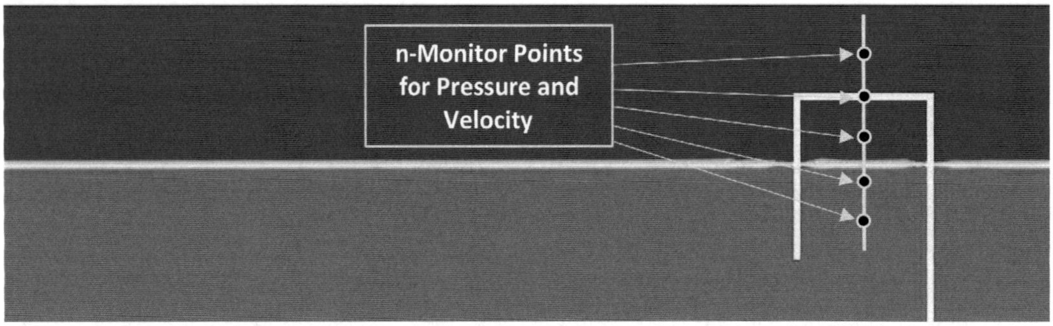

Figure 9. Monitor points (n-number) for Pressure and Velocity of OWC chamber

3. Results and discussion

3.1 Grid Independence and validation of the NWT numerical model

To validate the produced NWT model, present study was wave heights generated H (mm) was compared with the NWT model by AHS Weerakoon et al. [33] study experimental NWT model. Referring to

IOP Conf. Series: Materials Science and Engineering 1294 (2023) 012015 doi:10.1088/1757-899X/1294/1/012015

Figure 10 the comparison is illustrated. To assess the mesh's insensitivity within this study, the wave height profile was acquired to attain a solution that is independent of the mesh. To achieve this objective, as previously illustrated in Figure 05 and Figure 06, a fine grid is employed in the region encompassing both phases (air and water). To maintain mesh quality and ensure the grids' maximum aspect ratio remains below 20, it is necessary to reduce cell heights at the common interface between the two fluids. Various grid sizes are employed to verify the mesh independence of the numerical simulation, as demonstrated in Figure 10. In Figure 10, 'Y' represents cell heights in meters at the interface between the two phases, and experimental findings are included to validate the numerical outcomes.

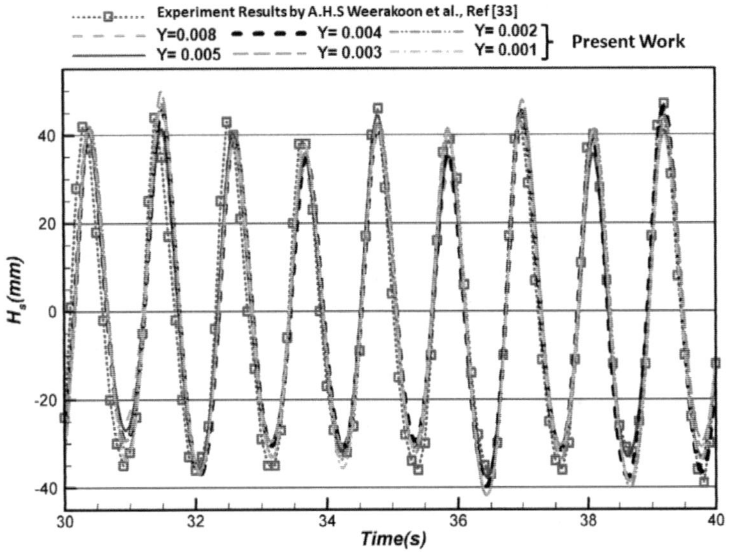

Figure 10. Validation and mesh independence of the simulation.

As indicated in Figure 10, results from cell heights below 5 mm are equivalent to Y=0.005 m. Consequently, reducing cell sizes below 5 mm does not alter the results but escalates computational expenses by increasing the number of grids. Figure. 11 provides the count of grids generated for different cell heights, and for the study's simulation, a grid count of 96,574 at Y=0.005 m is selected.

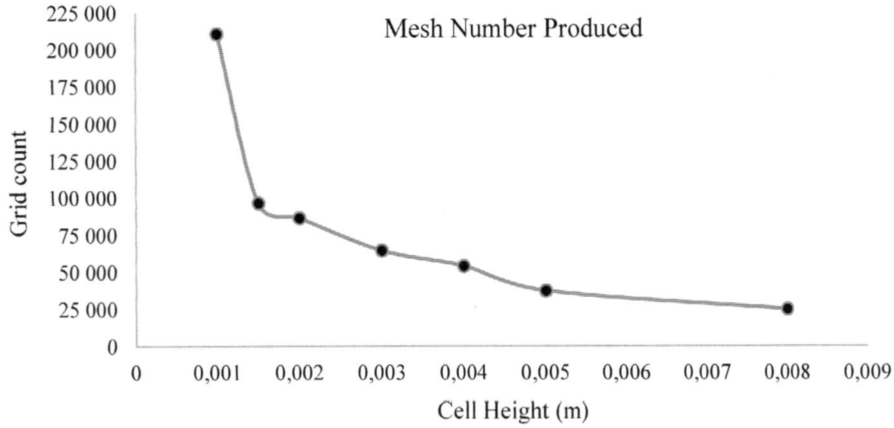

Figure 11. The number of grids produced for different cell heights.

3.2 Numerical Results with Changing the Front Wall Angle

At the end of series of simulations at each angle of front wall, pressure and velocity variations were identified. The chamber front wall being geometrically altered with 10° angle gaps starting from reference case of 0 and reaching 80. At each of these cases, average values of air velocity, ai pressure and chamber air power potential per wave meter front evaluated using CFD simulations. The wall angles from 0 to 30 divided as 1st half and 40 to 80 angle wall changes taken as 2nd half of analysis. On the 1st half (50%) of angles which provide the considerable pneumatic power potential inside the air chamber analyzed initially. At the reference case, 0° angle (vertical wall, 90° to SWL) the velocity initially experiencing a minimal increase due to slower air response inside the OWC chamber, eventually reaching 7 to 10 m/s with a sinusoidal waveform indicating successful wave energy capture. Pressure at the outlet starts low but steadily rises as wave energy accumulates, consistently staying positive (40 to 60 kPa) and peaking at 61.8 kPa. The power output per unit area initially starts low and follows a sinusoidal pattern like velocity, with a maximum of 605.08 W/m and an average of 123.72 W/m of power.

With 10° angle, velocity oscillating between positive and negative values, eventually forming a sinusoidal pattern, indicating successful wave energy capture with a maximum velocity of 8.15 m/s and an average velocity of 4.25 m/s. Examining the pressure graph (In Figure. 12. a) at the chamber outlet reveals consistently positive values, reaching a maximum of 61.8 kPa, attributed to the accumulation of wave energy within the chamber. The power output, peaking at 537.11 W/m, highlighting the impact of velocity fluctuations on power output with an average of 106.068 W/m. At 20° angle, the outlet velocity ranged from 4 to 8 m/s, reaching a maximum velocity of 7.13 m/s and an average velocity of 3.71 m/s. The pressure hits consistently positive values, peaking at 44.46 kPa, attributed to wave energy accumulation, with an average pressure of 12.43 kPa. Power output peaked 368.69W/m, concurrent with velocity fluctuations, and averaging 71.10 W/m. More opened wall at 30° angle, outlet velocity ranged between 3.71 m/s, and 6.22 m/s giving average velocity of 2.81 m/s. The pressure, peaking at 61.8 kPa, attributed to wave energy accumulation, with an average pressure of 7.25 pascals. Power output per unit area initially started low but rose over time, reaching a peak of 177.64W/m, correlated with velocity fluctuations, and averaging 32.21 W/m. The values of pressure and velocity sinusoidal variation for the 40s simulation time is illustrated in Figure 12 (a), and the chamber pneumatic power variation for 0 to 30 wall angle cases are shown in Figure 12 (b). Clear comparison is visible for each case.

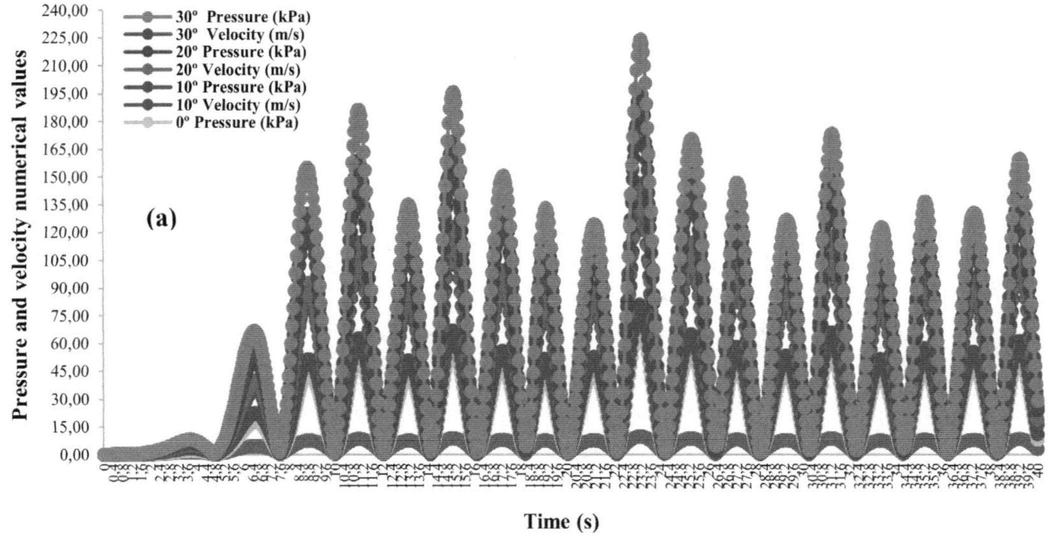

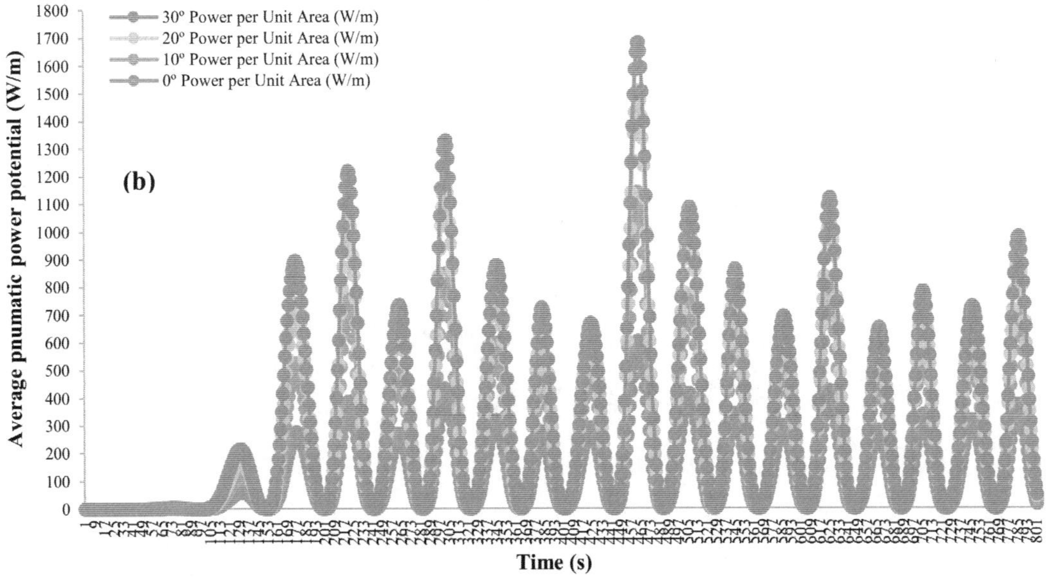

(b)

Figure 12. (a) Pressure and Velocity variations from 0° to 30° angle cases and (b) Chamber pneumatic power potential per meter of wave front in W/m

On the 2nd half (latter 50%) of wall angle cases varying from 40 to 80° can be clustered to a separate set, a contrast variation can be seen in average power values as the wall opening go pass beyond 30°. At both a 40° and 50° angle, the outlet velocity ranged from 1 to 6 m/s and 1 to 4.5 m/s, reflecting airflow dynamics and sinusoidal waveforms signifying successful wave energy capture, reaching maximum velocities of 5.33 m/s and 4.06 m/s, and average velocities of 1.59 m/s and 1.27 m/s, respectively. The pressure remained consistently positive, peaking at 61.8 kPa and 13.06 kPa, with average values of 2.56 kPa and 7.25 kPa, reflecting wave energy accumulation and efficient utilization. Power output per unit area initially started low but increased over time, peaking at 99.34W/m and 53.05W/m, with average values of 7.94 W/m and 4.16 W/m. These variations are visible in the Figure 13(a) and (b).

Simulations conducted at 60° and 70° wall angles, inside the OWC chamber, air velocities reached 4.35 m/s and 4.0 m/s, with average speeds of 2.07 m/s and 1.78 m/s. These velocities followed sinusoidal waveforms, reflecting wave energy capture and efficient utilization. Pressure values like previous cases remained consistently positive, peaking at 19.92 kPa and 15.59 kPa, with average values of 4.44 kPa and 3.11 kPa. Power output per unit area of wave front initially lower but increased with simulation time, reaching maxima of 102.73W/m and 52.38 W/m, with average values of 14.42 W/m and 8.19 W/m. In simulations at 80° velocity at the outlet ranged from approximately 0 to 2.5 m/s. The maximum velocity reached 2.34 m/s, with an average velocity of 1.07 m/s. Pressure values ranged from 1 kPa to 18 kPa with a maximum of 17.58 kPa. Power output per unit area started gave a maximum of 26.55 W/m, with an average of 3.48 W/m. These values of air velocities and pressures are illustrated in Figure 13(a) and (b) for comparison and contrast.

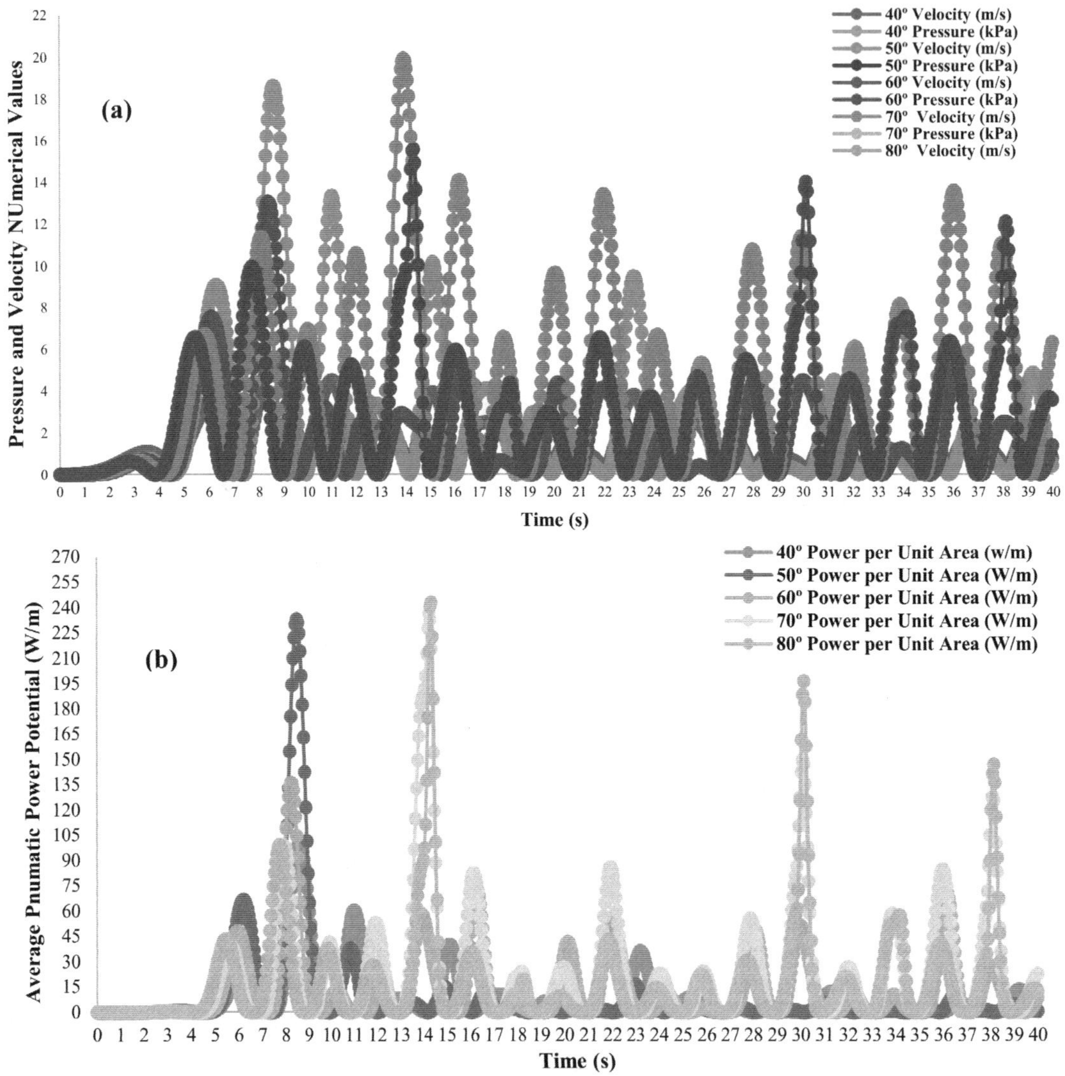

Figure 13 (a) Pressure and velocity variations from 40° to 80° angle cases and (b) Chamber pneumatic power potential per meter of wave front in W/m.

With these values the 1st 4-cases from 0 to 30° angles shown higher pressures and mainly high pneumatic power potential per unit area of wave front. Compared to the 5-case angles of 40 to 80°, which only yields considerably lower average power values. The summary of results is illustrated in Table 3. Here, the maximum values of pressures and power potentials per unit area changes did not account in summary of results. Mainly because during the simulation run, wave interference causes wave amplitude and wave lengths to superimpose, and some instances there may occur wave resonance conditions resulting extremely larger pressures and power fluctuations. To avoid these undesirable effects during analysis, the average velocities, average pressures, and average powers are illustrated in Figure 14 (a), and how the maximum velocity values varied compared to average velocities and pressures are illustrated in Figure 14 (b), for all the 9-cases.

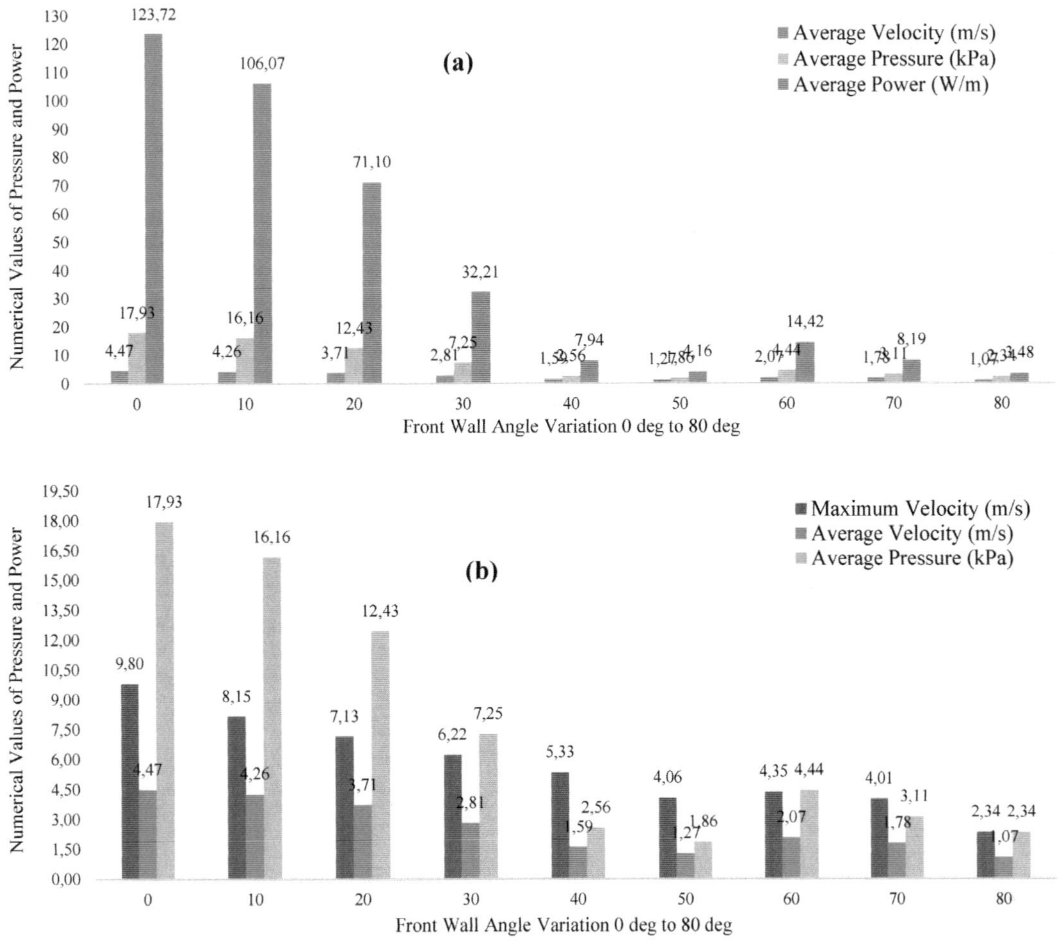

Figure 14. (a) Average pressure, velocity, and power variation inside OWC chamber for studied 9-wall angles, (b) Maximum velocity and average power and pressure variations for the 9-wall angles.

4. Conclusions

This paper presents the front wall angle variation effects on the (OWC fixed type) power potential of the air chamber. The major variation can be seen that the reference case at 0° angle produces the maximum pneumatic power potential unit are of wave front with a 123.72 W/m. And the least produced by 80° case with 3.48 W/m. A gradual power potential decrement is visible (Figure 14a) from reference wall angle to 40° angle. In between each 10° space reduction varied as 17.65, 34.97, 38.89, 24.27 and 3.77 W/m in values, in each of these stages the reduction accounts 14.27%, 39.97%, 54.7% 75.36% and 47.55% in order. Largest drop in power seen between 30 to 40° angle variation. Lowest variation seen between 0 to 10° angle locations. Beyond 30° the OWC chamber power potential diminishes exponentially and can be concluded these wall angles are not suited for OWC chamber designs, modifications, or optimization purposes. Due to wave resonance phenomena and two incident (incoming and outgoing) waves being super-positioned during collision, the chamber pressure and power potential values at 60° position reaches higher than 40 and 50° cases deviating from the pattern, and again for 70 and 80° positions these values drop with a non-linear patter which was witnessed between 0 to 50° angles.

Figure 12 illustrates the pressure and velocity variations for the 1st half (50%) of front wall angle changes are clustered. From 0° to 30° angle considered relatively smaller and the wall exerted wave pressures are higher, due to these angles making the wall closer to normal hence the wave loading can be considerable on the wall. This causes the wave chamber pressures to increase by 20 to 30% compared to mid-span of wall angles. The highest pressures can be seen at 0° case and lowered with at 50° case, interestingly not at the 80° case, but only with a 0.2 kPa's (Figure 14 (a) and (b)). The main reason is the wave front energy being accumulated with sharp angle at 0-degree. Thus, the air chamber compressing power of the heaving motion of the wave gets increased like a liquid column entrapped pressurized in a cringer. With lower degrees of front wall angle, say the wall is perpendicular (approximately) with SWL the wave energy transferred in to the OWC chamber becomes higher, thus the compressibility increases inside a confined space facing the chamber dead wall. It is clear (In Figure. 12.b) when the chamber front wall tilted further away from the horizontal angles, the wave front energy transmitted in to the OWC chamber does not receive adequate constant cross sectional area ratio in 3-D space ($height\ x\ width\ x\ breadth$) inside the OWC air chamber because the front wall is further expanded (from 0 to 10, 20, 30...etc.) form the mean reference vertical hinge point. The pneumatic (air) power potential which works as the energy source for an air turbine, or any PTO method device, on which the useful amount of work being absorbed shows gradual decrement after 0° and further. As the front wall angle is changed it can be concluded having a constant cross section (A) to height (h) ratio is essential to increase the pneumatic (air) power potential inside the air chamber of the OWC. This condition should remain as much as possible in a uniform manner to extract useful power when an air turbine is mounted with OWC, thus the efficiency of power generation can be increased by approximately 10 to 20% with having a 0° angle of front wall without any appendages fitted with into the wall.

Funding: This project has received funding from the European Union's Horizon 2020 research and innovation program under the Marie Skłodowska-Curie grant agreement No. 861079 ("Next MGT").

Reference

[1] Zhang Y, Zhao Y, Sun W and Li J 2021 Ocean wave energy converters: Technical principle, device realization, and performance evaluation. *Renewable Sustainable Energy Rev.* **141**, 110764.

[2] Falcão A F O and Henriques J C C 2016 Oscillating-Water-Column Wave Energy Converters and Air Turbines: A Review, *Renewable Energy*, **85**, 1391-1424.

[3] Cui Y, Liu Z, Zhang X, Xu C 2019 Review of CFD Studies on Axial-Flow Self-Rectifying Turbines for OWC Wave Energy Conversion, *Ocean Eng.* **175**, 80-102.

[4] Zhou Y, Ning D, Liang D and Cai S 2021 Nonlinear Hydrodynamic Analysis of an Offshore Oscillating Water Column Wave Energy Converter, *Renewable Sustainable Energy Rev.* **145**.

[5] Doyle S and Aggidis G A 2019 Development of Multi-Oscillating Water Columns as Wave Energy Converters. *Renewable and Sustainable Energy Rev.* **107**, 75-86.

[6] Nguyen H P, Wang C M, Tay Z Y and Luong V H 2020 Wave Energy Converter and Large Floating Platform Integration: A Review. **213**.

[7] Weerakoon A H S, Kim B H, Cho Y J, Prasad D D, Ahmed M R and Lee Y H 2021 Design Optimization of a Novel Vertical Augmentation Channel Housing A Cross-Flow Turbine and Performance Evaluation as a Wave Energy Converter, *Renewable Energy*, **180**, 1300-1314.

[8] Ning D, Wang R Q, Chen L F and Sun K 2019 Experimental Investigation of a Land-Based Dual-Chamber OWC Wave Energy Converter. *Renewable and Sustainable Energy Rev.* **105**

[9] Teixeira P R F, Davyt D P, Didier E and Ramalhais R 2013 Numerical Simulation of an Oscillating Water Column Device Using a Code Based on Navier–Stokes Equations, *Energy*, **61**

[10] Hong D C, Hong S Y and Hong S W 2004 Numerical Study of the Motions and Drift Force of a Floating OWC Device, *Ocean Eng.* **31**

[11] Gomes R P F, Henriques J C C, Gato L M C and Falcão A F O 2012 Hydrodynamic Optimization of an Axisymmetric Floating Oscillating Water Column for Wave Energy Conversion, *Renewable Energy*, **44**
[12] Nunes G, Valério D, Beirão P, Sá da Costa J 2011 Modelling and Control of a Wave Energy Converter, *Renewable Energy*, **36**
[13] Mustapa M A, Yaakob O B, Ahmed Y M, Rheem C K, Koh K K and Adnan F A 2017 Wave Energy Device and Breakwater Integration: A Review. *Renewable and Sustainable Energy Rev.* **77**
[14] Weerakoon A H S, Young H S, Kim W K and Lee Y H 2020 *Novel Tidal Energy Harnessing System Utilizing Quadruple Bi-directional Turbine Arrangement.*
[15] He F and Huang Z 2014 Hydrodynamic Performance of Pile-Supported OWC-Type Structures as Breakwaters: An Experimental Study, *Ocean Eng.* **88**, 618-626.
[16] Paixão Conde J M and Gato L M C 2008 Numerical Study of the Air-Flow in an Oscillating Water Column Wave Energy Converter. *Renewable Energy*, **33**
[17] Ali T H, Nikseresht A H and Hayati M 2021 Numerical Analysis of Hydrodynamic Performance of a Dual-Chamber Oscillating Water Column, *Energy*, **221**
[18] Brito-Melo A, Gato L M C and Sarmento A J N A 2002 Analysis of Wells Turbine Design Parameters by Numerical Simulation of the OWC Performance. *Ocean Eng.* **29**
[19] Setoguchi T and Takao M 2006 Current Status of Self-Rectifying Air Turbines for Wave Energy Conversion. *Energy Conversion and Management*, **47**
[20] Weerakoon A H S, Lee Y H and Assadi M 2023 Wave Energy Convertor for Bilateral Offshore Wave Flows: A Computational Fluid Dynamics (CFD) Study. *Sustainability*
[21] Thakker A and Hourigan F 2005 Computational Fluid Dynamics Analysis of a 0.6m, 0.6 Hub-to-Tip Ratio Impulse Turbine with Fixed Guide Vanes. *Renewable Energy,* **30**
[22] Thakker A and Abdulhadi R 2008 The Performance of Wells Turbine Under Bi-directional Airflow. *Renewable Energy*, **33**
[23] Renzi, Emiliano, Michele S, Zheng S, Jin S and Greaves D 2021 Niche Applications and Flexible Devices for Wave Energy Conversion: A Review. *Energies* **14**
[24] Falcão A F O and Henriques J C C 2019 The Spring-Like Air Compressibility Effect in Oscillating-Water-Column Wave Energy Converters: Review and Analyses. *Renewable and Sustainable Energy Rev.* **112**, 483-498.
[25] Dean R G and Dalrymple R A 1991 Water Wave Mechanics for Engineers and Scientists. Advanced Series on Ocean Engineering, *World Scientific Publishing Company.*
[26] Dao B V and Penzien J 1982 Comparison of Treatments of Non-linear Drag Forces Acting on Fixed Offshore Platforms, *Appl. Ocean Res.* **4**
[27] McCormick M E 2013 Ocean Wave Energy Conversion. *Courier Corporation.*
[28] Vyzikas T, Deshoulières S, Giroux O, Barton M and Greaves D 2017 Numerical Study of Fixed Oscillating Water Column with RANS-type Two-phase CFD Model, *Renewable Energy*, **102**, Part B
[29] Chen J, Wen H, Wang Y and Wang G 2021 A Correlation Study of Optimal Chamber Width with the Relative Front Wall Draught of Onshore OWC Device, *Energy*, **225**
[30] Hayati M, Nikseresht A H and Ali T H 2020 Sequential optimization of the geometrical parameters of an OWC device based on the specific wave characteristics, *Renewable Energy,* **161**,386-394
[31] Elhanafi A, Macfarlane G, Fleming A and Leong Z 2017 Scaling and Air Compressibility Effects on a Three-dimensional Offshore Stationary OWC Wave Energy Converter. *Applied Energy*, **189**, 1-20
[32] Prasad D D, Ahmed M R, Lee Y H and Sharma R N 2017 Validation of a Piston Type Wave-maker Using Numerical Wave Tank, *Ocean Eng.* **131**, 57-67.
[33] Weerakoon A H S, Kim B H, Cho Y J, Prasad D D, Ahmed M R and Lee Y H 2021 Design Optimization of a Novel Vertical Augmentation Channel Housing a Cross-flow Turbine and

Performance Evaluation as a Wave Energy Converter, *Renewable Energy*, **180**, 1300-1314.

[34] Mohsin S I and Al-Faruk A 2021 Numerical Simulation of an Oscillating Water Column Device and Investigating the Effects of Lip Submergence on Velocity and Pressure.

[35] Mia M R, Zhao M, Wu H, Palmer H 2022 Numerical Simulation of a Stationary Offshore Multi-chamber OWC Wave Energy Converter, *Ocean Eng.* **265**

[36] Nguyen H P, Wang C M, Tay Z Y and Luong V H 2020 Wave Energy Converter and Large Floating Platform Integration: A Review. *Ocean Eng.* **21**

[37] Zheng S, Zhang Y, Iglesias G 2019 Coast/Breakwater-integrated OWC: A Theoretical Model. *Mar struct.* **66**

[38] Ning D, Zhou Y, Zhang C 2018 Hydrodynamic Modelling of a Novel Dual-chamber OWC Wave Energy Converter. *Applied Ocean Research*, **78**, 180-191

[39] Mahnamfar F, Altunkaynak A 2017 Comparison of Numerical and Experimental Analyses for Optimizing the Geometry of OWC Systems, *Ocean Eng.* **130**, 10-24.

[40] Shalby M, Elhanafi A, Walker P and Dorrell D G 2019 CFD Modelling of a Small–scale Fixed Multi–chamber OWC Device, *Applied Ocean Research*, **88**, 37-47.

[41] Bouali B and Larbi S 2017 Sequential Optimization and Performance Prediction of an Oscillating Water Column Wave Energy Converter, *Ocean Eng.* **131**, 162-173.

[42] Xu C and Huang Z 2019 Three-dimensional CFD Simulation of a Circular OWC with a Nonlinear Power-Take-off: Model Validation and a Discussion on Resonant Sloshing Inside the Pneumatic Chamber *Ocean Eng.* **176**, 184-198.

[43] Moñino A, Quirós C, Mengíbar F, Medina-Lopez E and Clavero M 2020 Thermodynamics of the OWC Chamber: Experimental Turbine Performance under Stationary Flow, *Renewable Energy*, **155**, 317-329.

[44] Wang C and Zhang Y 2022 Performance Enhancement for a Dual-chamber OWC Conceived from Side Wall Effects in Narrow Flumes, *Ocean Eng.* **247**

[45] Rodríguez A A M, Silva R C and Ilzarbe J M B 2022 The influence of oblique waves on the hydrodynamic efficiency of an onshore OWC wave energy converter, *Renewable Energy*. **183**, 687-707

[46] Kush D, Gandhi M and Banerjee 2023 Numerical Validation of Power Take-Off Damping in an OWC Chamber and Modifications for Increased Efficiency. In: Banerjee, J., Shah, R.D., Agarwal, R.K., Mitra, S. (eds) Recent Advances in Fluid Dynamics. *Lecture Notes in Mechanical Engineering*. Springer, Singapore

IOP Conf. Series: Materials Science and Engineering 1294 (2023) 012016 doi:10.1088/1757-899X/1294/1/012016

Dynamic response of a floating dock under corrosion-induced accidents

T Zahi*, X Wen and M C Ong

Department of Mechanical and Structural Engineering and Materials Science
University of Stavanger, NO-4036, Stavanger, Norway

* Correspondence: 268048@uis.no (Terry Zahi)

Abstract. Dynamic responses of a floating dock under corrosion-induced accidents are studied using a numerical method. The numerical model is proposed to calculate the dynamic responses of the floating dock during operations. It includes a six-degree-of-freedom (6-DOF) model, a hydrostatic force model, a hydrodynamic force model, and a hydraulic model. The effects of the corrosion-induced holes on the stability of the floating dock are investigated and the results show that the maximum pitch and roll angles are 0.18° and 0.69° respectively when there is one hole located at one tank. The maximum pitch and roll angles become 0.42° and 2.04° respectively when there are two holes located at different tanks. The results indicate that situations involving more than one corroded hole result in large roll and pitch angles, which ultimately increase the risk of the vessel capsizing. This analysis not only emphasizes potential hazards but also presents an opportunity for the maritime sector to enhance safety, operational efficiency, and environmental responsibility.

1. Introduction

A floating dock as illustrated in Figure 1 is a specialized pontoon. It plays a crucial role in the maritime industry, serving as essential infrastructure for vessel maintenance, repairs, and cargo handling operations [1]. It is equipped with chambers and has a cross-sectional shape resembling the letter "U." The structure is designed to accommodate varying water levels and provides a stable platform for carrying out these activities. These ballast tanks allow for the adjustment of the floating dock's draft by adding water into them or removing water from them and maintaining the desired trim and heel as the same time [2]. Typically, the process of raising the dock involves pumping out the ballast water, while submerging it is achieved by allowing seawater to flow into the tanks by gravity. The flow rate of the ballast water going into or coming out of each tank is controlled by a valve specifically assigned to that tank. The floating dock is gaining popularity as a solution to address the shortage of dock space in diverse areas of ocean engineering because it eliminates the need for ground yard space [3]. However, they can be subjected to accidental conditions, such as extreme weather events or unexpected equipment failures, which can significantly impact their stability and structural integrity.

Figure 1. Vessel undergoing maintenance on a floating dock [4].

There are two accidents related to the corrosion of the ballast tanks. In one accident, a floating dock sank during a docking operation due to excessive trim caused by the failure of ballast valves following corrosion. This excessive trim submerged the aft vent pipe, leading to rapid flooding of compartments and subsequent sinking of the floating dock at the berth [5]. Another case involved a floating dock sinking due to inaccurate positioning of keel blocks. When the floating dock emerged from the water, several keel blocks punctured the pontoon deck, causing flooding in multiple ballast tanks, rapid loss of buoyancy, and eventual sinking [6].

The aim of this paper is to evaluate the dynamic behaviors of a floating dock under corrosion induced accidents. This evaluation is rendered using a numerical model that has been created to replicate the dynamic operations of the floating dock. This model incorporates several components, including a six-degree-of-freedom (6-DOF) model, a hydrostatic force model, a hydrodynamic force model, and a hydraulic model. The outcomes of this study will provide valuable insights for the maritime industry regarding the maintenance strategies necessary to ensure the safety of floating docks.

2. Numerical model

2.1. Specifications of a floating dock system

Figure 2 shows the floating dock system, where a floating dock, a vessel, mooring ropes connecting the dock and the vessel, mooring lines and the docking blocks on the top deck are shown. The floating dock system is described within a global coordinate system called *OXYZ*. The origin of this coordinate system is positioned at the centre of the dock's keel. The *X*-axis represents the direction along the keel, from the aft side to the fore side. The *Y*-axis is from the starboard to port direction, while the *Z*-axis points vertically upward from the bottom of the dock.

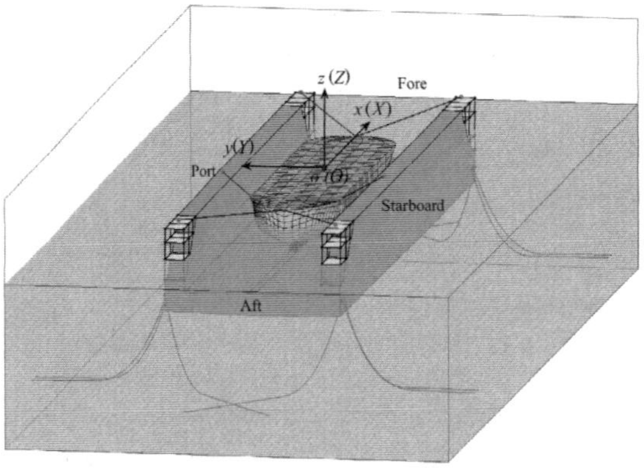

Figure 2. Illustration of a floating dock system.

A body-fixed coordinate system named *oxyz* initially coincide with the global coordinate system. The body-fixed coordinate system moves with the floating dock. It will become different from the global coordinate system as the dock moves. Table 1 shows the specifications of the floating dock system. The Details of the ballast tanks are shown in Figure 3, where the number and the maximum volume of each ballast tank are given.

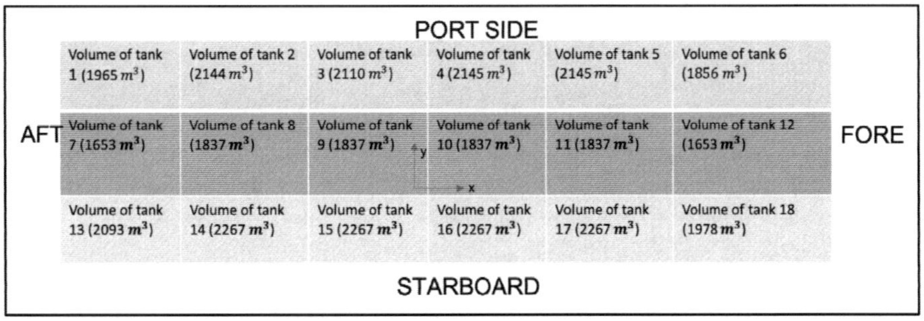

Figure 3. Details of the ballast tanks.

Table 1. Specifications of a floating dock system

Dimension of dock $L_d \times B_d \times H_d$	$168.48\ \mathrm{m} \times 39.8\ \mathrm{m} \times 18.2\ \mathrm{m}$
Mass of dock	5.1782×10^6 kg
Initial *X* of the dock's CoG	-0.435 m
Initial *Y* of the dock's CoG	0.093 m
Initial Z of the dock's CoG	5.497 m
Dock's mass moment of inertia I_{11dock}	$9.56 \times 10^8\ \mathrm{kg \cdot m^2}$
Dock's mass moment of inertia I_{22dock}	$1.026 \times 10^{10}\ \mathrm{kg \cdot m^2}$
Dock's mass moment of inertia I_{33dock}	$1.096 \times 10^{10}\ \mathrm{kg \cdot m^2}$
Dimension of vessel $L_v \times B_v \times H_v$	$95.217\ \mathrm{m} \times 20\ \mathrm{m} \times 8\ \mathrm{m}$
Mass of vessel	5.1292×10^6 kg
Initial *X* of the vessel's CoG	-0.435 m
Initial *Y* of the vessel's CoG	0.093 m
Initial *Z* of the vessel's CoG	13.09 m
Vessel's mass moment of inertia $I_{11vessel}$	$2.234 \times 10^8\ \mathrm{kg \cdot m^2}$
Vessel's mass moment of inertia $I_{22vessel}$	$2.968 \times 10^9\ \mathrm{kg \cdot m^2}$
Vessel's mass moment of inertia $I_{33vessel}$	$2.968 \times 10^9\ \mathrm{kg \cdot m^2}$
Density of seawater	1025 kg/m^3
Gravitational acceleration	9.81 m/s^2

2.2. 6-DOF model

The motions of the floating dock are updated using a 6-DOF model. The 6-DOF model includes the dock's translational motion equations in the global coordinate system and the dock's rotational motion equations described in the body-fixed coordinate system. The translational motion equations are given in Eq. (1) based on Newton's Second Law.

$$\frac{\mathrm{d}^2 \boldsymbol{X}_{CG}}{\mathrm{d}t^2} = \boldsymbol{m}^{-1} \sum \boldsymbol{F}_G \tag{1}$$

where $\mathbf{X}_{CG} = (X_{CG}, Y_{CG}, Z_{CG})$ represents the CoG of the floating dock, $\boldsymbol{m}$ represents the mass matrix of the floating dock. $\boldsymbol{F}_G$ is denoted as the external force vector applied to the CoG. The dock's angular velocity vector is modelled using Eq. (2) [7].

$$\frac{\mathrm{d}\boldsymbol{\omega}_B}{\mathrm{d}t} = \boldsymbol{I}^{-1} \left[\sum \boldsymbol{M}_B - \boldsymbol{\omega}_B \times (\boldsymbol{I}\boldsymbol{\omega}_B) \right] \tag{2}$$

where $\boldsymbol{I}$ represents the inertial tensor of the floating dock, $\boldsymbol{M}_{\mathrm{B}}$ is the moment vector and $\boldsymbol{\omega}_{\mathrm{B}} = (\omega_{B1}, \omega_{B2}, \omega_{B3})$ is the dock's angular velocity vector. The rotational angles of the floating dock are computed using Eq. (3)

$$\begin{cases} \frac{\mathrm{d}\phi}{\mathrm{d}t} = (\omega_{B2}\sin\gamma + \omega_{B3}\cos\gamma)/\cos\psi \\ \frac{\mathrm{d}\psi}{\mathrm{d}t} = (\omega_{B2}\cos\gamma - \omega_{B3}\sin\gamma) \\ \frac{\mathrm{d}\gamma}{\mathrm{d}t} = \omega_{B1} + (\omega_{B2}\sin\gamma + \omega_{B3}\cos\gamma)\tan\psi \end{cases} \tag{3}$$

The yaw pitch and roll angles are denoted as ϕ, ψ and γ are the yaw, respectively. The interaction between the dock and the vessel is addressed in an approximate way. The dock and the vessel are considered as a single rigid body with hybrid mass and mass moments of inertia.

2.3. Hydrostatic force model

The buoyancy forces of the dock and vessel, as well as the gravitational forces of water in the ballast tanks are considered as the hydrostatic forces. They are all calculated using the Archimedes' law. To assess the submerged volumes of the floating dock and the ballast tanks, a strip theory is employed, where the three-dimensional (3D) structure is divided into two-dimensional (2D) sections. By integrating the hydrostatic forces along the longitudinal direction of the floating dock and the ballast tanks, the overall hydrostatic loads on the 3D structure are calculated. The submerged regions of the 2D sections are represented by a few boundary points, enabling the determination of their area, first moment of area, and second moment of area. Using the section areas, the displaced water volume of the floating dock, as well as the volume of ballast water, can be computed by the integration along the longitudinal direction.

The height of the water level in a ballast tank is calculated using a secant iteration method of a single point in Eq. (4).

$$h^{(n+1)} = h^{(n)} - \frac{h^{(n)} - h_{\mathrm{pre}}}{V^{(n)} - V_{\mathrm{pre}}}(V^{(n)} - V) \tag{4}$$

In Eq. (4), V is the water volume in a ballast tank, h_{pre} is the water level height in the previous time step and V_{pre} is the water volume in the water level h_{pre}.

2.4. Hydrodynamic force model

The added mass of the floating dock is calculated by forming the 2D results along the longitudinal direction of the dock. The 2D results are based on the added mass and mass moments of inertia of 2D plate. The 3D correction given by the aspect-ratio formula of Pabst [8]. Table 2 shows the results of the floating dock's added mass and mass moments of inertia.

Table 2. Added mass and mass moments of inertia.

Motion	Formula	value
Heave	$\mathrm{m}_{33\mathrm{added}} = \frac{1}{8}\rho\pi\mathrm{B}^2\mathrm{L}\Psi(\mathrm{B/L})$	9.4604×10^7 kg
Roll	$\mathrm{I}_{11\mathrm{added}} = \frac{1}{256}\rho\pi\mathrm{B}^4\mathrm{L}\Psi(\mathrm{B/L})$	$4.6830\times 10^9\ \mathrm{kg\cdot m^2}$
Pitch	$\mathrm{I}_{22\mathrm{added}} = \frac{1}{96}\rho\pi\mathrm{B}^2\mathrm{L}^3\Psi(\mathrm{B/L})$	$2.2378\times 10^{11}\ \mathrm{kg\cdot m^2}$

The damping coefficients are modelled based on the mass matrices of the dock, a damping ratio of 5% and natural frequencies of the heave, roll and pitch. According to the DNV recommended practice, when the damping of the system is unknown, a damping ratio of 5% is suggested [9]. The natural frequencies are calculated using Eq. (5) [9].

$$\omega_{\text{heave}} = \sqrt{\frac{C_{33}}{m_{33}}}, \quad \omega_{\text{roll}} = \sqrt{\frac{C_{44}}{I_{11}}}, \quad \omega_{\text{pitch}} = \sqrt{\frac{C_{55}}{I_{22}}} \tag{5}$$

where m_{33}, I_{11} and I_{22} are the total mass, mass moments of inertia of the floating dock, and C_{33}, C_{44} and C_{55} are the hydrostatic restoring coefficients in heave, roll and pitch motions, respectively.

2.5. Modelling of the ballast water system and the corrosion holes

The ballast system of the floating dock can be divided into six groups and each group can be illustrated in Figure 4. There are six ballast pumps that individually control three ballast tanks positioned from port to starboard, and two types of pipelines with diameters of 400mm and 600 mm. Each ballast tank is equipped with its own butterfly valve, and the main pipes feature inlet and outlet valves used for ballasting and de-ballasting operations, respectively.

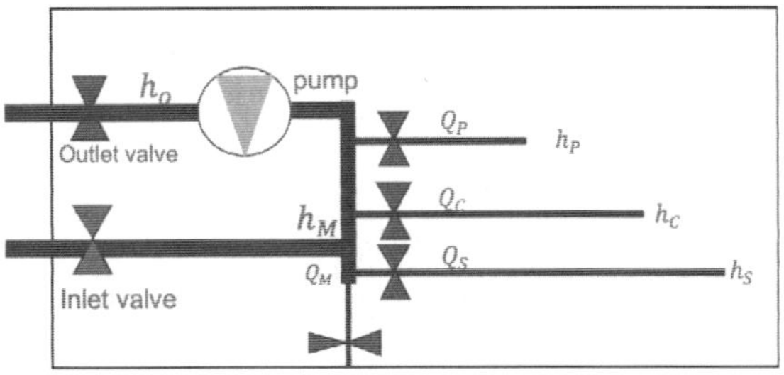

Figure 4. Description of one group of the ballast water system during the de-ballasting operation.

The numerical modelling of the ballast water system is based on the pressure difference characteristics of different elements, i.e., pipe, valve and pump. The difference of water head when the water is passing the element is calculated in Eqs. (6) – (9).

$$h_O - h_{\text{out}} = \lambda_{\text{M}}|Q_{\text{M}}|Q_{\text{M}} \tag{6}$$

$$h_{\text{P}} - h_{\text{M}} = \lambda_P|Q_{\text{P}}|Q_{\text{P}} \tag{7}$$

$$h_{\text{C}} - h_{\text{M}} = \lambda_{\text{C}}|Q_{\text{C}}|Q_{\text{C}} \tag{8}$$

$$h_{\text{S}} - h_{\text{M}} = \lambda_{\text{S}}|Q_{\text{S}}|Q_{\text{S}} \tag{9}$$

h_{M}, h_{P}, h_{C}, h_{S}, and h_{o} are the water heads at right sides of pump, port, centre, starboard and outlet (inlet) valves in Figure 4, and h_{out} is the water head at the left side of outlet valve. Q_{M}, Q_{P}, Q_{C} and Q_{S} are the flow rates in main, port, centre and starboard pipes. The coefficients λ_{M}, λ_{P}, λ_{C} and λ_{S} are given by the $K_V$values of the butterfly valves, as reported by Wen et al. [10].

Table 2. K_V value at different hole diameter for the butterfly valves [10]

Diameter [mm]	40	50	65	80	100	125	150	200	250	300	400
K_V [m³/hour/bar]	53	133	240	410	665	900	1800	3550	7350	9100	10500

The K_V values depend on the diameter of the corrosion hole and are given by experimental measurements [11]. λ can be written as Eq. (10).

$$\lambda = \frac{1}{\mathrm{g}(\mathrm{KV}/36000)^2} \quad [\mathrm{s}^2/\mathrm{m}^5] \tag{10}$$

The water head difference when the water is passing through pumps is calculated using Eq. (11).

$$h_O - h_{\text{M}} = h_O - \lambda_{\text{pump}}|Q_{\text{M}}|Q_{\text{M}} \tag{11}$$

where $h_0 = 21.25$ m is the pump's total water head when $Q_M = 0$, and $\lambda_{\text{pump}} = 20\ \mathrm{s}^2/\mathrm{m}^5$ is the pump coefficient. A continuity equation between the main pipe and the branch pipes is modelled in Eq. (12).

$$Q_{\mathrm{M}} = Q_{\mathrm{P}} + Q_{\mathrm{C}} + Q_{\mathrm{S}} \tag{12}$$

For the ballasting operation, the solution is given by taking $h_0 = 0$ and $\lambda_{\mathrm{pump}} = 0$. The flow rate through the hole is then given by Eq. (13)

$$Q = \frac{(h_{out} - h)}{\sqrt{\lambda_{hole}|h_{out} - h|}} \tag{13}$$

Where λ_{hole} is hole coefficient. After the flow rate of a ballast tanks is obtained, the ballast water volume is updated using Eq. (14).

$$\frac{\mathrm{d}\alpha_j}{\mathrm{d}t} = \frac{Q_j}{V_{\max,j}} \tag{14}$$

where α_j is the volume fraction of the water in j^{th} ballast tank, $V_{\max,j}$ is the total volume of jth tank and Q_j is the corresponding flow rate.

3. Results and discussions

3.1. Case description

The maximum pitch and roll angles of the floating dock are investigated. The dock is initially docking a vessel on top without heel and trim at the draught of 3.5m. The dock will tilt when there is corrosion hole located at one of two ballast tanks. This present study involves two scenarios: one is that a corrosion-induced hole with different diameters occurs in one single ballast tank during the ballasting operation and another is that two ballast tanks have corrosion-induced holes of 300mm each.

The sensitivity study of section number for the dock and the ballast tanks, and the convergence study of the water levels in ballast tanks were conducted by Zhang et al. [12]. According to the convergent results of these sensitivity studies, the dock's section number is chosen as 150 and the section number of each ballast tank is selected as 20 in the present study. In a dock's section, 18 points are used while those of port, centre and starboard tanks are 8, 8 and 6, respectively.

3.2. Dynamic behaviours

To guarantee the dock's performance under these unexpected circumstances, a time-step sensitivity analysis is to be conducted. Figure 5 shows the time-step sensitivity study of the draught, pitch and roll angles during the corrosion induced accident in Tank No.1, where the hole diameter is 50mm. Three time-steps of 1s, 0.5 s, and 0.25 s are examined to assess the convergence of the results. The results obtained using these three different time steps appeared the almost same, a time step of 0.5s is chosen to proceed the further analysis to achieve a balance between the computational time and the temporal resolution. As can be seen in Figure 6, the draught becomes larger as the weight of the ballast water increases. The dock continues to tilt with final heel and trim angles of 0.55° and 0.14° respectively.

Figure 6 shows the roll and pitch angles for the scenarios of one corrosion hole with different diameters located at Tank. No.01. The roll and pitch angles of different hole diameters have the same final convergent results. Table 5 presents the time taken to fill up the ballast tank through the corrosion hole in Tank No. 01. The time is recorded when the flow rate through the hole decreases to $0.5 \times 10^{-3}\ \mathrm{m^3/s}$. From the table, the time increases with a decreasing hole diameter. For a hole diameter of 300mm, the dock tilts to the maximum heel and trim in about half an hour.

Table 3. Time taken to fill up the ballast tank through the corrosion hole in Tank No. 01.

Hole diameter [mm]	Duration time [hour]
50.00	21.34
100.00	4.39
150.00	1.63
200.00	0.83
250.00	0.40
300.00	0.32

IOP Conf. Series: Materials Science and Engineering 1294 (2023) 012016 doi:10.1088/1757-899X/1294/1/012016

Figure 7 shows the maximum pitch and roll angle for the scenarios of one corrosion hole with a diameter of 300mm located at different ballast tanks. A large maximum pitch angle corresponds to the hole located at the tanks near the aft or fore, and a large maximum roll angle corresponds to the hole located at the tanks near the port or starboard. It can be attributed to the moment due to the extra weight of the ballast water flow into the tank through the corrosion hole. The maximum roll angles of the scenarios of one corrosion hole located at the starboard tanks are always larger than those at the port tanks. The reason is that the total volumes of the starboard tanks are larger than those of the port tanks, as shown in Figure 3. Figure 8 shows the volumes of the ballast water in Tanks No.02 and 014 for the scenarios of one corrosion hole located at Tanks No.02 and 014, respectively. The volume of the ballast water in Tank No. 14 is much larger than that in Tank No.02, which can also explain the reason of the different maximum roll angle in Figure 6(b).

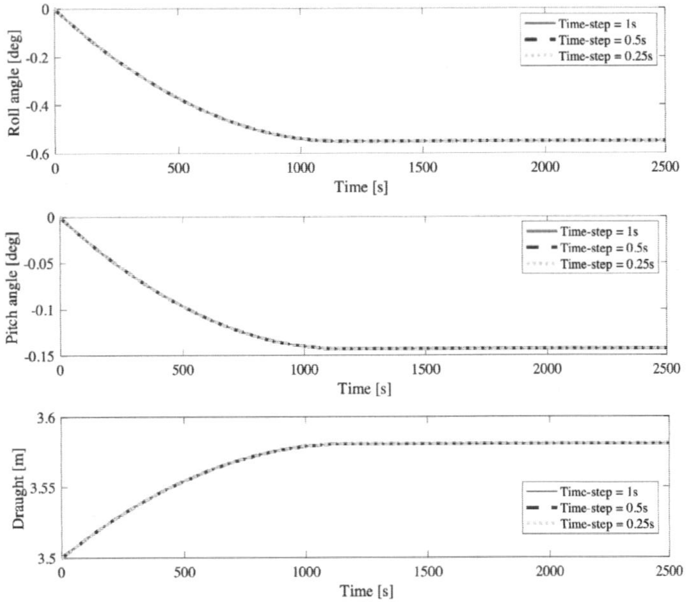

Figure 5. Time-step sensitivity study of the roll, pitch angles and draught during the corrosion induced accident in Tank No. 01, where the hole diameter is 50 mm.

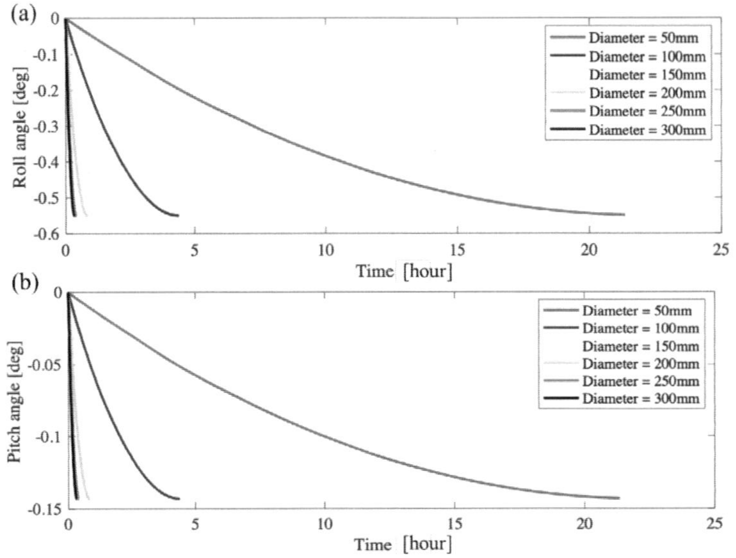

Figure 6. Roll and pitch angles for the scenarios of one corrosion hole with different diameters located at Tank. No.01.

0.14	0.09	0.03	0.03	0.09	0.14
0.17	0.11	0.03	0.03	0.11	0.17
0.18	0.12	0.04	0.04	0.11	0.17

(a) Maximum pitch angle [°]

0.55	0.58	0.54	0.56	0.58	0.55
0.00	0.00	0.00	0.00	0.00	0.00
0.72	0.75	0.71	0.68	0.71	0.69

(b) Maximum roll angle [°]

Figure 7. Maximum pitch and roll angles for the scenarios of one corrosion hole with a diameter of 300 mm located at different ballast tanks.

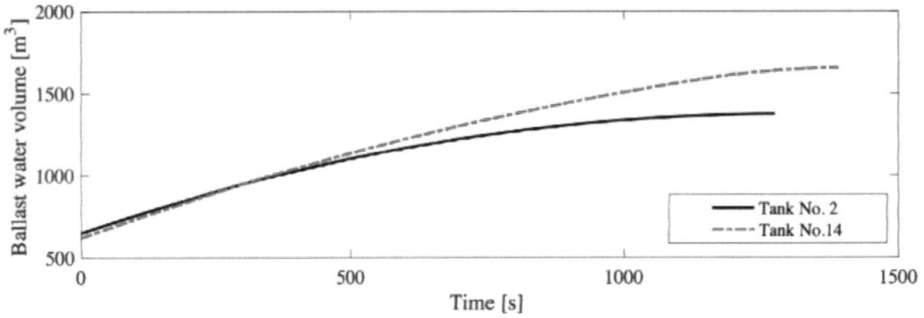

Figure 8. Volumes of the ballast water in Tanks No.02 and 014 for the scenarios of one corrosion hole located at Tanks No.02 and 014, respectively.

Figures 9 and 10 shows the maximum pitch angles for the scenarios of two corrosion holes with a diameter of 300 mm located at different ballast tanks. The distributions follow the principle as discussed for one corrosion hole cases. A large maximum pitch angle corresponds to the holes located at the tanks near the aft or fore, and a large maximum roll angle corresponds to the holes located at the tanks near the port or starboard. The maximum pitch and roll angles for two corrosion hole cases are 0.42° and 2.04° respectively.

tank No.	1	2	3	4	5	6	7	8	9	10	11	12	13	14	15	16	17	18
1	0.14	0.30	0.21	0.13	0.06	0.00	0.35	0.27	0.19	0.12	0.05	0.00	0.31	0.24	0.17	0.11	0.05	0.00
2	0.30	0.09	0.15	0.07	0.00	0.06	0.29	0.21	0.14	0.06	0.00	0.06	0.25	0.19	0.12	0.06	0.00	0.05
3	0.21	0.15	0.03	0.00	0.06	0.12	0.18	0.13	0.06	0.00	0.06	0.11	0.17	0.11	0.06	0.00	0.05	0.10
4	0.13	0.07	0.00	0.03	0.14	0.21	0.12	0.06	0.00	0.06	0.13	0.19	0.11	0.06	0.00	0.06	0.11	0.17
5	0.06	0.00	0.06	0.14	0.09	0.32	0.06	0.00	0.06	0.13	0.21	0.28	0.05	0.00	0.06	0.12	0.19	0.25
6	0.00	0.06	0.12	0.21	0.32	0.14	0.00	0.05	0.12	0.19	0.27	0.35	0.00	0.05	0.11	0.17	0.24	0.31
7	0.35	0.29	0.18	0.12	0.06	0.00	0.17	0.32	0.22	0.14	0.06	0.00	0.42	0.32	0.22	0.13	0.06	0.00
8	0.27	0.21	0.13	0.06	0.00	0.05	0.32	0.11	0.16	0.07	0.00	0.06	0.33	0.24	0.15	0.07	0.00	0.06
9	0.19	0.14	0.06	0.00	0.06	0.12	0.22	0.16	0.03	0.00	0.07	0.13	0.22	0.15	0.07	0.00	0.07	0.13
10	0.12	0.06	0.00	0.06	0.13	0.19	0.14	0.07	0.00	0.03	0.15	0.22	0.13	0.07	0.00	0.07	0.15	0.22
11	0.05	0.00	0.06	0.13	0.21	0.27	0.06	0.00	0.07	0.15	0.11	0.33	0.06	0.00	0.07	0.16	0.24	0.33
12	0.00	0.06	0.11	0.19	0.28	0.35	0.00	0.06	0.13	0.22	0.33	0.17	0.00	0.06	0.14	0.22	0.32	0.42
13	0.31	0.25	0.17	0.11	0.05	0.00	0.42	0.33	0.22	0.13	0.06	0.00	0.18	0.39	0.27	0.16	0.07	0.00
14	0.24	0.19	0.11	0.06	0.00	0.05	0.32	0.24	0.15	0.07	0.00	0.06	0.39	0.12	0.19	0.09	0.00	0.08
15	0.17	0.12	0.06	0.00	0.06	0.11	0.22	0.15	0.07	0.00	0.07	0.14	0.27	0.19	0.04	0.00	0.09	0.16
16	0.11	0.06	0.00	0.06	0.12	0.17	0.13	0.07	0.00	0.07	0.16	0.22	0.16	0.09	0.00	0.04	0.18	0.26
17	0.05	0.00	0.05	0.11	0.19	0.24	0.06	0.00	0.07	0.15	0.24	0.32	0.07	0.00	0.09	0.18	0.11	0.41
18	0.00	0.05	0.10	0.17	0.25	0.31	0.00	0.06	0.13	0.22	0.33	0.42	0.00	0.08	0.16	0.26	0.41	0.17

Figure 9. Maximum pitch angles (°) for the scenarios of two corrosion holes with a diameter of 300 mm located at different ballast tanks.

tank No.	1	2	3	4	5	6	7	8	9	10	11	12	13	14	15	16	17	18
1	0.55	1.44	1.34	1.26	1.18	1.12	0.68	0.64	0.60	0.57	0.54	0.51	0.02	0.02	0.02	0.01	0.01	0.01
2	1.44	0.58	1.41	1.36	1.31	1.26	0.69	0.66	0.63	0.61	0.59	0.57	0.02	0.01	0.01	0.01	0.01	0.01
3	1.34	1.41	0.54	1.26	1.24	1.22	0.59	0.58	0.57	0.56	0.56	0.55	0.02	0.02	0.02	0.02	0.02	0.02
4	1.26	1.36	1.26	0.56	1.34	1.35	0.58	0.58	0.59	0.59	0.60	0.61	0.01	0.01	0.01	0.01	0.02	0.02
5	1.18	1.31	1.24	1.34	0.58	1.54	0.57	0.59	0.61	0.63	0.66	0.68	0.01	0.01	0.02	0.02	0.02	0.02
6	1.12	1.26	1.22	1.35	1.54	0.55	0.51	0.54	0.56	0.60	0.64	0.67	0.01	0.02	0.02	0.02	0.02	0.02
7	0.68	0.69	0.59	0.58	0.57	0.51	0.00	0.00	0.00	0.00	0.00	0.00	0.83	0.78	0.73	0.68	0.64	0.61
8	0.64	0.66	0.58	0.58	0.59	0.54	0.00	0.00	0.00	0.00	0.00	0.00	0.81	0.78	0.75	0.72	0.69	0.67
9	0.60	0.63	0.57	0.59	0.61	0.56	0.00	0.00	0.00	0.00	0.00	0.00	0.72	0.71	0.70	0.69	0.68	0.67
10	0.57	0.61	0.56	0.59	0.63	0.60	0.00	0.00	0.00	0.00	0.00	0.00	0.67	0.68	0.69	0.70	0.71	0.72
11	0.54	0.59	0.56	0.60	0.66	0.64	0.00	0.00	0.00	0.00	0.00	0.00	0.67	0.69	0.72	0.75	0.78	0.81
12	0.51	0.57	0.55	0.61	0.68	0.67	0.00	0.00	0.00	0.00	0.00	0.00	0.61	0.65	0.68	0.73	0.78	0.84
13	0.02	0.02	0.02	0.01	0.01	0.01	0.83	0.81	0.72	0.67	0.67	0.61	0.72	1.90	1.79	1.67	1.56	1.47
14	0.02	0.01	0.02	0.01	0.01	0.02	0.78	0.78	0.71	0.68	0.69	0.65	1.90	0.75	1.87	1.78	1.71	1.64
15	0.02	0.01	0.02	0.01	0.02	0.02	0.73	0.75	0.70	0.69	0.72	0.68	1.79	1.87	0.71	1.71	1.68	1.66
16	0.01	0.01	0.02	0.01	0.02	0.02	0.68	0.72	0.69	0.70	0.75	0.73	1.67	1.78	1.71	0.68	1.68	1.71
17	0.01	0.01	0.02	0.02	0.02	0.02	0.64	0.69	0.68	0.71	0.78	0.78	1.56	1.71	1.68	1.68	0.71	2.04
18	0.01	0.01	0.02	0.02	0.02	0.02	0.61	0.67	0.67	0.72	0.81	0.84	1.47	1.64	1.66	1.71	2.04	0.69

Figure 10. Maximum roll angles (°) for the scenarios of two corrosion holes with a diameter of 300 mm located at different ballast tanks.

4. Conclusion

The dynamic responses of a floating dock under corrosion-induced accidents are studied using a numerical method. The numerical model is proposed to calculate the dynamic responses of the floating dock during operations. It includes a six-degree-of-freedom (6-DOF) model, a hydrostatic force model, a hydrodynamic force model, and a hydraulic model. The effects of the corrosion-induced holes on the stability of the floating dock are investigated and the results show that the maximum pitch and roll angles are 0.18° and 0.69° respectively when there is one hole located at one tank. The maximum pitch and roll angles become 0.42° and 2.04° respectively when there are two holes located at different tanks. The results indicate that situations involving more than one corroded hole result in larger roll and pitch angles, which ultimately increase the risk of the vessel capsizing. This analysis not only emphasizes potential hazards but also presents an opportunity for the maritime sector to enhance safety, operational efficiency, and environmental responsibility.

References

[1] Wankhede A, 2021, Dry Dock, Types of Dry Docks & Requirements for Dry Dock, last modified 9 January 2021. https://www.marineinsight.com/guidelines/dry-dock-types-of-dry-docks-requirements-for-dry-dock/#Types_of_Dry_Dock. Accessed 14 March 2023.

[2] Yoon K, Kim H-J, Yeo S, Hong Y, Cha J and Chung H 2020 Ballasting plan optimization for operation of a 2D floating dry dock, *Structural Engineering and Mechanics.* 74(4), 521–532.

[3] Shan X, Yu Q, Tian J 2009 Risk management of mooring operation of floating dock," Journal of Tianjin University Science and Technology. Technol. **42**, 335-339.

[4] ASRY (2023) Arab ship building and repair yard company, available at: https://www.asry.net (accessed Sep. 01, 2023)

[5] Morra T 2011 The Evolutionary Development of Floating Dry Docks, Master's Thesis, East Carolina University, East Carolina.

[6] El-Maadawy M, Moustafa M M, El-Kilani H S and Tawfiq A A 2018 Structural safety assessment of a floating dock during docking operation, Port-said Engineering Research Journal. **22**(2), 32 - 39.

[7] Schaub H and Junkins J L 2003 Analytical Mechanics of Space Systems. American Institute of Aeronautics and Astronautics (AIAA), Inc., Reston, VA, pp. 23,77-123.

[8] Pabst W 1930 Theory of the landing impact of seaplanes,NTRS – NASA Technical Report. Available at: https://ntrs.nasa.gov/api/citations/19930094836/downloads/19930094836.pdf.

[9] Det Norske Vertias (DNV) (2012), DNV Recommended Practice DNV-RP-C205:Environmental conditions and environmental loads.

[10] Wen X, García Conde A, Zhang J, and Ong M C 2023 Numerical modelling of a floating dock and its automatic control, in Proceedings of the 42nd International Conference on Offshore Mechanics and Arctic Engineering. American Society of Mechanical Engineers, Melbourne, Australia, June 11–16, 2023. (Submitted: OMAE2023-102873).

[11] Eurotorc, Butterfly valve engineering data, Nov. 12, 2022. https://www.eurotorc.com/butterfly-valve-engineering-data.html. (accessed Sep. 16, 2023).

[12] Zhang J, Wen X, Ong M C 2023 Development of a Floating Dock Numerical Model and the Ballast Water Distribution Strategy. Proc. ASME 2023 42nd Int. Conf. Ocean, Offshore and Arctic Eng. **5**: Ocean Eng.. Melbourne, Australia. June 11–16, 2023. V005T06A068. ASME. https://doi.org/10.1115/OMAE2023-102996

IOP Conf. Series: Materials Science and Engineering 1294 (2023) 012017 doi:10.1088/1757-899X/1294/1/012017

Numerical study on the lifting operation of a gravity-type fish cage

A Gjerde *, X Wen and M C Ong

Department of Mechanical and Structural Engineering and Materials Science
University of Stavanger, NO-4036, Stavanger, Norway

* Corresponding author: andr.gjerde@stud.uis.no

Abstract The dynamic behaviours of the lifting operation of a gravity-type fish cage under calm sea conditions are investigated in this study using an extended position-based dynamics (XPBD) method to obtain the structural deformations of the aquaculture nets. The original XPBD is improved to accurately predict the tensions of the aquaculture nets by applying correction forces. The present XPBD is validated by comparing the experimental results of a flexible horizontal net. The time-step sensitivity is verified for the case of the lifting operation of the fish cage. Results show that the lifting force increases rapidly resulting from the weight of the sinkers at the bottom of the side net. The maximum tension of the net is located at the net ropes connected to the centre point of the bottom net. The structure of the bottom net should be enhanced for safety lifting operations.

1. Introduction

Fish cages are the most used equipment in aquaculture [1-3]. As technology advances, the number of different types of fish cages used in aquaculture are becoming less [4]. The gravity-type fish cages have become the more frequently used as a fish cage for the aquaculture in many countries such as Norway, China, and Japan. The gravity-type fish cages usually consist of aquaculture nets made of nylon or polyethylene (PE), high-density polyethylene (HDPE) floating collars and sinker tube, concrete centre sinker at the bottom of the net and related mooring system. All the materials can be produced in the manufactories and transported to specific sea sites to be assembled. The usage of the gravity-type fish cages can help reduce the capital cost for the industrial-scaled fish farming. Another important way to increase to the harvest and the profits of the industrial-scaled fish farming is to ensure the health and the good quality of the fish by monitoring the sea lice and conducting the de-lice operations [5]. One type of de-lice operations is to pump the fish into the de-lice machines and clean the fish by fresh water. Before the de-lice of the fish, the aquaculture nets need to be lifted and the fish will gather for pumping. The extra oxygen supply is required when the fish is gathering. The lifting operation is conducted by a crane installed on a vessel, which is very similar to the harvest operation. The lifting operations of a gravity-type fish cage are important for the fish farming and the dynamic response of the fish cage during the lifting operation is investigated in the present study.

A numerical method is adopted to study the dynamic response of the fish cage during the lifting operation in the present study. For numerical simulations, there are a few approaches and methods that can be used, such as finite element method (FEM) [6], OrcaFlex [7], and mass-spring model [8]. Rui

Dou [6] used the FEM method together with a panel model, a mass model and Morison model [2] to model and analyse a semi-submersible fish cage. Cifuentes and Kim [7] calculated the current load acting on a fish cage using a Morison-force model applied at instantaneous positions of equivalent-net modelling using Orcaflex [7]. Lee et al. [8] used the mass-spring model to simulate the flexible structures' behaviour to understand the movements and design an appropriate system. A new efficient method called extended position-based dynamic (XPBD) method [9-10] was proposed to simulate the cloth dynamics in games. It is an implicit method and more robust compared with the explicit method of mass-spring model. The method will be used in the present study to simulate the dynamic behaviors of the flexible fish cage.

The present study focuses on the dynamic analysis of the lifting operations of a gravity-type fish cage using the XPBD method. The present paper is organized as follows. The fish cage specifications and the XPBD method is given in Section 2. Section 3 presents a validation case of a flexible horizontal net, and the results and discussions of the dynamic process of the lifting operations of the fish cage. The conclusion is summarized in Section 4.

2. Numerical model

2.1 Fish cage specifications

Figure 1 displays the full-scale structure of a gravity-type fish cage. The cage has a circumference of 160m and a total height of 35 m. It consists of two main parts: a 30 m high side net and a 5 m high bottom net. The fish cage includes thousands of panels which are enclosed by net ropes. The net ropes are constructed using PE lines. For the side net, each net rope has a length of 1.25 m. The bottom net has 8 net ropes from the centre to the outer boundary and 128 net ropes at the outer boundary. Inside the net panels, there are small inner nets made of nylon with square mesh patterns. These inner nets have a diameter of 2.85 mm and a length of 25 mm. There are side net sinkers installed at the bottom of the side net and centre sinker installed at the centre point of the bottom net. Table 1 shows the specifications of the gravity-type fish cage, including the dimensions, materials, and key features.

During the lifting operation of the fish cage, the centre point of the bottom net is lifted from the lowest point to a specific position. The present lifting speed is 0.5m/s, as shown in Figure 2, chosen based on the mechanism of the lifting crane. The crane starts at 0s and accelerates to the normal lifting speed in 10s. The crane totally stops at 100 s.

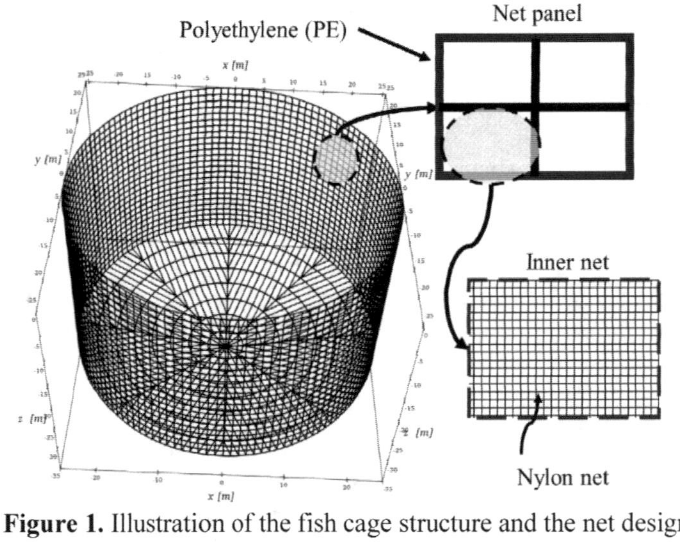

Figure 1. Illustration of the fish cage structure and the net design.

Table 1. Specifications of the gravity-type fish cage.

	Value	Unit
Cage circumference	160	m
Cage height	35	m
Side net height	30	m
Bottom net height	5	m
Side net sinkers' mass	80	kg/m
Bottom net sinker's mass	200	kg
Polyethylene density	960	kg/m^3
Nylon density	1140	kg/m^3
Elastic modulus polyethylene	1.5e9	Pa
Net rope diameter (PE)	10	mm
Inner twine diameter (nylon)	2.85	mm
Length inner net	25	mm
Solidity Sn	0.2	-

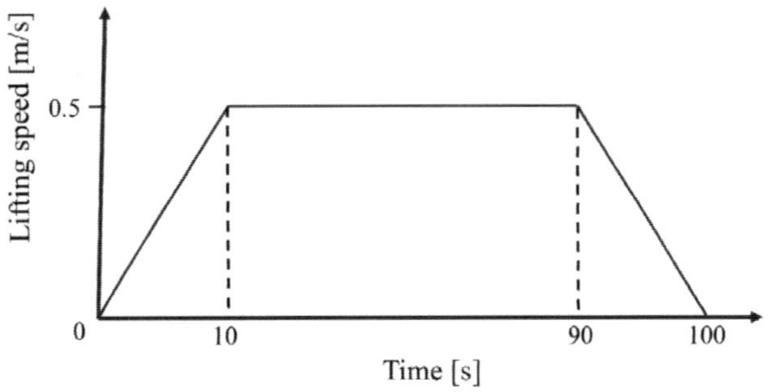

Figure 2. Description of the lifting speed during the lifting operation of the fish cage.

2.2. XPBD method

The extended position-based dynamics (XPBD) is an implicit method and serves as a numerical simulation technique used for simulating complex dynamic behavior of deformable objects, making it a valuable tool for investigating the fish case behavior and optimizing the design parameters. XPBD was developed as an extension of the position-based dynamics (PBD) approach, introducing distinct advantages in terms of stability, easy implementations, and computational efficiency. The details of the original XPBD simulation loop are shown in Algorithm 1.

Algorithm 1: original XPBD simulation loop	
1:	predict position $\tilde{\boldsymbol{x}} \leftarrow \boldsymbol{x}^n + \Delta t \boldsymbol{v}^n + \Delta t^2 \boldsymbol{M}^{-1} \boldsymbol{F}_{ext}$
2:	**for all** constrains **do**
3:	compute $\Delta\lambda$ using Equation (7)
4:	compute Δx using Equation (6)
5:	**end for**
6:	update positions $\boldsymbol{x}^{n+1} \leftarrow \tilde{\boldsymbol{x}} + \Delta\boldsymbol{x}$
7:	update velocities $\boldsymbol{v}^{n+1} \leftarrow \frac{1}{\Delta t}(\boldsymbol{x}^{n+1} - \boldsymbol{x}^n)$

The method first updates the positions of the nodes based on the external forces $\boldsymbol{F}_{ext}$ using Newton's second law. Then, the positions of the nodes need to be corrected according to the inner force resulting from every constraint. For a single constraint, the changing of the positions of these two nodes is given by evaluating Equation (1).

$$\begin{aligned}\boldsymbol{\Delta x}_1 &= +\Delta\lambda m_1^{-1}\boldsymbol{q}\\ \boldsymbol{\Delta x}_2 &= -\Delta\lambda m_2^{-1}\boldsymbol{q}\end{aligned} \tag{1}$$

The changing of Lagrange multiplier λ is given by Equation (2).

$$\Delta\lambda = \frac{-C}{m_1^{-1} + m_2^{-1} + \tilde{\alpha}} \tag{2}$$

where $\tilde{\alpha} = \frac{1}{k\Delta t^2}$ and $C = l - l_0$.

Compared with the mass spring model, XPBD can provide a quick result for the net dynamics in an implicit way. However, XPBD has its inevitable error for calculating the tensions of the constraints. It is because the masses of the nodes are involved when evaluating the Lagrange multiplier change. It is against the physics intuition that the elastic force is only determined by the stiffness and the constraint. To correct the results of XPBD, the term of $m_1^{-1} + m_2^{-1}$ in the denominator of Equation (2) should be removed. However, the simulation loop as shown in Table 1 will become unstable if $m_1^{-1} + m_2^{-1}$ is removed. To maintain the stable simulation loop and correct the results of XPBD at the same time, an XPBD correction is proposed by applying extra forces acting on the two nodes of the constraint, where the expression is shown in Equation (3)

$$\begin{aligned}\boldsymbol{F}_1 &= +\gamma\boldsymbol{q}\\ \boldsymbol{F}_2 &= -\gamma\boldsymbol{q}\end{aligned} \tag{3}$$

where γ is evaluated by Equation (4).

$$\gamma = \frac{-C(m_1^{-1} + m_2^{-1})}{(m_1^{-1} + m_2^{-1} + \tilde{\alpha})\tilde{\alpha}\Delta t^2} \tag{4}$$

The new simulation loop is shown in Algorithm 2. Before predicting the position, the correction forces $\boldsymbol{F}_{cor}$ are calculated for every constraint. It will increase the computational cost when using the same time step and the same structure configuration. However, a small-time step close to the mass-spring model as shown in Equation (3) is required to ensure a close prediction accuracy, which reduces the advantage of an implicit method of XPBD. The present modified XPBD not only enhances the precision of predictions and but also significantly lowers computational expenses. As a result, the previously incurred drawback of introducing correction force calculations is effectively mitigated.

Algorithm 2: modified XPBD simulation loop
for all constrains **do**
compute γ using Equation (4)
compute correction force $\boldsymbol{F}_{cor}$ using Equation (3)
end for
predict position $\tilde{\boldsymbol{x}} \leftarrow \boldsymbol{x}^n + \Delta t\boldsymbol{v}^n + \Delta t^2 M^{-1}[\boldsymbol{F}_{ext} + \boldsymbol{F}_{cor}]$
for all constrains **do**
compute $\Delta\lambda$ using Equation (2)
compute Δx using Equation (1)
end for
update positions $\boldsymbol{x}^{n+1} \leftarrow \tilde{\boldsymbol{x}} + \boldsymbol{\Delta x}$
update velocities $\boldsymbol{v}^{n+1} \leftarrow \frac{1}{\Delta t}(\boldsymbol{x}^{n+1} - \boldsymbol{x}^n)$

An in-house code is developed based on the MATLAB platform. More functions of the present code will be added in the future. The comparison between the original and modified XPBD methods will be given in our further study.

3. Results and discussions

3.1. Validation study

The validation case is based on Lee et al.'s [8] experimental test of a flexible horizontal net, as shown in Figure 3 in its top view. The net consists of 214 nodes and 424 line elements. The twine length is 100 mm in the inner net and 141.4 mm on the boundaries. The twine diameter is 0.4 mm. Additionally, the net incorporates three sinkers with masses: F_1 = 1.5 kg (middle), F_2 = 0.5 kg (left), and F_3 = 0.7 kg (right). The specifications of the flexible horizontal net are shown in Table 2. During the experiment, the four corner nodes were fixed. The three sinkers are dropped from the initial positions and reached their final positions as shown in Figure 4(a). During the calculations, the velocities of each node are damped into 99% at every 0.5 ms to approach the final convergent position of the flexible horizontal net. Figure 4(b) presents the results of the final position of the flexible horizontal net obtained using three time-steps of 0.025 ms, 0.05 ms and 0.1 ms. The results of the three time-steps exhibit a good agreement and they show good consistent with the experiment result, which confirms the feasibility of the present XPBD method used to simulate the flexible fish cage. The validation is based on the visual comparison because of the limited data from the experiment of Lee et al.'s [8].

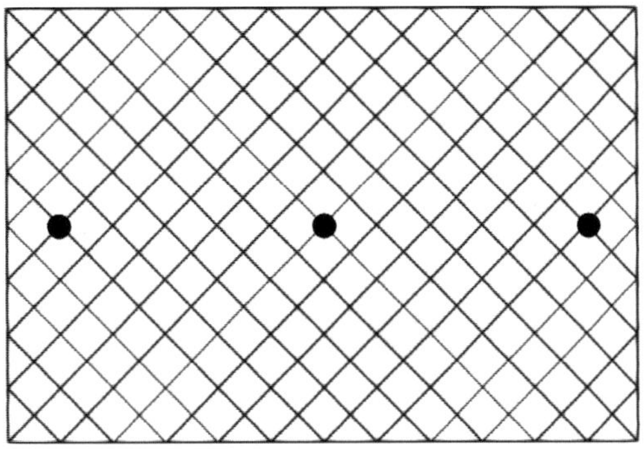

Figure 3. Initial position of the flexible horizontal net (top view).

Table 2 Validation case specifications.

Parameters	Values
Net Dimensions (meshes)	12x8
Node count	214
Element count	424
Stiffness	15.000 N/m
Mass	0.0015 kg
Twine diameter	0.4 mm
Half mesh size	100 mm
Sinker 1's mass	1.5 kg
Sinker 2's mass	0.5 kg
Sinker 3's mass	0.7 kg

IOP Conf. Series: Materials Science and Engineering 1294 (2023) 012017 doi:10.1088/1757-899X/1294/1/012017

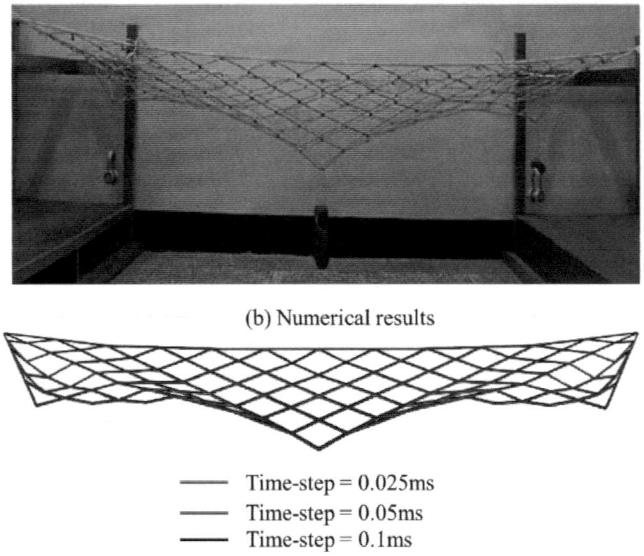

Figure 4. Comparison between the experiment result of Lee et al. [8] (a) and the numerical results of the present XPBD (b).

3.2. Dynamic process of the lifting operation

Before the lifting operation, the initial position of the fish cage is calculated using the similar way of the validation case. Figure 5 shows a time-step sensitivity study of the lifting force acting on a fish cage in a still water. Three time-steps of 0.25ms, 0.5ms and 1ms are examined in this analysis. The results of all the tested time-steps exhibit remarkable agreement. Consequently, the time-step of 1ms is chosen as the preferred time-step for subsequent calculations to reduce the computational cost.

Figure 6 shows the time-step sensitivity studies of the tensions of the fish cage in XZ- and YZ-planes at 50s. There is a slight reduction in tension on one side of the cage during the initial phase of the lifting operation, specifically at the time-step of 0.25ms in XZ-plane. The overall tensions remain consistent across all time-steps, while some variations in tension are observed during the lifting of the net. Looking at the results of the time-step of 0.25ms in Figure 6, the maximum tension reaches 1400N.

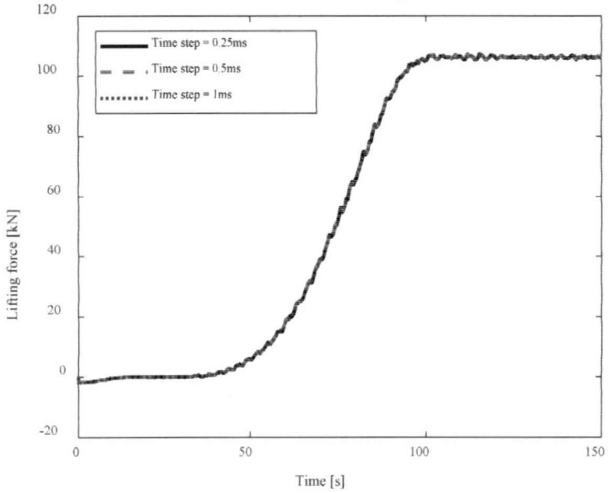

Figure 5. Time-step sensitivity study of the lifting force acting son the fish cage.

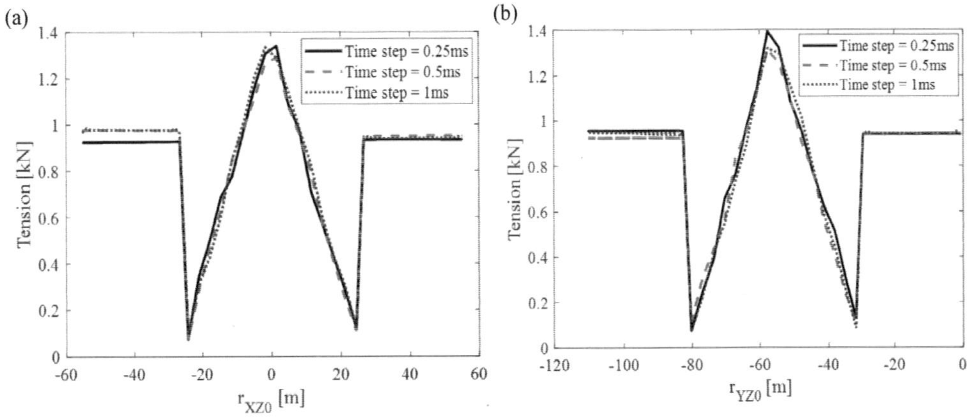

Figure 6. Time-step sensitivity study of the tensions of the fish cage in XZ and YZ-planes at 50 s.

As shown in Figure 5, the lifting force is small in [0 s, 50 s] and quickly increases over a short duration of [50 s, 100 s]. The force keeps the same value as the crane stops operation. The reason for the rapid increase of the lifting force is that the bottom net starts to support the weight of the sinkers at the bottom of the side net in [50 s, 100s]. Figures 7 shows the comparisons of tensions in XZ- and YZ-planes at different time during the lifting operation. The time instants shown in this figure are chosen as 50 s, 75 s, and 100 s. As can be seen, the tension on the bottom net has a significant increase in [50 s, 100 s]. The maximum tension can reach 12.5 kN, which is almost 13 times of the value of the side net in Figure 7. The reason can be attributed to the special structure of the bottom. As can be seen form the deformations and tensions of the fish cage inside and top views at different time in Figure 8, the maximum tension of the net is observed to be located at the net ropes connected to the centre point of the bottom net. Only eight net ropes support the sinkers' weight, while for the side net, there are 128 net ropes supporting the total sinkers' weight. It can be concluded that the net ropes of the bottom net experience the maximum tension during the lifting operation of a fish cage. The structure of the bottom net, especially for the eight net ropes, should be enhanced.

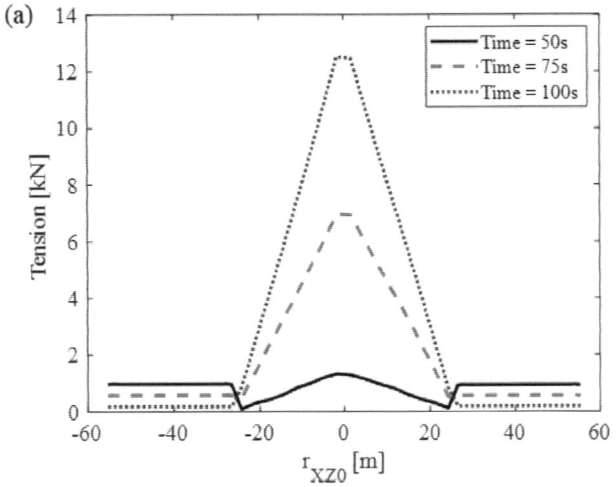

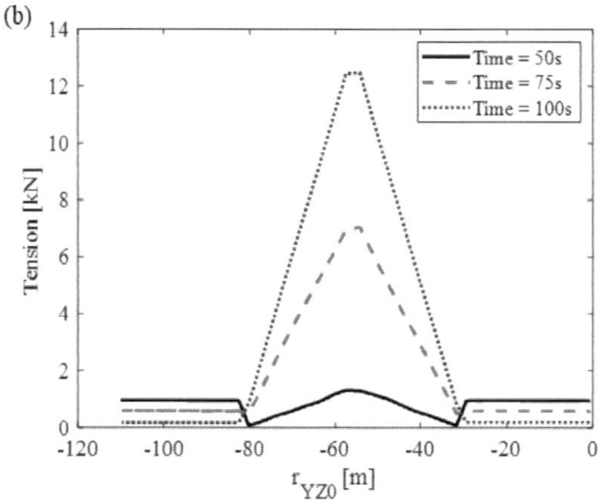

Figure 7 Comparisons of tensions in XZ- and YZ-planes at different time during the lifting operation.

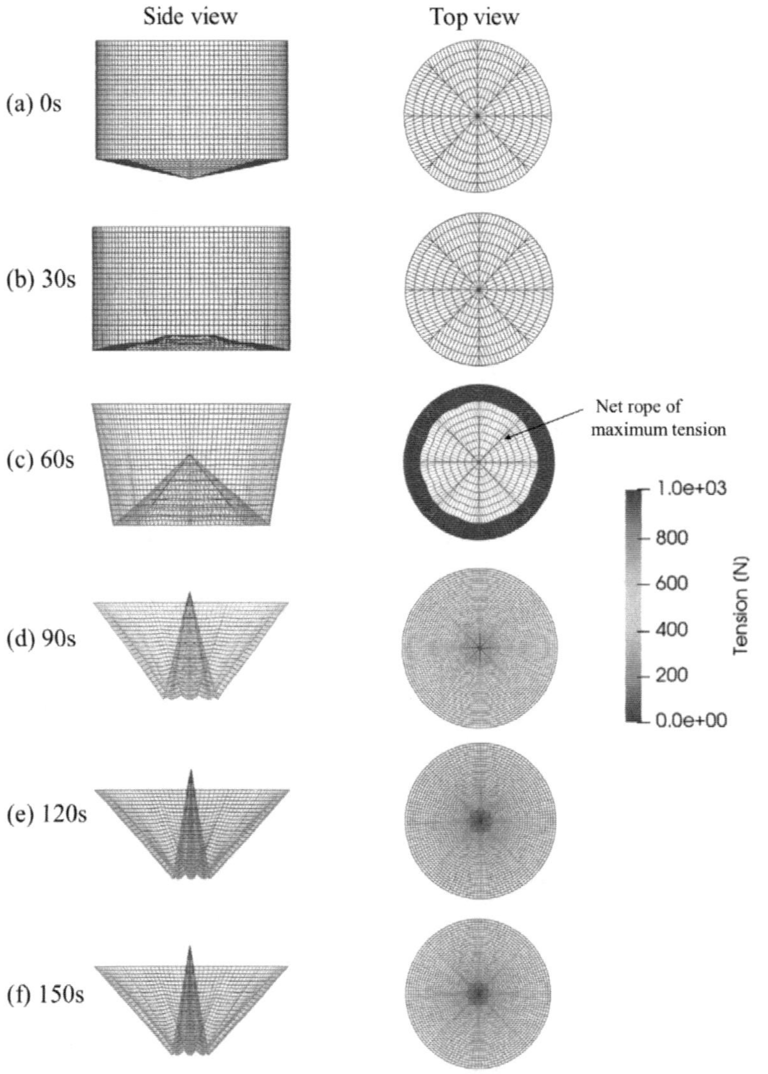

Figure 8 Deformations and tensions of the fish cage at different time.

4. Conclusion

The dynamic behaviors of the lifting operation of a gravity-type fish cage under calm sea conditions are investigated in this study. The structural deformations of the aquaculture net are obtained using an improved XPBD method. Correction forces are applied to the aquaculture net to predict the accurate tensions. The present XPBD is validated by comparing the experimental results of a flexible horizontal net. The time-step sensitivity study is conducted for the lifting operation of the fish cage. Results show that the lifting force increases rapidly resulting from the weight of the sinkers at the bottom of the side net. The maximum tension of the net is located at the net ropes connected to the centre point of the bottom net. The structure of the bottom net should be enhanced for safety lifting operations. In future study, the structure of the bottom net will be investigated in detail under failure conditions of the net ropes.

Acknowledgment. This study was supported by the Research Council of Norway through the project "Unleashing the sustainable value creation potential of offshore ocean aquaculture" (Project number: 328724).

References

[1] Huguenin J E 1997 The design, operations and economics of cage culture systems. *Aquacul. Eng.* **16**, 167–203.
[2] Guo Y C, Mohapatra S C and Guedes Soares C 2020 Review of developments in porous membranes and net-type structures for breakwaters and fish cages. *Ocean Eng.* **200**, 107027.
[3] Sievers M, Korsøen Ø, Warren-Myers F, Oppedal, F, Macaulay G, Folkedal O and Dempster T, 2021 Submerged cage aquaculture of marine fish: a review of the biological challenges and opportunities. *Rev. Aquacult.*, **14**(1), 106-119.
[4] Beveridge, M.C., 2008. *Cage Aquaculture*, Third ed. Oxford, UK. Blackwell Iowa.
[5] Berstad A J, Tronstad H, Sivertsen S A and Leite E 2005 Enhancement of design criteria for fish farm facilities including operations. Proc. 24th Int. Conf. Offshore Mech. Arctic Eng. **41960**, 825-832.
[6] Dou R 2018. Numerical modelling and analysis of a semi-submersible Fish-cage, *Master's thesis*, Norwegian University of Science and Technology (NTNU).
[7] Cifuentes C and Kim M H 2017 Numerical simulation of fish nets in currents using a Morison force model. *Ocean Syst. Eng.* **7**(2), 143-155.
[8] Lee C W, Lee J H, Cha B J, and Kim H Y 2005 Physical modelling for underwater flexible systems dynamic simulation. Ocean Eng. **32**(3-4), 331-347.
[9] Müller M, Heidelberger B, Hennix M and Ratcliff J 2007. Position based dynamics. J. Visual Commun. Image Represent. **18**(2), 109-118.
[10] Miles M, Müller M and Chentanez N 2016. XPBD: position-based simulation of compliant constrained dynamics. *Proc. 9th Int. Conf. Motion in Games*. October 2016, P. 49 – 54.

IOP Conf. Series: Materials Science and Engineering 1294 (2023) 012018 doi:10.1088/1757-899X/1294/1/012018

Optimization of bend stiffeners of a suspended inter-array power cable between two floating offshore wind turbines

D Liu[1], I Ahmad[2], M J Janocha [1,*], P Nystrøm[2] and M C Ong[1]

[1] Faculty of Science and Technology, University of Stavanger, Norway
[2] IKM Ocean Design AS, Norway

[*] Coresspondence: marek.j.janocha@uis.no

Abstract. The concept of a suspended inter-array power cable assumes that the cable floats within the water column instead of being laid on the seabed. This setup requires additional equipment, such as buoyancy modules or buoys, to achieve the desired cable configuration. The implementation of buoyancy modules introduces abrupt changes in stiffness between the cable sections with clamped-on buoyancy modules and bare cable sections. Large stiffness variations can negatively impact cable bending, causing excessive curvature and fatigue damage. In order to form a smooth transition in stiffness between the buoyancy sections and the bare cable, bend stiffeners can be equipped. The study aims to optimize the bend stiffener design for a representative suspended power cable between two floating offshore wind turbines (FOWTs). The inter-array power cable system is simulated in OrcaFlex. Two parameters, including the outer diameter and the length of the bend stiffener, are adjusted to generate different cases. Eight environmental conditions are applied to the dynamic analysis of the cases. The fitness factor approach is used as a criterion to assess the overall performance of different bend stiffener designs. Adjusting the outer diameter of bend stiffeners clearly influences the maximum effective tension and the minimum bending radius by changing the stiffness profile of the bend stiffener and its submerged weight. In the investigated range of parameters, increasing the overall length of bend stiffeners is found to be less effective than adjusting the outer diameter of the bend stiffener.

1. Introduction

Offshore wind is one of the most abundant renewable energy resources. In deep water areas, floating offshore wind turbines (FOWTs) are necessary to utilize offshore wind potential. A significant effort is currently dedicated to lowering the levelized cost of energy generated by FOWTs. One of the areas with considerable potential for optimization and cost savings is the design of the power cable system. The inter-array power cables are used to connect individual wind turbines in the wind farm. There has been a growing interest in applying inter-array power cables in a suspended configuration. In the suspended configuration, the power cables are submerged instead of being laid on the seabed. Such a technical solution can reduce the overall necessary length of the power cables, resulting in lower investment costs and a decreased inter-array energy loss [1]. For a suspended power cable system, ancillary equipment such as buoyancy modules or buoys is necessary to provide the desired geometry and maintain the cable suspended through the water column. One of the technical challenges of using ancillary equipment is maintaining an acceptable minimum bending radius of the cable. The stiffness can change abruptly in sections where the flexible, dynamic power cable is equipped with buoyancy modules. Such significant stiffness differences between different sections of the cable affect the bending behavior in an unfavorable way, leading to excessive cable curvature and accumulation of fatigue damage. In order to form a smooth transition in stiffness between the buoyancy sections and bare cable, bend stiffeners can be equipped, as shown in Figure 1. Bend stiffeners were found effective in preventing excessive bending of suspended power cables [1]. Results reported by Yang et al. [2] suggest that employing bend stiffeners in the power cable system of a wave energy converter (WEC) can reduce the maximum curvature response by 85% in the first element near the WEC and by 90% near the export hub.

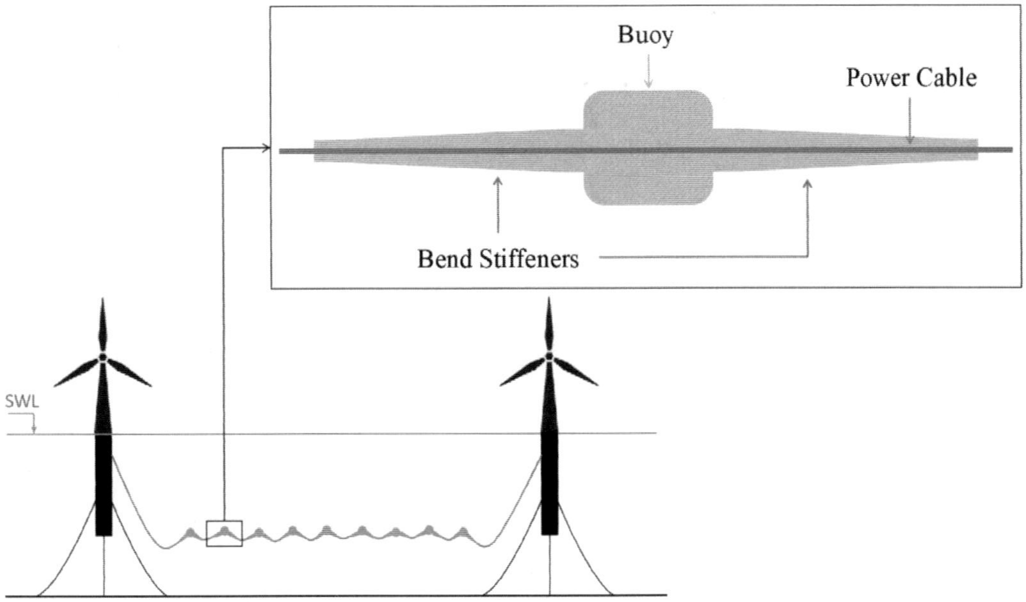

Figure 1. Power cable with buoy attached to bend stiffeners.

Bend stiffeners were successfully applied to dynamic power cables of floating WECs before being adopted in inter-array power cables of FOWTs. Thies et al. [3] investigated mechanical loads and the associated failure modes of dynamic power cables connected to a floating WEC. They observed that the attachment/hang-off points experienced the most mechanical failures attributed to fatigue damage. Thies et al. [3] suggested a dedicated bend stiffener can mitigate fatigue failure near the attachment points. These findings were confirmed by Yang et al. [2], who analyzed power cables' dynamic motions and fatigue life for WEC applications. They found that fatigue life near the cable connection points was significantly lower. Rentschler et al. [4] presented a design approach for dynamic inter-array cable systems based on a genetic algorithm. One of the key parameters influencing the effectiveness of a bend stiffener is the end diameter of the bend stiffener. Their parametric study revealed that the bend stiffener is unable to limit the bending when the end diameter is close to the cable's outer diameter, leading to excessive fatigue damage values. Schnepf et al. [5] investigated a suspended power cable with different configurations between a FOWT and a Floating Production Storage and Offloading Unit (FPSO). In their study, the cable was observed to move upwards into regions with strong currents when the buoys' distribution was concentrated on the midspan of the cable. When subject to current in the transverse direction, the cable turned almost into a horizontal catenary. Schnepf et al. [5] reported that this led to large tension and a large departure angle at the hang-off points. An extensive range of departure angles needs to be considered in the design process of bend stiffeners. The feasibility study on suspended inter-array power cables between two FOWTs was conducted by Schnepf et al. [6]. The lowest values of the minimum bending radius were reported at connection points between the cable and bend stiffeners attached to the buoys. Ahmad et al. [1] investigated the dynamic response of fully suspended inter-array power cable configurations between two FOWTs. Their study found the bend stiffener limited the bending radius of the cable at the buoy-supported section to a radius greater than 2.6 m, which is the lowest MBR in their study.

To the best of the authors' knowledge, no prior research has been conducted to optimize the bend stiffeners attached to the buoys of suspended power cables, nor has there been a fatigue analysis of the cables with varying parameters of bend stiffeners. The present study aims to fill this gap by investigating the influence of the bend stiffeners with different parameters on suspended power cables between two FOWTs.

The present paper is organized as follows: A detailed description of the methodology employed in the study is presented in Section 2. The numerical setup and the properties of the investigated inter-array cable and ancillary equipment are provided. Section 3 discusses the parametric study's results investigating the effects of bend stiffeners on the fatigue life of suspended power cables. Furthermore, an optimization method is proposed to aid in the design of bend stiffeners for suspended power cable applications. The summary of the main findings of the study is given in Section 4.

2. Methodology and Numerical Setup

Details of the methodology and numerical setup in the study are presented in this section. The properties of OC3-Hywind FOWT and suspended inter-array power cable are provided. The variable parameters of bend stiffeners are listed. The environmental conditions and wind farm layout of the present study are presented. The procedure of the case study is outlined. All simulations in the present study are conducted with OrcaFlex version 11.3.

2.1 Wind Farm Layout and Environmental Conditions

The modelling of OC3-Hywind FOWT follows the definition by Jonkman [7]. The layout of the investigated configuration of two FOWTs and the directions of environmental loadings considered in the present study are shown in Figure 2. The details of environmental conditions are listed in Table 1, including wind, wave and current. Table 1 presents the rated loading condition for OC3-Hywind FOWT and extreme loading conditions with a return period of 50 years based on metocean data from the Visund field in the Norwegian Sea.

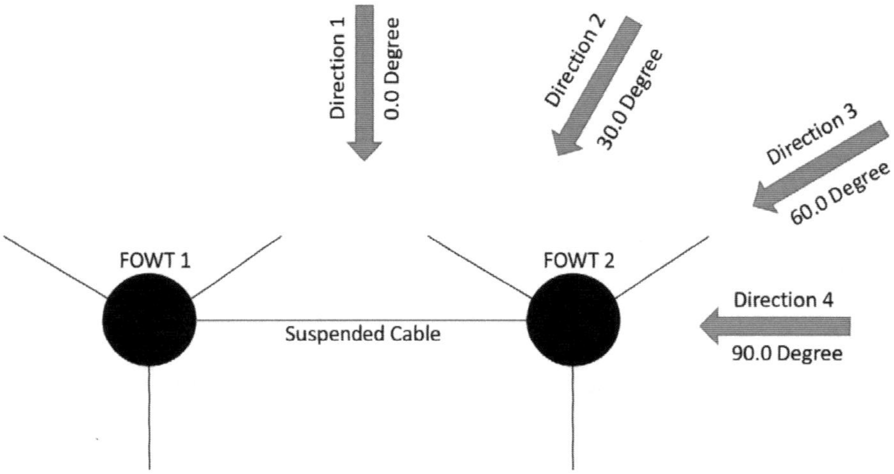

Figure 2. Two FOWTs Setup.

Table 1. Environmental conditions [1].

Wind Velocity at hub height (m/s) Type: NPD Spectrum	11.4	37.45
Wave Type: JONSWAP	H_s: 6 m T_p: 10 s γ: 2.872	H_s: 13.2 m T_p: 15.1 s γ: 2.639
Current velocity at sea surface (m/s)	0.486	1.070

Notes: H_s: Significant wave height; T_p: Peak wave period and γ: Peak enhancement factor

2.2 Suspended inter-array power cable setup

The power cable properties in the present study are the same as those used in the study by Beier et al. [8]. The previous study [8] showed that for similar suspended configurations, the differences between the dynamic response of cables in the Start-of-Life (SOL) and End-of-Life (EOL) states are small. The marine growth effects are not explored in the present study.

The buoys' properties are the same as in the study by Ahmad et al. [1]. The configuration of the suspended inter-array power cable is selected based on the best configuration reported in the study of Beier et al. [8], who investigated the optimal buoy arrangement on suspended power cable between two OC3 FOWTs. An identical two turbine' configuration is used in the present study. In the adopted configuration, the power cable length is 1240 m with ten buoys attached, and buoy spacing is 100 m.

2.3 Bend stiffeners setup

A set of geometric and material parameters used as variables in the present optimization study are summarized in Table 2 and Table 3. The outer diameter of the bend stiffener varies linearly along the bend stiffener length. The density of bend stiffeners is assumed to be constant with varying Young's Modulus and Poisson ratio. The dimension of one case is shown in Figure 4. All cases are set up based on practical engineering and physics. Table 4 shows all configurations of bend stiffeners' outer diameter.

Table 2. Bend stiffeners' geometry and mass parameters.

Parameter	Unit	Min	Max	Step
Length (BSL)	m	2.4	3	0.6
Outer diameter close to the buoy (OD_a)	m	0.42	0.54	0.04
Outer diameter away from the buoy (OD_b)	m	0.20	0.28	0.04
Inner diameter (ID)	m	0.16	0.16	0
Material density	ton/m^3	7.85	7.85	0
Poisson ratio	-	0.4	0.5	0.1

ton/m^3: ton per cubic meter

Table 3. Bend stiffeners' Young's Modulus.

Properties	Unit	Type No.1 (TPU S-60D)	Type No.2 (Polyamide 11)	Type No.3
Young's modulus	kPa	$186 \cdot 10^3$	$320 \cdot 10^3$	Figure 3

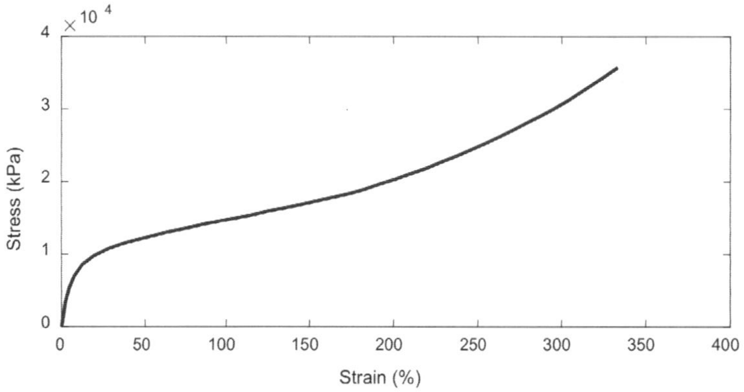

Figure 3. Stress-strain relationship of A bend stiffener (adopted from Ahmad et al. [1])

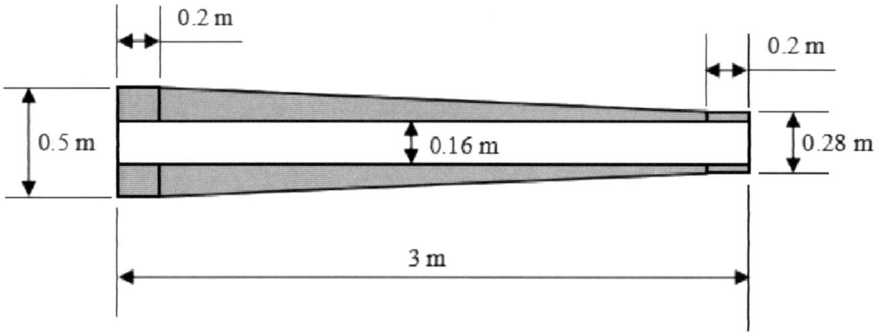

Figure 4. Dimension of A bend stiffener (BSL: 3 m, OD: 0.28-0.5 m, ID: 0.16 m)

Table 4. Outer diameter of bend stiffeners

OD	1	2	3	4	5	6	7	8	9	10	11	12
OD_a	0.20	0.24	0.28	0.20	0.24	0.28	0.20	0.24	0.28	0.20	0.24	0.28
OD_b	0.42	0.42	0.42	0.46	0.46	0.46	0.50	0.50	0.50	0.54	0.54	0.54

2.4. Case study procedure

The optimization procedure used in the present study is shown in Figure 5. A summary of acceptance criteria for the power cable is listed in Table 5. Equations (1) – (3) are applied to normalize the minimum bending radius (MBR) and minimum breaking load (MBL). The fatigue life performance of the power cable is more sensitive to MBR than to MBL. In order to reflect that, more emphasis should be given to MBR in fitness factor calculation than to MBL. In the present study, the fitness factor weights for MBR and MBL are 2/3 and 1/3, respectively. The optimization goal is to minimize the fitness factor by adjusting the geometry and structural parameters of the bend stiffener investigated in the present study.

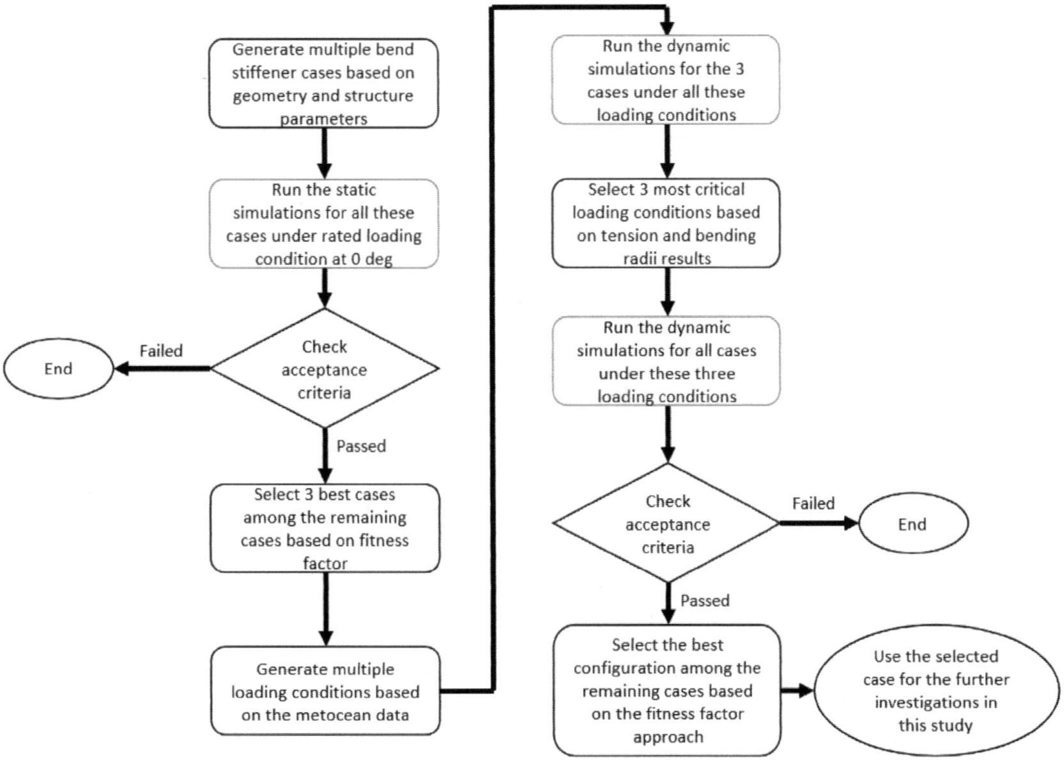

Figure 5. Bend stiffener case study procedure.

Table 5. Acceptance criteria (based on the study by Beier et al. [8]).

Criteria	Unit	Limit
Minimum bend radius (MBR)	m	2.5
Minimum breaking load (MBL)	N	90000
Minimum seabed clearance	m	10
Minimum sea surface clearance	m	30

$$Normalized\ MBR = \frac{2.5\ m}{MBR} \cdot \frac{2}{3} \quad (1)$$

$$Normalized\ MBL = \frac{MBL}{90\ kN} \cdot \frac{1}{3} \quad (2)$$

$$Fitness\ Factor = Normalized\ MBR + Normalized\ MBL \quad (3)$$

3. Results and discussion

3.1. Static analysis

The static analysis results are presented in Table 6. Surface clearance of less than 30 m caused most failure for the bend stiffeners of 1.8 m and 2.4 m lengths. For the bend stiffeners 3 m long, maximum effective tension exceeding MBL led to most failures. The successful cases in Group 1 are used to perform the static analysis with different Young's Modulus and Poisson's Ratio. No failure occurred in Group 2 and Group 3.

Table 6. Static analysis results.

Group	Young's Modulus [kPa]	Poisson's Ratio	Length of bend stiffener (BSL) [m]	Total tested cases	Successful cases	Causes of failure			
						MBL (> 90 kN)	MBR (< 2.5 m)	Surface clearance (< 30 m)	Seabed clearance (< 10 m)
1	Stress-strain curve (Figure 3)	0.5	1.8	12	1	0	0	11	0
		0.5	2.4	12	5	2	0	5	0
		0.5	3	12	5	6	0	1	0
2	186e3	0.4, 0.5	1.8	1*2	2	0	0	0	0
		0.4, 0.5	2.4	5*2	10	0	0	0	0
		0.4, 0.5	3	5*2	10	0	0	0	0
3	320e3	0.4, 0.5	1.8	1*2	2	0	0	0	0
		0.4, 0.5	2.4	5*2	10	0	0	0	0
		0.4, 0.5	3	5*2	10	0	0	0	0

According to static analysis results, the geometry parameters of the bend stiffener are dominant factors affecting effective tension, bending radius and vertical excursion of the power cable. The submerged weights of bend stiffeners with different dimensions are listed in Table 7. Successful cases in static simulation are marked in green. All the successful cases have suitable submerged weight to keep the cable at the proper depth. Figure 6 shows the vertical position of bend stiffeners with the same length of 2.4 m and the same inner diameter of 0.16 m.

Table 7. Submerged Weight of Bend Stiffeners.

Submerged weight per BS (ton)	OD1	OD2	OD3	OD4	OD5	OD6	OD7	OD8	OD9	OD10	OD11	OD12
BSL = 1.8 m ID = 0.16	2.92	3.28	3.68	3.39	3.77	4.19	3.90	4.29	4.73	4.44	4.85	5.30
BSL = 2.4 m ID = 0.16	4.21	4.72	5.29	4.87	5.41	5.99	5.58	6.14	6.74	6.33	6.91	7.55
BSL = 3.0 m ID = 0.16	5.42	6.08	6.81	6.27	6.96	7.71	7.17	7.89	8.68	8.14	8.89	9.71

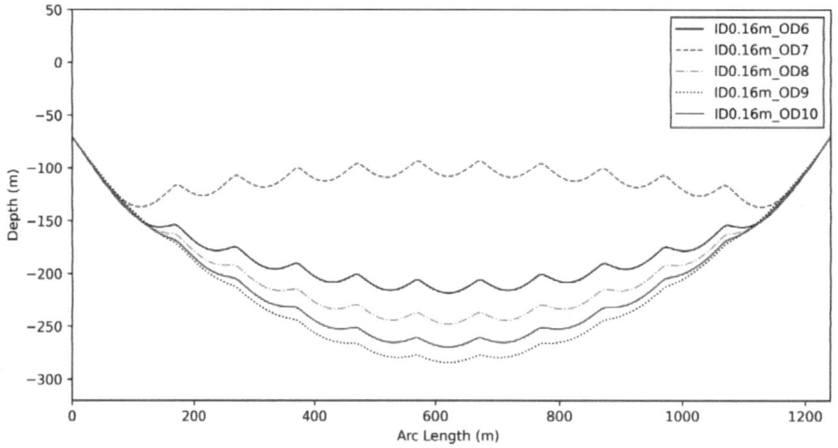

Figure 6. Vertical excursions for different outer diameter. (BSL: 2.4 m)

Figure 7 shows the normalized values and fitness factor of the successful cases in static analysis. Four cases with the lowest fitness factor were selected for the dynamic simulation under all environmental conditions.

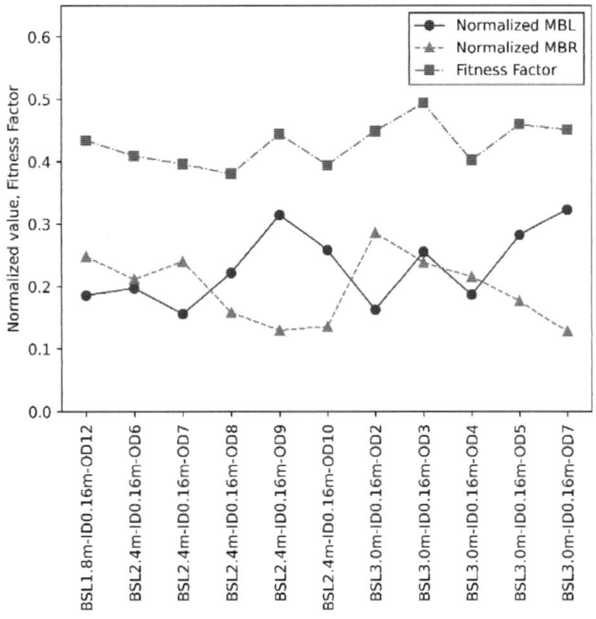

Figure 7. Fitness Factors

3.2. Dynamic Analysis

The dynamic analysis results of selected 4 cases are listed in Table 8 and Table 9. The fitness factor of each case is listed in Table 10. A colour scale is used to show the performance of different cases, where green colour indicates better performance. In the present parametric space, the optimal effective tension and minimum bending radius areas generally do not overlap. This means that it is not possible to achieve the best performance in both MBL and MBR at the same time. The fitness factor is applied to determine the overall performance of the bend stiffeners by accounting for both the MBL and MBR effects. The BS1 case (BSL2.4m_ID0.16m_OD8) performs best among the four selected bend stiffeners.

Table 8. Maximum effective tensions obtained from dynamic simulations.

		Maximum Effective Tension (kN)			
Environmental Condition	Load Direction	BS1	BS2	BS3	BS4
Extreme	1	63.21	75.34	44.25	52.79
	2	66.49	81.59	45.66	54.57
	3	71.71	90.74	49.08	58.05
	4	73.27	94.56	51.24	59.88
Rated	1	61.61	72.20	43.19	51.76
	2	62.60	73.23	44.03	52.62
	3	64.39	75.63	44.97	53.70
	4	64.14	75.42	45.45	53.80

Notes: BS1: BSL2.4m-ID0.16m-OD8; BS2: BSL2.4m-ID0.16m-OD10;
BS3: BSL2.4m-ID0.16m-OD7; BS4: BSL3.0m-ID0.16m-OD4.

Table 9. Minimum bending radii obtained from dynamic simulations.

Environmental Condition	Load Direction	Minimum bending radius (m)			
		BS1	BS2	BS3	BS4
Extreme	1	9.89	10.88	6.67	7.60
	2	9.48	10.22	6.32	7.49
	3	8.83	9.34	5.59	7.16
	4	8.42	8.94	5.55	7.08
Rated	1	10.15	11.63	6.72	7.61
	2	10.25	11.70	6.65	7.66
	3	10.23	11.60	6.56	7.67
	4	10.21	11.55	6.47	7.62

Table 10. Fitness Factor

Environmental Condition	Load Direction	Fitness Factor (-)			
		BS1	BS2	BS3	BS4
Extreme	1	0.403	0.432	0.414	0.415
	2	0.422	0.465	0.433	0.425
	3	0.454	0.515	0.480	0.448
	4	0.469	0.537	0.490	0.457
Rated	1	0.392	0.411	0.408	0.411
	2	0.394	0.414	0.414	0.412
	3	0.401	0.424	0.421	0.416
	4	0.401	0.424	0.426	0.418

The three most critical environmental conditions are selected based on **Table 8** and **Table 9**. These environmental conditions are applied to the other successful cases from static analysis. **Table 11** shows the summary of dynamic analysis results. The maximum effective tension exceeding 90 kN becomes the only cause of failure in dynamic analysis.

Table 11. Dynamic Analysis Summary.

Directional Environmental Condition	Total Tested Cases	Successful cases	Causes of Failure			
			MBL (> 90 kN)	MBR (< 2.5 m)	Surface clearance (< 30 m)	Seabed clearance (< 10 m)
Ext_30deg	11	7	4	0	0	0
Ext_60deg	11	6	5	0	0	0
Ext_90deg	11	5	6	0	0	0

3.3 Effect of Two Parameters

3.3.1 Submerged weight and outer diameter (OD). Varying the outer diameter of the bend stiffener can change the submerged weight of the bend stiffener, which significantly influences the vertical excursion of the power cable based on the static analysis results. The positive correlation of submerged weight, vertical excursion, MBL, and MBR is clearly visible in Table 12. The vertical excursion and maximum effective tension along the cable are presented in Figure 8 and Figure 9. A larger vertical excursion means a larger pulling length, which causes higher MBL and larger MBR.

Table 12. Dynamic Results with Varying OD (Density: 7.85 ton/m^3).

Case BSL2.4m_ID0.16m	Density (ton/m^3)	Submerged weight (ton)	MBL (kN)	MBR (m)	Vertical excursion max. (m)	Vertical excursion min. (m)
OD6	7.85	5.99	63.656	7.352	-67.584	-219.301
OD7	7.85	5.58	51.244	5.554	-67.556	-144.771
OD8	7.85	6.14	73.272	8.424	-67.601	-249.357
OD9	7.85	6.74	137.775	10.841	-67.685	-286.780
OD10	7.85	6.33	94.559	8.941	-67.633	-271.862

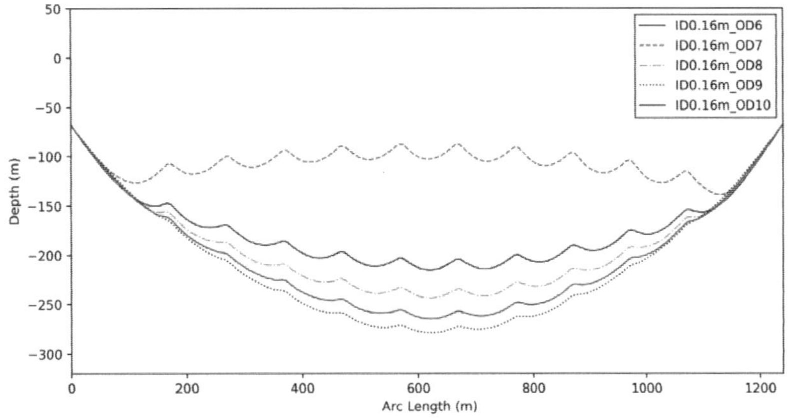

Figure 8. Vertical excursions with varying OD (Density: 7.85 ton/m^3).

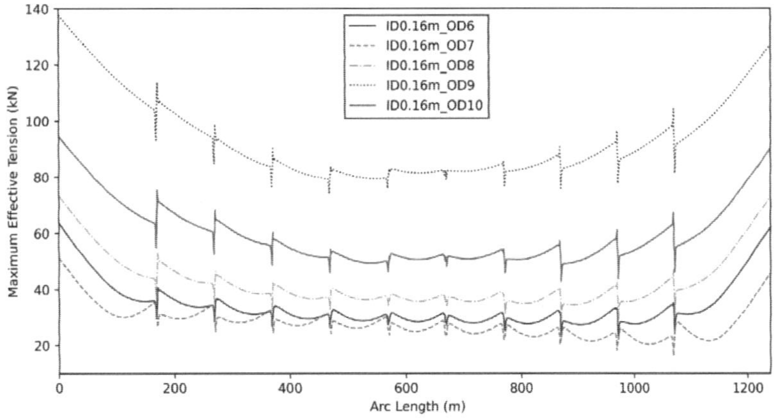

Figure 9. Maximum effective tension with varying OD (Density: 7.85 ton/m^3).

Vertical excursion, MBL, and MBR of the additional four cases are provided in Table 13, which have the same length, inner diameter, and submerged weight as "BSL2.4m_ID0.16m_OD8". Figure 10 and Figure 11 show the vertical excursion and maximum effective tension along the cable. The results show that bend stiffeners with different ODs, but the same submerged weight have similar vertical excursions, MBL, and MBR. The different submerged weight caused by varying OD is the intrinsic factor, which significantly affects the cable's vertical excursion, MBL, and MBR. It is necessary to keep the submerged weight of the bend stiffener in a proper range for the design of the bend stiffener.

Table 13. Dynamic Results with Varying OD (Submerged Weight: 6.14 ton).

Case BSL2.4 m _ID0.16 m	Density (ton/m^3)	Submerged weight (ton)	MBL (kN)	MBR (m)	Vertical excursion max. (m)	Vertical excursion min. (m)
OD6	8.02	6.14	70.328	7.662	-67.595	-242.106
OD7	8.54	6.14	70.189	7.600	-67.595	-241.861
OD8	7.85	6.14	73.272	8.424	-67.601	-249.357
OD9	7.23	6.14	77.242	8.876	-67.607	-255.620
OD10	7.64	6.14	78.363	7.948	-67.609	-257.147

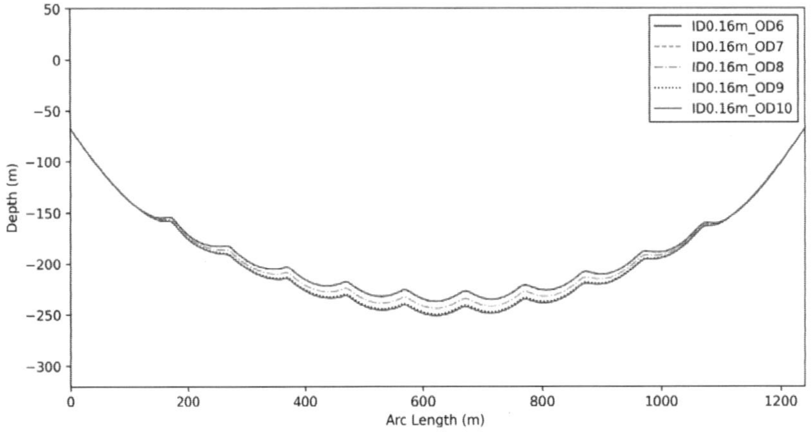

Figure 10. Vertical Excursions with varying OD (Submerged Weight: 6.14 ton).

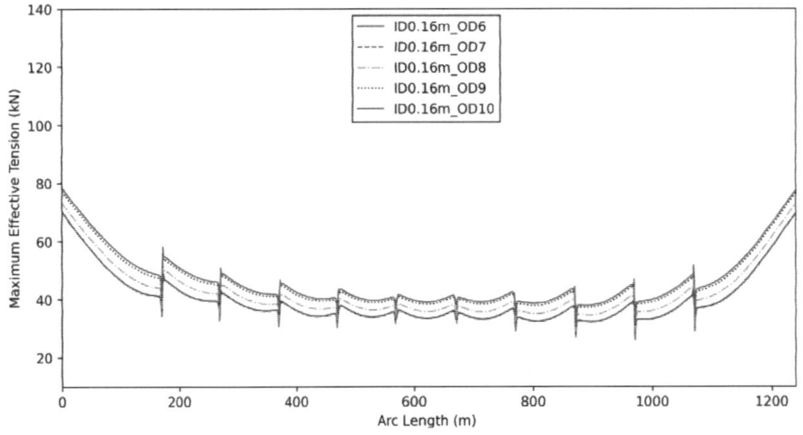

Figure 11. Maximum effective tension with varying OD (Submerged Weight: 6.14 ton).

3.3.2 Length of Bend Stiffener (BSL). The positive correlation between MBL and MBR is apparent, as shown in Figure 12. The positive effect of increasing BSL on MBR is limited to a small range. Bend stiffeners of 2.4 m length have larger MBR than those of 3.0 m length when $60\ \mathrm{kN} < MBL < 90\ \mathrm{kN}$. This finding can explain why three of the four best cases selected have 2.4 m length.

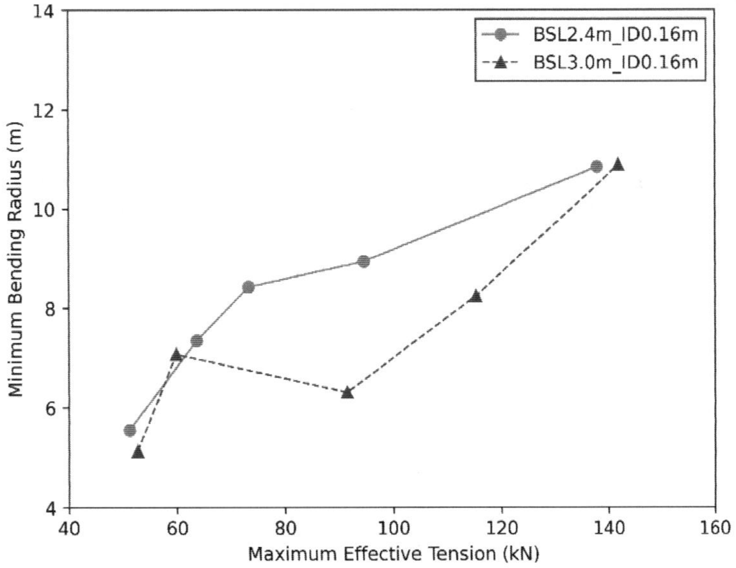

Figure 12. Relation between MBL and MBR under EL4

4. Conclusions

The mechanical properties of bend stiffeners applied to a suspended inter-array power cable system are investigated numerically. Different bend stiffeners with varying outer diameters and lengths are analyzed. The most suitable bend stiffener is determined based on the maximum effective tension, minimum bending radius, and fitness factor of both static and dynamic analysis. The influence of several parameters on the suspended inter-array power cable is investigated. The main conclusions in the present study are listed as follows:

1. The bend stiffener of length 2.4 m, inner diameter 0.16 m, and outer diameter 0.24-0.50 m is the best configuration in this study, which has the best balance between the MBL and MBR.
2. The outer diameter of the bend stiffener influences MBL and MBR by changing the submerged weight of the bend stiffener. The submerged weight of the bend stiffener grows when the bend stiffener is thicker, which leads to higher MBL, higher MBR and larger vertical excursion.
3. MBR will increase when BSL becomes larger, which positively influences the design. At the same time, the weight of the bend stiffener will increase with BSL, leading to larger maximum effective tension (MBL), which is undesirable from the design perspective. The positive effect of increasing BSL (3 m) on MBR is limited to a small range in the present study. Bend stiffeners of 2.4 m length have larger MBR than those of 3.0 m length when $60\ \mathrm{kN} < \mathrm{MBL} < 90\ \mathrm{kN}$.

References

[1] Ahmad I B, Schnepf A and Ong M C 2023 An Optimisation Methodology for Suspended Inter-Array Power Cable Configurations between Two Floating Offshore Wind Turbines, *Ocean Eng.* **278**, 114406.

[2] Yang S-H, Ringsberg J W and Johnson E 2018 Parametric Study of the Dynamic Motions and Mechanical Characteristics of Power Cables for Wave Energy Converters. *J. Marine Sci. Technol.* **23**(1), 10–29.

[3] Thies P, Johanning L and Smith G 2011 Assessing Mechanical Loading Regimes and Fatigue Life of Marine Power Cables in Marine Energy Applications. *Special Issue Proc. of the Institution of Mechanical Engineers, Part O, J. Risk and Reliab.* **226**, 18–32.

[4] Rentschler M U T, Adam F and Chainho P 2019 Design Optimization of Dynamic Inter-Array Cable Systems for Floating Offshore Wind Turbines. *Renewable Sustain. Energy Rev.* **111**, 622–35.

[5] Schnepf A, Lopez-Pavon C, Devulder A, Johnsen Ø and Ong M C 2022 Suspended Power Cable Configurations for Floating Offshore Wind Turbines in Deep Water Powering an FPSO. *Proc. of OMAE2022. Volume 8: Ocean Renewable Energy, 2022.*

[6] Schnepf A, Lopez-Pavon C, Ong M C, Yin G and Johnsen Ø 2023 Feasibility Study on Suspended Inter-Array Power Cables between Two Spar-Type Offshore Wind Turbines. *Ocean Eng.* **277,** 114215.

[7] Jonkman J, 2010 Definition of the Floating System for Phase IV of OC3. Report No. NREL/TP-500-47535, 979456.

[8] Beier D, Schnepf A, Van Steel S, Ye N and Ong M C 2023 Fatigue Analysis of Inter-Array Power Cables between Two Floating Offshore Wind Turbines Including a Simplified Method to Estimate Stress Factors. *J. Marine Sci. Eng.* **11**(6), 1254.

[9] Kvitrud A, Løland A H 2018 Observed Wave Actions on Norwegian Semi-Submersible and TLP Decks. *Proc. of the ASME 2018 37th International Conference on Ocean, Offshore and Arctic Engineering. Volume 1: Offshore Technology. Madrid, Spain. June 17–22.*

MATT'S PATROL: A Generalized framework to track marine litter path, fare and toll

J R. Jao[1,*], E A. Vallar [2], J R. Jao[4] , M Era [3] and M J Lañada [3]

[1]Department of Software Technology, College of Computer Studies, De La Salle University, 2401 Taft Ave, Malate, Manila, 1004 Metro Manila, Philippines
[2]Department of Physics, College of Science, De La Salle University, 2401 Taft Ave, Malate, Manila, 1004 Metro Manila, Philippines
[3]Department of Sociology and Behavioral Sciences, De La Salle University, 2401 Taft Ave, Malate, Manila, 1004 Metro Manila, Philippines
[4]Management and Organization Department, Ramon V. del Rosario College of Business, De La Salle University, 2401 Taft Ave, Malate, Manila, 1004 Metro Manila, Philippines

[*] Correspondence: jazzie.jao@dlsu.edu.ph

Abstract. Attention is given to the mapping of marine litter pathways due to their negative impact on the environment, society, and economy. Many models have been created to examine ocean dynamics and particle tracking systems. This study aims to contribute to these existing models by introducing a framework called "MATT'S PATROLL: MArine liTTer'S PATh, faRe, and tOLL." This framework examines three pillars needed for the management of marine litter, the computational, socio-ecological, and economic arms. The Lagrangian particle tracking model is used to model the spatio-temporal behavior of litters on the surface of the ocean. The "faRe" component describes the direct and indirect effects of meso-particles on sea and land populations while they wait to accumulate on shorelines and other settlements. The "tOLL" component examines the long-term ecological and socio-economic costs on ecosystems in land and water, as well as the damage and loss in gross revenue from the marine sector when marine litters penetrate habitats and livelihoods. To illustrate the framework, artificially generated litter data around Visayas region is used to simulate path, fare, and toll calculations.

1. Introduction

The term marine litter refers to any solid waste materials that have been discarded, disposed of, or abandoned in the marine and coastal environment and this can include a variety of materials such as plastics, wood, metals, glass, rubber, clothing, paper, and more [1]. The most prevalent form of marine litter is plastic waste, which is predicted to reach 12 billion metric tons in landfills or natural environments by 2050 and could potentially populate the oceans with at least 5 trillion particles if current management and human actions continue as they are. This shows that knowledge on litter behaviour is crucial for effective waste management [2] and preservation of marine life as it can also threaten the underwater ecosystem [3]. Various studies have shown that plastic debris has disrupted the feeding patterns of marine animals [3], resulting in a disturbed aquaculture and livelihood processes. According to [4] plastic can also serve as a breeding ground for dangerous microorganisms that can inflict harm on organisms that consume them. The consumption of plastic debris in oceans may result in the transfer of these pathogenic microorganisms, particularly when the creatures that consume them can act as sources and carriers of infectious diseases. Since plastics tend to remain in the marine environment for a longer period due to their durability and slow degradation, the possibility of these materials in various oceans around the globe is highly likely [5]. This litter accumulations impose economic, environmental, human health, and aesthetic problems [6] on communities and fisherfolks. In general, marine litter path is a well-established problem statement that are already solved using several computational solutions. For example, studies of [7–9] involves the development of models to trace the origin of debris found at coastal locations and applied this to the East Asian area focusing on the west coast of Japan. Using a forward and backward-time particle-tracking model, the researchers successfully identified the origin of the debris that was found on the west coast of Japan and to estimate the quantity of debris arriving from specific sources. [10] used a Lagrangian modeling method to examine the movement patterns of neutrally buoyant waste materials in the upper ocean layer of the coastal region of South-East Bay of Bengal.

However, despite the existing mathematical tools available for use, marine litter tracking is not a straightforward and requires specific assumptions [11. 12] asserts that floating marine litter tracking heavily relies on the monitoring and prediction of coastal water transport. The accumulation of continental outflows commonly occurs in the coastal region, which makes it a crucial area to observe. Nevertheless, the surface ocean dynamics in coastal areas are intricate because of numerous forces acting simultaneously over various time and space scales. In terms of model inputs, reliable data on coastal and marine litter count is relatively scarce and often they are only complete or at least available on places with well-established automated monitoring and data storage systems [6]. Marine litter is becoming a growing issue in the Philippines and its consequences have already been felt [13]. According to research, the Philippines is one of the top twenty countries with the highest amount of mismanaged plastic waste and is responsible for 0.28-0.75 million metric tons of plastic waste entering the ocean [14]. Despite the increasing concern about marine litter pollution in the Philippines, there are still only a few studies examining this problem. [6] frames marine litter as a complex and multi-dimensional challenge requiring a non-singular solution. This can be done through collaborations between multidisciplinary fields of study. For example, aside from predicting pathways of marine litter to prevent accumulation, understanding the socio-economic impact of plastics is also essential in shaping future international negotiations on its design, production, reuse, and reprocessing. Further, research on the sources, fate and consequences of marine litter is still in its infancy [15]. With this, study presents the following objectives below:

(1) Provide a unified framework of marine litter management with computational, sociological, and economical components to track and quantify marine litter life cycle from source to sink in the long-term basis.
(2) Simulate the path of several marine litter particles using the Lagrangian ocean dynamics framework in PARCELS [16] library in Python.
(3) Apply the simulation of path tracking to a specified marine area around a specified area in the island of the Philippines.

As marine litter caters several compartments, the study at hand is focused only on inspecting the path of meso-particles with a scale of 5-2.5 cm, since numerical description requirement for micro and nano particles populating the marine ecosystem requires different domain. Moreover, since marine litter particles encompass a wide range of types, the study is only focused on evaluating plastic debris and a static hydrodynamic data from 2020 and interpolated over a wide range of days to accommodate the tracking mechanism of the model. While it is generally accepted that plastic waste found in the ocean primarily comes from sources on land, recent studies indicate that sea-based sources may also play a significant role in contributing to the problem [17]. However, accurately modelling the distribution of litter on a global scale is difficult due to the uncertainty surrounding the location and amount of plastic waste that needs to be adequately managed and is entering the ocean. Our study explicitly examines waste sources originating from land.

2. Methodology

2.1. MATT's PATROLL

Marine litter management involves a multi-stage process that requires the use of multidisciplinary techniques. This study presents a generalized framework to tackle the issue by identifying key processes: Path, Fare, and Toll. The framework begins by modeling the problem using OGCMs and particle tracking models. Oceans play a crucial role in transporting materials, so predicting their patterns is essential in managing the impact of the marine litter life cycle. The "path" component of the framework aims to establish this information. To achieve this, multidimensional hydrodynamic data and particle tracking systems may be used to understand ocean processes and resulting transports at different time and spatial scales. It is recognized that predicting accurate ocean circulation and particle dynamics is challenging without problem simplification. Therefore, the development of sophisticated models is discouraged at the first run. While ocean dynamics are not entirely predictable, other features like temperature, elevation, pressure, and wind data may be included to simulate predictions with more realistic contexts. This study uses a Lagrangian particle tracking approach to model transport and surface current fields with particle dimension ignored to reduce the complexity of dynamics brought by volume and shape profile. The litter path is afterward assessed by a user-defined spatiotemporal frame and is quantified by setting an origin or location where observations are made, with a threshold value indicating the "radius of coverage" for litters surrounding the area. This radius is relative and may be defined according to the following factors: litter type and amount, uncertainty of ocean data, and limited monetary budget for ocean drive cleanups.

As there may be a set of locations where marine litter is expected to pass, the particles may be simplified by using a clustering algorithm to map these locations into a representative point in space, used as a reference location where litters are expected to be found. The identification of this clustering algorithm may depend on the expected path of waste and the following targets: (1) the representative points should not over-generalize the cluster of points, and it should also not over-specify the litter locations, and (2) the point should refer, as much as possible, to only a specific type of waste in cases when transported litters are of mixed-waste type. This specific requirement may be complex as different types of waste have varying weight, size, volume, and shape contributing to their dynamics in ocean currents and other variables. Dealing with marine plastic pollution is a complex issue that spans across various sectors. As such, it is crucial to develop customized strategies that can facilitate informed decision-making. The context used in marine fare identification phase serves as a guiding principle in formulating valuation models, risk assessment, and mitigation scenarios in litter management. For example, the lack of action in marine litter prevention has direct and indirect costs [18] that should be calculated after identifying the tendency to participate of affected community which may depend on several factors and actors. While marine litter is generally perceived to be unfavourable [19] some social groups have lower willingness to participate due to lack of incentives and behavioral differences. When examining the impact of tourism on an area, it is important to first evaluate its economic significance. This involves determining whether the area is a popular destination in its own right or just a passageway to other destinations.

To accurately assess the risks and consequences of tourism, it's recommended to classify the "fare" component into four categories: marine, tourism and recreation, ecosystem, and economy. The last category considers the indirect effects of marine litter caused by transport paths being redirected to vulnerable ecosystems that weren't factored into the simulation. To reduce the negative impact of marine litter on society, it's essential to have a policy mix that considers public support and trade-offs between short-term actions and long-term prevention policies. This is especially important in situations where integrated solutions are needed. A comprehensive approach that explores the root causes of litter, including the top contributing factors, is crucial. Such an approach will help to identify the most effective measures to address the issue and prevent the negative consequences of marine litter. [20]. When adopting a multidisciplinary approach towards the reduction of marine litter, it is imperative to have a comprehensive understanding of the community's attitudes towards this issue, as well as the economic benefits that can be gained from the reduction of marine waste on beaches. Such knowledge is critical to comprehend the decline in welfare benefits resulting from marine debris. The findings may support evaluation of costs and benefits of different public policies and management strategies, including beach clean-ups and integrated policy/management actions. In essence, the quantification of these evaluated measures and risks can be effectively determined by utilizing a base valuation model such as the Marine Litter Valuation Model. The model considers the direct use-value, indirect use-value, option value, bequest value, and existence value, as well as the exposure of marine litter, and the concentration of marine litter at the minimum [21]. On the other hand, when it comes to the "toll" component, it considers the long-term effects of marine litter transport and accumulation. This phase's valuation may differ from the "fare" component as certain costs are not realized until a specific period. To identify such scenarios, one needs to look at the long-term behaviour of litter and how it is likely to settle in particular locations compared to others. Additionally, it is assumed that marine debris encountered can be of different types and sizes other than plastic. Therefore, the accumulation patterns and life cycle of these litters vary greatly and may contribute to how the long-term valuation model is perceived.

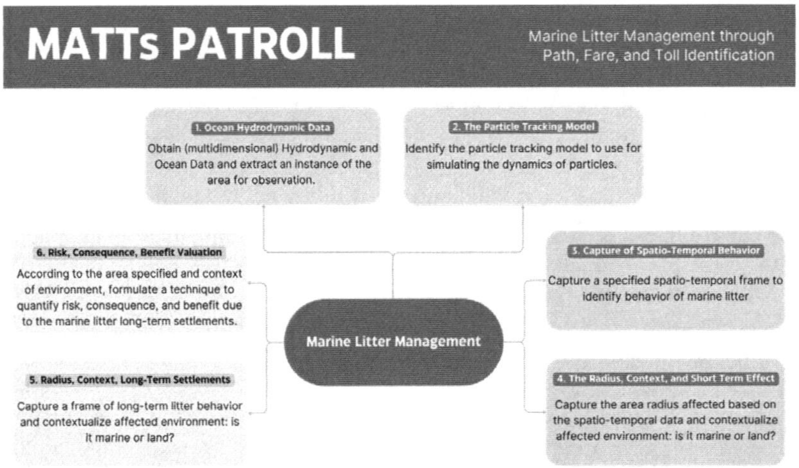

Figure 1. (a) The framework of MATTs PATROLL for Marine Litter Management through Path, Fare and Toll identification.

The particle trajectories are modelled using the solution of the following equation [22]

$$\vec{x}_{t+\Delta t} = \vec{x}_{t+\Delta t} + \int_{t}^{t+\Delta t} \{\vec{u}_c(\vec{x}_t, t)\}\Delta t + R\sqrt{2K_h \Delta t}$$

where $\vec{x}_t$ is the position of a particle represented as a vector with x-y components at time t, $\vec{u}_c(\vec{x}_t, t)$ is the ocean current. The study employs a daily interval and $\Delta t = 12$ hours. The time integration is done using the Fourth Order Runge-Kutta advection scheme of [23]. During the simulation, several particles that are pushed ashore along the coast and remain within a specified zone for more than 12 consecutive hours will be treated as long-term litters floating around this area. The litter is tracked within a 300-day period and spatio-temporal captures are examined according to a maximum radius specified around selected coastal locations. The OceanParcels v2.1.5 global framework enables customizable particle tracking simulations using outputs from ocean circulation models to track marine litter. The study uses a two-dimensional hydrodynamic data (U, V) such that the U is the eastward seawater velocity and V is the northward seawater velocity, referring to the horizontal speed of water moving towards the east and towards the north, respectively. These two variables describe the speed and direction of sea water, necessary for modelling litter under the assumption of known flow environments. The meridional and zonal ocean data from Copernicus Marine global ocean reanalysis is used based on assimilation of observations and interpolation of the Arakawa C native grid with 1/12 degree and 50 vertical levels. The dataset consists of daily mean 2D fields from January 1, 2021 to January 1, 2022. The field set input in Ocean Parcels is an area cropped around (N, S, E, W) = (15.143828754476074, 8.352073931417541, 125.60136840796869, 117.74763010348647) covering the greater Visayas region of the Philippines. It is highlighted that this study does not provide a reliable count of meso-litter particles as they are artificially simulated. Figure 2 shows the location of these particles at day 0 of the simulation. The approximate weight of plastic, accounted as one of the factors determining the path of particles, is determined using the 2014 statistics report from [24] which details the average weight of different plastic products. The reason behind artificially simulating the particles is the lack of information on the quantity of marine plastic debris in many parts of the Philippines [4]. The following assumptions are set in the simulation of the meso-litter path:

1. The interaction of particles with the hydrodynamic field does not cause significant modification to the latter.
2. The significant contributor to the displacement of particles is only the hydrodynamic field; thus, wind vectors and other factor do not contribute to its motion.
3. The particle-particle interaction effect is not significant, that is, a group of particles cannot cause significant path and dynamics modification to another group of particles.
4. Since the simulation uses hydrodynamic data defined on C-grids, particles will not be retained on land areas where there is an undefined velocity data.

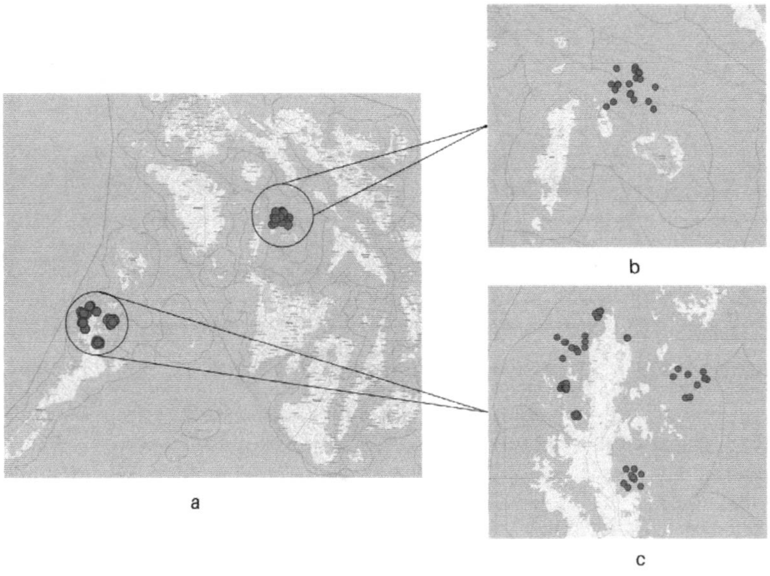

Figure 2. (a) Distribution of Simulated Marine Meso-litters around the vicinity of (b) El Nido-Calabugdong Island, and (c) Romblon Island.

2.2. *The design of marine litter fare and toll*

After path identification, it is expected to have several displacement locations. A clustering technique called Density-Based Spatial Clustering of Applications with Noise (DBSCAN) [25] algorithm is employed to reduce the number of data points through (1) partitioning group of particles and (2) calculating a centroid. DBSCAN is an unsupervised machine learning algorithm that works by identifying clusters in space based on their density and is more effective in handling spatio-temporal data [26]. The model obtains densities of displaced litter in a single latent variable represented as a pair of latitude-longitude positions and considers dense regions as clusters and the surrounding areas with low density as noise. To detect the cluster's borders, the algorithm looks for a decrease in density, which indicates the transition zone between the high-density core and the low-density surroundings such that the cluster's boundary can be thought of as the area where the density starts decreasing. Figure 3 shows the result of DBSCAN algorithm for selected simulation days where highly-dense collection of litter are treated to be of one cluster. These cluster movements serve as the basis in calculating a representative point, the one closest to the centroid of the cluster, used in plotting the litter path in a chosen spatio-temporal frame.

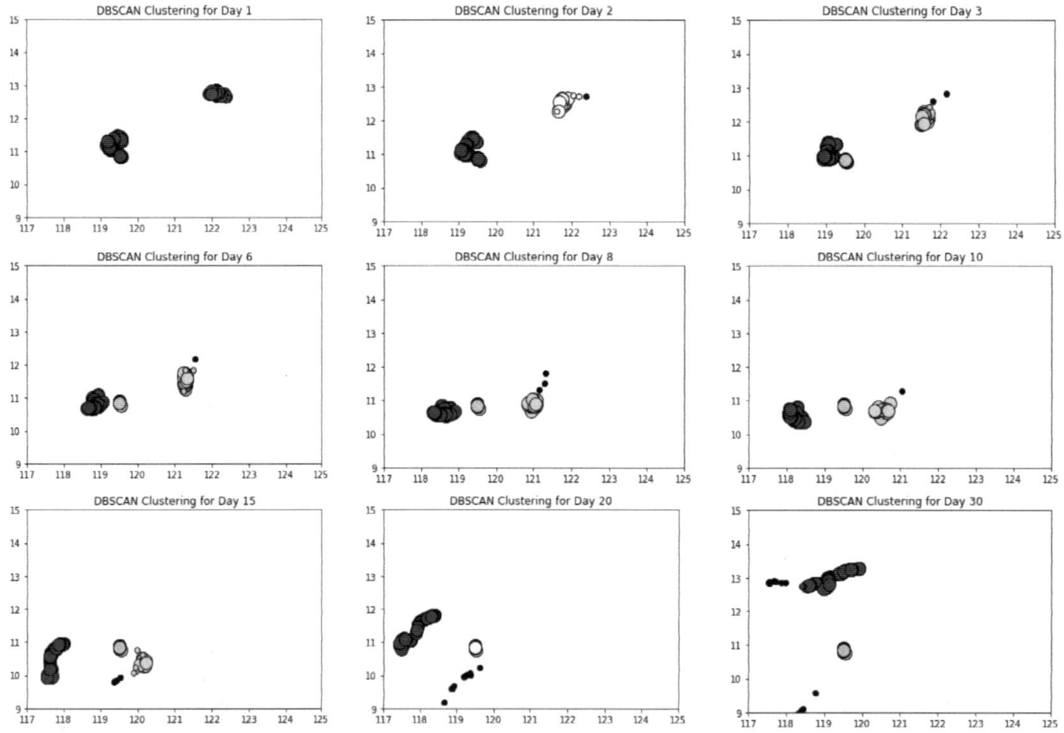

Figure 3. Day 1-6,10,15,20 Result of DBSCAN Clustering Algorithm.

Determining the precise quantity and location of meso-litter on the surface of the ocean can be quite a complex task. To simplify the situation, it is common practice to identify representative points that can help generalize the distribution of debris as a dense collection of particles pivoted as a single point. By using these points, it becomes possible to capture the spatio-temporal behavior of litter and determine its overall transport path. Additionally, setting a radius around these selected areas can help identify the affected communities, making it easier to calculate the cost brought by this transport of litter.

3. Results and Discussions

A scatter plot of the captured position of particles in time is lacking when it comes to showing the density of particles situated in space. To have a clearer observation of the change in particle density in a specific area for a period time, kernel density map is used as illustrated in **Figure 4**. The simulated plastics are tracked forward in time near El Nido, Taytay, and Romblon islands, carrying the following assumptions: their weights, shape, and chemical composition are negligible in their dynamics on the field data itself.

Day 1 to Day 8 shows a significant change in the transport of marine litters turning in the westward direction due to the high zonal velocity, v.

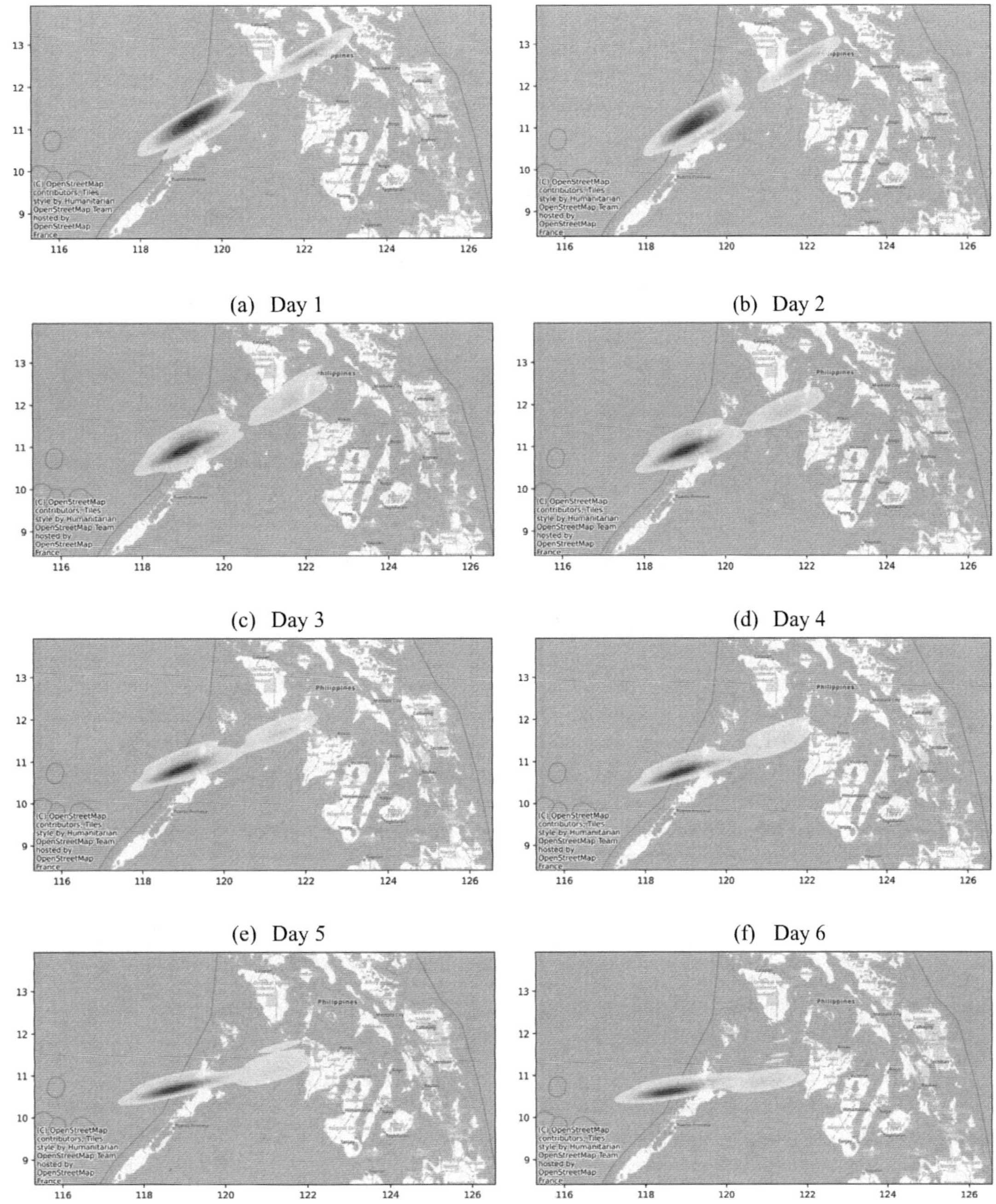

(a) Day 1 (b) Day 2

(c) Day 3 (d) Day 4

(e) Day 5 (f) Day 6

(g) Day 7 (h) Day 8

Figure 4. Simulation of Meso-litters Displacement from Day 1 to Day 8 (January 1, 2021-January 9, 2021), daily 12-hour average.

Figure 5 shows that over the next 20 days, ocean velocity fields will continue to direct particles towards the northern part of the region. As a result, any litter present in the vicinity of Palawan Island will likely be carried towards the region surrounding southern Luzon. This movement will occur gradually over time, and the concentration of dense particle collections is expected to remain intact. At day 50, the highly dense particle collection starts to traverse the upper part of the Philippines, while other particles remain on the vicinity of Palawan. It is possible to prevent and manage the consequences of the theoretical scenario by conducting ocean clean-up drives before the predicted location of litter. To avoid this issue, communities near Palawan can be identified to gauge their willingness to reduce marine debris by implementing measures such as limiting single-use plastics, enforcing strict waste management practices by local government units and penalizing those who do not comply with environmental regulations. These values can be expressed as either positive (supporting prevention practices) or negative (not supporting prevention practices) in the marine "fare," which can be evaluated later in the valuation model. To determine the affected communities while evaluating the marine "fare," the representative points from DBSCAN are used to obtain a convex hull that indicates the approximate affected areas, as shown in Figure 6. The area of the polygon may be used to assess the extent of marine litter transport during the specified period.

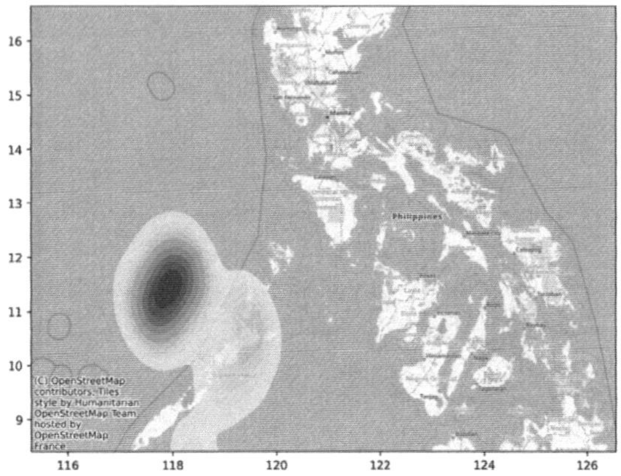

(a) Day 20 Density of Displaced Particles

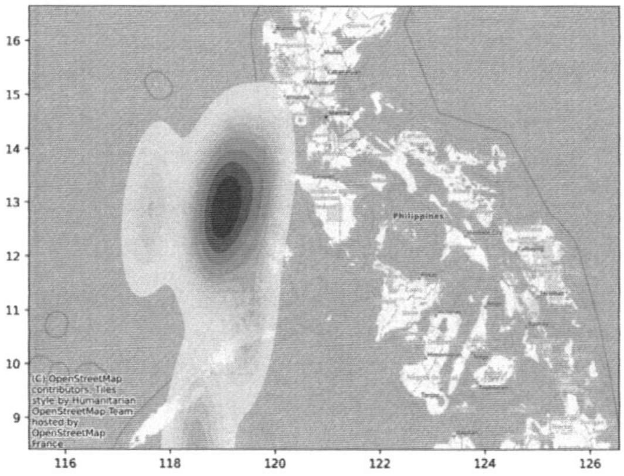

(b) Day 30 Density of Displaced Particles

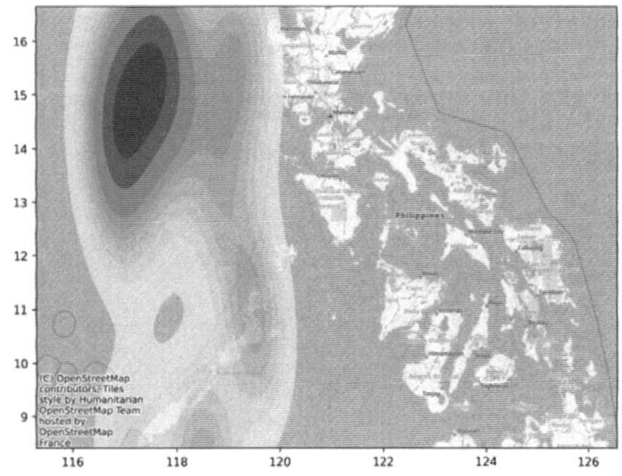

(c) Day 50 Density of Displaced Particles

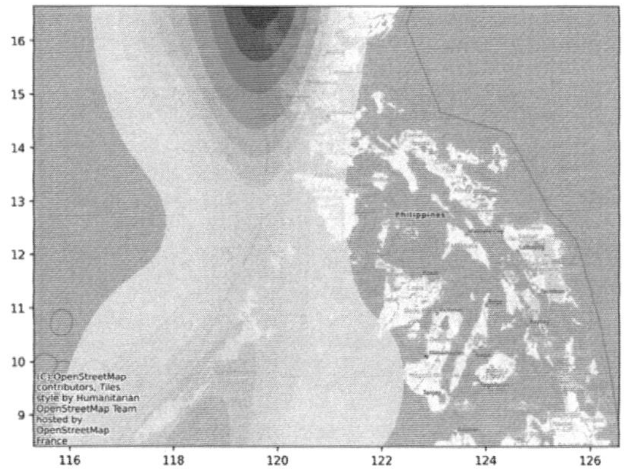

(d) Day 300 locations of All particles

Figure 5. Simulation of meso-litters displacement for Day 20, 30, 50, daily 12-hour average.

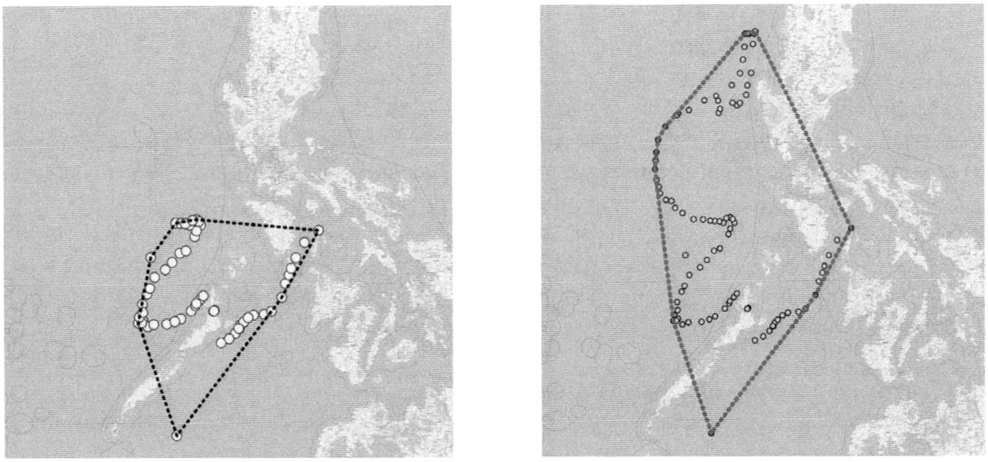

(a) Day 30 of displaced particles using the representative points from DBSCAN.

(b) Day 300 of displaced particles using the representative points from DBSCAN.

Figure 6. The convex hull of displaced particles using the representative points from DBSCAN.

The simulation study examined the long-term effects of marine litter by focusing on the period of 300 days and beyond. The results indicated that despite the passage of time, the amount of litter in the Palawan region did not decrease and instead showed a significant accumulation. These findings underscore the importance of carefully evaluating the impact of marine litter in specific coastal areas of Palawan. Quantifying the risks and consequences associated with the buildup of litter can aid in this evaluation. It is essential to consider the socio-economic costs, such as the negative effects on tourism and the livelihood of the local community, during the summer months when Palawan is a popular tourist destination. Incorporating these costs in the valuation model formulation will provide a more comprehensive understanding of the impact of marine litter in the region.

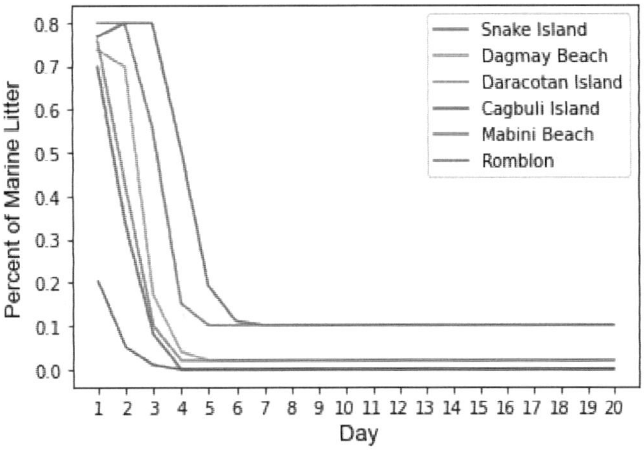

Figure 7. The percent of marine litter observed within a 10-km radius of several observation locations from Day 1-10, daily 12-hour average.

Determining the extent of marine litter involves carefully selecting six observation points: Snake Island, Dagmay Island, Daracotan Island, Cagbuli Island, Mabini Beach, and Romblon Island. To evaluate the amount of litter within a given threshold radius, a buffer area with a radius of 10 km is created for each point. Figure 7 portrays the percentage of marine debris occupying different locations within the radius. Based on the research, litter points adjacent to Romblon Island typically migrate away quickly, while debris around Snake Island remains longer. The transportation of litter impacts not only the source but also the surrounding communities in Mindoro, Tablas, and Semirara Island, as well as the southern part of Palawan. Controlling marine litter in the short-term is crucial as it tends to oscillate and return to the coastal regions, negatively impacting community perception, marine ecosystems, and livelihoods. Therefore, socio-economic and legislative measures are necessary to establish a sustainable, closed-loop management system.

4. Conclusions and Recommendations

Using a model that tracks particles via the Lagrangian method, the research investigated the coastal accumulation and trajectory of artificially generated marine litter originating from five coastal regions. The transport patterns were categorized based on ocean hydrodynamic data and dependence on particle weight. Litter coming from Romblon is seen to have faster rate of litter displacement and Palawan Island tends to have greater tendency to accumulate them in longer amount of time. In long-term basis, the litter source assumed to be situated in these five locations, are expected to damage even those locations which are assumed to not discharge any significant marine litter. This shows that the existence of marine litter in vulnerable and open areas, may be brought along those areas wherein good waste management is being practiced. Some debris in Palawan tends to be retained on the coasts from which it originated. In conclusion, it is imperative to recognize that marine litter management is not a problem that can be tackled by a single entity or location. Rather, it is a global issue that requires collective action from individuals, governments, and organizations around the world. Responsibility for the actions is shared, and different entities must work together to conserve both marine and land ecosystems. While the study offers a valuable framework for addressing marine litter, its limitations must be acknowledged. Specifically, it focuses on only a portion of the plastic life cycle in the ocean and does not account for factors such as litter size and the effects of wind and ocean currents. Nevertheless, the study remains a valuable resource for governments, non-profit organizations, and the public to take meaningful steps towards reducing marine litter. For future work, it is recommended to validate the simulation obtained by comparing it with known particle trajectory data in region of Visayas.

References

[1] Sivadas S K, Mishra P, Kaviarasan T, Sambandam M, Dhineka K, Murthy M V R, Nayak S, Sivyer D and Hoehn D 2022 Litter and plastic monitoring in the Indian marine environment: A review of current research, policies, waste management, and a roadmap for multidisciplinary action *Mar. Pollut. Bull.* **176** 113424.

[2] Liubartseva S, Coppini G, Lecci R and Creti S 2016 Regional approach to modeling the transport of floating plastic debris in the Adriatic Sea *Mar. Pollut. Bull.* **103** 115–27

[3] Edgar G J, Stuart-Smith R D, Willis T J, Kininmonth S, Baker S C, Banks S, Barrett N S, Becerro M A, Bernard A T F, Berkhout J, Buxton C D, Campbell S J, Cooper A T, Davey M, Edgar S C, Försterra G, Galván D E, Irigoyen A J, Kushner D J, Moura R, Parnell P E, Shears N T, Soler G, Strain E M A and Thomson R J 2014 Global conservation outcomes depend on marine protected areas with five key features *Nature* **506** 216–20

[4] Abreo N A S, Macusi E D, Cuenca G C, Ranara C T B, Andam M B, Cardona L C and Arabejo G F P 2015 Nutrient Enrichment, Sedimentation, Heavy Metals and Plastic Pollution in the Marine Environment and its Implications on Philippine Marine Biodiversity: A Review *IAMURE Int. J. Ecol. Conserv.* **15**

[5] Barnes D K A, Galgani F, Thompson R C and Barlaz M 2009 Accumulation and fragmentation of plastic debris in global environments *Philos. Trans. R. Soc. B: Biol. Sci.* **364** 1985–98

[6] Chassignet E P, Xu X and Zavala-Romero O 2021 Tracking marine litter with a global ocean model: where does it go? Where does it come from? *Front. Mar. Sci.* **8** 667591
[7] Kako S, Isobe A, Kataoka T and Hinata H 2014 A decadal prediction of the quantity of plastic marine debris littered on beaches of the East Asian marginal seas *Mar. Pollut. Bull.* **81** 174–84
[8] Isobe A, Uchida K, Tokai T and Iwasaki S 2015 East Asian seas: A hot spot of pelagic microplastics *Mar. Pollut. Bull.* **101** 618–23
[9] Pilechi A, Mohammadian A and Murphy E 2022 A numerical framework for modeling fate and transport of microplastics in inland and coastal waters *Mar. Pollut. Bull.* **184** 114119
[10] Declerck A, Delpey M, Rubio A, Ferrer L, Basurko O C, Mader J and Louzao M 2019 Transport of floating marine litter in the coastal area of the south-eastern Bay of Biscay: A Lagrangian approach using modelling and observations *J. Oper. Oceanogr.* **12** S111–25
[11] Schneider F, Parsons S, Clift S, Stolte A and McManus M C 2018 Collected marine litter — A growing waste challenge *Mar. Pollut. Bull.* **128** 162–74
[12] Lentz S J and Fewings M R 2012 The wind- and wave-driven inner-shelf circulation *Annu. Rev. Mar. Sci.* **4** 317–43
[13] Cruz L L B and Shimozono T 2021 Transport of floating litter within Manila Bay, Philippines *Marine Pollution Bulletin*
[14] EY S 2018 Modeling of Small Sea Floaters in the Central Mediterranean Sea: Seasonality of At-- Sea Distributions *Oceanogr. Fish. Open access J.* **7**
[15] Woods J S, Verones F, Jolliet O, Vázquez-Rowe I and Boulay A-M 2021 A framework for the assessment of marine litter impacts in life cycle impact assessment *Ecol. Indic.* **129** 107918
[16] Kehl C, Nooteboom P D, Kaandorp M L A and Sebille E van 2023 Efficiently simulating Lagrangian particles in large-scale ocean flows — Data structures and their impact on geophysical applications *Comput. Geosci.* **175** 105322
[17] Lebreton L, Slat B, Ferrari F, Sainte-Rose B, Aitken J, Marthouse R, Hajbane S, Cunsolo S, Schwarz A, Levivier A, Noble K, Debeljak P, Maral H, Schoeneich-Argent R, Brambini R and Reisser J 2018 Evidence that the Great Pacific Garbage Patch is rapidly accumulating plastic *Sci. Rep.* **8** 4666
[18] Mcilgorm A and Rule M 2009 *Understanding the Economic Benefits and Costs of Controlling Marine Debris in the APEC Region*
[19] Athens N T U, Aretoulakei E, Ponis S, Plakas G and Agalianos K 2021 Marine plastic littering: A Review of socio economic impact *J. Sustain. Sci. Manage.* **16** 276–300
[20] Grilli G, Andrews B, Ferrini S and Luisetti T 2022 Could a mix of short- and long-term policies be the solution to tackle marine litter? Insights from a choice experiment in England and Ireland *Ecol. Econ.* **201** 107563
[21] Lee J 2015 Economic valuation of marine litter and microplastic pollution in the marine environment: An initial assessment of the case of the United Kingdom
[22] Seo S and Park Y-G 2020 Destination of floating plastic debris released from ten major rivers around the Korean Peninsula *Environ. Int.* **138** 105655
[23] Dormand J R and Prince P J 1980 A family of embedded Runge-Kutta formulae *J. Comput. Appl. Math.* **6** 19–26
[24] Statista-Research 2014 *Average weight of polyethylene terephthalate bottles as of 2014, by type (in grams)*
[25] Ester M, Kriegel H-P, Sander J and Xu X 1996 A Density-Based Algorithm for Discovering Clusters in Large Spatial Databases with Noise 226–231
[26] Khan K, Rehman S U, Aziz K, Fong S and Sarasvady S 2014 DBSCAN: Past, Present and Future *Fifth Int. Conf. Appl. Digit. Inf. Web Technol. (ICADIWT 2014)* 232 8

Effect of a rotational damper on a moored and articulated multibody offshore system in waves

Qi Zhang, Ould el Moctar, Changqing Jiang*

Institute of Ship Technology, Ocean Engineering and Transport Systems, University of Duisburg-Essen, Duisburg 47057, Germany

*E-mail: changqing.jiang@uni-due.de

Abstract. Motions and loads of an articulated and moored floating platform consisting of multiple bodies in waves are investigated through numerical analysis. The wave–structure interaction (WSI) problem is solved using a high-fidelity viscous–flow solver that couples nonlinear rigid body motions, multibody interactions with an internal connection, and mooring dynamics. The study focuses on two modular floating structures (MFSs) connected by a flexible joint, with and without a rotational damper, and positioned using four symmetrical mooring lines. Multibody responses and the corresponding loads acting on the mooring lines and hinged joints are analyzed. Our results reveal that the influence of the damper on heave motions is less significant. Notably, the presence of the rotational damper has a noticeable impact on pitch motions between the two hinged MFSs. Introducing a rotational damper on the flexible joint effectively dampens the highly dynamic pitch motions while not imposing additional loads on the flexible joints.

1. Introduction

As the global population continues to grow, the availability of natural resources and land space is increasingly constrained. To address this challenge, the concept of very large floating structures (VLFS) has been proposed to alleviate land space limitations. They possess the potential to fulfill a diverse array of functions, such as supporting marine resource development, facilitating oil and gas exploration, enabling offshore tourism, and facilitating fish farming. Additionally, VLFSs can contribute to expanding land space for airports and offer opportunities for future generations. The prototype for VLFSs can be traced back to the "sea station" concept proposed by Edward R. Armstrong [1]. Various concepts and applications of VLFSs for offshore and coastal use have been reviewed and summarized by Lamas-Pardo et al. [2]. Jiang et al. [3] provide a comprehensive overview of the state-of-the-art advancements in modeling hydrodynamics and hydroelasticity for VLFSs.

Due to their substantial size, VLFSs often experience significant bending moments and shear forces. Consequently, it is crucial to carefully evaluate their hydroelastic behavior in waves during the design process [4]. In contrast to traditional VLFSs, the concept of modular floating structures (MFSs) has emerged. MFSs are constructed from multiple interconnected modules, offering advantages such as reduced fabrication, transportation, and installation costs [5]. MFSs have even been developed as fundamental building blocks for artificial islands within the Space@See project [6]. For modularized VLFSs, one of the pivotal technologies is the connection design, which can employ various articulation techniques, including hinged, prismatic, cylindrical, and screw joints. Additionally, these techniques can be categorized into rigid and flexible joints. A novel design presented by [7] introduces the flexible-base hinged connector (FBHC), comprising a hinged joint and two flexible bases, to reduce connection loads in VLFSs. Figure 1 provides examples of common applications of hinged joints in offshore floating bodies.

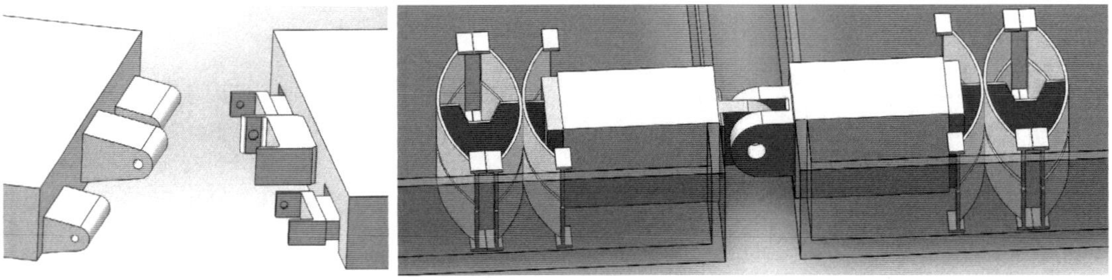

Figure 1. Typical articulated connectors for VLFSs [3]. (reproduced with permission from Elsevier).

Based on our review of research on MFSs, it becomes evident that the potential flow theory approach has been widely utilized. In these studies, connections are conventionally modeled as independent linear springs with six degrees of freedom [8]. For instance, Jiang et al. [9] assessed the potential flow theory solver's ability to simulate an articulated and moored floating system consisting of multiple bodies. Their findings indicated that a satisfactory agreement between simulations and measurements was attained, particularly when wave frequencies significantly differed from the bodies' natural frequencies. Nevertheless, discrepancies emerged when dealing with wave frequencies near the natural frequencies of the bodies. This deviation can be attributed to the restricted incorporation of flow viscosity or intense nonlinear free surface flows within the potential flow theory. To enhance accuracy in such scenarios, potential flow theory solvers often require the introduction of additional damping mechanisms.

In previous work, we developed and validated a mooring–viscous flow solver to consider viscous-flow effects, nonlinear wave-structure interaction as well as mooring dynamics, focusing on a single body [10]. Subsequently, we extended our investigation to include two different connections, namely rigid joints and flexible joints, between two modular floating structures to study the motions and loads of an articulated and moored floating platform consisting of multiple bodies in waves [11, 12]. In this paper, we introduce a novel aspect by presenting an analysis of a rotational damper integrated into an articulated and moored floating multibody platform subjected to waves. Our study primarily examines the influence of the rotational damper on the wave-induced motions and loads experienced by the MFSs. The comprehensive analysis encompasses various aspects, including the characterization of multibody motions, the evaluation of forces within the mooring lines, and the assessment of forces and moments acting on the connection joints. Ultimately, we present our results and derive conclusions based on our findings.

2. Theoretical background

This section gives a brief overview of used numerical methods, including the Navier-Stokes equations governing fluid dynamics, a lumped-mass model addressing mooring line dynamics, and a forward algorithm simulating the nonlinear rigid body motions of the MFSs. For a more comprehensive description, refer to Jiang [13]; Jiang and el Moctar [11].

2.1. Fluid dynamics

The flow in the present study is assumed to be incompressible, viscous and Newtonian, where the governing continuity and momentum conservation equations are written as follows:

$$\nabla \cdot \mathbf{v} = 0 \tag{1}$$

$$\frac{\partial \mathbf{v}}{\partial t} + \nabla \cdot (\mathbf{v}\mathbf{v}) = \nu \nabla^2 \mathbf{v} - \frac{1}{\rho} \nabla p_d \tag{2}$$

$$p_d = p - \rho (\mathbf{g} \cdot \mathbf{r}) \tag{3}$$

where $\mathbf{v}$ denotes the velocity vector, ∇ is the gradient operator; ν and ρ are the kinematic viscosity and density; p_d represents the dynamic part of pressure p. And $\mathbf{g}$ denotes the gravity vector and $\mathbf{r}$ is the position vector.

For hydrodynamic applications with free surface, the effective density and kinematic viscosity are expressed in terms of the volume fraction α using the Volume-Of-Fluid (VOF) method [14]:

$$\rho = \alpha \rho_w + (1 - \alpha) \rho_a \tag{4}$$

$$\nu = \alpha \nu_w + (1 - \alpha) \nu_a \tag{5}$$

where subscripts w and a stands for water and air, respectively. The existing velocity field converts phase fractions at each time step, the distribution and the development of the free surface is then estimated using the extended VOF formulation [15]:

$$\frac{\partial \alpha}{\partial t} + \nabla \cdot (\mathbf{v}\alpha) + \nabla \cdot (\mathbf{v}_r \alpha [1 - \alpha]) = 0 \tag{6}$$

where $\mathbf{v}_r$ represents the velocity normal to the interface, with its magnitude being proportional to the instantaneous velocity, symbolizing the artificial compression on the free surface.

2.2. Multibody dynamics

The behavior of multibody dynamic without joints is a extension of the formulation of nonlinear rigid body motions [16]:

$$\mathbf{H}(\mathbf{q}) \ddot{\mathbf{q}} + \mathbf{C}(\mathbf{q}, \dot{\mathbf{q}}) \dot{\mathbf{q}} = \tau \tag{7}$$

where $\mathbf{H}$ is the matrix of mass, $\mathbf{C}$ is the matrix of body force and τ is the external force vector. When kinematic constraints are existing, the motion equation transforms as follows:

$$\mathbf{H}(\mathbf{q}) \ddot{\mathbf{q}} + \mathbf{C}(\mathbf{q}, \dot{\mathbf{q}}) \dot{\mathbf{q}} = \tau + \tau_c \tag{8}$$

where τ_c represents the constraint force. Although this force is unknown, it possesses a crucial property that enables us to either calculate its value or eliminate it from the equation [17].

2.3. Mooring dynamics

The mooring dynamics are approached by a lumped-mass model [18]. This approach involves lumping the effects of mass, inertial reactions, and external forces at a finite number of nodes along the mooring line. The motion equation for each node i is formulated as follows:

$$(\mathbf{m}_i + \mathbf{a}_i) \ddot{\mathbf{r}}_i = \mathbf{W}_i + \mathbf{B}_i + \mathbf{D}_i + \mathbf{T}_i + \mathbf{C}_i \tag{9}$$

where $\mathbf{m}_i$ denotes the mass matrix of node i; $\mathbf{a}_i$, the corresponding added mass; $\mathbf{W}_i$, the net buoyancy; $\mathbf{T}_i$, tensile force; $\mathbf{B}_i$, the interaction with sea bed; $\mathbf{C}_i$, a numerical internal damping force; $\mathbf{D}_i$, the hydrodynamic drag force. Then the second-order ordinary differential equations is transformed into a first-order system of differential equations. Subsequently, it can be solved using the second-order Runge-Kutta integration scheme via a constant time step.

2.4. Coupling method

The developed coupling method consists of a mooring model used to account for mooring dynamics, and a mechanical model representing an articulated multibody floating system. The coupling technique involves connecting each solver module with the rigid body motion module. Forces induced by mooring and kinematic constraints at joints directly impact the equations governing rigid body motion. The interaction between flow dynamics and rigid body motion is achieved through an iterative process. The forces and moments acting on rigid bodies in the inertial coordinate system are expressed as:

$$\mathbf{F}_c + \mathbf{F}_m + m\mathbf{g} + \mathbf{F}_p + \mathbf{F}_v = \mathbf{F} \tag{10}$$

$$\mathbf{M}_c + \mathbf{M}_m + \mathbf{M}_p + \mathbf{M}_v = \mathbf{M} \tag{11}$$

where subscripts c and m are the constraint and mooring components, respectively. Subscripts p and ν are the pressure and viscous components. The pressure and viscous components of fluid forces and moments are derived from the aforementioned Navier-Stokes equations Eqs. (1) to (3).

The forces and moments induced by the constraint are calculated using the equations of motion given in Sect. 2.2. Considering that constraint forces do not perform any work along any direction of velocity freedom that complies with motion constraints:

$$\begin{bmatrix} \mathbf{F}_c \\ \mathbf{M}_c \end{bmatrix} \cdot \dot{\mathbf{q}} = 0 \tag{12}$$

Here, the mooring-induced forces and moments are computed via the lumped-mass model introduced in Sect. 2.3. The mooring solver (Π) uses the boundary values of rigid body solver (Ω) from the previous iteration or time step to calculate its values for the next iteration or time step:

$$\mathbf{X}_\Pi^{n+1,\, i+1} = \Pi\left(\mathbf{X}_\Omega^{n+1,\, i}\right) \tag{13}$$

$\mathbf{X}$ is the vector space containing boundary values, n represents the time steps and i denotes the iterations. Then the updated $\mathbf{X}_\Pi$ is used to compute $\mathbf{X}_\Omega$ for the next iteration or time step:

$$\mathbf{X}_\Omega^{n+1,\, i+1} = \Omega\left(\mathbf{X}_\Pi^{n+1,\, i+1}\right) \tag{14}$$

For further details of the coupling between fluid dynamics, mooring dynamics and nonlinear rigid body motions, see [10, 11, 13, 16].

3. Test case description

To study the effects of the rotational damper on connection joints of an articulated and moored floating system in waves, the present test cases consist of two boxes representing the floating modules articulated by flexible connection joints. Table 1 provides specific information about the floating box, including the position of its Center of Gravity (CoG) above the keel. For more comprehensive information, please refer to the model test report [19].

Table 1. Particulars of the tested floating box in waves.

Breadth	Length	Depth	Draft	CoG	K_{zz}	K_{yy}	K_{xx}
0.643 m	0.643 m	0.2143 m	0.1286 m	0.1246 m	0.2625 m	0.24 m	0.24 m

In our previous study [11], discrepancies were noted between the simulated and measured motions in pitch. Specifically, the simulated pitch responses of both floating bodies exceeded

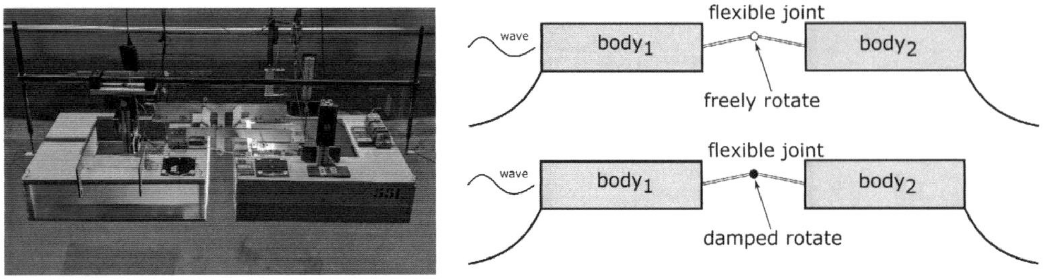

Figure 2. An overview of the experimental setup within Space@Sea project [19], along with a perspective view of the current configuration.

the measured values. This discrepancy can be attributed to several factors. Firstly, the hinged joint used during the experiment is more complex than the numerical representation. In the experimental setup, the joint had to incorporate load cells, as shown in Fig. 2, which limited its location during the physical test. This difference in joint configuration between the experiment and our simulations contributed to the discrepancies. Furthermore, the physical model test revealed that the manufactured joint was relatively inflexible, and the rotational hinge damping in the experiments was not accounted for in our previous simulations. As an extension, this study focuses on evaluating the effect of a rotational damper on the hydrodynamic responses of a flexibly connected multibody floating platform. By incorporating this damper into our simulations, we aim to address its hydrodynamic effects.

4. Results and discussion

To assess the impact of a rotational damper, this section analyzes multibody motions and the corresponding loads on the joints. Figure 3 illustrates the comparison of heave motions between two bodies connected using a flexible joint, without and with a rotational damper. We consider a head wave with a wave period of $T = 1.81$ s and an amplitude of $\zeta = 0.021$ m. The floating body closer to the wave maker is denoted as the front body (Body 1, $B1$), while the one farther away from the wave maker is labeled as the rear body (Body 2, $B2$). The subscript d indicates results obtained from the flexible joint integrated with a damper ($B1_d$ and $B2_d$). It is evident that heave motions are hardly affected by the added damper. When a floating structure is moored with a soft mooring system, heave motions are generally dominated by its hydrostatic stiffness and inertia [20].

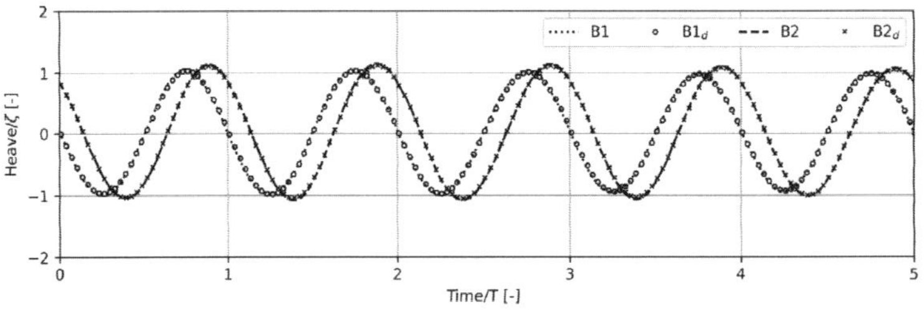

Figure 3. Comparative heave responses of the front body (B1) and the rear body (B2), where ζ is the wave amplitude and T is wave period.

However, significant discrepancies are evidenced in the pitch motions between the flexible joint without rotational damping and the one with rotational damping, as illustrated in Fig. 4.

Specifically, the pitch motions of both $B1$ and $B2$ without rotational joint damping are greater than those obtained from the flexible joint with rotational damping.

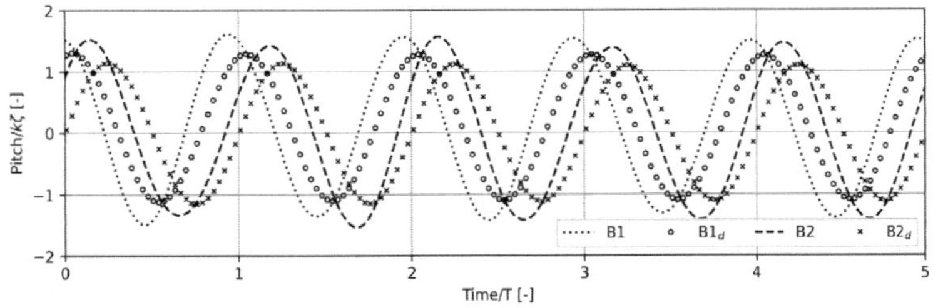

Figure 4. Comparative pitch responses of the front body (B1) and the rear body (B2), where T is wave period, ζ is the wave amplitude, and k is the wave number.

Figure 5 illustrates the associated longitudinal forces F_x and vertical forces F_z acting on the flexible joint without and with a rotational damper, respectively. Remarkably, the longitudinal and vertical hinge forces of the flexible joint without any rotational damping closely match those obtained from the flexible joint with rotational damping. This is mainly because the forces affecting the hinge joint are primarily governed by surge and heave motions, which are relatively similar for both of the flexible joints under consideration. Furthermore, strong nonlinearities are evident in the vertical joint forces for both cases.

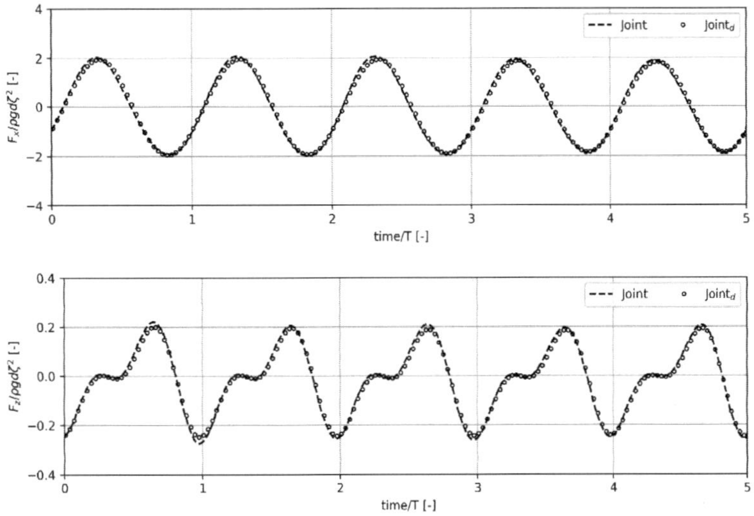

Figure 5. Comparative horizontal (top) and vertical (bottom) forces acting in the flexible joints, where g is the acceleration of gravity, ρ is water density, d is the length of body, and ζ is the wave amplitude.

5. Conclusions

This paper investigates the hydrodynamic performance of a rotational damper integrated into an articulated and moored floating multibody platform. Two types of flexible joints, a rotational freely flexible joint, and a rotational damped flexible joint, were employed to connect two floating modules. The system was positioned using four symmetric mooring lines. A coupled mooring-joint-viscous flow solver was utilized to predict the motion and load responses of the configured concepts, considering fluid dynamics, mooring line dynamics, and nonlinear rigid body motions. The primary focus was on the influence of the damper-based hinge joint on wave-induced motions and loads.

The results indicate that a rotational damper integrated into the flexible joint had a minimal impact on the translational motions of the system. However, notable differences were observed in pitch motions, where the pitch motions of a rotational damped flexible joint were smaller than those of the rotational freely flexible connected joint for both front and rear bodies. Overall, the hinge forces of the two types of flexible joints were comparable. This finding provides clear evidence to MFSs designers that a rotational damper can be installed to improve the stability of articulated and moored floating multibody platform without significantly altering the forces and moments acting on the system compared to those without a rotational damper.

To draw more generalized conclusions, additional sea states, such as irregular waves, would need to be considered. However, simulating irregular waves can be computationally expensive. As these approaches model highly resolved spatial flow features with nonlinear and coupled equations in the time domain, making it challenging to perform a large number of long-time irregular wave simulations. To address this challenge, various frameworks for efficient irregular wave simulations have been developed [21–24], and this will be the focus of our future work.

Acknowledgments

The authors also acknowledge the German Research Foundation (DFG: 448471847).

References

[1] Armstrong E R US Patent No 1511153, 1924 Sea station
[2] Lamas-Pardo M, Iglesias G and Carral L 2015 A review of very large floating structures (vlfs) for coastal and offshore uses *Ocean Engineering* **109** 677–690
[3] Jiang C, Xu P, Bai X, Zhao Z, el Moctar O and Zhang G 2023 A review of advances in modeling hydrodynamics and hydroelasticity for very large floating structures *Ocean Engineering* **285** 115319
[4] Kashiwagi M 2000 Research on hydroelastic responses of vlfs: Recent progress and future work *International Journal of Offshore and Polar Engineering* **10**
[5] Ren N, Zhang C, Magee A R, Hellan Ø, Dai J and Ang K K 2019 Hydrodynamic analysis of a modular multi-purpose floating structure system with different outermost connector types *Ocean Engineering* **176** 158–168
[6] Flikkema M and Waals O 2019 Space@ sea the floating solution *Frontiers in Marine Science* 553
[7] Shi Q, Xu D and Zhang H 2018 Design of a flexible-base hinged connector for very large floating structures *International Conference on Offshore Mechanics and Arctic Engineering* vol 51203 (American Society of Mechanical Engineers) p V001T01A015
[8] Jiang C, el Moctar O and Schellin T E 2021 Hydrodynamic sensitivity of moored and articulated multibody offshore structures in waves *Journal of Marine Science and Engineering* **9** 1028
[9] Jiang C, el Moctar O and Schellin T E 2022 Capability of a potential-flow solver to analyze articulated multibody offshore modules *Ocean Engineering* **266** 112754
[10] Jiang C, el Moctar O, Paredes G M and Schellin T E 2020 Validation of a dynamic mooring model coupled with a rans solver *Marine Structures* **72** 102783
[11] Jiang C and el Moctar O 2023 Extension of a coupled mooring–viscous flow solver to account for mooring–joint–multibody interaction in waves *Journal of Ocean Engineering and Marine Energy* **9** 93–111
[12] Jiang C, Xu P, el Moctar O and Zhang G 2023 Analysis of a moored and articulated multibody offshore system in steep waves *Journal of Offshore Mechanics and Arctic Engineering* **145** 041401
[13] Jiang C 2021 *Mathematical modelling of wave-induced motions and loads on moored offshore structures* Ph.D. thesis Dissertation, Duisburg, Universität Duisburg-Essen, 2021
[14] Hirt C W and Nichols B D 1981 Volume of fluid (vof) method for the dynamics of free boundaries *Journal of computational physics* **39** 201–225
[15] Rusche H 2003 *Computational fluid dynamics of dispersed two-phase flows at high phase fractions* Ph.D. thesis Imperial College London (University of London)

[16] el Moctar O, Ley J, Oberhagemann J and Schellin T 2017 Nonlinear computational methods for hydroelastic effects of ships in extreme seas *Ocean Engineering* **130** 659–673
[17] Featherstone R 2014 *Rigid body dynamics algorithms* (Springer)
[18] Hall M and Goupee A 2015 Validation of a lumped-mass mooring line model with deepcwind semisubmersible model test data *Ocean Engineering* **104** 590–603
[19] Thill C and Raghu S P 2018 Model tests report D4.3 Space@Sea Tech. rep. Delft University of Technology
[20] Jiang C, el Moctar O, Schellin T E and Paredes G M 2021 Comparative study of mathematical models for mooring systems coupled with cfd *Ships and Offshore Structures* **16** 942–954
[21] Paulsen B T, Bredmose H and Bingham H B 2014 An efficient domain decomposition strategy for wave loads on surface piercing circular cylinders *Coastal Engineering* **86** 57–76
[22] Baquet A, Kim J and Huang Z 2017 Numerical modeling using cfd and potential wave theory for three-hour nonlinear irregular wave simulations *International Conference on Offshore Mechanics and Arctic Engineering* vol 57632 (American Society of Mechanical Engineers) p V001T01A002
[23] Gatin I, Vukčević V and Jasak H 2017 A framework for efficient irregular wave simulations using higher order spectral method coupled with viscous two phase model *Journal of Ocean Engineering and Science* **2** 253–267
[24] Wiley W, Tran T T, Boerner T, Weston C and Wang L 2023 An efficient three-dimensional cfd-based numerical wave tank for a wave energy converter in extreme irregular waves *ASME 2023 42nd International Conference on Ocean, Offshore and Arctic Engineering* (American Society of Mechanical Engineers Digital Collection)

IOP Conf. Series: Materials Science and Engineering 1294 (2023) 012021 doi:10.1088/1757-899X/1294/1/012021

A parametric study on the initial transverse stability of suspension ships

J Han[1,*], and M Murai[2]

[1] Faculty of Engineering, Yokohama National University
[2] Faculty of Environment and Information Sciences, Yokohama National University

E-mail: han-jialin-pk@ynu.ac.jp

Abstract. Suspension ships are a novel type of ships that utilize a suspension system as an integral part of their structure. One critical aspect of a suspension ship is its stability characteristics. The parameters that affect the initial transverse stability of these ships include the ship's mass ratio, loaded height of suspension, placement location height of suspension, and beam. It is found that designing a suspension ship in monohull configuration is more difficult than in a catamaran configuration. Special attention is required when the static mass ratio is below 2 for both monohulls and catamarans.

1. Introduction

Suspensions employed in land vehicles are known for their ability to improve traction and isolate shocks, thereby improving ride comfort. Similarly, adoption of suspension systems in ships aims to achieve similar objectives. However, due to the more complex ocean wave surface compared to land road profiles, designing of ship suspensions presents considerable challenges.

A suspension ship, which differs from conventional ships having rigid-body structures, comprises of a suspended body, a suspension system, and a hull unit. A suspended body, also known as a cabin or an upper deck, rests on a suspension system consisting of mechanical linkages, springs, and dampers. This system guides the relative movement between the suspended body and the hull unit. The hull unit provides buoyancy force and may consist of one or multiple hulls. From this point forward, these elements will be referred to as the cabin, suspension, and hull.

In the past two decades, several suspension ships have been developed as illustrated in Fig.1. Suspension controllers are employed to minimize the motion of the cabin. There are variations in the suspension systems and control methods.

A. Nauti-craft 2 play. The hydro-pneumatic system of the Nauti-craft 2 enables active counteraction of heave, pitch and roll movements by shifting fluids between cylinders [1].

B. Martini 1.5. The servo-yachts, previously known as Velodyne-Marine, utilized air suspensions and developed the suspension ship Martini 1.5. The cabin remained stable while advancing at a speed up to 30 knots [2].

C. Proteus. The first 100-ft long technology demonstrator of wave adaptive modular vehicle (WAM-V) developed and tested by Marine Advanced Research in 2009 [3]. The suspension linkages make use of leaf springs, slider, and rocker arms.

Figure 1. Different designs of suspension ships.

D. WAM-V 16. Autonomous surface vehicles incorporating suspension technology utilize air springs, hinges, and ball joints to operate on the water's surface and adapt to various sea conditions [4].

E. Wave Harmonizer 7. The ship's suspension adopted helical springs, pangtograph, and Watt's linkage. Direct Current (DC) motors are utilized to provided the desired force to suppress the motion of the cabin [5].

F. Wave Harmonizer 6. The suspension design was similar to Wave Harmonizer 7. The control system comprised a wave energy harvesting mode [6] and a motion reduction mode [7].

2. Design principle of suspension ships

Typically, a floating rigid body possesses six degrees of freedom (DOF). When it is joined to another rigid body, the total DOF can increase up to twelve. Suspension linkages of ships should be designed to constrain motions that could endanger the ship while allowing for other motions to be manipulated to enhance the ship's ride comfort.

2.1. Concept of suspension ships

A nine DOF suspension ship is examined in this research. The relative surge, sway, and yaw motions between the cabin and the hull should be constrained by the suspension linkages, while the heave, pitch, and roll motions of the cabin should be separated from those of the hull. Therefore, the surge, sway, and yaw motions depict the motions of the entire ship, while the heave, pitch, and roll motions relate to both the cabin and the hull.

A two-dimensional dynamic sketch of this suspension ship is presented in Fig.2. The cabin pitches around the center of gravity of the sprung mass G_s, while the hull pitches about the center of gravity of the unsprung mass G_u. Typically, the mass supported by the suspension and the mass below the suspension are termed as sprung mass M_s and unsprung mass M_u, respectively. The center of gravity of the entire ship is denoted as CoG, and the center of buoyancy is represented by CoB. K indicates an imaginary point on the baseline. In an initial state of static equilibrium, these five points align in a hypothetical vertical line. The initial trim angle and heel angle are both considered zero.

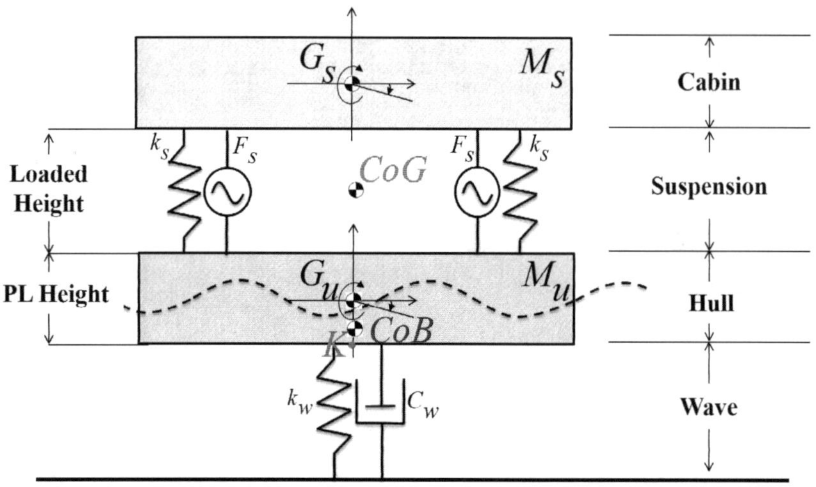

Figure 2. Dynamic sketch of suspension ships.

2.2. Initial transverse stability

The primary requirement in ship design is the initial transverse stability of a ship, which determines its ability to resist capsizing or heeling to one side when subjected to external forces.

Transverse stability is evaluated by examining the ship's metacentric height (GM). CoG denotes an imaginary point through which the entirety of the ship's weight presumably concentrates. The metacenter (M) is a theoretical point where the vertical line, which crosses CoG and CoB, intersects with the vertical line of buoyant force acting when the ship is inclined. If the GM value is positive, indicating that the point M is above CoG, the ship will tend to return to its upright position when tilted, which is considered a stable condition. However, if the GM value is negative, the ship will tend to continue tilting, which is unstable and can result in the ship capsizing.

Designing a suspension ship with appropriate initial transverse stability is essential for ensuring its safety at sea. The factors to consider are the ship's mass ratio of the unsprung mass and sprung mass, the loaded height of the suspension, the placement location of the suspension, and the beam of the ship. It is important to ensure the ship adheres to proper stability requirements to prevent accidents and maintain crew safety.

2.2.1. Mass ratio The mass ratio, q, is defined as the ratio of the sprung mass to the unsprung mass. It is expressed as

$$q = \frac{M_s}{M_u} \tag{1}$$

The distribution of mass across the cabin, the suspension, and the hull determines the CoG's location. When examining land vehicles, the suspension's mass is considered not as a single parameter but an additional mass to both the suspended body and the hull. The suspension system is represented by springs and dampers, which are massless.

Accurately determining the sprung and unsprung masses is crucial for modeling and simulating the ship's motion responses. In physical models, suspension linkages are either connected to the cabin or the hull. While it is possible to measure the mass of each suspension component, determining to which amount that the sprung mass or the unsprung mass should account for the suspension's mass remains challenging. The sprung mass of a physical model can be estimated through a free decay test [7]. The sprung mass can be estimated by utilizing the dampened natural frequency. This results in a dynamic mass ratio, which is differentiated from the static mass ratio.

The cabin-to-hull mass ratio, referred as the static mass ratio, is defined as

$$p' = \frac{M_{cabin}}{M_{hull}} \tag{2}$$

whereas the M_{cabin} and M_{hull} represent the overall measured mass of the cabin and hull, respectively.

The mass ratio inevitably changes during the cargo loading and offloading, as well as during passenger boarding and alighting. In this study, it is assumed that the fluctuations in mass ratio will not lead to alterations in the trim and heel of either the cabin or the hull.

2.2.2. Loaded height The loaded height, H_L, as depicted in Fig.2, indicates the vertical distance between the cabin and the hull when the suspension is stationary, and the ship attains static equilibrium.

The loaded height encompasses the allowable travel distance and the solid height of the suspension. The peak-to-peak vertical distance of the suspended cabin can travel when the hull is at rest represents the allowable travel distance. It determines the maximum motion amplitude that can be compensated by the suspension system. The solid height is the vertical distance between cabin and the hull while the cabin remains at its lowest point and the hull is stationary.

When the suspension is fully compressed, suspension bottom out happens, causing the cabin to hit the hull. The impacts of suspension bottom out on suspension ships is similar to the hard landings on aircraft by resulting in significant vertical acceleration spikes on the cabin which can cause damage and injuries.

The parameters affecting the loaded height vary among suspension designs. For this study, the considered parameter is the spring stiffness k_s. At a certain M_s and free length L_f of the spring, the height under loads varies according to changes in k_s, while the solid height is assumed constant.

2.2.3. Placement location height The height of the suspension's placement location, H_{PL}, shown as PL Height in Fig.2, indicates the vertical distance from the baseline to the placement location of the suspension.

This parameter has a significant bearing on the suspension design and the selection of ship configurations. All current suspension ships, shown in Fig.1, are catamaran with entirely suspended cabins. The suspension is located on the top deck of the hull, with its placement height equal to that of the hull. However, for potential future designs, the suspension may be positioned lower than the top deck of the hull or even inside the hull, resulting in varying placement heights.

2.2.4. Beam of ship It is widely recognized that the beam of the hull and the overall beam of the ship play a crucial role in determining its initial transverse stability. Additionally, the shape of the hull is also a key factor. However, the hull's shape is not highly relevant to this study like other factors and will be explored in future research.

3. Parametric investigation on a suspension model

A suspension design, along with the proposed initial assembly, is presented. A parametric investigation into the initial transverse stability is carried out to identify feasible ship designs that possess a positive GM.

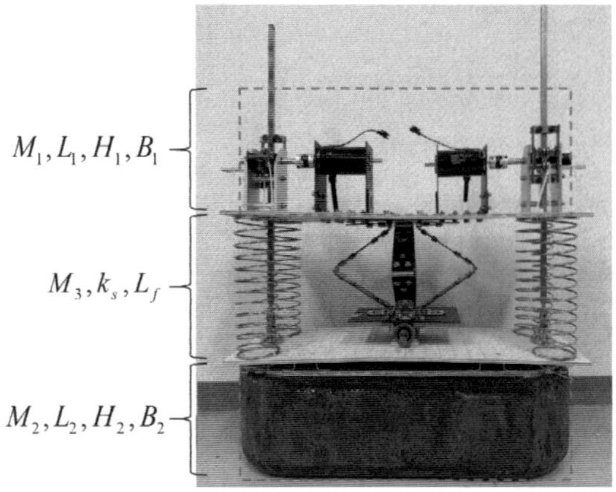

Figure 3. Initially assembly of a suspension model.

Table 1. Design specifications of a suspension model

Item	Value
mass of cabin M_1	2.5 kg
mass of suspension M_3	2.6 kg
mass of hull M_2	1.3 kg
spring stiffness k_s	235 N/m
free length of spring L_f	0.22 m
Diameter of spring	0.05 m
Dimension of cabin L_1, H_1, B_1	L0.5 m H0.20 m B0.15 m
Dimension of hull L_2, H_2, B_2	L0.5 m H0.15 m B0.15 m

3.1. Metacentric height GM

It is known that GM of a floating body can be calculated by

$$\mathrm{GM} = \mathrm{KB} + \mathrm{BM} - \mathrm{KG} \tag{3}$$

where, KB is the vertical distance from K to CoB, KG is the vertical distance from K to CoG, and BM is the distance from CoB to the metacenter M.

3.2. Suspension prototype

An initial assembly of a suspension design and its specification are presented in Fig.3 and Table 1, respectively. The cabin and hull are considered as blocks with the dimensions provided. The block's center of gravity is assumed to be at its geometric center, significantly simplifying the calculation the GM for the initial layout.

Let the water density ρ be 1025 kg/m^3, the draft of ship d can be obtained as

$$d = \frac{\frac{M_1+M_2+M_3}{\rho}}{L_1 \cdot B_1} = 0.0833 \text{ m} \tag{4}$$

Then, KB is

$$\mathrm{KB} = \frac{d}{2} = 0.0416 \text{ m} \tag{5}$$

The second moment of waterline area about longitudinal axis I_L is estimated by

$$I_L = \frac{1}{12} L_2 \cdot B_2^3 = 1.4062 \times 10^{-4} \text{ m}^4 \tag{6}$$

Then, BM is

$$\text{BM} = \frac{I_L}{\frac{M_1+M_2+M_3}{\rho}} = 0.0225 \text{ m} \tag{7}$$

The loaded height of the suspension is given as

$$H_L = L_f - \frac{M_1}{4k_s} = 0.2173 \text{ m} \tag{8}$$

The vertical distance from the center of gravity of the hull, suspension, and cabin to the point K on the baseline can be obtained as

$$KG_1 = H_2 + H_L + \frac{H_1}{2} = 0.4673 \tag{9}$$

$$KG_3 = H_2 + \frac{H_L}{2} = 0.2587 \tag{10}$$

$$KG_2 = \frac{H_2}{2} = 0.0750 \tag{11}$$

Then, KG is

$$\text{KG} = \frac{M_1 \cdot KG_1 + M_2 \cdot KG_2 + M_3 \cdot KG_3}{M_1 + M_2 + M_3} = 0.3029 \text{ m} \tag{12}$$

Therefore, the GM of the initial assembly is

$$GM = -0.2387 < 0 \quad \text{unstable} \tag{13}$$

This demonstrates that the initial assembly of the prototype suspension ship is not stable. The main reasons are

- the mass of the cabin and the suspension components are substantial, leading to a high center of gravity and a large GM;
- the loaded height of the suspension is considerable, also contributing to a large KG;
- the placement location of the suspension is so high that causes a large KG;
- the beam of the hull are narrow, resulting in a small KB and BM.

3.3. Parameter tuning

The aim of tuning the parameters is to identify design proposals that possess satisfactory initial transverse stability, and more importantly, to comprehend the impact of the parameters.

Hereafter, the suspension mass and the ship's total mass remain constant. The static mass ratio p' is adjusted within a range of $0.1 \sim 0.9$, while the spring stiffness is modified from soft to hard. The suspension is relocated from the bottom to the top deck of the hull. The free length of the spring is reduced to a certain value if permissible. The hull's beam is increased up to 2.5 times its original size. Catamaran configuration is taken into account with three transverse distances between the two hulls. Changes in GM along with adjustments to these parameters are illustrated from Fig.4 to Fig.7.

IOP Conf. Series: Materials Science and Engineering 1294 (2023) 012021 doi:10.1088/1757-899X/1294/1/012021

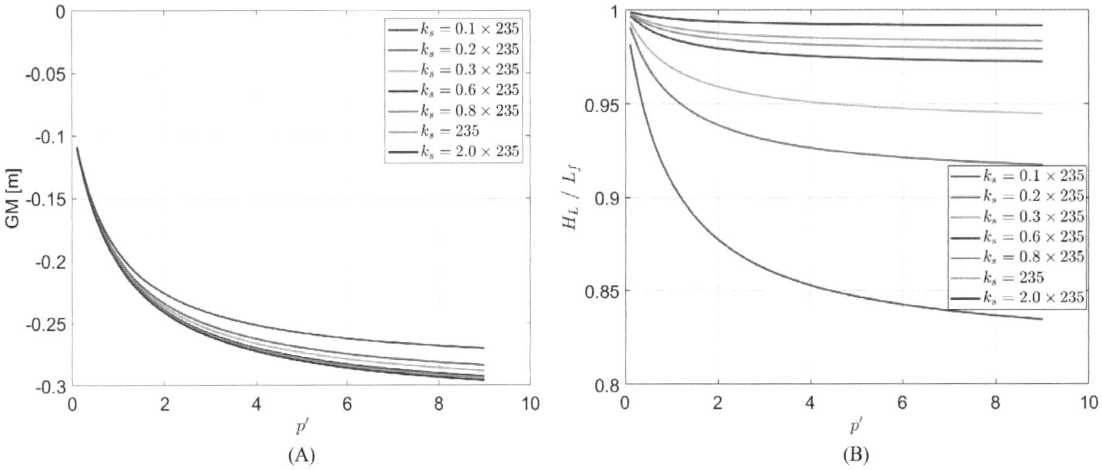

Figure 4. GM and H_L vary against the static mass ratio p' and spring stiffness k_s.

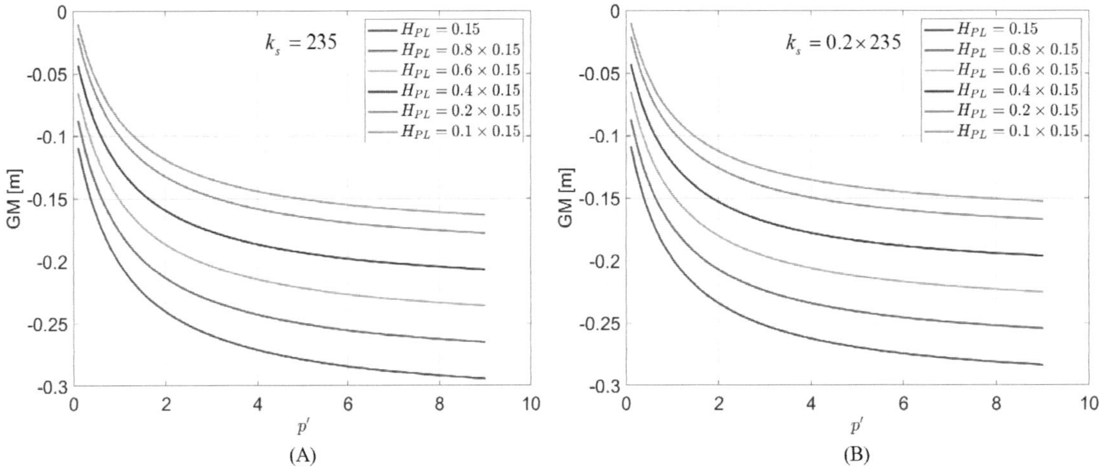

Figure 5. GM vary against the static mass ratio p' and placement location height H_{PL}.

3.3.1. static mass ratio, spring stiffness The GM that changes along with spring stiffness at different static mass ratio p' is illustrated in Fig.4(A), with the corresponding loaded height shown in Fig.4(B). The dimensionless loaded height is acquired via dividing the free length of spring L_f.

As shown in Fig.4(A), altering the mass distribution between the cabin and hull, as well as modifying the stiffness of the springs did not attain initial stability. A lighter cabin and softer springs are found effective in improving stability. In regions where the static mass ratio is less than 2, the GM increases sharply with a reduction in p'. The impact of changing spring stiffness, however, decreased.

Fig.4(B) demonstrates that although the static mass ratio increased by approximately 8 times, the loaded height of suspension H_L decreased by less than 20%. At static mass ratios lower than 2, H_L decreases significantly, particularly when k_s is below half of its original value. This suggests the need for a softer spring stiffness and potentially a shorter free length of spring.

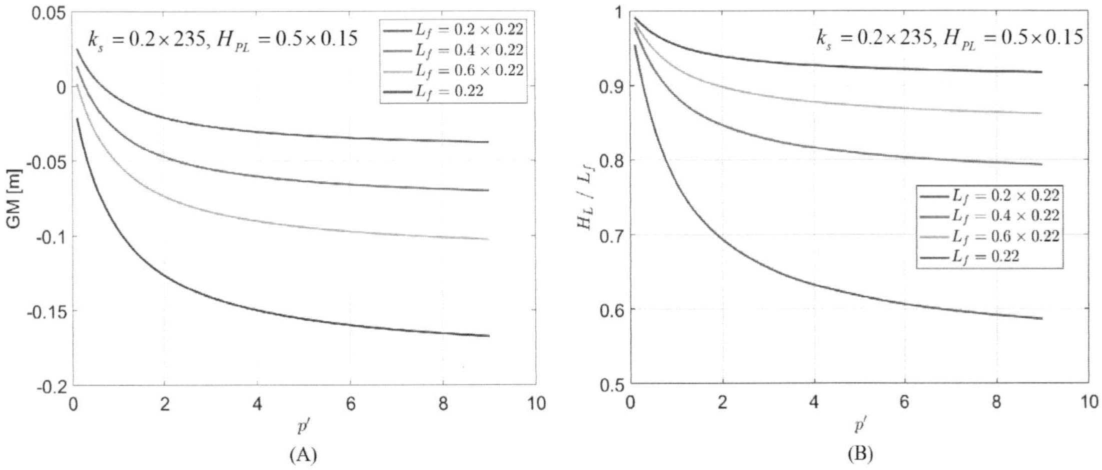

Figure 6. GM and H_L vary against the static mass ratio p' and free length of springs L_f.

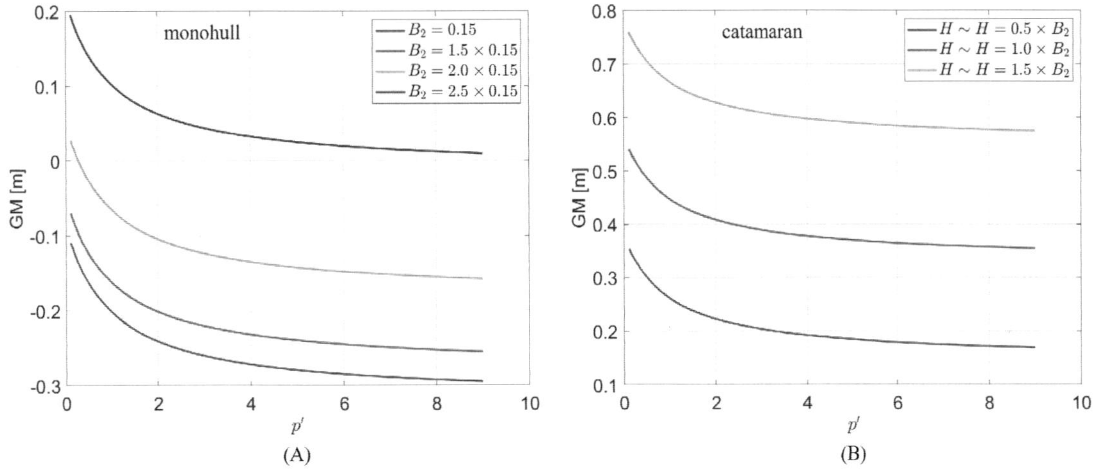

Figure 7. GM vary against the static mass ratio p' and beam.

3.3.2. placement location height The GM varies with the placement location height of suspension at different static mass ratios p' and spring stiffness is illustrated in Fig.5.

The initial transverse stability remains unobtained by lowering the placement location to one tenth of the hull height, even when the spring stiffness is reduced to one fifth of its original value.

It is observed that a lower placement location height results in increased stability. With smaller p', the increase in GM achieved by decreasing H_{PL} becomes less significant. It should be noted, however, that tuning H_{PL} has a greater impact than that of k_s.

3.3.3. free length of springs The GM at various free lengths of springs for different static mass ratios p' is depicted in Fig.6(A), and the associated loaded height is shown in Fig.6(B). The suspension's placement location is now halfway and the adopted spring stiffness is one fifth of the initial value.

The initial transverse stability, GM > 0, is achieved only within a narrow parameter range, where p' and L_f are both small. From a stability perspective, a shorter spring length is preferable. Nevertheless, it should be noted that the loaded height might changes significantly with variations in p', particularly for $p' < 2$. This could result in a high occurrence of suspension bottoming out. Therefore, it is essential to consider the trade-off between the free length and stiffness of the spring.

3.3.4. beam The GM that alters with changes in beam at varying static mass ratios p' is illustrated in Fig.7(A), and alterations based on the transverse distance between the two hulls of a catamaran can be observed in Fig.7(B). The monohull suspension ship is used as a half model to build the catamaran. The transverse distance between the two hulls is used as a tuning parameter. The decks of the two hulls are fixed by crossbeams, and so are the cabins. The mass of the crossbeams is neglected.

Monohull initial transverse stability, where GM > 0, is achieved when the beam is 2.5 times its original size. For a catamaran, the initial transverse stability can be achieved despite the fact that the transverse distance between the two hulls is only half of the beam of one hull. It demonstrated that the most dominating parameter that influences the initial transverse stability is the beam.

This suggests that implementing suspensions for a monohull is incredibly challenging. In addition, this explains why all the current suspension ships are in a catamaran configuration.

4. Conclusions

Initial transverse stability is a crucial factor in the design of suspension ships. Various parameters need to be considered when calculating GM, such as the mass ratio, loaded height of suspension, placement location height of suspension, and the ship's beam. A parametric investigation was carried out to analyze the impacts on GM. The following conclusions can be drawn from the analysis:

1. A reduced mass ratio results in a increased GM. The initial transverse stability displays significant variation when the mass ratio of the cabin and hull is less than 2. It is imperative to pay close attention if the static mass ratio falls within this range;
2. A shorter loaded height is preferable for a larger GM. However, the allowable travel distance must be taken into account to prevent bottom out. If springs are utilized, the spring stiffness and free length must be carefully balanced.
3. A lower placement location height results in a larger GM. Although, the impact becomes weaker with the reduced static mass ratio, it exerts more impact compared to the spring stiffness.
4. A larger beam produces a larger GM. The beam of the hull for a monohull and the overall beam of a catamaran are the dominating parameters that influence the initial transverse stability.

Achieving initial transverse stability for a monohull with a fully suspended cabin was found challenging. However, suspending only a portion of the cabin to improve ride comfort could be a more practical solution which could be one of the future studies.

Acknowledgments

This research was supported by JSPS KAKENHI Grant Numbers 21K14363.

IOP Conf. Series: Materials Science and Engineering 1294 (2023) 012021 doi:10.1088/1757-899X/1294/1/012021

References

[1] Nauti-Craft marine suspension technology, https://www.nauti-craft.com/marine-suspension/ (Accessed: 15 September 2023)
[2] Servo-yachts, https://www.servo-yachts.com/press, (Accessed: 15 September 2023)
[3] Conti U and Gundersen M 2011 Second Generation Design of Wave Adaptive Modular Vessels (WAM-V®): A Technical Discussion of Design Improvements. *Proceedings of* 11th *International Conference on Fast Sea Transportation* pp 772-776
[4] Wave Adaptive Modular Vehicle®, https://wam-v.com/, (Accessed: 15 September 2023)
[5] Han J, Kanno S, Mochizuki A, Kitazawa D, Maeda T and Itakura H 2019 Study on Attitude Control of a Cabin-Suspended Catamaran by Using a Double-Loop Control System *Proceedings of the ASME 2019 38th International Conference on Ocean, Offshore and Arctic Engineering* (Glasgow: June 9–14, 2019) vol(9) 95827.
[6] Han J, Maeda T, Itakura H and Kitazawa D 2022 Experimental study on the wave energy harvesting performance of a small suspension catamaran exploiting the maximum power point tracking approach *Ocean Engineering* vol(243) 110176.
[7] Han J, Maeda T, Itakura H and Kitazawa D 2023 Experimental study on the motion reduction performance of a small suspension catamaran exploiting an active skyhook control strategy *Ocean Engineering* vol(281) 114642.

Numerical analysis of centrifugal water pump impeller under varying loads

L Yadeta[1 *], H G Lemu[2] and A K/M Tadese[1]

[1] Jimma Institute of Technology, Jimma University, Ethiopia
[2] Faculty of Science and Technology, University of Stavanger, Stavanger, Norway

[*] Correspondence: letamchengrju@gmail.com

Abstract. Centrifugal pumps are widely used to move liquids from one place to another at varying pressure and flow rate, and to an elevated height for irrigation, water supply plants, and steam power plants. The load variations often lead to unpredictable performance of the impeller. The objective of this study is to investigate the performance of a three-stage centrifugal water pump impeller under varying loads using numerical methods. To conduct the investigation, a water pump used by Jimma town water treatment station, which operates with head of 191 m and discharges 439.2 m^3/hr water was used as a case study. Computational fluid dynamics analysis of the pump was conducted using ANSYS 19.0 CFX to get the pressure distribution at different flow rates (60%, 100% and 140% of the design flow rate). The fluid pressure was then used to make static stress analysis of the impeller employing one way fluid structural interaction analysis for three different materials. The finding of the study shows that the fluid pressure on the impeller has great impact on the impeller performance causing high stress and deformation of the impeller.
Keywords: Centrifugal pump, computational fluid dynamics, fluid-structure interaction, Navier Stokes equation, numerical analysis

1. Introduction

Pump is a mechanical device that applies energy to move liquids from one place to another at increased pressure, flow rate and to an elevated height. Irrigation, fire-fighting services employ pumps of different sizes [1]. Pumps and turbo-pumps are used in many technological areas and cover a wide range of applications such as thermal power generation, nuclear, propulsion, marine and water supply [2]. The pumps can be divided into two general categories, namely dynamic pumps and displacement pumps. In a dynamic pump, such as a centrifugal pump, energy is added to the pumping medium continuously and the medium is not contained in a set volume. The energy, in a displacement pump such as a diaphragm pump, is added to the pumping medium periodically while the medium is contained in a set volume. The pump is driven by a prime mover that is either an engine or an electric motor. The capacity of a pump is defined based on the pressure head (in meters) and the maximum delivery flow rate at a specific speed of the shaft [3]. Centrifugal pumps are one of the dynamic pumps which consists of a set of rotating vanes enclosed within a casing and is used to impart energy to a fluid through centrifugal force. Centrifugal pump is a device mainly used for transporting liquid from lower level to higher level. Centrifugal pumps convert mechanical energy from a motor to energy of a moving fluid, where some of the energy goes into kinetic energy of fluid motion, and the rest is converted to potential energy that is represented by a fluid pressure or by lifting the fluid against gravity to a higher level [4][5].

Centrifugal pumps are widely used for irrigation, water supply plants, steam power plants, sewage, oil refineries, chemical plants, hydraulic power service, food processing factories and mines, because of their suitability in practically any service. In pumps the mechanical energy is converted into hydraulic energy [6]. The two main components of centrifugal pump are impeller and casing. Centrifugal pump moves liquid by rotating one or more impellers inside a volute casing. The liquid is introduced through the casing inlet to the eye of the impeller where it is picked up by the impeller vanes. The rotation of the impeller at high speeds creates the centrifugal force that throws the liquid along the vanes, causing it to be discharged from it's out-side diameter at a higher velocity. This velocity energy is converted to pressure energy by the volute casing prior to discharging the liquid to the system [7]. The second part of centrifugal pump is stationary element comprising the casing, bearings and seals. A shaft is used to support rotating components or to transmit power or motion by rotary or axial movement [7]. Mechanical component fails due to either a decrease in its strength or an increase in load acting on it. Failure of a component affects the performance of a pump, causing either a reduction in its efficiency or its complete breakdown. In industrial applications, the probability of occurrence of critical problems such as component damage and pump failure is high due to heavy loads and the demand for continuous operation of the pump. In such cases, the entire plant will have to be shut down until the pump is either brought back to service or replaced. In order to prevent huge economic losses incurred due to such a shut-down, a pump must function reliably under specified operating conditions [8]. Centrifugal water pump components are subjected to different loading conditions. It is also subjected to corrosion and corrosion assisted fatigue due to the operating or working environment.

In order to perform numerical analysis of centrifugal pump impeller, numerical calculations were performed by solving the 3D Navier-Stoke´s and energy equations using the commercial code ANSYS CFX [9]. In the same study, steady-state analysis was performed using the κ-ω based shear stress transport (SST) model, which proved to give relatively accurate predictions in fluid machine analysis. Using an FSI analysis, it is possible to see the Coriolis forces and centrifugal forces generated by the impeller. FSI analysis is a multiphysics interaction between fluid flow and a solid structure. There are two ways to do an FSI analysis (1) one-way coupling and (2) two-way coupling. Benra et al. [10] employed one-way FSI coupling, in which the CFD results were transferred to the structural analysis, while in two-way FSI, the deformations in the structure obtained from the transfer of the CFD results are fed back to the CFD environment. The objective was to determine the deviation in the flow created as a result of the structural deformation. The study by Gu et al. [11] demonstrated the effect of FSI on rotational forces. The study indicated that fluid pressures significantly affect von-Mises stress.

Kobayashi et al. [12] investigated a mixed-flow pump with an unshrouded impeller by one-way coupled FSI, and the distribution of stress on the impeller was obtained. Piperno et al. [13] concluded that to ensure the safe operation of a rotating structure for every flow rate, the analysis of stress and deformation on the impeller in the unstable operation region needs to be solved by the consideration of FSI. Kang and Kim [14] pointed out that structural safety needs to be evaluated because the impeller receives fluid pressure load and centrifugal force during the operation. Even though different researches were conducted, many of them focused on centrifugal compressor impellers and less study was conducted on the FSI stress analysis of centrifugal pump impeller. This study focuses on one-way FSI stress analysis of three stage centrifugal water pump impeller. It is case study of Jimma town water treatment station, which is located at Boye and pumps water with head of 191 m and 439.2 m^3/hr discharges.

2. Methodology

Actual design data about the specification of a given centrifugal water pump was collected from the Jimma town water treatment station located at Boye. Based on the collected data, the 3D model of the centrifugal water pump shaft, impeller and return guide vane was prepared by using ANSY 19.0 geometry modelling and SolidWorks modelling software. Impeller was modelled using Vista CPD Design to model impeller blades from 1D to 3D geometry.

In this study based on recommended number of impeller, in this study six blades impellers were chosen. In this study Complex 3-D Computational Fluid Dynamics analysis was performed to get the pressure distribution of the impellers by using ANSYS 19.0 software. ANSYS CFX was used to perform steady state CFD analysis of the three-stage centrifugal pump. One-way partitioned FSI analysis was used to determine the stress of the three-stage pump where the fluid pressure on the structure will be transferred to the solid solver. The fluid flow is modelled as incompressible flow using water as a working fluid. k-SST omega turbulence model was used by considering wall of solid is modelled as no-slip condition. The principles of conservation law governed by fluid dynamics are [15]:

$$\frac{\partial u_i}{\partial x_i} = 0 \tag{1}$$

$$\rho(\frac{\partial u_i}{\partial t} + u_j \frac{\partial u_i}{\partial x_j}) = -\frac{\partial p}{\partial x_j} + \frac{\partial}{\partial x_j}(\mu \frac{\partial u_i}{\partial x_j} - \overline{\rho \acute{u}_i \acute{u}_j}) \tag{2}$$

Where u_i is velocity vector, p is pressure scalar, ρ is density, i and j is tension notations, $\overline{\rho \acute{u}_i \acute{u}_j}$ is apparent stress turbulence tensor, μ is the dynamic viscosity. The k–ω based SST model accounts for transport of turbulence shear stress and highly accurate prediction of the onset of the amount flow separation under adverse pressure gradients. The unknown turbulent viscosity μ_t is determined by solving two additional transport equations for turbulent energy k and for the turbulence frequency ω. These two equations can be written as:

$$\frac{\partial(\rho\kappa)}{\partial t} + \frac{\partial}{\partial x_j}\left(\rho k u_j\right) = \frac{\partial}{\partial x_j}\left[\left(\mu + \frac{\mu_t}{\sigma_k}\right)\frac{\partial k}{\partial x_j}\right] + p_k - \beta'_{\rho k\omega} + p_k b \quad (3)$$

$$\frac{\partial(\rho\omega)}{\partial t} + \frac{\partial}{\partial x_j}\left(\rho\omega . u_j\right) = \frac{\partial}{\partial x_j}\left[\left(\mu + \frac{\mu_t}{\sigma_\omega}\right)\frac{\partial \omega}{\partial x_j}\right] + \alpha\frac{\omega}{\kappa}p_k - \beta_{\rho\omega^2} + p_k b \quad (4)$$

The impeller is modelled in the blade frame, and the diffuser and return guide vane is in the fixed frame of the reference and both of them are related each other through the frozen rotor. The boundary conditions given in Table 1 were applied:

Table 1. CFD simulation boundary conditions

Boundary condition	Value
Working fluid	Water
Temperature	25 ^{0}C
Inlet pressure	0.3 bar
Speed	1480 rpm
Turbulence intensity	5%
Discharge flow rate	73.2 kg/s, 122 kg/s, 170.8 kg/s
Turbulence model	κ-ω SST

To simulate the pump under different flow rate operating conditions, the outlet mass flow rate was set to flow rates 0.6 Qd = 73.2 kg/s, Qd = 122 kg/s and 1.4 Qd = 170.8 kg/s design flow rate. Three dimensional incompressible Navier Stokes equations were solved with ANSYS 19.0-CFX Solver. In addition, a no-slip flow condition is applied on the walls (on the blade, hub and shroud). Smooth wall function was selected.

Unstructured meshes in CFX meshing was used to reduce the amount of time spent generating meshes, simplifying the geometry modeling and mesh generation process. Apart from its higher memory requirement unstructured mesh is also preferable for complex geometry. In this study unstructured mesh tetrahedral cells were used because the fluid domain in the pump has complex geometry. The geometry and the mesh of a six bladed pump impeller and diffuser with return guide vane domain were generated with CFX meshing as shown in Figure 1(a)-(d). Un- structured meshes with tetrahedral cells were used for the domain of impeller and diffuser with return guide vane as well as for the inlet and out-let casing of fluid domain.

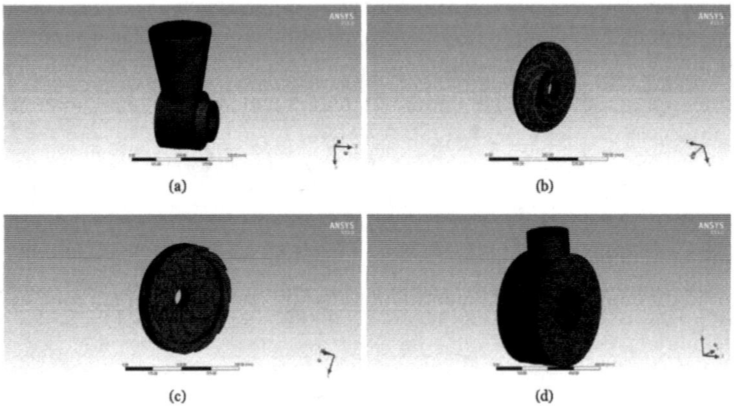

Figure 1. CFD Fluid domain mesh

To assure accuracy of simulation mesh independence in the flow domain must be determined before starting the CFD studies. In this study, the calculation domain was divided into unstructured grids by CFX Meshing. Seven mesh element size with the same numerical settings were analyzed to select appropriate mesh number of elements to carry out the CFD analysis of three stage centrifugal pump. As it can be observed from Figure 2 grid size minimally influences the numerical results, and the overall difference is within 1%. So, for the analysis of the three-stage centrifugal pump 2 353 607 elements were selected considering time and computational cost.

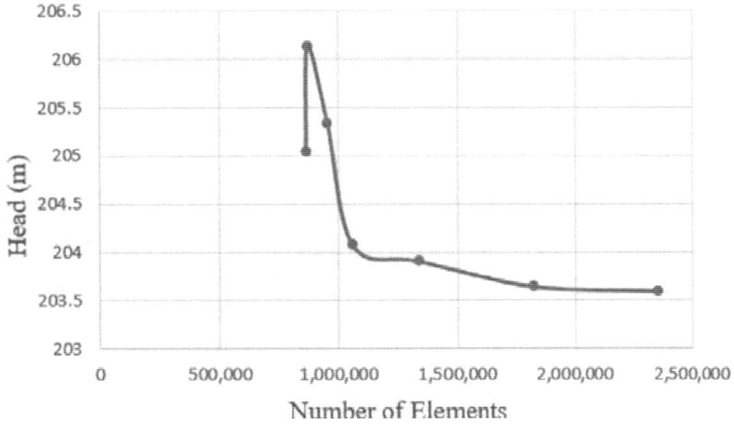

Figure 2. Mesh independency test

For centrifugal pump impeller the following materials was selected. Pump impeller operates in corrosive environment. The three selected material for impeller stress analysis are G-Cu-Sn10 Copper Tin Alloy, EN-GJL-250 Grey Cast Iron and G-X6CrNiMo18-10 stain- less steel.

3. Discussion of results

With the aid of CFD, the complex internal flow in the three stage centrifugal pump was predicted quite well. In this study, a steady state solution with κ-ω SST turbulence model was used in ANSYS-CFX for analysis of three stage centrifugal pump CFD simulation. From the manufacturers experience pumps may operate other than BEP of the pump which operates at nominal operating conditions. In this study the analysis of three stage centrifugal pump at off design condition was carried out at different flow rates 60% Qd, Qd and 140% Qd design flow rate. The analysis of the three-stage pump at different flow rate was carried out to see the impact of different flow condition on stress performance of an impellers. The CFD simulation result of the pump is shown in Figure 3.

The total head of the pump at the design flow rate of 439.2 m^3/hr obtained from the CFD analysis of the three-stage pump is 203.597 m. The head rise obtained through CFD analysis was over predicted the head rise than the design head of the pump which was 191 m. The over prediction of the CFD result is because leakage was not considered due to its computational cost and complexity in CFD simulation, additionally surface roughness of impeller, diffuser with return guide vanes was assumed to be smooth which in turn increases head of CFD when neglected. Figure 3 shows CFX simulation result of flow rate versus head. As it can be observed from Figure 3 as the flow rate increases the head of the pump decreases and vice versa.

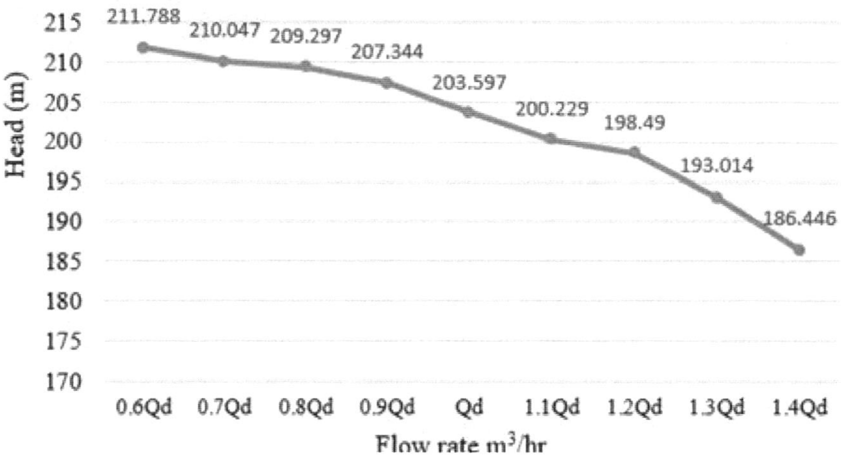

Figure 3. Flow rates versus head

3.1 Pressure distribution

From CFD simulation of the pump pressure distribution of the pump was obtained. Total pressure on pressure side of the blade is more than that of suction side. The difference of pressure from the pressure side to the suction side of the impeller blade is increasing from leading edge to trailing edge of the blade. The total pressure patterns are varying along the span of the impeller. Low total pressures are observed near hub of the impeller. Figure 4 shows pressure distribution of the CFD simulation of the pump at design flow rate Qd = 439.2 m^3/h. The pressure rise increases from the first stage impeller to the subsequent stage impellers of the pump.

As it can be observed from Figure 4(a), the maximum total pressure for the first stage impeller at design flow rate (Qd) is 0.6962 MPa and the minimum total pressure is -0.181 MPa. Similarly, as it is shown in the Figure 4(b) and (c) maximum pressure of second and third stage impeller is 1.330 MPa and 1.916 MPa respectively while minimum total pressure is 0.5127 MPa and 1.150 MPa. As it is indicated in Table 2, for 0.6Qd flow rate the maximum total pressure for the first stage impeller is 0.7633 MPa and the minimum total pressure is -0.09933 MPa. Similarly, the maximum pressure of second and third stage are 1.386 MPa and 1.973 MPa respectively while the minimum total pressure of second and third stage is 0.5365 MPa and 1.155 MPa respectively.

For 1.4Qd flow rate (Table 2), the maximum total pressure for the first stage impeller is 0.6686 MPa and the minimum total pressure is -0.1267 MPa. Similarly, the maximum pressure of the second and third stage impellers are 1.261 MPa and 1.794 MPa respectively, while the minimum total pressure of second and third stage is 0.5154 MPa and 1.096 MPa respectively. Generally, as it can be observed from Table 2, at different flow rates of the pump, the pressure of the fluid increases as the fluid flows from the suction side to the discharge side at each stage of the pump. In all operating conditions, the global maximum pressure was observed at the third stage impeller. Additionally, the pressure of the pump decreases as the flow rate of the pump increases and vice versa.

IOP Conf. Series: Materials Science and Engineering 1294 (2023) 012022 doi:10.1088/1757-899X/1294/1/012022

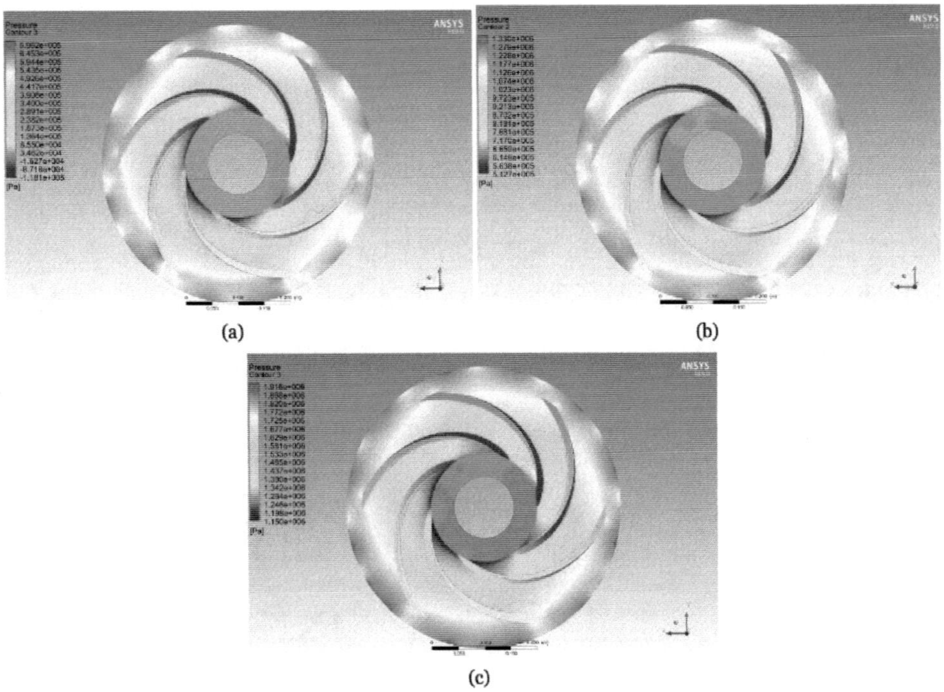

Figure 4. Pressure distribution in impeller (a) Stage I (b) Stage II (c) Stage III

Table 2. Pressure distribution

Flow Rate (m^3/hr)	Pressure distribution (MPa)					
	Stage I		Stage II		Stage III	
	Max	Min	Max	Min	Max	Min
0.6Qd	0.7633	-0.099331	1.386	0.5365	1.973	1.155
Qd	0.6962	-0.1181	1.330	0.5127	1.916	1.150
1.4Qd	0.6686	-0.1267	1.261	0.5154	1.794	1.096

3.2 FSI stress analysis in the impeller

Numerical stress analysis of the impellers of the three-stage centrifugal water pump was performed. The pressure distribution on the surface of the impeller was imported to FEA module to perform FSI stress analysis of the impeller at 0.6Qd, Qd and 1.4Qd flow rates of the pump operation. In this study, factor of safety was analyzed only for the third stage impeller because the third stage impeller experiences maximum pressure in all flow rate conditions which leads to higher stress and deformation.

3.2.1 FSI stress analysis at 0.6Qd flow rate: FSI stress analysis was analyzed at 0.6Qd flow rate of the pump operation for three different materials. In the first stage impeller, maximum equivalent stress for G-Cu-Sn10 Copper Tin Alloy is 50.521 MPa and minimum equivalent stress is 0.00399 MPa (Table 3). The data in the table also shows that FSI stress analysis result for EN-GJL-250 Grey Cast Iron and G-X6CrNiMo18-10 impeller materials is 49.676 MPa and 50.057 MPa respectively. Similarly minimum equivalent stress for both material is 0.0044873 MPa and 0.0052831 MPa.

As the pressure of water increases from first stage to the second stage, it is expected that equivalent stress of the second stage increases. As it can be observed from Table 2, the maximum equivalent stress of the second stage impeller of the pump is for G-Cu-Sn10 Copper Tin Alloy is 110.61 MPa and minimum equivalent stress is 0.094794 MPa. Similarly maximum equivalent stress of the second stage impeller for EN-GJL-250 Grey Cast Iron and G-X6CrNiMo18-10 is 108.33 MPa and 109.36 MPa respectively. Additionally, minimum equivalent stress for both materials are 0.083661 MPa and 0.0720061 MPa respectively.

IOP Conf. Series: Materials Science and Engineering 1294 (2023) 012022 doi:10.1088/1757-899X/1294/1/012022

Table 3. FSI equivalent stress of impeller at 0.6Qd flow rate

Materials	Equivalent stress (MPa)					
	Stage I		Stage II		Stage III	
	Max	Min	Max	Min	Max	Min
G-Cu-Sn10 Copper Tin Alloy	50.521	0.00399	110.61	0.094794	171.32	0.10044
EN-GJL-250 Grey Cast Iron	49.676	0.0044873	108.33	0.083661	167.72	0.12008
G-X6CrNiMo18-10	50.057	0.0052831	109.36	0.0720061	169.4	0.088268

Generally, as it can be observed that the stress in the impeller increases from the first stage to the third stage because as the number of stages of the impeller increases, the pressure of the water rises which results in high pressure on the surface of the impeller as the pump stage increases. It can be concluded that the third stage impeller is critical impeller as it experiences high stress.

3.2.2 FSI stress analysis at Qd flow rate: This section discusses results of FSI stress analysis in the impeller at design flow rate operation of the pump. Like previous case, FSI stress analysis was performed for three different materials of the impeller. Results obtained from the analysis are illustrated in Figure 5 and summarized in Table 4. As it is shown in the table, for first stage impeller, maximum equivalent stress for G-Cu-Sn10 Copper Tin Alloy is 50.802 MPa and minimum equivalent stress is 0.018 MPa. Similarly, maximum equivalent stress of the first stage impeller for EN-GJL-250 Grey Cast Iron and G-X6CrNiMo18-10 is 48.1 MPa and 48.51 MPa respectively. Additionally, minimum equivalent stress for both materials is 0.025 MPa and 0.013 MPa. From second stage impeller, FSI stress analysis for G-Cu-Sn10 Copper Tin Alloy maximum equivalent stress is 108.98 MPa while minimum equivalent stress is 0.128. Similarly, as it is shown in Table 4, maximum equivalent stress of the second stage impeller for EN- GJL-250 Grey Cast Iron and G-X6CrNiMo18- 10 is 105.93 MPa and 107.05 MPa respectively.

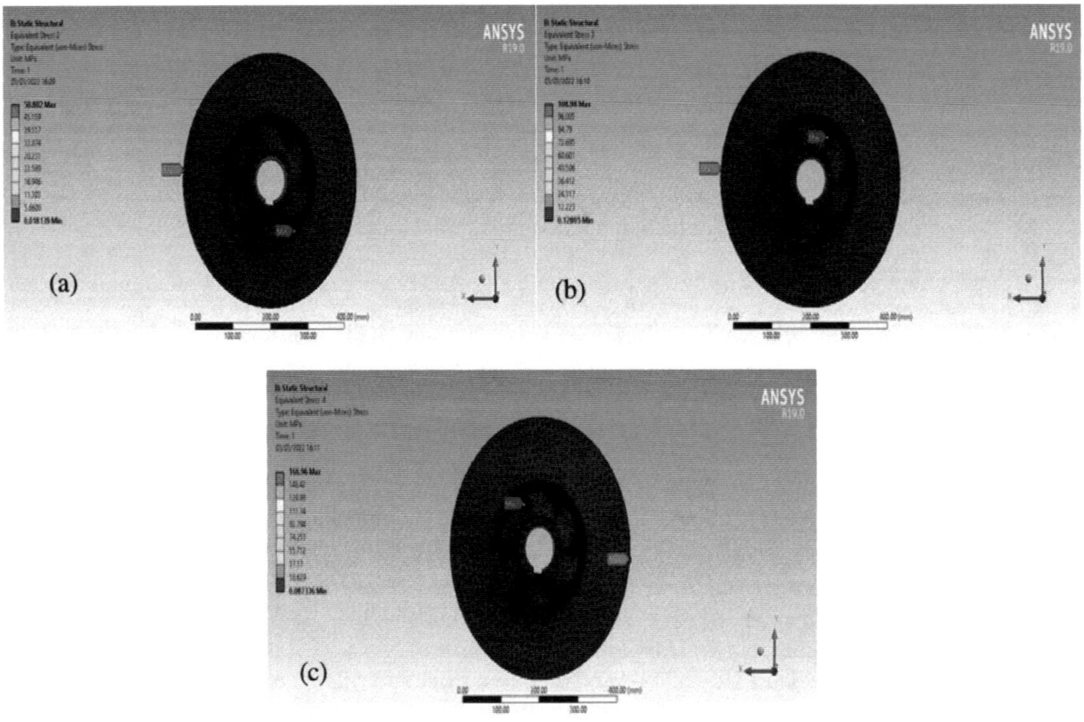

Figure 5. Equivalent stress for G-Cu-Sn10 Qd flow rate (a) Stage I (b) Stage II (c) Stage III.

Also, minimum equivalent stress for both material is 0.094 MPa and 0.115 MPa respectively. Compared to first stage impeller, the equivalent stress for the three materials has been increased. In the case of third stage impeller FSI stress analysis, equivalent stress increased for all the three materials as compared to the first and second stage impeller. As it can be observed from Table 4, the maximum equivalent stress of the third stage impeller of the pump for G-Cu-Sn10 Copper Tin Alloy is 166.96 MPa and minimum equivalent stress is 0.087 MPa. Similarly maximum equivalent stress of the third stage impeller for EN-GJL-250 Grey Cast Iron and G-X6CrNiMo18-10 is 161.34 MPa and 163.05 MPa respectively. Additionally, minimum equivalent stresses for both materials are 0.12671 MPa and 0.10297 MPa respectively. Generally, the stress in the impeller increases from the first stage to the third stage because the pressure of the water rises as the pump stage increases, which results in high pressure on the surface of the impeller.

Table 4. FSI equivalent stress of impeller at Qd flow rate

Materials	Equivalent stress (MPa)					
	Stage I		Stage II		Stage III	
	Max	Min	Max	Min	Max	Min
G-Cu-Sn10 Copper Tin Alloy	50.802	0.018	108.98	0.128	166.96	0.087
EN-GJL-250 Grey Cast Iron	48.097	0.025	105.93	0.094	161.34	0.127
G-X6CrNiMo18-10	48.509	0.013	107.05	0.115	163.05	0.103

3.2.3 FSI impeller equivalent stress at1.4Qd flow rate: The FSI stress analysis results for three stage impeller is presented in Table 5. As shown, maximum equivalent stress of the first stage impeller of the pump for G-Cu-Sn10 Copper Tin Alloy is 49.251 MPa and minimum equivalent stress is 0.012 MPa. Similarly maximum equivalent stress of the first stage impeller for EN-GJL-250 Grey Cast Iron and G-X6CrNiMo18-10 stainless steel are 48.366 MPa and 48.764 MPa respectively. Also, the minimum equivalent stress for both materials are 0.0174 MPa and 0.020 MPa respectively.

On the other hand, the equivalent stress in the second stage impeller for G-Cu-Sn10 Copper Tin Alloy is 103.96 MPa and minimum equivalent stress is 0.096 MPa. Similarly, as it can be observed from Table 5 that maximum equivalent stresses of the second stage impeller for EN-GJL-250 Grey Cast Iron and G- X6CrNiMo18-10 are 101.78 MPa and 102.78 MPa respectively. Additionally, minimum equivalent stresses for both materials are 0.0476 MPa and 0.092 MPa.

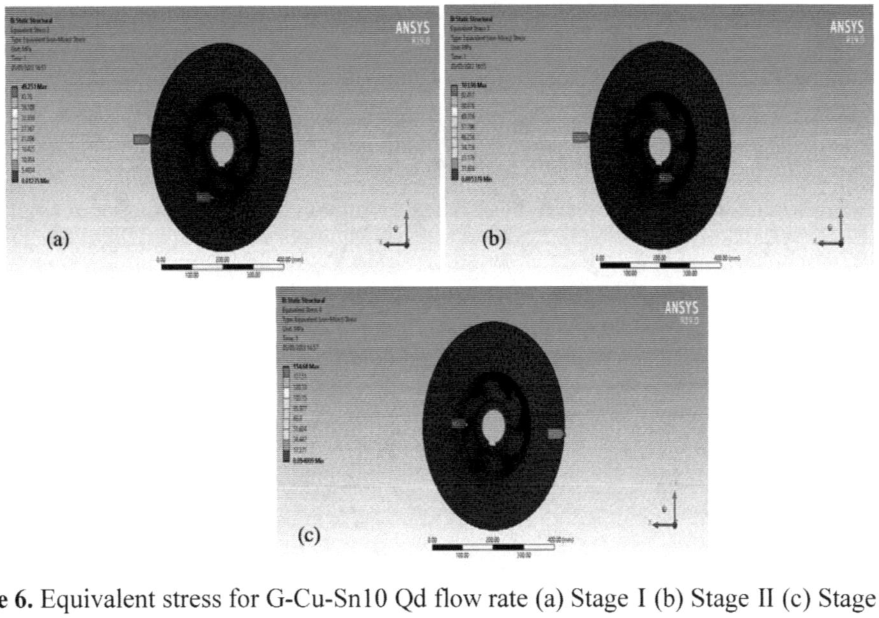

Figure 6. Equivalent stress for G-Cu-Sn10 Qd flow rate (a) Stage I (b) Stage II (c) Stage III.

At 1.4Qd flow rate operation of the pump, the equivalent stress at third stage is greater than that of the first two stage impellers. As it can be observed from Table 5, maximum equivalent stress of third stage impeller of the pump for G-Cu-Sn10 Copper Tin Alloy is 154.68 MPa and minimum equivalent stress is 0.094 MPa. Similarly, as it indicated in Table 4.20 maxi- mum equivalent stress of the third stage impeller for EN-GJL-250 Grey Cast Iron and G- X6CrNiMo18-10 is 151.21 MPa and 152.86 MPa respectively. Also minimum equivalent stress for both material is 0.122 MPa and 0.076 MPa. Generally, the stress of the impeller increases from the first stage to the third stage because of the fact that as the stage of the impeller increases the pressure of the water rise which results in high pressure on the surface of the impeller as the pump stage increases.

The study shows that in all operating conditions of the pump minimum stress was observed near the hub on the suction side of the pump. Research indicated that in centrifugal pump impellers, the minimum pressure exists at the suction side of the impeller. The lowest result of pressure at the suction side of the impeller results in the lowest equivalent stress value at the suction side of the impeller. As it can be observed from the above stress analysis of the impeller at all flow rate conditions i.e at 60% Qd, Qd and 1.4 Qd flow rate the maximum equivalent stress was observed on the pressure side at the trailing edge of the pump. The result agrees with previous research conducted [16].

Table 5. FSI equivalent stress of impeller at 1.4 Qd flow rate

Materials	Equivalent stress (MPa)					
	Stage I		Stage II		Stage III	
	Max	Min	Max	Min	Max	Min
G-Cu-Sn10 Copper Tin Alloy	49.251	0.012	103.96	0.095	154.68	0.094
EN-GJL-250 Grey Cast Iron	48.366	0.017	101.78	0.048	151.21	0.122
G-X6CrNiMo18-10	48.764	0.020	102.78	0.091	152.86	0.076

3.3 FSI Total deformation of impeller

3.3.1 FSI Total deformation of impeller at 0.6Qd flow rate: From FSI stress analysis of impeller total deformation of impeller at different stages of the pump was carried out. In this study total deformation analysis was carried out only for the third impeller because third stage impeller experiences high pressure and stress. Maximum total deformation of third stage impeller of the pump for G-Cu-Sn10 Copper Tin Alloy is 1.4609 mm and minimum total deformation is 0.037053 mm. Similarly total deformation of the third stage impeller for EN-GJL- 250 Grey Cast Iron and G-X6CrNiMo18-10 is 1.5 mm and 1.3959 mm respectively. Additionally, minimum deformation for both material is 0.035975 mm and 0.038336 mm.

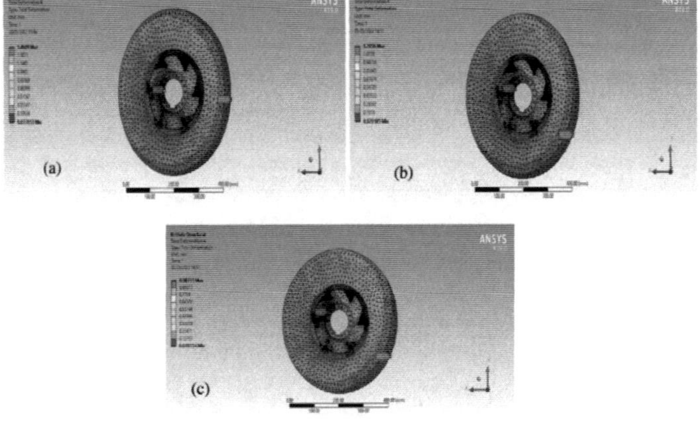

Figure 7. Total deformation for G-Cu-Sn10 impeller material at (a) 0.6Qd flow rate, (b) Qd flow rate and (c) 1.4Qd flow rate

Table 6. FSI Equivalent stress of impeller at 0.6Qd, Qd and 1.4Qd flow rate

Materials	Total deformation (mm)					
	0.6Qd		Qd		1.4Qd	
	Max	Min	Max	Min	Max	Min
G-Cu-Sn10 Copper Tin Alloy	1.4609	0.037053	1.2056	0.020185	1.4661	0.037053
EN-GJL-250 Grey Cast Iron	1.5	0.035975	1.2556	0.021183	1.5	0.019502
G-X6CrNiMo18-10	1.3959	0.038336	1.2065	0.020871	1.4425	0.01894

3.3.2 FSI Impeller total deformation at Qd flow rate: As it can be observed from Table 3 above the maximum total deformation of third stage impeller of the pump for G-Cu-Sn10 Copper Tin Alloy is 1.2056 mm and minimum total deformation is 0.020185 mm. Similarly, from Table 4.19 total deformation of the third stage impeller for EN-GJL-250 Grey Cast Iron and G-X6CrNiMo18-10 Stain- less steel is 1.2556 mm and 1.2065 mm respectively. Additionally, minimum total deformation for both material is 0.021183 mm and 0.020871 mm respectively.

3.3.3 FSI impeller total deformation at 1.4Qd flow rate. Total deformation of the third stage impeller at 1.4Qd flow rate of the pump discussed. Results obtained from total deformation is presented in Table 3 above. As it can be observed from the Figure 7 maximum total deformation of third stage impeller of the pump is for G-Cu-Sn10 Copper Tin Alloy is 1.46609 mm and minimum deformation is 0.037053 mm. Similarly total deformation of the third stage impeller for EN-GJL-250 Grey Cast Iron and G-X6CrNiMo18-10 is 1.5 mm and 1.4425 mm respectively. Also, minimum total deformation for both material is 0.019502 mm and 0.01894 mm. In all cases of the operating conditions of the pump maximum impeller deformation was observed at the trailing edge of the impellers in all the three materials of the pump. This is due to the fact that the pressure inside the impeller increases from the leading edge of the pump to the trailing edge of the pump hence it creates high stress and deformation at the trailing edge of the impeller of the pump.

3.4 FSI safety factor of impeller at different flow rate:

Stage III FSI safety factor at 0.6Qd flow rate: In this study, the safety factors were analysed only for the third stage impeller. This is due to the fact that the impellers have the same geometry, and from the three impellers arranged in series on the shaft, the third stage impeller experiences high stress and deformation. As it can be observed from Figure 8(a), the safety factor at 0.6Qd flow rate of the pump for G-Cu- Sn10 Copper Tin Alloy is around 0.76, while for EN-GJL- 250 Grey Cast Iron and G-X6CrNiMo18-10, the safety factors 0.98 and 1.24 respectively. Factor of safety for G-Cu-Sn10 Cop- per Tin Alloy and EN-GJL-250 Grey Cast Iron is less than one which means the impeller fail be- fore design life of the impeller. Factor of safety for G-X6CrNiMo18-10 is 1.24 which indicates that the impeller is safe because the safety factor is greater than one.

Stage III FSI safety factor at Qd flow rate: As it can be observed from the Figure 8(b), the safety factor for G- Cu-Sn10 Copper Tin Alloy is around 0.78, while for EN- GJL-250 Grey Cast Iron and G-X6CrNiMo18-10 the safety factors are 1.02 and 1.3 respectively. Factor of safety for G-Cu-Sn10 Cop- per Tin Alloy is less than one, which means the impeller fails before its design life. The safety factor for G-X6CrNiMo18-10 is 1.29 which is also greater than one implying that the impeller is safe.

Stage III FSI safety factor at 1.4Qd flow rate: Plot of the safety factor distribution for this case is given in Figure 8(c), where it is observed that the value for G- Cu-Sn10 Copper Tin Alloy is 0.76. For EN-GJL-250 Grey Cast Iron and G-X6CrNiMo18-10, the minimum safety factor values are 0.98 and 1.24 respectively. It indicates that the safety factor for G-Cu-Sn10 Copper Tin Alloy and EN-GJL-250 Grey Cast Iron are less than one, which means the impeller fails before its design life. On the other hand, the G-X6CrNiMo18-10 material is safe because its safety factor is 1.24 which is greater than one.

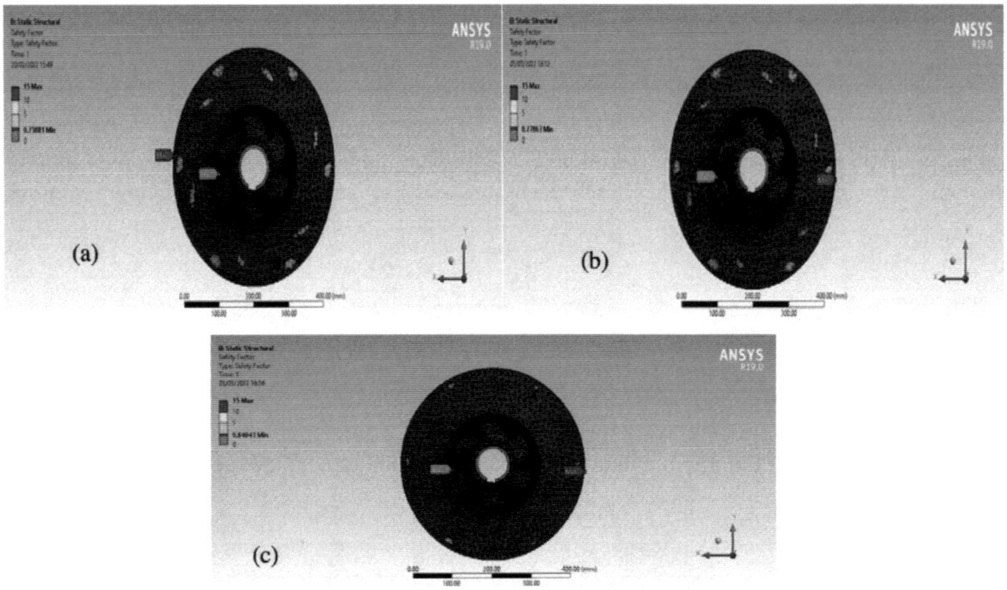

Figure 8. Stage III safety factor for G-Cu-Sn10 impeller material at (a) 0.6Qd flow rate, (b) Qd flow rate and (c) 1.4Qd flow rate

4. Conclusion

To investigate the effect of the fluid pressure on centrifugal water pump impeller, CFD analysis of a three-stage centrifugal pump was carried out using ANSYSS CFX at different flow rates of the pump ranging from 0.6Qd to 1.4Qd flow rates. In the analysis, a steady state solution with k- SST turbulence model was used in ANSYS-CFX for analysis of the centrifugal water pump. The result shows that as the flow rate increases the head of the pump decreases and as the flow rate of the pump decreases the head developed by the pump increases. The head rise obtained through CFD analysis was over predicted head rise than the design head of the pump which was 191 m. This is due to the fact that leakage was neglected in the analysis of the pump during the analysis and also surface roughness of the impeller, diffuser with return guide vanes was assumed to be smooth.

Numerical stress analysis of the impellers of the three-stage centrifugal water pump was performed, in which it was observed that the pressure distribution on the surface of the impeller varied from suction side to the pressure side of the impeller. The pressure of the fluid does not only vary along the surface of the impeller but also it varies along different stages of the pump. From the finding of the study, the third stage experiences higher deformation and stress than the first and the second stage impeller. The results obtained from the impeller analysis shows that the impeller stress increases as the flow rate decreases because the pressure acting on the wall of the impeller increases as the pump flow rate decreases.

Acknowledgement: The authors would like to acknowledge the support provided INDMET project funded by NORHED II program (grant nr. 62862) to cover the publication cost of this work.

References

[1] Cherkassy V M 1985 Pumps, Fans, Compressor, Mir publishers,.

[2] Benturki M, Dizene R, Ghenaiet A 2018 Multiobjective optimization of two-stage centrifugal pump using NSGA-ii algorithm, *J. Appl. Fluid Mech.* **11**(4) 929- 943.

[3] Khaing C C, Lynn A Z and Nyi N 2019 Design of multistage centrifugal pump impeller for high head applications, *Int. J. Latest Technol. Eng., Manage. Appl. Sci. (IJLTEMAS)* **8**, 115-120.

[4] Rajanand M P 2016 Design & analysis of centrifugal pump impeller by FEA, *Int. Res. J. Eng. Technol. (IRJET),* **03**, 420-428.

[5] Iratkar G G and Gandigude A U 2017 Structural analysis, material optimization using FEA and experimentation of centrifugal pump impeller. *Int. J. Adv. Res. Ideas. Innov. Technol.*, **3**(4), 97-105.

[6] Singh V R, Zinzuvadia M, Sheth S M 2017 Parametric study and design optimization of centrifugal pump impeller-a review, *Int J Eng Res Appl.* **1**, 507-515.

[7] Tamin M N, Hamzah M A 2017 Fatigue failure analysis of a centrifugal pump shaft, In: *Fail. Anal. Prev., Ed. Aidy Ali, IntechOpen,* 1-14, Rijeka, Croatia,.

[8] McKee K K, Forbes G L, Mazhar M I, Entwistle R and Howard I M 2011 A review of major centrifugal pump failure modes with application to the water supply and sewerage industries, *In: Asset Manage. Council (ed), ICOMS Asset Manag. Conf.* (Gold Coast: May 16 2011)

[9] Bardina J E, Huang P G, Coakley T 1997 Turbulence modeling validation, *28th AIAA Fluid Dynamics Conference, AIAA- 1997-2121.*

[10] Friedrich-Karl B, Hans Josef D, Ji P. Sebastian S and Bo W 2011 A comparison of one-way and two-way coupling methods for numerical analysis of fluid-structure interactions, *J. Appl. Math.*, 1-16.

[11] Gu Y, Pei J, Yuan S, Xing L, Stephen C, Zhang F and Wang X 2018 Effects of blade thickness on hydraulic performance and structural dynamic characteristics of high-power coolant pump at overload condition, *Proc. Inst. Mech. Eng., Part A: J. Power Energy,* **232**, 992-1003.

[12] Kobayashi K, Ono S, Harada I and Chiba Y 2010 Numerical analysis of stress on pump blade by one-way coupled fluid-structure simulation. *J. Fluid Sci. Technol..* **5**, 219–234.

[13] Piperno S, Farhat C and Larrouturou B 1995 Partitioned procedures for the transient solution of coupled aroelastic problems Part I: Model problem, theory and two-dimensional application. *Comput. Methods Appl. Mech. Eng.* **124**, 79–112.

[14] Kang H S and Kim Y J 2016 A study on the multi-objective optimization of impeller for high-power centrifugal compressor. *Int. J. Fluid Mach. Syst.* **9**, 143–149.

[15] Suh S-H, Kyung-Wuk K, Kim H-H, Cho M T and Yoon I S 2015 A study on multistage centrifugal pump performance characteristics for variable speed drive system, *Procedia Eng.* **105**, 270-275.

[16] Ji P, Shouqi Y and Jianping Y 2014 Dynamic stress analysis of sewage centrifugal pump impeller based on two-way coupling method *Chin. J. Mech. Eng.* **27**(2), 369-375

IOP Conf. Series: Materials Science and Engineering 1294 (2023) 012023 doi:10.1088/1757-899X/1294/1/012023

Design optimization of small fishing vessel structures: A case study

G L Putra[1, *], F Rizqillah[1], A Yatim[1], and Y Liu[2]

[1] Universitas Indonesia, Indonesia
[2] Norwegian University of Science and Technology, Norway

[*] Correspondence: gerrylp@eng.ui.ac.id

Abstract. The cost of vessel operations poses a significant challenge for fishermen in Indonesia, with 60% - 70% of their operational expenses being allocated to fuel. To address this issue and decrease fuel consumption in fishing vessels, one possible solution is to construct lighter boat designs. Through careful optimization of the construction process and the selection of suitable materials, it becomes feasible to achieve such a lighter design, leading to reduced operational and construction costs. The optimization process involves considering various frame spacing variations, like 400 mm, 500 mm, and 600 mm, and utilizing materials such as mild steel and AH32. Additionally, optimizing the plate thickness ensures that the vessel's stress and deformation remain within the limits set by the rules. As a result of this optimization, a weight reduction of up to 20% is observed.

1. Introduction

One of the primary concerns, particularly for fishermen in Indonesia, revolves around the expenses associated with operating their boats. The reason for this financial burden is that a significant portion, around 60% - 70%, of their operational costs, is allocated to fuel [1]. The situation becomes even more challenging due to the consistent increase in fuel prices in Indonesia, which further escalates the operating expenses for fishermen [2]. To mitigate the impact of rising fuel costs, one practical approach is to design boats with lower resistance, thereby reducing the power required for operation. Achieving this goal entails constructing lighter fishing boats through meticulous optimization of boat construction and the careful selection of appropriate materials. By employing these optimization techniques and choosing suitable materials, it becomes entirely feasible to develop a fishing boat design that is significantly lighter and more fuel-efficient.

Presently, the construction of fishing boats in Indonesia heavily relies on wooden materials. Nevertheless, the accessibility of wood for boat building is progressively dwindling, and not all wood varieties can serve as viable raw materials in traditional boat construction [3]. Consequently, numerous wooden boats are witnessing a decline in their overall quality because of the scarcity of suitable wood types, both in terms of species and size adequacy [4].

In order to achieve the most efficient and effective boat construction in terms of strength and weight, there is a need for research focused on optimization methods and a comparative analysis of different materials used in the process. Several studies have been conducted in this area, exploring various aspects of structural optimization and material selection. For instance, one study delves into the use of genetic algorithms for structural optimization in the engine room [5]. Another research project investigates stiffener optimization in plates utilizing Finite Element Analysis (FEA) with Ansys [6]. Moreover, there are studies that explore structural optimization using evolutionary algorithms, considering multiple variables such as plate thickness, longitudinal & transverse member scantlings, and spacing between members [7].

Furthermore, research has been conducted on material selection with a focus on achieving lightweight designs while considering environmental life cycle assessment [8]. Additionally, there are studies employing genetic algorithms to optimize structural elements such as plate thickness, plate layout, stiffener type, and material selection, all aimed at achieving cost-effectiveness [9]. Lastly, one study is dedicated to the optimization of plate thickness in the hull construction of small fishing vessels previous research [10] solely concentrated on optimizing plate thickness, considering the material's yield strength as the optimization constraint. Unfortunately, the outcomes of that optimization fell short of meeting the deformation limit constraints [11]. To address this issue, the present research seeks to enhance boat construction optimization by incorporating various frame spacing variations and employing two types of steel materials: mild steel and AH32. Moreover, the optimization of plate thickness will now consider deformation limits to ensure compliance with existing rules and regulations. The ultimate objective is to discover an optimal construction that is both lightweight and in full adherence to the required rules and regulations. The novelty of this research lies in its comprehensive approach, considering multiple variables including frame spacing, material types, and deformation limits, with the aim of achieving a lightweight yet fully compliant construction.

2. Method

The optimization method consists of three essential processes. First, the determination of the type of material to be used in the construction. Second, the determination of various frame spacing variations to be applied to the ship's structure. And third, the optimization of plate thickness to achieve the most optimal design. The combination of these three processes is expected to result in a lighter ship construction while still complying with applicable rules and regulations.

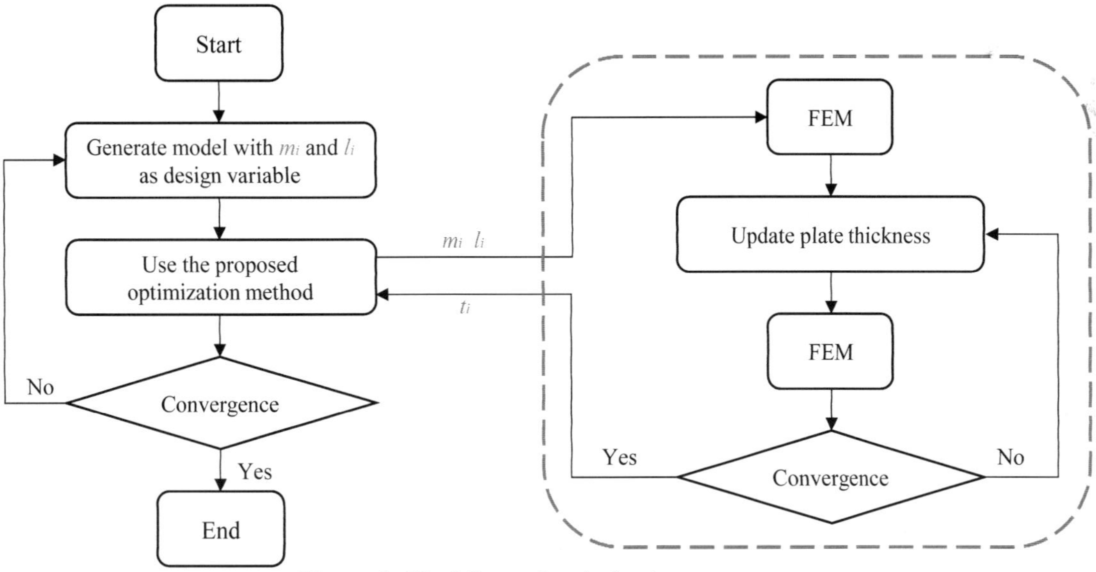

Figure 1. Workflow of optimization process

As depicted in Figure 1, the approach involves a step-by-step application of various types of materials, along with different frame spacing combinations. Subsequently, the focus shifts to optimizing the plate thickness while adhering to predefined constraints. This optimization process is performed iteratively, repeating the defined constraints until the desired state of convergence is attained. By following this systematic method, the research aims to achieve an optimal design for the small fishing vessel, considering both the materials used and the frame spacing variations, ultimately leading to a well-optimized and efficient construction.

2.1. *Update of plate material type*

In the context of this study, a meticulous material selection process is applied to determine the most suitable plate material, guided by a specific objective function that prioritizes the minimization of construction weight. The shipbuilding industries are driven by a strong incentive to design lightweight vessels due to the manifold advantages it offers. By opting for lighter materials, manufacturing costs are reduced significantly, resulting in enhanced cost-effectiveness. Additionally, this emphasis on lightweight design aligns with the global commitment to reducing carbon emissions, making it an eco-friendly approach. As a result, the integration of optimized plate materials in ship construction not only benefits the industry financially but also contributes to mitigating the environmental impact by curbing CO_2 emissions.

2.2. *Update of frame spacing*

The design under consideration is intended for a fishing vessel hull with a 5 gross tonnage, where the distance between the frames is irregular, with a maximum of 1000 mm and a minimum of 500 mm. For the structural analysis, a model is utilized that encompasses the main deck, building upon the groundwork laid by previous research [10]. To tailor this model to the specific requirements of the study, additional modifications are made employing Ansys software. These modifications result in the creation of a construction model with varying configurations of frame spacing. Notably, three configurations are examined, as shown in Figure 2, with frames spaced at intervals of 400mm, 500mm, and 600mm, respectively. These different configurations are crucial to understanding how frame spacing impacts the overall structural performance and optimization goals of the fishing vessel.

Upon completion of the modelling process, three construction models were derived, each featuring distinct frame spacing variations. These models will undergo comprehensive analysis using Ansys Static Structural. The final constructed models are as follows:

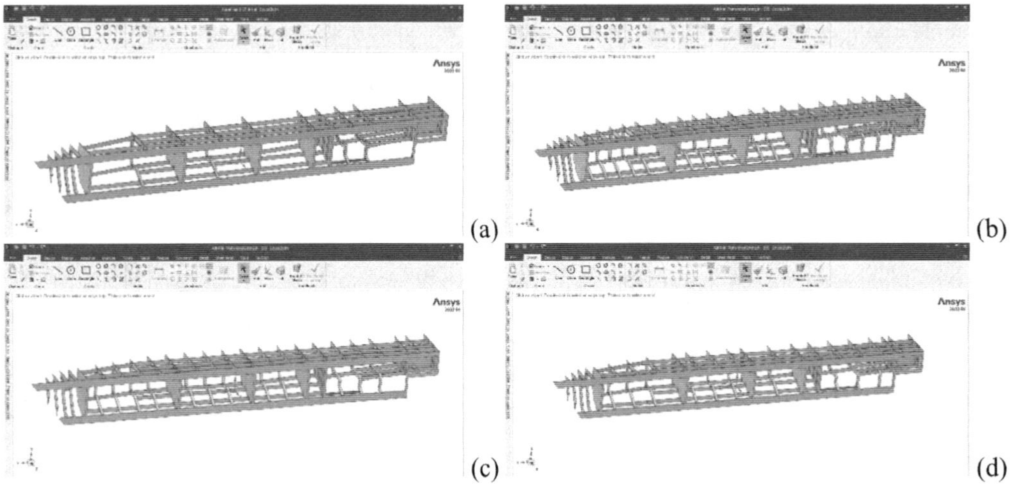

Figure 2. FE models of the fishing vessel (a) Initial, (b) 400 mm, (c) 500 mm, and (d) 600 mm.

2.3. *Update of plate thickness*

Size optimization is a method employed to enhance the dimensions of an object, aiming for optimal results. In the realm of structural engineering, one application of size optimization pertains to optimizing plate thickness. This process relies on equations that consider the interplay between plate thickness, axial force, bending moment, and stresses [9]. The plate thickness t update process is carried out by using Equations 1 and 2. Equation 1 is derived from the expression of the maximum stress σ_{max}^{a}, while Equation 2 relates to the maximum deformation w_{max}^{a}. Following these equations, the recommended optimal plate thickness is determined at step a+1 when the maximum stress reaches the strength limit, σ_{perm}. This iterative process continues until the maximum stress value approaches the strength limit σ_{perm}[12]. Through this iterative approach, the plate thickness is continuously adjusted, ultimately leading to an optimized design where the maximum stress is closely aligned with the strength limit.

$$t^{a+1} = t^{a} \sqrt{\frac{\sigma_{max}^{a}}{\sigma_{perm}}} \tag{1}$$

$$t^{a+1} = t^{a} \sqrt[3]{\frac{w_{max}^{a}}{w_{perm}}} \tag{2}$$

3. Optimization model

3.1. FEM model
In this analysis, a uniform element size of 22 mm has been employed, resulting in a total of 620,379 elements and producing stress values that tend to converge, as shown in Figure 3. This choice is rooted in several considerations, including the capabilities of the computer, computational time requirements, and the intricate shape of the object under study. Furthermore, the meshing technique adopted is the quadratic tetrahedral mesh. This mesh type is preferred due to its ability to accurately represent and model objects with complex and intricate geometries, ensuring precise and reliable results in the analysis [13].

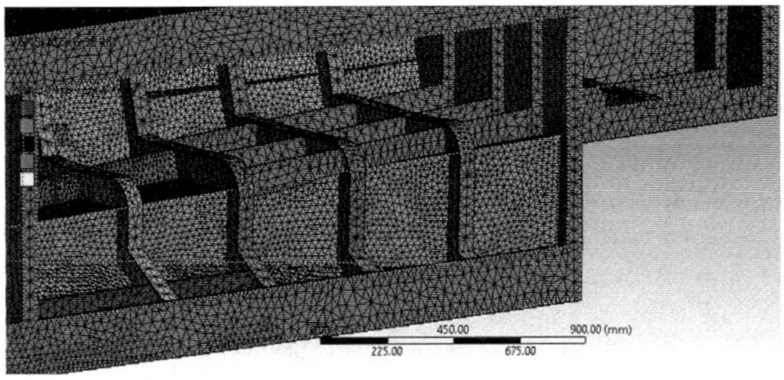

Figure 3. The mesh result for model

3.2. Design variables
This study focuses on three key design variables to optimize the ship's structure: the choice of material, frame spacing, and plate thickness. Specifically, two distinct materials were proposed, MS (Mild Steel) and AH32, with the primary difference lying in their yield strength characteristics as shown in Table 1. Furthermore, frame spacing varies on 400 mm, 500 mm, and 600 mm. Each frame spacing model presents unique structural considerations.

The last variable of interest is plate thickness, a critical factor that significantly influences both the strength and weight of the ship's construction. It's important to note that among these variables, material type and frame spacing are considered discrete variables, while plate thickness is treated as a continuous variable. These design considerations pave the way for a comprehensive optimization process that seeks to strike the right balance between structural integrity, weight efficiency, and performance.

Table 1. Material properties

Material Type	Yield Strength (MPa)	Young's Modulus (MPa)	Density (kg/m^2)	Poisson Ratio
Mild Steel (NS)	235	200,000	7800	0.3
AH32	315	200,000	7800	0.3

3.3. Design constraints

To identify the most optimal plate, several constraints must be satisfied initially. The primary constraint is related to the material's strength limit. According to IACS rules, the maximum allowable stress limit is determined by the material's yield strength (R_{eH}), as expressed in Equation 3. This analysis involves the use of a total of 15 plates, as seen in Figure 4. Furthermore, in accordance with the rules outlined by [14], there are constraints governing the deformation limits of a plate. The allowable deformation limits for a plate, whether using steel or aluminum materials, as indicated in Equation 4, are determined by the span or distance between frames (ι) and values provided in Table 2.

$$\sigma_{perm} = 0.8R_{eH} \tag{3}$$

$$f = \frac{l}{value} \tag{4}$$

Table 2. Values for determination of permissible deflection.

Structural element	Value	
	Steel	Alu
Slamming (shell)	500	700
Shell, Bhd, tank	400	560
Stringer, decks, ramps	200	280

Additionally, the optimization is bounded by practical considerations to facilitate material sourcing and to prevent the plate from becoming excessively thin. Consequently, constraints on the minimum plate thickness and the permissible number of thickness variations are imposed, as described in Equations (5) and (6).

$$t_{min} = 2\ mm \tag{5}$$

$$\text{Number of thickness variations} = 3 \tag{6}$$

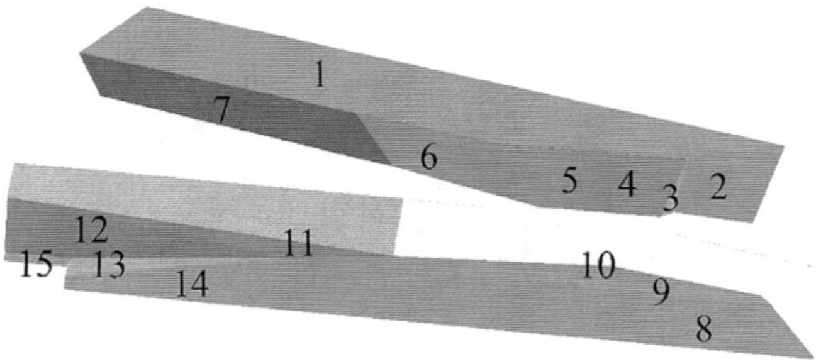

Figure 4. Plate number

4. Results and discussion

In fishing vessels, transverse strength takes precedence over longitudinal strength due to the vessel's relatively short length, which necessitates the use of transverse construction. Consequently, the primary emphasis is placed on ensuring robust transverse strength. Therefore, the results presented below are directly related to the transverse forces acting on the vessel.

When a vessel carries a given load, it experiences tension in response to external forces, including the cargo load and hydrostatic pressure. In the design of ship construction, it is crucial to ensure that the stress occurring in the ship's plates remains below the limit of the ship's material capabilities, particularly its yield strength, as explained previously. Therefore, the constraints associated with structural optimization are closely tied to the properties of the materials utilized in construction.

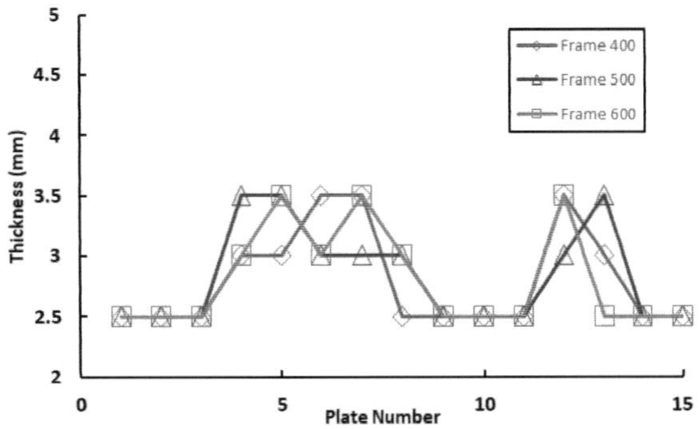

Figure 5. The optimal plate thickness of mild steel & AH32

Figure 5 depicts the results of the optimization process. Post-optimization, the plate thickness does not quite reach the minimum required thickness, despite the von-mises stress levels remaining well below the material's yield strength as shown in Table 3. This phenomenon can be attributed to the presence of deformation limits affecting steel materials with variable frame distances. Interestingly, both AH32 and mild steel materials converge to the same thickness following optimization. From these results, both MS and AH32 show similar outcomes, so there is no significant difference when using either of these materials. This similarity is due to the resulting thickness values being the same. AH32 would have a different impact compared to MS when its stress-bearing capacity is thinner than MS.

Table 3. Max stress & deformation.

Material	Frame spacing (mm)	Initial		Optimum	
		Max stress (MPa)	Max deformation (mm)	Max stress (MPa)	Max deformation (mm)
MS	400	61.86	0.65	93.34	0.99
	500	61.92	0.59	100.60	1.21
	600	62.14	0.66	106.39	1.43
AH32	400	61.86	0.65	93.34	0.99
	500	61.92	0.59	100.60	1.21
	600	62.14	0.66	106.39	1.43

As shown in Figure 6, the optimization results for both mild steel and AH32 materials demonstrate a weight reduction of 21.3% for the 400 mm frame, 21% for the 500 mm frame, and 20.8% for the 600 mm frame. Meanwhile, it's worth noting that an increase in construction mass can provide additional strength [15]. Consequently, When the deformation-to-weight ratio value is smaller, and this value is obtained from the maximum deformation and the total weight produced in various frame spacing variations, it signifies that a relatively minor increase in weight can significantly enhance the structure's strength, resulting in improved overall performance and safety.

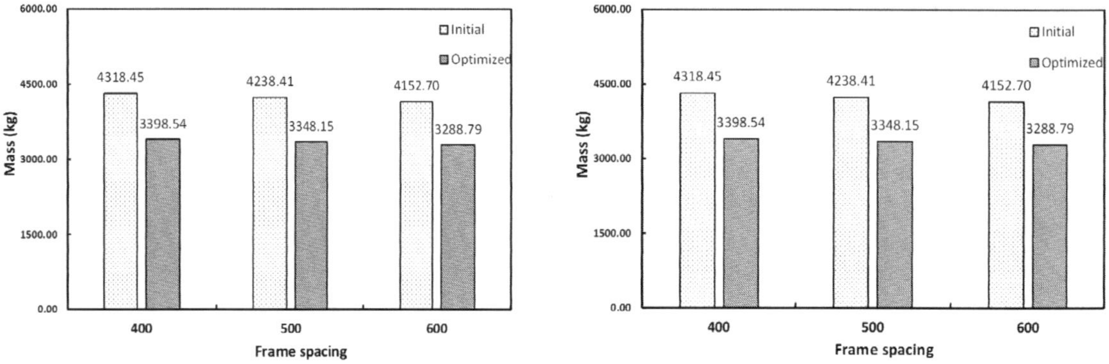

(a) (b)

Figure 6. The Hull structure's weight (a) Mild steel and (b) AH32.

Analyzing the deformation-to-weight ratio in Table 4, it becomes apparent that the optimal choice is the 400 mm frame, as indicated by the yellow colour. However, if the aim is to achieve the lightest construction possible, then the 600 mm configuration is the preferred choice. Furthermore, in this construction, the difference between using mild steel and AH32 does not yield a significant impact. This is because the maximum stress that occurs in most frame configurations remains below the yield strength of both materials. Additionally, given that both materials have the same density, the resulting weight remains consistent for the same thickness. Therefore, when considering the cost of material procurement, it is recommended to opt for mild steel, which tends to be more cost-effective than AH32.

Table 4. Strength to weight ratio

Material type	Frame spacing (mm)	Deformation to weight ratio (mm/kg)	
		Initial	Optimum
MS	400	0.00015	0.00029
	500	0.00014	0.00036
	600	0.00016	0.00043
AH32	400	0.00015	0.00029
	500	0.00014	0.00036
	600	0.00016	0.00043

5. Conclusions

Based on the research results of the structural optimization, which considered frame spacing and plate thickness while taking into account yield strength and deformation constraints outlined in the rules, the following conclusions can be drawn to address the research objectives:

1. The optimization results in a weight reduction of 21.3% for the 400 mm frame, 21% for the 500 mm frame, and 20.8% for the 600 mm frame for both mild steel and AH32 materials.
2. Frame spacing exerts an impact on the weight of the construction, resulting in a difference of 80 to 90 kg. In the initial condition, the weight of the construction is 4318 kg for the 400 mm frame, 4238 kg for the 500 mm frame, and 4152 kg for the 600 mm frame.
3. When evaluating the deformation-to-weight ratio, the most favourable combination is the 400 mm frame. However, if the goal is to achieve the lightest possible construction, the 600 mm configuration is a viable choice.

In the future, optimization of frame and stiffener sizes will be needed to achieve a more optimal design. This optimization process involves carefully adjusting the dimensions of the frames and the types and placements of stiffeners within the ship's hull.

Acknowledgements
This publication has been sponsored by the Norway-ASEAN Consortium in Risk Management for Safer and Sustainable Oceans (NESS) and is supported by the Innovation Funding Program for the Year 2023 under the Product Development Program (P4) Scheme No: PKS-138/UN2.INV/HKP.05/2023.

References

[1] Simbolon F. Nelayan 2022, 60%-70% Biaya Operasional Nelayan Melaut Adalah BBM. Kontan.co.id. Available at: https://nasional.kontan.co.id/news/nelayan-60-70-biaya-operasional-nelayan-melaut-adalah-bbm

[2] Lathif AN 2022 Biaya Operasional Tinggi, Ratusan Kapal Nelayan di Pati Tak Bisa Melaut. iNewsJateng.id. Available at: https://jateng.inews.id/berita/biaya-operasional-tinggi-ratusan-kapal-nelayan-di-pati-tak-bisa-melaut

[3] Kurniawati V, Novita Y, Kusumanti I. 2011 Tingkat Pemanfaatan Kayu Pada Pembuatan Gading-Gading di Galangan Kapal Rakyat UD. Semangat Untung Desa Tanah Beru, Bulukumba, Sulawesi Selatan. *Buletin PSP*. **19**(3),219–28.

[4] Nasution P, Hutauruk RM. 2016 Analisis Konstruksi Kapal Nelayan Tradisional di Kabupaten Bengkalis Provinsi Riau. *J. Perikanan dan Kelautan*. **21**, 7–17.

[5] Kitamura M, Nobukawa H, Yang F. 2001 Application of a genetic algorithm to the optimal structural Design of a ship's engine room taking dynamic constraints into consideration. *J Mar Sci Technol.* **5**,131–46.

[6] Alinia MM. A 2005 Study into optimization of stiffeners in plates subjected to shear loading. *Thin-Walled Struct.* **43**(5), 845–60.

[7] Sekulski Z. 2014 Ship hull structural multiobjective optimization by evolutionary algorithm. J. Ship Res. **58**, 45–69.

[8] Poulikidou S, Schneider C, Björklund A, Kazemahvazi S, Wennhage P, and Zenkert D. A 2015 Material selection approach to evaluate material substitution for minimizing the life cycle environmental impact of vehicles. *Mater Des*. **83**, 704–12.

[9] Putra GL, Kitamura M. 2021 Study on optimal design of hatch cover via a three-stage optimization method involving material selection, size, and plate layout arrangement. *Ocean Eng.* **219**, 108284.

[10] Christofer. Studi Optimasi Struktur Kapal Ikan 5 Gross Tonnage Dengan Material Baja. [Depok]: Universitas Indonesia; 2023.

[11] Germanischer Lloyd. *Rules for Classification and Construction*. DNV 2012. Available at: https://www.dnv.com/rules-standards/gl-rules-guidelines.html

[12] Putra GL, Kitamura M, Takezawa A. 2019 Structural optimization of stiffener layout for stiffened plate using hybrid GA. *Int. J. Naval Arch. Ocean Eng.* **11,** 809–18.

[13] Cifuentes AO, Kalbag A. 1992 A performance study of tetrahedral and hexahedral elements in 3-D finite element structural analysis. *Finite Elem. Anal. Des.* **12**(3-4), 313–8.

[14] Biro Klasifikasi Indonesia. Rules For Classification and Construction. Vol. 2. 2022.

[15] Pratama EZ. Evaluasi Kekakuan dan Kekuatan Material Komposit terhadap Beban Statis pada Body dan Chassis Widya Wahana V. Institut Teknologi Sepuluh Nopember; 2015.

IOP Conf. Series: Materials Science and Engineering 1294 (2023) 012024 doi:10.1088/1757-899X/1294/1/012024

Combining machine learning methods and data augmentation for misaligned journal bearings design

K Arvanitis* and P G Nikolakopoulos

Machine Design Laboratory, Department of Mechanical Engineering and Aeronautics, University of Patras, 26504 Patras, Greece

* Corresponding author e-mail: up1067325@upnet.gr

Abstract. Shaft misalignment is one of the most common defects observed in rotating systems and has a substantial effect on dynamic behaviour, stability, and lifetime. Aim of this study is the binary identification of misalignment using five Machine Learning techniques: Logistic Regression, K-Nearest Neighbours, Support Vector Machines, Decision Tree and Random Forest. Nevertheless, the limited quantity of provided data points coupled with the substantial imbalance between the aligned and misaligned cases necessitated the implementation of oversampling and data augmentation methods. The utilization of SMOTE-LOF for oversampling the minority class, alongside the adoption of a Conditional Tabular GAN for the generation of synthetic data points yielded substantial outcomes. The application of SMOTE with the Local Outlier Factor on the original dataset achieves the oversampling of the minority class by using additional synthetic data, while the LOF factor overcome the noise problem. After the SMOTE-LOF implementation, new synthetic samples of the minority class are added to the dataset, eliminating the imbalance, however the 'cleaned' dataset's small number of samples could lead to poor performance of the Machine Learning techniques. Hence, using a Generative Adversarial Network to expand the dataset is requisite, and since the experimental data is tabular, utilizing a Conditional Tabular Generative Adversarial Network is ideal for this case of study. The results of the current study elucidate the dataset that, among the augmented datasets, exhibits the best quality score when contrasted with the original dataset. Furthermore, the dataset that performs the best on the Machine Learning algorithms is determined.

1. Introduction

Journal bearings are prevalent engineering components in almost all types of rotating machinery. The identification of defects observed in rotating systems has already attracted several researchers with remarkable results. More specifically, a lot of articles have been examining the diagnosis of misalignment on journal bearings, since it is one of the most severe defects that influences the bearing behavior and consequently the rotor bearing dynamics. Traditional methods such as finite element, finite difference, meshless methods are already being utilized to solve the Reynolds equation for the pressure distribution.

On the other hand, as computer science continues to advance, there has been a notable surge in the utilization of machine learning applications within the field of mechanical engineering. However, within the domain of Tribology, machine learning applications still constitute a new era of research [1], and their growth is steadily extending across various facets of the field.

Gertzos et al. [2] presented a CFD analysis of hydrodynamic lubricated bearings lubricated with a Bingham fluid. Nuruzzaman et al. [3] investigated the pressure distribution and load capacity of a journal bearing using the finite element method (FEM), while Wenjie Zhou et al. [4] proposed a novel numerical method for studying the lubricated characteristics of journal bearings, including the FEM and mixed method. Moreover, J.W. Lund [5] reviewed the dynamic coefficients for fluid film journal bearings, and Zhou et al. [6] proposed a method to find the bearing's static equilibrium position. In addition, Yiyang Fu et al [7] calculated the pressure development of journal bearings by using inhomogeneous and homogeneous grid in the Finite Difference Method (FDM).

Since misalignment is having a significant impact in the performance of the journal bearing, many researchers have studied misaligned journal bearings, both experimentally and analytically. Jang and Khonsari [8] presented a literature review of the characteristics of the misaligned journal bearings, while Singh and Sinhasan [9] investigated the effects of small misalignment on the dynamic characteristics of an ungrooved big end bearing utilizing the FEM method. Furthermore, Nikolakopoulos and Papadopoulos [10] studied the nonlinear behavior of misaligned journal bearings, while Jun Sun et al [11] presented a hydrodynamic lubrication analysis of misaligned journal bearings due to shaft deformation. Experimental work of axial misalignment effect on the seizure load of journal bearings was executed by Zhang et al [12].

Furthermore, experimental studies of misaligned bearings were conducted by Huber et al. [13] and Bouyer and Fillon [14]. Huber et al. investigated the misaligned plain circular bearing characteristics such as pressure and oil flow under a static load, while Bouyer and Fillon evaluated the performance of a plain journal bearing submitted to a misalignment torque.

The field of artificial intelligence (AI) applied to journal bearing lubrication is relatively new, and although some research has been published there remains ample room for forthcoming contributions. Rosenkranz, Marian et al. [15] presented an overview of the various applications of artificial intelligence (AI) and machine learning (ML) in tribology. Machine learning techniques have been applied for journal bearings' misalignment monitoring. Saridakis et al. [16] proposed a misalignment diagnosis approach that uses artificial neural networks. The proposed model uses values of journal bearing performance characteristics, such as eccentricity, attitude angle and film thickness and it trains an ANN so that reliable identification is performed. Also, Nikolakopoulos and Kokkinidis [17] made use of ANNs to solve the problem of misaligned hydrodynamic journal bearing. Firstly, the Reynolds equation was solved using finite difference methos (FDM) for different operating and misalignment conditions. The results was used as an input to train four ANNs, one for each design parameter. After the networks were tested for several operational characteristics, the results were compared with the results of the finite difference method. As an outcome, the force and the torque can be predicted with maximum error of 5% with less computational cost than the FDM. Furthermore, the application of ANNs to calculate the hydrodynamic forces and the dynamic of the rotor was investigated by Kornaev et al. [18]. Nevertheless, especially in tribology, collecting extensive datasets can be costly, time-consuming, or even infeasible. Datasets that suffer from imbalance and limited data availability represent common scenarios and frequently result in poor performance of machine learning algorithms.

Oversampling technique is widely used in order to handle the imbalance of datasets, however in this case the data distribution is ignored, and the risk of overfitting is increased. Aiming at this problem, Chawla et al. [19] proposed the Synthetic Minority Oversampling Technique (SMOTE). Zhenya Wang et al [20] proposed a method using SMOTE and CNN to deal with imbalanced bearing data, while Xinqian Hu et al. [21] utilized a variation of SMOTE method in order to enhance the variety of fault samples for bearing fault diagnosis. Additionally, data augmentation is an imperative technique in engineering problems that suffer from limited data availability. Data augmentation techniques provide a pragmatic solution by synthetically expanding the dataset through various transformations. Generative Adversarial Networks (GANs) are a popular class of data augmentation models introduced by Ian Goodfellow et al. in 2014 [22] and has few applications on fault diagnosis of rotating systems in case of limited datasets. Wentao Mao et al. used a GAN to improve the generalization ability of the fault diagnosis model of a rolling bearing, and Diwang Ruan et al [23] investigate the bearing fault diagnosis by optimizing a CNN and a GAN collaboratively. Additionally, Cheng Peng et al. [24] substantiate the enhancement of the rolling bearing fault diagnosis method as a result of the GAN utilization.

The aim of this study is the identification of journal bearing misalignment within the context of a limited and imbalanced dataset. The integration of the Synthetic Minority Oversampling Technique (SMOTE) and Conditional Tabular Generative Adversarial Network (CTGAN) is used to effectively address the constraints presented by the limited and imbalanced journal bearing dataset provided by previous work of Saridakis et al. [16]. Then, five machine learning algorithms are utilized: Logistic Regression, K-Nearest Neighbors, Support Vector Machines, Decision Tree and Random Forest. The outcomes of this work elucidate the optimal quantity of generated synthetic samples necessary for the augmented dataset to exhibit the best quality score compared to the original dataset. Additionally, the performance of the machine learning algorithms is taken into consideration.

2. Theory

2.1. Misalignment on journal bearings

Ideally, in a journal bearing, the axes of both shaft and bush are parallel when installed and remain so during the operation under the imposed load and speed. Nevertheless, in practice, this ideal condition rarely exists, and the shaft tends to suffer from some degree of misalignment while rotating inside its bushing. The primary effect is considerable reduction in the minimum thickness of the lubricant film, which is responsible for protecting the surfaces from direct contact.

Substantially, reduced minimum film thickness also alters the entire pressure and temperature fields. In fact, the maximum values of pressure and temperature in misaligned bearings are much higher than those in the aligned ones. To counteract this effect a slight defect in the minimum film thickness area is sometimes added to the bearing geometry to prevent contact. Misalignment can also result from an individual or combination of the following factors: asymmetric bearing loading, elastic deflection of the shaft, under an imposed under load or its own weight., thermal distortion of the shaft, distortions caused by bearing housing support, distortions caused by bearing housing supports., and errors due to installation and assembly defects.

The governing equation is the Reynold's equation, which can be derived through two distinct approaches: either from Navier-Stokes equations (conservation of mass and momentum) or from equilibrium of an element of the fluid. The simplified equation considering isoviscous lubricant with steady film thickness and unidirectional velocity approximation [25] is,

$$\frac{\partial}{\partial x}\left(h^3\frac{\partial p}{\partial x}\right)+\frac{\partial}{\partial y}\left(h^3\frac{\partial p}{\partial y}\right)=6U\eta\frac{dh}{dx} \tag{1}$$

The most important effect of bearing misalignment is the drastic alteration of the protective film thickness in both the circumferential and axial directions. Therefore, the Reynolds equation should be modified to allow the variation of film thickness in both directions, meaning that an analytical solution is not available. This needs to be carefully taken into consideration in the analysis. Figure 1(a) shows a typical section of the journal and the bearing for the aligned case.

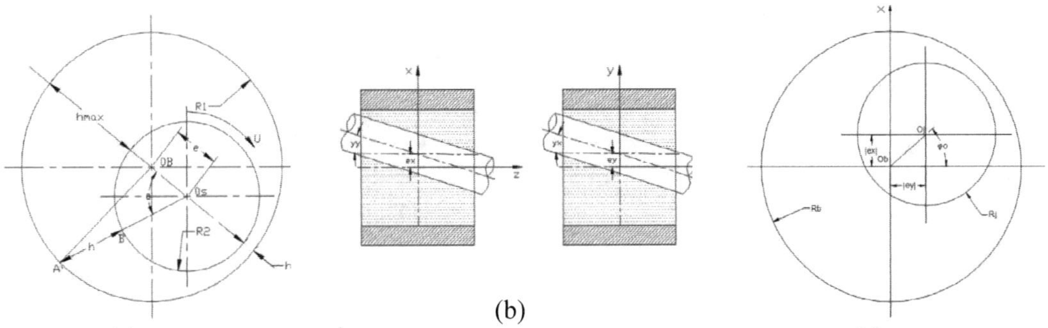

(a) (b) (c)

Figure 1. Geometry of aligned and misaligned journal bearing [17].

The sections in the x-z and y-z planes of a misaligned journal bearing are shown in Figure 1(b), while Figure 1(c) provides a representation of the geometry in the x-y plane for z=0.

The film thickness prediction plays a significant role in misaligned journal bearings. Nikolakopoulos and Papadopoulos [10] mentioned the following expression describing the film thickness with provision for misalignment:

$$h(\theta, z) = c + e_0 \cdot \cos\theta + z \cdot (yy \cdot \cos(\theta + \varphi_0) + yx \cdot \sin(\theta + \varphi_0)) \tag{2}$$

Another useful parameter, which indicates the bearing lubrication regime, is the minimum film thickness h_{min}, which occurs at one of the two ends of the bearing. The minimum film thicknesses for misaligned and worn journal bearings are defined by the following expressions, as mentioned by Saridakis et al. [16]:

$$h_{min}|_{z=L/2} = c - \sqrt{(e_0 \sin(\varphi_0) - \psi_y L/2)^2 + (e_0 \cos(\varphi_0) + \psi_x L/2)^2 + c(\delta_0 - 1 - \cos(\varphi_0))} \tag{3}$$

$$h_{min}|_{z=-L/2} = c - \sqrt{(e_0 \sin(\varphi_0) - \psi_y L/2)^2 + (e_0 \cos(\varphi_0) - \psi_x L/2)^2 + c(\delta_0 - 1 - \cos(\varphi_0))} \tag{4}$$

2.2. Synthetic Minority Oversampling Technique (SMOTE)

Imbalanced data typically refers to a condition in which a number of data samples in a certain problem is not equally distributed. This disproportion in distribution occurs when the number of data samples representing a class is significantly lower than those in other classes, thereby leading to the underrepresentation of one or more classes in the dataset. These underrepresented classes are referred to as the minority or positive class, while others are called the majority or negative class. Generally, there are two traditional techniques for handling class imbalance: undersampling and oversampling. Undersampling involves removing instances of the majority class while in oversampling, instances of the minority class are duplicated.

Another canonical technique is Synthetic Minority Oversampling Technique (SMOTE). The SMOTE [19] is based on the idea of the nearest neighbor algorithm (kNN) and assumes that a synthetic data sample can be interpolated between members of the minority class in "feature space" (as opposed to "data space"). Concretely, the SMOTE algorithm chooses k nearest neighbors for each data point based upon the similarities in feature vectors and creates a synthetic sample at a random point on the line between each data point and chosen neighbor. As a result, the original data obtained using SMOTE are used to synthesize new minority data that are different from the original ones, thereby alleviating the impact of overfitting on the minority class, which is a common issue caused by random oversampling. This technique boosts class membership whilst introducing enough variety to encourage classifier generalization and grow minority class decision regions.

The minority class is over-sampled by taking each minority class sample and introducing synthetic examples along the line segments joining any/all of the k minority class nearest neighbors (usually five nearest neighbors are used). Depending upon the amount of over-sampling required, neighbors from the k nearest neighbors are randomly chosen. Synthetic samples are generated in the following way (Figure 2): Take the difference between the feature vector (sample) under consideration and its nearest neighbor. Multiply this difference by a random number between 0 and 1 and add it to the feature vector under consideration. This causes the selection of a random point along the line segment between two specific features. This approach effectively forces the decision region of the minority class to become more general. The synthetic examples cause the classifier to create larger and less specific decision regions, rather than smaller and more specific ones that may lead to overfitting.

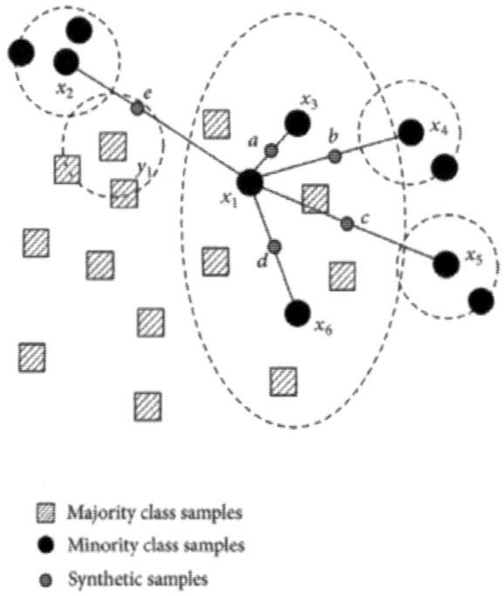

Figure 2. Example of generating synthetic data with SMOTE [19].

However, SMOTE can produce synthetic minority data samples considered as noise, which is also part of the majority classes. Therefore, an improvement of the technique by adding the Local Outlier Factor (LOF) has been introduced. [26] The proposed method is called SMOTE-LOF and produces better accuracy, AUC and f-measure than the SMOTE. An outlier is the odd point in a dataset. It is also known as abnormalities, discordances, deviations or anomalies, while noise can be defined as the wrong label (class noise) or errors in the attribute value (attribute noise). Figure 3 shows noise samples in a dataset. Typically, most outliers are noise, however, sometimes they are correct data. Therefore, out of a number of outliers detected, the ones identified as noise are derived from synthetic minority class samples generated by the SMOTE. The LOF value is the mean ratio of mean reachability distance of each minority sample divided by the mean reachability distance of each k-neighbor.

$$lrd(x) = -\frac{k}{\sum_{o \in N_k(x)} reach\_dist(x, o)} \tag{5}$$

$$lof(x) = -\frac{\sum_{o \in N_k(x)} \frac{lrd(o)}{lrd(x)}}{k} \tag{6}$$

Where:

- lof(x): Local Outlier Factor of object x in dataset o
- lrd(o): Local Reachability Density of objects in dataset o
- lrd(x): Local Reachability Density of object x

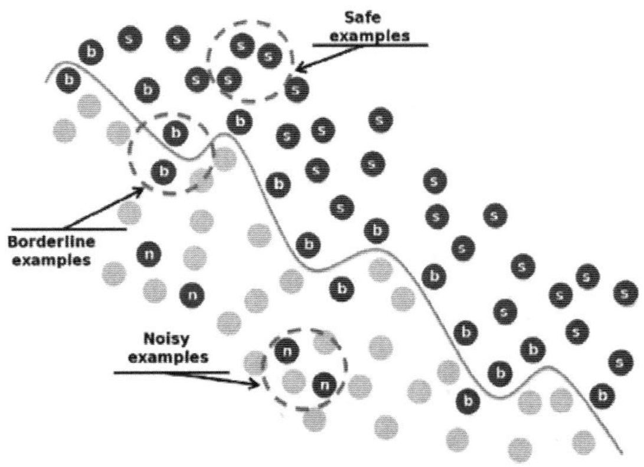

Figure 3. Noise samples in a dataset [26].

2.3. Synthetic Conditional Tabular Generative Adversarial Network (CTGAN)

Generative Adversarial Networks [22], or GANs for short, are an approach to generative modeling using deep learning methods, such as convolutional neural networks. Generative modeling is an unsupervised learning task in machine learning that involves automatically discovering and learning the regularities or patterns in input data in such a way that the model can be used to generate or output new examples that plausibly could have been drawn from the original dataset.

The basic principle of GANs, illustrated in Figure 4, is inspired by two-player zero-sum game, in which the total gains of two players are zero, and each player's gain or loss of utility is exactly balanced by the loss or gain of the utility of another player. GANs often comprise a generator and a discriminator that learn simultaneously. The generator tries to capture the potential distribution of real samples and generates new data samples. The discriminator is often a binary classifier, discriminating real samples from the generated samples as accurately as possible. Both the generator and the discriminator can adopt the structure of currently popular deep neural networks. The optimization process of GANs is a minimax game process, and the optimization goal is to reach Nash equilibrium, where the generator is considered to have captured the distribution of real samples.

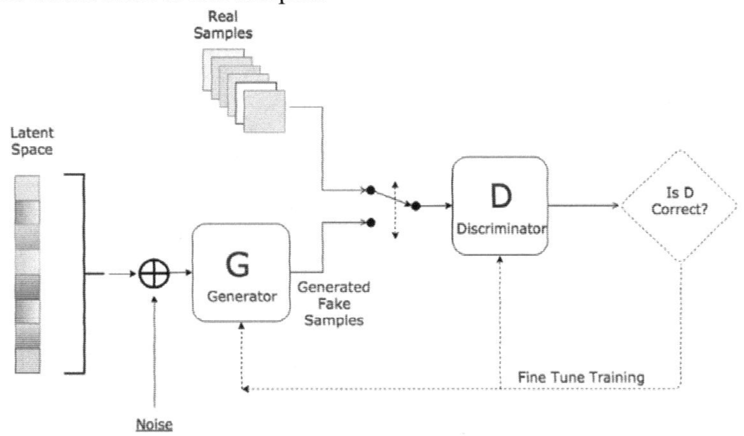

Figure 4. The structure of a Generative Adversarial Network (GAN) [27].

The discriminator evaluates the quality of the data created by the generator [27]. It attempts to predict the source of the sample, using samples that come from two sources:

- Real data instances. The discriminator uses these instances as positive examples during training.
- Fake data instances created by the generator. The discriminator uses these instances as negative examples during training.

The discriminator connects to two loss functions, the discriminator one and the generator one (Figure 5). During its training, it ignores the generator loss and just uses the discriminator loss. Generator loss is used during generator training.

During discriminator training:

- The discriminator classifies both real data and fake data from the generator.
- The discriminator loss penalizes the discriminator for misclassifying a real instance as fake or a fake instance as real.
- The discriminator updates its weights through backpropagation from the discriminator loss through the discriminator network.

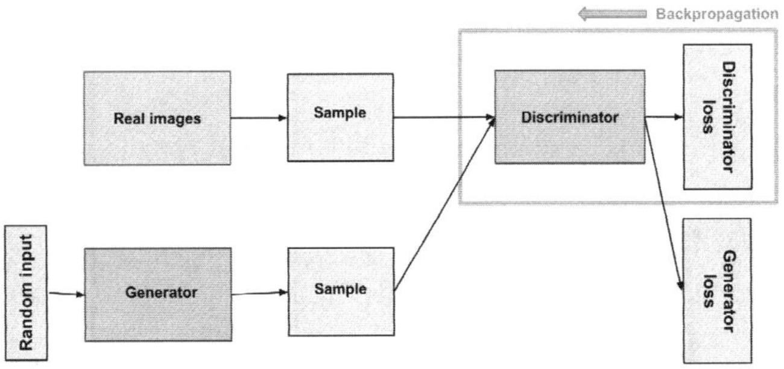

Figure 5. Backpropagation in discriminator training [27].

The generator learns to map a latent space to the distribution of the data it aims to reproduce, so that when fed with a noise vector from the latent space, it predicts a sample from the estimated distribution [27]. The generator is evaluated by the discriminator, meaning that its goal is to create data samples that are similar to those in the original data set. The discriminator penalizes the generator for producing implausible results. The generator output is connected directly to the discriminator input. Through backpropagation, the discriminator's classification provides a signal that the generator uses to update its weights.

Generator training requires tighter integration between the generator and the discriminator than discriminator training requires. The portion of the GAN that trains the generator includes: random input (noise), generator network, which transforms the random input into a data instance, discriminator network, which classifies the generated data, discriminator output., and generator loss, which penalizes the generator for failing to fool the discriminator.

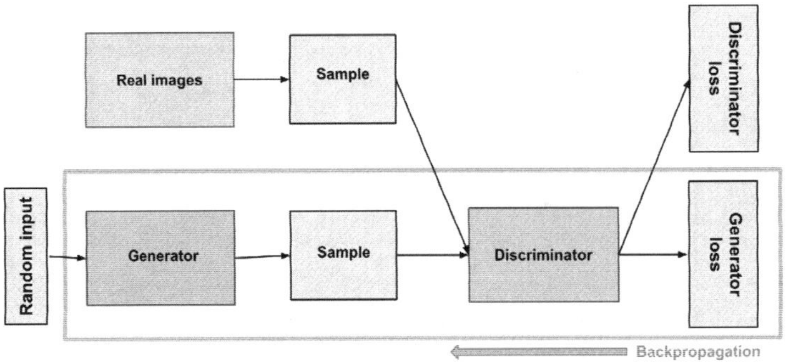

Figure 6. Backpropagation in generator training [27].

To train a neural net, we alter the net's weights to reduce the error or loss of its output. In GAN, however, the generator is not directly connected to the loss that we're trying to affect. The generator feeds into the discriminator net, and the discriminator produces the output we're trying to affect. The generator loss penalizes the generator for producing a sample that the discriminator network classifies as fake.

This extra chunk of the network must be included in backpropagation. Backpropagation adjust each weight in the right direction by calculating the weight's impact on the output – how the output would change if the weight is changed. However, the impact of the generator weight depends on the impact of the discriminator weights it feeds into. So, backpropagation starts at the output and flows back through the discriminator into the generator.

Meanwhile, the discriminator must not change during generator training. So, the generator is trained with the following procedure [27]: sample random noise, produce generator output from sampled random noise, get discriminator 'Real' or 'Fake' classification for generator output, calculate loss from discriminator classification, backpropagate through both the discriminator and generator to obtain gradients, and use these gradients to change only the generator weights.

Similarly, discriminator is constant during generator training phase. Otherwise, the generator would be trying to hit a moving target and might never converge. It's this back and forth that allows GANs to solve generative problems that would otherwise be insoluble. The discriminator will try to get better at distinguishing fake and real data, while the generator will progressively get better at generating data that is closer to the original.

On the other hand, for a GAN, convergence is often a fleeting, rather than stable, state. As the generator improves with training, discriminator performance declines because it can't easily tell the difference between real and fake data. In case that the generator succeeds perfectly, then the discriminator has a 50% accuracy, meaning that it flips a coin to make its prediction. As a result, discriminator feedback gets less meaningful over time. To better understand the minimax optimization in GANs, one needs to first answer the following question: What is the proper notion of equilibrium in GAN zero-sum games? In other words, what should be the optimality criteria in the GAN's minimax optimization problem? A classical notion of equilibrium in the game theory literature is the Nash equilibrium, a state in which no player can improve its individual gain by choosing a different strategy. As a well-established result, for a generator G expressive enough to reproduce the distribution of observed samples, Nash equilibrium exists for that generator function producing the data distribution [28].

Most commonly, GANs are used for image or even audio/speech generation. However, there are plenty of Real-world domains that are often described by "tabular data", i.e., data that can be structured and organized in a table-like format. As a standard, features (sometimes called "variables" or "attributes") are represented in columns, whereas observations (or "records") correspond to the rows. Additionally, real-world data usually comprises both numeric and categorical features. Numeric features (also called "continuous") are those that encode quantitative values, whereas categorical (also called "discrete") represent qualitative measurements. CTGAN (Conditional Tabular Generative Adversarial Network) was conceptualized to partially "capture" this heterogeneity of real-world data and compared to other architectures such as WGAN and WGAN-GP, has proven to be more robust and generalizable for a variety of datasets [29]. The "conditional" part of CTGAN refers to the fact that it can also generate synthetic data that is conditioned on certain input variables, such as gender or age.

Contrary to image data, where pixel values normally follow a Gaussian-like distribution, continuous features in tabular data are often non-Gaussian. Moreover, they tend to follow multimodal distributions (as depicted in Figure 7), where probability distributions have more than one mode, i.e., they present distinct local maxima (or "peaks").

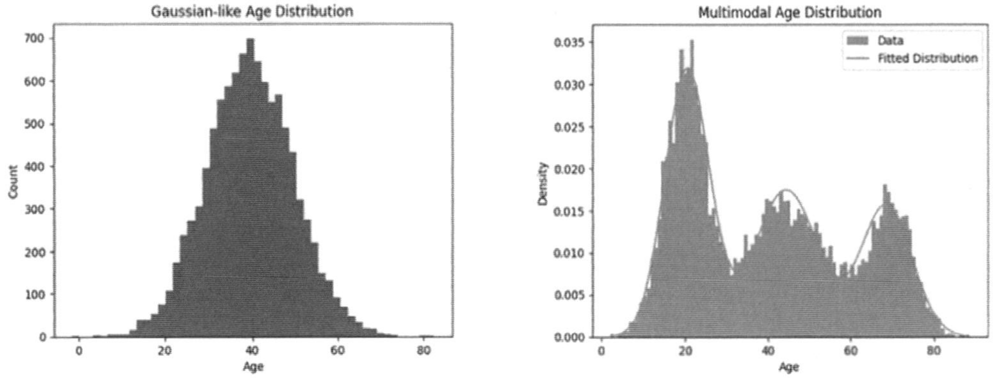

Figure 7. Gaussian versus multimodal distribution [29].

To capture these behaviours, CTGAN uses mode-specific normalization. Using a VGM (Variational Gaussian Mixture) model, each value in a continuous feature is represented by a one-hot vector indicating its sampled mode and a scalar that represents the value normalized according to that mode (Figure 8). An example of mode-specific normalization is presented in Figure 8. The VGM finds three modes, namely n_1, n_2 and n_3. C_{ij} represents a value in a feature j (e.g., j="Age"), for which ρ_3 is picked. C_{ij} is therefore represented by a one-hot vector $\beta_{i,j} = [0,0,1]$ indicating the third node, and a scalar $\alpha_{i,j} = \frac{c_{i,j}-n_3}{4\varphi_3}$, where φ_3 represents the standard deviation of the third mode (ρ_3).

The representation of a row become the concatenation of continuous and discrete columns.

$$r_j = a_{1,j} \oplus \beta_{1,j} \oplus \cdots a_{N_c,j} \oplus \beta_{N_c,j} \oplus d_{1,j} \cdots \oplus \beta_{N_d,j} \tag{7}$$

Where $d_{i,j}$ is one-hot representation of a discrete value.

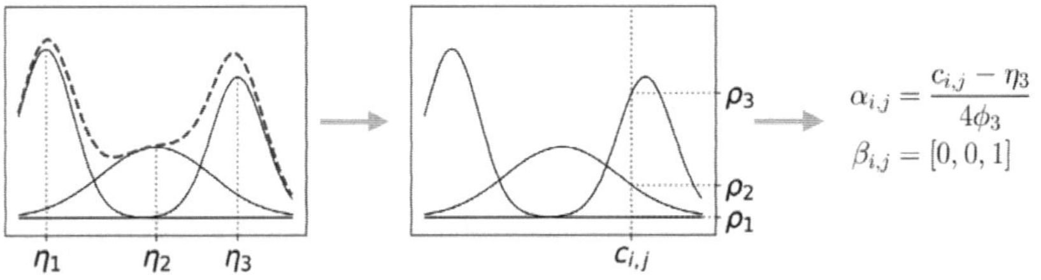

Figure 8. Example of mode-specific normalization [29].

On the other hand, CTGAN aims to address essentially two main challenges introduced by categorical features. The first challenge is the sparsity of one-hot-encoded vectors in real-world data. While the generator outputs the probability distributions over all possible categorical values, the original "real" categorical values are directly encoded in a one-hot-vector. These are easily distinguishable by the discriminator by comparing the distribution sparseness between real and synthetic data. The second challenge is the imbalance associated to some categorical features. If some categories of a feature are underrepresented, they cannot be learned adequately by the generator. If we were concerned with predictive modelling or classification tasks, data oversampling could be a solution to alleviate this issue. However, since the goal of synthetic data generation is to mimic the properties of the original data, this is not an option.

CTGAN introduces a Conditional Generator to deal with the challenges imposed by imbalanced categories which usually lead to GAN's infamous mode collapse. Specifically, the goal is to resample efficiently in a way that all the categories from discrete attributes are sampled evenly (but not necessary uniformly) during the training process, and to recover the (not-resampled) real data distribution during test. Let k* be the value from the i*th discrete column D_{i*} that has to be matched by the generated samples $\hat{r}$, then the generator can be interpreted as the conditional distribution of rows given that particular value at that particular column, i.e. $\hat{r} \sim \mathbb{P}_G(row|D_{i*} = k^*)$. For this reason, it is named Conditional Generator and the GAN is referred as Conditional GAN. Figure 9 illustrates an overview of a CTGAN model. With training-by-sampling, examples are conditioned on the possible values of categorical features, sampled according to their log-frequency [29].

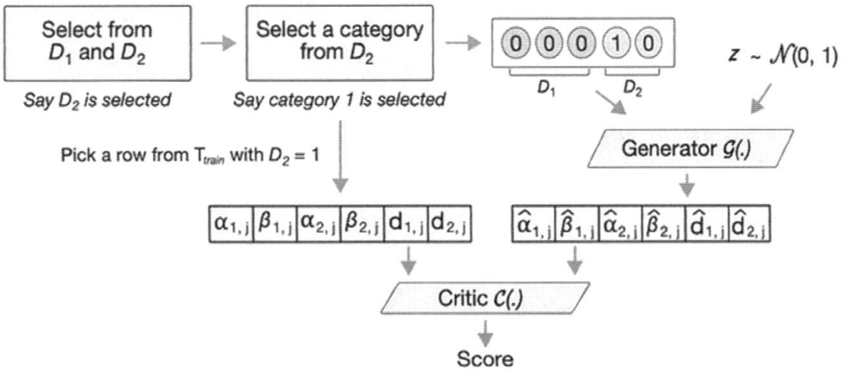

Figure 9. Overview of the CTGAN model.

The key to training a GAN and evaluate its performance over time are the loss functions, a set of formulas that indicate each network how to improve after each training iteration (or epoch). The discriminator and the generator each have their own loss values. Epoch after epoch, these networks learn by trying to minimize their loss function.

Loss functions are an area of active research, and many approaches have been proposed for different uses. In CTGAN, the custom loss functions formulated by SDV [30] for the purposes of creating synthetic data are used.

$$L_D = \frac{1}{m}\sum_{i=1}^{m}[D\left(x'^{(i)}\right) - D(x^{(i)})] \tag{8}$$

$$L_G = \frac{1}{m}\sum_{i=1}^{m}\left[D\left(x'^{(i)}\right)\right] + H \tag{9}$$

Where:

- x: real data
- x': synthetic data
- D(x): Discriminator's output given the real data
- D(x'): Discriminator's output given the synthetic data

The discriminator is learning to produce low values if the data is synthetic and high values if it is real. The range of these values is dependent on linear transformations and dimensions of the input data. When CTGAN performs well, both the generator and discriminator loss will eventually stabilize when they have run through enough epochs [30].

3. Results and Discussion

Using the results of the previous work, the primary objective of this current study is the identification of misalignment of a journal bearing. The 13×1331 dataset from [16] was used for implementation of a variety of Machine Learning algorithms., while a binary column named 'Misaligned' was added in order to determine if the bearing is aligned or not. The journal bearing was judged to be aligned if

IOP Conf. Series: Materials Science and Engineering 1294 (2023) 012024 doi:10.1088/1757-899X/1294/1/012024

dimensionless misalignment angles were less than 0.05. It is mentioned that wear depth is used as a training input, but it is not a target variable, nevertheless.

Table 1. Dataset used in this study (first 5 rows).

Record index	f	e	hmin0	hminL	PSIx th	PSIy th	D0 th	Misaligned
1	72.073559	0.371509	0.628491	0.628491	0.0	0.00	0.0	0
2	72.077757	0.371411	0.652290	0.604727	0.0	0.05	0.0	0
3	72.057533	0.371207	0.675995	0.580942	0.0	0.10	0.0	1
4	72.012865	0.370896	0.699544	0.557163	0.0	0.15	0.0	1
5	71.943629	0.370480	0.722856	0.533414	0.0	0.20	0.0	1

Two primary issues were identified through the application of a preliminary analysis to the data:

- Limited dataset. The Machine Learning algorithms would not perform well given the 1331 experimental data samples.
- Imbalanced dataset. Aligned cases outnumber the misaligned cases by a wide margin, making the Machine Learning models prone to overfitting.

This work introduces a new model for the identification of misalignment, particularly when faced with limited and imbalanced dataset. Initially, the imbalance within the minority class is addressed through the application of SMOTE, coupled with the LOF mechanism to mitigate noise interference. Subsequently, the resampled, 'cleaned' dataset is further augmented with synthetic samples using a CTGAN, thereby ensuring a sufficient volume of data samples. Finally, the expanded dataset is subjected to various Machine Learning algorithms for predictive modeling. A visual representation of this workflow is depicted in Figure 10. The Aligned – Misaligned cases on the resampled and the cleaned dataset are presented, where resampled is the dataset after the SMOTE implementation and the cleaned is the one after the SMOTE-LOF implementation. It is clear that the SMOTE method oversampled the minority class (aligned case), and as an outcome the number of aligned cases is equal with the misaligned ones. Then, the 258 generated outliers are erased using the LOF, and the misaligned cases subsequently outnumber the aligned cases by a small margin.

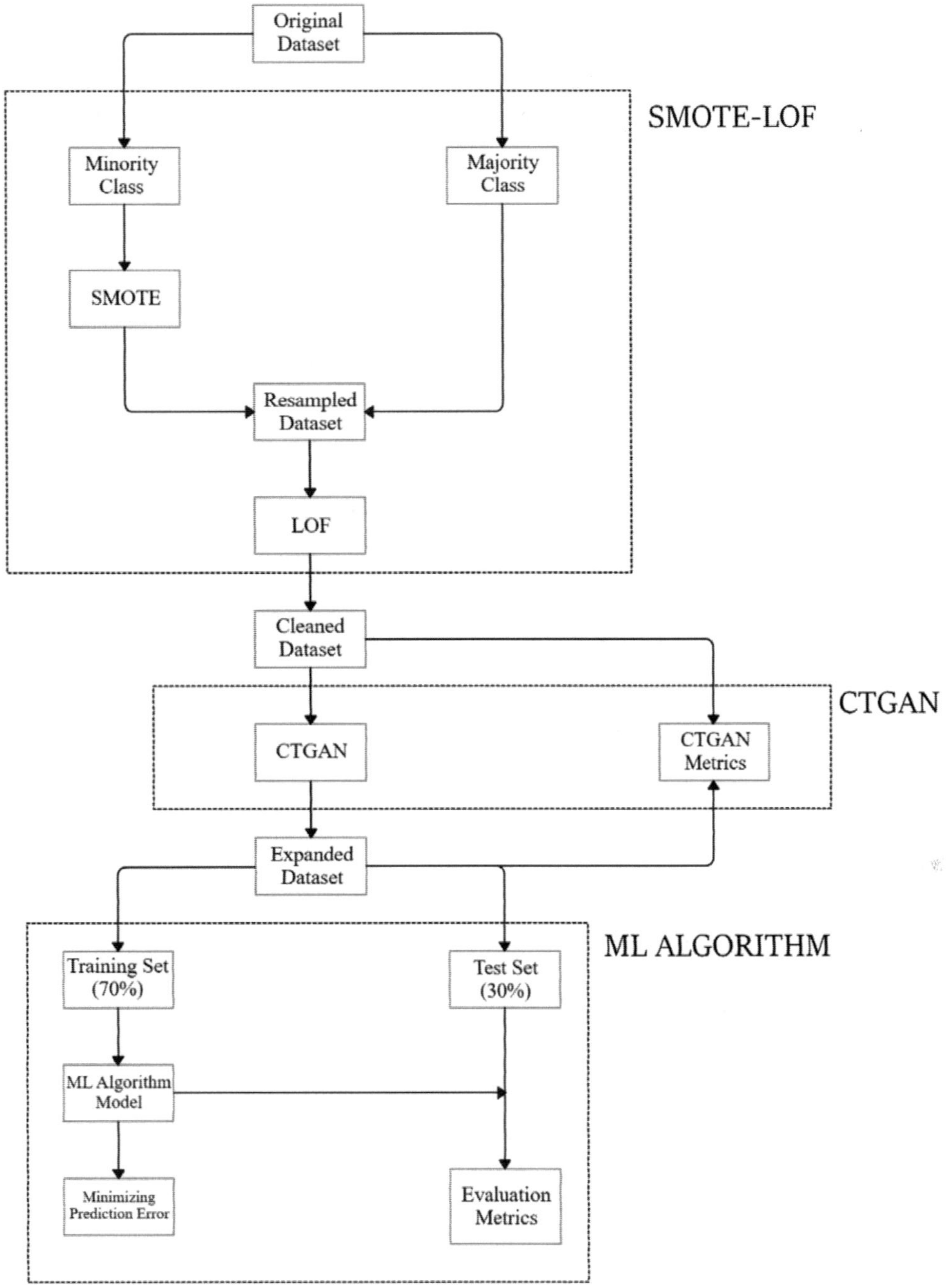

Figure 10. Flowchart for this study.

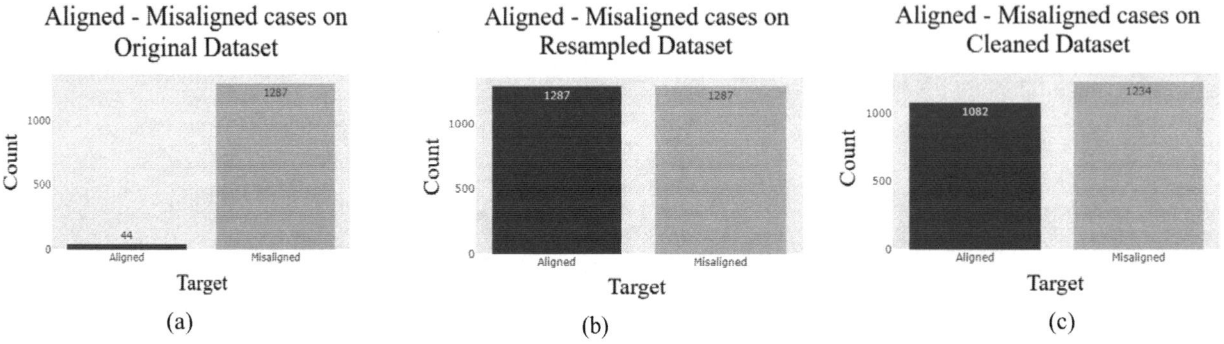

Figure 11. Aligned-Misaligned cases on the original, resampled and cleaned dataset.

Additionally, some pairwise relationships of the cleaned dataset variables, with the noisy samples highlighted, are presented as well.

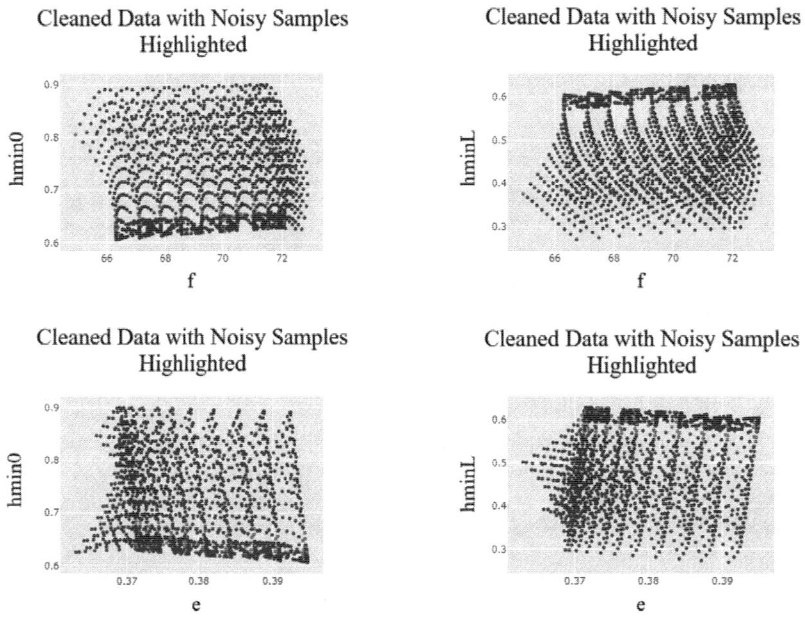

Figure 12. Cleaned data with noisy samples highlighted.

Then, using the SDV library in Python, a CTGAN model is created and fit to the cleaned dataset and new synthetic data is generated. It is noted that all the features of the input data is numeric; the "Misaligned" column is erased and then added to the generated dataset, depending on the values of the nondimensional misalignment angle. The CTGAN model is based on the GAN-based Deep Learning data synthesizer which was proposed in [29].

In CTGAN, the mode-specific normalization is invented to overcome the non-Gaussian and multimodal distribution. A conditional generator and training-by-sampling are designed to deal with the imbalanced discrete columns. Finally, fully-connected networks and several recent techniques are utilized to train a high-quality model.

Define the following notations:

- $x1 \oplus x2 \oplus \ldots$: concatenate vectors x1, x2, . . .
- $\text{gumbel}_\tau(x)$: apply Gumbel softmax with parameter τ on a vector x

- $leaky_{\gamma}(x)$: apply a leaky ReLU activation on x with leaky ratio γ
- $FC_{u\to\upsilon}(x)$: apply a linear transformation on a u-dim input to get a υ-dim output

Tanh, ReLU, softmax, BN are used for batch normalization, as well as drop for dropout.

Since columns in a row do not have local structure, fully-connected networks are used in generator and discriminator to capture all possible correlations between columns. Specifically, two fully-connected hidden layers are utilized in both generator and discriminator. In generator, batch-normalization and Relu activation function are used. After two hidden layers, the synthetic row representation is generated using a mix activation functions. The scalar values αi is generated by tanh, while the mode indicator βi and discrete values di is generated by gumbel softmax. In discriminator, leaky relu function and dropout on each hidden layer are used.

Finally, the conditional generator G(z, cond) can be formally described as:

$$\begin{cases} h_0 = z \oplus cond \\ h_1 = h_0 \oplus ReLU(BN(FC_{|cond|+|z|\to 256}(h_0))) \\ h_2 = h_1 \oplus ReLU(BN(FC_{|cond|+|z|+256\to 256}(h_1))) \\ \hat{\alpha}_i = tanh\left(FC_{|cond|+|z|+512\to 1}(h_2)\right), \quad 1 \leq i \leq N_C \\ \hat{\beta}_i = gumbel_{0.2}\left(FC_{|cond|+|z|+512\to m_i}(h_2)\right), \quad 1 \leq i \leq N_C \\ \hat{d}_i = gumbel_{0.2}\left(FC_{|cond|+|z|+512\to |D_i|}(h_2)\right), \quad 1 \leq i \leq N_d \end{cases} \tag{10}$$

For the discriminator, the PacGAN framework is used with 10 samples in each pac to prevent mode collapse. The architecture of the critic (with pac size 10) C(r1, . . . , r10, cond1, . . . cond10) can be formally described as:

$$\begin{cases} h_0 = r_1 \oplus \ldots \oplus r_{10} \oplus cond_1 \oplus \ldots \oplus cond_{10} \\ h_1 = drop(leaky_{0.2}(FC_{10|r|+10|cond|\to 256}(h_0))) \\ h_2 = drop(leaky_{0.2}(FC_{256\to 256}(h_1))) \\ C(\cdot) = FC_{256\to 1}(h_2) \end{cases} \tag{11}$$

The model is trained using WGAN loss with gradient penalty. Adam optimizer is used with learning rate $2*10^{-4}$. Using the SDV library in Python, new samples of data are generated. In this study we train the CTGAN for 1000 epochs, while both the generator and the discriminator have two fully connected hidden layers with 256 neurons each. Also, the batch size (the number of samples used in each step) is set to 500.

Consequently, several generated synthetic samples are added to the original dataset, resulting to the new, expanded dataset. The performance of the Machine Learning algorithms is evaluated for 3000, 4000, 5000 and 8000 synthetic samples. SDV library in Python provides a variety of metrics to evaluate the quality of the generated samples compared to the original dataset. For this study, quality report as well as visualizing the synthetic data are used. Additionally, the distribution of generator and discriminator loss per epoch is presented. The quality report provides an overall quality score, column shapes and column pair trends. Column shapes is a property that looks at the overall shape the columns to see if the synthetic data matches the real data, while the column pair trends between two columns describe how they vary in relation to each other, for example the correlation. The higher the score, the more the trends are alike. The results of this study for 4000 generated samples are presented in the Table 2.

Table 2. CTGAN metrics for each expanded dataset.

Metric	Overall Quality Score	Column Shapes	Columns Pair Trends
3000 Generated Samples	89.86%	91.64%	88.08%
4000 Generated Samples	87.34%	89.8%	84.89%
5000 Generated Samples	93.47%	91.2%	95.73%
8000 Generated Samples	88.07%	89.93%	86.21%

It is noticed that the expanded dataset of 5000 generated samples has the best overall quality score and column pair trends, while the column shapes value is above 90%. Then, we plot the column shape for each feature of this specific dataset.

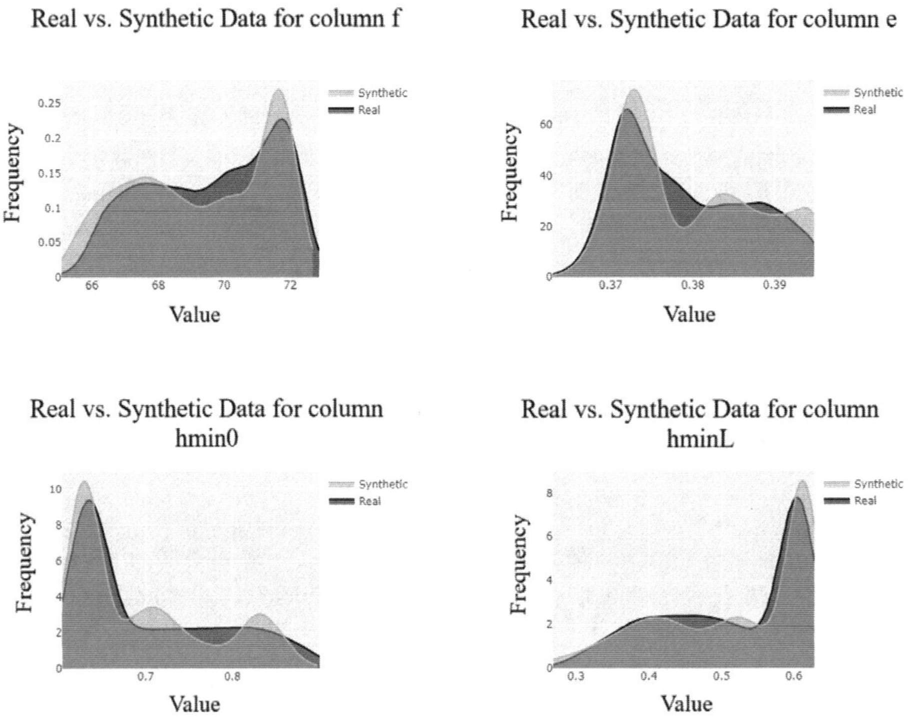

Figure 13. CTGAN Column shape for each feature of the expanded dataset of 5000 synthetic samples.

It is observed that the data distribution of the synthetic data samples is satisfying similar to the original ones, with some noticeable inconsistencies especially at the distribution of column "f". Finally, the generator and the discriminator loss are plotted for each epoch.

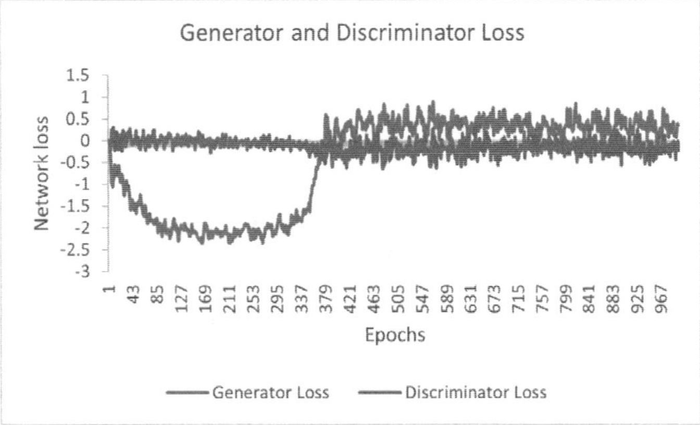

Figure 14. Generator and Discriminator loss.

As mentioned beforehand, generator and discriminator loss stabilization over time indicates a good performance of the CTGAN. From the plot it is observed that after approximately 400 epochs both generator and discriminator loss eventually stabilize and get less noisy, suggesting that CTGAN performs well. It is noteworthy that the selection of 1000 epochs for runtime proved judicious, as both the discriminator and generator loss reached a state of stabilization at an extended number of epochs.

After the data augmentation, the machine learning techniques are applied to each new extended dataset, utilizing the sklearn library in Python.

Table 3. Accuracy and AUC of each ML algorithm on the augmented datasets.

ML algorithm	Logistic Regression		KNN		SVM		Decision Tree		Random Forest	
	Accuracy	AUC	Accuracy	AUC	Accuracy	AUC	Accuracy	AUC	Accuracy	AUC
3000 Generated Samples	86.52%	0.88	88.78%	0.89	90.16%	0.9	85.58%	0.85	89.09%	0.88
4000 Generated Samples	89.39%	0.9	91.50%	0.92	90.61%	0.91	87.28%	0.87	91.03%	0.91
5000 Generated Samples	89.98%	0.9	90.66%	0.91	89.48%	0.9	86.79%	0.86	90.16%	0.9
8000 Generated Samples	85.75%	0.87	87.01%	0.87	87.53%	0.88	82.10%	0.81	87.04%	0.87

4. Conclusions

This work resulted in some important conclusions regarding the application of data augmentation and Machine Learning methods to the tribological design of misaligned journal bearings. The rarity of comparable studies in scientific literature not only amplifies the significance of the findings, but also initiates the exploration of this new field of research. To sum up, the application of the SMOTE-LOF eliminated the imbalance of the original dataset. Then, four (4) augmented datasets were generated by utilizing a Convolutional Tabular GAN (CTGAN), each one consisted of 3000,4000,5000 and 8000 synthetic data samples accordingly. Finally, the Machine Learning algorithms were utilized on each expanded dataset. For this study, a variety of ML techniques were chosen to be utilized on the augmented datasets, in order to determine which one is the ideal for this kind of problem. Furthermore, the application of multiple ML algorithms on the augmented datasets affords a more comprehensive assessment of the datasets' overall performance, thereby facilitating the selection of the most suitable one. More specifically, the following are observed:

- The SMOTE technique eliminates the original dataset's imbalance effectively, while the LOF erases a considerable number of noisy samples.

- The augmented dataset consisted of 5000 synthetic data samples has the best performance and quality compared to the original dataset.
- A good performance of the CTGAN is indicated by the stabilization of both the discriminator and generator loss after an extended number of epochs.
- The augmented dataset consisted of 4000 synthetic samples exhibits the superior performance on the Machine Learning algorithms.
- KNN and SVM are the algorithms with the best performance on the augmented datasets.

Therefore, in the context of monitoring applications, it is judicious to opt for the generation of 4000 synthetic samples, since the augmented dataset exhibits superior performance when subjected to Machine Learning techniques. As a generic conclusion, the proposed methodology can effectively be employed for the identification of the misalignment of journal bearings, especially when dealing with limited and imbalanced data. This is a very helpful tool when collecting a sufficient amount of experimental data is not feasible.

References

[1] Marian M and Tremmel S 2021 Current trends and applications of machine learning in tribology—A review, *Lubr.* **9**(9), 86, doi: https://doi.org/10.3390/lubricants9090086.

[2] Gertzos K P, Nikolakopoulos P G and Papadopoulos C A 2008 CFD analysis of journal bearing hydrodynamic lubrication by Bingham lubricant, *Tribol. Int.* **41**(12), 1190 - 1204, doi: https://doi.org/10.1016/j.triboint.2008.03.002.

[3] Nuruzzaman D M, Khalil M K, Chowdhury M A, and Rahaman M L 2010 Study on Pressure Distribution and Load Capacity of a Journal Bearing Using Finite Element Method and Analytical Method, *International Journal of Mechanical & Mechatronics Engineering IJMME-IJENS.* **10**(5).

[4] Zhou W, Wang Y, Wu G, Gao B, and Zhang W 2022 Research on the lubricated characteristics of journal bearing based on finite element method and mixed method, *Ain Shams Engineering Journal.* **13**(4), doi: https://doi.org/10.1016/j.asej.2021.11.007.

[5] Lund J W 1987 Review of the Concept of Dynamic Coefficients for Fluid Film Journal Bearings, *Journal of Tribology.* **109**(41), doi: https://doi.org/10.1115/1.3261324.

[6] Zhou W, Wei X, Wang L, and Wu G 2017 A superlinear iteration method for calculation of finite length journal bearing's static equilibrium position, *Royal Society Open Science.* **4**(5), doi: https://doi.org/10.1098/rsos.161059.

[7] Fu Y, Jin J, and Chen X 2018 Numerical analysis of Journal Bearing based on Finite Difference Method with Inhomogeneous Grid, *Journal of physics.* **1064**, doi: https://doi.org/10.1088/1742-6596/1064/1/012059.

[8] Jang J and Khonsari M 2015 On the Characteristics of Misaligned Journal Bearings, *Lubricants.* **3**(1), doi: https://doi.org/10.3390/lubricants3010027.

[9] Singh D, Sinhasan R, and Pal R 1989 Performance Characteristics of an Ungrooved Big-End Bearing with Misalignment, *Tribology Transactions.* **32**(2), doi: https://doi.org/10.1080/10402008908981883.

[10] Nikolakopoulos P G and Papadopoulos C A 1994 Non-linearities in misaligned journal bearings, *Tribology International.* **27**(4), doi: https://doi.org/10.1016/0301-679x(94)90004-3.

[11] Sun J and Changlin G 2004 Hydrodynamic lubrication analysis of journal bearing considering misalignment caused by shaft deformation, *Tribology International.* **37**(10), doi: https://doi.org/10.1016/j.triboint.2004.05.007.

[12] Liang X, Yin Z, and Dong Q 2019 An experimental study of axial misalignment effect on seizure load of journal bearings, *Tribology International.* **131**, 476–487, doi: https://doi.org/10.1016/j.triboint.2018.11.014.

[13] Huber M, Strzelecki S, and Steinhilper W 1998 Theoretical and Experimental Determination of the Performances of Misaligned and Statically Loaded Cylindrical Journal Bearings, *Tribology at Work: Proceedings of the 5th International Tribology Conference in Australia.*

[14] Bouyer J and Fillon M 2002 An Experimental Analysis of Misalignment Effects on Hydrodynamic Plain Journal Bearing Performances, *Journal of Tribology*. **124**(2), doi: https://doi.org/10.1115/1.1402180.

[15] Rosenkranz A, Marian M, Profito F J, Aragon N, and Shah R 2020 The Use of Artificial Intelligence in Tribology—A Perspective, *Lubricants*. **9**(1), doi: https://doi.org/10.3390/lubricants9010002.

[16] Saridakis K M, Nikolakopoulos P G, Papadopoulos C, and Dentsoras A J 2012 Identification of wear and misalignment on journal bearings using artificial neural networks, *Proceedings of the Institution of Mechanical Engineers, Part J: Journal of Engineering Tribology*. **226**, doi: https://doi.org/10.1177/1350650111424237.

[17] Kokkinidis K A and Nikolakopoulos P G 2022 Simulation of Misaligned Journal Bearings Using Neural Networks, *Tribology in industry*. **44**(1), doi: https://doi.org/10.24874/ti.1078.03.21.10.

[18] Alexey Kornaev, Nickolay Kornaev, Kornaeva E, and Savin L 2017 Application of Artificial Neural Networks to Calculation of Oil Film Reaction Forces and Dynamics of Rotors on Journal Bearings, *International Journal of Rotating Machinery*. **2017**, doi: https://doi.org/10.1155/2017/9196701.

[19] Chawla N V, Bowyer K W, Hall L O, and Kegelmeyer W P 2002 SMOTE: Synthetic Minority Over-sampling Technique, *Journal of Artificial Intelligence Research*. **16**, doi: https://doi.org/10.1613/jair.953.

[20] Wang Z, Liu T, Wu X, and Liu C 2023 A Diagnosis Method for Imbalanced Bearing Data Based on Improved SMOTE Model Combined with CNN-AM, *Journal of Computational Design and Engineering*. **10**(5), doi: https://doi.org/10.1093/jcde/qwad081.

[21] Hu X, Man J, Yang H, Deng J, and Liu Y 2022 An Improved Metalearning Framework to Optimize Bearing Fault Diagnosis under Data Imbalance, *Journal of Sensors*. **2022**, doi: https://doi.org/10.1155/2022/1809482.

[22] Goodfellow Ian J, Pouget-Abadie Jean, Mirza Mehdi, Xu Bing, Warde-Farley David, Ozair Sherjil, Courville Aaron, Bengio Yoshua 2014 Generative Adversarial Nets, doi: https://doi.org/10.48550/arXiv.1406.2661

[23] Ruan D, Song X, Gühmann Clemens, and Yan J 2021 Collaborative Optimization of CNN and GAN for Bearing Fault Diagnosis under Unbalanced Datasets, *Lubricants*. **9**(10), doi: https://doi.org/10.3390/lubricants9100105.

[24] Cheng P, Zhang S, and Li C 2022 A Rolling Bearing Fault Diagnosis Based on Conditional Depth Convolution Countermeasure Generation Networks under Small Samples, *Sensors*. **22**(15), doi: https://doi.org/10.3390/s22155658.

[25] Stachowiak G W and Batchelor A W 2016 Engineering tribology, *Amsterdam: Butterworth-Heinemann*.

[26] Asniar, Maulidevi N U, and Surendro K 2021 SMOTE-LOF for noise identification in imbalanced data classification, *Journal of King Saud University - Computer and Information Sciences*. **34**, doi: https://doi.org/10.1016/j.jksuci.2021.01.014.

[27] Google Developers 2019 Overview of GAN Structure | Generative Adversarial Networks, https://developers.google.com/machine-learning/gan/gan_structure

[28] Farnia Farzan and Ozdaglar Asuman 2020 Do GANs always have Nash equilibria, *International Conference on Machine Learning*.

[29] Xu L, Skoularidou M, Cuesta-Infante A, and Veeramachaneni K 2019 Modeling Tabular Data using Conditional GAN, *33rd Conference on Neural Information Processing Systems (NeurIPS 2019), Vancouver, Canada*.

[30] SDV 2022 Demystifying the CTGAN Loss Function | Synthetic Data Modeling · sdv-dev/SDV · Discussion #980, *GitHub*, https://github.com/sdv-dev/SDV/discussions/980.

IOP Conf. Series: Materials Science and Engineering 1294 (2023) 012025 doi:10.1088/1757-899X/1294/1/012025

Material characterization of EN AC-Al Si12CuNiMg alloy in stress relaxation and creep conditions

G G Sirata * **and K Waclawiak**

Silesian University of Technology, Department of Materials Technologies, Poland

* Corresponding author e-mail: goftila.sirata@polsl.pl

Abstract. Al−Cu−Mg alloys are widely used in various applications, especially engine components, where it is subjected to elevated temperatures during operation. Prolonged exposure of these components to elevated temperatures gradually leads to creep and relaxation phenomena. This study aims to investigate the effects of temperature on stress relaxation and creep behavior of the aluminum-silicon alloy EN AC-Al Si12CuNiMg. Furthermore, this study evaluates the effect of temperature on the rheological model parameters of this alloy. Experimental tests were conducted to investigate the stress relaxation and creep behavior of this alloy at different temperatures of 150°C, 250°C, and 350°C. The rheological model parameters were then determined from the obtained experimental data. The Standard Linear Solid Model (SLSM) was used to fit the stress relaxation and creep curves obtained from the experiment. Overall, the results indicate that increasing temperature significantly affects the stress relaxation and creep behavior of the investigated alloy, with higher temperatures leading to increased relaxation and creep rates. Additionally, the rheological model parameters of the alloy were found to vary with temperature.

1. Introduction

Aluminum alloys are commonly used in the automotive and aerospace industries due to their favorable properties, such as a high strength-to-weight ratio, ease of fabrication, good thermal conductivity, and good corrosion resistance [1–4]. Among these alloys, aluminum-silicon alloys are commonly used to manufacture combustion engine components, such as pistons and cylinder heads, which typically operate at elevated temperatures [5, 6]. The main challenge for these components while operating at elevated temperatures is the time-dependent deformation phenomenon. Deformation occurs gradually when these components operate at elevated temperatures for a long time. Understanding and analyzing these phenomena is essential for designers during material selection for such applications.

Stress relaxation and creep tests are two of the most widely used experimental approaches for determining the time-dependent behavior of materials when exposed to elevated temperatures [7, 8]. Stress relaxation is a mechanical behavior in which the material produces a certain strain at a specific temperature and initial stress. This phenomenon usually occurs when a material is subjected to elevated temperatures for a prolonged duration at constant strain [9]. During the experimental investigation, the temperature and total strain of the material are kept constant while the stress gradually decreases over time. Under the action of time, the elastic strain inside the material gradually transforms into the process of plastic strain [4]. During the relaxation test, stress reduction is measured as a function of time. Therefore, when evaluating structural response to thermally induced strains, it is essential to evaluate the stress relaxation in the structure of metal alloys. Creep is a time-dependent deformation behavior of a material under constant stress or load conditions at elevated temperatures. All materials deform first elastically and then plastically once the stress has exceeded the yield stress. However, creep can occur even at stress levels below the elastic limit under the effect of temperature and constant load over a long period [10].

Over the last few decades, numerous studies and models have been developed to characterize creep damage under various creep conditions. These studies have focused on the mechanisms of creep deformation and the factors that affect the creep life of the material. For example, the Larson-Miller model [11] is a widely used method for predicting the creep life of materials under high temperatures and constant stress levels. It considers the combined effects of time, temperature, and stress on creep deformation. However, the model has a limitation in its ability to accurately predict creep conditions because some other factors, such as material composition, variable loads, and variable temperature, that affect the creep life of the material have not been considered. Furillo et al. [12] modified the Larson-Miller parameter by incorporating additional factors, such as microstructure and alloy composition, to predict the creep life of materials more accurately. Considering these factors, the modified model achieved a better correlation between predicted and experimental creep-life results. This suggests that the original Larson-Miller model did not adequately account for these factors, highlighting a limitation in its accuracy.

Similarly, Pavlou [13] proposed a non-linear cumulative creep damage model that considers the damage history and loading effect to describe the creep deformation behavior of metallic materials under variable stresses and temperatures. However, this model may not accurately predict the creep deformation of metallic materials with shorter lifespans under variable temperature conditions, mainly when high temperatures are initially applied under varying temperature conditions [14]. Batsoulas [15] proposed a non-linear model for creep damage accumulation that considers loading sequence damage and creep strain. However, contradictions were found in the model, making it difficult to establish a correct creep damage accumulation rule for the strain under the same loading conditions. Hu, et al. [16] proposed a new creep damage assessment model based on Batsoulas' model. The model considers the loading process and the material's creep behavior under various conditions. However, the proposed model lacks clear physical significance, which is essential for its practical implementation.

Similar to other metallic materials, the primary factors affecting the stress relaxation and creep behavior of aluminum alloys are loading conditions and operating temperature. According to the studies by Mansoor and Shahid [17], Lyu et al. [18], and Hamasaki et al. [19], operating temperature generally leads to enhanced stress relaxation in aluminum alloys. Mansoor and Shahid [17] studied the stress relaxation behavior of polycrystalline aluminum of different grain sizes, annealed at 500°C, and aged at room temperature for six months. The results show that stress relaxation is affected by aging time. Lyu F. et al. [18] examined the influence of temperature levels on the stress relaxation and aging behavior of the Al-Zn-Mg alloy AA7B04-P. The study proved that increasing temperature increases stress relaxation with a predominant dislocation creep mechanism. Hamasaki et al. [19] conducted stress relaxation tests at 25, 100, 200, and 300°C for the AA5182-O aluminum alloy sheet to observe the temperature and strain rate dependence on its flow stress and stress relaxation. The results showed that the dominant relaxation mechanism is grain boundary slip due to the influence of temperature.

Other studies also reported that the operating temperature plays an important role in influencing the creep behavior of aluminum alloys. The two important creep parameters, creep strain and creep strain rate, tend to increase with temperature, as reported by the studies conducted by Yu W. et al. [20], Zhang et al. [21], and Golshan et al. [22]. Zhan et al. [23] investigated the creep behavior of 2A14 aluminum alloy at different temperatures by conducting a creep test, and the results show that the creep strain and creep strain rate increase with temperature. Another study by Zhang et al. [21] found that the creep resistance of cast aluminum alloys decreases with increasing temperature, particularly between 200°C and 300°C. Golshan et al. [22] analyzed the effects of temperature on the creep behavior of the AlSi12CuNiMg aluminum alloy. They found that an increase in temperature resulted in a decrease in the creep lifetime and an increase in the minimum strain rate. Zhan et al. [23] studied the creep behavior of the Al-Cu-Mg alloy and constitutive models that describe the high-temperature creep behavior of the alloy. They reported that the creep behavior of the Al-Cu-Mg alloy depends on temperature. These studies showed the significant impact of temperature on stress relaxation and creep behavior of different aluminum alloys. However, the variation of stress relaxation and creep model parameters with temperatures has not been studied so far. Understanding the variation of stress relaxation and creep model parameters with temperature is essential for accurately predicting the behavior of different aluminum alloys under different temperature conditions.

The aim of this article is to evaluate the influence of temperature on stress relaxation and creep behavior as well as on the constitutive model parameters of the EN AC-Al Si12CuNiMg alloy.

2. Materials and methods

2.1. Material and specimen preparation

In this study, experimental investigations were conducted on a cast EN AC-Al Si12CuNiMg alloy. Each prepared specimen had a gauge length of 60 mm and a diameter of 10 mm. The chemical composition of the investigated aluminum alloy was determined using optical emission spectroscopy (OES) and is given in table 1. The specimens for stress relaxation and creep experiments were prepared by stir casting method. Figure 1 shows the prepared specimens for the stress relaxation and creep tests.

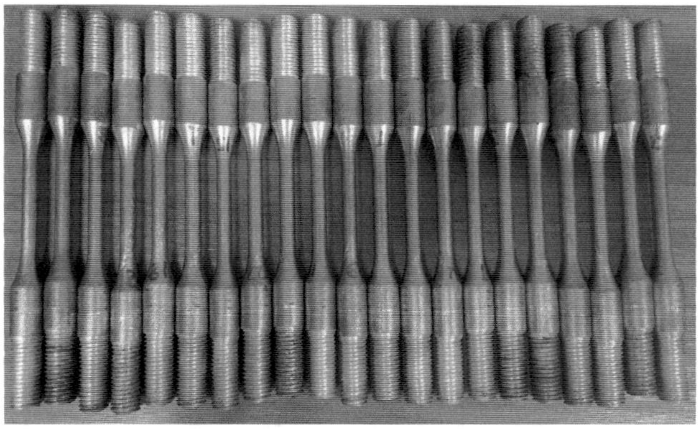

Figure 1. Relaxation and creep test specimens.

Table 1. The determined chemical composition of the investigated alloy in weight percentage.

Fe	Si	Cu	Mg	Zn	Mn	Ti	Ni	Al
0.4	11.4	1.27	1.24	0.18	0.18	0.04	1.48	rest

2.2. Experimental procedures

The stress relaxation and creep tests were conducted on a universal testing machine (Z100) manufactured by Zwick/Roell. The testing machine was fitted with a furnace and universal 3-zone furnace controller, as indicated in Figure 2. A 3-zone universal furnace controller is used to control the furnace temperature and specimen surface temperature via six thermocouples. The Zwick testXpert software was used to monitor and control the load application and temperature through a digital closed-loop control system. The stress and creep tests were carried out at temperatures of 150, 250, and 350°C. To study the effect of these temperatures on the stress and creep properties of the alloy, an initial stress level equivalent to 50% of the alloy's yield strength was used. The authors carried out the study of the mechanical properties of the alloy at each corresponding temperature [24] as presented in Table 2.

Figure 2. Stress relaxation and creep test set-up.

Table 2. Mechanical properties of the studied alloy [24].

Temperature, °C	Young's modulus (E), GPa	Yield strength, MPa	Ultimate strength, MPa	Fracture strain, %
150	82	100	125	0.6
250	55	90	115	0.8
350	55	60	65	1.9

Initially, the specimen was installed and aligned in the center of a furnace, then heated for about one hour until steady state temperature was reached. Throughout the process, heat was controlled using a 3-D universal furnace controller through thermocouples. Once the target temperature was attained, 50N of preload was applied to ensure no slippage occurred during the testing process. The relaxation tests were conducted in accordance with ASTM E328-21 [25]. During the relaxation test, the load was applied at a strain rate of 2 × 10-4 sec-1 until initial stress was imposed. Once the initial stress level was reached, the strain was kept constant while the stress was relieved until the end of the test. Throughout the process, the load was monitored as a function of time, and the test was generally continued for about 24 hours. During the creep test, the load was applied until the initial stress was reached, which was kept constant for about 45 hours. The extensometer was used to measure the elongation of the specimen.

2.3. Rheological model of stress relaxation and creep behavior of EN AC-Al Si12CuNiMg alloy

The Standard Linear Solid Model (SLSM) is a constitutive model commonly used in the field of engineering to describe the mechanical behavior of materials under different loading and temperature conditions. It is based on the assumption that the material can be represented as a linear combination of springs and dashpots. It is commonly used to characterize the stress relaxation and creep behavior of materials [26, 27]. This model assumes that the material is a linear combination of a spring element and a dashpot element, with the spring element representing the elastic response of the material and the dashpot element representing the viscous response of the material. It is usually represented by two different forms: the Maxwell form and the Kelvin form. Figure 3 illustrates the differences between the Maxwell form and the Kelvin form of the standard linear solid model [cite]. The Maxwell form consists of a spring element and a dashpot element arranged in series, while the Kelvin form consists of a spring element and a dashpot element arranged in parallel. For this study, the Kelvin form of the SLSM.

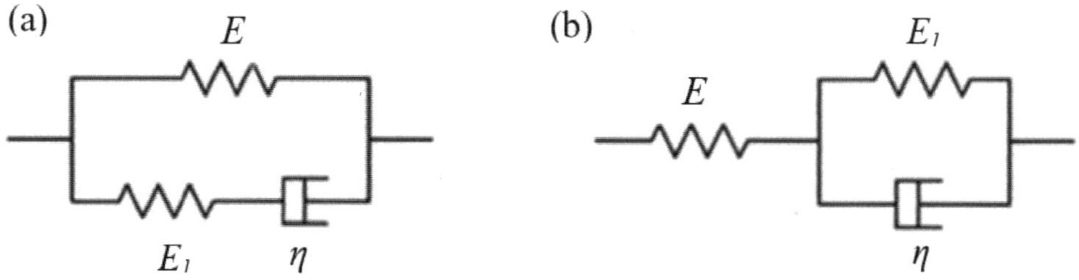

Figure 3. The three parametric standard linear solid model: (a) the Maxwell form of SLSM and (b) the Kelvin form of SLSM [27].

Based on a standard linear solid model, the stress relaxation behavior of the material can be described using either equation (1) or equation (2):

$$\sigma(t) = E\varepsilon_1\left(\frac{E_1}{E_1+E}+\frac{E}{E_1+E}\exp\left(-\left(\frac{E_1+E}{\eta}\right)t\right)\right) \tag{1}$$

$$\sigma(t) = E\varepsilon_1\left(\frac{H}{E}+\left(1-\frac{H}{E}\right)\exp\left(\frac{-t}{n}\right)\right) \tag{2}$$

Where, ε_1 stands for the constant strain during relaxation, $E = E_e$ stands for the relaxation modulus at the beginning of stress relaxation, E_1 stands for the relaxed modulus that describes how much the relaxation modulus drops from the beginning of stress relaxation to the end, η, stands for viscosity which describes the material's ability to flow under applied stress, and H stands for the relaxation modulus as time approaches to infinity ($t \to \infty$) and n is a time constant. The model parameters of equation 1 can be related to the model parameters of equation (2), which can be described by equations (3) and (4).

$$\frac{1}{H} = \frac{1}{E}+\frac{1}{E_1} \tag{3}$$

$$n = \frac{\eta}{(E+E_1)} \tag{4}$$

The creep strain, as per the SLSM model, can be described as a function of time by using either equation (5) or (6).

$$\varepsilon(t) = \frac{\sigma_1}{E}\left[\frac{E}{H}+\left(1-\frac{E}{H}\right)\exp\left(-\frac{H}{nE}t\right)\right] \tag{5}$$

$$\varepsilon(t) = \frac{\sigma_1}{E}\left(1-\frac{E}{H}\right)\left[1-\exp\left(-\frac{H}{nE}t\right)\right]\exp\left(-\frac{H}{nE}t\right) \tag{6}$$

where σ_0 is the constant stress during the creep test, the other parameters are the same as the relaxation parameters.

3. Results and Discussion

3.1. Stress relaxation curves

Figure 4 shows the stress relaxation curves obtained at different temperatures and the fitted curves achieved through the SLSM of equation (1). The relaxation curves indicate a good agreement between experimentally obtained results, and the SLSM curve fits with coefficients of determination (R2) values more than 0.99. The value of the stresses is expressed as a function of time. The initial stress level used at this temperature was about 50% of the yield stress (50, 45, and 30 MPa) at each corresponding temperature. As clearly indicated in Figure 4, relaxed stress increases with increasing temperatures. For example, at 150°C, the relaxed stresses are 23 MPa, indicating that the material is able to release 46% of its initial stress. As the temperature increases to 250°C, the relaxed stresses increase to 33 MPa, with a temperature drop of 73%. Finally, at 350°C, the relaxed stresses reach 27 MPa, with a temperature drop of about 90%. The most significant drop in the stress level occurs within approximately 2 hours of loading. Depending on the level of stress relief, the entire relaxation process can be categorized into three distinct stages. In the initial phase, stress exhibits rapid relaxation; however, in the subsequent phase, stress demonstrates a slower relaxation rate. After a long period of relaxation, the stress eventually approaches a steady state where the relaxation rate becomes almost constant in the third phase. In the first stage of relaxation, the relaxation behavior is non-linear and decreases with the duration of the test as it progresses. In the second stage of stress relaxation, the material develops compressive stress, and the stress relaxation rate decreases significantly. Finally, in the third stage, the stress relaxation rate becomes almost constant.

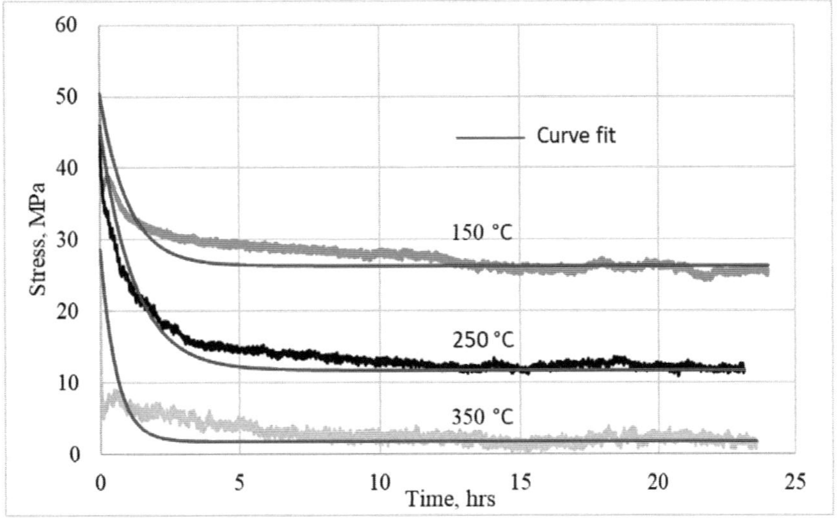

Figure 4. Experimental stress relaxation curves and fitted curves under different temperatures.

3.2. Creep curves

Figure 5 shows the obtained creep-strain curves of the investigated alloy. These curves were obtained at different temperatures of 150, 250, and 350°C with the corresponding 50% of the yield stress of the material, 50, 45, and 30 MPa, as presented in Table 2. As clearly shown in Figure 5, the creep strain increases as the temperature increases. For example, at a time of 10 hours and a temperature of 150°C, 250°C, and 350°C, the creep strain values are 0.09%, 0.19%, and 0.48%, respectively. This indicates the significant effect of temperature on the creep behavior of the investigated alloy. It also shows that the material is more vulnerable to creep deformation at higher temperatures. There are similarities between the plotted creep curves and typical creep curves defined by two stages: the primary stage, where the creep rate decreases, and the secondary phase, where the creep rate becomes constant [28, 29]. As stated in reference [30,31], the creep rate in the primary creep stage is initially high due to the initial formation and movement of dislocations. It is followed by a rapid decrease as the dislocations accumulate and their mobility decreases. As shown in Figure 6, the creep strain curve corresponding to a temperature of 350°C shows higher creep rates than the creep rate associated with temperatures of 250°C and 150°C. This suggests that a higher creep temperature results in higher creep strain.

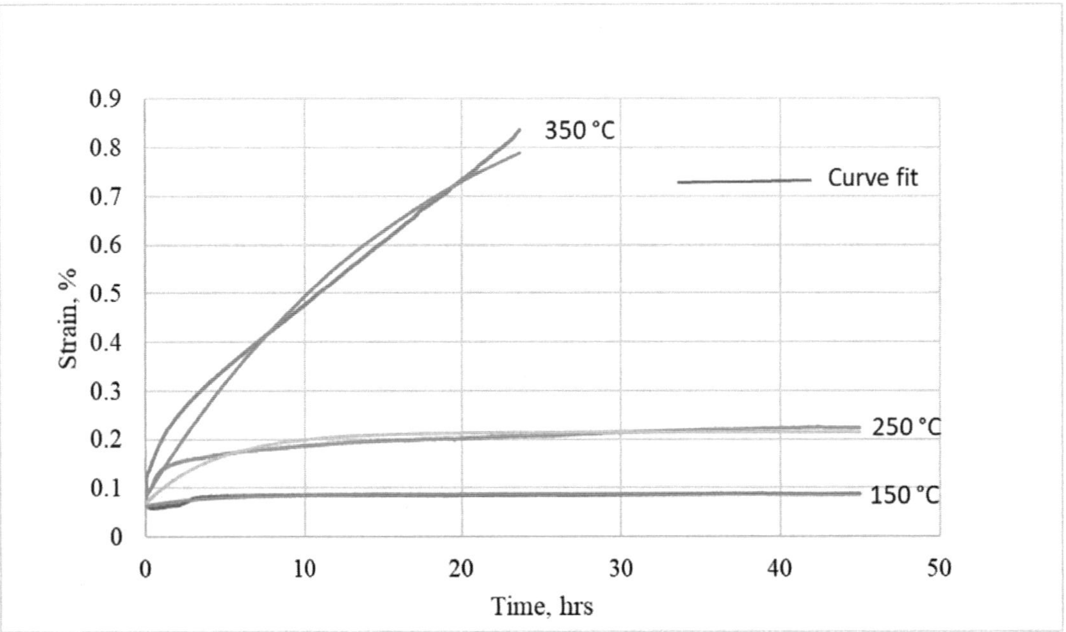

Figure 5. Creep curves under different temperatures.

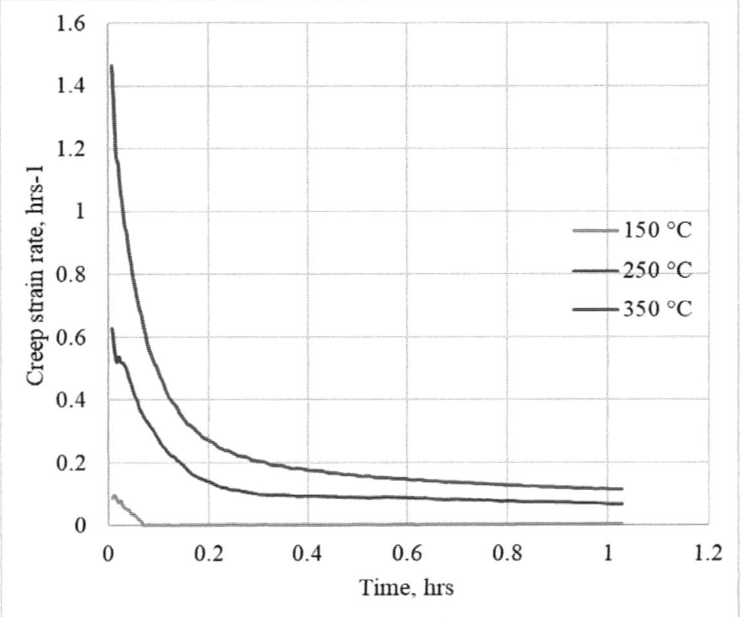

Figure 6. Creep rate curves of the primary stage under different temperatures.

3.3. The standard linear model of solids (SLSM) parameters determination

The test result indicates that all parameters of the SLSM relaxation model vary significantly with temperature. For example, the relaxation modulus at the beginning of relaxation (E) decreases as the temperature increases. When the temperature increased from 150°C to 250°C, it decreased by about 7%. However, a significant reduction was observed when the temperature increased from 150 to 350°C, with its value reduced by approximately 32%. For the relaxation modulus E_1, there is no significant change in temperature from 150°C to 250°C. However, when the temperature increases from 150 to 350°C, it decreases by 45%.

Additionally, the viscosity decreases significantly as the temperature increases. The viscosity values decreased by 30 to 70% as the temperature increased from 150 to 250 C and 350°C. This shows that temperature significantly affects the relaxation model parameters of the investigated alloy. The chart in Figure 7, which displays the relaxation model parameter values as a function of temperature, clearly illustrates the trend of decreasing values as the temperature increases.

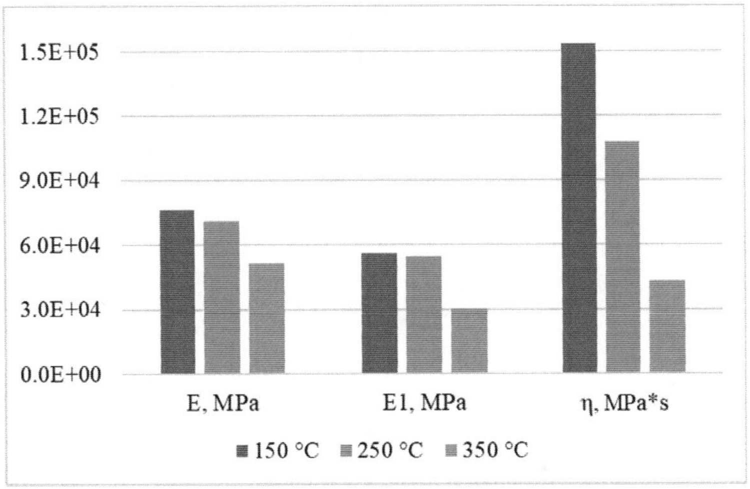

Figure 7. Relaxation model parameters of the investigated alloy at different temperatures.

The creep model parameters of the studied alloy are influenced by temperature. Figure 8 illustrates how these parameters change with temperatures. The three standard model parameters (E, E_1, and η) all decrease with increasing temperature, with E_1 experiencing a more significant reduction than the other parameters. Specifically, as the temperature increased from 150°C to 250°C, the value of E_1 decreased by approximately 85%. This reduction becomes even more noticeable as the temperature increases from 150°C to 350°C, with E_1 decreasing by approximately 95%. Since E_1 measures the material's resistance to creep deformation, these substantial decreases in this parameter indicate that the alloy's ability to resist creep deformation decreases at elevated temperatures. This means that the alloy may experience more plastic deformation and structural changes at elevated temperatures. As the temperature increases, the thermal energy within the material increases and promotes the mobility of dislocations and the diffusion of atoms [31]. The enhanced mobility of dislocation and diffusion of atoms are likely attributed to these changes in creep behavior.

In stress relaxation and creep, model parameters (E and η) decrease as temperature increases by almost the same values. However, E_1 drops significantly as the temperature increases from 150 to 350°C during creep rather than during stress relaxation. This indicates that the increase in temperature has a greater impact on E_1 during creep compared to stress relaxation.

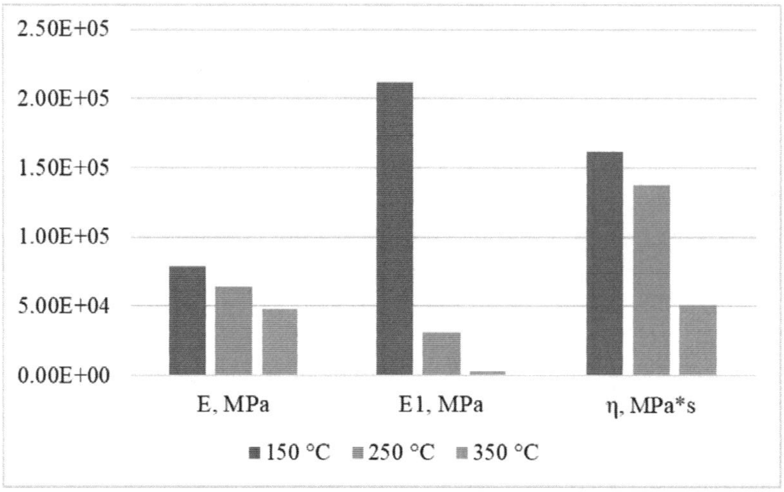

Figure 8. Creep model parameters of the investigated alloy at different temperatures.

4. Conclusion

In this study, stress relaxation and creep behavior of EN AC-Al Si12CuNiMg alloy were studied at different temperatures: 150, 250, and 350°C. To do so, the stress relaxation and creep tests were carried out. By fitting the stress relaxation and creep curves using the standard linear solid model (SLSM), the model parameters were determined for each test temperature. The results showed that the model parameters vary with temperature. The fitted curves showed good agreement with the experimental data, indicating that the determined SLSM model parameters can define the investigated alloy's stress relaxation and creep properties under different temperatures.

In the case of stress relaxation, the stress level of the alloy decreases rapidly in the initial phase. The most significant reduction in stress levels occurs around 2 hours after loading. The rate of relaxation also increases with temperature. Unlike stress relaxation, creep deformation does not show any reduction; however, it increases with temperature.

The trends of creep and relaxation rates are almost similar. In the initial phase, the rapid change in relaxation and creep rates is attributed to the rearrangement of dislocations within the alloy. As the dislocations gradually rearrange and realign, the rates of relaxation and creep decrease, leading to the second phase, where the rates become almost constant. This indicates that the material has reached a more stable state of stress relaxation and creep behavior. By reaching this steady state, the material has balanced the rearrangement of dislocations and the applied stress. Relaxation rate and creep strain rate increase with temperature, leading to accelerated deformation of the alloy.

References

[1] Sáenz de Argandoña E, Galdos L, Ortubay R, Mendiguren J and Agirretxe X 2015 Room temperature forming of AA7075 aluminum alloys: W-temper process *Key Eng. Mater.* **651** 199–204.

[2] Wang R, Ram D, Stauder B, Gutiérrez R F, Gariboldi E and Poletti M C 2022 Stress Relaxation during Artificial Aging of an AlSi7Cu0. 5Mg Cast Alloy *Crystals* **12** 1168.

[3] Polmear I 2005 *Light alloys: from traditional alloys to nanocrystals* (Elsevier) P437.

[4] Puspitasari P, Soepriyanto O R, Sasongko M I N and Dika J W 2018 Mechanical and physical properties of aluminium-silicon (Al-Si) casting alloys reinforced by Zinc Oxide (ZnO) (MATEC Web of Conferences: EDP Sciences) vol 204 p 05003.

[5] Lee J A and Chen P S 2002 Aluminum-silicon alloy having improved properties at elevated temperatures and articles cast therefrom. U.S. Patent 6,399,020.

[6] Lee J A 2003 Cast aluminum alloy for high temperature applications (132nd TMS Annual Meeting and Exhibition).

[7] Lee J 2012 Elevated-temperature properties of ASTM A992 steel for structural-fire engineering analysis PhD Thesis, University of Texas at Austin, USA.

[8] Mishra S, Yadava M, Kulkarni K N and Gurao N P 2018 Stress relaxation behavior of an aluminium magnesium silicon alloy in different temper condition, *Mech. Mater.* **125** 80–93.

[9] Costa Junior P R, Moura Neto C de and Wade D A 2014 Evaluation of a 7050-TAF aluminum alloy submitted to creep age forming, *Mater. Res.* **17** 603–11.

[10] Vattur Sundaram M 2014 Low Temperature Creep/Relaxation Behaviour of PM Steels under Static Load, *Master's Thesis* Chalmers University of Technology, Sweden.

[11] Larson F R and Miller J 1952 A time-temperature relationship for rupture and creep stresses *Trans. Am. Soc. Mech. Eng.* **74** 765–71.

[12] Furillo F T, Purushothaman S and Tien J K 1977 Understanding the larson-miller parameter *Scr. Metall.* 11.

[13] Pavlou D G 2001 Creep life prediction under stepwise constant uniaxial stress and temperature conditions *Eng. Struct.* **23** 656–62.

[14] Zhao X, Niu X, Song Y and Sun Z 2022 An investigation of the non-linear creep damage accumulation of different materials: Application of a novel damage model *Fatigue Fract. Eng. Mater. Struct.* **45** 530–45.

[15] Batsoulas N D 2009 Creep Damage Assessment and Lifetime Predictions for Metallic Materials under Variable Loading Conditions in Elevated Temperature Applications *Steel Res. Int.* **80** 152–9.

[16] Hu X, Ye W, Ma X and Song Y 2019 A new creep damage assessment method for metallic material under variable load conditions at elevated temperature *Fatigue Fract. Eng. Mater. Struct.* **42** 2725–37.

[17] Mansoor M and Shahid M 2018 Stress relaxation behavior of annealed aluminum-carbon nanotube composite *Sci. Eng. Compos. Mater.* **25** 1015–9.

[18] Lyu F, Li Y, Shi Z, Huang X, Zeng Y and Lin J 2020 Stress and temperature dependence of stress relaxation ageing behaviour of an Al–Zn–Mg alloy *Mater. Sci. Eng. A* **773** 138859.

[19] Hamasaki H, Morimitsu Y and Yoshida F 2017 Stress relaxation of AA5182-O aluminum alloy sheet at warm temperature *Procedia Eng.* **207** 2405–10.

[20] Yu W, Zhan L, Xu Y, Chen K, Yang Y, Xu L, Peng N, Ma B, Liu C and Chen Z 2022 Temperature-dependent creep aging behavior of 2A14 aluminum alloy *J. Mater. Res. Technol.* **19** 1343–54.

[21] Zhang S, Zhang Y, Chen M, Wang Y, Cui Q, Wu R, Arola D and Zhang D 2018 Characterization of mechanical properties of aluminum cast alloy at elevated temperature *Appl. Math. Mech.* **39** 967–80.

[22] Golshan A M A, Aroo H and Azadi M 2021 Sensitivity analysis for effects of heat treatment, stress, and temperature on AlSi12CuNiMg aluminum alloy behavior under force-controlled creep loading *Appl. Phys. A* **127** 1–18.

[23] Zhan L, Wu X, Wang X, Yang Y, Liu G and Xu Y 2018 Effect of process parameters on fatigue and fracture behavior of Al-Cu-Mg alloy after creep aging *Metals* **8** 298.

[24] Sirata G G, Wacławiak K and Dyzia M 2022 Mechanical and Microstructural Characterization of Aluminium Alloy, EN AC-Al Si12CuNiMg *Arch. Foundry Eng.* **22**.

[25] ASTM E328-21 2008 Standard Test Methods for Stress Relaxation for Materials and Structures (West Conshohocken, US: Annual Book of ASTM Standards).

[26] Plaseied A and Fatemi A 2008 Deformation response and constitutive modeling of vinyl ester polymer including strain rate and temperature effects *J. Mater. Sci.* **43** 1191–9.

[27] Lin C-Y 2020 Alternative form of standard linear solid model for characterizing stress relaxation and creep: Including a novel parameter for quantifying the ratio of fluids to solids of a viscoelastic solid, *Front. Mater.* **11**.

[28] Beddoes J and Mohammadi T 2010 Comparison of stress relaxation and creep strain rates for the superalloy IN738LC *J. Strain Anal. Eng. Des.* **45** 587–92.

[29] Zhan L H, Li Y G, Huang M H and Lin J G 2011 Comparative study of creep and stress relaxation behavior for 7055 aluminum alloy *Adv. Mater. Res.* **314** 772–7.

[30] Zheng J-H, Jin Y, Xu L, Fan C, Song W and Chen Y 2023 Comparative Study of Creep and Stress Relaxation Behaviour during Ageing of 7050 Aluminum Alloy *Metals* **13** 778.

[31] Naumenko K and Altenbach H 2016 *Modeling high temperature materials behavior for structural analysis* (Springer) Vol. 112.

Effect analysis of wooden fence width on wave transmission by SWASH model

T Mai[1,*], H T Dao[2], T T A Ngo[3], H H Pham[1], Y Liu[4]

[1] Hanoi University of Civil Engineering, Vietnam
[2] Hanoi University of Natural Resources and Environments, Vietnam
[3] Thuyloi University, Vietnam
[4] Norwegian University of Science and Technology, Norway

[*] Correspondence: trimc@huce.edu.vn

Abstract. Wooden fences, a permeable structure, have become a nature-based solution for supporting traditional structures to restore mangrove forests along the Mekong deltaic coasts. Even though prior studies have explored a number of hydraulic functions of these fences, an in-depth investigation into the influence of fence width on wave dissipation and damping is needed to consider. Therefore, this paper employs a numerical approach to thoroughly examine the impact of fence width on wave damping. The findings illustrate the substantial role of fence width in governing the dissipation of incoming waves. The correlation between the transmission coefficient and the fence width is established. This relationship also concludes that the larger the fence thicknesses, the lower the transmission coefficients. Notably, the study also identifies that the transmission coefficient experiences a slight decline beyond a certain width threshold.
Keywords: wooden fence, wave transmission, Mekong Delta, mangroves, relative fence thickness.

1. Introduction

The Mekong deltaic coast is known as one of the largest deltas in the world, with a rich diversity of ecosystems, including mangroves. Mangrove is the natural defense and protection for the coast because it morphs massively and naturally coastal sediment along the coast. Furthermore, mangrove forest provides a lively environment for local animal and livelihood communities.

However, due to climate change and sea level rise, the mangrove belt along the coast in the Mekong Delta provinces has been seriously degraded [1,2]. As a result, the retreatment of the mangrove belt has led to an increasing rate of coastal erosion. There have also been a number of hard/gray construction solutions applied, such as hollow pillar breakwaters and centrifugal concrete pile walls for breaking waves, to protect and prevent the coast from erosion. However, these solutions are oddly expensive and harmfully environmental. Additionally, gray structures often break the equilibrium of morphodynamics, such as wave, flow, and sediment transport.

On the other hand, planting young mangrove trees for gradual restoration and maintaining the mangrove belt is a nature-based solution currently applied to these damaged areas [3]. Furthermore, building soft walls, e.g., wooden fences, for waves and currents reduction and sediment acceleration, gives an important job in the work of supporting young mangroves. This solution also supports the old mangrove forests at the side already protected by the sea dikes.

In order to widely apply wooden fences for wave load reduction to mangrove forests, determining the influence of the wooden fence and wave parameters on the wave-damping efficiency by formulas or relational lines is essential today. In the context of designing wooden fences, Dao et al. (2020, 2021) [4,5] studied wave transmission over different fence thicknesses ranging from 1.4 m to 3.6 m, scaled from laboratory data. This study approached a numerical model to assess the relationship of wave transmission coefficient ($K_T = H_T/H_I$, with H_T and H_I are the transmission and incident wave heights, respectively) and the relative thicknesses with wave height (*B*/*H*). However, the result only presents wave height reduction occurred when low incoming wave heights impact the fences. For a similar context of thicknesses, wave reduction over the mangrove's width was studied in the previous research, i.e., Phan et al. (2019) [6] for laboratory study and Dao et al. (2018, 2021) [4,7] for numerical studies. However, this study only focused on the assessment of transmission for the non-linearity of waves. In more recent studies, Shu et al. (2023) [8] studied wave energy reduced by artificial bamboo fences in the laboratory scales, concluding the effect of wave transmission on the wave steepness and the non-linearity, but the fence's thickness on wave transmission is still hindered. Mancheno et al. (2023) [9] and Qu et al. (2023) [10] applied the open-source non-hydrostatic model to study wave attenuation over emergent rigid cylinders, which consider wooden fences and vegetations, respectively. While Qu et al. (2023) focused on using a numerical model (NHWAVE) to get a further understanding of vegetation densities and width on wave transmission, Mancheno et al. (2023) then used the SWASH model to study the effect of distance between cylinders on wave reductions.

Further understanding of relationships between fence widths and wave transmissions still needs to be carried out since a fence's design should be carefully considered for both efficiency and construction cost. Therefore, in this study, the main objective is to identify the relationship between wave transmission and the thickness of the fence relative to wavelength. An open-source model, SWASH, developed by Delft University of Technology, is used to simulate wave-fence interaction and assess the wave transmission coefficient with relative fence thicknesses. This paper will present the results of research to evaluate the effect of fence width on wave propagation through wooden fences through a mathematical model.

2. Methodology

2.1 SWASH model

SWASH model is an open-source model based on the nonlinear shallow water equations [11] that can simulate the multidimensional time-domain model for non-hydrostatic, free surface flow. The SWASH model can also consider shallow processes in the near-shore, such as breaking, reflection, and diffraction [12–15]. SWASH model is also applied for further research on wave-runup and overtopping [16] and prediction of wave-runup on a coral-reef bed [17]. Further study on the improvement of wave height prediction in the near-shore is also carried out by Umesh, P and Behera, M (2021) [18].

The government equations used to simulate waves traveling through a section perpendicular to the shore are as follows:

$$\frac{\partial u}{\partial x}+\frac{\partial w}{\partial z}=0 \tag{1}$$

$$\frac{\partial \eta}{\partial t}+\frac{\partial}{\partial x}\int_{-d}^{\eta} u dz=0 \tag{2}$$

$$\frac{\partial w}{\partial t}+\frac{\partial uw}{\partial x}+\frac{\partial ww}{\partial z}+\frac{1}{\rho}\frac{\partial (P_{nh})}{\partial z}+\frac{\partial \tau_{zz}}{\partial z}+\frac{\partial \tau_{zx}}{\partial x}=0 \tag{3}$$

$$\frac{\partial u}{\partial t}+\frac{\partial uu}{\partial x}+\frac{\partial wu}{\partial z}+\frac{1}{\rho}\frac{\partial (P_h+P_{nh})}{\partial x}+\frac{\partial \tau_{xz}}{\partial z}+\frac{\partial \tau_{xz}}{\partial x}=0 \tag{4}$$

where *x* and *z* are the horizontal and vertical directions in Decade coordinate, respectively; u and w are the horizontal and vertical particle velocities, respectively; η is the water elevation; *t* is the time; P_h and P_{nh} are the hydrostatic and non-hydrostatic pressures, respectively; the bottom stress follows a quadratic friction law and expressed as:

$$\tau_b = c_f \frac{U|U|}{\eta + d} \tag{5}$$

where u is the average depth velocity and c_f is the friction coefficient based on Manning:

$$c_f = \frac{n^2 g}{d^{1/3}} \tag{6}$$

In this study, a crossed profile was set in Figure 1, defined with a slope to create a breaking point for waves before entering the horizontal zone. Note that the profile mimics the gentle slope (approximately 1:1000) found in the Mekong deltaic coast. Wave inputs were set at the west side of the profile. At the east side, a 75m-sponger layer was set to absorb most of the reflection from both short- and longwaves. A spatial resolution (Δx) of 0.05, which is approximately $L_p/100$ (with L_p is the peak wavelength), was applied. The background viscosity of 3.10^{-4} was put in the model. A constant friction coefficient of 0.02 was also added. Other settings were set as default [11].

An assumption is that the bamboo poles are rigid in the SWASH model simulation. The vegetation implementation model was used in order to simulate wooden fences with three main parameters of a cylinder, including density (number of cylinders per area), diameter, and bulk drag coefficient [4,7]. Irregular wave characteristics include significant wave height and peak period applied at the boundary (left side of the profile, Figure 1). When waves come into contact with the fence (at x = 125 m, Figure 1), transmission wave height (H) can be calculated as:

$$\frac{H}{H_0} = \frac{1}{1+\tilde{\beta}x} \tag{7}$$

where, $\tilde{\beta}$ is the dissipation parameter expressed by Suzuki et al. (2019) [19] after implementing a horizontal component of wave forces to a cylinder from the study of Mendez and Losada (2004) [20]:

$$\tilde{\beta} = \frac{1}{3\sqrt{\pi}} \overline{C_D} D N H_0 k \frac{sinh^3(k\alpha H_f) + 3\,sinh(k\alpha H_f) + cosh^3(k\alpha H_f) - 3\,cosh(k\alpha H_f) + 2}{[sinh(2kd) + 2kd]\,sinh\,(kd)} \tag{8}$$

in which αH_f (m) is the submerged depth of the fence and k is the wave number. As seen in equation (8), transmission wave height is influenced by the three main parameters mentioned above. In Figure 1b, wooden fence settings in the model are expressed with horizontal and vertical orientations representing for vertical and horizontal forces, respectively. Note that the orientation of the fence is the showcase for interpretation of equation (8).

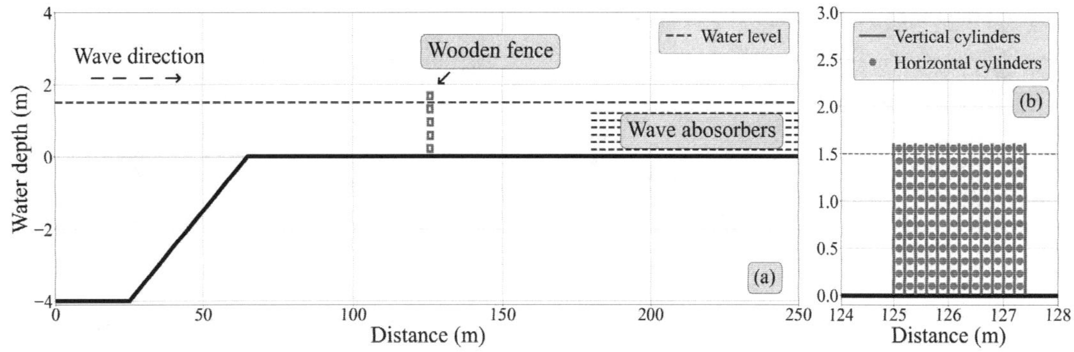

Figure 1. (a) Simulation profile in SWASH; (b) Wooden fence formulation with vertical (green line) and horizontal cylinders (red dot).

2.2. Field data and scenarios

In this study, wave and wooden fence characteristics were mimicked from field measurements in Nhat Mat, Bac Lieu, Vietnam [21] (Figure 2). Waves, including wave heights and wave periods, were obtained from measurements in front and behind the fence (Figure 2b), while only fence thicknesses were measured with two values of 0.8 and 1.6 m.

Due to the high dissipation environment for waves during propagation to the shore, the wooden fence was set at x = 125 m, far enough from the breaking point, as shown in Figure 1. Additionally, sea dikes in the field create unpredictable reflections due to longwaves. Therefore, sponger layers were set instead of dike to reduce as much complexity as possible. Other parameters include a cylinder diameter (D) of 0.02 m, a density of 636 cylinders per m^2 equal to about 80% porosity, and a bulk drag coefficient inherited from Dao et al. (2020) 2.0 for every scenario. Following Dao et al. (2020) [5], the relation between cylinder gap, s, and diameter D is calculated as 1.0, and it is suitable for this study to use because of the similar application. Also, only one fence height was used, 1.6 m. To assess the thickness dependency, several fence widths were set, ranging from 0.8 to 2.4 m. However, note that only one density was used to necessarily reduce the complexity of the assessment.

Table 1. Simulation scenarios.

Wave height (H_s, m)	Wave period (T_p, s)	Water depth (d, m)	Fence width (B, m)
0.39	5.57	1.53	0.8, 1.0, 1.2, 1.4, 1.6, 1.8, 2.0, 2.4
0.42	6.37	1.46	0.8, 1.0, 1.2, 1.4, 1.6, 1.8, 2.0, 2.4
0.52	7.31	1.58	0.8, 1.0, 1.2, 1.4, 1.6, 1.8, 2.0, 2.4
0.66	8.41	1.44	0.8, 1.0, 1.2, 1.4, 1.6, 1.8, 2.0, 2.4

As can be seen in Table 1, four different wave heights, wave periods, and water depths were applied to the input boundary. For each fence thickness, four-wave characteristics were used. All scenarios are presented in Table 1.

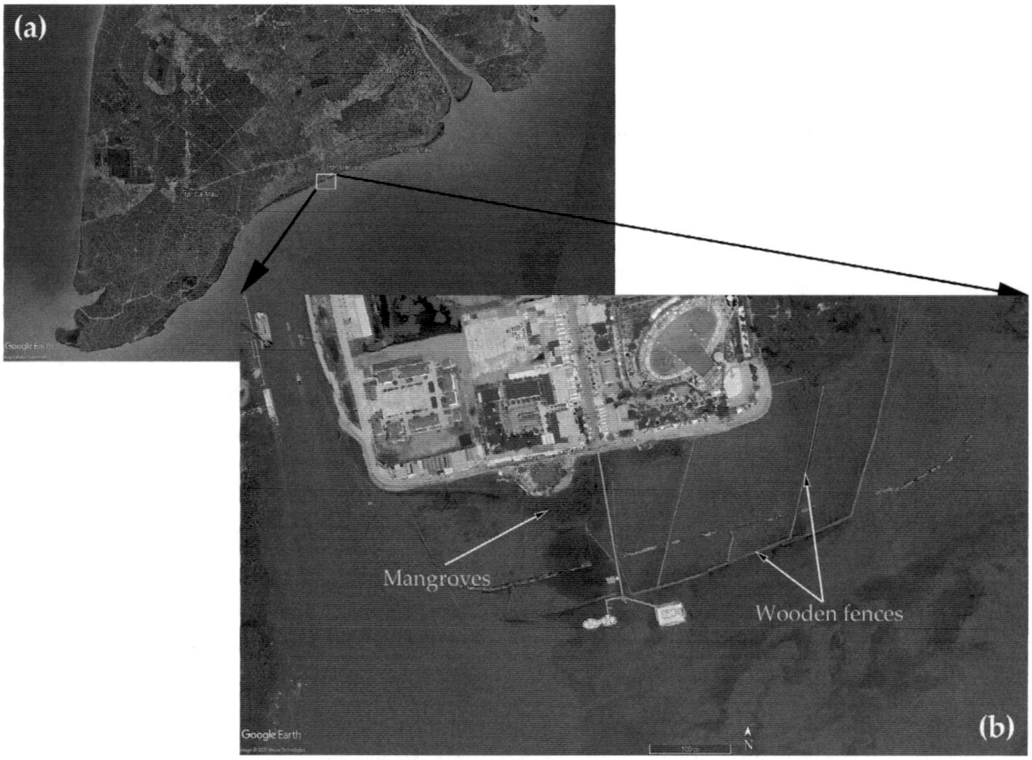

Figure 2. Fence location where field measurements were obtained (Dao et al. 2021) [4]: (a) Mekong Delta; (b) Nha Mat, Bac Lieu, Mekong.

IOP Conf. Series: Materials Science and Engineering 1294 (2023) 012026 doi:10.1088/1757-899X/1294/1/012026

3. Results and discussions

In this section, wave heights and surface elevations are presented as a part of the result of wave transformation over the profile. Next, the relationships between wave transmission coefficient and the relative fence thickness with wavelength and wave height are shown to assess the roles of width on wave reduction.

3.1 Wave transformation over the profile

Figure 3 presents standing wave heights in different scenarios. Wave propagation in each case illustrates different stages of a wave during its traveling to the near-shore. In the showcases in Figure 3, breaking waves appear at the early stage when water depth reduces from 4 m to about 2 m at $x = 50$ m to 80 m. However, the higher the incoming waves, the more wave breaking, as shown in Figures 3a&b. Smaller waves presented in Figure 3c&d slightly break after a short delay since these waves enter the shallow area lately.

Due to the appearance of the wooden fence at $x = 125$ m (Figure 3e), wave heights in all showcases are reduced by about 50% of their energies after reflection. Also, most wave energies are absorbed by the sponger layers. Therefore, transmission waves seem to have minor impacts from reflections at the east side.

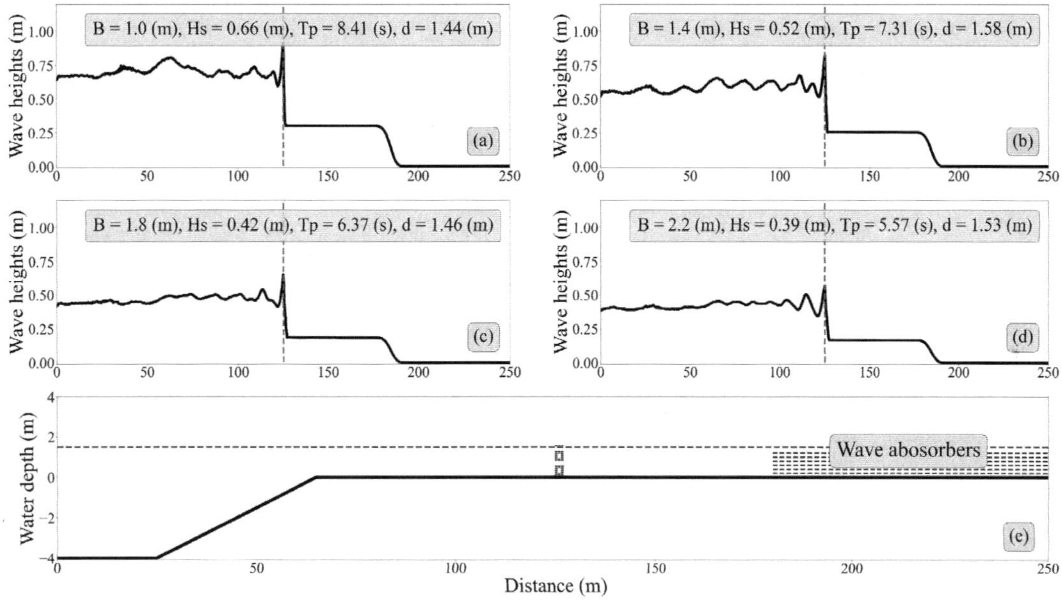

Figure 3. Standing waves over the fences.

Additionally, surface elevations and wave spectrum at locations in front and behind the fence of all showcases in Figure 3 are presented in Figure 4. For each case, surface elevations behind the fence are significantly reduced for larger waves (Figures 4b&d) but slightly decreased for smaller waves (Figures 4e & g). Yet, the porosity of wooden fences was set at about 85%, and the wave damped by them is highly presented. For example, all transmission wave spectrums (blue line) in the showcases (Figure 4a,c,f,h) are much lower than incoming waves (red line).

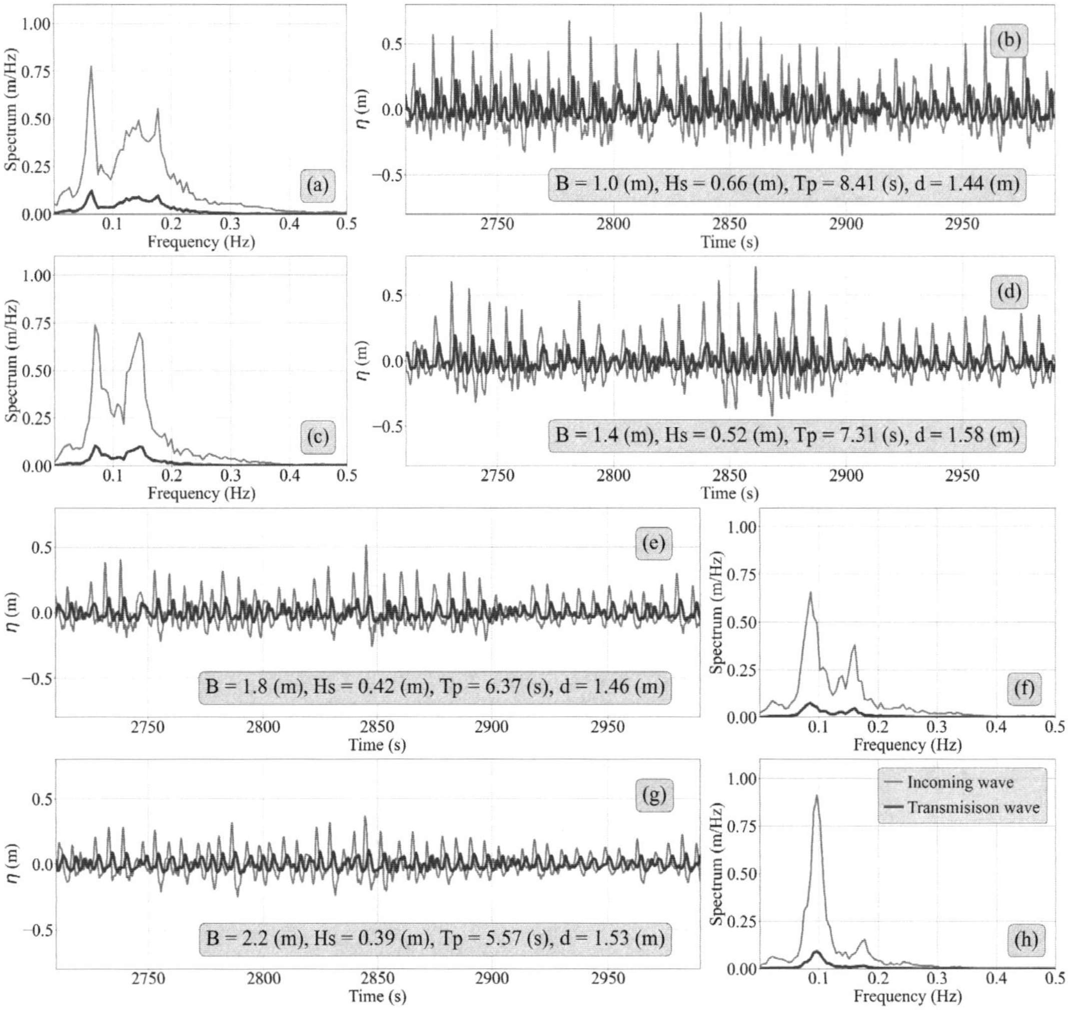

Figure 4. Surface elevation (b, d, e, g) and wave spectrum (a, c, f, h).

3.2 Wave transmission and fence thicknesses

In this section, the influence of fence thickness is presented in Figure 5 and Figure 6. The wave transmission coefficient (K_t) is plotted against two dimensionless parameters, which are expressed as:

$$K_t = \frac{H_t}{H_i} \tag{9}$$

where H_t and H_i are transmission and incoming wave heights, respectively.

In Figure 5, the relative fence thickness with incoming wave height is plotted against the transmission coefficient, and this plot is compared with the previous study [4]. The scatters from Figure 5 illustrate the reduction trend of incoming waves, indicating that larger waves tend to be damped more significantly and even more for the larger fence thickness.

The transmission coefficients in this study (red circles) value from 0.3 to 0.5 with the B/H_i from 1 to 4. These scatters interestingly match the data from Dao et al. (2021) [4] for the same range of B/H_i. The gap of K_t between the two studies can be interpreted by the difference of porosities applied. Furthermore, the fit line (dashed line) from this study is well-fit with Dao et al. (2021) [4] showing an influence of fence thickness on incoming wave height. The two scatters and the fit-line are interestingly matched at a *B/Hi* value of about 4. This can be explained by the fact that the larger thicknesses dissipate most of the wave, and at a certain point, wave reduction reaches the same proportion in Dao et al. (2021) with lower porosity. The larger the thickness is, the higher the wave is damped.

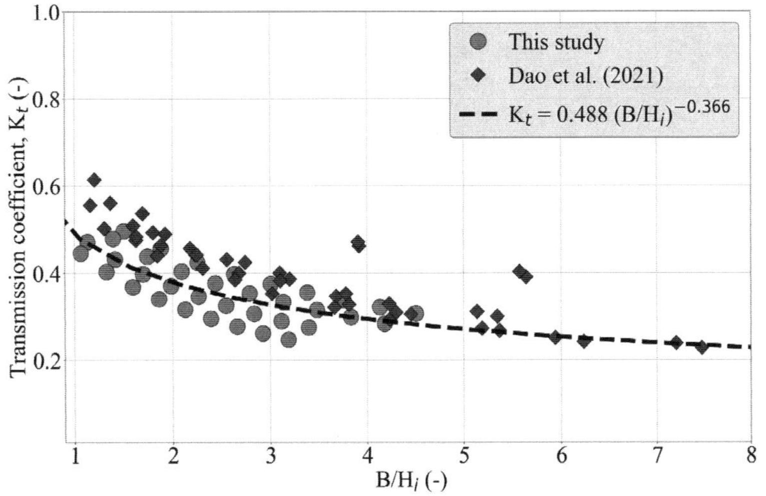

Figure 5. Relative fence thickness with incoming wave height and transmission coefficient with additional data from Dao et al. (2021) [4].

Since the influence of thicknesses on incoming wave heights is shown in Figure 5, the impact of this characteristic on wavelength, directly calculated from wave period and water depth, is indicated in Figure 6. Since water depths for every scenario are merely similar, it is acceptable to ignore the depth effects. In Figure 6, the larger waves (higher period) tend to be damped more than smaller waves, such as K_t ranges from 0.3 to 0.5 for B/L_i from 0.02 to 0.12.

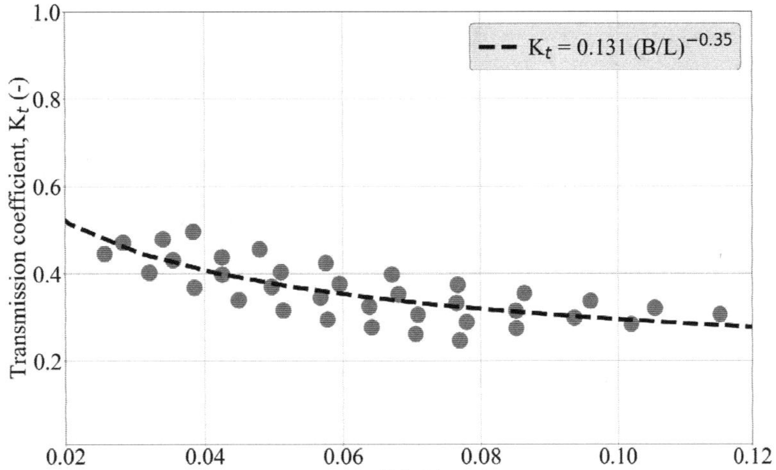

Figure 6. Relative fence thickness with wavelength and transmission coefficient.

Both results in Figures 5 and 6 have power fit lines that give an idea for wooden fence designs. As a result, the optimization for fence thicknesses can be indicated based on those two power expressions.

4. Conclusions

This study indicates the influence of fence thickness on wave damping, and the efficiency of wave reduction based on different widths can be indicated from the simulation results.

The SWASH model has been applied in previous studies, such as Dao et al. (2018, 2021) [4,7], showing a capability for simulating wave-fence interactions. The results show that wave heights are reduced from 30% to 70% of heights after contacting wooden fences. The large range of wave reduction also illustrates the dependency of wave damping on fence thicknesses. Additionally, the relations of wave transmissions and the relative fence thicknesses are presented as a decreased trend. In detail, the larger wave reduction often appears with the high value of the dimensionless parameters B/H_i and B/L_i. These results conclude that waves tend to be highly damped due to larger fence width. Note that other influencing parameters, such as water depth and reflection, are made as minimum as possible.

Furthermore, the other stages of wave-fence interactions, including wave reflection and dissipation, are left out due to the complexity of the fence's characteristics. For example, the high reflection often occurs for higher density or lower porosity, resulting in the difficulty of predicting the dissipation rate inside a wooden fence.

References

[1] Anthony EJ, Brunier G, Besset M, Goichot M, Dussouillez P and Nguyen VL 2015 Linking rapid erosion of the Mekong River delta to human activities. *Sci Rep* **5** 14745.

[2] Báo Tài nguyên và Môi trường (Vietnamese) 2018 Đồng bằng sông Cửu Long: Xói lở bờ biển bủa vây. n.d.

[3] Mazda Y, Magi M, Kogo M and Hong PN 1997 Mangroves as a coastal protection from waves in the Tong King delta, Vietnam. *Mangroves and Salt Marshes* **1** 127–35.

[4] Dao HT, Hofland B, Suzuki T, Stive MJF, Mai T and Tuan LX 2021 Numerical and small-scale physical modelling of wave transmission by wooden fences.

[5] Dao HT, Hofland B, Stive MJF and Mai T 2020 Experimental assessment of the flow resistance of coastal wooden fences. Water (Basel) **12**:1910.

[6] Phan KL, Stive MJF, Zijlema M, Truong HS and Aarninkhof SGJ 2019 The effects of wave non-linearity on wave attenuation by vegetation. *Coastal Eng.* **147**:63–74.

[7] Dao T, Stive MJF, Hofland B and Mai T 2018 Wave Damping due to Wooden Fences along Mangrove Coasts. *J Coast Res* **34**:1317–27. https://doi.org/10.2112/JCOASTRES-D-18-00015.1.

[8] Shu A, Zhu J, Cui B, Wang L, Zhang Z and Pi C. 2023 Coastal wave-energy attenuation by artificial wooden fences deployed for mangrove restoration: an experimental study. *Front Mar Sci* **10**:1165048.

[9] Mancheno AG, Reniers A and Suzuki T 2023 Optimising the wave attenuation of bamboo fences using the numerical wave model SWASH. *J. Coastal Hydr. Struct.* **3**, Paper 27.

[10] Qu K, Lan GY, Sun WY, Jiang CB, Yao Y, Wen BH, et al. 2021 Numerical study on wave attenuation of extreme waves by emergent rigid vegetation patch. *Ocean Eng.* **239**:109865.

[11] Zijlema M, Stelling G and Smit P. 2011 SWASH: An operational public domain code for simulating wave fields and rapidly varied flows in coastal waters. *Coastal Eng.* **58** 992–1012.

[12] Smit P, Zijlema M and Stelling G. 2013 Depth-induced wave breaking in a non-hydrostatic, near-shore wave model. *Coastal Eng.* **76** 1–16.

[13] Zijlema M and Stelling GS. 2005 Further experiences with computing non-hydrostatic free-surface flows involving water waves. *Int J Numer Methods Fluids* **48** 169–97

[14] Zijlema M. Modelling wave transformation across a fringing reef using SWASH. ICCE 2012: Proc. 33rd Int. Conf. Coastal Eng.g (Santander, Spain, 1-6 July 2012)

[15] Rijnsdorp DP, Smit PB and Zijlema 2012 M. Non-hydrostatic modelling of infragravity waves using SWASH. *Coastal Engineering Proceedings* **1** 27.

[16] Henderson CS, Fiedler JW, Merrifield MA, Guza RT and Young AP 2022 Phase resolving runup and overtopping field validation of SWASH. *Coastal Eng.* **175** 104128.

https://doi.org/https://doi.org/10.1016/j.coastaleng.2022.104128.

[17] Liu Y, Liao Z, Fang K and Li S. 2021 Uncertainty of wave runup prediction on coral reef-fringed coasts using SWASH model. *Ocean Eng.* **242** 110094. https://doi.org/https://doi.org/10.1016/j.oceaneng.2021.110094.

[18] Umesh PA and Behera MR. 2021 On the improvements in nearshore wave height predictions using nested SWAN-SWASH modelling in the eastern coastal waters of India. Ocean Eng. **236**:109550. https://doi.org/https://doi.org/10.1016/j.oceaneng.2021.109550.

[19] Suzuki T, Hu Z, Kumada K, Phan LK and Zijlema M 2019 Non-hydrostatic modeling of drag, inertia and porous effects in wave propagation over dense vegetation fields. *Coastal Eng.* **149**, 49 – 64, https://doi.org/10.1016/j.coastaleng.2019.03.011

[20] Mendez FJ and Losada IJ 2004 An empirical model to estimate the propagation of random breaking and nonbreaking waves over vegetation fields. *Coastal Eng.* **51**:103–18.

[21] Mai T, Dao T, Ngo A and Mai C 2019 Porosity Effects on Wave Transmission Through a Bamboo Fence. International Conference on Asian and Pacific Coasts, Springer 1413–8.

IOP Conf. Series: Materials Science and Engineering 1294 (2023) 012027 doi:10.1088/1757-899X/1294/1/012027

Reuse of decommissioned offshore steel components for new buildings: A case study

L V Valen[1*], S C Siriwardane[2], F Bjørheim[2] and S Røland [1]

[1]Sweco Norge AS, Storgata 136, 3915 Porsgrunn, Norway
[2]Department of Mechanical and Structural Engineering and Material Science, Faculty of Science and Technology, University of Stavanger, Stavanger, Norway

[*] Contact Author: larsvegar.valen@sweco.no

Abstract. Reuse process of structural steel members, which are extracted from decommissioned offshore platform, for new apartment building is presented in this paper. An index is introduced to assess the technical feasibility of the reuse process of offshore steel components to steel buildings. The applicability and significance of the reuse process is demonstrated by a real case study. Steel columns of a new four-storey apartment building are replaced by selected members from recently decommissioned offshore platform in this case study and hence feasibility, cost benefits and environmental advantages are investigated. Utilization ratios of selected pre-used columns shows that those columns have more than sufficient capacities to withstand the loads. This indicates that considered decommissioned module of offshore topside is applicable for similar size of new buildings. The results indicate significant environmental benefits such as reduction of the CO_2 emission, low global worming potentiality, and etc. The reusability index of the considered case study is 51.5%, which shows marginal feasibility of reuse, and this can be increased, if the reuse process is industrialized and streamlined by the offshore decommisioning companies. Three ways are proposed to industrialize and streamline the reuse process in the latter part of this paper.

1. Introduction

Construction industry is taking countless measures to reduce the carbon footprint as world is currently facing environmental and climate challenges. Literature has reported that one third of the CO_2 emissions are produced by the construction industry [1-3]. Production of structural steel requires quite lot of energy which emit approximately three billion tons of CO_2 each year [4]. This amount is approximately 8% of all CO_2 which is being released by human beings in the world. Researchers gave a high attention for hydrogen breakthrough ironmaking technology to decarbonize the common steel production process [4] and this technology is not yet fully established as it may increase the production cost. Therefore, construction industry should pay attention about other alternatives to reduce the CO_2 emission. Reuse of steel members of decommissioned structures is another way to reduce CO_2 emission. Use of structure or part of it after its first use (i.e pre-used) is defined as "Reuse" in this paper and this differ from the "Recycling" process. Life cycle of steel commonly has five main phases such as raw materials, steel production, manufacturing, service life and recycling. The phase 1, 2 and 4 are energy demanding and release lot CO_2. Therefore, these phases can be bypassed by reusing the pre-used steel for new steel buildings.

There are methods and European recommendations have recently been developed to reuse steel, which were taken from decommissioned steel building, for single-storey onshore buildings [4-6]. However, there is a doubt that sufficient amount of pre-used steel can be obtained from only decommissioned onshore structures. Another source of pre-used steel supply is decommissioned offshore oil and gas platforms. Majority of offshore platforms are reaching to end of design life and decommisioning is now required (i.e currently 12% of offshore installations, which is 88 oil platforms) [7,8]. Around 2000 offshore oil and gas platforms in the world are planned to be decommissioned by year 2040 [9] and this could be a big source of supply for reuse process/industry.

Repurposing of steel members is an important process when reusing offshore steel in component level. This process requires different attention on the feasibility assessment (i.e reusability assessment) which includes comprehensive degradation assessment, remaining capacity/strength analysis and re-design procedures. Reusing offshore steel for new onshore buildings may provide economic benefits, in addition to above discussed environmental benefits. However, there are lack of studies to check the feasibility and quantify both environmental and economic benefits, when reusing offshore decommissioned steel members/structures for new onshore buildings.

Therefore, main objective of this paper is to investigate technical feasibility, advantages, and disadvantages, when using the reclaimed steel members from decommissioned offshore platforms to new onshore buildings. A real-life case study is performed to quantitatively assess the feasibility/technical reusability and environmental benefits. This case study only focusses analysis and design of new apartment building with reclaimed steel, not on designing the building for the purpose of being reused. The paper first discusses the reuse methodology. The details and the results of the case study is presented. Comparison of the results and discussions are shown in the latter part of the paper.

2. Reuse methodology

Generally accepted reuse practice is required to streamline and industrialize the supply of reuse offshore steel in the building industry. This practice should include all the stages/operations from pre-decommissioning assessment of the donor platform to delivery on new building construction site, based on a reusability assessment. The reusability assessment consists of three major aspects, such as feasibility (i.e. technical aspects), cost benefits and environmental impacts. The major operations of the reuse process are decommisioning, demounting, handling & manipulation, separation & cleaning, redesigning & repurposing, modification, quality check, and geometric verifications. The flowchart of the reuse process is shown in Figure 1. The more details of each operation and related conceptual framework are presented in recent publication of the authors [8,10].

2.1. Reusability index

The technical feasibility of the reuse process can be assessed by a recently proposed approach, which is based on reusability index [5,8]. This index involves assessing and quantifying above mentioned individual operations (i.e reusability performance categories) of reuse process based on own feasibility. The reusability index r is defined for a single member as,

$$r = \sum \rho_i w_i \tag{1}$$

where ρ_i is the feasibility of each performance categories, which is graded from zero (impossible) to unity (very easy), w_i is the weighting factors of the individual performance categories. Generally, quantitative assessment of contributing parties, which are involving in each operation of the process, should be required for determining relevant values for ρ_i and w_i. The grades of ρ_i are divided by 20% increments and differences between each grade are clearly tabulated in the original article published by Technical Research Center of Finland [5]. The proposed values for weighting factors, w_i of each operation of the reuse process are presented in the Section 2.2.

The reusability index for entire structure (R) is defined as [5,8],

$$R = \frac{\sum m_i r_i}{\sum m_i} \tag{2}$$

where m_i is the weight of the structural member and r_i is the reusability index of ith structural component calculated as defined in equation (1). The final reusability index is case-dependent.

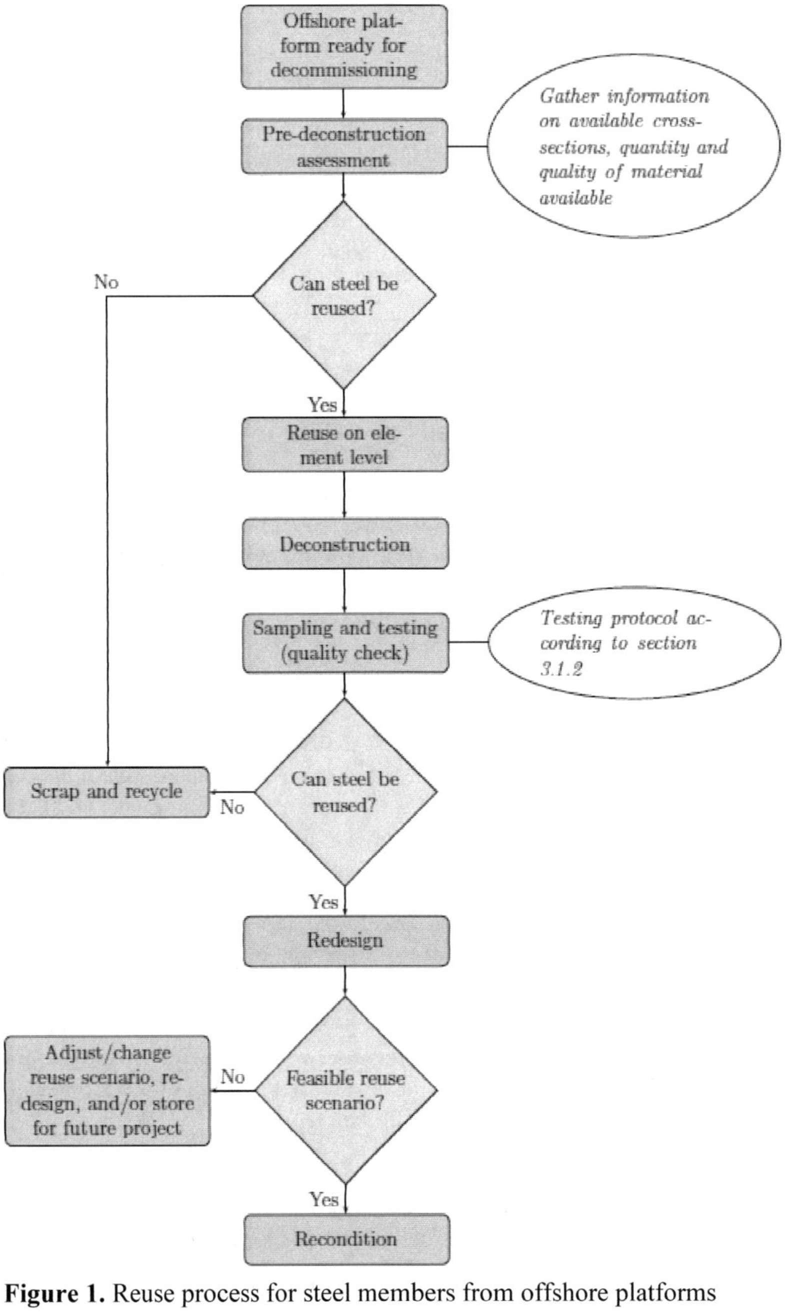

Figure 1. Reuse process for steel members from offshore platforms

2.2. Proposed weighting factors

The weighting factor w_i ranks the significance of each operation (i.e performance category) of reuse process and sum of w_i of each operation is equal to unity. Offshore steel structures are generally subjected to harsh marine environment. Therefore, quality check operation/performance category has a higher potential weightage than other operations specially when consider the reuse process of offshore steel. The modification and repurposing performance categories should be weighted lower than other categories in the process as reusing element/member level (i.e. beams and columns) requires less effort and planning than structure or module levels. The proposed weighting factors for reusability assessment offshore steel are shown in Table 1.

Table 1. Proposed weighting factors

Performance category	Weighting factor (w_i) %
Deconstruction	25
Handling and manipulation	10
Separation and cleaning	10
Redesigning	10
Repurposing	5
Modification	5
Quality check	30
Geometry check	5

3. Case study: Reuse of offshore steel for new apartment building

3.1 Considered apartment building

An apartment building, which is part of Nybyen housing project, is chosen by Sweco AS for this case study. It is a four-storey building which consists of 11 separate apartments. The overall length, width and height of the building are 23m, 12m and 12m. The building is designed by concrete floors supported by steel columns as shown in Figure 2. This type of building is very common in Norway. Therefore, results (i.e. feasibility assessment, environmental benefits and cost aspects) of this case study can be applicable for many other apartment buildings in Norway and this is one of the major reasons to select this building for this case study. The objective of this case study is to check the feasibility of replacing all the steel columns of original design by decommissioned offshore steel columns and hence compare/discuss the environmental and economic benefits, pros and cons. The floor height is around 2.7 m and 69 steel columns with steel grade S355 are used in the original design. The total weight of the new columns is 5220kg. The different types of cross sections used for the columns are shown in master's thesis of the first author [10].

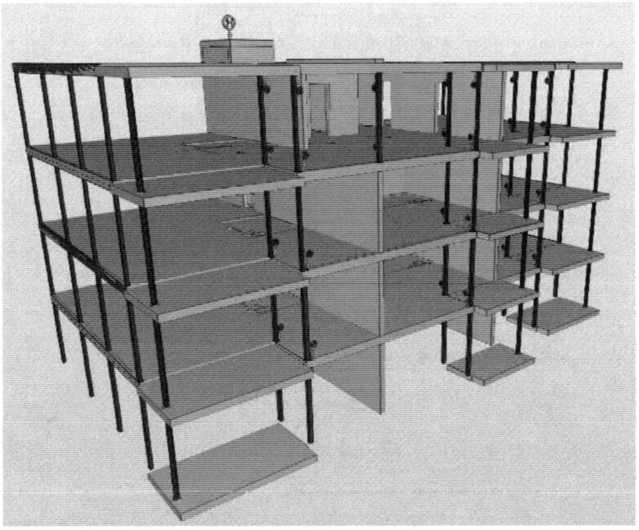

Figure 2. 3D model of considered apartment building

3.2 Structural analysis of original building/design results

The dead, imposed, wind and snow loads were calculated, and ultimate limit state analysis of the building is performed by finite element method employed structural analysis package, FEM design [11]. The steel columns are subjected to interaction of axial force and biaxial bending. Therefore, individual buckling (stability) checks were performed using the equations 6.61 and 6.62 which is given in clause 6.3.3 in the Eurocode 3 [12]. Hence, corresponding utilization ratios (UR) of the steel columns are obtained for the all the columns in each floor. The obtained UR for most critical columns (i.e. which has a UR more than 0.7) are assigned an ID and shown in Table 2.

Table 2. Internal forces and utilization ratios of critical columns of original design

Column ID	Floor	Column type	N_{Ed} (kN)	$M_{y,Ed}$ (kNm)	$M_{z,Ed}$ (kNm)	UR (%)
1	2	RHS200x100x10	594.4	12.81	3.56	85.3
2	2	RHS200x100x10	850.1	18.21	0.56	98.3
3	2	RHS200x100x10	729.0	0.79	5.49	81.3
4	2	RHS200x100x10	697.3	0.90	4.87	76.7
5	2	RHS200x100x10	438.4	1.64	7.79	67.5
6	2	RHS200x100x10	595.9	1.48	13.83	73.8
7	3	RHS150x100x10	429.3	17.84	17.46	93.1
8	3	RHS150x100x10	607.5	11.33	14.95	97.9
9	3	RHS150x100x10	504.8	2.43	5.93	73.8
10	3	RHS150x100x10	474.4	1.19	4.79	68.2
11	3	RHS150x100x10	443.1	1.22	12.86	71.7
12	4	SHS100x100x10	273.8	8.14	11.85	83.3
13	4	SHS100x100x10	388.3	4.62	9.64	88.0
14	5	SHS100x100x10	126.8	8.17	11.07	76.6

3.3 Availability check of decommissioned steel

The topside of an oil platform is commonly constructed by smaller modules which can be fabricated in the shipyard and transported to the site for further assembly. The steel components of one of the similar modules of decommissioned platform (Figure 3) has been considered for reuse purposes in this case study. The platform was operated in shallow water and the topside of it was supported by a bottom fixed jacket. Mainly the I type of cross sections are used for girders, transverse deck beams and stiffeners. Members with H type of cross sections have been used for supporting the decks and all the joints are welded. These types of members with H cross sections were selected for reusing in this apartment

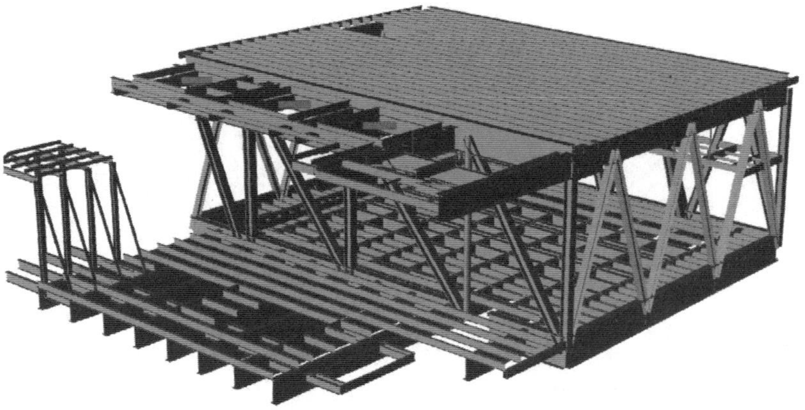

Figure 3. Decommissioned steel module of offshore donor platform

building based on the reduced capacities due to interaction of axial force and biaxial bending. The selected cross sections are universal columns (UC) 305×305×97, UC 356×406×235, UC 356×406×287 and UC 356×406×393. The original drawings and inspection/maintenance history of these members are available, while material documentations are not available. The material of the columns is grade 40EE and strengths are shown in British Standard BS 4360 [13]. The geometric details of the steel members and corresponding cross-sectional properties are given in the master's thesis of the first author [10]. The class of the cross-sections were classified based on guidelines given in Eurocode 3 [12] and results are shown in Table 3. Design resistance of the cross-sections for uniform compression ($N_{c,Rd}$) was calculated following the guidelines in Eurocode 3 as shown in Table 3.

Table 3. Material, geometrical properties and cross-section capacities of selected columns

	UC305x305x97	UC356x406x235	UC356x406x287	UC356x406x393
Steel grade	40EE	40EE	40EE	40EE
Maximum t (mm)	15.4	30.2	36.5	49.2
Yield strength (N/mm^2)	260	245	245	240
Flange class	2	1	1	1
Web class	1	1	1	1
Class of cross section	2	1	1	1
$N_{c,Rd}$ (N/mm^2)	3055.8	6976.7	8563.3	11442.3

3.4 Degradation status of the selected members

Accumulated fatigue damages of selected members are negligible since the selected members are extracted from a module of the topside. Therefore, corrosion is the main degradation mechanism which should be considered for the selected members. To take account intact to worst-case of corrosion degradation, three different uniform corrosion scenarios/cases were considered such as Case 1. no corrosion, Case 2. uniform corrosion along whole cross-section, and Case 3. Uniform corrosion along parts of the cross-section (which provides eccentricity of the centroid and additional bending moment). Considered three corrosion scenarios are shown in Figure 4.

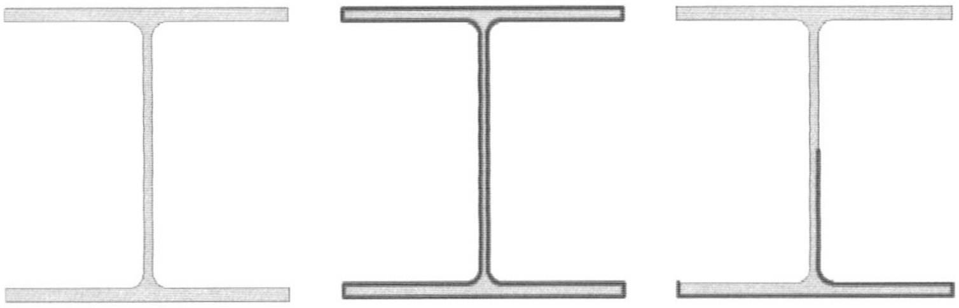

(a) Case 1 (b) Case 2 (c) Case 3

Figure 4. Corrosion cases

The service life of the considered module is 25 years (i.e. life span t =25 years). The coating life of the members is considered as 10 years (i.e t_{pt} = 10 years) for corrosion Case 2 and t_{pt} is assumed to be zero for Case 3 to simulate the worst case scenario. Time dependent thickness reduction due to uniform corrosion can be calculated using below function [14].

$$w(t) = A(t - t_{pt})^B \quad (3)$$

where $w(t)$ is the depth of corrosion in millimeters and t is age of the structure. The A and B are the parameters which depend on the corrosive environment and the type/grade of steel. The values for A

and B are given in previously published article [14]. The corresponding cross sectional details and properties are shown in Table 4 for all 3 corrosion cases.

Table 4. Geometrical properties of the UC305x305x97with different corrosion cases

	Case 1	Case 2	Case 3
Depth of corrosion $w(t)$ (mm)	0	1.0	1.5
Cross section area (mm^2)	12345	10559	11462
Change of centroid e_z (mm)	0	0	11
Change of centroid e_y (mm)	0	0	2

3.5 Performance category assessment

The ranking of the performance categories (i.e operations of the reuse process) are determined based on the data and criteria description is presented in original article published by Technical Research Centre of Finland [5]. The feasibility values of ρ_i is selected to obtain the most conservative reusability index and the results are shown in Table 5.

Table 5. Proposed ranking of the performance categories

Category	ρ_i	Comment
Deconstruction	0.5	Welded connections, moderately difficult access
Handling and manipulation	0.6	Manipulation by crane, forklift and small lifting devices
Separation and cleaning	0.4	Separation and cleaning using hand tools
Redesigning	0.8	Detailed drawings and maintenance history available
Repurposing	0.4	Some remanufacturing required
Modification	0.6	Columns are applicable for reuse without too much work
Quality check	0.4	Material documentation not available, require laboratory testing
Geometry check	0.8	Simple straightness testing using lasers

3.6 Life cycle assessment

One Click LCA [15] software was used for life cycle assessment (LCA) of this case study. The global warming potential (GWP) of this construction project was calculated by above software using environmental product declaration (EPD) according to Norwegian standard which provides the method for greenhouse gas calculations for buildings [16]. This method takes account every source of CO_2 emissions for building materials throughout their life cycle.

The GWP of the steel columns was assessed. The EPD of original RHS columns were taken from steel manufacturer, Tibnor [17]. The generic EPD for H-sections was utilized with reuse option/function for universal beams (UB) which were selected from decommissioned module. The total steel masses for both cases were given as an input to these assessments. It was assumed that only columns with SHS and RHS cross-sections replaced by UC 305×305×97. The circular columns in balconies with CHS 139.7×8 were not replaced with reuse columns as it is required to replace by similar shape and size due to aesthetic appearance/aspects. The transport distances were based on locations of Tibnor (i.e new steel manufacturer) and Aker Solution shipyard in Stord (i.e., location of decommissioned offshore module) relative to the apartment building location at Sandefjord, Norway. It was assumed that mode of transport is onshore by lorries. The LCA results are shown in Section 4.3.

4. Results

This section presents the capacities and utilization ratios of the selected members for reuse, results of reusability index and LCA. Comparison of capacities and design optimization with related discussion are made at the latter part of this section.

IOP Conf. Series: Materials Science and Engineering 1294 (2023) 012027 doi:10.1088/1757-899X/1294/1/012027

4.1 Capacities of reused columns

The RHS columns used in original design (Table 2) were replaced by selected UC columns in frame element model made by FEM-Design programme. The corresponding steel grade is also changed to 40 EE and ultimate limit state design checks were performed according to Eurocode 3 [12]. Additional bending moments due to eccentricities of corroded case 3 were imposed for ULS design checks. Hence, corresponding UR of the steel columns are obtained for the all the columns in each floor. Analysis results shows that UC305×305×97 cross section gives the highest URs and obtained URs for each degradation cases are shown in Table 6.

Table 6. Utilization ratios of selected UC305x305x97 columns with different corrosion cases

Column ID	Floor	N_{Ed} (kN)	$M_{y,Ed}$ (kNm)	$M_{z,Ed}$ (kNm)	Utilization ratio (UR) %		
					Case 1	Case 2	Case 3
1	2	594.4	12.81	3.56	33	42	40
2	2	850.1	18.21	0.56	38	49	47
3	2	729.0	0.79	5.49	39	36	34
4	2	697.3	0.90	4.87	27	34	32
5	2	438.4	1.64	7.79	19	24	22
6	2	595.9	1.48	13.83	28	35	32
7	3	429.3	17.84	17.46	29	33	32
8	3	607.5	11.33	14.95	31	38	35
9	3	504.8	2.43	5.93	21	26	25
10	3	474.4	1.19	4.79	19	24	23
11	3	443.1	1.22	12.86	22	27	25
12	4	273.8	8.14	11.85	18	21	19
13	4	388.3	4.62	9.64	19	23	22
14	5	126.8	8.17	11.07	12	14	12

4.2 Calculated reusability index

The reusability index was calculated for selected columns from equation (1) by following the guidelines given Section 2. The proposed weighting factors in Table 1 and ranking values of performance category given in Table 5 were used to calculated final reusability index as shown the Table 7.

Table 7. Reusability index calculation

Category	ρ_i	w_i	$\rho_i\, w_i$
Deconstruction	0.5	0.25	0.125
Handling and manipulation	0.6	0.10	0.06
Separation and cleaning	0.4	0.10	0.04
Redesigning	0.8	0.10	0.08
Repurposing	0.4	0.05	0.02
Modification	0.6	0.05	0.03
Quality check	0.4	0.30	0.12
Geometry check	0.8	0.05	0.04
$r = \sum \rho_i\, w_i$			0.515

IOP Conf. Series: Materials Science and Engineering 1294 (2023) 012027 doi:10.1088/1757-899X/1294/1/012027

4.3 Life cycle assessment results
The LCA was performed as described in Section 3.6. The input parameters of LCA and GWP results are shown in Table 8. Figure 5 shows the distributions of GWP for new and reused steel columns respectively.

Table 8. Input parameters and LCA results

Category	New steel	Reused steel
Mass (*kg*)	5220	14928
Assumed transport distance (*km*)	185	430
Global warming potential, GWP (*kgCO2*)	14838	667

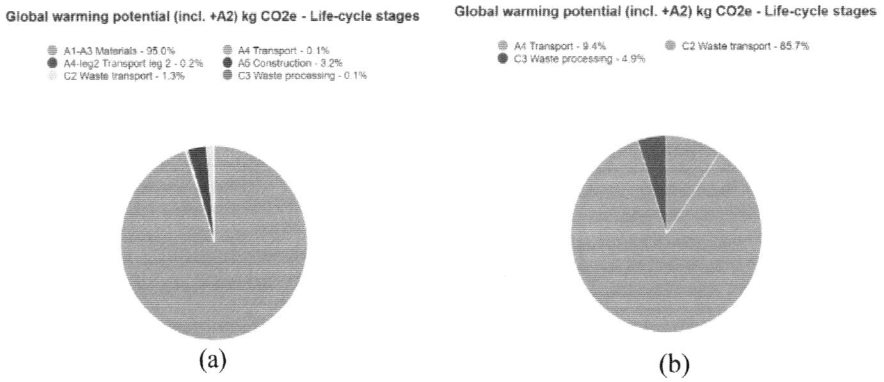

Figure 5. LCA results (a). GWP for original RHS columns, (b). GWP for reused UC305x305x97

5. Comparison and discussions
Comparison of the results between new steel and reused steel applications are made in this section and related discussions are presented.

5.1 Design optimization
Utilization ratios of selected reuse columns in Table 6 (i.e max UR is 47%, whereas new steel shows max UR as 98.3%) shows that those columns have a more than sufficient capacity to withstand the loads. This indicates that considered decommissioned module of offshore topside is applicable for new buildings which are same size, and larger buildings than considered in this case study. The capacities of the degraded/corroded columns shows that remaining capacities are sufficient to withstand the load though there is possibility of some corrosion degradation of the reuse columns/members. The degraded columns should be treated and protected against development of further corrosion.

This case study results shows that selecting suitable columns from a larger pool of steel components from decommissioned oil platforms is beneficial to choose most suitable columns which gives higher UR. Design optimization can be achieved in this way and, leads to huge cost reduction than new steel. Alternatively, change of structural configuration/architecture (i.e., conceptual design) can be done to suit the capacity of pre-used (reused) steel columns. The changes would include changing span length and height of columns, changing the load paths and etc which could change the modification performance category in the reusability index.

5.2 Feasibility/reusability
The reusability index (*r*) describes the degree of practicality/feasibility of reuse of decommissioned steel members for this case study. However, this always depends on the performance categories and weighting factors. The *r* for considered case study is 51.5% as shown in Table 7 and this hardly indicate a profitable

reuse operation. The drawback should be identified and address to increase the reusability index. Low-scoring performance categories with corresponding largest weighting factors (i.e. deconstruction, quality checks and etc) are most potential for improvement.

5.3 Environmental and economic perspectives
Significant environmental benefits with regards to global warming potential (GWP) can be achieved when reuse the decommissioned steel components for this case study as shown in Table 8. More specifically, 5220 kg of new steel columns yields emission of 14838 kg of CO_2. The total mass of 14838 kg yields 667kg of CO_2. This provides GWP of 2.84 $kgCO_2/kg$ and 0.04 $kgCO_2/kg$ for new and reused steel respectively. The GWP associated to re-manufacturing and construction were neglected by the LCA software in this case study. The GWP associated with number of additional emissions in the reuse process such as decommissioning, dismantling, storage and etc, should be considered in LCA. Therefore, it would be beneficial to have both EPD, which specifically made for reused steel from offshore topsides and EPD, which took accounts all the emissions throughout all the service life of the new steel from production stage. It may be sensible to focus on the fact that emissions due to remelting during recycling process of the decommissioned steel can be omitted when reusing the steel.

Calculating cost benefits behind the reuse process is currently an impossible task as there is no marketplace for it. A full cost breakdown structure (CBS) should be created to determine the accurate cost for reuse steel in this case study by involving all the parties in the supply chain. Hence, cost and potential savings can be calculated.

6. Conclusions
An index has been introduced and modified in this paper to quantify the reusability of pre-used steel from offshore topside module to new building. A real-life case study was performed to observe the applicability and significance of the reuse process.

The case study results shows that reusability index is 51.5% which make room for improvement of the reuse process with regards to economical and sustainability aspects. The improvement can be achieved by changing the decommisioning process more industrialize and streamline the supply of the reused offshore steel for the building construction industry such as i) facilitate the shipyard to be able to de mounting the topsides in way the members are reusable afterwards, ii) incorporate stock list in the steel supply chain that specialize in the refurbishing process, and have capacity for quality and geometric checks, and iii) establish a competitive logistic network with storage and transportation capacities for reused steel. Therefore, it is beneficial to have above three facilities in companies which responsible for decommisioning activities and own the shipyards. This study concludes that decommissioned steel has a sufficient capacity for large number of building construction projects and there would be a market for reused offshore steel in building construction industry if above three changes are implemented.

Development of generalized guideline/framework for inspection and testing for remaining capacities of degraded/damaged members and development of EPD for reused offshore steel are recommended for future studies.

References
[1] Fujita M, Fujita T, Iwata M, Kenmitsu T, Kimura U, Koiwa K, Midorikawa M, Okazaki T, Takahashi S, Tanaka and Wada, M 2023, Japanese efforts to promote steel reuse in building construction, *J. Struct. Eng., ASCE*, **149** (1), 04022225.
[2] Tingley D D, Cooper S and Cullen J 2017, Understanding and overcoming the barriers to structural steel reuse, a UK perspective, *J. Cleaner Prod.*, **148**, 642-652.
[3] Barth F, Jacobsen J 2023, How do we upcycle steel, *website of Nordic Circles*, *https://www.nordiccircles.com/upcycling.*
[4] Coelho A, Pimentel R, Ungureanu V, Hradil P, Kesti J 2020, European recommendations for reuse of steel products in single-storey buildings, *PROGRESS 2017-2020*, *European Convention for Constructional Steelwork*, Portugal,
[5] Hradil P, Talja A, Ungureanu V, Koukkari H, Fülöp L 2017, Reusability indicator for steel framed

buildings and application for an industrial hall, *Proc EUROSTEEL 2017*, Ernst & Sohn, (Copenhagen, Denmark, September 13-15, 2017)

[6] Hradil P, Fülöp L, Ungureanu V 2019, Reusability of components from single-storey steel-framed buildings, *Steel Constr. Des. Res.* **12** (2), 91-97.

[7] OSPAR Commission 2019, *Inventory of Offshore Installations*, ODIMS, https://odims.ospar.org/en/search/?search=inventory.

[8] Siriwardane S C, Pavlou D 2023, Reuse of Offshore Steel Components for Buildings: Opportunities, Challenges and a Framework *Proc. of 33rd Int. Ocean Polar Eng. Conf.* (Ottawa, Canada: June 19 –23, 2023) Paper Number ISOPE-1-23-131.

[9] Vidal P, 2022 Decommissioning of offshore oil and gas platforms: A systematic literature review of factors involved in the process *Ocean Engineering.* **255**, [Online]. Available: https://www.sciencedirect.com/science/article/pii/S0029801822008101.

[10] Valen L V 2023 *Reuse of structural steel components from decommissioned offshore structures* Master's Thesis, Uiversity of Stavanger, Norway.

[11] Veteläinen J 2022 *FEM design User Manual* StruSoft AB https://strusoft.com/wp-content/uploads/2022/07/FEM_English_Manual_English-1.pdf

[12] EN 1993-1-1, Eurocode 3 2005 *Design of steel structures - Part 1-1: General rules and rules for buildings*, European Committee for Standardization, Brussels, Belgium.

[13] BS 4360 1990 *British Standard Specification for Weldable structural steels*, British standard institution, Loandon, UK.

[14] Aeran A, Siriwardane S C, Mikkelsen O, and Langen I 2017 A framework to assess structural integrity of ageing offshore jacket structures for life extension *Marine Struct.*. **56**, 237–259.

[15] One Click LCA. [Online]. Available: https://www.oneclicklca.com/

[16] NS 3720, 2018 *Method for greenhouse gas calculations for buildings* Standard Norge, Lysaker, Norway.

[17] Tibnor 2023 Kaldformede hulprofiler. [Online]. Available: https://www.epd-norge.no/epder/bygg/stal-armering-aluminiumskonstruksjoner/kaldformede-hulprofiler-1

W-beam guardrail crash analysis and evaluation using finite element method

Y S Lerabo and H S Habte[*] (0000-0002-3517-9835)

Addis Ababa Institute of Technology, Addis Ababa University, Ethiopia

*Correspondence: haileleoul.sahle@aait.edu.et

Abstract: Guardrails are highly recommended to be erected in the median and side corner of a highway road to secure vehicle accidentally loss it track, supports the safety of the vehicle driver and passenger from catastrophic accidents. The design and constructions of different guardrails are done confirming standard to retain their performance that related to energy absorption capacity. Besides, damaged guardrails that compromise their performance are seen in highways due to poor maintenance or repair. This study was conducted to investigate the structural integrity of W-beam guardrail damage. An impact test which leads to non-linear material deformation behavior was simulated using commercially available software, ABAQUS. The geometrical models of the guardrail were modeled adopting AASHTO standard, and a representative Jeep truck used to impact the guardrail at 90o with 42.26km/hr speed. The non-conforming W-beam guardrail was modeled considering a loss connection of supporting leg with the W-beam. Results of the study showed that the structural integrity of the damaged guardrail is affected as the internal and plastic energy was decreased from the W-beam guardrail without damage. More, the damaged W-beam was displaced more, by 14.15%, to the restricted area as compared to the W-beam guardrail without damage.

Keywords: Crash simulation, non-conforming W-beam guardrail, dynamic explicit FEA.

1. Introduction

Road traffic accident (RTA) is one of the causes for human life threatening problems for different citizens especially in developing counties and also leads to traffic jamming on highway and results for huge economic losses. A report on RTA has shown that the death of rate pedestrians followed by passengers and drivers dynamically increased.

To reduce the risks of drivers and pedestrian, different road side guardrail are used. The guardrails usually design and constructed following regulation set by countries that their design and manufacturing categorized under flexible, semi-rigid or rigid in relation to the road location risk. Standard risk management is used test the quality of the guardrail taking an evaluation criterion; structural adequacy, occupant risk, and Post-Impact vehicular trajectory [1 - 3].

In the study of the performance or innovation on guardrail, researcher follows experimental, analytical and numerical simulation for guardrail performance tests. Besides, the experimental investigation of guardrail crashworthiness performance required complex test set-up and costly operation investment owing to the destructive operation; hence number researcher used FEA using commercial software like ABAQUS and LS-DYNA to simulate the complex nonlinear impact behavior of guardrail with different vehicle types. Martin and Wekezer [4] conducted a validation for numerical approach for full-scale crash impact to supplement the highway roadside safety hardware (guardrail) test, where the research result shows that the nonlinear explicit dynamic FE code using LS DYNA supplements the experimental test. And more, the finite element method for road safety guardrail, W-beam guardrail, crash test adopting the European standard EN 1317 supports to comply as alternative option of test standard proposed by the AASHT, Teng et al. [5]. Paweł and Sebastian [6] used FEA code on LS-DYNA for simulating the road guardrail subjected to a bus crash to investigate the effective material modeling of the guardrail. They have considered three zones guardrail model, their material properties characterized by different constants to comply with the real one, and their results shows a close match to the experimental ones. Ren and Vesenjak [7] conducted a nonlinear elastic-plastic dynamic crash simulation for designing a new roadway guardrail and result closely matches to the experimental one.

Nevesa et al. [8] in their investigation on the performance of flexible and rigid road guardrails to vehicle impact according to test TB11 from EN 1317 standard showed that the flexible guardrail is much safer than the concrete guardrail. Amirarsalan et al. [9] have investigated the effective impact variables for both flexible and rigid barriers, the hight of the barriers yield as the main parameter to influence the efficiency of the barriers. Teng et al. [5] develop a finite element analysis for W-beam guardrail using Euroapean standard EN 1317 which the research was validated with experimental tests. The same authors identified the effect of different space for W-beam barrier using numerical simulation apporach [10].

However, roadside guardrails can be observed with non-confirming construction due to poor construction processes or poor maintenance provision after they are damaged or long term of operations. As can be seen from Figure 1 (a), the w-beam supports the leg bend and loss of the connections with the beam, which potentially leads to sagging for the vehicle during post collision period. And in Figure 1(b), the joint connection between these two different types of guardrails is not constructed using the conforming guardrail assembling and installation procedure [2].

(a) (b)

Figure 1. W-beam guardrail without (a) proper maintenance and (b) construction

For W-beam guardrail which is very popular among the flexible guardrail category, countries are following their supervision and maintenance procedure for maintaining the good conditions of the guardrails. Following the site supervision and the status of the damage conditions, damage category between low to high severity, maintenance and/or replacement recommended. It is evident that with the low rate for maintaining or replacing the damaged guardrails in urban and rural streets, this condition promotes a catastrophic accident when it crash happened at these areas once again.

The objective of this study is to investigate the crashworthiness performance of damaged deformable guardrails using nonlinear finite element analysis adopting a numerical simulation on vehicle crash on deformable guardrail. For the numerical simulation a W-beam representing the deformable guardrail, guardrail and a Pickup, Jeep, use to represent the crashing body.

2. Method and Material

2.1 Method

According to AACRA's standard, level 3 test model for guardrail structural adequacy for urban area was adopted. MASH 2016 manual test level 3 recommended a pickup truck with the weight of 2.27 Ton impact the guardrail at impact angle 25° with maximum speed 100 k/hr [12, 13]. The guardrail structural adequacy and occupant risk was considered as main evaluation criteria which were done by evaluating the deformation of the guardrail and energy absorption.

The deformable guardrail material and geometrical data were adopted from the AASHTO. This study considered two cases to represent damaged deformable guardrail, where the first case considered the deformable guardrail losses it support column connection with ground, and the second considered a lose connection between beam to beam in longitudinal segment arrangement. The selected model of damage W-beam guardrail represents the mild sever damage (Category 2) that observed in field when a proper maintenance failed to be done after a damage occur on the guardrail [15].

Since the standard longitudinal segment W-beam guardrail length is 30m which required minimum five segments for performance test, this study consider two segments to represent the second damaged guardrail. Common guardrail evaluation conditions are related with holding a vehicle at collision period, proper workable space after crash and safety of the occupant at crash period. The evaluation criteria are summarized with set safety levels related with the impact severity, containment level and deformation of the guardrail [5, 10].

2.2 Material

The study considered the standard flexible/deformable guardrail, W-beam, which is constructed and assembled in urban and rural dangerous vehicle pathways to protect vehicle not to overpass their track to the other side.

The specification of the W-beam guardrail specification was adopted from AASHTO. This guardrail type is developed using special steel structures, like MIT 180. The beam has a W-shape profile with an overall dimension of (310 x 3 x 4,128) mm and the support leg is designed with a C shape profile with an overall dimension of (150 x 5 x 1,820) mm, length of 1120 mm is expected to earthen, which the support covering material considered as non-deformable and so the covered leg was ignored in the CAD model. For assembling the beam with the support leg, a steel connector with an overall dimension (150 x 5 x 350) mm is used. Figure 2 illustrates a single segment W-beam guardrail assembled on three legs whose center-to-center distance is 1,905.0 mm.

The study considers a special steel material AASHTO MIT 180 which is one of the standard materials adopted in manufacturing the W-beam guardrail. The elastic material behavior is depicted in Table 1 while to capture the plastic behavior of the special steel structure, a Johnson-Cook material model was adopted, and its constants are listed in Table 2.

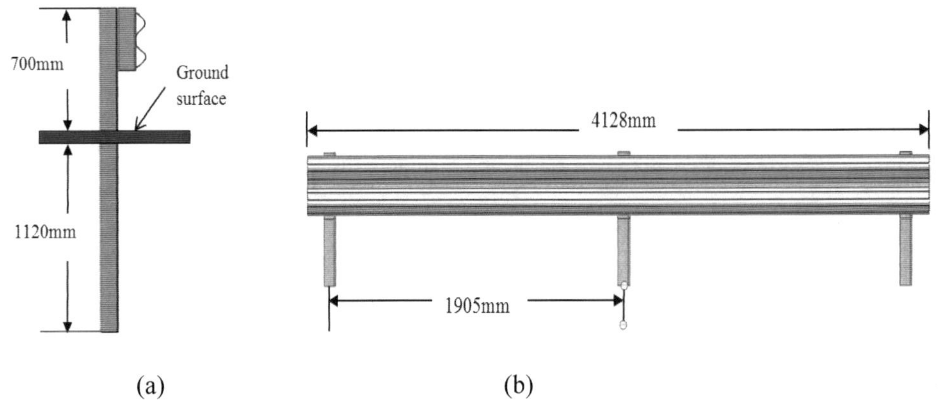

Figure 2. Single segment W-Beam guardrail (a) side view and (b) front view

Table 1. Material behavior of W-Beam guardrail parts [14]

Material	Density (t/mm3)	Modulus of elasticity (MPa)	Yield strength (MPa)	Poisson's ratio
Special steel MIT 180	7.813e-9	2.00e3	419.0	0.29

3. Finite element analysis

This study used a non-linear dynamic explicit FEA considering the deformation process following the impact load will be elastic-plastic. For the selected FEA approach, the guardrail parts were modeled as deformable bodies and the vehicle as rigid body. Parts of the guardrail were modeled as shell considering the thickness is small compared with other geometrical data. Adopting Test 3 of MASH standard, the full longitudinal arrangement was modeled using eight segments to get 30m length [11] for the targeted guardrail impact test. Considering the connecting parts, bolts and nuts, to connect segments one to the other and to the spacer and support are firm and tight, there specific model types were ignored in this study since their effect is not the interest of the study. Hence, Figure 3 illustrates the CAD model of 30m long W-beam guardrail as a single instant, part for the finite element analysis.

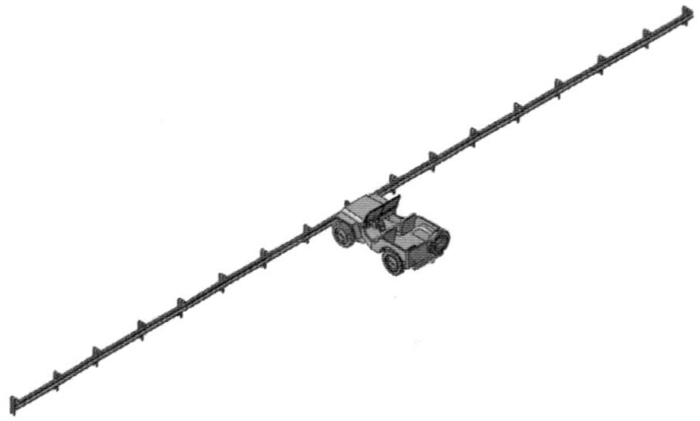

Figure 3. W-beam segments with Jeep truck impact orinted at 90°.

The first and second conditions for damaged W-beam guardrail are depicted in Figure 4, which the condition consider for support leg losses its connection to ground, hence a model of W-beam where a support leg removed was used to represent the damage.

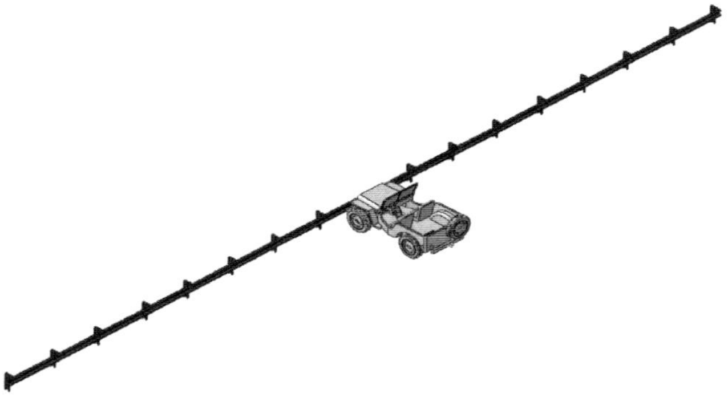

Figure 4. W-beam guardrail with single support leg removed, support damage.

For representing the impacting body, a medium scale vehicle for impact test, Jeep, was used (Figure 5). Model effective impact height and front bumper geometry (1,440 mm) meets the interest of the research as suggested in MASH. The model was imported from free-access CAD models [3]. The mass of the Jeep used is 2.27 tonne.

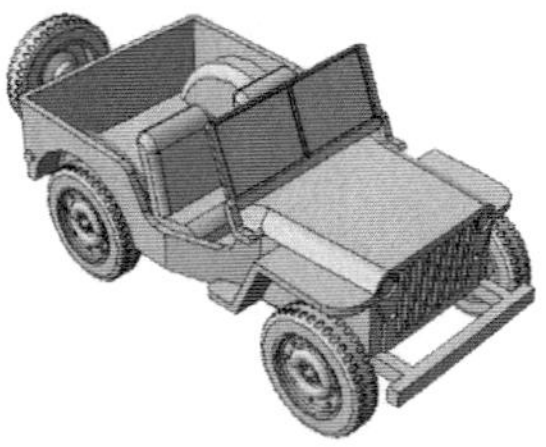

Figure 5. Jeep truck to represent vehicle model.

The crash simulation targeted an elastic-plastic material behavioral change. Hence, the elastic and plastic behavior of the selected material was defined in ABAQUS. To model the plastic behavior of the W-beam guardrail material, Johnson-Cook non-linear material modeling equation was adopted. Its material constants are listed in Table 2. For the Johnson-Cook model, the effect of temperature was neglected for this study.

$$\sigma = [A + B\varepsilon^n].\left[1 + Cln\left(\frac{\varepsilon^{\cdot}}{\varepsilon_0}\right)\right].\left[1 - \left(\frac{T-T_0}{T_m-T_0}\right)\right] \quad (1)$$

Where: σ: stress , A: Yield stress, B: Hardening constant, C: Hardening coefficient, n: Hardening exponent, $\varepsilon^{\cdot}$: strain rate, ε: strain, T_m: Melting temperature, T_0 : Transition temperature, m: Thermal softening exponent.

Table 2, Johnson-Cook Parameters for Special steel [14].

Material property	Symbol	Value	Unit
Yield Stress	A	792	MPa
Hardening constant	B	510	MPa
Hardening exponent	n	0.26	
Thermal softening exponent	m	1.03	
Melting temperature	T_m	1793	K
Transition temperature	T_{room}	293	K

The time increment T_{cr} was estimated by Eqn. (2) [2]

$$T_{cr} = \frac{l}{\sqrt{c_d}} = l * \sqrt{\frac{\delta}{E}} \tag{2}$$

Where: c_d: Material property, l: the smallest element dimension, E - Young's modulus, δ - Density

Using Eqn. (2), the minimum time increment was estimated to get reliable simulation result, where the geometrical values of mesh element and material data of the W-beam was used. The time increment for analyzing the W–beam under dynamic impact model yields approximately 2.62E-6. The total of 0.08 step time was used for the dynamic explicit. The increment time was chosen automatically and the stable increment estimator was selected globally. The maximum time increment was selected as unlimited with a time scaling factor of 1.

To define the interaction between deformable self-surface contact and contact between rigid and the deformable bodies, a general contact was defined. The penalty algorism was selected to specify the coefficient of friction, μ = 0.35, and hard contact to assert zero relative motion between surface after contact. The bolt and nut joints to connect guardrail in serious were not considered as a single entity for the guardrail rather a tie constraint was adopted for the connections.

The deformable bodies were discretized with reduced shell element, S4R. Fine mesh size was focused on the areas where the rigid body in contact first and to the area where the rigid bodies complete displaced for defined analysis period. The mesh sizes for deformable bodies were optimized to achieve the mesh convergence that would supports to get reliable results. As it can be seen in the Figure 6, the fined element size was 15 mm seen in center of the guardrail and the larger element size was 210 mm. The rigid body was discretized with R3D4 solid element, and the minimum element size was 100 mm.

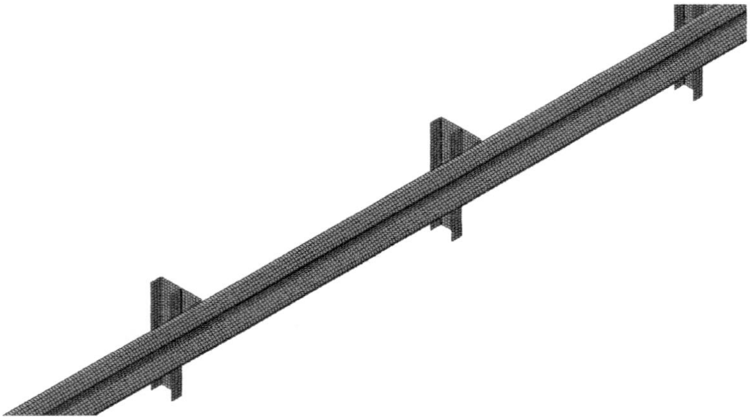

Figure 6. Discretized W-beam segment

The boundary conditions were defined using a connection between the W-beam support with the ground and the impact inertia condition. With the consideration of that 1120 mm of the support length are embedded and casted, an Encastered constraint condition. Supporting legs lower nodes were constrained with a rigid point to act as a rigid connection, see Figure 7.

For the impacting inertia, zero angle impact orientation was used for the study. The Jeep truck was targeted to impact the W-beam in full front face with a vehicle speed of 42.26 km/hr which is 90° vehicle impact speed of 100 km/hr. The vehicle impact speed magnitude is adopted from MESH standard. Figure 8 shows the arrangement of the vehicle arrangement and velocity vector of truck impact direction.

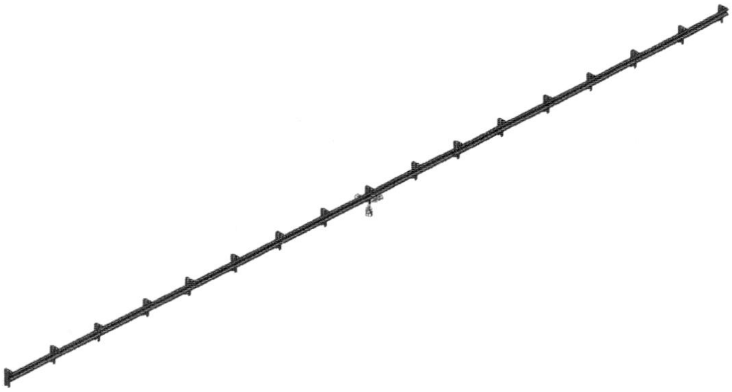

Figure 7. Fixed boundary condition to support leg applied with Rigid point constrained

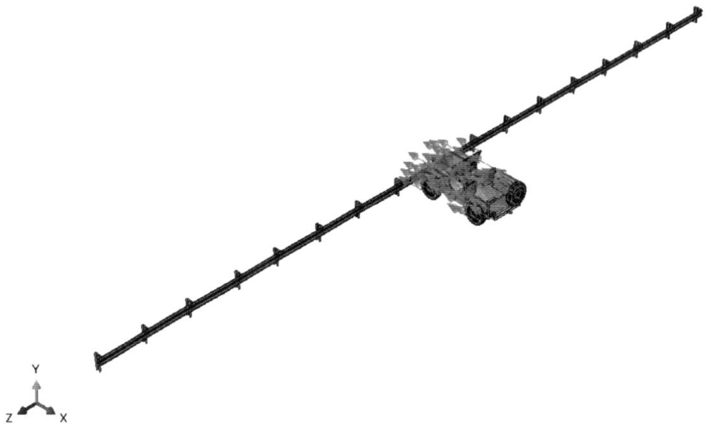

Figure 8. Impact speed orientation of the Jeep truck

4. Result and discussion

W-beam guardrail is one of the popular guardrails observed in dangerous roads or highways. In this study, W-beam guardrail was modeled using AASHTO design standard to simulate and study crash behavior due vehicle crash. Guardrail with single support damage (condition of removed support due previous crash or poor maintenance) was considered as a damaged W-beam guardrail model.

Herewith, results of energy balance that assure the reliability of crash simulation on the W-beam guardrail, results of internal and potential energy, displacements of W-beam guardrail with and without damages results are discussed.

4.1. Energy balance

Figure 9 illustrates the energy balance for the W-beam guardrail impacted with vehicle oriented at 90 degree and supported to check the reliability of the dynamic explicit FEA in simulating the crash behavior of the W-beam. Figure 9 (a) shows, W-beam guardrail without damage yielding the total energy been constant during impact period. The kinetic energy of the structure decreases gradually with time as the internal energy of the structure increases gradually. This condition supports to validate the simulation imitate crash behavior. Further, Figure 9(a) shows the artificial energy of the structure is in minimum range (<1%) compared to total energy that supports to justify minimum element distortion [5].

And Figure 9 (b) shows the simulation result of W-beam guardrail with SSupport damage in relation to the energy relationships. It illustrates the total energy constant across the impact period while the kinetic and internal energy illustrate inverse relations that justify the change of kinetic energy due to the vehicle crash with energy absorption by the W-beam guardrail in the system. Further, the same figure shows the artificial energy is in very low range compared to the total energy.

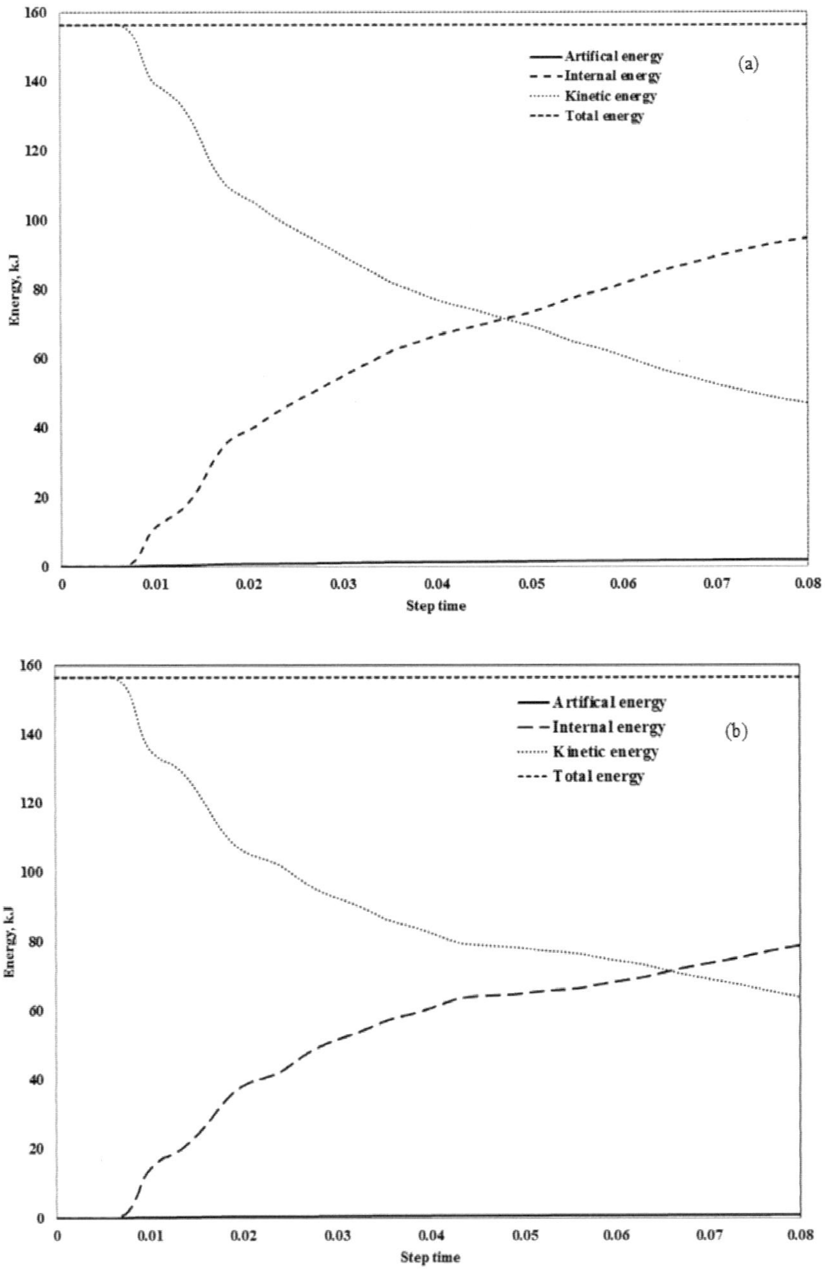

Figure 9. Energy balance for W-beam guardrail (a) without and (b) with damage

4.2. Internal and plastic energy

The internal and plastic energies for W-beam guardrail with and with damages were retrieved from the crash simulation to identify the energy absorption capacity of the guardrail. Results shows, for both models, energy absorption process through the crashing period, the internal energy been developing in conversion to the kinetic energy, and the plastic energy related with the permanent deformation of the material through the crashing. The results of damaged W-beam guardrail for internal and plastic energies are minimum compared to guardrail without damage, depicting the damage compromise its efficiency in relation to energy absorption capacity.

4.3. Displacement

The displacements of W-beam guardrails with and without damages were identified and illustrated in Figure 11. As it can be seen from Figure 11(a), the maximum displacement, 402.9 mm, of the W-beam guardrail without damage is observed close to the vehicle crash direction. While the maximum displacement for damaged ones identified with 459.9 mm, see Figure 11 (b). With the consideration of the identified displacements for both models that lies in the W1 Level of working width (W1 ≤ 0.6 m), the W-beam guardrail with damage displaced more, 14.15%, compared to W-beam guardrail without damage.

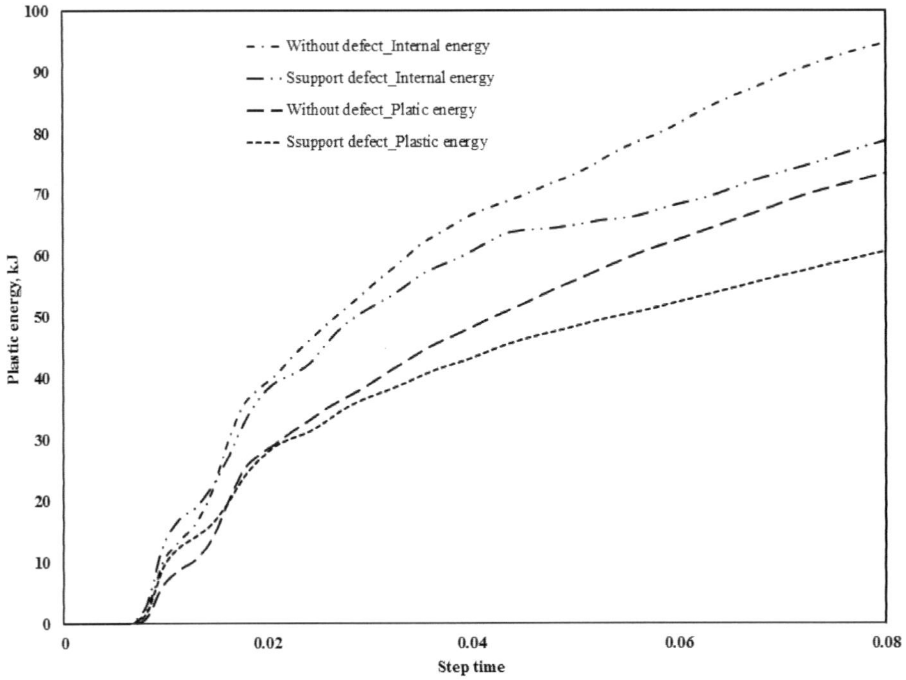

Figure 10. Internal and plastic energies results of the W-beam guardrails without and with damages.

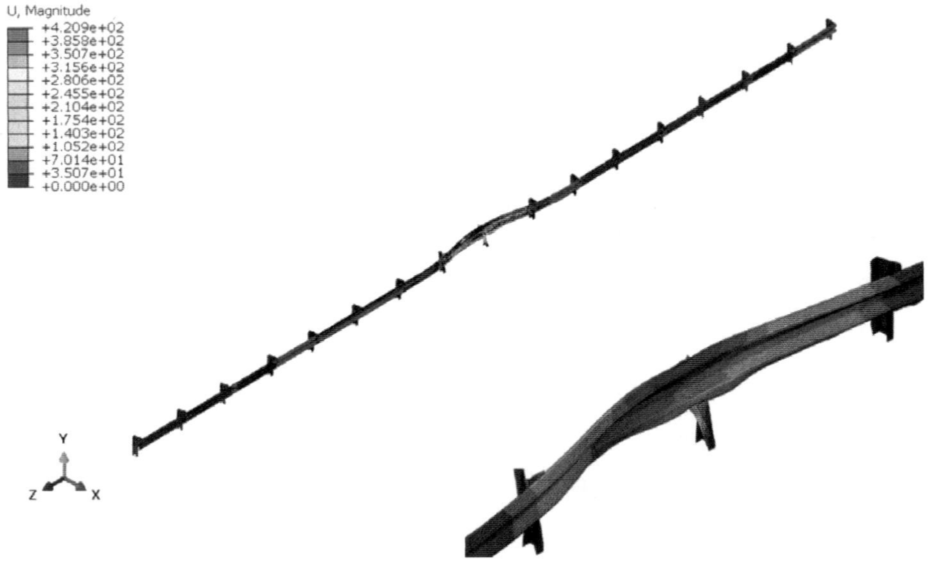

(a)

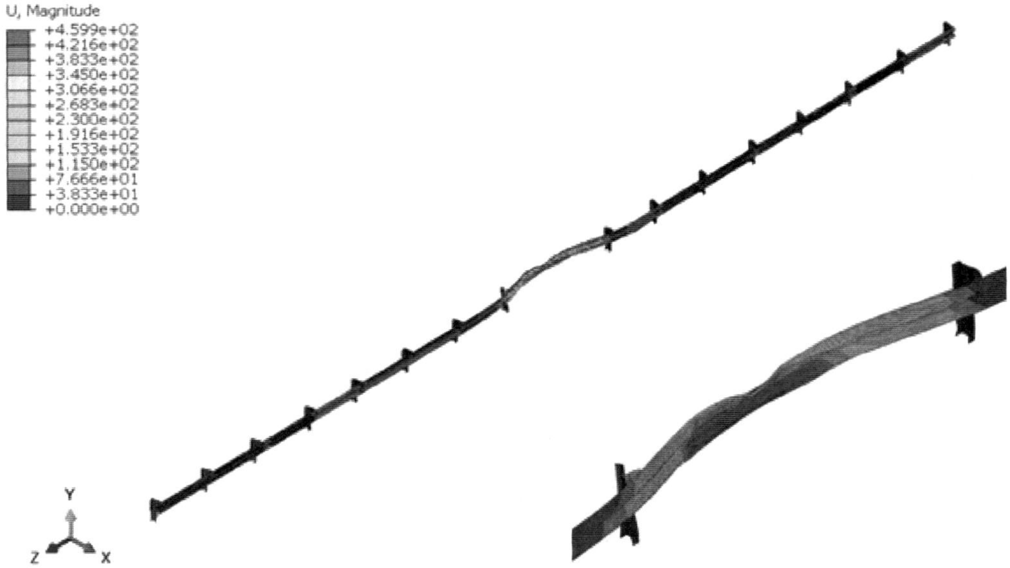

(b)

Figure 11. Overall displacement of W-beam guardrail (a) without and (b) with damage

4.4. Force versus time

Figure 12 shows the force versus time relation for guardrail without damage and with damage. As it can be seen from the graph, reaction force in relation to impact period shows a zigzag curve that shows the point of reaction exercise a variable load that reflect the variable internal load condition, stretching and bending. Further, the variable reaction loads asserts that the guardrail possess a plastic deformation that support for better energy absorption in impact period.

The identified variable loads of reaction on the guardrails depicted that the pick loads recorded from guardrail without damage is higher than the pick loads recorded for guardrail with damage. Besides, for both conditions, it is observed that the variable load is declining in its magnitude as the impact time is increasing, which confirm the kinetic energy is decreasing. Further, the figure shows the average force of guardrail without defect is higher than with defect (>40%) that indicate the defect free guardrail supports to resist the crash compared with the defected ones.

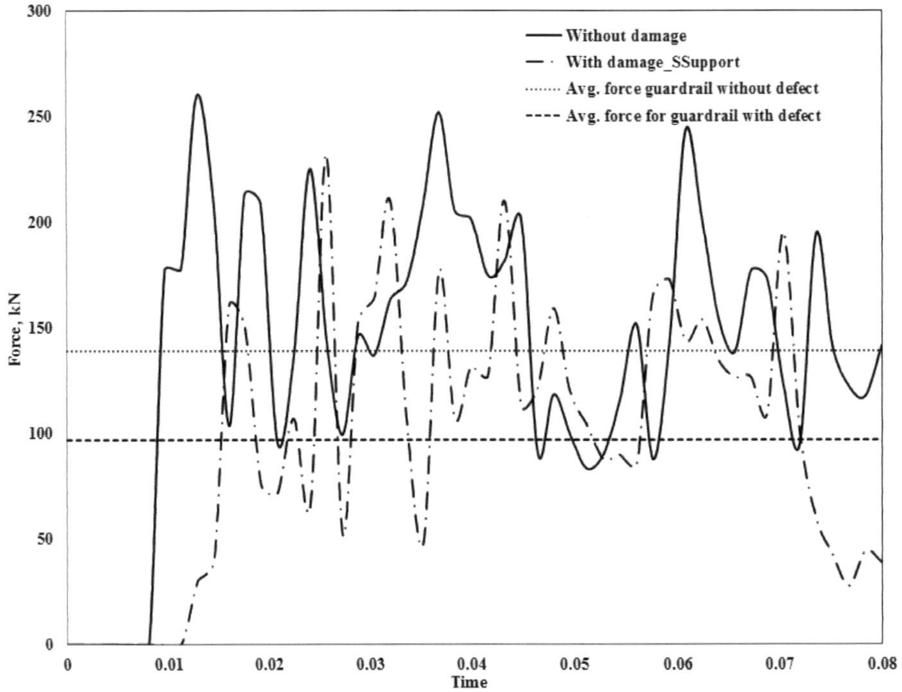

Figure 12. Force versus time for guardrail without and with defect

5. Conclusion

A non-linear finite element analysis was done to analyze the crash of Jeep truck on W-beam guardrail which is modeled using AASHTO design standard. For the crash test, for guardrail without and with damage, the standard MASH parameters for vehicle speed, impact angle, and vehicle inertia load and front bumper width was adopted.

The energy balance results show that the simulations for crash tests are reliable to investigate other performance criteria. The internal and plastic energies conversions for W-beam guardrail without damage shows higher compared to the damaged one, hence the energy absorption capacity can be considered to be compromised when damage on the guardrail processed. And more, the total displacement of the W-beam guardrail with damage increase by 14.15% compared with guardrail without damage, limits its future use under safe level of working width. Further, the average reaction force of the guardrail without damage (defect) is greater than 40% to guardrail with defect, that comply a lower resistant to impact.

Reference

[1] Michie J D, Calcite L R, and Bronstad M E 1971 Guardrail performance and design, National Cooperative Highway research Program Report 115, *Southwest Research Institute, San Antoni, Texas.*

[2] Ross H E , Sicking D L, Zimmer R A, and Michie J D 1993 Recommended Procedures for the Safety Performance Evaluation of Highway Features, *National Cooperative Highway research Program Report 350, National Academy Press, Washington, D.C.*

[3] https://obexsystems.com/about/en-1317-specification/ accessed August 28, 2023.

[4] Martin O S and Wekezer J W 1998 Crash impact analysis of the G2 guardrail, *FAMU-FSU College of Engineering Computer Impact Simulation Laboratory, TRB National Research Council*, Annual Meeting.

[5] Teng T L, Liang C L, and Tran T T 2016 Development and validation of a finite element model for road safety barrier impact tests, *Simul.: Trans. Soc. Model. Simul. Int.* **92**(6), 565-578.

[6] Dziewulski P and Stanisławek S 2019 The impact of forming processes on road barrier strength, *AIP Conf. Proc.* **2078**, 020003, https://doi.org/10.1063/1.5092006

[7] Ren Z, Vesenjak M 2005 Computational and experimental crash analysis of the road safety barrier, *Eng. Fail. Anal.* **12**, 963 – 973.

[8] Nevesa R R, Fransplassb H, Langsethb M, Driemeiera L., Alvesa M 2018 Performance of Some Basic Types of Road Barriers Subjected to the Collision of a Light Vehicle, *J. Braz. Soc. Mech. Sci. Eng.* 40, 274, https://doi.org/10.1007/s40430-018-1201-x

[9] Molan A M, Rezapour M andd Ksaibati K. 2019 Modeling traffic barriers crash severity by considering the effect of traffic barrier dimensions, *J. Mod. Transport.* 27**,** 141–151.

[10] Teng T L, Liang C L, and Tran T T 2015 Effect of various W-beam guardrail post spacing and rail heights on safety, *Advances in Mechanical Engineering* 7(11)**,** 1–16.

[11] AASHTO Standard Specifications for Transportation Materials and Methods of Sampling And Testing, 26th Edition, 2007.

[12] Dobrovolny C S, Arrington D R, Schulz N D, Xavier C B 2015 *Design and Finite Element Analysis of a MASH 31-inch W-Beam Guardrail System for Placement on 3H:1V Sloped Terrain Configuration (2014 WV-62)*, Test report 12-602951-00001, Texas A&M Transportation Institute, Texas.

[13] Dassault Systemes. Abaqus 6.14 Online Documentation: 9.3.2 Definition of the stability limit. Dassault Systemes, http://abaqus.sys.kth.se/v6.14/books/gsa/default.htm

[14] Plaxico C A, Hacket R M, and Uddin W, 1997, Simulation of a vehicle impacting a modified thrie-beam guardrail. *Trans. Res. Record*, **1599**(1), 1–10. https://doi.org/10.3141/1599-01

[15] Fitzgerald W J 2008 W-Beam Guardrail Repair: A Guide for Highway and Street Maintenance Personnel Report FHWA-SA-08-002 Federal Highway Administration Office of Safety, United States. https://rosap.ntl.bts.gov/view/dot/49224

IOP Conf. Series: Materials Science and Engineering 1294 (2023) 012029 doi:10.1088/1757-899X/1294/1/012029

An approach for fatigue life assessment of a road bridge based on measured corrosion and actual traffic loading

M T Gebremeskel, A T Cruz and N D Adasooriya*

Department of Mechanical and Structural Engineering and Material Science, Faculty of Science and Technology, University of Stavanger, Stavanger, Norway

* Contact Author: mudiyan.n.adasooriya@uis.no

Abstract. Fatigue and corrosion play a significant role in the aging of steel bridges, especially in a marine environment. An absence of fatigue strength curves for details which are exposed to corrosive environments, realistic traffic load models and accurate damage accumulation models increase the susceptibility of fatigue damage of steel bridges. In this paper, remaining fatigue life of a road bridge is assessed based on the measured corrosion wastage and more realis-tic traffic loads. Corrosion fatigue is also studied as it is one of the main factors that reduce the fatigue life and recently proposed fatigue strength curve for structural details exposed to corrosive environments is utilized for this study. A non-linear fatigue damage accumulation model is used for life assessment apart to the conventional damage model, Miner's rule. The calculated fatigue lives are compared and discussed in quantitative manner to identify the most conservative remaining life and corresponding methods.

1. Introduction

Bridges are subjected to multiple recurring traffic loads that may significantly fall below their structural resistance limit. Structural damage accumulated continually over a period of time results in localized and cumulative failure processes known as fatigue. It is well known that all bridges are subjected to environment assisted cracking (EAC) which includes three major types such as stress-corrosion cacking (SCC), hydrogen damage (HD) and corrosion fatigue (CF) due to different environment, structural and metallurgical factors [1-7].

The steel bridges are subjected to different aggressive corrosive environment, and it causes the time dependent loss of protective coating and material loss due to corrosion. When the steel is exposed corrosive environment and alternative cyclic stresses, then steel is subjected to corrosion fatigue. It is obvious that corrosion fatigue is significant compared to the damage from fatigue and corrosion [1-4,6-8,11]. The corrosion affects the material loss, and it changes the cross-sectional properties of the members due to changes of the surface roughness, irregularities, corrosion pits. The deterioration process caused by EAC will affect the integrity of the bridges by the degradation of material strengths. There may be a reduction in remaining fatigue life of the steel bridges due to changes in structural behavior and stiffness [11]. Fatigue is still less understood regarding the cause of formation and failure mechanism with respect the structural details even though fatigue is considered as one of the most critical form of failures of steel structures. The aging of bridges needs the attention as the replacement of the bridges are not economically viable and fatigue pays a significant role in their service life. Meanwhile much research shows that corrosion and fatigue are the main mechanism for deterioration.

There are a quite lot of research work have been done to study the interaction effect of corrosion and fatigue. Creating improved technologies for capturing the fatigue phenomena and undertaking a reliable assessment of the fatigue damage are necessary for steel bridges [5,8-14].

Main objective of this this paper is to spot and assess the importance of using an accurate fatigue assessment procedure which consists of accurate models for stress evaluation, S-N curves, and fatigue damage accumulation. The applicability and significance of the recently proposed generalized formula of the S-N curve is confirmed by applying to a case study road bridge. Further, a conventional method predicted fatigue lives are compared with recently developed assessment procedures which are utilized to study remaining life of the bridges.

2. Recently proposed fatigue damage models, strength curves and frameworks

2.1. Fatigue strength curve for structural details exposed to corrosive environments

The fatigue performance of steel components of these structures is affected by the exposed environment. Determination of fatigue strength of structural joints and constructional elements in corrosive media or corrosive environment has been studied. The recently proposed fatigue damage model aimed the fatigue strength of structural joints and constructional elements in corrosive environments and modification has been done by considering the design fatigue strength curves of constructional details and detail classes [9-11].

The recently proposed formula for fatigue strength of a corroded detail is shown as follows,

If $\Delta\sigma_{cor} \geq \Delta\sigma_{D,cor}$,

$$\Delta\sigma_{cor} = \Delta\sigma_D \left[N_{f,LCF}^{c} N_{f,CAFL}^{1/m}\right] N_R^{(-c-1/m)} \quad where\ c = \frac{log\left[\frac{\Delta\sigma_D}{\Delta\sigma_{D,cor}}\right]}{log\left[\frac{N_{f,CAFL}}{N_{f,LCF}}\right]} \tag{1}$$

If $\Delta\sigma_{cor} \leq \Delta\sigma_{D,cor}$

$$\Delta\sigma_{cor} = \Delta\sigma_{D,cor}\left[N_{f,CAFL}^{-\acute{c}}\right] N_R^{\acute{c}} \quad where\ \acute{c} = \frac{log\left[\frac{\Delta\sigma_{D,cor}}{\Delta\sigma_{L,cor}}\right]}{log\left[\frac{N_{f,CAFL}}{N_{f,VAFL}}\right]} \tag{2}$$

where $\Delta\sigma_D$ is the stress range at the fatigue curve slope changing point, corresponding to the $N_{f,CAFL}$ cycles. The $-{}^{1}/_{m}$ is the slope of the fatigue strength curve uncorroded details, where m is equal to 3 when $\Delta\sigma \geq \Delta\sigma_D$, equal to 5 when $\Delta\sigma_D \geq \Delta\sigma > \Delta\sigma_L$ and infinite when $\Delta\sigma \leq \Delta\sigma_L$, where $\Delta\sigma_L$ is the fatigue endurance limit of the detail corresponding to $N_{f,VAFL}$. $\Delta\sigma_{D,cor}$ is the stress range for $N_{f,CAFL}$ cycles at the intersection of the two corroded fatigue curves. $N_{f,CAFL}$ is the number of cycles to fatigue failures of uncorroded details at the intersection point [11].

The corrosive environment dependent parameters have been determined conservatively for corrosive environments such as urban and maritime environments and tabulated in the Table 1 and Table 2 [9-11].

Table 1. Parameters used in proposed fatigue strength curve of corroded details

Parameter	Constructional details in Eurocode	Constructional details in DNV code
$N_{f,LCF}$	10^4	10^4
$N_{f,CAFL}$	$5x10^6$	10^7
$N_{f,VAFL}$	10^8	10^8
$\frac{\Delta\sigma_L}{\Delta\sigma_D}$	0.549	0.631

Table 2. Mean values and conservative values used in fatigue strength curve of corroded details

Corrosion parameters	Constructional details in Eurocode				Constructional details in DNV code			
	Mean value	Conservat -ive value	Mean value	Conservat -ive value	Mean value	Conservat ,ive value	Mean value	Conservat -ive value
$\frac{\Delta\sigma_{D,cor}}{\Delta\sigma_D}$	0.497	0.308	0.641	0.536	0.46	0.27	0.61	0.50
$\frac{\Delta\sigma_{L,cor}}{\Delta\sigma_L}$	0.356	0.175	0.518	0.40	0.356	0.175	0.518	0.40

2.2. Non-linear fatigue damage model based only on S-N curve parameters

A recently proposed non-linear fatigue damage model, which take loading sequence effect more accurately than Miner's rule, is adapted in this paper to calculate cumulative fatigue damage by utilizing nominal and/or hot-spot approach [12]. The model is applicable for several engineering fields. The equation illustrates an effective and simple fatigue damage model which has been developed based on fatigue damage evolution curve. It gives a damage at a particular level.

$$D_i = 1 - \left[1 - \frac{n_i}{N_i}\right]^{\delta_i} = 1 - \left[1 - \frac{n_{(i+1),eff}}{N_{i+1}}\right]^{\frac{\delta_{i+1}}{\mu_{i+1}}} \tag{3}$$

where D_i is the damage at load level i when exposed to a given stress amplitude (or range) σ_i for n_i cycles. $n_{(i+1),eff}$ is the effective number of cycles corresponding to σ_{n+i} at level $i+1$. N_i and N_{i+1} are the cycle to failure number of cycles which can be taken from *S-N* curves given in design standards and regulations. The μ_i and δ_i are model parameters that depend on *N* and stress levels.

$$\delta_i = \frac{-1.25}{lnN_i} \tag{4}$$

$$\mu_i = \left(\frac{\sigma_{i-1}}{\sigma_i}\right)^2 \tag{5}$$

2.3. Life assessment framework/guidelines

The life assessment framework is recently proposed to be applied for existing steel bridges [7-11]. The newly introduced formula of S-N curve for corroded details and Miner's rule were used to find the remaining fatigue life which exposed to corrosive environment or EAC. In this study, previously proposed framework was modified by adapting recently proposed non-linear damage model mentioned in section 2.2 in-stead of Miner's rule. There are five steps of the framework. First step is to do the structural analysis to replicate the current state of the bridge. Critical element is found by using structural analysis and it's necessary to determine the stress spectrum of the identified critical element. The remaining fatigue life is calculated by using conventional approach and newly proposed method [7-11]. The current state of the structural details is categorized as corroded or uncorroded or /and in a corrosive environment.

3. Case study bridge: Description and its current state

The considered bridge was constructed in 1937 which is in Strand Municipality, Norway. The view of the bridge is shown in Figure 1. The bridge was partially destroyed by floods and rehabilitated and rebuilt in 1942 with modifications as per today. There are indications that bridge has exposed to corrosive environment and subjected to increased load cycles.

Figure 1. Considered road bridge

3.1 Geometrical information and material properties

The considered bridge has two simple spans of non-composite sections, each with an equal length of 19.5m and an end span of concrete T- beams with 12.7m in length as shown in Figure 2. There are two lane single carriageway and superstructure is supported by two pillars. The non-composite section consists of a reinforced concrete deck. The main steel girder consists of three evenly spaced rolled sections, designated as DIP 95 as shown in Figure 3. The total width of concrete deck has a total width of 5.82m with an average depth of 190mm as shown in the Figure 3.

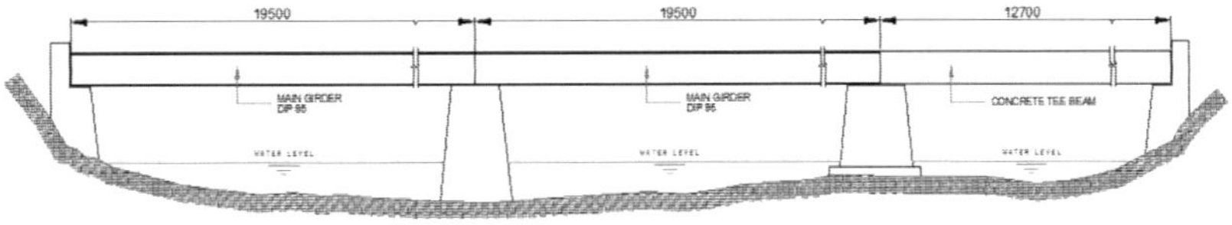

Figure 2. Longitudinal section of bridge

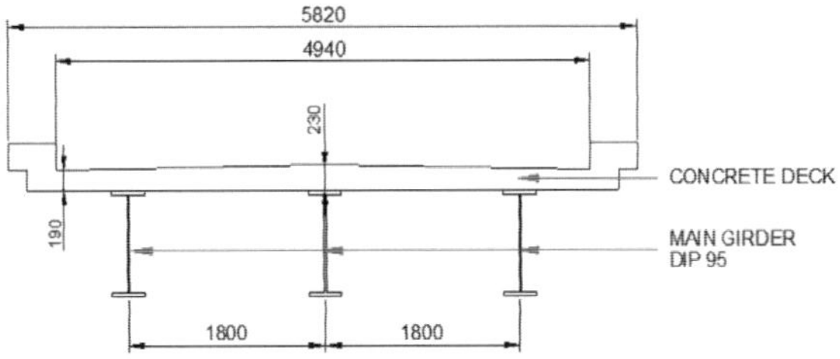

Figure 3. Cross sectional view and main girder steel beams (DIP 95)

The DIP 95 is doubly symmetric I shape plate girder beam with depth, flange width, web thickness and flange thickness 950mm, 300mm, 19mm and 36mm respectively. The cross-sectional properties such as gross area, 2nd moment of area about major axis and elastic section modulus about major axis are 390.55 cm^2, 572953 cm^4 and 12062 cm4 respectively. The grade of the steel is S275, where characteristic yield strength and modulus of elasticity 275 MPa and 200GPa respectively.

3.2 Degradation, damage and defects description

The bridge inspection reports, and current visual inspection results provides of coating loss and corrosion in the bridge girders as shown in Figure 4. It is reported that the bridge is exposed to uniform/patch corrosion and those are in the mid-span of the exterior girder and bottom surface of the top flange. In general, it is found that bottom of the top flange, bottom and top surface of the bottom flange have exposed to corrosion as shown in Figure 5. Visible cracks were not identified during the visual inspection of any steel portion of the bridge. A maximum of 4mm uniform corrosion is recorded in the midspan of the exterior girder. The thickness reduction of the uniform corrosion changes the structural stiffness by changing cross sectional properties. Therefore, degraded DIP 95 cross section at mid span is shown in Figure 5.

Figure 4. Bridge girders with corrosion

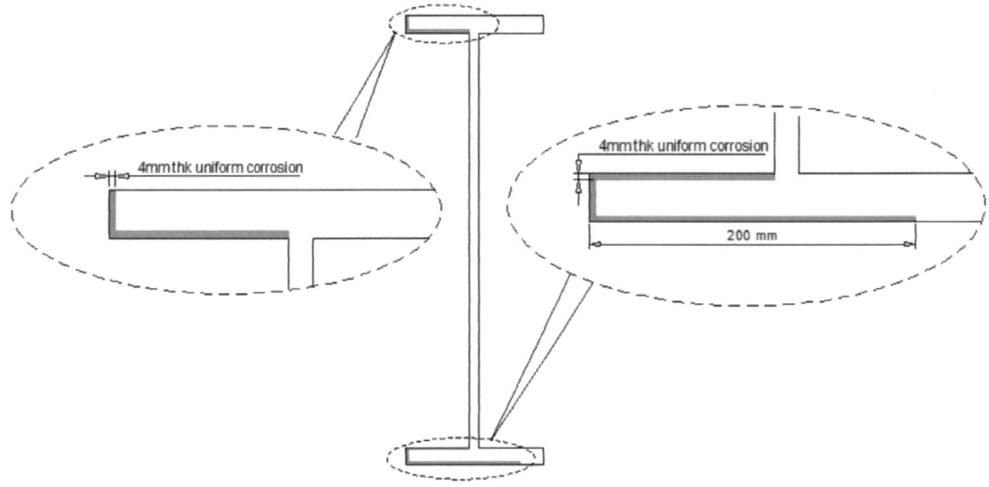

Figure 5. Corroded DIP 95 cross section at midspan of exterior girder with about 4mm thickness reduction at the top and the bottom surfaces.

4. Structural analysis of the bridge

4.1 Load models

Eurocode defines five different fatigue load models [13]. Fatigue load model 4 (FLM 4) is used to calculate the total fatigue damage accumulated through the design service life of the bridge as recommended for road bridges [13]. The total number of lorries crossing the bridge per year is 125 000 which was taken from Euro-code based on traffic category 3 which is main roads with low flow rates of lorries. The axle spacing, equal axle loads and percentages of heavy traffic for five lorries were taken from the Table 4.7 of Eurocode by considering medium distance road category [14].

In addition to Eurocode FLM 4, an alternative fatigue load model (AFLM) pro-posed for bridges in Norway [15] is used for fatigue life assessment. The AFLM consists of 5 lorries in FLM4 and additional six other light vehicles. The load and axle spacings of additional vehicles are shown in Table 3.

Table 3. Additional vehicle details of AFLM

Vehicle type	Axle spacing	Axle load
Combi	2.5	12
		12
Sedan	2.9	16
		16
Station	3.0	19
		19
SUV/Minivan	3.1	21
		21
Pickup/Van	3.2	29
		29
Tractor/Smaller trucks	3.5	64
		64

Probabilities for the six remaining vehicle types have been derived from the distribution of registered cars in Norway. Based on the annual average density of traffic (AADT) and the route's location, public road administration given traffic data estimates that 5% to 20% of all vehicles are heavy vehicles. FLM4 has an 8.56% share of heavy vehicles based on the expected AADT and the predicted number of trucks. FLM4 Between 8 and 15% of the five FLM4 cars on the bridge is likely to be ac-counted for by the AADT by categorizing to three scenarios, such as Scenario 1: AFLM with 4% heavy vehicles, Scenario 2: AFLM with 8.56% heavy vehicles, and Scenario 3: AFLM with 15% heavy vehicles. Probability of occurrence of different scenarios are shown in Table 4. The total number of vehicles crossing the bridge per year is 365 000.

Table 4. Probability of occurrence for the scenarios AFLM (1) One lane loaded (2) Both lanes loaded

Vehicle type	Traffic percentage of AFLM (%)		
	Scenario 1	Scenario 3	Scenario 3
Combi (1)	10.56	10.0584	9.35
Sedan (1)	10.56	10.0584	9.35
Station wagon (1)	9.12	8.6868	8.075
SUV/Minivan (1)	7.92	7.5438	7.0125
Pickup/Van (1)	6.00	5.715	5.3125
Tractor/Smaller trucks (1)	3.84	3.6576	3.40
Lorry 1 FLM4 (1)	0.80	1.712	3.00
Lorry 2 FLM4 (1)	0.20	0.428	0.75
Lorry 3 FLM4 (1)	0.60	1.284	2.25
Lorry 4 FLM4 (1)	0.10	0.214	0.375
Combi (2)	10.56	10.0584	9.35
Sedan (2)	10.56	10.0584	9.35
Station wagon (2)	9.12	8.6868	8.075
SUV/Minivan (2)	7.92	7.5438	7.0125
Pickup/Van (2)	6.00	5.715	5.3125
Tractor/Smaller trucks (2)	3.84	3.6576	3.40
Lorry 1 FLM4 (2)	0.80	1.712	3.00
Lorry 2 FLM4 (2)	0.20	0.428	0.75
Lorry 3 FLM4 (2)	0.60	1.284	2.25
Lorry 4 FLM4 (2)	0.10	0.214	0.375

4.2 Stress evaluation

The current bridge has two spans, and a single-span bridge was modelled. The bridge's behavior during its service life was done by the structural behavior under moving load. The influence line analysis was done for each single passage of vehicle and hence bending moments are established based on the influence line during the passage of each lorry, which will result in a stress history at the mid-span where severe corrosion is reported. Fatigue assessment for the current bridge classified as a safe life method adopting the recommended value for partial factor for fatigue (γ_{mf}) is 1.35. A 3m with of one notional lane was considered and therefor width of remaining area is 1.94m. For assessing the local effect with notional lanes, the single lorry is placed on the span cantered in the notional lane. One lane loading is analysed and obtained load distribution factor to exterior and interior girders are 0.55 and 0.14 respectively. The calculation of the nominal stress ranges of exterior girder involves multiplying the highest response by the distribution factor and the dynamic amplification factor of 1[12,16]. The result is then divided by the related section modulus of DIP 95 (corroded/uncorroded condition). The obtained stress ranges are shown in Table 5.

Table 5. Nominal stress ranges

Vehicle Type	ΔM_i (kNm)	Uncorroded state		Corroded state	
		$\Delta\sigma_i$ (MPa)	$\Delta\sigma_R$ (MPa)	$\Delta\sigma_i$ (MPa)	$\Delta\sigma_R$ (MPa)
Combi (1)	56.10	6.05	8.16	6.60	8.91
Sedan (1)	73.04	7.87	10.63	8.60	11.60
Station wagon (1)	73.04	7.87	10.63	8.60	11.60
SUV/Minivan (1)	94.60	10.20	13.76	11.13	15.03
Pickup/Van (1)	130.00	14.01	18.91	15.30	20.65
Tractor/Smaller trucks (1)	281.60	30.35	40.97	33.14	44.74
Lorry 1 FLM4 (1)	453.20	48.84	65.94	53.34	72.01
Lorry 2 FLM4 (1)	707.85	76.29	102.99	83.31	112.47
Lorry 3 FLM4 (1)	856.35	92.29	124.60	100.79	136.06
Lorry 4 FLM4 (1)	657.25	70.84	95.63	77.35	104.43
Lorry 5 FLM4 (1)	694.65	74.87	101.07	81.75	110.37
Combi (2)	56.10	6.05	8.16	6.60	8.91
Sedan (2)	73.04	7.87	10.63	8.60	11.60
Station wagon (2)	73.04	7.87	10.63	8.60	11.60
SUV/Minivan (2)	94.60	10.20	13.76	11.13	15.03
Pickup/Van (2)	130.00	14.01	18.91	15.30	20.65
Tractor/Smaller trucks (2)	281.60	30.35	40.97	33.14	44.74
Lorry 1 FLM4 (2)	453.20	48.84	65.94	53.34	72.01
Lorry 2 FLM4 (2)	707.85	76.29	102.99	83.31	112.47
Lorry 3 FLM4 (2)	856.35	92.29	124.60	100.79	136.06
Lorry 4 FLM4 (2)	657.25	70.84	95.63	77.35	104.43
Lorry 5 FLM4 (1)	694.65	74.87	101.07	81.75	110.37

5. Fatigue life assessment

5.1 Detail categories and fatigue strength curves

The main girder cross section is DIP 95 and it is a rolled I section. The main girder is subjected bending and therefore the bottom flange subjected to tensile stresses. If the fatigue stresses of the bottom flange are due to in-plane bending, the detail category 160 with constructional detail number 2 can be chosen according to Eurocode 3 [17]. The corresponding detail category-based S-N curve is shown Figure 6. This S-N curve is used for fatigue life assessment of uncorroded details of the girder.

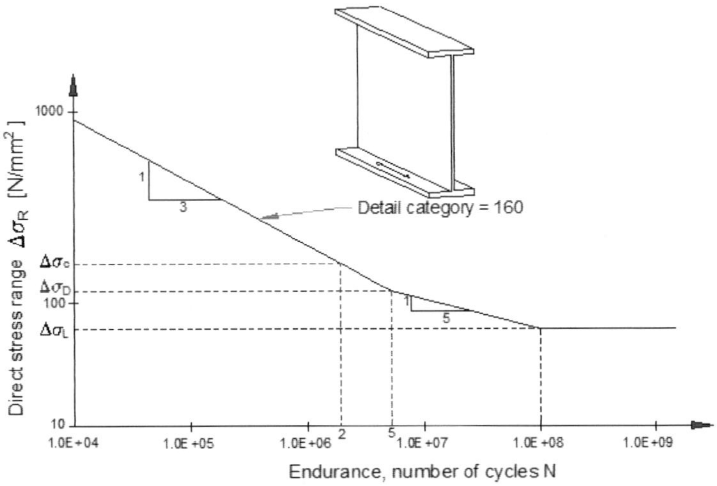

Figure 6. Design S-N curve for detail category 160 which is used for uncorroded DIP 95 girder

Fatigue strength curve for structural details exposed to corrosive environments, which is corroded DIP 95 girder, was obtained from the recently proposed formula [9-11] which formulation was discussed in the section 2.1. The S-N curve used for corroded details are shown in Figure 7, where $\Delta\sigma_D = 0.737 \times 160$ MPa.

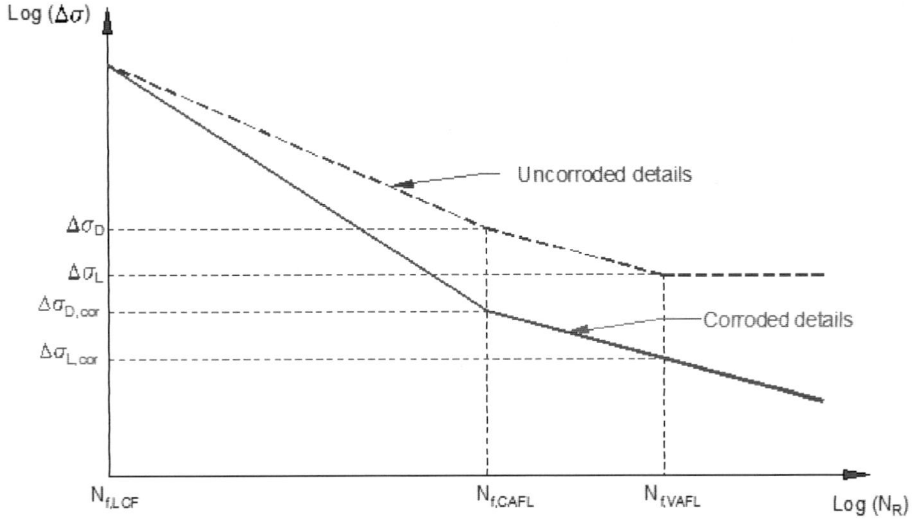

Figure 7. S-N curve used for corroded DIP 95 girder

5.2 Fatigue life calculation

Fatigue life of the critical location of the girder was calculated by four different methods [19-23].

- Method 1: Life assessment using stress ranges due to FLM4, both corroded and uncorroded S-N curves and Miner's rule.
- Method 2: Life assessment using stress ranges due to AFLM, both corroded and uncorroded S-N curves and Miner's rule.
- Method 3: Life assessment using stress ranges due to FLM4, both corroded and uncorroded S-N curves and non-linear damage model described in section 2.2 [19-22].

- Method 4: Life assessment using stress ranges due to AFLM, both corroded and uncorroded S-N curves and non-linear damage model described in section 2.2 [19-22]. All the calculated fatigue lives of the bridge are shown in Table 6.

Table 6. Fatigue lives of the bridge girder

Fatigue load model	Remaining fatigue life (years)			
	Miner's rule		Non-linear damage model [Section 2.2]	
	Uncorroded member	Corroded member	Uncorroded member	Corroded member
FLM4	284	94	143	85
AFLM-Scenario 1	681	146	343	108
AFLM-Scenario 2	318	109	160	90
AFLM-Scenario 3	181	93	91	82

5.3 Results and discussion

Stress analysis results shows that that bridge exposed to alternative stresses and vulnerable to corrosion fatigue as there are reported corrosion damage and it is in corrosive marine environment. Table 6 shows that there is significant change of fatigue lives depend on the method that used for analysis and life assessment. A 49.65% and 9.57% difference can be seen when comparing Method 1 and 3 for both uncorroded and corroded members respectively. There are 49.63%, 49.69% and 49.72% differences when compare Method 2 and 4 for Scenario 1, 2 and 3 respectively for uncorroded members. Similarly, 26.03%, 17.43% and 11.83% differences can be observed when compare Method 2 and 4 for Scenario 1, 2 and 3 respectively for corroded members. Comparison of all the methods has numerous uncertainties associated with fatigue life evaluation methods that may lead to different outcomes.

6 Conclusions

Remaining fatigue life evaluation of a specific road steel bridge was performed by using, (i) two different fatigue load models (i.e. FLM4 and AFLM), (ii) simulating structural degradation based on the reported corrosion damage (i.e. stress analysis using reduced cross section and using recently proposed S-N curve of corroded de-tails), and (iii) using conventional Miner's damage rule and recently proposed non-linear damage accumulation model. The fatigue lives calculated based on above methods were compared and discussed.

Results shows that Method 4-scenario 3 gives the lowest remaining fatigue life which is 82 years under the condition that there will not be any increase of loading, and/or any change of current corrosion state and corrosive environment. This indicates the using scenario 3 (i.e. AFLM with 15% heavy vehicles) of the discussed alternative fatigue load model with non-linear fatigue damage accumulation model provides conservative prediction to remaining fatigue lives for both corroded and uncorrected states. The differences of calculated fatigue lives between each method, emphasize the need for precise S-N curves which represent degraded strength state of the critical detail, realistic load models and an accurate fatigue damage accumulation model for aging bridges to implement conservative judgement remaining life. In addition, the actual traffic load measurements can be considered to provide more accurate response ranges.

Consideration could be given to a comprehensive examination of the bridge's cur-rent condition to propose strategies for life-extension and bridge maintenance. In addition, measurements of the actual traffic load can be undertaken to provide more accurate bridge response ranges.

References

[1] Adasooriya N. D, Pavlou D and Hemmingsen T 2020 Environment-assisted corrosion dam-age of steel bridges: a conceptual framework for structural integrity, *Corros. Rev.* **38**(1), 49-65.

[2] Nada Al-Eidan 2015 What are the most dangerous types of corrosion and does the dangerous vary depending on the type of metal?, [Researchgate post].

[3] Gibson stainless & specialty INC. 2017 Corrosion types and prevention, Available at: https://www.gibsonstainless.com/types-of-corrosion.html (Accessed: 2023.10.20)

[4] Revie R.W: 2008 *Corrosion and corrosion control an introduction to corrosion science and engineering.* John Wiley & Sons.

[5] Al-Emrani M and Mustafa A 2014 *Fatigue design of steel and composite bridges. Göteborg*: Chalmers Reproservice.

[6] Griggs-Jr F: Eads Bridge at St. Louis, 2017 *Structure Magazine*, 18-22.

[7] Kreislova K, and Geiplova H 2012 Evaluation of corrosion protection of steel bridges, *Proce-dia Eng.* **40**, 229-234.

[8] Kulicki J M, Prucz Z, Sorgenfrei D. F, and Mertz D R 1990, Guidelines for evaluating cor-rosion effects in existing steel bridges, *Transportation Research Board*, NRC, Washing-ton, D.C.

[9] Adasooriya N D, Pavlou D, and Hemmingsen T 2020, Fatigue strength degradation of cor-roded structural details: A formula for S-N curve, *Fatigue Fract Eng Mater Struct.*, **43**(6), 1199-1213.

[10] Adasooriya N. D, Pavlou D, and Hemmingsen T 2020, S-N curve for riveted details in corro-sive environment and its application to a bridge, *Fatigue Fract Eng Mater Struct.*, **43**(4), 721-733.

[11] Adasooriya N D 2020, Structural integrity of steel bridges: Environment-assisted cracking. *PhD Thesis,* University of Stavanger, Norway.

[12] Aeran A 2019, Life extension of offshore structures: A conceptual framework and fatigue damage models. *PhD Thesis,* University of Stavanger, Norway.

[13] NS-EN 1991-2:2003+NA:2010. Eurocode 1: Actions on structures part 2: Traffic loads on bridges.

[14] Sandviknes J S S 2021, Environment-assisted fatigue of steel bridges: A conceptual frame-work for structural integrity/life assessment. *Master's thesis*, University of Stavanger, Norway.

[15] Sæstad A, Ersdal G and Siriwardane S. C 2016, Comparison of deterministic and probabilistic fatigue assessment methods: A case study of a road bridge, *Proc. IABSE Congress Challenges in Design and Construction of an Innovative and Sustainable Built Environment*, Sept. 21-23, Stockholm, Sweden, 275-283.

[16] Aeran A, Siriwardane S.C, Mikkelsen O, and Langen I 2017, An accurate fatigue damage model for welded joints subjected to variable amplitude loading. *In IOP Conf. Se-ries: Mater. Sci. Eng.* **276**, 012038. IOP Publishing.

[17] NS-EN 1993-1-9:2005 + NA: 2010.Eurocode 3: Design of steel structures – part 1-9 Fatigue

[18] Siviero E. and Pavan R 018, Assessment of existing steel bridges: codes and standard, *IOP Conf. Series. Mater. Sci. Eng.* **419**(1).

[19] Siwowski T 2015, Fatigue assessment of existing riveted truss bridges - case study, *Bull. Pol. Acad. Sci. Tech. Sci.* **63**(1), 125-133.

[20] Siviero E and Pavan R 2018, Assessment of existing steel bridges: codes and standard. *In IOP Conf. Series: Mater. Sci. Eng.*, **419**, 012006. IOP Pub-lishing.

[21] Oehme P 1989, Damage analysis of steel structures. *In IABSE Proc.*, 139–189.

[22] Ye X.W, Su Y.H, and Han J 2014, A state-of-the-art review on fatigue life assessment of steel bridges. *Math. Probl. Eng.* **2014**, https://doi.org/10.1155/2014/956473.

IOP Conf. Series: Materials Science and Engineering 1294 (2023) 012030 doi:10.1088/1757-899X/1294/1/012030

Non-linear buckling analysis of ship hull stiffened panels

E Hacihamud[1*], K Dahl[2], O Gabrielsen[2], H G Lemu[1] and S C Siriwardane[1]

[1] Department of Mechanical & Structural Engineering and Materials Science, University of Stavanger, Norway

[2] Department of floating structures, DNV AS, Stavanger, Norway

*Contact author: enesalhh@gmail.com

Abstract. Stiffened panels are crucial in maritime applications, where buckling is a primary structural failure concern. Typically, assessing the buckling capacity of these panels relies on established guidelines like DNV-RP-C201 and NS-EN 1993-1-5. Alternatively, non-linear finite element methods, guided by DNV-RP-C208, can be used. However, there's a significant research gap in calibrating buckling capacities for large ship-hull stiffened panels. This paper bridges that gap by employing non-linear finite element analysis to evaluate the buckling capacity of ship hull stiffened panels and compare the results with DNV-RP-C201. It establishes a capacity benchmark by thoroughly comparing results for a plated panel. Subsequently, it develops a validated model using ABAQUS, introducing material and geometric nonlinearities, including a mid-stiffener span imperfection ranging from 0.06% to 1% of the stiffener span length for 99.7% calibration. The validated non-linear model then assesses the impact of holes on buckling capacity under uniaxial and gravity loads, revealing a clear correlation between reduced buckling capacity.

1. Introduction

Stiffened panels find widespread applications across various industries, with a particular emphasis on offshore and maritime sectors. Their popularity stems from their exceptional capacity to withstand heavy loads while maintaining a favorable stiffness-to-weight ratio. Given that buckling represents a prevalent failure mode in stiffened panels, extensive research efforts have been dedicated to its study over the years. This research encompasses analytical approaches, experimental testing, and the utilization of non-linear finite element (NLFE) methods.

Several standards such as DNV-RP-C201 [1] and NS-EN 1993-1-5 [2] provide design guidelines against failure by buckling. Based on recommendations of these and similar standards, it is possible to predict failure capacities of plated structures, though the scope of these standards may not be included in some cases, where instead NLFE can be used to determine the buckling capacity. DNV-RP-C208 [3] is a standard developed by NLFE as a recommended practices for this last case.

As real stiffened panels have geometrically imperfections introduced from fabrication methods and tolerances, introducing equivalent geometric imperfections in order to predict the buckling capacity correctly. The DNV-RP-C208 [3] recommended practice provides guidance to geometric imperfection amplitudes. It is concerned with identifying the characteristic resistance of a structure or section of a structure to meet the DNV criteria for ultimate strength in DNV recommended practices DNV-RP-C201 [1].

Conducting non-linear buckling analysis using recommendations of DNV-RP-C208 [3] does not imply that the aim is to replace the determination of structural buckling resistance compared with the

IOP Conf. Series: Materials Science and Engineering 1294 (2023) 012030 doi:10.1088/1757-899X/1294/1/012030

traditional standards, but to address the cases that are not within the limitations of the standards. It is crucial to consider the statistical variation of the different parameters, when using NLFE method to determine buckling resistance. This can be achieved by comparing the FE capacities with experimental capacities if physical testing could be performed.

In the past years, a significant amount of research has been conducted on the development and formulation of the ultimate limit strength for buckling capacities. Paik et al. [4] derived sets of such formulations for the steel plate elements that are under four load components such as compression/tension, edge shear, and lateral pressure loads assuming that the plates are simply supported along all edges. Cho et al. [5] used a simplified numerical approach to detect the structural behavior under combined loads. A parametric study was then done to determine the ultimate strength of the stiffened plates under different cases of loading. The results were then studies using regression approach to find the strength formulations, which provided acceptable results with DNV (Plate ultimate limit state) standard and ABAQUS predictions. The aim of Cho's study [5] was to develop formulas that predict the ultimate strength of stiffened plates under the influence of combined axial compression, transverse compression, shear force and lateral pressure loadings.

Ozguc et al. [6] adopted a simple design formula to evaluate the buckling strength of stiffened panels considering both the welding-induced residual stresses and the geometrical imperfections. Formulas were proposed for use in investigation of ship panel geometries using NLFE analysis and validate the results. For instance, Zhang [7] focused on panels under compression and developed a semi-analytical formula to predict ultimate strength under compression (i.e. buckling capacity). Recently, Ozguc [8] utilized the NLFE code named ADVANCE ABAQUS, where an imperfection sensitivity work of a stiffened deck panel on an FPSO vessel is additionally accounted for. In-plane bi-axial compression was explored/studied in two orthogonal directions in the case study. The obtained buckling capacities of the stiffened panels are compared with DNV PULS (Panel Ultimate Limit State) buckling code. The capacities estimate from ADVANCE ABAQUS and DNV PULS code are found to be highly similar.

Though there are several research studies, there is a need for more studies on achieving the calibration of compression of buckling capacities which are obtained by the NLFE method and analytical buckling capacities given in the traditional standards. This is especially important for complex geometries, such as a large ship hull stiffened panels with openings and other irregularities where nonlinearities and geometric deformation are difficult to predict and those have a significant effect on the buckling capacity. However, buckling capacity analysis using NLFE method has few challenges such as (i) introducing applicable geometrical nonlinearities, (ii) identifying issues with material nonlinearity and evaluating the consequences of large strains, (iii) defining boundary conditions that represent the real problem appropriately and (iv) introducing a combined local and global imperfections.

This study aims to apply Nonlinear Finite Element Analysis (NLFE) to assess the buckling capacity in a specific design scenario, specifically, a stiffened panel with substantial openings. Traditional methods do not adequately address this situation. To begin the paper, a stiffened panel with a geometry that can calculate buckling capacity analytically (i.e., using DNV-RP-C201) is chosen. Finite element modeling (FEM) is then conducted for this selected stiffened panel geometry, incorporating nonlinear material properties and accounting for geometric imperfections. The FEM is modified by adjusting the patterns and amplitudes of imperfections, thus yielding buckling capacities. Finally, significant openings are introduced into the nonlinear finite element model of the stiffened panel, and the reduced buckling capacity of the panel due to these large openings is computed.

2. Finite element model description of ship hull stiffened panel

A portion of ship hull stiffened panel (Figure1(a)), which is a geometry commonly used in North Sea, is selected for non-linear buckling analysis. Geometry, material properties, applied loads and boundary conditions, which were used for finite element model description of the selected portion, are described in this section.

IOP Conf. Series: Materials Science and Engineering 1294 (2023) 012030 doi:10.1088/1757-899X/1294/1/012030

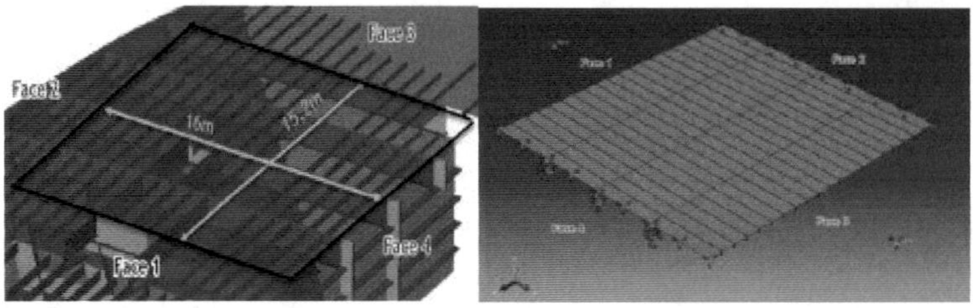

Figure 1. a) Real ship hull stiffened panel model b) FE ship hull model

The selected portion of ship hull stiffened panel, which is a rectangular panel of 15.2 m × 16 m, consists of 19 equally spaced bulb flats stiffeners and three equally spaced girders as shown in Figure 1(a). The geometrical data of the considered panel are shown in Table 1.

Table 1. Geometrical data of FE model

Parameters	Values
(a) Plate	
Plate thickness	15 mm
Plate length	16000 mm
Plate width	15200 mm
(b) Stiffeners geometrical properties	
Profile type	Bulb flats
Number of stiffeners	19
Stiffener length	15200 mm
Stiffener Span	3800 mm
Stiffener spacing	800 mm
Stiffener height	240 mm
Stiffener web thickness	10 mm
Stiffener Bulb width	39.5 mm
Stiffener Bulb thickness	28.8 mm
(c) Girders geometrical properties	
Profile type	Girder T
Number of Girders	3
Girder length	16000 mm
Girder spacing	3800 mm
Girder web height	930 mm
Girder web thickness	12.5 mm
Girder Flange width	450 mm
Girder Flange thickness	30 mm

The material of the panel is S235 and the non-linear material behavior is considered for the analysis. True stress versus strain relation, given by DNV recommended practice, is used for the analysis and non-linear properties of S235 were selected from the same code of practice for thickness less than 16 mm [3].

The boundary conditions and constraints under consideration are illustrated in Figure 1 and summarized in Table 2. To elucidate, a force is applied along the x-axis to Face 1, permitting unrestricted translation in that direction. In contrast, displacement in the y-direction is restricted to prevent any out-of-plane movement of the plate. Moreover, the plate is assumed to be firmly affixed to a girder at Face 1, thereby constraining rotation around the z-axis (Rz). Face 2, 3, and 4 are assumed to be immovable, as they are connected to both pillars and vertical plates. Displacement in the Z-direction at Faces 1, 2, and 3 is allowed to mimic transverse deformation due to Poisson's effect. Finally, translational displacement along the x-direction at Faces 2 and 4 is unimpeded, as it aligns with the loading direction.

Table 2. FE model boundary conditions

Face	Constrained degrees of freedom
Face 1	Y, Rz
Face 2	Y, Rx, Ry, Rz
Face 3	X, Y, Rx, Ry, Rz
Face 4	Y, Z, Rx, Ry, Rz

3. Non-linear analysis of stiffened ship hull panel using finite element method

Finite element analysis for buckling capacities of geometrically imperfected stiffened ship hull panel is presented in this section in detail. General purpose Finite Element analysis (FEA) program ABAQUS [9] was utilized for this non-linear analysis and the Riks method, which is generally used to simulate geometrically non-linear unstable behavior of structures, was employed. Details of the FE mesh, the considered assumptions the geometrical imperfection and the obtained buckling capacities are presented in this section.

3.1. FE Mesh and assumptions

To conduct the FE mesh, S4R which is a linear quadrilateral shell element was used because shell elements enable specification of the membrane stress components within each element easier. Moreover, the nodes of the element are located at the mid-thickness of each element resulting in no element mesh assigned to the thickness layers [10]. The mesh was generated using the auto meshing tool of ABAQUS. A mesh sensitivity study was performed by few different mesh samples to decide acceptable mesh size to ensure the accuracy of the stress concentrations. Finally, a finite element model meshed with 49703 elements with an average aspect ratio of 1.45 was selected by considering both convergence of the response and computational power.

Furthermore, the assumption used in the analysis are

(i) perfectly constructed panel before introducing geometrical imperfections,

(ii) straight and continuous stiffeners throughout the girders, which are involved with their full moment capacity on the stiffeners- girders crossing point,

(iii) No buckling in the girders because of the compression stress in the stiffener's direction. This assumption presumes that the girder's function is only to reduce the buckling length of the longitudinal stiffeners and

(iv) perfect contact and stress distribution between the different components of the panel.

3.2. Geometrical imperfection of the panel/model

Perturbations in the geometry was employed to introduce imperfections to the model, while the imperfections were defined by applying linear superposition of multiple eigenmodes. Then the analysis of the eigenmode was done on the ideal structure. Geometrical imperfections were imported using ABAQUS´s built-in features by superposing the weighted mode shapes. Thereafter, the displacements of the nodes from the eigenmode value were multiplied by scale factor to determine the magnitude of the nodes´ coordinate alteration that would provide the desired geometrical imperfections for the non-linear analysis model [9].

The DNV recommended practice [3] provides guidelines to generate geometrically imperfected shapes/patterns based on eigenmodes of the panel. The imperfections patterns recommended by DNV-RP-C208 are divided into global and local imperfections. The local form of imperfections are the out of plane deformation of the plane plate between transverse stiffeners as shown in Figure 2 (left), while the right-hand side figure shows the longitudinal out of plane deformation along the direction on the longitudinal stiffener that are defined as the global imperfections. The recommend imperfection amplitude/values are show in Table 3.

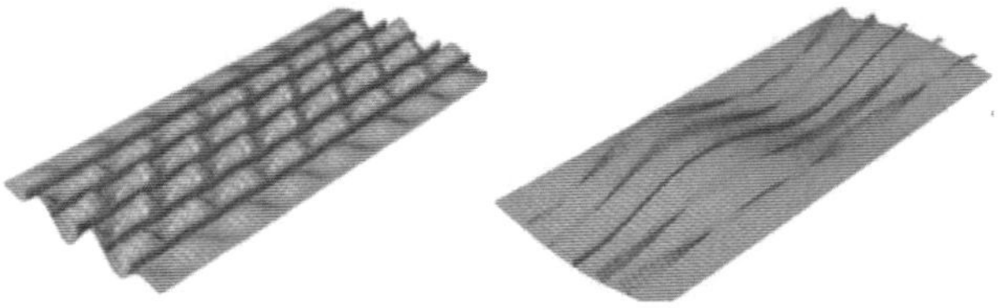

Figure 2. DNV-RP-C208 given local imperfection pattern (left), global Imperfection pattern (right)

Table 3. DNV-RP-C208 imperfection amplitude recommendation [3]

Component	Shape	Magnitude	Imperfection
Longitudinal stiffener girder webs (global imperfection)	Bow	L/400	9.5 mm
Plane plate between stiffeners (Local imperfection)	Eigenmode	S/200	4.0 mm

Note: L stands for length of the girder and S is spacing between transverse stiffeners.

Eigenvalue analysis is performed and three eigen modes patterns were studied as two different cases to determine their patterns and magnitude that can result with the buckling capacity of the calibration benchmark. The eigen modes that provide imperfection patterns similar to the patterns recommended by DNV recommended practice [3] are chosen. Figure 3 shows the pattern of the global imperfection, which is identical for both cases. This imperfection represents the curved bow shape along the stiffener in between the girders.

However, displacement magnitude is not identical at each mid-stiffener span along the panel. Thus, a displacement average amplitude was determined at the middle of the relevant spans indicating the maximum and minimum displacements of the relevant spans. This also helps to have a more realistic case of global geometrical imperfections distribution and magnitude.

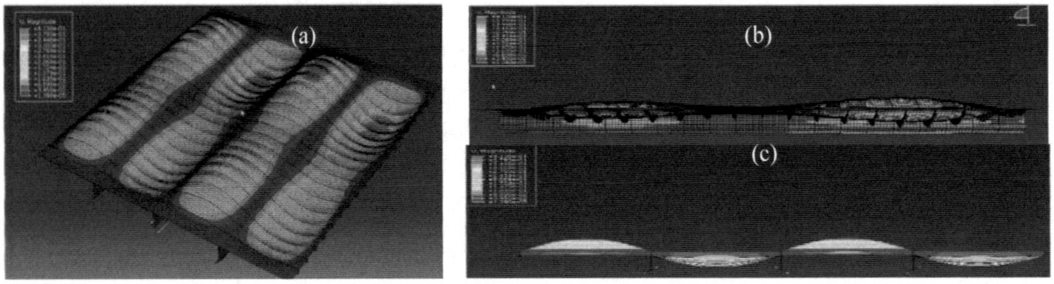

Figure 3. Global imperfection pattern (a) Plate (b) Stiffener side view (c) Girder side view

Figure 4 depicts the local imperfection patterns, which shows the displacement magnitude in the eigenmode at the nodes in the middle plate between the stiffeners. The unequal displacement reveals a more realistic local imperfection distribution. The difference between the two patterns is shown in Figure 3 and Figure 4, where pattern 1 depicts displacements that have sine function shape with the displacement altering to the opposite direction in every sequential stiffener span.

IOP Conf. Series: Materials Science and Engineering 1294 (2023) 012030 doi:10.1088/1757-899X/1294/1/012030

While in pattern 2 (Figure 5), the displacement change direction in every two sequential stiffener span so that the first two stiffener spans have an opposite displacement direction from the last two stiffener spans.

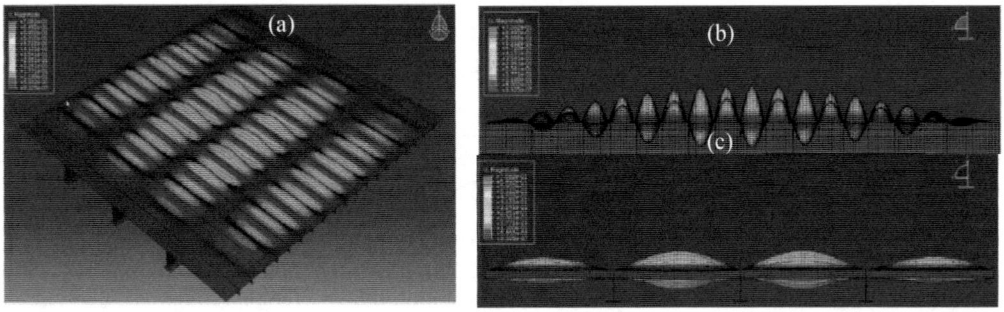

Figure 4. Local imperfection pattern-1 (a) Plate (b) Stiffener sideview (c) Girder side view

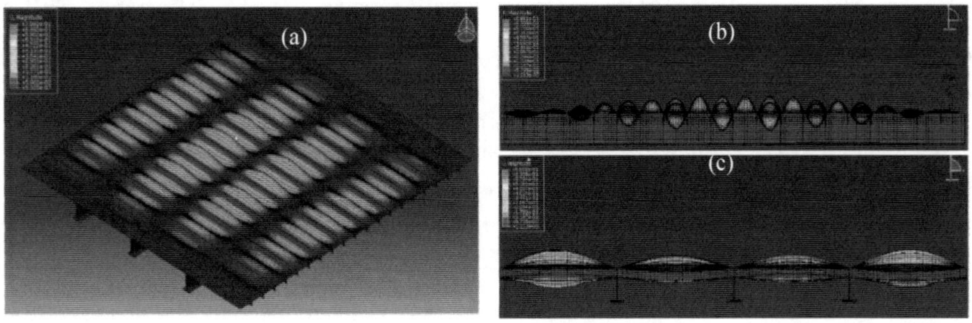

Figure 5. Local imperfection pattern-2 (a) Plate (b) Stiffener sideview I Girder side view

The nodal imperfection, which is the final scaled nodal displacement introduced to the non-linear analysis model from the global and local eigen mode patterns, was obtained using a relation between the scale factor and the nodal displacement as shown below equation.

$$\text{Nodal imperfection} = U_L \times F_L + U_g \times F_G \tag{1}$$

where nodal imperfection is the final scaled node displacement due to the non-linear analysis model from the global and local eigen mode patterns, U_L is the nodal displacement due to the local pattern, U_g is the nodal displacement due to the global pattern, F_G is Global pattern scale factor and F_L is local pattern scale factor.

The DNV recommended practice [3] provides a minimum mid bow imperfection for the global pattern and mid plate in between stiffeners imperfection as shown in Table 3. Due to the properties of the model geometry properties, the nodes of the global and local patterns obtained from the eigenmode have a wide range of nodal displacement. To apply a scale factor representing a realistic final imperfections pattern in the non-linear analysis model, a study was conducted that can also help to determine the suitable nodal displacement average from each pattern and to determine the scale factor.

The trail scale factor (SF) are found using equation (1), so when multiplied with $U_{L,av,max}$ and $U_{g,av,min}$ the resulting maximum nodal imperfection would be close to U = 13.5 which is the recommended according to DNV-RP-C208 [3].

For each case, three trail Scale Factor (SF) have been found. Two additional SFs were applied to case 1 upon considering the results that will be discussed later. The SF case that results in average of imperfection amplitudes (U_{av}) within the acceptable range results in a buckling capacity close to the benchmark buckling capacity was obtained according to the standards. This was chosen as a calibrated case for the stiffened panel under compression loads. The considered trial SFs are shown in Table 4.

Table 4. Applied scale factors

Case	Imperfection pattern	$U_{av,max}$ (mm)	$U_{av,min}$ (mm)	SF1	SF2	SF3	SF4	SF5
1	Global	0.380	0.1098	47.5	63.63	95	85	77.5
	Local 1	0.410	0.126	18.18	18.18	18.18	18.18	18.18
2	Global	0.380	0.1098	8.9	63.63	95		
	Local 2	0.372	0.128	31.67	18.81	18.18		

3.3. Buckling capacities of imperfected panels

Three non-linear buckling analyses were performed for each scale factor at the combined cases. The obtained results are plotted as force (MN) vs. displacement (mm) of the nodes on Face 1, which is the face where the load is applied. Then the results are compared with DNV-RP-C201, which is hereafter termed as "calibration force". The load versus displacement obtained from FE non-linear analysis are shown in Figure 6 only for both local and global buckling combined cases. In order to establish a calibration capacity, the stiffened panels buckling capacity is determined according to DNV-RP-C201 [1] and considered as a benchmark to calibrate the non-linear finite element model buckling capacity.

ABAQUS combines the local and the global imperfection patterns into one patten of imperfections this is then used in the non-linear analysis. Three non-linear buckling analyses were performed for each scale factor: (1) only local pattern, (2) only global pattern, and (3) the combined case of imperfection pattern. To determine the imperfection case that results in values closest to the reference case provided by DNV-RP-C201, the results are represented as force (MN) vs. displacement (mm) of the nodes in Face 1. The results are shown in Figure 6, Table 5 and Table 6, which are discussed in the next chapter.

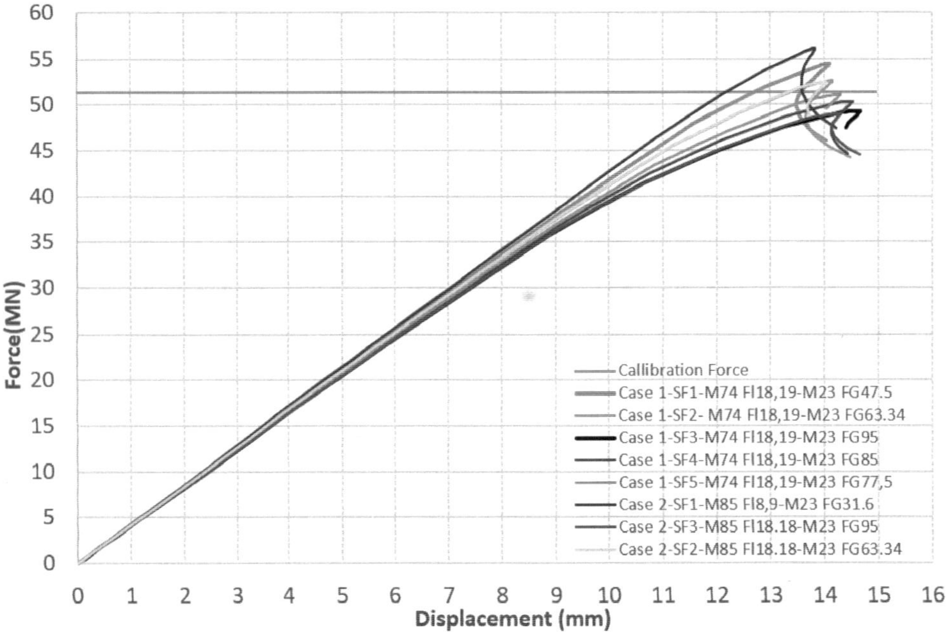

Figure 6. Comparison of the combined patterns cases, nonlinear buckling analysis results

Table 5. Results summary: Nodal Imperfection magnitude of combined imperfection pattern

Case	Imperfection Pattern	Nodal imperfection magnitude of combined imperfection pattern (mm)				
		SF1	SF2	SF3	SF4	SF5
1	Global (M23)[1)] Local1 (M85)[2)]	5.2-25.6	2.3-31.63	2.29-43.55	2.29-39.75	2.29-36.9
2	Global (M23) Local 2 (M74)[3)]	4-11.78	2.4-31.17	2.3-42.89	_	_

Note:
1) M23 is the global pattern eigen mode number.
2) M85 is Local pattern 1 eigen mode number.
3) M74 is Local pattern 1 eigen mode number.

Table 6. Results summary of comparison with reference case

Case	Imperfection Pattern	Diversion to calibration capacity (%)				
		SF1	SF2	SF3	SF4	SF5
1	Global (M23) Local1 (M85)	5.7	+2.3	-4.1	-2.3	-0.58
2	Global (M23) Local 2 (M74)	8.9	+2.09	-4.68	_	_

4. Comparison and discussion

4.1. Comparison of non-linear buckling capacities

To choose the best factors with acceptable imperfection magnitude range along the panel, an educated guess to start with factors that would result in combined imperfection magnitude around double the recommendation of DNV-RP-C208 [3]. For case 1, the chosen scale factor resulted in an imperfection magnitude that range between 5.2 - 25.6 mm along the whole panel. Whereas, for case 2 the range of imperfection magnitude is 4-11.8 mm. Case 2 imperfection range is close to the recommended values in DNV-RP-C208 which is (0 - 13.5 mm) for the investigated panel.

Both those cases gave buckling capacities with a 5.7- 8.9% over estimation of the buckling capacities for this panel. This can be justified by the difference in the shape of the patterns because the global and the local imperfection patterns recommended by DNV-RP-C208 consider an equal imperfection magnitude along the whole panel. The eigen mode analysis of this panel did not provide global or local modes with equal imperfection magnitudes at the mid-spans.

It can be seen from Table 5 that the average imperfection that ranges from SF2 to SF3 are very close in case 1 and 2. This is done to determinate the local imperfection pattern case that gives a buckling capacity closer and under the calibration capacity. As shown in Table 6 and Figure 6, local pattern 1 in case 1 provides a capacity 0.58% closer to the calibration capacity than local pattern 2. This is considering the imperfection contribution of the global pattern, which is the same for both cases. For this reason, case 1 is chosen to continue with the calibration process with SF4 and SF5.

As depicted in Figure 6, SF3, SF4 and SF5 for case 1 and SF3 for case 2 provide estimates of buckling capacity that is lower than the calibration capacity. However, SF3 for case 2 and case 1 provide an underestimate of the capacity of 4.1% and 4.68% while introducing 15% higher imperfection magnitude than SF5 case 1. Also, in case 1, SF4 represents a larger underestimation of the buckling capacity with higher imperfection magnitude than SF5. SF5 in case 1 is thus chosen for calibration cases because it provide only an underestimation of 0.58% of the calibration buckling capacity.

IOP Conf. Series: Materials Science and Engineering 1294 (2023) 012030 doi:10.1088/1757-899X/1294/1/012030

4.2. Calibration of the FE model

The largest imperfection range (31.4 - 37 mm) is concentrated in the middle of the spans 5A,5B, 5C and 5D as illustrated in Figure 7. There is a lower concentration of high imperfections at 7A, 7B, 7C and 7D and 15A, 15B, 15C and 15D due to the global and local patterns of imperfection that intersect with their highest magnitudes at those locations. The highest nodal imperfection of 36.9 mm at middle of the mentioned stiffener spans represents 1% of the stiffener span length. This is higher than the 13.5 mm or 0.355% of the stiffener span length (i.e 3800 mm) which is given in the DNV-RP-C208 [3] recommendation for the case where the maximum imperfections from both patterns intersect. However, the global and the local patterns of imperfection used in the DNV-RP- C208 have equal amplitude of imperfection along the panel. The case was not the same because of the panel geometrical properties and the eigenmode analysis provided global and local patterns of imperfection with different magnitudes along the panel. The von Mises stress distribution along the panel is shown in Figure 8.

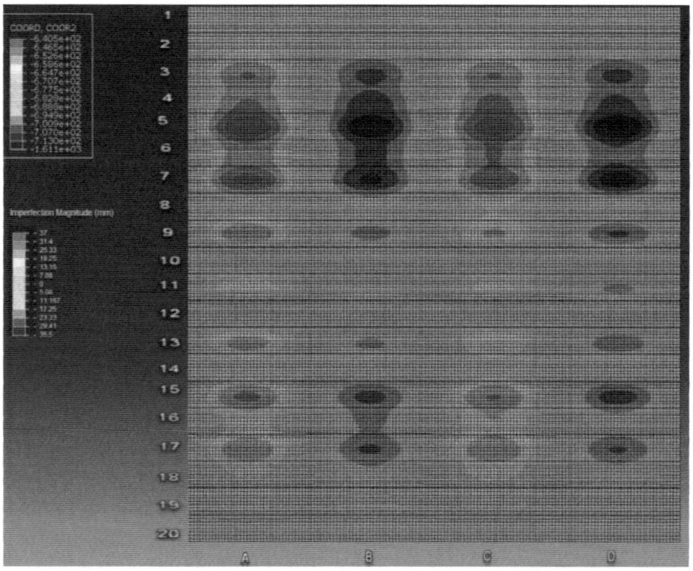

Figure 7. Calibrated case imperfection magnitude and distribution

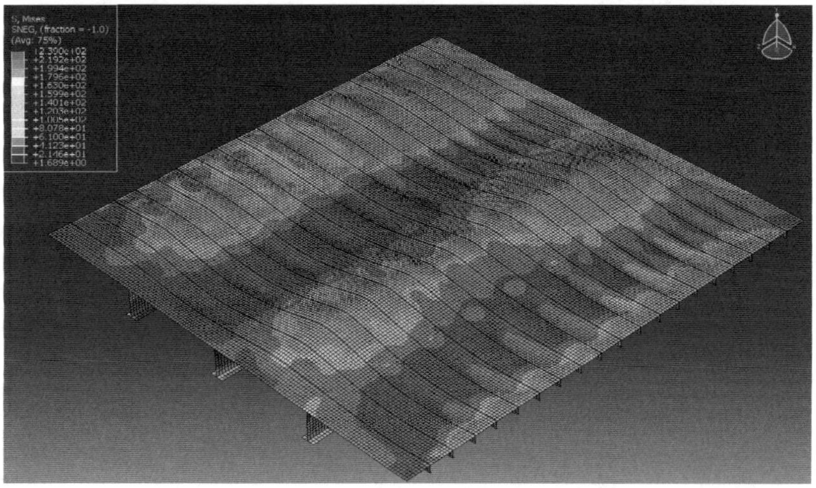

Figure 8. Von-mises stress distribution (MPa)

4.3. Parametric study on effect of holes in the panels on buckling capacity

To investigate the effect of holes on the buckling capacity of the model, the several spans were removed as a first step before conducting non-linear buckling analysis. The same imperfection patterns and magnitudes were imposed on the model. Shapes and location of holes 1 and 2 are shown in Figure 9.

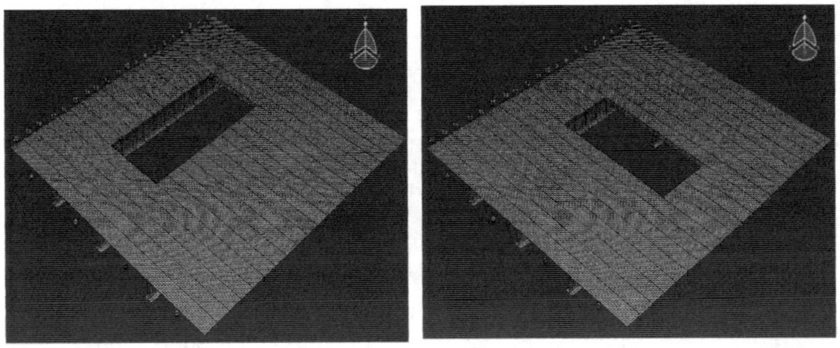

Figure 9. Calibrated model- Hole 1 (left), Calibrated model Hole 2 (Right)

Results of the analysis in Figure 10 and Table 7 display that the percent reduction of the buckling capacity for the calibrated model for hole 1 is almost equal to the span cross sectional reduction, though the area reduction in hole 1 is along spans B and C (Figure 9). The area reduction is related to the area reduction of one stiffener span. This is also the case with hole 2, where the hole is along 10 plates between stiffener spans (stiffener spacing) and one stiffener span (A). The cross-sectional reduction was the governing factor where 50% reduction resulted in 47.8% reduction in the buckling capacity of the model, and 20% reduction resulted in 19.6% buckling capacity reduction.

Table 7. Effects of holes on buckling capacity of the calibrated model

	Area reduction		Results	
	Surface (%)	Stiffener span CS (%)	Buckling capacity (MPa)	Reduction of buckling capacity (%)
Hole 1	10	20	137	-19.6
Hole 2	12.5	50	88.7	-47.8

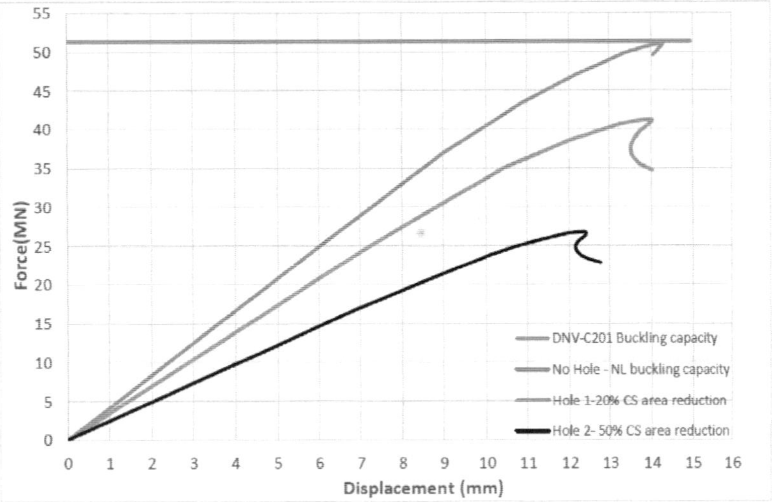

Figure 10. Effects of holes on buckling capacity of the calibrated model

5. Conclusions

The purpose of this article is to explore the application of a non-linear finite element method as outlined in DNV-RP-C208 [3]. This method is used to assess the buckling capacity of a ship hull stiffened panel. The goal is to then compare and adjust the non-linear buckling capacity with the buckling capacity determined through analytical formulas found in the conventional code DNV-RP-C201 [1]. To achieve this adjustment, the study examines the impact of combinations of one global and two local imperfection patterns, each with various imperfection magnitudes. The selected imperfection patterns closely resemble those recommended in DNV-RP-C208.

The model introduces a mid-stiffener span imperfection magnitude ranging from 0.06% to 1% of the stiffener span length. The aim is to achieve a calibration level of 99.4% with respect to the buckling capacity defined in the DNV-RP-C201 standard. It's noteworthy that the 1% mid-span imperfection magnitude exceeds the DNV's recommended 0.355% of the stiffener span length. This deviation is due to the imperfection patterns derived from eigenmode analysis, which did not provide uniform imperfection magnitudes along the mid-spans of the panel, as assumed in the DNV-RP-C208 recommended patterns. This discrepancy is attributed to the geometric properties and boundary conditions of the panel, as well as the assumption of perfect contact and stress distribution among stiffeners, plates, and girders, which collectively contribute to the larger imperfections introduced for calibration purposes.

With the calibrated non-linear model, the study investigates the influence of holes on the buckling capacity under both uniaxial and gravity loads. The findings demonstrate a reduction in non-linear buckling capacity associated with the reduction in the cross-sectional area of stiffeners and plates caused by the presence of holes in the plate.

As a suggestion for future research, it is recommended to conduct an experimental study on the buckling behavior of a scaled-down version of the same structure under identical loads and boundary conditions. This experimental study would serve to validate the results obtained in this analysis.

References

[1] DNV Recommended Practice 2010 *Buckling strength of plated structures* DNV-RP-C201, dated October 2010.
[2] NORSOK standard N-004 2013 *Design of steel structures*, 3rd ed, Norwegian Petroilum Industry.
[3] DNV Recommended Practice 2021 *Determination of structural capacity by non-linear finite element analysis methods* DNV-RP-C208, Det Norske Veritas, Amended September 2021.
[4] Paik J K and Thayamballi A K 2003 *Ultimate Limit State Design of Steel Plated Structures*. Chichester, England; Hoboken, NJ: J. Wiley.
[5] Cho S R, Kim H S, Doh H M and Chon Y K 2013 Ultimate strength formulation for stiffened plates subjected to combined axial compression, transverse compression, shear force, and lateral pressure loadings *Ships Offshore Struct.*, **8**(6), 628–637.
[6] Ozguc O, Das P K and Barltrop N 2007 The new simple design equations for the ultimate compressive strength of imperfect stiffened plates *Ocean Eng.* **34**, 970–986.
[7] Zhang S 2016 A review and study on ultimate strength of steel plates and stiffened panels in axial compression *Ships and Offshore Struct.* **11**(1), 81-91.
[8] Ozguc O 2020 Assessment of buckling behavior on an FPSO deck panel Polish Maritime Research, 27, 50–58.
[9] Michael S 2009 *ABAQUS/Standard User's Manual* Version 6.9 Providence, RI: Dassault Systèmes Simulia Corp.
[10] Cook R D, Malkus D S, Plesha M E 1989 *Concepts and Applications of Finite Element Analysis* (4th ed.). New York: Wiley.

IOP Conf. Series: Materials Science and Engineering 1294 (2023) 012031 doi:10.1088/1757-899X/1294/1/012031

Spatial and temporal crack initiation prediction in steels based on residual stress tensor distribution monitoring

E Mangiorou[1], T Damatopoulou[1], S Angelopoulos[1], P Vourna[1], A Ktena[2] and E Hristoforou[1,*]

[1]Laboratory of Electronic Sensors, National TU of Athens, Athens, Greece
[2]National Kopodistrian University of Athens, Athens, Greece

[*] Correspondence: hristoforou@ece.ntua.gr

Abstract. We have developed a method and the corresponding instruments, able to provide the spatial and temporal determination of crack initiation in ferromagnetic steels and their welds. The method is based on the monitoring of the differential magnetic permeability μ_r along the length of the steel, since a sigmoid monotonical dependence of μ_r on residual stresses has been determined in the past. Thus, the magnetic stress calibration (MASC) curves can inform on the localized residual stresses. Therefore, the temporal stress monitoring at different areas of interest can inform about the status of residual stresses. Then, the tendency of the residual stress value can inform about reaching the positive (tensile) or negative (compressive) yield point of the steel. As an example, a quasi-parabolic temporal dependence of μ_r offers the tendency to determine the time, or the time window, required to reach yield point. This way, the temporal prediction of the crack initiation is provided for the given point in space, since reaching the yield point results in crack initiation.

1. Introduction

Structural health monitoring is an important issue in steel production and manufacturing, as well as in steel structures in several applications [1]. Among several parameters, residual stress monitoring is a crucial one, determining the threshold, after which nano-cracks start being initiated, propagating along the material, leading to failure. This threshold is the yield point of the steel under examination.

The standard method to determine the residual stress component in a given direction on the surface of the steel is the X-Ray Diffraction, using the Bragg-Brentano set-up, XRD-BB for short [2]. In a similar way, the bulk residual stresses are determined by neutron diffraction [3].

The current state of the art in stress monitoring is either the strain gauges [4] or the drill hole method [5]. Strain gauges are able to monitor stresses at a given point or area, where they are set, while drill hole method is a rather destructive technology.

For this reason, our team has long time ago initiated work in stress monitoring in steels, using contactless, non-destructive technology, based on the measurement of the differential magnetic permeability on the surface or the bulk [6], the non-linear magnetoacoustic properties [7] and the Barkhausen noise [8] of the steel under test. The comparison between magnetic properties and residual stresses determined by XRD-BB for surface or neutron diffraction for bulk in the same direction resulted in a monotonic sigmoid function, the so called Magnetic Stress Calibration curve (MASC) that allows for stress determination in these materials. Furthermore, a Universality law has been determined, permitting the measurement of stresses in an unknown type of steel [9].

Bearing in mind to predict the area and the time of a crack initiation, the correlation between magnetic properties and residual stresses has been used. In the following chapters the proof of principle of the method to provide the spatial and temporal prediction of crack initiation is illustrated. For this reason, a Duplex steel has been employed, that has a vast variety of important applications.

2. Monitoring the surface magnetic permeability

The 10 cm X 15 cm X 2 mm selected Duplex steel was cut into two identical parts and then it was autogenously welded, in order to generate a variety of residual stresses from the negative yield point (compressive stresses) up to the positive yield point (tensile stresses).

Then, the XRD-BB method was used to determine the residual stresses across the weld in three different lines, vertical to the welding line. Consequently, the magnetic differential permeability was determined by using a surface permeability sensor illustrated in Figure 1.

The sensor is based on the magnetostrictive delay line principle and on the magnetic circuit balance in the presence of the steel under test at a well-defined position [10]. Two pulsed current conductors, symmetrically set above and below a magnetostrictive ribbon or wire, are used to host pulsed current of the same amplitude, transmitted simultaneously above and below the magnetostrictive ribbon or wire.

In the absence of the steel surface under test, the magnetic fields from the two conductors are counterbalanced inside the magnetostrictive material and therefore no pulsed microstrain is generated along the length of the magnetostrictive ribbon or wire. Therefore, no pulsed voltage output is detected at the search coil.

In the close presence of the magnetic surface under test, the balance of the magnetic field inside the ribbon or wire breaks, resulting in the generation of an elastic pulse that propagates in the ribbon or wire, subject to be detected by the search coil as a pulsed voltage output. The higher the permeability of the steel under test, the higher the response of the search coil.

Arranging an electromagnetic yoke to magnetize the steel under test, as illustrated in Figure 2, the change of the permeability is depicted in the amount of the microstrain generated at the area of the cross section of conductors and magnetostrictive element. Thus, the dependence of the voltage output on the magnetic field on the under test steel corresponds to the steel magnetic permeability dependence on the applied field. Thus, a precise determination of magnetic permeability of the steel under test can be realized. Figure 3 depicts the maximum magnetic differential permeability dependence on the applied field by the electromagnetic yoke. Apart from that, using a permanent magnet yoke, as illustrated in Figure 4, can result in the determination of the permeability at a given biasing field, which is also of importance for fast quality tests in steels.

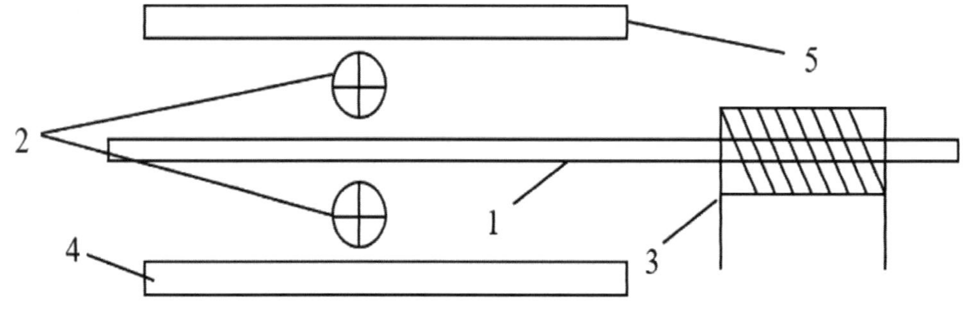

Figure 1. The surface magnetic permeability sensor: (1) magnetostrictive delay line, (2) pulsed current conductor, (3) search coil, (4) steel under test, (5) steel sample counteracting on the response of the steel under test.

In the experimental results provided in this work, the maximum differential permeability of the steel under test was monitored, compared with the corresponding residual stress at the given point and direction of the steel under stress. Residual stress was monitored using the X-Ray Diffraction in the Bragg-Brentano mode (XRD-BB). The dependence of the maximum magnetic permeability at a given point on the corresponding residual stress is illustrated in Fig. 5. A monotonic dependence of such dependence can be observed. These measurements illustrated that the yield point of the Duplex steel is 605 MPa.

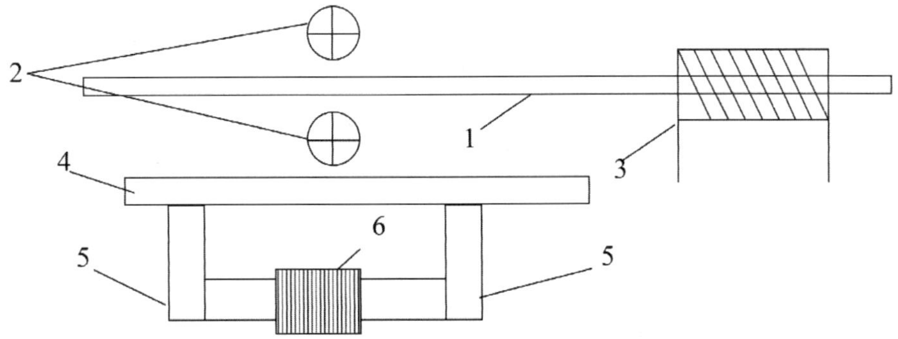

Figure 2. The surface permeability sensor with electromagnetic yoke magnetizing the steel under test: (1) MDL, (2) pulsed current conductors, (3) search coil, (4) steel surface under test, (5) & (6) soft ferromagnets and excitation coil respectively, to form the biasing electromagnet.

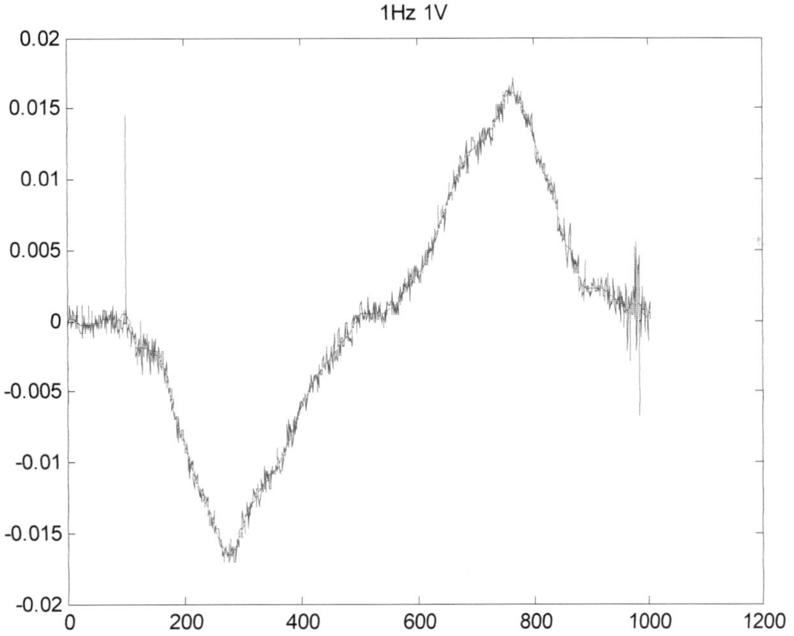

Figure 3. The response of the search coil under sinusoidal excitation current transmitted in the electromagnetic yoke, illustrating a correspondence of the surface permeability of the steel under stress on the applied field. X axis, field in A/m. Y axis in V.

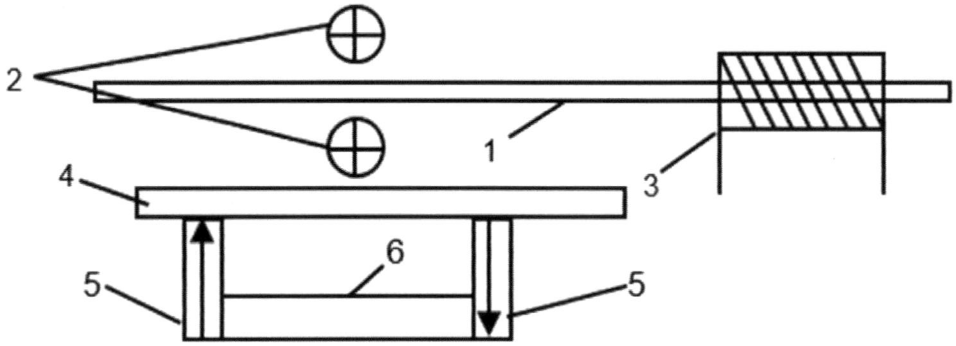

Figure 4. The surface permeability sensor with permanent magnet yoke magnetizing the steel under a given magnetic field: (1) MDL, (2) pulsed current conductors, (3) search coil, (4) steel surface under test, (5) permanent magnets, (6) soft magnetic bar for magnetic circuit closure.

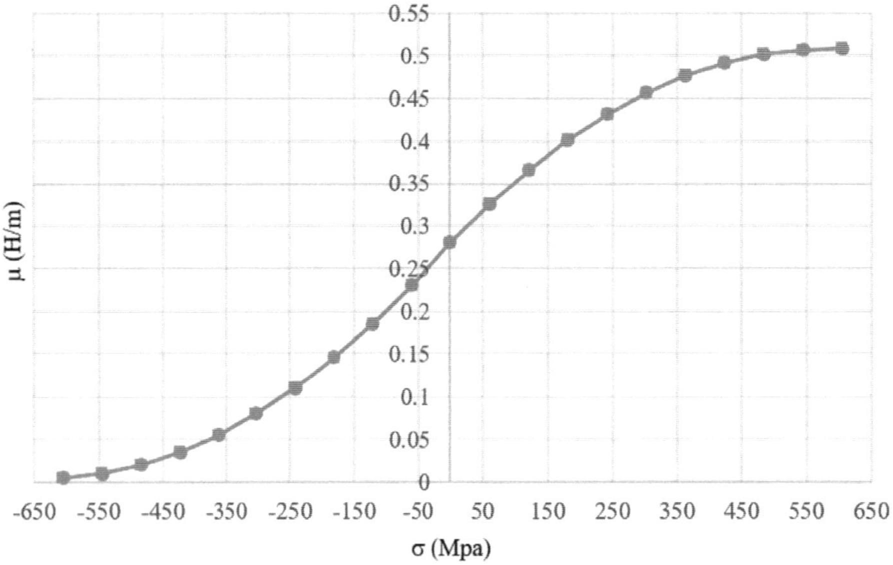

Figure 5. The dependence of the maximum differential permeability of the steel under test on the monitored residual stresses.

3. Determination of the crack initiation time

Having determined the MASC for the given Duplex steel, the initiation of experiments for the prediction of crack initiation was realized. The first set of experiments was realized by using a fatigue Instron instrument. The fatigue, namely repetitions of cyclic loads, was realized for the case of high cyclic fatigue, namely loading with stresses significantly smaller than the yield point of steel under test. In fact, the pre-loaded pressure was equal to 110 MPa. The actual residual stress was determined by means of monitoring the response of the surface permeability sensor. Bearing in mind that the frequency of loading was 1 kHz and the excitation field of the electromagnetic yoke was at a frequency of 1 Hz, the monitored permeability and the corresponding residual stress was averaged every 1 second or one thousand cycles. Figure 6 illustrates the dependence of residual stresses on the fatigue cycles at a given point of the Duplex steel. Apparently, an inversely proportional dependence of the residual stresses (and the maximum differential permeability) on the fatigue cycles is observed. Reaching the yield point corresponds to more than 10 million cycles.

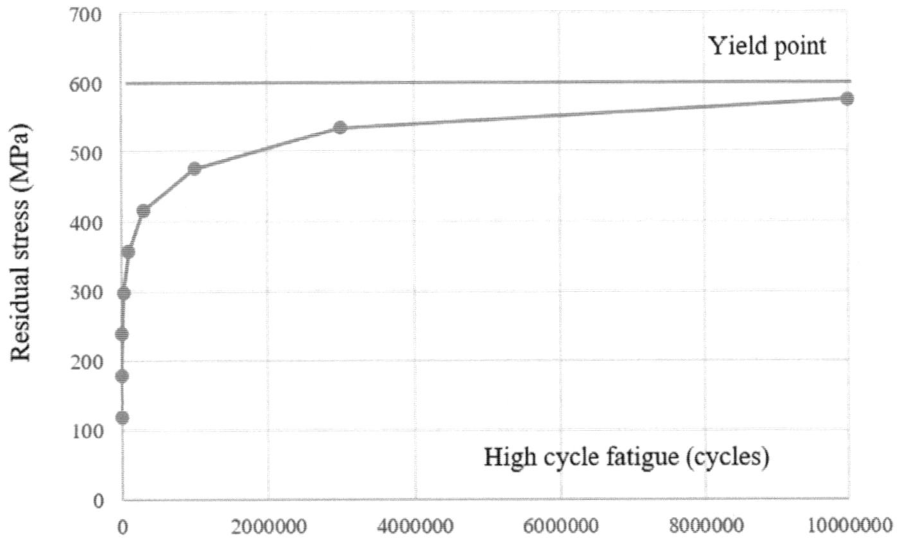

Figure 6. The dependence of residual stresses on the fatigue cycles of the Duplex steel in the high cycle fatigue mode, using 110 MPa load.

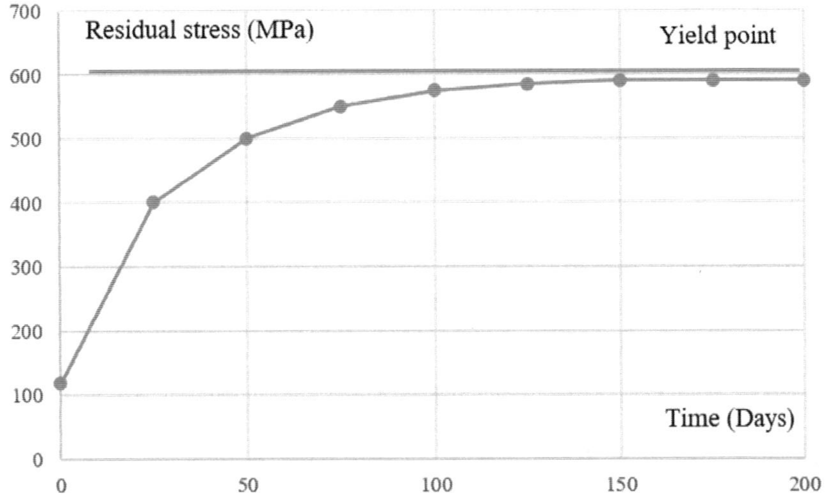

Figure 7. The dependence of residual stresses on days of applying fatigue cycles of the Duplex steel in the high cycle fatigue mode, using 50 MPa load.

Following the above-mentioned experiment, another high cycle fatigue test was realized, under 50 MPa load with the same cyclic load frequency of 1 kHz. The dependence of the residual stresses due to the monitoring of the maximum differential permeability on time is illustrated in Figure 7.

In both cases, a tendency of the yield point approach is observed, allowing for the determination of the proposed algorithm: the continuous monitoring of residual stresses with respect to time, determines a mathematical approach to the yield point, and therefore to the crack initiation. Such function can be determined from the first cycles of loading. Therefore, the projection of the function in time determines the time window, where the yield point is reached, being equivalent to the crack initiation.

Apart from this detailed measurement of residual stresses on the surface of the under test steel, requiring time for detecting the residual stress at a given point of the steel, there is also the need of fast characterization, even if the permeability does not correspond to its maximum value. For this reason, the surface permeability sensor with the permanent magnet yoke, as illustrated in Figure 4 or even the sensor of Figure 1, can be employed. Running such a test on top of a steel surface with more or less uniform response is illustrated in Figure 8. The pulsed voltage output of the sensor is amplified 100 times. Such a sensor can be used for fast monitoring along the length of the steel under test.

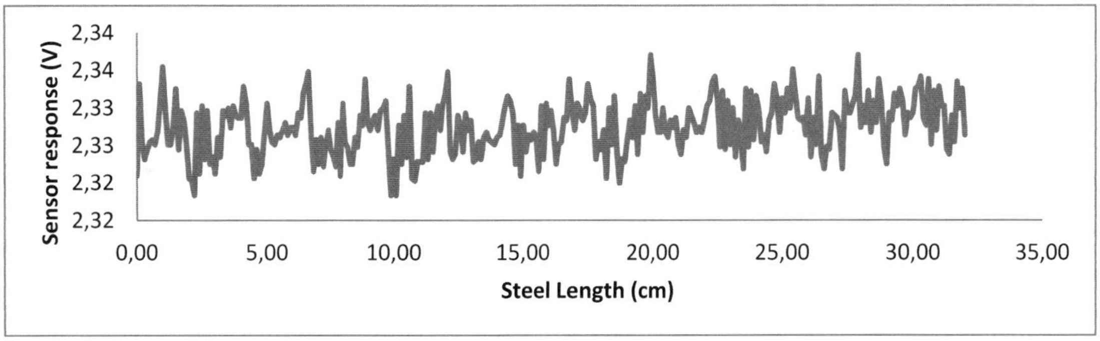

Figure 8. Fast permeability tests on top of the under-test steel using the sensor of Figure 5.

4. Conclusions

The ability to determine the time for the crack initiation in a point or area of a steel is described in this paper. It has been proven for the case of high cyclic fatigue under 110 MPa and 50 MPa stress, with yield point in the order of 600 MPa. Such a high cycle fatigue resulted in a well predictable way of reaching the yield point, thus resulting in a nano-crack initiation.

Acknowledgements: This research has been co-financed by the European Union and Greek national funds through the Operational Program Competitiveness, Entrepreneurship and Innovation, under the call RESEARCH-CREATE-INNOVATE (project code: T2EDK-02521).

References

[1] Li, Z.; He, J.; Teng, J.; Wang, Y. 2016 Internal stress monitoring of in-service structural steel members with ultrasonic method, *Mater.,* **9,** 223. https://doi.org/10.3390/ma9040223.

[2] Kriegner D., Matej Z., Kuzel R. and Holý V., 2015 Powder diffraction in Bragg-Brentano geometry with straight linear detectors, *J. Appl. Cryst.* **48**, 613, https://doi.org/10.1107/S1600576715003465.

[3] Tomota Y., Lukas P., Harjo S., Park J-H., Tsuchida N., Neov D., 2003 In situ neutron diffraction study of IF and ultra low carbon steels upon tensile deformation, *Acta Mater.*, **51**(3) , 819.

[4] Pang, C., Lee, GY., Kim, Ti. *et al.* 2012 A flexible and highly sensitive strain-gauge sensor using reversible interlocking of nanofibres. *Nat. Mater.,* **11**, 795, https://doi.org/10.1038/nmat3380.

[5] Hill E.J., Robertson J., Uvarova Y. 2015 Multiscale hierarchical domaining and compression of drill hole data, *Comput. Geosci.,* **79**, 47.

[6] Hristoforou, E., Ktena, A., Vourna, P., Argiris, K., 2018 Dependence of magnetic permeability on residual stresses in alloyed steels, Open Access, *AIP Adv.*, **8**(4), 047201.

[7] Hristoforou E. and Reilly R.E., 1991 Nonuniformity in amorphous ribbon delay lines after stress and current annealing, *Jpn. Appl. Phys.,* **69**, 5008.

[8] Augustyniak B., Piotrowski L., Chmielewski M., Kosmas K., Hristoforou E., 2010 Barkhausen noise properties measured by different methods for deformed armco samples, *IEEE Trans. Magn.,* **46**, , 544.

[9] Hristoforou E., Vourna P., Ktena A., Svec P., 2016 On the universality of the dependence of magnetic parameters on residual stresses in steels, *IEEE Trans. Magn.,* **52**(5), , 7362189.

[10] Hristoforou E., Kosmas K., Kollar M., 2008 Surface magnetic non-destructive evaluation using permeability sensor based on the MDL technique, *J. Electr. Eng.,* **59**, 90.

Experimental assessment of cracked tubular joints repaired with crack-deflecting holes and weld-toe grinding

S Riise, M Atteya, O Mikkelsen and G Ersdal*

Faculty of Science and Technology, University of Stavanger, Norway

* Corresponding author: gerhard.ersdal@uis.no

Abstract. Tubular joints in offshore structures for energy production are exposed to cyclic loading and may experience fatigue cracking. Presently, the inspection method commonly used for such structures will primarily detect whether a member is flooded, which is an indication of through thickness cracks. As a result, repair methods suitable for repairing through thickness cracks that are cost-effective and quick to implement are needed. One such repair method could be crack arrest by hole drilling. However, the validity of such a repair method is at present not sufficiently proven. Hence, this paper presents the results of an experimental fatigue test of pre-cracked tubular joint repaired with crack-deflecting holes combined weld-toe grinding. One tubular double T joint was tested and underwent three-phases comprising of a pre-cracking phase, a repair phase and a post-repair testing phase. The through-thickness fatigue crack was achieved by subjecting the intact test joint to cyclic axial loading within the high-cycle fatigue domain. In the repair phase, crack-deflecting holes were drilled in front of each crack tip to arrest the crack ends. In addition, the weld-toe behind these holes were grinded to enhance the fatigue life of the repaired joint. The performances of the repaired test joint were then investigated in the last phase by subjecting it to further cyclic loading within the high cycles fatigue domain. In the test, the crack was successfully arrested in the crack-deflecting holes. Furthermore, the repair method indicated an apparent enhancement of the fatigue endurance of the cracked tubular joint when subjected to stress ranges within the high-cycle fatigue domain.

1. Introduction

Offshore jacket structures fabricated from steel tubular are widely used as offshore substructures for oil and gas exploration and are also increasingly used as substructures for offshore wind turbine. Figure 1, shows a typical offshore jacket structure. These fixed structures are exposed to a harsh environment with considerable cyclic wind and wave loadings in a corrosive environment, which can lead to fatigue cracking. Consequently, fatigue is one of the critical design criteria for such structures [1-4].

Although every effort is made in the design stage to avoid fatigue cracking in the offshore structures' service life, this may not be achieved. In addition, numerous offshore installations are ageing and have already passed their design life. Therefore, adequate inspection and repair are vital to maintain safe operating conditions and safely extend the life of the existing offshore structures. In most cases, the repair is difficult and expensive and may result in significant downtime with consequential loss [5].

Figure 1. Heimdal platform, typical offshore jacket structure (photo by Simen Riise)

An extensive number of repair methods exist, spanning from weld repair to mechanical fixing or removal and replacement of structural elements [6]. A similarity among these techniques is the necessity for heavy equipment, extensive planning, and substantial resource allocation, all of which contribute to high expenditure. Hence, there is a lack of economically temporary repair methods for tubular joints subjected to fatigue cracks. A well-known temporary repair method for plated structures subjected to fatigue cracks is crack tip holes and crack deflecting holes [7]. This method has been used with success in aeronautical industries and bridge engineering. It comprises drilling a through-thickness hole in the crack tip or crack tip proximity to arrest the crack in the hole, thereby reducing the stress intensity at the crack tip. Another cost-effective repair method proven to increase fatigue life is weld toe grinding [5], which implies profiling the weld toe to reduce the notch stress. The advantage of both these methods is that they do not require heavy equipment and are relatively easy to perform, which means lower costs.

The objective of the present investigation was to assess the effect of repairing a cracked tubular joint with crack-deflecting holes in combination with weld toe grinding, where the crack has been formed under high cycle fatigue conditions.

2. Specimen construction and fatigue test method

2.1. Specimen design

The joint design and fabrication are equal to the specimen tested by [8], thus enabling a comparison of the results. One DT-joint specimen was fabricated with the geometry shown in Figure 2.

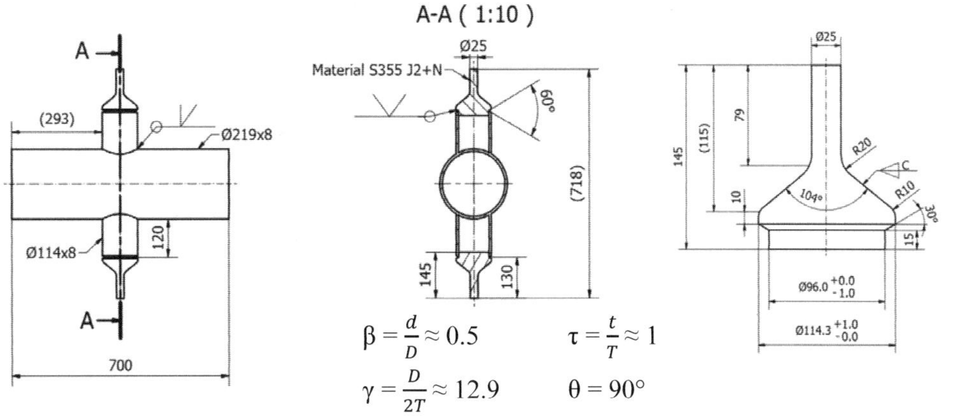

$$\beta = \frac{d}{D} \approx 0.5 \qquad \tau = \frac{t}{T} \approx 1$$

$$\gamma = \frac{D}{2T} \approx 12.9 \qquad \theta = 90°$$

Figure 2. DT-joint Geometry [8]

NDT testing was conducted in compliance with the requirements stipulated in DNV-OS-C401, including 100% visual inspection, 100% magnetic particle testing and 100% ultrasonic testing. No weld improvement techniques were applied before the pre-cracking of the specimen. A picture of the DT-joint as fabrication and installed in the fatigue testing rig is shown in Figure 3.

Figure 3. Specimen as-fabricated and installed in fatigue rig, respectively

2.2. Materials

The brace and chord were both made of hot-rolled, seamless pipes formed from normalised structural steel satisfying the grade requirements S355 G15+N. The cones were machined from a 120mm shaft of steel S355 J2+N.

A total of eight tensile tests was performed of the steel used in the fatigue test. The specimens were constructed according to ASTM A370 [9]. Four 3/4" specimens were cut out from the brace pipe and four 1" specimens were cut out from the chord pipe. The stress strain curve for one representative specimen is shown in Figure 4. This specimen had a tensile stress at yield of 423MPa (with a mean value for all tests at 406.5MPa) and an ultimate tensile stress of 544MPa (with a mean value for all tests at 528MPa).

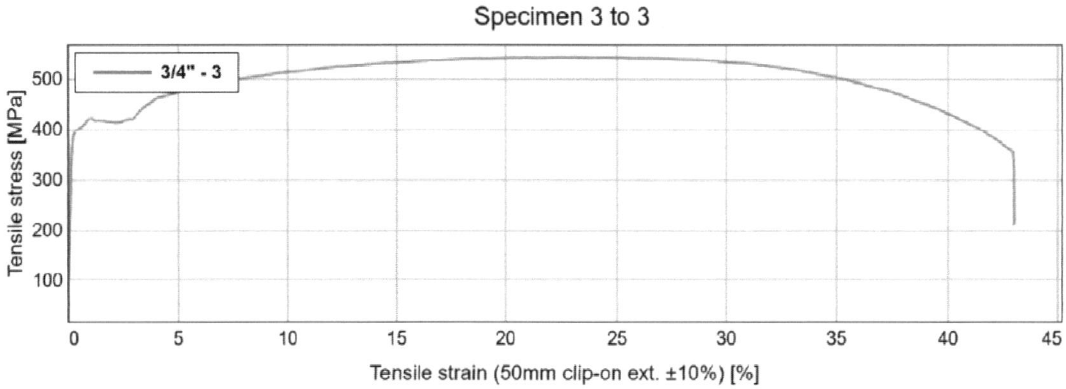

Figure 4. Stress strain curve for the steel applied in the tests.

2.3. Instrumentation

Linear electric strain gauges with grid lengths of 3mm and a nominal resistance of 120 were installed at critical locations on the DT joint to measure the stress distribution in the chord, detect and map crack propagation, and acquire the stress concentration factors. As the strain gauges are sensitive to temperature, thermocouples PT100 were glued to the specimen to constantly measure the temperature of the specimen and correct the strain readings.

In each quadrant of the specimen, six strain gauges were attached. The strain gauge layout within a single quadrant is depicted in Figure 5, featuring two SGs at the saddle point to capture the maximum hot spot stress with the aid of the linear extrapolation method described in DNV-RP-C203 [10]. The first row of strain gauges nearest to the weld toe were placed in extrapolation point A, as specified in DNV-RP-C203. Meanwhile, the strain gauges farthest from the weld toe, located at the saddle position, were positioned at extrapolation point B.

- Point A at chord surface normal to the weld toe = $0.2\sqrt{rt} = 4.4mm$
- Point B at chord surface normal to the weld toe = $2\pi R \frac{5}{360} = 9.6mm$

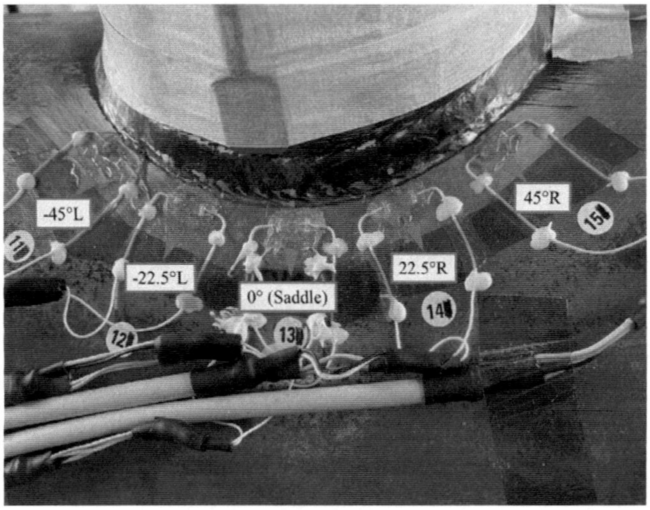

Figure 5. Strain gauge layout at quadrant 1

IOP Conf. Series: Materials Science and Engineering 1294 (2023) 012032 doi:10.1088/1757-899X/1294/1/012032

2.4. Fatigue testing condition

The specimen was tested under constant amplitude sinusoidal loading, in which sole axial force was applied through the brace. The entire fatigue testing process was carried out under load control, maintaining a positive R-ratio and a constant frequency between 3 to 10 Hz in air at room temperature. Before initiating the fatigue test, several preparatory steps were performed to ensure reliable results. Among these, the clamping stresses were recorded, and the reading frequency of the strain gauges was determined through a convergence study to ensure all peaks and valleys were captured during testing. The clamping stresses were decided to not be included in the testing and the strain gauge reading frequency was set to 300 Hz.

3. Result of fatigue test

3.1. Overview

The complete fatigue testing scheme of the DT joint aimed to evaluate the fatigue endurance of a cracked DT joint repaired with crack deflecting holes combined with weld toe grinding when degraded under high cycle fatigue conditions. To investigate this, one tubular DT-joint was tested through a three-phased testing scheme comprising:

1. **Pre-cracking:** the specimen was subjected to cyclic tension load under HCF conditions until a through-thickness crack was achieved.
2. **Repair:** two repair methods were performed on the specimen: weld toe grinding at the cracked quadrant and crack deflection holes at the crack tip proximity.
3. **Post-repair testing:** the repaired specimen was subjected to further cyclic loading under HCF conditions to evaluate the performance of the repair method. As a clear, indication of increased fatigue endurance was achieved the load was then significantly increased to break the specimen.

3.2. Pre-cracking phase

3.2.1. Stress Distribution. The stress distribution in the intact specimen was recorded as the pre-crack testing commenced and as expected for DT joints subjected to axial load the highest strain range was found in the saddle position. Furthermore, it can be seen in Figure 6 that the stress range is gradually decreasing towards the crown. It is also apparent in Figure 6 that stress ranges in all four quadrants are quite similar and symmetric around the saddle. This indicates symmetrically placed strain gauges and a lower probability of significant eccentricities in the specimen or test setup.

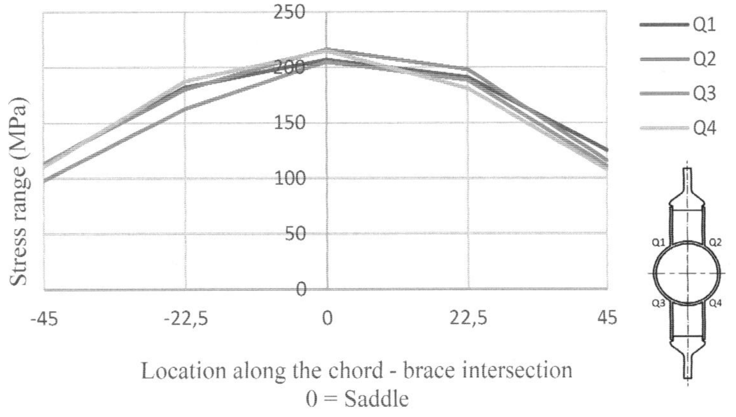

Figure 6. Stress range distribution at strain gauge closest to weld along chord-brace intersection ($\Delta P = 40$ kN)

3.2.2. SCF Calculation. In the pre-cracking phase, the stress concentration factors in the four saddle positions in the intact specimen were calculated based on the recorded HSS. The maximum SCF recorded under an axial load range of 40 kN was 20.45 located in Q2 and the average SCF of the four saddle positions was 19.89. Figure 7 below shows the SCF in the four quadrants. Furthermore, it shows the measured location of the strain gauges at a perpendicular distance from the weld toe along the chord surface and the ideal SG location when using linear extrapolation according to DNV is marked with the red dashed lines. Note that the SG positioned in saddle-b (in Figure 7) at Q3 was not measured after installation. Hence, the average SG placement in saddle-b from the other three quadrants is plotted.

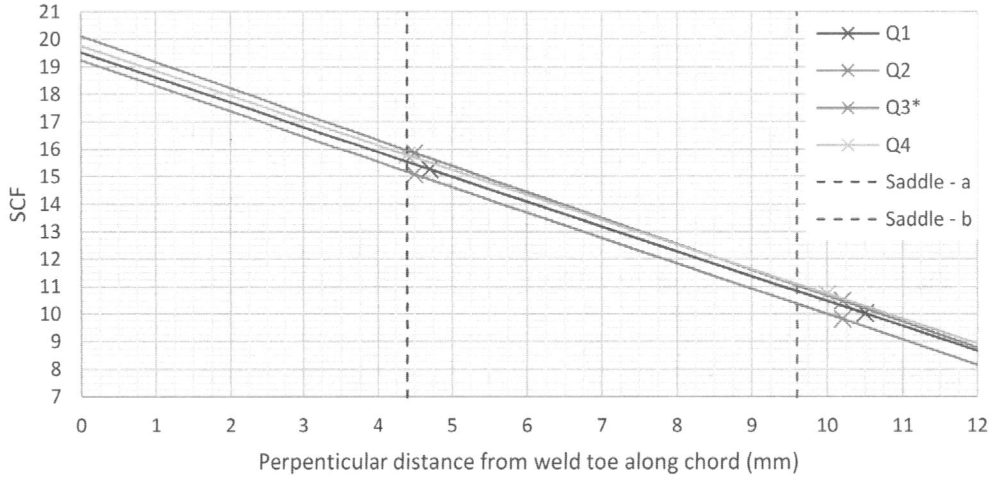

Figure 7. SCF at the four quadrants from the experimental work, with ΔP = 40 kN.
*The SG positioned in saddle-b at Q3 was not measured after installation. Hence, the average SG placement in saddle-b from the other three quadrants is plotted.

3.2.3. Pre cracking test loading. The load range and R-ratio maintained throughout the pre-cracking stage were 40kN and 0.1, respectively. Initially, the frequency was set to 6Hz but reduced to 3Hz after an 11% loss in the hot spot stress range at saddle point in Q3 at 300k cycles to ensure documentation of critical events. The maximum load reach in the testing (Pmax) was 45kN, while the minimum (Pmin) was 5kN. Stress distribution along the brace chord intersection and hot spot stresses at saddle positions were recorded. HSS recordings and load cell readings facilitated the calculation of SCFs.

3.2.4. Result of pre-cracking. The objective in the pre-cracking phase was to achieve a through-thickness crack, which could be observed with a 100% loss in hot spot strain readings. It was decided to stop the test at 1.515 million cycles as the strain gauges at the saddle and -22.5° degrees in Q3 had passed 90% and 96% drop in strain range, respectively. Hence, it was assumed that a through-thickness crack was achieved between the two strain gauges marking N3. The surface crack length was measured to 73mm, spanning between strain gauges at -45° to the strain gauge at +22.5°. Based on the strain readings the remaining three quadrants seemed to not have undergone any significant damage. A summary of the fatigue life in Q3 is shown in Table 1.

Table 1. Fatigue testing results of the pre-cracking stage

Fatigue life	Number of cycles	Remark
N1	342 000	15% drop in the measured hot-spot strain*
N2	410 000	7.5mm surface crack length
N3	1 515 000	observation of 100% loss in the hot-spot strain readings**

*The hot-spot strain is measured by linear extrapolation from two strain gauges placed perpendicular to the weld toe.
**Two strain gauges lost over 90% of the strain range. Therefore, 100% strain range loss between these strain gauges is assumed.

3.3. Repair of the cracked DT-joint

3.3.1. Locating the crack tip. For crack-deflecting holes to increase fatigue endurance, it is vital to determine the exact location of the crack tips to place the holes in a favourable spot. Several steps were carefully carried out to validate the crack tip's location. First, the crack tips were located with the strain gauge readings and a microscope. Then MPI with fluorescent was conducted to validate the position. Finally, MPI with fluorescent was performed repeatedly with the specimen subjected to a static tensile load of 30kN to open the crack. As there was a good agreement on the crack tip position in all three inspections, the crack tip position was concluded to span from -38° to +21° in Q3.

3.3.2. Hole drilling and weld toe grinding. The specimen was repaired by drilling 18mm holes in the crack tip proximity. Positioning of the holes was made in alignment with Atteya [8], by centring the holes 12° ahead of the crack tip and shifting them 15mm perpendicular from the weld toe. As the crack spanned from -38° to +21°, the holes were placed centred at -50° and +33°. The objective of the crack-deflecting holes was to deflect the crack away from the weld toe, which holds significant stresses and arrests the crack in the holes.

In addition to the crack deflecting hole repair, weld toe grinding was performed in Q3 behind the holes. The repair was performed with a rotating "balled-nosed tree" carbide burr with a maximum diameter of 8mm and was used to grind the weld toe in compliance with BS 7608 [11]. The weld toe was further enhanced by sanding the weld toe in a direction perpendicular to the weld toe with a 240-grit paper, which produced a surface roughness of 58.5μm [12]. Post repair, the emerged weld toe profile exhibited an average depth and radius of 0.5mm and 3.4mm, respectively, which aligns with the DNV C203 and BS 7608 specifications.

3.3.3. Instrumentation in the repaired quadrant. After the repair, a set of three new strain gauges was installed around each crack deflecting hole in the repaired quadrant to monitor existing and new crack development. The strain gauge layout in Q3 is shown in Figure 8.

3.4. Performance of the repaired DT

3.4.1. Post-repair test Loading. In the final phase of the testing program, the performance of the repaired DT-Joint was tested by subjecting the specimen to additional cyclic stress. As the crack induces greater SCF while propagating, the load range was reduced to ensure that the stress range remained within the bounds of high cycle fatigue throughout the entire post-repair testing. A load range of 27kN was selected, maintaining an R-ratio of 0.1, resulting in a minimum load (Pmin) of 3kN and a maximum load (Pmax) of 30kN. The frequency was initially set at 5Hz but was later increased to 10Hz once the strain measurements stabilised and showed no indication of further crack propagation or initiation. The strain gauge readings stabilised after achieving a through-thickness crack between the crack-deflecting holes.

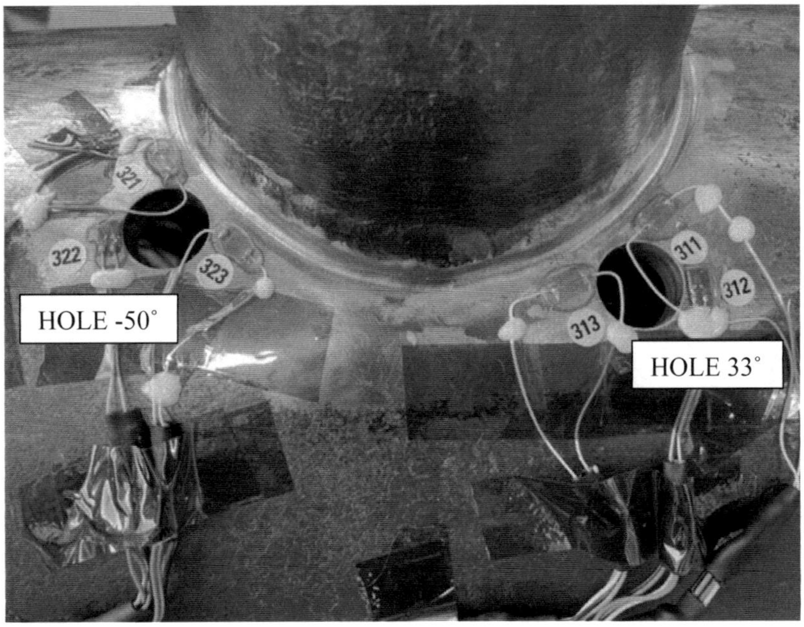

Figure 8. Strain gauge layout in repaired quadrant

3.5. *Result of post-repair testing.* A full through-thickness crack was achieved between the holes after 930 000 cycles, and the testing continued for another 2 585 000 cycles before test termination giving a total of 3 515 000 cycles in the post-repair testing. A summary of the fatigue life in the post-repair testing is shown in Table 2.

Table 2. Repaired specimen fatigue life

Fatigue Life	÷50° Hole	+33° Hole
Ni1	540 000	900 000
Ni2	910 000	930 000
Run out	3 515 000	

Ni1 surface crack into the hole; Ni2 though-thickness crack into the hole

3.6. Discussion

As there was no sign of crack initiation or further crack development, the conclusion that the repair had successfully increased the fatigue endurance in the specimen was drawn. Thus, the testing was stopped after a total of 3 515 000 cycles in the post-repair testing phase. However, to study further crack initiation and propagation it was decided to break the specimen by continuing the test at a significantly higher load range. The new load range chosen gave stresses in the borderline between High and low cycle fatigue.

3.7. Braking specimen load

3.7.1. *Test loading when breaking the specimen.* The new load range selected was adjusted to 54 kN, maintaining an R-ratio of 0.1. This load range gave a maximum load (Pmax) of 60 kN and a minimum load (Pmin) of 6 kN. Since the new stress range fell into the low cycle fatigue regime, events were anticipated to occur at a considerably accelerated rate. Hence the frequency was reduced back to 3 Hz.

IOP Conf. Series: Materials Science and Engineering 1294 (2023) 012032 doi:10.1088/1757-899X/1294/1/012032

3.7.2. Results from breaking the specimen. Previous testing of equal specimens has shown reverse coalescence, a phenomenon where the crack initiates in the weld toe beyond the crack-deflecting hole (crown side) and propagates back into the hole [8]. This behaviour does not align with the expected behaviour before the work of Atteya [8], where the crack initiates at the hole.

As per Atteya's discovery, the reverse coalescence phenomenon was present at both the crack-deflecting holes in the specimen tested in this experimental investigation, as shown in Figure 9.

Figure 9. Crack initiation in front of SG crown-side at +33° hole

The testing was terminated after 260 000 cycles. At this point, the crack had propagated back into both holes and extended over a total length exceeding 1.5 multiplied by brace diameter marking N4. The fatigue life stages are summarized in Table 3.

Table 3. Fatigue life during breaking the specimen

<table>
<tr><th>Fatigue Life</th><th>÷50° Hole</th><th>+33° Hole</th></tr>
<tr><td>N1*</td><td>208 000</td><td>93 000</td></tr>
<tr><td>N2*</td><td>250 000</td><td>145 000</td></tr>
<tr><td>N4</td><td colspan="2">260 000</td></tr>
</table>

N1 Crack initiation beyond the hole 15% drop in strain range*
N2 Crack entering hole reversely*

4. Evaluation of the repair's effectiveness

4.1. Comparison with existing tests

An empirical approach to evaluating the efficacy of the repair is by comparing the number of cycles until N3 with those leading to N4. In this paper the ratio between N4 and N3 is defined as dR, Equation (1). From previous testing of similar tubular joints and test conditions without any repair the maximum recorded dR is about 30% [8, 13]. If the post repair fatigue test would reach a dR of 100% without any further crack initiation the repair would have fully reinstated the as-built fatigue life.

$$dR = \frac{N4 - N3}{N3}\% \tag{1}$$

The specimen was subjected to a lower cyclic stress range in the post-repair phase than in the pre-cracking phase. Therefore, a direct comparison to the pre-crack fatigue endurance would not be possible. Nevertheless, the cycle count to reach N3 under the post-repair stress range was estimated by shifting the SN curve to the N3 result from the pre-cracking phase, as shown in Figure 10. Subsequently, from this shifted S-N curve, the projected cycle count to achieve N3 under the stress range employed in the post-repair phase could be derived. Eventually, this enabled comparing the number of cycles in the post-repair phase and the estimated number of cycles to N3. The procedure employed to calculate dR is summarised as follows:

1. Pre-crack result ($\Delta P = 40\text{kN}$):
   a. $\sigma_{\text{nom}}(14.16 \text{ MPa}) \cdot \text{SCF}(19.4) = \text{HSSr}(274.7 \text{ MPa})$
   b. Number of cycles to N3 = 1 515 000
2. Shifting the SN curve to N3 from the pre-crack phase:
   a. New log(a) = 13.50
   b. m = 3
3. Estimate of the Number of cycles to N3 under post-repair load ($\Delta P = 27\text{kN}$):
   a. $\sigma_{\text{nom}}(9.19 \text{ MPa}) \cdot \text{SCF}(19.4) = \text{HSSr}(178.24 \text{ MPa})$
   b. Estimated number of cycles to N3 = $10^{\log(a) - m \cdot \log(\Delta\sigma)} = 5\ 546\ 000$
4. Number of cycles in the post-repair phase ($\Delta P = 27\text{kN}$):
   a. Number of cycles to run out = 3 515 000
5. Calculating dR = 3 515 000/ 5 546 000 = 63%

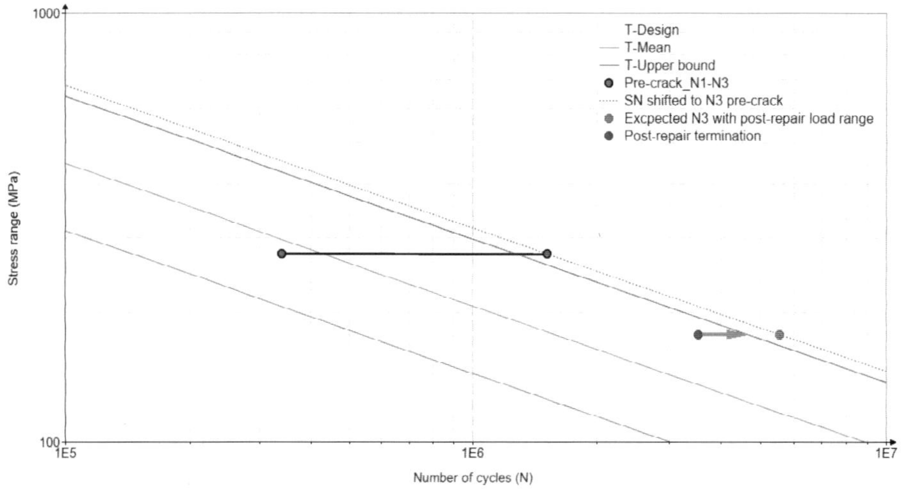

Figure 10. Standard SN curve and SN curve shifted to pre-crack result

The dR ratio of 63% derived from the approach employed in this section is about 100% higher than the greatest dR number for similar specimens tested without repair. Consequently, the conclusion can be drawn that drilling crack-deflecting holes combined with weld-toe grinding has successfully increased fatigue endurance. It should also be emphasised that N4 has not been reached under these loading conditions, but it was decided to stop the testing due to time limitations and that sufficient data

IOP Conf. Series: Materials Science and Engineering 1294 (2023) 012032 doi:10.1088/1757-899X/1294/1/012032

was achieved. Furthermore, there was no sign of crack initiation or development at the point of test termination. It should be emphasised that this approach is very conservative, as the post repair testing results is compared to a number of cycles laying above the upper bound of the SN curve.

4.2. *Comparing recorded stress levels with SN-curve*

4.2.1. Determining cycle count and stress range. In the post-repair testing, the specimen underwent 3 515 000 cycles until defined as a run-out. However, it took 930 000 cycles to achieve a through-thickness crack between the crack deflecting holes, in which the crown side strain gauge (SG with highest strain values) location where the new crack was expected to occur was subjected to a variable stress range as the crack propagated.

To evaluate the accumulated damage during these first 930 000 cycles and its relevance for the fatigue life of this location, an equivalent stress range for these cycles was calculated, and the resulting damage was established based on the SN curve. This damage was then used further to determine the equivalent number of cycles this part of the test represents, correcting for the stress level at this location after through thickness between the crack deflecting holes. The total number of cycles was then determined to 2 721 000.

The stress range recorded in post repair testing after a full through thickness crack between the holes was 250 MPa. However, only the directional strains are captured by the strain gauge therefore a finite element model was created to determine the factor to obtain the principal stresses [14]. Based on this analysis an amplification factor of 1.08 was determined which then gives a stress range of 269 MPa.

4.2.2. *Comparison with SN curve*

The basis for the comparison is 2.7 million cycles with a stress range of 269 MPa. Considering this result, it is immediately evident that the fatigue endurance in the post-repair testing is significantly higher than the upper bound of the T curve, which sets the basis for discussing the weld-toe grinding effect.

Both DNV-RP-C203 and BS 7608 provide an improved SN curve for welded connection subjected to weld toe grinding, as presented in Figure 11. By comparing the number of cycles from the design T curve with the improved SN curves, it is possible to derive an improvement factor of the weld toe grinding from DNV and BS for the given stress range of 269 MPa.

- Improved fatigue life according to DNV = 1.6
- Improved fatigue life according to BS = 2.0.

The reason for deriving these improvement factors, as opposed to simply extracting the number of cycles from the SN curves, is to enable the consideration of the fatigue endurance during the pre-cracking phase in the comparison. By utilising the SN curve shifted to the pre-crack result, the following number of cycles is expected:

- No improvement from weld-toe grinding: 1 616 000 cycles
- Improvement from grinding according to DNV: 2 636 000 cycles
- Improvement from grinding according to BS: 3 227 000 cycles
- **Number of cycles to run out in post-repair testing: 2 721 000 cycles**

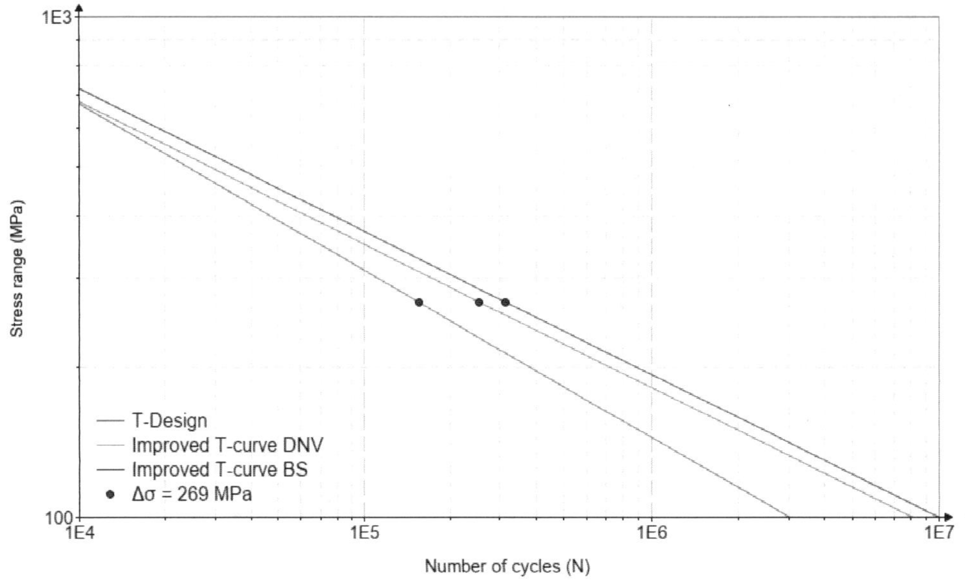

Figure 11. Improved SN curve from weld-toe grinding

It is evident from these considerations that the weld-toe grinding successfully increased the specimen's fatigue endurance. The number of cycles in the first attempt of the post-repair testing excided the number of cycles expected according to DNV when improved with weld toe grinding.

5. Conclusion

The performance of crack deflection holes as a repair method in the high-cycle fatigue domain has been investigated in this work, where crack deflection holes has been drilled on the chord-side in the vicinity of the crack tip. In addition, weld toe improve by grinding have been implemented.

Based on this work it is indicated that crack deflection holes combined with the weld-toe grinding managed to arrest the crack initiation in post-repair testing for high cycle fatigue loading. This is assumed to be a result of the weld-toe profile improvement, the improved surface roughness from the grinding, the sanding at the weld-toe and the reduced stress concentration due to the holes compared to an unrepaired cracked tubular joint.

The load range was increased in a second post-repair test to a high-stress range, leading to plastic strain and stress amplitudes and subsequent crack initiation and failure. In this high-stress range test, the weld toe-grinding had a limited enhancement on the specimen's fatigue life under such strain amplitudes.

Given the considerable variation and substantial uncertainties associated with the fatigue phenomenon, a single test cannot definitively establish a general understanding of the fatigue performance. However, in light of existing testing and the findings within this work, the effect of the studied repair method appears very promising.

6. Further work

As the experimental results indicated a significant fatigue life enhancement of tubular joints repaired with crack-deflection holes combined with weld-toe grinding, future research should focus on providing more empirical evidence on the method by investigating the effects of the repair on:

- Different joint configurations, e.g. T-joints, K-joints, and Y- joints
- Joints with different diameter and wall thickness ratios
- Joints subjected to different cyclic loading modes, e.g. In-plane and out-of-plane bending
- Joints in seawater with cathodic protection and free corrosion

Considering the crack initiation beyond the holes and the reverse coalescence phenomenon in the post-repair testing, future work should also focus on optimizing the cut-out repair, as already stated by Atteya [8]. An optimal hole shape, size and placement would decrease the stress concentration even further, making the repair method efficient for tubular joints subjected to higher nominal stress ranges and more extensive degradation.

Acknowledgement: The opinions expressed herein are those of the authors and they should not be construed as reflecting the views of any organization associated with the author.

References

[1] Almar-Næss A 1985 Fatigue handbook : offshore steel structures. 1985, Trondheim: Tapir.
[2] Atteya, M et al. 2021 Experimental and numerical study of the elastic SCF of tubular joints. *Mater.* **14**(15), 4220.
[3] Ersdal, G., J.V. Sharp, and A. Stacey, *Ageing and Life Extension of Offshore Structures: The Challenge of Managing Structural Integrity*. 2019, Newark: Newark: John Wiley & Sons, Incorporated.
[4] Lotsberg I 2016 *Fatigue Design of Marine Structures*. Cambridge: Cambridge: Cambridge University Press.
[5] Tubby P 1989 *OTH 89 370 Fatigue performance of repaired tubular joints.*
[6] Sharp J V and G Ersdal 2021 *Underwater inspection and repair for offshore structures*, Hoboken: John Wiley & Sons, Inc.
[7] Atteya M et al. 2020 *Crack arresting with crack deflecting holes in steel plates.* in *ASME 2020 39th Int. Conf. Ocean, Offshore Arctic Eng.***3,** https://doi.org/10.1115/OMAE2020-19358
[8] Atteya, M., 2023 Crack arresting using cutout repair in tubular offshore joints, *PhD thesis,* University of Stavanger (to be published), Norway.
[9] ASTM, *A370 – 19 Edition September 2019*. 2019, ASTM INTERNATIONAL.
[10] DNV, *DNV-RP-C203 - Fatigue design of offshore steel structures*. 2019, DNV AS.
[11] British Standard, *BS 7608:2014+A1:2015 - Guide to fatigue design and assessment of steel products*. 2015, BSI Standards Publication.
[12] Upmold. *Surface finish sandpaper grit chart*. 2017; Available from: https://upmold.com/surface-finish-sandpaper-grit-chart/.
[13] HSE, *OTH 92390 Background to new fatigue design guidance for steel joints and connections in offshore structures*. 1999.
[14] Riise S 2023 Experimental assessment of cracked tubular joints, *Master thesis,* University of Stavanger, Norway.

Statistical evaluation of tubular joint SCFs from test data and finite element analyses

G Ersdal [*], M Atteya and O Mikkelsen

Faculty of Science and Technology, University of Stavanger, Norway

[*] Correspondence: gerhard.ersdal@uis.no

Abstract. Stress concentration factors (SCF) derived from both FEA and measurements from laboratory experiments are associated with uncertainties. To understand these uncertainties, SCF values can be determined from a series of tests or possibly also several SCF-values estimated from converged finite element analysis. The challenge is to determine a recommended value for the SCF based on these tests or analysis, ensuring that the resulting SCF along with the SN-curve meets the code required 2.3% exceedance probability. In this paper, structural reliability analysis (SRA) methods are employed to model the combined uncertainty of the parameters representing the SN-curve and the measured SCFs from experimental test joints. Based on this analysis, the SCF that, when paired with the design SN-curve, provides the minimum fatigue life is established. From this analysis, a method to provide characteristic values of the SCF is suggested. Finally, FEA of the same detail, with and without modelling the weld, using both first and second-order elements are performed. From these analyses the SCF is estimated using linear extrapolation and direct extraction. Based on these findings and their comparison to the characteristic SCF, a suggested method to extract SCFs from FEA is provided.

1. Introduction

The design of tubular joints against fatigue failure is normally based on the 'hot-spot' stress derivation then using it with the T-curve, as found in standards like DNV RP C203 [1]. The T-curve correlates the 'hot spot' stress range at the joint intersection with the number of cycles to develop a through-thickness crack in the stressed member. The design T-curve is assumed to have a normal distribution in logarithmic scale, and can be expressed as:

$$\log N = \log A - m * \log \Delta S \tag{1}$$

where N is the number of cycles, $\log A$ is the curve intercept, m is the inverse slope of the curve and ΔS is the stress range. The value of $m = 3$ if $N \leq 10^7$ cycles and $m = 5$ if $N > 10^7$ cycles are applicable for fatigue curve in air environment.

SN-curves are based on linear regression by the least mean square error of selected experimental data points of a range of tubular joint types, loading types and loading ratios. It is well known that there is a significant variation in the resulting fatigue lives and more or less identical tests can be performed gaining different number of cycles to failure. These uncertainties are normally assumed to be related to issues like inherent randomness in material surface and quality, the quality of the welding and associated presence of crack-like defects from the welding, the magnitude of the stress level, stress increase due to local joint geometry and stress increase due to weld geometry. In practical engineering the inherent randomness in material and weld quality is assumed included into SN-curves, determined by several laboratory experiments. Hence, the stress ranges from these experiments are plotted against the number of cycles until occurrence of a through-thickness crack in the specimen and these are plotted in a log-log scale diagram. A regression analysis can then be used to determine the best approximation for a mean value of these data. In addition, the standard deviation (SD) for the $\log A$ parameter is established and as a lower bound the design SN-curve is determined 2 SD below the mean curve, giving a maximum of 2.3% chance of a fatigue crack at the calculated value of cycles. The data used to establish the design T-curve, along with the mean and design T-curves for in air conditions, are shown in Figure 1, based on HSE [2].

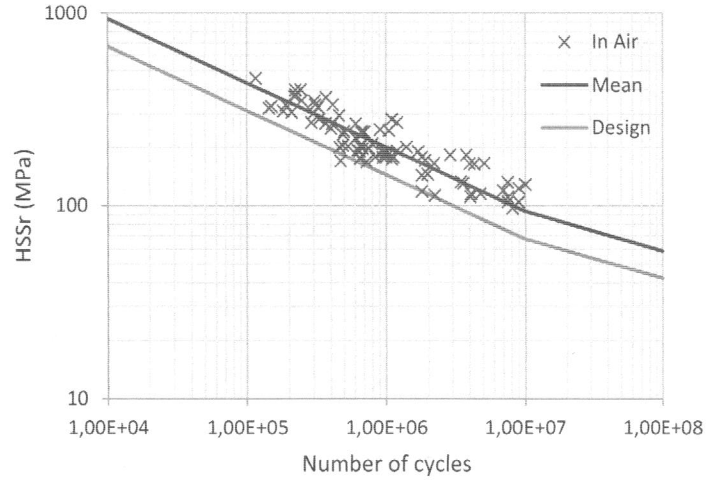

Figure 1. T-curve and data, based on HSE [2] where HSSr is the hot spot stress range

The T-curve used in standards for design of offshore structures are reported to have a mean value of the *logA* of 12.92 and a standard deviation (SD) of the *logA* value of 0.23. The design S-N T-curves is taken as the mean value of the *logA* minus two standard deviations of *logA*.

To create data-points for the SN-curve, careful measurements of the cyclic stress magnitude (S) is needed, along with accurate determination at which number of cycles (N) the specimen is experiencing a through-thickness crack. The cyclic stress magnitude is often determined by the hot-spot stress (HSS) range, which is again often determined by the stress concentration factor (SCF) multiplied by the nominal stress in the brace. The SCF is the ratio between the HSS and the nominal stress in the specimen away from the joint and can be found in design codes, typically providing the Efthymiou stress concentration factors (SCF) for the fatigue life estimation of tubular joints [3][4]. These are often assumed to be slightly biased to the safe side. Hence, many engineers seeking fatigue life optimization are using finite element analysis (FEA) or measurements direct from laboratory tests to obtain unbiased SCFs for tubular joints, both for design optimization and particularly for life extension of existing structures. However, stress concentration factors (SCF) both based on FEA and measurements from laboratory experiments are associated with uncertainties. For laboratory experiments these uncertainties may be a result of the individual geometry of the weld profile, the placement of the strain gauges, the strain measurements and the SCF extrapolation methodology. For an FEA such uncertainties are related to modelling of the weld profile, element type, mesh density, SCF extrapolation methodology and other numerical inaccuracies.

In this work we will investigate the combined uncertainty of the inherent randomness in material and weld quality and the stress concentration factor (SCF). One method of investigating uncertainties is the use of structural reliability analysis (SRA), for example by the use of the so-called first-order reliability method (FORM). By this method it is possible to model the uncertain parameters by their assumed distribution functions and determine a so-called design point providing a given probability of limit state exceedance (where the limit state in this case is the SN-curve being the assumed formula for when a through-thickness crack is expected). Standard practice is to use the mean value of the measured SCFs and this study is essentially to check if the mean value is the correct choice.

Further, the same joint has been numerically modelled by several methods using finite element analysis (FEA), as previously presented by Atteya et al [5]. These simulated SCFs will be compared with what is found to be the "recommended" SCF to determine some initial recommendation for determining SCFs from FEA.

2. Recent experimental work and numerical studies used as basis for the statistical evaluations

As part of PhD research project [6], the HSSs and their associated SCFs were measured for four double T tubular joints (DT1-DT4), with each joint having 4 quadrants (Q1-Q4) providing a total of 16 quadrants where the SCFs could be determined. The joint configuration and the arrangement of strain gauges is shown in Figure 2.

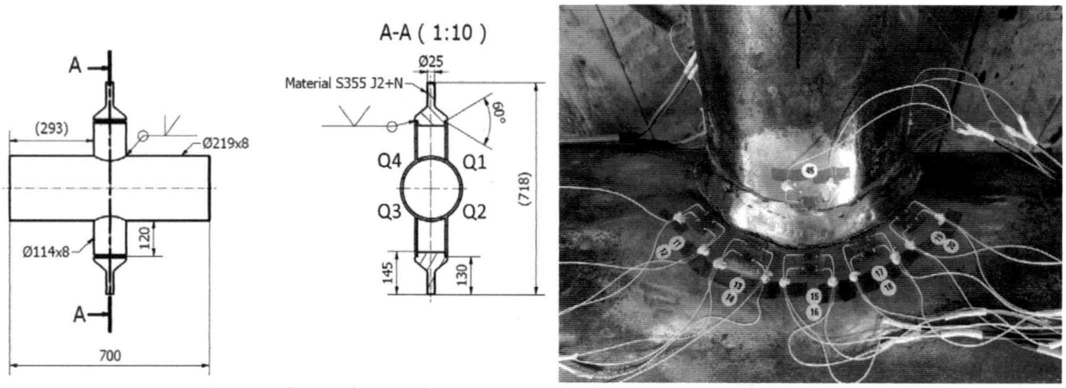

Figure 2. Joint configuration and arrangement of strain gauges for Q2 (quadrant 2)

The SCFs were determined by standardised methods and found to be as shown in Table 1, as also reported in [5]. The observed variations in results were assumed to be due to deviations in actual weld profiles, in the precision of actual coordinates and orientation of strain gauges as well as the selected extrapolation method.

Table 1. SCFs determined by standardised methods for each quadrant of the four test joints.

	Test joints				
	DT1	DT2	DT3	DT4	All
Quadrant 1	18.80	19.09	17.18	21.19	
Quadrant 2	18.22	20.67	22.24	23.30	
Quadrant 3	18.96	20.05	19.96	18.65	
Quadrant 4	17.76	21.60	*no reading*	20.40	
Mean value	18.43	20.36	19.79	20.88	19.87
SD	0.55	1.06	2.54	1.93	1.72
CoV	0.03	0.05	0.13	0.09	8.7%

The distribution of the SCF's at the saddle point from the experimental work shown in Table 1 were plotted in a normal distribution probability paper, as shown in Figure 3, and found to fit reasonably well the normal distribution.

By bootstrapping these data, it was indicated that the 95% confidence interval for the mean value of the SCF was [19.17, 20.63] and for the standard deviation [1.2, 2.0], providing a coefficient of variation (CoV) in the range of 6–10%.

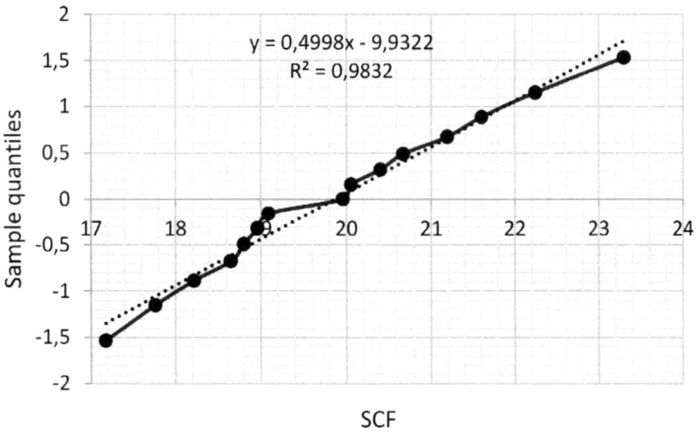

Figure 3. SCF values from experimental work plotted in a normal distribution paper, indicating a mean value of the SCF of 19.9 and a standard deviation of 2.

After these initial evaluations were performed, the precision in the strain gauge locations was studied in more detail at eight locations on the saddle of the chord side, using two specimens from a total of four specimens [5]. The variation between actual and average strain gauge locations serves as useful indicator of the overall accuracy of the SCF for the specimen. The actual strain gauge's locations to the weld toe were measured to an accuracy of ±0.1 mm, then the measured points and their strain readings were used to extrapolate for the SCFs linearly. Figure 4 shows the distribution of strain gauges at the intended "a" and "b" locations and the extrapolation lines to the weld toe to estimate the hot-spot stress and associated SCF value for each pair of points. The results show that the location of the strain gauges at point "a" were scattered without evidence of any systematic error, as indicated by Figure 4. However, for the strain gauges at point "b" there is a scatter around a systematic mean shift of 1.1 mm from the intended "b" location, as indicated by Figure 4.

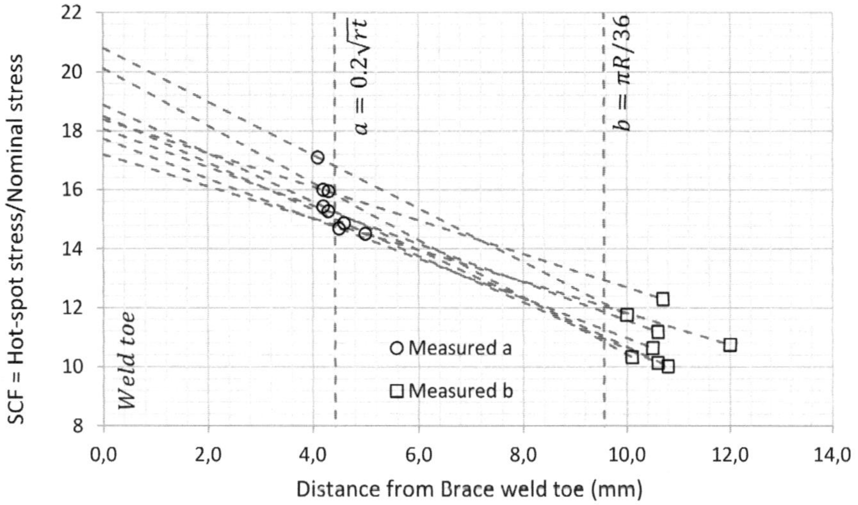

Figure 4. Experimentally measured SCFs sensitivity

The systematic shift in mean value can be explained by the length of the strain gauge carriers. The carrier's length was 5 mm while the distance between point "a" and "b" was only 5.1 mm, which made it challenging to manually position the strain gauges spot on. After strain gauges installation, the average distance between "a" and "b" was approximately 6.3 mm.

The actual extrapolated SCFs for the four quadrants on the two test joints, prior to correction and after correction, are shown in Table 2.

Table 2. SCFs at test joints after correcting for the location of the strain gauge positions.

	Test joints – uncorrected and corrected (*)					
	DT1	DT2	All	DT1*	DT2*	All *
Quadrant 1	18.80	19.09		18.07	18.39	
Quadrant 2	18.22	20.67		17.73	20.12	
Quadrant 3	18.96	20.05		18.50	18.89	
Quadrant 4	17.76	21.60		17.19	20.80	
Mean value	18.43	20.36	19.4	17.87	19.55	18.71
SD	0.55	1.06	1.29	0.55	1.11	1.21
CoV	3%	5%	6.6%	3%	5.7%	6.5%

For the SCFs corrected for the strain gauge location, the average from these eight points is 18.71 with a standard deviation of 1.21, providing a CoV of 6.5%. This should be compared to the uncorrected average for the same two specimens, providing a mean SCF of 19.39, a standard deviation of 1.29 and a CoV of 6.6%. As reported in [5], the difference between the mean value between these two samples can be assumed to be due to the systematic inaccuracy in the location of strain gauge b, while the variation in the results from the eight points is a combination of other inherent uncertainties (such as weld geometry, material properties and measurements). This implies that it could be argued that the shift in strain gauge location introduces a bias of 3.7% to the safe side, which can also easily be seen to be reasonable by a geometric evaluation of the mean values of locations and stresses at "a" and "b".

$$\frac{S_a - \frac{S_b - S_a}{x_b - x_a} \cdot x_a}{S_a - \frac{S_b - S_a}{x_{b\_corrected} - x_a} \cdot x_a} = \frac{16 - \frac{11 - 16}{9.56 - 4.42} \cdot 4.42}{16 - \frac{11 - 16}{10.66 - 4.42} \cdot 4.42} = 1.04 \quad (2)$$

where $S_a - \frac{S_b - S_a}{x_b - x_a} \cdot x_a$ is the extrapolation formula for the stress at the hot-spot, S_a and S_b the stress reading at locations a and b respectively (average values of S_a and S_b is used), x_a and x_b is the intended locations of strain gauge a and b respectively and $x_{b\_corrected}$ is the corrected mean location of strain gauge b.

The data in Table 2 can be plotted in a normal probability paper as shown in Figure 5, still to some extent indicating that the data follows a normal distribution. The shift in the mean value is obvious and the slope is showing signs of being similar.

For the entire sample, the average SCF was found to be 19.87, with a standard deviation of 1.72 and CoV of 8.7%. To adjust for the strain gauge location, assuming the same trend for the two remaining specimens as for the two tested, the mean value of the SCF is set to 19.87/1.037 = 19.16, while there were not found any obvious reason to reduce the CoV from 8.7% [5]. It is also assumed that the data still can be modelled by a normal distribution function.

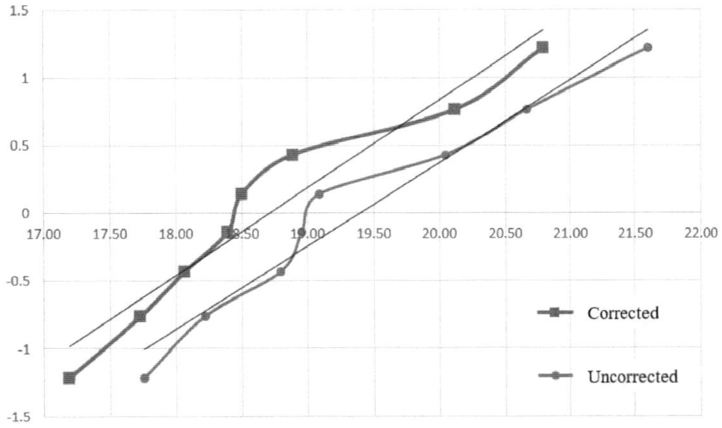

Figure 5. Corrected and uncorrected SCF values from the two experimental test joints plotted in a normal distribution paper, indicating a change in the mean value for the SCF from 19.4 to 18.7.

In Atteya et al [5], the SCFs were also determined by different finite element analysis (FEA). The variation of the SCFs for models with different weld profiles and different extrapolation methods is shown in Table 3. The "smallest" accepted weld profile was the shortest acceptable weld leg length as per fabrication specifications and the "largest" was the most extended acceptable weld leg length as per fabrication specifications [5].

Table 3. SCFs for different weld profiles based on unaveraged principal stresses with 1st and 2nd order brick elements.

	Direct Extraction from $0.1\sqrt{rt}$ Location		Linear Extrapolation	
Weld profile	1st Order	2nd Order	1st Order	2nd Order
No Weld	-	-	20.2	22.2
Idealised smallest accepted weld profile	20.1	19.8	18.5	20.4
Idealised largest accepted weld profile	18.3	20.3	19.1	21.0

The mesh discretization with convergence study and the use of extrapolation method is reported in [5]. With the mean SCF determined above, only two of the models are underpredicting the SCF. However, the further studies will determine if the mean value of SCF are the correct SCF to be used in design fatigue analysis.

3. Probabilistic modelling of uncertainty in experimental SCFs

To study the combined uncertainty of the inherent randomness in material properties and weld quality and the stress concentration factor (SCF), the problem is investigated by a first order reliability method (FORM) analysis [7]. A simple limit state is established for the problem illustrating the probability of the number of cycles determined from the SN-curve based on $logA$ and SCF as stochastic variables exceeding the number of cycles determined from characteristic values of $logA$ and SCF. The purpose is to determine the characteristic value of the SCF providing a limit state exceedance probability of 2.3% (as per the definition of the fatigue limit state).

The limit state function is described as:

$$g(logA, SCF) = [logA_{ch} - m \cdot \log(SCF_{ch} \cdot \Delta\sigma_{nom})] - [logA - m \cdot \log(SCF \cdot \Delta\sigma_{nom})] \quad (3)$$

where $logA_{ch}$ is the characteristic value of the slope of the SN-curve as found in standards and SCF_{ch} is the characteristic value of the stress concentration factor (SCF) determined by the analysis. The stochastic variables are modelled as shown in Table 4.

Table 4: Description of stochastic variables

Variable	Description	Modelling	parameter 1	parameter 2
$logA_{ch}$	Characteristic $logA$	Fixed	12.46	
m	Inverse slope of SN curve	Fixed	3	
SCF_{ch}	Characteristic SCF	Fixed	Determined by the analysis	
$\Delta\sigma_{nom}$	Nominal stress range	Fixed	10 MPa	
$logA$	the SN curve intercept	Normal distributed	Mean=12.92	SD=0.23
SCF	Stress concentration factor	Normal distributed	Mean=19.16	SD=1.67

Based on these parameters, a FORM analysis was performed for the given limit state, providing a 2.3% exceedance probability ($\beta = 2$) of the limit state function for a SCF_{ch} of 19.9. The design points in the FORM analysis are $logA_{DP} = 12.5$ and $SCF_{DP} = 20.45$. The design points can be argued to be the values that should be used in a design calculation. However, since $logA$ is given in design codes as 12.46 the SCF should be adjusted to take this $logA$ value into account.

$$SCF_{adj} = \frac{10^{\frac{\left(12.46 - \left(logA_{DP} - m \cdot log(SCF_{DP} \cdot \Delta\sigma_{nom})\right)\right)}{m}}}{\Delta\sigma_{nom}} = 19.9 \quad (4)$$

The limit state function with a "characteristic" SCF of 19.9 is illustrated in Figure 6, with a circle of radius 2 illustrating the exceedance probability 2.3% to indicate the design point for the limit state. The sensitivity to changes in the two stochastic parameters is shown in Figure 7 and clearly indicate that the uncertainty in $logA$ by far supersedes the uncertainty in the SCF.

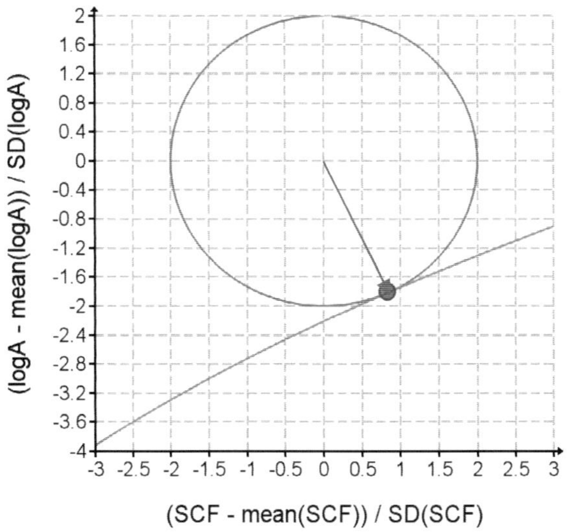

Figure 6. Graphical representation of the limit state function with a characteristic SCF of 19.9

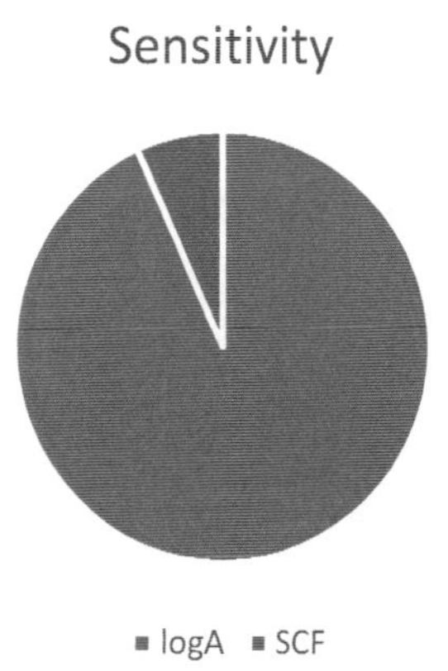

Figure 7. Sensitivity of the two stochastic parameters in the limit state

An alternative to the above FORM analysis would be an inverse first-order reliability method (IFORM) analysis as developed by Winterstein et al [8]. By an IFORM analysis any combinations of $logA$ and SCF that gives 2.3% probability of exceedance (probability of fatigue crack according to the SN-curve) can be found approximately. In these analyses both $logA$ and SCF are modelled as normal distributed in accordance with the data given above. However, as the limit state includes a term $\log(SCF)$, this term needs to be modelled as a log-normal distribution. Hence, the $logA$ and SCF variables are transformed into a standard normalised U-space by the following transformation:

$$logA = SD_{logA} \cdot U_{logA} - \mu_{logA} \tag{5}$$

$$SCF = 10^{[SD_{\log\_SCF} \cdot U_{SCF} + \mu_{\log\_SCF}]} \tag{6}$$

where μ_{logA} is the mean value of $logA$, SD_{logA} is the standard deviation of $logA$, U_{logA} is $logA$ transferred to a standard normalized parameter, $\mu_{\log\_SCF}$ is the mean value of $log\ (SCF)$, $SD_{\log\_SCF}$ is the standard deviation of $log\ (SCF)$ and U_{SCF} is the SCF transferred to a standard normalized parameter.

By defining the parameters in the U-space in such a manner that they define a circle with a radius of $\beta = 2$ (indicating a 2.3% exceedance probability), the contour line of allowable combinations of $logA$ and SCF are found.

At the minimum value of $logA$, the mean value of SCF is found and, hence, it could be argued that a characteristic value of $logA$ should be combined with the mean value of SCF. However, if the number of cycles according to the SN-curve is calculated for all allowable combinations of $logA$ and SCF, the set of parameters ensuring the lowest fatigue life with exceedance probability of 2.3% can be found. This minimum $logN$ is found to be 5.565 for the variable pair $logA = 12.51$ and $SCF = 20.63$. This SCF is then adjusted to account for the fact that the $logA$ in codes are 12.46:

$$SCF_{ch} = \frac{10^{\left(\frac{12.46-5.565}{3}\right)}}{\Delta\sigma_{nom}} = 19.9 \tag{7}$$

The "contour line" for values of logA and SCF that fulfils the required exceedance probability of 2.3% are shown in Figure 8, with marking of the design point pair of variables pair $logA = 12.51$ and $SCF = 20.63$ (black circle) in addition to the adjusted pair of variables pair $logA = 12.46$ and $SCF =$

19.9 (red square, slightly off the contour line indicating 2.3% exceedance probability). In addition, the closest value on the contour line with $logA = 12.46$ and $SCF = 19.3$ is indicated (black cross).

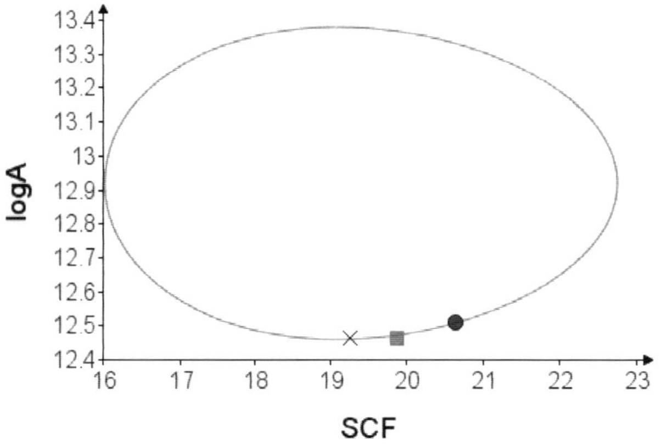

Figure 8. Contour line for logA and SCF providing a 2.3% exceedance of SN-fatigue limit state, indicating the SCF at the logA and the minimum logN and the adjusted SCF to account for standardised logA.

The argument for using the adjusted value of 19.9 is, hence, to ensure that the minimum value of $logN$ is obtained, even if this set of parameters provides an exceedance probability that is slightly less than 2.3%. The alternative pair of variables that are on the contour line ($logA = 12.46$ and $SCF = 19.3$) in contrast does not provide a $logN = 5.565$. Hence, the pair of values recommended here are a substitute set of parameters for the design values providing the same $logN$ with the boundary condition given by the standardised $logA$.

The recommended SCF of 19.9 could possibly be estimated by two simplified methods:

1) the mean value of 95% confidence level from the bootstrap, which adjusted by the bias of 3.7% also provides a value of 19.9,
2) the mean value of the measured SCFs plus 0.443 times the standard deviation of the SCFs, as presented in Table 4, which result in a SCF of 19.9.

However, this example is insufficient to indicate whether any of these ways to determine the recommended SCF can be used as a rule. Nevertheless, these approaches appear to be reasonable estimates until more data becomes available.

4. Implication of the probabilistic analysis on the choice of FEA and model

The findings of the probabilistic analysis can be used to evaluate the FEA models and extrapolation methods presented in Atteya et al [5]. As shown in Table 3, the variation in the SCF results is depending both on the element type used (first-order or second-order), the modelling of the weld and the extrapolation method. In Table 5, which is similar to Table 3, the FEA models and extraction methods that produced SCFs lower than what has been found to be the recommended SCF of 19.9 is greyed out.

An analysis neglecting the weld profile combined with linear extrapolation will provide SCFs on the safe side, but it is worth noticing that second-order elements are overestimating the stress by approximately 10%. The direct extraction method is in most cases providing acceptable results, except from first-order elements with the idealized largest accepted weld profile. The linear extrapolation method is providing acceptable results only for second-order elements.

Table 5. SCFs for different weld profiles based on unaveraged principal stresses.

Weld profile	Direct Extraction from $0.1\sqrt{rt}$ Location		Linear Extrapolation	
	1st Order	2nd Order	1st Order	2nd Order
No Weld	-	-	20.2	22.2
Idealised smallest accepted weld profile	20.1	19.8	18.5	20.4
Idealised largest accepted weld profile	18.3	20.3	19.1	21.0
Efthymiou			22.4	

There seems to be some system for the linear extrapolation cases where second-order elements are providing higher SCFs than first-order elements. In addition, in both cases the idealised largest accepted weld profile is providing larger SCFs than the idealised smallest accepted weld profile. The results from the direct extraction method are less systemised.

5. Conclusion

In this paper, a method to determine the SCF that, with the SN-curve given in standards, are expected to provide the code-recommended 2.3% fatigue limit state exceedance. The recommended SCF seems to fit with the upper mean value with 95% confidence interval. In addition, it is shown that $\mu_{SCF} + 0.443 \cdot SD_{SCF}$, where μ_{SCF} is the mean value of the measured SCFs and SD_{SCF} is their standard deviation, may also be used.

Based on the experimental SCF measurements from this study (corrected for strain gauge location), it seems like the SCF predicted by this method will underestimate the experienced SCF in 5 of 15 cases. Hence, this may be a middle route that can be taken in some cases. However, due considerations to the savings in design cost versus the possible significant increase in operational costs, should significant repairs be needed, should be made.

It is important to note that there is no definitive way to prove that measurements and related uncertainties discussed here are representative of similar work conducted in other laboratories. Nevertheless, it is argued that the issues discussed in this paper hold broader significance in a general sense.

A general recommendation regarding FEA to determine SCFs is difficult to give based on this one example, but an indication that an FE model without the weld included in the model seems to be a reasonable choice and is likely the easiest to model. Alternatively, reliable SCF estimates can also be found by using second-order elements in combination with linear extrapolation.

References

[1] DNVGL (2016). Recommended practice DNVGL-RP-C203 Fatigue design of offshore steel structures. DNVGL, Høvik, Norway
[2] HSE Health and Safety Executive *Background to New Fatigue Guidance for Steel Joints and Connections in Offshore Structures* report OTH 92 390, February 1999.
[3] Efthymiou M *Development of SCF formulae and generalised influence functions for use in fatigue analysis* In Recent Development in Tubular Joints Technology; Anugraha Centre: Egham, UK, 1988.
[4] ISO *Petroleum and Natural Gas Industries—Fixed Steel Offshore Structures* ISO/FDIS 19902:2020(E); International Organization for Standardization, Geneva, Switzerland, 2020.
[5] Atteya M, Mikkelsen O, Wintle J and Ersdal G 2021 Experimental and Numerical Study of the Elastic SCF of Tubular Joints, *Mater.* **14**, 4220 https://doi.org/10.3390/ma14154220
[6] Atteya M 2023 Crack arresting using coutu repair in tubular offshore joints, *PhD Thesis*, University of Stavanger, Norway (to be published).
[7] Melchers R E 1999 *Structural Reliability Analysis and Predictions*, John Wiley & Sons Ltd.
[8] Winterstein S R, Ude T C, Cornell C A, Bjerager P and Haver S 1993 Environmental parameters for

IOP Conf. Series: Materials Science and Engineering 1294 (2023) 012033 doi:10.1088/1757-899X/1294/1/012033

etreme respnse: Inverse FORM with omission factors *Proc of ICOSSAR-93* (Insbruck, Austria, 9-13 August 1993).

IOP Conf. Series: Materials Science and Engineering 1294 (2023) 012034 doi:10.1088/1757-899X/1294/1/012034

Availability of fresh water in cold climate regions

A D Awong[1] and O T Gudmestad [1,2*]

[1] UiT, The Arctic University of Norway
[2] University of Stavanger, Stavanger, Norway

*Corresponding author: otgudmestad@gmail.com

Abstract: Challenges related to the availability of fresh water in cold climate regions are discussed with emphasis on possible degradation of the quality of drinking water. The permafrost in this region has been consistent in making sure there is little drainage in the ground. This is important, as fresh water is mainly supplied through melting water from snow and ice in the many winter months. Climate change and warmer weather plays a role in heating up the permafrost and increasing the rate of the melting snow. This has an impact on both the amount of water and potentially the drinking quality. In certain cold region areas, fresh water sources are located not far from the shore. With increased shorline erosion due to warming climate, the fresh water supply is in danger of being contaminated by salt water during large storm events. Furthermore, human activities such as road transport and sewage contamination do also have a significant impact on the water's quality. A bowtie diagram will be used to illustrate the different risk factors and suggestions on how to maintain the quality of fresh water supplies in cold climate areas. The primary focus areas will be Isdammen, near Longyearbyen on Svalbard and Utqiaġvik (formerly Barrow), Alaska.

1. Introduction

The access to clean fresh water is one of the most important requirements for a society. Although there are ways to handle situations where the water quality is poor, the preference is to use clean water from the natural environment. The upgrading of the water quality is typically handled by filtering the water or by adding chemicals to ensure bacteria are destroyed. Some societies (for example in the Middle East) are, however, lacking access to clean fresh water so desalination of seawater is necessary [1, 2]. The process is, however, very costly and requires much energy. Smaller and vulnerable societies cannot afford such expensive solutions. This applies to most smaller societies located in cold climate regions. [3, 4]. In these areas the access to clean fresh water is under pressure. The water quality, however, are normally carefully checked and the water is filtered, and it should not be necessary for the population to boil the water to ensure the quality.

According to the Government of Norway, the temperature within the Arctic has been increasing twice as fast as the global average [5]. The pressure on the clean freshwater resources of drinking quality in cold climate regions comes partly from heating of the permafrost in the ground whereby the water locked in shallow dams could disappear [6]. The water quality could also gradually worsen because of increased human activity. In the worst situation, erosion of the shoreline and increased storm surges and wave-heights due to melting of the sea ice (the fetch length for generation of large waves does increase) [7], could cause saltwater to penetrate freshwater reservoirs. Desalination may not be an economic option, another freshwater reservoir must be identified and established as the drinking water reservoir for the society.

In this paper we will illustrate the challenges by discussing the clean freshwater availability situation in Utqiaġvik (formerly Barrow), Alaska and Longyearbyen, Svalbard, Norway. Emphasis will be on Longyearbyen, located at 78.25o N as this is the location best known by the authors.

2. Permafrost as a barrier to ensure the integrity of a fresh water reservoir

The ground in the Arctic is usually frozen throughout the year. This permafrost holds a significant quantity of frozen water. We call it permafrost when the ground's temperature stays below zero degrees Celsius continuously for minimum two years in a row [8, 9]. The permafrost may go deep below the surface. During summer, when the snow has melted, the top layer of the soil may begin to thaw, but it can still be frozen solid lower down [4]. As the permafrost ensures that when the ice and snow melts, it does not drain the water away, so water can stay there for a longer period. Even though the top layer may melt, the deep permafrost works as a protective ice shield, preventing water drainage [8, 10]. There are multiple factors that may contribute to melting the permafrost. The main factors are ground and surface warming from climate change. These contribute to the melting and increase of the temperature of the permafrost. Another factor is changes in snow cover, which can also be responsible for the development of this phenomenon [11, 12]. The conditions of the hydrologic environment are changing as the results of the thaw in the permafrost [6]. Figure 1 shows how a warmer climate can affect lakes in permafrost regions. If we compare the left lakes on sketches shown in Figure 1a) and b), we see that the water is starting to drain away more on b) because the permafrost is not keeping it in place. This will reduce the amount of water available in the lake (the reservoir). In the pictures on the right, we see the opposite. On Figure 1b) the permafrost has melted some, but there is still ice left keeping it from draining too much. Some of the melting water goes into the lake and increases the amount of water present. We see here that we can use the state of the lakes as an indicator of how much impact the increased temperature has in certain places.

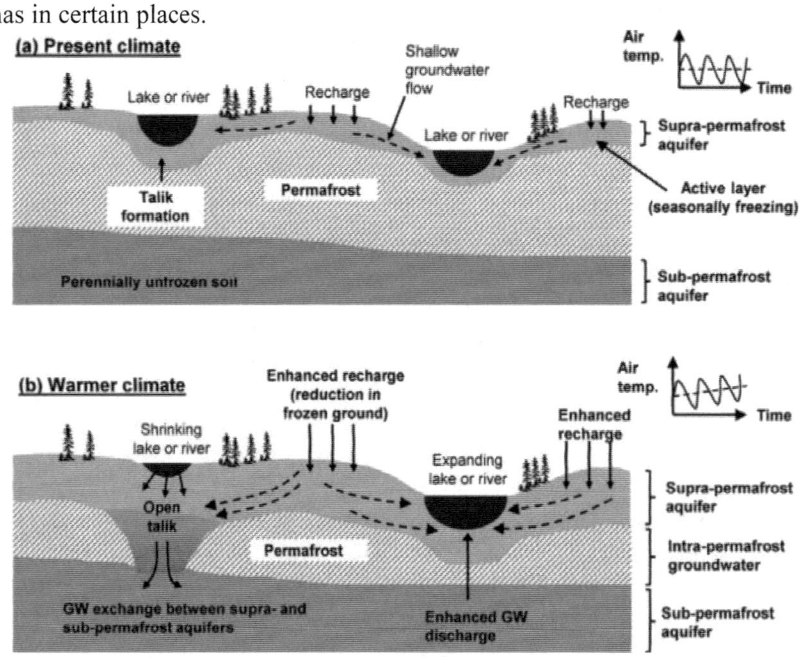

Figure 1. Evolution of the water level in a lake when the permafrost is melting [6].

3. Degradation of water quality due to human activities

Degradation of the quality of fresh water due to human activities is of considerable concern in cold climate as contaminated run-off from human activities will reach the reservoirs more easily than in tempered regions and nothing will disappear into the ground and be filtered, [3, 4]. All contamination is accumulated over the winter season; melting snow where the winter's pollution has accumulated will run-off onto the nearest lake. It should also be noted that foam is used more extensively to extinguish fires in cold climate than elsewhere (as water will freeze to ice when used during the winter), and toxic pollution from the fire-fighting foam may reach the freshwater reservoirs [15].

4. Seawater penetration into a freshwater reservoir

A concern is freshwater reservoirs located close to the shoreline. With decreasing ice cover in the autumn months, the open fetch length will increase causing larger storm surges and generating waves. This cause accelerated shoreline erosion and flooding [6]. Freshwater reservoirs located closer to the waterfront, therefore run an increased risk of being contaminated with salt-water, making the water in the reservoir non-drinkable and unfit for use in most civil activities, as for example for use as firewater during freezing temperature, see also [3, 4].

5. Case study number 1: Utqiaġvik (formerly Barrow), Northern Alaska

5.1 Climate at Utqiaġvik

Utqiaġvik is the northernmost town in North America at 71 °N. The population is about 4.500. The warm season in Utqiaġvik (Figure 2) starts in June and lasts to mid-September. During this period, the average daily highest temperature is 2 °C. The average daily highest temperature is in July with 8 °C [9]. In January, the average daily highest temperature is -21 °C, temperatures down to – 40 °C are not uncommon.

During the year, the sky is cloudy most of the time, partly clouded or mostly clear only 30% of the time in June. The climate is mostly dry with a maximum average monthly rainfall during July and August of 2-3 cm. During the winter, the monthly snowfall is in average 5 cm. Snow can be expected during the months September to May. The wind has an average velocity of 5 to 6 m/s year around [13].

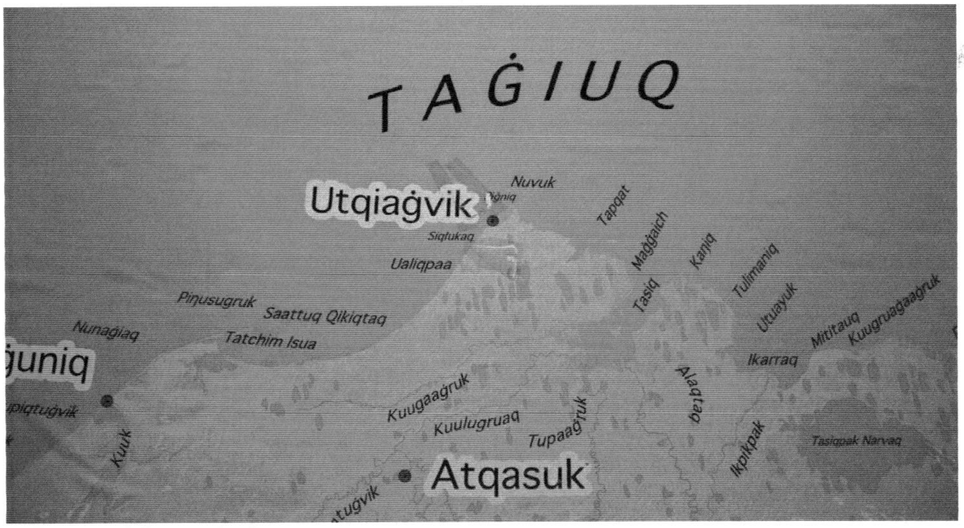

Figure 2. Map showing Inupiaq language place names shows the city of Barrow as Utqiaġvik at the Iñupiat Heritage Centre. (Map prepared by Marc Lester / Alaska Dispatch News)

5.2 Fresh water reservoirs in Utqiaġvik.

In the area, there are many lakes and freshwater ponds, several of these are located very near to the coastline that is exposed to heavy erosion [14] and salt-water intrusion. The reservoir providing fresh water to the town, used to be the Imikpuk Lake, however, the lake is contaminated, and the Isatkoak

IOP Conf. Series: Materials Science and Engineering 1294 (2023) 012034 doi:10.1088/1757-899X/1294/1/012034

Lagoon presently provides the fresh water for the town (Figure 3), although this Lagoon is prone to seawater flooding in case of large storm surges [14]. It is located at the end of the airport runway.

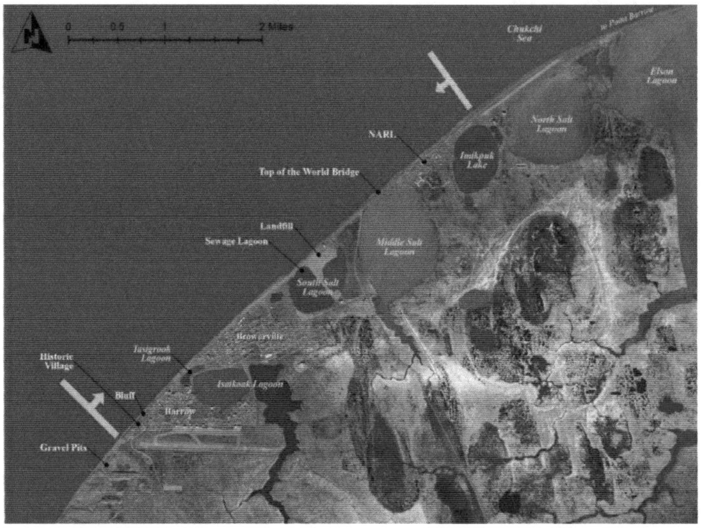

Figure 3. Satellite photo of Utqiaġvik (formerly Barrow), Northern Alaska. [14]. Note the location of the Isatkoak Lagoon providing drinking water to the town and the Imikpuk Lake previously providing the fresh water to the town.

The previous freshwater reservoir, the Imikpuk Lake was declared as unfit for drinking water as it was documented in 2017 that the levels of PFAS (per- and polyfluoroalkyl substances, released as a result of historical firefighting activities) exceeded the U.S. Environmental Protection Agency's (EPA) Lifetime Health Advisories (LHA). The contamination followed the response to two airplane crashes that occurred near Site 5 Airstrip in the 1970's [15]. The fresh water supply is a constant worry for the people living in the town and all efforts are made to keep the present drinking water reservoir free of contamination.

6. Case study number 2: Longyearbyen, Svalbard, Norway

6.1. Climate at Svalbard

6.1.1 Air Temperatures. Svalbard has an Arctic climate, but it has recently warmed significantly due to global warming. According to Norwegian Meteorological Institute, the average temperature in Longyearbyen during December, January, and February was -16° Celsius in the period 1961 to 1990. From 1991 to 2002, the temperature increased to -12 °C, which is 4 °C higher over the same period. In addition, the temperature has risen by 5°C in December and January in other parts of Svalbard [13]. It should be noted that the increase during the winter months is the highest. This will most likely lead to a severe impact on snow cover and the permafrost in the area as well as to provide more rain in future years.

6.1.2 Ground temperature at Svalbard. Svalbard has permafrost, but the ground temperature is rising. The graph below, Figure 4, shows the change in ground temperature in an area on Svalbard from 1998-2017 [16]. It shows a steady rise during the period. The temperature is measured at a depth of 10 m. Closer to the surface, it could be higher. When the temperature in the air increases, the ground temperature will increase even more. The measurements in the graph were taken in a cold area of the archipelago. In Longyearbyen, the temperature in the ground is higher than this [16]. This means that the permafrost in Longyearbyen in the future may be in jeopardy.

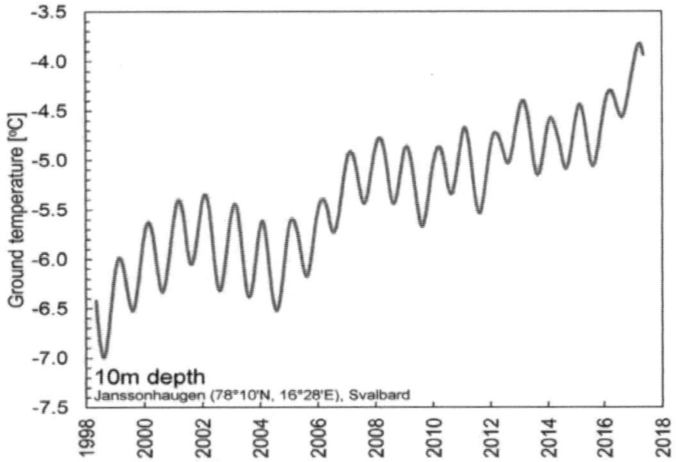

Figure 4. Ground temperature at Janssonhaugen Svalbard (near Longyearbyen) [17].

6.2 Fresh water availability

In Svalbard, freshwater is taken from Isdammen (see discussion below) during most of the year. However, during summer months, water is taken from the river running through the Longyearbyen valley and origination from the glaciers to the south of the valley. During winter, this river is frozen. The access to safe drinking water is under pressure [18, 19].

6.3 Isdammen (The Ice-dam reservoir)

Isdammen is located in Svalbard, east of Longyearbyen and is operated by the local government, Figures 5 and 6. The lake is around 2,5 km long and 0,6 km wide located at an elevation of 14 m above mean sea level. A 4 m high dam keeps the water in place. Its coordinates are 78° N and 15° E. According to the Norwegian Water Resources and Energy Directorate (NVE), the bottom of the lake is made of stone and earth and has an overflow of concrete [19]. There are numerous sources of water that lead to the lake. Some of the water supply comes from the Endalen watershed. Endalen watershed is composed of various elements, such as rainwater, snow, and ice melt. It also has a moderate mining presence, [20]. This is also true for other watersheds in the area. The lake is next to a cliff which also supplies it with melting water in addition to releasing more sediments into the water.

In 1960, the dam was created to provide Longyearbyen with more fresh water. Before this, the locals got some of their water from the watersheds leading to Isdammen and from other watersheds, [21]. The lake supplies the region with drinking water in winter, spring and autumn and is the town's only drinking water source in winter.

A road has also been constructed leading to the dam and beyond, Figure 7. At the outlet of the dam to the north, a pump house is built. The water is piped to nearby Longyearbyen. Instead of being buried underground, the pipeline is placed atop of the ground because of the permafrost [19]. A map showing Adventdalen in 1896 is added (Figure 8). This year was prior to the industrialization of Longyearbyen (prior to mining activities), and the map clearly shows a considerable loss of glaciers.

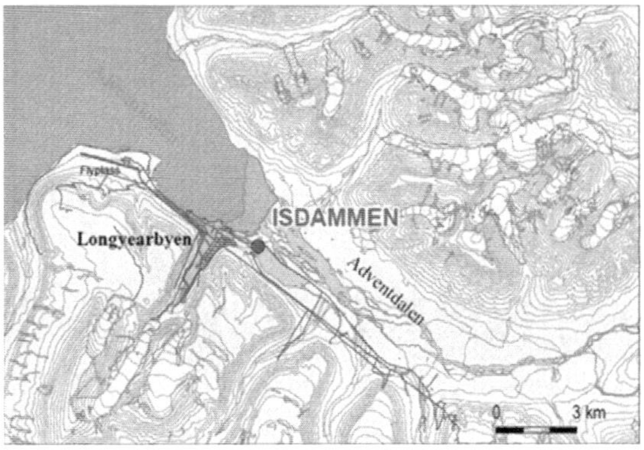

Figure 5. Location of Isdammen, Svalbard

Figure 6. Longyearbyen (to the right) and Isdammen (to the left). Photo by Helena Nynås, NVE

Figure 7. The road passing Isdammen (seen to the right). Photo: O. T. Gudmestad, August 2023.

IOP Conf. Series: Materials Science and Engineering 1294 (2023) 012034 doi:10.1088/1757-899X/1294/1/012034

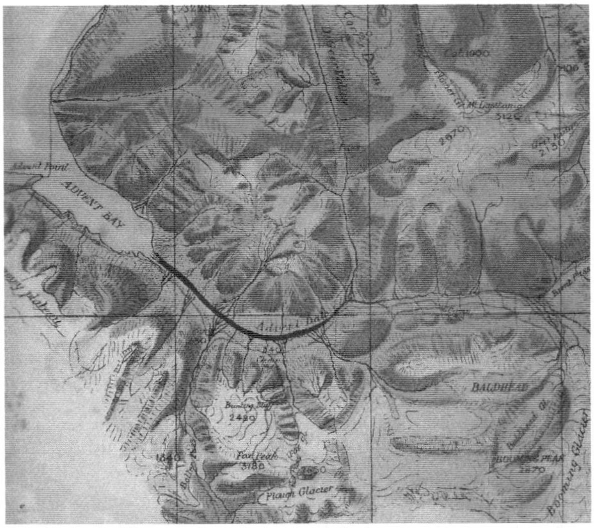

Figure 8. Sketch map of part of Spitsbergen (the main island of the Svalbard Archipelago). From a survey of Sir W. Martin Conway, summer of 1896. (Map in Privat ownership). The map shows a considerable loss of glaciers compared with present glaciers, compare with Figure 5.

7. Qualitative risk analysis

7.1 Bow-tie diagram

A Bow-tie analysis has been conducted to identify causes and consequences, The causes of an (unwanted) event can be mitigated by prevention barriers/ controls, while the consequences of the event happening, can be reduced, or avoided by barriers/ controls providing mitigation or recovery, see Figure 9. It must be noted that there might be escalation factors influencing the barriers/controls to enhance the effects of the prevention barriers or to strengthening the mitigation and recovery controls.

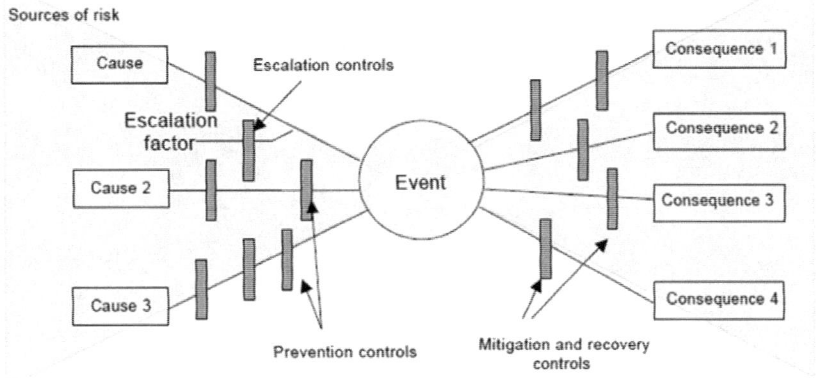

Figure 9. Basic elements of a bow-tie analysis

To analyse a bow-tie diagram, we need to identify the top event. In this case, it is "Degradation of drinking water sources in the Arctic," See Figure 10. This event represents a serious issue and there are many causes that can lead to this event. In Section 7.2 causes and consequences are discussed as well as prevention and mitigation barriers. Figure 10 does not show any escalation controls; however, physical attention and daily control of the drinking water source represents a recommended control measure reducing the top event to occur and reducing the consequences of a top event.

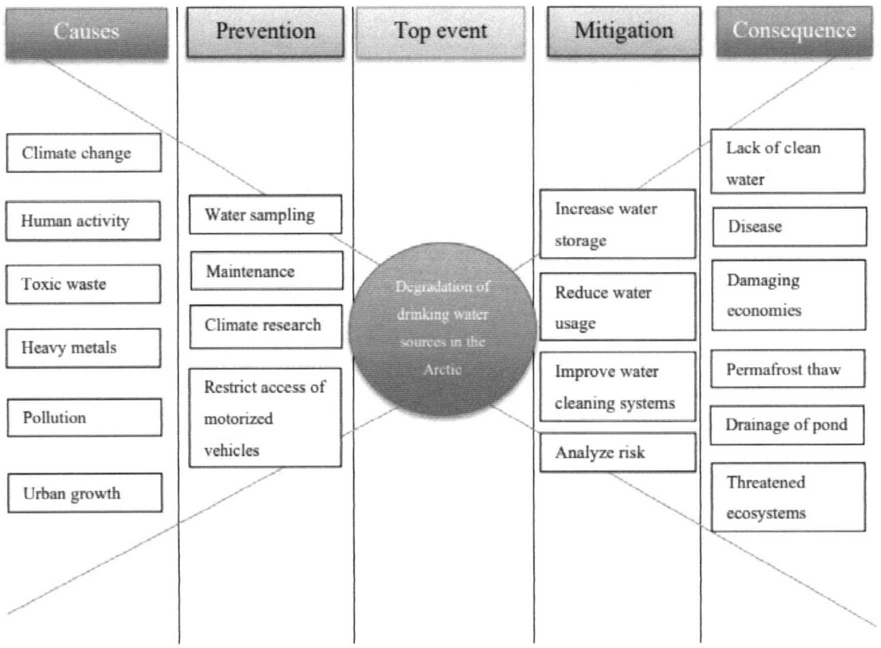

Figure 10. Bow-tie diagram of undesirable event "Degradation of drinking water sources in the Arctic."

7.2 The qualitative risk evaluation

7.2.1 Causes. The primary causes of degradation are climate change and human activity. The warming of the environment causes more snow and ice to melt and it heats up the permafrost. This causes more sediments to get into the water systems, some of which may be heavy metals that cause illness when consumed. In addition to this, when the permafrost is gone, the water may even drain away, [10], [22].

Humans impact areas where they stay. In general, we can think of pollution made by vehicles and different types of waste, from trash to factory waste products to construction waste, [23]. These pollutants may all enter the water. At Isdammen, there is a road right next to the reservoir, which can lead to emissions from cars and other vehicles to enter the water. All these factors can lead to negative consequences.

7.2.2 Consequences. The consequences are all negative, however some are more impactful than others. First, clean water may be a lacking resource in the Arctic, especially during winter. Although it is possible to melt snow and ice with heat to get water, it is energy consuming and not very efficient. That is why lack of a clean water resource can become a serious issue. Either if it is because of pollutants in the water or because there are insufficient volumes of water. Pollutants in the water may cause diseases and are costly to remove. The economy may take a toll from degrading drinking water. This can be through having to heat up snow/ice for water, having to import water or having to improve the water filtering systems or other parts of the water distributing infrastructure. Lastly, the ecosystems in the Arctic can be affected by the warmer weather and changes that can come from variations in snow cover and reduction of permafrost. The reduction of permafrost may be the most influential factor on how the water conditions are, and especially the amount of water. As seen before, in a lot of arctic regions, the permafrost ensures that the water does not drain away, [9]. A lot of arctic lakes are shallow [10]. Isdammen on Svalbard is around 4 meters deep, meaning it is susceptible to increased drain in the ground. Increased temperature will also increase the amount of water supplied to the lake from melting

IOP Conf. Series: Materials Science and Engineering 1294 (2023) 012034 doi:10.1088/1757-899X/1294/1/012034

snow and ice, so there may become a balance between water supply and loss. It is uncertain how it will be in the future. The topography of the area is also something that can affect how much water is in the lakes. Isdammen has a cliff on one side and downward slope on the other side. Increased melting on the cliff-side will cause more sediments to enter the water, maybe making it murkier and shallower. There is also a fear that the cliff-side will become unstable and cause mudslides, [16]. The sea is also not far away. Having the sea nearby will affect how the permafrost melts, as the salt in the water reduces the heat needed to melt ice.

7.2.3 Prevention. To prevent some of the consequences will be difficult. Climate change has already started and the temperature has risen significantly in the Arctic. This means that there in a lot of Arctic regions will be thawing of permafrost, and lakes will be affected, [8, 10]. One thing that is done in order to try and prevent the consequences is climate research and other research, which gives us answers that can help reduce the speed of degradation and hopefully keep the lakes and water supply in good condition. In addition to this, water sampling is being done to monitor the condition of the water. This helps being ahead of the top event and seeing if the water's condition degrades. Isdammen is currently being monitored regularly [20]. Maintenance is also important, for instance on dams and other infrastructure, such as pipes and pumps. To help keep the water as fresh as possible, restricting the access of motorized vehicles is also an option. This is especially the case around Isdammen, where a road is right next to the lake. On Svalbard, a lot of people use snowmobiles. Right now, these are prohibited from driving on the lake, so some measures are in place. What can be done if these measures were not sufficient?

7.2.4 Mitigation. To mitigate the effect of the top event, there are some things that can be done. First, the water storage can be increased by building a higher dam. Higher temperatures will lead to more melting which in turn will lead to more water that can be stored. This is the case in many Arctic regions. One thing that can be done if there is water shortage is to reduce the amount of water used. This is at least doable in a short term, but if a place is going to grow, there should be an abundance of resources, including water. The water filtering system and other infrastructure can also be improved. At least in cases where the biggest issue is pollutants in the water. In some cases, the only thing you can do is monitor and analyse how the top event unfolds and try to mitigate the worst consequences. This is especially the case when we talk about permafrost thaw, which seems inevitable from global warming [24]. Because this is a relatively new phenomenon and we are only able to study the causes of it now, research must be done continually to know which consequences it will lead to. Change in snow cover and higher temperatures will make room for more biodiversity and a change in the biome growth, [25], [26]. The changing ecosystem will also be an unknown factor that needs to be monitored. If there is more plant life, maybe there will be more algae in the water that makes it undrinkable.

8. Conclusions

Drinking water in the Arctic is affected by many different sources. They can each cause different consequences, and some are easier to prevent and mitigate than others. Global warming and change in temperature are the hardest to do something with. It influences a lot of different conditions. It affects snow- and ice cover, precipitation and most importantly permafrost.

The permafrost has done a lot to make sure that the conditions in the ground in the Arctic has stayed relatively the same over thousands of years. Now, signs say that it is starting to thaw. This can have a large impact on water sources, either by causing more drainage, making more pollutants or sediments enter the water or make whole cliff-sides enter the water.

This can be the case at Isdammen, Svalbard, Norway. At Isdammen, there are already signs that the water contains pollutants. That is why monitoring and researching the area is important to prevent the water from becoming undrinkable. In Arctic areas in general, it is important to monitor drinking water supply and make sure that there is sufficient volume and of drinkable quality.

The situation in Utqiaġvik (formerly Barrow), Alaska is in addition complicated due to the erosion of the shoreline with probability of saltwater intrusion as well as the potential for contamination due to the closeness to the airstrip. How can we prevent the degradation of drinking water in the Arctic? First, improving infrastructure such as water filtering, pumps and dams are necessary if the degradation comes from increased amounts of pollutants in the water. More pollutants will enter with more melting snow and ice. If the degradation comes from reduced amounts of water, it is harder to prevent it. Improving the infrastructure can mitigate as well as prevent degradation. If one water source disappears or becomes unusable, there may be a need to consider other sources. As is the case in Longyearbyen, which in summer uses a different water source than in winter. If temperature increases, maybe the town will not need to rely on Isdammen as much in winter.

References

[1] Elimelech M and Phillip W A (2011). The future of seawater desalination: Energy, technology, and the environment. *Sci.*, **333**, NO 6043. DOI: 10.1126/science.1200488.
[2] Spiritos E and Lipchin C (2013). Desalination in Israel. In: Water Policy in Israel, ed. Becker, pp. 101 – 123. Springer. The book is number 4 of the books in the series Global Issues in Water Policy.
[3] Lane K, Stoddart A K and Gagnon G A (2018). Water safety plans as a tool for drinking water regulatory frameworks in Arctic communities. Environmental Science and Pollution Research, Vol. 25, pp. 32988–33000
[4] Medeiros, A.S., Wood, P., Wesche, S.D., Bakaic, M: & Peters, J.F. (2017) Water security for northern peoples: review of threats to Arctic freshwater systems in Nunavut, Canada. *Reg Environ Change* **17**, 635–647. https://doi.org/10.1007/s10113-016-1084-2
[5] Regjeringen.no. (2022, 06.07). Miljøvern og miljøutfordringer i nordområdene. Miljøvern og miljøutfordringer i nordområdene - regjeringen.no
[6] Sinitsyn A and Kvitsand H (2019) Sustainable drinking water supply in Longyearbyen and climate change: present challenges and future opportunities. https://www.researchgate.net/publication/337256966_Sustainable_drinking_water_supply_in_Longyearbyen_and_climate
[7] Gudmestad O T (2020). Technical and economic challenges for Arctic Coastal settlements due to the melting of the Arctic ice and the permafrost". http://www.wreconf.org/ Published in IOP Conference Series: Earth and Environmental Science EES), *IOP Conf. Ser.: Earth Environ. Sci.* **612** 012049, 2020.
[8] Woo M-K (2012) Permafrost hydrology. Springer Science & Business Media.
[9] Yang D and Kane D L (2021) Arctic hydrology, permafrost, and ecosystems. Springer.
[10] Swanson D K (2022) Permafrost and Draining Lakes in Arctic Alaska. https://doi.org/10.3389/frym.2022.692218
[11] Cheng G and Jin H (2013) Permafrost and groundwater on the Qinghai-Tibet Plateau and in northeast China. Hydrogeology Journal, 21(1), 5.
[12] Osterkamp, T. (2007). Characteristics of the recent warming of permafrost in Alaska. Journal of Geophysical Research: Earth Surface, 112(F2).
[13] Weather Spak. Climate and Average Weather Year-Round in Barrow, (Internet page, no date). https://weatherspark.com/y/163/Average-Weather-in-Barrow-Alaska-United-States-Year-Round
[14] U.S. Army Corps of Engineers Alaska District, (2019, November). Barrow Alaska Coastal Erosion Feasibility Study. Economic Analysis Technical App. https://www.poa.usace.army.mil/Portals/34/AppendixCBarrowEconAppendix11052019.pdf
[15] Naval Arctic Research Laboratory Utqiaġvik (Barrow), Alaska (2017) Airstrip Site (Site 5). Imikpuk Lake Drinking Water Investigation. https://pacific.navfac.navy.mil/Portals/72/Northwest/Documents/Final%20NARL%20Barrow%20PFAS%20Imikpuk%20Lake%20Sampling%20Fact%20Sheet%2C%20August%202017.pdf
[16] Mortensen, M., Lanes, L., & Rommetveit, A. (2017, 12. juni). Dramatisk smelting av permafrosten på Svalbard. NRK (Troms og Finnmark). https://www.nrk.no/tromsogfinnmark/dramatisk-smelting-av-permafrosten-pa-svalbard-1.13555731
[17] Meterologisk Institutt, MET, (2022). Svalbard: Størst endringer i vinterklima. Svalbard: Størst endringer

IOP Conf. Series: Materials Science and Engineering 1294 (2023) 012034 doi:10.1088/1757-899X/1294/1/012034

i vinterklima (met.no)
[18] Longyearbyen Lokalstyre, (2023, 17.08) Arendalsuka: Sikker vannforsyning er under press.
[19] NVE, (2015, 16.08.2021). Isdammen, Svalbard. Dam: Isdammen, Svalbard (nve.no)
[20] Cold region science. The water supply to Longyearbyen. Available at: https://www.coldregionscience.com/endalen
[21] Knudsen E and Yri H T (2010). Teknisk industrielle kulturminner i Longyearbyen med omegn (in Norwegian).
[22] Wang G, Hu H and Li T (2009). The influence of freeze–thaw cycles of active soil layer on surface runoff in a permafrost watershed. *J. Hydrol.*, **375**(3-4), 438-449.
[23] Holm, P.A., (2022, 21. november). Fyllekjørte ut i Isdammen - nå er dommen falt. Svalbardposten. Fyllekjørte ut i Isdammen - nå er dommen falt (svalbardposten.no) (in Norwegian)
[24] Osterkamp T (2007). Causes of warming and thawing permafrost in Alaska. Eos, Transactions American Geophysical Union, 88(48), 522-523.
[25] Brittain J, Schartau A and Svenning M (2020). Biologisk mangfold i ferskvann på Svalbard: kunnskapsgrunnlag, påvirkninger og forslag til framtidig overvåking. Oslo (NVE Rapport, 13(2020), 97 (in Norwegian).
[26] Strauss J, Abbott B W, Hugelius G, Schuur E, Trea, C, Fuchs M, Schädel C, Ulrich M, Turetsky M, and Keuschnig M, (2021) Permafrost. Recarbonizing global soils–A technical manual of recommended management practices: Volume 2–Hot spots and bright spots of soil organic carbon, 130.

IOP Conf. Series: Materials Science and Engineering 1294 (2023) 012035 doi:10.1088/1757-899X/1294/1/012035

Transitioning towards renewable energy and sustainable storage solutions at remote communities in the Arctic, Case study of Flatey, Iceland

M R A Hjallar[1,*], E D Víðisdóttir[2], O T Gudmestad[1,3,*]

[1] UiT, The Arctic University of Norway
[2] Orkubú Vestfjarða, Ísafjörður, Iceland
[3] University of Stavanger, Stavanger, Norway

*Corresponding authors: mhj015@post.uit.no and otgudmestad@gmail.com

Abstract: The need for transitioning towards renwable energy and sustainable storage solutions is particularly challenging for remote communities in the Arctic, located far away from the electricity grid. This paper explores the potential for use of renewable energy on the remote island of Flatey, Iceland, which currently relies on two diesel aggregates for power. The primary goal is to assess the feasibility of transitioning from fossil fuels to renewable energy sources, such as solar or wind power, to meet the island´s variable energy demand while reducing its environmental impact. With a year-around population of 5 and with a considerably increased energy consumption during vacation times, due to more population, Flatey´s annual energy consumption is ~ 209.000 kWh, peaking in July at ~ 25.000 kWh. This fluctuation requires an adaptable and resilient energy infrastructure. The paper examines the viability of Flatey as a self-sufficient renewable energy provider. The study considers the island´s energy requirements, consumption patterns, and geographical constraints, while also evaluating technical, economic, and social factors that may influence renewable energy adaption. This paper, thereafter, investigates the feasibility of achieving energy self-sufficiency on the small island of Flatey. Different energy storage options is considered, focusing on battery storage, underground solar power/energy storage, and hydrogen storage.

1. Introduction

The availability of energy at smaller remote societies is challenging. The remoteness restricts the possibility to ensure connection to the national electric grid and the costs prohibits the society to build separate electricity generating systems based on renewable energy. Therefore, coal or diesel is normally used to support the electricity plants at these locations. It shall be noted that use of diesel gives rise to emittance of less greenhouse gases than coal. The emission of CO_2 per million Btu is according to [1] 74 kg for diesel while it is 96 kg for coal. Switching to natural gas, the emission is 53 kg per million Btu. In some remote societies the use of compressed gas (normally methane) is encouraged, [2]; larger societies are using LNG (Liquefied Natural Gas); however, the use of LNG requires building of very costly regasification facilities. For any of the different fuel sources, emissions associated with production and transportation must be added.

A shift to renewable energy sources with very limited emission of greenhouse gases is much encouraged. In this paper, the transition to use of renewable energy at a remote location is discussed by considering a case study. The motivation for presenting the paper is to investigate the feasibility of the transition from being a society basing the energy on fossil fuel to a society where the emissions are reduced compared with present situation. It should be noted that the sources of energy used at remote locations is of considerable discussion worldwide. The International Renewable Energy Agency (IRENA) and the Natural Resources Canada (NRCan) initiated the "Global Initiative for Transitioning Remote Communities to Renewable Energy" [3], in 2021. Reference is also made to [4], where Renewable Energy Alternatives for Remote Communities in Northern Ontario, Canada, are considered. In reference [5], a cost analysis is carried out for remote communities when hydro, solar and wave energy are available. To identify renewable energy alternatives where hydropower or wood is not available (like on Flatøy) or where the construction costs of wave energy plants will be too onerous, is a large challenge, therefore, the case of providing the barren Flatøy with renewable energy represents the ultimate challenge.

The location considered is Flatey, a remote island with a very famous history, one of the main cultural centres on Iceland, the location where one of the most important old Icelandic sagas, the Flateyjarbók [6], was written during the period 1387 to 1394. The book tells the history of the Norwegian kings and is a very important document for Norway and Iceland.

Flatey is a small island located in the centre of Breiðarfjörður in the north-western part of Iceland (Figures 1 and 2). The island is around two kilometres long and one kilometre wide and the landscape is, as the name indicates, almost completely flat, Figure 3. Until the latter part of last century the population on the island was growing and the island was in relation to its size heavily populated. Due to social change, change in the labour market and settlement condensation, the population has dwindled down and as of today there are only around 6 to 10 persons with permanent year-round residence on the island. The habitation on the island is seasonal, in the summertime considerably more persons live or spend time on the island. Some permanently move to vacations homes, some spend a few days at the island's hotel and others spend the day exploring the island. The main industry on the island today is tourism and the main attraction on the island is the cultural history and old village that mostly is in its original shape along with the unspoiled nature and wild bird life.

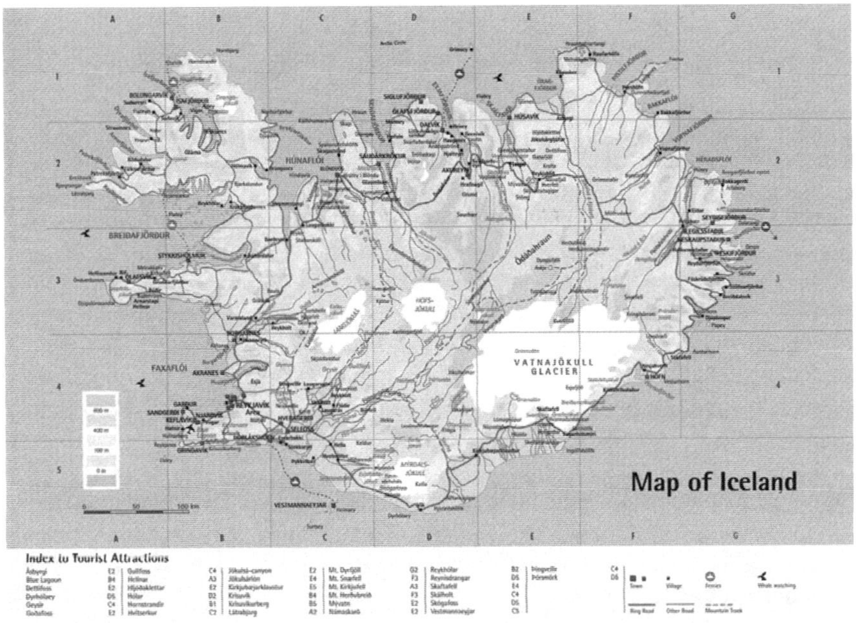

Figure 1. Map of Iceland. Note the location of Flatey in Breiðarfjörður to the North West.

The climate on Flatey is like Arctic climate in general. Flatey experiences a distinct climate characterized by relatively cool temperatures and moderate precipitation. The yearly average temperature in the district is 2,13 °C, which is approximately -1,04 °C lower than the average temperatures recorded across Iceland. In terms of precipitation, Flatey receives an average of approximately 59,75 mm of rainfall throughout the year. The precipitation in form of rain, is spread over a total of 144,3 rainy days, which amounts to approximately 39.71% of the year. This indicates that rain is a common occurrence on Flatey, with a considerable portion of the year experiencing wet weather. In contrast, the total snowfall precipitation in January 2020, which had the highest snowfall of that year, amounted to only 114 mm. This suggests a comparatively low occurrence of snowfall on Flatey [7].

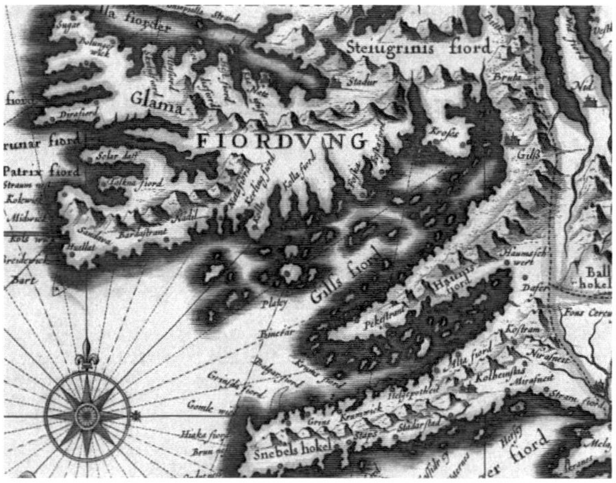

Figure 2. Clearly marked old map of Flatey (Platey) [8].

Figure 3. Map of Flatey [9].

Flatey is not connected to the national grid in Iceland and grid connections to the island are not viable due to technical, economic, geographical, and climatic limitations. Due to this fact all the island needs to be self-sufficient when it comes to power production. All power production on the island today is done using two diesel aggregates. With increased awareness of the polluting effects of burning fossil fuels along with rising oil price and rapid price reduction of renewable energy, all further development on Flatley's power system is set towards a greener system. In Iceland 98% of electricity production

IOP Conf. Series: Materials Science and Engineering 1294 (2023) 012035 doi:10.1088/1757-899X/1294/1/012035

comes from renewable energy sources, where hydro power and geothermal power are the building pillars. That leaves around 2% of the electricity generation to the burn of fossil fuels [10]. The goal is to further reduce this ratio as well as the Icelandic government has set a goal to increase the diversity of energy production where the focus has been on integrating both solar and wind power on larger scales than existing in the power system to this day [11].

The burning of fossil fuels for power production has many disadvantages other than the emission of CO_2 and other greenhouse gases, the maintenance cost is relatively high and the continuous diesel transportation to the island is both expensive and time-consuming. For the power system on Flatey where connection to the national grid is not an option, solutions of local sustainable energy production are set in focus. The potential local renewable energy sources are wind, solar, tidal, and green hydrogen. The option of tidal power has been researched in the area around the island, but the results were disappointing as the current measured is too slow for power production, or around 1 m/s. The use of hydrogen as fuel instead of the diesel has been researched but the results show that this would not be cost efficient. Leaving the first two options, wind, and solar power. Those options are the most feasible and despite high upfront cost for those renewable energy sources, the maintenance costs are low. These two power sources will be discussed further in this paper. For longer perspective, also the question of energy storage must be discussed as relying on wind and solar power results in using the existing diesel aggregates during high stable weather pressures wintertime, when there is no wind and limited daylight or when the wind velocities are high (at around 25 m/s) such that the wind turbine must close. The Flatey power generation project could be seen as a case study for the introduction of renewable power at an isolated location and its findings could be useful for other similar communities around Iceland or at other cold climate locations.

2. Present energy production and energy demand on Flatey

The annual energy production on Flatey has been around 210.000 kWh for the last years, that corresponds to burning over 60 thousand litres of diesel. Of those, around 50 thousand litres are used directly for heating of houses on the island, [12]. The electricity is produced by two 30kW diesel aggregates on automatic control and depending on the energy demand, either one or both machines are in operation. Due to the variating population on Flatey, the energy consumption varies drastically between seasons and, as in other bigger power systems, there are as well large variations in power demand within each day.

During low season in the winter the energy consumption during this period is quite stable at around 15.000 kWh per month. This changes with the increased population on the island and the number of visitations to the island during the summertime causing a large increase in the power demand. The power demand reaches over 24.000 kWh per month in this season, which lasts from June to August and in this period the maximum power demand per one day does reach up to 765 kWh. During these high peak energy consumption months, the variations within each day get larger and the maximum power demand reaches around 60kW, that happens at and around dinner time. Meaning that the power system on the island must be able to produce up to around 60 kW to cover the demand when it peaks, although the power demand is much lower at most times.

The required power generation supplied by the diesel aggregates is highly controllable, and the production can easily and rapidly be either increased or decreased in accordance with the power demand in the system at any time. When it comes to the generation of the renewable solar and wind power they are not as controllable or reliable since they do not generate energy when the sun is not shining, or the wind is not blowing, even not during the situations where the wind is excessively strong. Because of this, the introduction of these power sources into the power generation system will increase fluctuation in the system. With increased fluctuation in a power system, a rather large and easily accessible reserve capacity is needed to act as backup, balancing out the fluctuations caused by the renewable source [10]. In this case it would therefore be necessary to keep the diesel aggregates in the system as reserve capacity at first and later, they could possibly be replaced by a greener option, batteries or potentially hydrogen.

Before the integration of the variating renewable power sources to the system takes place, a comparison of the power demand on the island and the power production from the power sources needs to be done. If the curves of power demand and production are in line with each other, on both a longer time scale like seasons and on a shorter time scale like when during the day when the demand and production tends to peak, that shows that the power source is suitable for the selected power system. Of concern is also situations when the energy production is low. The potential for energy storage and full self-sufficiency based on energy storage is in this respect, of importance.

3. Solar power availability

3.1 Power demand vs. power production

As Flatey is located at latitude 65.3° North the winters are long, and the summers are relatively short. In these conditions, close to the Arctic Circle the power production of solar panels cannot be expected to be even throughout the year. During the winter, when the sun rises low above the horizon the production from the solar panels is low, but during the summer when the sun is high above the horizon and barely sets during the night, the production is high.

In the case of Flatey these large seasonal variations in the solar power production are not necessarily negative since the most important factor is how the power output from the solar panels and the power demand in the system correspond. In Flatey these two curves are correspondent, as both power demand and production peak during the summer.

The power demand is high when the sun rises high, and the days are the longest during the summer months at that time the solar panels have large power output. The power output from the panels diminishes as the days gets shorter but simultaneously the power demand on the island gets much lower as the number of persons staying and visiting the island decreases drastically.

Figure 4 shows the power demand in Flatey in the years 2018 and 2019 which can represent a typical year in Flatey with regards to the number of persons living on and visiting the island. From the graph the power demand is quite stable during September to May and that the curve rises significantly during June to August with a peak in July. This is completely in line with the tourist season on the island.

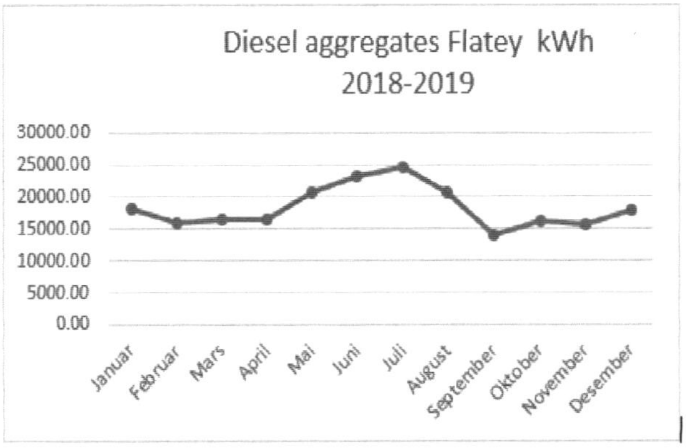

Figure 4. Power demand in Flatey (kWh), analysis of actual data

When researching the power demand in a power system, it is important to look at the variation of power demand on several different time scales. The daily power demand variations in the power system in Flatey are different during the different seasons as well. During winter the power demand is rather stable around the clock, varying with around 3kW per hour. In summer the power demand is not only larger in general, but the daily fluctuations are much larger as well. The average days during high season in June, July, and August the power demand peak occurs at or around dinner time and the hourly variations between night and day reach up to 17 kW. These variations need to be considered when it comes to estimating the available power in the system at each instant and the overall feasibility integrating solar power to the system.

3.2 Solar power in Iceland

Solar power has not, yet, been utilized on a large scale in Iceland. The utilization is mostly limited to regions that are not connected to the grid such as vacation homes and measurement stations that do not have a large power demand. There is although one solar power project on a larger scale installed in Iceland as 17 kW of solar power has been installed on one of IKEA's buildings in Reykjavík. The project is a research project for solar power in Iceland and there the panels are installed in two different angles, 50 panels are placed on a flat roof at a 20° angle from the horizontal and 13 panels are at a 90° angle on the wall of the building. Regarding the production per solar panel on the roof and wall of IKEA for the first two years of operation, the vertical panels on the wall produce more power when the sun rises low above the horizon during winter, from September to February and when the sun rises higher the production increases and the panels at 20° position on the roof exceed the ones on the wall by quite a bit [13].

When comparing the solar radiation data from Reykjavík (IKEA's location) to the solar radiation data from the nearest weather measurement station to Flatey, located at Gufuskálar on Snæfellsnes peninsula, the results are quite similar. The annual solar radiation at Gufuskálar is shown on Figure 5. The annual radiation is measured to 750 kW/m^2 while at IKEA´s location it is measured to 777 kW/m^2, but since Reykjavík is south of Gufuskálar the solar radiation is slightly more evenly spread out there. Indicating that at Gufuskálar the production of panels in a vertical position would be higher for a longer time period compared to in Reykjavík [14].

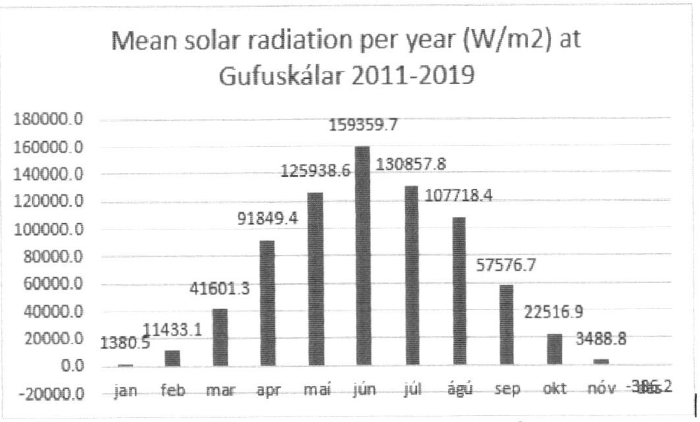

Figure 5. Mean solar radiation at Gufuskálar in W/m^2 from 2011-2019 [12]

The solar radiation data shown on Figure 5 is from a station located 70 km away from Flatey and the data has been accumulated from the year 2011. It must be noted that solar radiation devices from the meteorological office are from the horizontal placed panels and that the solar panels are most likely not going to have the same angle as well as more snow accumulates on a horizontal surface than on an angled one. The data does include some errors since it shows negative radiation for December but that cannot happen [15].

It can be seen from the solar radiation data at Gufuskálar that the monthly and seasonal radiation variations are drastic. Very little radiation is measured during the darkest months from October to March but as the sun rises higher and higher, the radiation increases with a significant peak in June.

Previously, it was believed that solar panels would perform better in warmer climates. However, more recent studies have shown that they perform better in colder climate. According to a study conducted by [16], the efficiency of solar panels decreases by 0,.5 %/°C increase in temperature.

Therefore, a temperature increase of 20 °C would reduce the efficiency of solar panels by up to 10%. They exhibit optimal efficiency at -5 °C, making them particularly suitable for Arctic climates.

3.3 Installation of solar power in Flatey

The accumulation of snow on the solar panels does affect the production a lot, especially in cold climate where the temperature can be expected to be below zero a large part of the year like in this case. This needs to be considered when the optimal angle for solar panels is decided on every specific location.

In Flatey the first step toward a greener power system is to be installation of solar panels on an old building once operating as a fish factory. There are several different options when it comes to the placement of the panels on the building, the two most feasible ones are either on a part of the roof or a large south-south-east (SSE) facing wall. The building, on the other hand, is old and closer inspection of the roof would need to be done in order to decide if it can support the panels without too expensive renovations. To get a better idea of the different scenarios regarding the placement and the angle the solar panels are installed at, calculations are done using an online calculator established by the European Commission [17].

The panels installed 2023 have an efficiency of 19,8% and each module has the maximum power of 335 W at standard test conditions. In the calculations the before mentioned wall of the old fish factory is taken as an example, it is 80 m^2 and is estimated to fit 40 Jinko solar panels. The calculated annual energy output from these 80 m^2 is just below 9000 kWh. This calculation refers to panels in a 90° position on a wall facing SSE in Flatey. When changing the angle that the panels are installed in is performed and the results shows that an optimal angle at this location is according to these calculations to be 47° from the horizontal. The energy output curve has the shape of the normal distribution. The annual energy output is calculated to be quite a lot higher compared to the 90° installation, or around 12.700 kWh. That is close to 40% higher. The output during the darkest months is similar in both cases but the curve is much steeper in the latter case and the output during summer is much larger. In both cases the energy output peaks in May with just above 1250 kWh for the 90° system and just above 2000 kWh in the 47°system.

Since the power demand in Flatey is much higher during summer and the main challenge for the system is to be able to meet that peaking demand, it would be wise to install the solar panels at an angle that provides the most during summer. As stated earlier the energy production from the diesel aggregates is around 210.000 kWh per year and this first installation of solar panels on the island will only account for a small amount of the total production. But it is a step in the direction to a greener power system in Flatey and will act as a research project for the actual potential of solar power production on the island.

3..4 Economics

The electricity cost from solar photovoltaics (PV) has reduced significantly in the last decade. As the cost reduces, PV becomes a more viable and competitive option. An economic analysis will be necessary, considering the Net Present Value of the economical return compared to the investments and the operating costs. The discount rate selected will be important driver for the result of the economic analysis. When considering the costs of PV it is also important to know the estimated lifetime of the panels.

Maintenance cost is very low compared to the start-up costs of installing the solar panels. As of today, it does only account for about 1% of the start-up cost but with continuing cost reduction of the panels, that share will increase [13]. Solar panels require very low maintenance and that especially in PV systems that have no moving parts as expected in this case, the only predictable maintenance is washing and visual inspection of all parts once or twice a year.

4. Wind power availability

The wind power potential in Flatey looks promising although further research would need to be performed by setting up a measuring mast, measuring both wind speed and direction so that the feasibility of the installation of a wind turbine can be determined more accurately. The online available wind data from the Icelandic meteorological office shows two drastically different results.

The older data from a small weather station in Flatey shows that the mean wind speed in the years between 1953 and 1978, is only 4.8 m/s which is rather low for wind power production [11].

The data available from a newly prepared wind atlas from the Icelandic Meteorological office from a location just outside of Flatey shows that the mean wind speed is just over 10 m/s with the prevailing wind direction from east-north-east [19]. The power production of wind turbines is controlled directly by the wind speed at any given moment. Typically wind turbines start producing power at a so-called cut-in wind speed around 4m/s, the power output does then increase with increasing wind speed up to around 13 m/s where it reaches the maximum output. A typical cut-off frequency of the turbines is than often around 25 m/s [20]. With a mean windspeed of just over 10 m/s along with the power production characteristics of a wind turbine, installing wind power to the power system looks promising.

4.1 Installation of wind turbine on Flatey

The power demand in Flatey, shown on Figure 4 above, is rather stable during winter at just over 16 kW per month. Installation of around 25 kW of wind power to act as a base power in a green power system in Flatey could therefore be an exciting option.

4.2 Other factors that affect the feasibility of a wind power project

When researching the feasibility of a wind power project on a specific site, the data from the wind atlas along with detailed data on the topography, surface roughness and obstacles around the turbine siting to calculate the estimated power production. On Flatey, which is a small and flat island, as the name indicates, a wind turbine would likely be located close to the shoreline, where roughness is low and no, or few obstacles are in its way. A study to find the optimal location would be required and as large areas both on Flatey and the smaller islands close to it are protected areas due to archaeological and cultural remains as well as the unique wild bird life the available locations are limited. The potential wind turbine must be installed on a location where it does not disturb the persons permanently living on the island, the wild bird life and where it does not shun away the tourists, which will be challenging. When calculating the feasibility of a wind power project the mean values for the specific wind parameters do not, as well as with solar power projects, require more detailed information on the wind characteristics, for example in which seasons and what time of the day the wind speeds tend to be the strongest.

The curve of the estimated power production from the wind turbine and the power demand on the island need to be compared, as well as comparison to the power production curves from other power sources in the system need to be considered. As the different power sources in the system need to be coordinated in the most efficient manner.

In the case of the power system in Flatey the production curve for wind power opposes the production curve for solar power. The production from the wind turbines is estimated to be largest during the winter months when the wind speed tends to be higher, and the production of solar power is largest during summer. This shows that the combination of these two power sources in the same power system potentially should be efficient.

4.3 Other factors at this site

The small community in Flatey relies on the tourism as the main industry and income, where the main attractions are the unspoiled nature and wild bird life. These facts need to be considered when it comes to choosing the size and the location for a wind turbine on the island. Infrastructure on the island is very limited as well, with only one road leading from the harbour to the old village. Large constructures like the building of new roads on the island are not likely to be accepted or well-liked by the islanders or people that have strong relation to the island and its community.

Therefore, the choice of a turbine needs to be smart and with these special circumstances in mind. The turbine would need to be easily installed without the use of large cranes or heavy machinery. Smaller turbines as predicted to suite well in Flatey must be identified.

5. Energy storage options

5.1 Storing solar power in battery banks

The use of battery storage in Photovoltaic systems (PV-systems) has its drawbacks [18]. Research has been limited when it comes to stationary batteries for larger facilities, but more prevalent in the realm of portable batteries – primarily electric car batteries. Additionally, energy can only be stored for a short duration and cannot be stored seasonally, for instance from spring to winter. In comparison to other storage methods, battery storage for PV-systems experiences greater energy loss, including heat loss and reduced efficiency from an inability to extract enough energy from the Suns radiation. Moreover, a considerable number of batteries must be connected in series to meet the necessary energy demand for a larger demand.

On the other hand, there are numerous advantages to using battery storage for PV-systems. Old, non-stationary batteries, such as used electric-car batteries, can be repurposed, which is both cost-effective and sustainable. Car batteries have greater size and weight requirements than stationary batteries, which are more flexible in terms of form factor. This makes it possible for larger stationary facilities with higher energy-demand, to connect old car batteries in series. Furthermore, battery storage for PV-systems requires less space than hydrogen storage.

5.2 Underground solar power storage

Borehole thermal energy storage (BTES) is a viable method of storing solar power, utilizing the ground as a storage medium for supplied heat. This system is particularly suitable for seasonal storage of solar energy. Ground source heat exchangers (BHE) are an essential component of BTES and are available in two main types: U-tube and coaxial wells.

U-tubes typically comprise either a single or double pipe circuit (double U-tubes), and wells with U-tubes require fill-material to enhance the heat transfer between the pipe/tube and the surrounding bedrock. Coaxial wells, on the other hand, feature two coaxial pipes, with downward or upward flow interacting with the bedrock. Subsequently, the flow direction is reversed to charge and extract heat. Effective contact between the outer pipe and the surrounding rock can enable a coaxial heat exchanger to facilitate efficient power transmission between the working fluid and the surrounding rock. The suitability of Flatey for BTES is contingent on the geology and physical properties of the bedrock, such as thermal heat capacity and heat conductivity, which are crucial factors in determining its compatibility with the system. Without detailed knowledge of Flatley's geology, it is challenging to ascertain whether it is a suitable candidate for BTES implementation [21, p. 37].

5.3 Storing power by using hydrogen

There are two main methods of hydrogen production: grey hydrogen and green hydrogen. In this context, the focus will be on green hydrogen, which is typically produced through electrolysis - the process of splitting water molecules into hydrogen and oxygen gas. The resulting hydrogen can then be used in fuel cells, where it reacts with oxygen to produce heat and electricity.

However, storing hydrogen presents a significant challenge due to its high volumetric density. To reduce its volume, hydrogen is typically stored under high pressure, with typical storage pressures ranging from 350.000 to 700.000 hPa. Such high-pressure storage requires additional safety measures to prevent leaks or explosions, necessitating more secure storage tanks. Additionally, storing hydrogen in liquid form requires extremely low temperatures or high pressures, as hydrogen evaporates at -253 °C.

When hydrogen and oxygen gas react in a fuel cell, they release heat or electricity. The efficiency of a fuel cell delivering electricity is typically 40-60%, significantly higher than that of solar power (η = 20%) or wind power with similar average efficiency. By utilizing both the electricity and heat generated by the fuel cell, an efficiency of 85-90% can be achieved [22]. So, in order to produce green hydrogen, the only required regular input is water. The economics of using hydrogen for power storage, must, however, be further evaluated.

6. Review of the various options for providing renewable energy and energy storage at Flatey

Based on the information given in Sections 3, related to solar power, Section 4, related to wind power and to Section 5, related to energy storage, a summary of the options is presented in Table 1. The table assesses availability, installation difficulty, operational complexity, and maintenance requirements.

Table 1 Comparison of different methods to produce and store renewable energy.

Energy resource/ Storage option	Availability	Installation difficulty	Operational complexity	Maintenance requirements
Solar power	Available in abundance during the summer period when the demand of energy is highest. Limited availability during winter months, location is 65,3°N.	Low difficulty. A gradual investment in solar panels is possible according to available budget. The initial costs of the first panels and the required electric system are relatively high	The operation of solar panels is considered to have low complexity. However, there are frequent reports of fires associated with solar panels and panels should be separated by distance to ensure redundancy.	Maintenance requirements are medium. It is necessary to clean the panels from dirt and from snow. Furthermore, regular checks of the safety of the electrical connections are needed.
Wind power	Available most of the year, noticeably better during fall and winter months than during the summer period	Relatively high difficulty. Installation of wind turbines is a complex task that requires careful planning to ensure that the bird population is protected. Also, it is a challenge to bring heavy construction equipment to the location. Smaller wind turbines rather than one big will ensure redundancy.	The operation of wind turbines is highly complex and requires knowledge of mechanical systems as well as monitoring of weather to ensure that the turbines shall not be overloaded in strong winds. During periods with atmospheric icing, the turbines should also be stopped to avoid that ice is thrown onto buildings or other infrastructure	Maintenance requirements are large and complex. The access to the nacelle involves climbing stairs that could be taken care of by the local population, while the checking of blades will normally involve specialists. Maintenance costs might run very high and cause considerable downtime.
Use of fuel	Available all year, however, diesel is not a renewable option, although ammonia might be considered as renewable if produced by renewable sources	Diesel generators are installed and should be kept available as backup. To retrofit the diesel aggregate for use of ammonia is considered and the costs of refurbishing the generator to run on ammonia might be affordable	As ammonia is extremely toxic, extreme care must be taken to handle ammonia.	Engines running on ammonia might be susceptible to unfamiliar wear and tear compared with diesel fuel. Maintenance assistance might have to be called to large costs and possible delays where the aggregate is down.
Battery storage	Us of older car batteries can be considered. The storage time and capacity will be limited though.	Installation of used batteries is quite easy. Beware not to put "all eggs in one basket"	Operations of the system is somewhat complex as many batteries are required. Note that the batteries will have different level of wear and tear. The potential for battery fires must not be underestimated.	Maintenance requirements are medium, damaged batteries must be taken out and the power bank must be checked daily.
Underground energy storage using boreholes	The drilling of borehole for energy storage requires knowledge about the geology. On Iceland, the temperature in the ground might be favourable for this solution.	The energy provided during times of abundant solar power or wind power could be transferred to hot water that can be stored in a borehole. This will require drilling of borehole. Heavy equipment for	The operation of such system should be feasible by the persons living on Flatey.	The maintenance requirements would be relatively low and well known for a technically interested person.

		drilling of the borehole is required		
Production of hydrogen for storage	A production system for hydrogen might be outside the financial resources of the Flatey community	A suitable hydrogen production system will have to be assembled locally. The need for careful workmanship to ensure no leaks makes the operation complex.	Operation of hydrogen facilities require extreme careful handling of all equipment. Any leak can cause a huge explosion	Maintenance of hydrogen facilities is outside the capability of the residents. Maintenance personnel must be hire in at large costs.

7. Conclusions and recommendations

The paper reviews feasibility of achieving energy self-sufficiency on the small island of Flatey using a combination of solar and wind power in combination with necessary energy storage; the storage being a most crucial part of the system. Different energy storage options are considered, focusing on battery storage, underground solar power/energy storage, and hydrogen storage. While battery storage has its drawbacks, such as limited storage duration and energy loss, repurposing old electric car batteries can provide a cost-effective and sustainable solution. The paper also discusses the potential of underground solar power storage using Borehole Thermal Energy Storage (BTES) and the challenges of implementing this system without detailed knowledge of Flatley's geology. Additionally, the use of hydrogen production through electrolysis is discussed, also pointing to the challenges of storing hydrogen due to its potential for explosions is discussed.

The discussion presented shows that utilization of both solar and wind power on Flatley's power system is promising, with relatively high mean wind speeds, according to the wind atlas data, and long days during summer when the sun rises high above the horizon. The combination of these two power sources at this location fits well, since the power curves of the two sources are quite different, with high wind power production during winter and high solar power production during summer. The power demand in this system has quite large seasonal variations with a lower stable demand from September to May and a higher demand during summer, from June throughout August. This fact plays a role when it comes to determining what amount of renewable power will be viable to install into the power system.

In a system without any batteries to store excess power production it may not be viable to install renewables with a maximal production higher than the demand at each season or time period. As the power system in the case of Flatey varies so much between seasons it can be questioned whether it would be viable to aim for a system completely driven by renewable sources as the peak power demand is high and only last for short periods of time. That is given that the diesel aggregates already operating in the system will remain there so that no further investment will be needed. Use of ammonia produced from renewable energy, rather than on diesel fuel, is considered in Table 1.

The possibility of energy storage to ensure full energy self-sufficiency based on renewable sources has been discussed in the paper and should be further evaluated.

References

[1] U.S. Energy Information Administration. Environment, (2023, September 7). Carbon Dioxide Emissions Coefficients. https://www.eia.gov/environment/emissions/co2_vol_mass.php. (Accessed: 2023-09-10).

[2] ClarkeEnergy, Compressed and liquified gas as a fuel for gas engines. https://www.clarke-energy.com/applications/compressed-and-liquified-gas-as-a-fuel-for-gas-engines/. (Accessed: 2023-09-10).

[3] IRENA, International Renewable Energy Agency (webpage, 2022, November). Building Knowledge of the Energy Transition in Remote Communities. https://www.irena.org/News/articles/2022/Nov/Building-Knowledge-of-the-Energy-Transition-in-Remote-Communities

[4] Arriaga M, Canizares C A and Kazerani M (2020). Renewable energy alternatives for remote communities in Northern Ontario, Canada. *IEEE Trans. Sustain. Energy*, **4**(3). 10.1109/TSTE.2012.2234154

[5] Robertson B, Bekker J and Buckham B (2020). Renewable integration for remote communities: Comparative allowable cost analyses for hydro, solar and wave energy. *Appl. Energy*, **264**, 114677. https://doi.org/10.1016/j.apenergy.2020.114677

[6] Flateyjarbók, (1390). Old Norse manuscript prepared on Flatey by the priests Jón Þórðarson and Magnús Þórhallsson.

[7] Weather and Climate - The Global Historical Weather and Climate Data, (2020). https://tcktcktck.org/iceland/westfjords/flatey/january-2020. Accessed 2023.09.10).

[8] Blaeu W, (1643-50). Tabula Islandiae Auctore Georgio Carolo Flandro. https://sanderusmaps.com/our-catalogue/antique-maps/europe/northern-europe/old-antique-map-of-iceland-by-w-blaeu-26533. (Accessed: 2023-09-10).

[9] Icelandic Times, Flatey, Breiðarfjörður. Editorial. https://icelandictimes.com/flatey-breidafjoerdur/. (Accessed: 2023-09-

10).
[10] Orkustofnun, (2020). OS-2020-T012-01: Installed capacity and electricity production in Icelandic power stations in 2019. [online] Available at: https://orkustofnun.is/gogn/Talnaefni/OS-2020-T012-01.pdf. (Accessed: 2023-09-10).
[11] Government of Iceland. Ministry of Industries and Innovation, (2018). *A Sustainable Energy Future: An Energy Policy to the year 2050.* (Accessed: 2023.09.10). Available at: https://www.stjornarradid.is/library/01--Frettatengt---myndir-og-skrar/ANR/Orkustefna/201127%20Atvinnuvegaraduneytid%20Orkustefna%20A4%20EN%20V4.pdf
[12] Sólbergsson, Sölvi, FAO Energy Division Orkubú Vestfjarða, (email to Víðisdóttir on 8th of July 2020) FAO Energy Division Orkubú Vestfjarða, Iceland.
[13] International Energy Agency, (n.d.). Variability and Predictability of Large-Scale Wind Power. Available at: https://studylib.net/doc/18356795/variability-and-predictability. (Accessed: 2023-09-10).
[14] Þrastarson, S., (2019). *Nýting sólarorku á Íslandi með sólarsellum*. University of Iceland. MSc Thesis. Available at: https://skemman.is/bitstream/1946/34488/1/N%c3%bdting%20s%c3%b3larorku%20%c3%a1%20%c3%8dslandi%20 me%c3%b0%20s%c3%b3larsellum%20-%20Sindri%20%c3%9erastarson.pdf. (Accessed 2023.09.10).
[15] Veðurstofa Íslands, (n.d.). *Ársmeðaltöl fyrir stöð 210 – Flatey.* Available at: https://www.vedur.is/Medaltalstoflur-txt/Stod_210_Flatey.ArsMedal.txt (Accessed: 2023-09-10].
[16] Gullbrekken L, Bunkholt N S, Grynning S, Bellman M and Kvande T, (2019) Efficiency of BIPV system – Field study in Norwegian climate. IOP Conf. Ser.: *Earth Environ. Sci.* **352** 012033
[17] European Commission, (Internet). Photovoltaic geographical information system. https://re.jrc.ec.europa.eu/pvg_tools/en/#PVP. (Accessed: 2023-09-10).
[18] NREL, (2018). Best Practices in PV System Operations and Maintenance. 3rd edition. National Renewable Energy Laboratory. Available at: https://www.nrel.gov/docs/fy19osti/73822.pdf. (Accessed: 2023-09.10).
[19] Danish Wind Industry Association, (2003). *Weibull Distribution Plotter Programme.* (Accessed: 2023.09.10). http://xn--drmstrre-64ad.dk/wp-content/wind/miller/windpower%20web/en/tour/wres/weibull/index.htm
[20] Li G, and Zhi J, (2016). *Analysis of Wind Power Characteristics*. [online] Available at: https://www.sciencedirect.com/topics/engineering/rated-wind-speed (Accessed: 2023-09.10).
[21] Heen A, (2018). Sesonglagring av solvarme til bygningsoppvarming: et case-studie (Master's thesis, Norwegian University of Life Sciences, Ås).
[22] Horne H and Hole J (2019). Hydrogen i det moderne energisystemet. http://publikasjoner.nve.no/faktaark/2019/faktaark80772019_12.pdf. Accessed 2023.09.10).

IOP Conf. Series: Materials Science and Engineering 1294 (2023) 012036 doi:10.1088/1757-899X/1294/1/012036

Impact of and solutions to effects of climate changes for Longyearbyen, Svalbard, Norway

S N Winther[1,*] and O T Gudmestad[1,2]

[1] UiT, The Arctic University of Norway
[2] Faculty of Science and Technology, University of Stavanger, Stavanger, Norway

Corresponding author: susannanw@hotmail.com_copy to otgudmestad@gmail.com

Abstract: Climate changes forces us to make significant mitigation and adaptation measures. As temperature rises and the environmental conditions changes, a variety of challenges occur. Across the whole globe, harming consequences are already being experienced. The Arctic region is particularly vulnerable to changes in Earth's climate system, especially because of the albedo effect, and the region is already heavily impacted. Primarily through the melting of ice, both sea ice and glaciers, permafrost thawing and changing precipitation patterns. Longyearbyen, Svalbard, is one examples of a society having to change as a direct result of global warming. This paper focuses on the challenges Longyearbyen will be facing and possible adaption methods and solutions. Because of environmental changes, the risk of natural hazards will increase, infrastructure will get damaged and traditional engineering methods will be impossible to implement. This is threatening to human lives as well as the habitat and survival of mammals, birds, and plants. It will be essential to find ways to predict and limit the effects of climate challenges, by protecting people and infrastructure from them. This will require innovation, adaption and risk-taking. By investigating the climate challenges facing Longyearbyen and possible ways to address them, this paper emphasizes the urgency of tackling the effects of climate changes in the Arctic in order to protect the communities in the Arctic region.

1. Introduction

The term climate change is used to describe the long-term variation of typical weather patterns, regionally or globally. Throughout the history, the Earth's climate has continually changed and gone through cycles with periods of warming and cooling [1, p. 81]. These climate variations are caused by various natural factors [2], and such processes occur slowly. However, large, and relatively rapid variations during the later decades are ongoing, indicating a change happening at a faster rate due to influence from human activities [1, p. 84]. The temperature is rising most rapidly in the Arctic, and Longyearbyen is one of the places on Earth where the average temperature is rising fastest. Further warming will amplify the melting of ice, thawing of permafrost and change of precipitation patterns, which will lead to severe consequences [3]. This paper will investigate how climate changes may affect the Arctic town Longyearbyen located at 78.25° North and how solutions to these challenges possibly can be tackled. Firstly, it will discuss why an increasing risk of natural hazards such as avalanches, landslides, and floods, necessitates major measures. Then, several adaption challenges regarding infrastructure and technical solutions will be explored. Effects of climate change on terrestrial and marine ecosystems are not part of this study, however, these effects will also cause major changes to the landscape.

2. Climate change

2.1 The globe's energy budget

The Earth's average surface temperature is maintained by a balance in the energy budget. When incoming short-wave solar radiation reaches the Earth's atmosphere, some is absorbed by gases, some is reflected into space by clouds or aerosol particles, and the rest passes through to the surface where it again is either reflected or absorbed [4]. Bright surfaces such as snow and ice reflect solar radiation very efficiently. In addition, there is outgoing radiation. The Earth's surface emits long-wave radiation, and the atmosphere also emits radiation, however, because of greenhouse gases this radiation is emitted both towards space and the Earth's surface, which is what causes the greenhouse effect. If the incoming and outgoing radiation is not balanced, the temperature increases or decreases which in turn eventually affects the climate [1, p. 25].

Human activities influence the global surface temperature. Most significantly by adding to the natural concentration of greenhouse gases in the atmosphere, primarily through the burning of fossil fuels. Furthermore, emission of methane, either caused by human activities or by leakage of methane from melting permafrost is of large concern. The greenhouse gas effect of methane is suggested to be 25 times more powerful than the emission of CO_2 [5]. This creates an enhanced greenhouse effect where more heat is trapped in the lower atmosphere leading to warming of the surface [1, p. 28].

2.2 Amplification of climate changes in the Arctic

Arctic amplification refers to the Arctic being the fastest warming region on the planet. An analysis from 2022 shows that since 1979 the temperature in the Arctic has risen nearly four times faster compared to the rest of the planet [6], as shown in Figure 1.

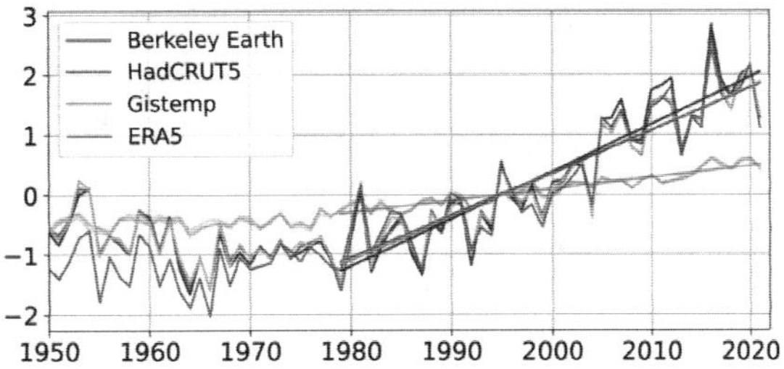

Figure 1. A comparison between the global mean (light colours) and the Arctic, 66.5 to 90° N (darker colours) average temperatures for the years 1950 to 2021 [6].

Legend: Berkeley Earth (non-profit research organization), Had CRUT5: the Met Office Hadley Centre/Climatic Research Unit version 5.0.1.0, Gistemp: NASA's Goddard Institute for Space Studies Surface Temperature version 4, ERA5: Data produced by the Copernicus Climate Change Service.

There are many processes contributing to the Arctic amplification. One of the main ones is the sea-ice albedo feedback. This process allows more heat to enter the Arctic system. As it gets warmer in the region, sea ice melts. When this happens, more of the Arctic Ocean gets exposed to solar radiation resulting in more warming as the dark ocean surface absorbs the radiation instead of reflecting it. This in turn leads to even more melting and warming. Another process involved in the Arctic amplification is the thawing of permafrost. When permafrost thaws, organic matter such as methane and carbon dioxide is released, leading to more greenhouse gases entering the atmosphere [7]. This in turn contributes to global warming, and permafrost thawing therefore makes up a feedback loop. What happens in the Arctic does not only lead to harming consequences locally, but it also affects the rest of the globe.

3. Longyearbyen, Svalbard, Norway

Longyearbyen is the administrative centre of Svalbard. The town is on the main island Spitsbergen and has approximately 2400 citizens [8]. The geography of Longyearbyen is characterized by mountains, glaciers and fjords as can be seen in Figure 2. In addition, Arctic tundra with limited vegetation and permafrost soil is widespread. The Arctic wildlife is diverse and consists of species that have adapted to the harsh environment, and Longyearbyen has a large amount of rare and endangered plant species. The town also holds the location of the Global Seed Vault, a facility that serves as a backup storage for seeds from around the world. Currently, the vault holds over 1.2 million seed samples from more than 6000 plant species [9].

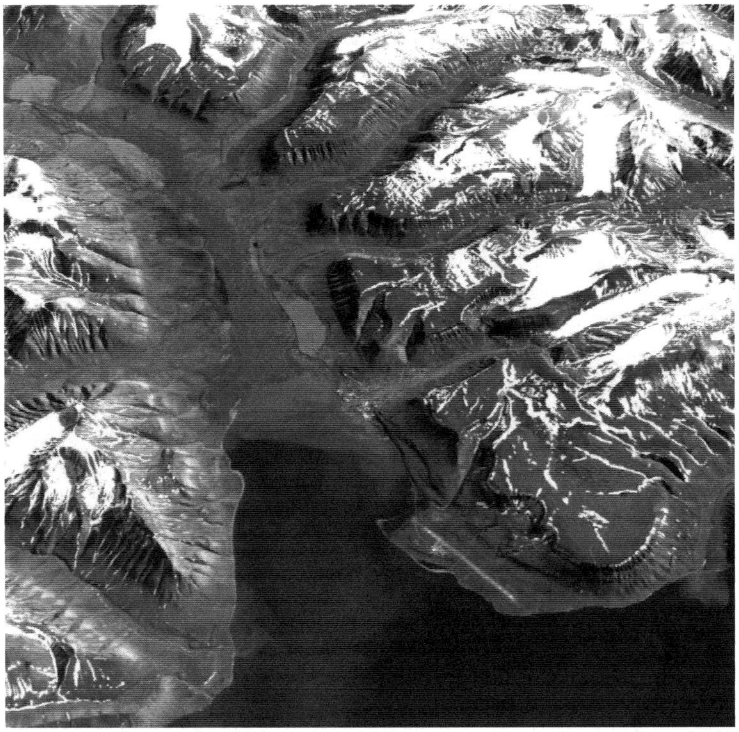

Figure 2. Image of part of Spitzbergen with Longyearbyen. Image from NASA [10].

3.1 Climate changes in Longyearbyen

Historical statistics from the weather station at Svalbard Airport give some insight in the weather trends Longyearbyen is experiencing. The station was established in 1964 and measures temperature, precipitation, snow depth and wind. Figure 3 illustrates the significant upward trend in temperature. This rapid warming causes increased melting of glaciers, permafrost thawing and precipitation, as well as more frequent extreme weather events.

When glaciers are melting in Longyearbyen, more water is added to the hydrological cycle. The most important effect of this in Longyearbyen is the increased wetness of the ground. Higher temperatures in Longyearbyen also increases the thawing of permafrost which makes the soil unstable. A monitoring station placed nearby Longyearbyen (at some higher location though) is indicating an increase in the ground temperature of about 1.8 °C over the period from 1998 to 2023 [11]. The trend of the annual mean ground temperature in Longyearbyen is heading towards crossing the zero degrees mark. This would mean the end of permafrost in Longyearbyen. Over time, the permafrost will gradually thaw more and more until it all eventually is gone. The thawing will change the terrain drastically, with respect to the effects of landslides and vegetation.

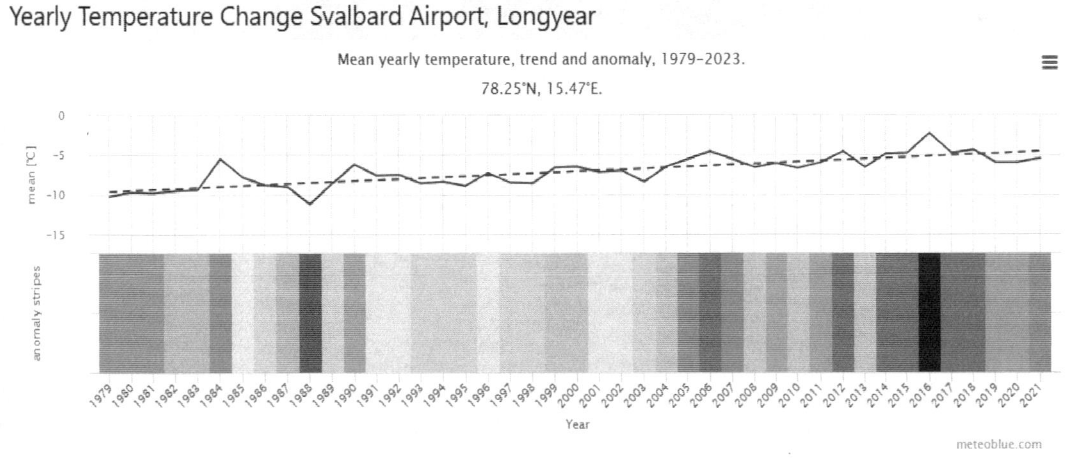

Figure 3. Yearly temperature change at Svalbard Airport, Longyearbyen, 1979 – 2021 [12].

The weather statistics also reveals that the amount of precipitation in Longyearbyen is increasing, which is an effect of the rising temperature [12]. Climate models predict that rain will be the dominating form of precipitation in the Arctic in the future. A larger part of snowfall will melt before reaching the surface because of the atmosphere being warmer. This will be particularly noticeable in autumn and spring when temperatures are close to zero degrees. It will get wetter also because warmer air has a higher capacity of holding moisture, leading to heavier rainfall [13]. Such changes in the weather pattern can lead to more permafrost thawing.

During recent years, temperature records and events of damaging avalanches and serious floodings have been linked to the changing climate. Other consequences concerning infrastructure are also present. Measures such as building of avalanche hindering barriers have been implemented to prevent and reduce damage of climate change in Longyearbyen [14].

4. Natural hazards caused by the effects of the warming climate

The changing climate in Longyearbyen will change the landscape as we know it, making it more likely to be affected by natural hazards such as avalanches, landslides, and floodings. Longyearbyen has experienced such situations in recent years. Mass movements (snow and ground slides) and accumulation of water are extremely dangerous and damaging. Therefore, it is important by the engineering society to identify ways to prevent, protect, and monitor these naturally triggered dangers.

4.1 Avalanches

In December 2015, eleven houses in Longyearbyen were taken by an avalanche that was released from the mountainside of Sukkertoppen. The incident resulted in several people getting injured and two people losing their lives [15], Figure 4. Because of Longyearbyen's isolated location, doctors and a surgical team had to be sent by plane from Tromsø to assist with the injuries, a trip that takes over an hour. Additionally, wounded people had to be transported to Tromsø. Homes and other buildings were left in ruins and the whole city contributed with search, rescue, and recovery. In 2017 another avalanche hit the town causing major damage. Both events have been linked to climate change as weather conditions were untypical for the season [14].

Figure 4. Damaged houses after the avalanche December 2015 [15] (photo: Geir Barstein)

The wetter climate and melting glaciers are huge avalanche triggers as the terrain becomes less stable. In order to understand the behaviour of snowpacks and avalanches, it is crucial to know what type of snow is present and its properties. However, this can be difficult, especially as the nature of avalanches have started changing due to global warming. As it gets milder and wetter, it is likely that the prevalence of wet-snow avalanches will increase in Longyearbyen. Such avalanches have a completely different dynamics than the dry avalanches that by far are the most common today. Higher levels of water in the snowpack makes it denser and weakens the bonds and layers, creating instability. Avalanches can also be very difficult to forecast and changes in weather patterns such as temperature, precipitation, and wind, expose new areas to avalanche risk [14]. For example, if wind direction and speed changes, that will affect where snow accumulates and thus potential avalanche risk areas.

4.2 Landslides

Permafrost thawing, more precipitation as rain and less sea ice contributes to more erosion and ground instability. This increases the probability of landslides. Due to the permafrost layer in Longyearbyen, the ground has the potential to become extremely wet. This is because the water does not have anywhere to escape as soon as it reaches down to the frozen ground. Therefore, the groundwater flow is highly restricted by permafrost. A higher amount of water increases the mass of the melted ground material and therefore also the gravitational driving force. Additionally, water can elevate pore-water pressures in the ground to the threshold of failure.

Another potential trigger for landslides is coastal erosion due to loss of the permafrost. Furthermore, as the sea level increases, more of the coast gets eroded, which in turn makes the terrain more vulnerable. So far, this has not been a significant issue in Longyearbyen, although a recent sea-cliff erosion failure occurred at Forkastningsfjellet northwest of Adventfjorden [16], see Figure 5. A larger failure could generate waves that would be of grave concern in Longyearbyen. Note that in other Arctic regions such as Siberia, coastal erosion represents a visible challenge [17].

Figure 5. Sea-cliff failure of approx. 300,000 m^3 of rock at Spitzbergen, not far from Longyearbyen, August 2016. (Photo: J. C. Nygaard [16]).

Figure 6 is a map showing Longyearbyen's risk picture, which considers landslides and floods. The red zones indicate the highest risk and shows areas where the probability of a landslide occurring is once every hundred years. However, keep in mind that the IPCC report [3] has signalled that extreme weather events have lower return intervals than thought, a more updated version would possibly suggest that the risks are somewhat more frequent than this figure says. While most of the infrastructure is placed outside the risk areas, we can find a few buildings and some roads even inside the red zones. A landslide will clearly have the potential to cause major destruction and threaten the safety of people.

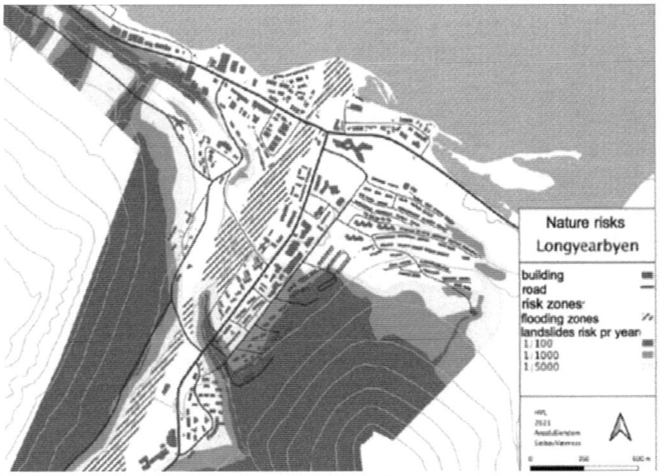

Figure 6. Areas of natural risk in Longyearbyen. Locations of highest flooding and landslide risk [18].

4.3 Flooding

Climate changes also increase the risk of flooding in Longyearbyen. When glaciers melt and permafrost thaws, large volumes of water get released. Additionally, it is expected that extreme weather events such as heavy rain or snowfall could become more frequent [14]. This will lead to increased runoff. It is most likely to happen in areas where permafrost and ice are close to the surface, and water cannot be absorbed into the ground or reach the river, resulting in the ground quickly getting fully saturated with water. If the amount of accumulated water is large enough, this can lead to flash floodings in downstream areas. One of the largest concerns is the Longyear River getting flooded. The river stretches through the Longyear valley, passing the town, before it runs out into Adventfjorden. Figure 6 reflects this risk of the river being flooded as water flow increases, by the marked flooding zones along the river.

Some years ago, the entrance to the Global Seed Vault (Figure 7), experienced several floodings mostly due to permafrost thawing, but also because of increased amounts of precipitation [19]. In the summer of 2020, Longyearbyen experienced record high temperatures which caused water from a melting glacier to flood a mine outside the town [20]. These events were reminders of how the climate change in Longyearbyen affects resources and societal functions.

Figure 7. The Svalbard Global Seed Vault [21] (Image Ambrosetti Tonatiuh)

IOP Conf. Series: Materials Science and Engineering 1294 (2023) 012036 doi:10.1088/1757-899X/1294/1/012036

5. Solutions to mitigate the effects of climate change

Because of Longyearbyen's isolated location and limited population, it is important to take great measures to reduce the severity of events when they occur. When time is crucial, transportation to other cities and lack of necessary equipment or capacity can be fatal. As the risk of avalanches, landslides and floods increases, it also becomes important to prevent destruction because of financial reasons. Such events usually require major clean-up work and reconstruction. This is both time-consuming and expensive, meaning frequent occurrences would be less manageable in the long run. Additionally, areas that today are considered safe may in the future lie in the path of new potential run-out zones, which can be very difficult to anticipate and therefore weaken the safety.

5.1 Technical solutions

There has been taken various measures in Longyearbyen to protect the infrastructure from damage, and there are comprehensive building regulations for the town. New buildings in Longyearbyen are being built on steel pillars that are drilled far down in the ground, through the permafrost and into the underlying bedrock [17]. Today, facilities are also lifted higher above the ground than traditionally to further separate the permafrost from the warmth radiating out of the buildings. Additionally, there have been placed sensors to measure the temperature and moisture of the ground, so that it can be better monitored.

5.1.1 Avalanche barriers. Longyearbyen has an advanced alert system that updates citizens about the daily avalanche situation. This local warning system, that uses data from sensors and local observers, was a direct result of the avalanches in 2015 and 2017 [14]. Evacuations happens regularly and many people have been permanently relocated to avoid risk zones. There has also been placed massive protection barriers in many areas. For example, numerous avalanche fences have been built on Sukkertoppen to protect structures that lie nearby (Figure 8). The situation prior to building the barrier is clear from the photo in this figure, i.e. no protection existed against an avalanche. At the upper and lower foundation, piles made of steel are drilled into the ground before additional rock anchors are secured into the bedrock to make sure all tension forces are accounted for [20]. Avalanche fences reduces the speed and intensity of the flow, as well as it traps snow and debris from the avalanche.

Figure 8. Avalanche fence barrier, Longyearbyen. (Photo: Maria Philippa Rossi/UNIS [14].

5.1.2 Monitoring. In Longyearbyen the permafrost layer is being monitored to better assess the risk of landslides, and there has been taken measures to flood-proof the Longyear River [14]. After the floodings of the Global Seed Vault, a major reconstruction project was conducted outside the vault. A new waterproof entrance was built, drainage ditches was added to carry away meltwater, and the ground surrounding was stabilized by getting artificially frozen and protected by a 20 meters high frozen wall, [19], as shown in Figure 9, which can be compared with Figure 7 that was taken at the inauguration of the Vault. When the vault was first built, it was not anticipated that such efforts would be needed.

F

Figure 9. For protection of the entrance to the Svalbard Seed Vault, a 20 m high frozen wall is created to stabilize the permafrost [19]. (Photo: Thomas Nilsen, Barents Observer)

As the conditions in Longyearbyen are getting warmer and wetter, it is crucial to be able to predict the behaviour of natural hazards. It will be important to further advance monitoring systems in the future, so they can give more precise information about when and where a potential avalanche or landslide will hit. Frequent evaluations and risk assessments should be done so that mitigation measures can be implemented [22], and the safety of the citizens can better be ensured.

5.1.3 The roads. Roads are highly exposed to the thawing permafrost. Potholes and cracks can become a frequent issue as the ground becomes unstable and shifts. Repairing or replacing are expensive processes, so there are strict regulations for road building in Longyearbyen. The roads are built on a specific fibre cloth that separates the masses from the underlying raised gravel bed, so that heat radiation which can further contribute to permafrost thawing is reduced [23]. The Svalbard Airport runway has experienced unevenness due to thaw subsidence. In 1998 a major, expensive reconstruction was carried out, however, it was only partly successful and is today regularly subject to re-pavement reparations.

5.1.4 Piping. In Longyearbyen, piping (water and sewage) that are buried in the ground are prone to break or rapture when the properties of the ground change. This can lead to leakage of pollutants and emissions of hazardous gases. Additionally, it is very costly and challenging to install pipelines in the ground in Longyearbyen because of the changing permafrost conditions. Because of the vulnerability of pipes running below the surface, the major piping systems in Longyearbyen are found above the ground. This includes water supply, sanitation, and district heating [23]. Above the ground, these systems can be made more flexible and adjustable. This reduces the risk of damage, as well as it makes repairing and maintaining easier. The pipes should be insulated to avoid freezing as well as avoiding outward radiation. It is important that there is no leakage leaving the pipes or the ground exposed. Additionally, the terrain must be levelled before construction can begin, Figure 10.

Figure 10. Piping over the ground on permafrost [24].

6. Conclusions

Climate change is drastically changing the Longyearbyen. The warming will impact the human communities, infrastructure, and landscape in the region. Retreating glaciers, changing weather patterns and permafrost thawing, as well as more frequent extreme weather events are all causing major challenges for Longyearbyen. For example, by increasing the risk of natural hazards such as avalanches, landslides, and floods. Additionally, essential structures and facilities, including buildings, roads, and pipelines, are getting more vulnerable. The survival of mammals, birds and plants will also be affected, as well as people's health and well-being. These challenges threaten the sustainability of the community. Extensive measures have already been taken. However, for the future of the town it will be important to continue coming up with new and improved ways to prevent, protect and restore. Because of Longyearbyen's isolated and northerly location, natural hazards are particularly critical. By reducing the severity of such events, we can make sure the consequences are restrained to a manageable level.

This can be done by relocating absolutely all structures to areas that are not at high risk, building more protection barriers and continuously improving monitoring systems to give early warnings. Infrastructure damage is also a concerning issue as buildings, roads and pipelines are particularly exposed to destruction when the ground loses its stability.

The mitigating effects of the activities undertaken to protect the buildings and the population of Longyearbyen are considered effective as improved safety against the effects of permafrost melting have been implemented in all building projects. The effects of increased precipitation are taken care of by ensuring drainage and use of material that can resist water and by constructing avalanche barriers that can resist large avalanches to form. There have been made many innovative improvements and solutions, however, we must make sure these can serve long-term and preferably permanent.

References

[1] Houghton J T 2015 *Global Warming*: The Complete Briefing. Cambridge University Press.

[2] University of Washington, College of the Environment. Climate Impacts Group: Climate Variability. https://cig.uw.edu/learn/climate-variability

[3] Intergovernmental Panel on Climate Change. (2021, August 9). Climate change widespread, rapid, and intensifying - IPCC. Retrieved from Newsroom: IPCC: https://www.ipcc.ch/2021/08/09/ar6-wg1-20210809-pr/

[4] NASA Science. (2010). The Earth's Radiation Budget. Retrieved from NASA Science: https://science.nasa.gov/ems/13_radiationbudget

[5] United States Environmental Protection Agency, EPA. Global Methane Initiative, Importance of Methane. https://www.epa.gov/gmi/importance-methane

[6] Rantanen M, Karpechko A Y, Lipponen A, Nordling K, Hyvärinen O, Ruosteenoja K, Vihma T and Laaksonen A 2022 The Arctic has warmed nearly four times faster than the globe since 1979. Retrieved from https://www.nature.com/articles/s43247-022-00498-3

[7] Flanagan E 2021. The Global Carbon Budget and Permafrost Feedback Loops in the Arctic. Retrieved from https://www.thearcticinstitute.org/global-carbon-budget-permafrost%20feedback-loops-arctic/

[8] Visit Svalbard (Online). Longyearbyen. Retrieved March 26, 2023, from Destinations: Visit Svalbard: https://en.visitsvalbard.com/visitor-information/destinations/longyearbyen

[9] Svalbard Global Seed Vault. (Online). Safeguarding seeds for the future. Retrieved March 31, 2023, from https://www.seedvault.no/

[10] NASA, Image of part of Spitzbergen with Longyearbyen. https://eoimages.gsfc.nasa.gov/images/imagerecords/8000/8577/longyearbyen_AST_2003193_lrg.jpg

[11] The Norwegian Meteorological Office, Cryo. https://cryo.met.no/en/home. Selected figure: https://cryo.met.no/index.php/en/permafrost/janssonhaugen.

[12] Meteoblue (Online). Climate Change Svalbard Airport, Longyear. Retrieved May 1, 2023, from Meteoblue: History & Climate: https://www.meteoblue.com/en/climate-change/svalbard-airport%2C-longyear_svalbard-and-jan-mayen_6296766

[13] Bintanja R, Van Der Wiel K, Van Der Linden E, Reusen J, Bogerd L, Krikken F and Selten F 2020 Strong future increases in Arctic precipitation variability linked to poleward moisture transport. Retrieved from https://www.science.org/doi/10.1126/sciadv.aax6869

[14] Albrechtsen E, Indreiten M and Wickström S 2022 Look to Longyearbyen for climate adaption. Retrieved from UNIS - The University Centre in Svalbard: https://www.unis.no/news/look-to-longyearbyen-for-climate-adaptation/

[15] Ylvisåker L N 2020 Skredet som endra ein by. Retrieved from NRK. https://www.nrk.no/tromsogfinnmark/xl/fem-ar-etter-snoskredet-pa-svalbard-19.- desember-2015-_-slik-har-det-endra-longyearbyen-1.15293520

[16] Kuhn D, Redfield T F, Hermanns R L, Fuchs M, Torizin J and Balzer D 2019 Anatomy of a mega-rock slide at Forkastningsfjellet, Spitsbergen and its implications for landslide hazard and risk considerations. *Norwegian J Geology* **99**, 41-61. https://dx.doi.org/10.17850/njg99-1-03.

[17] Gudmestad O T 2020 Technical and economic challenges for Arctic Coastal settlements due to the melting of the Arctic ice and the permafrost. In: *IOP Conf. Ser.: Earth Environ. Sci.* **612** 012049

[18] Van Nes A., Sæbø M, Værness K and Reksten C 2022 Strategies for densification of Longyearbyen: Spatial, morphological, geological, and social analyses of the arctic town Longyearbyen in Norway. *Proc. 13th Space Syntax Sympos.* Available at: https://www.hvl.no/globalassets/hvl-internett/arrangement/2022/13sss/452sabovaerness.pdf (Accessed: 20.10.2023)

[19] Nilsen T 2018 Thawing permafrost makes big trouble for world's northernmost town. T Retrieved from he Barents Observer: https://thebarentsobserver.com/en/arctic/2018/10/thawing-permafrost-troubles-longyearbyen (Accessed: 20.10.2023).

[20] Nilsen T 2020 Climate change hits back, Svalbard coal mine flooded by melting glacier. Retrieved from The Independent Barents Observer. https://thebarentsobserver.com/en/industry-and-energy/2020/07/coal-mine-flooded-melting-glacier (thebarentsobserver.com)

[21] Turne, C 2011. Doomsday Vault: Svalbard Global Seed Vault, ICON. Available at https://www.iconeye.com/architecture/doomsday-vault-svalbard-global-seed-vault.

[22] Hellum Ø S, Jonsson A, Kanstad S B, Lande E J and Nerland Ø 2018. The Challenges of Mitigation Measures in Longyearbyen Svalbard. Retrieved from https://www.researchgate.net/publication/332447708_the_challenges_of_mitigation_measures_in_Longyearbyen_Svalbard

[23] Longyearbyen Lokalstyre (2020, December 15). Lokal norm - tekniske anlegg, Longyearbyen (Norwegian). Retrieved from https://lokalstyre.custompublish.com/getfile.php/4457532.2046. qwspm7twqpzkmm/Lokal+norm+for+tekniske+anlegg+i+Longyearbyen.pdf

[24] SwissEduc. Glaciers online 2017 J. Alean · M. Hambrey · Impressum. Retrieved (20.10.2023) from: https://www.swisseduc.ch/glaciers/svalbard/longyearbyen/index-en.html?id=12

IOP Conf. Series: Materials Science and Engineering 1294 (2023) 012037 doi:10.1088/1757-899X/1294/1/012037

Dealing with abnormalities and deviations to enhance resilience in engineering Assets: A critical review from human factors and decision-making perspectives under complex operational contexts

M Bakhshandeh* and J P Liyanage

Faculty of Science and Technology, University of Stavanger, Norway

*Correspondence: mohammad.bakhshandeh@uis.no

Abstract. With the growing scale of industrial demands, complexities, and uncertainties around asset engineering and operations due to advanced technology utilization, digitalization, sustainability, new operating models, etc., the sensitive role of abnormalities and deviations towards human safety, systems security, reliability and resilience of engineering assets and industrial systems are becoming even more significant for modern industrial sectors as well as societies in general. In these contexts, the abilities of operators to capture and sense-make early signals that emerge from engineering assets and systems need more attention since it enables them to enhance critical situation awareness (SA) during complex operations. This calls for proactive solutions that can integrate core data with operator knowledge using suitable logical approaches, particularly in a period where there is growing recognition that asset data can provide strong support for engineering and operational decisions in demanding contexts. Based on an ongoing research project, this paper sheds light on abnormalities and deviations; two specific attributes that should be better understood. The purpose is to explore how to capitalize them at very early sense-making stages to enhance situation awareness and thus resilience of dynamic and complex engineering assets and systems. Through a critical review of the current state of knowledge, together with industrial observations, this paper studies these core concepts in detail with due attention to the critical need of so-called priory contextual knowledge and hybrid contextual decision solutions. This R&D work explores proactive possibilities to mitigate inherent potentials for unwanted events and incidents to enhance resilience in the era of digital twins and cyber-physical systems, where complex technologies and operational demands generate new conditions for asset performance.

Keywords: Complex systems, Industrial assets, Digitalization, Abnormalities, Deviations, Contextual knowledge, Advanced situation awareness, Early-sense deviations, Operational risk.

1. Introduction

With increasing reliance on complex dynamic systems, our lives, jobs, properties, and environment are becoming more complex, interconnected, and interdependent [1]. Decades of evidence from a range of industrial sectors have proven that unwanted events and incidents happen continuously in various industrial contexts challenging our conventional comprehension, despite technological and operational advances. Thus, being able to think and act early is arguably a critical attribute to be possessed by industrial assets and systems owners, and as Lane [2] emphasizes a small cause in current contexts can make a big difference, for instance in terms of reliability, safety, and security. Throughout the history of industrial evolution, unwanted events have left a range of unresolved questions and issues, which today and tomorrow can become the main concerns due to rapid and fast-tracked growth. With reference to investigation reports about major accidents such as the Fukushima Daiichi Nuclear Power Plant disaster, Deepwater Horizon blowout, Boeing Max-8 crashes, etc. (see for example, [3,4,5], etc.), one can argue that major industrial accidents have indicated the critical role of early warning-signals, pre-accident details, and non-attended data towards the mitigation of exposed risks. For example, detailed analyses of more recent Fukushima Daiichi Nuclear Power Plant [6,7,8], and Deepwater Horizon blowout accidents [9,10], unveil a range of abnormalities and deviations that had given some form of early indications prior to the disaster that had not been duly attended. Arguably, such early indications of abnormalities and potential deviations are vital information of modern complex assets that can be used smartly to inform operators early and promptly about potential conditions that can lead to unwanted events and incidents.

Thus, we argue that the clarity as well as critical roles of systems abnormalities and deviations still largely remain unexplored domains in terms of mitigating systems' inherent risks and major accident potentials. Hence, they are reviewed, and their potential is explored critically in detail in terms of enhancing resilience of modern complex assets and systems.

2. Abnormality and deviation concepts in complex systems: A critical review

Complex systems are formed when relationships between different layers in an industrial ecosystem such as political, organizational, environmental, natural, technical, operational, engineering, etc., are entangled and converged. These layers interact dynamically, and the effects of these numerous interactions propagate throughout the system over time sometimes leading to unfamiliar system behavior [11]. Unique features of complex systems make it difficult to acquire causal explanations of various behavior patterns in a highly dynamic system. Interestingly, Lane [2] asserts that we should change our attitudes about complex systems and that we should learn how to make future states more predictable in specific ways. Complex systems by nature are neither completely predictable nor fully controllable and hence refocus should be made on elements that make the system more fragile and susceptible to perturbations, and at the same time be sensitive to issues that may affect system's well-being [2,11]. However, it can be argued that complex systems naturally possess a wide range of abnormalities despite rules, controls, and standards. We argue that Abnormalities are a natural property of complex industrial assets and systems, they appear explicitly or implicitly as a characteristic of a complex system and can emerge from multiple layers or sources within its context. In addition, changing patterns of behaviors can create a degree of freedom for agents and components to alternate in complicated manners in a dynamic environment [12], resulting in generating various forms of deviations. The inability to understand abnormalities and deviations, lack of awareness of influential sources, lack of knowledge of their consequences, and lack of early preparedness and resolutions to handle them, etc. can expose complex systems to critical conditions. In order to avoid this, boundaries and critical points, so-called tipping points, can be defined according to the features of the system of interest, which can signal if the system behaves in an unexpected manner. According to Burton et al. [1], defining such boundaries and tipping points pre-requires identification of objectives, goals, functions, values, and scope of the system in focus.

From a systems deviation perspective, Kjellén [13] emphasizes on two specific aspects, namely variables and norms. He argues that a deviation occurs when a value of the system falls outside of defined boundaries. Boundaries can be interpreted as planned and intended norms, or constraints and markers, built into the physical system. According to Scheffer et al. [14], in complex dynamic systems, deviations often cover a period when variables start gradually crossing the pre-set boundaries and evolve into a fuzzy manner. Subsequently, the stability of the system can gradually decline, and variables can show unusual fluctuations indicated as early signals [14], yet the recovery chance is relatively high due to slow early transition path. When the system has passed a certain threshold, sudden changes may occur, making the interpretation of the pattern of change, propagation path, and recovery efforts more difficult. Eventually, the system can transit into an uncontrollable mode. Figure 1 illustrates some scenarios related to deviation path profiles in four possible modes.

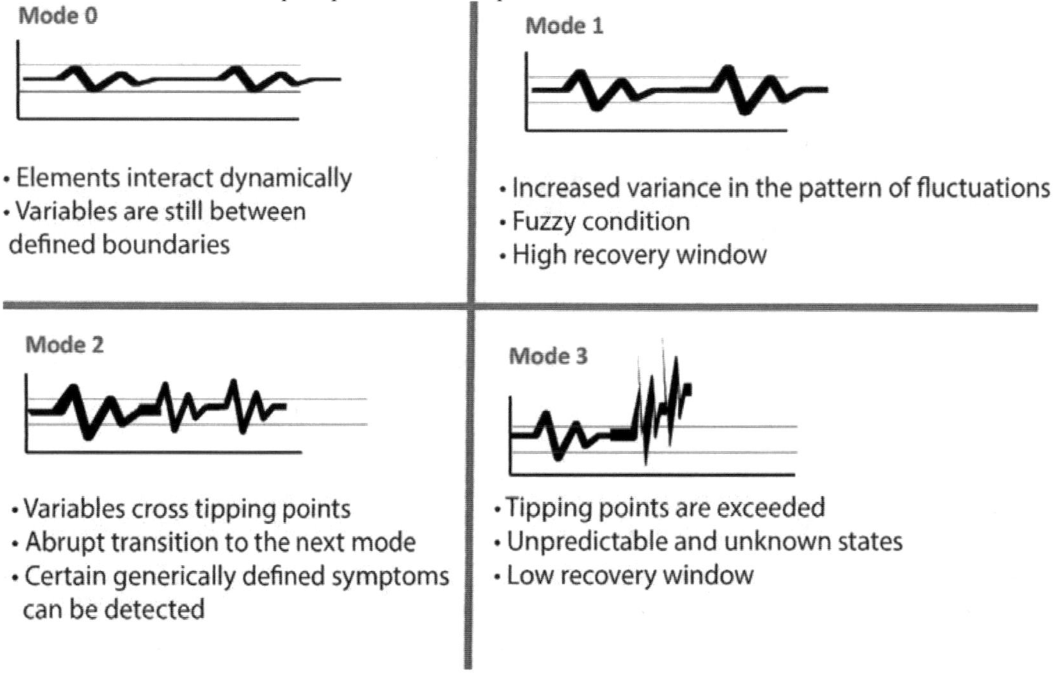

Figure 1. Some scenarios on Deviation path profiles due to inherent abnormalities.

3. Role of knowledge and situation awareness in sensing early- signals

Evidence from serious accidents shows that it is not easy to predict deviation transition paths as the state of the system may show only a few changes before any major deviation occurs [14]. If deviations grow over time to a higher level of severity (for example, Mode-3), activating alarm signals to the operator, then the available capacity to execute counteractions to compensate can become exhausted. If the system is loosely coupled with a higher degree of non-linear interactions, then abnormalities can even propagate through the systems in various paths resulting in complex deviation patterns and unpredictable outcomes. In many industrial contexts, human operators have to monitor the system constantly to locate and define highly likely causes of deviations. They should carry out appropriate remedial actions to mitigate consequences based on prior knowledge, experience, checklist, and pre-defined procedures. Such situations need several sequential search actions through several sources of information which naturally stress operators, especially when alarms are triggered, which may even cause automation bias and lead to erroneous decision-making. When it comes to sensing deviation signals, its potential impact on the level of Situation awareness (SA) of human operators, plays quite a critical role.

Endsley [15] describes SA in terms of three attributes namely: perception of elements, understanding of current situations, and prediction of future states. Bakhshandeh and Liyanage [16] argue however that the current definition of SA cannot address all concerns associated with complexity in a highly dynamic environment, and therefore it is a must to shift the boundaries of knowledge to acquire a higher level of SA, called Advanced situation awareness (Ad-SA).

As Baker [17] emphasizes, linear thinking when dealing with complexity can easily result in blinding humans to a few critical causes and consequences in a dynamic environment. Therefore, any specific knowledge creation process should be able to provide operators with a relatively more comprehensive basis and support, which can cover more engaging layers and even potentially hidden *black swan events*. This creates an increasing focus on the utility of data and data-driven solutions, and particularly

IOP Conf. Series: Materials Science and Engineering 1294 (2023) 012037 doi:10.1088/1757-899X/1294/1/012037

automated solutions. However, as Cilliers [11] points out, interpretation of complex systems data to generate relevant deeper knowledge cannot purely be left to an automated system, even though human capabilities in understanding complex behavior in a highly dynamic environment, particularly during early stages are yet far from being perfect. Human operators should still be designed into the decision-making loops in complex industrial processes and operations in a smart manner, since operator experience and tacit knowledge arguably play pivotal roles in terms of safety, security, reliability, and overall systems resilience. This indeed underlines the necessity of a hybrid approach that can capture deeply embedded meanings of abnormalities and deviations in complex industrial contexts to support operational decisions.

From such a perspective, it is argued here (See Figure 2) that a synergy of Human-Machine intelligence is an ideal approach. The underlying need is to generate necessary hybrid intelligence that can in turn enable human operators to make informative decisions by enriching their level of awareness and knowledge of dynamic situations. Such hybrid intelligence is a key factor in creating a specific knowledge type, so-called contextual knowledge. As Brézillon and Pomerol [18] delineate, contextual knowledge is more focused on the relevant knowledge for a given situation, and hence arguably is also quite critical for Advanced SA (Ad-SA). Moreover, contextual knowledge can enhance the human cognitive model for reasoning and decision-making in real-time contexts [19], which will make human operators more focused, ignoring irrelevant data or noise-information, when searching for effective counteraction options. However, as Beyerlein and Lin [20] imply, human knowledge alone is not comprehensive enough to deal with all complex conditions in a dynamic environment.

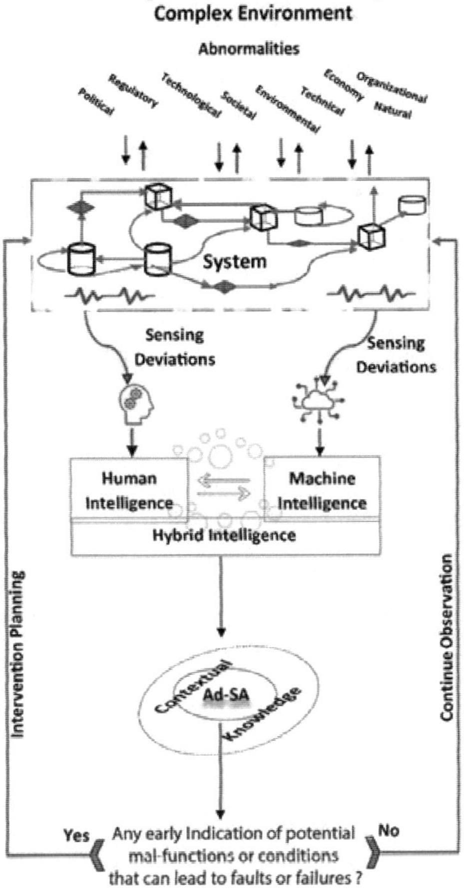

Figure 2. Fusion of Human-Machine Intelligence to generate contextual knowledge and Ad-SA to support early interventions.

From an industrial perspective, there are many efforts underway to develop suitable models to integrate information and interface them with existing knowledge to create better mental pictures for human operators for the purpose of making complex systems more understandable [21]. The central argument here is that as the complexity increases it is essential to explore a new class of models that can capture critical information relatively earlier and connect that with existing knowledge enabling new paths to deal with uncertainties of complex industrial systems proactively enhancing system resilience. This paper sheds more light on this elaborating on a Hybrid contextual decision solution.

4. Towards a Hybrid Contextual Decision Solution

From the point of view of a Human operator who is interacting with a complex and highly dynamic system, it is argued here that a Hybrid contextual decision solution is an increasing necessity in many modern industrial contexts. If adopted smartly, it can help operators deal with deviations, starting from relatively early stages of complex transition processes. The main purpose of such a solution is to provide the basis to enhance the SA (i.e., Ad-SA) of operators through a deeper contextual knowledge with a deeper understanding of ongoing system behaviors, early sense-making of deviations where the potential causes are rooted, and awareness of what their potential propagation paths can be.

However, it is often discussed that the emergent properties of modern complex systems represent a big challenge for predicting future states. But, following Cilliers [11], the presence of these properties does not pose a significant challenge to uncovering the causality. Moreover, the notion of abnormality does not convey the message that a complex system is a chaotic phenomenon, on the contrary, it is acknowledged to have a robust structure. The underpinning central issue is that inherent abnormalities and deviations in a system can be rooted in deep layers and develop into known or unknown paths with potential negative consequences. As Woods [22] points out temporal characteristics of such deviations can provide useful information about the root causes and alternatives for counteraction plans to operators. As a matter of fact, the process should be represented so that operators gain the abilities to track deviations that grow, spread, and subside [22]. Similar views seem to have triggered new efforts in developing semantic modeling of dynamic complex systems to assist humans in making sense of the elements in their interactive systems environment.

Over the years, various mathematical models, qualitative models, and algorithms have been developed based on artificial and computational techniques to create decision support systems (DSS) to assist humans with abductive reasoning, understanding the current situation, and predicting future states based on existing knowledge and experience [23]. However, from a more human operator perspective, Stanton [24] argues that the ways of presenting information based on such new solutions in modern times are heavily affected merely by technological capability rather than human performance or what operators actually need when dealing with perturbations. Therefore, Bordasch and Göhner [23] suggest that additional functionalities need to be added to existing models with specific considerations on human factors in real operational contexts. For instance, recent developments in remote operations in offshore oil and gas sectors have clearly indicated the need to provide operators with not only all real-time data relevant to field operations but also with solutions to create right cognitive maps to enhance SA (Ad-SA) during continuous operations. In such highly interactive and complex contexts, integrating data, utilizing rule-based technologies, using hybrid physics-based and data-driven models in combination with operational experience, modernizing operating procedures, etc. are industrial considerations that can better represent the demanding context and the dynamics of complex systems to help developing right cognitive models and contextual knowledge for advanced SA (Ad-SA). Moreover, ideally, any such contextual decision approach for supporting operators should also take into account human cognitive abilities, capacities, and limitations in dealing with complex data with diverse characteristics. The level of human sensitivity towards different stimuli can vary from person to person. The ability to detect right cues among irrelevant ones, sensitivity to interpret occurring events, and thus to ensure effective perceptual judgment, etc., purely based on cognitive skills and abilities in emerging complex contexts will be quite challenging [16]. Hence, in modern industrial contexts where human operators' work context and task characteristics become relatively more demanding, an ideal solution has to be hybrid in nature where algorithms and applications are integrated with operator experience, based on representative data, logic, and rules, reasoning processes, and interpretation and visualization support to enrich the contextual knowledge and situation awareness.

Based on the aforementioned reflections (and also deviation path profiles in Fig. 1), Figure 3 illustrates two competing scenarios: *alarm-based scenario* and *improved scenario*, for the purpose of sense-making of deviations in complex industrial contexts.

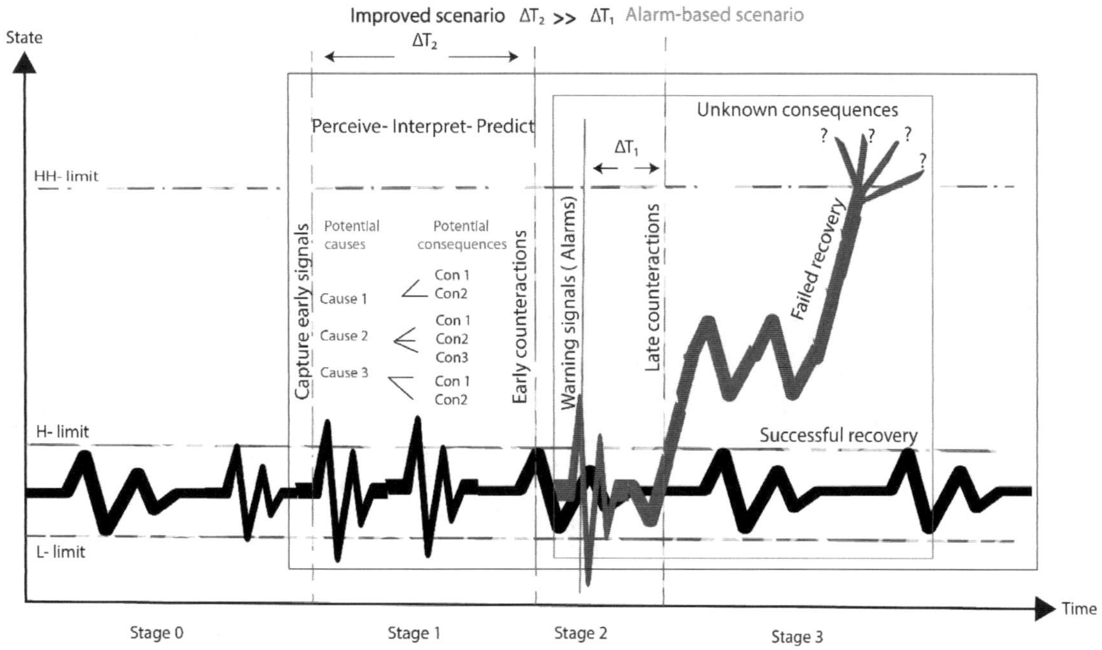

Figure 3. The basis for enabling better system resilience through a hybrid contextual decision solution.

The traditional scenario (ΔT1) is mostly based on an *alarm-based solution* where operators naturally have a limited time window (*Stage-2* + *Stage-3*) to become aware of the situation, detect all root causes, predict possible consequences, and take appropriate counteractions. Many such instances follow pre-defined routines and procedures, with well-known practical limitations and challenges in terms of alarm management as well as operators' cognitive workload [22,32]. Due to a higher degree of interactivity between system elements and the unpredictable behavior patterns of complex systems, considering all possible scenarios under stressful alarm situations is a real challenge both in terms of cognitive demands and execution of right mitigating actions. Thus, in very demanding, highly dynamic, and complex contexts that can be characterized by stress-inducing alarm patterns and late counteractions, a failed recovery tends to be the highly probable outcome as many major industrial accidents (e.g., Chernobyl disaster, Fukushima Daiichi Nuclear Power Plant disaster, Deepwater Horizon blowout, Boeing Max-8 crashes, etc.) have underlined. Therefore, to tackle these problems, human operators need to be attentive and resourceful to be able to maintain a proper level of awareness of what is changing in a given context in modern industrial systems. Any approach to capture the most likely scenarios in advance allowing operators to prioritize them according to an acceptance criterion, is of immense benefit, particularly in emerging contexts. This is the basis for the *improved scenario* (ΔT2) where operators are given the ability to work within an extended intervention window *(Stage-1* + *Stage-2* + *Stage-3)* so that they acquire possibilities to perceive relevant information early, interpret potential causes, predict potential consequences in real-time, and prioritize their interventions appropriately. As Figure 3 illustrates, there is a relatively larger window of time (ΔT2 >> ΔT1) for operators to sense deviations and take proactive measures. Once an early deviation signal has been captured and communicated by a technical solution, human operators can trace the potential causes of the early signal and take necessary actions, for instance, to prevent the system from reaching a more critical state where an alarm system would naturally be triggered. Such an early sense-making approach, characterized by capturing of early signals, more time to perceive-interpret-predict, and implementation of early counter actions, can naturally enable effective use of knowledge, time, and capacity, leading to successful recoveries, thus preventing costly downtime, damage to equipment, and potential safety hazards. As aforementioned, a whole range of major accidents inclusive of the Fukushima Daiichi Nuclear Power Plant disaster, Deepwater Horizon blowout, Boeing Max-8 crashes, etc., as well as many other unwanted events and incidents that occurred in recent years, have already indicated the practical utility of such early sense-making efforts towards critical operational decisions.

5. Conclusion

The paper emphasized that abnormalities and deviations are inherent characteristics of complex industrial systems, and deviations occur during interactions across pre-defined boundaries. Abnormalities may not be desirable, but they do not necessarily indicate failures or malfunctions until some form of deviation begins to occur. At best, tipping points can be pre-defined, according to matching criteria and expected values, to detect such deviations much early and subsequently to track the potential propagation paths ahead of time. Abilities for sense-making of an industrial system's early signals play an increasingly pivotal role by enabling operators to plan intervention measures and to deal with unwanted conditions effectively ensuring resilience. Through a critical review of these core concepts with respect to modern industrial contexts, this paper underlined the necessity for priory contextual knowledge and hybrid contextual decision solutions to mitigate inherent potentials for unwanted events and incidents. Reflections here are based on a critical review of the current state of knowledge and various industrial observations. The real practical utility of such solutions is clearly evident with respect to recent cases of unwanted events and incidents, and importantly more so for emerging complex industrial systems in the era of digital twins and cyber-physical systems.

References

[1] Burton S and Mcdermid JA Garnett P Weaver R 2021 *Safer Complex Systems: An Initial Framework* The University of York England.
[2] Lane SN 2011 The tipping point: How little things can make a big difference *Geography* **96** 34-38.
[3] International Atomic Energy Agency (IAEA) 2015 The Fukushima Daiichi Accident *IAEA* Vienna.
[4] British Petroleum (BP) 2011 Deepwater Horizon accident investigation report In: Bly M (Ed): Diane Publishing.
[5] The House Committee on Transportation and Infrastructure 2020 *Final Committee Report: BOEING 737 MAX* [Online] Available at: *https://transportation.house.gov*
[6] Perrow C 2011 Fukushima and the inevitability of accidents *Bulletin of the Atomic Scientists* **67** 44-52.
[7] Hollnagel E and Fujita Y 2013 The Fukushima disaster–systemic failures as the lack of resilience *Nucl Eng Technol* **45,** 13-20.
[8] Khan AH Hasan S and Sarkar MAR 2018 Analysis of possible causes of Fukushima disaster *International Journal of Nuclear and Quantum Engineering* **12,** 72-77.
[9] Reader TW and O'Connor P 2014 The Deepwater Horizon explosion: non-technical skills, safety culture, and system complexity *J Risk Res* **17,** 405-424.
[10] Oudhuis M and Tengblad S 2018 BP and Deepwater Horizon: A Catastrophe from a Resilience Perspective In: Tengblad, S and Oudhuis M (Eds) *The Resilience Framework: Organizing for Sustained Viability* Springer Singapore 71-87.
[11] Cilliers P 2005 Knowing Complex Systems In Richardson KA (Ed) *Managing organizational*

complexity: Philosophy, theory, and application Information Age Publishing (IAP) 7-20
[12] Poon L and Grebogi C 1995 Controlling complexity *Phys Rev Lett* **75** 4023-4027.
[13] Kjellén U 1984 The deviation concept in occupational accident control—I: Definition and classification *Accid Anal Prev* **16** 289-306.
[14] Scheffer M Bascompte J Brock WA Brovkin V Carpenter SR, Dakos V, Held H, Van nes EH Rietkerk M and Sugihara G 2009 Early-warning signals for critical transitions *Nature* **461** 53-59.
[15] Endsley MR 1995 A taxonomy of situation awareness errors In: R Fuller N Johnston and N McDonald (Eds) *Human factors in* aviation *operations* Ashgate Publishing **3** 287-292.
[16] Bakhshandeh M and Liyanage JP 2022 Advanced Situation Awareness, Human Vigilance, and Sensitivity in Complex and Dynamic Industrial Systems: Perspectives Towards Enhancing Systems Resilience Under Digitalization Contexts In: *Proceedings of the 32nd European Safety and Reliability Conference (ESREL 2022)* Research Publishing Services.
[17] Baker AC 2010 Shifting landscape: Differences that Spawn New Knowledge. In: Richardson KA and Tait A (Eds) *Complexity and Knowledge Management Understanding the Role of Knowledge in the Management of Social Networks* Information Age Publishing 109-122.
[18] Brézillon P and Pomerol JC 1999 Contextual knowledge and proceduralized context *Proceedings of the AAAI-99 Workshop on Modeling Context in AI Applications,* Orlando Florida USA *AAAI* Technical Report.
[19] Burov Y 2021 Knowledge-Based Situation Awareness Process Based on Ontologies In: *5th International Conference on Computational Linguistics and Intelligent Systems CEUR-Workshop Proceedings* 413-423.
[20] Beyerlein M and Lin J 2010 Participation and complexity in collaborative knowledge generation In: Richardson KA and Tait A (Eds) Complexity and Knowledge *Management: Understanding the Role of Knowledge in the Management of Social Networks* Information Age Publishing 153-173.
[21] Wickens CD 2000 The trade-off of design for routine and unexpected performance: Implications of situation awareness In: Endsley MR and Garland DJ (Eds) *Situation awareness analysis and measurement.* Lawrence Erlbaum Associates Inc 211-225.
[22] Woods DD 1994 Cognitive demands and activities in dynamic fault management: abductive reasoning and disturbance management In: Stanton NA (Ed) *Human factors in alarm design* CRC Press 63-92.
[23] Bordasch M and Göhner P 2013 Fault prevention in industrial automation systems by means of a functional model and a hybrid abnormity identification concept *IECON 39th Annual Conference of the IEEE Industrial Electronics Society* IEEE 2845-2850.
[24] Stanton NA 1994 *Human factors in alarm design* CRC Press
[25] Skilton M and Hovsepian F 2018 *The 4th industrial revolution* Springer.
[26] Zhu P Liyanage JP Kumar R and Panesar S 2021 Decision quality related to Emergency shout down system in the oil & gas industry: Influences from data & information flow *International Journal of Decision Sciences, Risk and Management, Inderscience Enterprises Ltd* **10** (1/2) 131-159.
[27] Zhu P and Liyanage JP Jeeves S 2019 Data-driven failure analysis of Emergency shutdown systems in oil & gas industry: Evaluation of detectability from failure progression perspective *J. Qual. Maint. Eng.* Emerald **26** (1) (online DOI: 10.1108/JQME-03-2018-0022).
[28] Liyanage JP 2012 Hybrid Intelligence through business socialization and networking: Managing complexities In: the digital era Cruz-Cunha MM Putnik GD (Eds) *Handbook of Research on Business social networking: Organizational, managerial, and technological dimensions* IGI-Global 567-582.
[29] Raza J and Liyanage JP 2010 Managing hidden systems threats for higher production regularity using intelligent technological solutions: A case study *Eur J Ind Eng* **4** (2) 249-263.
[30] Raza J and Liyanage JP 2009 Application of intelligent technique to identify hidden

abnormailities in a system: A case study from oil export pumps form an oil production facility *J Qual Maint Eng* **15** (2) 221-235.

[31] Liyanage JP 2008 Managing industrial assets and complex technologies in the digital era: Human and Organizational integration, Guest Editorial, Special issue on 'Managing industrial assets and complex technologies in the digital era: Human and organizational integration' *Int J Technol Hum Interact* IGI-Publishing i-iv.

[32] Rothenberg DH 2009 *Alarm management for process control: a best-practice guide for design, implementation, and use of industrial alarm systems* Momentum Press USA.

IOP Conf. Series: Materials Science and Engineering 1294 (2023) 012038 doi:10.1088/1757-899X/1294/1/012038

Ship emission projections based on time series forecasting model for sustainable shipping in the strait of Malacca and Singapore

K-H Ten[1], H-S Kang[1,2,*], K-Y Wong[1], C-L Siow[1,2], C-H Ong[3], K-C Hoo[4], Y Liu[5]

[1] Faculty of Mechanical Engineering, Universiti Teknologi Malaysia, Johor Bahru, Malaysia
[2] Marine Technology Center, Institute for Vehicle System & Engineering, Universiti Teknologi Malaysia, Johor Bahru, Malaysia
[3] Azman Hashim International Business School, Universiti Teknologi Malaysia, Johor Bahru, Malaysia
[4] Software and Advanced Technology Group, Intel Corporation, Penang, Malaysia
[5] Department of Mechanical and Industrial Engineering, Norwegian University of Science and Technology, Oslo, Norway

[*] Correspondence: kanghs@utm.my

Abstract. As maritime activities continue to play a pivotal role in global trade, concerns over ship emissions' environmental impact have intensified. This study presents detailed projection of ship emissions in Strait of Malacca and Singapore (SOMS), based on Automatic Identification System (AIS) data. By harnessing the rich AIS dataset, emission profiles were developed out of the ship activities data. To envision a sustainable maritime future, we incorporate some possible scenarios around the region combined with a time series forecasting model to project the future conditions of ship emission in SOMS. By analysing the conditions in each scenario, essentials for shaping intelligent systems for efficient maritime traffic can be discovered. Our analysis considers evolving factors such as various ship properties, operational modes, and trajectories. The results provide insights for policymakers, industry stakeholders, and environmental planners seeking to mitigate the local maritime sector's carbon footprint. This study signified the value of AIS data-driven approach to facilitate regional strategist in confronting resolutions for greener maritime operation, aligning with the transition to intelligent and sustainable practices in the maritime industry within the SOMS.

1. Introduction

Moving forward with the synergies of Sustainable Development Goals (SDGs) in combating climate change, maritime authorities were also committed to extend their effort into the fulfilment of net zero emissions by 2050. In the course of transitional phase of technical advancement to completely resolve the greenhouse gases (GHGs) emissions from ships, an institutional platform should be in place for collectively monitoring and analysis of the maritime carbon footprint relative to the ambient air quality. For this platform is not only for awareness raising to the society, but also to enhance the surveillance of maritime activities in conjunction with emissions tracking, AIS data were predominantly utilized for the assessment from environmental aspects, especially being an enabler for the use of machine learning model such as neural network. With the study of historical ship emissions over certain regions, it is also possible to project a logical trend of emissions that is likely to be foreseen with certain conditions applied. Relatively, SOMS is a critical region where ship traffic is dense and complex that is prone to congestions and form basin of emissions yet lacking the tracks of maritime carbon footprint in this region that many have concerned.

As the major conduit linking the international East-West shipping routes, SOMS is estimated to have enabled the transit of one-third of global trades annually, corresponded to 15 millions of oil barrel [1]. Aside from bringing prosperities to many countries that benefits from this shipping routes, the heavy marine traffic of about 120,000 vessels annually in the compact area of SOMS also brought about challenges, such as marine environmental and safety hazards, for the littoral countries to deal with. Upon the development of maritime safety awareness and environmental protection over the years, the partnerships of Tripartite Technical Experts Group (TTEG) between the members of Indonesia, Singapore and Malaysia are still continuing to address the latest maritime issues in SOMS, including further enhancement in ship reporting and routing system [2]. This suggested that the conventional Traffic Separation Scheme (TSS) alone has become overwhelmed to accommodate the current traffic size while ensuring a regulated traffic flow along the strait, leading to unwelcomed accumulation of emissions.

In order to maintain efficient traffic flow with minimum environmental impact in SOMS, it is essential to ensure an effective reporting of navigational status of transiting ships to the Vessel Traffic Service (VTS) authorities for in situ assessment and make necessary control. Besides upholding the mandatory reporting procedure (STRAITREP) in SOMS, AIS can be more reliable as a track record of ships trajectories with various instantaneous navigational parameters, such as speeds, headings, etc. While the improved ship routing system is expected to provide optimal routing alternatives for ships based on the distance to destination, available voyage time, and the instantaneous traffic conditions, to prevent disruptive traffic system and deteriorate shipping schedule. For such a complex problem, the effectiveness of the VTS in handling SOMS maritime data and feedback with appropriate aids is in concern. On the other hand, projections of maritime emissions in SOMS may help to reveal piecewise information about the ship traffic from bottom-up approach, demonstrating the outcomes of different scenarios.

Despite the ambitious targets to reduce GHGs, the trend of CO_2 emissions from international shipping was reportedly showing bad signs, with 5% rebound to 706 Mt CO_2 in 2022 since the dip in 2020 pandemic time [3]. To investigate whether the seemingly adverse trend will be replicated in SOMS even with decarbonization measures in place, this paper presents the projection of emission using a time series forecasting model based on the available AIS data from SOMS. The preceding computation of shipping emissions are conducted based on different scenarios to assess the effectiveness of different measures implementation against the maritime emission in SOMS. The results of this paper are expected to provide insights about AIS data-driven emission profile and future projections, as well as assessing the mechanism of each measure with regards to the resulted carbon footprint in compliance with the requirements.

2. Related works

The AIS-based emission estimation methodologies are commonly known as bottom-up approach or activity-based approach, which literally indicates the consideration of various activity parameters into the inference of emission factors (EF) rather than direct derivation from fuel consumptions that may be more susceptible to uncertainties in macro-estimation of emissions. Notwithstanding the dependency of fuel oil consumption (FOC) variable on the ship operational condition, considering the real harsh meteorological conditions in the ocean will be more coherent with the parameters recorded by AIS data, thus forming the legacy for weather routing [4]. Thanks to the varieties of parameters recorded by AIS, the data could also be served as indicator for different performance indexes of the overall ship traffic operations, determining the tradeoff between different technologies or policies towards decarbonized shipping. For instance, Dettner and Hilpert [5] attempted to model the resultant CO_2 emissions from

Northern European shipping if E-methanol is used as the main substitute for conventional fuel-powered ships, and has demonstrated its potential to reduce CO_2 emission by up to 50% by 2030 while abiding with the energy demands of the ships.

As such, for more detailed understanding of the environmental impact from maritime transportation, trend projections based on current the emission trend computed from AIS-based methodologies can provide much direct implications to reason the likelihood of the worst conditions in future with all variables being examined. In order to construct a reliable projection model, the predictive element is required to sustain precisions over a time series of period, where machine learning methods are often used to handle big data for processing the prediction part [6]. With the prevailing AIS-based emission estimation, the emission data pivoted with spatiotemporal and other navigational data can be further processed to form the elements for projection to visualize future emission trend [7]. Likewise, with consideration of scenarios like policies implementation and technological maturity, the most optimal alternatives could be reverse engineered from the projections. To incorporate different scenarios, certain level of parameter alteration can be done during the computation of input data, initializing another iteration of modified runs that will show a different projections than the baseline scenario [8].

Apparently, the overall emission trend from freight transportations can be envisioned by various projections including transport demands, energy consumption and finally, associated emissions. However, a national extent of projection may not be intended to project a precise trend of emission [9], but its value is to compare the outlook of different policies enforcement, as policies are often implemented over a macro scale of area. For instance, the common policy scenarios are implementation of Emission Control Area (ECA) and restriction use of conventional fuel oil, together with other conditions, like Business-As-Usual (BAU) and High-Growth (HiG) traffic conditions, to set the range of results to best- and worst-case scenarios [10]. Such qualitative study has to be carefully considered as the resultant differences of scenarios may unintentionally be deviated by external factors, such as the ship navigational behaviors in ECA, increasing traffic in narrow waterway, etc. Alternatively, a systematical investigation with discrete scenario design for certain parameters like fuel ratio, speed mode, ship properties and route selection can be considered for more precise reasoning [11].

Following the necessity of machine learning algorithm to handle big data, many studies had proposed varieties of approaches to assess and forecast the shipping emission based on the carbon footprint quantified from maritima data, majorly AIS data. As a general example, Paternina-Arboleda et al. [12] employed different machine learning algorithms, including Artificial Neural Network (ANN), linear regression, gradient descent, and AutoML-TPOT, as attempts to examine the best models for their predictions of SO_2 emission from ships based on AIS data. Beyond that, there were also other suggested algorithm worth to be used, such as Long Short-Term Memory (LSTM) [13], Recurrent Neural Network (RNN) [14], Gated Recurrent Units (GRU), Convolutional Neural Network (CNN) [15], etc. ANN, being a much robust generic algorithm in handling big complex data, can be utilized directly based on varieties of input data, to study the correlations of parameters on resulting different size of emissions. With that, Bilgili and Buğra Çelebi [16] concluded that the effect of harsh meteorological conditions on emission rates is seemingly greater than route distance.

Beside direct projection on emissions, computation of other variables would also be useful for indirect deduction or provides insights to the underlying factor of excess emissions. A comprehensive ship activity modelling is also able to calculate the ship power consumption and other derivatives as a preceding process for forecasting the attainable carbon footprint [17]. As such, numerical methods were employed to predict ship power in order to discover the substantial demand for FOC under different navigational conditions by means of data-driven machine learning methods [18]. Studies by Ren et al. [19] also reveal the consistency of utilizing AIS data for FOC estimation in way of predicting ship emissions, compared to direct reference on daily fuel oil measurement data. Similarly, Kaklis et al. [4] also demonstrated FOC estimation by a deep-learning model (SplineLSTM) to reflect the associated emissions of a ship. Correspondingly, ship speed as the primary parameter that dictated the FOC of ships constituting major part of ship emissions is to be optimized to maintain efficient operational condition [20].

Nevertheless, the operational factors of the ship navigation are not to be neglected during the midst of exploiting the primary substitute of cleaner fuels, as we expect to resolve the GHG emissions entirely. Given the stagnant condition for existing stage, some focus should be directed to the operational viewpoint of decarbonization like route-based action plan, which also required some efforts to pull together interest of funding to coordination span across jurisdictions for setting up smooth and safe voyage line [21]. For SOMS being the only channel to connect maritime from the East and West, it is important to sustain the maritime activities in the waterway without much interference with one another [22] . In that sense, reducing the energy demand works on another way in lowering the ship emissions, such as route planning [23], weather routing [24], just-in-time arrival [25], Speed optimization [26], etc. Fundamentally, the shipping network need to be established with explicit routes identification [27] so as to facilitate aforementioned strategies.

3. Methodology

The AIS data-driven emission estimation is processed based on a well-established framework of Methodologies for Estimating air pollutant Emissions from Transport (MEET) [28]. This framework facilitates micro-scale of derivation from ship properties and navigational parameters to the resulted emissions for every trajectory of the voyage. The subsequent future projection of ship emissions required a machine learning method to assimilate the underlying patterns of emission temporally. The extracted information through certain logical interpretation of curve fitting and trend modelling by the machine learning model allow more practical predictions of the future emission forming the trend of projections.

3.1. Data preparation

The AIS data are collected within a one-month period in September 2019, corresponded to a total of 4,552 ships activities that were captured in SOMS with range of 1.0°N to 4.5°N in latitude and 102°E to 106°E in longitude. The extent of ship data covers the open sea areas in South China Sea, significant port waters in Malaysia, Singapore and Indonesia, and some inland waterways, suggesting a diversified yet rational inference of the local maritime emissions. The information that is maintained for the process of emission estimation includes MMSI, date and time as timestamp, longitude and latitude of ship position and speed over ground (SOG). Besides, ship properties information such as ship type and gross tonnage are lookup from web-based database. The attributes of ships that are used to segregate data for specific energy demand variables are ship type and operational mode [29].

During data cleaning process, inappropriate status or extreme outliers were removed from the dataset. Data segregation is subsequently carried out based on the ship types and modes of operation based on the methodology demonstrated by Ten et al. [29]. Taking into consideration that the parameters applicable for each specific ship type may varied, every AIS data of specific MMSI were pivoted with normalized ship type, namely dry bulk, liquid bulk, general cargo, container, passenger, Ro-Ro, fishing, and others. Furthermore, the range of ship speed for classification of operational modes by different ship type were also being determined as provision for load factor assignment to compute the energy demand of each ship during their voyage.

3.2. Emission profile

The fundamental formulation of emission estimation for every voyage, $E_{Trip,i,j,m}$ in this study is based on derived equation justified in the report by Trozzi [28]. Owing to limited access to fuel-related information, this paper handles the derivation of emissions based on energy demand viewpoint, which imply the energy required to arrive at each trajectory with the required time and speed with consideration of possible tonnage, nominal power rating and associated load factor at specific operational modes.

In order to compute the trend of ship emissions along its voyage, the nodal emission product on each trajectory were computed to form an emission profile following the voyage captured by the AIS data along SOMS. Consequently, all ships will form their own specific emission profile corresponding to their ship type, geographical locations, and operational mode. For the context of being practical for making subsequent projections, influences of the granularity of data were being concerned as projection may be biased. Time series were constructed in daily and hourly units for comparison of the effect on projections.

Aside from the principal parameters of navigational behavior that determine the energy demand, other attributes are normalized to streamline the computational process of the projections. For instance, P_e and EF that is determined by the engine type and fuel type respectively are prorated into generic parameter for all ship type classifications with probabilistic approach based on the percentage of each engine-fuel type found from investigation of each particular ship type stated in Trozzi [28] reports. Correspondingly, this study only considered the dominating pollutant, carbon dioxide (CO_2) emissions as indicator of the corresponded shipping emissions in SOMS.

3.3. Emission projection

Based on the resulted profile computed from the AIS data, this study used the Prophet Forecasting Model, as attempt to make subsequent forecasts of the emissions for the maritime activities in SOMS. This model basically uses temporal component as regressor and is effective in decomposing trendline by different level of time series [30]. As there are different extents of projections, this study made reference to the guideline by EEA [31] to construct an appropriate projection methodology to adapt the context of this study, as shown in Figure 1. Taking the activity data as the fundamental basis for estimating the shipping emissions, the projection serves as an indicator of the impact brought about by the operational variables of the ship. While other influential variables, such as ship properties, engine-fuel type and retrofit technologies, correlated to emissions were not enumerated as aforementioned.

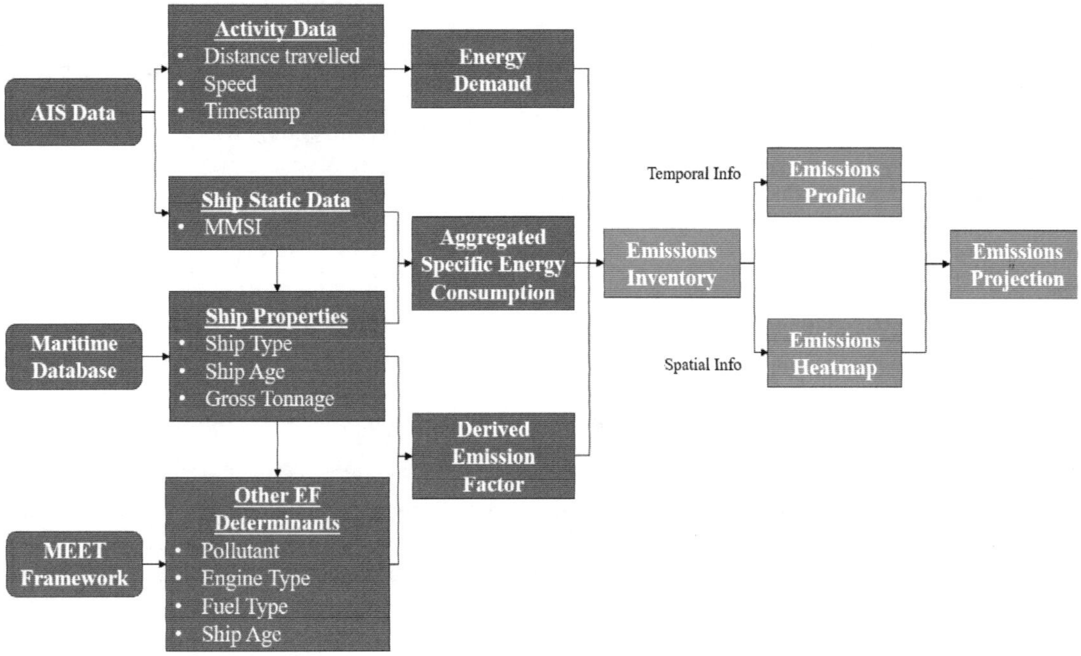

Figure 1. Schematic chart of the feature engineering for emission estimation as provision of emission projection.

The AIS data-driven energy demand as a derivative of ship activity data is the focal point of the basis for detailed assessment of emission projection. These AIS provided information are stratified with respect to the ship type classification, to segregate the distinct behaviours or traffic patterns in the region of SOMS and ease to subsequent assessment of the ship operational mode impact on emissions.

3.4. Scenario formulation

Without considering hypothetical scenario such as measure implementation and technological advances, the projections based on a one-month data is barely feasible for short-term projections. Hence, this study also considered several scenarios of possible measure implementations in SOMS that may enforce compliances of operational limit for ship navigating in this area to reinforce the regional decarbonization efforts, as shown in Table 1.

Table 1. Scenario formulation and application criteria.

Scenario	Parameter Change	Application Criteria	Ref
BAU	-	-	[10]
Speed Limit	Speed < 12 knot	Apply to ship type: Container, Dry Bulk, Liquid Bulk, General Cargo	[32]
HFO banned	CO_2 EF < 524 g/kWh	HFO EF substituted with other cleaner fuel	[33]
JITA	Energy demand - 15%	Apply to ship type: Container, Dry Bulk, Liquid Bulk, General Cargo	[34]

BAU: Business As Usual; HFO: Heavy Fuel Oil; JITA: Just-in-time Arrival

Considering the projection of emissions under existing circumstance, business as usual (BAU) scenario is considered the baseline other scenario analysis as the reference for assessment [10]. This is essentially applied to all scenario analysis where assumptions of specific conditions are modified on top of the BAU case. The scenarios considered are basically a soft demonstration of the possible measures being enforced to certain categories of ships where applicable, as shown in the application criteria. Nevertheless, the measures would be executed more deliberately in real cases, to which the policies would cover this part.

Speed limit in SOMS is actually undefined by the officials, but 12 knot is the only restriction that apply to Very Large Crude Carrier (VLCC) and deep draught vessels when navigating in certain areas of high risk [32]. In the context of this study, cargo carrying ships (container, dry bulk, liquid bulk and general cargo) are considered under this restriction, whereas the remaining ships (passenger, Ro-Ro, fishing and others) are restricted under 16 knots abiding to the speed limit scenario.

HFO is conventionally used as the primary source of fuel for maritime vehicles, to which global authorities were going against due to its high sulfur content and other pollutants emissions as a result of burning it. On the other hand, Marine Diesel Oil (MDO) was the main substitute for HFO during the time when alternative fuels were not yet well developed. Taking into account that before a green fuel can take over as the dominance of maritime power source, this study considered about substituting HFO with MDO as a progressive step towards maritime decarbonization.

JITA is conceptually an ideal operational measure to be achieved by active coordination of optimal shipping schedule and port operations to streamline the entire process of shipping. However, the optimal speed for each ship, subjected to respective schedule and ship properties, would turn into improvement of operational modes unproportionally. Owing to different typical hoteling durations reported [35] due to varying schedule arrangement and port handling methods, this study apply the scenario of JITA based on the change in energy demand. It is reported that by the best case of JITA, the energy demand corresponded by fuel consumption of the ships can be reduced up to 15% if operating at optimal speed [34].

4. Results and discussions

4.1. Emission profile

For every CO_2 emission instance pivoted to the trajectorial date, the emission profile by ship type under different scenarios are illustrated in Figure 2. It can be noticed there is no significance differences of profile pattern despite additional formulations on respective trajectories. Only the extents of CO_2 generation are being reduced by 8.07%, 3.14% and 15% with the implementation of speed limit, HFO banned, and JITA. On the other hand, an obvious upward trend can be noticed over the timeframe of the emission profile, which could be a misleading intuition that could possibly be due to skewing in data distribution and unfavourable aggregation of data. When comparing the CO_2 emission profile by level of time series in daily and hourly as shown in Figure 3, it can be noticed that this trend is smoothen when temporal data is granularized into hours. Hence, by this perception, it is believed that the hourly emission profile would generate more realistic emission projection.

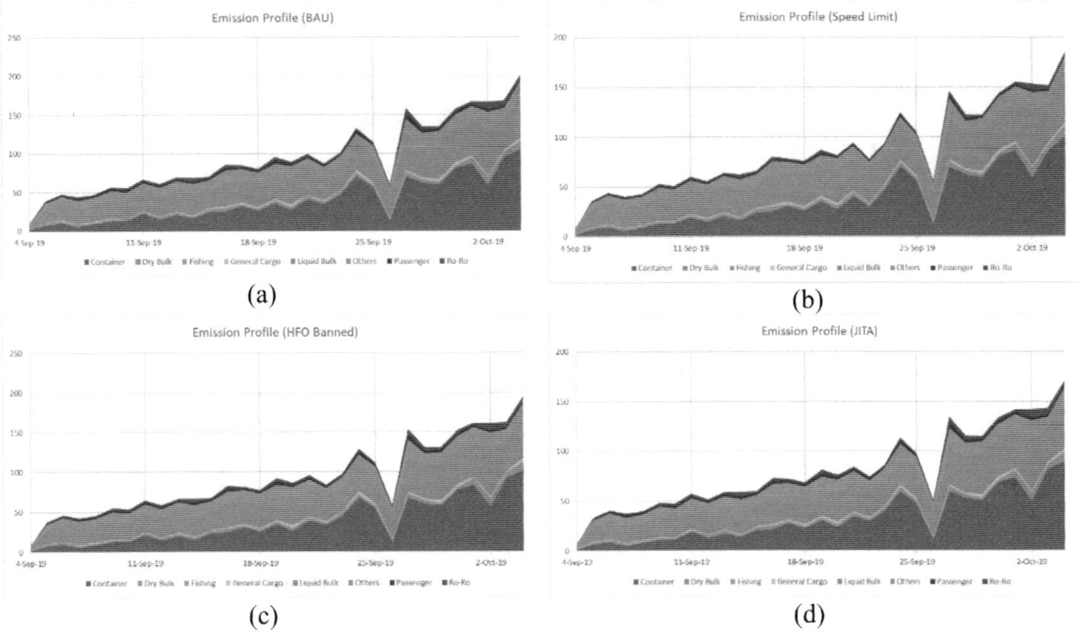

Figure 2. CO_2 emission profile (in tonnes) by ship types under scenario of (a) BAU, (b) Speed Limit, (c) HFO Banned, and (d) JITA.

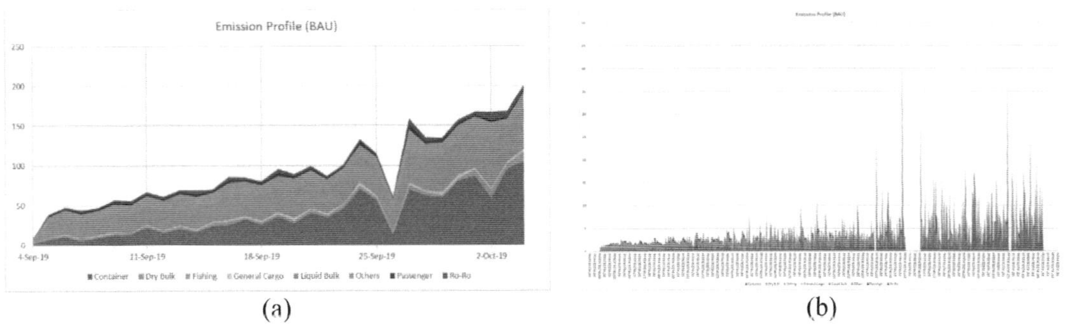

Figure 3. CO_2 emission profile (in tonnes) under scenario of BAU by level of time series in (a) Daily, and (b) Hourly.

IOP Conf. Series: Materials Science and Engineering 1294 (2023) 012038 doi:10.1088/1757-899X/1294/1/012038

4.2. Emission projection

As presented in Figure 4, the daily and hourly emission profile generated projections apparently diverged with one another. Following the trend of each data series, the way Prophet Forecasting model fits the trendline curve to make subsequent projections are seemingly coherent considering this is only a short-term projection. It is also worth to note that there is an uncertainty interval projected along with the projection to indicate the level of confidence for these projections made. The uncertainty interval ranged around 4.64 tonnes emission on average for the hourly emission projection under BAU scenario. While it would be irrational to concede the daily emission projection, considering that the CO_2 emission would be doubled within a period of one-month with a peak of 200 tonnes of CO_2 emission per day.

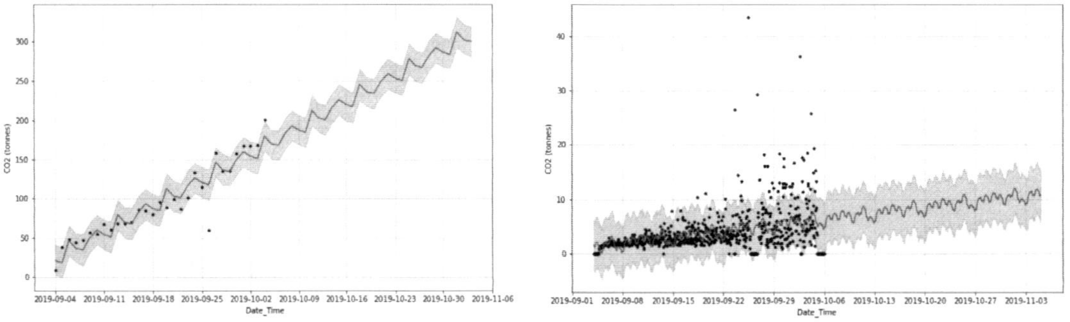

(a) (b)

Figure 4. Overall CO_2 emission profile (in tonnes) under scenario of BAU based on data granularity of (a) Daily and (b) Hourly.

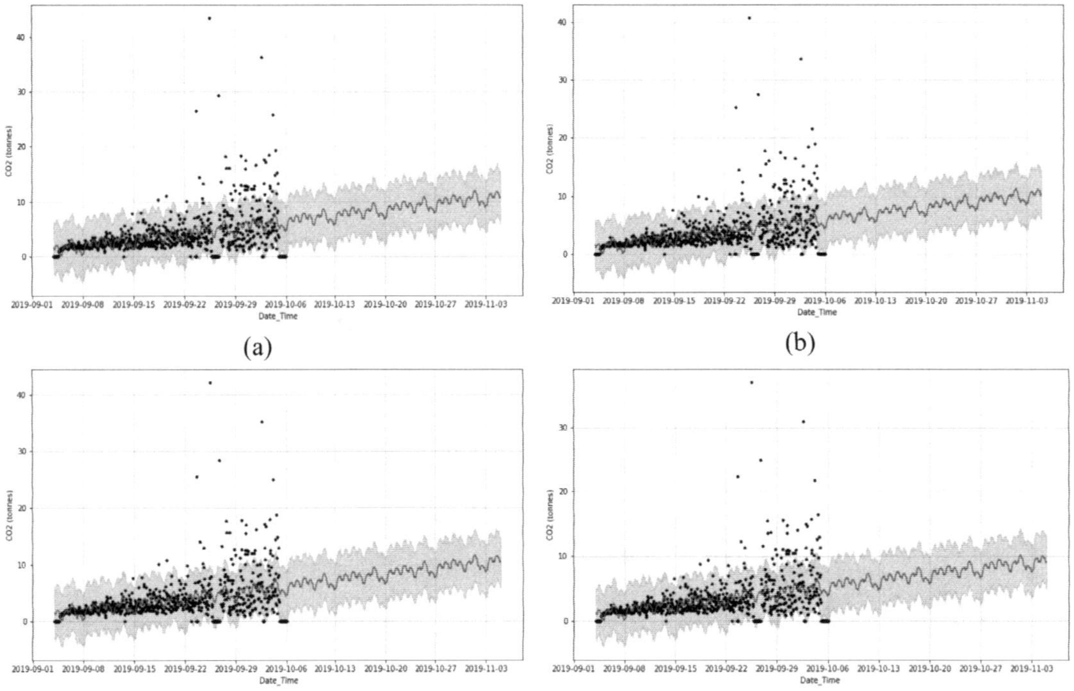

(a) (b)

(c) (d)

Figure 5. Overall CO_2 emission projection (in tonnes) under scenario of (a) BAU, (b) Speed Limit, (c) HFO Banned, and (d) JITA.

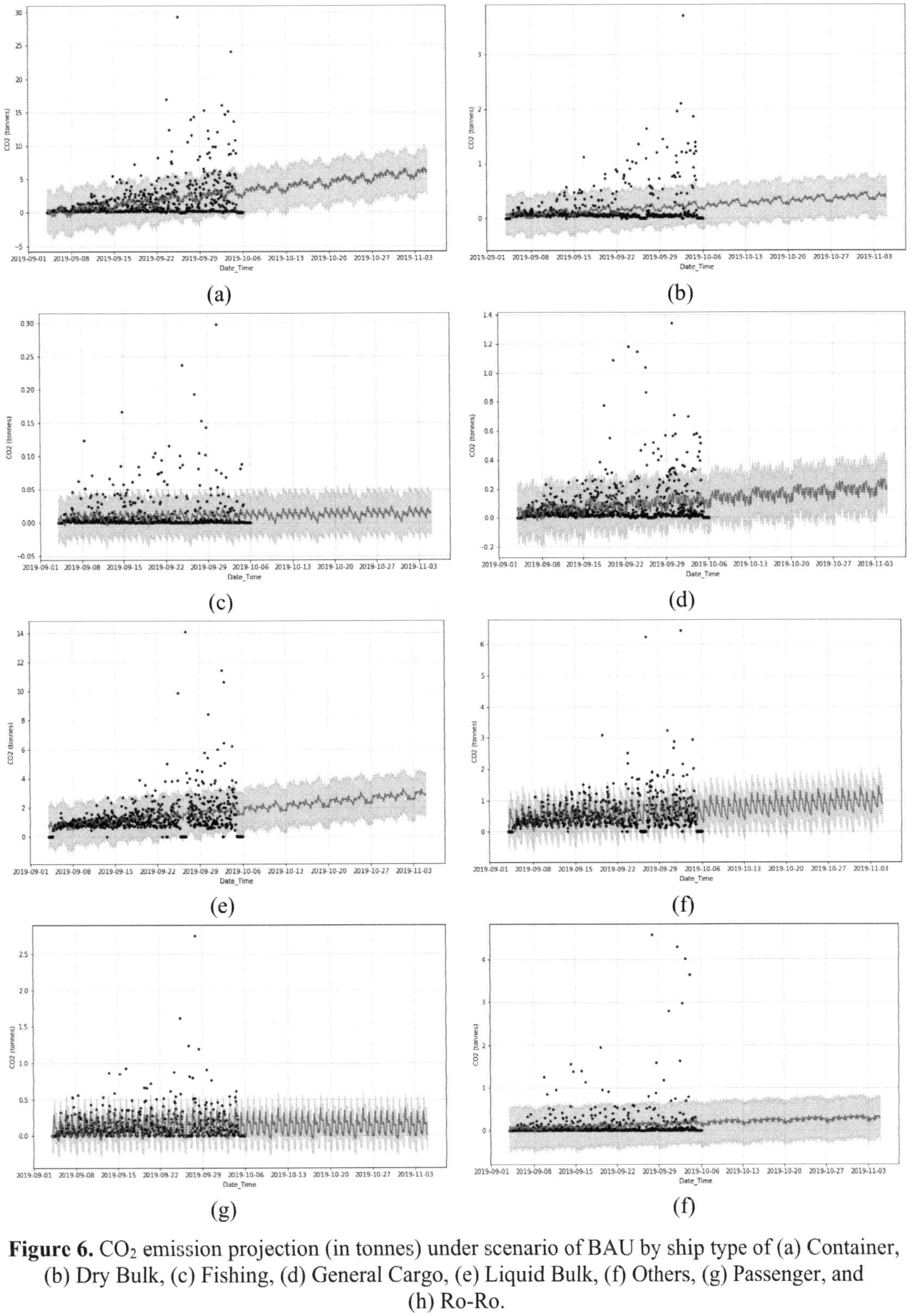

Figure 6. CO_2 emission projection (in tonnes) under scenario of BAU by ship type of (a) Container, (b) Dry Bulk, (c) Fishing, (d) General Cargo, (e) Liquid Bulk, (f) Others, (g) Passenger, and (h) Ro-Ro.

Accordingly, hourly emission profile is computed with the parameters of other scenarios, as illustrated in Figure 5. Surprisingly, the total emission projected by speed limit, HFO banned, and JITA respectively attained reduction of 6.06%, 3.06% and 55.32% respectively compared to the BAU scenario. Both speed limit and HFO banned were showing declining effect on decarbonizing maritime, whereas JITA can further leverage the decarbonization if fully established. Nevertheless, projection in each of the scenario were showing consistent uncertainty interval, indicating a steady trend and consistent pattern of the timeframe of AIS data.

Correspondingly, hourly projection was also considered for further stratified projections, to which overall emission projections in SOMS may not be practical enough either, as the Prophet Forecasting model only considered time series regression. For that, emission profile for each ship type under the scenario of BAU were computed to iterate independent projections through all ship types. It can be noted that the shipping emissions in SOMS were mainly originated from container and liquid bulk, to which the projections showed both ship types may have greater potential of bringing about additional emissions that burden to local ambient atmosphere, considering the overall global shipping capacity will continue to grow in near decades. While other ship types, including dry bulk, fishing, general cargo, passenger and Ro-Ro are less likely to bring about significant impact on the carbon footprint in SOMS, with projected emissions below 1 tonne.

5. Conclusion

The shipping capacity in SOMS are showing an increase and so do the carbon footprints. Considering the maritime decarbonization was showing stagnant progress while waiting for a promising solution, this paper demonstrated projection of emission with the incorporation of progressive measures implementation to reduce local shipping emissions. The overall outcome in estimating CO_2 emissions generated by marine traffic in SOMS from this paper is showing a critical condition, where lack of active measure was being in place despite maritime carbon footprints can be seen in growth. In the scenario of BAU, 92 tonnes on average of CO_2 emission are found. While the scenario analysis implied that up to 15% of reduction of shipping emission could be achieved by operational measures. Despite the good sign shown, a more detailed assessment of measures needs to be done to eliminate any possible trade off for such implementations to ascertain the value for worthwhile effort.

Based on the assessment through different scenario, it is discovered that JITA is the most effective measures to optimally reduce emissions operationally, notwithstanding the challenges to establish an effective traffic management framework to realize JITA. While using the best case of JITA, the projected emission could be leveraged to more than 50%. However, this study only considered the projections by utilizing pattern extraction in time series, but it may need to be technically validated on the deliverables required for such a forecast trend. Owing to limited timeframe of AIS data collection, it is recommended for a thorough projection model being developed based on AIS data of at least one year, in order to form solid basis for long-term projection.

In account to the lacking information about maritime carbon footprint in SOMS, the emission profile developed from the AIS data to make subsequent emission projection provides insights to local industrial players and global maritime authorities on the urgent need of focus on expediting decarbonization measures. Moreover, a deep focus on the value of AIS data for being an enabler for initiating emission profile and emission projection, there are plenty of other machine learning models, such as ANN, LSTM, deep learning models, etc. worth for trying to explore a much comprehensive model for better precision of analytics or artificial intelligence application. With well utilization of AIS data, intelligent systems are expected to be exploitable to foster the sociotechnical environment in maritime industry.

References

[1] RFC. Revolving Fund – 40 years of Excellence. 2021; Available from: https://rfcsoms.org/wp-content/uploads/2022/06/mobile/index.html.

[2] Benardy Ferdianshay, R.S. Indonesia, Singapore, and Malaysia Discuss Shipping Safety. 2023; Available from: https://en.antaranews.com/news/290250/indonesia-singapore-and-malaysia-discuss-shipping-safety.

[3] IEA. CO2 emissions from international shipping in the Net Zero Scenario, 2000-2030. 2023; Available from: https://www.iea.org/data-and-statistics/charts/co2-emissions-from-international-shipping-in-the-net-zero-scenario-2000-2030-3.

[4] Kaklis D, Eirinakis P, Giannakopoulos G, Spyropoulos C, Varelas T and Varlamis I 2022 A big data approach for Fuel Oil Consumption estimation in the maritime industry, in *IEEE Eighth Big Data Service and (BigDataService)*, 39-47.

[5] Dettner F and Hilpert S 2023 Modelling CO_2 emissions and mitigation potential of Northern European shipping, *Transp. Res. Part D Transp. Environ.* **119**: 103745-103759.

[6] Sim S Park J H and Bae H 2022 Deep collaborative learning model for port-air pollutants prediction using automatic identification system, *Transp. Res. Part D Transp. Environ.* **111**: 103431-103450.

[7] Li Y et al. 2023 Research on the carbon emissions traceability inventory and multi-horizon prediction of ship carbon emissions: a case study of Tianjin Port, *Front. Mar. Sci.* **10**, 1174411.

[8] Lamboll R D et al. 2021 Modifying emissions scenario projections to account for the effects of COVID-19: protocol for CovidMIP, *Geosci. Model Dev.* **14**(6), 3683-3695.

[9] Yan S, de Bruin K, Dennehy E and Curtis J 2021 Climate policies for freight transport: Energy and emission projections through 2050, *Transport Policy*. **107**, 11-23.

[10] Geels C et al 2021 Projections of shipping emissions and the related impact on air pollution and human health in the Nordic region, *Atmos. Chem. Phys.* **21**(16), 12495-12519.

[11] Jing D, Dai L, Hu H, Ding W, Wang Y and Zhou X 2021 CO_2 emission projection for Arctic shipping: A system dynamics approach, *Ocean Coastal Manage.* **205**,105531-105544.

[12] Paternina-Arboleda C D, Agudelo-Castañeda D, Voß S and Das S 2023 Towards of in **15**(16), 12171.

[13] Wang K et al 2023 A novel GA-LSTM-based prediction method of ship energy usage based on the characteristics analysis of operational data, *Energy*. 128910-128939.

[14] Wang Y, Watanabe D, Hirata E and Toriumi S 2021 Real- Management of Vessel Carbon Dioxide Emissions Based on Automatic Identification System Database Using Deep Learning, *J. Mar. Sci. Eng.* **9**(8), 871.

[15] Guo T and Xie L 2022 Research on Ship Trajectory Classification Based on a Deep Convolutional Neural Network, *J. Mar. Sci. Eng.* **10**(5), 568.

[16] Bilgili L and Buğra Çelebi U 2020 Estimation of ship flue gas emissions in dynamic operational conditions with ANN, *Proc. Inst. Mech. Eng., Part M: J. Eng. Marit. Environ.* **235**(2), 432-447.

[17] Guo B, Liang Q, Tvete H A, Brinks, H and Vanem E 2022 Combined machine learning and physics-based models for estimating fuel consumption of cargo ships, *Ocean Eng.* **255**, 111435-111450.

[18] Lee J B, Roh M I and Kim K S 2021 Prediction of ship power based on variation in deep feed-forward neural network, *Int. J. Nav. Archit. Ocean Eng.* **13**, 641-649.

[19] Ren F, Wang S, Liu Y, Han Y and Prabowo A R 2022 Container Ship Carbon and Fuel Estimation in Voyages Utilizing Meteorological Data with Data Fusion and Machine Learning Techniques, *Math. Probl. Eng.* 1-21.

[20] Bassam A M, Phillips A B, Turnock S R and Wilson P A 2023 Artificial neural network based prediction of ship speed under operating conditions for operational optimization, *Ocean Eng.* **278**, 114613-114624.

[21] CMD. Dual Actions: National and Route-Based Plans to Tackle GHG Emissions from

International Shipping. 2022; Available from: https://www.gcformd.org/_files/ugd/95aa28_2666a77dc9e84b89ae7735302ad8235c.pdf.
[22] MSC. 50 Years of Cooperation in the SOMS by Malacca Strait Council of Japan. 2019; Available from: http://www.cm-soms.com/uploads/2/75/CF12 20-204.3.2 2050 20years 20of 20Cooperation 20(by20MSC).pdf.
[23] Li H and Yang Z 2023 Incorporation of AIS data-based machine learning into unsupervised route planning for maritime autonomous surface ships, *Transp. Res. Part E Logist. Transp. Rev.* **176**, 103171-103202.
[24] Borén C, Castells-Sanabra M and Grifoll M 2022 Ship emissions reduction using weather ship routing optimisation, *Proc. Inst. Mech. Eng., Part M: J. Eng. Marit. Environ.* **236**(4), 856-867.
[25] IMO 2020 Just In Time Arrival Guide - Barriers and Potential Solutions. London, UK: GloMEEP Project Coordination Unit, International Maritime Organization (IMO)
[26] Andersson P and Ivehammar P 2017 Green approaches at sea – The benefits of adjusting speed instead of anchoring. *Transp. Res. Part D Transp. Environ.* **51,** 240-249.
[27] Liu L et al 2023 Data-driven framework for extracting global maritime shipping networks by machine learning, *Ocean Eng.* **269**, 113494-113512.
[28] Trozzi C 2010 Emission Estimate Methodology for Maritime Navigation, in *Task Force on Emission Inventories and Projections*, Techne Consulting srl: Convention on Long-range Transboundary Air Pollution.
[29] Ten K H et al. 2022 Assessment of Ship Emission Inventory in Strait of Malacca and Singapore based on Automatic Identification System Data. in *IEEE International Conference on Industrial Engineering and Engineering Management (IEEM)*.
[30] Taylor S J and Letham B 2018 Forecasting at Scale, *The American Statistician.* **72**(1), 37-45.
[31] EEA, EMEP/EEA 2019 Air Pollutant Emission Inventory Guidebook Copenhagen K, Denmark, European Environment Agency
[32] MPA. Safety of Navigation in the Singapore Strait. 2006; Available from: https://www.mpa.gov.sg/docs/mpalibraries/circulars-and-notices/port-marine-circulars/-pc06-20.
[33] Gilbert P et al. 2018 Assessment of full life-cycle air emissions of alternative shipping fuels, *J. Cleaner Prod.* **172**, 855-866.
[34] Merkel A, Kalantari J and Mubder 2022 A Port call optimization and CO_2-emissions savings – Estimating feasible potential in tramp shipping, *Marit. Transport Res.* **3**, 100054-100064.
[35] Wärtsilä 2019 Port Call Optimisation paves the way towards a higher port performance. Available from: https://www.wartsila.com/insights/article/port-call-optimisation-paves-the-way-towards-a-higher-port-performance.

IOP Conf. Series: Materials Science and Engineering 1294 (2023) 012039 doi:10.1088/1757-899X/1294/1/012039

Systems dynamic model of wear evolution for sheaves used in oil and gas hoisting operations

S Barua* **and I El-Thalji**

Department of Mechanical and Structural Engineering and Material Science, University of Stavanger, Stavanger, Norway.

* Corresponding author: b.soumya@stud.uis.no

Abstract. The global offshore drilling contracted fleet consisted of 378 jack-ups, 68 semisubmersibles and 73 drillships, according to Westwood Global Energy Group, as of September 2023. In most regions of the world, rig activity has picked up from increased operator demand and pushed the marketed utilization for jack-ups from 76% to 85%, semis from 60% to 80% and drillships from 82% to 90%. The utilization of a drilling rig depends on various factors, mainly drilling demand, rig capability and specifications, and rig efficiency and reliability. Worn sheave grooves can pull down the efficiency of a drive by 8% accelerate the wearing of ropes reduce rope life, increase maintenance costs, and the need for more frequent rope replacements. Wear depth for sheaves is typically measured as the depth of wear in the groove of the sheave, which indicates the extent to which the sheave has worn down over time. Research on surface wear of wire rope caused by the contact between the wire rope and the sheave has rarely been carried out. Upgrading Archard's Wear model is needed to provide a better estimation of wear depth for sheaves. Moreover, the wear coefficient shall be determined. Therefore, the purpose of this paper is to model the wear evolution caused by the sliding contact between the wire rope and the sheave. To achieve this purpose, case sheaves were purposefully selected, studied and modelled using both the analytical and simulation modelling approaches.

1. Introduction

Oil and gas drilling operations involve complex machinery and equipment subjected to extreme conditions that can result in significant wear and tear. Sheaves, as type of pulleys, play a critical role in these operations by supporting and guiding cables or ropes used for lifting heavy loads, such as drill pipes and casings, as illustrated in Figure 1. In general, the travelling block is connected to the drawworks by a cable that ascends to the crown block at the top of the derrick from the drawworks (situated on or close to the main platform). The friction and mechanical stress experienced by sheaves from continuous rotation with loads during drilling operations lead to wear and eventual degradation of their performance. As a result, timely maintenance and replacement of sheaves are essential to ensure safe and efficient drilling operations.

The wire and the rope drive, often known as the sheave, are a wire rope drive's main part. However, the ropes are generally more prone to failure since they have a smaller cross-sectional area and are often subjected to fretting, fatigue, and abrasive wear [1]. Due to this, the wire rope has received the majority of attention in the tribological studies supporting wire rope drives [2,3] while the sheaves have often received far less significance.

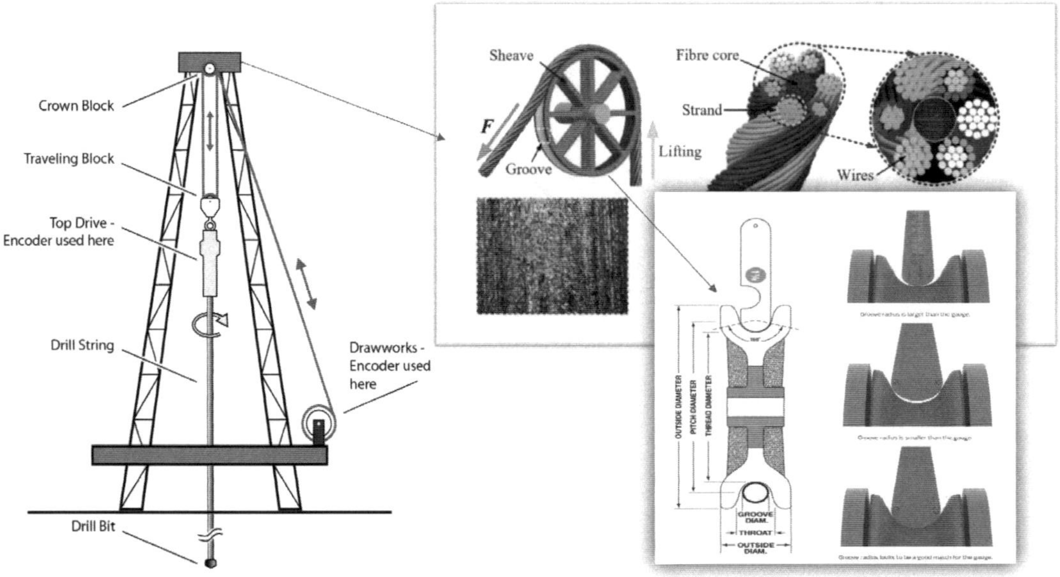

Figure 1. Sheave in drilling rig.

The sheave can also be worn out as abrasive wear act as failure mechanism in the sheave groove. The most common method to monitor the sheave health today is limited to visual inspections and manual groove measurements. The condition monitoring or predictive maintenance techniques are not applied yet. Thus, the purpose of this paper is to build a model that mimics the wear evolution caused by the sliding contact between the wire rope and the sheave. To achieve this purpose, case sheaves from the real-world were purposefully selected, studied and modelled using both the analytical and simulation modelling approaches. The developed model upgraded the Archard's Wear model [4] to provide a better estimation of wear depth for sheaves and estimated the wear coefficient for sheaves.

In the following section, the case studies and simulation models are presented. In section 3, the main results of the analytical and simulation models are presented and discussed, followed by some conclusions about the most sensitive parameters in sheave wear evolution process.

2. Materials and methods

In this section, the general wear theory and contributions toward rope and sheave wear are reviewed. Then, the analytical model to estimate the wear depth in sheaves are presented, together with the selected case study. Finally, the system dynamics model that combine the analytical model with actual operating conditions are presented.

2.1. Wear models for sheaves

In order to detect any signs of wear in the grooves that could lead to rope abrasion, pinching, or bird-caging, regular measurements of sheave groove are taken using two distinctive features – Groove depth and Groove radius. While groove depth of describes the vertical distance between the groove's deepest point and a reference plane (Figure 2), measuring groove radius ensures that sheave groove does not become too wide (overworn) or too narrow (aggregates rope pinching).

There are six distinct kinds of wear that can be characterized: abrasive; adhesive; fatigue; impact by erosion; corrosive; and electrical-arc-induced wear [5]. Corrosion and fretting are two additional typically highlighted wear modes. However, according to Hong et al., the industry's main wear mechanisms are abrasives, fatigue, and adhesives [6]. Metal-to-metal sliding contact during rope-sheave interaction is dominantly abrasive, while at the beginning the wear is adhesive. It is possible to model abrasive and adhesive wear using Archard's wear equation [4], while to model fatigue wear,

(a)

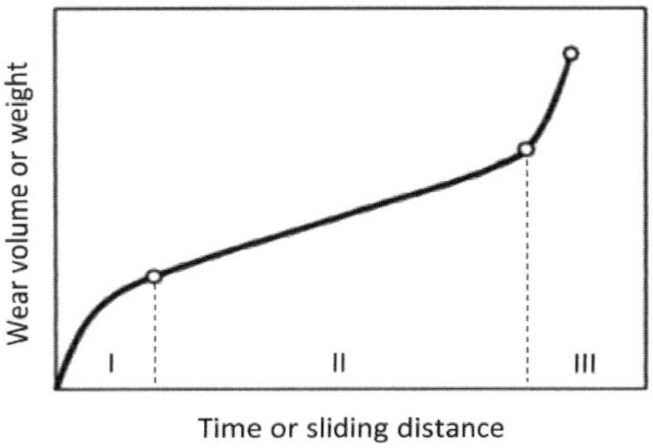

(b)

Figure 2. a) Measurement of groove depth (Source: Case Company) and b) Typical Wear curve with I. Transient II. Steady-state and III. Wear-out stage of wear phenomenon [4]

Coffin-Manson-type relation of fatigue fracture is more suitable [7]. Archard's wear equation calculates the volume of wear per unit sliding distance (q) depending on Normal Load (F), and the hardness of the softer surface (H), then Archard's wear equation is given as –

$$q = \frac{K \times F}{H} \quad (1)$$

where, K is a dimensionless constant. The wear coefficient (k) can be calculated by dividing the K over H (k = K/H).

In terms of wear depth (h), the Archard's equation can be utilised by considering the sliding distance (s) and contact pressure per unit area (p), then the wear depth equation is given as following:

$$h = k \times p \times s \quad (2)$$

Archard's seminal work in 1953 laid the foundation for understanding wear in sheave and rope interactions. He posited that wear is proportional to applied load and inversely proportional to the hardness of the softer surface. This equation has been widely used in engineering practice to estimate wear rates [8]. He also introduced the concept of a running-in stage, during which wear rates tend to decrease to a constant value after initial higher wear [9].

Building upon Archard's model, Varenberg's recent work extended the understanding of wear by considering the running-in stage. Varenberg incorporated surface topology data retrieved from the bearing ratio curve, also known as the Abbott-Firestone curve, into a modified Archard's equation. This modification allows for more accurate wear measurements during the running-in period and steady-state wear [4].

Hutchings and Shipway [5] emphasized that wear in sheave and rope systems is governed by various factors, including mechanical stresses, thermal and chemical properties, and surface interactions. Their research highlights the complex interplay between these factors and how they evolve over time as the contact progresses. According to them, there is more than one wear mechanism at work when metals slide against each other, and as the sliding conditions change, the relative value of each mechanism changes.

Oksanen et al. conducted extensive wear mechanism analyses in sheave grooves. Their work revealed deformation tongues from plastic deformation, crack propagation, and contact pressure-induced wear. They noted that initial contact with wire rope occurs at groove flanges, leading to diagonal cracks and deep pits in the material [10,11].

Hamblin and Stachowiak contributed to understanding environmental effects on wear, finding that dry conditions lead to adhesive wear and corrosion, while wet conditions accelerate corrosion wear and surface pitting [12].

Kato [7] underscored the multiparameter-sensitive nature of wear. Wear is influenced by numerous contact factors, including material properties, stress, contact angle, sliding speed, suspension stiffness, lubrication, and environmental conditions. Kato's research highlighted that metal-to-metal wear is dominantly adhesive at the beginning, and then the wear mode changes to abrasive wear as the microstructure and micro geometry change forms a rubbing texture on the contact surface.

Kulinaowski et al. addressed the significance of the friction coefficient in sheave and rope wear, particularly in realistic industrial conditions. Their research focused on measuring the operational friction coefficient of belt/drum drives under varying speeds and loads, highlighting the impact of friction on wear [13]. Hrabovský et al. developed a methodology to determine the friction coefficient in sheave grooves, noting variations based on load, lubrication, and groove type [14].

Shi et al. used finite element analysis to explore the relationship between wrap angle, pretension, contact force, and wear [15]. Guo et al. introduced the concept of the "global dynamic wrap angle," providing insights into how rope behaviour changes as it approaches and unwinds from the wrap angle [16]. Olszyna et al. examined the impact of wear on the geometrical parameters of sheave grooves, emphasizing the role of pulley positioning and rope design in influencing wear patterns [17].

Nabijou and Hobbs studied the effect of rope core type on friction, finding that it did not significantly affect the effective friction coefficient [18]. Studies by Goto and Amamoto as well as Xiangdong et al. explored wear behavior under varying loads and dynamic conditions. Goto and Amamoto's experiments revealed the transition from mild to severe wear and "quasi-mild wear"[19]. Xiangdong et al. observed the inverse relationship between friction coefficient and sliding distance and the evolution of wear debris [20].

Liu et al. introduced an automated machine vision-based system for wear detection in sheave grooves, providing a non-contact and precise method for assessing wear in industrial settings [21]. Ge and Zhang et al. investigated the frictional behavior of ropes under different conditions, considering parameters such as velocity, pressure, and tension force [22, 23]. Their research contributed to a better understanding of the complex relationship between these factors and friction.

In summary, the literature reviewed here underscores the multidimensional nature of wear in sheave and rope systems, emphasizing the importance of factors like load, friction, surface topology, and environmental conditions. These factors are accounted for in the dimensionless wear coefficient, k, in Archard's wear equation. Hence, the value of k will be different from case to case.

2.2. Analytical Model

In order to formulate the mathematical model that estimates the wear depth per revolution, the contact pressure against sheave groove (Figure 3), sliding distance and wear coefficient are required, which can be formulated as-

$$\frac{\text{Wear depth}}{\text{Circumference of sheave groove tread}}$$

$$= \frac{\text{Wear coefficient}}{\text{Hardness of sheave groove}} \times \text{Peak pressure against groove} \times \text{Relative sliding length per revolution}$$

2.2.1 Properties of wire rope and hoisting system. Circumference of sheave groove, contact pressure and total sliding distance depend on the contact's geometrical and material properties. In Table 1, the rope diameter, grade, cross-section, breaking load, tension, and number of falls are listed. Moreover, the sheave properties are listed such as hoisting capacity, sheave efficiency, sheave bending pitch diameter, and sheave hardness.

We have considered (1) Properties of wire rope against which the sheaves are in a relative sliding motion, (2) the properties of the hoisting system itself, and (3) the load and sliding distance experienced by the sheaves (Load-level data).

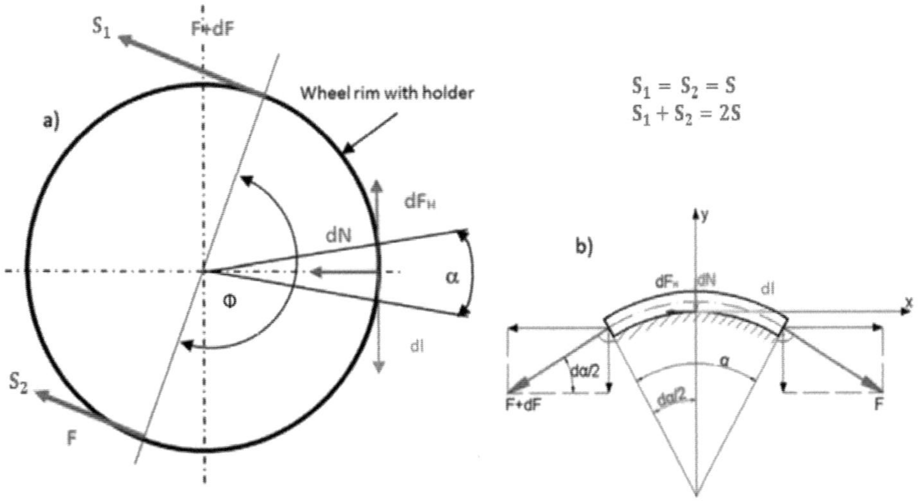

Figure 3. (Left) Schematic diagram of rope mechanics bent over sheave, (Right) Mechanics of rope power transmission on the elementary section of the sheave [24]

Table 1. Rope and Sheaves' properties.

Properties	Symbol	Value with unit
Nominal Rope Diameter	d	50.8 mm
Rope Grade (specific material grade or quality of a wire rope)	R_m	1960 MPa
Minimum Breaking Load (load endured by a rope without breaking)	F_u	1760 kN
Metallic Cross-Section (collective cross-sectional area of the metallic elements/wires)	A	1159 mm^2
Nominal Cross-Section (rope's overall cross-sectional area)	A_n	2027 mm^2
Nominal Rope Tension (pulling force applied to the wire rope)	S	72.1 kN
Rope falls (number of wire rope segments supporting the load)	n_f	16
System hoisting capacity	F_{max}	11122 kN/1250 short tons
Sheave bending pitch diameter (where the rope wraps around the sheave)	D_s	1735 mm
Hardness of sheave groove (toughness of sheave interior to resist wear and deformation)	H	2453 MPa/250 HB
Sheave efficiency (a measure of effectiveness to transmit power to wire rope)	η_s	98%
Kinetic friction coefficient	μ_k	0.1
Peak pressure coefficient	ψ_p	5
Coefficient of Nonlinear Elastic Modulus	C	434618649 MPa
Exponent of the Non-linear Elastic Modulus	β	3
Tangential Elastic Modulus (at 20% load)	$E_t^{20\%}$	102675 MPa

The circumference of sheave groove tread (O) which is the entire area of the sheave groove tread that comes into contact with the rope (in mm^2), can be calculated by,

$$O = \pi \times (D_s - d) \tag{3}$$

2.2.2 Load and contact pressure: To estimate the Peak pressure against Groove (p), which is the highest level of pressure encountered by the wire rope when it comes into contact with the groove of the sheave, the following four properties are needed: Peak pressure coefficient (ψ_p), Nominal rope tension (S),

Sheave bending pitch diameter (D_s), and Nominal Rope Diameter (d). Thus, the Peak pressure against Groove (p) is given as:

$$p = \frac{\psi_p \times 2 \times S}{(D_s - d) \times d} \tag{4}$$

The Peak pressure coefficient (ψ_p) is the ratio of peak pressure to nominal pressure experienced by the system. As specified in [26], we will use the peak pressure coefficient value of 5 in this model. However, total rope tension is the product of the Nominal rope tension (S) acting on each side of the sheave (Figure 2), which can be deduced the following equation:

$$S = \frac{F}{n_f} \tag{5}$$

Here actual hook load F refers to the total of actual load carried and the weight of the attachments (such a trolley or lifting tackle) linked to it and is calculated as -

$$F = \varphi \times \text{System hoisting capacity } (F_{max}) \tag{6}$$

Here, Φ is the load percentage rating of the hoisting system and in our case, it is 10.4%.

2.2.3 Total Sliding Distance: The second variable in the wear depth equation is the relative sliding length per revolution (δ_O), which is the distance the wire rope travels relative to the sheave or pulley during a full rotation and can be calculated as:

$$\delta_O = \frac{\text{Nominal rope tension (S)} \times \text{Sliding distance at groove tread (s)}}{\text{Nominal Cross–Section } (A_n) \times \text{Actual tangential E–modulus } (E_t)} \tag{7}$$

The nominal rope tension (S) is already previously calculated and the nominal cross section area, which is the overall cross-sectional area (mm^2), including metallic and non-metallic components, can be easily calculated by:

$$A_n = \frac{\pi d^2}{4} \tag{8}$$

However, to estimate the sliding distance at groove thread (s), which is the linear distance the rope travels along the groove tread during each cycle (in mm), the sliding arc angle (α) is required, which is an angular distance covered by the rope as it slides along the sheave's groove tread (in radians). the sliding arc angle (α) by itself depends on the Sheave efficiency (η_s) and Kinetic friction coefficient (μ_k). Sheave efficiency (η_s) measures power transmission effectiveness taking into account friction and deformation losses. As per industry norm, we will consider 98% efficiency for the sheaves in our model. Kinetic friction coefficient (μ_k) represents friction between the sheave and wire rope. As specified by Timoshenko, the metal-to-metal friction is kinetic, and deduced that the value of this kinetic friction coefficient is 0.1 [27]. Therefore, the Sliding arc angle (α) can be estimated by:

$$A \text{ (rad)} = \frac{1}{\mu_k} \times \ln\left(\frac{1}{\eta_s}\right) \tag{9}$$

Thus, the sliding distance at groove tread (s):

$$s = \alpha \times \frac{D_s - d}{2} \tag{10}$$

To estimate the relative sliding length per revolution (δ_O) still the actual tangential E-modulus (E_t) is required. The actual tangential E-modulus measures wire rope's modulus of elasticity in the tangential direction, using the following equation:

$$E_t = \beta \times \sigma \times \left(\frac{C}{\sigma}\right)^{\frac{1}{\beta}} \tag{11}$$

Tangential Elastic Modulus (E_t) is a quantitative indicator of a material's stiffness within the plastic deformation range, calculated based on the gradient of the stress-strain curve. Its value can vary depending on the specific measurement location. Exponent of the Non-linear Elastic Modulus (β) is a parameter that shapes the curve representing the relationship between stress and strain in the wire rope. Due to geometrical considerations of wire ropes, value of 3 has been selected as the exponent of non-linear elastic module [25]. In some cases, the modulus of elasticity may not remain constant but may vary with the applied load. The Coefficient of Nonlinear Elastic Modulus (C) accounts for the wire rope's nonlinearity in the relationship between stress and strain.

In general, the actual load carried by the hoisting system is 20% of its' maximum capacity and thus the Coefficient of Nonlinear Elastic Modulus is derived from:

$$C = \sigma_{20\%} \times \left(\frac{E_t^{20\%}}{\beta \cdot \sigma_{20\%}}\right)^{\beta} \tag{12}$$

Rope stress, σ is the ratio of the Internal force (Nominal Rope Tension, S) and the total metallic cross-sectional area (A).

2.2.4 Wear coefficient and wear depth: Assuming a set value of wear coefficient k_w and recalling the values of peak pressure against groove and relative sliding distance per revolution from Equations 4 and 7 respectively, the amount of relative radial wear in the groove for a single revolution can be retrieved using the Archard's wear equation as-

$$h/O = \frac{k_w}{H} \times p \times \delta_O \text{ (mm/rev.)} \tag{13}$$

Ton-km and $\text{Ton}^2 - \text{km}$ are two important parameters that are logged by rig systems for fatigue calculations, and we can apply these parameters in deriving the average load and sliding distance. Ton-km represents the movement of one metric ton of cargo over a distance of one kilometer, while $\text{Ton}^2 -$ km measures the square of the weight of goods (in metric tons) multiplied by the distance travelled.

$$\frac{Ton^2\text{−km}}{\text{Ton−km}} = \text{Average Load in Metric Tons} \tag{14}$$

$$\text{Total Sliding Distance} = \frac{\text{Ton−km}}{\text{Average Load}}$$

Thereby, the accumulated travelling length,

$$Z_{tot} = \frac{\text{Ton−km}}{\text{Average Load}} \tag{15}$$

Recalling the circumference of the sheave tread from Equation (3), we get the total number of sheave revolutions as -

$$n = i \times \frac{Z_{tot}}{O} \tag{16}$$

where, *i* is the Peripheral speed of a sheave, referring to the average velocity difference between the two falls as they rotate over a specific sheave which increases linearly as alternating increments occur between the traveling and the crown block. The peripheral speed of the rope determines the friction and wear of a particular sheave experience.

Table 2. Sheave number vs. peripheral speed chart

Crown block sheaves:	CB1	CB2	CB3	CB4	CB5	CB6	CB7	CB8
Peripheral speed (i):	2	4	6	8	10	12	14	16
Traveling block sheaves	TB1	TB2	TB3	TB4	TB5	TB6	TB7	TB8
Peripheral speed (i):	1	3	5	7	9	11	13	15

Finally, total wear depth in the groove can be determined from the value of wear per revolution by the sheave h/O multiplied by the total number of sheave revolution *n*, as shown in Equation (17) as –

$$h = n \times h/O \tag{17}$$

2.3. Case studies and Parameter Extraction
Based on the analytical model for wear evolution in sheaves, there are several sensitive parameters that shall be extracted in a practical manner, particularly, sliding distance and the wear coefficient.

The drilling operations may take several years to start and may go through breaks due to various reasons such as maintenance intervals, hot/warm/cold stacked conditions, etc. The key challenge in systematically estimating wear of a running system is to accurately derive the total number of revolutions while taking into account all the breaks and pauses. Original Equipment Supplier (OEM) supplied the values ton-km and ton^2-km from fatigue calculations which is vital to apply the devised mathematical model for finding accumulated wear.

The wear coefficient value of a specific case can be retrieved through several ways: (1) Analysis of design parameters and historical measurements, (2) Laboratory tests with similar loading conditions and wear environment and/, or (3) Finite Element Analysis or other numerical analysis tools, and (4) Experiential extraction by collecting the wear depth and the sliding distance. The last way was possible in this case study since there were two sets of data (As-born and one measurement logged after 8 years) available, with no measurements at the early stage of the wear duration. This resulted in the wear curve following a linear trend while the transient or running-in state not being represented in the wear curve.

As the change in the wear depth is measured over specific period and given the ton-km data, the specific wear coefficient could be extracted by taking the average of the individual sheave wear coefficients.

2.4. System dynamics simulation model
Estimating the wear evolution means that the wear depth will be calculated over time. Thus, a dynamic model that estimate wear depth every time unit (e.g, every hour) based on the analytical model is needed.

System dynamics modelling perfectly provide the change of specific variable over time in a time plot. It is almost same as putting the analytical model in a loop that calculate the wear depth value at each time unit over a specific time period. It is important to mention that when wear depth increases over time, the wear rate increases and that exponentially increase the future wear depth. This reinforcing phenomenon can be considered in system dynamics models, as shown in Figure 4.

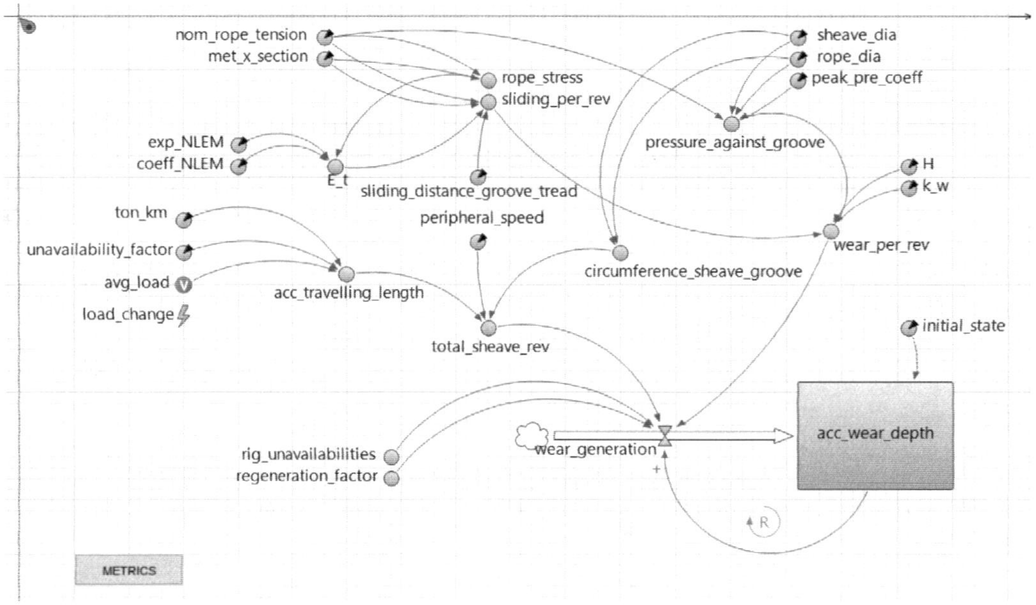

Figure 4. System dynamics simulation model for sheave wear depth.

AnyLogic simulation software was used to incorporate the mathematical model into a system dynamic simulation model of wear phenomena. The parameters and equations used in the analytical model are simply placed in the dynamic model through parameters and dynamic variables respectively, while variable 'avg_load' was used to determine the average load from fatigue calculations and the event 'load_change' fluctuates the average load over a normal distribution. The simulation model was detrimental in terms of simplifying lengthy calculations, producing wear curve and dynamically view wear progression, testing the effect of variable parameters such as wear regeneration factor, load and sliding distance, and finally predicting the amount of wear.

Since numerical and design intensive simulation software are time-consuming and require a wide range of data as well as prior training, it was decided that AnyLogic would be a better option to proceed, since its fits the purpose and a model could be developed without requiring any additional data. At first, the current model was developed by introducing parameters 'rig_unavailabilities' and 'unavailability factor' in order to include the non-operational periods during the timeline. The simulation model made is simpler to derive the wear values in an accurate manner. Additionally, it was easier to change and add different parameters such as load, regeneration factor, etc. However, predicted wear in the future would are based on the condition that the operation will run continuously, without any stoppage or shutdowns.

3. Results and discussion

In this section, three results' categories are presented. First, the individual and average wear coefficients are extracted and presented. Second, the wear depth time plot is presented considering the operational and non-operational periods over 10 years. Third, the effect of the load variation is presented and discussed. The calculated values display the highest amount of wear for the Fastline sheaves whereas deadline sheaves experience no wear. This pattern pertains to the mechanics of wear in multi-sheave rope drives. As power is transferred from the input sheave to the output sheave, the peripheral speed increases to compensate for the drop in diameter and keep the wire rope at a constant linear velocity throughout the drive system, resulting in no rotation in Deadline sheaves while Fastline sheaves rotates at highest speed. The inconsistent distribution of logged wear along the sheaves, as well as occurrence of wear in the Fastline sheaves suggest that the positions of the sheaves were interchanged at some stage of the wear period during periodic maintenance programs.

The wear curve derived from analytical model resembles an almost linear trend (Figure 5). This behavior is expected since the transient state of wear was accounted for due to lack of measurement points. However, since Archard's wear equation is valid for long wear periods, the wear curve determined from the calculations can be viewed as realistic, even if it only displays the steady-state wear progression.

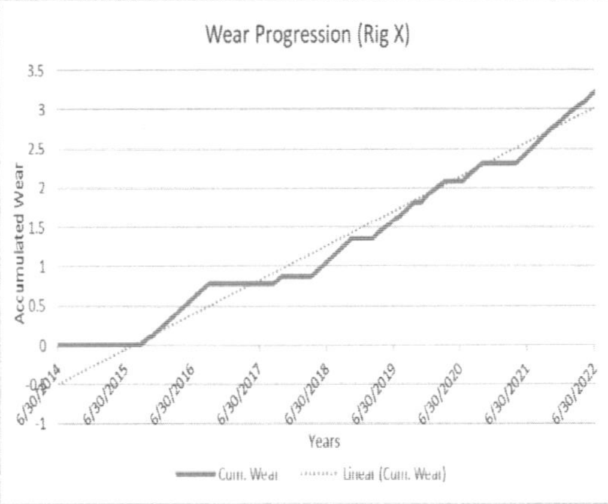

Figure 5. Projected Wear Curve for Sheave 5 in Rig X.

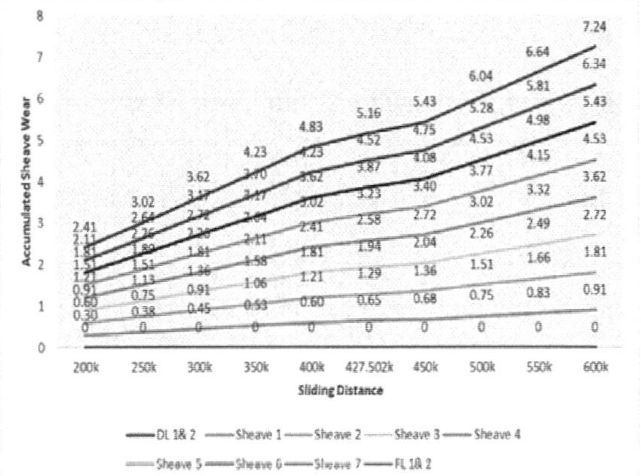

Figure 6. Average Load vs. Sheave Wear

It is visible from the Archard's wear model that wear depth is directly proportional to the load applied and total sliding distance. However, according to Hutchings and Shipway [5], while direct proportionality between sliding distance and wear is true (Figure 6), sheave wear is generally proportional to load applied for up to a point (Figure 7), usually until preliminary surface layers are worn off.

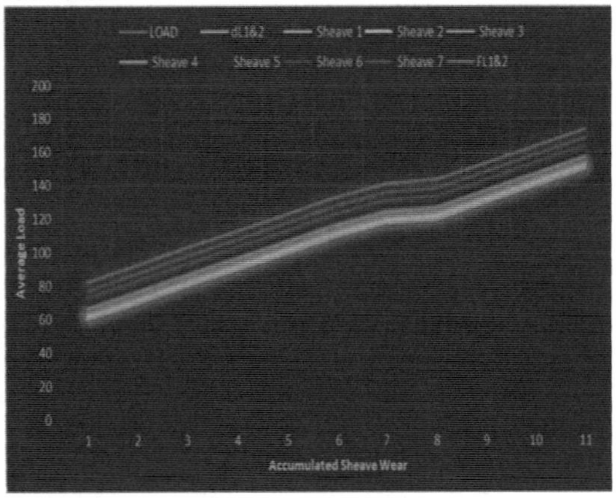

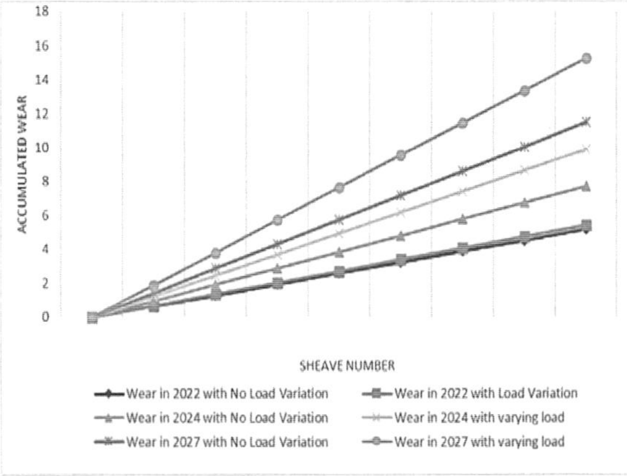

Figure 7. Sliding Distance vs. Sheave Wear

Figure 8. Predicted trend of Sheave Wear

The AnyLogic simulation software was useful to predict the amount of wear to be accumulated in any given number of years in the future (Figure 8). The data from the fatigue calculations provided the value of the average load carried by the crown block (Figure 9). However, using Anylogic it was possible to display the effect of load fluctuating over a given range (Figure 10), which is logical in terms of real-life operational scenario. Lastly, higher amount of wear resulted from fluctuating loads shed more light on the importance to consider real-life loading conditions and AnyLogic's capabilities to simulate them (Figure 11).

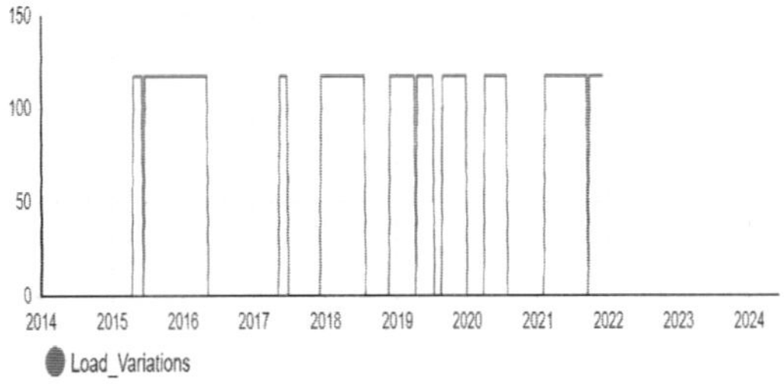

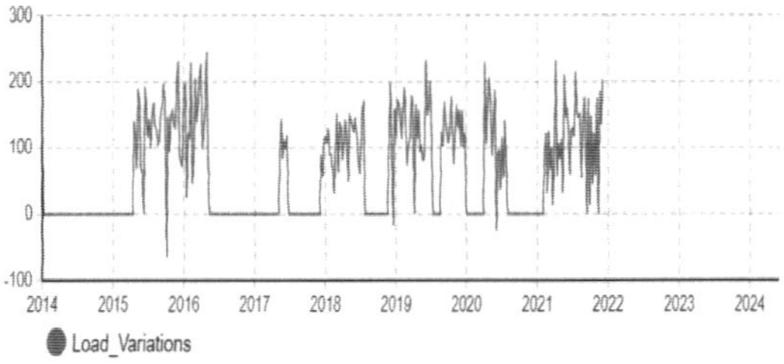

Figure 9. Average load throughout wear period

Figure 10. Fluctuating average load

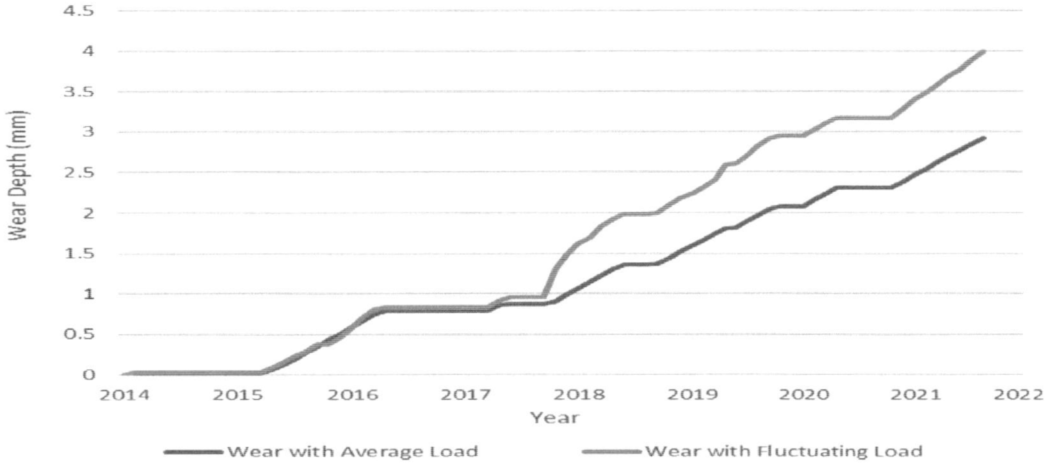

Figure 11. Wear with Average and Fluctuating Load

4. Recommendations for future work

Future studies should focus on improving data collection processes, particularly by incorporating multiple measurements over time on fixed reference points along the sheave's circumference. Techniques like non-contact laser profilometry and atomic force microscopy can be employed to measure wear based on surface roughness [5,28]. Most importantly, to address the limitations of the current methodology, future research should explore numerical modelling using tools like Ansys or Abacus [11]. This approach can provide a more comprehensive understanding of rope-sheave interactions and may lead to a more accurate estimation of sheave wear by considering both groove depth and groove radius changes.

5. Conclusion

The analytical model presented in this paper upgrades the general wear model to be more customised for sheave wear considering the sheave mechanics, geometrics, and properties. Then, the study was able to combine the real-world measurements with the analytical model to estimate the wear coefficient in a practical manner. It can be concluded that the wear coefficient is the most sensitive factor that should be carefully estimated and extracted. The system dynamic model enabled us to get the time plot of the wear depth and also model the third stage of the wear curve, known as the wear-out stage, which depends on the reinforcing phenomenon of the wear rate.

The research brings attention to several critical considerations and limitations in the development of the wear model and the overall study. Firstly, the model assumes uniform abrasive particle geometry and consistent material removal behavior, which may not align with real-world scenarios where irregular particle shapes and variable material removal are common. Additionally, the intricate nature of rope-sheave interactions presents a challenge, leading to the exclusion of groove radius measurements in favour, which has the potential to increase the accuracy of the projected wear. The study also makes an assumption regarding the maintenance strategy for sheave orientation (sheave orientations were changed but were not logged in records) due to limited operator verification. Furthermore, the research does not account for the influence of accumulated wear on wear rates, as quantifying the effects of wear debris remains a challenge. Lastly, the absence of measurements at the start of the wear period means that the model cannot represent the transient wear state, resulting in a linear wear curve trend. These limitations highlight opportunities for future research to delve deeper into these complexities and improve the accuracy of wear predictions in sheave systems.

References

[1] Feyrer K 2007 Wire Ropes: Tension, Endurance, Reliability, Springer

[2] Ridge I M L, Chaplin C R and Zheng J 2001 Effect of degradation and impaired on wire rope bending over sheave fatigue endurance. *Eng. Fail. Anal.* **8**(2), 173–87.

[3] Cruzado A, Hartelt M, Wäsche R, Urchegui M A and Gómez X 2010 Fretting wear of thin steel wires Part 1: Influence of contact pressure. *Wear* **268**(11–12), 1409–16.

[4] Varenberg M 2022 Adjusting for running-in: Extension of the archard wear equation. *Tribol. Lett.* **70**(2). https://doi.org/10.1007/s11249-022-01602-6

[5] Hutchings I and Shipway P 2017 Tribology: Friction and Wear of Engineering Materials. Butterworth-Heinemann, 2nd Ed., Elsevier.

[6] Hong W, Cai W, Wang S and Tomovic M 2018 Mechanical wear debris feature, detection, and diagnosis: A review. Chinese J. Aeronaut. **31**(5), 867–82.

[7] Kato K. *Classification of Wear Mechanisms/Models*. John Wiley & Sons, Ltd eBooks [internet]. 2014. p. 9-20. Available at: https://onlinelibrary.wiley.com/doi/abs/10.1002/9780470017029.ch2

[8] Archard J F and Hirst W 1956 The wear of metals under unlubricated conditions Proceedings of the Royal Society of London 2 236 (1206)397–410

[9] Archard J F 1953 Contact and Rubbing of Flat Surfaces, *J. Appl. Phys.* **124**(8)

[10] Oksanen V, Andersson P, Valtonen K, Holmberg K and Kuokkala V 2013 Characterization of the wear of nodular cast iron rollers in contact with wire ropes, *Wear,* **308**(1–2) 199-205

[11] Oksanen V Valtonen K Andersson P Vaajoki A Laukkanen A and Holmberg K 2015 Comparison of laboratory rolling–sliding wear tests with in-service wear of nodular cast iron rollers against wire ropes. *Wear,* **340–341** 73–81

[12] Hamblin MG and Stachowiak G 1995 Environmental and sheave material effects on the wear of roping wire and sheave. *Tribol. Int.* **28**(5), 307–15.

[13] Kulinowski P Kasza P and Zarzycki J 2021 *Identification of the operating parameters of the friction drum drive in industrial conditions* Eksploatacja I Niezawodność 31 23(1) 94–102

[14] Hrabovský L Fries J Kudrna L and Gaszek J 2022 Determination of the coefficient of friction in a pulley groove by the indirect method. *Coatings* **12**(5) 606.

[15] Shi X Pan Y Ma X 2017 Modeling and analysis of the rope–sheave interaction at traction interface, *J. Appl. Mech.* **84**(3)

[16] Guo Y Zhang D Feng C and Liu Y 2016 Dynamic creeping behaviors between hoisting rope and friction lining in friction transmission*, J. Vibroeng.* **18**(8)5010–28

[17] Olszyna G Sioma A Tytko A 2014 Laser Measurement System for the Diagnostics of Mine Hoist Components/Laserowy System Pomiarowy Do Diagnostyki Elementów Górniczych Wyciągów Szybowych, Arch. Mining Sci. **59**(2) 337–46

[18] Nabijou S Hobbs R E 1995 Frictional performance of wire and fibre ropes bent over sheaves, *J. Strain Anal. Eng. Des.* **30**(1) 45–57

[19] Goto H and Amamoto Y 2003 Effect of varying load on wear resistance of carbon steel under unlubricated conditions, *Wear* **254**(12) 1256–66

[20] Xiangdong C Peng Y Zhu Z Lu H Tang W and Zhang X 2022 Sliding Friction and Wear Characteristics of Wire Rope Contact with Sheave under Long-Distance Transmission Conditions *Mater.* **15**(20) 7092

[21] Liu S Ma D and Liu G 2022 Wear Detection System for Elevator Traction Sheave, J. Shanghai *Polar* Jiaotong University (Sci.) **27**(5) 706–14

[22] Ge S 1992 The friction coefficients between the steel rope and polymer lining in frictional hoisting. *Wear* **152**(1) 21–9

[23] Zhang Q Peng Y Zhu Z Tang W Chen G Zhao X 2021 Studies on friction and wear characteristics of wire rope used in multi-layer winding hoist during inter-circle transition under dry friction. *Proc. Inst. Mech. Eng.* **236**(1) 90–104

[24] Krešák K Peterka P Ambriško L and Mantič M 2021 Friction lining coefficient of the drive friction pulley, *Eksploatacja I Niezawodność* **23**(2) 338–45

[25] Jordan D W Smith P 1999 Nonlinear ordinary differential equations: An Introduction to Dynamical Systems. Oxford University Press USA

[26] Bhushan B 2001 Modern Tribology Handbook: Materials, coatings, and industrial applications
[27] Timoshenko 2007 Engg Mechanics Revsd 4E Sie Tata McGraw-Hill Education
[28] Blau P J 1996 Needs and challenges in precision wear measurement Oak Ridge National Lab. (ORNL), Oak Ridge, TN USA Jan

Service and inventory model for maintenance workshop in the short cycle operation region: Agent-based simulation approach

M Shalfawi *, **I El-Thalji** and **A Turkyilmaz**

Faculty of Science and Technology, University of Stavanger, Norway.

* Correspondence: mf.shalfawi@stud.uis.no

Abstract. Tools used in the North Sea region has a high cycle and come very often to maintenance workshop to be checked, repaired if needed and prepared for the next operations. The availability of spare parts at the workshop plays a significant role to keep the flow time as short as possible and meet such high-cycle operations. Supplying spare parts from the best-cost countries to the North Sea region with about one-year lead time makes the situation more critical and the economic order quantity and reorder point need to be found. However, having inventory at the workshop and ordering a batch of spare parts increases operational expenditures. Moreover, frequent supplies increase the environmental impacts of shipping CO2 and spare scrap rates, whereas repairing the used spare parts and reusing them again can offer a more sustainable solution. Therefore, the purpose of this paper is to develop a simulation model that can quantify the cost and benefits of reusing repaired spares compared to supplying newly built spares from the best-cost countries. To achieve this purpose, a case study has been implemented on a specific maintenance workshop within the North Sea region and the entire tool repair and spare part supply operations are conceptualized and modelled with the help of the simulation modelling approach. Two scenarios have been simulated: (1) the maintenance workshop fully depends on supplying newly built spares from the best-cost countries with no inventory stock or spare reuse, and (2) the Maintenance workshop is primarily dependent on repaired spares with an optimal level of spares stock. The simulation results, from the studied case, support the second scenario where a repair path cycle is introduced to the maintenance workshop, as a 78% reduction in lead times, a 116% improvement in worker utilization, a 73% reduction in crowding levels, a 52% reduction in scrap rate, and a potential profit increase of roughly three million NOK (20%). Therefore, it can be concluded that a local repair service is required to keep maintenance workshops in high-cycle regions at high-performance levels.

1. Introduction

Spare part inventory management is full of critical trade-off decisions where you need to ensure the availability of parts and effective cycle service level whilst limiting the investment in inventories [1]. Kosanoglu found that the unavailability of spare parts accounts for 80% of all system downtime [2]. On the other hand, spare parts inventories can tie up large amounts of cash and incur significant costs due to write-offs, duplication and insurance. Schultz claims that a firm's ability to meet short lead times, without excessive inventories, requires short manufacturing cycle times [3]. Reducing a firm's average cycle time without costly increases in capacity or expedited engineering processing times can often be accomplished through an inventory of spare components. Maintaining an inventory of spare components decreases the mean time to repair resulting in increased machine availability and reduced cycle times. Determining the appropriate investment in spare component inventories requires a tradeoff between increasing inventory carrying costs and improvements in service performance. Inventory-service models are capable of predicting and analyzing average cycle times, departure variability, and average inventory levels in order to enable managers to determine the best allocation of spare inventories for machines at a given workstation.

Several studies have shown that maintenance workshop services have better cycle time and net profit using a stocking strategy over the just-in-time strategy due to their unexpected demand variability or lead time variations [4, 5]. Sha Zhu highlights that an uneven distribution of maintenance tasks over time is an important cause of intermittency in spare parts demand, and this intermittency complicates spare parts inventory control severely [6]. Therefore, the stocking strategy by itself has several policies depending on whether it is time-based, historical demand based or real-time demand based: (1) Fixed Order Interval (FOI) where orders occur at fixed time intervals, regardless of the current inventory level (2) Lot-for-Lot (L4L) where replenishment order is for the exact quantity required to satisfy the current demand, (3) Reorder Point (ROP) or Reorder and Quantity (R,Q) that sets a predetermined inventory level at which a replenishment order is triggered, (4) The Min-Max policy (s,S) that When the inventory level reaches the minimum threshold, a replenishment order is placed to bring the stock back up to the maximum level, (5) Demand-driven replenishment policies which utilize real-time demand data to trigger replenishment orders [7]. The selection of a specific reordering policy depends on factors such as the demand pattern, lead time, cost considerations, desired service levels, and the organization's inventory management goals. It is common for organizations to implement a combination of reordering policies tailored to different spare parts or equipment based on their specific characteristics and requirements. Maintenance workshops tend to utilize the ROP or Min-Max policies as demand variability or long lead time to get the spare parts, which might be not optimal when it comes to investment in inventories. Therefore, there are three main approaches to reducing the lead time by utilizing the repaired spare parts instead of totally ordering newly built spares: (1) Repair and Stock (RS) where spare parts are repaired or refurbished as needed and then stocked in inventory, (2) Stock and Queue (SQ) spare parts are stocked in inventory without repair or refurbishment and waiting to be repaired by an external supplier when it is needed, (3) Stock and Serve (SS) spare parts are stocked in inventory without repair or refurbishment, and waiting to be repaired internally when it is needed. Utilizing the repaired spare parts has various economic and environmental impacts, such as reduced CO2 emissions due to the logistical operations involved in procuring new build parts through the reuse of what could have been scrapped.

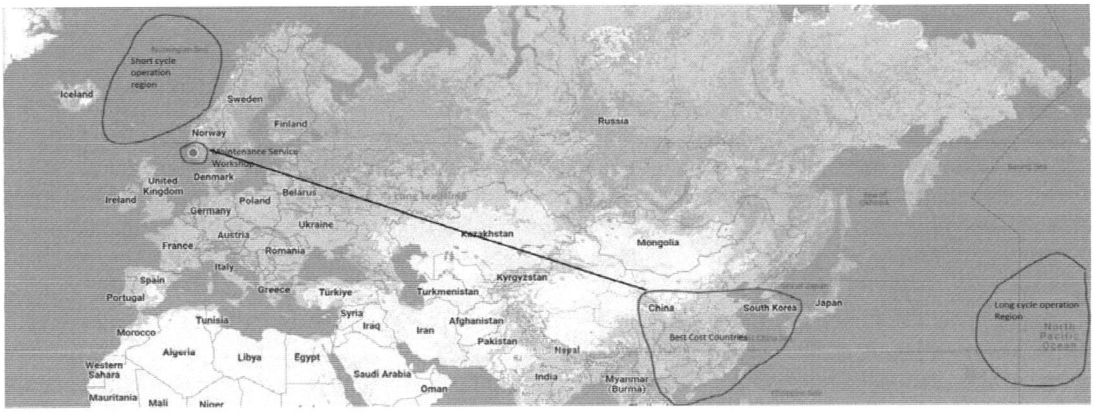

Figure 1. Maintenance logistics for Short Cycle Operation Region.

The spare parts inventory is further challenging for maintenance services in short-cycle operation regions, e.g. North Sea, compared to the long-cycle operation regions, e.g. Pacific Ocean, as the maintenance service workshops have limited time to repair the tools and send them for the next operations, especially when their best cost countries (BCC) are quite far, as illustrated in Figure 1.

The stocking strategy with the repairing spare parts approach can potentially solve such a challenge. Thus, the purpose of the paper is to explore the potential cost and benefits of applying the Reorder Point policy with the Stock and Serve (SS) approach for maintenance service workshops in short-cycle operation regions. It means we are going to study the effect of spare parts inventory policies and levels on the average maintenance workshop cycle time, departure variability, and associated cost and cost-saving parameters.

To achieve this purpose, a case study has been implemented on a specific maintenance service workshop within the North Sea region and the entire tool repair and spare part supply operations are conceptualized and modelled with the help of the simulation modelling approach. Two scenarios have been simulated: (1) the maintenance workshop fully depends on supplying newly built spares from the best-cost countries with no inventory stock or spare reuse, (2) the Maintenance workshop mainly depends on repaired spares with an optimal level of spares stock. The tools and critical parts are classified based on criticality, where the most critical tool and its five most critical parts were considered in the study. In this paper, we first present the literature, methods and data sources that provide the theoretical background and modelling inputs of this study. Then in section 3, the results from the simulation model are presented and discussed. We end this paper with conclusions regarding how cost-beneficial the Reorder Point policy with the Stock and Serve (SS) approach is for maintenance service workshops in short-cycle operation regions.

2. Materials and Methods

2.1. Replenishing spare parts strategies

Replenishing spare parts strategies can be categorized into three main approaches: Repair and Stock (RS), Stock and Queue (SQ), and Stock and Serve (SS). In the RS strategy, spare parts are repaired or refurbished as needed and then stocked in inventory to fulfil future demands. When a spare part is depleted from inventory, the repair process is initiated to replenish the stock. This strategy aims to strike a balance between repair costs and inventory levels. In the SQ strategy, spare parts are stocked in inventory without repair or refurbishment. When a spare part is depleted, a replenishment order is placed with the supplier. The lead time for the replenishment order creates a queue or waiting time before the spare part is available for use. This strategy prioritizes fast response times by relying on supplier availability. In the SS strategy, spare parts are stocked in inventory without repair or refurbishment, similar to the SQ strategy. However, instead of relying on external suppliers, the spare parts are serviced internally. This means that the organization has the capability to quickly address and replace depleted spare parts from its own inventory. The SS strategy offers greater control over lead times and response times. The selection of the most appropriate replenishment strategy depends on various factors, including the criticality of the spare parts, their lead time, the cost and availability of repair services, the organization's repair capabilities, and the desired balance between inventory costs and response times. However, the Lot-for-Lot policy may not be suitable for all scenarios. If there are significant costs associated with each order or if lead times are long, frequent ordering may not be practical or cost-effective. In such cases, other reordering policies like the Economic Order Quantity (EOQ) or Min-Max may be more appropriate.

In certain areas of the North Sea, particularly in more mature fields where the infrastructure is well-established, drilling operations can be considered a relatively short cycle. These operations may involve drilling exploration wells, appraisal wells, or development wells in fields with known reserves. However, in other areas of the North Sea, such as deepwater or frontier exploration regions, drilling operations can be more complex and time-consuming, falling into the long-cycle category. These operations often require specialized equipment, extensive planning, and longer drilling durations due to challenging environmental conditions, deeper water depths, or the exploration of unconventional resources.

2.2. The Case Study and Simulation Model Development

In this section, the case study will be presented together with the entire simulation model building. The case study is about a maintenance workshop that serves several tools operated in the North Sea. Each tool may be used for several operations and holds over 50 different components. Therefore, the inventory of spare parts is crucial for this workshop, especially due to the high operational cycle of these tools (they should be fixed and go back to operation). On average, the workshop receives five to ten tools every month. The workshop has a designed process, illustrated in Figure 2, where it receives the tool, checks if it is used or not, disassembles the tool and inspects the component. If the tool just needs standard replacement components, the replacement work will be performed and followed by cleaning and preservation to be ready for upcoming operations. If the tool needs to be repaired and the workshop is not able to perform that, it will be replaced with a new component from the stock. It is clear, the current maintenance service process is highly dependent on the spare parts stock, its safety stock, reorder point, batch size and lead times to BCC. Therefore, the enhanced maintenance service process is proposed, illustrated in Figure 3, to enable the dispatch of tools within seven working days from their receipt. This is done by introducing the local repair path. So, the failed components within the serviced tools will be replaced by spares in the stock, however, the stock will be filled by both repaired spares from previous jobs, and the newly ordered spares from BCC. In this case, the stock will be more dependable, and spare purchases and spare scraps will be reduced. It is worth mentioning that the failed components which will be repaired, are batched three units together and then sent to the local repair service for increased cost savings.

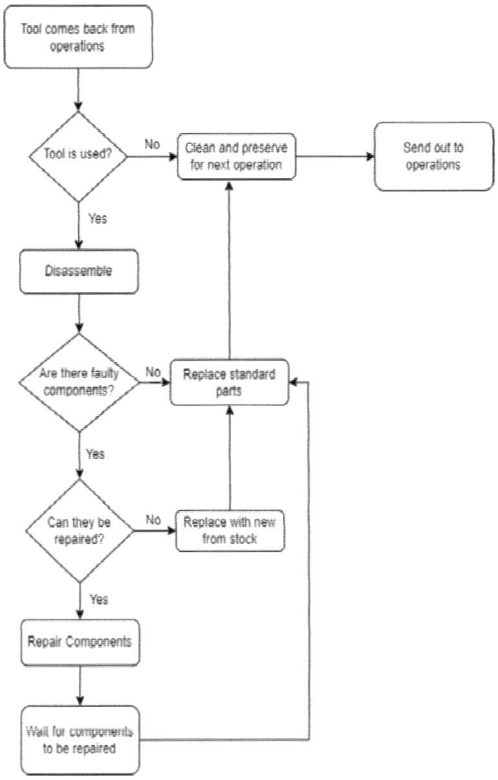

Figure 2. Process flowchart for the current scenario [8]

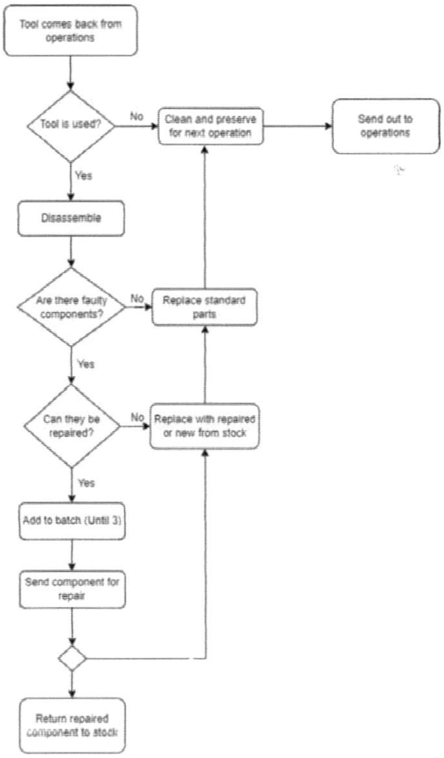

Figure 3. Process flowchart for the enhanced scenario with repair path [8]

IOP Conf. Series: Materials Science and Engineering 1294 (2023) 012040 doi:10.1088/1757-899X/1294/1/012040

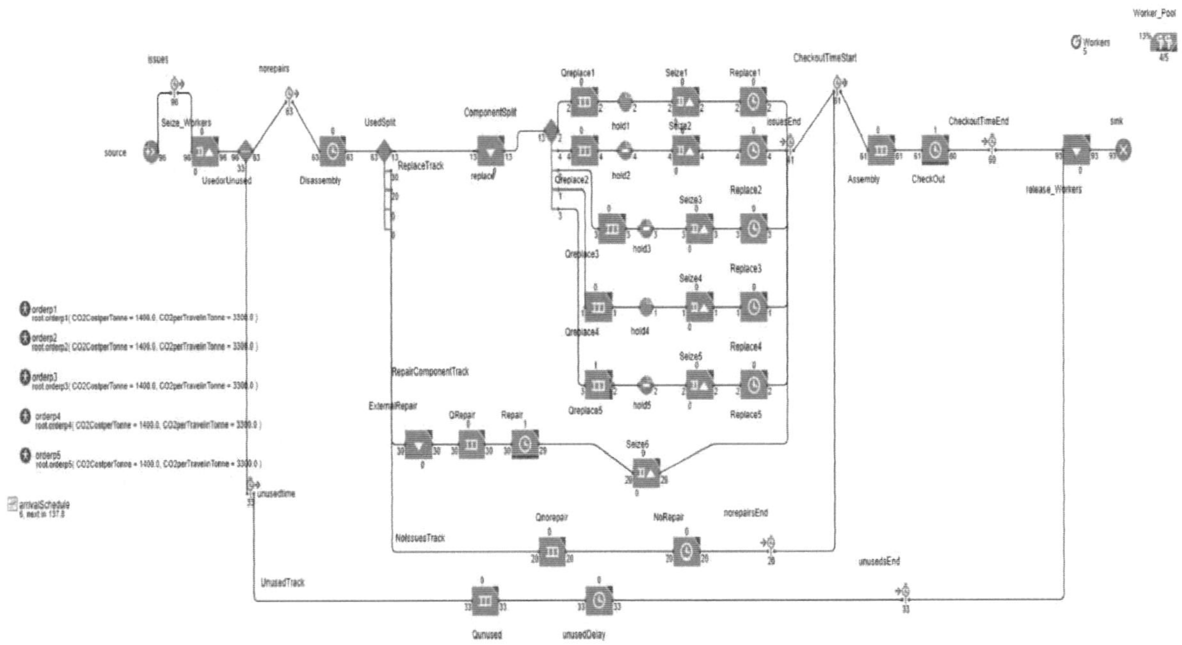

Figure 4. Simulation model for the current scenario [8]

IOP Conf. Series: Materials Science and Engineering 1294 (2023) 012040 doi:10.1088/1757-899X/1294/1/012040

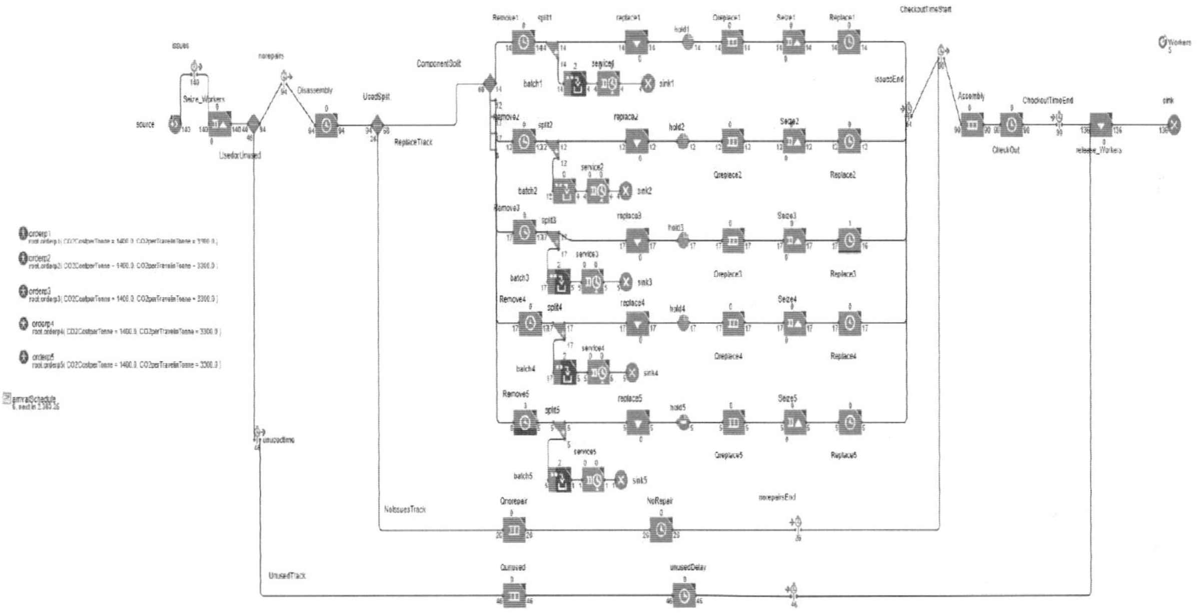

Figure 5. Simulation model for the enhanced scenario [8]

In summary, there are four paths in this process, as seen in Figure 4: unused tools that require only cleaning and preserving, tools with no issue as nothing was detected after the disassembly task, tools that need external repair and tools that require replacement work and will be locally performed. However, the replacement path is further split into five sub-paths based on the tools' criticalities (frequency and consequences). The simulation model for the enhanced process, shown in Figure 5, is almost similar to the current scenario model, except that it splits each replacement path (five sub-paths) into two tracks, one for the tool by itself and the second for the repair path for the failed component that will be batched in threes, repaired and sent to stock. The simulation model in both scenarios considered the worker capacity of the workshop (5 workers) and spare parts inventory capacity. Therefore, the "seize and release" functions are presented in the simulation model. Besides, the "hold" function is used to hold the replacement work once the spare parts stock is empty.

The simulation model has two more agents, one is the spare part inventory or stock and the second is the spare parts logistics from BCC. In Figure 6, the spare part inventory is modelled using system dynamics and state machine libraries in the Anylogic software.

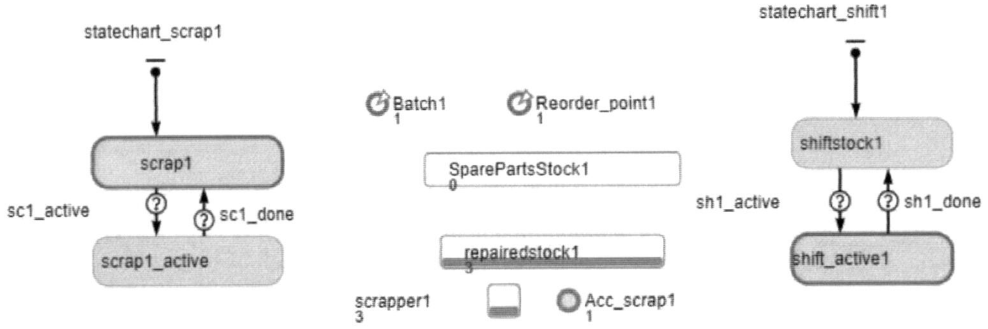

Figure 6. Spare parts inventory modelled as an agent. [8]

The spare parts logistics is modelled as a discrete event process, where the order enters the process and after waiting in the order queue the supply process starts and takes on average around 365 days. The order is triggered by the state chart (shown in Figure 7, left side) when the condition of the spare parts stock goes below a certain level. Once the spare parts arrive at the workshop, the ordering state will receive a message to indicate that the spare parts are available. The simulation time unit is hours, and the simulation time interval is 26000 Hrs.

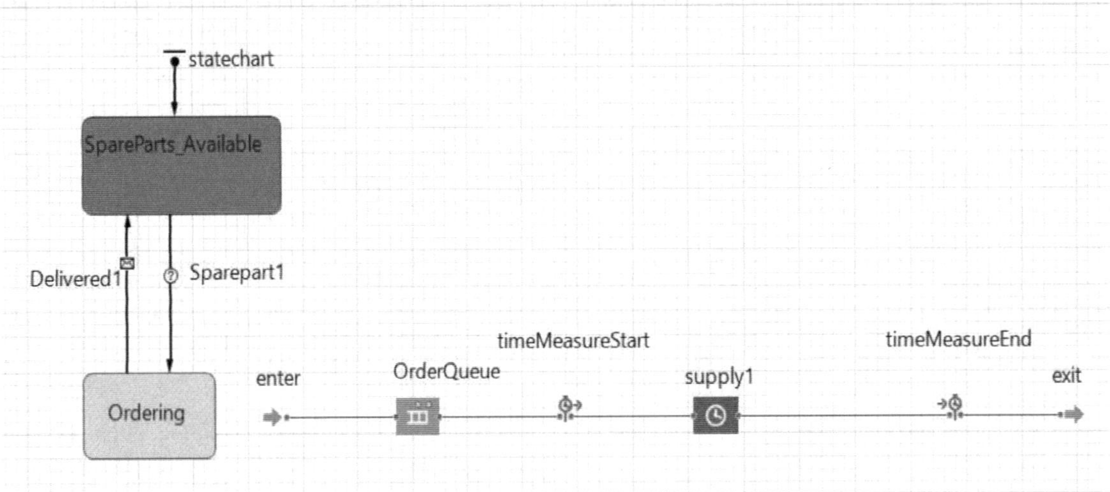

Figure 7. Spare parts logistics modelled as an agent [8]

3. Results and Discussion

The simulation model provides an estimation of several key performance indicators such as the number of served tools (which is a direct indicator for the revenue), average lead time (please remember the workshop seven days target), average worker utilization, workshop crowding and spare parts scrap rate. In Table 1, the key performance indicators, obtained by the simulation model, are summarized. The total served tools increased in the enhanced scenario as the workshop had short lead and flow times and was able to receive and repair more tools. The service lead time (from receiving the tool until it is ready for the next operation) was reduced to 193 hours, which is almost near to the target of seven days. There are around 5% of the tools still take more than 500 hours and further enhancement is required. The workshop crowding has also enhanced and instead of having on average 5 or 6 tools at once, it has on average 1 to 2 tools. The scrap rate has reduced and more components are repaired.

Table 1. Key performance indicator comparison between current and enhanced scenarios.

	Current scenario	Enhanced scenario with repair path	Difference
No. of tools served	207	243	17% increase
Average lead time in Hrs	888.65	193.09	78% reduction
Average worker utilization	0.73	1.58	116% increase
Tools with a lead time of less than 500 hrs	20%	90%	70% improvement
Average workshop crowding (number of tools in the workshop)	5.33	1.59	73% reduction
Scrap rate	21	10	48% reduction

The lead time histograms in Figures 8 and 9 illustrate the mean time for the workshop to prepare the tools ready for the next operation. It is clear that depending on the spare parts availability, some tools must wait more than 500 hours and some need in total 5500 hours. The enhanced scenario has shortened the lead time in general, however, there is a room for further enhancement to handle some cases where the lead time is still 3000 and 5000 hours.

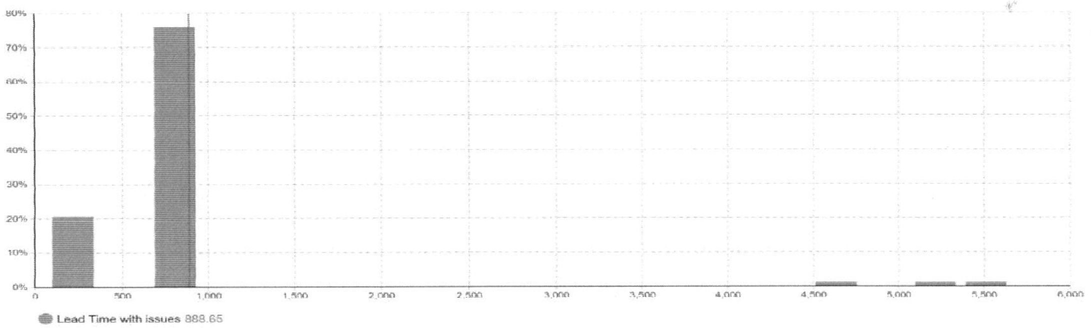

Figure 8. Lead time histogram for the current scenario.

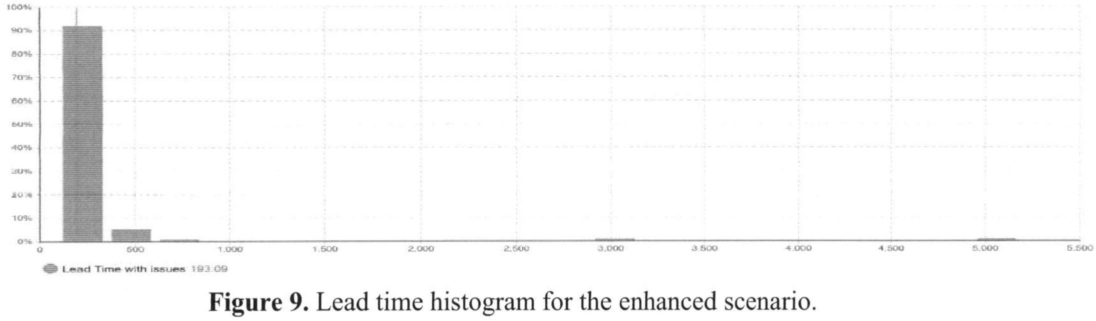

Figure 9. Lead time histogram for the enhanced scenario.

Figures 10 and 12 illustrate how the five workers are utilised over time and it is clear that the enhanced scenario has increased the utilization rate. However, the utilisation rate for the enhanced scenario is still low, which is 1.58 out of five, as shown in Figure 13.

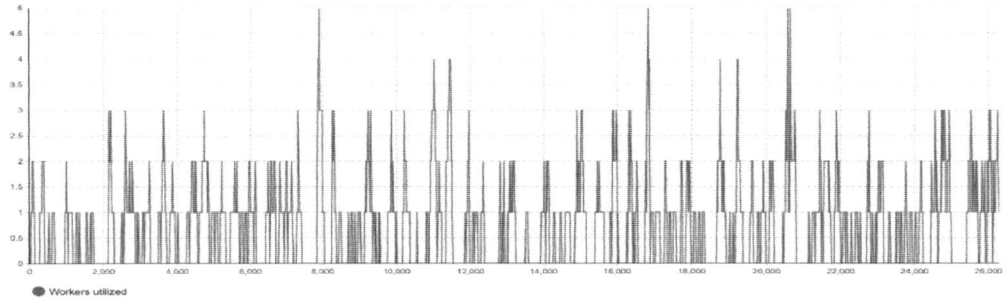

Figure 10. Worker utilization per hour for the current scenario.

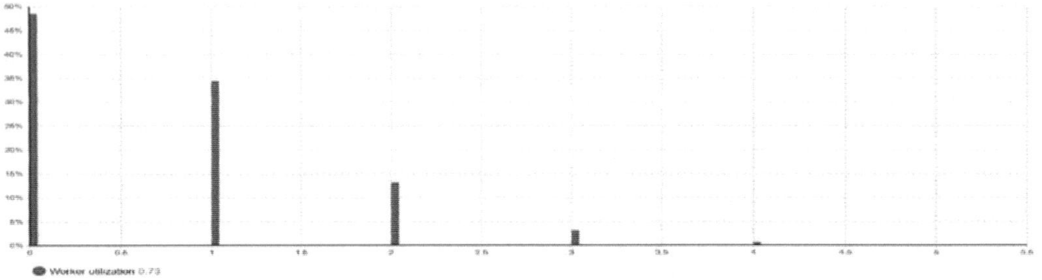

Figure 11. Average worker utilization for the current scenario. The worker utilization is 0.73 workers out of 5 workers.

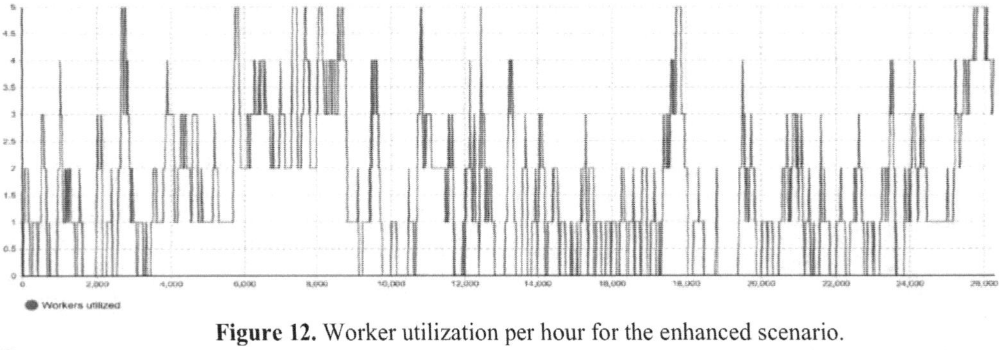

Figure 12. Worker utilization per hour for the enhanced scenario.

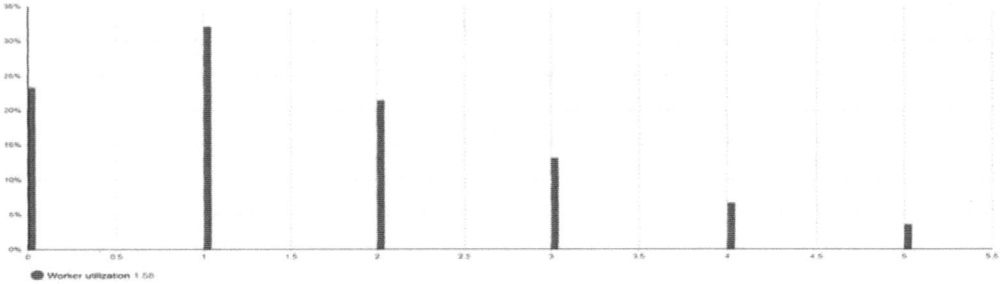

Figure 13. Average worker utilization per hour for the enhanced scenario. The worker utilization is 1.58 workers out of 5 workers.

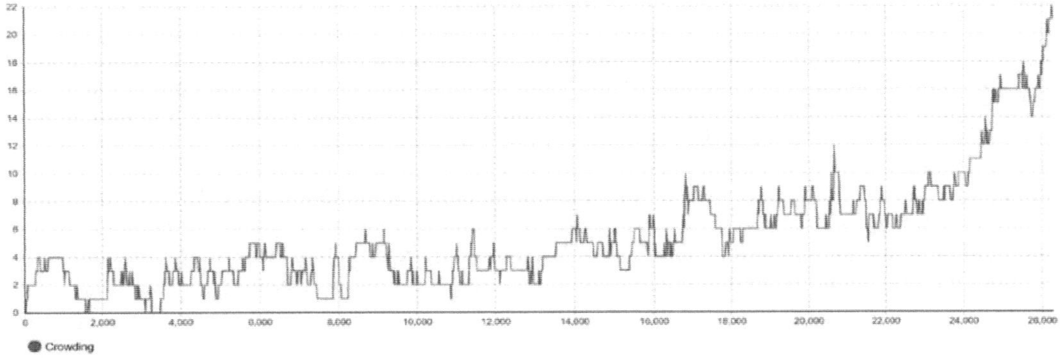

Figure 14. Crowding per hour for the current scenario.

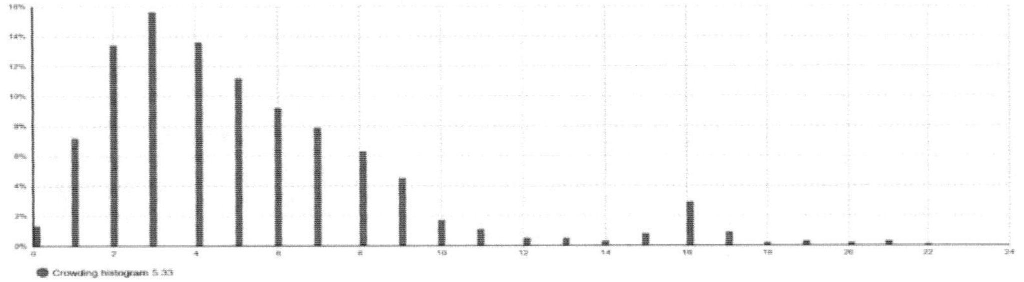

Figure 15. Average crowding for the current scenario.

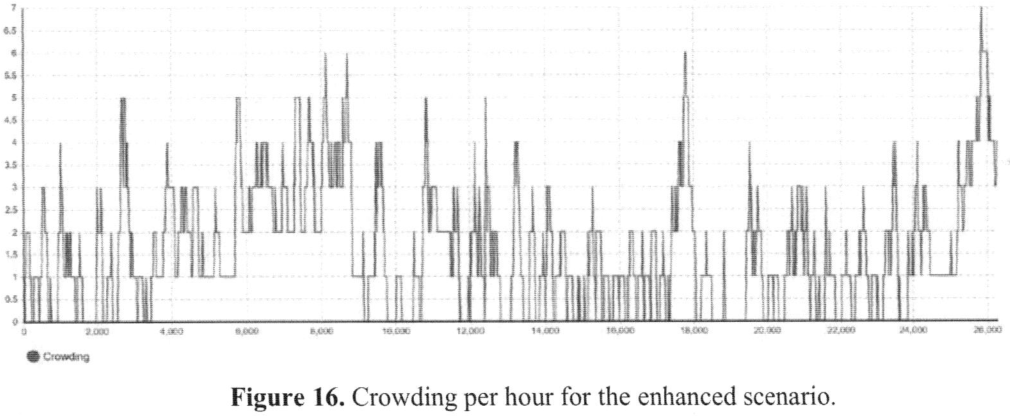

Figure 16. Crowding per hour for the enhanced scenario.

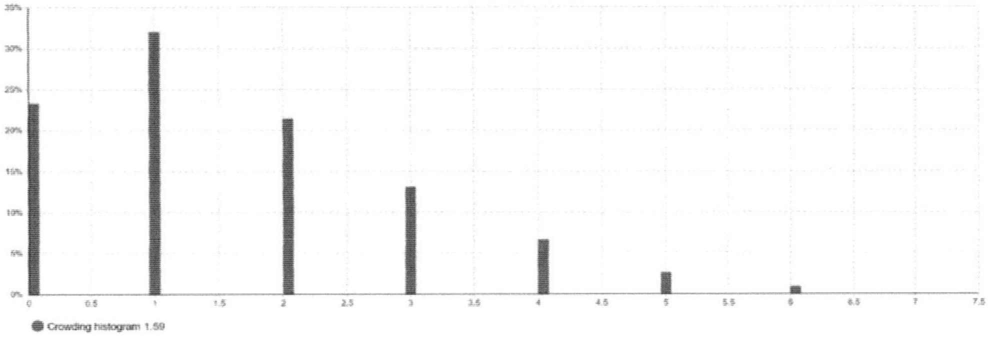

Figure 17. Average crowding per hour for the enhanced scenario.

Workshop crowding here refers to the number of tools that are simultaneously served in the workshop. In Figure 14, the number of tools in the workshop for the current scenario is increasing over time, whereas the crowding for the enhanced scenario fluctuates around two tools at once, as shown in Figure 16, allowing for easier navigation and less physically strenuous tasks, where it develops a safer, and more reliable work environment. Future research can be done to enhance operational efficiency and tool life-cycle management. One area is the development of alternative future-proofing strategies or improvements to the repair-path cycle stock inputs. This exploration can identify a leaner approach to finding optimal stock values that minimize inventory levels. Additionally, investigating advanced stock management techniques, such as dynamic stock optimization using real-time data, holds promise for improving the model's performance. Integrating advanced optimization algorithms like Particle Swarm Optimization and Simulated Annealing could further enhance the effectiveness of the inventory management system.

4. Conclusion

The simulation results for the study case support the conclusion that a local repair service is required to keep maintenance workshops in high-cycle regions at high-performance levels. The repair path cycle that was proposed for the current maintenance workshop has several benefits for the workshop, the region and the globe. For the workshop, 78% reduction in lead times, a 116% improvement in worker utilization, a 73% reduction in crowding levels, a 52% reduction in scrap rate, and a potential profit increase of roughly three million NOK (20%). For the region, the repair path cycle creates jobs for local repair and reduces the scrap rate. For the globe, the repair path cycle reduces the maintenance logistics between Norway and BCC and the associated CO_2 emissions.

References

[1] Van der Auweraer S and Boute R 2019 Forecasting spare part demand using service maintenance information *Int. J. Prod.n Econ.* **213** 138–149
[2] Kosanoglu F Turan H H Atmis M 2018 A simulated annealing algorithm for integrated decisions on spare part inventories and cross-training policies in repairable inventory systems. *In: Proc. Int.Conf. Comput. Ind. Eng* 1-14
[3] Schultz C R 2004 Spare parts inventory and cycle time reduction. *Int. J. Prod. Res.* **42**(4), 759-776
[4] Li X Zhang K Tan Z 2015 Spare part management in a testing workshop *2015 Winter Simul. Conf. (WSC)* 3170–3171
[5] Saeed M F Osman F 2022 Modeling and simulation for inventory management of repairable items in maintenance systems. *Simul.* **98**(11), 1013–1037
[6] Zhu S van Jaarsveld W Dekker R 2020 Spare parts inventory control based on maintenance planning *Reliab. Eng. Syst. Safety* **193,** 106600
[7] Tagaras G Vlachos D 2001 A periodic review inventory system with emergency replenishments *Management Science* **47**(3), 415–429
[8] Shalfawi M 2023 Modelling of Maintenance ServiceWorkshop and Inventory Operations for a Short Cycle Operational Region University of Stavanger Norway

Internal clearance behaviour in healthy and faulty bearings

I El-Thalji[*] and Y Khattab

Faculty of Science and Technology, University of Stavanger, Norway.

[*] Corresponding author's email: idriss.el-thalji@uis.no

Abstract. Internal clearance is a critical parameter in bearing design and operation because it affects the bearing's performance. It can be considered as a special type of the looseness fault. Several studies indicate that rolling bearings with large internal clearances showed a short predicted fatigue life. However, the time waveform and spectral features of bearing with internal clearances are partially studied and require further studies. To address this gap, the purpose of this paper is to study the effect of internal clearance on the dynamic response of healthy and faculty rolling bearings in both time waveform and spectral representations. Four experiments are designed and performed: (1) rigid bearing with no defect, (2) rigid bearing with bearing defect, (3) bearing with high internal clearance, (4) bearing with high internal clearance and bearing defect. The results indicate an increase in the overall vibration level with high internal clearance is introduced, however, there is an increase in the overall vibration level when high internal clearance and bearing defect are both presented. The internal clearance negatively affects the impact amplitude of the bearing defect, which also affects the spectral pattern at the natural frequency zone.

1. Introduction

Rolling bearings are critically important components in various industries, playing a crucial role in their machineries. Rolling bearings are indeed one of the main monitored components as they can provide valuable insights into the condition of the entire equipment in which they are installed. Internal clearance, also known as radial internal clearance, is an important parameter in the design and operation of rolling bearings. It refers to the space or gap between the rolling elements (balls or rollers) and the inner and outer raceways of a bearing, illustrated as Ci in Figure 1. High internal clearance in a bearing can result in higher levels of vibration, as it allows for more free movement of the rolling elements which causes vibrations, especially at high speeds or under dynamic loads.

There are several mathematical models used to estimate the effect of internal clearance on the bearing vibration response such as Tiwari et al. [1] Lioulios and Antoniadis [2], Harsha [3]. Ambro et al. [4], Changqing and Qingyu [5] Upadhyay et al. [6]. Xu et al. [7]. However, when the authors were testing bearing faults using the fault simulator rig at the University of Stavanger, a strange result was observed and motivated us to experimentally explore the effect of internal clearance on the bearing vibration response under healthy and faulty conditions. The issue was observation when we started with a set of bearings (internally rigid, almost no clearance) and later we had to use another set of bearings as we ran out of the first set of bearings. The second set of bearings was a bit internally loose which provided unexpected time wave and spectral features. Thus, the author decided to explore this issue further and perform four experiments and report the results: (1) rigid bearing with no defect, (2) rigid bearing with

IOP Conf. Series: Materials Science and Engineering 1294 (2023) 012041 doi:10.1088/1757-899X/1294/1/012041

bearing defect, (3) bearing with high internal clearance, (4) bearing with high internal clearance and bearing defect.

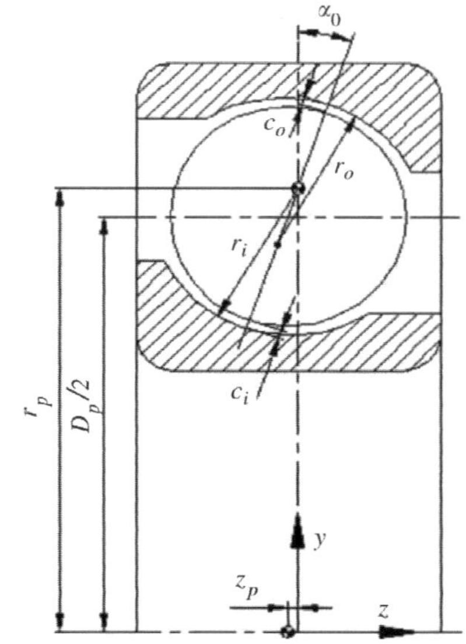

Figure 1. Bearing cross-section with internal clearance [1].

In the following section, the experimental set up is presented. In section 3, the main results of the experiments are illustrated and discussed, followed by some conclusions about the effect on internal clearance under healthy and faulty conditions and how to detect it.

2. Materials and Methods

The aim of this experiment is to study the internal clearance effect under healthy and faulty bearing conditions. Under healthy bearing conditions, two types of bearings are considered, one has high internal clearance (internally loose) and the second has no internal clearance (internally rigid), as shown in Figure 1. Under faulty conditions, the same type of bearings were used and an artificial defect was introduced to the bearing outer race, of 2 mm size.

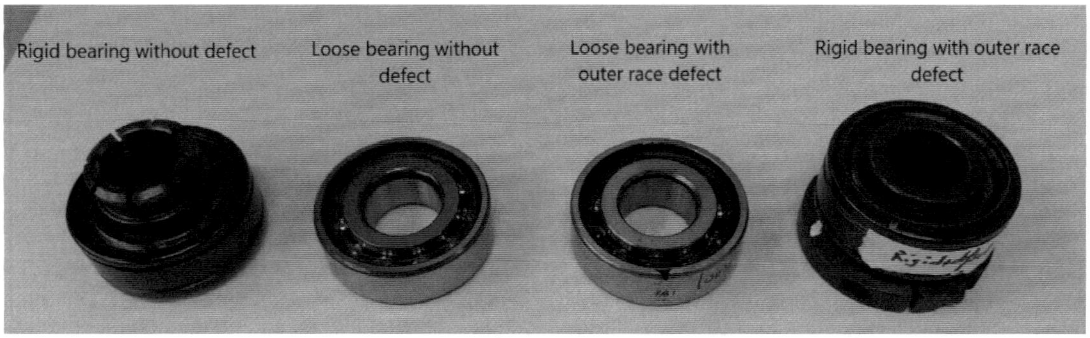

Figure 2. Bearings used in the experiments.

The bearing fault simulator rig at the University of Stavanger, shown in Figure 3, is used to run four experiments. The bearing fault simulator is a simple rotating set-up where a motor is driving a shaft carried by two bearings. Each bearing housing is monitored with three accelerometer sensors to collect vibration measurements. The sensors are connected with the SKF IMX8 monitoring module and connected to SKF cloud and @ptitude software for getting time waveform and frequency plots. All experiments are performed at a rotating speed of 20 Hz.

Figure 3. Bearing fault simulator rig with healthy rigid bearing (no internal clearances).

Figure 4. Bearing with high internal clearance.

Figure 5. Rigid bearing with defect.

Figure 6. Bearing with high internal clearance and defect.

Table 1. Experiment setup

	Experiment 1	Experiment 2	Experiment 3	Experiment 4
Rotating speed in Hz	20	20	20	20
No. rolling elements	8	8	11	11
Clearance effect			Yes	Yes
Bearing fault		Yes		Yes
Overall description	Rigid bearing without internal clearance or defect	Rigid bearing with defect	Bearing with high internal clearance and without defect	Bearing with high internal clearance and defect

3. Results and discussion

In this section, three results categories are presented: the overall vibration level in g, time waveform plots and spectral plots for each experiment. The overall vibration values, presented in Table 2, indicate the effect of internal clearance under healthy and faulty conditions. First, the overall vibration value for the internally loose bearing (experiment 3), which is 0.275 g, is clearly higher than the overall vibration value for the internally rigid bearing (experiment 1), which is 0.084g. Second, comparing the overall vibration values for experiment 2 (0.726 g) and experiment 4 (0.389 g), interestingly indicates that internal clearance diminishes the effect of the bearing defect. This diminishing impact phenomenon of bearing defect can be explained as the space between rolling elements and races being internally loose, the chance and impact severity are getting lower and therefore the overall vibration values are getting lower as well.

Table 2. Overall vibration value for the performed experiments.

Experiment setup	Vibration level (RMS value) [g]
Rigid bearing without internal clearance or defect	0.084
Rigid bearing with defect	0.726
Bearing with high internal clearance and without defect	0.275
Bearing with high internal clearance and defect	0.389

3.1. Time waveform analysis

To explain further the effect of internal clearance under healthy and faulty conditions, the time waveform analysis is required. The time waveform shows what physically happened during each experiment. In Figure 7, the time waveform for experiment 1 is presented, where hardly a specific pattern can be captured, and amplitudes are of low values (g values between 0 and 0.35). Comparing the time waveform of experiment 2 (Figure 8) with experiment 1 (Figure 7), a clear increase in the amplitudes can be observed (g values between 0 and 3) and a clear impact pattern is observed (impact pulses with almost fixed time interval between them). There are around 60 impact pulses that appeared in one second of the time wave presented in Figure 8. Comparing the time waveform in Figures 7 and 8 provides a good understanding of the effect of bearing defects, especially when the bearing is internally rigid.

To understand the effect of internal clearance, the time waveforms from experiment 1 (Figure 7) and experiment 3 (Figure 9) shall be compared. The time waveform in Figure 9 shows a clear increase in the amplitudes, i.e. the g values are between 0 and 1 compared to the g values in Figure 7 which are between 0 and 0.35. Moreover, there are impact pulses, even though, there is no bearing defect. It can be physically explained as the internal clearness offers a space for the rolling element to vibrate and then hit the race surfaces. However, the hits are not strong like the hits between the rolling element and the defect surface. It can be concluded that internal clearance generates impacts pulses even under healthy bearing conditions.

To understand the effect of internal clearance under faulty bearing conditions, the time waveforms of experiment 3 (Figure 9) and experiment 4 (Figure 10) shall be compared. The time waveform in Figure 10 shows a clear increase in the amplitudes, i.e. the g values are between 0 and 1.5 compared to the g values in Figure 9 which are between 0 and 1. Moreover, the duration of the length of the impact pulses is longer in Figure 10 compared to Figure 9. However, the amplitudes of impact pulses due to the bearing defect are much weaker compared to amplitudes in experiment 2 (Figure 8), which indicates that internal clearance diminishes the effect of the bearing defect. There are around 20 impact pulses that appeared in one second of the time wave presented in Figure 10, compared to 60 impact pulses that appeared in Figure 8. Therefore, it can be concluded that internal clearance affects the impact force severity and passing frequency (rolling elements are not picking the shaft speed as expected).

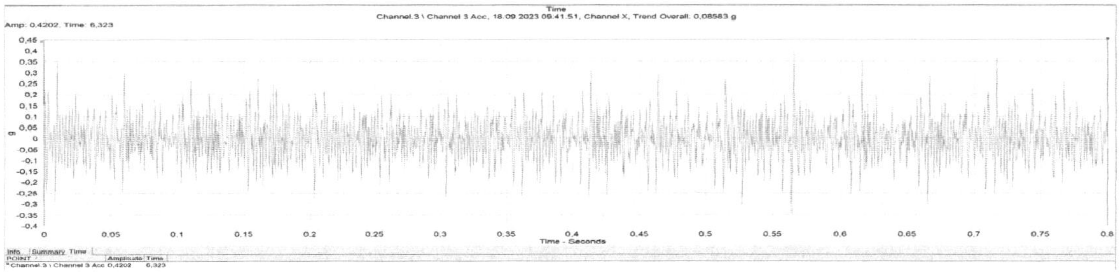

Figure 7. Time waveform for Experiment 1: rigid bearing without defect.

IOP Conf. Series: Materials Science and Engineering 1294 (2023) 012041 doi:10.1088/1757-899X/1294/1/012041

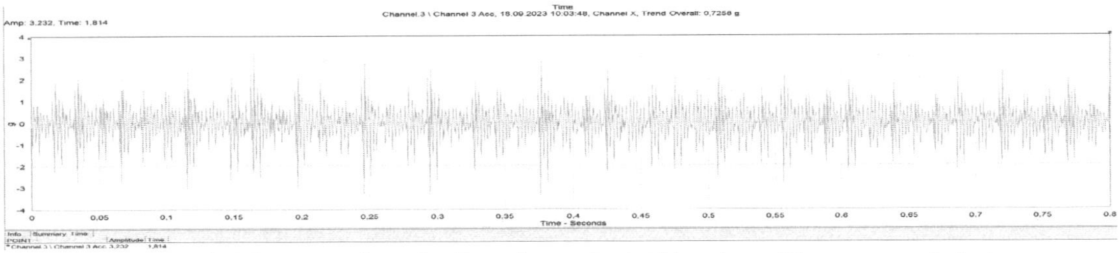

Figure 8. Time waveform for Experiment 2: rigid bearing with outer race defect.

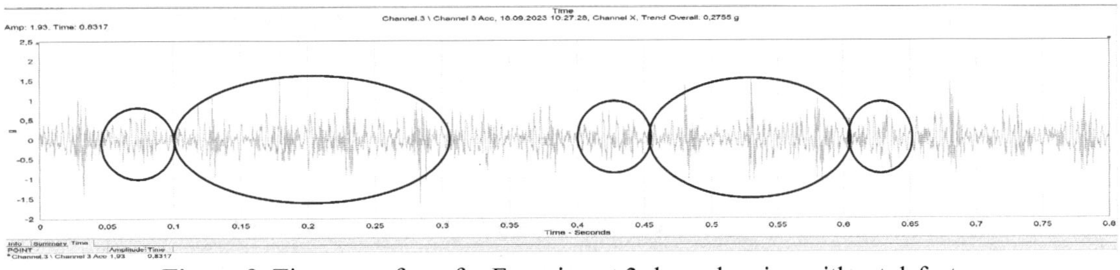

Figure 9. Time waveform for Experiment 3: loose bearing without defect.

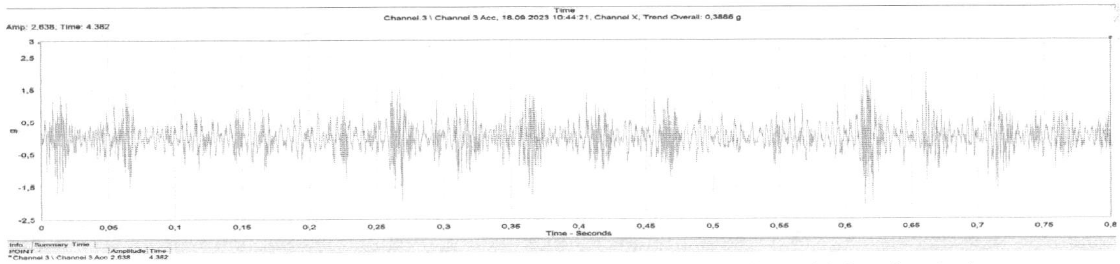

Figure 10. Time waveform for Experiment 4: loose bearing with bearing fault.

3.2. Spectral Analysis

The internal clearance made the task of detecting bearing fault using time waveform much harder as the impact and rolling element passing frequency were reduced. Frequency domain analysis might provide a better understanding of what happened from a spectral point of view. In Figure 11, the frequency plot for all four experiments is illustrated. There are three main frequency zones that we are going to study to understand the effect of internal clearance on the spectral features: (1) Fundamental frequency zone which is related to the rotating speed (1X) and its harmonies (2X,3X, etc.), (2) bearing fault frequency zone, which is related to the passing frequency of rolling element, (3) bearing natural frequency zone. At the fundamental frequency zone, in the first and second experiments, which are for the internally rigid bearing, there were no bins or peaks shown. However, tiny peaks at 1x (20Hz), 2x (40Hz), 3x(60Hz), and 4x(80Hz) can be seen for experiments 3 and 4, where the internally loose bearings are tested. Having peaks at the fundamental frequency zone might indicate that internal clearance generates shaft imbalance and misalignment. At the bearing fault frequency zone, the internal rigid bearing with fault (experiment 2), shows high amplitude peaks compared to all other experiments. Interestingly, in experiment 4 where a bearing defect exists like in experiment 2, the amplitudes of the peaks are much weaker. At the bearing natural frequency zone, the internal rigid bearing with fault (experiment 2), shows high amplitude peaks compared to all other experiments. It is clear that internal clearance excites the natural frequency of around 800-1000 Hz, however, the excitation is very weak. A closer look into these frequency zones has been given in Figure 11.

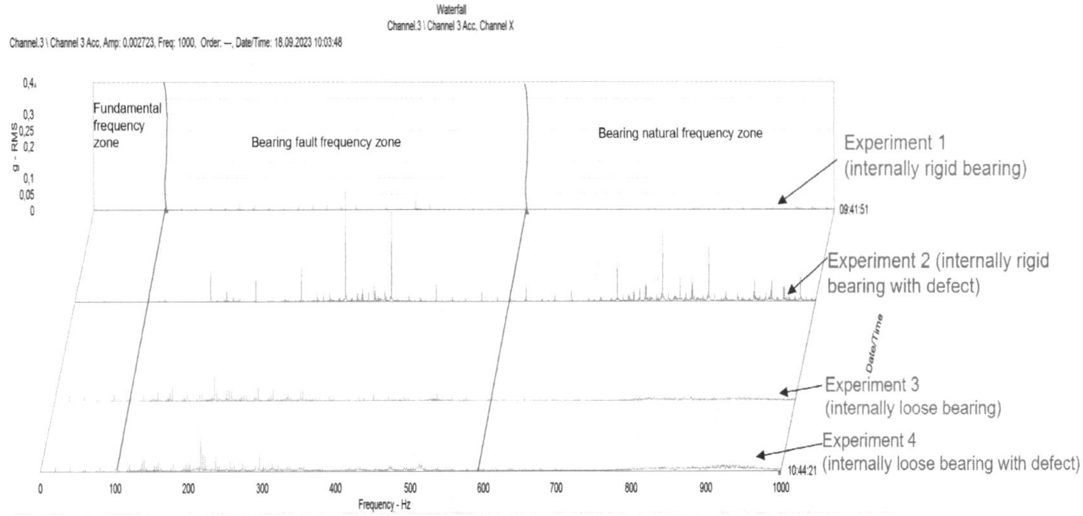

Figure 11. Waterfall plot for the performed experiments.

At the fundamental frequency zone, the internally rigid bearings (experiments 1 and 2) show no bins or peaks up to 100 Hz, as shown in Figure 12. The clear bins appear at 180 and 240 Hz for experiment 2 which might be related to the bearing defect. Remember, we have an outer race-bearing defect, its passing frequency can be theoretically estimated as (rotating speed*number of rolling elements) which is in this case equal to 160 Hz (20 Hz*8). This equation assumes that the rolling elements pick totally the shaft speed, which is not true. There is a relative speed factor for outer race fault of 40-100%. Thus, the practical bearing fault frequency can be estimated to range between 64 Hz and 160 Hz. However, the observed peaks are not within this estimated range. Looking at the time waveform of the impact pulses, a clear amplitude modulation exists, which is usually responsible for generating harmonics and sidebands.

The internally loose bearings (experiments 3 and 4) show bins or peaks up at fundamental frequency (20Hz) and its harmonics 40, 60, 80, 100, 120, 140, 160, 180, 200, 220, and 240 Hz. However, the highest amplitude is at 220 Hz. The estimated outer race fault frequency for bearings in experiments 3 and 4 is a range of 88-220 Hz (20 Hz*11 rolling elements). It can be concluded that internal clearance generates spectral peaks from 2x to 12x. The only difference between an internally loose bearing and an internally loose bearing with a defect is the amplitude of these frequency peaks, as shown in Figure 13.

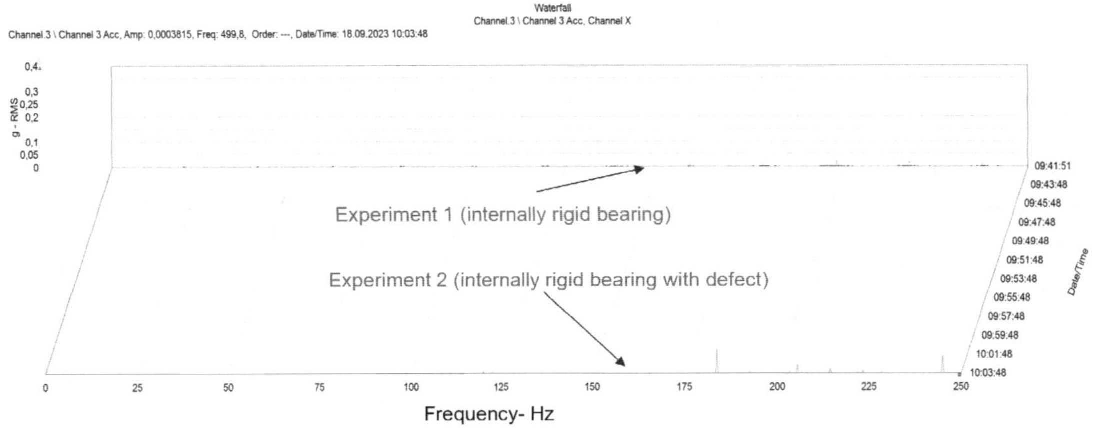

Figure 12. Waterfall plot for the fundamental and bearing fault frequencies zones, rigid bearing with and without bearing fault.

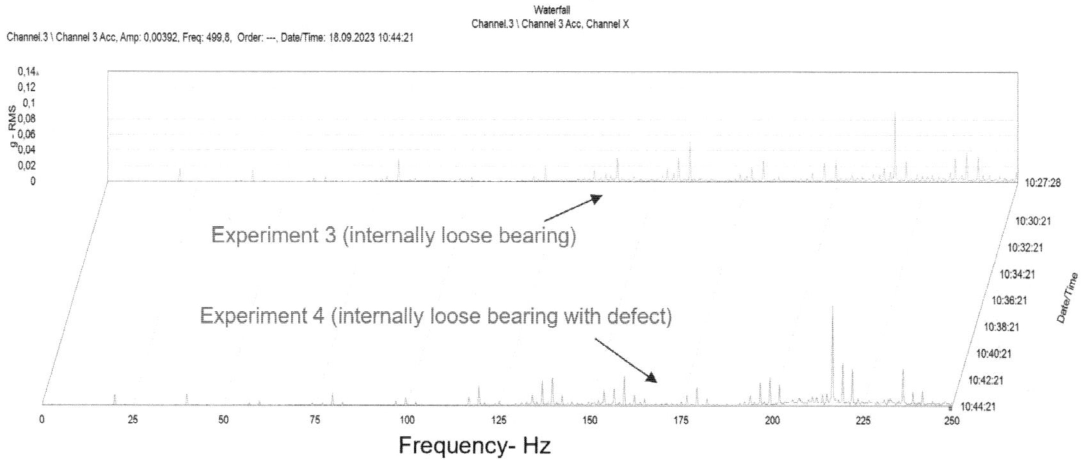

Figure 13. Waterfall plot for the fundamental and bearing fault frequencies zones, loose bearing with and without bearing fault.

At the bearing natural frequency zone (700-1000 Hz), the internally rigid bearings (experiments 1) show no bins or peaks, as shown in Figure 14. However, there are clear bins at the natural frequency zone in the case of experiment 2 (Figure 14), as the impacts of the bearing defect on the bearing defect area excited the bearing natural frequency. For internally loose bearing (experiment 3), it is expected to have no excitation at the natural frequency zone, same as the case of experiment 1. However, there is a clear excitation in Figure 15, from 750 Hz to 1000 Hz. It physically means that the rolling element vibration due to the internal clearances generates minor impact events that lightly excite the bearing natural frequency. In the case of experiment 4, the excitation at the natural frequency zone (750-1000 Hz) is a bit higher compared to the one in experiment 3, however, it is much lower compared to the one in experiment 2. It physically means that bearing defect under internally loose conditions excites the natural frequency, however, the effect of bearing defect is much higher under internally rigid conditions.

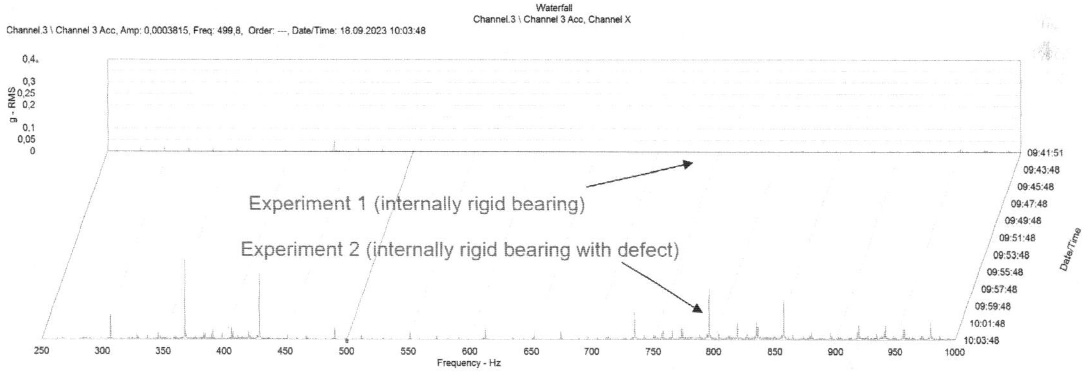

Figure 14. Waterfall plot for the natural frequencies zones, internally rigid bearing with and without bearing fault.

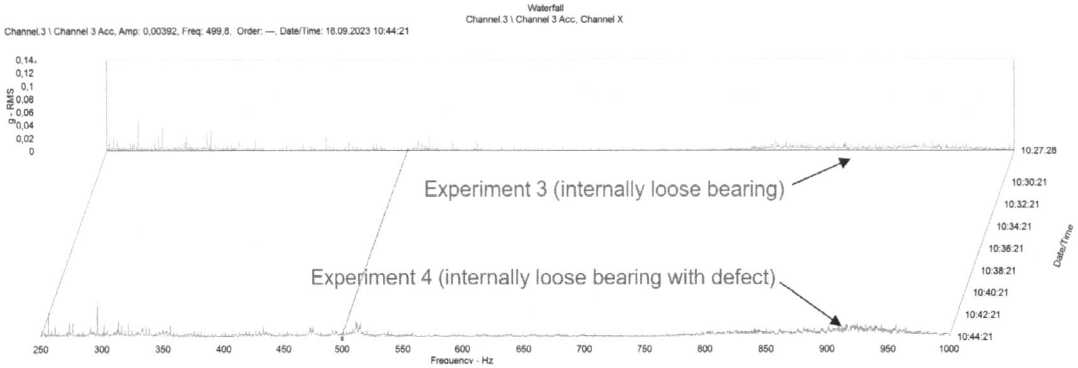

Figure 15. Waterfall plot for the natural frequencies zones, internally loose bearing with and without bearing fault.

4. Conclusion

The performed experiment in this paper shows the effect of internal clearance on the bearing vibration, its time wave and spectral features. Moreover, it explains the effect of internal clearance on bearing defect detection. The internal clearance generates impacts that vary in their amplitudes and interval (time between impacts), several peaks from 2x-12x and wide peaks at the natural frequency zone. The internal clearance diminishes the impact of bearing defects and makes them harder to detect. This effect of internal clearance acts as a smoothening agent for the defect impact, especially since the overall vibration values are much lower compared to the internally rigid bearing. It will be interesting to study the effect of internal clearance over time and observe if the bearing is going to deteriorate faster than the internally rigid bearing.

References

[1] Tiwari M, Gupta K and Prakash O 2000 Effect of radial internal clearance of a ball bearing on the dynamics of a balanced horizontal rotor. *J Sound Vib* **238**, 723–756.

[2] Lioulios A N and Antoniadis I A 2006 Effect of rotational speed fluctuations on the dynamic behaviour of rolling element bearings with radial clearances. *Int J Mech Sci* **48**, 809–829.

[3] Harsha S P 2006 Nonlinear dynamic response of a balanced rotor supported by rolling element bearings due to radial internal clearance effect. *Mech Mach Theory* **41**, 688–706.

[4] Ambrozkiewicz B, Syta A, Gassner A, Georgiadis A, Litak G and Meier N 2022 The influence of the radial internal clearance on the dynamic response of self-aligning ball bearings. *Mech Syst Signal Process* **171**, 108954.

[5] Changqing B and Qingyu X 2006 Dynamic model of a ball bearings with internal clearance and waviness. *J Sound Vib* **294**(1-2), 23 – 48.

[6] Upadhyay S H, Harsha S P and Jain S C 2006 Analysis of nonlinear phenomena in high speed ball bearings due to radial clearance and unbalanced rotor effects. *J Sound Vib* **296**, 23–48.

[7] Xu M, Han Y, Sun X, Shao Y, Gu F and Ball A 2022 Vibration characteristics and condition monitoring of internal radial clearance within a ball bearing in a gear-shaft-bearing system. *Mech Syst Signal Process* **165**, 108280.

IOP Conf. Series: Materials Science and Engineering 1294 (2023) 012042 doi:10.1088/1757-899X/1294/1/012042

Optimization of the routing problem using the physics-based electromagnetism-like algorithm

J R Jao[1,*] and E A Vallar [2]

[1]Department of Software Technology, College of Computer Studies, De La Salle University, 2401 Taft Ave, Malate, Manila, 1004 Metro Manila, Philippines
[2]Department of Physics, College of Science, De La Salle University, 2401 Taft Ave, Malate, Manila, 1004 Metro Manila, Philippines

[*] Correspondence: jazzie.jao@dlsu.edu.ph

Abstract. Waste collection is considered as a first objective in planning a sustainable waste management, and provision of substantial collection coverage in urban and rural areas should be satisfied before investing in more sophisticated infrastructure. Therefore, good waste collection service is essential to upgrade further the next levels of a municipal solid waste system. One of the ways to enhance performance on waste collection is to plan effective routing of trucks. The study formulated a Waste Collection Arc Routing Problem to model the waste collection routing problem of Parang, Marikina City. The study revealed that the EM-like with local search and selection strategy mechanism displayed significant minimization abilities of distance values, and the capacity consideration of truck is important to attain better efficiency and cost savings of route.

1. Introduction

Municipal Solid Waste Management (MSW) has been considered to be one of the challenging sectors to manage [1–3] due to its complex-natured behavior [4], quick growth of waste amount brought about by development, and high management costs [5] both in developing and developed countries. To be specific, developing countries are faced with greater magnitude of problems in inefficient poor segregation, collection, storage, treatment and disposal practices [6] which can be associated, in general, with lack of enabling system made up of legal, regulatory, and economic instruments, including sufficient financing, staff capacity, and stakeholder awareness [3,7]. According to [8], these waste management problems are evident in the Philippines as the country grows its population increases its household consumption with 0.03 percent from 2019 to 2020 [3] and expands its urbanization with 0.13 percent from 2010 to 2015.

The process of MSW management can be grouped into four stages of waste life cycle: generation, collection and transportation, transformation, treatment, and final disposal, where each one depends upon different factors in society and are vertically integrated, that is, changes to one part of the system impact the entire value chain. Waste collection and transportation in the Philippine-context is defined to be the act of removing waste from the source or common bin points in a given area, where waste segregation is expected to be done at source with corresponding labels before collection, and that waste trucks may be heterogenous to satisfy the storage of a given type of waste [8]. Local Government Units (LGUs) are given the responsibility to conduct this mandate. Some of them have implemented mandatory segregation at source by a "no segregation, no collection" city ordinance, like in Parang, Marikina City, in line with scheduling mechanisms. Collection can be done in the following ways: door-to-door collection, stationary collection through the Materials Recovery Facility (MRFs), and mobile waste collection through waste collection vehicles. Approximately 2 billion people worldwide do not have access to waste collection and around 3 billion lacked controlled waste disposal facilities in developing countries. Waste collection is considered as first and basic objective in planning a sustainable waste management [3] and provision of substantial collection coverage in urban and rural areas should be satisfied before investing in more sophisticated infrastructure. Negligence of this aspect has been shown to overthrow waste treatment despite the availability of treatment facilities as waste is not properly collected and disposed. Therefore, good waste collection service is essential to upgrade further the next levels of a MSW system like waste segregation, recycling, and recovery. In fact, [9] calls for efficient and effective solid waste collection by system analysis and operations optimization. One of the ways to enhance performance on waste collection is to plan effective routing of trucks.

Significant amount of research on collection route optimization have been conducted through the years. Commonly, this problem has been formulated as a Vehicle Routing Problem in two different ways: Node Routing Problem and Arc Routing Problem. The first one is when waste is to be collected along street segments, and the latter is when waste is serviced from each house, considered as a node. A distinct instance of ARP called Capacitated Arc Routing Problem, first introduced by [10] is the one relevant to this study. It requires to find a set of routes for vehicles with limited capacities, situated at a source called the depot, such that the total demand serviced on each route does not exceed the vehicle capacity while the total cost is minimized. The study specifically uses the single-objective ARP formulation to minimize the distance taken by the waste truck. Other measures like fuel consumption and Greenhouse Gas (GHG) Emissions are also calculated using the Comprehensive Modal Emission Model (CMEM) where the load and speed of the collection vehicle are included in the predictor variables. VRP is one of the intractable problems in combinational optimization that can only be solved with exact methods of solution when the problem size is small. This motivates the use of EM-like algorithm, an approximation algorithm to solve the waste collection routing problem formulated as a single-objective optimization problem where a distance function is minimized. The waste collection is done in the road network located in the boundary of Parang, Marikina City.

2. Background

Waste collection routing can be modeled as an NRP or ARP. The use of NRP is common in handling problems with large containers that are based in relatively small number of collection points, while ARP is useful when it comes to servicing a continuous collection demand like in door-to-door and household collection cases. The main objective of both problems is in general similar, that is, to minimize some cost function. However, few literatures have only dealt with general routing applications in waste collection since the use of different vehicles usually enables the separation of the two types of VRP mentioned. Many authors have based their research on increasing the efficiency of waste collection and transport processes because they were identified as being the most expensive functional elements in the entire waste management process. Some focused on shortening routing distance, minimization of collection expenses and number of collection vehicles. Lack of scientific and technological interventions in route selection can result in poor and costly planning of collection systems [11] so various studies have been employing: (1) mathematical optimization, grouped as either exact methods or approximate methods (2) GIS-based approaches using QGIS and ARCGIS, and (3) specialized software in solving the VRP [9]. Exact methods to solve the CARP were only proven to be accurate at few instances, like the study conducted by [12], having networks of 200 nodes and 300 edges. This has caused the area of approximate algorithms to emerge as method to generate relatively good quality solutions at a reduced computation time by sacrificing the guarantee of global optimality. The CARP was mostly solved alone by constructive heuristics in the early stages of its emergence. But lately, recent researchers have opted on using it as only a part of a more sophisticated algorithm for generating initial solutions like metaheuristics. Some of the works that employed constructive heuristics include the work of [10, 13],

[14], and [15]. The Path-Scanning algorithm is the commonly used method. It was proposed by [10] and it constructs routes one by one. It starts from a node called the depot and a route is constructed through repeatedly adding a task, which is treated as an arc in this case. Note that different directions of servicing tasks lead to different solutions since the line is treated an arc. The criteria are the arc that will be added is nearest from the current end of route (i.e. the head of last arc) among all the other arcs that do not go over beyond the truck capacity. In case there is more than one of this selection, tie-breaking rules are given to aid the decision.

For the metaheuristics approach in waste collection routing problem, there have been generous amount of works conducted through the years. [16] integrated an evolutionary algorithm Chaotic Particle Swarm Optimization in GIS in aiming to maximize the MSW collection at Danang City. The implementation of the algorithm obtained better total collected waste amount than the currently applied method in the said city. A multi-objective model of the waste bin collection in selected rural areas of Spain was analyzed by [17], where Tabu Search algorithm is used in minimizing cost of transport and improvements to the level of service, that is, the frequency of waste collection at each point over the planning period. The model is compared to another variant of Genetic Algorithm (GA) and showed better and denser set of solutions. [18] formulated a mathematical model in an oriented graph to minimize total costs with respect to time and capacity constraints of a waste collection arc routing problem. The total instance was 1,467 vertices and 3,529 edges. Input parameters (operation costs, vehicle capacities, waste collection demand). were obtained from demographic database and waste management expert estimates. The input data distances and times were defined prior to calculation, that is, distance of the end node of one arc to the initial node of the second arc. Using GA with local search the solution is found in the form of demand arc sequences for each route of each vehicle for all days throughout the entire planning horizon. One key component of the algorithm is the use of a fitness function which was used for checking feasible and unfeasible solutions. The study also suggested that stopping criterion of the algorithm for waste collection can depend on the preferred total number of iterations, running time of algorithm, or combinations of the previous ones. The proposed algorithm showed better performance than those already operating routes and a potential savings of around 7 % were computed.

Originally, EM-like is formulated for continuous optimization problems. An interesting approach to its usage on discrete optimization problems is done by several authors. [19] applied the algorithm to an optimization task of discrete nature called single machine scheduling problem. In order to do this, Random-Key (RK) procedure with genetic operations are done to encode the feasible solutions. By performing ANOVA and Duncan Test on the obtained function values, the study concluded that compared to the standard genetic algorithm, the attraction-repulsion mechanism has better performance. [20] also used the EM-like for the same problem except that Tabu Search instead of GA operators are used to improve the search process. It can be concluded that across different algorithms, the EM-TS garnered the lowest values. Similar positive results appeared on [21]. They considered a hybrid EM-like with scatter search and its application on Resource Constrained Project Scheduling Problem (RCPSP). Moreover, the force and charge calculation are modified such that Euclidean distance is removed and the charge values now takes [-1,1] values. The previous formulation only takes [0,1] values. The results of the comparison between other algorithm's performance in terms of solving benchmark instances on Project Scheduling Problem (PSP) show that the EM-like is superior and is effective in optimizing the given cost function, except for the results obtained from [22]. A study by [23] was conducted to hybridize the EM-like for solving the Capacitated VRP using a new local procedure iterated swap procedure. The solution yields a sequence of nodes, such that the initial is the depot, that satisfies given constraints of tour time/length limit and vehicle capacity. After testing the efficiency of the algorithm using 14 classical instances, the hybrid EM-like is found to be competitive compared PSO and ACO. In conclusion, the EM-like is shown to be flexible in terms of being applied to wide range of problems: discrete optimization, continuous optimization, simulated data using test functions, training neural networks, and vehicle routing problems. It can be hybridized with other well-established algorithms to improve its efficiency or be used on its own and perform parameter modifications.

3. Methods

The study was conducted inside the administrative boundary of Barangay Parang, Marikina City located in National Capital Region, Philippines. The road segments that need to be visited are generated randomly. Each collection point is assumed to be a residential household that is assumed to consist of three adult people of random gender. The type of waste that these people produce is assumed to be of residual waste category. There are a total of 206 intersections and 203 road segments in the network data generated. **Figure 1** shows the area boundary of the study using satellite and non-satellite map. The red node represents the intersection point where the path of the outgoing street from the depot is heading towards, while the rest of the black nodes corresponds to the intersection points of the network boundary.

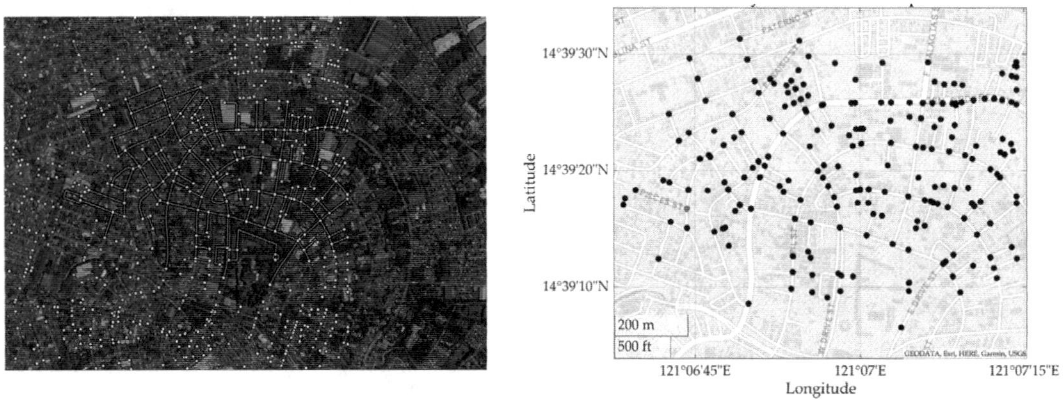

Figure 1. The streets and area boundary of the study.

3.1. The Network and Problem Design

The discussion of the network profile deserves careful design and discussion in this study as it will be the exploration medium of the EM-like algorithm. The shapefile data of the boundary area is extracted using OSM tools to render the edgelist information and its features. These features are composed of the street length, street speed limit, and the vector of coordinate points that span the street. Cleaning and the data include filling up the blank entries and adding additional column of information. The obtained CSV file is then loaded on MATLAB. The CSV file containing the edgelist will be modified, such that the two-way streets will be separated when plotted as a graph. Street network can be translated into a directed multigraph representation by first taking its undirected graph representation, as seen in **Figure 2**, and then duplicating the arc once when there exist in-out traffic directions. If there is only either in or out, then the arc is left as it is. The dotted ray in the same figure means that the road segment between nodes H, and I and I and J takes only one traffic direction. A strongly connected graph is created that will serve as the basis of generating random streets to service, such that every street and intersection is reachable given an arbitrary starting position.

The problem statement of the study is later defined to be a single optimization involving the minimization of the total distance taken by the truck while covering all the elements of the set of required streets. Two problem definitions are given below:

Definition 1. (The Baseline Scenario of ARP).
The baseline scenario of the ARP is the problem of finding a permutation of the set of required streets such that the following function is minimized:

$$f(\pi) = \sum_{i=1}^{O} l_i$$

where π is an arbitrary permutation of the required streets, l_i is the road length of the ith street from the constructed path, and O is the number of elements in this path constructed at every iteration such that it contains the sequence of edges or nodes to take to execute a feasible collection service.

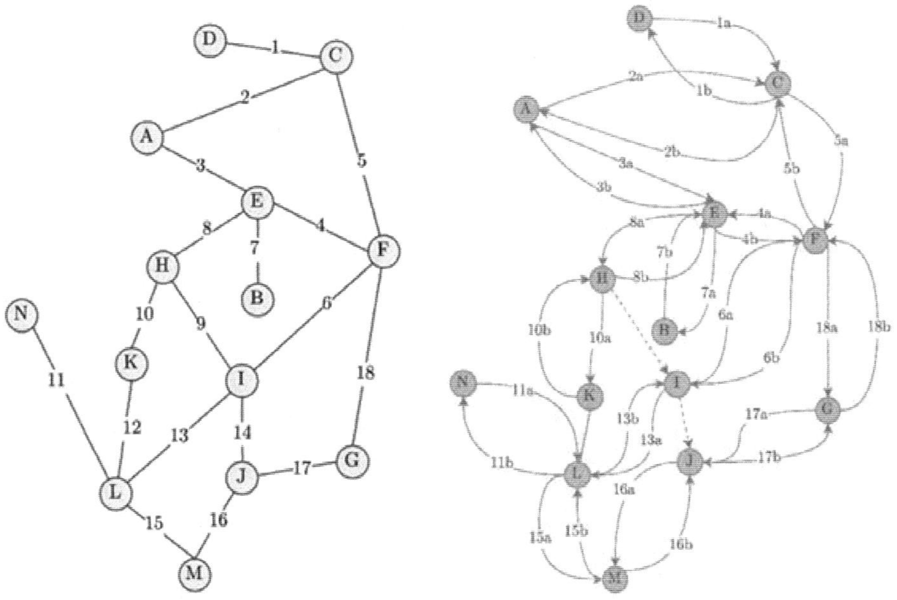

Figure 2. Undirected (Left) vs. directed (right) multigraph representation.

The calculation of fuel consumption of a waste truck traversing a street segment with length d, embodying a speed v, and carrying a load f, is done using CMEM [24] is used as seen in equation.

$$Fuel(v) = \lambda(FNV + w\gamma\alpha v + \gamma\alpha f v + \beta\gamma v^3)d/v \quad (1)$$

where: $\gamma = \xi/\kappa\psi$; $\alpha = g sin\theta + gC^r cos\theta$ and $\beta = 0.5C^d\rho A$.

The parameters of this equation are adapted from the study of [25] and are given in

IOP Conf. Series: Materials Science and Engineering 1294 (2023) 012042 doi:10.1088/1757-899X/1294/1/012042

Table 1.

In general, the waste collection route table will always contain street segments that are traversed more than once. This is because real-world data is inherently sparse. Moreover, due to its topology, some streets may only be accessed through a single road or detour is required due to traffic direction considerations, like if it's one-way or two-way. To prevent double counting the objective function value and collection service (load and time), equation (2) ensures that if a street appeared more than once in the path, the first traversal will have a non-zero value, otherwise zero for succeeding traversals. To calculate for the GHG emission, a carbon emission factor of 2.6391 kgs of $CO2$ per liter burnt is used.

$$x_{ij} = \begin{cases} 1, & if\ a_{ij}\ first\ appearance \\ 0, & if\ a_{ij}\ n^{th}\ appearance\ where\ n = 2,3,\ldots \end{cases} \tag{2}$$

Table 1. Parameters of the CMEM.

Symbol	Name	Unit	Value
F	Engine friction factor	kg/rev/liter	0.20
N	Engine speed	rev/s	34
V	Engine displacement	liters	7
A	Frontal surface area	m^2	7.6
C^d	Aerodynamic drag coefficients	-	0.55
C^r	Rolling resistance coefficients	-	0.009
r	Vehicle acceleration	m/s^2	0
w	Curb weight	kg	5,500
κ	Heaving value for diesel fuel	kJ/g	45
ε	Vehicle drive train efficiency	-	0.4
ϖ	Efficiency parameter for diesel engines	-	0.9
ξ	Fuel to air mass ratio	-	1
ψ	Conversion factor from grams to liters	-	737
ρ	Air density	kg/m^3	1.2041
g	Gravity	m/s^2	9.81

3.2. EM-like Algorithm

The EM-like being a population-based algorithm requires a set of initial solutions. A possible solution of the TSP as mentioned in the previous section can be regarded as a permutation of the set of cities or locations that requires service. The aim is to permute this set and determine what sequence of streets gives the minimum cost expense of waste collection. This permutation is embedded on the search agent of the EM-like, the charged particles. Since the population affects the effectiveness and efficiency of the algorithm [26], the candidate solutions, are heuristically initialized through declaring a canonical ordering and perturbing it to produce modified versions, where the canonical ordering follows the metric city block distance. The intuition behind is to be able to generate "good enough" permutation of streets that is ordered according to its proximity from the depot. Random generation of permutation is possible, and it is also considered as an alternative parameter of the population initialization process. After initializing population of solutions, the next step is to calculate their charges according to equation (3). This equation, unlike electrical charges, always assume that the particles are positive.

$$q_i = \exp\left[-n\frac{f(\mathbf{x}_i) - f(\mathbf{x}_{best})}{\sum_{j=1}^{m}[f(\mathbf{x}_j) - f(\mathbf{x}_{best})]}\right], \forall i = 1, \dots, m \tag{3}$$

Where n is the number of dimensions, in this context, the number of streets of the path, $f(\mathbf{x}_i), f(\mathbf{x}_j)$ the objective function value of particles i and j, and m is the number of initialized population of solutions. Next is to implement how new solutions are produced from the previous one. The EM-like is a population-based algorithm so it uses the presence and interactions of a collection of solutions to explore the search space. The mechanism of the attraction-repulsion of charged particles in physics is utilized to recalibrate the positions of particles according to their fitness value. This fitness value is embedded to the magnitude of their charge calculated in the previous section. The implementation is done by using equation (4). Notice that this is expressed in vector form. The traditional formulation of EM-like by [27] assumes that the particle can be placed in an $n-$dimensional space. As such, it is considered that each particle's position is composed of n −components that can be written as in equation (5).

$$\mathbf{F}_{ij} = \begin{cases} (\mathbf{x}_j - \mathbf{x}_i)\dfrac{q_i q_j}{\left|\mathbf{x}_j - \mathbf{x}_i\right|^2}, & \text{if } f(\mathbf{x}_i) > f(\mathbf{x}_j) \\ (\mathbf{x}_i - \mathbf{x}_j)\dfrac{q_i q_j}{\left|\mathbf{x}_j - \mathbf{x}_i\right|^2}, & \text{if } f(\mathbf{x}_i) \le f(\mathbf{x}_j) \end{cases} \tag{4}$$

$$\mathbf{x}_i = x_i^1\hat{\mathbf{x}}^1 + x_i^2\hat{\mathbf{x}}^2 + \cdots + x_i^n\hat{\mathbf{x}}^n \tag{5}$$

The two conditional statements on equation (4) signifies the need to specify which particle, between $\mathbf{x}_j$, $\mathbf{x}_i$ has "better" objective function value. Note that we are dealing with a minimization problem so the definition of a better solution in this case is the one with lower cost. The first statement means that if particle j has better (lower) value of f, it attracts the particle i and the direction of the displacement is towards the particle j, otherwise particle i gets attracted to it. Finally, the term in the denominator, is the Euclidean distance between particle i and j where the given result is a scalar quantity. After getting the resultant forces on each particle i due to the presence of other particles, the amount of movement resulting from the action of these forces is computed on every n component of each $f(\mathbf{x}_i)$.

4. Results and discussions

The EM-like algorithm is tested on a set of modified variables called the parameter scenarios detailed in **Table 2**. Each of this setting is compared in terms of solution results.

Table 2. Overview of Parameter Settings.

Scenario	Pairwise Distance	Local Search Improvement	Particle Movement Boundary
αCBL − o	City-block	Yes	[0,1]
αCBL − h			rand
αCBNL − o		No	[0,1]
αCBNL − h			rand
αRBL − o	Random	Yes	[0,1]
αRBL − h			rand
αRDNL − o		No	[0,1]
αRDNL − h			rand

Table 3. Results of different parameter settings on ARP using EM-like algorithm .

No	Scenario	Distance [m]		Cost [PHP]		GHG [g of CO2e]	
		Best	*Worst*	*Best*	*Worst*	*Best*	*Worst*
1	αCBL − o	12931	21129	335.2690	408.3794	16.0080	19.1097
2	αCBL − h	12934	20447	333.6144	404.0426	16.0758	19.5955
3	αCBNL − o	14979	20222	341.1939	392.2709	16.3717	19.9693
4	αCBNL − h	14984	22696	346.5605	416.1693	16.6292	18.8226
5	αRBL − o	22467	33476	411.4950	474.4036	19.4955	22.9103
6	αRBL − h	20522	35951	406.2944	477.4615	19.7450	22.7636
7	αRDNL − o	25249	35249	415.2007	476.7235	20.1936	23.2294
8	αRDNL − h	24901	36572	420.8440	484.1110	19.9228	22.8749

The $\alpha CBL - o$ obtained better distance value of 12931 compared to $\alpha CBL - h$ scenario's 12934. Both implemented a local search procedure. However, the corresponding cost of $\alpha CBL - h$ is better than that of $\alpha CBL - o$. This shows that minimizing the distance does not lead to minimization of the corresponding cost due to fuel consumption. This is because lack of capacity considerations forces the truck to be overfilled; thus, increased need to consume fuel. A total of 2000 iterations is made across all parameter scenarios as show in **Figure 3** - **Figure 6**.

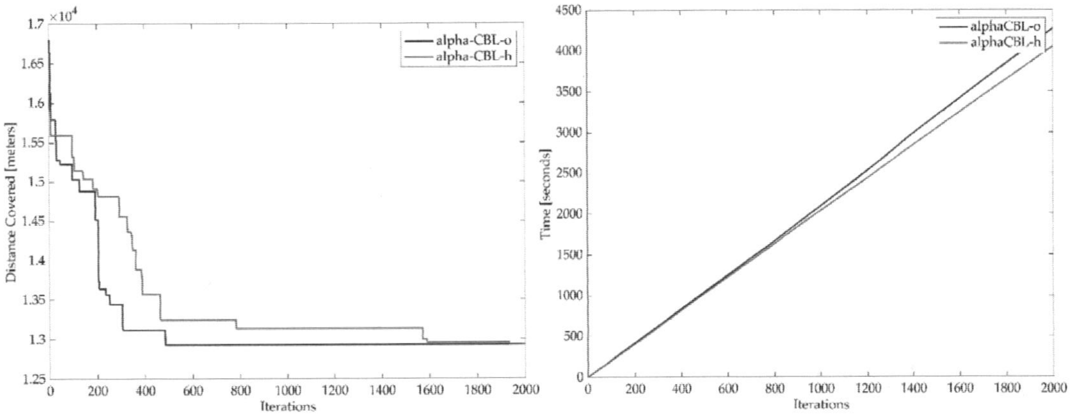

Figure 3. Distance and time minimization curve of $\alpha CBL - o$ vs. $\alpha CBL - h$ for 2000 iterations.

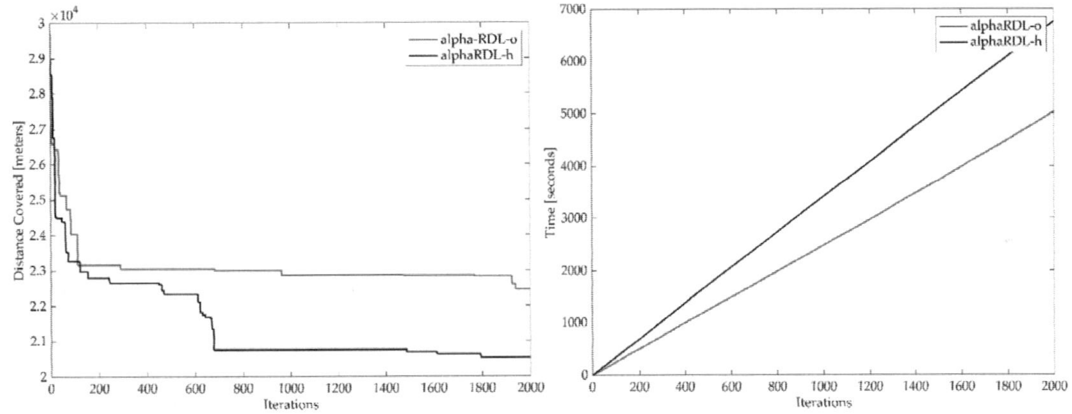

Figure 4. Distance and time minimization curve of $\alpha RDL - o$ vs. $\alpha RDL - h$ for 2000 iterations.

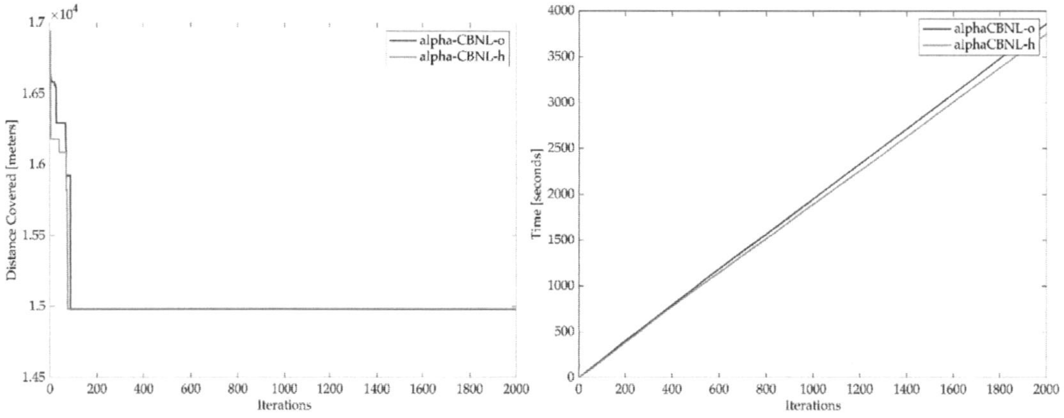

Figure 5. Distance and time minimization curve of $\alpha CBNL - o$ vs. $CBNL - h$ for 2000 iterations.

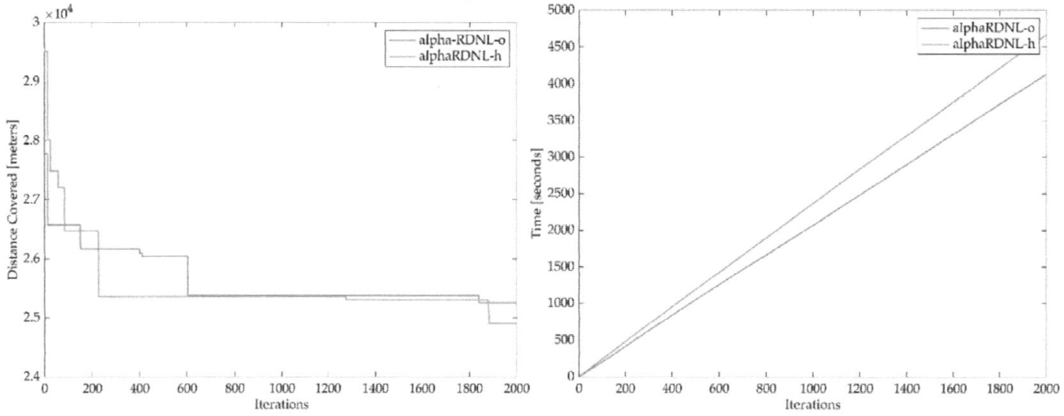

Figure 6. Distance and time minimization curve of $\alpha RDNL - o$ vs. $RDNL - h$ for 2000 iterations.

In general, the use of different particle boundary movement modified the search process of EM-like. The best distance value originated from the α-CBL-o scenario where the movement is restricted in [0,1]. The same scenario also displayed higher dissimilarity between the particles compared to the competing case of α-CBL-h. Dissimilarity indicates that the exploration is diverse and not redundant. This implies a good quality search as the aim of the EM-like is to explore diverse spaces and intensify the obtained good values. As shown in the comparison of the parameter scenarios in figure 5.9, the random permutation initialization scenario α-RD started from a large objective function value and although decreasing in value, its convergence to come close to the initial starting point of α-CD scenario appeared far-fetched. Therefore, it requires more iterations, leading to higher computational expense to eventually reach the range of values generated from the better initialization method implemented in the city-block based distance. This creates closer paths and spatially-informed configuration of feasible solution. Therefore, there is a speed of convergence issue with regards to the use of EM-like in this problem of VRP.

4.1. Metaheuristics for Maritime Operations

The study may also be applied in the problem of ship optimization wherein the medium of exploration is not described by a topological object represented by a street network. The Ship Routing Problem (SRP) involves assigning a fleet of known ships of different types to transportation requests. Although like the Vehicle Routing Problem (VRP), there are some distinct differences between the two. For instance, in VRP, vehicles must return to the depot, while ships are not required to return to the starting point. Additionally, certain ships cannot transport specific cargoes or enter ports due to physical or political/cultural limitations. Other variations include round-the-clock operation, differing start times (meaning some ships may have jobs at the start of the planning horizon), mandatory and optional cargoes, among other factors. When it comes to shipping companies, selecting the best route involves weighing the options of hub-and-spoke transportation or direct shipment. Factors such as the predicted volume of goods to be transported, the frequency of trips, and the available fleet size all play a role in the decision-making process. There exist many problems formulation of maritime shipping related problems. One example is the tramp shipping problem, wherein a company that has a fleet of ships with different sizes and fuel consumptions receives orders to load a full ship-load cargo at an origin port within a given time window and to transport it to a destination port. Therefore, timely decisions regarding spot orders and fleet assignments are crucial. The EM-like algorithm may be used to optimize such operations. This problem aims to maximize the overall profit of the company by subtracting the expenses incurred during empty sailings, such as transportation and fuel costs, from the total revenue generated.

Figure 7 below describes the solution representation of the tramp routing problem involving three distinct ships labelled as A, B, C and five cargoes. The simplest problem formulation involves a set of ships and cargoes that require service. The solution encoding may be represented as a 2D vector consisting of the shipping requests and ship allocations. At a port, ships either begin their journey to another port or terminate their journey if there are no further loading or unloading operations scheduled. The duration of each ship's voyage from one port to another is determined by the distance between the two ports and the speed of the ship. It is assumed that each port can accommodate only one ship for loading and unloading within a specified time. The distance lines may be generally computed using metric measures like the Euclidean distance and the vector $\vec{r}$ denotes the distance required to return.

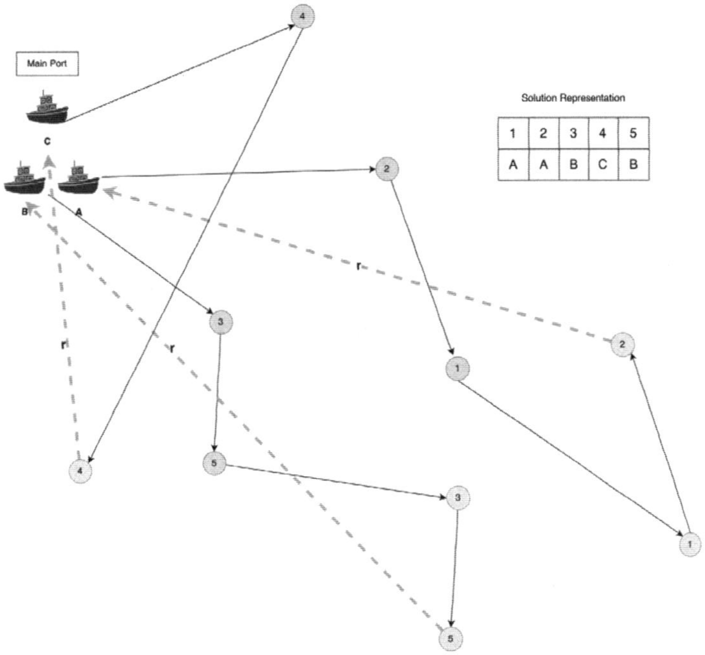

Figure 7. Problem and solution representation of the tramp routing problem

The method of local procedure and exploration of solutions in the EM-like will involve the different allocations of ships to cargoes using the heuristic procedure of equation (4). The attraction of cargoes will depend on the randomly assigned particle magnitude and the Euclidean distance specified between them. The algorithm's termination criteria may be based on the maximum iterations reached or a specific threshold of acceptable distance travelled given the constraints of the solution-finding process like immediate optimized route response and computational resources.

5. Conclusions

Based from the result of the simulation, modifying the boundary of particle movement makes significant difference in the search process of the EM-like; however, the study does not show any mathematical proof that it is indeed the particle movement boundary that causes the improvement of the search since how the permutations are modified per iteration, which is also crucial as the \ac{VRP} at hand is formulated as a permutation problem, is not detailed. It is only the solutions that are compared. However, based on the results, if the particle is set to move within a random boundary, while keeping other parameters constant, the solution obtained are better. There is an exception, however, since the αCBNL-o and αCBL-o displayed better values if the movement is only within [0,1]. Subjecting the algorithm with a local search procedure significantly improved the obtained values per iteration, with the highest improvement of 21%. This trend is evident to all the parameter scenarios. The study also recommends the implementation of capacity considerations in waste collection to ensure that the cost is also being minimized. Moreover, due to the minute difference and inconsistent results from the αCBNL-o and αCBL-h where particle boundary movement is modified, it is interesting for future work to study the behaviour of permutations across iterations between the two cases of the boundary of particle movement. For example, knowing how the permutations are re-arranged due to the interactions and modified movement can provide insight how well the mechanism obtain neighbour or diverse the decision variables.

Comparing these permutations may be done using metric such as Kendall-Tau and Spearman's Rho. To improve the speed convergence, it is suggested to modify parameters like random step length and particle boundary movement. It is also worth looking at the behaviour of the algorithm when the problem is formulated in multi-objective space or in terms of a penalty function to include the constraint of truck capacity. The multi-objective domain is more complex, so it is interesting to investigate how the algorithm re-formulation or modification is done to tailor the sophistication. In conclusion, the EM-like algorithm displayed significant minimization abilities of distance values. The traditional EM-like framework is used with local search and selection strategy implementation. It displayed 39 percent of solution improvement in the span of 71 minutes of computational completion time.

References

[1] Diaz L F 2017 Waste Management in Developing Countries and the Circular Economy *Waste Manage. Res.* **35**(1), 1-2.

[2] EPA 2020 *Best Practices for Solid Waste Management: A Guide for Decision-Makers in Developing Countries,* United States Environmental Protection Agency Office of Resource Conservation and Recovery. Available at: https://www.ccacoalition.org/resources/best-practices-solid-waste-management-guide-decision-makers-developing-countries.

[3] Group W B 2018 *Municipal Solid Waste Management: A Roadmap for Reform for Policy Makers* (1818 H Street NW Washington, DC 20433).

[4] Kolekar K A, Hazra T and Chakrabarty S N 2016 A Review on Prediction of Municipal Solid Waste Generation Models. *Procedia Environ Sci* **35,** 238–44.

[5] Fernández-Aracil P, Ortuño-Padilla A and Melgarejo-Moreno J 2018 Factors related to municipal costs of waste collection service in Spain. *J Clean Prod* **175,** 553–60.

[6] Agaton C B, Guno C S, Villanueva R O and Villanueva R O 2020 Economic analysis of waste-to-energy investment in the Philippines: A real options approach, *Appl Energ* **275,** 115265.

[7] Mmereki D, Baldwin A and Li B 2016 A comparative analysis of solid waste management in developed, developing and lesser developed countries *Environ Technology Rev* **5** 120–41

[8] Magalang A A 2013 Municipal Solid Waste Management in the Philippines, In: Municipal Solid Waste Management in Asia and the Pacific Islands, Challenges and Strategic Solutions,

Pariatamby A and Tanaka M (Eds.) *Environ Sci Eng,* 281–97, Springer. Available at: https://link.springer.com/book/10.1007/978-981-4451-73-4.

[9] Sulemana A, Donkor E A, Forkuo E K and Oduro-Kwarteng S 2018 Optimal Routing of Solid Waste Collection Trucks: A Review of Methods, *J Eng,* **2018,** 1–12

[10] Golden B L and Wong R T 1981 Capacitated arc routing problems, *Networks,* **11,** 305–15

[11] Tavares G, Zsigraiova Z, Semiao V and Carvalho M G 2009 Optimisation of MSW collection routes for minimum fuel consumption using 3D GIS modelling, *Waste Manage,* **29,** 1176–85

[12] Martinelli R, Poggi M and Subramanian A 2013 Improved bounds for large scale capacitated arc routing problem, *Comput. Oper. Res.* **40**(8), 2145 – 60.

[13] Benavent Enrique, Campos V, Corberán A and Mota E 1990 The capacitated arc routing problem. A heuristic algorithm, *Questtiio,* **14**(1-3), 107 -22.

[14] Pearn W L 1989 Approximate solutions for the capacitated arc routing problem

[15] Eydi A and Javazi L 2012 A novel heuristic method to solve the capacitated arc routing problem, *Comput. Oper. Res.* **16**(6), 589 – 600.

[16] Son L H 2014 Optimizing Municipal Solid Waste collection using Chaotic Particle Swarm Optimization in GIS based environments: A case study at Danang city, Vietnam *Expert Syst Appl,* **41,** 8062–74

[17] Gómez J R, Pacheco J and Gonzalo-Orden H 2013 A Tabu Search Method for a Bi-Objective Urban Waste Collection Problem

[18] Nevrlý V, Popela P and Šomplák R 2019 Heuristics for Waste Collection Arc Routing Problem. *Mendel – Soft Comput. J.* **25**(1), 15 – 22.

[19] Chang P-C, Chen S-H and Fan C-Y 2009 A hybrid electromagnetism-like algorithm for single machine scheduling problem *Expert Syst Appl* **36** 1259–67

[20] Sels V and Vanhoucke M 2014 A hybrid Electromagnetism-like Mechanism/tabu search procedure for the single machine scheduling problem with a maximum lateness objective *Comput Ind Eng* **67** 44–55

[21] Debels D, Reyck B D, Leus R and Vanhoucke M 2006 A hybrid scatter search/electromagnetism meta-heuristic for project scheduling *Eur J Oper Res* **169** 638–53

[22] Valls V, Ballestín F and Quintanilla S 2004 A Population-Based Approach to the Resource-Constrained Project Scheduling Problem *Ann Oper Res* **131** 305–24

[23] Yurtkuran A and Emel E 2010 A new Hybrid Electromagnetism-like Algorithm for capacitated vehicle routing problems *Expert Syst Appl* **37** 3427–33

[24] Demir E, Bektaş T and Laporte G 2012 An adaptive large neighborhood search heuristic for the Pollution-Routing Problem *Eur J Oper Res* **223** 346–59

[25] Lai D, Costa Y, Demir E, Florio A and Woensel T V 2021 The Pollution-Routing Problem with Speed Optimization and Uneven Topography *Arxiv* 2105.009229, Cornell University publication. Available at: https://arxiv.org/abs/2105.09229.

[26] Amodeo L, Talbi E and Yalaoui F (Eds.) 2018 *Recent Developments in Metaheuristics*, Springer. https://doi.org/10.1007/978-3-319-58253-5

[27] Birbil Ş İ and Fang S-C 2003 An Electromagnetism-like Mechanism for Global Optimization *J Global Optim* **25** 263–82.

IOP Conf. Series: Materials Science and Engineering 1294 (2023) 012043 doi:10.1088/1757-899X/1294/1/012043

Open material database for tensile test properties of additive manufacturing materials

A A Garcia-Granada *, **H. Rostro-González, J M Puigoriol-Forcada and G Reyes-Pozo**

Grup d'Enginyeria en Producte Industrial (GEPI), Institut Químic de Sarrià, Universitat Ramon Llull, Via Augusta 390, 08017 Barcelona, Spain

* Correspondeince: andres.garcia@iqs.url.edu

Abstract. In recent years, the investigation of material properties within additive manufacturing, also known as 3D printing, has gained significant research attention. The intricate interplay between numerous fabrication parameters and the resultant material properties of 3D-printed components has become crucial, particularly for enabling effective topology optimization. Considering this, we propose the establishment of an accessible open database. This repository stores a comprehensive collection of fabrication files corresponding to each distinct material and printer combination, accompanied by the outcomes of meticulous tensile testing. To support the research community, our initiative extends to the inclusion of material provider datasheets, facilitating comprehensive result comparisons. A standardized approach utilizing consistently applied strain rates is recommended, focusing on a compact dog bone specimen design. This pioneering attempt encompasses an expansive array of data derived from 25 distinct materials and 9 diverse printers, meticulously capturing the inherent variability within the samples. The database catalogues the complete spectrum of tensile test data, encompassing various essential measurements such as mass, and crucial material properties including Young's modulus, yield stress, fracture strain, and absorbed energy. These recorded metrics can be seamlessly correlated against density, manufacturing time, or cost parameters, enabling the generation of insightful plots and analysis. Through this collaborative effort, we aim to provide researchers with a robust foundation for informed decision-making and advancements in additive manufacturing.

1. Introduction

Additive manufacturing has emerged as a revolutionary technique, allowing the production of individual components without the reliance on traditional tooling methods such as molds for plastic injection or matrices for metal stamping [1,2]. For these materials you can find sdatabase such as Matweb [3], Campusplastics [4] and material selection software such as Granta Design [5]. Additive manufacturing is experiencing a rapid expansion in various applications, driven by the advent of novel printers, transformative technologies, and an evolving spectrum of materials [6,7]. Nevertheless, the successful utilization of additive manufacturing demands a comprehensive understanding of material properties, which exhibit notable variations contingent upon distinct manufacturing parameters [8]. For example, in the pursuit of optimal material properties for additive manufacturing, Garcia-Granada et al. [9] encountered challenges in implementing topology optimization. Their comparative study of two high-performance 3D printing machines revealed intriguing insights. The success of optimization was evident in the case of a Fortus 400mc employing Fused Deposition Modeling (FDM) technology, utilizing ABS-M30 material. However, the same level of optimization eluded the Polyjet Objet30 Prime printer, even with the utilization of Verowhite material. FDM is a trademarked term owned by Stratasys, a 3D printing company. Fused Filament Fabrication (FFF) is a generic term that describes the same technology.

In the realm of advanced manufacturing, FDM stands as a pioneering technique that holds immense promise for the fabrication of thermoplastic polymer composites. The synergy of thermoplastic polymers with composite materials presents a compelling avenue for achieving enhanced mechanical, thermal, and functional properties [10]. Domingo-Espin et al. [11] have made significant progression in unraveling the complexities of fused deposition modeling for polycarbonate parts. Through meticulous experimentation, they have discerned that the Young's modulus can fluctuate within a range of ±10%, a phenomenon attributed to the orientation during printing. Similarly, tensile strength exhibited variations of up to ±20%, while ultimate strain displayed a significant scatter, spanning from as low as 2% to a substantial 11.8%. On the tensile testing, a review of additively manufactured polymers is given in [12]. In this paper, it becomes evident that a significant gap exists in the realm of suitable standards designed to systematically evaluate the tensile properties of polymer components. This gap is particularly conspicuous when considering the intricate landscape of Fused Filament Fabrication (FFF), where the mechanical behavior is intricately governed by a multitude of printing parameters [13]. The inadequacy of existing standards becomes apparent in their inability to encompass the diverse array of influential factors that contribute to the mechanical characteristics of FFF-produced parts [14].

There is a need to document all sort of material properties for additive manufacturing such as thermal conductivity [15], viscoelastic properties [16], conductive properties [17].

The substantial diversity in printing parameters, printer models, and material properties is the focal point of our efforts in this work. To address this multifaceted challenge, we propose the creation of a comprehensive database that incorporates many of the key factors discussed here. This database is poised to serve as a foundational resource in the field of additive manufacturing.

Our initial iteration of this additive manufacturing database, denoted as Version 01, is now accessible through AMDB [18]. This repository stands as an invaluable asset, catering to the needs of both researchers and practitioners alike. It streamlines the exploration and integration of critical insights and data within the realm of additive manufacturing, offering a robust platform for advancing knowledge and innovation in this dynamic field.

2. Materials and methods

2.1. Geometry

Tensile tests are performed using a small geometry to minimise the use of material and manufacturing time. The geometric configuration is inspired by the ASTM_D638_V standard, with dimensions set at a thickness of 2 mm and width of 3.18 mm, culminating in a cross-sectional area of 6.36 mm². For precise specifications of an individual tensile specimen, refer to Figure 1a, which requires a material volume of 930 mm³. Length for strain calculation was calibrated using finite element analysis as shown in Figure 1c. For a displacement of $U = 1$ mm and $E = 2000$ MPa an average stress of $\sigma = 80$ MPa was observed in the cross-section area. Therefore, for deformation calculations ($\varepsilon = U/L$) we used the length $L = E*U/\sigma = 2000*1/80 = 25$ mm.

To streamline the process, an STL file encompassing four specimens has been curated and incorporated into the material database. This strategic optimization minimizes the required printing volume to 5541.19 mm³ (32.03 x 86.50 x 2), while material consumption remains confined to 3720 mm³ (4 x 930). The intricate details of the STL geometry earmarked for printing are meticulously outlined in Figure 1b, with slim 1.14 mm separations thoughtfully incorporated between specimens.

While the adoption of such diminutive specimens offers noteworthy advantages in terms of economizing both time and material resources, it is pertinent to acknowledge a limitation. Regrettably, due to their small size, the ability to investigate the implications of infill patterns is curtailed. This restriction is attributed to the preponderance of slicing utilized for generating outer surfaces, thereby warranting a judicious consideration of the experimental scope and its corresponding constraints.

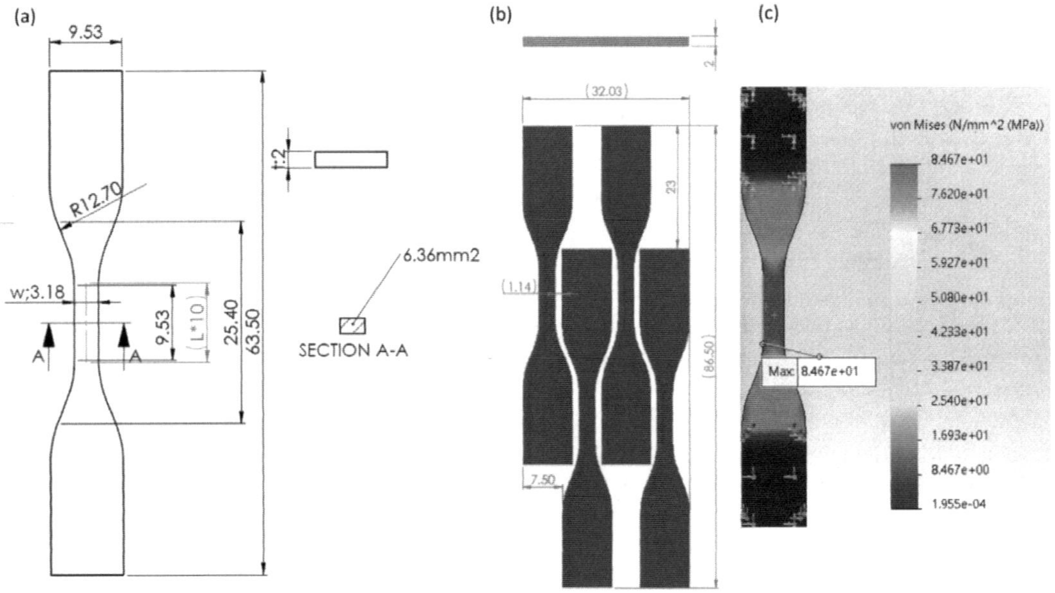

Figure 1. Geometry of (a) 1 tensile specimen and (b) downloaded STL for 4 tensile specimens and (c) length calibration.

2.2. 3D printers and slicing

The material database encompasses a diverse array of printers and slicing methods, accommodating a wide spectrum of choices. A primary premise of this database is to meticulously chronicle crucial details, including the specific material, printer model, slicing technique employed, and the associated manufacturing file. This comprehensive documentation is instrumental in fostering repeatability and ensuring the veracity of the results obtained.

In alignment with the scope of this paper, a specific subset of machines, slicing software, and materials has been judiciously considered, as outlined in Table 1. Each material, within the database, warrants a meticulous record encompassing the printer model utilized, the slicer employed, a visual representation of the slicing process, the manufacturing file itself, and, upon completion of testing, a minimum of four illustrative result images. The column run in material database makes reference to the specimen test run recorded in the tensile test machine.

By adhering to this meticulous documentation framework, the material database not only amplifies the value of this research but also lays a robust foundation for future investigations, augmenting the transparency, reliability, and transferability of findings across diverse additive manufacturing scenarios.

To exemplify the thorough documentation provided within the database, we present two snapshots showcasing the slicing process. Figure 2 captures a snapshot of Cura for Prusa, while Figure 3 displays a snapshot of Preform for Formlabs. These visual representations serve as illustrative instances, showing the intricate slicing strategies employed. The underlying intent is to facilitate direct comparisons, especially when different slicing software solutions are employed. By archiving these insights, the database delights a systematic approach to deciphering the impact of varying slicing methodologies, thus fostering a deeper understanding of the interplay between software choices and additive manufacturing outcomes.

Table 1. 3D printing machine, slicing and material

Technology	Slicing software	Machine	Materials	Run	Cost (€/min)	Cost (€/gr)
FFF / FDM	Cura 4.8.0	Ultimaker S5	PLA	135-138	0.02	0.1
			ABS	149-152	0.02	0.1
			TPU	145-148	0.02	0.1
			TPU+ABS	141-144	0.02	0.1
			PCL	186-189	0.02	0.1
		Prusa i3 mk3	PLA	115-119+121-126	0.01	0.1
			ABS	153-156	0.01	0.1
			PETG	191-194	0.01	0.1
			PLAFe	227-232	0.01	0.1
			PETG+PLA	196-199	0.01	0.1
		Ender 3v2	PHA210	238-240	0.01	0.1
			PHA190	234-237	0.01	0.1
	Axon 2	BFB 3000	PLA	131-134	0.01	0.1
			ABS	131b-134b	0.01	0.1
			PP0.5	204-207	0.01	0.1
			PP0.25	200-203	0.01	0.1
	Insight 14.4	Fortus 400mc	ABS-M30	241-244	0.30	1.0
	Cloudprint	Makerbot M.X	Nyloncarbon	176-179	0.10	0.25
SLA	Formlabs 3BL	Preform 3.25	FLEX80A	164-167+217-220	0.10	1.0
			BIO	180-183	0.10	1.0
			ELASTIC50A	168-172	0.10	1.0
			GREY	221-225	0.10	1.0
SLS	Sintratek Kit	Sintratec C1.2	TPU	233	0.02	0.1
Polyjet	Objet 30 Prime	Objet 9.2.8	Verowhite	127-130	0.10	0.25
			Med610	157-160	0.10	0.25
			Tangogrey	161-162+174-175	0.10	0.25

In Figure 2, the slicing process for FDM is depicted, showcasing an infill parameter configured at 100%. This deliberate choice is aimed at achieving the utmost density for the targeted layer height. Specifically, in the context of Draft mode, where layers are set at 0.2 mm thickness, the intricate composition involves the incorporation of 10 individual layers.

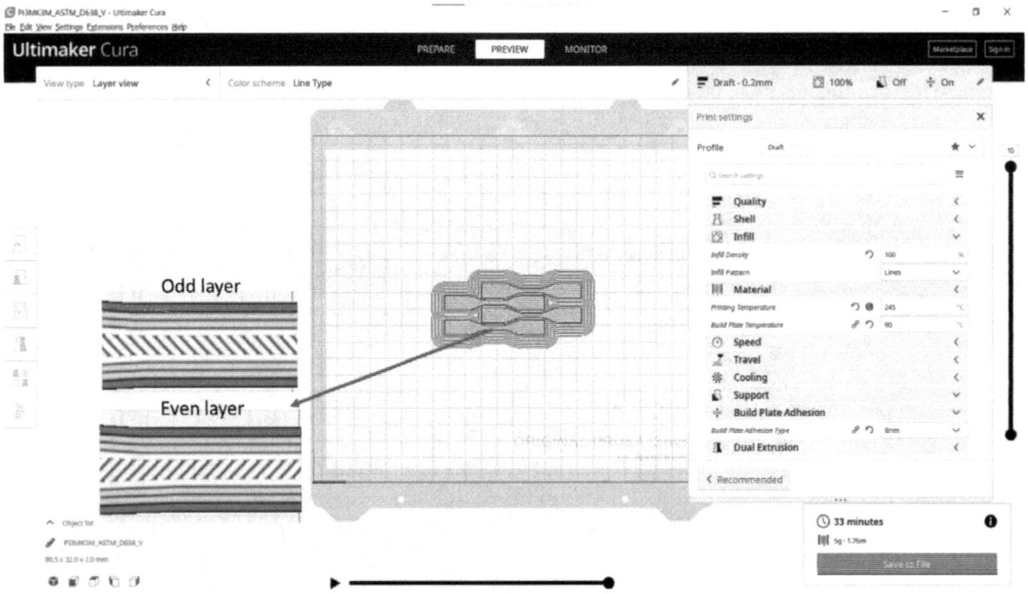

Figure 2. Slicing from Cura for Prusa with specimens manufactured in 33 minutes using 5 grams of material.

In Figure 3, the SLA slicing process is shown for Objet 30 Prime printer, encompassing 27 layers in total. Among these, 20 layers, each measuring 0.1 mm, constitute the part, while an additional 7 layers are dedicated to the base for secure platform attachment.

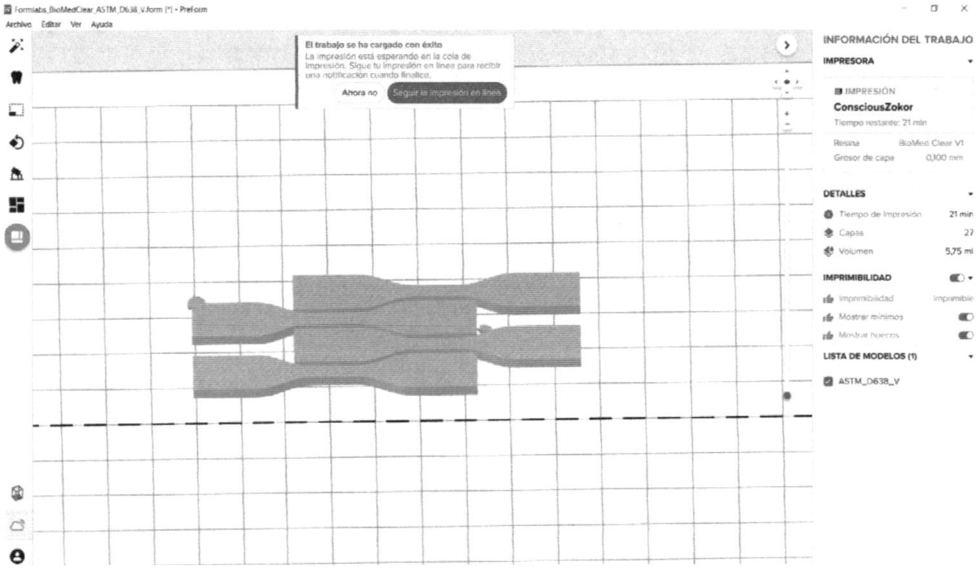

Figure 3. Slicing from Preform for Formlabs BioMed clear V1 specimens manufactured in 21 minutes using 5.75 grams of material.

2.3. *Time and cost*

For each individual setup involving four specimens, a multitude of adjustable parameters employs a discernible impact on manufacturing time. Delving into the realm of cost considerations, the calculations are meticulously derived from the operational experience garnered over the machine's working hours, aimed at spreading the acquisition cost across a seven-year utilization span. To illustrate, envision a machine operational for 500 hours annually. In this instance, the calculation unfolds as follows: 500 hours × 7 years = 3500 working hours. If the original machine cost is set to 7000€, this translates to a cost apportionment of 7000 € / 3500h = 2 €/h = 0.033 €/min. This column is shown in table 1 and can be changed in Excel file to be downloaded from the material database.

Encompassing these numbers, the equation further incorporates the inclusion of maintenance costs and energy expenditures, yielding the comprehensive cost per minute. Other overhead costs, including labour costs, are not considered because of the substantial differences among geographical regions. It is an attempt to standardize results for comparison purposes.

The essence of the database's design lies in its adaptability, offering users the latitude to configure their cost parameters in harmony with their distinct machine utilization patterns. This dynamic feature underpins the very essence of the database, placing the power to customize cost considerations squarely in the hands of individual users.

Expanding the ambit to material costs, the calculation includes the expenditure of acquiring material while considering the tangible utilization within a tank. On the realm of FDM filament, approximately 80% of material finds purpose, while a portion is forfeited to failed tests and material replacements. Similarly, for object's material, nearly 40% is lost during washing and material replacement. These empirical experiences intricately factor into the determination of real material costs. Therefore, the true cost of each material must be calibrated to reflect these practical realities, mirroring the authenticity of the user's experience.

2.4. Tensile test

Consistency and rigor establish the testing methodology for all tensile specimens, aligning with the established protocol outlined by ASTM D638: Standard Test Method for Tensile Properties of Plastics. Employing an MTS insight equipped with a 10 kN load cell, the testing transpires at a controlled displacement rate of 10 mm/min, meticulously capturing data at a frequency of 50 Hz. This data acquisition, amounting to 300 test points over a 1 mm displacement (1/25=4%) in 60 seconds, ensures a granular understanding of the mechanical response.

Each specimen's grip displacement, load cell force, and time are meticulously documented throughout the testing process, forming a comprehensive data repository. Stress and strain calculations are rendered feasible through engineering equations that presuppose low strain and a uniform cross-section. This foundational approach allows the derivation of these crucial mechanical properties with precision, laying the groundwork for insightful analyses and robust conclusions.

To calculate stress and strain, engineering equations assuming low strain and constant section. Therefore, stress is calculated as measured force divided by initial area $\sigma = \frac{F}{A}$ where σ is stress, F is force and A is cross section area. Strain is also calculated from measured displacement an initial calibrated length $\epsilon = \frac{U}{L}$ where ϵ is strain, U is measured displacement and L is calibrated length assumed to be constant during the test. Please note that changing in calculations L from 10 to 20 mm would imply to obtain a doubled Young's modulus. For all specimens the same L=25 mm was used even though some specimens showed early necking with deformation localised on small length. However, it was considered of interest to always compare the same force-displacement behaviour and therefore all tests used the same cross section and gauge length.

3. Results

The material database serves as a repository that stores an array of material tests, facilitating comprehensive comparisons across variables such as materials, printers, and costs. Essential mechanical properties like Young's Modulus, yield stress, ultimate stress, and energy can be juxtaposed for insightful analyses. Throughout all tests conducted, the highest recorded force reached 0.8 kN (approximately 125 MPa), which remained well within the range accommodated by the 10 kN load cell. However, certain soft materials, including Formlabs Flexible 80A and Objet TangoGray, exhibited a greater degree of noise in load measurements compared to the optimal capacity of the current load cell.

IOP Conf. Series: Materials Science and Engineering 1294 (2023) 012043 doi:10.1088/1757-899X/1294/1/012043

Despite encountering maximum loads of 0.04 kN (around 6.25 MPa) for these softer materials, the decision was made to retain the same load cell for the sake of maintaining uniformity across the comparisons.

Figure 4, shows a snapshot of the material database, illustrating the meticulous documentation of each printer and material. Each column is a material, printer and setup providing in each row material properties from literature and measured. This documentation encompasses slicing snapshots, downloadable manufacturing files, and material datasheets from providers, fostering seamless comparisons.

		MIN	MAX	acabado		1BFB-ABS	2BFB-PLA	3BFB-PPO_	4BFB-PPO_
				poca fuerz	0	1	2	3	4
L*(mm)	25			pendiente		BFB	BFB	BFB	BFB
w(mm)	3.18				CAD	ABS	PLA	PPO_25	PPO_5
t(mm)	2			Mat (g)	3.71	3.8	3.8	3.6	3.6
v(mm/min)	1			Sop (g)		0	0	0	0
PDF file	./ASTM_D638_V.pdf			Time (min)		16	16	19	19
STL file	./ASTM_D638_V.STL			Cost time (€/min)		0.01	0.01	0.01	0.01
				Cost mat (€/g)		0.1	0.1	0.1	0.1
				Cost sup (€/g)		0.1	0.1	0.1	0.1
	calcul	0.54 €	11.00 €	Cost (€)		0.54 €	0.54 €	0.55 €	0.55 €
not doc	datasheet	26	200000	E (MPa)		1681	3309	640	640
	datasheet	2	250	sy(MPa)		39	50	19	19
	datasheet	2	560	su(MPa)		40	53	25	25
	datasheet	5	600	ef(%)		5	6	600	600
	calcul	0.125	33.07692	ey(%)		2.320048	1.511031	2.96875	2.96875
	calcul	1.6536	12720	Fe1%(N)		106.9116	210.4524	40.704	40.704
	calcul	0.25	0.25	U1%(mm)		0.25	0.25	0.25	0.25
	calcul	15	15	t1%(s)		15	15	15	15
	calcul	12.72	1590	Fy(N)		248.04	318	120.84	120.84
	calcul	0.03125	8.269231	Uy(mm)		0.580012	0.377758	0.742188	0.742188
	calcul	1.875	496.1538	ty(s)		34.80071	22.66546	44.53125	44.53125
	calcul	12.72	3561.6	Fu(N)		254.4	337.08	159	159
	calcul	1.25	150	Uu(mm)		1.25	1.5	150	150
	calcul	75	9000	tu(s)		75	90	9000	9000
	Fabrication snapshot					./BFB_ABS	./BFB_PLA	./BFB_PPO	./BFB_PPO
	Fabrication file					./BFB_ABS	./BFB_PLA	./BFB_PPO	./BFB_PPO
	Fuente Datasheet					https://w	https://fila	https://3d	https://3d
	Resultados foto			24		./BFB_ABS	./BFB_PLA	./BFB-PPO	./BFB-PPO
	measured	1	16.84	m(gr)		3.36	4.97	3.49	3.49
	calcul	0.269542	4.539084	d(gr/mm3)		0.90566	1.339623	0.940701	0.940701
	measured	3.930818	2489.518	E(MPa)		1179.245	1965.409	589.6226	252.6954
	measured	0.786164	124.4759	E+(MPa)		58.96226	98.27044	29.48113	26.53302
	measured	0.786164	59.74843	sy(MPa)		23.58491	55.03145	23.58491	15.72327
	measured	0.117925	3.930818	sy+(MPa)		1.179245	2.751572	1.179245	3.930818
	measured	0.786164	55.03145	su(MPa)		22.01258	55.03145	26.72956	15.72327
	measured	0.117925	3.144654	su(MPa)		1.100629	2.751572	1.336478	0.786164

3DP-Materials | Control | Elastic Modulus vs € | Elastic Modulus vs density | Yield stress vs € | Yield stress vs density

Figure 4. Example of data on material database.

The material database further facilitates the generation of diverse graphs, exemplified in Figure 5, which plots yield stress against cost. With the added benefit of an available EXCEL file for the material database, the manipulation of material and printer costs becomes an effortless aim. This inherent flexibility empowers users to refine outcomes based on their unique considerations, thus enhancing the utility and adaptability of the material database.

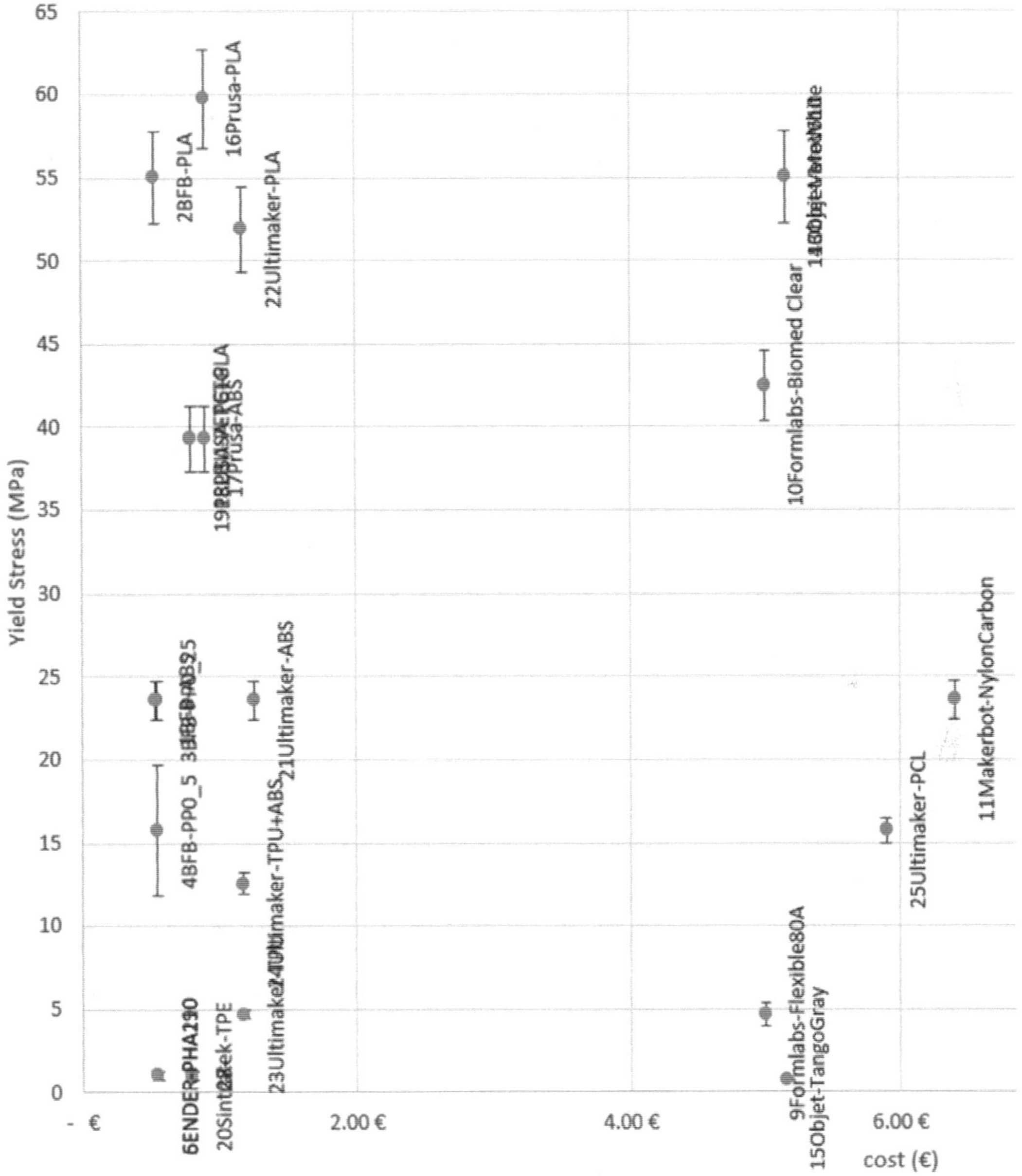

Figure 5. Example of chart showing measured yield stress versus cost.

The material database is producing results in a simplified model material based on a straight elastic line of Young's modulus E up to yield stress σ_y and another straight line to ultimate stress σ_u and fracture strain ε_f. Figure 6, shows 9 different tensile tests performed on Prusa-PLA material with the same fabrication file. Finally, a PRUSA PLA-model is obtained considering E = 2490±125 MPa, σ_y = 59.5±3 MPa, σ_u = 52.0±2.5 MPa and ε_f = 4.0±0.8 %. These error bars are also shown in Figure 5.

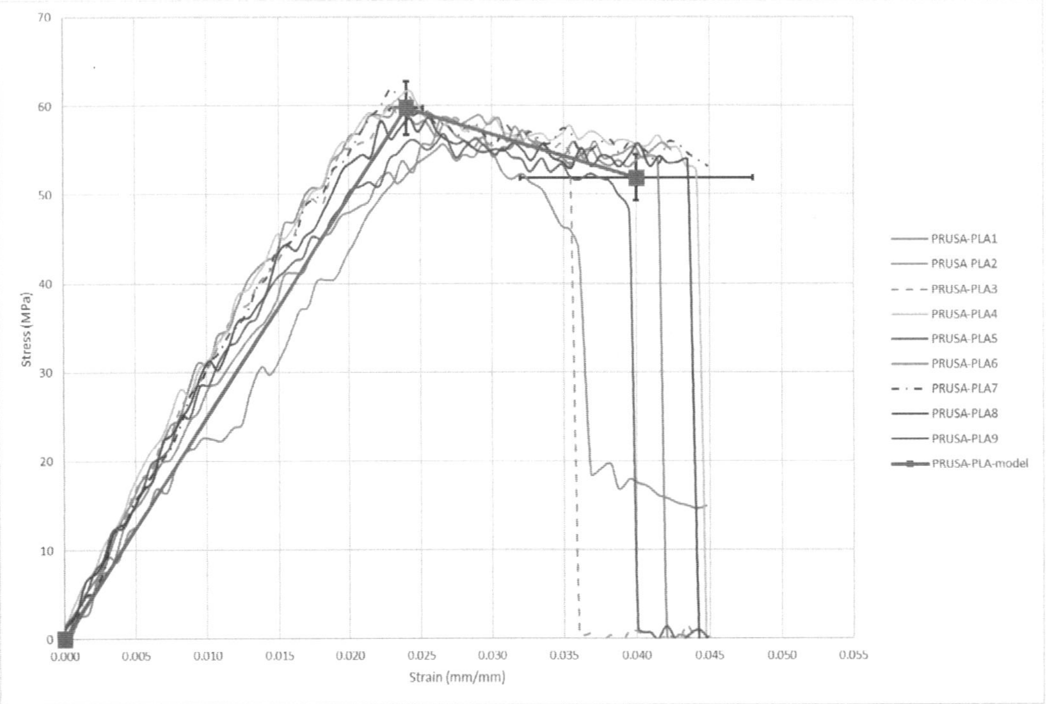

Figure 6. Stress strain curve for PRUSA PLA specimens to obtain variability.

After establishing an extensive overview of the fundamental parameters extracted from various sets of tensile curves, the subsequent step involves a comprehensive analysis of these parameters in relation to a range of relevant factors. These factors encompass density, cost, and variability, among others. Through this multi-faceted analysis, a deeper understanding emerges, shedding light on the intricate correlations and dependencies that underlie the mechanical behavior of the printed materials.

3.1. Elastic modulus

The elastic modulus, a fundamental metric, is derived by calculating the slope of the stress-strain curve. While there are distinct definitions for tangent and secant modulus within the realm of polymers, we opt for simplicity by employing a linear fit, illustrated in Figure 6, up to the yield stress.

Upon conducting this analysis across all specimens, the outcomes are synthesized into plots that offer insightful visualizations. These plots showcase the elastic modulus as a function of density in Figure 7, and as a function of cost in Figure 8. To elucidate the relationship between elastic modulus and density, reference lines are incorporated, demonstrating a proportional increase—when density doubles, elastic modulus also doubles.

Remarkably, within this framework, Ultimaker-PLA emerges as an exemplar, exhibiting a particularly favorable elastic modulus-to-density ratio. The same methodology is applied to cost considerations, revealing BFB-PLA's supremacy in the elastic modulus-to-cost ratio.

In essence, this systematic procedure not only determines the interplay between elastic modulus, density, and cost but also covers it within a visual narrative. These comparative insights serve as a compass, steering decisions towards materials that offer optimal trade-offs between elastic modulus, density, and cost parameters.

IOP Conf. Series: Materials Science and Engineering 1294 (2023) 012043 doi:10.1088/1757-899X/1294/1/012043

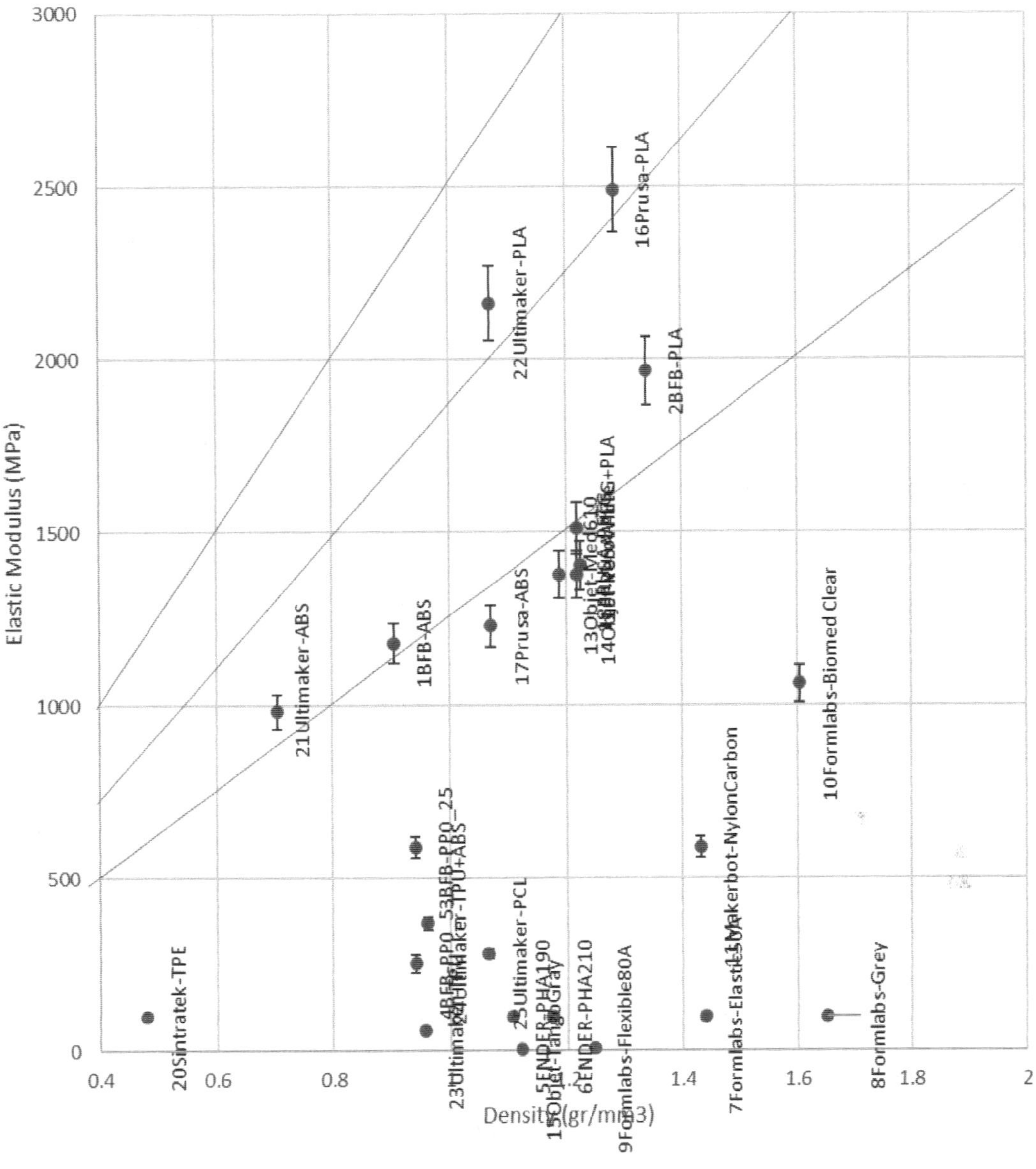

Figure 7. Elastic modulus versus density.

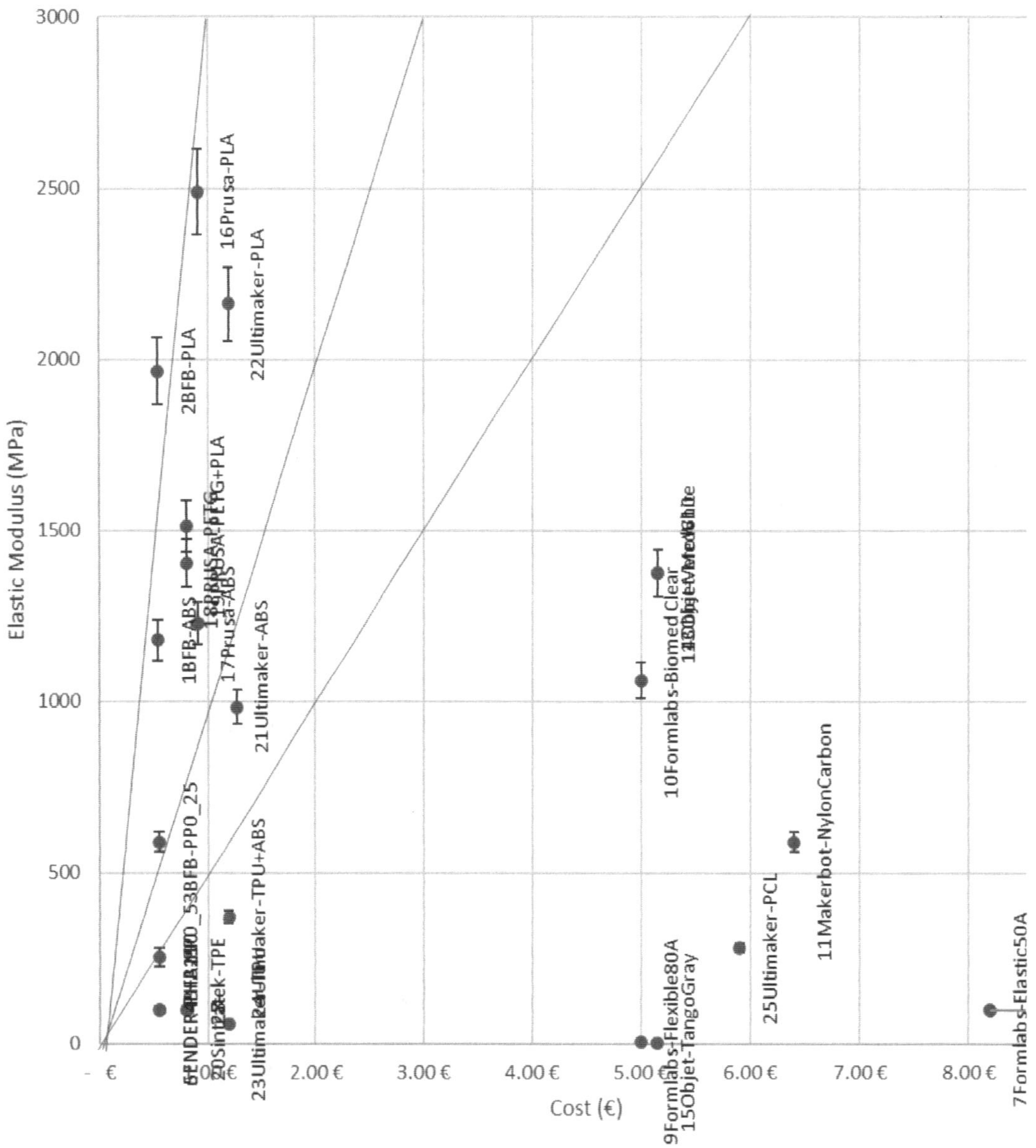

Figure 8. Elastic modulus versus cost.

3.2. Yield stress

Yield stress is deduced through the simplification of the stress-strain curve. As depicted in Figure 9, a discernible pattern emerges: Ultimaker-PLA boasts the most favorable yield stress-to-density ratio. Similarly, Figure 10 unveils another crucial insight: BFB-PLA emerges as the leader in the yield stress-to-cost ratio.

IOP Conf. Series: Materials Science and Engineering 1294 (2023) 012043 doi:10.1088/1757-899X/1294/1/012043

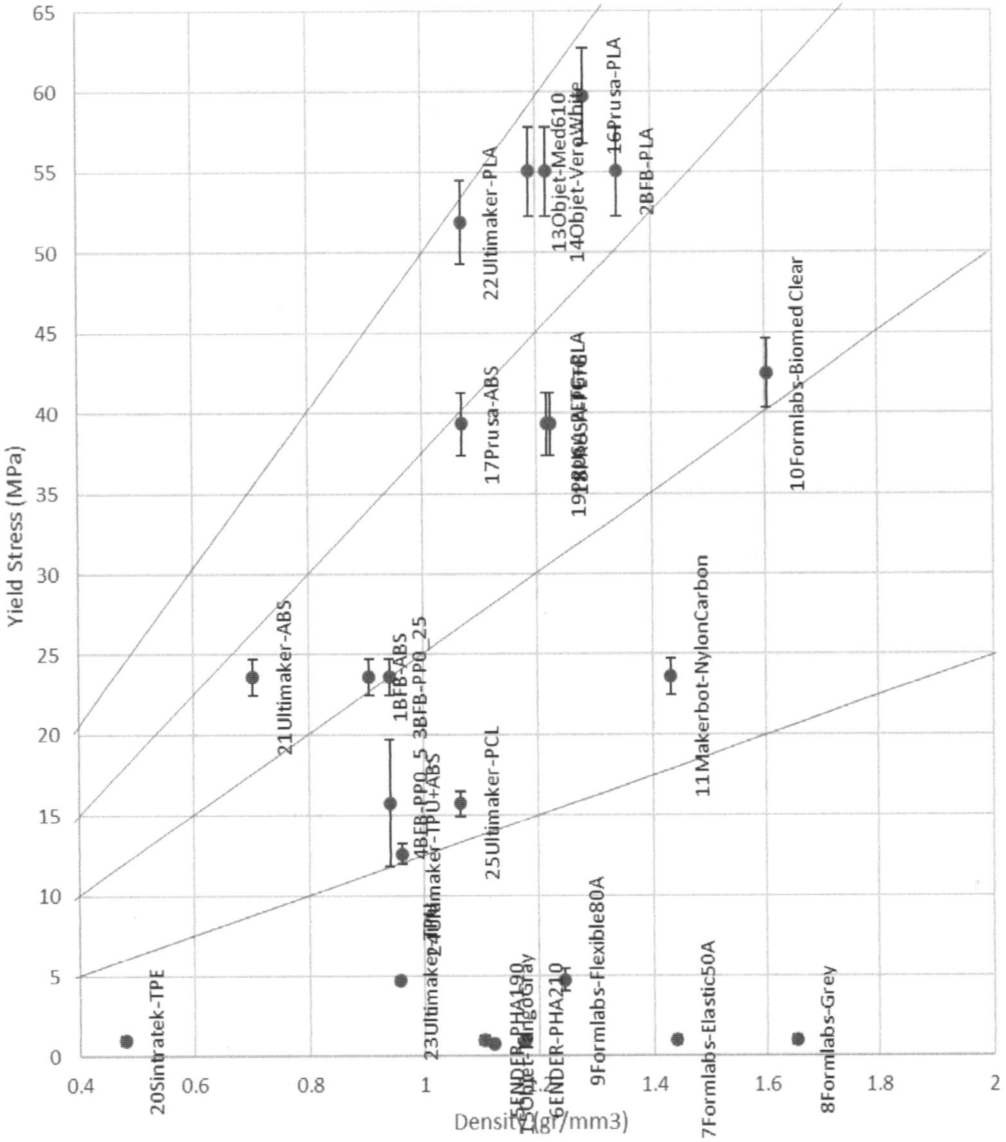

Figure 9. Yield stress versus density.

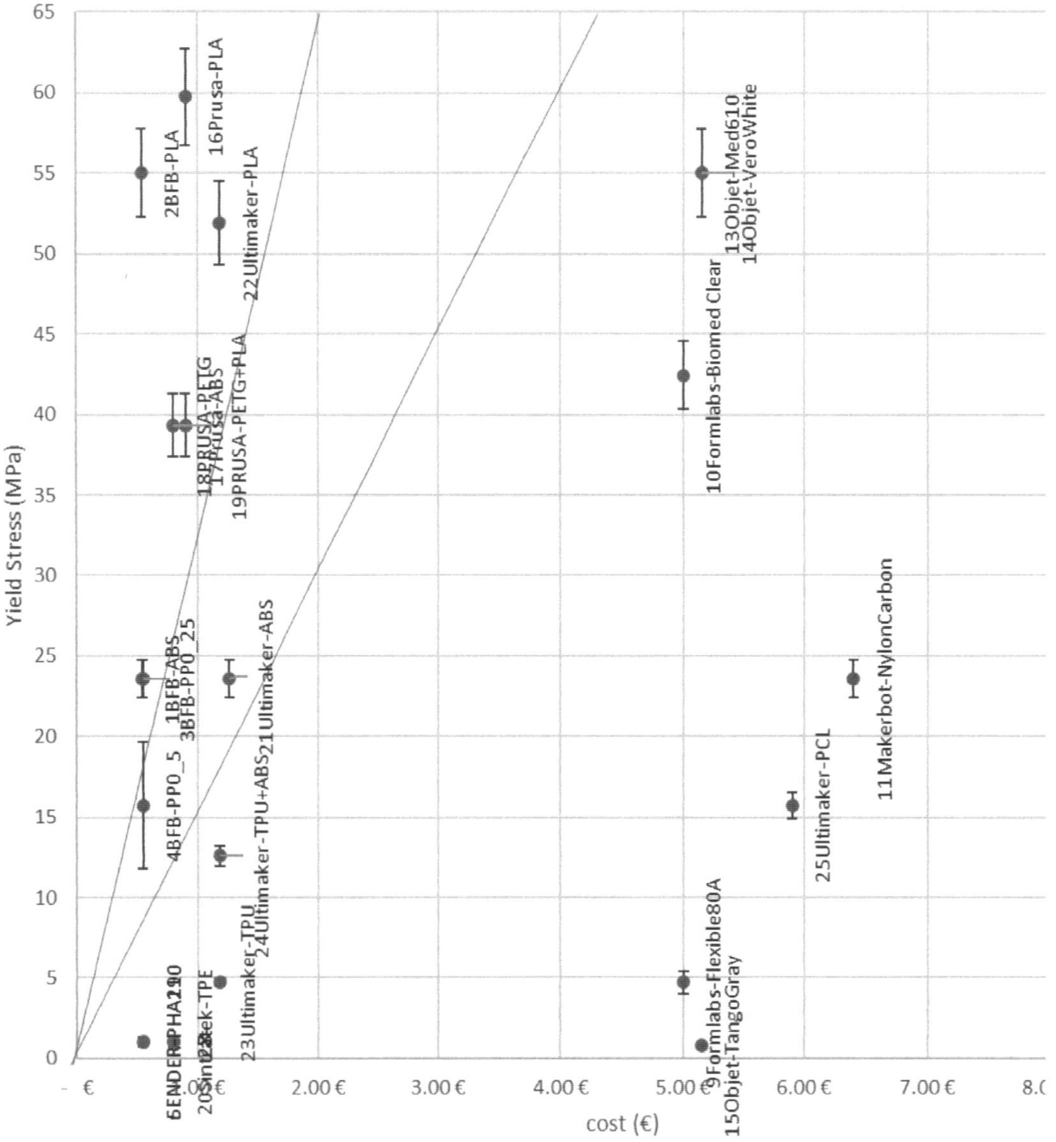

Figure 10. Yield stress versus cost.

3.3. *Energy*

Energy is computed by quantifying the area below the stress-strain curve, expressed in units of N/mm^2, mJ/mm^3, or MJ/m^3. To enable more meaningful comparisons, we normalize this energy by dividing it by the actual density of the component, resulting in units of mJ/g or J/kg. Notably, the highest energy yield is achieved by BFB-PP, a standout performer for both 0.5 mm and 0.25 mm layer thicknesses, as depicted in Figure 11. This elevated energy output can be attributed to the substantial deformation witnessed prior to fracture.

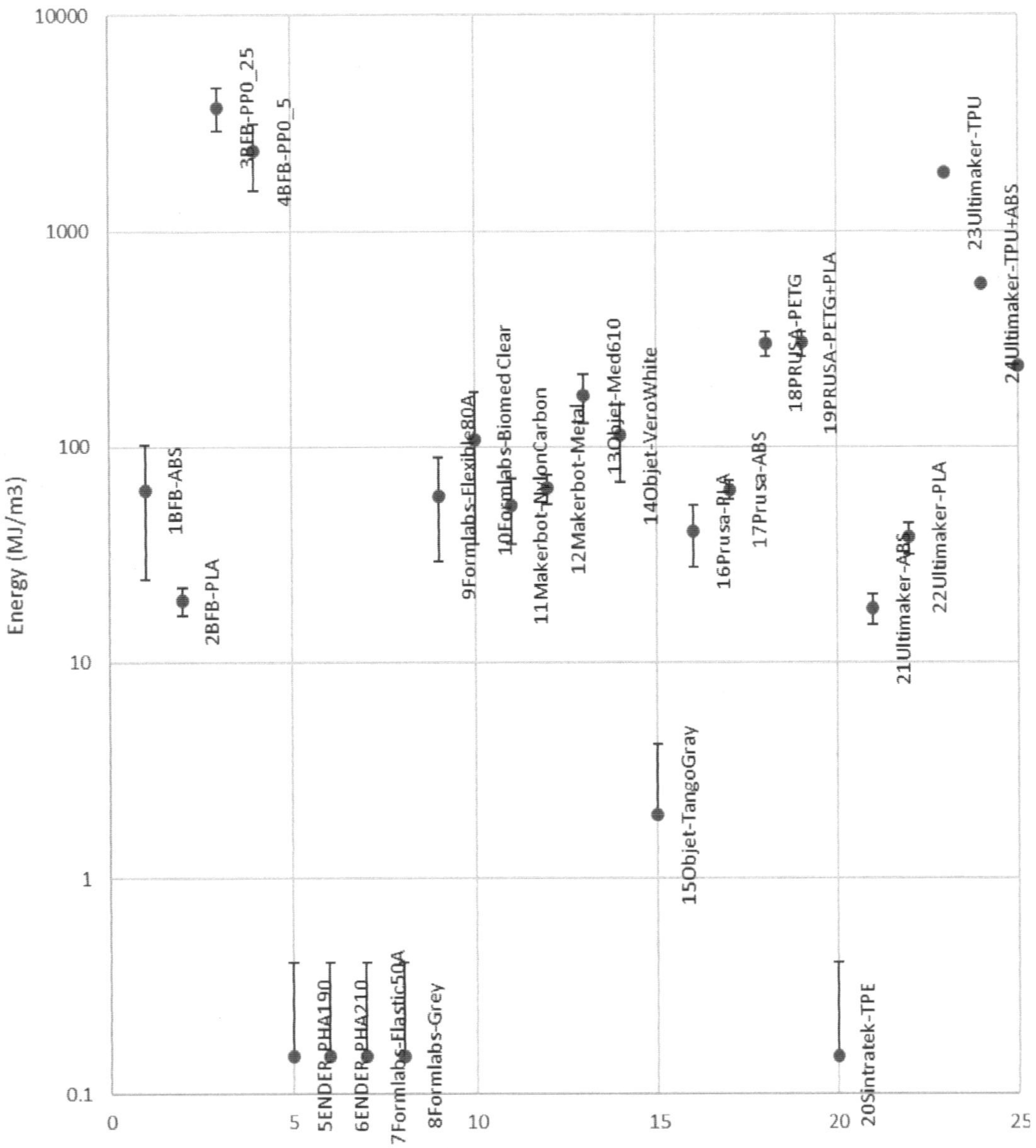

Figure 11. Energy for each material tested.

Interestingly, upon normalization by density, Figure 12 illustrates that there isn't a substantial shift in the observed trends. This suggests that, even when considering the influence of density, BFB-PP retains its position as the leader in energy-to-mass ratios. This stability underscores the robustness of BFB-PP's performance across varying contexts, further solidifying its position as an exceptional energy-absorbing material.

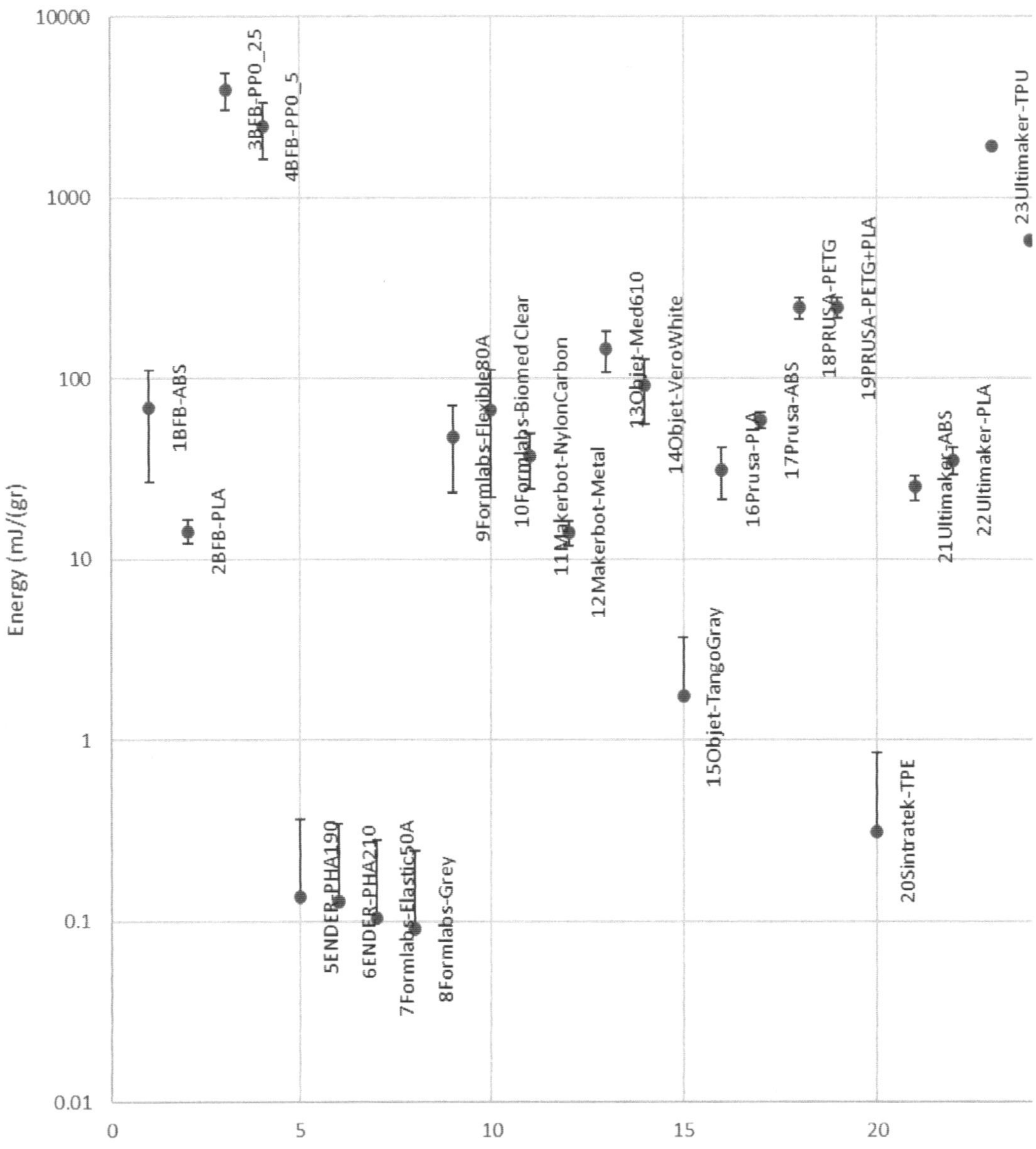

Figure 12. Energy for each material tested normalised for weight.

3.4. *Stress strain for simplified models*

Concluding our analysis, stress-strain relationships for all materials are visually represented using simplified models and a logarithmic scale, enclosed in Figure 13. The congruence with our previous findings is immediately apparent: Prusa-PLA prominently exhibits the highest Young's Modulus and yield stress, mirroring the observations in Figure 7. Similarly, the exceptional fracture strain of BFB-PP, as illustrated in Figure 11, is reaffirmed in this comprehensive depiction.

It is, however, prudent to acknowledge the unique case of Objet TangoGray and Formlabs-Elastic50A. Their values, although presented, warrant a note of caution as they should ideally be validated using a more suitable and smaller load cell. This consideration speaks to the inherent complexities that can arise when handling materials of distinct softness or flexibility and serves as a reminder of the nuanced approaches often required in their examination.

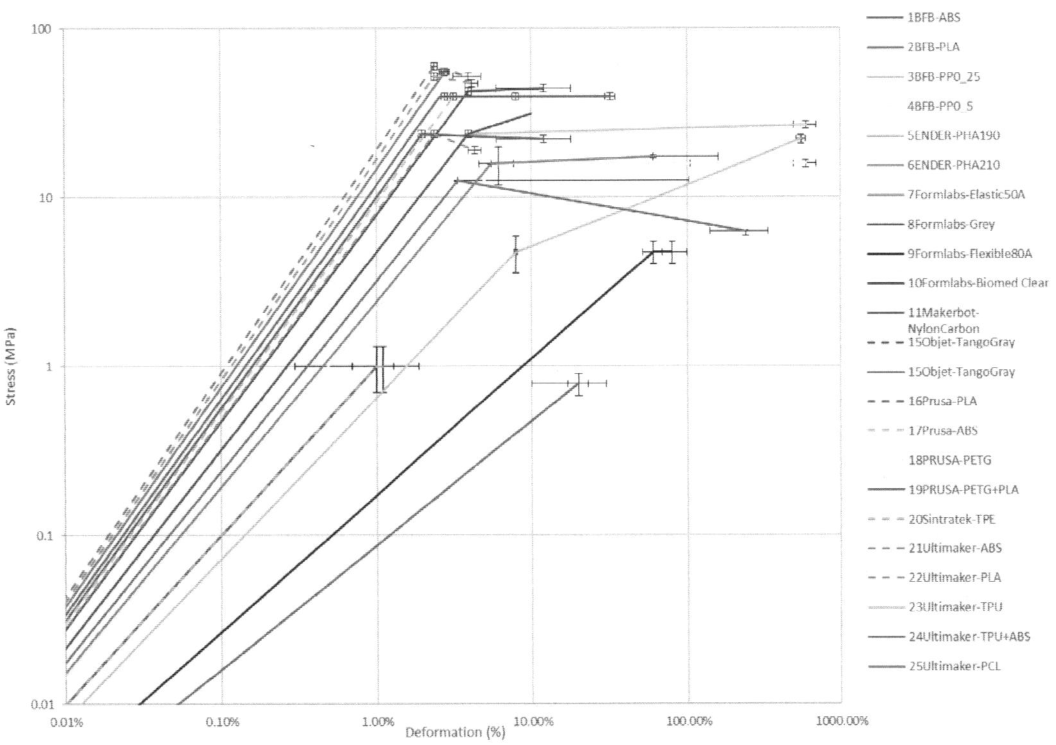

Figure 13. Stress strain for simplified models.

4. Conclusions

Concluding this study, we have constructed a comprehensive material database that establish the principle of reproducibility. In this repository, a minimum of four specimens per fabrication file have been meticulously tested in the lab and documented. The adoption of a smaller specimen design for these tensile tests has allowed for reasonable resource management in terms of material and manufacturing time. However, it's important to note that this choice does impose limitations on certain parameters, particularly regarding infill design for experiments.

The material database is thoughtfully designed to facilitate user interaction and adaptation. Available for download in Excel format, it offers the flexibility to adjust material selections and manufacturing costs, thereby aligning with individual user experiences and needs.

From our initial set of experiments, these valuable conclusions have emerged:

- The variation within Young's modulus, yield stress, and ultimate stress remains acceptable within a 10% range, but there's a noticeable 80% scatter in fracture strain, significantly impacting energy absorption.
- Among printed specimens, Ultimaker-PLA leads in the ratio of Young's modulus to density, followed by Prusa-PLA.
- In terms of the ratio of Young's modulus to IQS (industrial quality standards) cost, BFB-PLA takes the lead, trailed by Prusa-PLA.
- Yield stress to density ratio highlights Ultimaker-PLA as a frontrunner, with Objet-MED and Objet-VeroWhite trailing closely.
- The ratio of yield stress to IQS cost identifies BFB-PLA as superior, followed by PRUSA-PLA.
- BFB-PP emerges as the top contender for energy absorption, evident for both 0.5 mm and 0.25 mm layer thickness.

- The creation of simplified stress-strain curves, presented in logarithmic scale with error bars, provides a succinct yet comprehensive means of comparing all materials.

This database marks a foundational step towards establishing a repository for reproducible experiments, supported by consistent testing procedures. As we advance, this resource holds the potential to foster deeper insights, impulse innovation, and facilitate the development of new materials and processes within the realm of additive manufacturing.

References

[1] Ngo T D, Kashani A, Imbalzano G, Nguyen K T Q and Hui D 2018 Additive manufacturing (3D printing): A review of materials, methods, applications and challenges, *Composites, Part B* **143**, 172-196, doi:10.1016/j.compositesb.2018.02.012.

[2] Bhuvanesh K M and Sathiya P 2021 Methods and materials for additive manufacturing: A critical review on advancements and challenges, *Thin-Walled Struct.* **159**, doi:10.1016/j.tws.2020.107228.

[3] Matweb Material Database 2023, accessed 20 September 2023, https://www.matweb.com/,

[4] Campusplastics material information system for the plastics industry 2023, accessed 20 September 2023, https://www.campusplastics.com/

[5] Grantadesign material selector database 2023, accessed 20 September 2023, https://www.grantadesign.com/

[6] Bourell D, Kruth J P, Leu M, Levy G, Rosen D and Beese A M 2017 Materials for additive manufacturing, *CIRP Ann.* **66**, 659-681, doi:10.1016/j.cirp.2017.05.009.

[7] Calignano F, Manfredi D, Ambrosio E P, Biamino S, Lombardi, M, Atzeni E, Salmi A, Minetola P, Iuliano L and Fino P 2017 Overview on Additive Manufacturing Technologies, in *Proceedings of the IEEE*, **105**(4), 593-612, April 2017, doi: 10.1109/JPROC.2016.2625098.

[8] Kumar R, Kumar M and Chohan J S 2021 Material-specific properties and applications of additive manufacturing techniques: a comprehensive review, *Bull Mater Sci* **44**, 181. doi:10.1007/s12034-021-02364-y.

[9] Garcia-Granada A A, Catafal-Pedragosa J and Lemu H G 2019 Topology optimization through stiffness/weight ratio analysis for a three-point bending test of additive manufactured parts, in *IOP Conference Series: Materials Science and Engineering*, **700**(1), doi: 10.1088/1757-899X/700/1/012012.

[10] Penumakala P K, Santo J, and Thomas A 2020 A critical review on the fused deposition modeling of thermoplastic polymer composites, *Composites, Part B*, **201**, doi:10.1016/j.compositesb.2020.108336.

[11] Domingo-Espin M, Puigoriol-Forcada J M, Garcia-Granada A A, Llumà J, Borros S and Reyes G 2015 Mechanical property characterization and simulation of fused deposition modeling Polycarbonate parts, *Mater. Des.* **83**, 670–677, doi: 10.1016/j.matdes.2015.06.074.

[12] Sola A, Chong W J, Simunec D P, Li Y, Trinchi A, Kyratzis I and Wen C 2023 Open challenges in tensile testing of additively manufactured polymers: A literature survey and a case study in fused filament fabrication, *Polym. Test.* **117**, doi:10.1016/j.polymertesting.2022.107859.

[13] Fico D, Rizzo D, Casciaro R and Esposito C C 2022 A Review of Polymer-Based Materials for Fused Filament Fabrication (FFF): Focus on Sustainability and Recycled Materials. *Polymers (Basel).* Jan 24;**14**(3):465. doi:10.3390/polym14030465.

[14] Arifvianto B, Satiti B E, Salim U A, Nuryanti A and Mahardika M 2022 Mechanical properties of the FFF sandwich-structured parts made of PLA/TPU multi-material. *Prog Addit Manuf.* **7**, 1213–1223. doi:10.1007/s40964-022-00295-6.

[15] Dimla E, Rull-Trinidad J, Garcia-Granada A A, and Reyes G 2018 Thermal comparison of conventional and conformal cooling channel designs for a non-constant thickness screw cap, *J. Korean Soc. Precis. Eng.* **35**,(1), doi:10.7736/KSPE.2018.35.1.95

[16] Salazar-Martín A G, García-Granada A A, Reyes G, Gomez-Gras G, and Puigoriol-Forcada J M 2020 Time-Dependent Mechanical Properties in Polyetherimide 3D-Printed Parts Are

Dictated by Isotropic Performance Being Accurately Predicted by the Generalized Time Hardening Model, *Polymers (Basel)* **12**(3),678, Mar., doi:10.3390/polym12030678.

[17] Criado-Gonzalez M, Dominguez-Alfaro A, Lopez-Larrea N, Alegret N, and Mecerreyes D 2021 Additive manufacturing of conducting polymers: recent advances, challenges, and opportunities, *ACS Appl. Polym. Mater.* **3**(6), 2865-2883, doi:10.1021/acsapm.1c00252.

[18] AMDB Additive Manufacturing Data Base 2023, accessed 20 September 2023, https://meaagg.com/AMDB/AMDB.html

IOP Conf. Series: Materials Science and Engineering 1294 (2023) 012044 doi:10.1088/1757-899X/1294/1/012044

Bio-inspired design trends for sustainable energy structures

Y Regassa[1*], T Dabasa[1], G Amare[1] and H G. Lemu[2]

[1] Addis Ababa Science & Technology University, Sustainable energy center of Excellence, Addis Ababa, Ethiopia.
[2] University of Stavanger, N-4036 Stavanger, Norway

*Correspondence: yohannes.regassa@aastu.edu.et

Abstract. Researchers and engineers are constantly looking to nature for design inspiration as they work to create novel and effective energy conversion structures. Engineers are developing durable and resilient materials to survive extreme weather conditions and environmental issues using inspiration from the Nenuphar's durability in adverse aquatic environments and the Sunflower's adaptation to vary climatic situations. The analysis of new design ideas in renewable energy systems in this research was inspired by the Nenuphar (Water Lily) and the Sunflower. The Nenuphar's exceptional capacity to float and large leaf surface make it a model for enhanced water-based energy conversion systems. Researchers have been looking into novel concepts like floating solar farms and wave energy collecting platforms after being inspired by the Nenuphar's buoyancy and ability to efficiently use sunlight. Solar panel integration with water features can boost overall energy output while preserving precious land resources. The Sunflower, known for its phototropism and effective sun-tracking mechanism, offers suggestions for improving solar energy harvesting. Advanced heliotropic solar panels have demonstrated promising results in boosting energy production throughout the day by mimicking the Sunflower's ability to follow the trajectory of the Sun. These designs maximize energy outputs and boost overall system efficiency by maximizing solar exposure. Embracing these bio-inspired concepts can lead to the development of sustainable and efficient energy solutions, contributing to a greener and more sustainable future. Furthermore, the integration of smart technology and artificial intelligence is on the rise to optimize energy conversion structures.

Keywords: Bio-inspired, Energy conversion, Nenuphar, Sunflower, Sustainable solution

1. Introduction

Nature-inspired energy conversion design systems aim to provide sustainable solutions that minimize environmental impact [1] while optimizing energy production and utilization [2], [3]. These approaches can help address the challenges of climate change and promote a greener and more sustainable energy future [2]. Many researchers and engineers are actively exploring these concepts to create a positive impact on the planet [4].

Natural systems are self-optimizing, consequently structural engineers can learn a lot from them. Structural engineers are inspired by nature when designing and constructing a structure with a high degree of adaptability and little need for maintenance [5]. The global push towards renewable energy sources has led to innovative designs in energy conversion structures [6]. A 2MW integrated floating-wind turbine system was deployed as part of the Floatgen European research project on offshore wind energy [7]. It has been operating successfully since then, and it has paved the way for the development of other floating wind farms [7,8].

Many structural engineers have been driven and inspired to look outside of the box for creative problem-solving approaches by bio-inspired design [9]. Bio-inspired designer concur that biomimetics has not yet undergone a thorough investigation [4, 10]. Numerous engineering applications require the availability of lightweight structures with great energy absorption capability, as is well known [1, 4, 11]. Bio-inspired structures have been demonstrated to offer a considerable improvement in energy absorption capacity over conventional structures. These structures are inspired by various biological structures seen in nature. As a result, engineering fields have been using the biomimetic approach more frequently in recent years to design innovative, lightweight structures with outstanding energy absorption capability [1, 9]. Bio-inspired energy conversion structure design trends promise to revolutionize renewable energy technology and play a crucial role in combating climate change , the contribution of Bio-inspired design method is potential [12]. This review paper aims to provide a comprehensive design and production trends driven by nenuphar and sunflower structural features.

2. State of bio-inspired energy conversion structural design

Nature has served as a perpetual muse for architects and engineers, offering timeless lessons in efficient design and resource optimization. Sunflowers and water lilies, two iconic natural wonders, have inspired innovative structural design solutions that combine beauty with functionality. Bio inspired engineering design is the process of developing innovative and sustainable solutions for a range of problems, including energy technology, by taking cues from natural processes, patterns, and systems [13].

2.1 Biomimicry in structural design for renewable energy

Biomimetic design provides practical benefits and frequently leads to more durable as well as more ecologically friendly solutions by modeling performance and adaptability upon the principles of nature. In line with various energy conversion techniques, some of the prospective designs for renewable energy structures that were inspired by nature were thus outlined hereunder.

Bio-inspired Wind Turbine Design: Drawing inspiration from the aerodynamics of bird wings and the shape of tree leaves, wind turbine designs could mimic natural forms to improve efficiency and reduce noise [2,3, 9]. Studying the way birds and trees adapt to wind conditions can lead to more streamlined and silent turbine designs. In several industries, including renewable energy, biomimicry—the practise of using cues from nature to address human problems—has emerged as a powerful technique. In the design of wind turbines, biomimicry shows a lot of potential [4].

Researchers may create wind turbines that are more sustainable, effective, and environmentally friendly by researching and modelling the efficiency and flexibility of natural systems. Scholars are undertaking to examine how biomimicry might revolutionize wind turbine design, resulting in cleaner and more efficient energy-saving solutions. Birds have evolved to efficiently navigate through the air, minimizing energy consumption during flight. Wind turbine blades can take inspiration from the shape and structure of birds' wings to enhance aerodynamics. Many birds can fly over great distances because of the highly effective lift that their wings provide. The cambered form and swept-back leading edge of bird wings, as well as other aerodynamic characteristics, can be used to construct wind turbine blades [6]. This may help to increase the blades' effectiveness and cut down on energy use. The shape of a bird wing and a wind turbine blade are comparable, as seen in Figure 1. The shape of the bird wing is an airfoil with a cambered surface, which indicates that the top surface is more curved than the bottom surface. The force that keeps the bird flying is lift, which is helped by its form. The form of the wind turbine blade is also cambered.

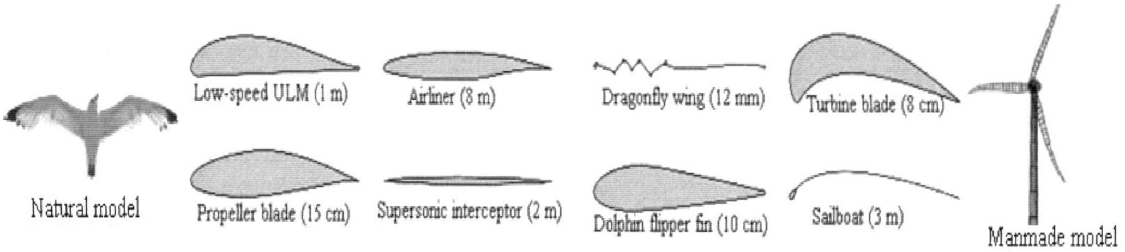

Figure 1. Bio-inspired wind turbine blade cross-section

By mimicking the flexible and adaptive wing movements, turbine blades can automatically adjust their angles and shapes to optimize energy capture under varying wind conditions. This biomimetic approach would lead to improved efficiency and stability of wind turbines, maximizing energy output. The need for the development of clean and renewable energy sources by world communities has never been more important. The use of palm tree leaves to design more durable wind turbine blades is one of the opportunities to develop adaptable engineering structures seen in Figure 2. Practitioners were inspired to develop flexible wind blades due to the fact that palm tree leaves have showed abilities to withstand strong winds and heavy rain without breaking.

Figure 2. Palm tree inspired flexible wind turbine blades.

Imitating the structures and textures of leaves, wind turbine designers can also reduce noise pollution caused by the blades. This biomimetic approach not only improves the overall aesthetic appeal of wind farms but also enhances public acceptance by mitigating the potential impact on nearby communities for generation of sustainable energy. Since palm tree leaves are made of a lightweight strong material and the foundation also anchored with many roots which can be an evidence for the state of wind turbine engineering design practices.

Engineers/Scientist has adapted solar panels to the efficient leaf structure and photosynthesis-enabling light-harvesting capabilities of leaves. Photovoltaic panels are analogous to green plant leaves in nature. To split water and produce chemical energy carriers, they absorb solar light and turn its energy into electricity (electrochemical energy). The chemical structure and function of biomolecules can be badly harmed by excessively heating leaves to temperatures above 40 to 45 ^{0}C. As a result, nature has created a number of adaptations that aid leaves in regulating temperature [14]. The first bionic photovoltaic panel was designed after typical tree leaf and twig structures observed in warm-climate regions and modelled as shown in Figure 3(a) and (b).

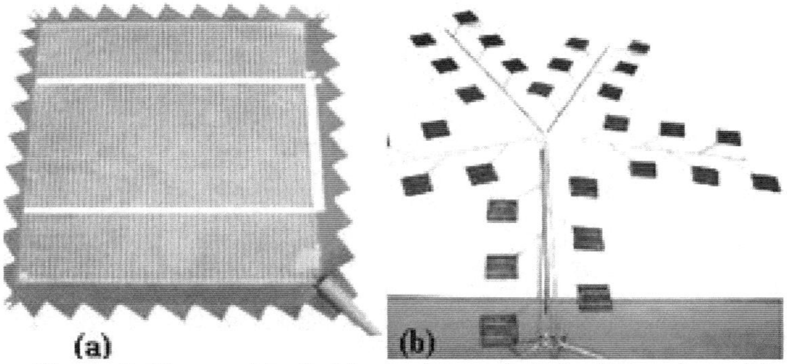

Figure 3. First model of a bionic photovoltaic leaf (a) and tree (b)

To increase light absorption and energy conversion efficiency, many researchers [2, 4, 11-12] have already begun to create solar panels that resemble leaf shapes. The mechanical and structural characteristics of the palm tree leaf-inspired solar PV structure depicted in Figure 4 that were investigated by [14], the designed prototype were tested with respect to mechanical and thermal properties, the design have provide a very acceptable outcome. Beside this the designed model was a lightweight moveable photovoltaic panel, which can be simply detached and re-assembled for installation at required place.

Figure 4. Palm leaf inspired photovoltaic structures.

There is an increasing interest in exploring alternative energy sources because of concerns about climate change and the need to reduce the dependency on fossil fuels. Accordingly, there is a greater need for clean energy sources that can convert energy effectively with the use of cutting-edge technological structures.

2.2 Nenuphar (water lilies) inspired renewable energy use structural design

Nenuphar is at the forefront of a cutting-edge technology model for renewable energy extraction, which includes wave solar and wind energy structures design and layout deigning. The actual model of nenuphar, if used for renewable energy conversion structure, will contribute to an effective use of land and geometries, especially for the wave energy source, which could significantly contribute to the world's electricity needs while reducing greenhouse gas emissions and optimising use of ocean surface area. So far, Nenuphar's model of a floating structure has created an innovative way of harnessing energy and turning it into electricity. Their design is based on the idea of "Nenuphar Islands," which are essentially collections of connected floating structures that were inspired by nenuphar pads. As shown in Figure 5 many lessons can be generated from the nenuphar front and back side orientations.

Figure 5. Nenuphar natural model for Bio-inspired design

Furthermore, being capable of imitating natural systems for structural design that adopt weather condition for the source of renewable energy conversion with optimal land usage might provide a more reliable and consistent energy supply. Due to the fact several practises are going to use the water's surface to develop renewable energy. Figures 6a and b depict a three-bladed floating horizontal axis wind turbine and Figure 6c a floating twin vertical axis wind blade that both create energy while floating over the water's surface as nenuphar did for photosynthesis process.

Figure 6. Floated wind turbine blades design and operation

Nenuphar's (water lilies) leaves provide an inspiration for innovative floating solar and wind turbines design have shown the possibilities to capture solar and wind energy while efficiently utilizing open space in the sea surfaces. This concept opens up new possibilities for energy conversion structures that can harness other renewable resources in offshore locations. Floating platforms could be adapted to accommodate solar panels, wave energy devices, or even hydro turbines, tapping into a diverse range of clean energy sources [15]. By exploiting the vast untapped potential of offshore regions, these floating energy platforms hold promise in significantly increasing renewable energy capacity. This idea may inspire the development of energy conversion systems with adaptive capabilities to adjust to shifting environmental conditions. Structures could, for instance, change their orientation to maximize energy capture or retract some parts during extreme weather. These adaptable structures would increase safety while simultaneously extending the life and increasing the effectiveness of renewable energy installations.

Water lilies, with their unique floating leaves, have long fascinated scientists and engineers due to their efficient use of sunlight and water surfaces. Natural water lilies have a distinctive natural system with a variety of engineering characteristics that can be imitated. Maintaining buoyancy is aided by the transpiration process, which involves water being absorbed by the roots and released through tiny holes called stomata on the surface of the leaf. The exchange of gases between the leaf and the surrounding water and air, such as carbon dioxide (CO_2) and oxygen (O_2), is facilitated by nenuphar evaporation system. There are renewable structures that mimic nenuphar platform, as seen in Figure 7, however many of those energy-conversion structures still require engineers and scientist to extract the actual natural working system of nenuphar for advanced product development. The large, flat leaves of nenuphar (water lilies) are able to float on water because of their high tensile strength and low density.

Figure 7. Bio-Inspired Mega project engineering practices

The choices of design for ocean wave, solar, and wind energy converter structures can be driven by nenuphar patterns and other natural phenomena. Apart from these merits, an additional hypothesized design factor for the nenuphar leaf's working mechanism includes the following areas:

1. Leaf Design for Optimal Sunlight Absorption: Water lilies have large, flat leaves that allow them to float on water while efficiently capturing sunlight for photosynthesis. Floating solar panels can be designed with similar large and flat surfaces to optimize sunlight absorption. This design would enable the panels to capture more solar energy throughout the day, increasing overall energy production.
2. Angle Adjustment Mechanism: Water lily leaves exhibit a unique angle adjustment mechanism that allows them to orient themselves to face the sun. Floating solar panels can incorporate a similar mechanism to adjust their tilt or orientation based on the sun's position. This feature ensures that the panels are always positioned optimally to receive maximum sunlight, further improving energy efficiency.
3. Natural Cooling System: Water lilies thrive on water bodies due to their natural cooling system. Their leaves remain partially submerged, enabling them to dissipate excess heat and maintain a cooler temperature. By integrating a similar cooling system into floating solar panel designs, the panels can operate at lower temperatures, which enhance their efficiency and prolong their lifespan.
4. Environmentally Friendly Material Choices: Water lilies are part of delicate aquatic ecosystems, making their ecological impact crucial to consider. Similarly, floating solar panels should be made from environmentally friendly materials that are non-toxic and have minimal ecological impact. Using sustainable materials for their construction ensures that they do not harm aquatic life or the surrounding environment.
5. Integrated Micro-habitats: Water lily leaves provide habitats for various microorganisms and small aquatic creatures. A biomimetic approach for floating solar panels could include integrating small, ecologically friendly habitats underneath the panels. These micro-habitats can foster biodiversity and contribute positively to the local ecosystem, supporting aquatic life and helping to maintain the ecological balance of the water body.
6. Floating Foundation and Stability: Water lilies are naturally designed to stay afloat on water surfaces without sinking. Similarly, floating solar panels can be engineered with efficient floating foundations and stability mechanisms. This design ensures that the panels remain securely in place, even during storms or varying water levels, preventing any potential damage to the panels or surrounding environment.

7. Biodegradable End-of-Life Management: Water lily leaves decompose naturally, returning nutrients to the water ecosystem. In the same spirit, floating solar panels should be designed with consideration for their end-of-life management. By incorporating biodegradable components or easily recyclable materials, the environmental impact of decommissioned panels can be minimized.

2.3 Sunflower inspired energy conversion structural design

Solar cells can harness the power of the sun to generate electricity. With the help of nature inspired design integration to efficient solar cell development, it is possible to produce energy without using coal or other fossil fuels. It has been observed that sunflower plants follow the sun as it moves from east to west. According to research, flower orientations increase vegetative biomass and warmth, attracting more pollinators. Researchers have been drawn to study sunflowers because of their fascinating natural design, which enables them to follow the sun all day. Sunflowers maximize their exposure to sunlight due to a mechanism called heliotropism, which also to increases the effectiveness of their photosynthetic process [16]–[18]. Since sunflowers are one of many plant species that exploit sunlight effectively to accelerate its growth [17] it can serve as an ideal method of bio-inspired design imitation.

The clean energy that would be generated from solar energy is an abundant available renewable resource. Nevertheless, because photovoltaic (PV) panels have low energy conversion efficiency, capturing this abundant energy source continues to be difficult for massive commercial use [19,20]. Energy conversion efficiencies for PV panels that are offered for sale range from 14% to 22% [19], [21,22]. Environmental variables like sun irradiation serve to aggravate this low efficiency even further. Depending on where a PV panel is located, different amounts of solar radiation impinge on its surface. PV panels really produce the most electricity when they are perpendicular to the direct solar radiation beam, where solar radiation is at its highest [21,22]. Figure 8 depicts the sunflower's process for tracking the sun using heliotropism model.

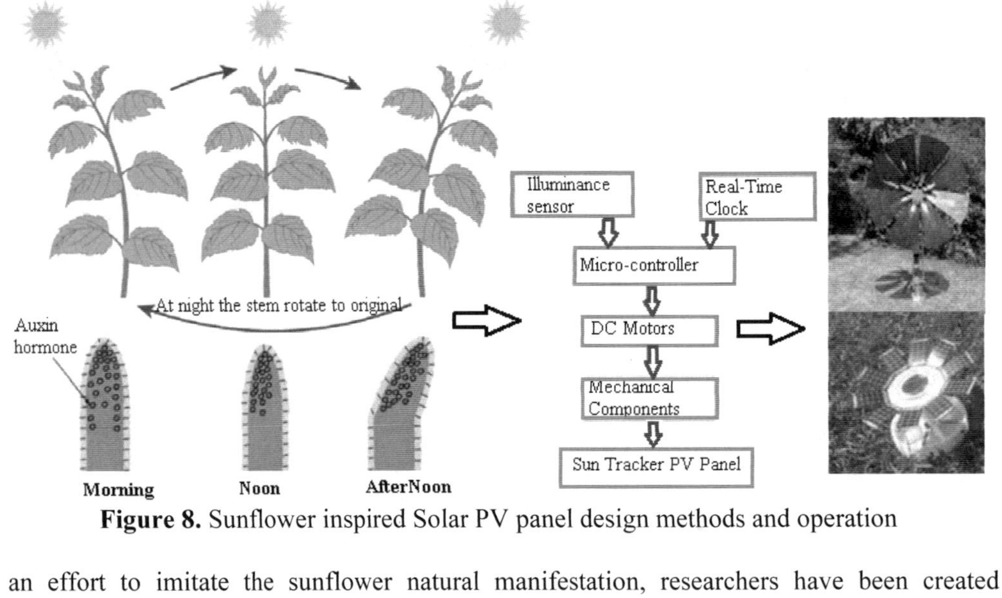

Figure 8. Sunflower inspired Solar PV panel design methods and operation

In an effort to imitate the sunflower natural manifestation, researchers have been created a manmade stem with unique components to support solar cells that can track sunlight, like sunflowers. These manmade stem-like materials can bend towards light, regulate themselves, and move into the proper position to collect solar radiation. Then, when the Sun moves, they make minor adjustments to stay put. As a result, they are able to collect 90% of the light that is available and is shining at a 75° angle on solar cells [23]. Beside this, sunflower energy systems include energy storage solutions inspired by the sunflower's capability to store energy through photosynthesis. Advanced batteries and

IOP Conf. Series: Materials Science and Engineering 1294 (2023) 012044 doi:10.1088/1757-899X/1294/1/012044

energy storage systems mimic the plant's energy conversion and storage processes, efficiently storing excess energy for later use [23, 24]. Nature continues to be a source of inspiration for innovative solutions to contemporary problems. The sunflower-inspired energy conversion structural design is a prime example of how biomimicry can revolutionize the renewable energy sector. By emulating the sunflower's ability to track the sun and efficiently convert energy, Designers are moving closer to a sustainable future where clean energy is abundant and accessible to all. As we continue to develop and refine these systems, we can look forward to a brighter, greener future for our planet.

3. Opportunities and challenges of bio-inspired energy conversion structural design

The need for effective and efficient energy conversion technology is urgently developing because of the world's tremendous energy crisis. In their search for novel solutions, scientists and engineers are looking to nature for ideas. Biomimicry, commonly referred to as bio-inspired energy conversion structure design, has a lot of potential for overcoming these difficulties. To create more effective, long-lasting and flexible energy conversion devices, researchers are modelling natural processes. But there are difficulties and complexity involved with bio-inspired design projects. Some of the observed and reviewed opportunities and challenges in line with nature inspired design were discussed as follows.

3.1 Opportunities by Bio inspired design

Structures for energy conversion that are designed using bio-inspiration have a wide range of possibilities for innovation, sustainability, and effectiveness. Engineers and scientists can develop novel and effective energy conversion systems by modeling nature's solutions[16]–[19], [24]. Listed below are a couple significant prospects for structural engineering design using bio-inspired energy conversion.

3.1.1. Energy Efficiency: Energy efficiency is a growing concern for the built environment because of increased urbanization. The built environment has an impact on people's health and comfort in addition to emitting greenhouse gases that damage the climate. In the recent years, there has been an increase in interest in the use of bio-inspired methods in the built. In comparison to what the engineering-materials industry is currently capable of, nature has created structures with properties and mechanisms that are considerably superior. The extraordinary efficiency of biological materials, including their exceptional properties that depend on fragile constituents, their high performance per unit mass, and their numerous functionalities in addition to their mechanical abilities, has been primarily attributed to their hierarchical organization to be accounted as a lesson for today's engineering design. Moreover bio-inspired designs offer the opportunity to significantly enhance energy conversion efficiency by learning from nature's time-tested solutions. This can lead to more sustainable and cost-effective energy technologies, helping to meet growing global energy demands while reducing greenhouse gas emissions [24-26].

3.1.2. Bio inspired design for unlimited source of Renewable Energy: Innovative thoughts have always been stimulated by nature to create new designs. Plants as well as animals offer a variety of structures having lower density, higher strength and high energy absorption that can inspire new design concepts with important characteristics [1, 6, 27, 29-31]. Bio-inspired structures have been demonstrated to offer a significant improvement over traditional counterparts for their applications in the energy sectors, by observing the critical functions of biological structures found in nature. The application of biomimicry principles to renewable energy technologies, such as solar panels and wind turbines, has the potential to revolutionize the energy sector sustainably with unlimited design options. Improved efficiency and adaptability can make these technologies more competitive and widely adopted for user demand [11, 27-29].

3.1.3. Nature as a Source of Inspiration for Innovation: Innovative thoughts have always been stimulated by nature to create new designs. Bio-inspired energy conversion structures can be integrated with biological systems. For example, artificial photosynthesis can be used to generate energy from sunlight, mimicking the efficiency of natural photosynthesis. This integration can revolutionize fields such as healthcare [32] and agriculture [25]. The study of natural systems serves as a wellspring of creative solutions to energy challenges [22]. By observing and learning from nature, researchers and engineers may unlock novel approaches and breakthroughs that lead to more sustainable and efficient energy conversion technologies Combining bio-inspired structures with existing technologies can create hybrid systems that leverage the strengths of both natural and engineered components. This synergy can lead to enhanced performance and versatility in various applications.

3.2 Bio inspired design challenges

To develop effective and sustainable energy conversion technologies, bio-inspired structural engineering design must overcome a number of constraints. These problems frequently involve several design, execution, and material considerations, making them complicated and multifarious. Following are some observed findings claimed as a challenge in designing renewable energy use structural engineering using bio-inspired methods.

3.2.1. Complexity and multidisciplinary: One of the primary challenges in bio-inspired energy conversion structure design is the intricate and multidisciplinary nature of the field. Nature's solutions have evolved over billions of years, resulting in complex, highly optimized systems. Likewise, the process of converting natural techniques into technological solutions for biomimetic designs may become a difficult interdisciplinary process [26, 31, 33]. By incorporating multiple kinds of mechanisms from many living things in search of superior solutions the design and implementation process complexity grows two three folds [29], [33-34]. Emulating natural systems requires expertise in biology, materials science, engineering, and other diverse disciplines. Bridging the gap between these fields can be a formidable task, demanding collaboration and communication between experts from different backgrounds [34].

3.2.2. Biocompatibility for engineering size design: Engineering design development must take biocompatibility into account, especially at the microscale where complex and delicate interactions with biological entities occur. Biocompatible engineering size design will be crucial in revolutionising healthcare and enhancing the lives of patients all around the world as technology keeps progressing. Even though nature provides a model for sustainable design in the actual translation of bio-system to engineering applications where bio-inspired structures interact as living organisms did, it will become impracticable in some areas like medical devices or biologically integrated systems that challenge to ensure prevalent biocompatibility. Systems of nature frequently function at scales that are substantially smaller than those of technical systems. Engineering systems that are compatible with biological systems may become challenging to develop as a result. Moving from small-scale laboratory prototypes to large-scale practical applications poses a substantial challenge. Scalability is crucial to make bio-inspired technologies economically viable and widely applicable. Maintaining efficiency, structural integrity, and cost-effectiveness when scaling up can be a formidable hurdle. To maintain effectiveness and functioning while maintaining the natural system for renewable engineering design is quite challenging [29, 33, 35-36].

3.2.3. Materials and manufacturing sustainability for bio inspired design model: The adoption of biomimetic approaches is constrained because conventional manufacturing processes that cannot accurately duplicate the complex biomaterials. Developing or finding materials that can replicate the properties of biological systems is a complex undertaking. Biomimicry often requires the creation of novel materials or the adaptation of existing ones. To maintain a similar design model to the imitated natural system, optimising manufacturing procedures for meeting the unique substances may prove challenging. In addition, the design and execution of bio-inspired energy conversion structures integrate sustainability of natural material manufacture into serious account. It is essential to ensure the sustainability for the manufacturing and disposal of bio-inspired renewable energy conversion structures to have as little impact on the environment as possible after their service life. Sustainability in materials and manufacturing is a key element of bio-inspired design, helping to produce more creative and eco-friendly solutions. This method addresses urgent environmental concerns while stimulating creativity and ingenuity by embracing the ideas of biomimicry, circular economy, and sustainable sourcing. Future developments in materials and manufacturing sustainability that will be advantageous to both people and the environment are on the horizon as we continue to explore the limitless potential of nature-inspired design [13, 15, 27-29].

However, bio-inspired designers and researchers have tried by providing unique solutions to a variety of technology and design challenges. In bio-inspired designs, there are a number of noticeable advantages that result from a smaller weight for the same amount of deflection, some of which can be explained as quantifiable values. Table 1 compares traditional and bio-inspired design parameters along with the advantages of using bio-inspired product design.

Table 1. Comparison of conventional and bio-inspired design parameters with BI product in use

Parameters	Conventional design (CD)	Bio-inspired design (BI)	Example
Weight reduction	Limited weight reduction achieved.	Significant weight loss while keeping the deflection the same.	A 20% reduction in wing weight over conventional designs was achieved by the Airbus A380's wing design, which was motivated by the structural concepts of bird bones [37].
Strength-to-weight ratio	Lower strength-to-weight ratio.	Higher strength-to-weight ratio.	The burr-inspired Velcro has a better strength-to-weight ratio than regular adhesives [38].
Material efficiency	Less efficient use of materials.	Improved material efficiency.	The lattice structure of the Tower of Paris, which was modelled by the natural plant stems, provided incredible durability while using less material [39].
Sustainability impact	Higher environmental impact.	Reduced environmental impact.	The kingfisher beak-inspired bullet train nose design lowered noise and energy consumption, making it more environmentally friendly [40].
Resilience	Less resilient.	Improved resilience to external forces.	Designs for biomimetic body armour modelled by fish scales inspiration for providing higher protection at reduced weight [41].
Adaptability	Less adaptable to changing conditions.	Enhanced adaptability to dynamic loads.	Improved energy capture and decreased structural fatigue are two benefits of wind turbine blades designed following humpback whale flippers mimic [42].
Cost savings	Higher production and operational costs.	Potential for cost savings.	The elephant-trunk-inspired Bionic Handling Aid boosts logistical efficiency and could lower costs [43].
Regulatory Compliance	May struggle to meet weight regulations.	Easier compliance with weight limits.	Bio-inspired design elements in the Boeing 787 DL, like lighter, more effective composite materials, aid in keeping to strict weight regulations [44].

Competitive Advantage	Potentially less competitive.	Competitive edge due to weight reduction.	Designing lighter, more aerodynamic cars using bio mimicry gives Formula 1 teams a significant advantage [45].

4. Research gaps

Numerous potential research gaps exist in the study of energy conversion structural design that takes inspiration from sunflowers and nenuphars (water lilies). These gaps can serve as a roadmap for future research and aid in the development of innovative, sustainable energy conversion systems. Potential research gaps are listed below.

1. *Productivity optimization details overlooked*: Explore methods to optimize the efficiency of energy conversion systems inspired by nenuphars and sunflowers. Investigate how design parameters, such as shape, size, and arrangement of nenuphars and sunflowers structural elements, impact energy capture and conversion efficiency which potential can be translated to machine learning.
2. *Material innovation for actual design lacks*: Develop and study new materials that mimic the mechanical properties and resilience of natural structures. Investigate the use of nenuphars and sunflowers materials for constructing mimicked energy conversion devices to enhance their performance and durability.
3. *Dynamic adaptation as natural system not implemented*: Research ways to create energy conversion systems that can dynamically adapt to changing environmental conditions, such as adjusting their orientation or configuration in response to varying sunlight, wind direction, or water flow.
4. *Efficient bio-inspired photovoltaic development challenged:* Explore the development of photovoltaic materials inspired by the light-absorbing properties of plant leaves and petals. Investigate their potential for improving the efficiency of solar panels, particularly in low-light conditions.
5. *Resilience to environmental stressors*: Investigate how natural structures like sunflowers withstand environmental stressors, such as wind, water, or pests. Apply this knowledge to design energy conversion systems that are more resilient and require less maintenance.
6. *Biological sensors and AI integration for bio-inspired design*: Explore the use of biological sensors or adaptations found in nenuphars and sunflowers to enhance the performance and safety of energy conversion systems. These research gaps highlight the diverse opportunities for advancing the field of energy conversion structural design inspired by nenuphars and sunflowers. Addressing these gaps can lead to the development of more efficient, sustainable, and environmentally friendly energy generation systems.

5. Conclusion

The application of bio-inspired design practises for the development and design of renewable energy structures is reviewed in this work in a systematic manner. Both the pros and cons of nature-inspired design have been examined, along with a possible list of subjects of study to be considered by those with appropriate expertise. The limited trail for sustainable energy conversation structure design was noted, but Nature gives a paradigm for intelligent, efficient designs that have been mostly disregarded or overlooked. The practise of mimicking nature is known as biomimicry, and it is something that humans have been doing for a very long time to find solutions to problems like hunger and shelter. The goal of biomimicry is to learn how nature lives and evolves so that it can be self-sustaining, not to imitate it in any way. The ability of nature to support life has required creativity. Through selection and testing throughout time, it has generated models, patterns, and solutions. Several of the challenges we now have are addressed in these solutions. Since 3-D printing employs materials that people are familiar with, including plastics, metals, and other materials, there may be an opportunity to use compatible elements to actualize the parts that were inspired by nature and manufactured for actual usage. On top of that, the incorporation of biological sensors and AI in bio-inspired design is now a promising innovation and technical frontier that has the potential to support engineering design and manufacturing of bio-inspired design in the renewable energy sector.

References

[1] N. S. Ha and G. Lu 2019 A review of recent research on bio-inspired structures and materials for energy absorption applications, *Compos. Part B Eng.* **181**,107496.

[2] I. Arrambide, I. Zubia, and A. Madariaga 2018 Critical review of o ff shore wind turbine energy production and site potential assessment, *Electr. Power Syst. Res,* **167**, 39–47, doi: 10.1016/j.epsr.2018.10.016.

[3] K. Mallard and V. Debusschere 2020 Multi-Criteria method for sustainable design of energy conversion systems, *Sustainability*, **12**(16), 6513; https://doi.org/10.3390/su12166513.

[4] E. Jamei and Z. Vrcelj 2021 Applied sciences biomimicry and the built environment, learning from nature's solutions. *Appl. Sci.*, **11**(16), 7514; doi.org/10.3390/app11167514.

[5] O. A. Oguntona and C. O. Aigbavboa 2017 Science direct science direct biomimicry principles as evaluation criteria of sustainability in the construction industry assessing the feasibility of using the heat temperature function n for demand forecast, *Energy Procedia*, **142**, 2491–2497.

[6] R. Guchhait 2023 Increasing growth of renewable energy a state of art, *Energies,* **16**(6), 2665; https://doi.org/10.3390/en16062665.

[7] R. C. Ramachandran, C. Desmond, F. Judge, and J. Serraris 2022 Floating wind turbines: marine operations challenges and opportunities, *Wind Energy Sci.* **7**(2), 903–924.

[8] P. Taylor, C. M. Wang, T. Utsunomiya, S. C. Wee, and Y. S. Choo 2010 Research on floating wind turbines: a literature survey, *IES J. Part A: Civil Struct. Eng.* **3** (4), 267-277.

[9] N. Pentelovitch and J. K. Nagel 2022 Understanding the use of bio-inspired design tools by industry professionals, *Biomimetics*, **7**(2), 63; https://doi.org/10.3390/biomimetics7020063.

[10] M. A. M. I lwich, T. H. S. Peck, O. L. G. A. S. Peck, T. H. S. Tegmaier, and H. E. P. Lanck, 2006 Biomimetics and technical textiles: solving engineering problems with the help of nature's wisdom, *American J. Bot.* **93**, 1455–1465.

[11] M. R. Chenaghlou, M. Kheirollahi, K. Abedi, and A. Akbari 2020 Inherent adaptive structures using nature-inspired compound elements, **6**, 1–13; doi.org/10.3389/fbuil.2020.561902.

[12] M. Rodrigues and C. G. Soares 2022 New wave energy converter design inspired by the nenuphar plant. *J. Mar. Sci. Eng.* **10**(11), 1612; https://doi.org/10.3390/jmse10111612.

[13] L. Ilieva, I. Ursano, L. Traista, B. Hoffmann, and H. Dahy 2022 Biomimicry as a sustainable design methodology-Introducing the 'biomimicry for sustainability framework, *Biomimetics*, **7**(2), 37; https://doi.org/10.3390/biomimetics7020037.

[14] M. Zähr, D. Friedrich, T. Y. Kloth, G. Goldmann, and H. Tributsch, 2010 Bionic photovoltaic panels bio-inspired by green leaves, *J. Bionic Eng.*, **7**, 284–293.

[15] C. Herrera et al., 2018 Structural design and manufacturing process of a low scale bio-inspired wind turbine blades, *Compos. Struct.*,**208**(13). DOI: 10.1016/j.compstruct.2018.08.061.

[16] P. Sun et al, Wang Zhang, Imran Zada, Yunxuan Zhang 2019 Improved energy efficiency in solar steam generation 3D-structured carbonized sunflower heads for improved generation, *ACS Appl. Mater. Interfaces,* **12**(2), 2171–2179.

[17] A. Lendlein, M. Balk, N. A. Tarazona, and O. E. C. Gould 2019 Bio perspectives for shape-memory polymers as shape programmable, *Active Mater.*, **20**(10), 3627–3640.

[18] X. Ling, M. I. Osoti, W. Zhang, Y. Wu, Q. Jin, and D. Zhang 2023 Bioinspired materials: From distinct dimensional architecture to thermal regulation Properties, *J. Bionic Eng.* **20**(3), 873-899.

[19] R. U. Sabran and A. C. Fajardo 2019 An improved sunflower-inspired solar tracking strategy for maximizing photovoltaic panel power generation, In: *IEEE 10th Int. Conf. on Humanoid, Nanotechnology, Information Technology, Communication and Control, Environment and Management (HNICEM)*, Baguio City, Philippines, 2018, pp. 1-6, doi: 10.1109/HNICEM.2018.8666379.

[20] Z. Usman, J. Tah, H. Abanda, and C. Nche 2020 A critical appraisal of PV-systems' performance, pp. 1–22 *Build.* **10**(11), 192; https://doi.org/10.3390/buildings10110192.

[21] S. Kurtz et al., 2013 Analysis of photovoltaic system energy performance evaluation method analysis of photovoltaic system energy performance evaluation method. http://www.osti.gov/bridge.

[22] A. Luceño-s and A. Mar 2019 Materials for photovoltaics: State of art and recent developments. *Int. J. Mol. Sci.*, **20**(4), 976.

[23] Bilge Baytekin, S Doruk Cezan, H Tarık Baytekin , Bartosz A Grzybowski 2018 Artificial Heliotropism and Nyctinasty Based on Optomechanical Feedback and No Electronic, *Soft Robot* **5**(1):93-98. doi.org/10.1089/soro.2017.0020.

[24] K. Koike, K. Fujii, T. Kawano, and S. Wada 2020 Bio-mimic energy storage system with solar light conversion to hydrogen by combination of photovoltaic devices and electrochemical cells inspired by the antenna-associated photosystem II, *Plant Signal. Behav.* **15**(3).

[25] Shashwat S. a, Kishor T. Zingre a, Niraj Thurairajah a, DEVS Kiran Kumar b, Krithika Panicker c, Prashant Anand c, Man Pun Wan 2023 A review on bioinspired strategies for an energy efficient built environment, *Energy build.,* **296,** 113382.

[26] Y. Wang, S. E. Naleway, and B. Wang 2020 Bioactive materials biological and bioinspired materials: Structure leading to functional and mechanical performance, *Bioact. Mater.* **5**(4), 745–757.

[27] C. Audibert, J. Chaves-jacob, J. Linares, and Q. Lopez 2018 Bio-inspired method based on bone architecture to optimize the structure of mechanical workspieces, *Mater. Des.*, **160**, 708–717.

[28] M. D. L. Á, O. Del, and K. Beermann 2023 Environmentally responsive materials for building envelopes: A review on manufacturing and biomimicry-based approaches, *Biomimetics*, **8**(1), 52; https://doi.org/10.3390/biomimetics8010052.

[29] R. T. Kapoor, M. Rafatullah, M. Qamar, and M. Qutob 2022 Review on recent developments in bioinspired-materials for sustainable energy and environmental applications, *Sustainability*, **14**(24), 16931; https://doi.org/10.3390/su142416931.

[30] H. Desing and R. Widmer 2022 How much energy storage can we afford? on the need for a sunflower society, aligning demand with renewable supply, *Biophys. Econ. Sustainability*, **7**(2), 1–15.

[31] A. Rani, R. Reddy, U. Sharma, P. Mukherjee, P. Mishra, and A. Kuila 2018, A review on the progress of nanostructure materials for energy harnessing and environmental remediation, *J. Nanostruct. Chem.* **8**:255–291

[32] D. Tripathi, K. Hajra, and D. Maity 2023 Recent advancement of bio-inspired nanoparticles in cancer theragnostic, *Nanotheranostics*, **4**(3), 299-322; https://doi.org/10.3390/jnt4030014

[33] M. C. Austin, D. Garzola, N. Delgado, J. U. Jiménez, and D. Mora 2020 Inspection of biomimicry approaches as an alternative to address climate-related energy building challenges: A framework for application in Panama, *Biomimetics*, **5**(3),40; https://doi.org/10.3390/biomimetics5030040

[34] L. L. Stevens, M. Fehler, D. Bidwell, A. Singhal, and D. Baumeister 2022 Building from the bottom up: A closer look into the teaching and learning of Life's Principles in biomimicry design thinking courses, *Biomimetics*, **7**(1), 25; https://doi.org/10.3390/biomimetics7010025

[35] S. Hayes, J. Toner, and C. Desha 2020 Enabling biomimetic place-based design at scale, *Biomimetics*, **5**(2), 21; https://doi.org/10.3390/biomimetics5020021

[36] K. Araque, P. Palacios, D. Mora, and M. C. Austin 2021 Biomimicry-based strategies for urban heat island mitigation: A numerical case study under tropical climate, *Biomimetics*, **6**(3), 48; https://doi.org/10.3390/biomimetics6030048

[37] J. Wang, Y. Li, G. Hu, and M.Yang 2019 Lightweight Research in Engineering: A Review, *Appl. Sci.* **9**(24); https://doi.org/10.3390/app9245322

[38] E I. Avgoulas and M P F. Sutcliffe 2016 A review of natural joint systems and numerical investigation of bio-Inspired GFRP-to-Steel joints, *Mater.* **9**(7), 566.

https://doi.org/10.3390/ma9070566

[39] E. Jamei, Z. Vrcelj 2021 Biomimicry and the built environment, learning from nature's solutions, *Appl. Sci.* **11**(16),7514; https://doi.org/10.3390/app11167514.

[40] J. Boaretto, M. Fotouhi, E. Tende, G F. Aver, V R R. Marcon, G L. Cordeiro, 2021 Biomimetics and Composite Materials toward Efficient Mobility: A Review Journal of *Compos. Sci.* **5**(22); http://dx.doi.org/10.3390/jcs5010022.

[41] M. Kamrul, J. Paul, P. E. Juan, H. Wang 2021 Biomimetic armour design strategies for additive manufacturing: A review, *Mater. Des.* **205**; doi:10.1016/j.matdes.2021.109730

[42] S.M.A. Aftab , N.A. Razak , A.S. Mohd Rafie , K.A. Ahmad 2016 Mimicking the humpback whale: An aerodynamic perspective, *Prog. Aerosp. Sci.* **(84)**, https://doi.org/10.1016/j.paerosci.2016.03.002.

[43] M. Rolf, K. Neumann, J.F. Queißer, R.F. Reinhart, A. Nordmann & J.J. Steil (2015) A multi-level control architecture for the bionic handling assistant, *Adv. Robot.* **29(**13), doi: 10.1080/01691864.2015.1037793.

[44] P., Bisma, M. I. Kittur, I. B, S. Kamangar, M. Hussien, and M. A. Umarfarooq 2022 Scientific advancements in composite materials for aircraft applications: A review, *Polym.* **14**(22), https://doi.org/10.3390/polym14225007.

[45] X. Castro, ZA. Rana 2020 Aerodynamic and Structural Design of a 2022 Formula One Front Wing Assembly, *Fluids,* **5**(4), 237. https://doi.org/10.3390/fluids5040237.

Assessment of awareness levels towards additive manufacturing best practices in transforming traditional manufacturing: A case study in Ethiopian context

T Lemma[1*], H G Lemu[2] and E M Gutema[1]

[1] College of Engineering and Technology, Wallaga University, Nekemte, Ethiopia;
[2] Faculty of Science and Technology, University of Stavanger, N-4036 Stavanger, Norway

* Correspondence: tekalign.lemma@ambou.edu.et

Abstract. The main objective of this study is to assess the awarness levels of acadamic staffs and industry experts towards the best practices of additive manufacturing (AM) in transforming traditional manufacturing (TM) in local context. The study used closed ended questionnaires with 1-5 liker scales to collect secondary data from the respondents. The collected data were analyzed using descriptive statistics (frequency and percentage) by using SPSS v20 software. The findings of this paper identified ten (10) AM best practice benefit factors (Resource consumption benefit factor, Waste management benefit factor, Pollution control benefit factor Material utilization benefit factors, Design optimization benefit factor, Manufacturing flexibility best practice factors, SC flexibility benefit factors, SC network improvement benefit factor, Cost reduction benefit factors, Ability of changing customer demand benefit factors) in transforming traditional manufacturing in local context. Among the eight AM best practices both academic staffs and industry experts have common awareness level on five factors. In addition, the findings of the study showed that 49% of academic staffs and 37% of industry experts have good awarness levels towards the identified AM best practices in transfornig traditional manufacturing. Based on the findings of this study conceptual framework was developed for future research work to quantitatively investigate the effects of AM best practices on traditional manufacturing performmance.

Keywords: Additive manufacturing, Pros of additive manufacturing, traditional manufacturing, best practices, transforming, awareness levels

1. Introduction

Additive manufacturing is the manufacturing process that joins materials to make objects based on 3D design data, usually in a layer-upon-layer manner and this technology was launched in the 1960s [1-3]. It is a means of rapid prototyping and manufacturing for components or products for actual end use, and it is a significant technology in the fourth industrial revolution [4, 5]. Around the world different types of AM technologies (processes) are used such as:- Stereo-lithography Apparatus (SLA), Fused Deposition

Modeling (FDM),Binder jetting (BJ), Selective Laser Sintering (SLS), Selective Laser Melting (SLM), and Electron Beam Melting (EBM). These technologies are classified into three categories base on the types manufacturing methods: liquid-based, solid-based, and powder-based. Among these, Liquid-based and powder-based systems dominate the industry today [6]. Different types of materials can be utilized in these AM processes such as: - Plastics, Resins, Rubbers, Ceramics, glass, concretes, and metals [7]. And AM technologies used these materials for different applications such as: lighter weight products, multilateral products, ergonomic products, efficient short production runs and fewer assembly errors. The benefits of AM have been studied and reported in different research papers. For example, according to [7] AM has the potential to reduce cost of production, logistics, inventories, and in the development and industrialization of a new product; reduce time to market, minimize materials and energy usage, and cut down waste [8]. And, AM could promote shorter, localized, collaborative, and more sustainable supply chains [9]; reduce raw material utilization and minimizes manufacturing and improve environmental pollution [10, 11].

Even if the review parts of this paper indicated the best practices of AM in transforming traditional manufacturing (TM), the attention given varies from continent to continent. As shown in Figure 1(a), for the past decade, 49% of the studies were conducted in Europe, 22% in North America, 16% from Asia, 11% in South America and 3% in Africa. In addition, the availability of AM system as shown in Figure 1(b) in the world indicated that 33% in North America, 26% in Europe, 24% in Asia and 17% in Central and South America. This indicated that still there is technological advancement difference and knowledge gap between developed and undeveloped countries like Ethiopia.

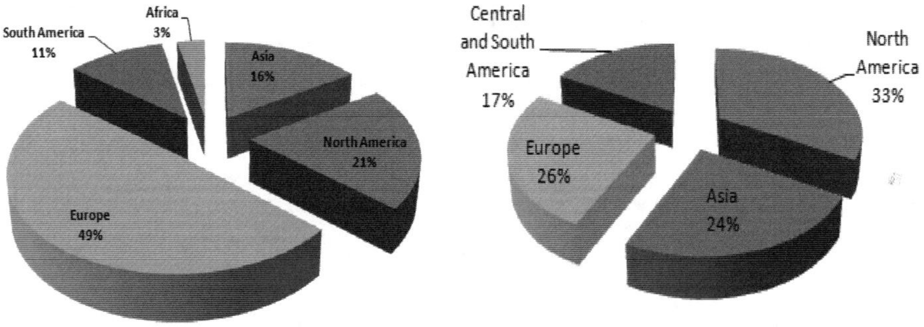

(a) (b)

Figure 1. (a) Published paper by continent, (b) AM installation per continent [12]

Observing the best practices of AM in transforming TM and the lower level of knowledge in undeveloped countries, this study aims to assess the awareness levels of industry experts and academic staffs in undeveloped countries in Ethiopia. The need of this study is twofold: - technically it helps to predict the acceptance levels for the implementation of the technology for the near future in the case country. In addition, by carrying out this study, Ethiopian academic staff and industry professionals might be inspired to accept this technology to offer solutions to numerous engineering and industry challenges.

2. Literature review

Additive manufacturing also known as direct manufacturing or 3D printing is a digital technology for producing physical objects layer by layer from a three-dimensional computer aided design (CAD) file, rather than through subtractive techniques (such as lathe and milling operations) or molding [13]. It was invented in the US in the 1980s and initially used as a method for producing rough physical prototypes of products. According to [10], AM production processes start with developing a 3D CAD model in solid modelling tools or scanning the object to generate its 3D CAD model with all its details and dimensions. In addition to 3D CAD, the initial three-dimensional data can be generated by computer tomography, magnetic resonance imaging, or using 3D digitizing systems [14]. To produce objects layer by layer, AM technology used different types of processes like powder bed fusion (laser or electron beam) by SLS, SLA and EBM), binder jetting by extrusion (injection) through indexing nozzles, etc. According to [15] the input materials used for these processes can be in the form of powders, filaments, liquids, or sheets.

Different studies illustrated the best practices of AM in transforming TM systems. The study in [10, 16] illustrated its environmental impacts of AM in three aspects: (1) Resource consumption, (2) waste management, and (3) pollution control.

According to the study in [9,17] AM has an advantaged in the areas of supply chain which are:
(a) shorter manufacturing time,
(b) reduced inventory,
(c) reduced production batch,
(d) lower transportation costs,
(e) less production waste, and
(f) better sustainable practices.

The article by Huang et al. [18] reported that AM flexibility provides the printing of products only when needed or at the time they will be utilized in production. And this transforms the production chain, it makes manufacturing on-demand, reduces inventories and waiting times and contributing to just-in-time practices. The study in [19] indicated the ability of AM to produce complex shapes, which are often impossible to create using TM methods. This study also illustrated AM's ability to convert raw materials directly into finished products and the device of AM are small and portable enough to fit on office desktop. In addition, 3D printers can create any shape or product without the need for any machine set-ups, which allows for levels of manufacturing flexibility that is hard to achieve in mass production. According to the study in [20] and [21] AM technology is growing rapidly with its unique features for producing an object without requiring any special, sophisticated tools or production lines. This unique feature of AM reduces the need for logistics, time from production to sale, and environmental impact [22]. The paper in [23] examined the impact of AM on companies' value creation on single areas in manufacturing sectors. And the result of their paper illustrated that AM changes time to market strategies, product variety and improve customer satisfaction. Based on this, the best practices of AM were summarized and illustrated in Figure 2.

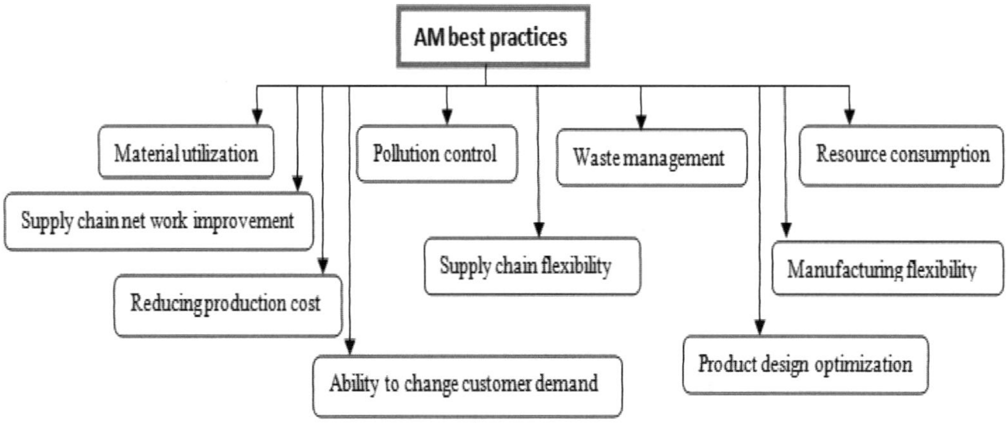

Figure 2. Categories of AM best practices.

3. Methodology

To meet the objective of this study, the authors used quantitative research approach. A survey study approach was used to collect primary data from the respondents, which comprised acadamic staffs from engineering disiplines (personnels) and industry experts such as production managers, supply chain and logistic experts, engineers and middle level workers…etc. The universities and case industries were selected by the use of convenient sampling, by considering the respondents' proximity, accessibility and availability. Among the 45 universities found in ethiopia, 7 universities were selected by considering their field of studies (specially engineering disciplines) and those who have Boarding schools. And from industry sectors 29 footwear and 3 plastic product manufcturing industies were considered for the case studies. To select the respondents for this study both simple random sampling and stratifies probability sampling techniques were used. The researchers were used stratified sampling techniques to create homogeneity between respondents. Accordingly, for this study general Yamane (1967) formula was used to calculate a precise sample size with 95% confidence level and p = 0.5 is assumed and given by the Equation (1)

Regarding this, the sample size of the respondents were 119 academic staffs and 251 industry experts. To collect data from the respondents' questionnaire was designed and structured based on the identified best practices of AM as illustrated in Figure 2. The questionnaire consisted of two parts. The first part has respondents profile, while the second part surveyed their awareness levels towards the identified 10 AM best practices. And the respondents were requested to respond to the questionnaires based on 1-5 liker scales (1= strongly disagree, 2 = disagree, 3 = neutral, 4 = agree, 5 = strongly agree). The prepared questionnaires were printed in hard copy and distributed to the respondents. Among the distributed questionnaires, 78.8% (219) were properly filled and returned for further analysis. And the collected data's were analyzed using descriptive statistics (frequency and percentage) with SPSS V 21 software.

$$n = \frac{N}{1+N(e^2)} \tag{1}$$

Where: n = sample size, N = Total population, e = margin of error 5%, and 1 = Probability of event occurring

Table1. Sample size determination

Organization	Total population	Sample size from each organization
Academic staffs (engineering professionals)	675	251
Footwear industry (designers, production managers, supply chain and logistic experts, engineers)	124	119
Plastic manufacturing industry (production manager and middle level workers)	42	
	Total sample size	370

4. Results and discussions

In this part of the study, the results of the survey assessments are presented and discussed. Both the awareness levels of academic and industry experts towards the identified 10 best practices of AM in transforming TM in local context was assessed and presented.

4.1 Findings and results about awareness level of academic staffs and industry experts

The environmental impacts of AM are illustrated in three aspects as: - (1) resource, (2) waste management and (3) pollution control related benefit factors. Thus, respondents are requested their awareness levels for

IOP Conf. Series: Materials Science and Engineering 1294 (2023) 012045 doi:10.1088/1757-899X/1294/1/012045

the three benefit factors in terms of material utilization, waste minimization, use of pollutant like cutting fluids, cast release components, and forging lubricants…etc.

Resource consumption benefit factor: The survey results in Figure 3(a) illustrated that 52% of the respondents from academic staffs have high awareness level about resource consumption benefit factor of AM. But 32% have very low awareness level or understanding about this benefit factor. And 14% of them have neither low nor high awareness level. Concerning industry expert's awareness level towards this factor, the survey findings illustrated that among the respondents only 32% have awareness about resource consumption benefit factors of AM. And 68 % (with 32% very low and 36% low response rate) have no awareness about this factor.

Waste management benefit factor: Concerning this factor, the result in Figure 3(b) illustrated that only 10% of the academic staffs have awareness about this factor. But almost 82% of the respondents are less aware (have low awareness) level (with 42% low and 40% very low response rate). The remaining 8% of the respondents are neither aware nor unaware towards this factor. Similarly, the findings in Figure 3(b) indicated that 28% (with 20% high and 8% very high response rate) industry experts have awareness about waste management benefit factors. But higher number of respondents that means 60% has no awareness about this factor. But 12% of the respondents are neither aware nor unaware about this factor.

Pollution control benefit factor: with regards to pollution control benefit factors, the findings in Figure 3(c) illustrated that 24% of the academic staffs have awareness about this benefit factor of AM (with 8% very high and 16% high response rate). But 52% of academic staff respondents have no awareness about the pollution control benefit factor of AM (with 16% very low and 36% low response rate). The remaining 24% have neutral opinion about this factor. On the other hands, 60% of the industry experts have no awareness about pollution control benefit factor of AM (with 28% low and 32% very low response rate). However, only 12% of them are aware about this factor. And 28% of the industry expert respondents are neither aware nor unaware about this factor.

These results indicated that the number of academic staffs and industry experts who were aware of waste management and pollution control benefit factors of AM was limited. This is because of non-availability of the technology in the industry and lack of exposure towards environmental related activities, knowledge gap and academic study in the areas of environmental related activities.

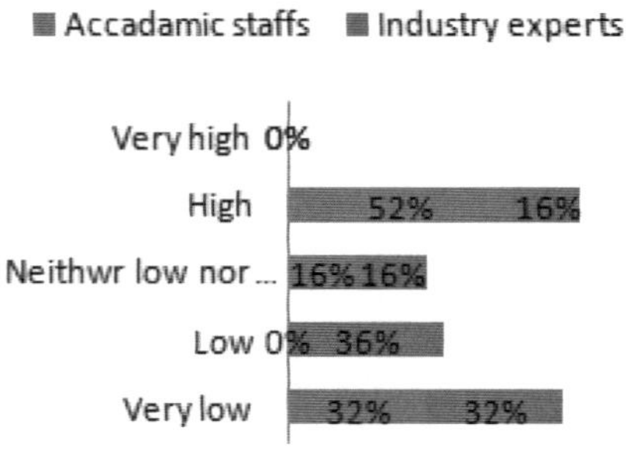

(a) Resource consumption benefit factor

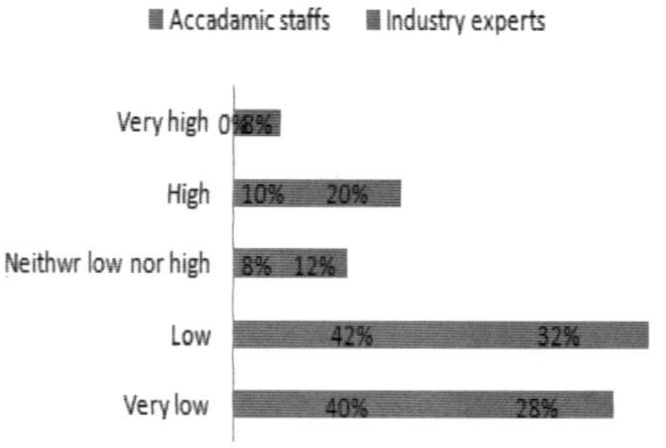

(b) Waste management benefit factor

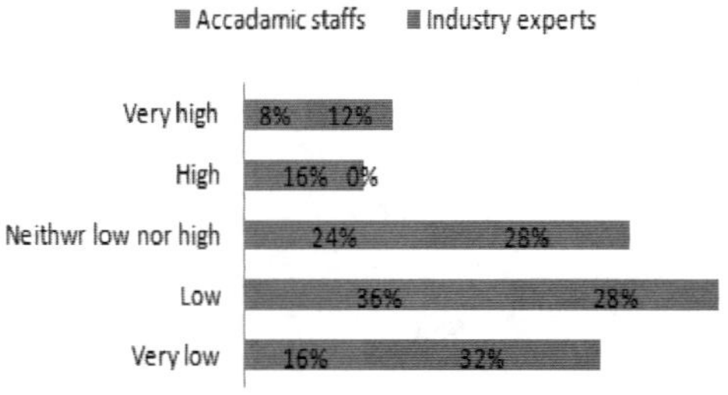

(c) Pollution control benefit factor

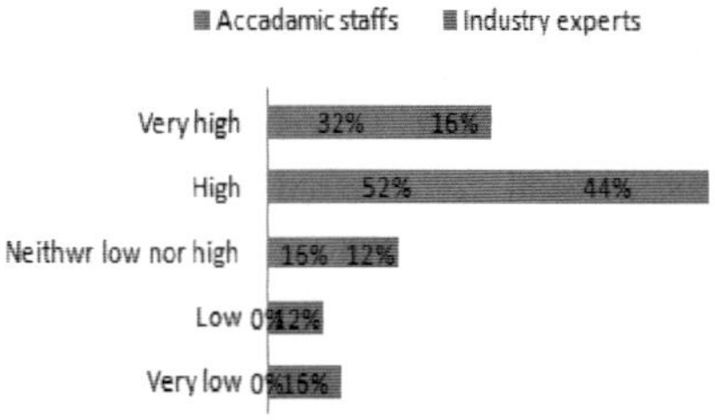

(d) Material utilization benefit factor

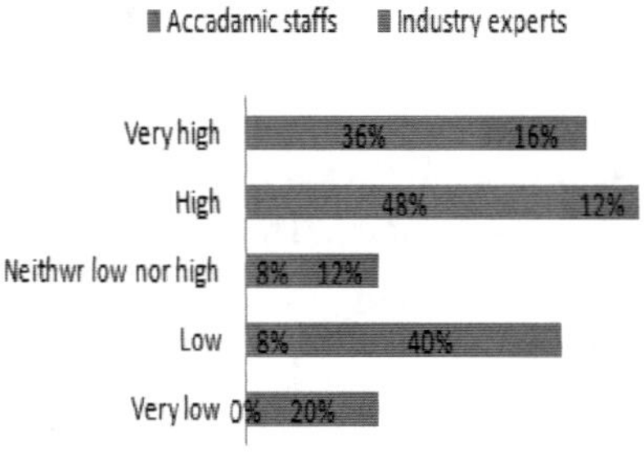

(e) Design optimization benefit factor

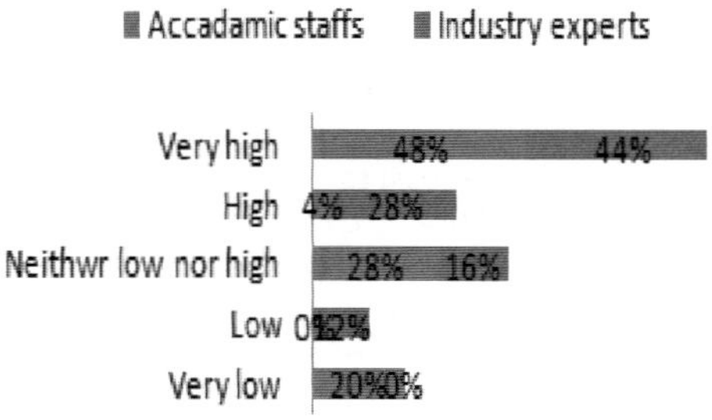

(f) Manufacturing flexibility benefit factor

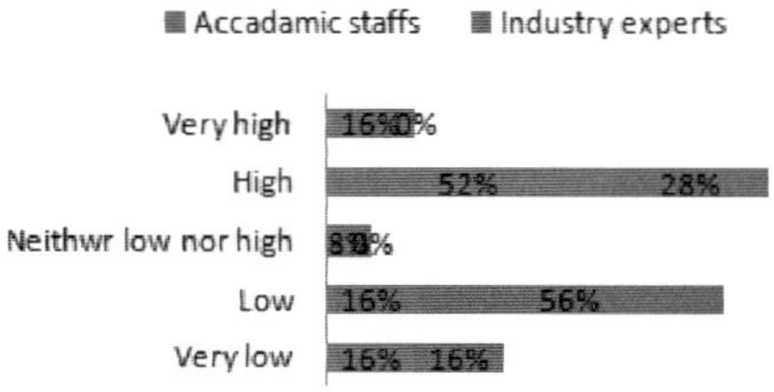

(g) Supply chain flexibility benefit factor

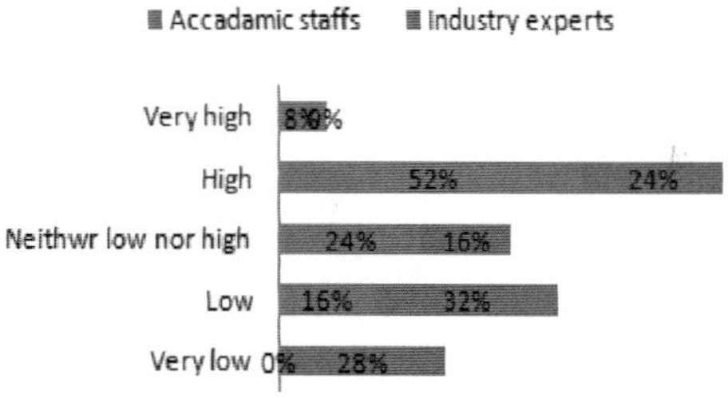

(h) Supply chain network benefit factor

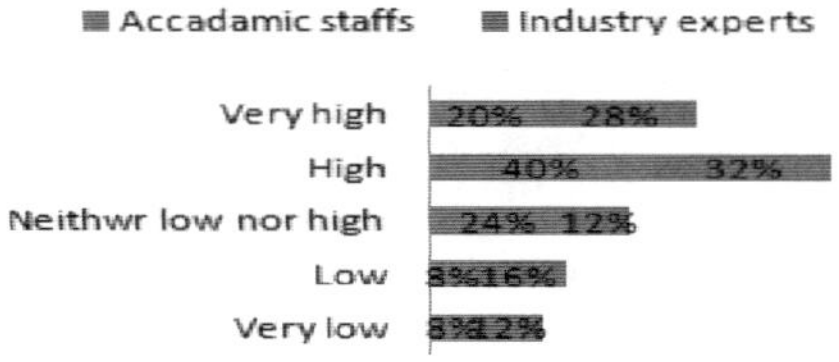

(i) Cost reduction benefit factor

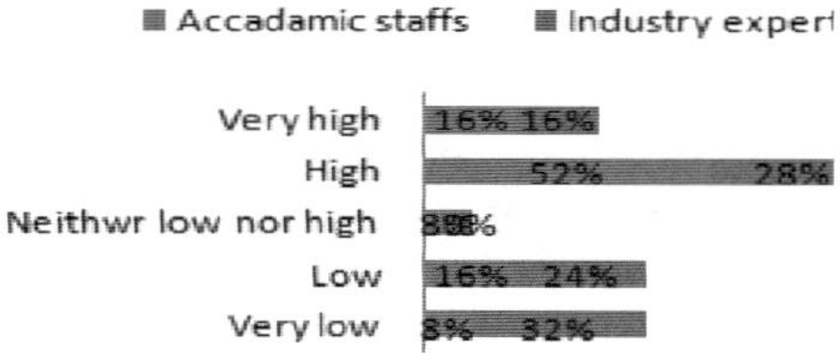

(j) Ability to change customer demand benefit factor

Figure 3. Respondents' response towards the survey factors.

In the areas of material utilization, design optimization and manufacturing flexibility best practice factors respondents were requested to respond their awareness level and perception. AM have best practices in the areas of novel materials and unique design solutions, mass reduction of components through highly efficient and lightweight designs, consolidation of multiple components and fabrication of complex parts without assembly process. It offers the ability to rapidly customize products at multiple scales and locations closer to consumers.AM allows production flexibility to create any shape or product without the need for any new machine set-ups and additional tooling.

Material utilization benefit factors: The findings in Figure 3(d) show that 84% of the academic staffs have high awareness level about the benefits of AM towards utilization of materials (with 52% high and 32% very high response rate). On the other hand, 16% of them have neutral perception concerning this factor. Similarly, majority of industry experts (60%) have heard and aware about material utilization benefit factors of AM (with 44% high & 16 % very high response rate). But 28% of them have no awareness (10% very low & 12% low response rate). In addition, 12% of them are neither aware nor unaware about this factor.

Design optimization benefit factor: In the areas of product design optimization benefit factor, as illustrated in Figure 3(e), majority of academic staffs (84%) have good awareness level (with 48% high and 36% very high response rate). But only 8% of them are unaware and neutral responses. With regards to this factor, the findings illustrated that 56% of industry experts have high awareness level (with 40% high and 16% very high response rates). On the other hand, 32% of them have no awareness about design optimization benefit of AM (with 12% very low and 20% low response rate). The remaining 12% of the respondents are neither aware nor unaware about this factor.

Manufacturing flexibility best practice factors: Among the academic staffs as the findings in Figure 3(f), illustrated, 52% have awareness about manufacturing flexibility best practice factors (with 4 % high and 48% very high response rate). And also 20% of them have no awareness about this factor. And 28% of the academic staff has neither aware nor unaware. On the other hand, the finding indicated that 72% of industry experts have awareness about manufacturing flexibility benefit factors (28% high and 44% very high response rate). And 12% of them have no awareness about this factor. In addition, 16% of industry experts have neither aware nor unaware about this factor. The highest level of awareness level of respondents may be from the skill developed in the industry through and teaching and learning experiences and though the comparison of the existing SM and AM.

Adaptation of AM improves SC network by changing product distribution network to the final consumers by promoting shorter, localized, collaborative production system. Reduce manufacturing and assembly steps which results in internal supply chain improvements and reduces the number of suppliers. In addition, the decentralized approach of AM eliminates long distances that products move within the supply chain and SC network is improved by reducing manufacturing time and inventory level. Furthermore, internal manufacturing SC is improved by minimizing suppliers, tools needed for production, by reducing number of labors and movement of components during manufacturing and stock of materials. AM supply chain shows greater mix and new product introduction flexibility with lower change over time and cost of new product introduction to the system.AM as one of the technologies that improves efficiencies of the entire SC from the cost of distribution to assembly and carry, all the way to the component itself. Based on these, respondents are requested their awareness level towards SC flexibility and SC network improvement benefit factors.

SC flexibility benefit factors: Figure 3(g) shows that 52% of academic staffs have awareness about SC flexibility benefit factors (with 44% high and 28% very high response rate), 32% of them have no awareness (with 16% low and very low response rate). On the other hand, 16% of the academic staffs' response is neutral. Among the industry experts, only 28% of them have high awareness about SC flexibility benefit factor. But 72% of them have no awareness about this factor (16% very low and 56% low response rate).

SC network improvement benefit factor: The findings in Figure 3(h) indicated that majority of academic staffs (60%) have awareness about the benefits of AM in improving SC network (with 52% high and 8% very high response rate). But 16% of the staffs have no awareness about this benefit factor, while the response of 24% of them are neutral. If we see the awareness of industry experts in the areas of AM supply chain improvement benefit factor, the findings illustrated that only 24% of them have high awareness and understanding. And 60% of the industry experts have no awareness about of this factor (with 28% very low and 32% Low respondent rate). In addition, 16% of them said that we are neutral about this factor.

The lower awareness level of industry experts in SC network and flexibility areas comes from knowledge gap in the areas of SC management and lack of training. But academic staffs' especially engineering professionals have got the awareness about these factors through training.

In AM technology each product can be customized to the customer's preferences. AM fulfill customer demand comes from continuously changing environment driven by changing customer needs, AM can respond to changing customer demands for new products and features. It reduces carbon footprint, logistics, transportation, distribution cost, etc. In this regard, both industry experts and academic staffs were requested their opinion and awareness towards cost reduction and Ability of changing customer demand benefit factors.

Cost reduction benefit factors: The findings in Figure 3(i) indicated that 60% of the academic staffs have good awareness level about cost reduction benefit factors (with 40% high and 20% very high response rate). But 16% of them are unaware about this factor (with 8% low and very low response rate). On the other hand, 24% of the academic staffs are neither aware nor unaware towards this factor. Similarly, the findings in this Figure confirmed that 60%of industry experts have aware about cost reduction benefit factor (32% high and 26% very high response rate). On the other hand, 28% of them have no awareness about this factor (12% very low and 16% low response rate). In addition, 12% of industry experts are neither aware nor unaware about this factor.

Ability of changing customer demand benefit factors: Concerning ability of changing customer demand benefit factors, the finding results in Figure 3(j), indicated that 68 % of academic staffs have awareness about this factor (with 52% high and 16% very high response rate). And also, the results under this factor indicated that 24% of the academic staffs have no awareness. In addition, 8% of them have neither aware nor unaware towards this factor. Similarly, the findings in relation to ability of changing customer demand benefit factors, 44% of industry experts have awareness (28% High and 16% very high response rate). But 56%) of them has no awareness about this factor (32% very low and 24% low response rate.

4.2 Findings and results on awareness level towards AM best practices

In this section of the analysis, the aggregate awareness level of academic staffs and industry experts are presented and compared. The results of the study in Figure 4 (a) and (b) illustrated the awarness level of the acadamic staffs and industry experts towards the best practices of AM in transforming TM. The results of Figure 12 (a) illustrated that 51% of industry experts have low awarnes about the identified 10 best practices of AM. And 37% of them are awared about the best practices of the identified AM. But the results in Figure 12(b) confirmed that only 37% of the acadamic staffs have low awarness level, while 49% of them are awared about the best practices of AM in transforming TM in local context. From this we can observe that there is awareness level gap between academic staffs and industry experts in local context. In other words, academic staffs are more aware than industry experts about the identified 10 best practices of AM, and this was our initial expectation because this technology has been around within the academia.

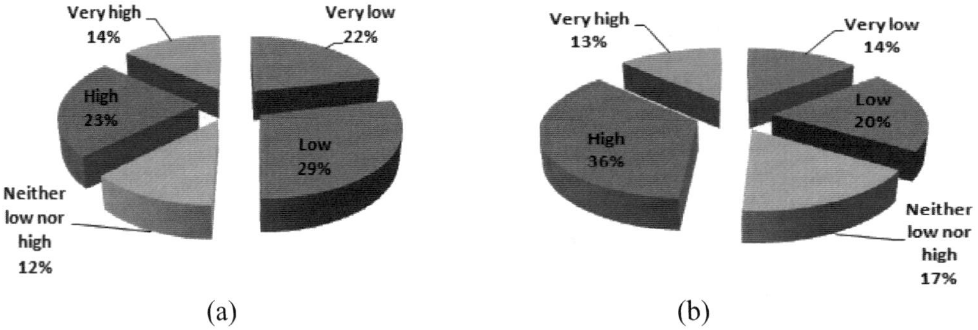

Figure 4. Aggregate awarness level (a) industry experts (b) academic staffs

5. Conclusion

The objective of this paper is to assess the awareness levels of industry experts and academic staffs towards the ten (10) best practices of AM in Ethiopian context. Knowing the level of awareness towards these best practices help to predict the acceptability and easy implementation of the technology. As a methodology, through literature review, the best practices of AM were identified and categorized into 10 factors, and then the awareness level of the industry experts and academic staffs were assesses with 1-5 liker scale open ended questionnaires.

The findings of this paper identified that both academic staffs and industry experts have common understanding (awareness level) on material utilization benefit factors, product design optimization benefits factors, cost reduction benefit factors, ability of changing customer demand benefit factors, and manufacturing flexibility benefit factors. But on the remaining factors there are awareness level gap between academic staffs and industry experts in local context.

Acknowledgement: The authors would like to acknowledge the support provided INDMET project funded by NORHED II program (grant nr. 62862) to cover the publication cost of this work.

Reference

[1] Saberi S, Mohd R, Zulkifli N, Megat M(2010), Effective factors on advanced manufacturing, technology implementation performance:a review, *J. Appl. Sci.*, **2**(10), 1229–1242
[2] ASTM Standard (2012), Standard Terminology for Additive Manufacturing Technologies, ASTM International, West Conshohocken, PA, **10**(4).
[3] Mellor, S. Hao, L. and Zhang, D. (2014) Additive manufacturing: a framework for implementation, *Int. J. Prod. Econ.* **149**, 194-201
[4] Rifkin J(2012) The third industrial revolution: How the internet, green electricity, and 3-D printing are ushering in a sustainable era of distributed capitalism, The World Financial Review. Available at: https://www.feelingeurope.eu/Pages/thirdindustrialrevolutionrifkin.pdf
[5] Wong V, Hernandez A(2012) A review of additive manufacturing, *ISRN Mech. Eng.* **5**(23) 1–10
[6] Bogue R(2013) 3-D printing: The dawn of a new era in manufacturing, *Assem. Autom.*, **33**(4), 307-311
[7] Radu G, Ines R, Florinda M, Bruna F, Helena C, Paulo P (2020), Impact assessment of additive

manufacturing onsustainable business models in Industry 4.0 Context, *Sustainability.* **12**, 7066

[8] Bourell D, Leu M, Rosen DW(2009) Roadmap for additive manufacturing: identifying the future of freeform processing, Austin, The University of Texas.

[9] Gebler M, Schoot U, Visser C(2014) A global sustainability perspective on 3D printing technologies, *Energy Policy*, **74,** 158-167.

[10] Allwood J, Ashby MF, Gutowski TG, Worrelln E(2011) Material efficiency: a white paper, *Resour. Conserv. Recycl.* **55**, 362 –381.

[11] Achillas, D. Aidonis, E. Iakovou, M. and Thymianidis, D(2015), A methodological framework for the inclusion of modern additive manufacturing into the production portfolio of a focused factory, *J. Manuf. Syst.* **37**, 328 –339

[12] Wohlers Report 20222 *3D printing and additive manufacturing: Globale state of the industry.* Available at https://wohlersassociates.com/wp-content/uploads/2022/05/2022_ExSum.pdf (Last accessed: 2023.10.18)

[13] Gardan J (2016), Additive manufacturing technologies: State of the art and trends, *Int. J. Prod. Res.* **54**(10), 3118–3132

[14] Gibson I, Rosen W,Stucker B(2009), Additive manufacturing technologies: Rapid prototyping to direct digital manufacturing, New *York: Springer*

[15] Manyika J, Chui M, Bughin J, Dobbs R,Bisson P, Marrs A (2013), Disruptive technologies: Advances that will transform life, business, and the global economy, 180, San Francisco, CA: McKinsey Global Institute.

[16] Achillas D, Aidonis E, Iakovou M, Thymianidis D (2015), A methodological framework for the inclusion of modern additive manufacturing into the production portfolio of a focused factory, *J. Manuf. Syst.* **37**, 328 –339.

[17] Attaran M(2017), The rise of 3-D printing: The advantages of additive manufacturing over traditional manufacturing,*Bus.Horiz.* **60**(5), 677–688.

[18] Huang Y, Leu M C, Mazumder J, Donmez A(2015), Additive manufacturing: current state, future potential, gaps and needs, and recommendations, *J. Manuf. Sci. Eng.***137**(1), 014001

[19] Cohen D, Sargeant M, Somers K (2014), 3-D printing takes shape, *J. Serv. Sci. Manag.* **10**(3), 89-99.

[20] Santos K, Loures E, Piechnicki F, Canciglieri O(2017), Opportunities assessment of product development process in Industry 4.0, *Procedia Manuf.* **11**, 1358-1365.

[21] Dimitrov D, Beer N,Hugo P, Schreve K(2014), Three dimensional printing, *In: Compr. Mater. Process.*, **10**, 217-250.

[22] Paris H, Mokhtarian H, Coatanéa E, Museau M, Ituarte I F (2016), Comparative environmental impacts of additive and subtractive manufacturing technologies, *CIRP Annals – Manuf. Technol.* **65**(1), 29–32.

[23] Kritzingera W, Steinwendera A, Lumetzberger S, Sihna W (2018), Impacts of additive manufacturing in value creation system, *Procedia CIRP.* **72,** 1518 – 1523.

Literature review on thin-walled and cellular structure designs for energy absorption

T Dabasa[1*], H G Lemu[2] and Y Regassa[1]

[1] Dept. of Mechanical Engineering, Addis Ababa Science and Technology University, Addis Ababa, Ethiopia
[2] Faculty of Science and Technology, University of Stavanger, N-4036 Stavanger, Norway

* Correspondence: tamana.dabasa@aastustudent.edu.et

Abstract. Bio-inspired structure is a topic of immense interest to researchers worldwide. In order to maximize energy absorption through biomimetic structures, this article presents bio-inspired structure particularly, thin walled and cellular structures thorough analysis of the interactions between experimental research and Finite Element Analysis (FEA) simulations. The study compiles the prior research on experimental investigations of thin-walled and cellular biomimetic structures in order to understand the significance of biomimetic structural energy absorption. These inventive works of nature serve as inspiration for these designs, which provide engineering solutions that excel in impact resistance and energy dissipation abilities. The study further highlights the mutual advantages of combining experimental research with FEA models, which enable a deeper understanding of the impact response and energy absorption mechanisms inherent in biomimetic structures, by exploring into recent developments in material science and design methodologies. The article emphasizes how important validations are in bringing experimental results in line with FEA predictions. Furthermore, the practical applications demonstrated in fields like aircraft engineering, automotive safety, and protections can serve as excellent examples of the paradigm-shifting potential of this method for boosting impact protection. This review proposes novel research avenues aimed at fully harnessing the potential of biomimetic architectures to enhance energy absorption, all while acknowledging and addressing the associated challenges.

Keywords: Biomimetic structures, Energy absorption, Thin-walled structure, cellular structures

1. Introduction

The ability of a structure to absorb and dissipate energy when subjected to an impact or external force is referred to as energy absorption. A system called the energy absorber transforms kinetic energy into another kind of energy [1]. Energy absorption must be taken into consideration during the engineering design process to reduce impact forces, increase structural safety, prevent collapses, protect equipment and products, and improve resilience, due to various structural elements found in biomimetic designs. Whether the design is inspired by plants, insects, or animals, it has consistently addressed one or more issues related to product/component optimization [2]. Modern scientists and engineers seek nature for optimal, economically viable, and long-term solutions.

Due to their extreme lightness and exceptional energy absorption capabilities, high energy-absorbing thin walled and cellular structure biomimetic designs are in high demand across a variety of technical sectors, including automobiles [3], aerospace [4], protective armours [5], biomedical fields [6] and civil engineering [7] due to their ultra-light and outstanding energy absorbing performance [8]. Consequently, several energy absorbers with different structures such as sandwich structures [9], plates [10], honeycomb [8], and foams [11] have been proposed in recent years. Various biological structures found in animals and plants could be used as inspiration to design novel structures that can sustain impacts generated during collision, exhibit excellent energy absorption capabilities and inspire the design of new energy absorbers. Further, a variety of materials have been used to manufacture bio-inspired structures, including ceramic, polymeric, metallic and fiber-reinforced composites and concrete [12]. It has always been possible for bio-inspired designs, whether they are based on plants or animals, to address one or more product/component optimization challenges [2].

Nowadays, the search for nature-inspired biomimetic is focused to develop best performing, most sustainable solutions. Two key methods are used to understand the performance of the solutions, particularly to investigate the mechanical properties and energy absorption of biomimetic structures: (1) experimental methods such as Drop-weight impact test [13], quasi-static compression test [14] and (2) FEA based simulations utilizing ABAQUS [8], ANSYS [15], and LS-DYNA [16]. Schematic images of some of the biomimetics testing approaches are shown in Figure 1.

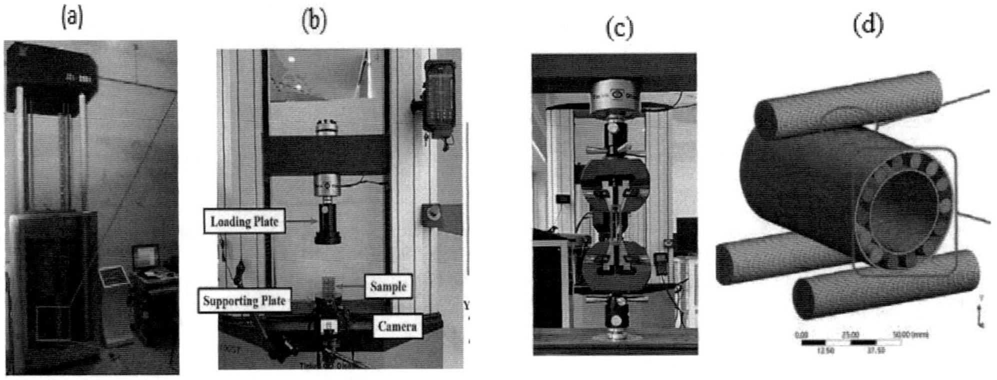

Figure 1. Schematic views of different testing techniques, (a) Drop-weight impact testing (b) Quasi-static compression test (c) Tensile test (d) Numerical simulation (FEA).

Observing the growing interest in use of biomimetic as nature-inspired design and analysis method, the objective of this article is to review recent research work on application of experimental and numerical research in this field. The review work is intended to lay the basis for continuing research in this direction.

2. Energy absorption criteria

Energy absorption (EA) quantifies the energy absorbed during compression and is calculated as the area under the compressive stress–strain curve, as depicted in Figure 2. Key parameters used for assessing energy absorption performance include EA, specific energy absorption (SEA), mean crush force (Pm), maximum crush force (Pmax), and crush load efficiency (CLE), which are crucial for evaluating crashworthiness [17].

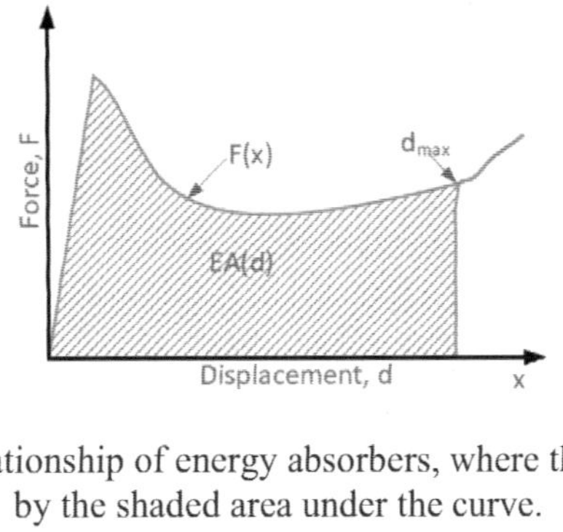

Figure 2. Force-displacement relationship of energy absorbers, where the absorbed energy is represented by the shaded area under the curve.

Energy absorption reflects the structure's capacity to absorb total impact energy, determined by integrating the crush force displacement curve (Equation (1)), while SEA measures the absorbed energy per unit mass of the structure (Equation (2)). As can be observed from the equation, higher SEA values correspond to higher elastic energy per unit mass, which indicates better energy-absorption capabilities. Pm (mean crushing force), on the other hand signifies the average compression force resisted during the total plastic deformation process and is calculated using Equation (3). In addition to calculation of the amount of energy absorbed, the uniformity of the crush-force displacement is also important to measure the crash load efficiency (CLE). This parameter is defined as the ratio of P_m to P_{max} (i.e., peak crushing force PCF), where P_{max} is the maximum of F(x), as given by Equation (4).

$$EA = \int_0^x F(x)dx \tag{1}$$

$$SEA = \frac{EA}{m} \tag{2}$$

$$P_m = \frac{EA}{x} \tag{3}$$

$$CLE = \frac{P_m}{P_{max}} \tag{4}$$

Where F(x) represents the instantaneous crush force at the effective crushing distance x and m is the total mass of the structure.

3. Methods studying biomimetic structure for energy absorption

The mechanical behaviour and energy absorption of biomimetic structures are explored using two essential techniques, which are highlighted in this section. It combines numerical simulations and experimental research (Figure 3) from previous studies to provide understanding of anticipated results.

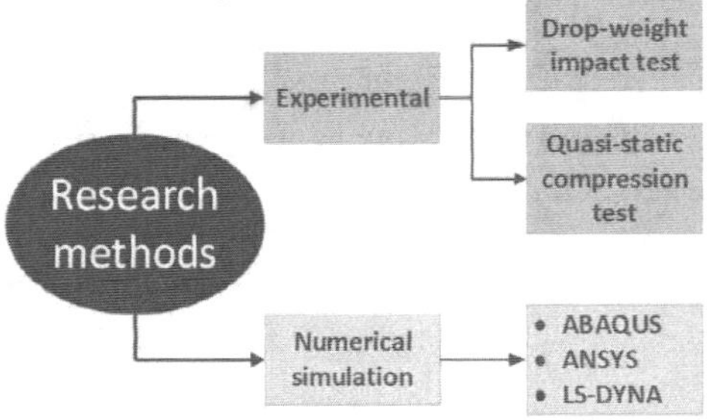

Figure 3. Methods for biomimetic structure for energy absorption.

3.1 Experimental analysis of biomimetic structures

Experimental analysis of biomimetic structures involves conducting physical tests and measurements on structures that are inspired by biological systems. These tests aim to evaluate the mechanical properties, performance, and energy absorption behaviour of these biomimetic structures under various conditions. The goal of experimental analysis is to gain insights into how these structures emulate natural systems and to assess their potential applications in engineering design and other fields. Bio-inspired structures for energy absorption experimental attempts made are particularly investigated for thin-walled structures and cellular structures.

3.1.1 Biomimetic thin-walled structure: Thin-walled structures have gained substantial attraction in industries such as automobile and aviation due to their lightweight, cost-effective, and high energy absorption properties [17]. Extensive research has focused on investigating the deformation modes [8] and energy absorption [18] of these structures. This exploration combines numerical simulations [19] and experimental studies like drop and quasi static test [20, 21]. Most common thin-walled structures are circular tubes like multi-cell tubes, multi-corner tubes, corrugated, and tapered tubes are thin-walled structure commonly used for energy absorption due to their high rigidity, strength, affordability, variety, and ease of manufacture [22].

Researchers have contributed significantly to the design evolution of energy-absorbing components within thin-walled tubes. Various strategies including structural modifications and diverse cross-sectional configurations such as circular, square [23], hexagonal [24] and non-convex multi-corner structures have been employed. Furthermore, hybrid structures [25], gradient thickness variations [21], and tapered tubes [22], as well as corrugated tubes [26] have been thoroughly examined. From previous studies, cross sectional configuration has a great effect on biomimetic structure. The impact performances of the multi-corner tubes with cross section of star-shaped were studied based on experimentally dynamic test and their numerical simulations were validated through impact tests on Aluminium ABAQUSS-tube samples. Due to its distinctive deformation mode, the S-tube's specific energy absorption exceeded that of polygon tubes (P-tube) [27]. As indicated in Figure 4 (a) lotus leaf inspired were designed for novel biomimetic hierarchical thin-walled structure (BHTS) design that has various advantages over conventional approaches to creating energy absorption structures [28].

Building on past work [29], bionic crossbeam and box designs, inspired by bamboo and cattail plants, significantly enhanced crashworthiness under quasi-static tests. In Figure 4 (b), designs reduced overall weight and crush deformation by 33.33% and 44.44%, respectively. Research into round and square cross sections of Al alloy thin-walled tubes [23] underscored the round section's superior energy absorption under quasi-static axial loading. Frusta and tapered tubes with varied cross-sectional forms have been extensively studied as energy absorbers due to their superior characteristics compared to circular tubes [22]. A novel thin-walled multi-cell tubular structure with a modified face-cantered cubic (MFCS) cross-section using experimental and numerical simulations in ABAQUS/Explicit 2020 software under quasi static were utilized. These studies showcased enhanced energy absorption, efficiency, and stable deformation [8]. A study in [30] investigated bionic tubes through drop weight tests, introducing three new tube types inspired by bamboo cross-sections. Notably, 8 out of 14 bionic specimens displayed greater SEA than cylindrical tubes, highlighting the efficacy of bionic design in innovations like bio-inspired aluminium honeycombs.

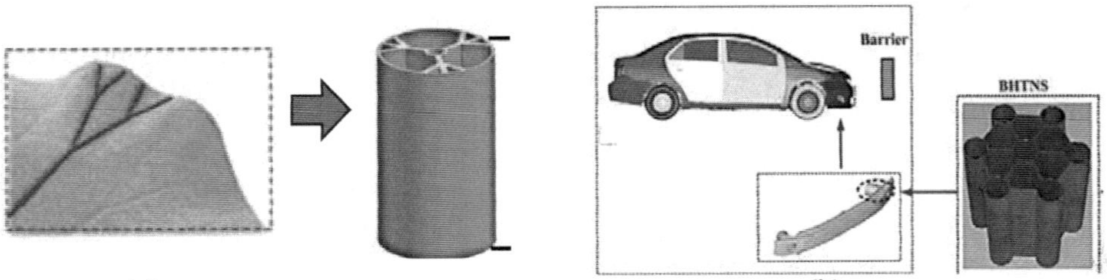

(a) (b)

Figure 4. (a) biomimetic structures inspired by a lotus leaf (b) Crash box of the vehicle with BHTNS inspired by the bamboo structure.

In another study [31] bionic thin-walled energy absorption structures inspired by bamboo's characteristics underwent comprehensive performance analysis. The quasi-static study conducted included specific energy absorption, peak crushing force, and load-carrying capacity behaviour under varying conditions. Bamboo-inspired honeycomb tubular structure was manufactured using the wire-cut electrical-discharge machining (WEDM), and subjected to a drop-weight impact test at an impact velocity of up to 4.4 m/s. It was concluded that the maximum specific energy absorption of the structures was higher than those of traditional metallic honeycomb structures (35 J/g) [32].

Use of foam-filled bionic structures are other ways to improve energy absorption. Horsetail bio-inspired multi-cell tubes (HBMTs) with foam were the subject of a numerical research using LS-DYNA, as well as an experimental lateral impact test with a 15 m/s impact velocity [33]. The foam density and wall thickness of the bio-inspired structure were discovered to have a substantial influence on their crashworthiness based on the examination of the novel design. When the bio-inspired construction was modelled with a 0.5 g/cm^3 foam density and a 3 mm tube thickness, the specific energy absorption was 1.6 kJ/kg. Another study [34] that was motivated by the cross-sections of bamboo and palms looked at the

energy absorption of multi-cell carbon fibre reinforced plastic (CFRP) and aluminium (Al) square tubes under quasi-static axial crushing.
Innovative configurations such as hybrid structures combining square and circular sections, were aimed to enhance energy absorption by introducing corners [25]. Bionic gradient thickness (BGT) tubes proposed for enhanced energy absorption underwent analysis, demonstrating improved deformation and energy absorption compared to uniform-thickness tubes [16]. Bamboo-inspired bionic design further enhanced energy absorption in thin-walled tubes attributed to gradient distribution of vascular bundles, nodes, and density. In order to determine the load-carrying and energy absorption capabilities of a thin-walled structure, researchers [35] studied it experimentally and numerically under quasi-static and dynamic loads. The structure was modelled after the biologic organism known as a balanus. The construction was deep drawn and consisted of an outer shell with a frusto-conical shape and an interior conical core with a hemispherical top. LS-DYNA was used to model the applied deep drawing process using a nonlinear finite element code. Additionally, it was demonstrated that the load supported by the balanus structure was larger than the total weight supported by the outer shell and inner core separately. At quasi-static strain rates, the mean force increase resulting from the interaction effect was around 5%, whereas at dynamic strain rates, it increased to nearly 26%.

3.1.2 Biomimetic cellular structure: Biomimetic cellular structures, inspired by nature, are designed for optimal energy absorption with applications in impact protection [36], automotive safety [19] and structural integrity [37]. Combining experiments and numerical methods, researchers are uncovering their performance potential. Cellular structures like foam-filled designs [19], lattice structure [38] and honeycomb [39] are most common structures used for energy absorption. Lattices have numerous superior qualities to foams and honeycombs, including light weight, high strength, energy absorption, and vibration reduction, all of which have received considerable attention [40]. Foam-filled straw structures have a great capacity for absorbing energy as shown in Figure 5. Based on two common straw constructions (Cornstalk and Reed), there are four different types of bio-inspired foam-filled structures as shown in Figure 5.

1) A bionic foam core with a central hole,
2) Four square holes,
3) Four taper holes, and
4) Two nodes.

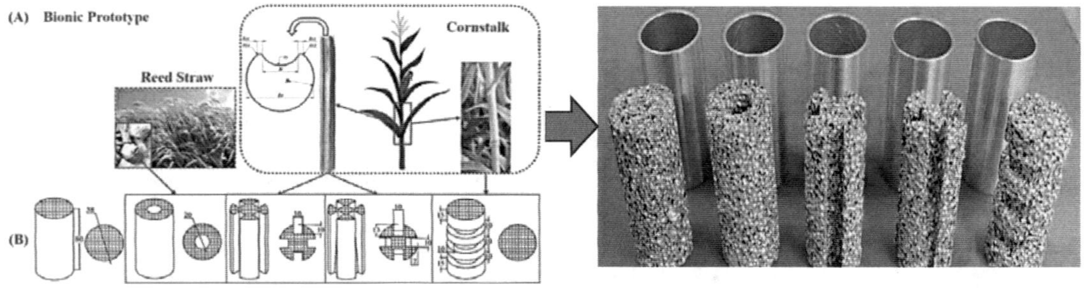

Figure 5. Bio-inspired aluminium foam-filled tubes

By using dynamic drop hammer and quasi-static compression, the energy absorption capacity and deformation mode of bionic structures were investigated [19]. According to the findings, 9 out of 12 bionic samples could absorb more energy than their matching fully filled counterparts. Exploring tortoiseshell-like composites [15], drop tests highlighted Bio-T composite's superior impact resistance compared to other structures (Bio-M, Bio-B, Bio N, HSP). Bio-T's arrangement at [0°/30°/0°] showed notably high load-bearing capacity prior to failure.

Uniquely shaped honeycomb with geometries influenced by natural shapes like the horseshoe, spiderweb, and pomelo peel are also used. Because they can survive high-stress impact situations, hooves have shown promising energy absorption properties and have similar porosity properties to human skin and have demonstrated promising energy absorption properties since they can withstand high-stress impact circumstances [41]. Horse-shaped aluminium honeycombs with different cross sections have been proposed based on triangular honeycomb, square honeycomb, hexagonal honeycomb and Kagome honeycomb to improve the energy absorption capacity and nonlinear explicit FEA code LS-DYNA was used [30]. In a work reported in [42] the unique microstructure of pomelo peel to create a novel hierarchical honeycomb structure was utilized. Spider-web hierarchy mimicking the porous honeycomb structure, crushing process was numerically simulated by LS-DYNA using a 400 kg impactor and a prescribed velocity of 15 m/s. The base material of the honeycomb was aluminium alloy 6060 T4.

Researchers [43] Hierarchical honeycombs are investigated to understand energy absorption characteristics compared to ordinary honeycombs in terms of their mechanical characteristics such as stiffness, and energy absorption and results reported that first-order spider-web hierarchical honeycomb's SEA increased by 62.1%, whereas second-order spider-web hierarchical honeycomb's SEA increased by 82.4%.

As shown in Figure 6 (a) researchers studied 3D printed cellular structures (spiral and honeycomb) using fused deposition modelling (FDM) technique with ABS plus material under compression for both experiential and numerical (ABAQUS), where crash tests showed spiral designs outperformed honeycomb in energy absorption [44]. Expanding on previous work, [45] developed shape-memory continuous fiber reinforced composite honeycomb structures (CFRCHSs) using fused filament fabrication (FFF) technique. The study highlighted the significant impact of honeycomb parameters to enhance compression behavior and energy absorption in 3D printed CFRCHSs.

Lattice structures are very good at absorbing energy [46] which indicated in Figure 6 (b). Lattice structure containing cubic cells, diamond cells, and re-entrant cells [47] conducted quasi-static test analysis, in which they investigated the failure process and stress-strain response of lattice structures as well as their dynamic load deformation behaviour. The lattice constructions clearly differed in their deformation behavior and energy absorption effect when subjected to low-speed and high-speed impact conditions.

Researchers in [46] looked at the degree to which uniform and graded lattice structures may absorb energy. According to the findings, the graded lattice structure absorbed 3.2 ± 0.1 MJ/m^3 of cumulative energy, while the uniform lattice structure absorbed 2.6 ± 0.2 MJ/m^3. For energy absorption, thin-walled structures filled or coated with various materials are being investigated [48]. The researchers developed novel hybrid structures thin-walled tubes filled with periodic lattice materials to increase energy absorption. As seen in Figure 6 (b), these hybrids showed a remarkable 115% improvement in impact energy absorption compared to the total of their individual components.

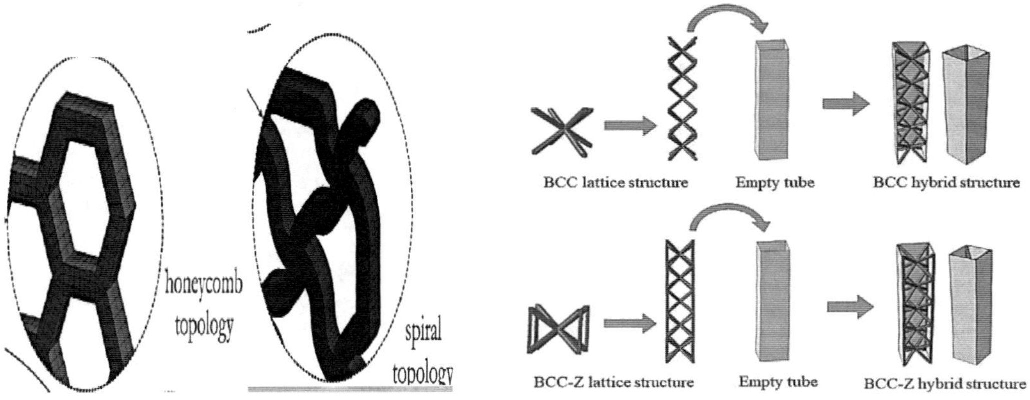

Figure 6. (a) Honeycomb and spiral structures (b)compressive behaviour of BCC and BCC-Z lattice.

3.2 Finite element modelling
The Finite Element Method (FEM) is vital in mechanical and structural engineering. Particularly, nonlinear FEM is valuable for large deformation plasticity problems, provides precise results compared to traditional methods. In various industries, FEA streamlines energy-absorbing structure design by replacing costly testing. These structures often collapse intricately under axial or multi-axial loads, challenging conventional analysis methods. Figure 7 illustrates an example of finite element model under impact testing which play a pivotal role in studying energy-absorbing structures [11].

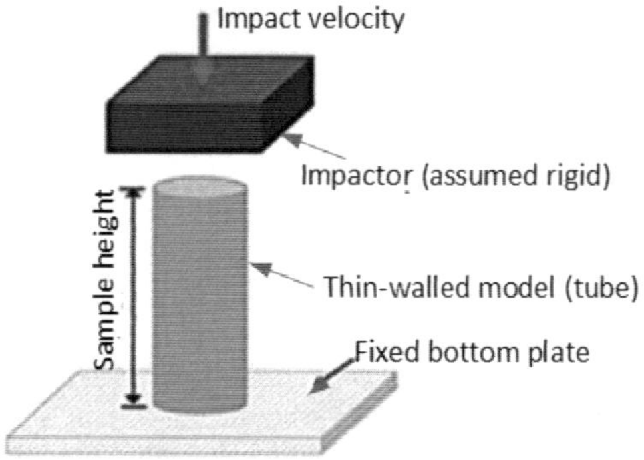

Figure 7. Finite element model under impact testing.

For meshing structures prior to simulations, choosing the right element size is essential to ensure accurate deformation and energy absorption analyses [17]. To visualize and mesh complex geometries, FEA tools such as Hypermesh, LS-Dyna utilising LS-Prepost file, and HyperView are helpful [16]. Lattice structures work best with the shrink wrap technique in Hypermesh [49]. Software like ABAQUS [50-52], ANSYS [6], and LS-DYNA [25,52] along with geometric models from SolidWorks and Catia meshed in HyperMesh [51] are used in FE models to imitate the behaviors of biomimetic structures. Quasi-static, dynamic, and impact simulations are all included in FEA, enabling thorough investigation of geometric parameters [18]. This includes analyzing PLA honeycombs with LS-DYNA, analyzing BHTS mechanical characteristics in ANSYS, and researching geometric parameter effects in ABAQUS. The paper evaluates 3D printed PLA honeycomb structures for sacrificial cladding solutions using experimental tests and LS-DYNA simulations. With a mesh size of 0.25 mm and second-order tetrahedral elements, Hypermesh uses the shrink wrap approach for meshing [49].

Quasi-static compression studies on selective laser melting (SLM)-printed specimens prepared in SolidWorks and simulated in ABAQUS were conducted on geometric parameter effects on energy absorption in HCFCC lattice structures [38]. The SEA of bio-inspired multi-cell conical tubes was observed to be higher than that of conventional designs, with a SEA of 46.2 kJ/kg, which is 1.3 and 1.8 times higher than that of conventional four-cell conical tubes, respectively [2]. The following is a representation of how the full finite element modelling of the simulation of biomimetic structures was evaluated, which were particularly used for simplicity of understanding.

3.2.1 Impact analysis: Nowadays, physical testing is becoming unnecessary because numerical simulations, such as LS-Dyna, ABAQUS, and ANSYS, can anticipate material and structure behavior under dynamic stresses precisely. The SEA of thin-walled constructions under dynamic axial loading in ABAQUS/Explicit with 4-node shell elements was observed to be 40% higher than that of P-tube [32]. ABAQUS/Explicit successfully simulates the energy absorption of bionic honeycomb tubular nested structures and showed good agreement with drop-weight experiment data. The development of nacre-inspired composite materials using impact testing and LS-DYNA computational modelling successfully prevents projectile perforation. Experimental impact tests and ABAQUS/Explicit simulations were used to examine the dynamic behaviour of thin-walled star-shaped cross-section tubes. When compared to square and circular multi-cell structures, hybrid structures can absorb up to 70.13% more energy, according to LS-DYNA numerical simulations

[25]. The finite-element model of BHTNS was established using ABAQUS/Explicit to simulate the drop-weight experiment (Figure 8).

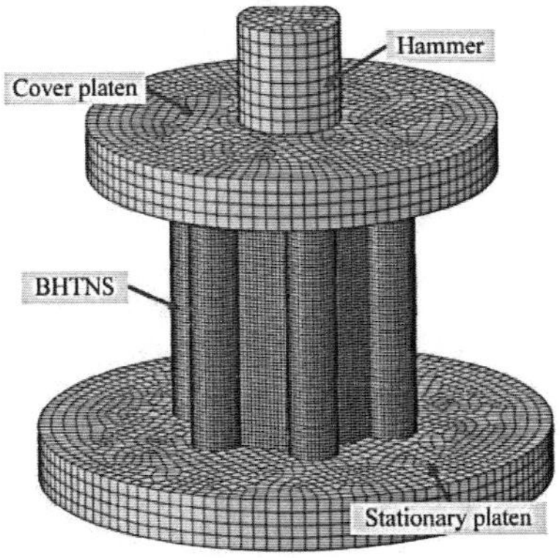

Figure 8. Finite modeling based on initial drop-weight experimental setups.

Aluminum alloy bio-inspired structures outperformed conventional honeycombs in terms of energy absorption. A 500 kg load and 10 m/s speed LS-DYNA investigation demonstrated their improved specific energy absorption [52]. Additionally, the ballistic impact resistance of bio-inspired composite scales and scale-like protective systems was evaluated numerically using LS-DYNA, with a focus on penetration resistance [53]. Using optimized hexagonal unit cell dimensions and density, ultralight bamboo-like structures with certain attributes were assessed using drop testing and Ansys simulations [15]. Energy absorption of lattice structures was shown to vary experimentally and numerically (using ABAQUS), with BLS A-6-0.75 and BLS B-6-0.2.5 showing slightly higher EA [48].

Researchers examined the mechanical durability and energy absorption of a bio-inspired honeycomb anti-collision pier using ABAQUS, selecting a central triangular section for enhanced performance [7]. Aluminum honeycombs with a horseshoe design that were inspired by nature were examined in LS-DYNA, and the results showed a considerable advantage in energy absorption, with a 63.9% increase in specific energy absorption [30]. A design inspired by trabecular bone demonstrated a 29% peak reduction and a 69% increase in energy absorption through successive buckling and collapse mechanisms. Using drop-hammer testing and Ansys simulations, a lightweight structure design inspired by the rostrum was optimized, yielding much greater total and specific energy absorption.

3.2.2 Deformation analysis: Numerical simulations via FEA predict material and structural deformation under various loads with quasi-static compression. Hybrid structures with external circle sections show improved deformation modes compared to inner circle sections [25]. Deformation and energy absorption analysis of bionic gradient thickness (BGT) tubes conducted in Ls-Dyna. Bionic circular tube (BCT) outperforms uniform-thickness circular tube (CT) in deformation and energy absorption, verified through quasi-static experiments [16]. LS-DYNA used to analyse horseshoe-shaped honeycombs, demonstrating improved plateau force and higher specific energy absorption, despite increased initial peak force [30]. Abaqus CAE 2017 simulations employed to study lattice structures' deformation and failure modes, revealing stress concentration sites [49].

Compression experiments are performed using bio-inspired energy-absorbing materials created utilising additive manufacturing with Beetle electron-DSM Somos 14120 resin. With a 15% greater compressive strength, 63% more deformation, and a staggering 115% increase in energy absorption, BEP outperforms HP. HP has an energy absorption of 154.80J and a specific energy absorption of 9.16 x 103 J/kg [54]. The geometrical characteristics are very important in the deformation mode. Conical corrugation tube (CCT), a unique tubular corrugated design resembling a coconut tree profile, was used to maximize energy absorption and reduce the initial peak crushing force. When compared to its circular straight tube and tapered tube equivalents, the CCT's initial peak force was significantly lower [55].

4. Optimization strategies for biomimetic energy absorption

Optimization strategies for energy absorption depend on problem complexity and design variables. Optimization objectives of biomimetic structure is based on specific energy absorption and peak crushing force that should be optimum solution [50].Common approaches include multi-objective optimisation with surrogate models for hybrid multi-cell structures [10],using Non-dominated Sorting Genetic Algorithm II (NSGA-II) for ship fender design to maximise SEA and minimise CFE Optimizing bionic structures inspired by bamboo with rate-dependent material models was implemented using Non-Sorting Genetic Algorithm 2 (NSGA2) in which 17% increase in energy absorption capacity was achieved [17]. Researchers in [17] investigated optimizing bionic tubes with different rib types to improve SEA. Using (NSGA-2), they determined the best-performing rib design, finding that the simple I-shape rib type P1 with 20 ribs yielded the highest energy absorption.

Improved crash performance and deformation mode of foam-filled ship fenders under quasi-static and dynamic stresses were the goals of [56]. Through simulation, six distinct fender models were evaluated. The examination criteria included crashworthiness factors including Specific Energy Absorption (SEA) and Crushing Force Efficiency (CFE). Finite element analysis was used to estimate crash responses and compare them to reference and experimental data. Height, foam density, thickness, and material were the four design parameters that were optimized. Non-dominated Sorting Genetic Algorithm II was used to increase SEA and decrease CFE. The Model 5 performed the best, showcasing its ability to replace traditional fender designs. Researchers [16] improved the BCT's (Bionic Composite Tube) crashworthiness design. They used multi-objective optimisation while taking the effect of load angle uncertainty into consideration. Peak force (PF) minimization and specific energy absorption (SEA) maximization were the two objectives. In comparison to the standard CT (Composite Tube), the ideal BCT showed a 9.3% increase in SEA and a 4% reduction in mass.

5. Validation of numerical models

Validating numerical models by comparing them with experimental or real-world data is crucial to ensure accurate representation of physical system behavior. Numerous studies have validated numerical results for both static and dynamic loading scenarios using experiments. For example, researcher in [32] studied axial crushing energy absorption of the BHTNS structure, with drop-weight impact experiments and computational models demonstrating reasonable agreement. LS-DYNA-based finite element modeling confirmed the impact performance of 3D-printed nacre-like composites, showing good alignment with experimental results. As shown in Figure 9, below comparison between experimental and numerical results were done on are:

a) Bionic gradient thickness (BGT) tubes [16],
b) Bionic honeycomb tubular nested structure (BHTNS) [52],
c) Hybrid tubes [25],
d) Bionic honeycomb thin-walled structure (BHTS) which filled the column in different ways inspired by the internal structure of the ladybeetle [53], and
e) Tube structure [57].

For the sake of comparison, the validation of numerical experiment with experimental work on different structures, reported in some research works are listed in Table 1. The error range obtained in these studies indicate that numerical research is well validated to be used instead of experimental research with significant cost benefits.

IOP Conf. Series: Materials Science and Engineering 1294 (2023) 012046 doi:10.1088/1757-899X/1294/1/012046

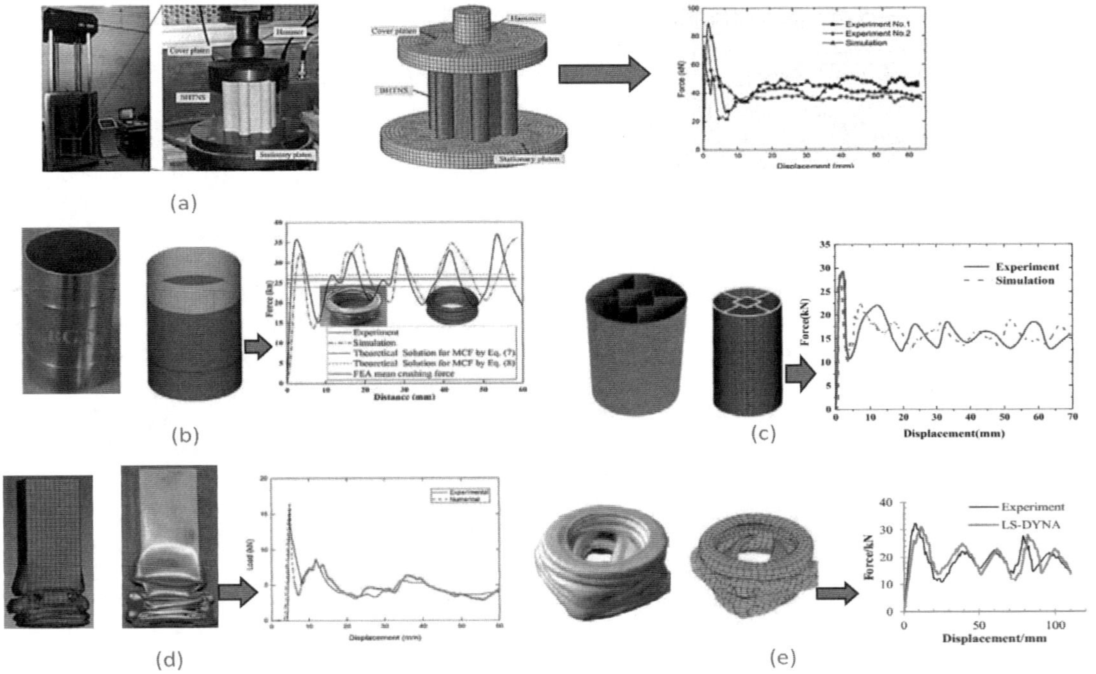

Figure 9. a comparison of experimental and numerical results for the following bionic structures: (a) bionic gradient thickness (BGT) tubes; (b) bionic honeycomb tubular nested structure (BHTNS); (c) hybrid tubes; (d) bionic honeycomb thin-walled structure (BHTS); (e) tube structure.

Table 1. Comparison of experimental and numerical results.

Structure	Methods	EA (J)	SEA (J/g)	PF(KN)	Ref.
BCT	Experiment	706.85	11.56	11.45	
	Numerical	673.82	11.38	10.81	[20]
	% error	4.9	1.61	5.92	
CS1	Experiment	-	9.74	29.65	[25]
	Numerical	-	9.29	30.14	
	% error	-	-4.62	1.65	
Foam filled fender	Experiment	133.569	-	21.9	[56]
	Numerical	133.637	-	22.4	
	% error	0.398	-	2.28	
Tube structure	Experiment	-	2.083	-	[57]
	Numerical	-	2.02	-	
	% error	-	0.903	-	
TPMS-CS structure	Experiment	669.97	-	66.07	[58]
	Numerical	669.97	-	65.92	
	% error	0.33	-	-1.34	

6. Conclusions

The finding of this study highlighted the significance of conducting impact and energy absorption analyses using FEA tools like ABAQUS, ANSYS, and LS-DYNA. Multi-objective optimisation approaches are recommended to handle trade-off problems and balance conflicting design objectives. Diverse biomimetic materials and structures are being actively investigated in research with the aim of enhancing energy absorption in a range of engineering applications. There are, nevertheless, considerable gaps in the literature. Although dynamic events frequently occur in biomimetic systems, the majority of recent study has been concentrated on quasi-static condition. To improve impact protection and crashworthiness, it is crucial to study the dynamic behaviour and energy absorption capabilities of biomimetic structures under high-energy situations.

Several key research gaps in the field of biomimetic structures and techniques of their analysis for energy absorption are noted for future work based on the literature review described in this article. Further research into alternative optimisation approaches is necessary despite the success of optimisation algorithms like genetic algorithms employing NSGA-II and particle swarm. Advanced optimisation techniques for the geometry and topology of structures, including their use in machine learning and artificial intelligence to find the best configurations suited to energy absorption needs, should also be investigated. Additionally, future study should also focus on the possible advantages of mixing several materials in one structure to maximise energy absorption, especially in figuring out how these materials interact with one another. In summary, this study highlights the promise of biomimetic structures in the field of energy absorption and impact prevention, while also identifying crucial areas for future research to address existing gaps and advance our understanding and capabilities in this field.

References

[1] Z. Zhao, X. Li, H. Jiang, X. Su, X. Zhang, and M. Zou 2023 Study on the mechanical properties and energy absorbing capability of polyurethane microcellular elastomers under different compressive strain rates, *Polymers.*,**15**(3), 778.

[2] A. Y. Patil, C. Hegde, G. Savanur, S. M. Kanakmood, and I. R. Fattah 2022 Biomimicking nature-inspired design structures-an experimental and simulation approach using additive manufacturing *Biomimetics,* **7**(4), 186.

[3] I. D. Wijegunawardana and W. R. de Mel 2021 Biomimetic designs for automobile engineering: A review, *IJAME*, **18**(3), 9029–9041,

[4] J. Yang, D. Gu, K. Lin, Y. Zhang, M. Guo, and L. Yuan 2022 Additive manufacturing frontiers laser additive manufacturing of bio-inspired metallic structures, *Chinese J. Mech. Eng. Addit. Manuf. Front.*, **1**(1), 100013.

[5] V. F. González-Albuixech, M. Rodríguez-Millán, T. Ito, J. A. Loya, and M. H. Miguélez 2018 Numerical analysis for design of bioinspired ceramic modular armors for ballistic protections, *Int. J. Damage Mech.,* vol. **28**(6) 815–837.

[6] Liang Meng, Hailong Liang, Hanchen Yu, Jingjing Yang, Fangzhi Li, Zemin Wang, and Xiaoyan Zeng 2019 The energy absorption and bearing capacity of light-weight bio-inspired structures produced by selective laser melting, *J. Mech. Behav. Biomed. Mater.*, **93**, 170–182.

[7] B. H. A. Pier, J. Wei, Y. Wang, J. Wang, X. Yao, and D. Wang 2022 Mechanical strength and energy absorption optimization of biomimetic honeycomb anti-Collision Pier, *Buildings*, **12**(11), 1941.

[8] G. Sun, D. Chen, X. Huo, G. Zheng, and Q. Li 2018 Experimental and numerical studies on indentation and perforation characteristics of honeycomb sandwich panels, *Compos," Struct.,***184**, 110–124.

[9] M. Xu, L. Pan, J. Chen, X. Zhang, and X. Yu 2019 The flexural properties of end-trabecular beetle elytron plates and their flexural failure mechanism, *Journal of Materials Science,* **54**(1)

[10] P. Wang, F. Yang, H. Fan, and G. Lu 2023 Bio-inspired multi-cell tubular structures approaching ideal energy absorption performance, *Mater. Des.,* **225**, 111495.

[11] J. Du, P. Hao, and L. Li 2021 Finite element analysis of energy absorption characteristics for biomimetic thin-walled multi- cellular structure inspired by horsetails, *Mech. Adv. Mater. Struct.*, vol. **29**, 6982–6993.

[12] R. Shao, C. Wu, Y. Su, Z. Liu, J. Liu, and S. Xu 2019 Numerical analysis on impact response of ultra-high strength concrete protected with composite materials against steel ogivenosed projectile penetration, *Compos. Struct.,* **220**, 861–874.

[13] S. Wang, J. Wang, and K. Komvopoulos 2023 Mechanical behavior of bamboo-like structures under transversal compressive loading, *Biomimetics,* **8**(1), 103.

[14] S. H. Siddique, P. J. Hazell, G. G. Pereira, H. Wang, J. P. Escobedo, and A. A. H. Ameri 2023 On the mechanical behaviour of biomimetic cornstalk- inspired lightweight structures, *Biomimetics*, **8**(1), 92.
[15] J. Wang, D. Hu, Z. Zhang, B. Pei, R. Xu, and X. Wu 2022 Anti-impact performance of bionic tortoiseshell-like composites, *Compos. Struct.,* **303**, 116315.
[16] J. Song, S. Xu, J. Zhou, H. Huang, and M. Zou 2021 Thin-walled structures experiment and numerical simulation study on the bionic tubes with gradient thickness under oblique loading, *Thin-Walled Struct.,* **163**(2), 107624.
[17] P. N. Panirani and J. Ghanbari 2022 Design and optimization of bio-inspired thin- walled structures for energy absorption applications, *Int. J. Crashworthiness*, **28**(1), 1-12.
[18] Ngoc San Ha 2020 Biomimetic structures and materials for energy absorption, *Euro. J. Appl. Eng. Sci*, **S**(3).
[19] J. Song, S. Xu, L. Xu, J. Zhou, and M. Zou 2020 Thin-walled structures experimental study on the crashworthiness of bio-inspired aluminum foam-filled tubes under axial compression loading, *Thin-Walled Struct.,* **155**, 106937.
[20] X. Yu, L. Pan, J. Chen, X. Zhang, and P. Wei 2019 Experimental and numerical study on the energy absorption abilities of trabecular honeycomb biomimetic structures inspired by beetle elytra, *J. Mater. Sci.,* **54**(3), 2193–2204.
[21] H. B. Rebelo, D. Lecompte, C. Cismasiu, A. Jonet, B. Belkassem, and A. Maazoun, 2019 Experimental and numerical investigation on 3D printed PLA sacrificial honeycomb cladding, *Int. J. Impact Eng.,* **131**, 162–173.
[22] N. S. Ha and G. Lu 2019 A review of recent research on bio-inspired structures and materials for energy absorption applications. *Composites Part B: Engineering*, **181**, 107496.
[23] A. Othman and A. E. Ismail 2018 Finite element modelling of AA6063T52 thin-walled tubes under quasi-static axial loading, *IOP Conf. Ser. Mater. Sci. Eng*., **342**(1).
[24] A. Ghazlan, T. Nguyen, T. Ngo, S. Linforth, and V. T. Le 2020 Performance of a 3D printed cellular structure inspired by bone, *Thin-Walled Struct.,* **151**, 106713.
[25] T. Chen, Y. Zhang, J. Lin, and Y. Lu 2018 Thin-walled structures theoretical analysis and crashworthiness optimization of hybrid multi-cell structures, *Thin Walled Struct.,* **142**, 116–131.
[26] N. S. Ha and G. Lu 2020 Thin-walled corrugated structures: A review of crashworthiness designs and energy absorption characteristics, *Thin-Walled Struct.,* **157**, 106995.
[27] W. Liu, Z. Lin, N. Wang, and X. Deng, 2016 Dynamic performances of thin-walled tubes with star-shaped cross section under axial impact, *Thin-Walled Struct.,* **100**, 25–37.
[28] L. Liu, L. Li, C. Guo, Y. Ge, Y. Chen, and L. Zhang 2023 The design of a biomimetic hierarchical thin-walled structure inspired by a lotus leaf and its mechanical performance analysis, *Materials*, **16**(11), 4116.
[29] T. Xu, N. Liu, Z. Yu, T. Xu, and M. Zou 2017 Crashworthiness design for bionic bumper structures inspired by cattail and bamboo, *Appl, Bionics Biomech*, **2017.**
[30] X. Yang, Y. Sun, J. Yang, and Q. Pan 2018 Thin-walled structures out-of-plane crashworthiness analysis of bio-inspired aluminum honeycomb patterned with horseshoe mesostructure, *Thin Walled Struct.,* **125,** 1–11.
[31] H. Zhang, Z. Huang, T. Li, C. Bao, and L. Zhang 2022 Multiple performance evaluation of bionic thin-walled structures with different cross sections considering complex conditions, J. *Environ. Public Health,* **2022.**
[32] D. Hu, Y. Wang, B. Song, L. Dang, and Z. Zhang 2018 Energy-absorption characteristics of a bionic honeycomb tubular nested structure inspired by bamboo under axial crushing, *Compos, Part B Eng,* **162**, 21–32.
[33] H. Yin, Y. Xiao, G. Wen, Q. Qing, X. Wu 2015 Crushing analysis and multi-objective optimization design for bionic thin-walled structure, *Mater. Des*, **87**, 825–834.
[34] E. Mahdi, D. Ochoa, A. Vaziri, and E. Eltai 2019 Energy absorption capability of date palm leaf fiber reinforced epoxy composites rectangular tubes, *Compos. Struct.,* **224**, 11100

[35] A. Tasdemirci, E. F. Akbulut, E. Guzel, F. Tuzgel, A. Yucesoy, and S. Sahin 2018 Thin-walled structures crushing behavior and energy absorption performance of a bio-inspired metallic structure: Experimental and numerical study, *Thin Walled Struct.,* **131**, 547–555.
[36] Shangqin Yuan, Chee Kai Chua, and Kun Zhou 2018 3D - printed mechanical metamaterials with high energy absorption, *Mater. Technol*, **4**(3).
[37] V. H. Carneiro, S. D. Rawson, H. Puga, J. Meireles, and P. J. Withers 2020 Additive manufacturing assisted investment casting: a low-cost method to fabricate periodic metallic cellular lattices, *Addit, Manuf.,* **33**, 10108.
[38] P. Wang, F. Yang, D. Ru, B. Zheng, and H. Fan 2021 Additive-manufactured hierarchical multi-circular lattice structures for energy absorption application, *Mater. Des.,* **210,** 110116.
[39] Y. Zhang, T. Liu, and W. Tizani, 2018 Experimental and numerical analysis of dynamic compressive response of nomex honeycombs, *Compos. Part B,* **148**, 27–39.
[40] Chen Pan, Yafeng Han and Jiping Lu 2020 Design and optimization of lattice structures: A review. *Appl. Sci*, **10**(18), 6374
[41] M. K. Islam, H. Wang, P. J. Hazell, M. A. Kader, and J. P. Escobedo 2022 Quasi-static response of horse hoof inspired biomimetic structures, *Mater, Today Proc.*, **59**(1),1168–1173.
[42] W. Zhang, S. Yin, T. Yu, and J. Xu 2019 Crushing resistance and energy absorption of pomelo peel inspired hierarchical honeycomb, *Int, J. Impact Eng*, **125,** 163–172.
[43] W. Zhang, S. Yin, T. X. Yu, and X. J. C 2019 Crushing resistance and energy absorption of pomelo peel inspired hierarchical honeycomb, *Int. J. Impact Eng.*,**125**, 163–172.
[44] M. Kucewicz, P. Baranowski, M. Stankiewicz, M. Konarzewski, P. Płatek, and J. Małachowski 2019 Modelling and testing of 3D printed cellular structures under quasi-static and dynamic conditions, *Thin-Walled Struct.,* **145**, 1-15
[45] C. Zeng, L. Liu, W. Bian, J. Leng, and Y. Liu 2021 Compression behavior and energy absorption of 3D printed continuous fiber reinforced composite honeycomb structures with shape memory effects *Addit. Manuf.,* **38**(6), 10184.
[46] D. S. J. Al-Saedi, S. H. Masood, M. Faizan-Ur-Rab, A. Alomarah, and P. Ponnusamy, 2018 Mechanical properties and energy absorption capability of functionally graded F2BCC lattice fabricated by SLM, *Mater, Des.*,**144**, 32.
[47] Zuhal Ozdemir, Everth Hernandez-Nava, Andrew Tyas, James A. Warren, Stephen D. Fay, Russell Goodall, Iain Todd, and harm askes 2016 Energy absorption in lattice structures in dynamics: Experiments, *Int. J. Impact Eng.*, **89**, 49–61
[48] E. Cetin and C. Baykasoğlu 2019 Energy absorption of thin-walled tubes enhanced by lattice structures, *Int. J. Mech. Sci.,* **157**, 471–484.
[49] Lucrezia Greco, Federica Buccino, Zhuo Xu, Laura Vergani, Filippo Berto, Mario Guagliano, Nima Razavi & Sara Bagherifard 2023 Design and analysis of energy absorbent bioinspired lattice structures, *J. Bionic Eng.,* **20**,1670–1686.
[50] K. Pajunen, P. Johanns, R. K. Pal, J. J. Rimoli, and C. Daraio 2019 Design and impact response of 3D-printable tensegrity-inspired structures, *Mater. Des.,* **182**, 107966.
[51] T. Van Le, A. Ghazlan, T. Ngo, and T. Nguyen 2020 Performance of a bio-mimetic 3D printed conch-lik structure under quasi-static loading, *Compos. Struct.,* **246**, 112433.
[52] D. Hu, Y. Wang, B. Song, L. Dang, and Z. Zhang 2019 Energy-absorption characteristics of a bionic honeycomb tubular nested structure inspired by bamboo under axial crushing, *Compos, Part B*, **162**, 21–32.
[53] P. Liu, D. Zhu, Y. Yao, J. Wang, and B. TQ 2016 Numerical simulation of ballistic impact behavior of bio-inspired scale-like protection system, *Mater. Des.,* **99**, 201–210.
[54] A. Ingrole, T. G. Aguirre, L. Fuller, and S. W. Donahue 2021 Materials Bioinspired energy absorbing material designs using additive manufacturing, *J. Mech. Behav. Biomed. Mater.,* **119**, 104518
[55] N. S. Ha, G. Lu, and X. Xiang 2018 International Journal of Mechanical Sciences High energy absorption efficiency of thin-walled conical corrugation tubes mimicking coconut tree configuration, *Int. J. Mech. Sci.,* **148**, 409–421

[56] F. Djamaluddin 2023 Finite element analysis and optimization of foam filled fender under quasi static and dynamic responses, *Front. Mech. Eng,* **9.**

[57] S. Pirmohammad and S. Esmaeili-Marzdashti 2019 Multi-objective crashworthiness optimization of square and octagonal bitubal structures including different hole shapes, *Thin-Walled Struct.,* **139**, 126–138.

[58] L. Liang, H. Zhu, D. Wei, Y. Wu, and W. Li, 2023 Energy absorption and multi-objective optimization of TPMS filled cylinder shell structures, *Electron. Res. Arch*, **31**(5), 2834-2854.

IOP Conf. Series: Materials Science and Engineering 1294 (2023) 012047 doi:10.1088/1757-899X/1294/1/012047

Process parameter modelling and optimization techniques applied to fused deposition modelling: A review

T Batu[1, 2*], H G Lemu[3], H G Shimels[4], and M Zebenay[5]

[1] Ethiopian Space Science and Geospatial Institute, Ethiopia
[2] Ethiopian Defence University, Ethiopia
[3] University of Stavanger, Norway
[4] Addis Ababa Science and Technology University, Ethiopia
[5] SSGI, Goddard Space Flight Center, USA

[*] Corresponding author's e-mail:- temesgen.batu@kiot.edu.et

Abstract. Manufacturing is the foundation of any industrialized country and involves making products from raw materials using various processes. Additive manufacturing (AM) was originally created as a method for swift prototyping, allowing the visualization, testing, and validation of a design prior to final production for end-users. FDM is the most commonly used additive manufacturing process for constructing products and prototypes. It encompasses numerous process parameters that impact the quality of manufactured products. Properly selecting these process parameters is crucial for producing products at a lower cost while enhancing mechanical properties, build time, and part quality, among other factors. Therefore, in the past, researchers have optimized the process parameters to achieve the desired product outcomes. In the present study, we provide an overview of FDM process parameters and review various design optimization methods. We present several experimental designs, such as the Taguchi method, response surface methodology, and design of experiments, as well as computational approaches like artificial intelligence, and machine learning.

1. Introduction

In any industry, manufacturing is a challenging sector due to its extraordinarily complex components. Among the recent advances in the manufacturing sector, additive manufacturing (AM) is the most recent method for fabricating complex components from 3D Computer-Aided Design (CAD) geometry, using the process illustrated in Figure 1 [1]. The AM process builds 3D objects or products layer by layer from a Computer-Aided Design (CAD) model, and this capability has experienced unprecedented growth as a manufacturing tool in some corporations. Combined with the advances in topology optimization, AM process aims to reduce mass or utilize materials where needed and has been successfully adopted in many engineering applications [2]. The main advantage with use of AM process is its ability to directly transform a computerized 3D model data into a finished product with no need for auxiliary tools. This facilitates the fabrication of parts having complex geometry that are difficult tfor conventional manufacturing methods [3].

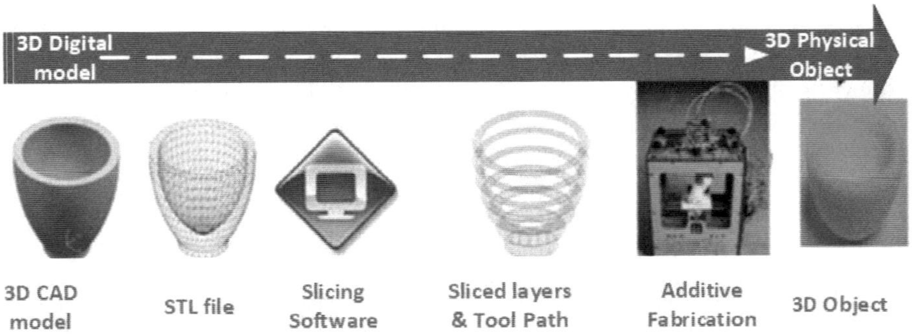

Figure 1. General additive manufacturing process flow (Adapted from [4])

Among the first forefront industries that utilized the amazing capabilities of additive manufacturing to transform their production are aerospace[5], medical[6], electronics[7], transportation, automotive[8], construction[9], healthcare monitoring[10] and sustainable energy generation [11]. There are different types of additive manufacturing techniques that have been developed recently. According to the American Society for Testing Materials (ASTM), AM has been classified into seven processes technologies as illustrated in Figure 2(a) [12,13].

From the various AM techniques, fused deposition modelling (FDM) process is one of the most popular and commonly used [14,15] in the industry. The main reason for its wider application are that FDM is cost-effective in manufacturing of thermoplastic parts and prototypes. Moreover, the wider availability of thermoplastic materials has made the lead times minimal and the process to be cheaper than other AM processes [12]. FDM was made commercially available in early1990s following the patent by the co-founder of Stratasys, Scott Crump in 1989 [16]. In the process, a continuous supply of thermoplastic filament is utilized for printing layers of materials to build the part. As illustrated in Figure 2(b), the material filament is heated to a semi-liquid phase upon heating the element inside the liquefying head, and this semi-liquid material is extruded through the nozzle on the printing platform.

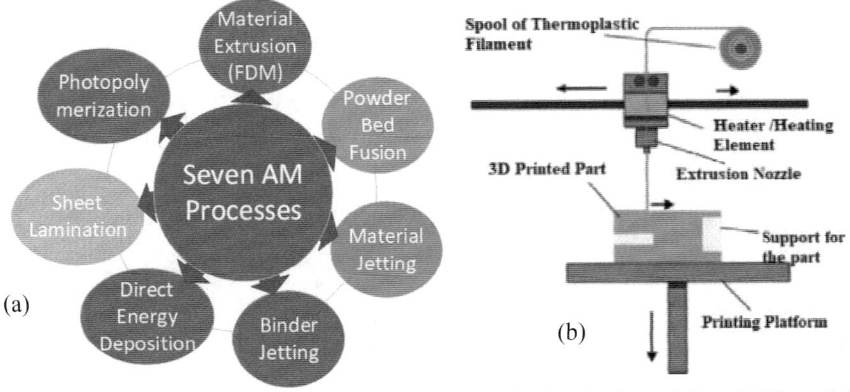

Figure 2. (a) Categorization of AM technologies (Adapted from[12]) and (b) Setup of FDM process [15,17]

The semi-liquid thermoplastic filament materials do not solidify immediately after extrusion from the nozzle onto the printing plate; instead, the material remains partially liquid and mergse during the construction of a specific layer, and then they solidify into a part that is built layer by layer at the prevailing ambient temperature [18]. As observed in Figure 2(b), support structures are essential in FDM process to ensure the stability of intricate designs. FDM offers versatile support options, including soluble support for complex geometries and break-away support for simpler ones. The requirement for support structures depends on the maximum overhang angle, which is influenced by material, layer height, and printer capabilities. Typically, most materials have a maximum overhang angle of around 45°, but advanced printers can exceed this without the need for support [19]. Among the main benefots of FDM, process simplicity, high-speed printing, and low cost can be mentioned. On the other hand, the disadvantages of the process include anisotropic properties, poor surface quality, layer-wise or stair case appearance, and limitation to access thermoplastic polymers because thermoplasticity is the essential property for a material to be 3D printed through FDM technique [15][20]. Since the mechanical characteristics and product quality of FDM-printed parts depend on the proper (or optimal) selection of process parameters, making the process suitable for mass production and more acceptable in the industries, finding the optimal process parameter combinations to improve the part quality and mechanical properties is very important [21]. The effect of the process parameters has a significant impact on the mechanical characteristics of the parts fabricated by FDM under static and dynamic loads [22]. The various common FDM process parameters, along with their descriptions are presented in Table 1.

Table 1 Process parameter of FDM and its description

Process parameter	Description	Reference
Layer thickness	Height of layers deposited after extrusion, determined by nozzle tip diameter and material	[14,15]
Build Orientation	Positioning of the part within the build platform in relation to X, Y, and Z directions.	[14,21].
Raster Angle/ Orientation	Angle of material deposition with respect to the X-direction on the build platform.	[14,15,21]
Air Gap	Distance between adjacent FDM-printed tool paths on a single layer.	[14,15,21]
Extrusion temperature	Temperature of thermoplastic filament materials inside the nozzle before extrusion.	[15,21]
Print Speed:	Speed of the nozzle tip in the XY plane during material deposition.	[15,21]
Infill pattern	Pattern used to form the internal structure of the FDM printed part (e.g., diamond, honeycomb).	[15,21]
Infill density/Interior infill	Solidity of the internal structure of FDM printed parts.	[15,21]
Nozzle diameter	Diameter of the extruder nozzle tip.	[15]
Raster width	Width of deposited beads along the extruder tool path.	[14,15,21]
Number of contours	Number of solid outer layers surrounding the internal infill pattern.	[14,15].
Contour width	Thickness of the outer contour layers.	[14,15]
Contour to contour Air gap	Distance between solid outer layers (contours).	[15]

The list in this table (Table 1) indicates that fabrication of components using FDM printing methods involve different processing parameters that play crucial roles in quality and performance of the printed parts. Accordingly, effects of these parameters on the quality of AM manufactured components have been investigated in several research works [14,15,21,23] Finding an optimum process parameter that improves the surface quality and mechanical properties (i.e., tensile strength, fracture toughness, plasticity, hardness, brittleness, and fatigue strength) of manufactured components is one of the research areas studied by researchers. To find an optimum amount of each parameters different optimization techniques were used. Thus, the goal of this survey is to summarize the optimization techniques that are applied in FDM AM methods. This review covers the period from 2000 to September 2023, and it offers well-organized information on objectives, process parameters studied, constraints, and optimization techniques for FDM process. This information can be valuable for researchers seeking to identify trends, challenges, and future directions in research gaps that exist in FDM process parameter optimization.

2. Highlights of studies on FDM process

In 2021, Dezaki, et al. [24] provided an overview of the research, development, and process optimization of FDM throughout history. They also presented an overview of the popular materials investigated to find out their features and mechanical properties in the FDM process. Similar work was also reported in [25], where a review was made to know the insights of one such AM process, i.e., FDM. Dey and Yodo

[21] conducted a comprehensive review of the current literature regarding the influence of parameters on part quality and process parameter optimization. In 2015, Mohamed et al. [14] conducted a comprehensive review of the research conducted up to that point to identify and enhance the process parameters in the FDM process. Future directions for FDM research in this domain are outlined focusing on the research on each quality characteristics (static and dynamic properties, material behavior, surface quality and build time). Quite recently, Rajan et al. [19] reported a review that explains the various 3D printing techniques used in FDM process as well as the polymer materials and polymer composites. The study also discussed the conducted investigations on different materials, parameters, properties, and the potential applications of FDM parameters. Kohad et al. [26] reviewed the identification of various FDM process parameters that affect the quality of the fabricated parts. They were also evaluated on two factors that influence fabricated parameters.

Even though there have been research achievements on FDM based AM process parameter optimization, the research directions for future development needs improvement. Some reviews about process parameter optimization in AM have been published, but mainly aiming at analyzing and discussing the general process parameter optimization work in[14,15,19] but gaps are observed on study of the modelling and optimization methods of process parameter in FDM as main study. The review also does not identify the objectives, design variables, and constraints that were used by the researchers. There is no doubt that the methods of modelling selecting or predicting an optimum process parameter for manufacturing high-quality components are a common problem for obtaining high quality product. Therefore, it is necessary to present a review of the process parameter modelling and optimization methods to summarize the existing achievements and give some future directions for further exploration in the finding of optimum process parameter for FDM based AM process. Besides, as far as the authors knowledge, similar review papers for process parameter optimization methods for FDM are neither under consideration nor have already been published elsewhere.

3. FDM process parameter modelling and optimization methods

Different optimization techniques have been used to obtain an optimal combination of all control variables to either minimize or maximize one or more desired outputs. These techniques are also used to identify the most influential parameters, the interactions between various parameters, and the extent to which an AM process parameter individually affects the output variables. Thus, this section presents experimental design approach and computational approach used for process parameter optimization of FDM based AM.

3.1. Experimental design approach

3.1.1. Taguchi method. The Taguchi method, a statistical approach also known as robust design, is often used in experimental design to enhance the quality of manufactured goods across various industries [27]. It minimizes the number of experiments needed, providing a systematic means to optimize designs in terms of quality and cost using orthogonal arrays of factors. Taguchi's primary goal is to reduce variability around target product properties through statistical experimental design, known as robust design. This method is extensively employed in AM research too to explore the impact of process parameters on additively manufactured components using this method. It integrates statistical and mathematical techniques to optimize performance traits, revealing effects with fewer experiments and focusing on controlling signal factors while mitigating noise factors that contribute to unpredictability [28].

Many studies used the Taguchi method along with analysis of variance (ANOVA) procedures to optimize the process parameters of FDM. These parameters include layer thickness [29–34], infill density [29,34] road width [31,32,35], speed of deposition [31,34], raster angle [30,32,33,36], air gap [30,32,33,36], raster width [30,36], slice height [36], deposition style [35], support style [35], print speed [29], direction of rotation [36], build orientation angle [36] and deposition orientation [35]. Most of the studies were conducted for ABS (acrylonitrile butadiene styrene) material with the goal of improving surface roughness [31,34,35] surface quality [32], dimensional accuracy [32,35], tensile strength [35], elastic performance [30,36], impact strength [34,36] build time [29,34] and production time [33]. Using the ANOVA, the researchers found significant parameters such as layer thickness[31,33] raster angle [36], raster width [36], and build orientation.

The Taguchi method was also employed for the optimization of FDM process parameters, namely extrusion and filling velocity, wire-width compensation and layer thickness of ABS material, to improve dimensional accuracy and reduce warpage deformation in printed parts [37]. Additionally, fuzzy logic was employed to optimize process parameters, including air gap, raster angle, layer thickness, raster width and orientation, aiming to enhance dimensional accuracy [38]. Gray Taguchi method was also utilized to optimize FDM process parameters, including wire-width compensation, extrusion and filling velocities, and layer thickness of ABS material to address dimensional errors and reduce warpage deformation in printed parts [39].

Other areas of application of the Taguchi method includes enhancing the quality of printed materials made from Polylactic Acid (PLA). For example, Maguluri et al [40] studied the influence of infill density, nozzle temperature, and printing speed on tensile properties using a Taguchi L8 array and obtained optimum values for each PLA specimens. Lokesh et al [41] examined the influence of various printing parameters on the mechanical properties of PLA specimens processed and tested as per ASTM standards through the Design of an Experiment (DOE), where the Taguchi approach by L9 orthogonal array was employed. To assess whether process variables have any significant effects, an ANOVA was performed. It was observed that layer thickness has more influence than build orientation and raster angle. Liu et al. [42] established a new theoretical model to reveal the distortion mechanism of PLA thin-plate parts in the FDM process, a theoretical model based on the theory of elastic thin plates in thermoelectricity. An experimental research methodology, utilizing the Taguchi method, was employed to design unique test specimens. Moreover, 81 test specimens were designed and prepared through the FDM process and measured by a portable 3D laser scanner. Two statistical analysis methods, the ANOVA and signal-to-noise ratio (S/N) were utilized to optimize process parameters and mitigate thin-plate part distortion. The experimental results indicated that the optimal process parameters can be obtained and that the proposed theoretical model was efficient.

In several studies, the Taguchi method was employed to optimize various process parameters in FDM. Santhakumar et al. [43] investigated the improvement of impact strength by optimizing FDM process parameters for polycarbonate material. They studied four key process parameters (1) layer thickness, (2) build orientation, (3) raster angle, and (4) raster width, and concluded using ANOVA analysis, that the most influential factor for impact strength in polycarbonate material is layer thickness. In their study, Anitha et al. [44] investigated the effects of parameters like layer thickness, road width, and deposition speed on FDM parts. They utilized an L18 orthogonal array for their experimental design. They found that layer thickness significantly affects the minimum surface roughness of FDM parts. Similarly, Alhubail et al. [45] investigated the effects of FDM variables, including layer thickness, air gap, raster width, contour width, and raster orientation, on surface roughness and tensile strength using Taguchi design (L32). Jiang et al. [46] explored the optimization of process parameters for FDM of Polyetheretherketone (PEEK) for biomedical implants, employing the Taguchi method to enhance tensile strength, print speed, layer thickness, temperature, and extrusion width. Finally, in [47], Gray Taguchi design of experiments was used to optimize FDM parameters leading to improvements in tensile strength, flexural strength, impact strength, and compression strength. Filippidis et al. [48] used a combined Taguchi with I optimality and grey analysis to determine the optimal combination: priming/retraction speed (8 mm/s), layer thickness (0.4 mm), temperature (190 °C) and nozzle speed (40 mm/s). The major influential factors for FDM process parameter optimization were layer thickness followed by printing speed and temperature.

3.1.2. Response Surface Method. Response surface methodology (RSM) integrates mathematical and statistical techniques to model and optimize processes. Its primary objective is to enhance responses influenced by various input parameters or factors. RSM employs the Design of Experiments (DoE) approach to collect sufficient data, enabling the establishment of a connection between controllable input parameters and the resulting outcomes[49]. This method is also used for the modelling and optimization of process parameter of FDM to improve the quality of printed parts.

In the research conducted by Wang et al. [50], RSM was used to optimize a multi-temperature parameter system for FDM by examining the effects of temperature conditions (nozzle, platform and environment temperature) on the tensile strength of carbon fiber/polylactic acid composite specimens using the constructed RSM model. The results showed that the RSM optimization method significantly improved tensile strength with a 3.2% gap compared to FDM results. Similarly, Mandge et al.'s research [51] utilized RSM to study the effect of process parameters on ABS specimen using RSM and validating the significance using ANOVA. The work aimed to overcome the shortcoming of the process while optimizing mechanical and thermal properties.

In the research conducted by Mohamed, et al. [52], they examined crucial FDM parameters using Q-optimal response surface methodology to investigate these parameters and how these parameters affect performance variables. To gauge the reliability of these models, ANOVA was employed to assess their adequacy and significance. This analysis led to the identification of optimal process parameter settings. To validate both the models and these optimized settings, a confirmation test was carried out. The results affirmed that the Q-optimal design approach shows promise as a method for optimizing FDM process parameters and validated the dependability of the developed models.

Equbal et al. [53] examined the performance of ABS P400 parts manufactured using FDM under compressive loading. They studied the effects of varying levels of three FDM process parameters: (1) raster angle, (2) air gap, and (3) raster width on three responses, including compressive stress, percentage deformation, and breaking stress of the fabricated parts. Experimental results were analyzed using ANOVA, response graphs, and 3D surface plots, revealing the anisotropic nature of ABS P400 parts and the significant impact of chosen process parameters on compressive properties. The relationships between process parameters and responses were complex, and predictive models for these responses were developed and validated through additivity tests. Multi-objective optimization using the desirability function approach was employed to discover the best combination of FDM process parameters, and the results were validated with a confirmatory test. The study provided valuable insights into the effects of critical FDM parameters on compressive properties beyond CS, including % D and BS of ABS P400 fabricated parts.

Comparison of Taguchi method with RSM was also reported in [54] using four parameters, namely (1) layer thickness, (2) extrusion temperature, (3) print speed and (4) raster width. In addition, their interaction terms are identified as control variables having three levels, where tensile strength and compressive strength are selected responses. L27 orthogonal array and face-centered central composite design (FCCCD) were used in the experiment using Taguchi and RSM, respectively. The S/N ratio and ANOVA were used to get the optimal FDM parameter combination as well as the main factors that influence the performance of the PLA samples. The results of the experimental results revealed that both Taguchi method and RSM succeeded in predicting better results compared with the original groups. In addition, the optimum combinations for tensile strength and compressive strength obtained from the RSM were 2.11% and 8.15% higher than the Taguchi method, respectively. Furthermore, in a study conducted by Panda et al. [55], they explored the influence of five crucial parameters: air gap, layer thickness, orientation, raster angle and raster width. Their investigation aimed to reveal the impact of these parameters on three specific outcomes of the test specimens:

(1) tensile strength,
(2) flexural strength, and
(3) impact strength.

The experiments were carried out using a central composite design (CCD), and they developed empirical models that established the relationships between each response and the process parameters. To verify the reliability of these models, ANOVA was employed for validation. In the end, a bacterial foraging technique was employed to recommend a theoretical combination of parameter settings that would yield strong performance across all the responses simultaneously

3.1.3. Factorial design. Full Factorial Design (FFD) is a classical optimization approach that employs statistical methods to demonstrate the relationship between variable parameters influencing the response through a linear regression model [56]. Information necessary for constructing the response model is gathered through experimental or simulation work, and the FFD method can be used to simultaneously investigate the effect of multiple independent variables and their interactions. Statistically, a FFD incorporates two or more variables within the experimental design, where each variable has distinct possible values or levels. The experimental units are displayed to all conceivable combinations of the levels over all variables. This approach facilitates the investigation of how each component influences the variables and the interactions among the factors affecting them [28].

Various factorial designs were utilized for modeling and optimizing process parameters in FDM printed parts, such as 2^3 [57], 3^2 [57], 2^5 [58], and 2^4 [59] full factorial designs. These designs encompassed parameters such as model temperature [57,60], layer thickness [57,58], part fill style [57], orientation [58], contour width [58], raster angle [58,59], part raster width [58], air gap [59–61], raster orientation [60], bead width [60], color [60], build orientation [61], raster width [59–61], build layer [61], part orientation [59], and build laydown pattern [61]. The primary objectives of these studies were to enhance tensile strength [59,60], surface roughness [57], reduce porosity [61], improve compressive yield strength [61], compressive modulus [61], compressive strength [60], support material volume [58], and build time [58] for FDM printed parts. Also for the 3D printed bottom housing part made from PLA, Haidiezul et al. [62] employed FFD to optimize shrinkage on the printed parts. The results of the optimization work demonstrated that the FFD significantly improved dimensional accuracyof the printed part. Gebisa and Lemu [63,64]conducted a FFD experiment on the high-performance ULTEM 9085 polymeric material. They analyzed the influence of five process parameters: air gap, contour number, raster width, raster angle and contour width. Their research findings indicated that only the raster angle had a substantial impact on the material's tensile properties [64], while both raster width and raster angle had the most effect on the flexural strength [63].

3.2. Computational approaches

3.2.1. Genetic Algorithm (GA). The GA is a computational method employed for optimizing process parameters within both conventional manufacturing industries [65] and in the realm of FDM. Specifically, it played a pivotal role in optimizing a myriad of process parameters. These encompassed layer thickness [66–69], orientation angle [66,67,69], raster angle [66,67,69], raster width [66,67,69], printing temperature [68], infill pattern [68], slice thickness [70], road width [70], liquefier temperature [70], and air gap [67,69,70]. The overarching aim was to minimize dimensional variability [67], reduce build time [71] , enhance accuracy [71], refine surface roughness [70], and meticulously control porosity [70] in FDM parts. Furthermore, this adaptable algorithm tailored its optimization strategies for various materials, including polymeric biocomposites [67], ABS [66], and copper-reinforced ABS [68]. Diverse GA types were harnessed, exemplified by the hybrid genetic algorithm [66] and the innovative three-step genetic algorithm-based approach [71]. Moreover, the GA was seamlessly integrated with different experimental design methodologies to optimize FDM process parameters. Notably, it harmonized with the Taguchi L9 technique [68] and was paired with RSMs [66,70] to achieve unprecedented precision.

3.2.2. Artificial Neural Network. The optimization of FDM process parameters for part quality improvement by using traditional methodologies will be costly and time consuming for the required level of precision. Thus, researchers give attention to an artificial neural network (ANN) method for process parameter modelling and optimization.

In their study, Giri et al. [72], employed critical process parameters as input variables for an ANN and utilized the number of contours to optimize and enhance the properties of 3D printed parts using FDM. The 3D printing material in this case was PLA. They trained the data sets and optimized them by utilizing ANN's function approximation. The ANN effectively forecasted the experimental data, achieving notably high correlation coefficients of $R = 0.9981$ for tensile strength, $R = 0.9984$ for build time, and R = 0.99837 for surface roughness, demonstrating the accuracy of the predictions. The root mean square error obtained for the three ouputs was 0.5543, 0.578 and 0.241. Furthermore, their research identified that the build orientation is the most crucial parameter for achieving optimal results in their study. Research from Lyu and Manoochehri[73] presented the study on the predictive model of FDM to optimize the dimensional accuracy of the process. Three process parameters, (1) extruder temperature, (2) layer thickness, and (3) infill density were considered. To achieve better prediction accuracy, three models were studied, (1) multivariate linear regression, (2) ANN and (3) SVR. These models were employed to characterize the complex relationship among the input variables and dimensions of fabricated parts. Based on the experimental dataset, it was observed that the ANN model performs better than the other two models. Correspondingly, the paper presented by Selvam et al. [74] characterized the influence of five manufacturing parameters on a part's ultimate tensile strength (UTS) and modulus of elasticity (E) experimentally, which was used to train an ANN. This ANN forms the basis of a capability profile that was shown to be able to represent the mechanical properties with RMSEP of 1.95 MPa for UTS and 0.82 GPa for E. They validated the capability profile and incorporated into a generative design methodology enabling its application to the design and manufacture of functional parts.

In addition to the above-discussed methods, there are also other methods used for optimizing process parameters of FDM. For instance, enabled teaching learning based algorithms [75] and particle swarm optimization [76–78] were employed to optimize the process parameters of FDM in order to enhance the quality of the manufactured parts.

4. Discussions

4.1. Observation of the trends and challenges in FDM parameter optimization

In today's competitive market, the quality of manufactured parts like dimensional accuracy, surface finish, manufacturing cost, mechanical strength, etc. is highly crucial to satisfy and attract customers. But as discussed earlier in this paper, the quality of parts produced by FDM based AM process highly depends on various process parameters. As a result, the ongoing research needs to focus on the continuous optimization of process parameters in the FDM process. Different process parameter modelling and optimization methods were applied by researchers during the development history of FDM based AM. In the present work, a review of methods used for AM process parameter optimization is presented.

To provide a review of optimization procedures for AM parameters, the applications of different optimization methods, such as the Taguchi method, genetic algorithms, artificial neural networks, response surface methodology, and factorial design, in the optimization process of FDM parameters are investigated. From this review, it is clearly observed that Taguchi method is mostly used and suggested for optimizing the process parameter of FDM process parameters. Taguchi helps to determine optimal sequence and ANOVA technique helps to determine which parameters are most significant and their percentage contribution. Taguchi methodology is widely used for the single and multi-optimization.

Even if different process parameter optimization methods were employed to determine optimal process parameters to improve aesthetics, build time, mechanical properties, surface finish and model material consumption, it's important to note that there are no universally ideal conditions that apply to all types of parts and materials [14]. It depends on the materials and indicates that for some materials, layer thickness is the critical parameter, and for others, another critical parameter do exist. So, future researchers must focus on finding the optimal condition for each material. With the same concept, there is no optimal condition for different mechanical properties. For instance, layer thickness is critical for flexural, tensile, and compressive strength. But for surface roughness and other properties, layer thickness is not the first critical parameter. Thus, identifying the optimal condition for each parameter for each property based on the applications area is expected. Also, it has been observed that there is no any method which considered all process parameters of FDM in optimization. So, for the future, it is expected that all parameters will be considered.

In comparison to components produced using traditional manufacturing processes like injection molding, FDM-manufactured parts typically exhibit lower mechanical properties and surface finish [14,79]. To enhance the quality and mechanical characteristics of parts produced through FDM, it is imperative to gain a comprehensive understanding of the interplay between material properties and process parameters.

Regarding material properties, the majority of studies have primarily concentrated on optimizing process parameters, utilizing the previously mentioned methods, to enhance the mechanical properties of ABS parts. However, there is a notable absence of published research articles that address the optimization of FDM process variables for the chemical, thermal and dynamic properties of FDM-fabricated parts using different materials. Consequently, there is a considerable need for extensive research in this domain in the future.

The majority of researchers have considered air gap, layer thickness, raster angle, raster width, and build orientations as input parameters or variables in their studies. While studying the effects of process parameters on the required response, air gap, layer thickness, and raster angle are important parameters to be considered. In this method, the majority of the researchers have done their work by considering the FDM parameters like air gap, layer thickness, and raster angle. In the end, they concluded that layer thickness is the most significant factor for build time, increasing impact strength, and minimizing surface roughness in FDM.

In the context of optimization, it's important to consider the physical constraints that impact the selection of optimal process settings in FDM. These constraints include limited choices for layer thickness, often confined to four specific values like 0.1270, 0.1778, 0.2540, and 0.3302 mm, which are determined by the nozzle diameter, leaving operators with no flexibility to choose other values. Additionally, each nozzle diameter is associated with its own range of raster widths. Furthermore, there is a restriction on the number of contours an operator can use when needed, and they must operate within predefined limits. These constraints have significant implications for FDM process optimization and should be taken into account in future research [14,15]. As a result, it becomes challenging to optimize the FDM process parameters in the presence of several such constraints. As a result, traditional Design of Experiments (DOEs) may face challenges in effectively addressing this type of problem. Therefore, new optimization techniques and mathematical modeling must be created in order to get around these restrictions and make the ideal parameters realistic and practicable in real-world applications.

FDM processes involve a multitude of interconnected parameters, often leading to complex, nonlinear relationships. Adjusting one parameter can have unexpected consequences on others, making it challenging to predict the overall impact on the final product's quality. This complexity is further exacerbated in high-dimensional parameter spaces, where optimization becomes resource-intensive and time-consuming. Engineers and researchers often grapple with finding the right balance between various parameters to achieve the desired results.

The use of computational methods for FDM parameter optimization can place significant demands on resources. Finite Element Analysis (FEA), Computational Fluid Dynamics (CFD), and machine learning algorithms require substantial computational power and time, limiting their applicability in environments with resource constraints. The iterative nature of optimization can also lead to extended production times, potentially affecting project schedules. Moreover, collecting and processing large datasets for machine learning-based optimization approaches can be daunting, requiring careful data management and analysis.

FDM encompasses a wide range of materials, each with its unique properties and behavior. This material variability poses a considerable challenge in optimization efforts. Different materials may demand distinct optimization approaches due to variations in melting points, viscosities, and thermal conductivities. Material properties can change over time due to factors such as humidity and storage conditions, necessitating adjustments to optimization parameters. Additionally, maintaining consistent material extrusion throughout the printing process is crucial, and any deviations can lead to variations in printed parts, making it harder to maintain optimized settings.

4.2. Future directions

As FDM technology continues to evolve, several promising avenues and innovations are shaping the future of FDM parameter optimization. These future directions aim to further enhance the efficiency, quality, and adaptability of FDM processes.

The integration of real-time monitoring and control systems is a crucial step in the evolution of FDM parameter optimization. By incorporating sensors and feedback mechanisms into 3D printers, manufacturers can dynamically adjust parameters during the printing process. This capability enables rapid response to variations in environmental conditions, material properties, or part geometry. Real-time monitoring can also detect anomalies or defects as they occur, allowing for immediate corrective actions. The adoption of Industry 4.0 principles, including the Internet of Things (IoT) and data analytics, will play a pivotal role in realizing this vision of adaptive and self-optimizing FDM systems.

The development of new materials specifically tailored for FDM applications is an exciting frontier. Innovations in materials chemistry and engineering are expected to yield materials with improved properties, such as enhanced strength, thermal resistance, or biocompatibility. These advanced materials will expand the range of applications for FDM, from aerospace components to medical implants. Moreover, material development will address the challenges of material variability in parameter optimization, as optimized settings may need to adapt to the unique characteristics of each material.

The integration of FDM into smart factories and Industry 4.0 environments is a transformative trend. Smart manufacturing systems will seamlessly connect FDM printers with other production equipment and databases, facilitating data-driven decision-making and process optimization. This integration will enable real-time tracking of production progress, quality control, and inventory management. Furthermore, it will allow for the exchange of optimization insights across the entire manufacturing ecosystem, fostering collaboration and knowledge sharing among manufacturers and researchers.

Research suggests that when it comes to mass production, FDM technology faces significant challenges that arise from both production times and costs, as well as the environmental impact associated with FDM when compared to other selective technologies like SLS and MJF [80]. Consequently, there may be a need to conduct a break-even point analysis for all materials, similar to what has been done for Direct Metal Laser Sintering [81], and compare it with traditional high-pressure die-casting and FDM with injection molding [82]. This analysis can help determine the maximum number of parts that should be manufactured using FDM, ensuring that production covers costs and assisting businesses in making informed decisions regarding pricing, investments, and profitability.

Future FDM parameter optimization approaches will likely incorporate multi-objective optimization techniques. Rather than focusing solely on a single optimization criterion (e.g., cost or quality), these methods will consider multiple objectives simultaneously. This will enable manufacturers to strike a balance between conflicting goals, such as minimizing production time while maximizing part quality. Multi-objective optimization algorithms will provide decision-makers with a range of Pareto-optimal solutions, allowing them to choose the one that best aligns with their priorities.

Future research in FDM should focus on the integration of Design for Manufacturing and Assembly (DFMA) principles into the context of Design for Additive Manufacturing (DFAM) and the concurrent optimization of process parameters. Researchers are encouraged to explore how the design freedom and reduced assembly constraints inherent in DFAM-informed designs can inform and enhance the selection and optimization of process parameters in FDM. Interdisciplinary collaboration between design and manufacturing teams, the incorporation of advanced computational tools and machine learning techniques, the development of materials tailored for DFAM-centric designs, and a focus on sustainability and environmental considerations are all promising avenues for further investigation. These directions hold the potential to revolutionize additive manufacturing by improving design flexibility, cost efficiency, and environmental sustainability [83].

Artificial intelligence (AI) and machine learning (ML) will continue to advance in the context of FDM parameter optimization. These technologies will become more sophisticated in predicting optimal parameters and identifying patterns in data, making them invaluable tools for manufacturers. As datasets grow and computational power increases, AI and ML models will provide more accurate and efficient optimization solutions.

5. Conclusions

Optimizing additive manufacturing process parameters is crucial for achieving cost-effective production while enhancing mechanical properties, build time, part quality, and more. In this paper, we examine several Design of Experiments (DOE)-based parameter optimization techniques applied to optimize process parameters in FDM. The methods considered encompass both experimental design and computational approaches for FDM process parameter optimization. Common methods include the Taguchi method, RSM, ANN, full factorial design method, among others, which are widely used in optimizing process parameters for FDM technique. This review work identified key FDM process parameters, including air gap, layer thickness, nozzle temperature, bed temperature, build orientation, raster width, and raster angle, that have been subjects of previous studies and possibly to be further investigated.

As AI techniques continue to mature, they are expected to remain attractive and powerful tools for optimizing process parameters in FDM. Thus, future works in this direction will focus on use of ML techniques in AI.

References

[1] Dehghanghadikolaei A 2018 Additive Manufacturing as A New Technique of Fabrication *J. 3D Print. Appl.* 3–4.

[2] Shi G, Guan C, Quan D, Wu D, Tang L and Gao T 2020 An aerospace bracket designed by thermo-elastic topology optimization and manufactured by additive manufacturing *Chinese J. Aeronaut.* **33** 1252–9.

[3] Gardan J 2016 Additive manufacturing technologies: State of the art and trends *Int. J. Prod. Res.* **54** 3118–32.

[4] Mendonca I E and Lorenzo S 2020 Sustainability analysis of the emerging technologies applied in manufacturing Presented by : Iván Esquivel Mendonça Faculty of Business Sciences.

[5] Kong L, Ambrosi A, Nasir M Z M, Guan J and Pumera M 2019 Self-Propelled 3D-Printed "Aircraft Carrier" of Light-Powered Smart Micromachines for Large-Volume Nitroaromatic Explosives Removal *Adv. Funct. Mater.* **29** 1–9.

[6] Murr L E 2020 Metallurgy principles applied to powder bed fusion 3D printing/additive manufacturing of personalized and optimized metal and alloy biomedical implants: An overview *J. Mater. Res. Technol.* **9** 1087–103.

[7] Khosravani M R and Reinicke T 2020 3D-printed sensors: Current progress and future challenges *Sensors Actuators, A Phys.* **305** 111916.

[8] Leal R, Barreiros F M, Alves L, Romeiro F, Vasco J C, Santos M and Marto C 2017 Additive manufacturing tooling for the automotive industry *Int. J. Adv. Manuf. Technol.* **92** 1671–6.

[9] Marchment T, Sanjayan J and Xia M 2019 Method of enhancing interlayer bond strength in construction scale 3D printing with mortar by effective bond area amplification *Mater. Des.* **169** 107684.

[10] Nasiri S and Khosravani M R 2020 Progress and challenges in fabrication of wearable sensors for health monitoring *Sensors Actuators, A Phys.* **312** 112105.

[11] Wang Y, Ahmed A, Azam A, Bing D, Shan Z, Zhang Z, Tariq M K, Sultana J, Mushtaq R T, Mehboob A, Xiaohu C and Rehman M 2021 Applications of additive manufacturing (AM) in sustainable energy generation and battle against COVID-19 pandemic: The knowledge evolution of 3D printing *J. Manuf. Syst.* **60** 709–33.

[12] Gibson I, Rosen D and Stucker B 2015 *Additive Manyfacturing Technologies: 3D Printing, Rapid Prototyping, and Direct Digital Manufacturing,* 2nd Ed., Springer..

[13] ISO/ASTM 2013 *Additive Manufacturing - General Principles Terminology (ASTM52900)*

[14] Mohamed O A, Masood S H and Bhowmik J L 2015 Optimization of fused deposition

modeling process parameters: a review of current research and future prospects *Adv. Manuf.* **3** 42–53.
[15] Jaisingh Sheoran A and Kumar H 2020 Fused deposition modeling process parameters optimization and effect on mechanical properties and part quality: Review and reflection on present research, *Mater. Today Proc.* **21** 1659–72.
[16] Crump S S 1991 Fast, precise, safe prototypes with FDM *Am. Soc. Mech. Eng. Prod. Eng. Div. PED* **50** 53–60.
[17] Dizon J R C, Espera A H, Chen Q and Advincula R C 2018 Mechanical characterization of 3D-printed polymers *Addit. Manuf.* **20** 44–67.
[18] Ngo T D, Kashani A, Imbalzano G, Nguyen K T Q and Hui D 2018 Additive manufacturing (3D printing): A review of materials, methods, applications and challenges *Compos. Part B Eng.* **143** 172–96.
[19] Rajan K, Samykano M, Kadirgama K, Harun W S W and Rahman M M 2022 Fused deposition modeling: process, materials, parameters, properties, and applications *Int. J. Adv. Manuf. Technol.* **120** 1531–70.
[20] Stansbury J W and Idacavage M J 2016 3D printing with polymers: Challenges among expanding options and opportunities *Dent. Mater.* **32** 54–64.
[21] Dey A and Yodo N 2019 A systematic survey of FDM process parameter optimization and their influence on part characteristics *J. Manuf. Mater. Process.* **3** 64
[22] Espín M D 2016 *Realitzada per* (Universitat Ramon Llull).
[23] Dandagwhal R D, Nikalje A M and Deore E R 2020 Effect of process parameters on additively manufactured parts using FDM process & material selection: A review *IOP Conf. Ser. Mater. Sci. Eng.* **810.**
[24] Lalegani Dezaki M, Mohd Ariffin M K A and Hatami S 2021 An overview of fused deposition modelling (FDM): research, development and process optimisation *Rapid Prototyp. J.* **27** 562–82.
[25] Sai P C and Yeole S N 2014 Fused Deposition Modeling *International Conference on Advances in Design and Manufacturing (ICAD&M'14)* pp 1–7.
[26] Kohad A, Dalu R and Student P G 2007 Optimization of Process Parameters in Fused Deposition Modeling: A Review *Int. J. Innov. Res. Sci. Eng. Technol. An ISO* **3297** 505–11
[27] Hamzaçebi C 2020 Taguchi Method as a Robust Design Tool *Quality Control - Intelligent Manufacturing, Robust Design and Charts.*
[28] Suniya N K and Verma A K 2023 A review on optimization of process parameters of fused deposition modeling *Res. Eng. Struct. Mater.* **9** 631–59.
[29] Mendonsa C, Naveen K, Upadhyaya P and Shenoy V D 2013 Influence of FDM Process Parameters on Build Time Using Taguchi and ANOVA Approach *Int. J. Sci. Res. ISSN (Online Index Copernicus Value Impact Factor* **14** 2319–7064.
[30] Lee B H, Abdullah J and Khan Z A 2005 Optimization of rapid prototyping parameters for production of flexible ABS object *J. Mater. Process. Technol.* **169** 54–61.
[31] Anitha R, Arunachalam S and Radhakrishnan P 2001 Critical parameters influencing the quality of prototypes in fused deposition modelling *Journal of Materials Processing Technology* vol 118 (Elsevier) pp 385–8.
[32] Nancharaiah T, Raju D and Raju V 2010 An experimental investigation on surface quality and dimensional accuracy of FDM components *J. Emerg. Technol.* **1** 106–11.
[33] Nancharaiah T 2011 Optimization of Process Parameters in FDM Process Using Design of Experiments *Int. J. Emerg. Technol. 2(1)* **2** 100–2.
[34] Sumalatha M, Malleswara Rao J N and Supraja Reddy B 2021 Optimization Of Process Parameters In 3d Printing-Fused Deposition Modeling Using Taguchi Method *IOP Conf. Ser. Mater. Sci. Eng.* **1112** 012009.
[35] Wang C C, Lin T W and Hu S S 2007 Optimizing the rapid prototyping process by integrating the Taguchi method with the Gray relational analysis *Rapid Prototyp. J.* **13** 304–15.

[36] Tharun M, Sumalatha M, Rao J N, G P, Cam M T and College V R S E 2017 Experimental Investigation of Impact Strength for Abs Plus FDM Parts Using Taguchi *Int. Res. J. Eng. Technol.* **4** 456–61.
[37] Zhang J and Peng A 2012 Process-parameter optimization for fused deposition modeling based on Taguchi method *Advanced Materials Research* vol 538–541 pp 444–7.
[38] Sahu R K, Mahapatra S S and Sood A K 2014 A Study on Dimensional Accuracy of Fused Deposition Modeling (FDM) Processed Parts using Fuzzy Logic *J. Manuf. Sci. Prod.* **13** 183–97.
[39] Sood A K, Ohdar R K and Mahapatra S S 2009 Improving dimensional accuracy of Fused Deposition Modelling processed part using grey Taguchi method *Mater. Des.* **30** 4243–52
[40] Maguluri N, Suresh G and Rao K V 2023 Assessing the effect of FDM processing parameters on mechanical properties of PLA parts using Taguchi method *J. Thermoplast. Compos. Mater.* **36** 1472–88.
[41] Lokesh N, Praveena B A, Sudheer Reddy J, Vasu V K and Vijaykumar S 2022 Evaluation on effect of printing process parameter through Taguchi approach on mechanical properties of 3D printed PLA specimens using FDM at constant printing temperature *Materials Today: Proceedings* vol 52 (Elsevier) pp 1288–93.
[42] Xinhua L, Shengpeng L, Zhou L, Xianhua Z, Xiaohu C and Zhongbin W 2015 An investigation on distortion of PLA thin-plate part in the FDM process *Int. J. Adv. Manuf. Technol.* **79** 1117–26.
[43] Santhakumar J, Maggirwar R, Gollapudi S, Karthekeyan S and Kalra N 2016 Enhancing Impact Strength of Fused Deposition Modeling Built Parts using Polycarbonate Material *Indian J. Sci. Technol.* **9** 1–6.
[44] Anitha R, Arunachalam S and Radhakrishnan P 2001 Critical parameters influencing the quality of prototypes in fused deposition modelling *Journal of Materials Processing Technology* vol 118 pp 385–8.
[45] Alhubail M, Alenezi D and Aldousiri B 2013 Taguchi-based Optimisation of Process Parameters of Fused Deposition Modelling for Improved Part Quality *Int. Jouranl Enggineering Res. Technol.* **2** 2505–19.
[46] Jiang C P, Cheng Y C, Lin H W, Chang Y L, Pasang T and Lee S Y 2022 Optimization of FDM 3D printing parameters for high strength PEEK using the Taguchi method and experimental validation *Rapid Prototyp. J.* **28** 1260–71.
[47] Nagendra J and Prasad M S G 2020 FDM Process Parameter Optimization by Taguchi Technique for Augmenting the Mechanical Properties of Nylon–Aramid Composite Used as Filament Material *J. Inst. Eng. Ser. C* **101** 313–22.
[48] Filippidis P, Papazetis G and Vosniakos G C 2020 Process parameter investigation for 3D printing of cellular structured parts *Procedia Manuf.* **51** 717–24.
[49] Myers R H, Montgomery D C and Anderson-Cook C M 2009 *Introduction* vol (5)2
[50] Wang Z, Li J, Wu W, Zhang D and Yu N 2021 Multitemperature parameter optimization for fused deposition modeling based on response surface methodology *AIP Adv.* **11.**
[51] Mandge V, Patel D M and Patil H G 2020 Experimental Investigation to optimise FDM process parameters for ABS material using RSM *REST J. Emerg. trends Model. Manuf.* **6.**
[52] Mohamed O A, Masood S H and Bhowmik J L 2016 Mathematical modeling and FDM process parameters optimization using response surface methodology based on Q-optimal design *Appl. Math. Model.* **40** 10052–73.
[53] Equbal A, Sood A K, Equbal M I, Badruddin I A and Khan Z A 2021 RSM based investigation of compressive properties of FDM fabricated part *CIRP J. Manuf. Sci. Technol.* **35** 701–14.
[54] Gao G, Xu F and Xu J 2022 Parametric Optimization of FDM Process for Improving Mechanical Strengths Using Taguchi Method and Response Surface Method: A Comparative Investigation *Machines* **10.**
[55] PANDA S K, PADHEE S, SOOD A K and MAHAPATRA S S 2009 Optimization of Fused

Deposition Modelling (FDM) Process Parameters Using Bacterial Foraging Technique *Intell. Inf. Manag.* **01** 89–97.
[56] Montgomery D C A S U 2017 *D esign and Analysis of Experiments, 10th Ed. Wiley.*
[57] Horvath D, Noorani R and Mendelson M 2007 Improvement of Surface Roughness on ABS 400 Polymer Using Design of Experiments (DOE) *Mater. Sci. Forum* **561–565** 2389–92,
[58] Pavan Kumar G and Regalla S P 2012 Optimization of support material and build time in fused deposition modeling (FDM) *Appl.Mech. Mater.* **110–116**, 2245–51.
[59] Rayegani F and Onwubolu G C 2014 Fused deposition modelling (fdm) process parameter prediction and optimization using group method for data handling (gmdh) and differential evolution (de) *Int. J. Adv. Manuf. Technol.* **73** 509–19.
[60] Ahn S H, Montero M, Odell D, Roundy S and Wright P K 2002 Anisotropic material properties of fused deposition modeling ABS *Rapid Prototyp. J.* **8** 248–57.
[61] Ang K C, Leong K F, Chua C K and Chandrasekaran M 2006 Investigation of the mechanical properties and porosity relationships in fused deposition modelling-fabricated porous structures *Rapid Prototyp. J.* **12** 100–5.
[62] Haidiezul A H M, Hazwan M H M, Soon Lee W, Gunalan, Fatin Najihah N and Fadhli I 2020 Full Factorial Design Exploration Approach for Multi-Objective Optimization on the (FDM) 3D Printed Part, *IOP Conf, Series: Mater, Sci, Eng,* **917**
[63] Gebisa A W and Lemu H G 2018 Investigating effects of Fused-deposition modeling (FDM) processing parameters on flexural properties of ULTEM 9085 using designed experiment *Mater,* **11** 1–23,
[64] Gebisa A W and Lemu H G 2019 Influence of 3D printing FDM process parameters on tensile property of ultem 9085 *Procedia Manuf.* **30** 331–8
[65] Leirmo T S and Martinsen K 2019 Evolutionary algorithms in additive manufacturing systems: Discussion of future prospects *Procedia CIRP,* **81,** 671–6,
[66] Tura A D and Mamo H B 2022 Characterization and parametric optimization of additive manufacturing process for enhancing mechanical properties, *Heliyon* **8** 1–8.
[67] Kumar R, Chohan J S, Singh S, Sharma S, Singh Y and Rajkumar S 2022 Implementation of Taguchi and Genetic Algorithm Techniques for prediction of optimal part dimensions for polymeric biocomposites in fused deposition modeling, *Hindawi, Int. J. Biomater.* **2022**
[68] Singh B, Kumar R and Singh Chohan J 2021 Multi-objective optimization of 3D printing process using genetic algorithm for fabrication of copper reinforced ABS parts, *Mater. Today: Proc.* **48,** 981–8.
[69] Chohan J S, Kumar R and Singh S 2022 Analysis of dimensional accuracy of fused filament fabrication parts using genetic algorithm and aguchi analysis, *Int. J. Mech, Eng.* **7,** 105–13.
[70] Arumaikkannu G, Uma Maheshwaraa N and Gowri S 2005 A genetic algorithm with design of experiments approach to predict the optimal process parameters for FDM *16th Solid Freeform Fabrication Symposium, SFF 2005* pp 150–61.
[71] Abdulhameed O, Mian S H, Moiduddin K, Al-Ahmari A, Ahmed N and Aboudaif M K 2022 A Multi-Part Orientation Planning Schema for Fabrication of Non-Related Components Using Additive Manufacturing *Micromachines* **13** 1–24.
[72] Giri J, Shahane P, Jachak S, Chadge R and Giri P 2021 Optimization of fdm process parameters for dual extruder 3d printer using artificial neural network, *Mater. Today: Proc.***43,** 3242–9.
[73] Lyu J and Manoochehri S 2018 Modeling machine motion and process parameter errors for improving dimensional accuracy of fused deposition modeling machines *J. Manuf. Sci. Eng. Trans. ASME* **140.**
[74] Goudswaard M, Hicks B and Nassehi A 2021 The creation of a neural network based capability profile to enable generative design and the manufacture of functional FDM parts, *Int. J. Adv. Manuf. Technol.* 2951–68
[75] Chohan J S, Mittal N and Kumar R 2020 Parametric optimization of fused deposition modeling

using learning enthusiasm enabled teaching learning based algorithm *SN Appl. Sci.* **2** 1–12.
[76] Dey A, Hoffman D and Yodo N 2020 Optimizing multiple process parameters in fused deposition modeling with particle swarm optimization *Int. J. Interact. Des. Manuf.* **14** 393–405
[77] Selvam A, Mayilswamy S, Whenish R, Naresh K, Shanmugam V and Das O 2022 Multi-objective optimization and prediction of surface roughness and printing time in FFF printed ABS polymer, *Sci. Rep.* **12** 1–12.
[78] Em T S and Kiong S C 2023 Optimization of Process Parameters for Polylactic Acid (PLA) of FDM Using Particle Swarm Optimization (PSO) *Res. Prog. Mech. Manuf. Eng.* **4** 205–17
[79] Mulay N N K A V 2018 Post processing methods used to improve surface finish of products which are manufactured by additive manufacturing technologies: A review *J. Inst. Eng. Ser. C* **99** 481–7
[80] Tagliaferri V, Trovalusci F, Guarino S and Venettacci S 2019 Environmental and economic analysis of FDM, SLS and MJF additive manufacturing technologies *Mater.* **12,** 4161.
[81] Atzeni E and Salmi A 2012 Economics of additive manufacturing for end-usable metal parts *Int. J. Adv. Manuf. Technol.* **62** 1147–55.
[82] Franchetti M and Kress C 2017 An economic analysis comparing the cost feasibility of replacing injection molding processes with emerging additive manufacturing techniques *Int. J. Adv. Manuf. Technol.* **88** 2573–9.
[83] Mäntyjärvi K, Iso-Junno T, Nemi H and Mäkikangas J 2018 Design for additive manufacturing in extended dfma process, *Key Eng. Mater.* **786** KEM 342–7.

IOP Conf. Series: Materials Science and Engineering 1294 (2023) 012048 doi:10.1088/1757-899X/1294/1/012048

The influences of international standards on structural performance and fracture behaviour of 3D printed long fiber composite structures

A D Akessa*, Y W Adugna, W M Tucho

Department of Mechanical and Structural Engineering and Materials Science, Faculty of Science and Technology, University of Stavanger, N-4036 Stavanger, Norway

* Correspondence: adugna.d.akessa@uis.no

Abstract. The material properties of the additive manufactured composite structures are usually done following international standards, that show some difference dimensionally. This study focuses on the impact-induced on the tensile properties of the 3D printed continuous carbon fiber reinforced composite part due to the standards applied for sample preparation. The specimens were fabricated by Markforged® Mark two 3D printing machine using carbon fiber as reinforcement and Onyx® as matrix material based on ASTM D638 and ASTM D3039-D3039M standards. The experimental results revealed that the specimens fabricated based on ASTM D638 showed a premature failure at the location where the straight gauge section of the specimen ends, and the curved transition regions begin due to stress concentration. The tests based on ASTM D3039-3039M standard showed better tensile strength and less stress concentration compared to ASTM D638. Fracture test with SEM reveals fiber breakage, debonding, and fiber pullout, which created cavities and voids between layers as the reasons for the tensile failure.

1. Introduction

Current developments in additive manufacturing (AM) have made popular manufacturing methods for, their low material waste, ease of manufacture, advantage of design freedom, and environmental friendliness. ASTM F2792−12a defines additive manufacturing as material joining to make an object from 3D model data, commonly layer upon layer, as opposed to the conventional machining method. [1,2]. According to ISO 17296-2 standard AM is grouped into seven based on the faction of machine parts as; material jetting, vat photopolymerization, powder bed fusion, binder jetting, Material extrusion, Directed energy deposition, and Sheet lamination. Additionally, the AM technique can be categorized based on the types of materials used and the deposition technique [3].

Additively manufactured parts have weaker when the load is applied at an angle to filament orientation than when applying load along its orientation, resulting, in AM parts being mostly used as a prototype than functional parts [4]. In the aircraft industry, the application of additive-manufactured parts especially composite parts is increasing because of low maintenance costs, weight, and better reliability of the parts printed from additive-manufactured composite [5]. The development of using composite fibers in the production of additive manufactured parts increases the functionality of the parts due to the improvement of the parts' properties [6]. Therefore, to have high-performance additive manufactured polymer products with high strength, reinforcing with high-performance reinforcement composite polymers is important [7]. Reinforcement of 3D-printed polymers can be done Using both long and segmented fiber reinforcement. Short fiber-reinforced additively produced parts have poor mechanical properties compared to continuous reinforced parts. The development of long fiber reinforcement 3D printed composites like Markforged Mark Two changed 3D printed composite parts into commercial products [8].

Though the start of AM technology lasted several decades, there is still a lack of understanding of the fundamental character of parts produced by AM technology. The study done to improve the mechanical properties of 3D printed composite parts is progressing as the core of research in 3D printing field. In tensile testing of AM composite parts, the cause of fiber fracture that occurs is induced by stress concentration due to the geometry of the specimen. Fiber fracture that occurs at an angle is coordinated fracture and propagates the matrix cracks from the fiber break points. The stress concentration causes the failure of many fibers at similar locations [9]. Another study done by Garrell M G et al. on stress concentration in tensile specimens produced based on ASTM D638 standard using finite element analysis concluded that the specimen failed at the end of the gauge area and where the curved regions begin, this is because the geometry difference which induces high stress at the location [10].

The properties like strength and structural efficiency of 3D printed composite parts are highly influenced by crystal structures, such as void content, geometry of the fiber, and harmony of both fiber and matrix ally. Based on the technique of digital imaged correlation, it is shown that crack tip strain has resulted ahead of the inter-layer fracture, reinforcement failure at different levels, and adherence to a short fiber connection. Different studies have been conducted to understand the properties of fiber-reinforced polymer parts. The failure behavior of 3D-manufactured long fiber reinforced polymer parts and stress concentration due to the geometry of the specimen has not been well understood either experimentally or numerically. Understanding the failure mechanism of 3D printed composite parts in designing and production of the parts [11].

The tensile properties of plastics and polymer matrix composite materials can be determined by ASTM D638-14 and ASTM D3039-3039M [12]. Whereas for characterization of tensile properties of conventional produced plastic parts that showed isotropic ASTM D638-14 was used. In contrast, 3D-printed reinforced plastic parts have shown anisotropic properties or transverse isotropic properties. Because of the anisotropic nature of the 3D material and tensile test specimen geometry, crack initiation at fillets results in premature failure. So, it is important to choose a different standard to reduce stress concentration [13]. ASTM D 3039/D 3039M is the standard method used with continuous or short fiber, to determine in-plane tensile properties of polymer matrix composite materials reinforced by high-modulus fiber. The standard method used straight tab and has less stress concentration compared to ASTM D638-14 standard [14].

The Markforged Mark Two 3D printing machine uses a dual nozzle system. The machine can print different materials like carbon fiber, Kevlar, glass fiber as reinforcement, and onyx and nylon as matrix material. Markforged Mark two 3D printing works in a similar way as material extrusion technology, where melted filament is extruded through a nozzle onto the 3D printer platform, by adding multiple layers to form the component [8]. However, since the machine is a newly developed technology (Markforged Inc., USA), its products are not well studied [13].

The composites' nature and process variation make 3D-printed composite parts not easy to predict mechanical and microstructural properties, because of the composite nature and process variation. In view of these factors, further study is needed to understand and predict the performance of continuous fiber-reinforced 3D-printed composite parts [15]. The aim of this work is to study the influence of international standards on the tensile properties of additively manufactured carbon-reinforced composite 3D printed parts.

2. Experimental methods

2.1. The material

Continuous carbon fiber and Onyx® materials supplied by PLM Group Norway AS were used as raw materials for the current study. Additively manufactured Carbon and Onyx as composite parts attain high strength and stiffness and they can be printed in different shapes [16,17]. Onyx is a combination of nylon-based thermoplastic and chopped carbon fiber made into filament. It was used as a matrix. The carbon as shown in Figure 1, is made from many micro-size continuous carbon filaments bundled together by sizing agent to form carbon filament as a reinforcement material [18]. The printing materials used in this work have a density of 1.4 gm/cm^3 and 1.2 gm/cm^3 of onyx and carbon respectively.

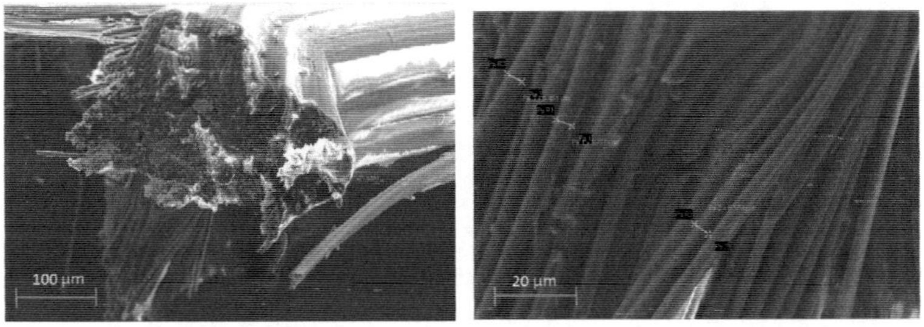

Figure 1. SEM image of carbon filament.

2.2. Model preparation and printing

The geometry of the sample was designed and converted to Stereolithography (STL) using Autodesk inventor 2023. The STL file is imported into the Markforged Eiger software for slicing, adjustment of printing parameters, material selection, etc. [19,20]. The test samples were manufactured by Mark Two desktop type 3D printing machine manufactured by Markforged. The machine has a printing volume of 0.584 m × 0.330 m × 0.355 m and has a pair of nozzles, through which fiber and material board on the printing platform. As shown in Table 1 the nozzle temperature for carbon is 252 $°C$ and that of Onyx is 277 $°C$. As Onyx filament passes through the nozzle during printing it melts, and the carbon fiber is heated to give adhesion to the matrix.

The printing parameters applied are shown in Table 1 for both matrix and fiber. Test specimens for tensile test were printed with orientations of 0° based on ASTM D638 and D3039-3039M standard method. For ease of reading, we refer ASTMD638-14 as S1 and ASMT D3039-3039M as S2, hereafter. Schematics that illustrate the geometry of the two standards are displayed in Figure 2. The geometry of the test sample manufactured based on S1 is rectangular while that of S2 is closer to the dog-bone type. The sample dimensions for both standard types are shown in Figure 2. To reduce experimental error, five test samples were fabricated using the same printing parameters for each standard as given in Table 1.

Table 1. Printing parameter used to create samples.

Parameters	Carbon fiber	Onyx
Orientation Angle	0°	±45°
Layer thickness (mm)	0.125	0.125
Number of layers	12	14
Infill density (%)	100	100
Roof and floor		4
Number of walls	2	
Fiber nozzle temp ($°C$)	252	
Plastic nozzle temp ($°C$)	277	

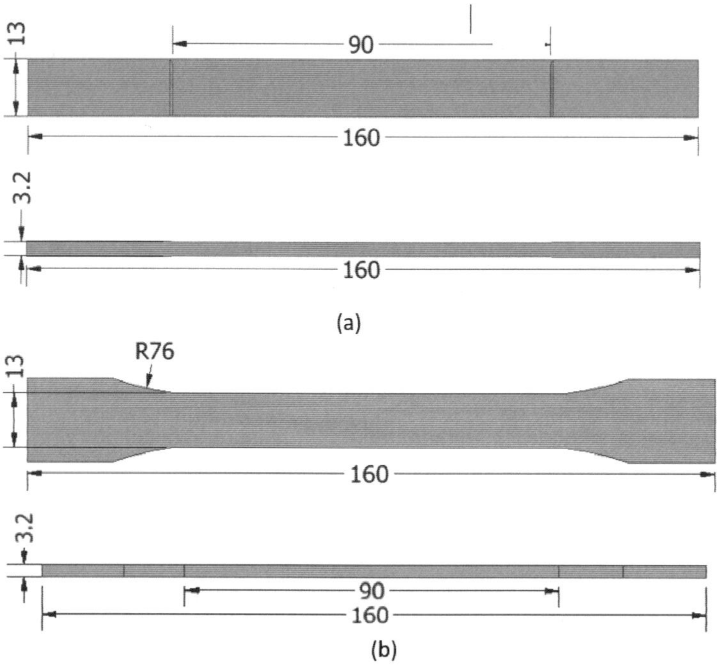

Figure 2. Tensile specimens manufactured by (a) S2 and (b) S1 standards.

Five samples were printed for each standard group (S1, S2) with the same amounts of carbon fiber onyx ratio, and stacking sequences as listed in Table 1. They were printed in unidirectional (0°) orientation of carbon fibers for both standard groups as displayed in Figure 3.

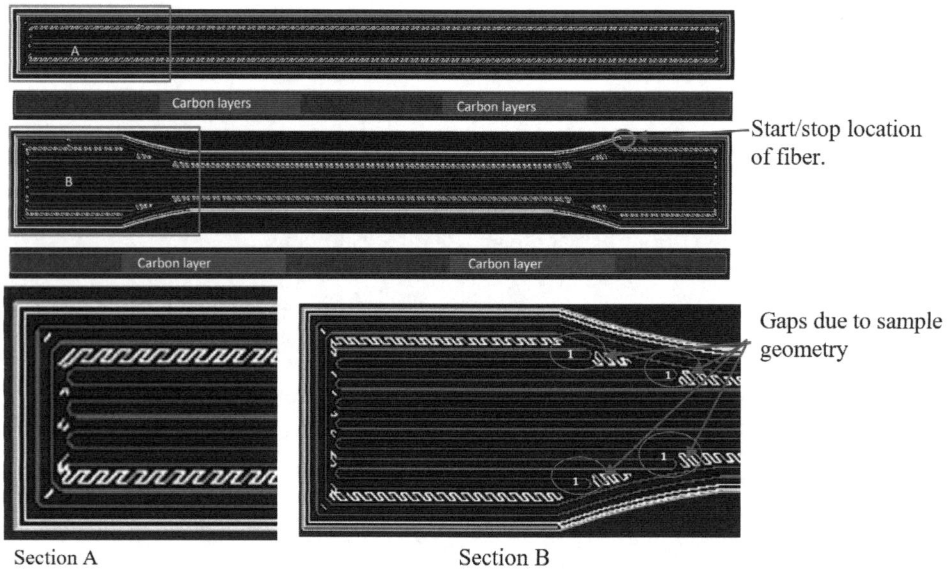

Figure 3. Fiber arrangement of S2 (A) and S1 (B) Specimen.

IOP Conf. Series: Materials Science and Engineering 1294 (2023) 012048 doi:10.1088/1757-899X/1294/1/012048

2.3Tensile test

Instron 5985 testing machine which has a load cell capacity of 250kN was used for tensile testing. Testing was done at room temperature and a strain rate of 1mm/min. The schematic drawing of the sample is seen in Figure 2.

2.4. Fracture properties

The fracture surface morphology of the tensile specimen was analyzed using scanning electron microscopy (SEM) (Gemini SUPRA 35VP ZEISS) equipped with EDAX energy dispersive X-ray spectroscopy (EDS).

3. Results and discussions

3.1. Tensile test

The effect of fiber stress concentration due to standard method selection on mechanical properties was studied under tensile load. Every test was done until the complete failure test sample. The tensile test results of the specimens were shown in Table 2 and the stain versus stress curve was also shown in Figure 5 (a) and (b).

Table 2. Tensile test result.

Standard type	Load (kN)	Strain (%)	Stress (MPa)	Modulus (MPa)
S1	10,51	4,79	245,51	6591,81
S1	10,61	4,26	247,96	7451,93
S1	11,8	4,76	275,75	7726,67
S1	10,17	4,12	237,61	7617,56
S1	11,24	4,55	262,67	7111,25
S1	11,5	4,7	268,75	7472,61
Average. *	*10,97*	*4,53*	*256,38*	*7328,64*
StDev. *	*0,64*	*0,28*	*14,89*	*416,66*
S2	12,06	5,58	281,7	8926,95
S2	12,04	5,87	281,22	8674,66
S2	13,93	4,52	325,49	8937,28
S2	12,2	4,82	285	9729,34
S2	13,09	5,22	305,78	9752,32
Average *	*12,66*	*5,20*	*295,84*	*9204,11*
StDev. *	*0,830*	*0,548*	*19,424*	*501,182*

The test results show brittle behavior as can be seen in Figure 5 (a) and (b). This shows that the specimens were influenced by the carbon fiber properties.

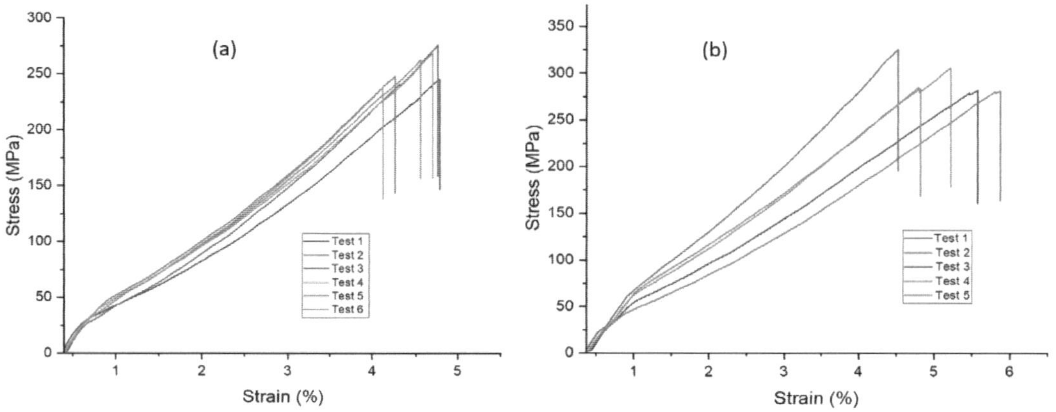

Figure 4. Tensile test stress vs. strain (a) S1 method, (b) S2 method

Both samples were printed with the same orientation, layer stacking, infill density, and percentage of carbon fiber filaments versus Onyx but with different standard methods as shown in Table 2. The samples tasted according to S2 Standard showed an average tensile strength of 295.84 MPa ±19 and elongation of 5.20 % ± 0.55, whereas the specimens produced according to S1 showed an average tensile strength of 256.38 MPa ± 14 and elongation of 4.28 % ± 0.28. The samples manufactured based on S2 standard have better strength and ductility compared to the samples produced based on the S1 standard. Using S1 standard for continuous reinforced composite 3D printed polymer has lower tensile strength and elongation values by 7.1% and 10.64% respectively, than that of S2. The S1 test result showed less deviation compared to S2.

S1 is a standard method used to arbitrate the tensile characteristics of plastic and composite parts that have isotropic properties. However, 3D-printed reinforced composite polymers have orthotropic/ transversely isotropic properties. Therefore, it is not easy to determine the properties of additively produced composite parts [13]. The result from the S1 standard shows lower strength, strain, and modulus of elasticity.

As shown in Figure 3 section B the fiber arrangement shows irregularity and gaps at the transition area from gauge to tab regions which were numbered 1 with a red circle, those gaps between adjacent layers increase the probability of failure because of high stress concentration. The start/end of filament shown at the start of the tab on the same part B also contributed to the premature failure of the samples manufactured based on S1 standard.

3.2. Fracture properties

Tensile tests after the failure of S1 and S2 were shown in Figure 4. In cases of the specimens fabricated based on S1, the failure occurred at the end of the gauge section. On the contrary sample from S2 standard fails in the gauge region of the samples as shown in Figure 4.

(a) 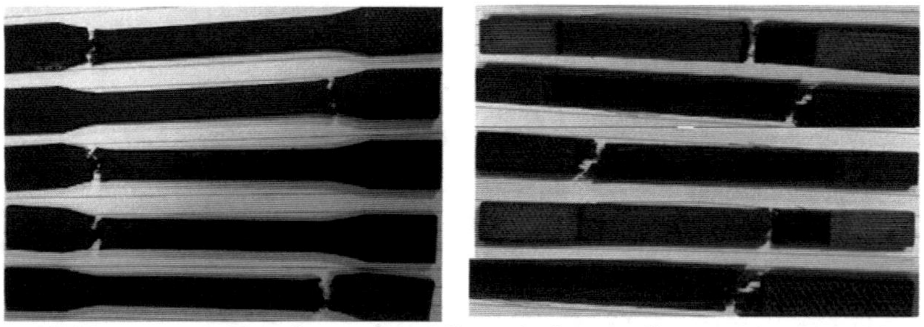 (b)

Figure 5. Fracture profile of materials failure during tensile test (a) S1, (b) S2

IOP Conf. Series: Materials Science and Engineering 1294 (2023) 012048 doi:10.1088/1757-899X/1294/1/012048

After tensile test of the sample, the structural analysis was done to characterize its failure property connected with damage to continuous fiber-reinforced 3D parts. It provides information that affects the strength as a result performance of 3D-printed composite parts. Cross-sectional images from SEM were displayed in Figure 6. to study tensile fracture parts at different resolutions. The fracture surface morphology Figure 6 (a & b) reveals matrix-dominated areas with voids and weak inter-bead adhesion as well as fiber debonding from matrix areas by poor bonding interfaces. In Figure 6 (c.) pulled-out carbon fibers can be seen which create voids in the matrix. While Figure 6 (d) shows inter-bead voids induced by a pattern of manufacturing. Figure 6 (e & f) also shows weak bonding between the onyx matrix and carbon fiber as well as pull out of carbon filament from the matrix. Figure (g) shows matrix-influenced areas, with vacant filament unglued from the matrix due to poor bonding interface, and dry fibers, with damaged carbon bundles and creates a cavity in the matrix parts [18].

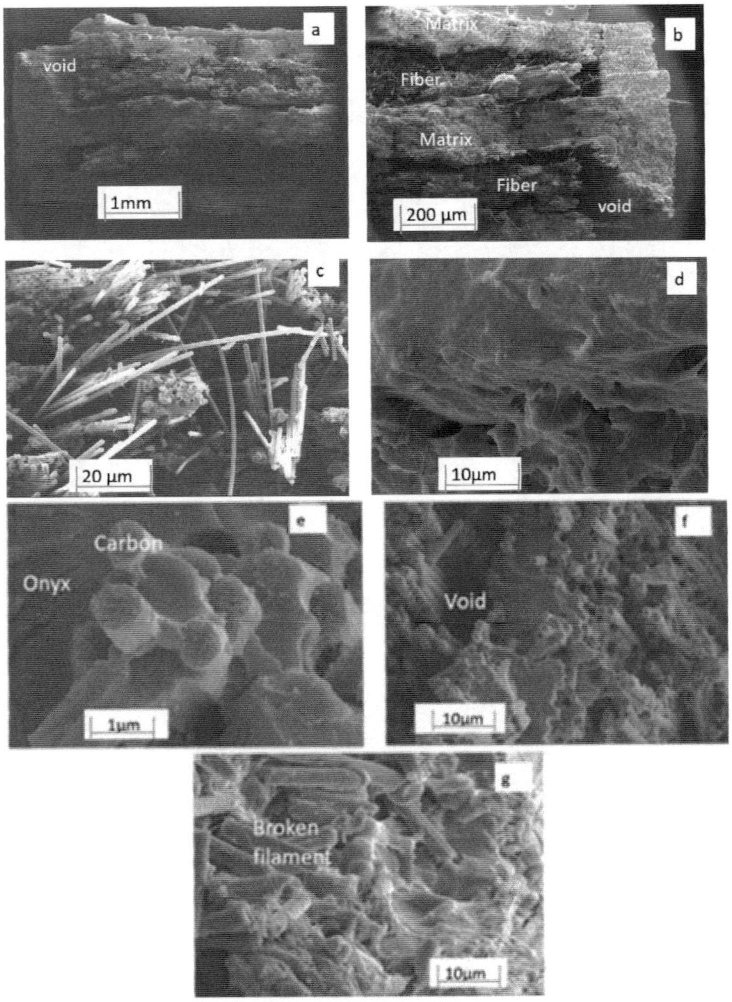

Figure 6. SEM images of the fracture surface

The engineering performance of additively manufactured long fiber reinforced polymer parts are influenced by printing defects such as inter-layer, inter-bead voids, and lack of fusion between matrix and reinforcement. As shown in Figure 3 section B the geometry of samples has a high contribution to the failure of 3D-produced parts.

In Figure 7, the energy in keV was shown on the conventional x-axis direction of the EDS, while the y-direction shows X-ray intensity. The "K" in the vertical axis implies that the values in the vertical axis

IOP Conf. Series: Materials Science and Engineering 1294 (2023) 012048 doi:10.1088/1757-899X/1294/1/012048

are multiplied by 1000. The source of Au/Gold is from the coating of carbon to make it conductive. Therefore, it was not considered in specifying element compositions.

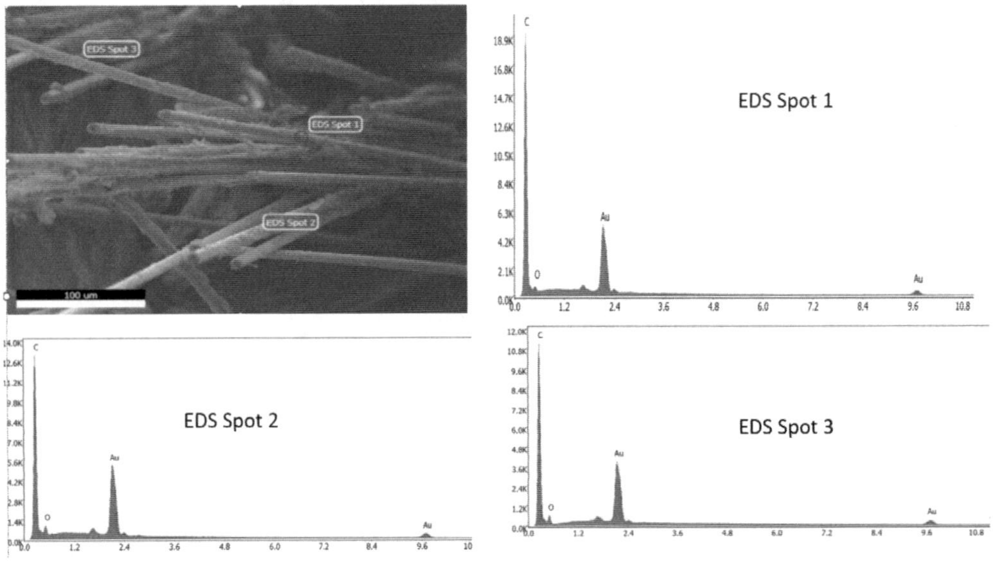

Figure 7. EDS elemental composition analysis of different carbon fiber filaments.

Table 3 shows the composition of carbon and oxygen molecules. The amount of carbon and oxygen in different filaments is different.

Table 3. The concentration of C and oxygen at different spots.

Spot	1	2	3
	Wt. %	Wt. %	Wt. %
C k	98.81	90.40	97.41
O k	1.19	9.60	2.59

4. Conclusion

In this work, the effect of international standards on tensile properties and the fracture morphology of long carbon fiber-reinforced polymer parts have been researched. The additively produced composite parts used carbon as reinforcement and Onyx as matrix materials.

In tensile testing of 3D printed parts, both ASTM D638 and ASTM D3039-3o39M were used as standards for producing and testing the parts. The test done based on the ASTM D638 standard shows lower results due to stress concentration induced by the specimen geometry. The fillet region of the specimens induces crack initiation that causes premature frailer.

The specimens produced based on the ASTM D3039-3039M standard have better tensile performance than the ASTM D638 standard. It has a straight tab and less stress concentration. Therefore, in tensile testing of 3D-printed composite parts, ASTM D3039-3039M has better performance.

A study of the failure characteristics of the printed composite part shows a lack of fusion of carbon fiber and onyx and individual fiber bundles together, so the synergetic effect of the carbon fiber bundle is lacking. It can be improved by using more printing temperature that enhances adhesion, denser composite, and minimizing voids.

IOP Conf. Series: Materials Science and Engineering 1294 (2023) 012048 doi:10.1088/1757-899X/1294/1/012048

References

[1] Croccolo D De, Agostinis M, and Olmi G (2013). Experimental characterization and analytical modelling of the mechanical behaviour of fused deposition processed parts made of ABS-M30, *Comput. Mater. Sci.*, **79**, 506-18.

[2] ASTM International 2013 ASTM F2792-12a. Rapid Manuf Assoc, 1-3.

[3] Standard ISO 17296-2, 2015 Additive manufacturing: Overview of process categories and feedstock, Geneva, Switzerland.

[4] Duty C, Failla J, Kim S, Lindahl J, Post B, Love L, and Kunc V (2017). Reducing mechanical anisotropy in extrusion-based printed parts. In: *2017 Int. Solid Freeform Fabrication Symp.*. University of Texas at Austin.

[5] Dantas F, Couling K, and Gibbons G J, 2020 Long-fibre reinforced polymer composites by 3D printing: influence of nature of reinforcement and processing parameters on mechanical performance. *Funct. Compos. Mater.* **1**(1), 1-12.

[6] Mori K I, Maeno T, and Nakagawa Y, 2014 Dieless forming of carbon fibre reinforced plastic parts using 3D printer, *Procedia Eng. 81*, 1595-1600.

[7] Sarvestani N A, van de, Werken N, Khanbolouki P, and Tehrani M, 2017, November 3D printed composites with continuous carbon fiber reinforcements, In *ASME and exposition* (Vol. 58356, p. V002T02A031). American Society of Mechanical Engineers.

[8] Adil S, and Lazoglu I, 2023 A review on additive manufacturing of carbon fiber-reinforced polymers: Current methods, materials, mechanical properties, applications, and challenges. *J. Appl. Polym. Sci.* **140**(7), e53476.

[9] Yamamoto G, Koizumi K, Nakamura T, Hirano N, and Okabe T, 2021 Tensile-strength-controlling factors in unidirectional carbon fiber reinforced plastic composites. *Composites, Part A: Appl. Sci. and Manuf.* **140**, 106140.

[10] Garrell M G, Shih A J, Lara-Curzio E, and Scattergood R O, 2003 Finite-element analysis of stress concentration in ASTM D 638 tension specimens, *J. Test. Eval.* **31**(1), 52-57.

[11] Yavas D, Zhang Z, Liu Q, and Wu D, 2021 Fracture behavior of 3D printed carbon fiber-reinforced polymer composites, *Comp. Sci. and Techn. 208*, 108741.

[12] American Society for Testing and Materials. (2014). *Standard test method for tensile properties of polymer matrix composite materials*. J. ASTM Int.

[13] Croccolo D De, Agostinis M, and Olmi G, 2013 Experimental characterization and analytical modelling of the mechanical behaviour of fused deposition processed parts made of ABS-M30, *Comput. Mater. Sci.* **79**, 506-518.

[14] ASTM Z, (2014). Astm D3039/D3039M. Annual Book of ASTM Standards, 1-13.

[15] Peng Y Wu Y, Wang K, Gao G, and Ahzi S, 2019 Synergistic reinforcement of polyamide-based composites by a combination of short and continuous carbon fibers via fused filament fabrication, *Compos. Struct. 207*, 232-239.

[16] Van Der, Klift F, Koga Y, Todoroki A, Ueda M, Hirano Y, and Matsuzaki R, 2016 3D printing of continuous carbon fiber reinforced thermo-plastic (CFRTP) tensile test specimens, *Open J. Compos. Mater.* **6**(01), 18.

[17] Brenken B, Barocio E, Favaloro A, Kunc V, and Pipes R B, 2018 Fused filament fabrication of fiber-reinforced polymers A review. *Addit. Manuf.* **21**, 1-16.

[18] Coelho A M G, Mottram J T, and Harries K A, 2015 Finite element guidelines for simulation of fibre-tension dominated failures in composite materials validated by case studies, *Compos. Struct.* **126**, 299-313.

[19] Yavas D, Zhang Z, Liu Q, and Wu D, 2021 Fracture behavior of 3D printed carbon fiber-reinforced polymer composites, *Compos. Sci. Technol.* **208**, 108741.

[20] Korkees F, Allenby J, and Dorrington P, 2020 3D printing of composites design parameters and flexural performance, *Rapid prototyping J.* **26**(4), 699-706.

Utilizing multiple criteria decision-making optimization methods to machine LM25Al/VC material under electro-discharge machining processes

M A Tolcha[1,2*] and H G Lemu[2]

[1]Jimma Institute of Technology, Jimma University, Ethiopia
[2]Department of Mechanical and Structural Engineering, University of Stavanger, 4036 Stavanger, Norway.

*Correspondence: mesay.alemu@ju.edu.et/ mesay.a.tolcha@uis.no

Abstract. In this work, sparking electro-discharge machining (EDM) process parameters are optimized by using multiple criteria decision-making (MCDM) methods. The work piece utilized for the investigation was LM25 aluminum alloy reinforced with Vanadium carbide (VC), processed through a stir casting technique. The EDM process parameters that have been considered in this work are by peak current, pulse-on-time, and pulse-off-time, with the effects of each variable and their combinations on the performance metrics of EDM like material removal rate, electrode wearing rate, and surface roughness. In this study, three MCDM methodologies were applied to evaluate EDM performance. Then, the obtained MCDM scores were compared using two different objective verification mechanisms. In this case, the VIKOR technique delivered the best-desired results relative to TOPSIS method, and factorial analyzing method. Also, the factorial method is simpler than the other methods, though it produced nearly identical results as the sophisticated MDCM method.

Keywords: Material removal rate, VIKOR, peak current, pulse-on-time

1. Introduction

The electro-discharge machining (EDM) process is a non-conventional and non-contact operation that is used in industry for high-precision products, especially in manufacturing industries such as the aerospace and automotive industries, communication, and biotechnology industries [1–3]. Most of those industries employed materials with superior properties like high corrosion and wear resistance, low cost, low weight and good strength. It is also appropriate in some high-strength and high-temperature applications. However, it is difficult to see the aforementioned properties in a single material, and to overcome these significant demands, composite materials were proposed by different researchers [4-8]. Aluminum alloys are an ideal material for different applications, like aircraft applications, cog wheels and shafts, rocket components, and regulating valve parts, while being supported by heat-refractory and hard materials.

Aluminum-vanadium carbide composites are providing superior properties which cannot be achieved by any existing monolithic or composite materials [9]. Vanadium carbide (VC) is the hard refractory material, and the cost of the reinforcement in comparison with other materials is very nominal [10]. VC particulate reinforcements help to improve the aluminum matrix in terms of hardness, heat refractory, strength, and wear resistance [11]. In fact, extremely high hardness will lead to machinability challenges. When it comes to the machining of difficult-to-machine and/or hard materials, the selection and optimization of the appropriate machining techniques are of prime importance. In addition to their unique characteristics, most modern materials need special manufacturing processes to enable them to be machined with ease [12]. EDM, a thermo-physically based material ablation technology, is a cutting-edge machining technique with a remarkable ability to machine extremely hard and brittle materials with complicated three-dimensional structures [13].

In this study, the focus is to increase material removal rate and minimize surface roughness for composite aluminum alloys with vanadium carbide (LM25Al/VC). The LM25Al/VC hybrid composite has been selected to analyze its machineability by EDM with the MCDM optimization method. The various EDM input parameters, such as peak current (I_p), pulse-on-time (t_{on}), plus-off time (t_{off}), and VC weight percentage of reinforcement materials, are considered in the analysis. It is also important to state that the influence of an increase in wt% of VC was investigated on performance characteristics during the machining of the aluminum alloy silicon composite [14]. It was found that the material erosion rate decreased, and the electrode wear increased with an increase in reinforcing quantity in the matrix phase [15]. Some researchers [16–19] have checked the feasibility of EDM in composite materials, and their results yield that the reinforcing constituents in the matrix obstruct effective sparking and reduce matrix phase erosion.

To improve the performance metrics of EDM, several sensitivity analysis and optimization approaches of the process variables were performed by previous researchers on various high-strength and harder materials [20–23]. The experimental findings indicate that positive polarity exhibited the maximum workpiece material removal rate accompanied by reduced tool erosion and surface waviness when compared to negative polarity [24]. An examination of electrically independent variables like pulse current, spark duration, and gap voltage was made for the Al7075/SiC/WS2 hybrid composite using the MCDM method. The results revealed that pulse current and discharging time had significantly varied all the performance metrics (MRR and tool erosion rate) in a linear manner with a gradual change in current levels [25]. EDM effectiveness, namely the quality of the machined holes, was measured through surface roughness and taperness. It was found that spark-off time had a large percentage contribution to hole quality metrics relative to discharge current, sparking duration, and flushing force [26]. Shao et al. [27] used dielectric fluid mixed with Al powder and conducted experiment in EDM machine, and to analyze the experimental results, response surface methodology (RSM) was used.

In real-world industry applications, many variables affect the manufacturing process. These variables can be analyzed by MCDM methods. In the same way, multi-objective approaches also have difficulties achieving many goals at the same time [28]. Various approaches, such as metaheuristics, Pareto-based strategies, and multi-model independent direction, are used to address improvement challenges. EDM process variables were optimized using different optimization techniques like Taguchi, TOPSIS, VIKOR, TLBO, RSM, and Grey relational analysis (GRA) [29, 30]. As a result, using various multi-standard methodologies may result in varied choice suggestions. However, most approaches gave some unique advantages and also had deficiencies of one sort or another. On the other hand, the choice of those models is based on the knowledge of researchers and professionals in the specific domain. In addition, the ML25Al/VC hybrid composite in the machining process is not yet fully understood due to the likely utilization of various materials for numerous non-conventional machining applications such as electrochemical machining, abrasive machining, chemical machining, etc. The problem is not only this,

IOP Conf. Series: Materials Science and Engineering 1294 (2023) 012049 doi:10.1088/1757-899X/1294/1/012049

but also there is very few literatures among the research done on the ML25Al/VC hybrid composite. Even it's possible to say, there are no works done on the ML25Al/VC under machining processes. While we summarized the advancements in using MCDM approaches, TOPSIS and the VIKOR method are considered in this work with different weighting metrics for genuine factories and place options based on alternative ranking and instructional recorded in terms of quality from a multifaceted perspective.

2. Materials and methods

2.1 Base metal and reinforcement

In this work, the LM25Al alloy is considered the base material. To enhance the material characteristics of the LM25Al alloy, vanadium carbide is used as reinforcement, as shown in Table 1. In addition, works related to utilizing vanadium carbide-reinforced aluminum matrix composites were scarcely available; hence, it is selected as reinforcement. VC tests often include an oxide debasement. It is produced by heating vanadium oxides with carbon to roughly 1000°C. Despite being thermodynamically stable, VC transforms to V2C at temperatures over 1000°C. VC particles are extremely strong at high temperatures and have a high oxidation capability. As a result, it is used for a variety of high-temperature applications. The metal network composite used in this investigation was joined using a mix-projecting method. To combine the composite, a mix-projection arrangement comprised of an obstruction stifle heater and a hardened steel stirrer was used. The stirrer assembly consisted of a stirrer connected to a variable-speed vertical boring machine with a range of 80 to 1400 rpm through a steel shaft. The stirrer had three edges that were spaced by 120°. Within this approach, the liquid aluminum mixture (LM25Al) is liquefied at a clear temperature near 780 °C in the silent heater (furnace), and then a warmed support material with 3% wt of VC is blended in with the liquid compound, and the mixture is mixed using the stirrer. To achieve consistent composite characteristics, the mixture was stirred at 800 rpm for 8 minutes. The improvement depends on the amount and uniformity of distribution of reinforcements, and the strength of the particle matrix boundary, and the mechanical properties of the matrix. Subsequently, to investigate the effect of 3 wt% VC in LM25Al alloy, brinell hardness as per ASTM E10 is 195±16 BNH obtained. When compared with the plain LM25Al alloy, the hardness of the synthesized composite increased up to 116%. This enhancement in hardness and strength value of LM25Al/VC is associated with the presence of VC as a load-bearing constituent in the ML25Al matrix during loading conditions. A similar phenomenon was reported in [26].

Table 1. Chimerical composition of base metals (ML25AL and VC)

Chemical composition	Cu	Mg	Si	Fe	Mn	Pb	Ni	Zn	Ti	Al
ML25 Al alloy	0.1	0.2-0.6	6.5-7.5	0.5	0.3	0	0.1	0.1	0.2	Bal.
Chemical composition of VC	Cu	Cr	Si	Fe	C	Pb	O	N	Al	V
	0.003	0.006	0.005	0.005	0.01	0.001	0.025	0.006	0.005	Bal.

Machining is carried out on the ML25Al/VC composite using different processing conditions. The PLUS Spark EDM PS 50 ZNC series machine is used for conducting experimental work, and the line diagram of the EDM process is elucidated in Figure 1. Copper electrodes and kerosene dielectrics are used in this study. The EDM process is influenced by both electrical and non-electrical parameters such as pulse on-time (t_{on}), pulse off-time (t_{off}), flushing pressure, polarity, peak current (I_p), electrode material, pulse interval, open discharge voltage, etc. It is difficult to examine the impact of all process parameters on performance characteristics at the same time due to different constraints. Thus, major influencing electrical parameters, namely I_p, t_{on} and t_{off} were considered in this work, and these parameters showed large deviations on the output characteristics [31]. The properties of the machine were that the connected load is 5 KVA and the capacity of flush pressuring is 196 kPa. The multi-design objective methods of ANOVA factorial, TOPSIS, and VIKOR are used in this research work to optimize the experimental trials. Twenty-seven runs are formulated by the design expert software, and the number of trials is purely dependent on the number of process parameters and the three ranges selected in this research work. The range for each variable is selected based on the preliminary trial runs on the ML25Al/VC composite and the capacity of the machine. It is important to state that when discharge current levels are less than 5 amps, there is not much change in workpiece material ablation. In addition, higher levels of current greater than 20 amps are not considered in order to facilitate stable machining and maintain parameter range consistency. To achieve the optimized parameters for pulse duration and spark interval, a wide range has been chosen based on the machine capacity. The range and levels of process variables are delineated in Table 2. The three replication runs of mid-level machining parameters are to measure the repeatability of the results. Tests are carried out in a randomized way, and performance metrics, namely material ablation rate and surface roughness, are considered. In addition, different combinations of process variables and their measured responses are delineated in Table 3.

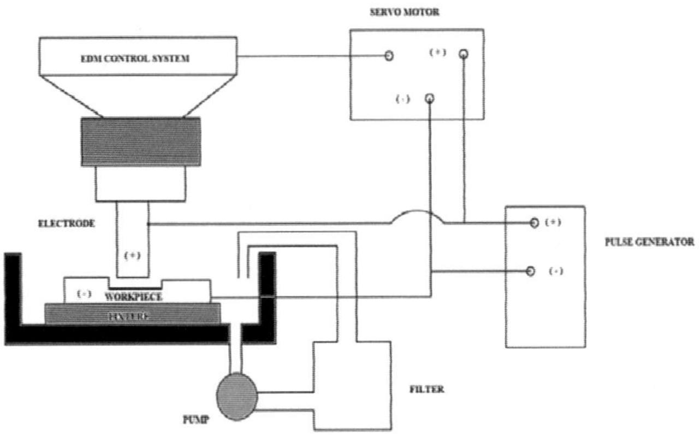

Figure 1. Schematic representation of EDM process

Table 2. Parameters and their levels

Parameters	Symbol	Level		
		-1	0	1
Pulse on time (μs)	t_{on}	100	350	500
Pulse off time (μs)	t_{off}	30	70	100
Peak current (Amps)	I_p	10	15	20

Table 3. Experimental trials and its responses

Expt. No.	Pulse on time (t_{on}) (μs)	Pulse off time (t_{off}) (μs)	Peak current (Amp)	Material removal rate (MRR) (g/min)	Surface roughness (R_a) (μm)
1	500	70	15	0.028248	4.67412
2	500	70	20	0.062556	4.256555
3	150	70	15	0.011592	4.751555

4	500	100	20	0.032388	6.11342
5	150	30	10	0.014544	4.861685
6	100	30	15	0.02454	4.431455
7	500	30	15	0.054792	3.85908
8	100	100	10	0.038892	5.26382
9	500	30	20	0.046284	3.337005
10	100	70	15	0.0183	5.176315
11	150	100	20	0.026544	4.106575
12	500	30	10	0.060648	5.32364
13	100	100	15	0.033708	5.106285
14	150	100	15	0.037692	4.596515
15	150	30	15	0.014784	5.50816
16	500	70	10	0.020928	4.99894
17	100	100	20	0.023832	4.61367
18	150	70	20	0.034224	4.62892
19	500	100	10	0.054612	4.822005
20	100	70	20	0.012696	5.205385
21	150	30	20	0.017868	4.818
22	150	100	10	0.01824	4.675
23	150	70	10	0.031824	4.29005
24	500	100	15	0.05568	4.3375
25	100	30	10	0.01314	4.7965
26	100	70	10	0.034668	5.0505
27	100	30	20	0.021144	4.671

The machinability of the process is calculated by determining material ablation rate, surface unevenness and electrode erosion rate. The formula used for measuring MRR

$$MRR = \frac{W_{in} - W_{fi}}{M_t} \text{ (gm/min)} \quad (1)$$

Where w_{in} is weight of the workpiece before machining, w_{fi} is weight of the workpiece after machining and M_t is machining time. The rate of material removal from the LM25Al/VC is calculated for each combination of parameters by considering loss in weight of the workpiece with respect to time taken to create a through hole. In this work, TOPSIS and the VIKOR are considered for further optimization and developing the models.

3. Multi criteria decision making techniques

3.1. TOPSIS

TOPSIS, known as technique for order of preference by similarity to ideal solution, is a multi-criteria decision analysis method. It compares a set of alternatives based on a pre-specified criterion. The various steps in performing the TOPSIS method are as follows:

Stage 1: Development of choice framework that comprising of all the data about the traits and the options of the interaction. Here, the potential options are addressed in line (i =1, 2......, m) and all credits connecting with every one of the choices are portrayed in the section (j =1, 2......, n).

Stage 2: Get the normalization choice framework utilizing the formulae given underneath: where, x_{ij} is the standardize choice. The normalized decision matrix grid can be addressed.

$$\bar{X}_{ij} = \frac{\mathrm{Xij}}{\sqrt{\sum_{\mathrm{i=1}}^{\mathrm{n}} (\mathrm{X_{ij}})^2}} \tag{2}$$

Stage 3: Acquire the weighted standardized choice network by assigning the heaviness of each. Property as: $\{w_j, j = 1, 2, \ldots\ldots, n\}$. It very well may be resolved utilizing the formulae given below

$$Wij = w_j * \bar{X}_{ij} \text{ , where } w_j \text{ is assigned value} \tag{3}$$

Stage 4: Decide the best (positive ideal) and most terrible (negative ideal) depending on the objective function.

Stage 5: Decide the partition measure for every elective utilizing the below relation

$$\lambda^+ = Max\ (Wij)_{i=1}^{n},\ \ \lambda^- = Min\ (Wij)_{i=1\ldots}^{n}, \quad \text{For beneficiaries} \tag{4}$$

$$\lambda^+ = Min\ (Wij)_{i=1}^{n},\ \ \lambda^- = Max\ (Wij)_{i=1\ldots}^{n}, \quad \text{For non beneficiaries} \tag{5}$$

Stage 6: Get relative closeness to the ideal arrangement utilizing the formulae given as Equation (5).

$$(s^+, s^-) = \sum_{j=1}^{n} \sqrt{(Wij - \lambda^{\pm})^2 + (Wij - \lambda^{\pm})^2} \tag{6}$$

Stage 7: Assign and positioning to all the overall closeness esteem in sliding request. Biggest or the smallest relative closeness esteem assigned as great execution of the other option as given in Equation (6) and results generated in Table 4.

$$P_i = \left(\frac{S^-}{S^+ + S^-}\right) \tag{7}$$

3.2. VIKOR

VIKOR means multi-criteria optimization and comprising solutions. The VIKOR technique focuses on positioning and picking an ideal substitute from a set of options and gives a compromise answer for an issue with clashing measures, hence coming to a last arrangement. It is a powerful apparatus for the advancement of machining activity, which comprises different reactions [32]. The proposed technique contains the accompanying advances:

Step 1: The different results obtained during the trial and error first and foremost changed over to a choice grid. In this step, all the criteria and output parameters are normalized irrespective of their units, and the values of the criteria lie between 0 and 1 by using Equation (2). And then determine the best and worst value, like $X_{ij}^{+} = \max(X_{ij})$ and $X_{ij}^{-} = \min(X_{ij})$.

Step 2: Standardize the choice of lattice amounts somewhere in the range of 0 and 1. The information standardization is done utilizing the accompanying conditions. The normalized choice grid can be composed as an underneath formulation

$$Y_i^+ = \sum_{j=1}^{n} W_j \left[(X_{ij})_{max} - (X_{ij})\right] / \left[(X_{ij})_{max} - (X_{ij})_{min}\right], \quad \text{for beneficial attribute} \tag{8}$$

$$Y_i^- = \sum_{j=1}^{n} W_j \left[(X)_{ij} - (X_{ij})_{min}\right] / \left[(X_{ij})_{max} - (X_{ij})_{min}\right], \text{non beneficial attribute} \tag{9}$$

IOP Conf. Series: Materials Science and Engineering 1294 (2023) 012049 doi:10.1088/1757-899X/1294/1/012049

Step 3: Register the gainful (ideal) and non-valuable (non-ideal) arrangements. These qualities can be determined utilizing the conditions indicated underneath.

$$R_i = Max^n of \left\{w_j \left[\left|(X_{ij})_{max} - (X_{ij})\right|\right] / \left[(X_{ij})_{max} - (X_{ij})_{min}\right]\right\} \tag{10}$$

Decide the utility measure and the lament measure by utilizing the accompanying conditions. Toward the end, the VIKOR file utilizes the condition given underneath (Equation (11)). Where is signifies the VIKOR file esteem i = 1, 2,..., m; likewise, addresses the heaviness of the most extreme gathering utility (in this work, v is 0.5).

$$Q_i = \frac{v((Y_i - Y_{i-min}))}{((Y_{i-max} - Y_{i-min}))} + (1 - v)\left(\left(\frac{((R_i - R_{i-min}))}{((R_{i-max} - R_{i-min}))}\right)\right) \tag{11}$$

Here Y_{i-max} is the maximum value of Y_i, and Y_{i-min} the minimum value of Y_i; R_{i-max} is the maximum value of R_i, and R_{i-min} is the minimum value of R_i.

Step 4: Ranking the acceptance choice with the cases, where case 1: $Q(a2) - Q(a1) \geq 1/(j-1)$ (acceptance advantage depending on the objective faction preference) and case 2: acceptance stability, likewise Q_j is ranking by Y_i and/or R_i with $v \geq 0.5$ in light of VIKOR file Q_i.The option with the biggest Q_i esteem is considered in this work as the best arrangement and result generated in Table 5.

Table 4. Optimization of EDM parameters using TOPSIS optimization technique

Exp. No.	Normalized decision matrix		Weighted normalized		Si+ (Ideal the best)	Si- (Ideal the worst)	Pi	Rank
1	0.15605	0.18815	0.11704	0.04704	0.14279	0.07052	0.3305	14
2	0.34559	0.17134	0.25919	0.04284	0.00925	0.21199	0.9587	27
3	0.06404	0.19127	0.04803	0.04782	0.21164	0.01371	0.0608	2
4	0.17893	0.24609	0.13419	0.06152	0.12808	0.08616	0.4021	16
5	0.08035	0.19570	0.06026	0.04893	0.19952	0.01756	0.0808	5
6	0.13557	0.17838	0.10168	0.04460	0.15790	0.05625	0.2626	12
7	0.30270	0.15534	0.22702	0.03884	0.03260	0.18042	0.8469	24
8	0.21486	0.21189	0.16114	0.05297	0.09995	0.11344	0.5316	21
9	0.25569	0.13433	0.19177	0.03358	0.06742	0.14643	0.6847	22
10	0.10110	0.20837	0.07582	0.05209	0.18430	0.02935	0.1373	7
11	0.14664	0.16531	0.10998	0.04133	0.14941	0.06516	0.3036	13
12	0.33505	0.21430	0.25129	0.05357	0.02150	0.20341	0.9044	26
13	0.18622	0.20555	0.13966	0.05139	0.12085	0.09219	0.4327	17
14	0.20823	0.18503	0.15617	0.04626	0.10380	0.10921	0.5127	20
15	0.08167	0.22173	0.06126	0.05543	0.19914	0.01456	0.0681	4
16	0.11562	0.20123	0.08671	0.05031	0.17329	0.04028	0.1885	9
17	0.13166	0.18572	0.09874	0.04643	0.16096	0.05291	0.2474	11
18	0.18907	0.18633	0.14180	0.04658	0.11811	0.09495	0.4456	18
19	0.30170	0.19410	0.22628	0.04853	0.03615	0.17872	0.8317	23
20	0.07014	0.20954	0.05260	0.05238	0.20744	0.01022	0.0469	1

21	0.09871	0.19394	0.07403	0.04849	0.18576	0.02909	0.1353	6
22	0.10077	0.18819	0.07557	0.04705	0.18411	0.03112	0.1445	8
23	0.17581	0.17269	0.13186	0.04317	0.12769	0.08581	0.4019	15
24	0.30760	0.17460	0.23070	0.04365	0.03022	0.18354	0.8586	25
25	0.07259	0.19308	0.05444	0.04827	0.20527	0.01472	0.0669	3
26	0.19152	0.20330	0.14364	0.05083	0.11683	0.09621	0.4516	19
27	0.11681	0.18803	0.08761	0.04701	0.17211	0.04216	0.1967	10

Table 5. Optimization of parameters utilizing VIKOR optimization technique

Exp. No.	Normalized decision matrix		Weight standardization		Y_i (Utility solution)	R_i (Regret solution)	Q_i	Rank
1	0.15605	0.18815	0.50489	0.12040	0.62529	0.50489	0.64742	14
2	0.34559	0.17134	0.00000	0.08280	0.08280	0.08280	0.00000	1
3	0.06404	0.19127	0.75000	0.12737	0.87737	0.75000	0.98497	26
4	0.17893	0.24609	0.44396	0.25000	0.69396	0.44396	0.64368	13
5	0.08035	0.19570	0.70656	0.13729	0.84385	0.70656	0.93196	23
6	0.13557	0.17838	0.55945	0.09855	0.65800	0.55945	0.70828	12
7	0.30270	0.15534	0.11426	0.04701	0.16127	0.11426	0.07147	2
8	0.21486	0.21189	0.34825	0.17350	0.52174	0.34825	0.46684	8
9	0.25569	0.13433	0.23946	0.00000	0.23946	0.23946	0.21302	6
10	0.10110	0.20837	0.65128	0.16562	0.81690	0.65128	0.87409	22
11	0.14664	0.16531	0.52996	0.06930	0.59926	0.52996	0.65033	15
12	0.33505	0.21430	0.02808	0.17889	0.20696	0.17889	0.14779	5
13	0.18622	0.20555	0.42453	0.15931	0.58385	0.42453	0.56192	16
14	0.20823	0.18503	0.36591	0.11341	0.47932	0.36591	0.45418	7
15	0.08167	0.22173	0.70303	0.19550	0.89853	0.70303	0.96268	25
16	0.11562	0.20123	0.61261	0.14965	0.76226	0.61261	0.81175	19
17	0.13166	0.18572	0.56987	0.11496	0.68483	0.56987	0.73247	17
18	0.18907	0.18633	0.41694	0.11633	0.53327	0.41694	0.52535	9
19	0.30170	0.19410	0.11691	0.13372	0.25062	0.13372	0.14059	4
20	0.07014	0.20954	0.73375	0.16824	0.90199	0.73375	0.98782	27
21	0.09871	0.19394	0.65764	0.13336	0.79100	0.65764	0.86304	21
22	0.10077	0.18819	0.65217	0.12048	0.77265	0.65217	0.84774	20
23	0.17581	0.17269	0.45226	0.08582	0.53808	0.45226	0.55476	11
24	0.30760	0.17460	0.10119	0.09009	0.19128	0.10119	0.07999	3
25	0.07259	0.19308	0.72722	0.13142	0.85864	0.72722	0.95647	24
26	0.19152	0.20330	0.41041	0.15429	0.56470	0.41041	0.53964	10
27	0.11681	0.18803	0.60943	0.12012	0.72955	0.60943	0.78941	18

The result obtained from the VIKOR optimization technique and portrayed in Table 5 fulfills the first acceptable advantage criteria from the file Q_i. Thus, VIKOR is a useful tool in MADM, mainly in situations where the decision-maker is not able or does not know in what way to express preference at the beginning of system design.

4. Result and discussion

The objective of this work is to estimate the optimized electrical discharge machining (EDM) parameters for maximum productivity and minimum surface integrity using the combined TOPSIS and VIKOR methods. The effect of three process parameters, namely pulse on time (t_{on}), pulse off time (t_{off}), and peak current (I_p), on responses such as *MRR* and R_a have been considered. The optimal conditions of the analysis were determined using the values obtained from TOPSIS and the VIKOR method, a multi-criteria decision-making method which determines the ranking list based on the particular measure of closeness to the ideal solution. From the design expert software analysis, it is found that the I_p=10Amp, t_{on}=150µs, t_{off} = 30 µs provides the lowest values for TOPSIS, and the I_p=15Amp, t_{on}=500 µs, t_{off}=30 µs provides the lowest values for VIKOR optimization techniques. The results were confirmed by conducting a conformation test and found to be comparable with the experimental results, as shown in Table 6. The findings of the study provide valuable information regarding TOPSIS and VIKOR optimization techniques, the responses, and confirm a higher *MRR* and minimum R_a. The VIKOR method provides a better compromise solution based on maximum *MRR* utility and minimum regret when compared with the TOPSIS method. However, it turns out that both the task of indicating the best alternative and ranking a set of decision options have similar results, but to a very limited extent, they are different, as shown in Table 6 in terms of errors.

Table 6. Conformation test as per the TOPIS and VIKOR Indexes

Predictor index	level of main effect for closeness solution	Predicted results	Experimental results for *MRR*	%Error
TOPSIS	I_p = 10 Amp, t_{on} = 150 µs, t_{off} = 30µs	0.0444861	0.014544	2.99421
VIKOR	I_p = 15 Amp, t_{on}= 500 µs, t_{off} = 30 µs	0.06220563	0.0547992	0.740643

In this work, optimization criteria are attempted for TOPSIS and VIKOR in design expert software to maximize MRR and minimize R_a. The dependent parametric condition with maximum composite desirability is viewed as the optimal level of parameters, as shown in Figure 2. As per the obtained results, I_p is the most vital parameter, followed by t_{on}. In this study, the obtained ideal parameter levels are I_p = 20 Amp, t_{on}=500 µs, and t_{off}=30 µs with desirability value of 0.553 and it is depicted in Figure 2. The combination of process parameters showing highest value of composite desirability are noted as the optimized parameter levels. Moreover, there is no significant rule that the composite desirability value ought to be near 1 [33]. It purely indicates how far the maximum and minimum values of the measured performances are from the optimized value of the responses.

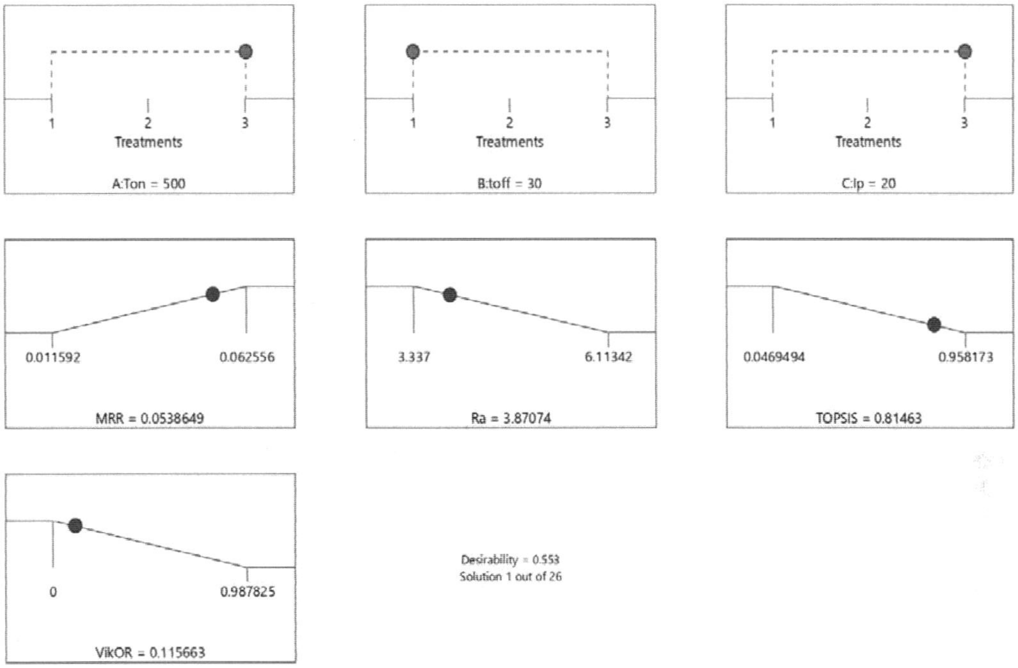

Figure 2. Optimized parameter levels for maximum MRR and minimum R_a

4.1. Factorial modelling and conformation

To assess performance characteristics for the EDM of LM25Al/VC, confirmation tests for the optimum parameters with their levels are done in Table 6. In the same way, factorial methods are used to build the machining characteristics into a mathematical and statistical numeric equation. The primary objective is to establish a relationship between the response and the spending process parameters.

$$TOPSIS = 0.3901 - 0.1264t_{on(1)} - 0.1508t_{on(2)} - 0.0294t_{off(1)} - 0.0544t_{off(2)} + 0.0101I_{p1} - 0.0001I_{p2} - 0.0590t_{on(1)}*t_{off(1)} - 0.1151t_{on(2)}*t_{off(1)} + 0.0026t_{on(1)} * t_{off(2)} + 0.1179t_{on(2)} * t_{off(2)} + 0.0762*t_{on(1)}*I_{p1} - 0.0403t_{on(2)} * I_{p1} + 0.0139t_{on(1)}*I_{p2} - 0.0254t_{on(2)}*I_{p2} \quad (12)$$

$$VIKOR = 0.5981 + 0.1372t_{on(1)} + 0.1547t_{on(2)} + 0.0291t_{off(1)} + 0.0604t_{off(2)} + 0.0017I_{p1} - 0.0042I_{p2} + 0.0538t_{on(1)}*t_{off(1)} + 0.1374t_{on(2)}*t_{off(1)} + 0.0049t_{on(1)} * t_{off(2)} - 0.1248t_{on(2)} * t_{off(2)} - 0.0826*t_{on(1)}*I_{p1} + 0.0237t_{on(2)} * I_{p1} - 0.0163t_{on(1)}*I_{p2} + 0.052t_{on(2)}*I_{p2} + 0.0500t_{off(1)} * I_{p1} -0.0247t_{off(2)} * I_{p1} \quad (13)$$

The equation contains all factor main effects and all factor-by-factor interactions to make predictions about the response for given levels of each factor. In this work, the high levels of the factors are designated by 2, and the low levels are assigned as 1 with a subscript in the equations.

5. Conclusion

In this work, the EDM process parameters are analyzed to machine the LM25 AL/VC composite material. Increasing I_p and t_{on} have produced a higher *MRR* and medium surface roughness. It has been indicated that I_p is the most crucial input parameter, followed by t_{on}. Using the VIKOR and TOPSIS techniques, the degree of impact of machining variables on the response parameters *MRR* and R_a have been analyzed. The prime conclusions are that TOPSIS and VIKOR can be used to find the ideal solution of EDM input parameters to machine LM25AL/VC composite material. The ideal solutions obtained at I_p=10 Amp, t_{on}= 150 μs, t_{off} = 30 μs provide the lowest values for TOPSIS, and the I_p = 15 Amp, t_{on}= 500 μs, t_{off} = 30 μs provide the lowest values for VIKOR optimization techniques. The models are valid with experimental results for selected responses. Therefore, VIKOR and TPOSIS can be used as prediction tools for EDM processes.

Funding: This research was done through the NORHED II project INDMET grant (grant nr. 62862) cooperation whose grant support is highly acknowledged. The authors will also gratefully acknowledge facility provided by Jimma University and University of Stavanger.

Reference

[1] Kumar A, Sharma R 2020 Multi-response optimization of magnetic field assisted EDM through desirability function using response surface methodology, *J. Mech. Behav. Mater.* **29**(1), 19–35. DOI: 10.1515/jmbm-2020-0003

[2] Gupta K, Gupta K 2019 Developments in nonconventional machining for sustainable production: A state-of-the-art review. *Proc. Inst. Mech. Eng., Part C, J. Mech. Eng. Sci.* **233**(12), 4213-4232.

[3] Sivaraman B, Padmavathy S, Jothiprakash P and Keerthivasan T 2016 Multi-response optimisation of cutting parameters of wire EDM in titanium using response surface methodology, *Appl. Mech. Mater.* **85**(4), 93–100. DOI: 10.4028/ www.scientific.net/ amm.854.93.

[4] Abbas N, Solomon G and Fuad M 2007 A review on current research trends in electrical discharge machining (EDM). *Int. J. Mach. Tools Manuf.* **47**(7), 1214-1228

[5] Sun Z, and Ahuja L 2010 Mechanical properties of vanadium carbide and a ternary vanadium tungsten carbide. *Solid State Commun.* **150**(6), 697–700

[6] Bodunrin M, Alaneme K and Chown L 2015 Aluminium matrix hybrid composites: a review of reinforcement philosophies; mechanical, corrosion and tribological characteristics, *J. Mat. Res. Tech.,* **4**(4), 434–445

[7] Yoo K, Kwon T and Kang S 2014 Development of a new electrode for micro-electrical discharge machining (EDM) using Ti(C,N)-based cerme, *Int. J. Precis. Eng.Manuf.* **15** (4), 609-616

[8] Selvarajan L, Narayanan S and Jeyapaul R 2015 Optimization of eDM hole drilling parameters in machining of MoSi2-SiC intermetallic/composites for improving geometrical tolerances. *J. Adv. Manuf. Syst.* **14** (4), 259–272. DOI: 10.1142/S0219686715500171.

[9] Kumar D, Mer S, Payal S and Kumar K 2021 Residual stress modeling and analysis in Aisi A2 steel processed by an electrical discharge machine. *Mater. Tehnol.* **56**(1), 65–72. DOI: 10.17222/mit.2021.325

[10] Ashebir D, Mengeshat A andSinha K 2022 An insight into mechanical and metallurgical behavior of hybrid reinforced aluminum metal matrix composite. *Adv. Mater. Sci. Eng.* https://doi.org/10.1155/2022/7843981

[11] Pradhan K and Biswas K 2008 Modeling of residual stresses of Edmed Aisi 4140 steel, *Proc. Int. Conf. Recent Adv. Mater., Processing and Characterization*, 49-55. A.P. India

[12] Kruth L, Stevens L and Froyen L 1995 Study of the white layer of a surface machined by die-

sinking electrical-discharge machining, *CIRP Annals,* **44**(1), 169–172, doi:10.1016/S0007-8506(07)62299-9

[13] Rajmohan T, Vinayagamoorthy R and Mohan K 2019 Review on effect machining parameters on performance of natural fibre-reinforced composites (NFRCs). *J. Thermoplast. Compos. Mater.* **32**(9), 1282-1302.

[14] Yadav V, Jain V and Dixit P 2002 Thermal stresses due to electrical discharge machining, *Int. J. Mach. Tools Manuf.* **42**(4), 877–888, doi:10.1016S0890-6955(02)00029-9

[15] Luzia O, Laurindo H, Soares C, Torres D and Mendes L 2019 Recast layer mechanical properties of tool steel after electrical discharge machining with Silicon powder in the dielectric. *Int. J. Adv. Manuf. Technol.* **103**(1-4), 15–28. DOI: 10.1007/s00170-019-03549-w

[16] Sundriyal S and Vipin S 2020 Study on the influence of metallic powder in near-dry electric discharge machining *J. Mech. Eng.* **66** (4), 243–253. DOI: 10.5545/sv-jme.2019.6475

[17] Sathish T 2019 Experimental investigation of machined hole and optimization of machining parameters using electrochemical machining, *J. Mater. Res. Technol.* **8**(5), 4354-4363

[18] Daneshmand S, Masoudi B and Monfared V 2017 Electrical discharge machining of Al/7.5% Al2O3 MMCs using rotary tool and Al2O3 powder, *Surf. Rev. Lett.* **24**(2), 1–17. DOI: 10.1142/S0218625X17500184

[19] Nagarajan P, Murugesan K and Natarajan E 2019 Optimum control parameters during machining of LM13 Aluminum alloy under dry electrical discharge machining (EDM) with a modified tool design. *Medziagotyra*, **25**(3), 270–275. DOI: 10.5755/j01.ms.25.3.20899

[20] Klocke F, Herrig T, Zeis M and Klink A 2018 Experimental investigations of cutting rates and surface integrity in wire electrochemical machining with rotating electrode, *Procedia CIRP*, **68** (7), 725-730

[21] Gudur P 2015, Effect of silicon carbide powder mixed EDM on machining characteristics of SS 316L material, *Int. J. Innov. Res. Sci. Eng. Technol.* **4**(9), 8133-8141. DOI: 10.15680/ijirset.2015.0404027

[22] Ablyaz, T R, Shlykov, E S, Muratov K R and Sidhu S S 2021 Analysis of wire-cut electro discharge machining of polymer composite materials. *Micromachines*, **12**(5), 571.

[23] Chow M, Yang D, Lin T and Chen F 2008 The use of SiC powder in water as dielectric for micro-slit EDM machining, *J. Mater. Process. Technol.* **195**(3), 160–170. DOI: 10.1016/j.jmatprotec.2007.04.130

[24] Fadhil S and Aghdeab S 2020 Effect of powder-mixed dielectric on EDM process performance. *Eng. Technol. J.* **38** (8), 1226–1235. DOI: 10.30684/etj.v38i8a.554

[25] Hosni J and Lajis A 2019 Experimental investigation and economic analysis of surfactant (Span-20) in powder mixed electrical discharge machining (PMEDM) of AISI D2 hardened steel. *Mach. Sci. Technol.* **24**(3), 398–424. DOI: 10.1080/10910344.2019.1698609

[26] Ehsan G, Pakseres H, Masoud A, Kamyar S and Ebadzadeh T 2016 Vanadium carbide reinforced aluminum matrix composite prepared by conventional, microwave and spark plasma sintering. *J. Alloy Com.* 688, **527** - 33. doi: 10.1016/ j.jallcom.2016.07.063.

[27] Shao M, Han Z, Sun J, Xiao C, Zhang, S and Zhao Y 2020 A review of multi-criteria decision making applications for renewable energy site selection. *Renewable Energy*, **157**, 377-403.

[28] Rouniyar K and Shandilya P 2021 Semi-empirical modeling and optimization of process Parameters on overcut during MFAPM-EDM of Al6061 alloy. *Proc. Inst. Mech. Eng. Part E J. Process Mech. Eng.* **235** (6), 1784–1796. DOI: 10.1177/09544089211015890

[29] Eyercioglu O, Cakir V and Gov K 2014, Influence of Machining Parameters on the Surface Integrity in Small-Hole Electrical Discharge Machining. *Proc. Inst. Mech. Eng. Part B J. Eng. Manuf.* **228**(1), 51–61. DOI: 10.1177/0954405413498730

[30] Suresh Kumar S, Uthayakumar M, Thirumalai Kumaran S, Parameswaran P, Mohandas E 2016; Electrical Discharge Machining of Al (6351) Alloy: Role of Electrode Shape, *Int. J. Mater. Prod. Technol.* **53**(1), 86–97. DOI: 10.1504/IJMPT.2016.076378

[31] Singh R, Ravindra R (2022), The electrical discharge machining process performance analysis for titanium alloy machining: Using TOPSIS and Taguchi technique, In: *Adv. Indu. Prod. Eng. Lecture Notes in Mechanical Engineering. Springer, Singapore.* 307–31

[32] Gangil M, Pradhan M 2018, Optimization the machining parameters by using VIKOR method during EDM process of titanium alloy. *Mater. Today Proc.*, 7486–7495

[33] Mohammadsadegh M, Seyed M, Mohammad K and Madjid T 2018 A hybrid desirability function approach for tuning parameters in evolutionary optimization algorithms. *J. of Meas.*, **114**(5), 417-427. https://doi.org/10.1016/j.measurement.2017.10.009

Comparative analysis of artificial neural network model and analysis of variance for predicting defect formation in plastic injection moulding processes

N D Dejene[1*] and D W Wolla[2]

[1]Faculty of Science and Technology, University of Stavanger, N-4036 Stavanger, Norway
[2] School of Mechanical and Industrial Engineering, Addis Ababa Institute of Technology, Addis Ababa, Ethiopia.

[*] Corresponding author: nd.dejene@stud.uis.no

Abstract. This study investigates the impact of plastic injection moulding process parameters on overflow defect formation. Experiments were conducted using a Taguchi L27 orthogonal array design. Multilayer Perceptron (MLP) artificial neural networks is explored and compared with ANOVA predictions. To assess model performance, the Root Mean Squared Error (RMSE) and the coefficient of determination (R2) is applied. The study considered temperature, speed, pressure, and packing force when constructing the MLP model using the back-propagation algorithm in Python. Results show that among the configured MLP neural networks, the 3-layer MLP architecture with sigmoid activation functions in hidden layers and a linear function in the output layer exhibited the lowest prediction error and the highest coefficient of determination. Comparative analysis reveals that the MLP neural network outperforms the ANOVA model, indicating superior prediction accuracy. The predicted outcomes from the ANN align well with experimental values, demonstrating the effectiveness of the ANN model in forecasting defect formation under specific process conditions. This research sheds light on the significance of process parameters and showcases the potential of MLP neural networks as a valuable tool in predicting and mitigating overflow defects in plastic injection moulding.

Keywords: Analysis of variance, Artificial neural network, Multilayer perceptron, injection moulding, defects.

1. Introduction

Plastic Injection moulding (PIM) is a technique that involves heating a polymer to a highly plastic state and forcing it to flow at high pressure into a mold cavity, in which it hardens [1–3]. After that, the molded portion is withdrawn from the hollow as shown in Figure 1. The technique yields discrete components that are close to the net form or near-net shapes. PIM is applied to produce complex features and intricate shapes. They are less expensive than other methods like 3D printing [4, 5]. By using the PIM method, a variety of thermoplastics, including ABS, Nylon, PLA, Polycarbonate, PEEK, Polyethylene, Polypropylene, Polystyrene, and Polyvinyl Chloride, are frequently utilized [6]. Fast-moving consumer goods, of which half are produced using plastic, are in high demand in this rapidly developing global economy [6, 7]. According to Maarif *et al.* [8], PIM now accounts for around 30% of the global production of products or components made of plastic. PVC is the most widely produced and reasonably priced polymer with the necessary chemical properties as well as mechanical and thermal stability [9–11]. Plastic Industries mold PVC to varieties of shapes according to the desired functions i.e., shoe sole is one of the applications.

Filling, holding, and cooling are the three fundamental stages of the injection moulding procedure. First, the cavity is filled with molten plastic material to form the product's desired shape. The pressure on the holding stage is then increased by packing additional material into the cavity. To solidify the molten materials, the temperature will be greatly reduced during the cooling phase. The last step ensures that the finished product of the moulding process is stable enough for ejection. If there are no flaws, the outcome of the injection moulding process is deemed satisfactory [12, 13].

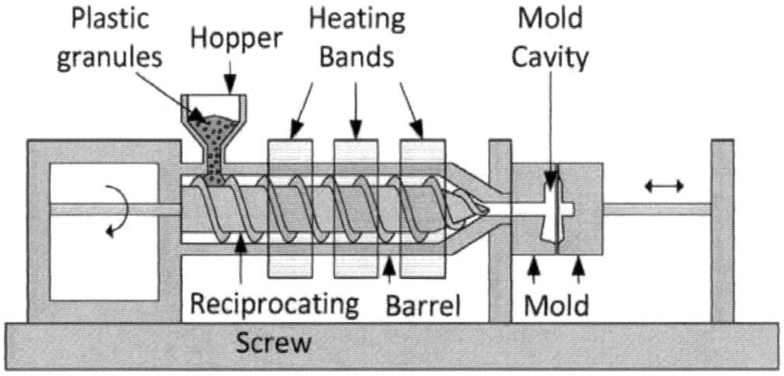

Figure 1. Plastic injection moulding machine in a simplified form [12].

Final products with defects are frequently the result of unforeseen behavioural changes in injection moulding parameters [14]. The common flaws detected on the final products of PIM are shrinkage and warpage, short-shot, and sink mark. This research addressed several defects in the process of shole sole injection moulding process as short-shot, weight variation, overflow, warpage, shrinkage, swelling, sink mark and black dot. However, based on the frequency and effects on the defects; this research addresses the overflow in the PIM. The different sorts of flaws are caused by different factors which influence the injection moulding process. An overflow defect due to different control process parameters involving in different stages PIM and causes incomplete parts or often causes weight variations. In PIM, process parameters need be managed to get the desired shape and minimum defects that may arise from different sources, including skills (man), machine, measurement, material, mold design, system parameters, and surroundings [11].

The process parameters such as flow velocity, melt temperature, heat input, metering length, hold pressure and holding duration are vital variables observed in this research. From the trends of scholars there have been several attempts to formulate the cause and effect of independent variables and the desired quality PIM final products. For instance, independent factors analysed to discuss the impacts on the microscopic moulding process and part quality of micro gears [15, 16]. Huang *et al.* [17], investigated the impacts of gate size, mold heat temperature, melting time, packing pressure, injection pressure and packing time to find the ideal warpage outcome of a molded plastic part on the quality characteristics. According to Aditya *et al.* [18], the mechanical and physical properties, dimensional or quantifiable features are the attributes of PIM quality characteristics.

The source of defects can be related to material, filling, packing, and cooling. Material-related defects are black specks and splays; filling-related are short fills, overflow, while packing and cooling associated defects such as sink mark voids, warpage, deformations, and others [11]. The cause effect analysis are the techniques used to identify the responsible factors for the formation of defects [19, 20] Except material-related flaws, the defects stated earlier are primarily formed due to poor selection of processing parameters. Massive research has been conducted from several angles to examine the effect of the PIM process parameters on the on the mechanical characteristics of molded parts and the incidence of moulding flaws [15]. The optimization of the involving process parameters is common in the current research to get better quality results of final components. Plastic pipe flaws were optimized to improve to the level of defects occurring in the process [21]. Similar techniques are followed to determine the influence of temperature on the tensile and flexural strengths of bidirectional kenaf/PLA composites and the optimal temperature. These demonstrates that including independent factors in PIM influences defect formation [22–24].

Numerous model-based and experimental research has been conducted on the impacts process parameters on the formation of several defects. Taguchi Method and Response Surface Method are generally preferred in experimental studies due to their effective combination of factors and cost. Predicting the impacts of process parameters on the quality characteristics of PIM are vital aspects of current research. For instance, the barrel temperature, holding time, mold temperature, holding pressure, and cooling time effects on the warpage formation are predicted by using machine learning, while their significancy performed by using Analysis of Variance (ANOVA) [25, 26]. Moreover, the correlation between PIM process parameters and particle size of kaolin clay affects strength, shrinkage and warpage polymer composite executed by ANOVA. In Chen *et al.* [27] study, the Taguchi technique and ANOVA are employed to identify the most important parameters to cause warpage during the moulding process. Then a numerical strategy with SolidWorks Plastics is used to optimize the process variables to lessen warpage. Analysis of the impacts of machine process parameters is vital to plan and design PIM process. The majority of current parameter analysis research, however, is focused on computer-aided engineering (CAE) or simulation, which has been shown to be insufficient for assessing complicated behavioural changes in the actual PIM process [8].

Artificial Intelligence (AI) has found diverse applications across various domains, encompassing tasks like calculating cutting forces, estimating tool lifespan, discovery new materials, implementing predictive maintenance, and forecasting process quality. Within the realm of AI techniques, Artificial Neural Networks (ANNs) stand out as one of the earliest and most widely adopted methods [28, 29]. ANNs, alongside other advanced machine learning approaches, have exhibited remarkable success in enhancing prediction accuracy [30]. In this study, we exploited the power of ANN to forecast the influence of crucial process parameters on the occurrence of overflow incidents in PIM processes. To achieve the experimental data was collected, and applied ANOVA and construct an ANN-based predictive model.

2. Experimentation and setups

2.1 Parameters selection

Among several defects the overflow was one of the significant challenges in the injection moulding process. The responsible process parameters were selected as an independent variable based on the cause-and-effect analysis conducted followed by 5-why. The independent variables are identified as temperature, injection pressure, injection speed and packing pressure. The dependent variable is overflow or flash out molten from the cavity. Melt temperature (T), injection velocity (S), injection pressure (P), and packing pressure (Pf) are recognized as the fundamental reasons for overflowing. These process parameters are independent variables and are responsible for the occurrence of overflows (R). The relationship between overflow and underweight is studied, and overflow becomes the cause of underweight.

2.2 Design of experiments

The Taguchi method orthogonal array L27 was used to conduct experiments. The level of temperature (160, 170, and 180 ^{O} C), injection speed (10, 15, and 20 cm/s), injection pressure (0.2, 0.4, and 0.6 bar) and packing pressure (0.1, 0.3 and 0.4 bar) based on the ranges of operating conditions stated by the manufacturer. The weight results are measured while the signal to noise ratio (S/N) is calculated as shown on Table 1. As measuring the overflow is challenging, the opposite measurement of the part left inside the mold weight is measured. The "Larger-is-better" method as shown on Equation (1) is used, and the S/N ratio methodology was used to reduce the overflows or increase the weight of the final part.

$$S/N = -10 \times Log10(\frac{1}{n}\sum_{i=1}^{n}\frac{1}{y^2}) \quad (1)$$

Table 1. Taguchi design of experimentation and response

Run	T(°C)	S (cm/s)	P (bar)	Pf (bar)	y(Kg)	S/N (dB)
1	160.0	10.0	0.20	0.10	0.200	-13.980
2	160.0	10.0	0.20	0.10	0.210	-13.560
3	160.0	10.0	0.20	0.10	0.220	-13.1530
4	160.0	15.0	0.40	0.30	0.240	-12.400
5	160.0	15.0	0.40	0.30	0.230	-12.770
6	160.0	15.0	0.40	0.30	0.260	-11.700
7	160.0	20.0	0.60	0.40	0.290	-10.7520
8	160.0	20.0	0.60	0.40	0.280	-11.060
9	160.0	20.0	0.60	0.40	0.270	-11.3730
10	170.0	10.0	0.40	0.40	0.280	-11.060
11	170.0	10.0	0.40	0.40	0.280	-11.060
12	170.0	10.0	0.40	0.40	0.270	-11.380
13	170.0	15.0	0.60	0.10	0.110	-19.1720
14	170.0	15.0	0.60	0.10	0.130	-17.7210
15	170.0	15.0	0.60	0.10	0.150	-16.480
16	170.0	20.0	0.20	0.30	0.170	-15.410
17	170.0	20.0	0.20	0.30	0.190	-14.430
18	170.0	20.0	0.20	0.30	0.180	-14.900
19	180.0	10.0	0.60	0.30	0.250	-12.0410
20	180.0	10.0	0.60	0.30	0.260	-11.7010
21	180.0	10.0	0.60	0.30	0.250	-12.0410
22	180.0	15.0	0.20	0.40	0.240	-12.370
23	180.0	15.0	0.20	0.40	0.250	-12.0410
24	180.0	15.0	0.20	0.40	0.230	-12.770
25	180.0	20.0	0.40	0.10	0.150	-16.490
26	180.0	20.0	0.40	0.10	0.160	-15.920
27	180.0	20.0	0.40	0.10	0.140	-17.080

As indicated on Table 2 and in Figure 2, the S/N is calculated, and these ratios offer valuable insights into the significance of each parameter for the formation of overflow are ranked.

Table 2. Response signal to noise ratios

Level	T	S	P	Pf
1	-12.32	-12.22	-13.64	-16.01
2	-14.67	-14.22	-13.33	-13.05
3	-13.62	-14.17	-13.64	-11.55
Delta	2.35	1.99	0.31	4.46

Rank	2	3	4	1

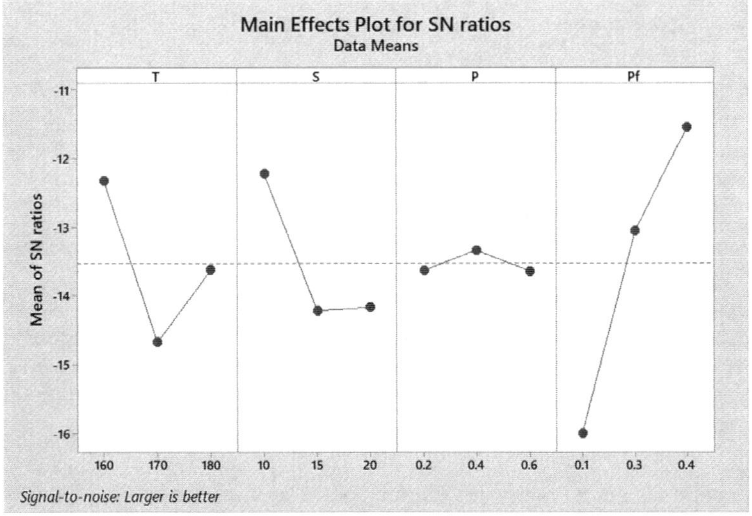

Figure 2. Signal- to- noise ratio: Larger is better result.

3. Results and discussions

3.1 Analysis of variance.

Based on the study of the experimental data, the ANOVA statistical approach is used to propose significant contribution of selected process parameters. ANOVA was used to assess the model's fit, indicating the effects of the model that were statistically significant at a 95% level of confidence (p-value < 0.05). The results of the multiple regression analysis used to fit the models to the experimental data, along with the values of the coefficients and the model summary statistics L27 designs, are displayed in Tables 3. The key selected parameters have statistically significant contribution for the formation of overflows. Moreover, it can be deduced that among the effective input parameters, the one with the higher F-value has a greater influence than the other parameters [31, 32]. Therefore, pressure is the most significant while temperature and packing pressure have slightly similar effects, and relatively speed has less contribution. Furthermore, as shown on Table 4, the computed regression coefficients R-sq. and R-sq. adjusted for overflow is 85.99% and 78.18% respectively. The result ensures a good match for the link between the process parameters and the examined quality requirements. The mathematical relationship for linking the overflow and the key process parameters was established as given in Equation (2).

Table 3. ANOVA results

Source	DF	Adj. SS	Adj. MS	F-Value	P-Value
T	2	0.015496	0.007748	23.77	0.000
S	2	0.007385	0.003693	11.33	0.001
P	2	0.016719	0.008359	25.65	0.000
Pf	2	0.015030	0.007515	23.06	0.000
Error	18	0.005867	0.000326		
Total	26	0.060496			

Table 4. The model summary

Residual std. deviation	R-sq.	R-sq. (adj.)	R-sq.(pred.)
0.0180534	90.30%	85.99%	78.18%

$$\text{Response (R)} = 0.21963 + 0.02593 \times T_{160} - 0.03185 \times T_{170} + 0.00593 \times T_{180} - 0.02185 \times S_{10} + 0.00370 \times S_{15} + 0.01815 \times S_{20} + 0.01815 \times P_{0.2} - 0.03519 \times P_{0.4} + 0.01704 \times P_{0.6} - 0.02852 \times PF_{0.1} + 0.02926 \times PF_{0.3} - 0.00074 \times PF_{0.4} \quad (2)$$

3.2 ANN model for response prediction.

In the present study, an ANN is composed of a few neurons connected by links. The feed forward neural network based on back-propagation is the best all-purpose model out of the different neural network types [33]. The network includes one response output, and four inputs: temperature, speed, pressure, and packing force. The TensorFlow library of python software package used for writing the programming code. To train and test the model; the data are partitioned by 70 and 30 % respectively. To identify the ideal architecture, several model iterations were conducted with different numbers of hidden layers, neurons, and activation functions for best training and testing developed models. Root mean square error (RMSE) is used in prediction, and R^2 values of the trained models as a primary selection criterion. The algorithm was trained using adaptive moment estimation (Adam) technique which combines the advantages of gradient descent (GD) and Stochastic gradient descent (SGD). The three layers as input layer, hidden layer, and output layer make up the usual type of ANN. An input unit layer is connected to a hidden unit layer, which is coupled to an output unit layer. The sigmoid activation function for hidden layer and linear function is chosen for output for training, and the MSE objective was set at 0.0001 to ensure that the MSE would not exceed this threshold. The selection of several hidden layers is executed by trial and error.

Several models are built, and some of the significant data are shown on Table 5. Equations (3)-(7) represent the feed-forward mechanism used by the MLP technique to estimate the output parameters based on the input parameters. The number of layers (i), sigmoid activation function S(X), input parameter (x), target value (y), output of the prediction (y_{pred}.), weight (w), bias (b), and error function (ε). The input value for the nth layer is multiplied by the weight value, and the bias value is added. The resulting value is entered into a sigmoid activation function, which computes a new value and transmits it to the next output layer. This method of training the model is continued until the computer error function between the output of the prediction model and the target value reduces to an acceptable level. Throughout the iterations 70/30 %, 15 second running time for each, 1000 epochs, 10 batch size and mse is utilized as metrics. So, that one hidden layer with 17 neurons provided minimum rmse, high R^2 value with demanded range of loss which is 0.00000612.

$$X = (x_1, x_2, x_3, x_4) = (T, S, P, Pf) \quad (3)$$

$$y_{train} = W_{i-1}X_{i-1} + W_iX_i + b \quad (4)$$

$$S(X) = \frac{1}{(1+e^{-x})} \quad (5)$$

$$y_{pred} = f(X_{test}) \quad (6)$$

$$\varepsilon = f(y_{test}, y_{pred}) = mse \quad (7)$$

$$R2 = r2\_score(y_{test}, y_{pred}) \quad (8)$$

The outputs of the hidden layer are used to construct the error function at the end of each feed-forward loop of the prediction model. When this error function returns a result, the weight values of the earlier layers are updated.

Table 5. The selection of neurons and hidden layer iterations

1st hidden	2nd hidden	RMSE	R^2	Loss (MSE)
7	0	0.02	0.82	0.00051462
8	0	0.02	0.83	0.00048113
9	0	0.01	0.96	0.00011662
10	0	0.02	0.87	0.00037283
11	0	0.02	0.90	0.00027857

12	0	0.02	0.87	0.00035588
13	0	0.02	0.88	0.00034911
14	0	0.02	0.82	0.00051478
15	0	0.02	0.86	0.00038781
16	0	0.01	0.97	0.00000975
17	0	0.01	**0.98***	0.00000612
18	0	0.01	0.94	0.000163909
19	0	0.01	0.95	0.000143263
20	0	0.01	0.94	0.000179964
21	0	0.01	0.94	0.000182297
9	1	0.05	0.03	0.002733573
	2	0.03	0.75	0.000704302
	3	0.01	0.96	0.000704302
	4	0.02	0.83	0.00038281

MLP artificial neural network model was used to predict the results, and the architecture is shown on Figure 3, which is built by using an online tool [34]. Three statistical performance evaluation metrics are utilized to evaluate the performance of the models: mean square error (mse), coefficient of determination (R^2) and root mean square error (rmse). The experimental results show that the 3-layer MLP architecture with sigmoid activation functions in each hidden layer and a linear function in the output layer has the lowest prediction error and the highest coefficient of determination among the configured MLP neural networks. Comparative analysis of performance findings shows that the MLP neural network has a lesser prediction error than the ANOVA model.

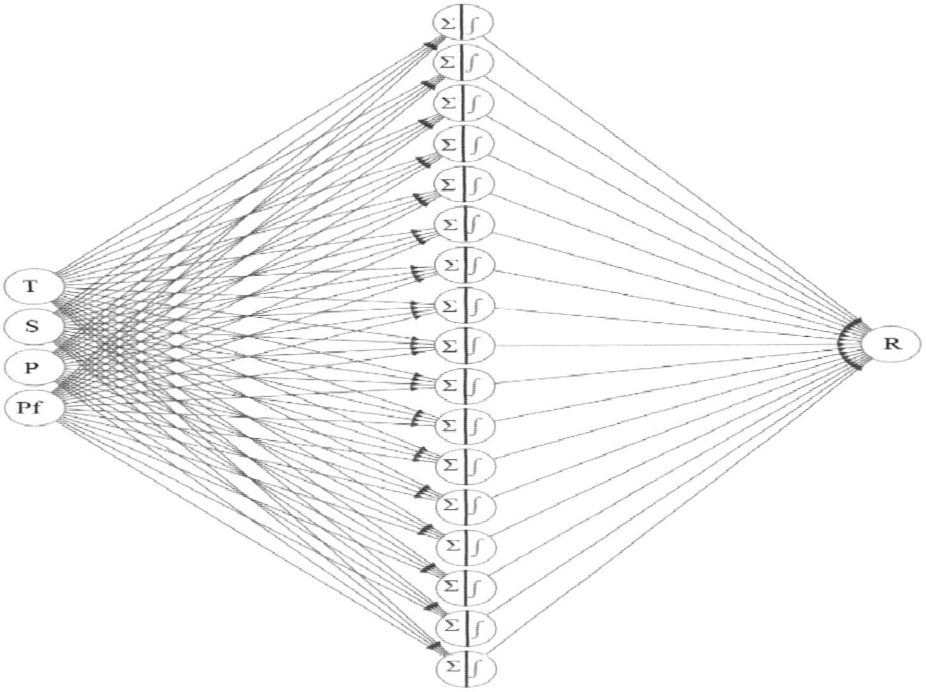

Figure 3. Multi-Layer Neural network architecture

The validation data set's regression coefficient (R^2), which was discovered to be 0.9693 and nearly one, shows a significant correlation between the experimental output and network predicted output. The result is given on Table 6, with the testing data.

Table 6. Experimental and ANN predicted results.

Run	T (°C)	S (cm/s)	P (bar)	Pf (bar)	Actual	Predicted	% Error
9	160	20	0.6	0.4	0.27	0.281	0.0407
17	170	20	0.2	0.3	0.19	0.171	0.0100
1	160	10	0.2	0.1	0.20	0.217	0.0850
25	180	20	0.4	0.1	0.15	0.157	0.0466
12	170	10	0.4	0.4	0.27	0.278	0.0296
10	170	10	0.4	0.4	0.28	0.278	0.0071
14	170	15	0.6	0.1	0.13	0.128	0.0154
2	160	10	0.2	0.1	0.21	0.217	0.0333

4. Conclusion

In this study the effects of plastic injection moulding process parameters on the formation of defect particularly overflow is studied. The study focused on the impacts of process parameters on overflow via comparison of ANN and ANOVA prediction and based on the findings; the following conclusion is observed.

- According to the ANOVA result, the selected factors temperature, speed, pressure, and packing force are vital process parameters with different levels of impacts on the formation of overflow or incomplete injection.
- ANOVA is better to assess the influence of each process on the formation of defects.
- The built MLP with 3 layers indicates better predicting capacity compared to ANOVA. The R^2 value of both ANOVA and ANN is 78.18 and 96.93% respectively.
- The ANN model has been demonstrated to be effective at predicting defect formation under process conditions.
- The drawback of ANN is that it needs more data to train the model compared to ANOVA prediction model.

Acknowledgement: This work is partially supported by INDMET project funded by NORHED II program (grant nr. 62862). The authors would like to acknowledge this support.

References

[1]. Attia U M, Marson S and Alcock J R 2009, Micro-injection moulding of polymer microfluidic devices, *Microfluidics and nanofluidics*. 7, 1-28.
[2]. Groover MP 2020, Fundamentals of modern manufacturing: materials, processes, and systems. *John Wiley & Sons*. **53**.
[3]. Fernandes C, Pontes A J, Viana JC and Gaspar-Cunha A 2018, Modeling and Optimization of the Injection-Molding Process: A Review. *Adv. Polym. Technol.* **37**(2), 429-449.
[4]. Khosravani M R, Nasiri S and Reinicke T 2022, Intelligent knowledge-based system to improve injection molding process, *J. Ind. Inf. Integr.* **25**, 100275.
[5]. Subramanyam B, Vineeta T V, Garre P, Bharadwaj V N and Shashank P S 2018, Comparative analysis of additive manufacturing over conventional manufacturing. In *IOP Conf. Ser. Mater. Sci. Eng.* **455**(1), 012102.
[6]. Ravikiran B, Pradhan D K, Jeet S, Bagal D K, Barua A and Nayak S 2022, Parametric optimization of plastic injection moulding for FMCG polymer moulding (PMMA) using hybrid Taguchi-WASPAS-Ant Lion optimization algorithm. *Mater. Today Proc.* **56**, 2411-2420.

[7]. Cooper T A 2013, Developments in plastic materials and recycling systems for packaging food, beverages, and other fast-moving consumer goods. *Trends in Packaging of Food, Beverages, and Other Fast-Moving Consumer Goods. (FMCG). Woodhead Publishing.* 58-107.

[8]. Maarif M R, Listyanda R F, Kang Y S and Syafrudin M 2022, Artificial Neural Network Training Using Structural Learning with Forgetting for Parameter Analysis of Injection Molding Quality Prediction. *Inf.* **13**(10), 488.

[9]. Lieberzeit P, Bekchanov D, and Mukhamediev M 2022, Polyvinyl chloride modifications, properties, and applications. *Polym. Adv. Technol.* **33**(6), 1809-1820.

[10]. Al-Oqla F M, Hayajneh M T and Fares O 2019, Investigating the mechanical thermal and polymer interfacial characteristics of Jordanian lignocellulosic fibers to demonstrate their capabilities for sustainable green materials. *Clean. Prod.* **241**, 118256.

[11]. Dejene N D and Gopal M 2021, The Hybrid Pareto Chart and FMEA methodology to Reduce Various Defects in Injection Molding Process. *Solid State Technol.* **64**(2), 3541-3555.

[12]. Khomenko M, Veligorskyi O, Chakirov R and Vagapov Y 2019, An ANN-based temperature controller for a plastic injection moulding system. *Electronics.* **8**(11), 1272.

[13]. Zhou H 2013, Computer modeling for injection molding. *Comput. Model. Inject. Molding Simulation, Optim. Contro.* doi: 10.1002/9781118444887.

[14]. Moayyedian M, Abhary K and Marian R 2016, The analysis of defects prediction in injection molding, *Int. J. Mech. Mechatronics Eng.* ***10***(12), 1883-1886.

[15]. Fei N C, Mehat N M and Kamaruddin S 2013, Practical applications of Taguchi method for optimization of processing parameters for plastic injection moulding: a retrospective review. *ISRN Ind. Eng.* **2013**, 1-11.

[16]. Zhao J, Mayes R H, Chen G E, Xie H and Chan P S 2003, Effects of process parameters on the micro molding process. *Polym. Eng. Sci.* **43**(9), 1542-1554.

[17]. Huang M C and Tai C C 2001, The effective factors in the warpage problem of an injection-molded part with a thin shell feature. *J. Mater. Process. Technol.* **110**(1), 1-9.

[18]. Darekar A M, Venkatesh T S, Patil B T and Yazad N 2015, Review of Optimization Aspects for Plastic Injection Molding Process. *An Int. J.: ESTIJ.* **5**, 2250–3498.

[19]. Sunadi S, Purba H H and Paulina E 2021, Overall Equipment Effectiveness to Increase Productivity of Injection Molding Machine: A Case Study in Plastic Manufacturing Industry. *ComTech Comput. Math. Eng. Appl.* **12**(1), 53-64.

[20]. Farahani S, Khade V, Basu S and Pilla, S 2022, A data-driven predictive maintenance framework for injection molding process. *J. Manuf. Process.* **80**, 887-897.

[21]. Tharazi I, Sulong A B, Muhamad N, Haron C H C, Tholibon, D, Ismail N F and Razak Z 2017, Optimization of hot press parameters on tensile strength for unidirectional long kenaf fiber reinforced polylactic-acid composite. *Procedia Eng.* **184**, 478-485.

[22]. Kumar B S and Balachandar S 2014, A study on the influence of hot press forming process parameters on Flexural Property of Glass/PP Based Thermoplastic Composites Using Box-Behnken Experimental Design. *Int. sch. res. notices.* **2014**.

[23]. Rathore A and Ujjainwala F 2017, Defects analysis and optimization of process parameters using Taguchi DoE technique for sand casting. Int. Res. J. Eng. Technol. **4**, 432-437.

[24]. Shahbudin S N A, Othman M H, Amin S Y M and Ibrahim M H I 2017, A review of metal injection molding-process, optimization, defects, and microwave sintering on WC-Co cemented carbide. *IOP Conf. Ser. Mater. Sci. Eng.* **226**(1), 012162.

[25]. Ahmed T, Sharma P, Karmaker C L and Nasir S 2022, Warpage prediction of Injection-molded PVC part using ensemble machine learning algorithm. *Mater. Today Proc.* **50**, 565-569.

[26]. Kowalczyk M 2014, Application of Taguchi, and ANOVA methods in selection of process parameters for surface roughness in precision turning of titanium. *Adv. Manuf. Sci.* **38**(2), 22–35, 2014.

[27]. Chen J, Cui Y, Liu Y and Cui J 2023, Design and parametric optimization of the injection molding process using statistical analysis and numerical simulation. *Processes*, **11**(2), 414.

[28]. Serin G, Sener B, Gudelek M U, Ozbayoglu A M and Unver H O 2020, Deep multi-layer perceptron-based prediction of energy efficiency and surface quality for milling in the era of sustainability and big data. *Procedia Manuf.* **51**, 1166-1177.

[29]. Wardah W, Khan M G, Sharma A and Rashid M A 2019, Protein secondary structure prediction using neural networks and deep learning: A review. *Comput. Biol. Chem.* **81**, 1-8.

[30]. Ahmed A N, Othman F B, Afan H A, Ibrahim R K, Fai C M, Hossain M S and Elshafie A 2019, Machine learning methods for better water quality prediction. *Introd. to Eng.* **578**, 124084.

[31]. Lillian S K, Charles E G 2008, Analysis of Variance: Is there a Difference in Means and What Does It Mean?," *J. Surg. Res.* **144**(1), 158–170.

[32]. Eren B, Yaqub M and Eyupoglu V 2019, A comparative study of artificial neural network models for the prediction of Cd removal efficiency of polymer inclusion membranes. *Desalin. Water Treat.* **143**, 48-58.

[33]. Choraś, M and Pawlicki M 2021, Intrusion detection approach based on optimized artificial neural network. *Neurocomputing.* **452**, 705-715.

[34]. LeNail, A 2019, NN-SVG: Publication-Ready Neural Network Architecture Schematics. J. Open Source Softw. **4**(33), 747.

Triply periodic minimal surfaces based functionally graded biomimetic scaffold fabrication via stereolithography

Y W Adugna[1, *], H G Lemu[1] and H R Hagland[2]

[1] Department of Mechanical and Structural Engineering and Materials Science, University of Stavanger, Stavanger, Norway
[2] Department of Chemistry, Biosciences and Environmental Engineering, University of Stavanger, Stavanger, Norway

* Correspondence: yosef.w.adugna@uis.no

Abstract. Triply Periodic Minimal Surfaces (TPMS), a class of intricate mathematical surfaces, have emerged as a promising framework for scaffold design due to their ability to replicate the complex geometries found in biological structures. Four TPMS structures, the Schwarz Diamond (D), Schwarz Primitive (P), Gyroid, and IWP (I-wrapped package) were designed for both uniform and graded density and additively manufactured through Stereolithography based additive manufacturing (AM) techniques using biomedical graded material. Two different mechanical tests, tensile and compression tests were examined on the TPMS structure to study their mechanical properties. The results showed that Schwarz D and IWP TPMS show greater tensile strength for both uniform and graded structures with 18.22 MPa and 14.41 MPa in uniform structures and 9.89 MPa and 9.23 MPa in graded structures of Schwarz D and IWP respectively. Uniform TPMSs show overall tensile strength over the graded TPMS. Compressive properties also show that Schwarz D and IWP TPMS have greater compressive strength in both uniform and graded TPMS, where overall graded structures show better strength over the uniform. Graded Schwarz D observed to have 100.68 MPa, and IWP TPMS has 99.57 MPa, and uniform Schwarz D has 33.94 MPa, whereas IWP TPMS shows 31.82 MPa compressive strength. Results reinforce the structure's suitability for scaffold applications, particularly in contexts demanding robust mechanical integrity. The application of SLA AM with biomedical-graded material strengthens the viability in areas like tissue engineering and regenerative medicine.

Keywords: Triple periodic minimal surface (TPMS), Tissue engineering, Biomimetic, Stress-strain curve

1. Introduction

Bone tissue engineering (BTE) is in great demand for the repair of defective bones due to an aging population and the increasing use of high-speed transportation, both of which are associated with an increase in bone diseases [1]. Diseased or injured bone tissue presents a significant requirement for bone replacements used in transplantation, making it the second most often transplanted tissue on an annual basis [2]. Access to bone replacements is a challenge and has been an area under intense research to find a solution to solve this issue, where Langer and Vacanti's 1993 Science article describing the properties and applications of biodegradable 3D scaffolds was pivotal in the development of the new field - tissue engineering [3].

The primary objective of tissue engineering is to imitate the inherent characteristics of natural tissue by facilitating tissue regeneration through the integration of a scaffold, cells, and bioactive molecules. One prominent benefit of bone tissue is its inherent capacity for self-repair, remodelling, and regeneration. Consequently, bone tissue engineering holds promise as a potential approach for addressing bone defects resulting from diverse causes [4].

The internal biological structure of bone tissue is characterized by a complex, porous, and curved composition, featuring irregular pore structure and uneven pore size and distribution. Additionally, the presence of internal communication pores facilitates the transportation of nutrients from the external environment to the interior of the system [5]. Presently, orthopedic implants like bone nails, bone plates, and intramedullary nails are not customized to match the specific bone characteristics of individual patients. Consequently, this lack of tailoring contributes to an elevated occurrence of complications during surgical procedures [6]. Furthermore, the presence of a significant mismatch in modulus between the bone and the implant results in stress shielding, which occurs because of the non-uniform distribution of stress at the interface between the bone and the implant. Thus, cellular materials are used to customize the characteristics of the implant and mitigate bone resorption around the implant caused by the stress shielding phenomenon [7].

Lattices represent a compelling subset of cellular materials and find extensive use in the design of body implants and scaffolds [8]. Triply periodic minimum surfaces (TPMS) are minimally shell-based sized surfaces with closed curves as boundaries, defined mathematically [9]. These surfaces serve as the foundation for TPMS scaffolds. Each unit cell is created by adding thickness to the surface, and the finished scaffold is cubically symmetric and has periodic repetitions of connected pores [10]. As an essential type of porous structure, functionally graded structures are an ideal option for producing lightweight, stress-resistant, and high-energy-absorbing components, which is also one of the reasons TPMS structures have attracted attention [11].

The design and pore geometry of scaffolds plays a significant role in the organization of cells and tissues in tissue-engineered constructs. The high biological and functional complexity of human tissues with specific micro-architecture and vascular networks require the use of more specific scaffold geometries made from multifunctional materials and structures [12]. Ideal 3D scaffolds should be highly porous, have interconnected pore networks, and have pores of a consistent and sufficient size for cell infiltration and migration [13]. As such, controlling porosity, pore size, surface area, and permeability, as well as manipulating strength and stiffness, is essential [14]. In general, a scaffold should provide a microenvironment that stimulates cell growth through cell-scaffold and cell-cell interactions, such as adhesion, differentiation, proliferation, migration, and maintenance [15, 16].

The formation of a porous structure inside the scaffold is one of the important elements in the manufacture of tissue engineering scaffolds both for uniform and graded structures [17]. Various conventional manufacturing techniques have been utilized to fabricate porous 3D scaffolds, such as solvent casting, fiber bonding, particulate leaching, phase separation membrane lamination, foaming, and moulding [18]. However, a limitation of all these approaches is the insufficient control over scaffold design, pore network, and pore size, resulting in the production of inconsistent and inefficient three-dimensional scaffolds [19]. Moreover, most of the traditional implants are solid structures, and their mechanical properties are unsatisfactory for the adjacent bone tissues, resulting in the occurrence of stress shielding between them [6]. On the other hand, the ongoing development of Additive manufacturing (AM) technology provides a promising method for fabricating structures with complex external geometries and well-defined internal connections i.e., in the creation of customized porous scaffolds [20].

Regardless of the material used, different AM technologies are employed to manufacture porous structures including selective laser melting (SLM), selective laser sintering (SLS), fused deposition modelling (FDM), digital light processing (DLP), stereo lithography apparatus (SLA) and others [21-25]. Most of the previous research preferred SLS and SLM. SLA AM technology is known for its efficiency, high dimensional accuracy, and speed. It offers unique ways to accurately control substrate architecture [26, 27]. Some of the studies on TPMS scaffolds manufactured through SLA, including Yu.

et. al., investigated the mechanical properties and energy absorption abilities of Schwarz P and Gyroid TPMS structures manufactured through SLA [28]. Chouhan *et. al* studied surface defects of Gyroid TPMS through varied thicknesses produced with SLA [29].

Previous investigations on TPMS have mainly focused on metallic additive fabrication, and the number of studies on SLA-manufactured TPMS structures is limited. In this work, four (4) different uniform and functionally graded TPMS structures (Schwarz P, Schwarz D, Gyroid, and IWP) were designed and fabricated by using the SLA AM technique, and their mechanical property is examined. A smooth variation in density was achieved in the graded TPMS structures by distributing the wall thickness linearly and continuously. The accuracy and manufacturability of SLA-fabricated TPMS specimens are investigated. Mechanical properties of both uniform and graded structures were examined through tensile and compression tests.

2. Materials and methods

2.1 Material

In this work, a medical-grade photocurable vat photopolymerization resin from Formlabs (Formlabs Inc., Somerville, MA, USA - Biomed Clear Resin (Product Code: FLBMCL01)) is used. BioMed clear resin is a photopolymer resin made of a mixture of methacrylic esters and photoinitiators. It is a light-curable polymer-based material designed for AM of medical grade, biocompatible, clear, and rigid parts for long-term surface contact. Table 1 and Table 2 provide Biomed clear material composition and properties according to the manufacturer's data [30].

Table 1. Biomed clear material composition

Material	Concentration, (wt.%)
Bisphenol A dimethacrylate	~50–70
2-hydroxyethyl methacrylate monomer(s)	7–10
Urethane dimethacrylate	25–45
Photoinitiator(s)	< 2

Table 2. Biomed clear properties

Material	Density, (g/cm^3)	Ultimate tensile strength (MPa)	Young's Modulus, (MPa)	Elongation (%)	Hardness, (Shore - D)
Biomed clear	1.09	52	2080	12	78

2.2. Structure design of TPMS

The Computer-Aided Design software Creo® Parametric (TM)™, based in Boston, MA, USA, was utilized to generate macro-lattice designs. The lattice tool of Creo® Parametric (TM) was utilized to generate various unit cells within part files. Subsequently, cell arrangements were formed from these part files and merged into part files to construct the overall intended structures. Mathematical representations of the surfaces are as follows:

$$\text{(i) Diamond: } \cos(\omega x)\cos(\omega y)\cos(\omega z) - \sin(\omega x)\sin(\omega y)\sin(\omega z) = c. \quad (1)$$

$$\text{(ii) Primitive: } \cos(\omega x) + \cos(\omega y) + \cos(\omega z) = c. \quad (2)$$

$$\text{(iii) Gyroid: } \cos(\omega x)\sin(\omega y) + \cos(\omega y)\sin(\omega z) + \cos(\omega z)\sin(\omega x) = c. \quad (3)$$

$$\text{(iv) IWP: } 2[\cos(\omega x)\cos(\omega y) + \cos(\omega y)\cos(\omega z) + \cos(\omega z)\cos(\omega x)] - [\cos(2\omega x) + \cos(2\omega y) + \cos(2\omega z)] = c. \quad (4)$$

Where, ω is used to define the unit cell length, and c controls the expansion of the surface in three dimensions. The unit cell design of the generated TPMS structure, constructed 3×3×3, is seen in Figure 1.

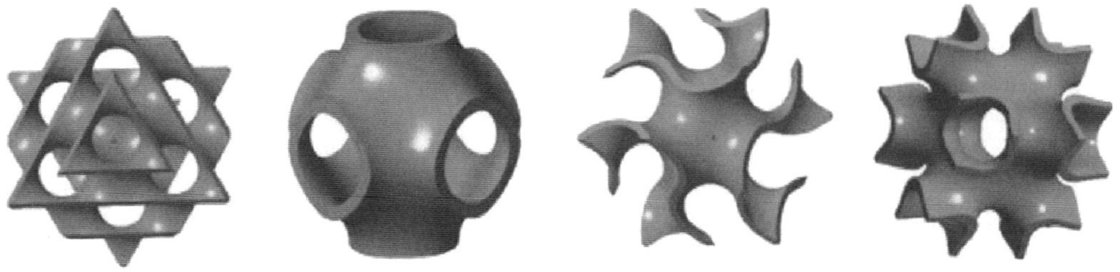

Figure 1. Unit cell design of TPMS structures (a) Schwarz D, (b) Schwarz P, (c) Gyroid and (d) IWP

For the tensile test specimen, the gauge length area of the specimen is considered for the generation of the TPMS structure. Whereas for the compression test, the whole geometry is considered. The lattices were constructed through repeating the unit cells through the part geometry. The uniform structure was designed to have a wall thickness of 0.3. The graded structure was built with a linear and continuous variation in wall thickness, ranging from 0.2 to 0.5 mm in the building's direction during the SLA printing process. Designed CAD models of the tensile and compression specimens are seen in Figure 2 and Figure 3, respectively.

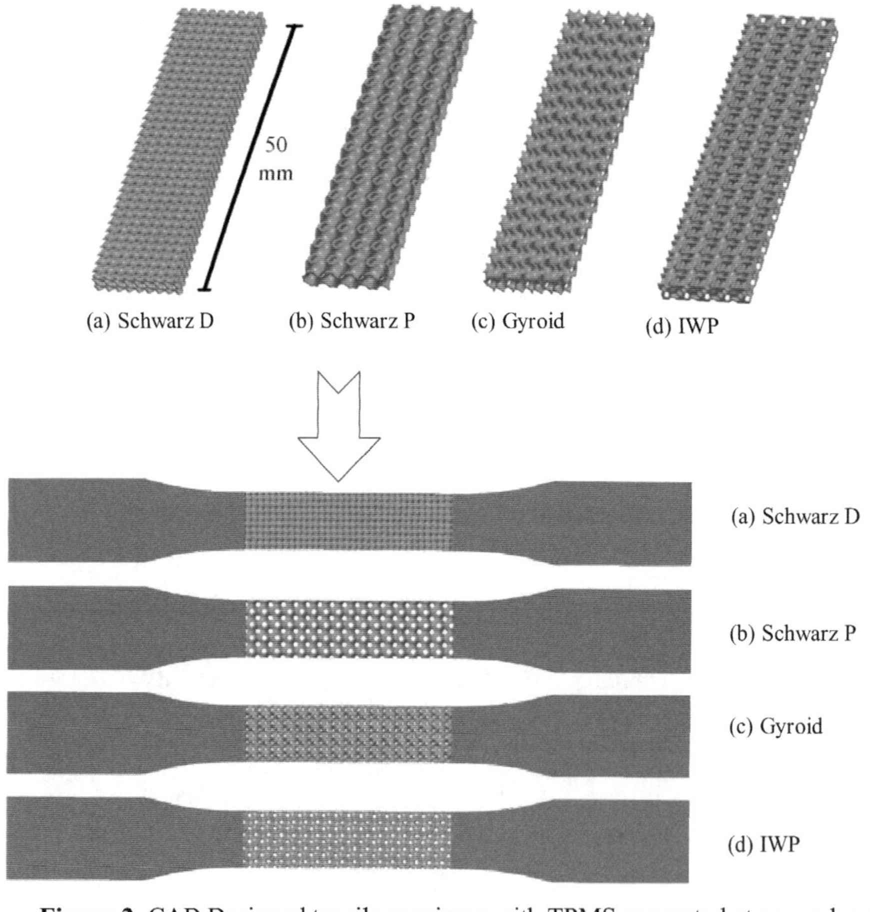

Figure 2. CAD Designed tensile specimen with TPMS generated at gauge length

IOP Conf. Series: Materials Science and Engineering 1294 (2023) 012051 doi:10.1088/1757-899X/1294/1/012051

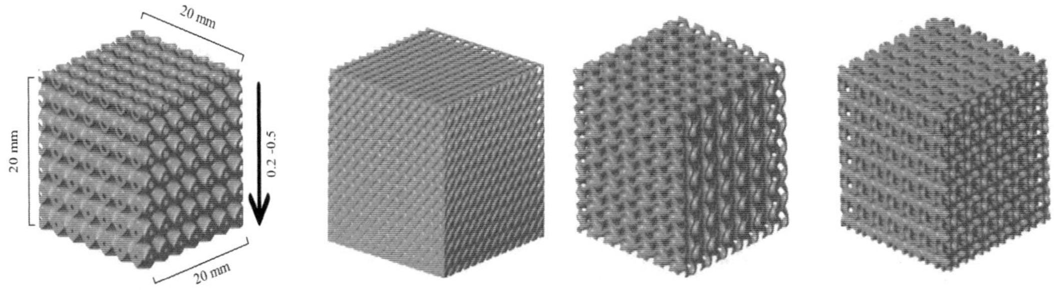

Figure 3. CAD-designed models of compression test specimens.

2.3. SLA 3D Printing

The 3D printer used was a Formlabs Form 3B+ (Formlabs Ohio Inc., Millbury, OH, USA) vat photopolymerization 3D printer with Formlabs Tank LT. The preparation of the necessary GCode and the slicing process was implemented on the Preform 3.16 software tool. Biomed Clear resin was 3D printed with 100 μm layer height for every case tested. The specimens were additively manufactured with a horizontal orientation on the build plate. After the 3D-printing process was completed, specimens were washed in a 99% pure isopropyl alcohol (IPA) solution for 15 min and dried for 5 min at room temperature and ambient conditions (this method was repeated twice), according to Formlabs specifications and guidelines. When the IPA cleaning process of the specimens was completed, a post-curing period of 60 min at 60 °C took place. Formlabs Form Wash and Form Cure (Formlabs Ohio Inc., Millbury, OH, USA) equipment have been used for the washing and the post-curing processes, respectively. Form 3B+ is equipped with a 405 nm UV laser. Three duplicates of each specimen type are manufactured for testing.

Table 3. SLA 3D Printing parameter

Parameter	Value
Slicing software	Preform v3.29
Layer height	100 μm
Build orientation	45^0
Wash time	15 min (99% IPA)
Dry time	5 min (99% IPA)

2.4. Mechanical testing

The tensile test specimen is prepared according to the ASTM D638 type IV standard. ASTM D638 type IV has a 50 mm gauge length. Thus, the total length of the gauge length (50 mm) is used for the generation of different TPMS structures. The tensile properties of additive manufactured tensile test specimens with TPMS structure were tested using Instron tensile testing machine that has a load cell capacity of 10 kN, at a speed of 2 mm/min, and a sampling frequency of 50 hz.

Compression test specimen is prepared according to the standard ASTM D-695. The initial design of the specimen has a cubic geometry with the dimensions of 20 x 20 x 20 mm^3. The Instron tensile testing machine with a load cell of 250 kN is used for a compression test. The test was set to terminate when the nominal strain reached 70% of the specimen geometry (14mm) and specimen fracture, respectively.

IOP Conf. Series: Materials Science and Engineering 1294 (2023) 012051 doi:10.1088/1757-899X/1294/1/012051

3. Results and discussion

3.1. Tensile test result

Before the production of specimens, it is crucial to consider a significant aspect in the context of the tensile test specimen, namely the occurrence of stress concentration at the solid-lattice transition, and failure near the grip section due to non-uniform stress at the interface. The presence of stress concentration resulting from the abrupt cross-sectional variation in the solid-lattice transition zone, which is deemed insignificant in the context of the compression test, can give rise to premature fracture of the components subjected to tensile loading, hence compromising the reliability of the test results. In the first design, the TPMS structure has a sudden change of cross section from porous to solid, but this clearly shows where the failure might happen. In the second design, grading of the transition between TPMS and the solid interface is introduced which significantly prevents the stress concentration, it was hypothesized that the lattices may be gradually graded as they make the transition into the solid zone, resulting in an augmented cross-sectional area of the material inside the region. The first design failed, and the specimens were manufactured using the second option. The designed specimens were successfully manufactured using the SLA AM technology. Figure 4 displays the as-manufactured tensile test specimen and the after-test broken specimen.

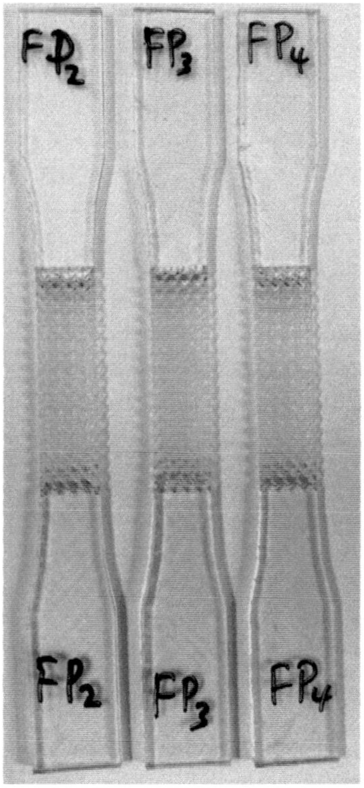

(a)

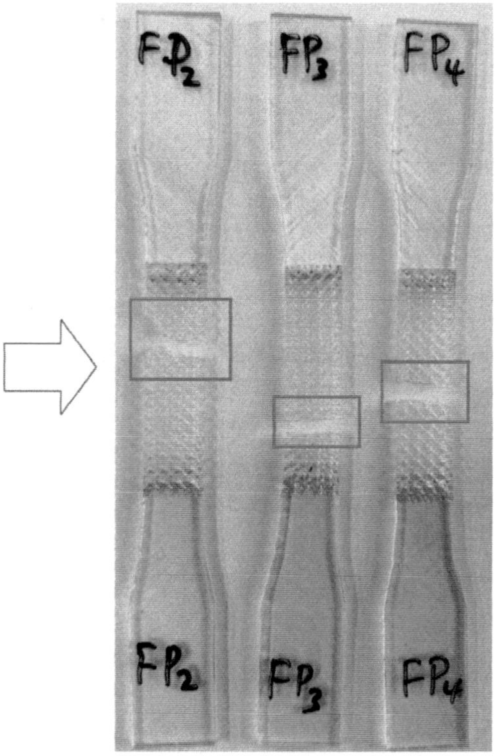

(b)

Figure 4. Specimen label (a) and After test specimen (b)

The tensile stress-strain curve for the AM'ed Uniform TPMSs is presented in Figure 5. The tensile response of some of the uniform TPMS constructions consists of a linear elastic zone, which is then followed by strain hardening in the same manner as the solid specimens i.e., Schwarz P and IWP TPMS. Whereas Schwarz D and Gyroid show fracture before reaching the strain hardening zone. An observable phenomenon characterized by a step-like pattern in stress decrease is evident before the occurrence of ultimate fracture, suggesting the occurrence of failure at the level of cells inside the TPMS structures. This can be seen clearly on the Uniform Schwarz P stress-strain curve (Figure 5b).

Schwarz D TPMS shows the highest mean tensile strength of 18.22 MPa, suggesting higher structural integrity and superior resistance to deformation under the applied tensile forces. IWP TPMS follows with a mean tensile strength of 14.41 MPa, showing substantial capacity to withstand loads. The intermediate level of strength could be attributed to the inherent geometric features and interconnectivity within the IWP structure. The Schwarz P shows the lower mean tensile strength, 6.67 MPa among all uniform TPMS structures.

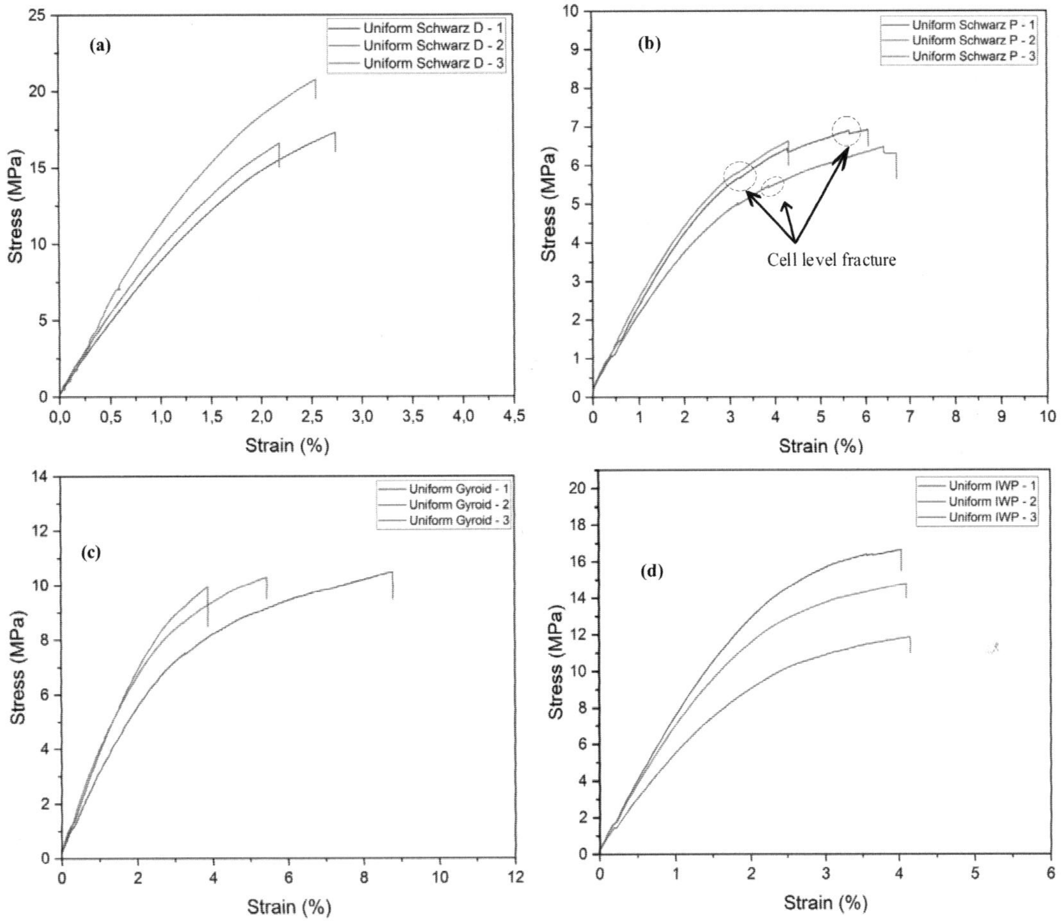

Figure 5. Stress-strain curve of Uniform TPMS Tensile test (a) Schwarz D (b) Schwarz P (c) Gyroid (d) IWP

All Uniform TPMS structures showed ductile behaviour during testing. Gyroid TPMS exhibited the highest mean elongation, reaching 3.7%. Next to Gyroid TPMS is Schwarz P. TPMS failed at 2.53% of mean elongation. Schwarz D TPMS is the lowest of all uniform TPMS with 0.89% mean elongation. It is noteworthy to mention that the observed reduction in elongation might be ascribed to the curing process that the material undergoes after production. In addition to the materials removed to generate the TPMS structure, the observed decrease in elongation can be attributed to the structural modifications that take place during the curing process.

The individual TPMS tensile stress-strain curve of the four Graded TPMS structures is given in Figure 6. Comparing the tensile strength of the graded TPMS structure, most of them show similarity except Gyroid. Schwarz D and IWP show closer highest tensile strength at 9.9 MPa and 9.22 MPa respectively whereas Schwarz P follows with 7.38 MPa of mean tensile strength. Gyroid demonstrates a comparatively lower mean tensile strength with 4.05 MPa compared to other graded TPMS structures.

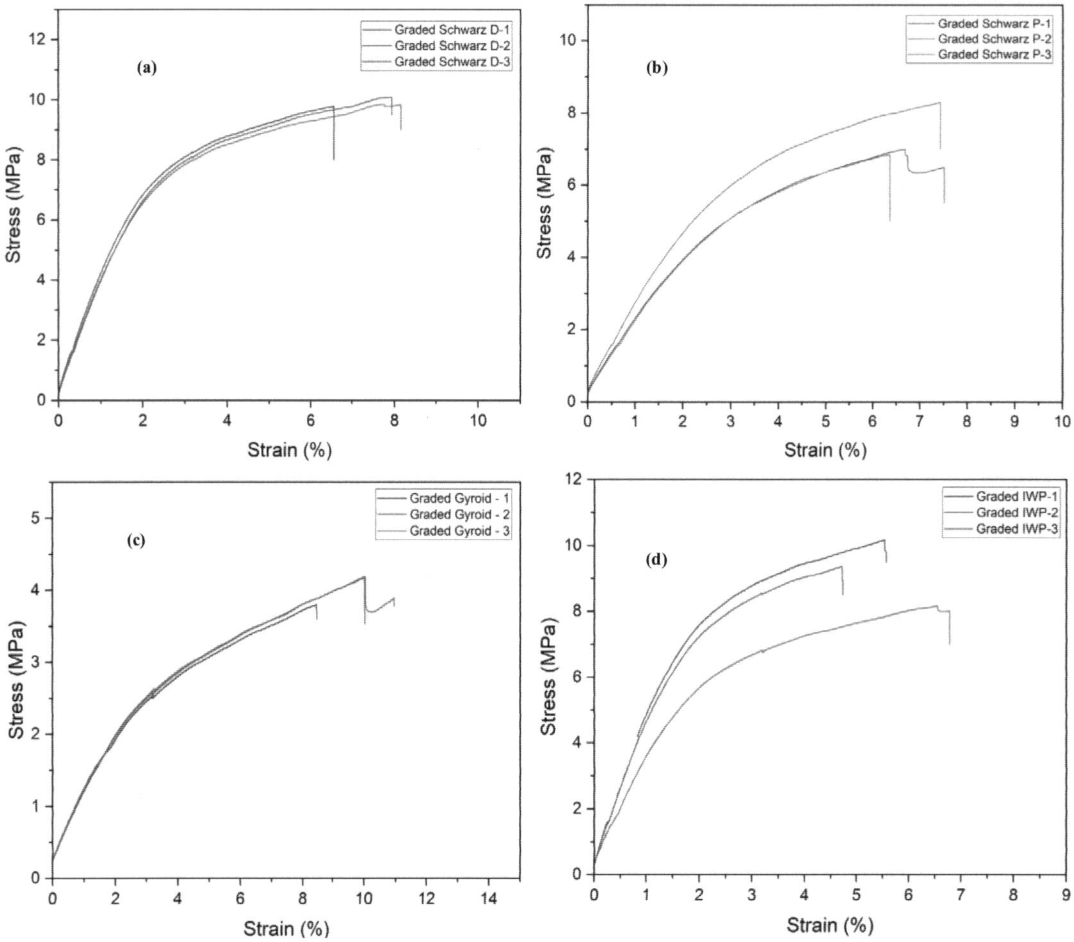

Figure 6. Stress-strain curve of Graded TPMS Tensile test (a) Schwarz D (b) Schwarz P (c) Gyroid (d) IWP

The graded TPMS constructions, much like the uniform structures, demonstrated ductile behaviour. Among all the Gyroid TPMS had the highest level of resilience against deformation, and greatest elongation, reaching 5.6%. The Schwarz D TPMS structure exhibited a significant elongation of 4.9%, in close succession. This observation implies a notable capacity for undergoing deformation while concurrently preserving structural integrity. Schwarz P and IWP TPMS structures had comparatively lower elongations, measured at 3.5% and 3.4% respectively. Although the ductility of these values is significantly lower in comparison to Gyroid and Schwarz D, they nonetheless fall within the acceptable range of mechanical performance for several applications. In contrast, the structures of Schwarz P and IWP TPMS exhibited comparatively lower mean elongations, measured at 3.5% and 3.4% respectively. The comparative results from the tensile properties of the uniform and graded TPMSs are summarized in Table 4 and Figure 7.

Table 4. Summarized tensile property of uniform and graded TPMS

TPMS geometry		Wall thickness (mm)	Max. Force (kN)	Tensile strength (MPa)	Elongation (%)	Elastic modulus (MPa)
Schwarz D	Uniform	0.3	758	18.22	0.89	1.1
	Graded	0.2 - 0.5	412	9.89	4.8	0.4
Schwarz P	Uniform	0.3	278	6.68	2.5	0.2
	Graded	0.2-0.5	307	7.38	3.5	0.2
Gyroid	Uniform	0.3	426.3	10.25	3.74	0.367
	Graded	0.2-0.5	161	4.05	5.6	0.1
IWP	Uniform	0.3	597	14.41	1.9	0.633
	Graded	0.2-0.5	383.6	9.23	3.4	0.433

The comparative tensile stress-strain curve of all structures, both uniform and graded TPMS considered for this study is seen in Figure 7. In both cases, Schwarz D TPMS shows greater mean tensile strength. The mean tensile strength of graded Schwarz D TPMS (18.22 MPa) is close to double for uniform Schwarz D (9.9 MPa). Schwarz P exhibits relevant mean tensile strength in both cases, where the graded tensile strength (7.39 MPa) is higher than the uniform (6.67 MPa). This implies that the graded structure enhances the tensile strength of Schwarz P TPMS. The mean tensile strength of uniform Gyroid TPMS is more than double (10.25 MPa) to the mean tensile strength of graded Gyroid TPMS (4.05 MPa). A similar trend is observed in IWP, the uniform shows the highest mean tensile strength (14.41 MPa) than graded TPMS (9.23 MPa). This further supports the trend observed across the other TPMS structures, where uniform structures tend to yield greater mean tensile strength.

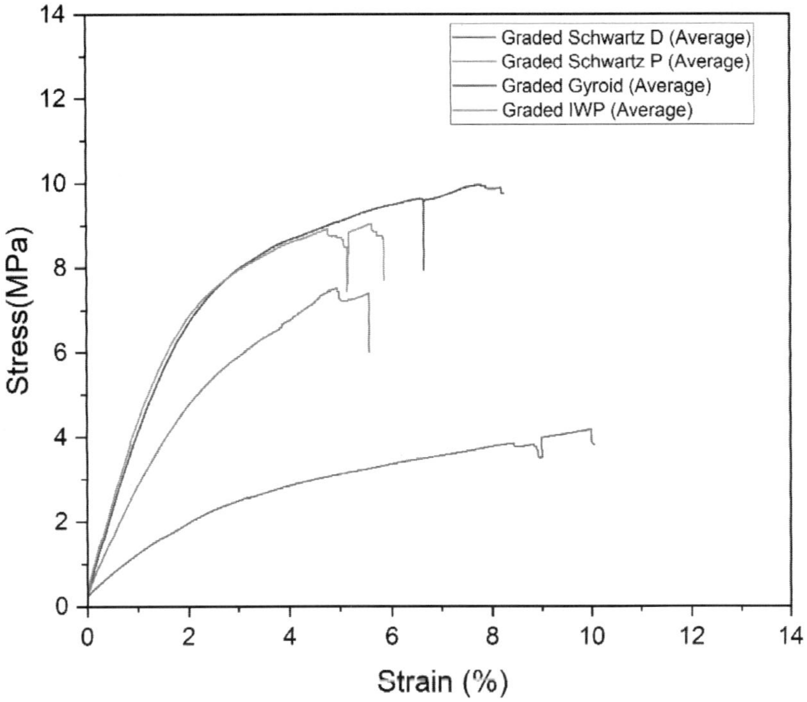

Figure 7. Stress-strain curve of TPMS structures under tension

In all of them, uniform TPMS structures show greater mean tensile strength except Schwarz P where the graded TPMS is higher. The observed variations in mean tensile strength among different TPMS structures can be attributed to their unique geometric properties and compositions. Uniform TPMS structures tend to exhibit higher tensile strength in most cases, as they maintain a consistent material distribution and are less prone to stress concentration. The variation in thickness results in variation of stress and material distribution in a graded TPMS can be considered as the major factor for comparatively less strength.

The analysis of ductility among the structures reveals compelling insights into their mechanical behaviour highlighting graded TPMS's superior elongation properties compared to their uniform counterparts. The graded Schwarz D TPMS displays a remarkable mean elongation of 4.87%, indicating its substantial capacity to undergo plastic deformation before failure. In contrast, the uniform Schwarz D TPMS exhibits a significantly lower mean elongation of 0.89%. This difference can be attributed to the change in the cross-section area of the graded TPMS which allows more elongation under stress. In Schwarz P TPMS a similar trend is observed. The graded Schwarz P TPMS shows a notable mean elongation of 3.5%, indicating its ability to undergo plastic deformation before failing. On the other hand, the uniform Schwarz P TPMS exhibits a slightly lower mean elongation of 2.53%. The graded Gyroid TPMS exhibits a substantial mean elongation of 5.6%. In contrast, the uniform Gyroid TPMS achieves a mean elongation of 3.75%, which, while still notable, falls short of the uniform structure's ductility. In the case of IWP TPMS, the graded structure displays a mean elongation of 3.4%, highlighting its capacity for plastic deformation. On the other hand, the uniform IWP TPMS exhibits a lower mean elongation of 1.9%. The observed increase in ductility in graded TPMS structures compared to uniform ones can be attributed to the graded structure's capacity to distribute stresses and strains more effectively. This leads to a higher degree of deformation before structural failure.

3.2. Compression test result

Compression results of tested TPMS structures are provided in Figure 8, Figure 9, and Table 5. Stress-stain curves for a typical lattice structure display three different regions: (1) an elastic compression region, where the stress rises linearly with strain during the elastic stage, illustrating a linear stress-strain relationship; (2) a plateau region following the elastic region, which is marked by a reduction in stress and load-carrying capacity because of wall yield, plastic deformation, buckling, and fracture of the lattice walls; and (3) a densification region, where collapsed walls and cells come into contact and cause a significant increase in stress [31].

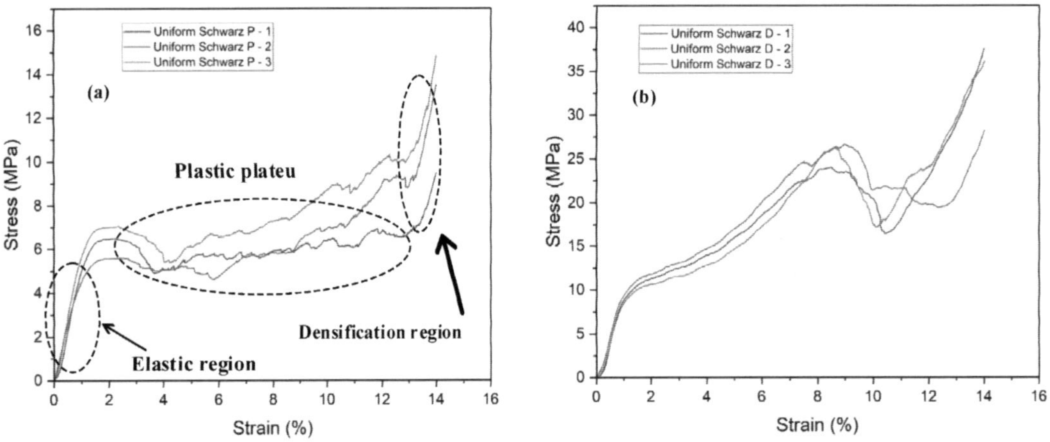

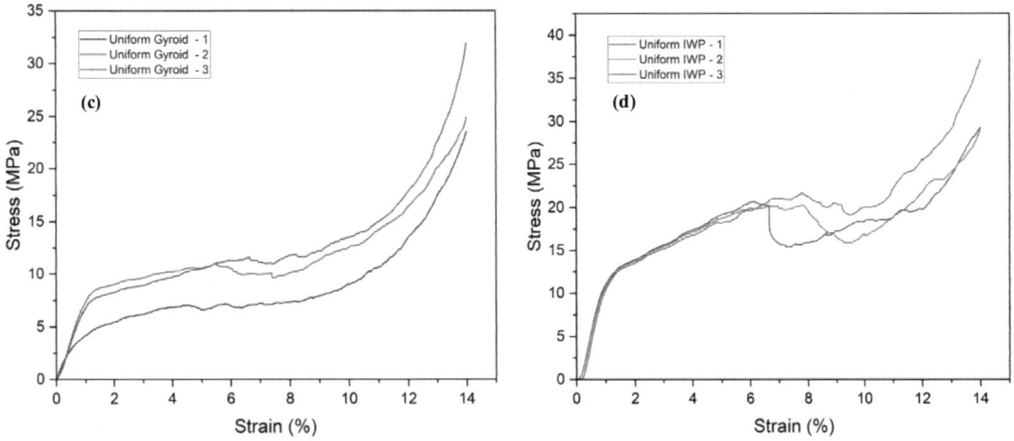

Figure 8. Stress-strain curve of uniform TPMS under compression (a) Schwarz D (b) Schwarz P (c) Gyroid (d) IWP

Figure 8 above displays the stress-strain property of uniform TPMS of Schwarz D, Schwarz P, Gyroid, and IWP where the elastic, plastic deformation and densification regions can be seen clearly. All four uniform structures exhibit elastic properties at lower strains. This is a characteristic common to many materials, wherein at low levels of stress, the material can deform elastically and return to its original shape once the stress is removed. This behavior is typically governed by the material's elastic modulus, which reflects its stiffness. As shown in the uniform Schwarz P stress-strain curve, the plastic collapse happened when the structure reached the first peak of the stress. Schwarz D demonstrates the highest mean compressive strength with 33.94 MPa, and IWP TPMS also closer with 31.82 MPa. This suggests that IWP TPMS is similarly adept at withstanding compressive loads. Gyroid has 26.78 MPa and Schwarz P with the lowest compressive strength 12.6 MPa. Schwarz P exhibits repeating ups and downs in the stress-strain curve, an indication of a layer-by-layer deformation pattern. It can be considered as every up and down attribute to each layer cell row deformation. Schwarz D, Gyroid, and IWP TPMS's stress-strain curve shows the structures deformed uniformly, an indication of consistent load bearing. The deformation behavior of the structures in the densification region is much like those of the solid parts and exhibits fast stress increment.

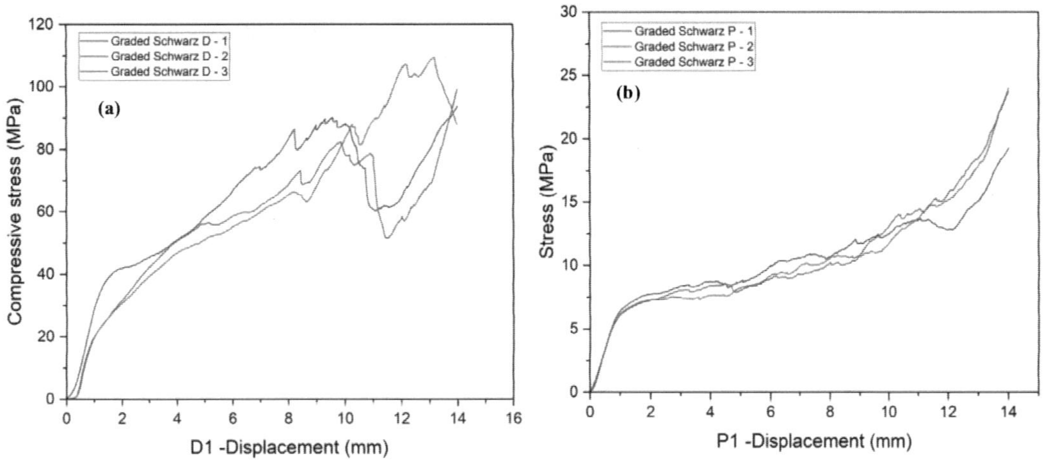

Figure 9. Stress-strain curve of Graded TPMS under compression (a) Schwarz D (b) Schwarz P (c) Gyroid (d) IWP

IOP Conf. Series: Materials Science and Engineering 1294 (2023) 012051 doi:10.1088/1757-899X/1294/1/012051

The graded TPMS structures stress-strain curve under compression is seen in Figure 9. Graded structures show overall greater compressive strength compared to the uniform TPMS. Schwarz D exhibits 100.68 MPa. The geometric arrangement and composition of Schwarz D in its graded form play a crucial role in achieving this high compressive strength. Like the uniform pattern, graded IWP TPMS also exhibits close mean compressive strength to the Schwarz D with 99.57 MPa. The fact that it approaches the performance of Schwarz D suggests that IWP when graded, can be a viable alternative for applications requiring high compressive strength. The Gyroid shows a mean compressive strength of 46.39 MPa. Whereas the Schwarz P has a comparatively lower mean strength of 22.3 MPa under compression. The summary of the different mechanical properties of both uniform and graded TPMS structures can be found in Table 5.

Table 5. Mechanical properties of Schwartz D, Schwarz P, Gyroid, and IWP TPMS Structures under compression

TPMS		Wall thickness (mm)	Max. Force (kN)	Compressive strength (MPa)	Compressive modulus (MPa)
Schwarz D	Uniform	0.3	13.58	33.94	233.98
	Graded	0.2 - 0.5	40.27	100.68	738.28
Schwarz P	Uniform	0.3	5.04	12.6	128
	Graded	0.2-0.5	8.92	22.3	171.7
Gyroid	Uniform	0.3	10.71	26.78	185.56
	Graded	0.2-0.5	18.56	46.39	194.08
IWP	Uniform	0.3	12.73	31.82	313.72
	Graded	0.2-0.5	39.83	99.57	684.97

Figure 10 shows the overall stress-strain curve under compression test for both uniform and graded TPMS structures. The compressive strength shows great variability between the two cases. The trend in both cases follows the same pattern in terms of highest to lowest mean strength. Schwarz D TPMS, whether graded or uniform, exhibits the highest mean compressive strength. Schwarz D's graded mean compressive strength (100 MPa) and IWP TPMS's graded mean compressive (99.47) are 3 times the strength in uniform of their respective (33.94 MPa for Schwarz D and 31.82 MPa for IWP). Schwarz P's graded mean compressive (22.3 MPa) is close to twice the strength exhibited in uniform (12.6 MPa). The Graded Gyroid (46.39) also displays greater mean strength to the uniform (26.78 MPa)

Overall grading of the TPMS structure according to their function enhances their compressive strength. These highlight the potential of graded TPMS materials for applications where high compressive strength is a critical requirement, such as in structural components subjected to heavy loads or in orthopaedic implants.

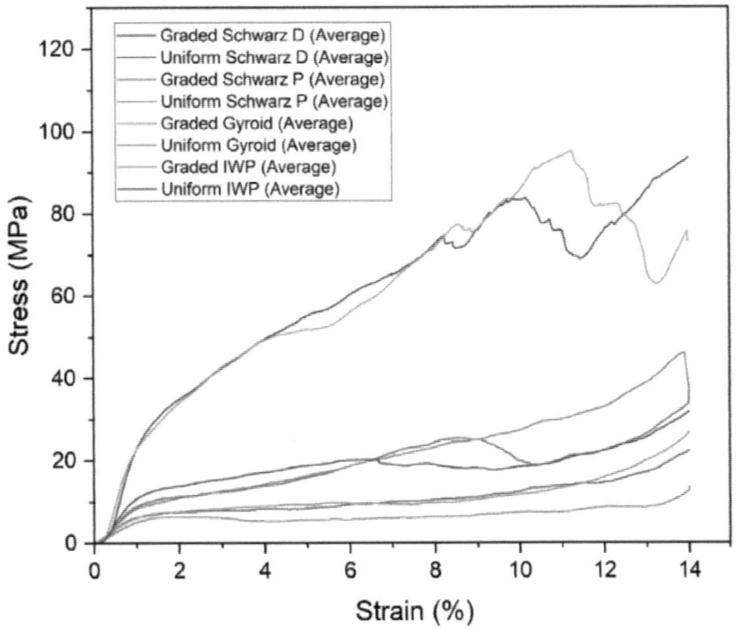

Figure 10. Stress-strain curve of TPMS structures under compression.

4. Conclusion

In this study, uniform and functionally graded four TPMS structures namely Schwarz Diamond (D), Schwarz Primitive (P), Gyroid and IWP were designed and successfully manufactured through SLA AM technology. The SLA AM technology proved to be highly effective in manufacturing complex structures with a good surface finish and accuracy. The mechanical properties of the structures were analyzed through tensile and compressive tests on the manufactured specimens. It is notable that both uniform and graded Schwarz D TPMS structures exhibited greater mean tensile strength compared to the other structures. Both uniform and graded Schwarz D TPMS show greater mean tensile strength, also greater compressive strength in both cases of uniform 18.22 MPa and graded structures with 9.89 MPa structures. IWP follows in tensile strength observed to have 14.41 MPa in uniform structure whereas the graded structure has 9.23 MPa. Additionally, both Schwartz D and IWP TPMS displayed higher compressive strength across all tests in both uniform and graded cases. Graded Schwarz D observed to have 100.68 MPa compressive strength, and IWP TPMS has 99.57 MPa, and uniform Schwarz D has 33.94 MPa, whereas IWP TPMS shows 31.82 compressive strength. This suggests that Schwarz D TPMS, in both uniform and graded forms, possesses the best mechanical properties, making it a promising candidate for applications demanding high strength and resilience. IWP TPMS structure also shows greater mean tensile and mean compressive strength in all cases following Schwarz D TPMS. The deformation behaviour of Schwarz P matches well with conventional lattices as per the data from stress-strain curve showing the typical compression property were an elastic compression region, a plateau region, and a densification region is observed. This study contributes to the growing interest in TPMS as a framework for designing scaffolds, leveraging their ability to mimic complex biological geometries. The successful additive manufacturing process using biomedical graded material further supports their viability in biomedical applications.

5. References

[1] Zhang L et al. 2020 Tailored mechanical response and mass transport characteristic of selective laser melted porous metallic biomaterials for bone scaffolds, *Acta Biomater.* **112**, 298-315.

[2] Collins M N et al. 2021 Scaffold fabrication technologies and structure/function properties in bone tissue engineering, *Adv. Funct. Mater.* **31**(21), 2010609.

[3] Langer R and Vacanti J P 1993 Tissue engineering. Science, **260**(5110), 920-926.

[4] Yu J, Xia H, and Ni Q Q 2018 A three-dimensional porous hydroxyapatite nanocomposite scaffold with shape memory effect for bone tissue engineering, *J. Mater. Sci.* **53**(7), 4734-4744.

[5] Xiong J et al. 2015 Advanced micro-lattice materials, *Adv. Eng. Mater.* **17**(9), 1253-1264.

[6] Zhang J et al. 2022 Design of a biomimetic graded TPMS scaffold with quantitatively adjustable pore size, *Mater. Des.* **218**, 110665.

[7] Yan C et al. 2015 Ti–6Al–4V triply periodic minimal surface structures for bone implants fabricated via selective laser melting, *J. Mech. Behav. Biomed. Mater.* **51**, 61-73.

[8] Al-Ketan O et al. 2020 Functionally graded and multi-morphology sheet TPMS lattices: Design, manufacturing, and mechanical properties, *J. Mech. Behav. Biomed. Mater.* **102**, 103520.

[9] Santiago R et al. 2023 Modelling and optimisation of TPMS-based lattices subjected to high strain-rate impact loadings, *Int. J. Impact Eng.* **177**, 104592.

[10] Dong Z and Zhao X 2021 Application of TPMS structure in bone regeneration, *Eng. Regener.* **2**, 154-162.

[11] Novak N et al. 2021 Development of novel hybrid TPMS cellular lattices and their mechanical characterisation, *J. Mater. Res. Technol.* **15**, 1318-1329.

[12] Blanquer S B and Grijpma D W 2021 Triply Periodic Minimal Surfaces (TPMS) for the Generation of Porous Architectures Using Stereolithography, *Computer-Aided Tissue Engineering: Methods and Protocols*, 19-30.

[13] Loh Q L and Choong C 2013 Three-dimensional scaffolds for tissue engineering applications: role of porosity and pore size.

[14] Verma R et al. 2022 Design and analysis of biomedical scaffolds using TPMS-based porous structures inspired from additive manufacturing, *Coatings*, **12**(6), 839.

[15] Ning L and Chen X 2017 A brief review of extrusion-based tissue scaffold bio-printing, *Biotechnol. J.* **12**(8), 1600671.

[16] Nguyen T L et al. 2011 A novel manufacturing route for fabrication of topologically-ordered porous magnesium scaffolds, *Adv. Eng. Mater.* **13**(9), 872-881.

[17] Huang Y et al. 2017 3D bioprinting and the current applications in tissue engineering, *Biotechnol. J.* **12**(8), 1600734.

[18] Yusong P, Qianqian S and Yan C 2013 Fabrication and characterisation of functional gradient hydroxyapatite reinforced poly (ether ether ketone) biocomposites, *Micro Nano Lett.* **8**(7), 357-361.

[19] An J et al. 2015 Design and 3D printing of scaffolds and tissues, *Engineering*, **1**(2), 261-268.

[20] Lu Y et al. 2020 Quantifying the discrepancies in the geometric and mechanical properties of the theoretically designed and additively manufactured scaffolds, *J. Mech. Behav. Biomed. Mater.* **112**, 104080.

[21] Li Y et al. 2018 Additively manufactured biodegradable porous iron. *Acta Biomater.* **77**, 380-393.

[22] Abou-Ali A M, Lee D W and Abu Al-Rub R.K 2022 On the Effect of Lattice Topology on Mechanical Properties of SLS Additively Manufactured Sheet-, Ligament, and Strut-Based Polymeric Metamaterials, *Polymers,* **14**(21), 4583.

[23] Sankineni R and Ravi Kumar Y 2022 Evaluation of energy absorption capabilities and mechanical properties in FDM printed PLA TPMS structures. Proceedings of the Institution of Mechanical Engineers, Part C, *J. Mech. Eng. Sci.* **236**(7), 3558-3577.

[24] Yang C et al 2023 Preparation and thermal insulation properties of TPMS 3Y-TZP ceramics using DLP 3D printing technology, *J. Mater. Sci.*, 1-16.

[25] M Li et al. 2023 Bioadaptable bioactive glass-β-tricalcium phosphate scaffolds with TPMS-gyroid structure by stereolithography for bone regeneration. *Journal of Materials Science & Technology*, **155**, 54-65.

[26] Melchels F P, Feijen J and Grijpma D W 2010 A review on stereolithography and its applications in biomedical engineering. *Biomaterials*, **31**(24), 6121-6130.

[27] Sun J, Binner J and Bai J 2019 Effect of surface treatment on the dispersion of nano zirconia particles in non-aqueous suspensions for stereolithography, *J. Eur. Ceram. Soc.* **39**(4), 1660-1667.

[28] Yu S, Sun J and Bai J 2019 Investigation of functionally graded TPMS structures fabricated by additive manufacturing, *Mater. Des.* **182**, 108021.

[29] Chouhan G et al 2023 Identification of Surface Defects on an SLA-Printed Gyroid Lattice Structure. *in International Conference on Research into Design*. Springer.

[30] Biomed Clear resin safety data sheet. [cited 2023, 06.10]; Available from: *https://formlabs-media.formlabs.com/datasheets/2001424-SDS-ENUS-0.pdf.*

[31] Austermann J et al. 2019 Fiber-reinforced composite sandwich structures by co-curing with additive manufactured epoxy lattices, *J. Compos. Sci.* **3**(2), 53.

Additive manufacturing best practices and its impact on supply chain performance through structural equation modelling: A case study in Ethiopian footwear industry

T Lemma[1*], E M Gutema[1], and H G Lemu[2]

[1]College of Engineering and Technology, Wallaga University, Nekemte, Ethiopia;
[2]Faculty of Science and Technology, University of Stavanger, N-4036 Stavanger, Norway

* Correspondence: tekalignlemma507@gmail.com

Abstract. The aim of this study is to analyze the effects of additive manufacturing (AM) best practices on supply chain (SC) performance. The study developed conceptual framework with four AM best practice as independent variables and five SC performance measuring factors as dependent variables. To test the developed model and hypothesis, data's has been collected from 129 respondents from 29 footwear industries in Ethiopia. AMOS v.23 is used to test model fit with confirmatory factor analysis (CFA) ($\chi 2$/df, RMSEA, CFI, GFI, RMR, Tucker-Lewis index (TLI)). Reliability analysis was carried out to test the reliability and stability of the research questionnaires. Correlation analysis was performed to test the strength and association between dependent and independent variables. And to test the developed hypothesis structural equation modelling (SEM) or paths analysis was used. The findings of the CFA result indicated that data fit the model reasonably well. The finds of correlation analysis indicated that, AM best practices have positive, strong and significant relationship with cost, time and reliability related factors of SC performance. But it has very weak and insignificant relationship with customer related factors. In addition, the findings revelled that AM best practices have negative and insignificant relationship with supplier related factors. On the other hands, the findings of SEM confirmed that the identified four AM best practices positively and significantly improve SC performance. And this study concludes that manufacturing flexibility, material utilization, environmental control and product design optimization of AM plays a vital role in improving SC performance.

Key words: - Additive manufacturing, supply chain, best practice, footwear industry

1. Introduction

The supply chain (SC) of the existing traditional manufacturing consists of several varied elements, and it makes the system more complex. And these complex systems generate disordered environment for the organization and increase its operational load for managing diverse actors [1, 2]. As reported in [3, 4], firms are exposed to a variety of operational risks such as supply chain disruptions due to increased levels of uncertainty and unpredictability that come from complexity and increased operational load. The negative effect of operational performance comes from an increased transaction costs (e.g., production, inventory, logistics, and communication), long and unreliable lead-times, difficulty in schedule attainment and inconsistent product quality [5,6], which all are parts of or comes from supply chain complexity [7]and it affects the SC performance.

In SC management, SC performance is measured through effectiveness and efficiency [8]. To enhance effectiveness and efficiency of SC performance, a disruptive technology like AM plays a vital role [9]. According to [10], AM implementation increases customer satisfaction by promoting rapid innovation and product design modifications. Additive manufacturing has been identified as enabling major change within the supply chain [11] by

- accelerating product development times [12],
- enabling on-demand production with short lead times [13],
- affording new distribution channels [14],
- changing market structures [15], and
- Supporting a wide variety of SC structures [16].

Additive manufacturing allowed flexibility of production in case of new product development and production, and this reduces the associated time and costs [17], it allows production of difficult and nonstandard product designs [18], and this directly influences cost and time-based performance capabilities of SC (i.e. quick and inexpensive introduction of new products to the market, short manufacturing lead times, or reductions in outbound costs). In addition, AM capability towards postponement of flexibility also influences cost, reliability and customer-oriented SC performance [19] by quickly responding to changes in market demand and customer requirements and increasing order fill rate by reducing inventory-holding costs and safety stocks. And adaptation of AM technology allows sourcing flexibility and enabling suppliers to operate efficiently at different levels of production volume and varieties [20], and this intern affects supplier-oriented SC performance by enabling suppliers to quickly modify or introduce new products into the market [20]. Similarly, the study by [21] illustrated that AM flexibility leads to customer satisfaction and minimization of inventory cost, which will improve SC customer-oriented and cost performance. In addition, according to [22] flexibility influences the efficiency of the SC through reliability of deliveries (reliability performance).

According to the study in [23], AM flexibility provides the printing of products only when needed or at the time they will be utilized in production. And this transforms the production chain, it makes manufacturing on-demand, it reduces inventories and waiting times and contributing to just in time practices. The decentralized approach of AM eliminates long distances that products move within the supply chain, reducing logistics and storage costs [24]. And it has the potential to change traditional structure of supply chains and location of manufacturing [25]. According to [26], AM technologies are opening new opportunities in terms of production paradigm and manufacturing possibilities. The paper illustrated that with AM technologies manufacturing lead times will be reduced, new designs will have shorter time to market, and customer demand will be met more quickly. In addition, it explores its impact on the supply chain, and investigates its transformative potential and impact on various industry segments. Even if the best practices of AM were illustrated in the existing studies, to our knowledge no studies were conducted to illustrate its effect on supply chain performance, especially in undeveloped countries. Thus, this study aims to study the effects of AM best practices on supply chain performance in context of Ethiopian footwear industry.

2. Supply chain performance and AM best practices

According to the works reported in [27, 28], SC performance represents a construct which measures and quantifies the efficiency and effectiveness of the SC processes to strengthening the market position of the firm. To evaluate the SC efficiency and effectiveness, SC management needs to assess its performance [29]. Efficiency aims to maximize the output of the firm with minimum input by reducing costs and waste (cost-related performance), but effectiveness seeks to achieve SC optimization by increasing customer satisfaction (service-related performances) [30]. According to [9], to enhance effectiveness, companies need to strive for innovation maximization in all possible areas and seeks to be as flexible and customer oriented as possible. Implementation of AM is one of the viable investments for promoting rapid innovation and product design modifications and resulting with the increased customer satisfaction, thus improves effectiveness and it contributes to flexibility advancements [14]. Firms' SC management improved performance through the effective use of resources and capabilities, and non-financial metrics are crucial in measuring and fostering the improvement of performance of contemporary SC [31].

Nowadays, to improve the performance of SC (effectiveness and efficiency); one of the industrial 4.0called AM technology plays a vital role. Existing studies illustrated manufacturing flexibility, material utilization, product design optimization and environmental factors as few of the best practices of AM in improving SC performance. It enables major change within the SC by producing complex shapes and forms without assembly, by minimizing waste of materials without the need for custom tooling, with low manufacturing process set up, and infinitely customizable [32, 33]. In practice, this implies on-demand production with short lead times [13] and highly flexible production. This can act to significantly reduce the need for buffer inventory of finished or semi-finished goods. AM can simplify the SC and only raw material provider, or suppliers are needed [34]. According to the study of [35], AM technology is growing rapidly with its unique features for producing an object without requiring any special, sophisticated tools or production lines. This unique feature of AM reduces the need for logistics, time from production to sale, and environmental impact [36]. In addition, AM machines have the ability to create any shape or product without the need for any machine set-ups, which allows for levels of manufacturing flexibility. According to[37]one of the crucial areas that AM can contribute are cost reduction and simplification of organizations' supply chains. And it revealed concrete benefits of AM in reducing lead-time and number of customers.AM adoption positively influences SC performance and as a result firm performance is improved [38].AM is one of the technologies that improves efficiencies of the entire SC from the cost of distribution to assembly, resulted for high level of customization, flexibility, possibilities in logistics management and potential for production cost savings.

3. Conceptual framework work and hypothesis

3.1. Research conceptual framework

In this section, a conceptual research framework was developed based on detail literature review and in consultation with footwear industry expert. As illustrated in Figure 1, this study considered AM best practices as independent variables and it is represented by four factors as: -

(1) Manufacturing flexibility (MF),
(2) Material utilization (MU),
(3) Environmental control (EC) and
(4) Product design optimization (PDO).

And SC performance is considered as dependent variables and measured in terms of cost related factors (CRF), time related factor (TRF), customer related factor (CURF), supplier related factor (SRF) and reliability related factor (RRF).

3.2 Research hypothesis

Thus, from the developed conceptual framework and literature review the following five hypotheses were developed and tested using structural equation modelling (SEM).

- **Hypothesis 1**: AM best practices positively and significantly affects customer related factor
- **Hypothesis 2**: AM best practices positively and significantly affects supplier related factor

- **Hypothesis 3**: AM best practices positively and significantly affects cost related factor
- **Hypothesis 4**: AM best practices positively and significantly affects time related factor
- **Hypothesis 5**: AM best practices positively and significantly affects reliability related factor.

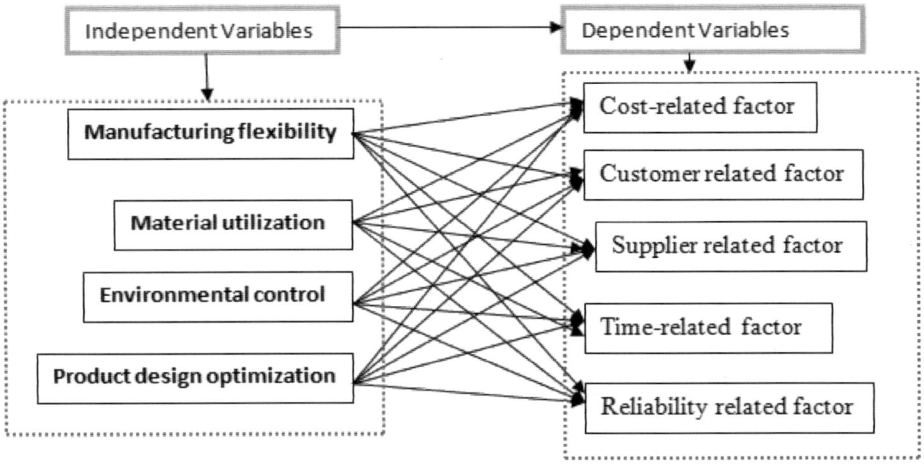

Figure 1. Conceptual framework developed from the review.

4. Methodology

In order to validate the developed research model, data for the constructs were collected using pre-coded standard questionnaire which is designed and prepared in 1 - 5 points Likert scales from 129 respondents. These scales were used to measure several constructs referring to the dependent and independent variables. The scale consists of 1=Strongly Disagree, 2=Disagree, 3=Neutral, 4=Agree and 5=strongly Agree. In data collection, 129 respondents from 29 footwear industries in Ethiopia were selected and participated in the study survey. Before the final paper was distributed, pilot test or pre-questionnaire test was conducted by distributing 25 papers. SPSS vr.20 and AMOS vr.23 were used to purify the data and analyze the confirmatory factor analysis (CFA). Though several different alternatives exist for selecting the model fit statistics or to conduct CAF, this study usedχ^2/df tests, RMSEA, CFI, GFI, RMR and Tucker-Lewis index (TLI) for testing model fit by adopting and as recommended in [39]. And also, to test the path analysis for hypothesis testing, SEM was used. In addition, correlation analysis was performed to study the direction and relationship between independent and dependent variables.

5. Analysis of findings

5.1 Reliability analysis

This study conducted reliability analysis before evaluating the research model to check the reliability of the questionnaires for each construct (variable), and then separately factor analysis (FA) was carried out for individual constructs. The researchers distributed 25 papers for pilot test before the final questionnaires were distributed. The reliability analysis results are shown in Table 1, which includes the number of items before and after some items were deleted and Cronbach's alpha (α) before and after the item was deleted. As per the result of the reliability analysis as indicated in Table1, the value of Cronbach's alpha (α) dropped to 0.732 and 0.901, which indicated higher or satisfactory reliability of the questioners.

5.2. Confirmatory factor analysis

Based on the results listed in Table1, the factor analysis for AM best practices variables i.e. manufacturing flexibility (MF), material utilization (MU), environmental control (EC) and product design optimization (PDO) indicated that the data fits the model well. However, after removing or deleting bad factor loading item and by conducting covariance between unobserved variables, the factor loading was improved and the items were retained as shown in Table 1. The FA model fit statistics result as indicated in the table (Table 1), which indicated a good model fit because.

- the data fit the model reasonably well to all the 9 constructs;
- CFI ranges from 0.963 to 1.00;
- TLI ranges from 0.886 to 1.01 and
- GFI ranges from 0.953to 1.00.

This indicates that the model fit for all AM best practice variables are good. On the other hand, concerning the factor analysis result of SC performance, variables CRF, TRF, CURF, SRF and RRF (TLI=0.959, CFI=0.993, RMSEA=0.118, GFI= 0.989, χ^2/df= 2.774; TLI=0.948, CFI=0.974, RMSEA=0.105, GFI= 0.967, χ^2/df= 2.425; TLI=1.00, CFI=1.00, RMSEA=0.000, GFI= 1.00, χ^2/df= 0.003; TLI=0.886, CFI=0.970, RMSEA=0.039, GFI= 0.950, χ^2/df= 5.477; TLI=0.950, CFI=0.990, RMSEA=0.025, GFI= 0.988, χ^2/df= 2.077) respectively were obtained which also indicated that the model fits well

Table 1. Model fit statistics result

	χ^2/df	GFI	TLI	CFI	RMSEA	RMR	# of items	Cronbach'salpha (α) Before item deleted	After item deleted
MF	3.217	0.953	0.945	0.972	0.132	0.036	5[5]	0.901	0.901
MU	1.771	0.984	0.969	0.991	0.078	0.036	5[5]	0.783	0.783
EC	0.591	0.998	1.01	1.00	0.000	0.009	5[4]	0.802	0.829
PDO	2.67	0.980	0.950	0.983	0.114	0.041	5[4]	0.788	0.816
CRF	2.774	0.989	0.959	0.993	0.118	0.018	4[4]	0.868	0.869
TRF	2.425	0.967	0.948	0.974	0.105	0.037	5[5]	0.833	0.833
CURF	0.003	1.00	1.00	1.00	0.000	0.000	6[5]	0.845	0.865
SRF	5.477	0.950	0.886	0.970	0.187	0.039	7[6]	0.879	0.891
RRF	2.077	0.988	0.950	0.990	0.092	0.025	5[4]	0.723	0.732

Figure 2 shows separate factor analysis for individual constructs. The FA result of manufacturing flexibility and material utilization variable as shown in Figure 2 (a) and (b) falls between 0.79 to 0.87 and 0.62 to 0.76 respectively after conducting covariance analysis and without removing any items from these factors. And, as shown in Figure 2 (c), FA result of product design optimization falls between o.55 to 0.89 after removing one item. The FA results for environmental related factor in Figure 2(d) shows that the factor loading falls between 0.67 and 0.9 after conducting covariance analysis and removing one lower factor loading items. On the other hand, the FA results of cost and time related factors as shown in Figure 2 (e) and (f) falls between 0.65 to 0.93 and 0.50 to 0.95 respectively without removing any items under these factors. And the factor loading of reliability related factors was improved and falls in between 0.50 to 0.76 as shown in Figure 2 (g) after removing one item. In addition, while analyzing factor loading of customer and supplier related factors, as shown in Figure 2 (h) and (i), their factor loading falls between 0.58 to 0.92 and 0.56 to 0.96 after removing one item from each factor.

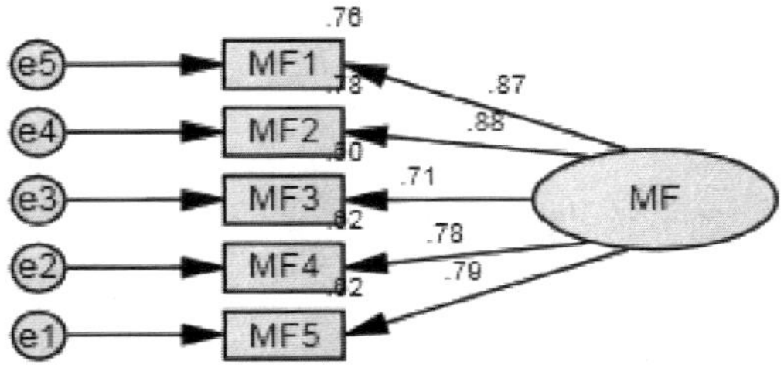

(a) manufacturing flexibility

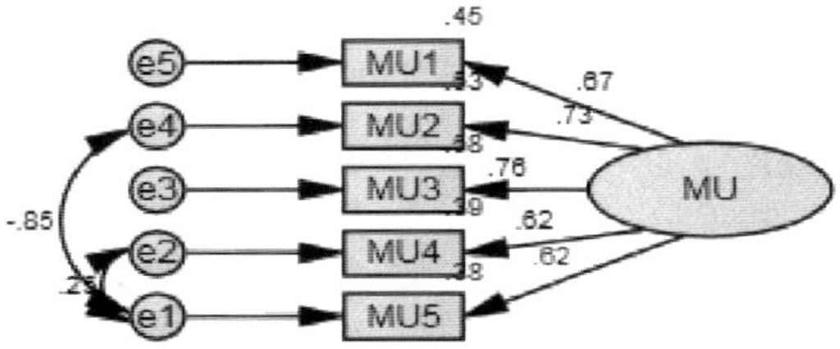

(b) material utilization

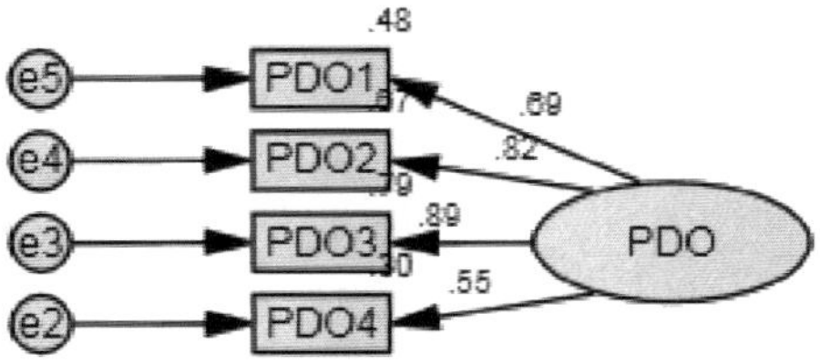

(c) product design optimization

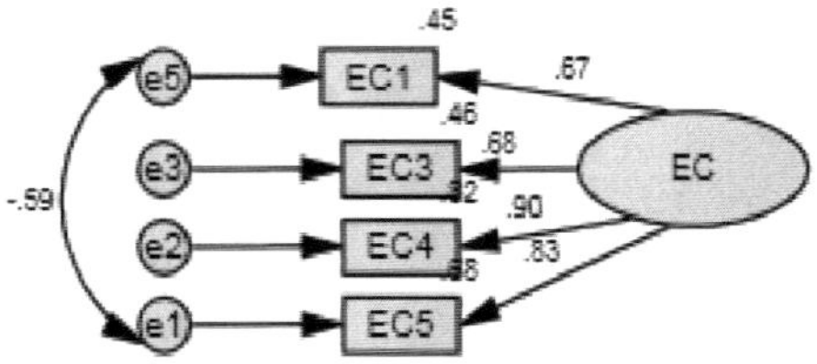

(d) environmental control factor

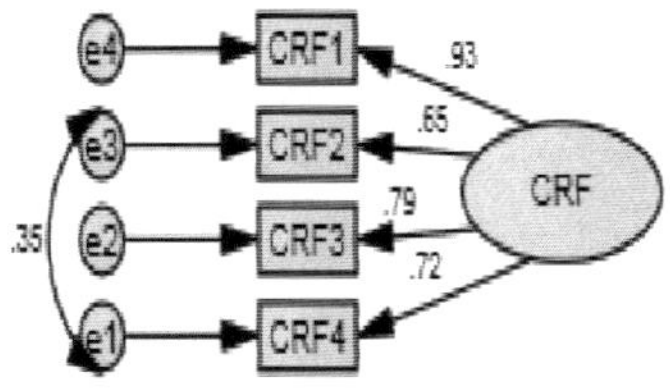

(e) Cost factor

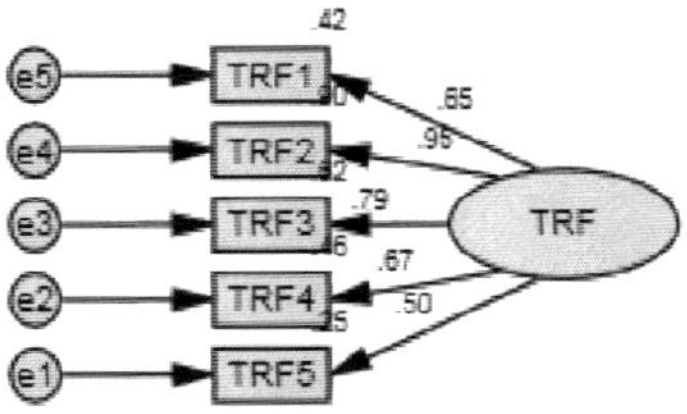

(f) Time factor

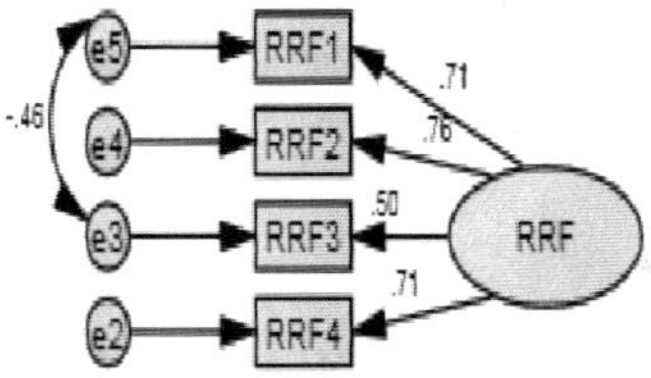

(g) Reliability factor

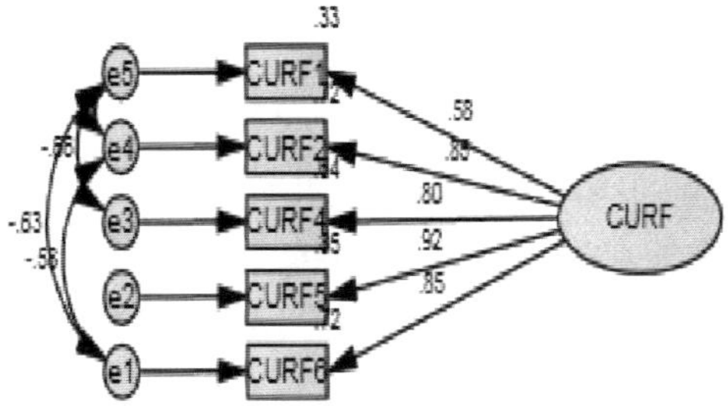

(h) Customer factor

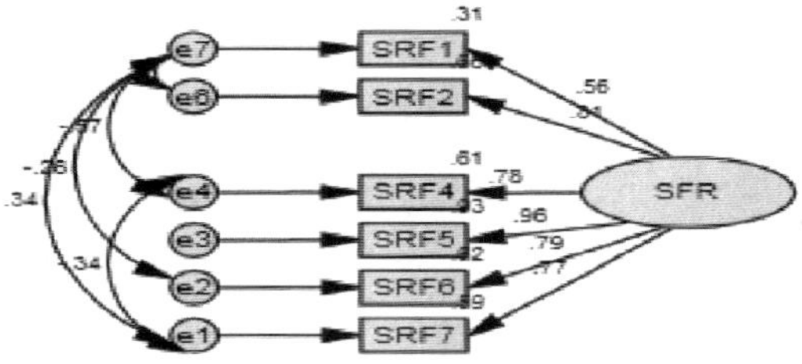

(i) Supplier factor

Figure 2. Factor analysis results of independent and dependent variable

5.3 Correlation analysis

This research paper used correlation analysis to examine the strength of the relationship between AM best practices and SC performance measuring factors. The result in Table 2 shows that AM best practices have positive, strong and significant relationship with cost related factor (r = 0.539, p= 0.000), time related factor (r= 0.613, p= 0.000) and reliability related factor (r= 0.634, p= 0.000). On the other hand, AM best practices have negative and insignificant relationship on supplier's related factors (r= -0.003, p= 0.969). In addition, the result in Table 2 indicated that AM best practices have very weak and insignificant relationship with customers related factors (r= 0.01, p= 0.912). From this result, we can say that implementation of AM technologies improves supply chain performance in terms of cost, time, and reliability. From this correlation results, unexpected results were obtained showing that AM best practices have negative, very weak, and insignificant relationship with suppliers and customer related factors. These unexpected results were obtained due to smaller number of sample size.

Table 2. Correlation analysis result

		AMR	CRF	TRF	SEF	CURF	RRF
AMR	Pearson Correlation	1	.539**	.613**	-.003	.010	.634**
	Sig. (2-tailed)		.000	.000	.969	.912	.000
	N		129	129	129	129	129
CRF	Pearson Correlation		1	.011	.456**	.490**	.218*
	Sig. (2-tailed)			.900	.000	.000	.013
	N			129	129	129	129
TRF	Pearson Correlation			1	-.445**	-.460**	.534**
	Sig. (2-tailed)				.000	.000	.000
	N				129	129	129
SEF	Pearson Correlation				1	.965**	.047
	Sig. (2-tailed)					.000	.595
	N					129	129
CURF	Pearson Correlation					1	.019
	Sig. (2-tailed)						.830
	N						129
RRF	Pearson Correlation						1
	Sig. (2-tailed)						
	N						129

**Correlation is significant at the 0.01 level, *Correlation is significant at the 0.05 level (2-tailed)

5.4 Structural modelling analysis

The developed hypothesis in Section 3.2 was tested using path analysis (SEM) and the results are shown in Figure 3, in which a positive and significant effects of AM best practices on SC performance cost related factor (β= 0.54, p=0.000) is observed. Hence hypothesis 1 is accepted. Similarly, the findings of the study illustrated that AM best practices have positive and significant effects on time related factor (β = 0.61, p = 0.000) and this also supports hypothesis 2. The results of SEM analysis also confirmed that SC performance customer related factor positively and significantly affected by AM best practices (β = 0.01, p = 0.000). Hence hypothesis 3 is supported.

On the other hands, material utilization, manufacturing flexibility, product design optimization and environmental best practices of AM significantly and positively influencing reliability related factor of SC performance (β = 0.63, p = 0.000) and this accepted hypothesis 4. The results of the analysis in Figure 3 also illustrated that the four AM best practices positively and significantly affects supplier related factor of SC performance measuring factor (β= 0.02, p=0.000). Thus, this also supported hypothesis 5.

These hypotheses testing results confirmed that all the identified AM best practices (material utilization, manufacturing flexibility, product design optimization and environmental factor) have positive and significant impacts in improving SC performance (efficiency and effectiveness). The result illustrated and supported or in line with the findings of existing studies. Through implementation of AM process, reliability of SC performance is improved by reducing buffer inventory of finished or semi-finished goods and improves customer-oriented performance through manufacturing flexibility without the need of tooling and new machine setups. On the other hands, adaptation of AM improves SC cost related performance by delivering raw material on time and logistic related costs. It improves customer satisfaction by promoting on demand production with shorter lead time, and high flexible production. It improves effectiveness of SC by reducing time from supplier to production, from production to sales and through reduction of carbon footprint.

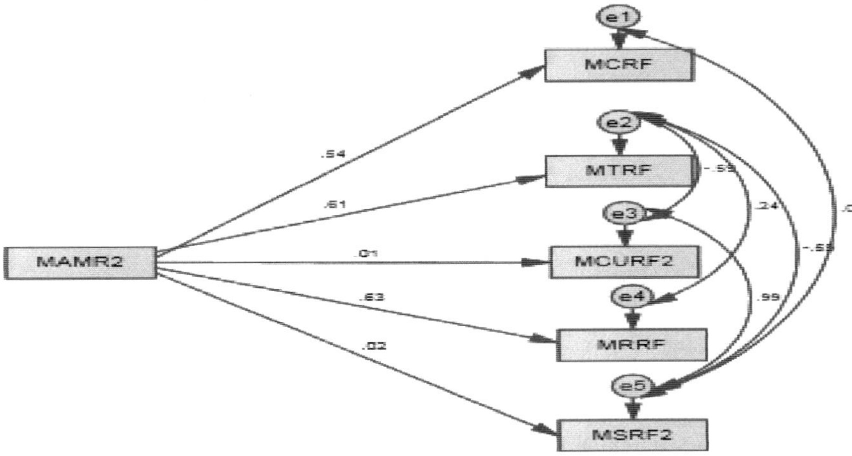

Figure 3.SEM path analyses

6. Conclusion

This study aims to analyze the effects of AM best practices on supply chain performance. The study identified four AM best practices and five SC performance measuring factors from the literature review. To test the developed conceptual framework and hypothesis, data have been collected through 1- 5 liker scale from 129 respondents from 29 footwear industries in Ethiopia. Through pilot study the reliability and validity of the questionnaires was tested. Correlation analysis was used to examine the strength of the relationship between independent factors (MF, MU, EC, and PDO) and dependent VARIABLES (CRF, TRF, CURF, SRF and RRF). And the model fits and the hypothesis were tested using confirmatory factor analysis and structural equation modelling respectively. And in conclusion the four AM best practices: - (a) manufacturing flexibility (MF), (b) material utilization (MU), (c) environmental control (EC) and (d) product design optimization will improve the performance of supply chain.

Acknowledgement: The authors would like to acknowledge the support provided INDMET project funded by NORHED II program (grant nr. 62862) to cover the publication cost of this work.

References

[1] Choi T, Krause D (2006), the supply base and its complexity: Implications for transaction costs, risks, responsiveness, and innovation, *J.Oper.Manag.*, **24**, 637–652

[2] De Leeuw S, Grotenhuis R, Van G (2013), Assessing complexity of supply chains: evidence from wholesalers. *Int. J. Oper. Prod. Manag.*, **33**, 960–980

[3] Bode C, Wagner S(2015), Structural drivers of upstream supply chain complexity and the frequency of supply chain disruptions, *J. Oper. Manage*, **36**, 215–228

[4] Birkie S, Trucco P (2020), do not expect others do what you should! Supply chain complexity and mitigation of the ripple effect of disruptions, *Int. J. Logist. Manage.*,**31**(1), 123-144

[5] Lorentz H, Toyli J, Solakivi T, Hälinen H, Ojala L(2012), Effects of geographic dispersion on intrafirm supply chain performance, *Supply Chain Manag.: An Int. J.*,**17**, 611– 626

[6] Lu G, Shang G (2017), Impact of supply base structural complexity on financial performance: Roles of visible and not-so-visible characteristics, *J. Oper. Manage*, **53**, 23–44

[7] Serdarasan S (2013), A review of supply chain complexity drivers, *Comput. Industr. Eng.*, **66**, 533–540

[8] Beamon B(1999), Measuring supply chain performance,*Int.J. Oper. Prod. Manage.*, **19**(3), 275-292

[9] Chan H, Griffin J, Lim J, Zeng F, Chiu, A (2018), The impact of 3C Printing Technology on the Supply Chain: Manufacturing and Legal Perspectives,*Int. J. Prod. Eco.*, **205**, 156-162

[10] Eyers D, Potter A, Gosling J, Naim M(2018), The flexibility of industrial Additive Manufacturing systems, *Int. J. Oper. & Prod. Manage*, **38**(12), 2313-2343

[11] Holmstrom J, Partanen J(2014), Digital manufacturing-driven transformations of service supply chains for complex products, *Supply Chain Manage.: An Int. J.*, **19**(4), 421-430

[12] Gibson I, Rosen D, Stucker B (2015), Additive Manufacturing Technologies – 3D Printing, *Rapid Prototyping and Direct Digital Manufacturing*, New York: Springer

[13] Petrovic V, Vicente H, Gonzalez J, Jorda F, Delgado G, Ramon B, Portoles G (2011), Additive layered manufacturing: sectors of industrial application shown through case studies, *Int. J. Prod. Res.***49**(4), 1061-1079

[14] Eyers D (2015), The Flexibility of Industrial Additive Manufacturing Systems. PhD Thesis, Cardiff University, UK

[15] Weller C, Kleer R, Piller F (2015), Economic implications of 3D printing: market structure models in light of additive manufacturing revisited, *Int. of Prod. Eco.*, **164**, 43-56

[16] Ryan M.J, Eyers D.R, Potter A.T, Purvis L, Gosling J (2017), 3D printing the future: scenarios for supply chains reviewed, *Int. J. Physic. Distr. Logistic.Manag.*,**47**(10), 992-1014

[17] Chan H, Griffin J, Lim J, Zeng F, Chiu A(2018), The impact of 3C Printing Technology on the Supply Chain: Manufacturing and Legal Perspectives, *Int. Prod. Eco*, **205**, 156-162

[18] Craeghs T, Bechmann F, BerumenS, Kruth J(2010), Feedback control of Layerwise Laser Melting using optical sensors, *Physics Procedia*, **5**(B), 505-514

[19] Nyman H, Sarlin P (2013), From Bits to Atoms: 3D printing in the context of supply chain strategies. In,*47th Hawaii Int. Conf. Syst. Sci.*, Hawaii

[20] Giffi C, Gangula B, Illinda P(2014), 3D opportunity in the automotive industry - Additive Manufacturing hits the road, A Deloitte series on Additive Manufacturing, Deloitte University Press. Available at: http://dupress.com/articles/additive-manufacturing-3dopportunity-in-

IOP Conf. Series: Materials Science and Engineering 1294 (2023) 012052 doi:10.1088/1757-899X/1294/1/012052

automotive/

[21] Lummus R, Duclos L, Vokurka R(2003), Supply chain flexibility: building a new model, *Global J. Flexible Syst. Manage.*, **4**(4), 1-13

[22] Chavosh A, Halimi A, Edrisi M, Hosseini S, Esferjani P(2011), A Model for Supply Chain Performance of Electronics Industry in Malaysia, I*n Proc. 5th Int. Conf. Social Sci.Humanit.*. Singapore: IACSIT Press, 387-391

[23] Huang Y, Leu M, Mazumder J, Donmez A (2015), Additive manufacturing: current state, future potential, gaps and needs, and recommendations, *J. Manuf.Sci. and Engineer.*, **137**(1), 014001

[24] Rubmann M, Lorenz M, Gerbert P, Waldner M, Justus J, Engel P, Harnisch M (2015), Industry 4.0 the future of productivity and growth in manufacturing, Boston *Consulting*, **l9**(1), 54-89,

[25] Cohen D, Sargeant M, Somers K (2014), 3D printing takes shape, *Int. J. Serv. Oper. Manage*, **10**(3), 89-99

[26] Mohsen A (2017), The rise of 3-D printing: The advantages of additive manufacturing over traditional manufacturing, *BUSHOR*-1399, **12**

[27] Li S, Ragu-Nathan B, Ragu-Nathan T.S, Rao S (2006), The impact of supply chain management practices on competitive advantage and organizational performance,*Omega*, **34**, 107-124

[28] Maestrini V, Luzzini D, Caniato F, Maccarrone P, Ronchi S(2018), Measuring supply chain performance: a lifecycle framework and a case study, *Int. J. Oper. Prod. Manage*., **38**(4), 934-956

[29] Gunasekaran A, Patel C, McGaughey R (2004) A framework for supply chain performance measurement, *Int. J. Prod. Econ.*, **87**, 333-347

[30] Balocco R, Miragliotta G, Perego A, Tumino A, (2011), RFId adoption in the FMCG supply chain: an interpretative framework, *Supply Chain Manage:An Int. J.*, **16**(5), 299-315

[31] Gimenez C, van der Vaart, T, van Donk D P (2012), Supply chain integration and performance: the moderating effect of supply complexity, *Int. J. Oper. Prod. Manage., 32(5), 583-610*

[32] Singamneni S, Yifan L, Hewitt A, Chalk R, and Thomas W (2019), Additive Manufacturing for the Aircraft Industry: A Review,*J. Aeron. Aerospac. Eng.*, **8**

[33] Bak D (2003), Rapid prototyping or rapid production? 3D printing processes move industry towards the latter, Assembly Automation, MCB UP Ltd, **23**,340-345

[34] Mellor S, Hao L, Zhang D 92014), Additive manufacturing: a framework for implementation, *Int. J. Prod. Econ.*, **149**, 194-201

[35] Dimitrov D, Beer N, Hugo P, Schreve K.(2014), Three Dimensional Printing, *In Comprehensive Mater. Process.*, **10**, 217-250

[36] Paris H, Mokhtarian H, Coatanéa E, Museau M, Ituarte I (2016), Comparative environmental impacts of additive and subtractive manufacturing technologies, *CIRP Annals – Manufact. Technol.*, **65**(1), 29–32

[37 Siavash H, Inigo F, Alireza J, Jia A, Chee K, Jan H (2020), Impact of Additive Manufacturing on Supply Chain Complexity, *Proc. 53rd Hawaii Int. Conf. Syst. Sci.*, 4505- 4514

[38] Mia D, Daniel R, Josip M(2019), Additive manufacturing: empirical evidence for supply chain integration and performance from the automotive industry. *Supply Chain Manage.: An Int. J.* **24**(5)*, 604 -21.*

[39] Kline R (2011), Principles and Practice ofStructural Equation Modeling, 3rd ed.,the Guildford Press: New York, NY, USA.

IOP Conf. Series: Materials Science and Engineering 1294 (2023) 012053 doi:10.1088/1757-899X/1294/1/012053

Fracture toughness analysis of a woven sisal-glass fiber hybrid composite

O K Gudeta[1*], M A Tolcha[2*], T D Badasa[3] and E G Koricho[4]

[1] Dire Dawa university, Ethiopia
[2] Jimma University, Ethiopia
[3] Addis Ababa Science and Technology University, Ethiopia
[4] Addis Ababa Science and Technology University, Ethiopia

[*] Correspondence: Email: Oda33kerreddu@gmail.com (O.K. Gudeta), mesay.alemu@ju.edu.et (M.A. Tolcha)

Abstract. This experimental work's goal was to investigate the mode-I fracture toughness behavior of a hybrid composite material made of woven sisal and glass fibers under mode-I loading circumstances to gauge the material's resistance to fracture behavior. The sisal plant was painstakingly stripped of its fibers. For the sisal fiber surface treatment, 5% NaOH was utilized. Next, treated fibers were weaved by hand to create the woven sisal fiber. Using the hand lay-up technique and the ASTM D5045-99 standard, compact tension specimens were created. The findings indicated that, under mode I loading circumstances, the mode I fracture toughness value of a specimen with a thickness of 5 mm has a (KIC) of 18.34 MPa. m. By determining a material's resistance to fracture under mode I stress conditions—a condition that is employed in a variety of engineering applications—it is a valuable parameter for developing and establishing preconditions for engineering applications. The manufactured hybrid specimens' performance and applications for various engineering materials for construction (such as tables, ceilings, and partitions), lightweight vehicle parts (like boards), and replacing synthetic materials with natural fibers in engineering applications were demonstrated by the results of the fracture toughness test.

1. Introduction

These days, the use of hybrid composite materials has expanded in the automotive, aerospace, nuclear, marine, biomedical, and other engineering domains for the following reasons: superior fatigue characteristics, high strength or stiffness for reduced weight, ability to alter fiber orientations, etc. (1). The reinforcement in hybrid composites is a combination of synthetic and natural fibers [2]. While both natural and synthetic fibers are excellent reinforcements for hybrid composites and have their own benefits, they also have disadvantages. For example, natural fibers absorb too much moisture and have poor mechanical properties, while synthetic fibers have issues with recycling, reusability, and biodegradability [3]. When natural and synthetic fibers are hybridized, performance is enhanced and maximizes the benefits of both natural and synthetic fibers while minimizing the negative effects of each kind of fiber [4, 5].

In addition, voids, discontinuities, inter-ply cracking, and fiber cracking are among the several issues that hybrid composite materials may encounter. Among these three issues, hybrid composite materials frequently experience mode I fracture toughness failure. Numerous elements will impact the failure process due to the intricate nature of the fracture mechanism [6]. It was discovered that hybrid composite materials are the greatest substitute for metals and have a wide range of applications [7]. The majority of study has recently concentrated on the fracture toughness of hybrid composite materials, as previously mentioned, because of their many benefits.

Since defects can occasionally arise during the processing, manufacturing, or servicing of a material or component, fracture toughness is a crucial mechanical attribute. Thus, it gauges a material's ability to withstand fracture and is applied in a variety of technical applications, such as chemical, mechanical, civil, and materials engineering. A material is more prone to ductile fracture if its fracture toughness is high. Materials with lower fracture toughness are more prone to brittle fracture [16].

This study describes the preparation of compact stress specimens for a fracture toughness test. During the test, software connected with universal testing equipment generates an estimate of the fracture length subsequent to the original crack. According to the criteria of ASTM standards, the test was developed on universal testing apparatuses, and results were acquired appropriately.

2. Materials and methods

The materials utilized for the experiment are shown in Figure 1. The materials used directly in the production of the hybrid composite sample made of sisal and glass fiber are woven sisal fiber, E-glass fiber, sodium hydroxide, polyester resin together with its hardener, a press machine, and a universal testing machine.

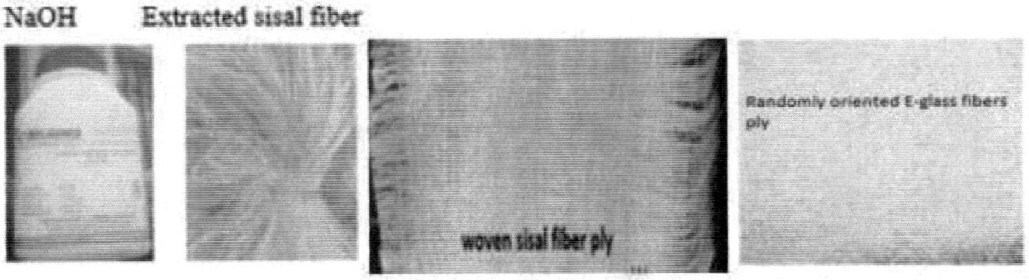

Figure 1. Basic materials.

2.1. Sisal and extraction process of sisal fibers

Sisal plant leaves were harvested by cutting them at the base and gathering them from the Ethiopian Rift Valley. After trimming, the leaves were divided into strips along their length. Hand labor was used to extract the fibers (Figure 2), which were subsequently dried in the sun after being cleaned of dust with pure water [9, 10].

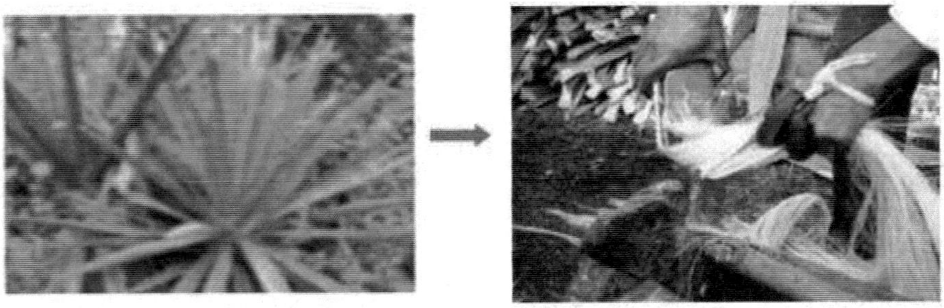

Figure 2. Extraction of sisal fibers.

2.2. Sisal fiber treated with sodium hydroxide

Surface modification techniques can be applied chemically or physically, depending on how the fiber surface is modified [12]. In order to improve the surface adhesion properties for the fiber-matrix interface, the fiber's shear strength, rigidity and stiffness, moisture absorption issues, and roughness of the fiber, waxy substances, pectin, lignin, and natural oils are removed from the external surface of the fiber cell wall through chemical treatment of natural fibers [13]. Aqueous sodium hydroxide is added to natural fiber to ionize the hydroxy group. For bleaching and cleaning plant fiber surfaces, sodium hydroxide is the most often used chemical [89]. Nevertheless, the characteristics of the fibers change and their connection weakens as the sodium hydroxide content rises.

$$\text{Sisal fiber - OH + NaOH} \rightarrow \text{fiber-O-Na} + H_2O \quad (1)$$

Four liters of distilled water and 400 g of NaOH pellet were combined at room temperature and left to soak for three hours. This was done based on a variety of literature sources as well as the positive outcomes of my trial work, which involved mixing one liter of distilled water with 100 g of sodium hydroxide, or a medium percentage. Next, to achieve a pH of neutrality, the soaked fibers were cleaned with pure water. In order to remove any remaining NaOH and achieve a neutral pH value, the soaked fibers were repeatedly rinsed with regular water. After that, they were left to dry in the sun for a whole day. The solution's moisture content rises because of the reaction between NaOH and air.

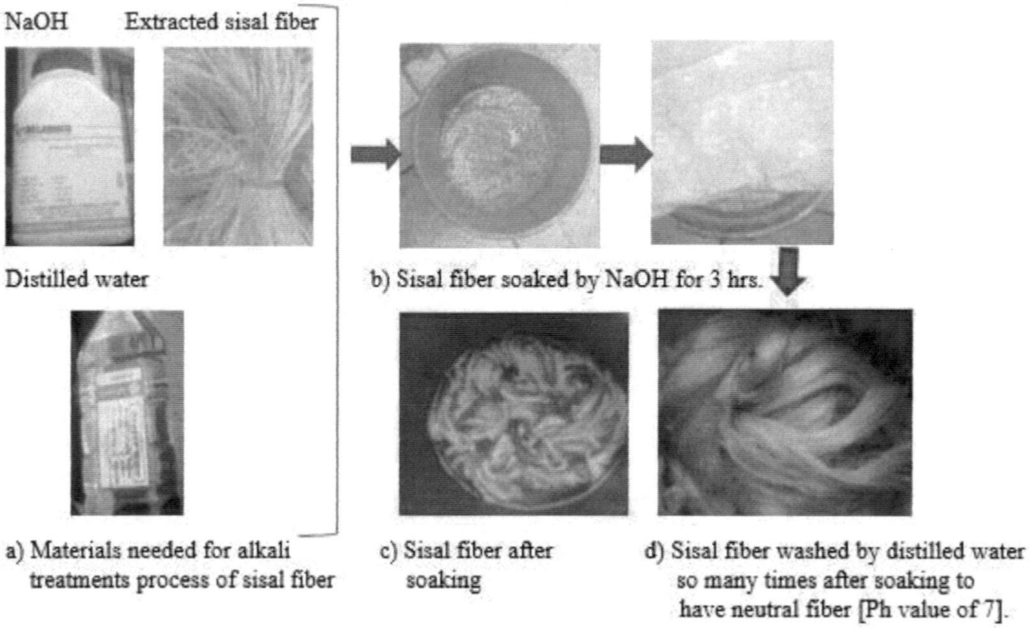

Figure 3. Alkali treatment of sisal fiber.

2.3. Preparation of woven sisal fiber

According to Zhang and Kelkar AD, the superior mechanical properties, conformability, and design flexibility of woven fabric textile structures make them a better choice for use as reinforcement in composite materials than other materials. In comparison to unidirectional laminates, the composite laminates made of woven fabrics exhibit superior impact resistance, more balanced properties, and good properties in mutually orthogonal directions [11, 14]. Consequently, bi-directional layers of treated sisal woven fabric with warp and weft in the perpendicular direction (00/900) were made when the fibers dried. The sample of woven fabric measured 38 cm in width and 38 cm in length (Figure 4).

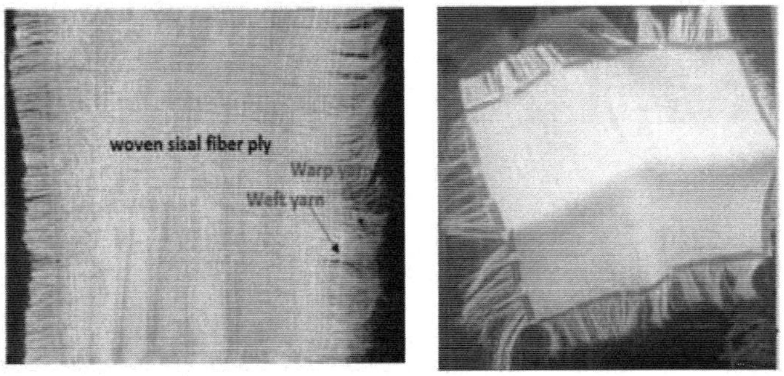

Figure 4. Manufactured woven sisal fiber ply.

2.4. Glass fiber

Common fiberglass fiber kinds, such as E-glass, are used in composite materials and offer good mechanical, insulating, and chemical resistance combinations. E-glass fiber exhibits superior strength and impact resistance, low thermal conductivity, low susceptibility to moisture, good electrical insulator qualities, high resistance to chemical agents, and dimensional stability at high temperatures when compared to other varieties of glass fiber [8]. They do, however, come with certain drawbacks, including low stiffness, a brief fatigue life, and extreme temperature sensitivity [15]. For this work, sisal fiber reinforcement in the form of E-glass fiber has been employed because of the previously described property. The supplier of randomly oriented E-glass fiber (Figure 5) is World Fiber Glass and Waterproofing Engineering plc, which is located in Addis Ababa.

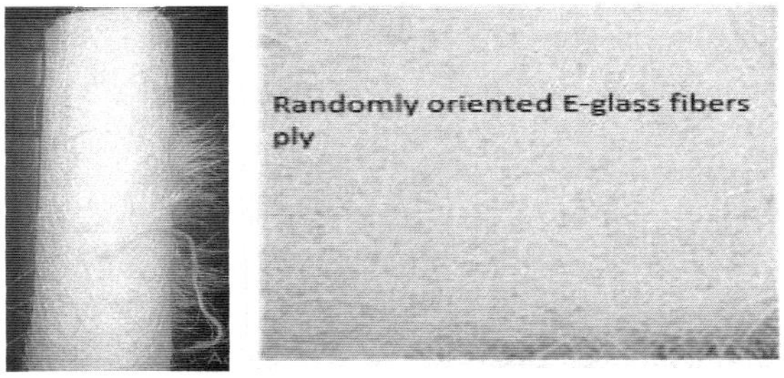

Figure 5. Randomly oriented glass fibers.

Glass fiber and sisal fiber were used to make the fibers in this case. The main data for E-glass, sisal fiber, and polyester resin in this study (fiber density and matrix densities) are derived from reviews of the literature. The rule of mixtures formula was used to generate the findings, which are evaluated and shown in Table 1 along with their values.

Table 1. Table fibers and matrix volume values.

Primary data		Calculated Results	
Parameters	Value	Parameters	Value
Sisal fiber ply weight (Wf-sisal)	71.25 g	Sisal fiber volume (Vf- sisal)	53.6 cm^3
Glass fiber ply weight (Wf-glass)	171.05 g	Glass fiber volume (Vf- glass)	66.6 cm^3
Total fiber weight	242.3 g	Total fiber volume -	120.2 cm^3
Sisal fiber density (ρf-sisal)	1.33 g/cm^3	Matrix volume (Vm)	121.5 cm^3
Glass fiber density (ρf -glass)	2.57 g/cm^3	Total fiber volume ratio (VF)	50 %
Matrix weight (Wm)	145.8 g	Matrix volume ratio (VM)	50 %
Matrix density (ρm)	1.2 g/cm^3	Total fiber weight ratio (WF)	62.4 %
		Matrix weight ratio (WM)	37.6 %

Layer (Ply) thickness	Woven sisal fibers ply thickness	0.35 mm
	Randomly oriented E-glass fibers ply thickness	0.25 m

2.5. Methods of hand lay-up

Glass fiber ply, woven sisal fiber ply, and an adequate amount of polyester resin mixture were added to the prepared mold using the hand lay-up fabrication process (Figure 6), beginning and finishing with layers of resin. One percent of the resin's volume was added to the catalyst when it was the resin was at room temperature for curing. The simplicity, adaptability, and inexpensive capital investment of hand lay-up procedures are their advantages. But the majority of it requires labor-intensive work [16]. In order to keep the matrix from adhering to the mold during the composite's separation, mold release is crucial. For this thesis project, paste wax (oil) is the most popular kind of releasing agent. Applying paste wax results in a smoother surface finish. Typically, paste wax is rapid, affordable, and easily accessible.

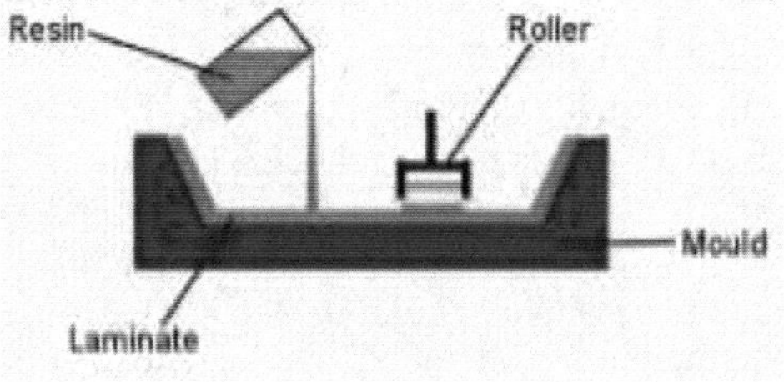

Figure 6. Hand lay-up method.

2.6. Hybrid composite plies and laminates

The component architecture of the bidirectional woven sisal and E-glass-reinforced polyester hybrid composite has been characterized as glass-sisal-sisal-glass (Figure 7) based on the reinforcement, matrix, and curing procedure. A 5 mm-thick hybrid specimen is created by weaving ten plies of sisal fibers and preparing two plies of randomly oriented E-glass fibers.

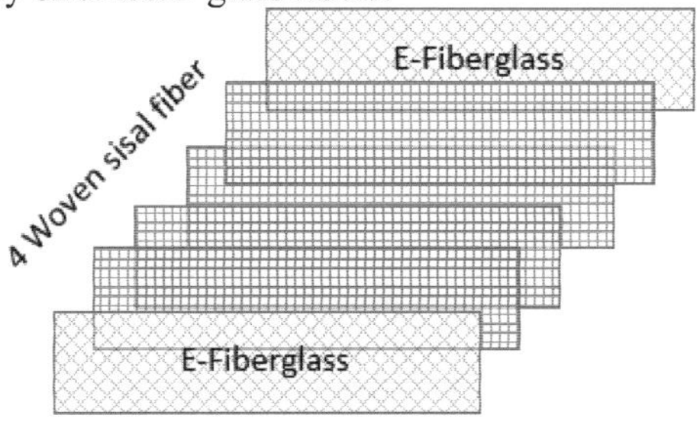

Figure 7. Hybrid plies configuration.

2.7. Fabrication process of a hybrid specimen

The flowchart for specimen preparation is shown in Figure 8. The composite plate is prepared by mixing polyester resin with a hardener, and the glass and sisal fiber plies are woven and stacked in a specified order on the mold surface. Next, the initial fiber ply is positioned. To create the extra resin, this layer is wetted with resin and then gently pressed with a roller. Finally, apply a compression pressure of 100 bar (10 MPa) to the mold using the hydraulic press machine, and then let it sit at room temperature for 12 hours to allow for consolidation.

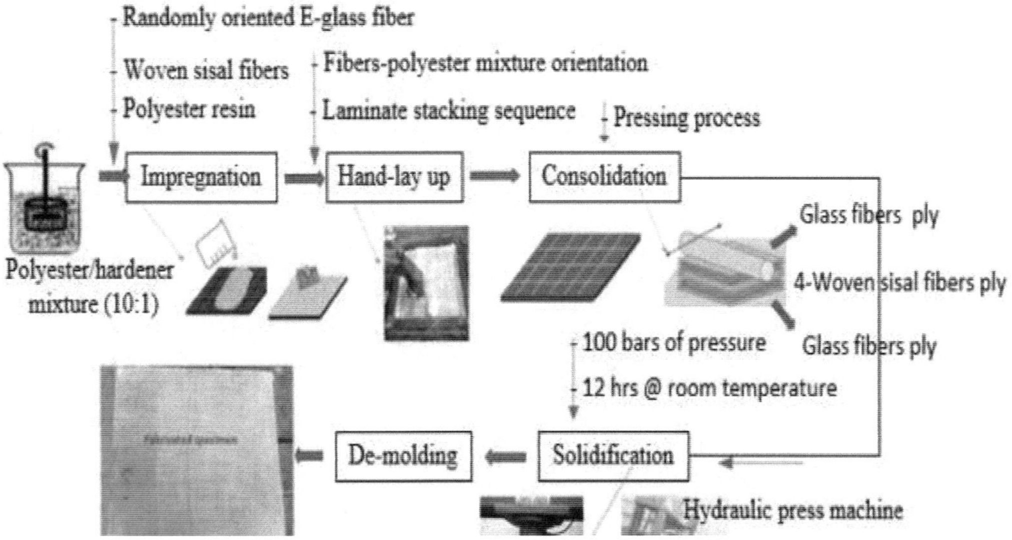

Figure 8. Specimen fabrication flow chart.

3. Testing conditions and experimental setup

3.1. ASTM dimensions requirements for band saws

Using a band saw, the prepared hybrid composite specimen was cut to the appropriate dimensions in accordance with the criteria for compact tension testing, which were carried out using a universal testing apparatus. The compact tension specimen for the rectangular test specimen was prepared in accordance with American Society of Testing Materials standards ASTM D5045-99; Figure 9 displays the sample's dimensions and configurations.

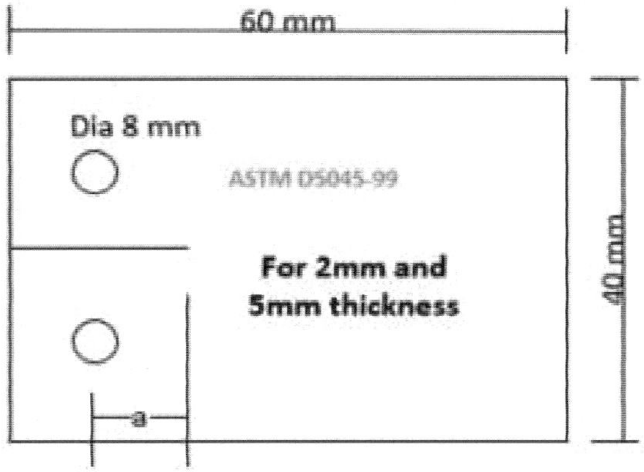

Figure 9. Dimensions and configuration of the specimen according to ASTM D5045-99.

3.2. Universal testing machine

A universal testing machine of model WP 310, with a test load capacity of up to 50 kN, a maximum piston stroke of 150 mm, and a crosshead speed of 6 mm/sec, was used to study compact tension fracture toughness tests at Bishoftu Defense College of Engineering. During testing, data from the machine is collected using the data gathering program. All specimens undergo continuous loading from the dynamometer load cell until the specimen fails the test. Figure 10 depicts the test setup.

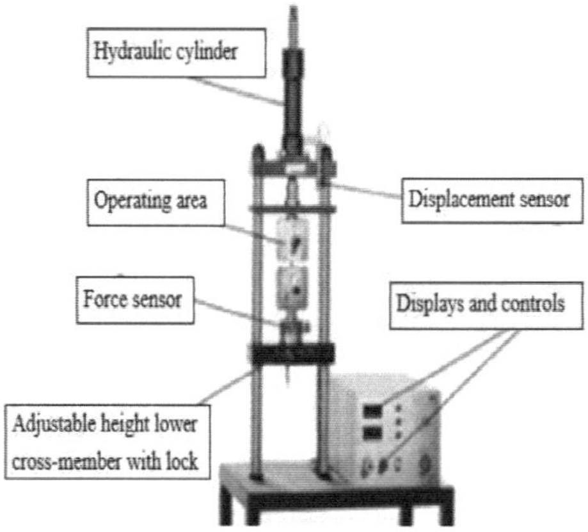

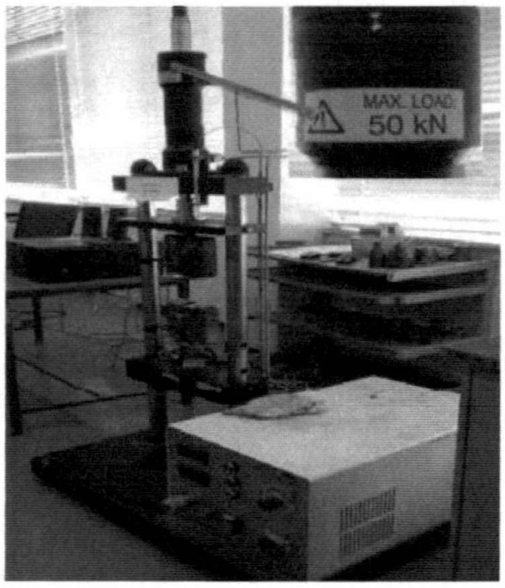

Figure 10. Universal testing machine.

3.3 Mode-I fracture toughness test (ASTM D5045-99) with compact tension

A compact tension specimen was constructed with a ratio of 0.5 between the crack width ('w') and length ('a') (a/w = 0.5). As per the ASTM D5045-99 standard, all dimensions are given, and the band saw blade's thickness is represented by the crack thickness (crack opening). A room-temperature universal testing equipment with a cross-head speed of 6 mm/sec was used to create and test five compact tension specimens for each test. The test's load and load line displacement were noted, and the loads' critical stress intensity factor was computed. Figure 11 displays the standard 5 mm produced compact tension test specimens.

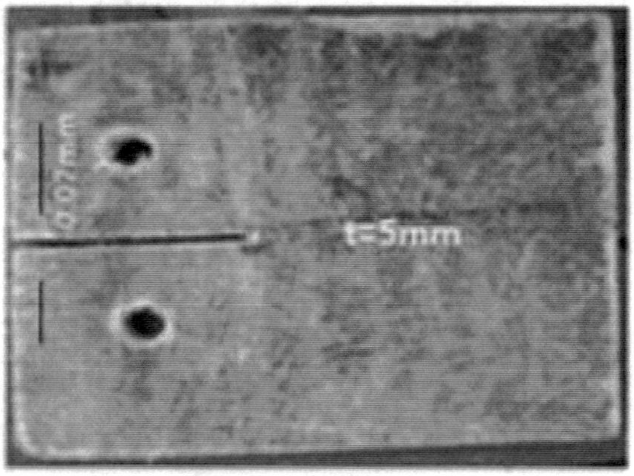

Figure 11. Produced compact tension specimen.

4. Result and discussion

4.1. Results of the Mode-I fracture toughness tests

For the compact tension test, five test samples made of hybrid composites of woven sisal and E-glass fiber were constructed (Figure 12). Graphs, which can relate various material properties used in fracture mechanics to demonstrate the fracture resistance of a certain composite material, have been used to summarize the experimental results of these specific experiments. In this experiment, the universal testing machine's integrated software automatically recorded loading and load progression, crack length, and crack advancement.

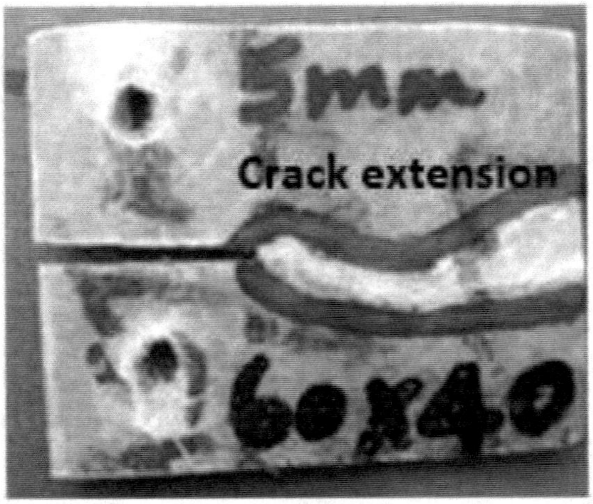

Figure 12. Test results for specimens with a thickness of 5 mm

4.2. Stress intensity factor and strain energy release rate

The compact tension specimen's stress intensity factor was calculated using ASTM D5045-99. Software generates crack and load advancement, and ASTM D5045-99 standard KIC or SIF calculations are used.

$$\text{KIC} = \frac{P}{h/w}\left(f\left(\frac{a}{w}\right)\right) \tag{2}$$

Where f (a/w = 0.5) is a geometrical factor obtained from standard, p is the applied load, h and w are the thickness and width of the hybrid specimen, respectively, and KIC is the critical stress intensity factor. The compliance calibration approach, which uses the concept of compliance as the ratio of deformation to applied load, was also used to determine the critical strain energy release rates. Figure 13(a)–(c) displays the outcomes.

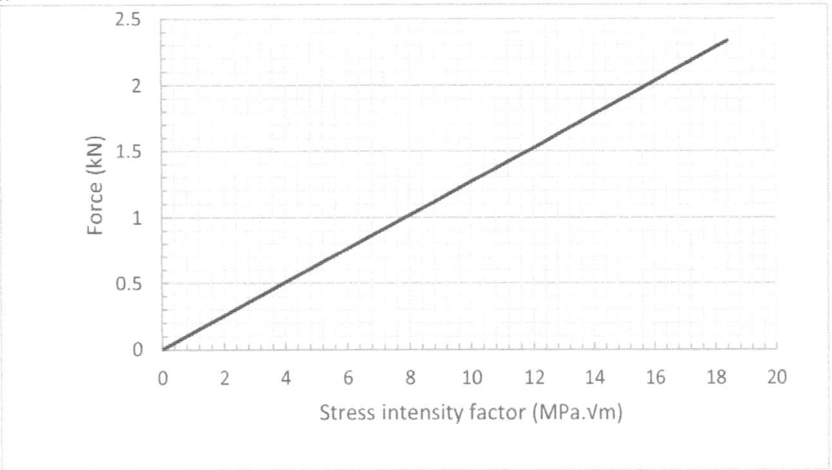

a) Force vs stress intensity factor

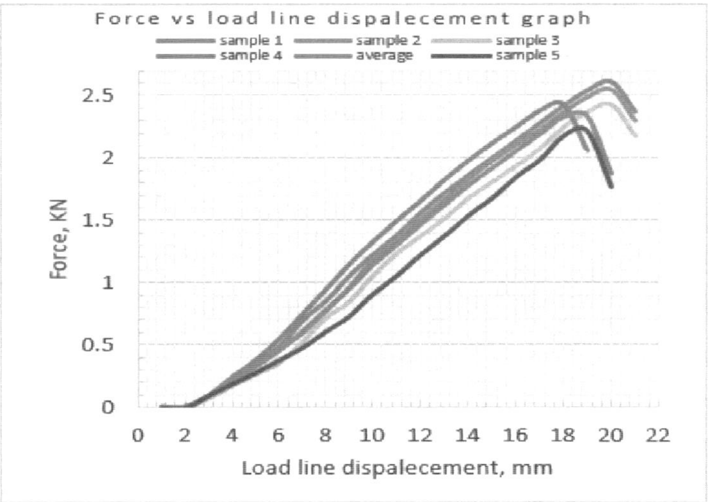

b) Force vs load line displacement

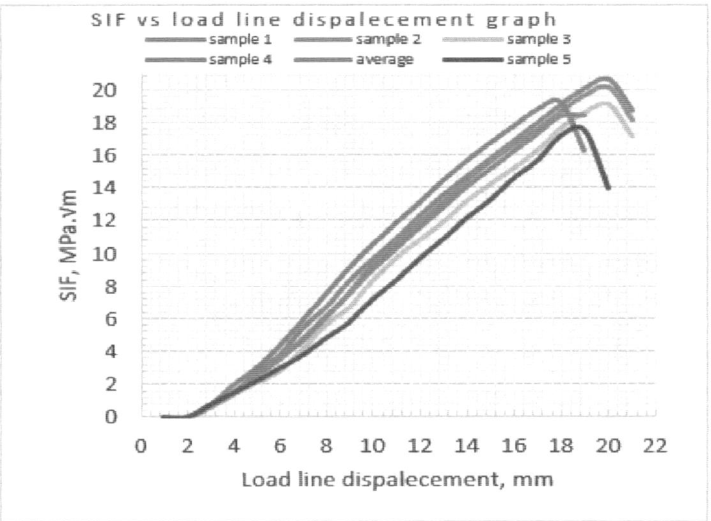

c) Stress intensity factor vs load line displacement

Figure 13. Compact tension test results.

Table 2 summarizes the specimens' average critical stress intensity factor and critical strain energy release rate (GIC), as indicated by the result graphs and the experimental results and analysis.

Table 2. Fracture toughness result.

Material	Specimen geometry	Crack extension (mm)	Max. force (kN)	Fracture toughness K_{IC}	Standard Deviation	G_{Ic} (KJ/m^2)	Standard deviation
SGHC-Compact Tension	Rectangular (5 mm thick)	11.55	2.328	18.34	1.301	13.57	1.485

According to the Yonas Tsegaye experiment on chopped sisal fiber composites, fracture toughness of 5.54 MPa was investigated through varying sisal fiber and epoxy ratios (30/70). Also, Z. Salleh studied mode I fracture toughness of kenaf-woven glass fibers and investigated a value of 252.1 MPa. √mm. Prem Kumar studied the fracture toughness of short banana fiber composites and investigated a fracture toughness value of 2.911 MPa. As shown from the results, the manufactured woven sisal-glass fiber hybrid composite is a tough material and is used for the replacement of some specific materials.

5. Conclusion

To enhance the mechanical and physical characteristics of both natural and synthetic fibers, hybridization of natural and synthetic fibers is crucial. Since defects can occasionally arise during the processing, manufacturing, or servicing of a material or component, fracture toughness is a crucial mechanical attribute. In other words, it gauges a material's ability to withstand fracture in a variety of engineering applications, such as chemical, mechanical, civil, and engineering materials.

Thus, the purpose of this research is to experimentally investigate the mode I fracture toughness behavior of hybrid composites made of woven sisal and E-glass fibers. The experimental work's findings demonstrated that a rectangular specimen with a thickness of 5 mm had a mode I fracture toughness value of KIC = 18.34 MPa. −m. Based on all the comparisons and results, the fabricated hybrid composite specimen is tough and is used to replace some specific materials. It also has a good fracture toughness, demonstrating the performance and applications of the materials in a variety of engineering applications.

Funding: This research was done through the NORHED II project INDMET grant (grant nr. 62862) cooperation whose grant support is highly acknowledged.

References

[1] T. Daniel Gay, 2003, Composite Materials Design and Applications, London: CRC Press LLC.
[2] Joseph R S, Wang J, Ong M C and Jakobsen J B 2017 Vortex induced vibration (VIV) effects of a composite, *Int. Res. J. Eng. Technol.* **3**(4).
[3] P.K. Mallick 2007, Fiber-reinforced composite materials, manufacturing, and design, Third edition, Michigan.
[4] Kedar S. Pandya, Ch. Veerraju and N.K. Naik 2011 Hybrid composites made of carbon and glass woven fabrics under quasi-static loading, *Mater. Des.* **32**(7), 4094 - 4099.
[5] P. S. Shivakumar Gouda 2011, Fracture Toughness of Glass-Carbon (0/90) Fiber Reinforced Polymer Composite: An Experimental and Numerical Study, *J. M. Mater. Charact. Eng.* **10**(8), 671-682.
[6] V. Santhanam 2013, Mode I Fracture Toughness of Banana Fiber and Glass Fiber Reinforced Composite, *Adv. Mater. Res.* **622–623**, 1320–1324.
[7] Shashank T.A. (2016, Evaluation of fracture toughness behavior of glass banana fiber reinforced epoxy hybrid composite, *Int. J. Innovations Eng. Res. Technol.* **3**(3).
[8] S¸ Aziye Arasan 2016, Fracture toughness of woven glass and carbon reinforced gybrid and non hybrid composite plate, *Polym. Compos.* **39**(3), 783 - 793.
[9] Harish K. Patel, 2017, The fracture toughness of natural fiber- and glass fiber-reinforced SMC, *Plast. Rubber Compos.* **46**(8), 355 - 364.
[10] Araya Abera (2018), Experimental investigation of fracture toughness for treated sisal epoxy composite, *AIMS Mater. Sci.* **5**(1).
[11] Prem Kumar Naik 2018, Mode I Fracture characterization of banana fiber reinforced polymer composite, *IOP Conf. Series: Mater. Sci. Eng.* **376**, vol. 376.
[12] Z. Salleh (2012, Fracture Toughness Investigation on Long Kenaf/Woven Glass Hybrid Composite Due to Water Absorption Effect, *Procedia Eng.* **41**, 1667–1673.
[13] Manickam Ramesh, 2016, Influence of fiber orientation and fiber content on properties of sisal-jute glass fiber-reinforced polyester composites, Journal of Applied Polymer Science.
[14] KJ. Wong and B F Yousif 2010, The effects of alkali treatment on the interfacial adhesion of sisal fibers fibers, *Proc. Inst. Mech. Eng. Part L: J. Mater.: Des. Appl.* **224**(3)
[15] D. Zhang 2014 Progressive damage and failure analysis of 3D textile composites subjected to flexural loading, University of Michigan.
[16] J A. Xian – Kui Zhu 2012, Review of Fracture Toughness Testing and Standardization, *Eng. Fract. Mech.* **85**, 1-46.

IOP Conf. Series: Materials Science and Engineering 1294 (2023) 012054 doi:10.1088/1757-899X/1294/1/012054

Artificial neural network model for predicting CO_2 heat pump behaviour in domestic hot water and space heating systems

F S Fadnes[1,2,*], R Banihabib[1] and M Assadi[1]

[1] Faculty of Science and Technology, University of Stavanger, Norway
[2] Norconsult AS, Norway

[*] Correspondence: fredrik.s.fadnes@uis.no

Abstract. The natural refrigerant, CO_2, possesses thermophysical properties that make it highly suitable for domestic hot water (DHW) production using heat pump technology. In this study, the development and validation of an artificial neural network (ANN) model that enables efficient design and control of a CO_2 heat pump is presented. The study employs experimental data from a CO_2 heat pump with a nominal heat capacity of 8 kW. The fully instrumented rig includes the heat pump and a pump rig designed to generate system temperatures representative of various space heat and DHW demands. A comprehensive dataset was generated through systematic variation of inlet temperatures and setpoints. The ANN provides predictions for outlet temperatures, heat production, and electricity consumption utilizing inlet flow rates, temperatures, and setpoints as inputs. These predictions are important for condition monitoring or in a smart operation management framework that determines optimal schedules for the machine.

1. Introduction

As nations and industries strive to meet global climate objectives, the adoption of efficient, standardized, and scalable solutions is vital. The ongoing global energy crisis has accelerated the need for a shift away from fossil fuels [1]. In 2022, nearly half of the energy demand in buildings was allocated to space heating and Domestic Hot Water (DHW), resulting in 2,400 Mt of direct CO_2 emissions, along with an additional 1,700 Mt of indirect emissions [2]. The transition from fossil fuels to renewable energy sources for space heating and DHW is a critical component of achieving a Net Zero Scenario. Although mature technologies such as heat pumps, bioenergy boilers, and district heating are available, their adoption needs to be accelerated [1].

While the demand for heating and cooling can vary based on numerous factors such as climate, geographic location, architectural design, building standards, and occupancy, the requirement for DHW remains relatively stable. As advancements in building design and retrofitting methods significantly reduce thermal energy needs, the demand for DHW remains unchanged [3] and in buildings without hydronic space and ventilation heating systems, there is still a DHW demand. In the US, 18% of home energy consumption is dedicated to DHW, while in the EU, it accounts for 14% [4]. Natural gas (41%), electricity (20%), district heating (13 %), renewables (13 %), petroleum products (11.5%) and solid fuels (1.5%) constitute the energy consumption for DHW in the residential sector in the EU [5]. Efficient and standardized DHW production can play a significant role in the green transition [6].

The heat pump is a green production technology [7] that is becoming increasingly popular in heating and cooling applications [8]. An established guideline for achieving an efficient heat pump system is to minimize the temperature lift between the evaporator and the condenser [9]. Consequently, this principle poses challenges when utilizing traditional heat pumps for DHW production, considering that a safe DHW temperature typically ranges between 60 and 70°C [10], to prevent legionella bacteria growth [11].

Moreover, there are concerted efforts within a growing segment of the heat pump industry to adopt refrigerants that mitigate environmental harm [12]. While heat pumps are generally favoured over electric heaters and fossil-fuelled boilers due to their efficiency and reduction of greenhouse gas emissions [13], the evolution of refrigerants has presented challenges. Early refrigerants were phased out due to their negative impact on the ozone layer. Similarly, hydrofluorocarbons (HFCs) are now being phased out due to their significant greenhouse gas contribution if released into the atmosphere [14]. Even the contemporary "low-GWP" alternatives, like hydrofluoroolefins (HFOs), come with potential issues. Their degradation products, such as trifluoroacetic acid, can pose environmental risks [15]. Natural refrigerants on the other hand, have been found to offer comparable, if not superior, performance to HFCs and HFOs, while significantly reducing the associated environmental drawbacks [16]. Among these, ammonia, propane, and CO_2 stand out as the commercially viable options, each presenting distinct advantages and challenges [17].

Heat pumps that operate with CO_2 as refrigerant have garnered significant interest as a sustainable and efficient alternative for DHW production [18]. Unlike the conventional heat pump processes, which reject heat at a constant temperature through condensation, CO_2-based heat rejection can occur in a supercritical state, releasing heat at a variable or "gliding" temperature. Such a process can reach the necessary temperatures for DHW production while maintaining standard heat pump efficiency [19]. Optimal performance, characterized by a high Coefficient of Performance (COP), is achieved by managing gas cooler pressures to ensure a low outlet temperature for the CO_2 gas; this parameter is significantly influenced by non-linear cooling curves near the critical point [20].

By distributing the heat rejection over multiple gas coolers, the CO_2 heat pump can be adapted for space heating. However, the need for a low gas cooler outlet temperature to achieve high process efficiency may be a challenge when combined with the temperature levels used in traditional space heating [21], which typically range between 30-60°C [22]. The operational conditions of space heating and DHW systems are closely related to the design of gas coolers and the efficiency of a CO_2 heat pump system. For an integrated CO_2 heat pump system with fixed external conditions, there is an optimum CO_2 gas cooler pressure that will maximize the system's COP [20].

While the use of CO_2 as a refrigerant is well-established in specific sectors, including DHW heating in Japan [23] and in supermarket refrigeration [24], its widespread adoption is still limited. Any reluctance to adopt CO_2 as a refrigerant primarily stems from a lack of familiarity, limited experience, and concerns about initial costs [25]. By investigating the challenges of integrating CO_2 heat pumps into a variety of systems, simplified approaches could provide industrial designers and end-users with the tools necessary to improve both design and operational efficiency. This, in turn, could facilitate a wider implementation of CO_2-based systems.

A standard thermodynamic analysis often requires computational efforts, the use of detailed physics-based equations, and a deep understanding of the internal components of the units and systems being modelled. Gathering this specific information can be time-intensive, costly, and sometimes unattainable [26]. Furthermore, many commercial systems have internal controllers whose operations are proprietary, which can complicate the modelling process. In contrast to physics-based modelling, the Artificial Neural Network (ANN), a data-driven approach, has garnered significant attention over the past decade, emerging as a cornerstone of the ongoing Artificial Intelligence revolution [27]. The ANN methodology efficiently captures the complex non-linear relationships between input and output parameters [28].

Related works and contribution: Since Lorentzen "reintroduced" CO_2 as an efficient and sustainable refrigerant in the early 1990s [19], the CO_2 heat pump process has garnered attention from both heat pump and refrigeration engineers as well as researchers. While the application of machine learning methods to describe the CO_2 heat pump process has been explored, it remains a subject with limited extensive literature.

Wu et al. utilized ANN to predict the performance metrics of a CO_2 air-conditioning system's gas cooler [29]. Trained with inlet conditions from both the CO_2 and air-side, the network achieved predictions of heat transfer, outlet temperatures and pressure drop across the gas cooler with deviations less than ±5% from actual measurements. A limitation, however, was its dependency on flow measurements from both gas coolers sides, which due to high cost, is uncommon in real-world heat pump systems [30]. Opalic et al. [30] explored ANN's potential to predict CO_2 mass flow rate and compressor power consumption in an industrial cooling system. Using easily measurable parameters like system temperatures and pressures as inputs, they constructed a model based on compressor polynomials. When tested in a real-world setting, the model achieved a Mean Average Percentage Error (MAPE) below 5%.

Xu et al. employed an ANN to enhance calculations when optimizing the performance of an air-source CO_2 heat pump [13]. Using heat rejection pressure, gas cooler outlet temperature, and compressor displacement as inputs, the ANN predicted the CO_2 mass flow rate, the heat transfer area of the gas cooler, the pressure ratio, electricity consumption, and heat production. In another study, Yin et al. combined Particle-Swarm-Optimization with a neural network to determine the optimal discharge pressure for a CO_2 heat pump. This method achieved a relative error close to 1.5% [31].

Combined DHW and space heating systems based on CO_2 heat pumps demand careful attention to both system design and operational control. Based on experimental data from a CO_2 heat pump rig at the University of Stavanger (UiS), this paper introduces an ANN-based model that effectively predicts the heat pump's performance under various operational conditions. The heat pump features two gas coolers, one for DHW production, the other for space heating. Using inlet conditions and process targets as input, the model predicts the performance and outlet conditions of the machine, relying only on externally measured parameters. This practical approach benefits industry professionals and building owners looking to gain better control of and improve their heat pump systems.

The paper is organized as follows: First a description of the CO_2 process, DHW production and space heating, and the experimental rig is given, before a discussion of the dataset, including the data collection, processing, and the model's structure. Then, the prediction results and their implications are presented and discussed. The paper concludes with a discussion of potential future research and development directions and a summary of the findings.

2. Methods and materials

2.1. The experimental rig at UiS

The experimental setup at UiS is designed to closely mimic a realistic system adaptable to various real-world applications. At its core is an industrial, small-scale CO_2 heat pump, enhanced with extra instrumentation, as illustrated in Figure 1 (a).

The CO_2 heat pump has an 8-kW nominal heating capacity, aligning well with the DHW demands of a medium-sized apartment building equipped with an accumulation system. The inverter-controlled compressor, a DORIN CD300H, offers a nominal cooling capacity spanning 2.2 kW to 5.2 kW. The experimental rig is designed for both DHW and space-heating, featuring two gas coolers connected in series. The first gas cooler (GC-HT) is designed for DHW production, while the second one (GC-LT) is for conventional space heating. A temperature setpoint can be set at each gas cooler. Due to their serial connection, the temperature configuration at the first gas cooler directly impacts the available heat at the second cooler. Care in the selection of setpoints is essential to guarantee optimal performance across both gas coolers. The system can be calibrated to represent DHW production, building space heating, or a combination of the two.

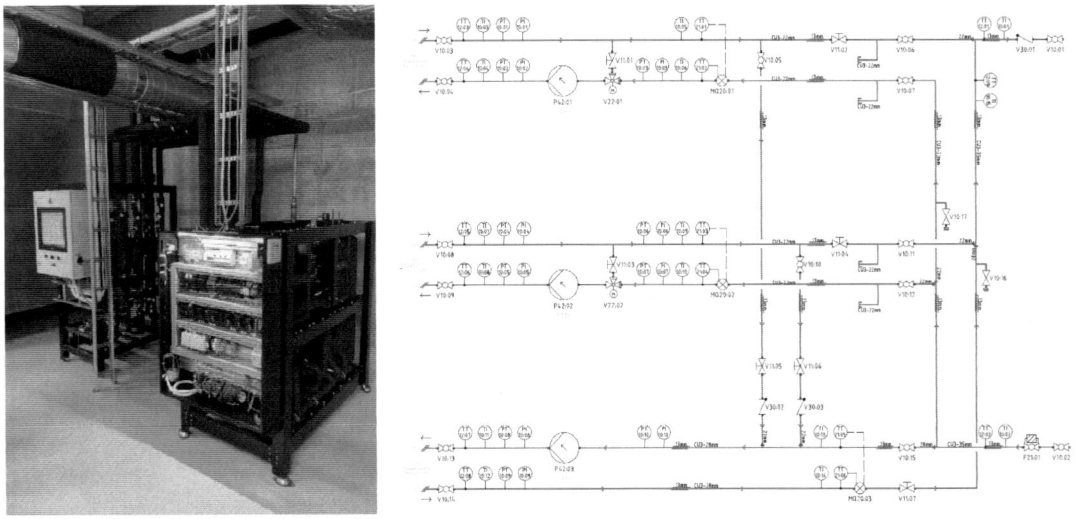

Figure 1. (a) Pump rig and CO_2 heat pump, (b) P&ID pump rig system.

The rig connects to the central cooling infrastructure at the UiS through a cold-water pipe. The temperature of the water this pipe varies between ca. +5°C and +20°C, depending on seasonal variations and the cooling requirements of the campus. The cold-water pipe serves as the heat source for the evaporator and the thermal sink for the gas coolers.

On the water side of the heat pump, each gas cooler and the evaporator are connected to their own dedicated circulation pump, as shown Figure 1 (b). Currently, the mass flow rate across each heat exchanger remains approximately constant. Shunt valves on the water side of the gas coolers enable adjusting inlet temperatures by mixing in hot water produced at the same gas cooler. This feature allows the user to quickly determine the gas cooler outlet CO_2 temperature's influence on available heat and process COP. The only uncontrollable variable in the configuration is the cold-water temperature, which essentially functions as the low-temperature heat source for the heat pump.

2.2. Model structure, Data capture and Processing

The primary objective of this work was to develop a model capable of swiftly predicting the performance of the CO_2 heat pump under varied operational conditions, with a focus on controllable setpoints and changing inlet values. The model was constructed to predict heat production at both gas coolers, the electricity consumption of the heat pump, and outlet temperatures for each heat exchanger. Inputs were flow and temperature to each heat exchanger, along with the heat production setpoint at the GC-LT. DHW was always prioritized with a setpoint of +60 °C at the GC-HT, and therefore this setpoint was not selected as model input.

To gather a comprehensive dataset, a systematic variation of the input parameters was conducted by adjusting the inlet temperatures for each gas cooler and the production temperature setpoint for GC-LT. This process was carried out over a month, from May 11th to June 10th, 2023. For every parameter combination, the system operated for at least 20 minutes, ensuring that it reached and maintained steady conditions. These variations aimed to produce a dataset that mirrors potential operational conditions for a combined building heating and DHW production heat pump. The GC-HT consistently maintained a production setpoint of +60 °C for DHW production, with inlet temperatures varying between +10 °C and +55 °C at 5 K intervals. Meanwhile, the GC-LT operated at setpoints of either +40 °C or +50 °C. Its inlet temperatures started at +10°C and rose in 5 K steps, stopping 5 K below the selected setpoint. This methodology resulted in 130 unique parameter configurations. Figure 2 provides a visual representation of the inlet temperature variations over a day.

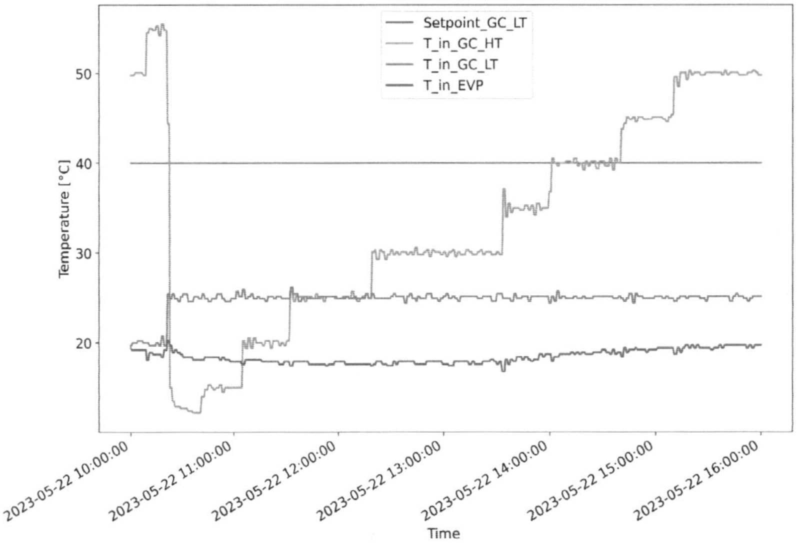

Figure 2. Example of variation of inputs to generate dataset.

In this configuration, the machine consistently prioritizes DHW production. When the GC-HT inlet temperature is low, most of the heat pump output serves DHW production, leaving only a small portion for space heating. This situation represents the heating cold mains water. With a higher GC-HT inlet temperature, less of the heat pump's capacity is available for DHW, and the rest is allocated to space heating. The amount of heat depends on the GC-LT water inlet temperature. Such a scenario can represent the heat pump heating a mix of cold mains water and higher temperature water from an accumulator tank without stratification.

The data were selected and downloaded manually from the digital control system of the rig. During preprocessing, the data were refined by removing outliers, noise cancellation, and resampling to a uniform 30-second interval to ensure consistency in later analysis and modelling.

2.3. ANN and Model Development

A data-driven model for the heat pump operation was developed using the collected data. ANNs were chosen for to their adaptability and ability to handle complex patterns and relationships in the data. Once the ANN model has been trained and set up, it offers a user-friendly tool that engineers or operators can utilize without requiring a deep understanding of its architecture. Users can enter the defined input parameters to model and instantly receive predictions of outlet temperatures and energy consumption. Within an integrated system, the ANN provides high-precision predictions, eliminating the need for potentially time-consuming recalculations. This approach facilitates real-time monitoring and adjustment of the heat pump system, potentially optimizing its performance and energy efficiency. The primary limitation of this method, within the presented framework, is the model's strong dependency on its training data. Values outside the model's training bounds may yield results with notable inaccuracies.

Figure 3 illustrates the architecture of the ANN, including inputs values and predictions. The model includes input and output layers, enhanced by three intermediate hidden layers. In each hidden layer, a designated number of artificial neurons receive input from the preceding layer. These inputs, multiplied by their associated weights, are combined, and then processed through an activation function. The training of an ANN involves the careful adjustment of these weights, which is crucial for efficiently converting system inputs into the desired outputs.

Prior to model training, determining an appropriate model configuration is essential, which includes defining hidden layers parameters such as layer count, the number of neurons per layer, and the activation functions. The training phase further involves the selection and adjustment of various parameters, such as the choice of optimizer and the calibration of the learning rate. These decisions are part of the hyperparameter tuning, which critical for the model's performance. In this work, the Bayesian optimization methodology was employed for hyperparameter tuning, resulting in the configuration shown in Figure 3 (a) that employs the Stochastic Gradient Descent optimizer at a learning rate of 0.37. The model was trained on 80% of the dataset, with the remaining 20% reserved for model validation. This reserved subset, commonly referred to as the test set in machine learning, remained untouched during training, thus facilitating an objective assessment of the model's validity.

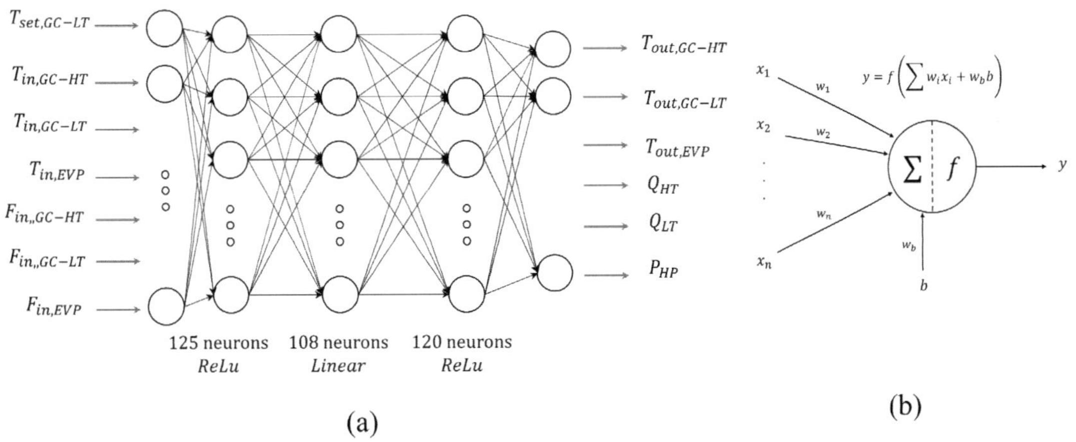

Figure 3. (a) ANN setup, (b) artificial neuron with inputs, weights, activation function, and output.

3. Results

Table 1 presents the error metrics for the test set. The table includes the maximum error, mean absolute error (MAE), root mean square error (RMSE), and mean absolute percentage error (MAPE) associated with the model's predictions across the dataset.

Table 1. Evaluation of model error over the unseen data in the test set.

	Max Error	MAE	RSME	MAPE [%]
$T_{out,GC-HT}$ [°C]	7.18	0.11	0.21	0.2 %
$T_{out,GC-LT}$ [°C]	4.43	0.10	0.17	0.3 %
$T_{out,EVP}$ [°C]	1.27	0.03	0.05	5.7 %
Q_{HT} [kW]	2.80	0.05	0.08	3.3 %
Q_{LT} [kW]	0.93	0.06	0.08	0.5 %
P_{HP} [kW]	1.04	0.02	0.03	2.4 %

The model consistently predicts accurately across all output parameters. The test set, derived randomly from the initial dataset during preprocessing, contains the distinct data points from training, leaving out the in-between values such as those that lie between the 5 K temperature intervals. Given the ANN's ability to interpolate, assessing the model's performance on these intermediate data points becomes essential. Furthermore, due to the random generation of the test set, presenting the model's results over a continuous period becomes challenging. To provide a more comprehensive validation, a secondary test was conducted. Over a 20-hour span, three months after collecting the initial dataset, the system was subjected to four distinct gas cooler inlet temperature configurations. The initial three configurations used temperatures that fell between those of the initial training, while the fourth used setpoints mirroring the temperatures from the original dataset. Details of the inlet temperature variations for this additional validation can be found in Table 2 and Figure 4.

Table 2. Inlet temperature setpoints for model evaluation.

	Inlet temperature HT GC [°C]	Inlet temperature LT GC [°C]
1	28	25
2	32	22
3	47	26
4	35	20

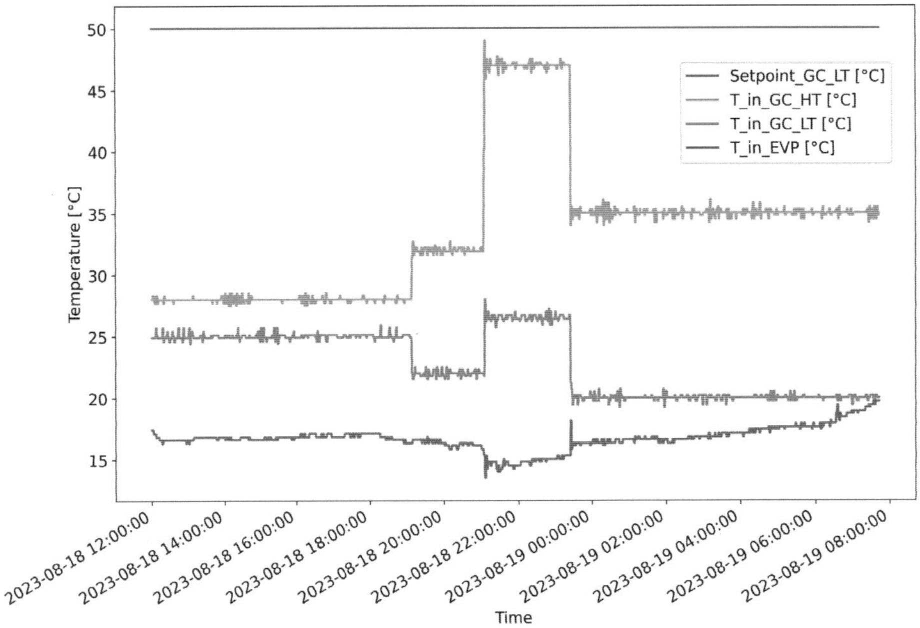

Figure 4. Visualization of inlet temperature during model evaluation.

Table **3** and Figure 5 document the error calculations and predictions when the trained model was applied to the data series from August 2023. The model's predictions for this secondary test set are more than acceptable, closely matching the results observed with the original test set. The most pronounced discrepancies occur when the gas cooler inlet temperatures change rapidly, leading the inlet temperature to oscillate before stabilizing in a new state.

Table 3. Evaluation of model error over the unseen data, August 2023.

	Max Error	MAE	RSME	MAPE [%]
$T_{out,GC-HT}$ [°C]	5.90	0.50	0.61	0.8 %
$T_{out,GC-LT}$ [°C]	2.62	0.29	0.42	0.7 %
$T_{out,EVP}$ [°C]	0.17	0.02	0.03	0.9 %
Q_{HT} [kW]	2.36	0.06	0.10	1.4 %
Q_{LT} [kW]	1.97	0.14	0.18	1.1 %
P_{HP} [kW]	0.17	0.02	0.03	0.9 %

Towards the end of the data series, a noticeable rise in error is evident in both the gas coolers' outlet temperatures and the heat production predictions. This variation can be attributed to the single unpredictable model input: the evaporator's inlet temperature. As illustrated in Figure 4, this temperature, labeled as T_in_EVP [°C], increases rapidly from +16 - 17 °C to around +20 °C, exceeding the temperature range used in the initial training. Figure 6 showcases conditions from the training stage with gas cooler inlet temperatures like those from August 2023. Within the red rectangle, the inlet temperature to the evaporator during training is documented. The temperature was relatively stable, within a range between 15-17 °C, and thus below the values used in the practical evaluation.

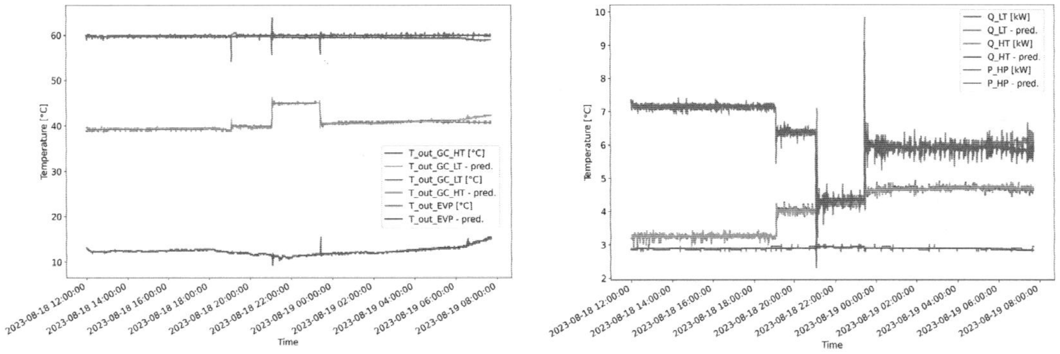

Figure 5. (a) Temperature predictions and, (b) heat and energy predictions, using the final model.

The visual validation emphasizes both the strengths and limitations of the ANN. It demonstrates its effective interpolation capabilities, while also revealing challenges encountered during extrapolation. From a design and operation standpoint, the model offers consistent accuracy for design evaluations and control operations, ensuring the training data aligns with realistic operational parameters. The model's occasional deviations, especially when inlet values are outside its familiar range, suggest caution in less-known conditions. However, these deviations might also underscore the model's potential in pinpointing operational anomalies, indicating its applicability in system malfunction detection.

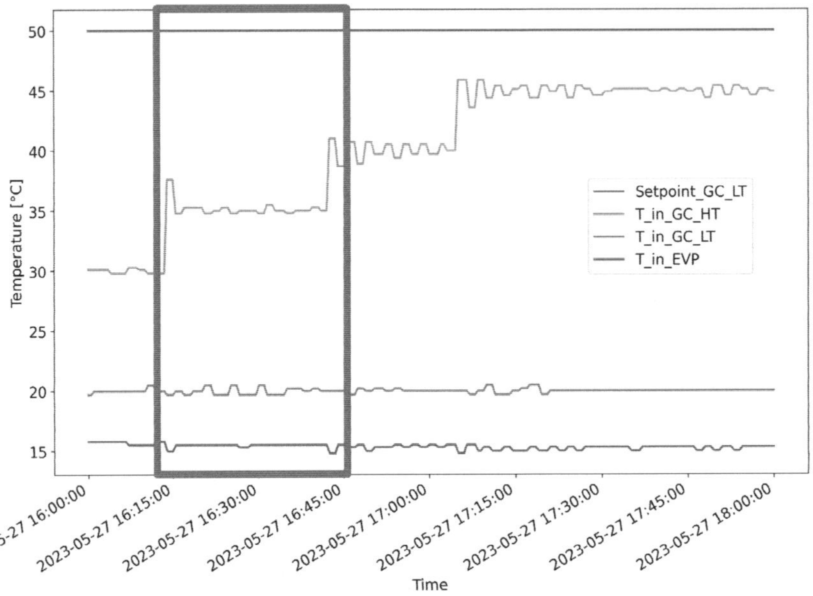

Figure 6. ANN training inputs corresponding to August 2023 data.

A notable limitation of this method is its dependency on large datasets for model development. The extent of data in this study is relatively limited. Up until now, the heat pump has operated only at its peak output, maintaining constant power consumption throughout the training phase. Additionally, only DHW production has been prioritized and the setpoint at GC-LT has been evaluated at two temperatures: +40 °C and +50°C. In different scenarios, the machine could operate without any heat extraction from GC-LT, adopt lower temperature setpoints for both gas coolers, and reduce the compressor frequency to operate at part-load conditions.

4. Further research and development activities

This research serves as a foundation for future developments by presenting a demo case. The model equips both users and designers with a robust tool for swift and precise analysis and control capabilities of the CO_2-heat pump. These are essential for achieving objectives like condition monitoring and the identification of optimum operation by set-point selection.

Fast et al. developed a Graphical User Interface (GUI) for an Artificial Neural Network (ANN) model for an industrial gas turbine [32]. Their Excel-based GUI allowed the plant operators, even those unfamiliar with ANNs, to evaluate the degradation of their gas turbine by comparing the predicted and the real-time data. Their aim was to automate this GUI for real-time condition monitoring. Inspired by Fast et al.'s work, Figure 7 showcases an initial GUI for the CO_2 heat pump investigated in this study. The GUI uses color coding to differentiate between inlet conditions (blue), controllable set-points (yellow), actual measurements (green), and predictions (red). The Coefficient of Performance (COP) is then estimated based on heat production and electricity consumption and compared against actual measurements.

This research may lead to numerous advancements. These range from data-driven demand predictions to system operation optimization influenced by dynamic elements, such as heat demand and electricity pricing. Additionally, there is potential for integration with energy storage, geothermal boreholes, and solar heating systems.

A CO_2 heat pump connected to a stratified storage system can play a significant role in a hybrid microgrid [33], which integrates electric and thermal loads and productions systems [34]. Managing decentralized energy systems like microgrids is becoming increasingly complex, given that energy supply is only partially controllable due to the variable nature of renewable sources [35].

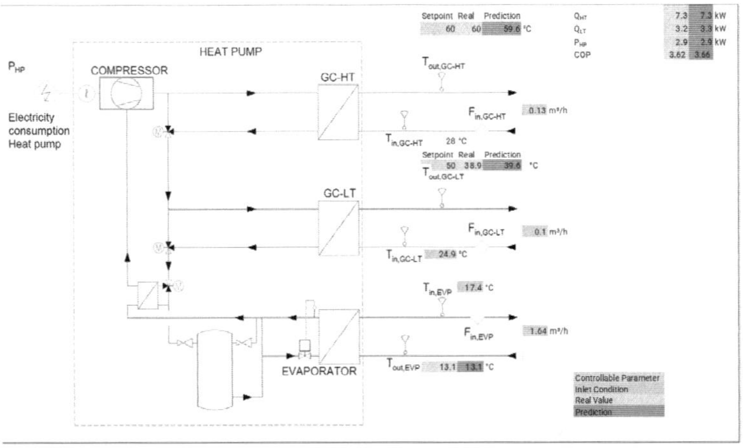

Figure 7. Example of GUI in Excel of the ANN model.

These systems often consist of multiple production units that need to be managed and optimized in a holistic manner. Therefore, the model and methodology presented in this study serve as an initial step toward developing a comprehensive sub-model for a CO_2 heat pump within a microgrid environment.

5. Conclusions

This paper has introduced the development of an ANN model tailored for a small-scale CO_2 heat pump. A profound understanding of the heat pump's behavior under variable inlet conditions provides designers and operators with insights essential for effectively managing these variations and building a robust energy system using the CO_2 heat pump. Employing a set of easy-to-measure parameters, the model predicts the heat pump operations, achieving a MAPE below 6% across all outputs in the test set, with the same or superior accuracy during the practical evaluation. The limitation of the methodology was documented when inlet parameters outside the bounds of the training set were used as input to the model, with predictions deviating from the real operation.

References

[1] IEA, Tracking Clean Energy Progress 2023, IEA, Paris, 2023. [Online]. Available: https://www.iea.org/reports/tracking-clean-energy-progress-2023,
[2] IEA, Heating, Paris, 2022. [Online]. Available: https://www.iea.org/reports/heating
[3] Ivanko D, Walnum H T, and Nord N 2020 Development and analysis of hourly DHW heat use profiles in nursing homes in Norway, *Energy Build.* **222** doi: 10.1016/j.enbuild.2020.110070.
[4] Fuentes E, Arce L and Salom J 2018 A review of domestic hot water consumption profiles for application in systems and buildings energy performance analysis, *Renewable Sustainable Energy Rev.* **81**, 1530-1547, doi: 10.1016/j.rser.2017.05.229.
[5] Sarabia-Escriva E-J, Soto-Francés V-M, Pinazo-Ojer J-M and Acha S 2023 Economic and environmental analysis of domestic hot water systems for single-family homes, *Energy Build.* **286**, doi: 10.1016/j.enbuild.2023.112925.
[6] Pomianowski M Z, Johra H, Marszal-Pomianowska A and Zhang C 2020 Sustainable and energy-efficient domestic hot water systems: A review, *Renewable and Sustainable Energy Rev.* **128**, doi: 10.1016/j.rser.2020.109900.
[7] Liu H, Katsura T and Nagano K 2022 An inverter-driven heat pump with a multi-tubular tube-in-tube heat exchanger for domestic hot water supply, *J. Build. Eng.* **54**, doi: 10.1016/j.jobe.2022.104694.
[8] You T and Wang F 2023 Green ground source heat pump using various low-global-warming-potential

refrigerants: Thermal imbalance and long-term performance, *Renewable Energy*, **210**, 159-173, doi: 10.1016/j.renene.2023.04.058.

[9] Wu J, Sun S, Song Q, Sun D, Wang D, and Li J 2023 Energy, exergy, exergoeconomic and environmental (4E) analysis of cascade heat pump, recuperative heat pump and carbon dioxide heat pump with different temperature lifts, *Renewable Energy*, **207**, 407-421, doi: 10.1016/j.renene.2023.03.028.

[10] Legionellaveilederen - Fordelingsnett til dusjer og andre aerosoldannende tappepunkter 2020 N.-N. I. o. P. Health.

[11] Toffanin R, Curti V, and Barbato M C 2021 Impact of Legionella regulation on a 4th generation district heating substation energy use and cost: the case of a Swiss single-family household, *Energy*, **228**, doi: 10.1016/j.energy.2021.120473.

[12] Abas N, Kalair A R, Khan N, Haider A, Saleem Z, and Saleem M S 2018 Natural and synthetic refrigerants, global warming: A review, *Renewable and Sustainable Energy Rev.* **90**,557-569, doi: 10.1016/j.rser.2018.03.099.

[13] Xu Y, Mao C, Huang Y, Shen X, Xu X, and Chen G 2021 Performance evaluation and multi-objective optimization of a low-temperature CO_2 heat pump water heater based on artificial neural network and new economic analysis, *Energy*, **216**, doi: 10.1016/j.energy.2020.119232.

[14] Brodal E and Jackson S 2019 A comparative study of CO_2 heat pump performance for combined space and hot water heating, *Int. J. Refrig.* **108**,234-245, doi: 10.1016/j.ijrefrig.2019.08.019.

[15] Dudita M and Kauffeld M. Environmental impact of HFO refrigerants &alternatives for the future. Open Access Government. (accessed.

[16] Mancinelli C, Manno M, Salvatori M and Zaccagnini A 2023 Thermal Energy Storage as a Way to Improve Transcritical CO_2 Heat Pump Performance by Means of Heat Recovery Cycles, *Energy Storage and Sav.*, doi: 10.1016/j.enss.2023.06.003.

[17] Ciconkov R and Refrigerants 2018 There is still no vision for sustainable solutions, *Int. J. Refrig.*, **86**,441-448, doi: 10.1016/j.ijrefrig.2017.12.006.

[18] Zhao X-X, He Y-J, Cheng J-H, and Zhang C-L 2023 Long-term performance evaluation of CO2 heat pump water heater under different discharge pressure control strategies, *Appl. Therm. Eng.*, **222**, doi: 10.1016/j.applthermaleng.2022.119918.

[19] Lorentzen G 1994 Revival of carbon dioxide as a refrigerant, *Int. J. Refrig.* **17**(5),292-301,doi: https://doi.org/10.1016/0140-7007(94)90059-0.

[20] Stene J 2004 Residential CO_2 Heat Pump System for Combined Space Heating and Hot Water Heating, Degree of Doktor Ingeniør, Department of Energy and Process Engineering, *NTNU - Norwegian University of Science and Technology Faculty of Engineering Science and Technology*.

[21] Cheng J-H, He Y-J, and Zhang C-L 2021 New scenario of CO_2 heat pump for space heating: Automatic mode switch between modified transcritical and cascade cycle in one system, *Appl. Therm. Eng.* **191** , doi: 10.1016/j.applthermaleng.2021.116864.

[22] Østergaard D S, Smith K M, Tunzi M and Svendsen S 2022 Low-temperature operation of heating systems to enable 4th generation district heating: A review, *Energy*, **248**, doi: 10.1016/j.energy.2022.123529.

[23] Maratou A, Lira J T, Jia H and Masson N 2012 CO_2 heat pumps in europe - market dynamics & legislative opportunities. *REHVA Journal*: Shecco publications.

[24] Karampour M and Sawalha S 2018 State-of-the-art integrated CO2 refrigeration system for supermarkets: A comparative analysis, *International Journal of Refrigeration*, **86**,239-257, doi: 10.1016/j.ijrefrig.2017.11.006.

[25] Market Data Forecast 2023 CO_2 Heat Pump Water Heater Market Research Report - Segmentation by Product (Air Source and Water Source); By Application (Residential, Commercial and Industrial) and Region - Industry Forecast of 2023 to 2028

[26] Deb C and Schlueter A 2021 Review of data-driven energy modelling techniques for building retrofit, *Renewable and Sustainable Energy Rev.* **144**, doi: 10.1016/j.rser.2021.110990.

[27] Abiodun O I, Jantan A, Omolara A E, Dada K V, Mohamed N A and Arshad H 2018 State-of-the-art in artificial neural network applications: A survey, *Heliyon*, **4**(11),00938, doi: 10.1016/j.heliyon.2018.e00938.

[28] Arat H and Arslan O 2017 Optimization of district heating system aided by geothermal heat pump: A

novel multistage with multilevel ANN modelling, *Appl. Therm. Eng.*, **111**,608-623, doi: 10.1016/j.applthermaleng.2016.09.150.

[29] Wu Z G, Zhang J Z, Tao Y B, He Y L and Tao W Q 2008 Application of artificial neural network method for performance prediction of a gas cooler in a CO2 heat pump, *Int. J. Heat Mass Transfer*, **51** (21-22), 5459-5464, doi: 10.1016/j.ijheatmasstransfer.2008.03.009.

[30] Opalic S M, Goodwin M, Jiao L, Nielsen H K, Pardinas A A, Hafner A, Kolhe M L 2020 ANN modelling of CO_2 refrigerant cooling system COP in a smart warehouse, *J. Cleaner Prod.* **260**, doi: 10.1016/j.jclepro.2020.120887.

[31] Yin X, Cao F, Wang J, Li M, and Wang X 2019 Investigations on optimal discharge pressure in CO2 heat pumps using the GMDH and PSO-BP type neural network—Part A: Theoretical modeling, *Int. J. Refrig.* **106**, 549-557, doi: 10.1016/j.ijrefrig.2019.04.027.

[32] Fast M, Assadi M and De S 2009 Development and multi-utility of an ANN model for an industrial gas turbine, *Appl. Energy*, **86**(1), 9-17, doi: 10.1016/j.apenergy.2008.03.018.

[33] Obara S, Nagano K and Okada M 2017 Facilities introduction planning of a microgrid with CO_2 heat pump heating for cold regions, *Energy*, **135**, 486-499, doi: 10.1016/j.energy.2017.06.154.

[34] Khare V and Chaturvedi P 2023 Design, control, reliability, economic and energy management of microgrid: A review, *e-Prime Adv. Electr. Eng. Electron. Energy*, **5**, doi: 10.1016/j.prime.2023.100239.

[35] Reynolds J, Ahmad M W and Rezgui Y 2018 Holistic modelling techniques for the operational optimisation of multi-vector energy systems, *Energy and Build* **169**, 397-416, doi: 10.1016/j.enbuild.2018.03.065.

IOP Conf. Series: Materials Science and Engineering 1294 (2023) 012055 doi:10.1088/1757-899X/1294/1/012055

Day-ahead optimal scheduling of micro gas turbine-based microgrid considering electricity and heating energy

Q Zhang[1,*] and M Assadi[1]

[1] Department of Energy and Petroleum Engineering, University of Stavanger, Stavanger, Norway

[*] Correspondence: q.zhang@uis.no

Abstract. Microgrids are considered as the crucial element in the integration of various distributed energy resources in buildings. They are capable of operating in both grid-connected and islanded mode and have shown immense potential in absorbing renewable energy. However, the widespread implementation of intermittent renewable energy sources, coupled with variable electricity pricing, has significantly increased the operation uncertainty of microgrids. This paper presents an analysis of the operation strategies of an integrated energy system that includes a micro gas turbine, a ground source heat pump, PV panels with the aim of meeting the heating and electricity demands of a commercial building. To facilitate this endeavor, a neural network model for micro gas turbines was developed with a focus on fast computation time and high accuracy in capturing off-design performance. Furthermore, mathematical models for ground source heat pump, PV panel were developed and validated using the Modelica language. Dymola optimization package was utilized to derive the day-ahead scheduling followed by one-hour intervals for the system, with the purpose of minimizing the electricity and heating costs associated with the system. The results demonstrate that the total costs could be reduced by approximately 51% during the analyzed period, indicating a promising avenue for cost savings in the system's operation.

1. Introduction

To address climate change, there is a growing global shift towards renewable energy sources as a sustainable and eco-friendly alternative to fossil fuels. However, the intermittent nature of renewable energy poses a significant obstacle to its consistent and reliable generation. One potential solution to tackle this challenge is the implementation of microgrid. A microgrid is a decentralized energy system that integrates a variety of distributed energy resources and energy storage techniques to meet users' demands, such as heating, cooling, and electricity [1]. It offers the flexibility to operate in both grid-connected and islanded configurations.

The MGT has captured considerable attention worldwide due to its compact size, high power-to-weight ratio, and fuel flexibility. As a dispatchable generation source, MGT can integrate with renewable energy sources, ensuring a reliable energy supply when renewable energy is not available. Extensive investigations have been conducted on MGT-based microgrids. Das et al. [2] examined the techno-economic aspects of an off-grid community's microgrid that consisting of PV, wind turbine, battery, and MGT, aimed to meet the electricity and heating demand. The findings revealed a significant reduction in the initial investment by utilizing waste heat from MGT. Furthermore, the proposed system, capable of meeting both electricity and heating demand, achieved a 40% reduction in CO_2 emissions compared to a system focused solely on electricity. Kumar et al. [3] proposed a microgrid that encompasses PV, battery, MGT, a desalination unit, and a gasifier. This hybrid energy system aims to meet the community's demand for electricity, clean water, and gas. Through techno-economic optimization, the researchers determined that the optimized system could generate 80,423 kWh of electricity and 34,778 MJ of heat annually. Moreover, the recovery of waste heat from MGT allowed the desalination unit to produce 9,516 liters of clean water per day.

Considering the intermittency of renewable energy and the continuously shifting user demands, implementing day-ahead scheduling in microgrids is paramount to efficiently manage energy supply and storage. Pallante et al. [4] conducted a study on day-ahead scheduling for a real office building equipped with a heating system, fan coil power supply network, and lighting network. MATLAB/Simulink was employed to create a physical model of the building. Taking into account weather conditions and hourly energy prices, the study determined optimal schedules for room temperature set points and flow water temperature set points. The results demonstrated significant potential cost savings, ranging from 10% to 28% across diverse scenarios. Reynolds et.al [5] developed a data-driven model for a small office building located in Cardiff, UK, using artificial neural networks. Genetic algorithm was employed to derive the day-ahead scheduling for the room temperature set-point. The study highlighted a significant 25% reduction in energy consumption achieved through optimized scheduling compared to a baseline heating strategy.

This paper examines a day-ahead scheduling strategy for a microgrid that integrates a MGT, PV system, and GSHP. Dynamic simulation models for the primary components have been developed and validated. The genetic algorithm has been employed for the dynamic scheduling. A comparison to a rule-based operation strategy emphasizes the advantages of the proposed approach.

2. System description

Within this study, we investigated a microgrid integrating a MGT, PV system, and GSHP to satisfy both the heating and electricity needs of an office building. The MGT, PV system, and grid power are utilized to address the electricity demand. Additionally, GSHP and the waste heat from the MGT was utilized to meet the heating demand. The schematic of the microgrid is shown in the Figure 1.

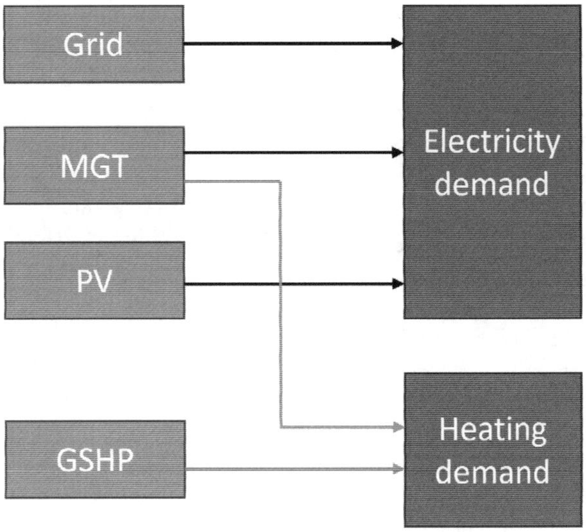

Figure 1. Schematic of the microgrid

The electricity and heating demand profiles for the office building is shown in the Figure 2.

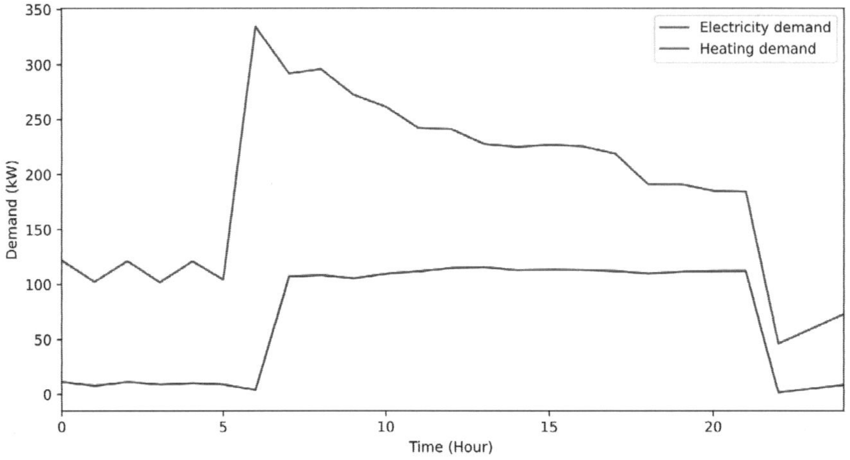

Figure 2. Electricity and heating demand for the office building (24 hours)

3. Modeling and validation

This study employs Modelica, an equation-based object-oriented language, to analyze the dynamic performance of the microgrid. The global Modelica community has made significant contributions by developing well-established open-source packages tailored for modeling and simulating microgrids and building energy systems. Some notable packages include IDEAS [6], Buildings [7], AixLib [8], and BuildingSystems [9].

3.1. Micro gas turbine

Assessing the off-design performance of MGT involves the use of detailed thermodynamic models, which have been extensively developed to analyze the influence of changing operational and environmental conditions. However, these detailed models come with a high computational load, making them impractical for operational optimization. Hence, this study introduces a data-driven model for MGT based on the real-world operation data, which can provide fast computational speed and accurate prediction. Inspired by the brain, an ANN models complex data relationships through weight and bias adjustments. The FFNN, one of the most common types of ANN, processes data unidirectionally from input to output layers. In this study, FFNN is utilized for modeling a MGT incorporating 4 inputs (ambient temperature, ambient pressure, air humidity, and power demand) and 2 outputs (fuel consumption and waste gas temperature). Optimizing internal model parameters, often referred to as hyperparameters, is a crucial aspect of developing effective ANN model. This encompasses parameters like the learning rate, the architectural aspects such as the hidden layer count, neuron number within each layer, as well as the activation functions and optimizer. These hyperparameters are of paramount importance in shaping the model's capacity to acquire knowledge from training data effectively and generalize this knowledge for accurate predictive outcomes. In this study, Table 1 presents an overview of the hyperparameters and their respective ranges.

Table 1: Hyperparameter and range

Hyperparameter	Range
Neurons in the hidden layers	[8 – 128]
Number of hidden layers	[1 – 4]
Activation function	[Relu, Elu]
Optimizer	[AdaGrad, RMSprop, SGD]
Learning rate	[1e-4, 1e-3, 0.01, 0.1]

The search for optimal hyperparameters is conducted via Bayesian optimization, with the results presented in Table 2.

Table 2. Parameter of ANN model for MGT

Parameter	Description
Number of hidden layers	3
Number of neurons of hidden layer 1	33
Activation Function of hidden layer 1	Elu
Number of neurons of hidden layer 2	95
Activation Function of hidden layer 2	Relu
Number of neurons of hidden layer 3	21
Activation Function of hidden layer 3	Relu
Cost function	Mean squared error
Optimizer	AdaGrad

Operation data obtained from Turbec T100, encompassing a variety of ambient conditions and power outputs, were harnessed in developing a data-driven model. The dataset was divided into training (70%), validation (10%), and testing (20%). Figure 3 displays a comparison between the predictions of the data-driven model and testing data that not utilized in the training process. The resulting MAPEs for fuel consumption and waste gas temperature were found to be 0.25% and 0.19%, respectively. More detailed information about modeling and validation of MGT can be found in our previous publication [10].

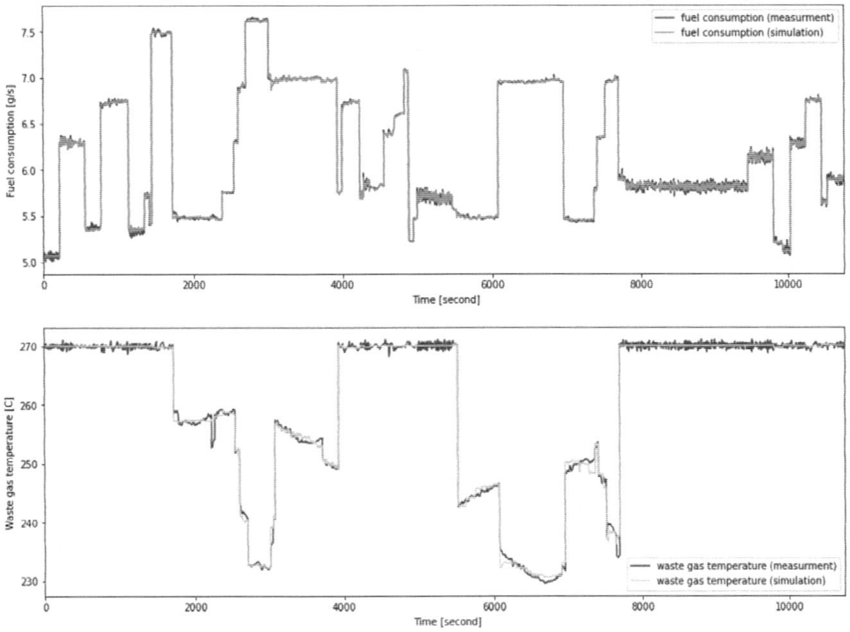

Figure 3. Validation of micro gas turbine

3.2. Ground source heat pump

In the ground source heat pump system, there are two essential components: the borehole heat exchanger and the brine to water heat pump. The borehole heat exchanger utilizes a series of vertical boreholes in the ground to aid in heat transfer. These boreholes are filled with a circulating fluid, commonly a mixture of water and antifreeze, that absorbs heat from the surrounding ground in winter and releases excess heat during summer. The brine to water heat pump is responsible for extracting or injecting heat by the circulating fluid, ensuring effective heating or cooling for the building or system.

The modeling of the borehole heat exchanger involves utilizing MoBTES [11], an open-source package developed using the Modelica language. The TRCM is employed to accurately model heat transfer within the borehole, emphasizing short-term accuracy. Additionally, the finite difference method is adopted to model heat exchange between the borehole and the ground. The accuracy of the borehole heat exchanger model is validated using real-world operation data from UPV, Spain [12]. The borehole heat exchanger consists of six single U-tube boreholes interconnected in a balanced parallel arrangement. Each borehole has a depth of 50 meters, and there is a 3-meter spacing between them. These six boreholes are organized in a rectangular grid layout of 2 by 3, covering a total area of 18 m^2. For additional details, refer to Table 3.

Table 3. Parameter of borehole heat exchanger model

Parameter	Description
Working fluid	Water
Tube outer diameter	0.016 m
Tube inner diameter	0.0127 m
Tube shank spacing	0.07 m
Borehole depth	50 m
Borehole diameter	0.150 m
Borehole number	6
Borehole distance	3 m
Pipe thermal capacity	$1.81\cdot10^6$ J/(m^3.K)
Pipe thermal conductivity	0.4 W/(m.K)
Grout thermal capacity	$3.2e^6$ J/(m^3.K)
Grout thermal conductivity	2.09 W/(m.K)
Ground thermal capacity	$3\cdot2e^6$ J/(m^3.K)
Ground thermal conductivity	2.09 W/(m.K)

A 4-day measurement data period (from September 13th to 16th, 2005) during the GSHP system's cooling mode operation has been selected for validation. The Figure 4 illustrates two boundary conditions for validation: the borehole heat exchanger 's inlet temperature and flow rate. The continuous stop-start intervals of the GSHP operation can also be observed through the flow rate of the borehole heat exchanger. The accuracy of developed model is established by comparing the predicted borehole heat exchanger outlet temperature with the measured data. The comparison result is represented in the Figure 4, indicating a good agreement between the model's predictions and the actual measurements. During the off periods of the GSHP system, some deviations are observed, however, these have no impact on the calculation of the GSHP system's heating capacity and energy consumption. Excluding these off periods, a statistical analysis yields a low MAPE of 2.81% when comparing the measured and simulated outlet temperatures.

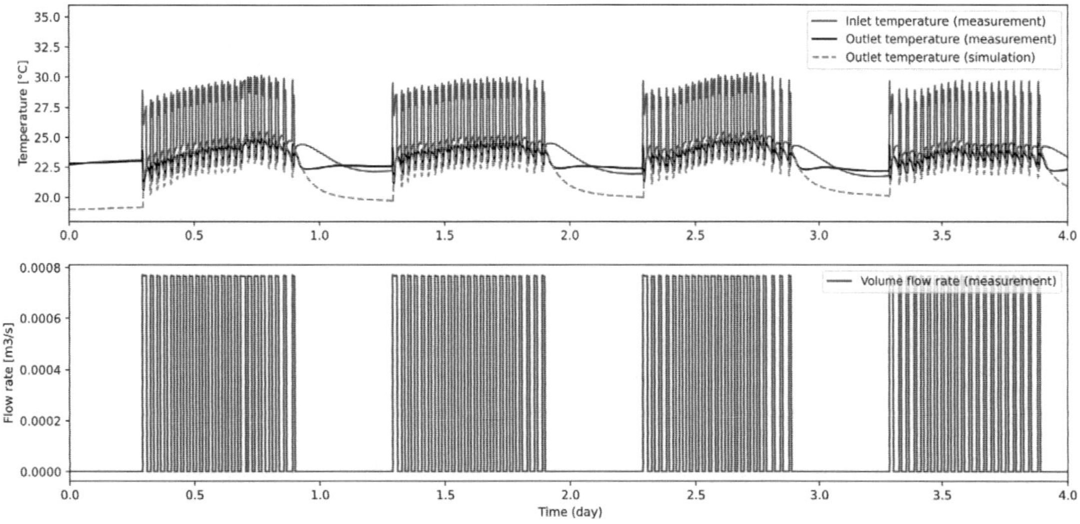

Figure 4. Validation of borehole heat exchanger model

For modeling the brine to water heat pump, a regression model based on performance data obtained from manufacturers is employed. This regression model considers key input parameters: source side brine inlet temperature, brine flow rate, load side water inlet temperature, and water flow rate. With this model, power consumption and heat capacity can be predicted. To demonstrate the model's accuracy, a validation study has been conducted using a 130 kW heat pump. In Figure 5 and Figure 6, we present a comparative analysis of the simulation results and the catalog data concerning the heating capacity and power consumption of heat pump. The visual representation clearly indicates a close match between the simulation results obtained from the regression model and the catalog data, with a relative error of less than 5% for both heating capacity and power consumption. Detailed information can be found in our previous publication [10].

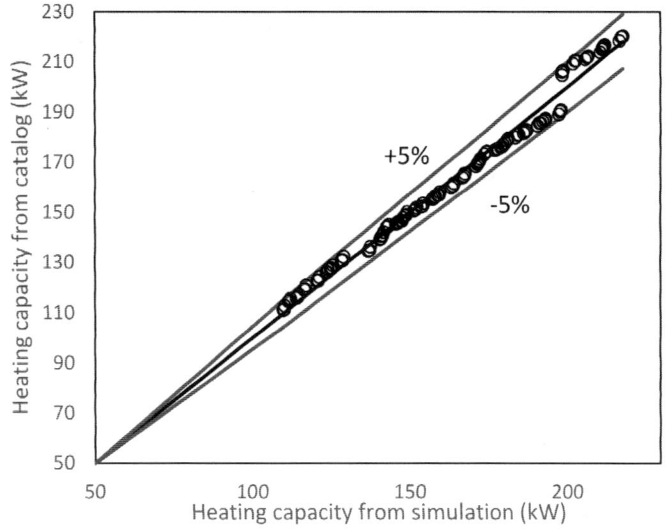

Figure 5. Validation of heating capacity of heat pump

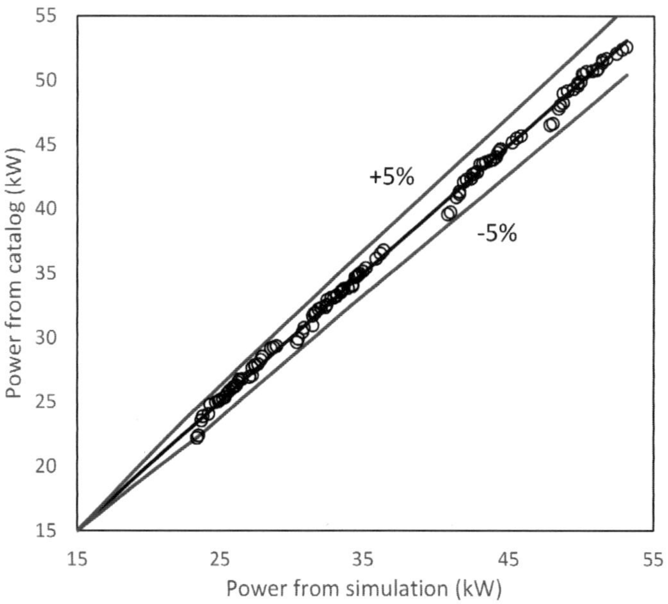

Figure 6. Validation of power consumption of heat pump

3.3. *PV panel*

AixLib, an open-source package developed in the Modelica language, is employed to model PV systems. This model enables accurate prediction of the nonlinear electrical behavior of PV systems, which are affected by variations in cell temperature and solar irradiance. The Single-Diode method [13] is employed to model the physical behavior of the PV cell, while an empirical approach is utilized to model the cell temperature based on the specific mounting types of PV.

To validate the model's accuracy, operation data collected from a PV system installed on the rooftop of the NIST Campus building is employed. The PV array is comprised of 312 Sharp NU-U235F2 modules, with a rated power output of 73.3 kW. Each panel is oriented southwards and fixed at a 10-degree tilt angle. One week of operation data, ranging from June 6, 2016, to June 13, 2016, is utilized to illustrate the model's accuracy. By utilizing solar irradiance, wind speed, and ambient temperature, the model can predict power generation from the PV system. Figure 7 demonstrates a good agreement between the predicted and measured power generation. The resulting MAPE is 6%.

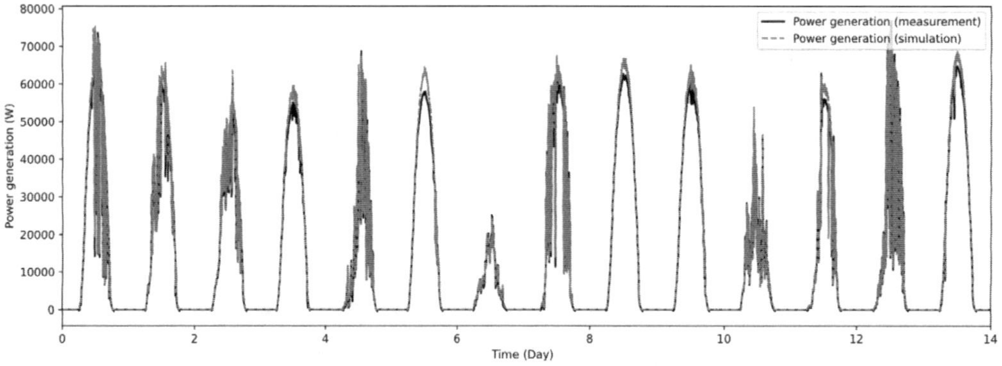

Figure 7. Validation of PV system model

4. Optimization

Day-ahead scheduling is a critical aspect of microgrid operations, enabling efficient energy resource management by planning generation and storage, considering variable load and energy prices. Day-ahead scheduling allows microgrids to harness the benefits of price fluctuations, redistributing user load to low-price periods. Furthermore, day-ahead scheduling promotes the integration of renewables like solar and wind, aligning with sustainability targets.

In this study, the optimization objective is to minimize operation costs, including both the cost of natural gas for the MGT and the cost of purchasing electricity. Calculating the operation cost involves utilizing the prices of electricity and natural gas. Figure 8 displays the hourly electricity prices. In this study, the cost of natural gas is set at 0.75 NOK/kWh.

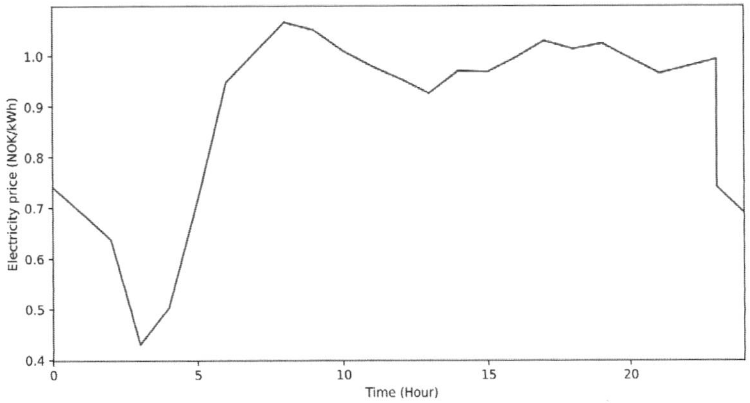

Figure 8. Electricity price for 24 hours

The operation cost can be represented as follows:

$$Cost = \int_{t=0}^{t=24} E_{grid} \cdot P_e dt + \int_{t=0}^{t=24} Fuel_{MGT} \cdot P_g dt$$

Where:
$Cost$ – operation cost of the microgrid (NOK)
E_{grid} – Electricity purchased from grid (kWh)
P_e – Price of electricity (NOK/kWh)
$Fuel_{MGT}$ – Fuel consumption of MGT (kWh)

The optimization variable in this study is the capacity and on-off status of the MGT. The optimization is carried out at one-hour intervals, which results in a 24-hour capacity time profile. This time profile is essential for effectively managing the energy generation and consumption patterns over the course of a day, considering varying demands and available resources. To achieve this, we employ the Dymola optimization package, a powerful tool that supports multiple optimization algorithms. In this case, the genetic algorithm is chosen as the optimization algorithm due to its capability to navigate and search through an extensive and complex solution space effectively. Genetic algorithms mimic the process of natural selection and evolution, making them adept at finding nearly optimal solutions in a computationally efficient manner.

5. Results and discussions

Day-ahead scheduling for the microgrid involved computing the capacity for the MGT, PV system, and purchased grid electricity for every hour. A comparison of performance was made by evaluating a baseline case utilizing rule-based operation.

5.1. Day-ahead scheduling of the microgrid

The genetic algorithm was utilized to determine the optimization variables—specifically, the MGT's capacity and its on-off status for each hour, illustrated in Figure 9. At the start of the day, MGT is switched on at 7 am, with a capacity of 108 kW. Shortly thereafter, MGT undergoes shutdown. In the afternoon, MGT operates for several hours. During the night, specifically at 8 pm, MGT is turned on.

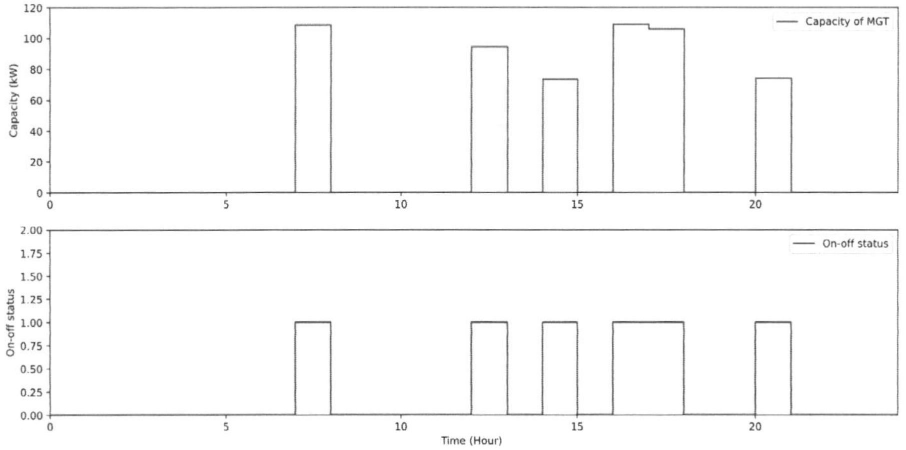

Figure 9. Optimization results for capacity and on-off status of MGT

The day-ahead scheduling results for the microgrid are depicted in Figure 10. The PV system contributes to electricity generation during the day's peak when solar irradiance is at its highest. MGT operations at 7 am, in the afternoon, and at 8 pm. Any additional demand is supplied through purchased electricity from the grid. Utilizing day-ahead scheduling, an operational cost of 3965 NOK is achieved.

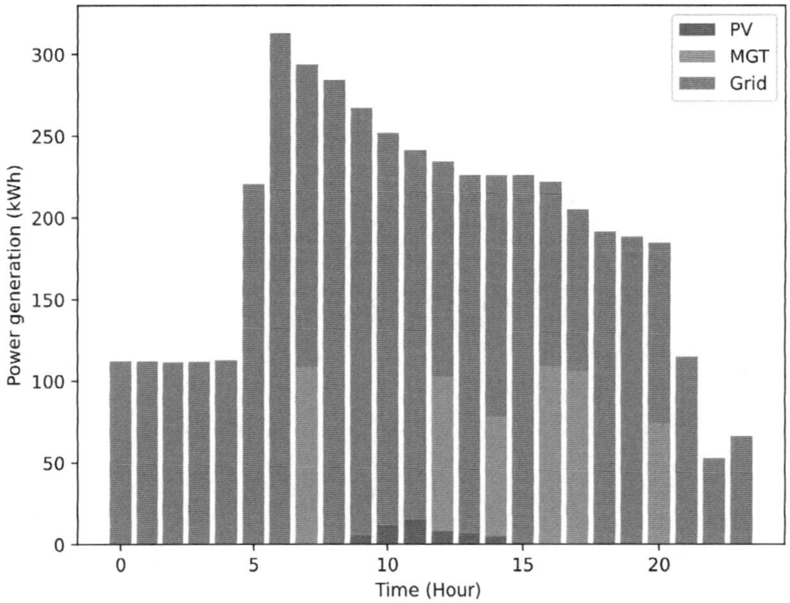

Figure 10. Day-ahead scheduling of the microgrid

5.2. Comparison with rule-based operation

A comparison was made between the day-ahead scheduling strategy and a rule-based operation approach for the microgrid. Under this rule-based strategy, the priority is to optimize electricity generation through the PV system. When the PV capacity proves insufficient, the MGT steps in to generate electricity and satisfy demand. If the MGT operates at its peak capacity and demand persists, grid electricity is acquired to meet the energy needs. Similarly, the waste heat produced by the MGT is utilized to meet the heating needs. If the waste heat proves inadequate, the GSHP operates to meet the remaining heating demand.

The rule-based operation strategy is applied to the microgrid, considering same electricity and heating demands. This approach results in an operational cost of 8116 NOK. Notably, the implementation of a day-ahead scheduling strategy showcases a significant 51% reduction in operational costs.

6. Conclusions

A day-ahead scheduling for the microgrid, integrating a MGT, PV system, and GSHP, is conducted to meet the heating and electricity demands of the office building. Key conclusions from this study can be outlined as follows:

- The models for the primary components within the microgrid were developed and validated. The MAPE for the borehole heat exchanger model was recorded at 2.81%, while the MAPE for the PV system model stood at 6%.
- The microgrid's operation cost was minimized by optimizing the capacity and on-off status of the MGT using a genetic algorithm, considering the provided demand profile and energy price. The optimal scheduling obtained resulted in a daily operation cost of 3965 NOK.
- A comparison was conducted between a rule-based operation strategy and a day-ahead scheduling strategy for the microgrid. The comparison revealed that the day-ahead scheduling strategy achieved a 51% reduction in operation costs compared to the rule-based operation strategy.

Abbreviation

ANN – Artificial neural network
CO_2 – Carbon dioxide
FFNN – Feed-forward neural network
GSHP – Ground source heat pump
MAPE – Mean absolute percentage error
MGT – Micro gas turbine
PV – Photovoltaic
TRCM – Thermal resistance and capacity model

References

[1] Gu W, Wu Z, Bo R, Wei L, Zhou G, Chen W and Wu Z 2014 Modeling, planning and optimal energy management of combined cooling, heating and power microgrid: A review, *Int. J. Electr. Power Energy Syst.* **54**, 26–37. https://doi.org/10.1016/j.ijepes.2013.06.028

[2] Das BK, Tushar MSHK and Zaman F 2021 Techno-economic feasibility and size optimisation of an off-grid hybrid system for supplying electricity and thermal loads, Energy, ***215***,119141. https://doi.org/10.1016/j.energy.2020.119141.

[3] Kumar N and Karmakar S 2022 Techno-economics of a trigeneration HRES; a step towards sustainable development, Renew Energy **198**:833–40. https://doi.org/10.1016/j.renene.2022.08.107.

[4] Pallante A, Adacher L, Botticelli M, Pizzuti S, Comodi G and Monteriù A 2020 Decision support methodologies and day-ahead optimization for smart building energy management in a dynamic pricing scenario, *Energy and Buildings,* **216**, 109963. https://doi.org/10.1016/j.enbuild.2020.109963

[5] Reynolds J, Rezgui Y, Kwan A S K and Piriou S 2018 A zone-level, building energy optimisation combining an artificial neural network, a genetic algorithm, and model predictive control, *Energy*, **151**, 729–739. https://doi.org/10.1016/j.energy.2018.03.113

[6] Jorissen F, Reynders G, Baetens R, Picard D, Saelens D and Helsen L 2018 Implementation and verification of the IDEAS building energy simulation library, *J Build Perform Simul* **11**:669–88. https://doi.org/10.1080/19401493.2018.1428361.

[7] Wetter M, Zuo W, Nouidui TS, and Pang X 2014 Modelica Buildings library, *J Build Perform Simul*, **7**:253–70. https://doi.org/10.1080/19401493.2013.765506.

[8] Müller D, Lauster M, Constantin A, Fuchs M and Remmen P AixLib - An Open-Source Modelica Library within the IEA-EBC Annex 60 Framework.

[9] Nytsch-Geusen C, Huber J, Ljubijankic M and Rädler J 2013 Modelica BuildingSystems – eine Modellbibliothek zur Simulation komplexer energietechnischer Gebäudesysteme. Bauphysik **35**,21–9. https://doi.org/10.1002/bapi.201310045.

[10] Zhang Q, Banihabib R, Fadnes F, Sazon T, Ahmed N and Assadi M 2023 Techno-economic analysis of a biogas-fueled micro gas turbine cogeneration system with seasonal thermal energy storage. *Energy Conversion and Management*, **293**, 117407. https://doi.org/10.1016/j.enconman.2023.117407

[11] Formhals J, Hemmatabady H, Welsch B, Schulte D, and Sass I 2020 A Modelica Toolbox for the Simulation of Borehole Thermal Energy Storage Systems, *Energies* **13**:2327. https://doi.org/10.3390/en13092327.

[12] Ruiz Calvo F and Montagud C 2014 Reference data sets for validating GS HP system models and analyzing performance parameters based on a five year operation period, *Geothermics*, **51**, 417 428. https://doi.org/10.1016/j.geothermics.2014.03.010

[13] Batzelis, E., & Papathanassiou, S. A. (2016). A method for the analytical extraction of the Single-Diode PV model parameters, *IEEE Trans. Sustainable Energy*, **7**(2), 504–512. https://doi.org/10.1109/tste.2015.2503435

IOP Conf. Series: Materials Science and Engineering 1294 (2023) 012056 doi:10.1088/1757-899X/1294/1/012056

Assessing impact of borehole field data's input parameters on hybrid deep learning models for heating and cooling forecasting: A local and global explainable AI analysis

N Ahmed[1,2,*], M Assadi[1], Q Zhang[1] and A A Ahmed[1]

[1] Department of Energy and Petroleum Engineering, University of Stavanger, 4021 Stavanger Norway
[2] US Pakistan Centre for Advanced Studies in Energy (USPCASE), National University of Sciences & Technology (NUST), H-12 Sector, Islamabad, Pakistan

[*] Correspondence: Naveed.ahmed@uis.no

Abstract. Achieving accurate performance forecasting of borehole heat exchanger is essential for optimizing ground source heat pump systems, enabling optimal control, and facilitating energy-efficient operations with enhanced sustainability of the built environment. This study aims to investigate and quantify the impact of model architecture, the number of input data sensors, and their accurate identification on multivariate hybrid deep learning models. Moreover, the significance of incorporating a recent development in deep learning to pay selective attention to the input data i.e., attention-based mechanisms in LSTM-CNN and CNN-LSTM architectures is also investigated. The significance of input parameters for the data-driven AI models is assessed through a significance interpretability analysis utilizing Explainable-AI local-method, namely Shapley Additive Explanations and global-explanation methods i.e., permutation feature importance method and Friedman statistical test. The findings highlight the efficacy of attention mechanisms in capturing temporal dependencies in LSTM-CNN-At and spatial patterns in CNN-LSTM-At, may not necessarily enhance their multistep forecasting capabilities for the borehole field data in comparison to LSTM-CNN architecture. The 24 hours ahead forecasting results show that the order of accuracy is LSTM-CNN> LSTM-CNN-At> CNN-LSTM> CNN-LSTM-At. The findings emphasize that by carefully designing the model layers, it is feasible to remove redundant borehole field sensors for data measurement while maintaining the forecasting accuracy of the hybrid data-driven models.

1. Introduction

The building sector's heating demands contribute significantly to global CO_2 emissions and consume a substantial amount of energy. To foster the transition towards safe and sustainable heating and cooling solutions on a global scale, borehole coupled heat pumps (BCHPs) play a vital role [1]. Ensuring accurate performance forecasting of borehole heat exchangers (BHEs) becomes crucial for optimizing ground source heat pump (GSHP) systems. By achieving precise forecasts, system design can be optimized, energy usage can be reduced, and overall system reliability can be enhanced, thus fostering the sustainability of the built environment [2]. The performance of GSHP systems is influenced by several factors, including mass flow rates, heat transfer fluid temperatures and variable weather conditions. Consequently, effective control measures encompassing both demand-side management and the geothermal heat source are necessary to minimize overall energy consumption [3]. Accurate forecasting of GSHP energy output becomes imperative to curtail energy waste and facilitate the development of reliable models for assessing the thermal characteristics of BHEs.

In current work, deep learning models are investigated due to their adaptability in handling complex, multivariable borehole heat exchanger systems. In such scenarios with dynamic data conditions, deep neural networks excel by learning intricate relationships directly from data, making them well-suited for real-time monitoring and forecasting of non-linear systems. Achieving accurate forecasting with data-driven machine learning models is influenced by various factors, including input variables, training algorithms, exogenous variables, and model architecture [4]. Hybrid deep neural networks like LSTM-CNN and CNN-LSTM are known for their complexity and tend to outperform simpler models (e.g., Long Short-Term Memory (LSTM), Convolutional Neural Networks (CNN) random forest, linear regression, decision trees) in delivering accurate predictions. However, the interpretability of these multi-layered networks poses a challenge as they often behave as "black-box" models, making it difficult to comprehend the rationale behind specific predictions [5]. In real-world heating and cooling applications, the interpretability of simpler models instills greater trust among users. Therefore, achieving a balance between performance and interpretability is crucial for developers when deploying deep learning models [6]. This research confronts the significant challenge of identifying key input variables that reliably forecast heating and cooling temperatures in borehole heat exchangers based on seasonal building loads.

Individual algorithms, such as LSTM and 1-D CNN, have demonstrated effectiveness in various cases [7–9]. However, the rising popularity of hybrid deep learning models, such as LSTM-CNN and CNN-LSTM, can be credited to their capability of capturing both temporal and spatial dependencies within input datasets, leading to enhanced accuracy in heating and cooling load forecasting. In the study by Zhang et al. [10], a CNN-LSTM model was developed to predict the outlet temperature of borehole heat exchangers by considering spatial-temporal features. The model was trained using soil temperature field, inlet and outlet water temperatures. The proposed model exhibited higher prediction accuracy, with an average coefficient of determination (R2) value that surpassed the ANN, LSTM, Simple-RNN, and CNN algorithms by 3.54%, 2.35%, 5.75%, and 4.60%, respectively. In another investigation, Yao et al. [11] introduced a combined approach that involved a differentiated CNN-LSTM for short-term heat demand prediction in district-level heating systems. The objective of this approach was to forecast heat demand half an hour ahead, utilizing solely historical heat data. In the realm of building load prediction, Explainable Artificial Intelligence (EX-AI) plays a crucial role in enhancing user trust and providing transparency for hybrid deep neural networks. Research conducted by Tsoka and fellow researchers [12] has highlighted the effectiveness of EX-AI methods in reducing the required number of input features for accurate load forecasting while preserving accuracy. Chung et al. [13] investigated the input parameters for EX-AI-based load forecasting in office buildings across different climate regions. By utilizing local feature importance methods of SHAP and LIME, the study revealed improved identification of essential input variables compared to global explainable methods.

Determining the optimal selection of sensors from the borehole field for deep machine learning forecasting models can be a challenging task due to the diverse influences of building characteristics, weather variables, and internal heat-gain profiles on building load variability. The extensive investigation of deep machine learning models has shed light on their performance; however, the influence of model architecture, the number and accurate identification of input sensors on the performance of hybrid algorithms for heating and cooling multistep forecasting remains challenging. This study aims to examine and quantify this impact while exploring the significance of incorporating attention-based mechanisms in LSTM-CNN and CNN-LSTM architectures. The effectiveness of attention mechanisms in capturing temporal dependencies in LSTM-CNN-At and spatial patterns in CNN-LSTM-At is analyzed, with a specific focus on identifying essential input parameters for the deep AI models using Explainable-AI methods at both local and global levels. The implications of this research are significant, as the proposed multistep prediction hybrid model (LSTM-CNN) can contribute to industry applications such as 24-hour ahead prediction of BHE output, demand-side management of heating and cooling, and building operations.

2. Methodology

This study aims to investigate the impact of model architecture, the number of input sensors, and their accurate identification on the performance of hybrid algorithms for multistep forecasting in ground source heat pump applications. The methodology, illustrated in Figure 1, involves a series of steps starting with data extraction from real borehole heat exchanger installations. The collected data undergoes cleaning and pre-processing before training and validating hybrid deep learning models to achieve the desired forecasting accuracy. The constructed hybrid deep models, including LSTM-CNN, CNN-LSTM, and attention-based models LSTM-CNN-At and CNN-LSTM-At, are trained, validated, and tested using varying numbers of independent input parameters in three different variations of dataset (Variation-1, Variation-2, Variation-3). The trained deep learning models are then utilized to predict the performance of borehole heat exchangers 24 hours in advance, ensuring efficient management of thermal loads. Furthermore, the importance of input parameters for the hybrid deep models is determined through an interpretability analysis using Local and Global Ex-AI techniques.

2.1 Pre-processing and cleaning of dataset

In the current investigation three variations of datasets with different numbers of input parameters were utilized to train and test the developed deep learning algorithms, as outlined in Table 1. These datasets consisted of essential parameters derived from real BHE installations, ensuring their relevance in training the AI models. Specifically, the dataset used in this study was sourced from a geothermal installation situated at the Gale Bullman Multi-purpose building in Missouri, USA [14]. It encompassed 144 boreholes, each with a depth of 122m, which were interconnected to a large-scale GSHP system. The collected data spanned approximately one year, recorded at 15-minute intervals. To ensure data quality, a thorough cleaning process was implemented, resulting in a dataset devoid of significant outliers. Any identified outliers were removed and replaced with interpolated values. Furthermore, supplementary weather data including wind speed (V_{wind}), solar radiation (R_{sol}) and ambient temperature (T_{amb}) were obtained from a reputable US weather service provider [15].

Table 1. Different variations of data with different number of input parameters

Datasets	Names of input-parameters	Input-parameters
Variation-1	T_{out}, T_{in}, T_{amb} , m^0	4
Variation-2	T_{out}, T_{in}, T_{amb}, m^0, R_{sol}	5
Variation-3	T_{out}, T_{in}, T_{amb}, m^0, R_{sol},V_{wind}	6

The identification and handling of outliers in these datasets were performed using three standard deviation method[16]. The effectiveness of the data cleaning procedure enhanced data quality with significantly reduced noise levels over time. To assess the accuracy of the hybrid models developed in this study, two commonly used evaluation metrics, Mean Squared Error (MSE) and Mean Absolute Percentage Error (MAPE), were employed and defined by the following equations:

$$MSE = \frac{1}{n}\sum_{i=1}^{n}(\hat{y}_i - y_i)^2 \quad (1)$$

$$MAPE = \frac{100}{n}\sum_{1=1}^{n}\left|\frac{\hat{y}_i - y_i}{y_i}\right| \quad (2)$$

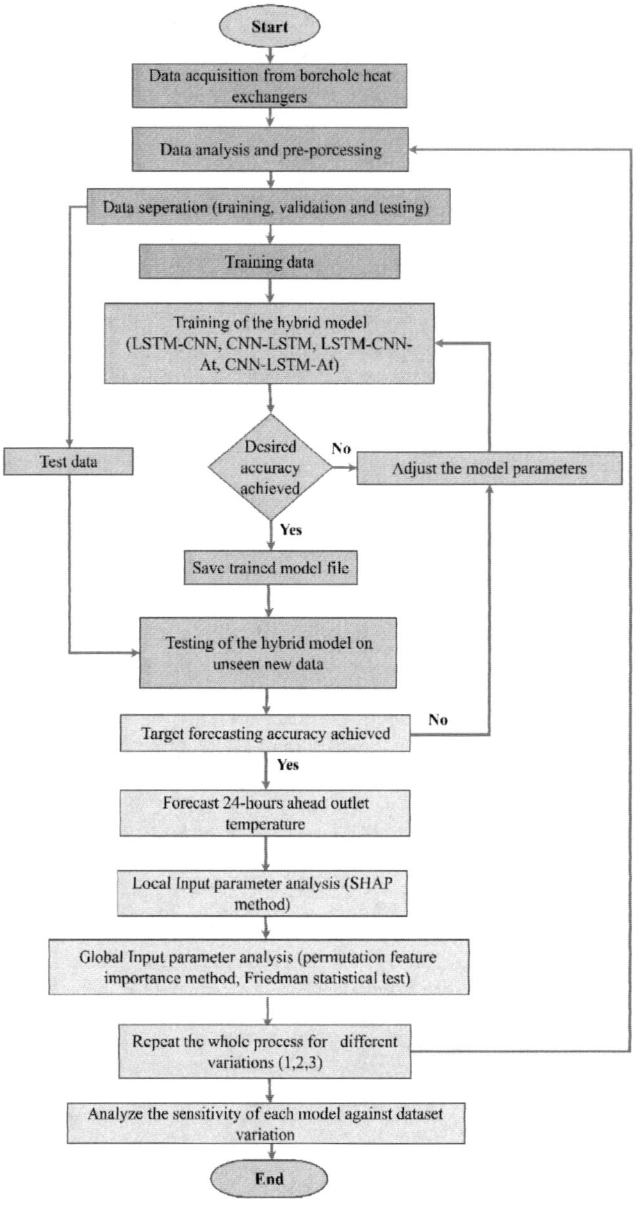

Figure 1. Overview of the implemented methodological approach for the current work

In the equations, n represents the total number of samples present in the dataset, y_i signifies the actual observed value at the ith time instance, and $\hat{y}_i$ denotes the corresponding predicted value

2.2 Feature importance analysis using local and global level methods

To determine the most crucial input parameters from BHE installations for accurate multistep forecasting of hybrid AI models and enhance their interpretability, three distinct methods are employed: a local-level method called Shapley Additive Explanations and two global-level methods, namely the permutation feature importance method and the Friedman statistical test. The SHAP method emplyed in the current work is a local interpretability approach providing insights into the contribution of individual input features (T_{out}, T_{in}, T_{amb}, m^0, R_{sol},V_{wind}) in generating predictions. In addition to the local-level analysis, the permutation feature importance method is used to assess the importance of input parameters on a global scale. This technique involved systematically permuting the values of all input parameters and evaluating the resulting influence on the model's forecasting performance. By measuring the change in model performance metrics (RMSE, MAPE), the relative importance of each input parameter was investigated in the overall model performance. Moreover, the Friedman statistical test is employed, which is a widely-used non-parametric test, to compare the ranks of the input parameters based on their influencing effect on the model's prediction capabilities. Collectively, these three methods provide a comprehensive insights into the significance and influence of input parameters from borehole installations on the deep learning hybrid AI models.

2.3 Hybrid deep learning models

In this study four distinct deep machine learning architectures are employed to predict the fluid outlet temperature of a borehole heat exchanger. These architectures are trained and tested using borehole field data, considering various numbers of input parameters, to forecast the outlet temperature for the next 24 hours. Each of these developed algorithms possesses its unique set of strengths and limitations, which are critical to consider. The LSTM-CNN architecture as shown in Fig. 2 (a), combines the Long Short-Term Memory network and Convolutional Neural Network layers. The LSTM layers capture temporal dependencies in the input sequence, while the CNN layers extract spatial features from the data. This architecture can effectively model both temporal and spatial patterns in the data, making it suitable for capturing complex relationships in time-series and spatial data. The LSTM-CNN architecture can be computationally expensive and may require a large amount of training data for optimal performance. Interpretability of the model can be challenging due to its complex nature. Whereas CNN-LSTM construction has the oppositely arrange layers combining the CNN layers with LSTM layers. The CNN layers extract spatial features from the input data, and the LSTM layers capture temporal dependencies. It is effective in capturing spatial patterns and temporal dependencies simultaneously. Similar to the LSTM-CNN architecture, this model may require a large amount of data for training and can be computationally intensive. Interpretability remains challanging due to the complexity of the model.

As illustrated in Fig. 2 (b), LSTM-CNN-At architecture arranges systematically LSTM and CNN layers with an attention mechanism. The attention mechanism function takes the LSTM-CNN layer's output as input and computes the attention weights to give more importance to certain time steps or features in the sequence The attention mechanism allows the model to selectively focus on relevant input features, enhancing its ability to capture important information. It can capture both temporal dependencies and spatial patterns while emphasizing important information.However, this architecture may require additional computational resources compared to the basic LSTM-CNN and CNN-LSTM models. Proper tuning of the attention mechanism and hyperparameters is essential for optimal performance. While the CNN-LSTM-At architecture combines CNN and LSTM layers with an attention mechanism. It leverages the spatial extraction capabilities of CNN layers and the temporal dependency modeling of LSTM layers, while also incorporating attention to highlight important features. Similar to other attention-based models, the CNN-LSTM-At architecture may require additional computational resources and careful hyperparameter tuning. The complexity of the model can make its interpretation challenging. The choice of architecture is made by taking into account the specific characteristics of the borehole heat exchanger dataset and the requirements of the forecasting task. This selection process is crucial as the considered architecture directly influences the model's capability to capture the inherent temporal and spatial patterns present in the measured BHE data.

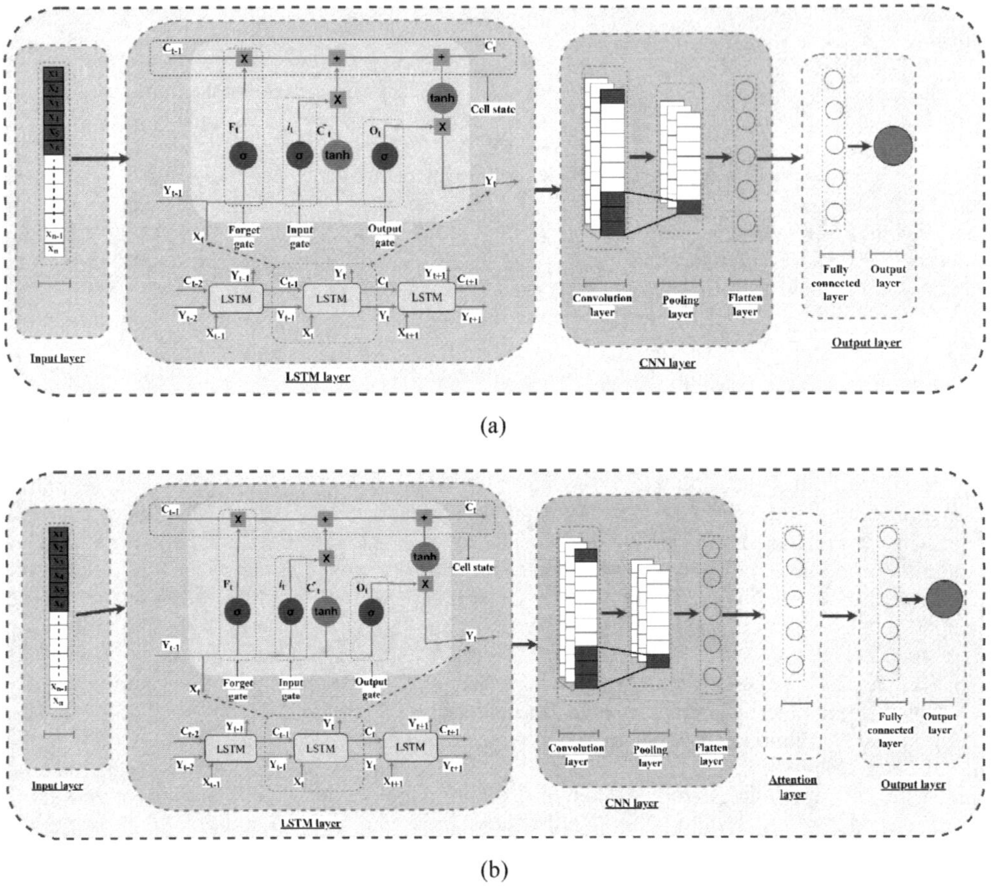

Figure 2. Architecture of hybrid model combining (a) LSTM and CNN (b) LSTM-CNN with attention mechanism.

3. Results and discussion

The analysis presented in this section provides a quantitative assessment of how the input parameters influence the multistep forecasting capabilities of four hybrid algorithms.

3.1 Effect on multistep forecasting accuracy

In this study, three different variations are considered, each utilizing a distinct number of input parameters during the developed model training. In variation-1, data from four input sensors (T_{out}, T_{in}, T_{amb}, m^0) is used. The hybrid deep learning algorithms are employed to predict the outlet temperature for the next 24 hours, and the predicted values were compared with the actual measured values, as depicted in Figure 3. The accuracy results as shown in figure 6 revealed that the LSTM-CNN model achieved the highest accuracy, followed by the LSTM-CNN-At, CNN-LSTM, and CNN-LSTM-At models, respectively. Both the LSTM-CNN and LSTM-CNN-At models performed comparably well, exhibiting MAPE values of 1.29% and 1.59%, respectively. These hybrid algorithms combine the strengths of LSTM layers in capturing temporal dependencies and long-term memory, along with the effectiveness of CNN layers in capturing spatial patterns within the data. However, the introduction of the attention mechanism did not lead to improved accuracy. Instead, it resulted in reduced ability to capture temporal and spatial patterns within the data, accompanied by a higher number of trainable parameters. The CNN-LSTM and CNN-LSTM-At models exhibited MAPE values of 1.62% and 2.53%, respectively. The reverse order of layers in these models prioritizes capturing spatial features first and then extracting temporal dependencies, which led to a decrease in forecasting accuracy.

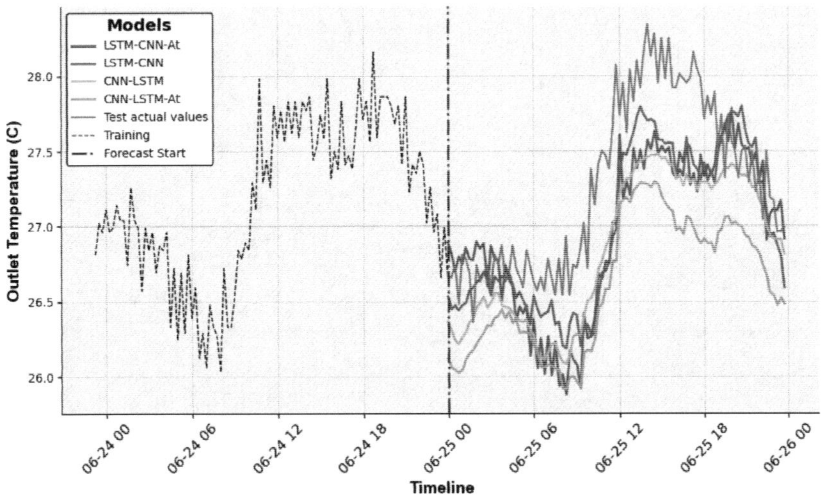

Figure 3. Performance of hybrid deep learning models on variation-1 for next 24-hour

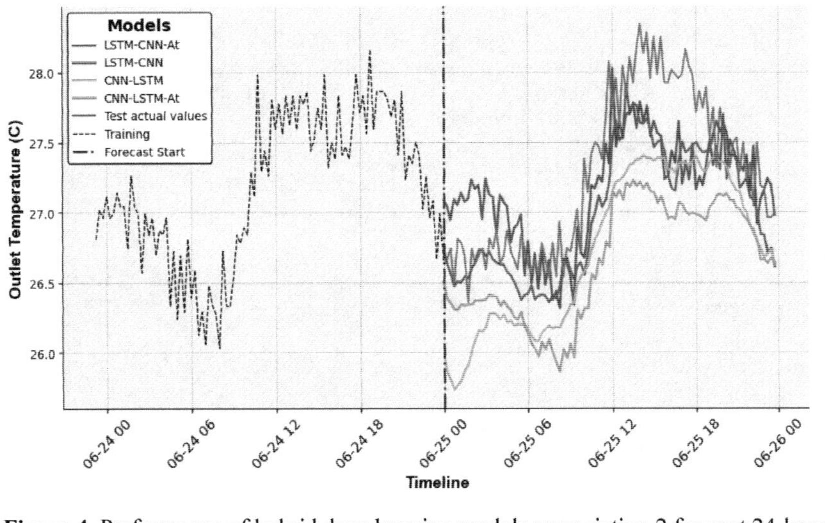

Figure 4. Performance of hybrid deep learning models on variation-2 for next 24-hour

In variation-2, an additional input parameter, solar radiation was included in the measured dataset of the borehole heat exchanger. This resulted in a total of five input parameters considered for the deep learning models (T_{out}, T_{in}, T_{amb}, m^0, R_{sol}). The results depicted in Figure 4 demonstrate that the inclusion of R_{sol} as an input feature has improved the forecasting accuracy of the hybrid algorithms, except for CNN-LSTM, for the next 24 hours. The order of prediction accuracy remains consistent with the previous variation, with LSTM-CNN achieving the highest accuracy, followed by LSTM-CNN-At, CNN-LSTM, and CNN-LSTM-At. Among these models, LSTM-CNN exhibits the lowest MAPE of 0.72%, indicating the highest level of accuracy. These findings suggest that the addition of solar radiation as an input parameter contributes to improved forecasting accuracy for most of the hybrid algorithms. However, CNN-LSTM-At performs relatively less accurately compared to other models.

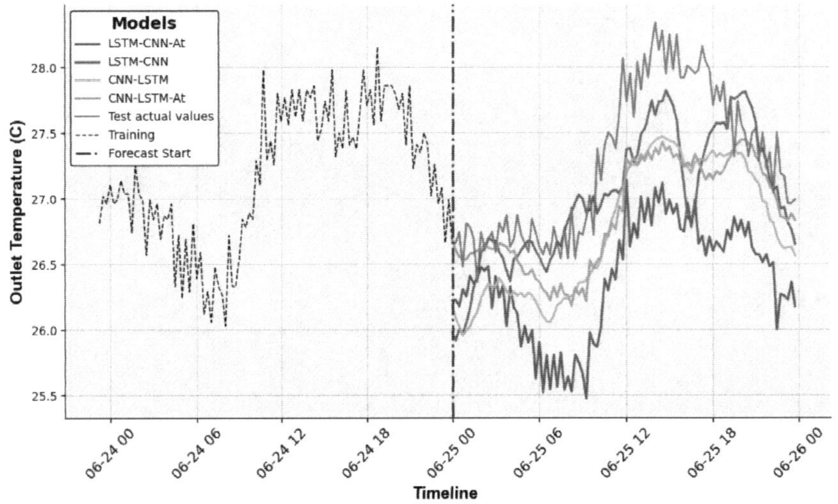

Figure 5. Performance of hybrid deep learning models on variation-3 for next 24-hour

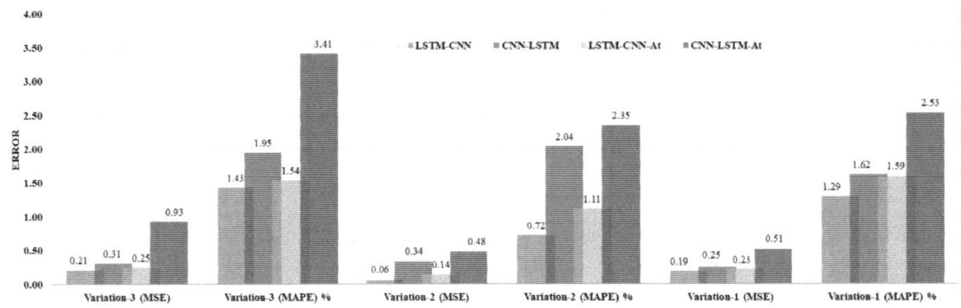

Figure 6. Performance evaluation (MAPE, MSE) of hybrid models under different input parameters

In variation-3, the number of input features considered for model training increased to six by including wind speed along with T_{out}, T_{in}, T_{amb}, m^0 and R_{sol}. The forecasting results presented in Fig. 5 demonstrate that the accuracy of CNN-LSTM increases, while the accuracy of LSTM-CNN, LSTM-CNN-At, and CNN-LSTM-At exhibit a decrease, with MAPE values of 1.95%, 1.43%, 1.54%, and 3.41% respectively. By conducting a comprehensive analysis of the forecasting results and the characteristics of the different hybrid models, valuable insights are gained regarding their suitability for predicting the outlet temperature of the borehole heat exchanger. Factors such as the number of input parameters, layer structure, and model complexities play a significant role in determining the performance of these models.

3.2 Feature importance analysis using global and local Ex-AI methods

In this section the critical input parameters for the data-driven AI models are identified through an interpretability analysis using Local and Global Explainable-AI techniques.

3.2.1 Local Ex-AI analysis. To evaluate the significance of input parameters of hybrid learning algorithms for borehole coupled heat pump systems, local-level explanations using the SHAP method was employed. This methodology leverages the Shapley value to quantify the contribution of each input measured sensor data (T_{in}, T_{amb}, m^0, R_{sol},V_{wind}) by assessing their impact on the output variable, T_{out}. Figure 7 presents the average impact of input parameters for Variation-1as measured by their corresponding SHAP values when employing the developed hybrid algorithms. The SHAP values provide insights into the range of average impact that each input parameter has on the output forecasting. Here only average impact results are discussed and the results indicate a consistent overall trend of relative average impact on the output across the considered hybrid algorithms. However, slight variations in the magnitude of their effects is observed. For the best-performing LSTM-CNN model, the order of importance of input parameters in terms of average impact is $T_{in} > T_{out} > m^0 > T_{amb}$, while for the least-performing CNN-LSTM-At model, the order is $m^0 > T_{amb} > T_{in} > T_{out}$.

In Variation-2 as depicted in Fig.8, the results demonstrate that the inclusion of solar radiation as an input parameter significantly impacts the forecasting accuracy and cannnot be disregarded. Depending on the architecture of the hybrid model, this inclusion can either lead to a decrease in accuracy (CNN-LSTM) or an improvement (LSTM-CNN, LSTM-CNN-At, CNN-LSTM-At). The order of average impact of input parameters for LSTM-CNN-At ($m^0 > T_{amb} > T_{in} > T_{out} > R_{sol}$), CNN-LSTM-At ($m^0 > T_{in} > R_{sol} > T_{amb} > T_{out}$), LSTM-CNN ($m^0 > R_{sol} > T_{in} > T_{out} > T_{amb}$), and CNN-LSTM ($T_{in} > T_{amb} > T_{out} > m^0 > R_{sol}$) highlights that m^0 and T_{in} have the most significant impact, while the impact of other parameters with reduced magnitude, such as R_{sol}, cannot be ignored.

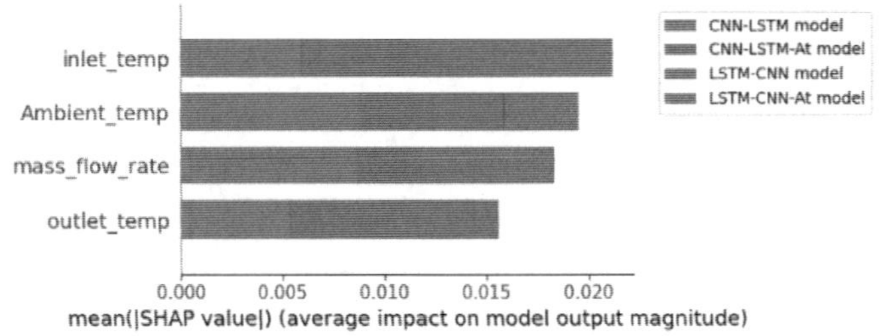

Figure 7. Quantifying input parameter impact on model outputs with SHAP values for Variation-1

Figure 9 illustrates the mean SHAP values for Variation-3 with the error analysis revealing that the inclusion of the sixth input parameter(V_{wind}), leads to a decrease in accuracy for all algorithms except CNN-LSTM, which shows an improvement. The order of parameter importance, based on the average impact value, for LSTM-CNN and LSTM-CNN-At is is $T_{in} > T_{out} > m^0 > R_{sol} > T_{amb} > V_{wind}$ and $T_{in} > T_{out} > T_{amb} > m^0 > R_{sol} > V_{wind}$, respectively. Overall, the results indicate that the addition of V_{wind} as an input parameter has the least significant impact on the forecasting accuracy of all hybrid algorithms, but its contribution cannot be overlooked. These results highlight the models' sensitivity to variations in the input parameters and emphasize the significance of obtaining precise measurements to ensure accurate forecasting of the outlet temperature.

3.2.2 Global analysis. Global interpretability of the hybrid deep learning models is investigated using the permutation feature importance method and Friedman statistical test. Fig. 10 shows global feature analysis using permutation feature importance method for three different variations. By randomly permuting the values of a feature in each variation and evaluating the model's performance on the permuted data, it was investigated how much the model's accuracy decreases. The greater the drop in accuracy, the more important that feature is for the model's multistep predictions. Overall results depict that the order of feature importance for variation-1($T_{in} > T_{out} > T_{amb} > m^0$), variation-2 ($T_{in} > T_{out} > T_{amb} > m^0 > R_{sol}$) and variation-3 ($T_{in} > T_{out} > T_{amb} > R_{sol} > V_{wind} > m^0$) very much depend on the model architecture. A higher feature importance score of inlet temperature for each variation suggests that this feature has a stronger influence on the model's predictions. The impact magnitude of other features varies depending on the architecture of hybrid model.

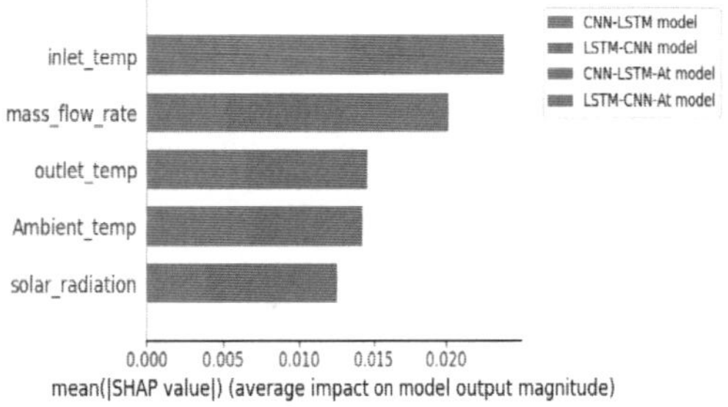

Figure 8. Quantifying input parameter impact on model outputs with local explainable SHAP values for Variation-2

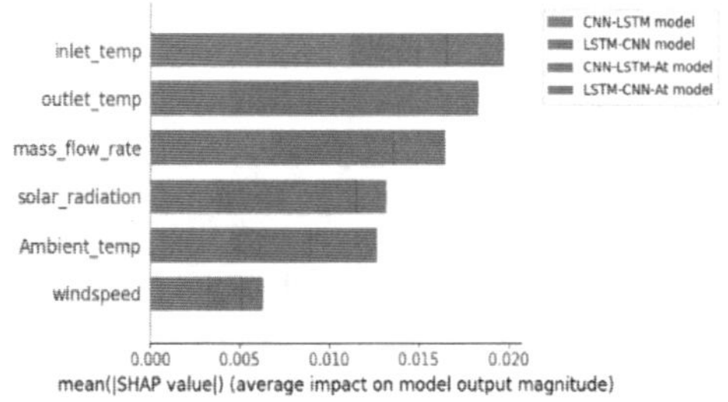

Figure 9. Quantifying input parameter impact on model outputs with local explainable SHAP values for Variation-3

To compare the performance of multiple hybrid models on the considered variations, a non-parametric test i.e., Friedman statistical test is used to determine whether there are significant differences among different sets of data. The Friedman test ranked the deep learning hybrid models based on their performance on multiple runs with various input data samples. Based on the average ranks obtained from the Friedman test, the performance order from best to worst as follows: LSTM-CNN model with an average rank of 1.33 (best performing), LSTM-CNN-At with a rank of 2.0, CNN-LSTM with a rank of 3.00, and CNN-LSTM-At with a rank of 3.66 (worst performing). The small differences in the average ranks indicate that the models' performance is quite comparable, and no statistically significant distinctions were found. The test results yielded an F-value of 5.80 and a p-value of 0.12. The p-value indicates that there is no significant difference in the performance of the four hybrid deep learning models at the chosen significance level of 0.05. Therefore, the null hypothesis cannot be rejected, suggesting no significant distinctions in the models' performance.

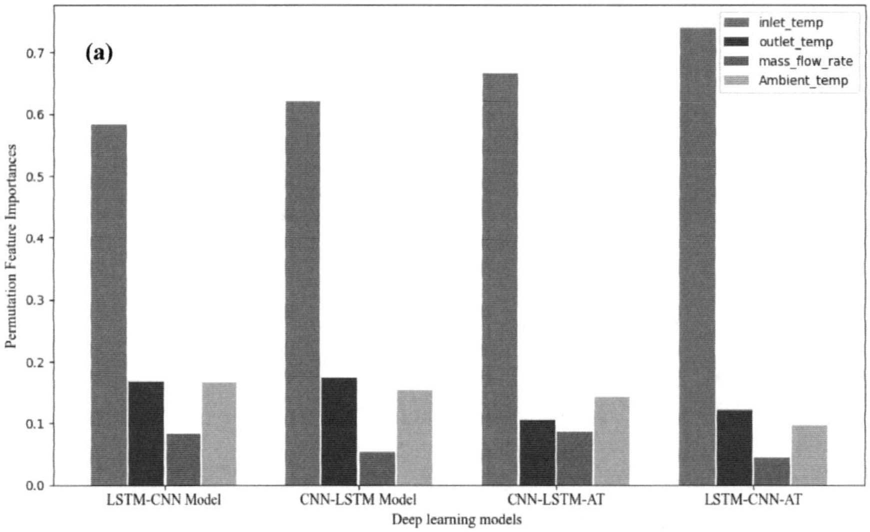

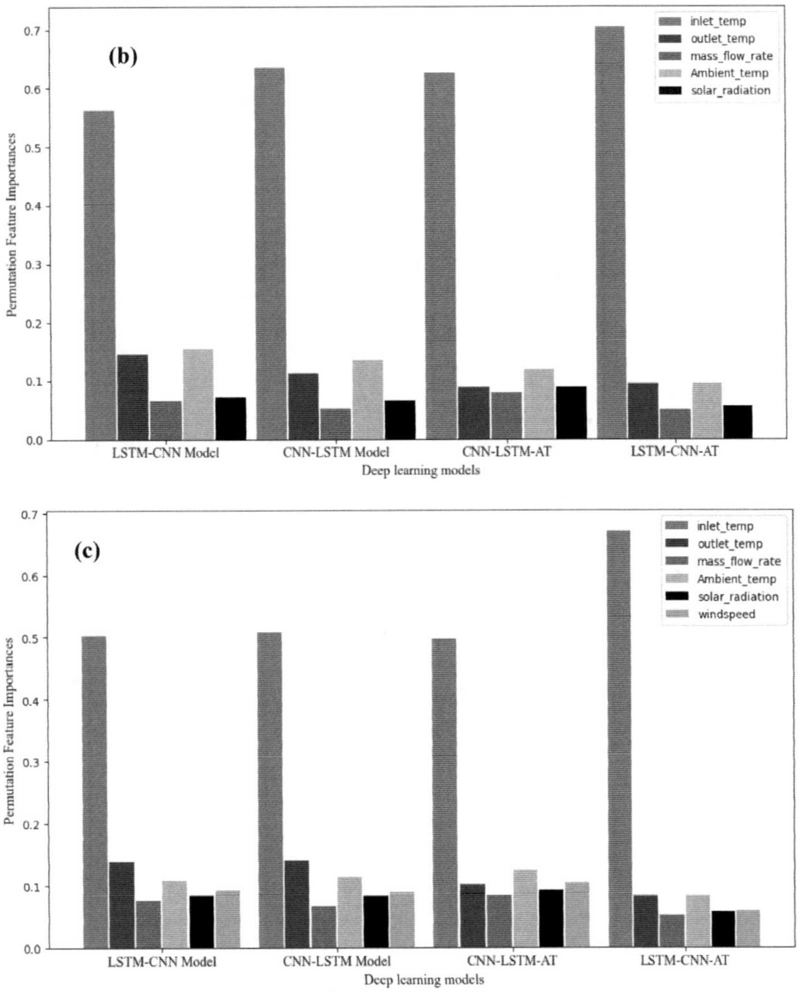

Figure 10. Global feature analysis using permutation feature importance method for (a) Variation-1, (b) Variation-2, (c) Variation-3

4. Conclusions

This study explores the impact of training algorithms, input parameters, model architecture, and exogenous variables on the multistep forecasting accuracy of data-driven models. It aims to identify key input sensors for deep learning-based load forecasting and investigates the significance of input parameters on 24-hour-ahead predictions for efficient thermal load management. The importance of input parameters is assessed using interpretability analysis with Ex-AI techniques, including SHAP analysis for local interpretability and the permutation feature importance method with the Friedman statistical test for global interpretability. The findings highlight the efficacy of attention mechanisms in capturing temporal dependencies in LSTM-CNN-At and spatial patterns in CNN-LSTM-At, may not necessarily enhance their multistep forecasting capabilities for the borehole field data in comparison to LSTM-CNN architecture. Local interpretability of the hybrid models by using SHAP method reveal that incorporating additional input features from BHE installations can improve the prediction accuracy.

However, it is important to consider the model architecture carefully, as there is a threshold beyond which adding more input parameters may not yield further improvements. Global interpretability of the hybrid deep learning models by using Permutation feature importance method show that the inlet temperature is the most influential parameter for all three variations and impact magnitude of other features varies depending on the architecture of hybrid model. Through this research, a comprehensive understanding of the interplay between model architecture, input sensors, and their identification is gained, enabling improved multistep forecasting performance in ground source heat pump applications.

Acknowledgments: We are indebted to the staff of the Department of Energy and Petroleum Engineering (IEP) at the University of Stavanger and for allowing to present this paper. The research leading to these results has received funding from the Norway Grants 2014–2021 via the National Centre for Research and Development in Warsaw.

References

[1] International Energy Agency (IEA) 2022 World Energy Oultlook, 1–524.
[2] Ahmed A, Assadi M, Kalantar A and Sliwa T A 2022 Critical Review on the Use of Shallow Geothermal Energy Systems for Heating and Cooling Purposes, *Energies,* 1–22.
[3] Ahmed N, Assadi M, Abbas A and Banihabib R 2023 Optimal design, operational controls, and data-driven machine learning in sustainable borehole heat exchanger coupled heat pumps : Key implementation challenges and advancement opportunities, *Energy for Sustainable Development*, **74**, 231–257, doi: 10.1016/j.esd.2023.04.004.
[4] Wang Z, Hong T and Piette M A 2020 Building thermal load prediction through shallow machine learning and deep learning, *Appl Energy*, **263**, 114683, doi: 10.1016/j.apenergy.2020.114683.
[5] Vishwarupe V, Joshi P M, Mathias N, Maheshwari S, Mhaisalkar S and Pawar V 2022 Explainable AI and Interpretable Machine Learning: A Case Study in Perspective," *Procedia Comput Sci*, **204**, 869–876, doi: 10.1016/j.procs.2022.08.105.
[6] Behzadi A, Holmberg S, Duwig C, Haghighat F, Ooka R and Sadrizadeh S 2022 Smart design and control of thermal energy storage in low-temperature heating and high-temperature cooling systems: A comprehensive review, *Renewable and Sustainable Energy Reviews*, **166**, 112625, doi: 10.1016/j.rser.2022.112625.
[7] Muzaffar S and Afshari A 2019 Short-term load forecasts using LSTM networks, *Energy Procedia*, **158**, 2922–2927, doi: 10.1016/j.egypro.2019.01.952.
[8] Jang J, Han J and Leigh S B 2022 Prediction of heating energy consumption with operation pattern variables for non-residential buildings using LSTM networks, *Energy Build*, **255**, 111647, doi: 10.1016/j.enbuild.2021.111647.
[9] Cai M, Pipattanasomporn M and Rahman S 2019 Day-ahead building-level load forecasts using deep learning vs. traditional time-series techniques, *Appl Energy*, **236**, 1078–1088, doi: 10.1016/J.APENERGY.2018.12.042.
[10] Zhang W, Zhou H, Bao X and Cui H 2023 Outlet water temperature prediction of energy pile based on spatial-temporal feature extraction through CNN–LSTM hybrid model, *Energy*, **264**, 126190, doi: 10.1016/j.energy.2022.126190.
[11] Yao F, Zhou W, Al Ghamdi M, Song Y and Zhao W 2022 An integrated D-CNN-LSTM approach for short-term heat demand prediction in district heating systems, *Energy Reports*, **8**, 98–107, doi: 10.1016/j.egyr.2022.08.087.
[12] Tsoka T, Ye X, Chen Y, Gong D and Xia X 2022 Explainable artificial intelligence for building energy performance certificate labeling classification, *J Clean Prod*, **355**, 131626, doi: 10.1016/j.jclepro.2022.131626.
[13] Chung W J and Liu C 2022 Analysis of input parameters for deep learning-based load prediction for office buildings in different climate zones using eXplainable Artificial Intelligence, *Energy Build*, **276**, 112521, doi: 10.1016/j.enbuild.2022.112521.
[14] Smith D C and Elmore A C 2018 The observed effects of changes in groundwater flow on a

borehole heat exchanger of a large scale ground coupled heat pump system *Geothermics*, **74**, 240–246, doi: 10.1016/j.geothermics.2018.03.008.
[15] Crossing V 2023 Global Forecast and History Data, Accessed: Jan. 05, 2023. [Online]. Available: https://www.visualcrossing.com/weather-data-editions#
[16] *Applied Statistical Modeling and Data Analytics* 2018, doi: 10.1016/c2014-0-03954-8.

IOP Conf. Series: Materials Science and Engineering 1294 (2023) 012057 doi:10.1088/1757-899X/1294/1/012057

Hydrogen safety considerations: Mitigating risks and securing operations in enclosed spaces

S Kleszcz1 * and M Assadi

Faculty of Science and Technology, University of Stavanger, Norway

* Correspondence: slawosz.a.kleszcz@uis.no

Abstract. The increased interest in hydrogen as a versatile and environmentally friendly energy carrier has led to its recognition as a potential solution for addressing environmental concerns and challenges associated with transitioning to alternative energy sources. However, working with hydrogen requires careful consideration of safety measures to ensure the safe handling, storage, and utilization of this highly flammable gas. This article focuses on the important safety considerations related to working with hydrogen in enclosed spaces and emphasizes the steps taken to minimize risks. Essential calculations were conducted to determine the concentration of hydrogen within a container during a leak. The findings revealed that, without proper ventilation and safety precautions, the concentration of hydrogen in the container can quickly reach explosive levels. To ensure safe operation, a secure system has been designed and described, which includes the implementation of various detection devices. Additionally, the article discusses the recommended course of action in the event of an explosion. In summary, this study offers a safety analysis of hydrogen leakage in a container. The suggested security system, incorporating detection devices and a safety valve, guarantees the safe functioning of the hydrogen supply system.

Nomenclature

Symbol	Description
p_1	fuel pressure [bar]
p_2	pressure in the container [bar]
p_b	pressure in gas bottle [bar]
p_e	critical flow pressure [bar]
T_1	fuel temperature [K]
T_2	temperature in the container [K]
T_e	critical flow temperature [K]
p_t	stagnation pressure [bar]
T_t	stagnation temperature [K]
D	diameter of fuel pipe [m]

IOP Conf. Series: Materials Science and Engineering 1294 (2023) 012057 doi:10.1088/1757-899X/1294/1/012057

$\dot{m}_{H_2}$	hydrogen mass flow [m/s]
γ_{H_2}	hydrogen isentropic expansion factor [-]
M	Mach number [-]
ρ_e	density at critical conditions [kg/m^3]
A_e	surface area [m_2]
v_e	critical velocity of fluid [m/s]
R_{H_2}	hydrogen gas constant [J/mol K]
a_e	speed of sound [m/s]
$\dot{V}_n$	volumetric flow [m^3/h]
V_c	volume of container [m^3]
c_{H_2}	concentration of hydrogen in the room [m^3/m^3]
q_{H_2}	amount of hydrogen added to the room [m3/h]
n	number of air volume changes per hour [1/h]
t	time [s]
c_1	concentration of hydrogen in the room at start [m^3/m^3]
c_2	concentration of hydrogen in the supply fluid [m^3/m^3]
A_f	flame surface area [m^2]
A_{in}	total internal surface area of the enclosure [m^2]
R_{cl}	external cloud radius [m^2]
p_{ex}	internal overpressure [m^2]

1. Introduction

Due to the impacts of fossil fuels on climate change the rapid development of renewable power generation has been observed. However, the impact of their inherent intermittency, fluctuation, and difficulty in prediction on the existing electric grid attracts more and more concern. For this reason, hydrogen started to being considered as an important energy store that can maximize the advantages of renewable and sustainable energies [1-2]. The core principle underlying this idea involves a new system that utilizes hydrogen for energy delivery. Such system includes the integration of hydrogen production, storage, transportation, distribution, and applications, as well as other aspects such as education, safety, codes, standards, and regulations. Effective, dependable, and swift monitoring for hydrogen leaks is crucial in guaranteeing the safety of hydrogen storage. Encountering high temperatures, hydrogen embrittlement damage, or external impacts can easily trigger leaks and the spread of high-pressure hydrogen, leading to potentially catastrophic explosions [3]. Unfortunately, hydrogen tends to leak and disperse easily [4], possesses a low ignition energy threshold, has a substantial potential for fuel explosions, and carries a significant amount of explosion energy [5].

The range for ignition volume fraction spans from 4% to 74%, while the explosion volume fraction lies within 18% to 59%. Furthermore, the minimum ignition energy required is just 0.02 mJ [6], which poses a certain level of risk when there is a leak.

Despite of it, hydrogen has an excellent safety record and is as safe for transport, storage and uses as many other fuels. Nevertheless, safety remains a top priority in all aspects of hydrogen energy. The hydrogen community addresses safety through stringent design and testing of storage and transport concepts, and by developing codes and standards for all types of hydrogen-related equipment. [7]

Up to this point, there has been extensive research conducted on the efficiency of ventilation. Cerchiara et al. [8] studied ventilation conditions related to a minor leak situation in which the hydrogen concentration did not exceed 2 vol% in a fuel cell room with a ventilation opening and a ventilation fan. Hyon at al [9] performed experiments and analyses on higher volumes of hydrogen leakage from household fuel cell rooms and reduced hydrogen concentrations through a variety of methods such as ventilation openings, ventilation fans, and automatic supply shutoff devices. Although the hydrogen concentration in a small hydrogen fuel cell room for home use can rapidly increase, a rapid reduction in the concentration of hydrogen with an appropriate ventilation system has been experimentally proven. In addition, natural and forced ventilation were studied as a function of the sizes, positions, leak quantities, and flow rates of vents [10-13]. In accordance with these findings, a new method was proposed for forced ventilation based on real-time sensing.

As hydrogen-based research and applications continue to advance, the need for detecting and monitoring hydrogen gas has grown significantly to prevent the risks associated with its leakage and potential explosions [14]. Due to hydrogen lacks color, odor, and taste, it remains imperceptible to human senses [15]. The aspect for a reliable use of hydrogen sensors is its functional safety. This aims to evaluate possible malfunctions and their consequences, especially from the electronic part of systems or equipment, in order to avoid unacceptable risks of relating physical injuries or damages to the human health [14]. Hence, sensitive, and reliable detection of hydrogen gas has become a major need [16].

This study presents a risk assessment needed to be conducted to evaluate the operational safety of hydrogen turbine operating in the container. The primary objective has been to prevent the accumulation of explosive gas mixtures in case of fuel gas leakage into the container. Using the worst-case scenario as a basis, calculations were performed to design an evacuation system. Further safety margins have been incorporated by considering additional safety factors in the chosen parameters to enhance the overall safety level.

2. Case description

In the pursuit of researching and developing gas turbines capable of utilizing hydrogen and hydrogen blends, a decision was made to establish a test rig within a container to increase its portability (Figure 1). Therefore, it became essential to conduct a risk assessment to thoroughly appraise the operational safety of this unit. The primary objective has been to prevent the accumulation of explosive gas mixtures in the event of fuel gas leakage into the container. To ensure that the hydrogen concentration does not exceed the ignition limit of 4%, a continuous ventilation of the container needs to be upheld through an open intake system, supported by a blower. Considering the worst-case scenario, represented by a pipe rupture event, the fuel system safety valve will be automatically triggered by fuel detections sensor, leading to its closure. This safety measure, coupled with the continuous evacuation process, is intended to uphold hydrogen concentrations within the container well below the threshold for potential explosion.

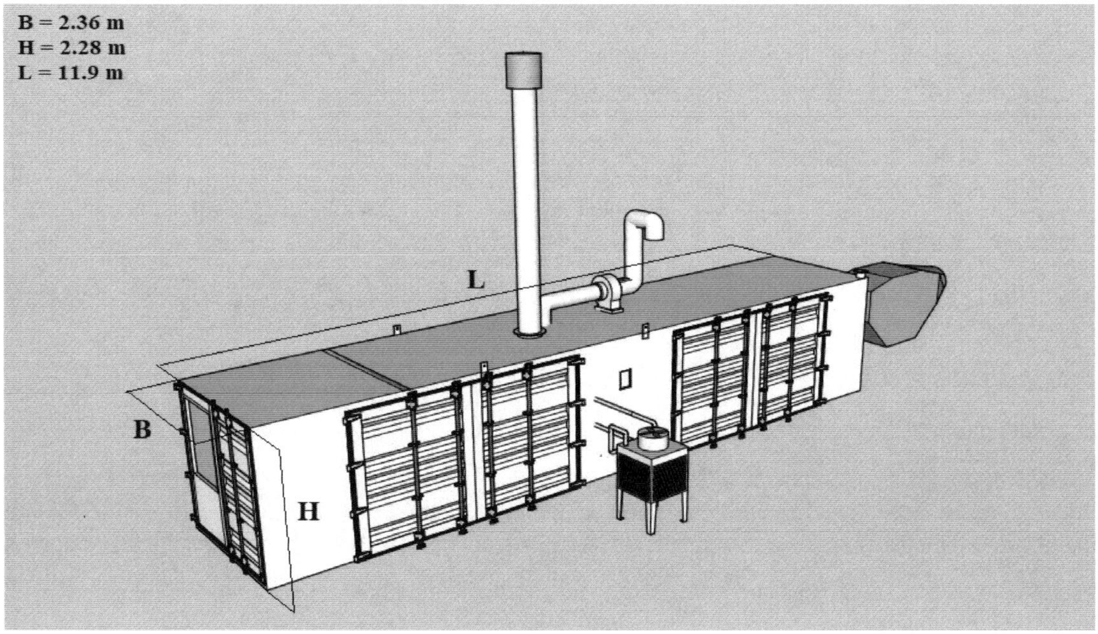

Figure 1. Steel container with basic dimensions.

2.1 Hydrogen supply system

To ensure the requisite fuel pressure within the combustion chamber a fuel delivery pressure of 10 bars (p_1) is required. This pressure accounts for the anticipated losses incurred due to pressure drops in both pipes and fuel valves. In the unfortunate event of pipe damage, the fuel would be released into the container, which maintains atmospheric pressure (p_2). Given that the fuel gas originates from a gas bottle pressurized at 200 bars (p_b), it is reasonable to assume that the gas parameters after reduction valve can be represented using the concept of stagnation state.

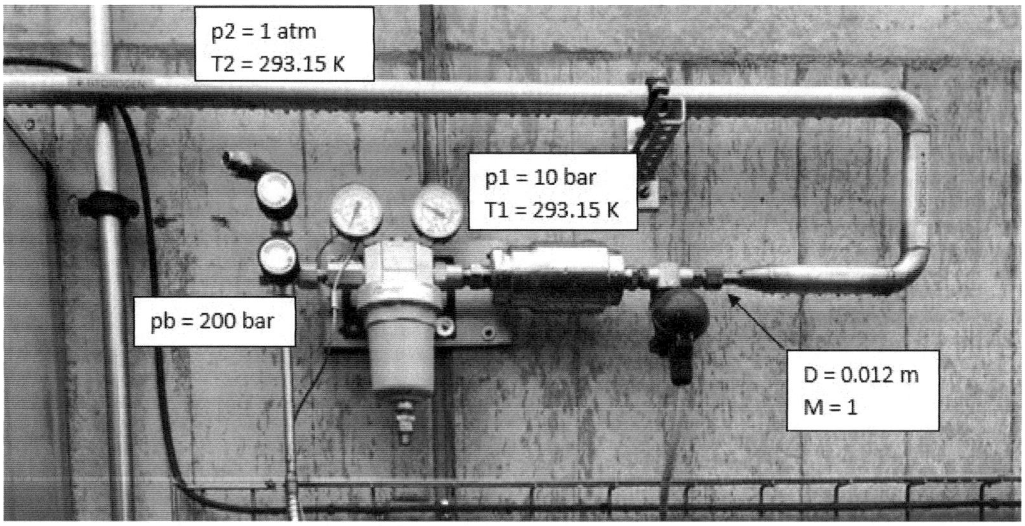

Figure 2. Photo of hydrogen supply system.

During chocked flow conditions, the velocity at the system narrowest juncture – situated immediately after the safety valve within the pipeline – aligns with critical flow, denoted as Mach = 1. This circumstance restricts the mass flow rate that can traverse the system. The diameter of this specific pipe is designated as D = 0.012 m. The principal parameters for calculations are indicated in Figure 2.

2.2 Critical temperature and pressure for chocked flow

Assuming ideal gas behaviour, steady state choked flow occurs when the downstream pressure falls below the critical value. The value for the critical pressure (1) and temperature (2) can be calculated by the following equations:

$$p_e = p_t\left(\frac{2}{\gamma_{H_2}+1}\cdot M^2\right)^{\frac{\gamma_{H_2}}{\gamma_{H_2}-1}} = 5.81\ bar \tag{1}$$

$$T_e = T_t\cdot\frac{2}{\gamma_{H_2}+1}\cdot M^2 = 244.3\ K \tag{2}$$

2.3 Mass flow through the pipe rapture

Applying the conservation of mass (3), relationships for Mach number (4), the speed of sound (5), and the ideal gas law (6), the convenient equation that apply for the mass flow rate through the pipe is given by (7):

$$\dot{m}_{H_2} = \rho_e A_e v_e = const \tag{3}$$

$$v_e = Ma_e \tag{4}$$

$$a_e = \sqrt{\gamma_{H_2}R_{H_2}T_e} \tag{5}$$

$$\mathrm{p}_e = \rho_e R_{H_2}T_e \tag{6}$$

$$\dot{m}_{H_2} = \rho_e A_e V_e = \left(\frac{p_e}{R_{H_2}T_e}\right)A_e\left(M\sqrt{\gamma_{H_2}R_{H_2}T_e}\right) = 0.0775\ \frac{kg}{s} \tag{7}$$

3. Leakage safety considerations

3.1 Ventilation of the container

Any structure that incorporates components utilizing hydrogen must possess appropriate ventilation. The standard air exchange rate (as stipulated for continuous ventilation designed for the container) should be 0.3 m³ of air per minute and per square meter of solid floor space (8), as outlined in reference [17]. Consequently, the hourly evacuation volume count (9) is formulated as follows:

$$\dot{V}_n = 11.9\ \mathrm{m}\cdot 2.36\ \mathrm{m}\cdot 0.3\ \frac{m^3}{min} = 8.43\ \frac{m^3}{min} = 505.8\ \frac{m^3}{h} \tag{8}$$

$$n = \frac{\dot{V}_n}{V_c} = \frac{505.8\,\frac{m^3}{h}}{45.64\ m^3} = 11.1\ \frac{1}{h} \tag{9}$$

It is assumed that number of the volume changes per hour is going to be 12 for continuous ventilation of the container.

3.2 Concentration of hydrogen in the container during leakage

The generic equation for hydrogen concentration can be calculated by equation (10). It is presupposed that the hydrogen is homogeneous mixture with air within the container:

$$c_{H_2} = \frac{q_{H_2}}{nV_c(1-e^{-nt})} + (c_1 - c_2)e^{-nt} + c_2 \tag{10}$$

Where c_{H_2} – concentration of hydrogen in the room when perfectly mixed (m^3/m^3), q_{H_2} – amount of hydrogen added to the room (m^3/h), n – number of volume changes per hour (h^{-1}), V_c – volume of the container (m^3), t – time (h), c_1 – concentration of hydrogen in the container at start (m^3/m^3), c_2 – concentration of hydrogen in the supply fluid (m^3/m^3).

The results derived from the solved equation (10) have been visualized in Figure 3. As showed on the graph, it becomes evident that the hydrogen contamination within the container surpasses 4% within a span of 2 seconds, rendering it to a combustible mixture.

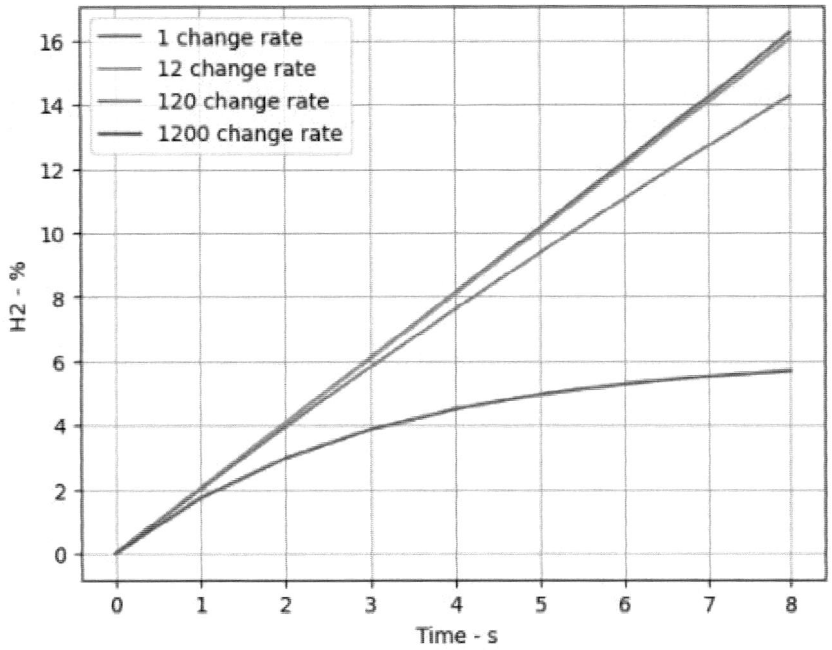

Figure 3. The hydrogen concentration in the container over time with ventilation.

It's clear that almost no ventilation system can guarantee safe operation in the event of pipe rapture. The approach presented above is employed to evacuate hydrogen from the container after successful halted leak, and in instances of minor leaks within the system, it hinders the build-up of combustible mixtures.

4. Security system

The most secure option involves interruption of the fuel supply in the event of a malfunction, but the solution should be adaptable without disrupting the device standard functioning. Therefore, the main protection will be a shut-off valve triggered by a peripheral device. Positioned on the supply line, a safety valve halts hydrogen flow when leakage is detected. The valve is controlled by the PLC controller, and it is triggered by signals from three different sources of gas leakage detection, pressure transmitter, mass flow meter and hydrogen sensor.

The maximum closing time t_{SC} of the safety valve (model Burkert type 6240) [18] is 0.1 second and the scan cycle speed t_{PLC} for PLC controller is 0.015 second.

IOP Conf. Series: Materials Science and Engineering 1294 (2023) 012057 doi:10.1088/1757-899X/1294/1/012057

4.1 The pressure transmitter

The pressure transmitter (model: WIKA E-10) [19] is continuously checking the pressure in the pipe. In case of pressure decrease in the fuel line closing signal for the safety valve is passed via PLC controller. The settling time for pressure transmitter t_{pt} is 0.002 second. To prevent unintentional stoppage of the turbine due to fluctuation of pressure measurement the signal from transmitter must be filtered. It is assumed to take in average 100 samples as security signal (11):

$$t_{av\_pt} = 100 \cdot 0.002 = 0.2\ s \tag{11}$$

According to this information the total time t_{1t} needed from detecting the leakage to closing the valve is equal to or less than sum of the times above (12):

$$t_{1t} = t_{sc} + t_{av\_pt} + t_{PLC} = 0.315\ s \tag{12}$$

4.2 Mass flow meter

The mass flow meter is controlling amount of gas delivered to the turbine. In case of pipe rapture, the value will differ significantly from the set value.

The typical Indication Response Time t_{mf} for this device (model Alicat 5000 SLPM) [20] is 0.01 second. Similarly, the signal should also be filtered (13) in this case:

$$t_{av\_mf} = 100 \cdot 0.01\ \mathrm{s} = 1\mathrm{s} \tag{13}$$

According to this information the total time t_{2t} needed from detecting the leakage to closing the valve is equal to or less than the sum of the times above (14):

$$t_{2t} = t_{sc} + t_{av\_mf} + t_{PLC} = 1.115\ s \tag{14}$$

4.3 Hydrogen sensor

The hydrogen sensor is measuring the concentration of the hydrogen inside the container.

The response time for this device t_{hs50} (model Oldham OLCT 100 XP) [21] is 30 second. Where t_{hs50} refers to the time it takes for the sensor to respond with an output signal that is 50% of the full value of the gas being detected (15). Havin t_{hs50} and knowing the concentration rate of hydrogen in the room over time (Figure 4) we can estimate the detection level of the sensor.

$$t_{res} = c_{H_2} * (1 - 0.5^{\left(\frac{t}{T50}\right)}) \tag{15}$$

where: c_{H_2} is concentration of hydrogen in the room, t is the time, and T50 is the response time of the device.

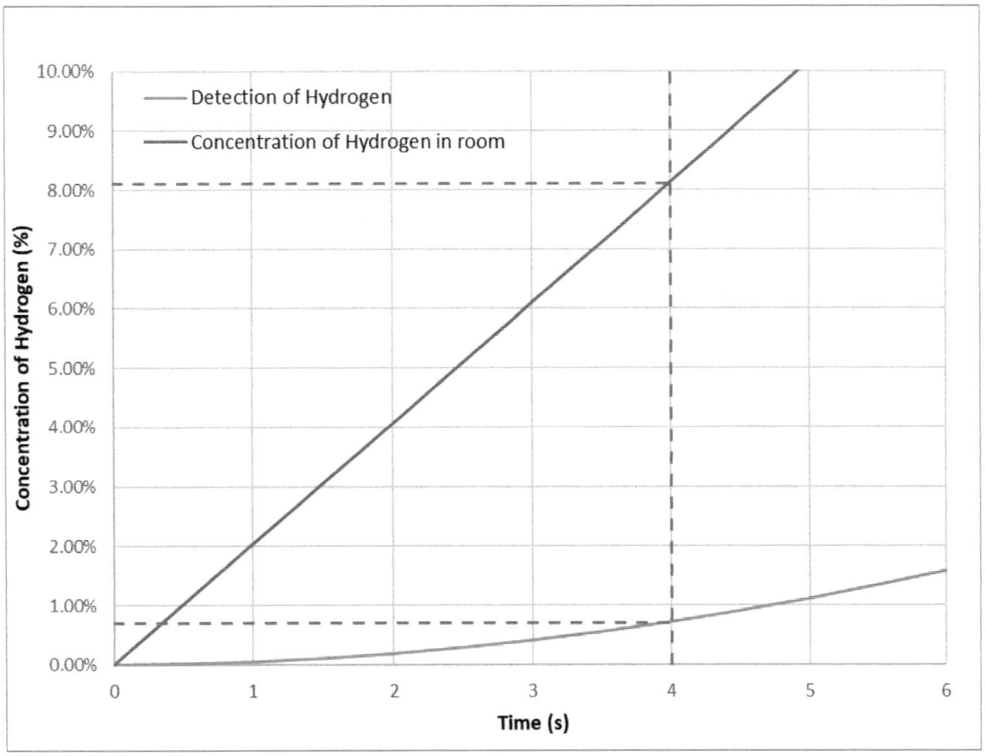

Figure 4. The response of detector over time during leakage.

Based on this data, the overall time t_{3t} required from leak detection to valve closure is either equal to or less than the cumulative times (16) indicated above.:

$$t_{3t} = t_{sc} + t_{hs90} + t_{PLC} = 4.115\ s \tag{16}$$

The sensor must be set on alarm at 17% of LEL.

4.4 The results

The 3-step secure solution for the gas leakage is presented in Figure 5. Each security step operates autonomously, and in the event of a power failure, the safety valve will autonomously close (valve position: normally closed). The performance of the hydrogen sensor is significantly influenced by its installation position, and this factor can lead to further delays in the triggering time. Consequently, the following paragraph examines the scenario of an explosion.

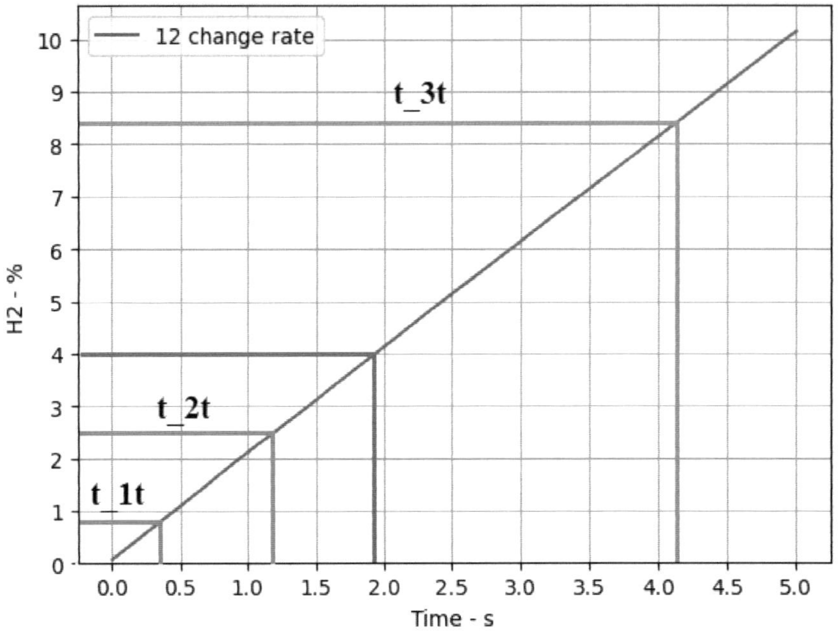

Figure 5. Hydrogen concentration in the container after activation of safety valve triggered by a) pressure sensor and mass flow meter (green line), b) hydrogen sensor (orange line).

Figure 6 displays the duration required for complete ventilation of the container when employing the hydrogen sensor as the triggering mechanism. After 200 seconds, it is anticipated that the air-hydrogen mixture will have reached a hydrogen concentration below the safety threshold of 4%.

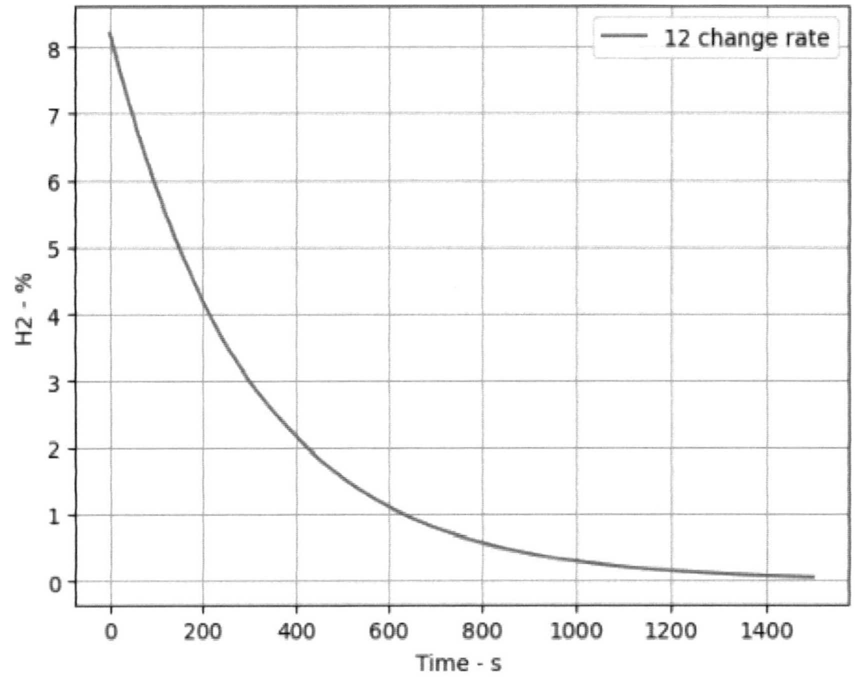

Figure 6. Concentration of hydrogen in time during ventilation process.

IOP Conf. Series: Materials Science and Engineering 1294 (2023) 012057 doi:10.1088/1757-899X/1294/1/012057

5. Explosion safety considerations

In the event of a hydrogen explosion, the emphasis was placed on constraining the highest level of overpressure generated by the explosion through careful sizing of the technical opening. The calculations for determining the vent size were derived from the referenced article [22], wherein the authors presented simplified formulas for estimating the highest peak pressure within a vented volume when a hydrogen explosion occurs. Steps for computing the Internal Overpressure (IO) are presented below:

1. Compute flame surface area for ignition at the centre of the enclosure (17) (A_f):

$$A_f = 0.25 \cdot A_{in} \tag{17}$$

where A_{in} is the total internal surface area of the enclosure (18):

$$A_{in} = 2 \cdot (L \cdot B + B \cdot H + H \cdot L) \tag{18}$$

2. Compute external cloud radius using Eq. (19)

$$R_{CL} = 0.5V^{0.3} \tag{19}$$

where V is the volume of the enclosure.

3. Compute G1 and G2 using Eq. (20) and (21). Select β1 and β2 from table A2 in the reference.

$$G1 = \left[\left(L_{eff}^{\beta 1}\right)^2 \left\{\left(\frac{A_f}{A_v}\right)^2 - 1\right\}\right] \tag{20}$$

$$G2 = \left[R_{Cl}^{\beta 2}\right]^2 \tag{21}$$

Where L_{eff} for ignition at the centre of the enclosure can be defined as:

$$L_{eff} = 0.5\, L$$

4. Compute internal overpressure using eq. 22. Select F1 and F2 from table A1 for hydrogen.

$$p_{ex} = (F1 \cdot G1) + (F2 \cdot G2) \tag{22}$$

Results for the overpressure calculations for different vents opening sizes are shown at Figure 7.

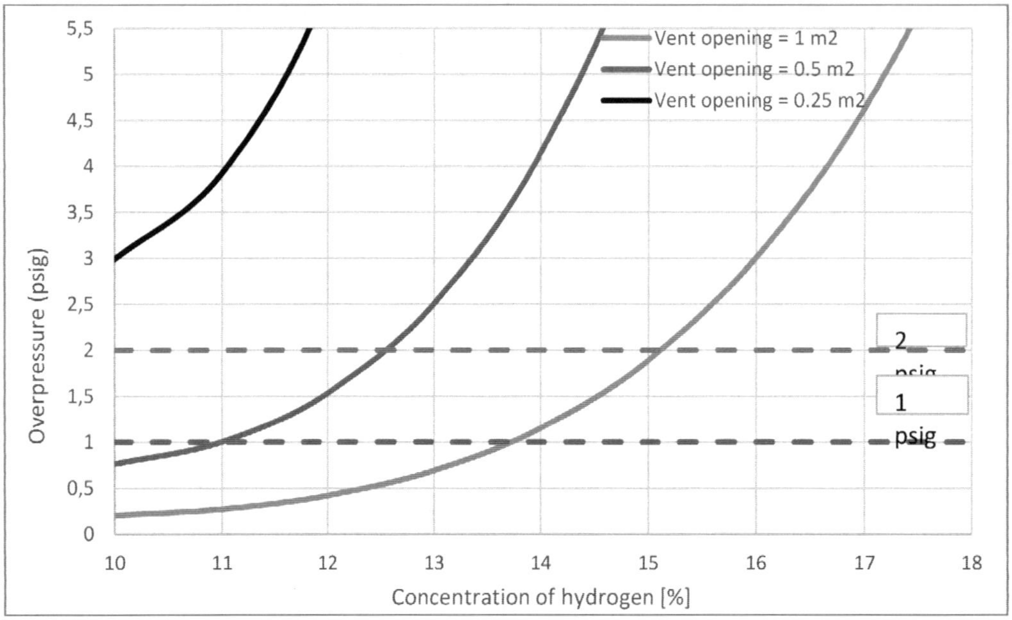

Figure 7. Peak pressure for different concentration of hydrogen with vent opening due to explosion.

Referring to Table 1, which outlines the correlation between damage and the resultant overpressure caused by an explosion, along with the information depicted on Figure 7, we can derive an estimation for the required opening size. As illustrated in Figure 6, if two security levels were to fail, the hydrogen concentration could potentially increase by up to 9%. Therefore, considering our approach, it was opted for an opening size of 1 m².

Table 1. Level of damage expected at specific overpressure values [23]

Overpressure (psig)	Expected Damage
0.04	Loud noise (143 dB), sonic boom glass failure
0.15	Typical pressure for glass failure
0.40	Limited minor structural damage
0.50-1.0	Windows usually shattered, some window frame damage.
0.70	Minor damage to house structures.
1.0	Partial demolition of houses, made uninhabitable
1.0-2.0	Corrugated metal panels fail and buckle. Housing wood panels blown in.
1.0-8.0	Range for slight to serious laceration injures from flying glass
2	Partial collapse of walls and roofs of house
2.0-3.0	Non-reinforced concrete of cinder block walls shattered

6. Summary and conclusion

In conclusion, maintaining continuous ventilation is imperative when dealing with hydrogen systems, particularly in scenarios involving minor leaks, to avert the formation of potentially hazardous mixtures. However, it's essential to integrate this ventilation approach with other security measures to comprehensively ensure the safety of the entire system operation. Nowadays modern electronic devices exhibit rapid response times, enabling the detection of gas leaks within seconds. The key consideration lies in selecting the appropriate device and strategically deciding upon the installation location. Additionally, each system should undergo preliminary testing with inert gases to fine-tune the sensors responses and guarantee accurate system operation. In the event of an explosion, damage can potentially be mitigated by strategically weakening the construction at specific points. In summary, it is feasible to create a secure working environment for hydrogen-related operations.

References

[1] Zhang F, Zhao P, Niu M and Maddy J 2016 The survey of key technologies in hydrogen energy storage, *Int. J. Hydrogen Energy* **41**, 14535–14552.

[2] Zainul A, Zafaranloo A, Rafiee A, Merida W, Lipinski W and Kaveh K 2020 Hydrogen as an energy vector, *Renewable and Sustainable Energy Rev.* **120**.

[3] Xianfeng C, Chuncheng Z and Yi L 2022 Research and development of hydrogen energy safety, *Emergency Management Science and Technology* **2**, 3.

[4] Hajji Y, Bouteraa M, ELCafsi A, Belghith A, Bournot P and Kallel F 2015 Natural ventilation of hydrogen during a leak in a residential garage, *Renewable and Sustainable Energy Rev.* **50**, 810-818.

[5] Borman G L, Ragland K W 1998 Combustion engineering, (NewYork: McGraw-Hill).

[6] Herath I, Madeley D, Bradshaw B. Hydrogen Fuel Cell taxi: Safety analysis experiences. 4th IET International Conference on System Safety (November 2009).

[7] Momirlan M, Veziroglu T N 2005 The properties of hydrogen as fuel tomorrow in sustainable energy system for a cleaner planet, *Int. J. Hydrogen Energy* **30**, 795–802.

[8] Cerchiara G M, Mattei N, Schiavetti M, Carcassi M N 2011 Natural and forced ventilation study in an enclosure hosting a fuel cell, *Int. J. Hydrogen Energy* **36**, 2478–2488.

[9] Hyon W J, Hongcheol L, Inju H, Heela J 2022 Safe Ventilation Methods against Leaks in Hydrogen Fuel Cell Rooms in Homes, *Energies* **15**, 5434.

[10] Matsuura K 2009 Effects of the geometrical configuration of a ventilation system on leaking hydrogen dispersion and accumulation, *Int. J. Hydrog. Energy* **34**, 9869-9878.

[11] Matsuura K, Nakano M, Ishimoto J 2010 Forced ventilation for sensing-based risk mitigation of leaking hydrogen in a partially open space, *Int. J. Hydrog. Energy* **35**, 4776-4786.

[12] Matsuura K, Nakano M, Ishimot J 2012 Sensing-based risk mitigation control of hydrogen dispersion and accumulation in a partially open space with low-height openings by forced ventilation, *Int. J. Hydrog. Energy*, 1972-1984.

[13] Matsuura K, Nakano M, Ishimoto J 2012 Acceleration of hydrogen forced ventilation after leakage ceases in a partially open space, *Int. J. Hydrog. Energy* **37**, 7940-7949.

[14] Hübert T, Boon-Brett L, Palmisano V, Bader M A 2014 Developments in gas sensor technology for hydrogen safety, *Int J Hydrogen Energy* **39**, 20474-20483.

[15] Spencer C L, Watson V, Hippler M 2012 Trace gas detection of molecular hydrogen H_2 by photoacoustic stimulated Raman spectroscopy, (PARS) *Analyst* **137**, 1384-1388.

[16] Buttner W J, Post M B, Burgess R, Rivkin C 2011 An overview of hydrogen safety sensors and requirements, *Int J Hydrogen Energy* **36**, 2462-2470.

[17] https://h2tools.org/bestpractices/ventilation, access on day: 13.10.2023.

[18] Technical documentation of sensor: https://www.burkert.com/en/type/6240, access on day: 13.10.2023

[19] Technical documentation of sensor: https://www.wika.com/en-in/e_10_e_11.WIKA, access on day: 13.10.2023.

[20] Technical documentation of sensor: https://store.alicat.com/products/m-5000slpm-d, access on day: 13.10.2023.

[21] Technical documentation of sensor: https://www.teledynegasandflamedetection.com/en-us/olct-100-olc-100-toxic-and-combustible-gas-detector, access on day: 13.10.2023.
[22] Sinha A., Wen J. X 2019 A simple model for calculating peak pressure in vented explosions of hydrogen and hydrocarbons, International journal of hydrogen energy 44.
[23] Frank P 1980 Loss Prevention in the Process Industries, Lees Vol. 1 (London and Boston: Butterworths).

IOP Conf. Series: Materials Science and Engineering 1294 (2023) 012057 doi:10.1088/1757-899X/1294/1/012057

Appendix

Table A1. F1 and F2 values for various hydrogen concentrations. Here E refers to the power of 10.

H2%	F1	F2
10	1.78E-05	1.04E-03
11	2.33E-05	1.52E-03
12	3.55E-05	2.57E-03
13	5.79E-05	4.61E-03
14	9.56E-05	8.29E-03
15	1.55E-04	1.46E-02
16	2.44E-04	2.47E-02
17	3.72E-04	4.02E-02
18	5.49E-04	6.30E-02
19	7.87E-04	9.52E-02
20	1.10E-03	1.40E-01
21	1.49E-03	1.98E-01
22	1.99E-03	2.75E-01
23	2.60E-03	3.72E-01
24	3.34E-03	4.92E-01
25	4.22E-03	6.38E-01
26	5.26E-03	8.12E-01
27	6.48E-03	1.02E+00
28	7.89E-03	1.25E+00
29	9.51E-03	1.52E+00
30	1.14E-02	1.82E+00

Table A2. Value of β1 and β2 used for various configurations for hydrogen.

	β1	β2
Ideal	0.243	0.243
obstacle-low congestion	0.243	0.243
obstacle-high congestion	–	–
Initial turbulence	0.5	0.243

IOP Conf. Series: Materials Science and Engineering 1294 (2023) 012058 doi:10.1088/1757-899X/1294/1/012058

Machine learning for underground gas storage with cushion CO_2 using data from reservoir simulation

J O Helland[1,*], H A Friis[1], M Assadi[1] & S Nagy[2]

[1] University of Stavanger, Stavanger, Norway

[2] AGH University of Science and Technology, Kraków, Poland

E-mail: *johan.o.helland@uis.no, jhel@norceresearch.no

Abstract. Underground natural gas storage (UNGS) is a means to store energy temporarily for later recovery and use. In such storage operations, carbon dioxide (CO_2) can be injected as cushion gas to improve the operating efficiency of the working gas and then be permanently stored in the same reservoir. A potential obstacle for widespread use of this technology is that the mixing of the different gases can lead to undesired CO_2 production. Herein, we use a two-component flow model to simulate injection and withdrawal periods of methane (CH_4) in idealized reservoirs containing CO_2. First, we simulate cases with a single well for both CH_4 injection and production. From 1200 simulations with systematic variation of reservoir temperature, porosity, permeability, height, and injection time, we find that the reservoir height and permeability have the most significant impact on the production time until the well stream reaches 1% mole fraction of CO_2. In another set of simulations, we investigate the impact of well spacing in seasonal gas storage scenarios with separate wells for CH_4 injection and production, while CO_2 injection occurs from a third well. Based on the simulated data we construct artificial neural networks (ANNs) that describe the relations between the varied input parameters and the production time of CH_4, well-block mole fraction and pressure. We conclude that trained and validated ANN models are useful tools to optimize important parameters for UNGS operations, including well positioning, with the aim at maximizing the amounts of delivered gas.

1. Introduction

Natural gases, e.g., methane (CH_4), are regarded as one of the cleanest energy resources among the fossil fuels [1]. In underground natural gas storage (UNGS) the gas is stored temporarily in salt caverns, reservoirs, or aquifers, before it is produced upon demand (e.g., for electrical heating in winter seasons) and to maintain a steady energy supply [1, 2]. Storage operations in an underground reservoir can benefit from using a cushion gas to improve the operating efficiency of the working gas by contributing with pressure support [2]. The compressible properties of carbon dioxide (CO_2) are an advantage for a cushion gas as it allows for large storage capacities of natural gas in the reservoir, while its expansion to lower pressure aids the gas production [3]. Another potential advantage is that injected CO_2 can be stored permanently in the reservoir after it served as a cushion gas for UNGS. However, in such storage strategies a disadvantage is that the mixing of gases in the reservoir can lead to undesired CO_2 production during continued seasons of injections and withdrawals.

In the literature, reservoir simulation studies of UNGS with cushion gas focus on demonstrating the advantageous density changes of CO_2 near its critical point in idealized reservoirs [3]; the mechanical responses to pressure build-ups in idealized, yet inclined reservoirs

[4]; and simulations of working gas quality caused by mixing with cushion gas [5]. While the entry of data-driven methods has proven useful in research on CO_2 storage [6], there are to date few data-driven methodologies demonstrated for UNGS with cushion CO_2 in reservoirs or aquifers. In the literature on UNGS, most data-driven approaches are concerned with describing the relations between gas deliverability (that is, the amount of gas that can be withdrawn per day [2]), the flowing bottom hole pressure (BHP) in the well, and reservoir pressure [7, 8]. On the other hand, Mann III & Ayala [9] developed an artificial neural network (ANN) model aimed at determining the optimum design of a storage facility, but their model did not include information on which gas was considered as cushion gas, nor the complete reservoir conditions, both of which are important for the behaviour of the gases and their mixtures.

Herein, we perform systematic reservoir simulations of UNGS at reservoir conditions with cushion CO_2 in idealized reservoir geometries, inspired by Oldenburg [3] and Ma et al. [4]. The set of simulations generates data that we use to develop ANN models that learn parameter relationships for seasonal storage of CH_4 with cushion CO_2. Our objective is twofold: First, we show that ANN models can predict production time until the CO_2 mole fraction exceeds 1% in the well block (that is, the grid block where the bottom hole of the well is located). Secondly, for seasonal storage with fixed injection and production periods, we develop ANN models that can describe the CO_2 mole fraction in the well block after each seasonal storage cycle. The prediction of these parameters under different well configurations and injection/production schedules contribute to the optimization of UNGS operation, with the aim at minimizing costly and undesired CO_2 production.

2. Methods

Reservoir simulation was used to generate data for seasonal CH_4 storage scenarios in idealized 2D reservoirs initially containing CO_2. Using the simulation results, artificial neural network (ANN) models were developed to describe relations between input parameters and calculated parameters, such as production time, well-block mole fraction and well-block pressure.

2.1. *Reservoir models*

For the reservoir simulations we use Matlab Reservoir Simulation Toolbox (MRST) [10] and its *Compositional* module [11]. MRST is free, open-source software written in Matlab and the MRST module *Compositional* includes some well-known solution strategies for compositional reservoir simulations. The Peng-Robinson equation of state (EoS) and Lohrenz-Bray-Clark correlation for viscosity [12], with industry standard corrections that account for mixing behaviour of gases, are default settings and are also used in the simulations carried out in this work. The compositional module of MRST has previously been benchmarked against the commercial reservoir simulator Eclipse 300, as well as AD-GPRS (the Stanford University research simulator) [11].

We use a two-component model with gravity to simulate various seasonal CH_4 storage scenarios in idealized, vertically aligned, rectangular 2D reservoirs, initially containing CO_2. 2D setups were used in this study, because 2D simulations are less computationally costly than 3D simulations, and hence it allows us to explore a more comprehensive set of parameter variations with simulations in shorter time. The aim here is thus not to develop ANN models that describe storage in a real reservoir, but rather to build confidence in ANN models as useful tools to describe relations between influential parameters in UNGS with cushion CO_2. All simulations assume the initial reservoir pressure is 6 MPa. During injection the pressure rises above the critical point where CO_2 exists as a supercritical phase. Hence, the simulations account for conditions where advantageous, large density changes of CO_2 occur with pressure [3]. The reservoirs have fixed length 1000 m, and we always assume vertical permeability is 10% of horizontal permeability, $K_v = 0.1K_h$, which is a reasonable assumption in non-fractured, laminated rock [13]. All simulations use vertical wells with radius 10 cm, while all other reservoir boundaries were sealed. Injection and production rates are specified at standard surface conditions (temperature 273.15 K and 1 atm pressure). For this setup, we carried out simulations with different grid block resolutions and maximal time steps and found that grid blocks of size $5 \times 1 \times 1$ m^3 (assuming depth of the 2D reservoir is 1 m) and maximal time step of 3 days were suitable for the investigations. These results were also validated against the commercial simulator GEM [14].

2.2. *Artificial neural network models*

To tailor the setup of feed-forward neural networks with backpropagation, data generated from the reservoir simulations were utilized. Due to its flexibility in design, it was decided to use *TensorFlow* and the high-level *Keras* library installed in a *Python* environment for the ANN model development. Keras is an open-source library that provides a Python interface for neural networks, and it acts as an interface for the TensorFlow library. Within this code framework we have high flexibility in determining the structure of the ANN and in training of the model.

When developing the optimal choice of the ANN structure, we started with a small number of hidden layers, and a small number of neurons in each layer. Then the networks were trained by systematically increasing the number of neurons and layers until no significant improvement in the results was observed. For each network the choice of activation functions, optimization method, and specifications of the learning rate were altered. We also investigated the deactivation of a fraction of neurons in each layer (called "dropout fraction") and the use of regularization weights in every layer based on a metric (for example L_2-norm) that will add a penalty to the total loss function. These latter functionalities are useful means to avoid underfitting and overfitting in ANN models.

Based on this experimentation, ANN models with two hidden layers and a relatively large number of neurons were developed. The model specifications also include the activation functions recified linear unit (*ReLu*) and tanh, whereas the use of dropout fractions did not improve the results. The ANN training also made use of a slightly declining learning rate and L_2-regularization. As optimization method we used Adam optimizer while mean squared error was used both as metric in the training and as loss function.

3. Results

We begin by describing two scenarios of UNGS with cushion CO_2 and present results from the corresponding reservoir simulations. These scenarios are outlined in Section 3.1.1 (Case 1) and Section 3.1.2 (Case 2). Then we proceed with presenting the corresponding ANN models for these cases and their performance.

3.1. *Reservoir simulations*

3.1.1. Case 1: One-well CH_4 injection & withdrawal cycle. Case 1 has one well in the upper left corner of the reservoir (3 m below the top seal, see Figure 1) which is used to simulate CH_4 injection followed by production with a rate of 0.015 m^3/s. Production ceases when well-block CO_2 mole fraction exceeds 1%, so that production time is different from injection time. The impact of reservoir thickness (vertical distance) H, horizontal permeability K_h, porosity ϕ, temperature T, and injection time t_{inj}, were studied by varying these parameters independently in different simulations. Figure 1 shows the distribution of CH_4 at a time during injection and production from one such simulation. The lighter gas, CH_4, tends to stay above CO_2 while a mixing zone of both gas components develops in between. Figure 2 shows the evolution of well-block pressure and well-block CH_4 mole fraction over time for this simulation. The first part of the production exhibits a time interval in which the well primarily produces CH_4, whereas at later times the well-block CH_4 mole fraction declines, leading to significant amounts of undesired

CO_2 production. We also note that the well-block pressure increases steadily during injection, but in the beginning of the production stage it decreases rapidly to enable flow toward the well from distant zones in the reservoir.

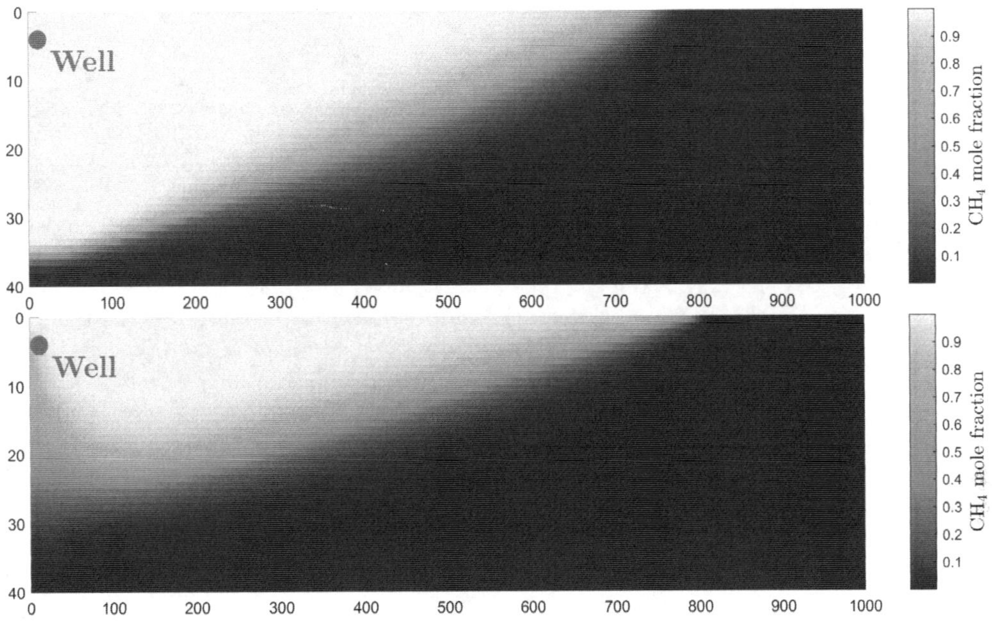

Figure 1. CO_2 and CH_4 distribution at 90 days (*top*) and 240 days (*bottom*) from a typical Case 1 simulation with $T = 323.15$ K, $H = 40$ m, $\phi = 0.2$, $K_h = 1000$ mD, and $t_{inj} = 180$ days.

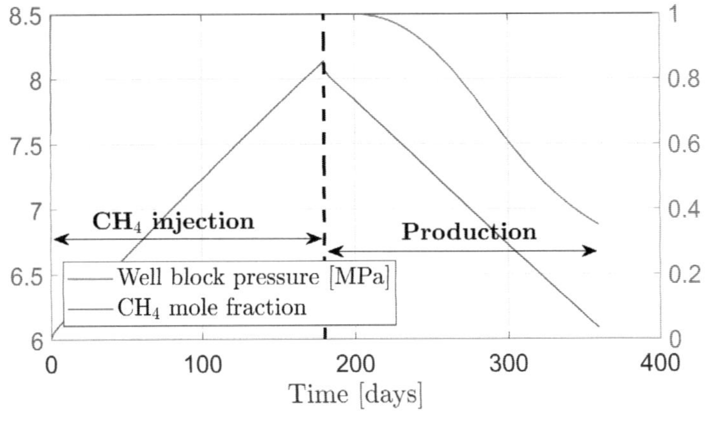

Figure 2. Pressure and CH_4 mole fraction in well block over time from a typical Case 1 simulation with $T = 323.15$ K, $H = 40$ m, $\phi = 0.2$, $K_h = 1000$ mD, and $t_{inj} = 180$ days.

Table 1 shows the range and resolution of the parameters varied independently in the simulations. Thus, a total number of 1200 ($= 4 \times 5 \times 4 \times 5 \times 3$) simulations were carried out with different parameter combinations. To identify the most influential parameters on the simulated well-block pressure and well-block mole fraction, a simulation with midrange parameters $T = 328.15$ K, $H = 60$ m, $\phi = 0.155$, $K_h = 700$ mD, and $t_{inj} = 180$ days was conducted. The midrange parameter simulation was then compared with new simulations where the midrange value of one parameter at a time was substituted with its maximum and

Table 1. List of parameter values varied independently in the MRST simulations.

Temperature T [K]	Reservoir thickness H [m]	Porosity ϕ [frac.]	Horizontal permeability K_h [mD]	Injection time t_{inj} [days]
313.15	30	0.11	100	150
323.15	40	0.14	300	180
333.15	50	0.17	500	210
343.15	70	0.20	900	–
–	90	–	1300	–

minimum value. Among the varied parameters, the results show that the reservoir thickness H and horizontal permeability K_h have the most significant impact on the well-block mole fraction, particularly in the interval between the minimum and midrange values, see Figure 3. Consequently, additional simulations were carried out with variation of H and K_h in this interval, so that the step size for the variation of these parameters is uneven, see Table 1.

From the set of simulations, we saved to file the five varied input parameters in addition to the maximum pressure, average reservoir pressure and average CH_4 mole fraction before and after production, well-block pressure after production, and production time (i.e., the time until well-block CO_2 mole fraction exceeds 1%).

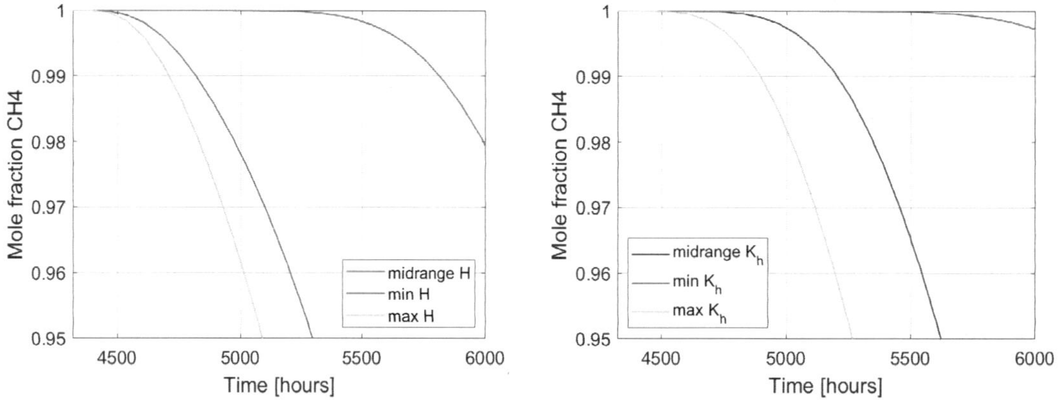

Figure 3. Well-block mole fraction of CH_4 during production from a Case 1 simulation with midrange parameter values compared to Case 1 simulations with maximum and minimum values of reservoir thickness H (*left*) and horizontal permeability K_h (*right*).

3.1.2. Case 2: Seasonal CH_4 injection & withdrawal cycles with continuous CO_2 injection using different wells. Case 2 explores the behaviour in seasonal storage cycles when CH_4 injection and production occur from two different wells, while continuous CO_2 injection occurs from a third well. This is a scenario for using the reservoir both as a facility for seasonal gas storage with cushion CO_2 and as a site for permanent CO_2 storage. In this case we set $H = 40$ m, $K_h = 1000$ mD, $\phi = 0.2$, and $T = 323.15$ K. The location of the CO_2 injection well is 37 m below the top seal at left boundary, the location of the CH_4 injection well is 3 m below the top seal at a distance $x = d_W$ from the left boundary, while the location of the production well is at the right boundary, 3 m below the top seal. This setup is illustrated in Figure 4 (top). The schedule in the simulations is as follows: (1) initial CH_4 injection with rate 0.025 m^3/s for 360 days, (2) 180 days with all wells closed which allows some gravitational segregation, and (3) three cycles of CH_4 injection (180 days) and withdrawal (180 days) with rate 0.015 m^3/s while continuous CO_2 injection occurs from the third well. We studied the impact of CO_2 injection rate ($Q_{CO2} = 0$, 0.005 and 0.010 m^3/s) and CH_4 injection well position ($d_W = 200$, 400, 600 and 800 m) by varying these parameters independently, which constitutes 12 ($= 3 \times 4$) simulations.

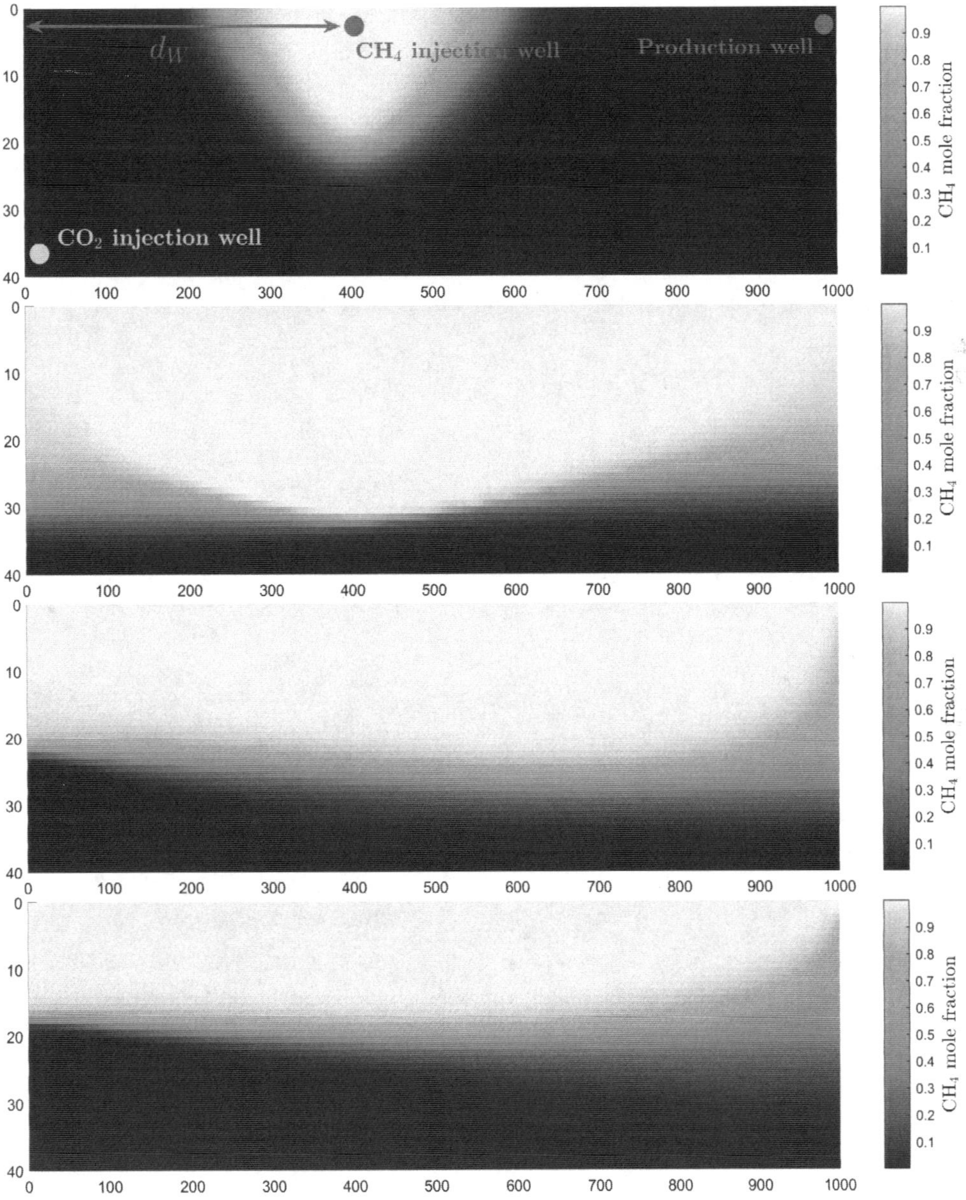

Figure 4. CH_4 distribution for Case 2 with $Q_{CO2} = 0.01$ m^3/s and $d_W = 400$ m at four time slots (ordered from top to bottom): during initial CH_4 injection (30 days), after the initial CH_4 injection (360 days), after the first seasonal production (900 days), and after the third seasonal production (1620 days).

Figure 4 shows images of the CH_4 distribution in the reservoir at four times, for the case with $Q_{CO2} = 0.010$ m^3/s and $d_W = 400$ m. After the initial CH_4 injection, the most significant CH_4/CO_2 mixing zones take place at the flanks. During CO_2 injection, the mixing zone at the left boundary diminishes while the mixing zone at the right boundary, where production occurs, expands slightly. Figure 5 shows the impact of Q_{CO2} on pressure and mole fraction in the production well-block, and the impact of injection well position d_W on the mole fraction in the production well-block. The impact of well position d_W on the pressure in the production well-block is insignificant and is not shown here. The results show, as expected, that a continuous CO_2 injection leads to an undesired pressure build-up over time which gets worse with injection rate and subsequent seasonal injection/withdrawal cycles. A significant increase in pressure could damage the reservoir rock, alter its permeability [4], and potentially lead to gas leakage. With respect to mole fraction, an increased CO_2 injection rate has a marginal positive effect in the first cycle, as it results in lower production of CO_2, while the subsequent seasonal cycles show more severe negative effects as the amount of CO_2 production increased significantly. Further, the simulations with varying well position shows that a decreased spacing between CH_4 injection and production wells leads to the least CO_2 production during seasonal cycles. Hence, the optimal setting is to use the same well for production and injection.

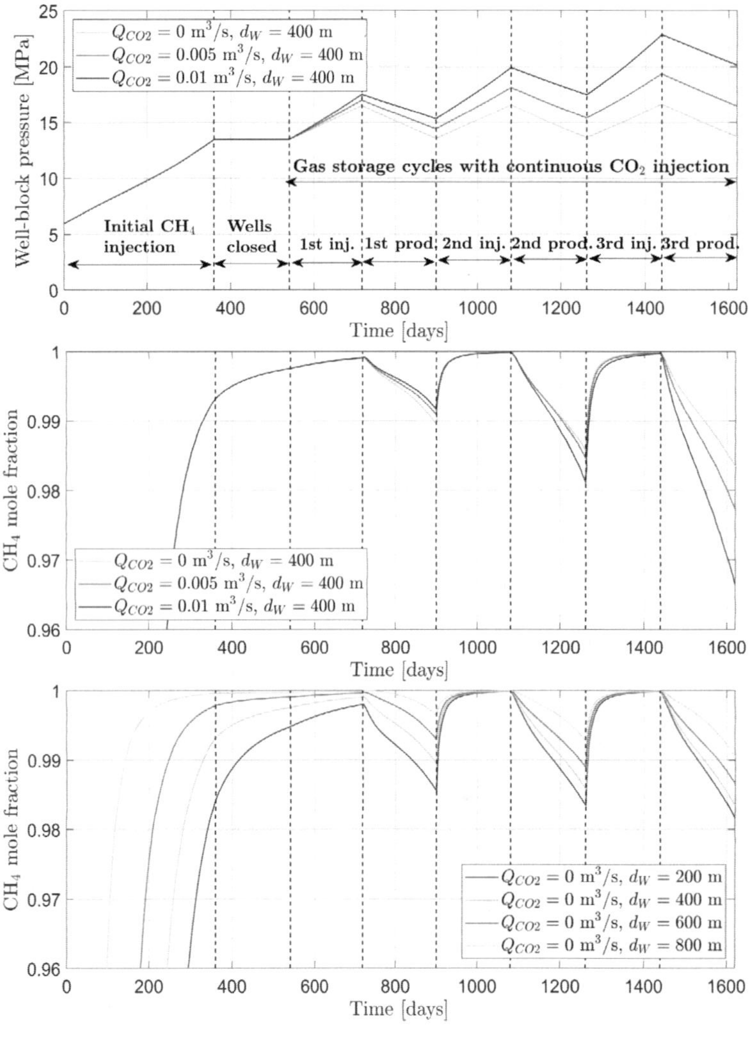

Figure 5. Well-block pressure (*top*) and well-block CH_4 mole fraction (*middle*) over time for Case 2 simulations with different CO_2 injection rates Q_{CO2} and fixed CH_4 injection-well position $d_W = 400$ m. *Bottom:* Well-block CH_4 mole fraction over time for Case 2 simulations with $Q_{CO2} = 0$ m^3/s and different d_w. Vertical lines (*black, dashed*) indicate the different injection/production schedules.

3.2. ANN results

ANN models for Case 1 and Case 2 were developed separately. The ANN model for Case 1 aims at predicting the production time until well-block CO_2 mole fraction exceeds 1%, while for Case 2 where the production time is fixed, the ANN model aims at predicting the CO_2 mole fraction in the production well-block after each seasonal withdrawal.

3.2.1. ANN model for Case 1. The ANN model for Case 1 takes as input the five varied parameters from the simulations (see Table 1). Before training the ANN model, it is crucial to normalize all the data to the range $[0, 1]$. Normalization is important for stability of the training, as the data varies significantly in magnitude between the parameters. We also shuffle the 1200 data sets from the simulations randomly, with possibility for replicating the shuffle if desired. After the shuffle, the first 75 data sets were taken as the validation set, the next 75 data sets as testing set, and the remainder (1050 data sets) as the training set. In the optimized ANN model structure for Case 1, the following number of neurons were obtained: 5 (input layer) – 42 (hidden layer) – 15 (hidden layer) – 1 (output). We used *ReLu* as activation function, a slightly declining learning rate, and L_2-regularizer penalty coefficient of 10^{-4}. The model was trained for 200 epochs. Figure 6 shows the ANN model's excellent performance on the training data, validation data, and test data sets.

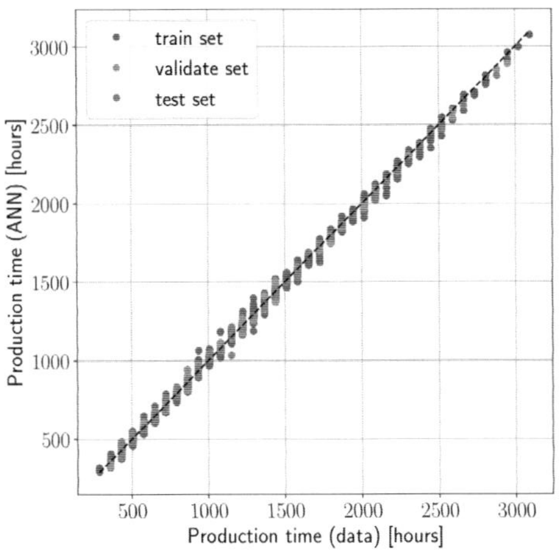

Figure 6. ANN calculation of production time until the well-block mole fraction exceeds 1% CO_2 versus corresponding data sets from training, validation, and testing.

3.2.2. ANN model for Case 2. For Case 2 only the CO_2 injection rate and the CH_4 injection-well position were varied. In addition, each simulation provided results for three seasonal injections and productions. Thus, for the Case 2 ANN model the injection rate (Q_{CO2}), well position (d_W), and the seasonal cycle number (1, 2 or 3) were considered as inputs. These yield 36 ($= 3 \times 4 \times 3$) data sets, all of which were used for training the ANN model. The output is the CO_2 mole fraction in the production well-block.

In the optimized ANN model structure for Case 2, the following number of neurons were obtained: 3 (input layer) – 21 (hidden layer) – 5 (hidden layer) – 1 (output). We used *ReLu*, tanh and *ReLu* as activation functions in the hidden layers and the output layer, a slightly declining learning rate, and L_2-regularizer penalty coefficient of 10^{-3}. This model was trained for 600 epochs. Figure 7 (left) shows the ANN model's excellent performance on the training data. To evaluate its performance on other data, we performed two new MRST simulations for different combinations of Q_{CO2} and d_W: $Q_{CO2} = 0.0075$ m^3/s, $d_W = 300$ m, and $Q_{CO2} = 0.0025$ m^3/s,

$d_W = 700$ m. Figure 7 (right) shows that the ANN model predicts the mole fraction at the end of each seasonal cycle very well for both simulations.

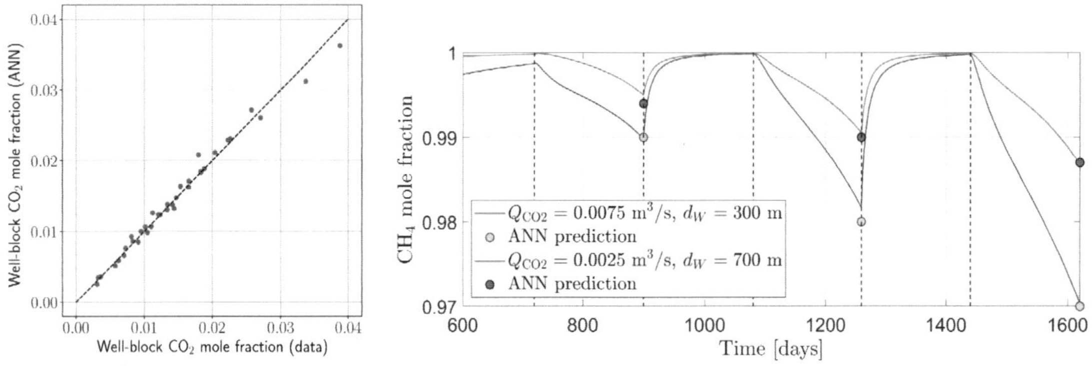

Figure 7. *Left:* ANN calculation of well-block CO_2 mole fraction for Case 2 versus corresponding data used in the training. *Right:* Performance of ANN model for well-block mole fraction when it is applied to two new simulations of Case 2.

Finally, the ANN model for Case 2 was expanded slightly by also taking pressure in the production well-block as output. For this problem, the optimal ANN structure included more neurons in each layer: 3 (input layer) – 27 (hidden layer) – 15 (hidden layer) – 2 (output). This time the L_2-regularizer penalty coefficient was set to 10^{-4}. Otherwise, the ANN specifications remained unchanged. This model was also trained for 600 epochs.

Figure 8 shows the ANN model's excellent performance on the training data set with respect to describing simultaneously the well-block CO_2 mole fraction and the well-block pressure accurately. The ANN model was also compared against the two new MRST simulations for validation. Figure 9 shows an excellent agreement between these simulations and the ANN model with respect to both pressure and mole fraction.

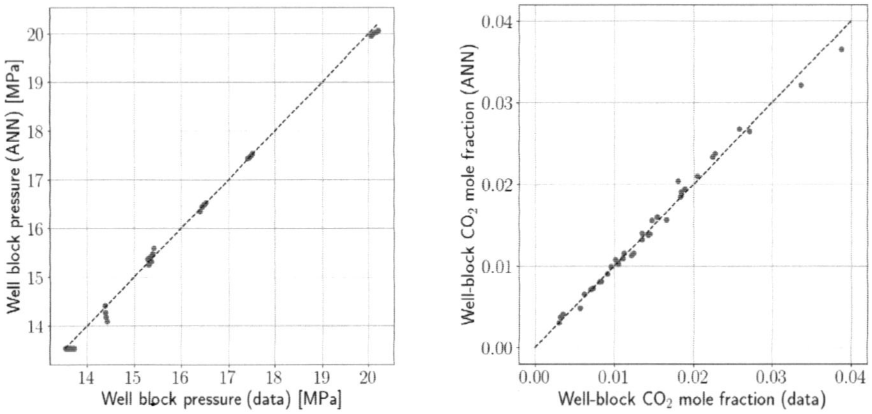

Figure 8. Simultaneous ANN calculation of well-block pressure (*left*) and CO_2 mole fraction (*right*) versus the data used in the training (from Case 2).

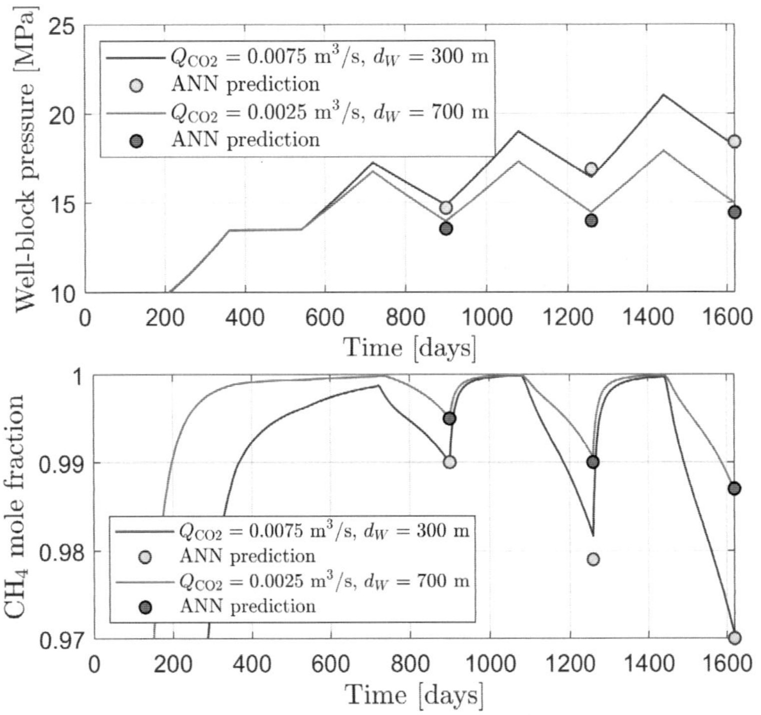

Figure 9. Performance of the ANN model for Case 2 that calculates both well-block CO_2 mole fraction and well-block pressure simultaneously. ANN predictions of well-block pressure (*top*) and well-block mole fraction (*bottom*) compared with two new test simulations of Case 2.

For the ANN model of Case 2 which takes both well-block mole fraction and well-block pressure as outputs, the RMSEs are 1.6×10^{-3} for mole fraction and 0.32 MPa for pressure when compared against the new test simulation with $Q_{CO2} = 0.0075$ m^3/s and $d_W = 300$ m, while the RMSEs are 3.37×10^{-4} (mole fraction) and 0.48 MPa (pressure) with respect to the other test simulation with $Q_{CO2} = 0.0025$ m^3/s and $d_W = 700$ m. In comparison, the ANN model that only takes well-block mole fraction as output obtains RSMEs of 9.95×10^{-4} and 6.6×10^{-4} for the first and second test simulation, respectively. Hence, accuracy is maintained in the ANN model when increasing the number of outputs, albeit at the expense of increasing the number of neurons in the hidden layers in the ANN. This shows promise in expanding the data-driven models with higher numbers of inputs and outputs, accompanied by more data for training and validation, to increase their predictive capability and utility value for seasonal UNGS with cushion CO_2.

4. Conclusions

Reservoir simulation and ANN models were used to explore seasonal cycles in UNGS with cushion CO_2. Our conclusions are as follows:

- To avoid CO_2 production, it is paramount to have a significant amount of CH_4 present in the reservoir before seasonal injection & withdrawal cycles begin.
- Decreased spacing between CH_4 injection & production wells leads to the least CO_2 production during seasonal cycles. Hence, the optimal setting is to use the same well for production and injection.
- Continuous CO_2 injection during UNGS has a marginal positive effect in the first cycle, while later cycles bear disadvantages with increased CO_2 production and undesired pressure build-up that worsen with increased CO_2 injection rate.
- Trained ANN models capture the behaviour of critical parameters that describes undesired

CO_2 production during UNGS. Hence, ANN models are useful tools to optimize UNGS operations.

Future work should supplement the data generated here with simulations of other schedules for CO_2 injection in conjunction with seasonal CH_4 injection and production. It will also be needed to expand the ANN models with other output parameters. For example, with the addition of well-block pressure and average reservoir pressure, the ANN model could contribute to establishing their relations to well deliverability [7, 8]. It is also desirable to explore deep-learning methods to describe the CH_4/CO_2 distribution in the reservoir, in particular the extent of mixing zones, potentially by utilizing methods developed for CO_2 storage that describe the migration of CO_2 plumes in sandstone formations [6].

Acknowledgments
This research was funded by EEA and Norway Grants, operated by The National Center for Research and Development, grant number NOR/POLNORCCS/AGaStor/0008/2019-00.

References

[1] International Energy Agency (IEA) Fuel & technologies, gas Available online URL `https://www.iea.org/fuels-and-technologies/gas`
[2] US Energy Information Administration (EIA) The basics of underground natural gas storage. natural gas report November 16, 2015. Available online URL `https://www.eia.gov/naturalgas/storage/basics/`
[3] Oldenburg C M 2003 *Energy & Fuels* **17** 240–246
[4] Ma J, Li Q, Kempka T and Kühn M 2019 *Energy & Fuels* **33** 6527–6541
[5] Sadeghi S and Sedaee B 2022 *Journal of Energy Storage* **46** 103885
[6] Chu A K, Benson S M and Wen G 2023 *Energies* **16** 246
[7] Ali A 2021 *Energy* **229** 120648
[8] Thanh H V, Zamanvad A, Safaei-Farouji M and Ashraf U 2022 *Renewable Energy* **200** 169–184
[9] Mann III A W and Ayala L F 2009 *Int. J. Modelling and Simulation* **29** 214–223
[10] Lie K A 2019 *An introduction to reservoir simulation using MATLAB/GNU Octave: User guide to the MATLAB Reservoir Simulation Toolbox* (Cambridge University Press)
[11] Lie K A and Møyner O (eds) 2021 *Advanced Modeling with the MATLAB Reservoir Simulation Toolbox* (Cambridge University Press) chap 8. Compositional simulation with the AD-OO framework, pp 324–374
[12] Lohrenz J, Bray B G and Clark C R 1964 *Journal of Petroleum Technology* 1171–1176
[13] Fanchi J R 2010 *Integrated Reservoir Asset Management* ed Fanchi J R (Boston: Gulf Professional Publishing) pp 49–69 ISBN 978-0-12-382088-4
[14] CMG 2015 CMG GEM User's Guide Computer Modeling Group Ltd., Calgary, Alberta, Canada

IOP Conf. Series: Materials Science and Engineering 1294 (2023) 012059 doi:10.1088/1757-899X/1294/1/012059

Numerical modelling of a high temperature borehole thermal energy storage system: Norway case study

A A Ahmed[1], M Assadi[1,*], R Gholami[1], N Ahmed[1,2]

[1] Department of Energy and Petroleum Engineering, University of Stavanger, 4021 Stavanger Norway

[2] US Pakistan Centre for Advanced Studies in Energy (USPCASE), National University of Sciences & Technology (NUST), H-12 Sector, Islamabad, Pakistan

[*] Corresponding author: mohsin.assadi@uis.no

Abstract. Global warming is threatening life on earth. Utilising renewable energy is considered as the most effective measure to minimise anthropogenic CO_2 emissions. High-temperature borehole thermal energy storage (BTES) systems have a world-wide potential to reduce energy consumption, increase energy utilisation of waste heat and provide efficient seasonal heat storage. Hybrid application of BTES and solar energy leads to net zero emissions. In this study HT-BTES is evaluated for seasonal thermal heat storage and recovery. To this end, a CMG STARS model was built and validated using the existing 100-wells BTES system in Norway. Then, the model was used to evaluate BTES thermal performance and thermal recovery efficiency. Sensitivity analysis was also conducted to study the dynamics of storage temperature in the BTES under different operating conditions, such as heat carrier flow rate, injection temperature, and charging period. Results of this case study show that the model-predicted temperatures during charging and discharging are in good agreement with the existing BTES system. In 5 years of operation, 35.5% of the heat injected into the BTES system was recovered, while the significant heat remained in the borehole region and lost to surrounding rock (64.5%). BTES was found very sensitive to flow rate, the charging period and injection temperature. Borehole depth has a minimal effect on BTES storage temperature at constant studied injection temperature.

1. Introduction

Anthropogenic climate change is threatening nature and society by altering temperatures and weather patterns. Floods, droughts, and wildfires are becoming more frequent in many parts of the world. Developing a sustainable energy strategy is imperative for protecting natural resources and ensuring future generations well-being. In response, most countries have joined an alliance to gradually reduce greenhouse gas (GHG) emissions as they strive to reach net zero emissions by 2050. Following the same energy transition policy, Norway is increasing its actions and responsibilities to tackle emissions despite having a low carbon-intensity energy system [1]. In its ambitious climate action plan, Norway aims to cut greenhouse gas emissions by at least 50% and towards 55 % by 2030 compared to 1990 emissions baseline [2]. That would result in less than 23 million tonnes of carbon dioxide equivalent emissions in 2030 compared with approximately 52 million tonnes in 1990 [3]. To achieve this ambitious goal, it requires electrification of various sectors, including transport, manufacturing and oil and gas extraction [1]. This in turn requires adequate, flexible and clean energy supply sources such as geothermal, wind and solar to reduce emissions. Solar power represents an almost insignificant share of renewable energy in Norway. However, the Norwegian Solar Energy market is expanding rapidly and is projected to increase from 358 MW in 2022 to 4943 MW by 2028 [4]. Furthermore, the intermittency of solar energy poses challenges to its broad application. However, seasonal borehole thermal energy storage (BTES), which includes technologies both for short- and long-term retention of heat, can create or improve the utility of solar energy.

The BTES system offers multiple opportunities as it offsets the mismatch between thermal energy production from solar and demand and reduces grid dependence [5]. The BTES are vertical boreholes that are piped in series or parallel to create concentric thermal zones (see Figure 1). Therefore, heat is stored in a rock environment near to the boreholes as a storage medium, and heat exchange occurs through vertical boreholes [6, 7]. It has the advantages of creating cold and warm zones as well as the possibility to switch to a reverse flow. Large solar thermal collector fields are promising technology for replacing fossil fuels in heating and cooling systems. So, BTES is a huge natural capacity to temporarily store this solar energy (harvested in summer) when the demand is low, and then recover it when the demand is high in the future. The recovered thermal energy can be used for heating buildings. The main drawbacks of BTES systems are their slow response time and high thermal losses. Heat transfer from the ground to the heat transfer fluid is relatively slow due to low heat transfer rates [8, 9].

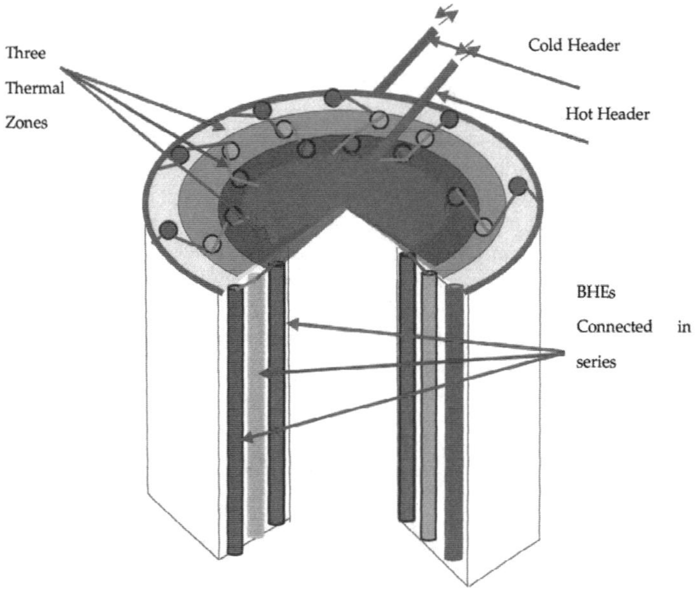

Figure 1. Schematic of Borehole Thermal Energy Storage (BTES and BHEs are connected in series).

This study, a simulation for BTES system, is performed to history-match an existing BTES system (GeoTermos Project) used to store heat in summer and provide space heating in winter for the Fjell School building (1000 m^2) located in Drammen, Norway. The GeoTermos consists primarily of a thermal storage facility consisting of a total of 100 wells, each with a depth of approx. 50 meters, which are organised as a cylinder with a diameter of 40 meters. Energy is supplied directly from solar collectors (150 m2) and solar cells mounted on roofs (900 m^2) and in façades (100 m^2) as shown in Figure 2. The solar cells supply electricity for an air source heat pump that uses natural refrigerant CO2. Direct heating of the school, by 30°C water-based underfloor heating system. Injection of heat started in April 2020. After 6 months of heat injection the rock temperature recorded in Sept 2021 reached <43°C. Figure 3 illustrates fibre for distributed temperature sensing installed in 11 of the BHEs.

The incorporation of additional sensors, particularly the mass flow rate probe, introduces significant implications both technically and economically. While the advantages of enhanced data acquisition and system monitoring are evident in our current study, the potential cost implications of such sensor implementations are acknowledged. To address this issue, a supplementary cost evaluation will be conducted for the installation of these sensors in our ongoing borehole field project at the University of Stavanger. This assessment will encompass an analysis of initial setup costs, ongoing maintenance expenses, and the potential economic benefits resulting from improved data quality and enhanced system efficiency facilitated by these sensors. The resulting assessment to be presented in our future works will enable the readers to gain a better understanding of the trade-offs and advantages associated with sensor deployment in BTES system design and operation.

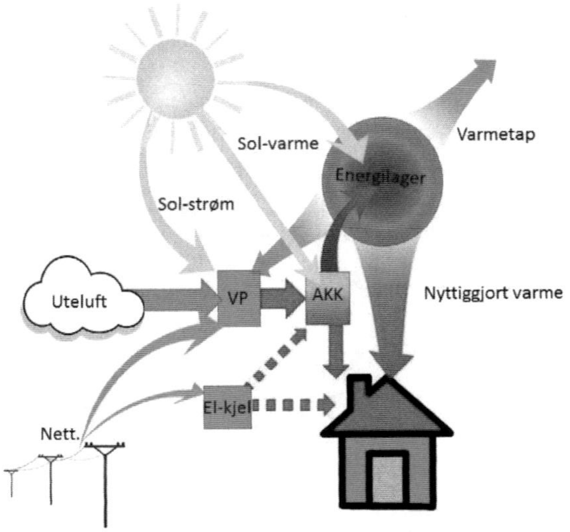

Figure 2. Schematic representation of the Fjell School BTES (GeoTermos) used in this study.

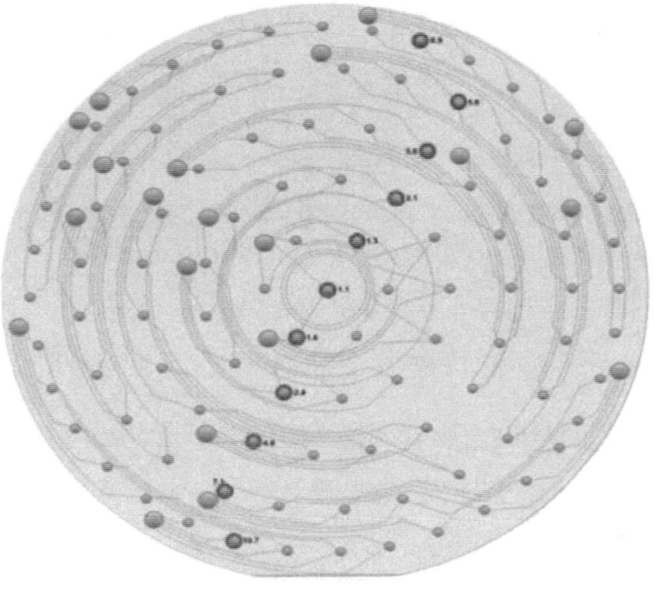

Figure 3. Fibre for distributed temperature sensing installed in 11 of the BHEs

The model was then used to investigate the effect of underground properties, depth of borehole exchangers, charging/discharge cycle time and discharge efficiencies to determine overall thermal efficiency. In addition, the impact of injected water temperature was explored. The underground thermal storage system is based on seasonal cyclic injection and production from BTES. The solar energy is used to directly charge the BTES or using a heat pump fed by electricity produced from solar. CMG STARS Software was used to model BTES using geological properties, heat transport in the subsurface, and well characteristics. Then, the model was used to run a sensitivity analysis to determine the impact of main operational parameters on the potential of storing solar energy, utilisation efficiency and overall performance.

2. Methodology

CMG-STARS simulator (Version 2022.10) is used to measure heat storage, thermal heat production, subsurface heat distributions and efficiency of BTES. CMG-STARS models the heat transfer from injected water to the rock matrix. A three-dimensional (3D) model was created to simulate heat transfer with 100 wells (50 to 55 m each) as shown in Figure 4. The model dimensions are 120x120x60 meters. A regular coordinate system having three dimensions (i, j, k) was applied to the model. High grid resolution (1 meter) was used for the 3D grid to provide an improved accuracy in the area of drilled wells and avoid numerical diffusion. The main input parameters used for the simulation are summarised in Table 1.

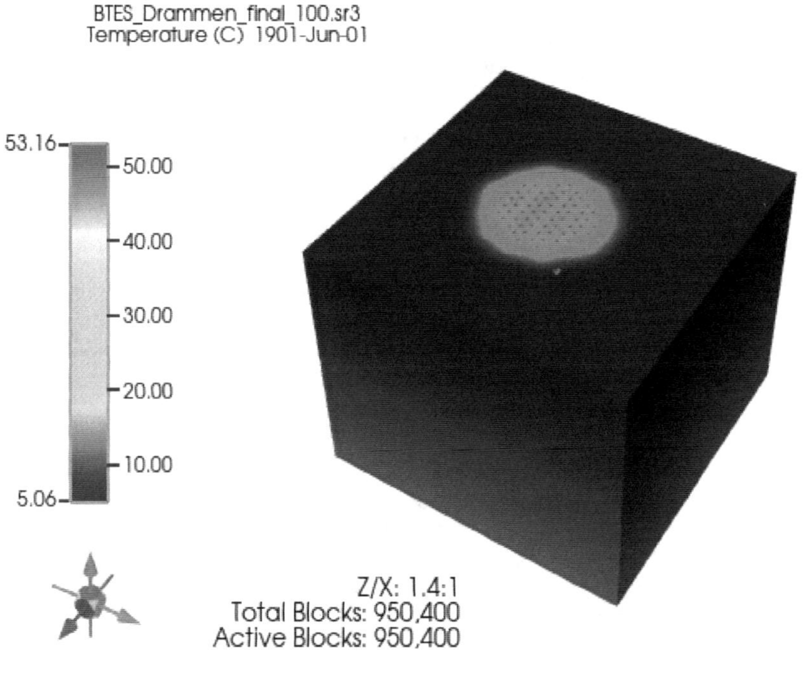

Figure 4. The 3D view of the numerical model

Table 1. Input parameters for HT-ATES CMG STARS modelling

Data	BTES
Length, I (120 cells), m	120
Width, J (120 cells), m	120
Thickness, K (60 cells), m	60
Porosity, %	15
Permeability, md	25
Undisturbed rock temperature, C	8
No. of wells	100
Well length, m	50-55
Distance between wells, m	4
Flow rate, m3/hr	4.2
Injection temperature (Charging), °C	60
Injection temperature (Discharging), °C	25
Rock thermal conductivity, W/mk	2.5
Seasonal cycle, month	6

2.1 Heating load

The seasonal BTES system studied is assumed to have a heating mode in the winter (discharge) and a charging mode in the summer. Charging and discharging loads in the simulations are simplified to a 6-month heating period in winter and a 6-month charging period in summer. A constant flow rate and injection temperature are used during summer and the same flow rate is used during winter for discharge. The extracted heat is assumed to directly heat the school by a 30°C water-based underfloor heating system. The maximum pumping rate is set to 4.2 m3/h with an injection temperature of 25°C. The undisturbed ground temperature is 8°C and the model was run for 10 years with 20 repeated cycles.

2.2 Thermal efficiency.

From the model, the extracted energy (Q_{ext}) for each 6-month heating period during the lifetime of the system is determined following Eq1:

$$Q_{ext} = C_p \, q \, \rho \, (T_{wh} - T_{inj}) \, \mathrm{n} \tag{1}$$

Here, q is the total pumping rate, T_{wh} is the temperature of the water that is extracted from the production well at the wellhead, T_{inj} is the temperature of the water that is injected in the injection well and n is a time increment. The injected heat during summer (Q_{inj}) for each 6-month charging period during the lifetime of the system is determined following Eq2.

$$Q_{inj} = C_p \, q \, \rho \, (T_{inj} - T_{in}) \, \mathrm{n} \tag{2}$$

IOP Conf. Series: Materials Science and Engineering 1294 (2023) 012059 doi:10.1088/1757-899X/1294/1/012059

Here, q is the total pumping rate, T_{inj} is the temperature of the water that is injected to the well at the wellhead, T_{in} is the temperature of the water that is fed to the heat exchanger. The thermal heat recovery efficiency (η_{th}) is defined by dividing the extracted energy by the energy that is loaded to ATES.

$$\eta_{th} = \frac{Q_{ext}}{Q_{inj}} \tag{3}$$

2.2 Sensitivity analysis and optimization.

The amount of energy that is provided by BTES varies with flow rates, ground temperature, and injected temperature. The effects of well depth on BTES thermal recovery efficiency were evaluated. Sensitivity analysis was extended to include well operation and control strategies. The injection temperature and cycle time were investigated to determine the optimum values that could result in better thermal recovery efficiency.

3. Results and discussions

In the following section the results of the simulation study are presented and discussed.

3.1 Model validation

To assess the validity of the model predictions, a history matching technique was employed, comparing the model outcomes with actual data obtained from the reference borehole field [10]. The validation procedure focused on a 2D cross-sectional representation of the BTES system, encompassing temperature distribution data from 11 wells equipped with temperature measurement sensors during the dynamic heat injection and extraction cycles. As illustrated in Figure 5, the temporal evolution of thermal distributions within the BTES system is presented, covering a 6-month heat injection cycle followed by a 6-month heat extraction cycle. Figure 6 reveals insights into the predicted reservoir temperature during the charging phase. It's noteworthy that the model projected a reservoir temperature of 45°C during this period, closely aligning with the measured temperature obtained from fiber optic sensors, which recorded a temperature just below 43°C. The variance between the modelled and actual temperature is less than 5%, well within an acceptable margin of error. While the pursuit of a perfect match between model and experimental data remains a goal, it is essential to consider the practical complexities and underlying assumptions inherent in the model such as variability in actual ground properties, changing dynamic behaviour of BTES system and the model assumptions. However, the existing level of error remains within an acceptable range. The alignment between the model's predictions and the measured data underscores the model's reliability in accurately simulating the behaviour of the BTES system, enhancing confidence in the results presented in the study.

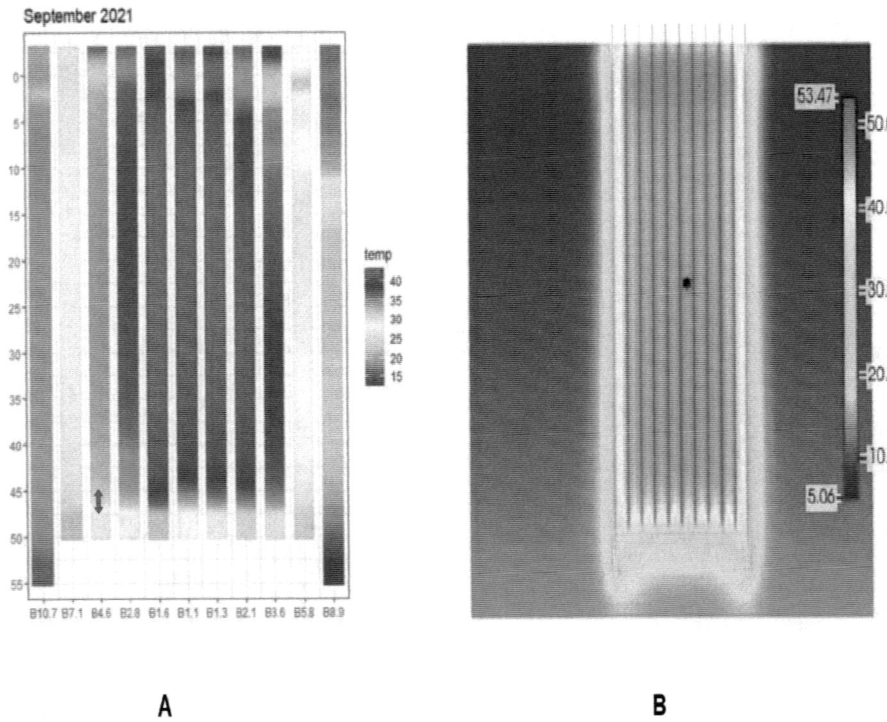

Figure 5. Spatial temperature distribution across 11 borehole exchangers, (A) actual data and (B) CMG STARS model.

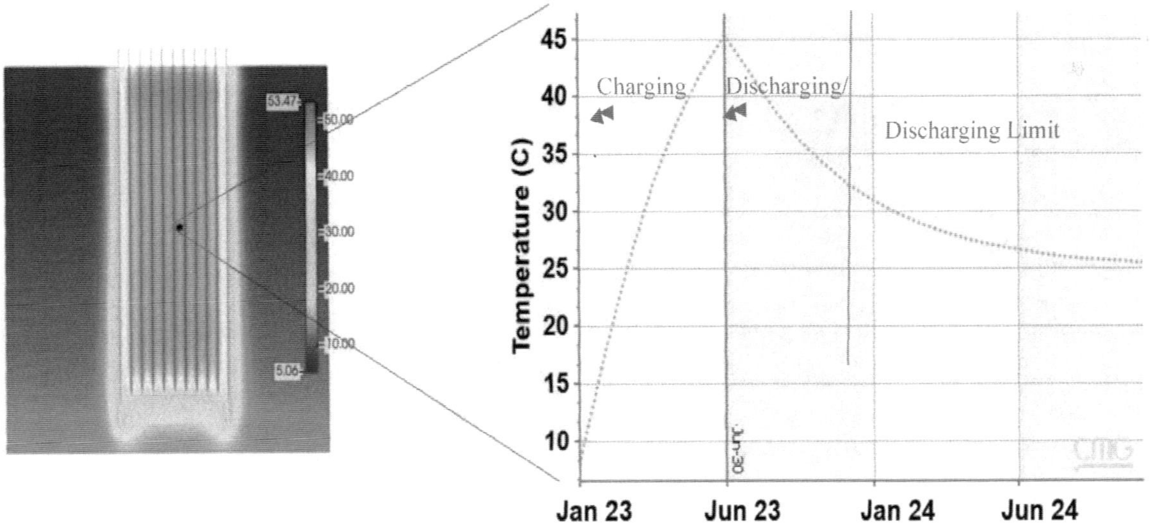

Figure 6. Reservoir temperature profile during charging and discharging cycles.

3.2 Temperature Distribution and Heat Extraction Efficiency

The evolution of thermal storage plumes with time is shown in Figure 7 for the cross section of BTES. The predicted temperature distributions are shown for charging and discharging periods. The BTES region shows a trend of increasing temperature with time. The average temperature at the centre of BTES reaches a maximum temperature of about 45°C. Heat distribution during the first cycle of heat

IOP Conf. Series: Materials Science and Engineering 1294 (2023) 012059 doi:10.1088/1757-899X/1294/1/012059

extraction (Figure 7, discharge t=6 months) shows that the temperature remains higher than the original reservoir temperature (8°C). That resulted in a low thermal recovery efficiency of only 30.9%. It could be explained by the fact that more than half of the heat injected is lost to heat up the rock matrix. As time passes, the heat loss to the rock reduces and the thermal efficiency increases to 35.3% after 5 years of operation (see Figure 7).

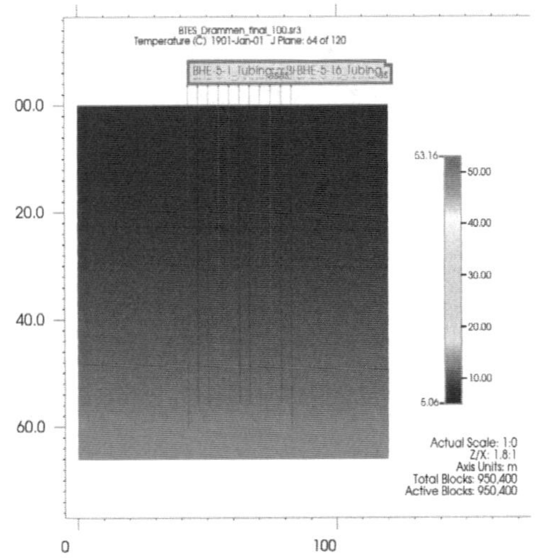

Charging: t = 0

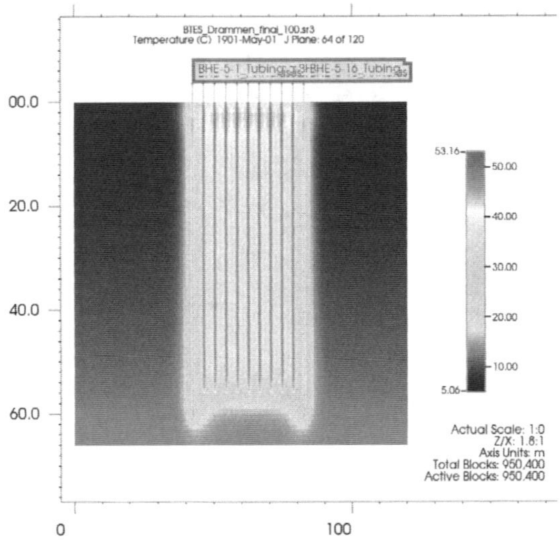

T = 4 months

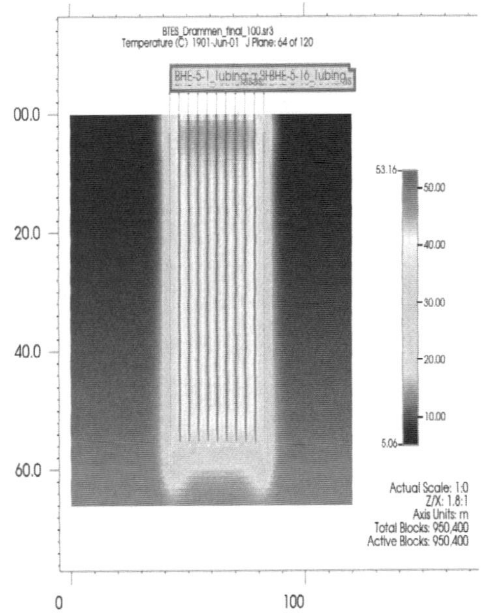

T = 6 months

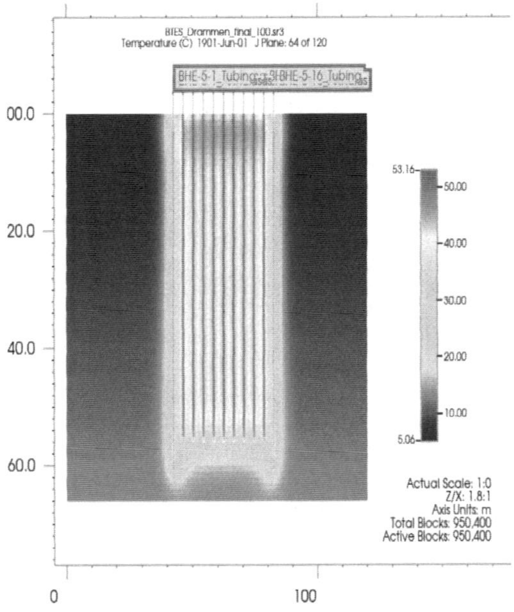

Charging: t = 0

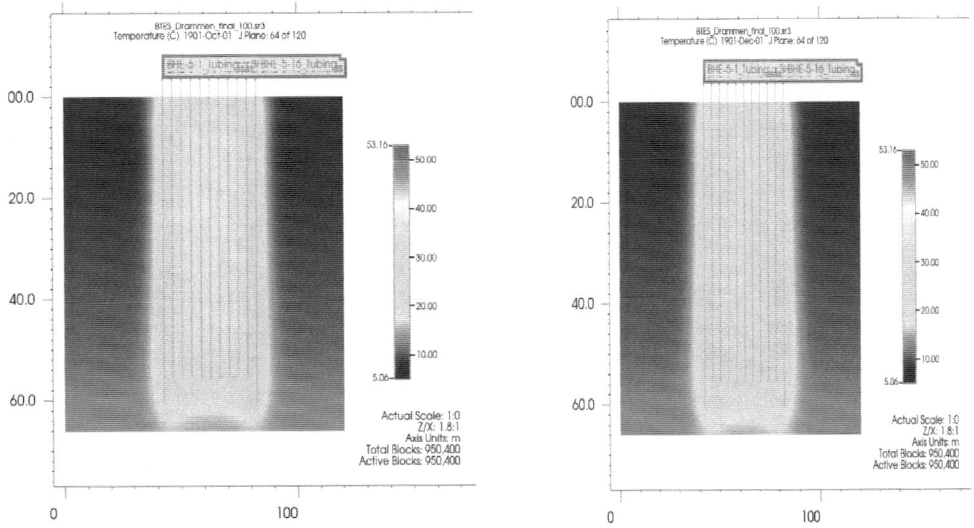

T = 4 months T = 6 months

Figure 7. Simulated temperature distribution at lateral cross-section across 11 BHEs. Results are shown for the ends of the charging (0, 2, 4, 6 months) and discharging (0, 2, 4, 6 months) for the first year of operation.

The CMG STARS BTES model is adopted in order to study the long-term effect (5 years) of a warm well's thermal performance. Figure 8 shows the repeated cycles of charging and discharging, in which heat carrier flow rate is kept constant to 4.2 m^3/h. Two cases for the initial charging period were evaluated, 6 months (case-1) and 15 months (case-2), while the discharging periods were fixed to 6 months for both cases. The first thing to note in Figure 8 is that the average BTES temperature reaches a quasi-steady state behaviour after approximately 2 cycles with slightly higher temperature from year to year for case-2. During the 1st cycle, the average BTES temperatures reach 45°C and 53°C for case-I and case-2, respectively. The temperature is observed to increase to around 50°C before declining to 47°C after 5 years for case-I. In Case-2, temperature almost levelled at 50°C for all the 5 years. However, in the winter, local temperatures almost levelled near the boreholes at 35°C for both cases, at the minimum temperature required for space heating.

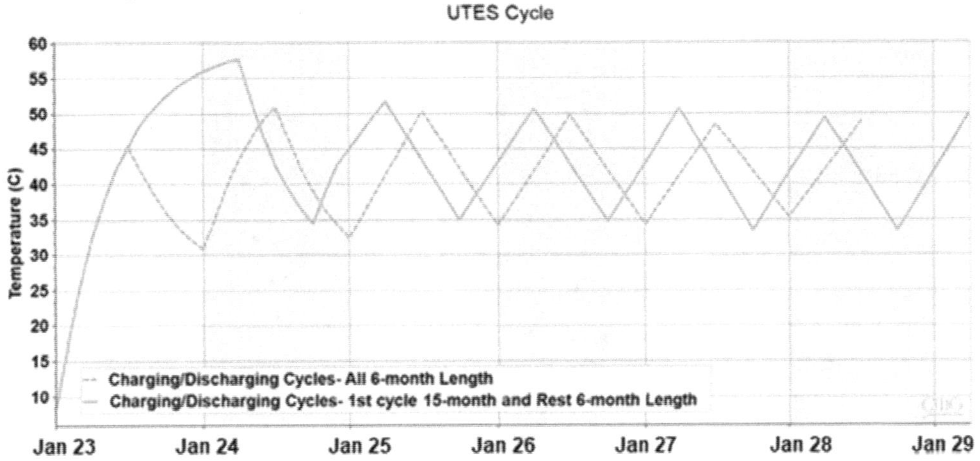

Figure 8. BTES temperatures for studied cases during charging and discharging over 5 years. Note: 6-month and 15-month charging for the 1st cycle are used, the following cycles fixed to 6-month for charging and discharging.

3.3 Sensitivity Analysis

3.3.1 Effect of heat carrier flow rate. Figure 9 shows the effect of the flow rate on the average BTES temperature. It is clear that as the flow rate increases from 1.05 m3/h to 4.2 m3/h, the average BTES temperature increases. However, at a flow rate higher than 4.2 m3/s no increase in BTES temperature was observed. The corresponding BTES efficiency is observed to increase with increasing the flow rate from 34 % and reached a plate of 36% at flow rate higher than 4.2 m3/h.

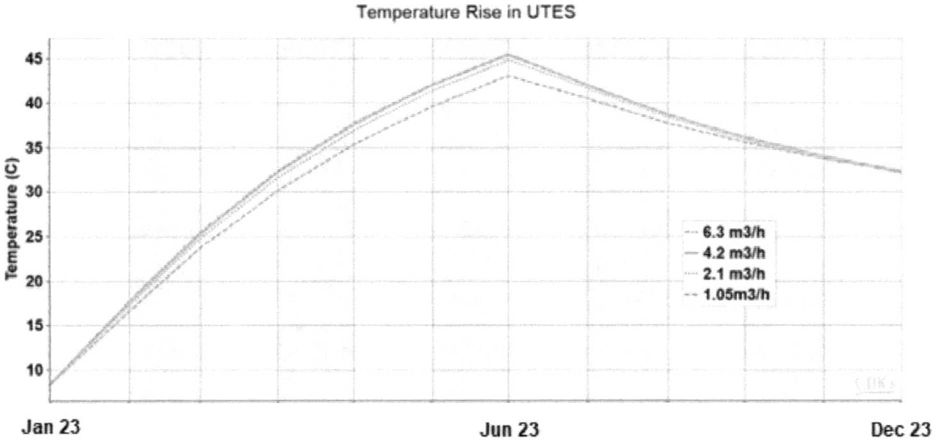

Figure 9. Impact of increased flow rate on average BTES temperatures.

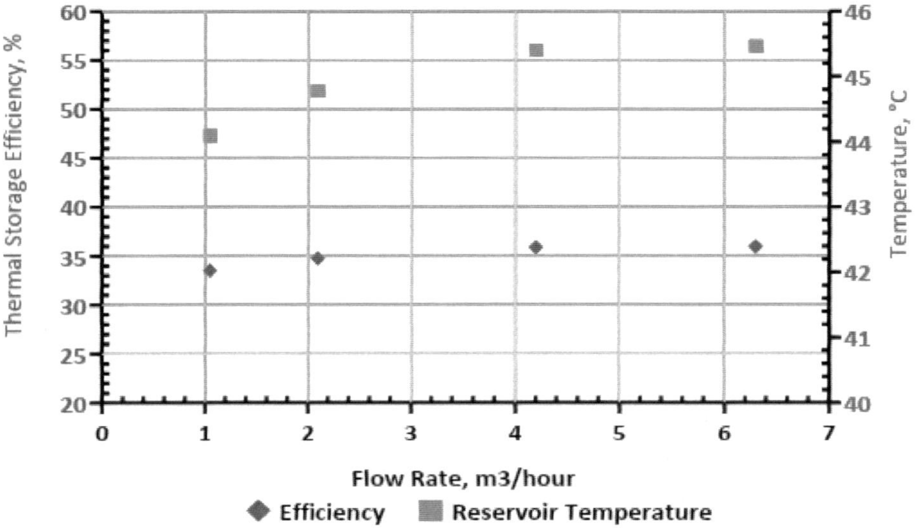

Figure 10. Impact of increased flow rate on average temperatures and efficiency of the BTES system.

3.3.2 Effect of heat charging period. In this section charging periods at constant injection temperature (60°C) were evaluated. As the charging period increased the BTES temperature increased and then almost levelled out at a longer charging time. Initially, BTES temperature increased by 12°C (from 33°C to 45°C) when the charging period was prolonged from 3 months to 6 months. However, only 4 to 3°C increments were observed when the charging period increased higher than 9 months. It's worth

IOP Conf. Series: Materials Science and Engineering 1294 (2023) 012059 doi:10.1088/1757-899X/1294/1/012059

noting that at higher BTES temperature the discharging period takes longer before reaching the discharging temperature limit (i.e., 30°C).

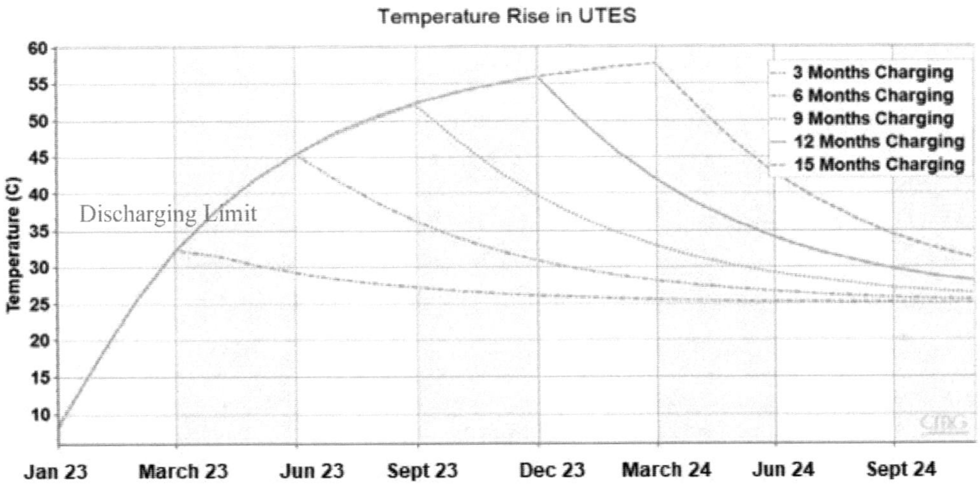

Figure 11. Impact of charging time on average BTES storage temperatures.

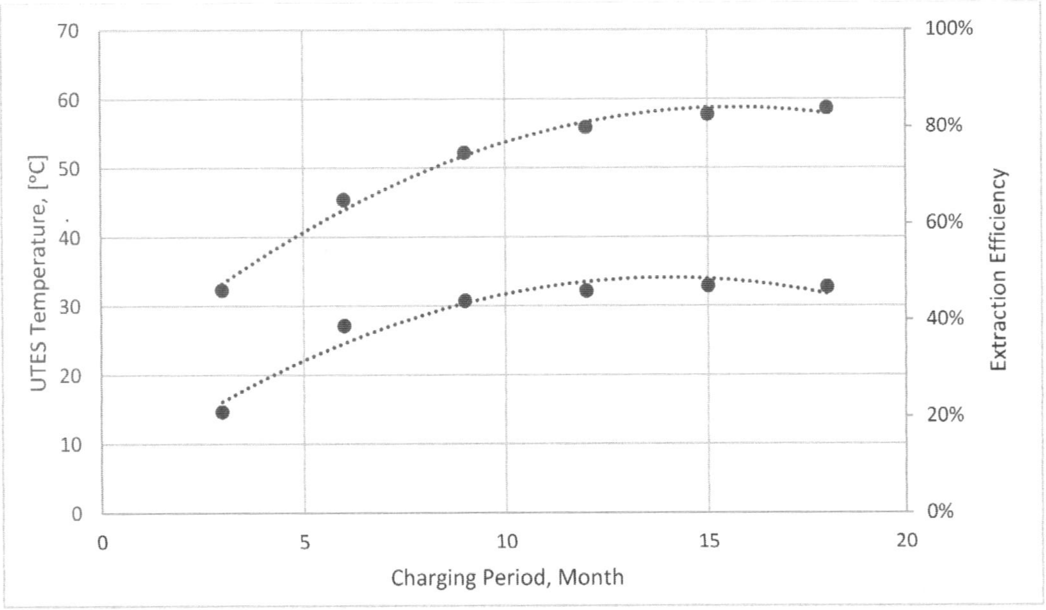

Figure 12. Impact of charging time on average BTES storage temperatures and thermal efficiency.

3.3.3 Effect of Borehole exchanger depth. Increasing BHE depth has been found to have a minimal effect on BHE charging up. As can be seen from figure 12, at constant charging fluid temperature, BTES temperature decreases with increasing depth.

IOP Conf. Series: Materials Science and Engineering 1294 (2023) 012059 doi:10.1088/1757-899X/1294/1/012059

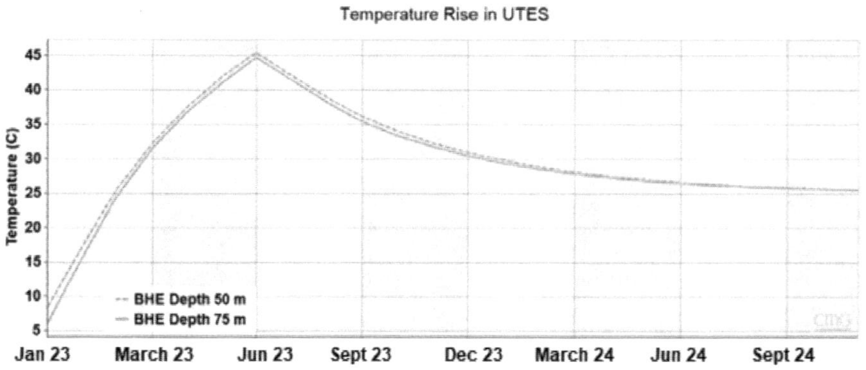

Figure 12. Impact of increased flow rate on average BTES temperatures.

3.3.4 Effect of injection water temperature

Figure 13 shows the influence of increasing injection temperature of heat carrier fluid. The BTES temperature steadily increases by increasing the injection temperature during charging up. BTES storage temperatures are found to increase linearly as injection temperatures increase (see figure 14). However, cooling down during discharge was faster at higher temperatures. As the temperature within the BTES increases, the rate of heat loss to the surrounding rocks also increases.

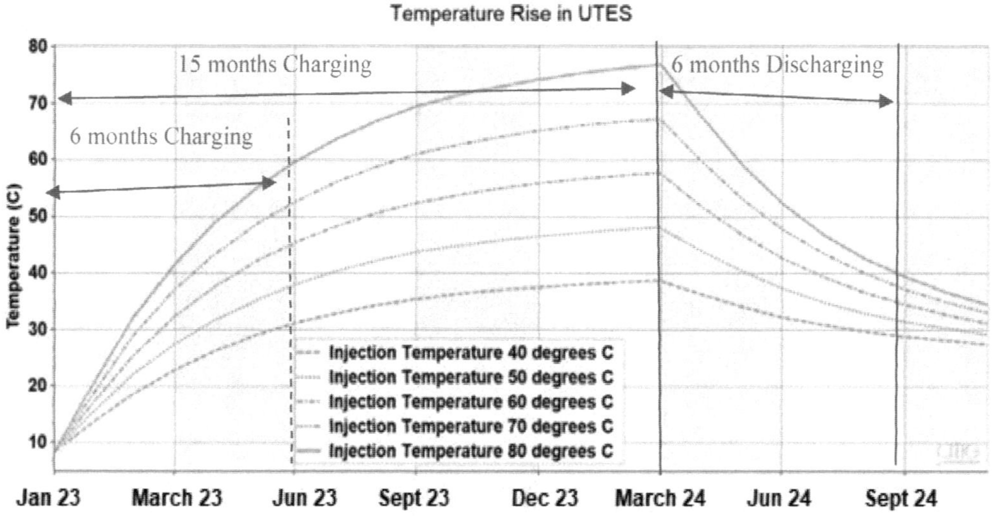

Figure 13. Impact of increased flow rate on average BTES temperatures after 15-months charging and 6 months discharging.

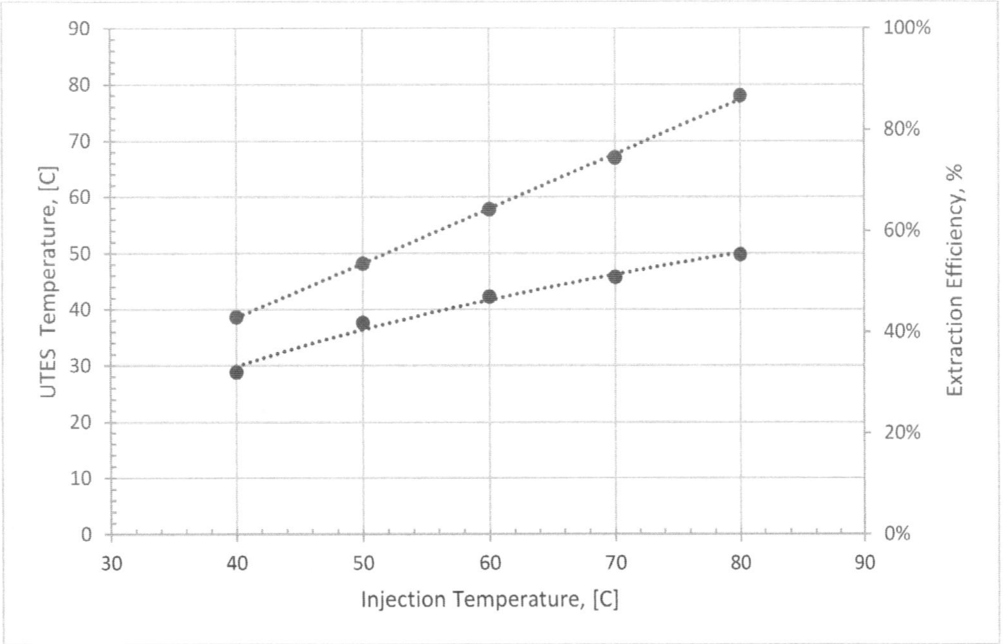

Figure 14. BTES storge temperature after 15 months charging and corresponding thermal efficiency.

4. Conclusions

In this paper, a large-scale BTES in Norway with 100 boreholes was modelled in CMG STARS. This model was validated by temperature distribution in 11 boreholes across the BTES measured by fibre optic temperature sensors. Following are the conclusions reached in this study:

- The simulated temperatures matched well with the measured data.
- The BTES storage temperature increases to a maximum temperature of 45°C, however during production, the temperature decreases to 32°C.
- In 5 years of operation, 35.5% of the heat injected into the BTES system was recovered, while the significant heat remained in the borehole region and lost to surrounding rock (64.5%).
- BTES was found very sensitive to flow rate, the charging period and injection temperature. Well depth has minimal effect at constant injection temperature.

ACKNOWLEDGEMENTS: We acknowledge POLNOR grand for funding the project. We are indebted to the staff of the department of energy and petroleum engineering (IEP) at the University of Stavanger and for allowing us to present this paper.

References

[1] Norwegian Ministry of Climate and Environment, 2021, 'Norway's Climate Action Plan for 2021–2030 available: Meld St. 13 (2020–2021) (regjeringen.no)
[2] Norway's new climate target: emissions to be cut by at least 55 % - regjeringen.no
[3] ENERGY TRANSITION NORWAY 2021, Available: Energy Transition Norway 2021 DNV
[4] Blackridge research and consulting, Norway solar power market outlook to 2028. Available: Norway Solar Power Market Outlook | Blackridge Research & Consulting
[5] Welsch, B., Göllner-Völker, L., Schulte, D.O., Bär, K., Sass, I. and Schebek, L., 2018, Environmental and economic assessment of borehole thermal energy storage in district heating systems. *Appl. energy*, **216**, 73-90.
[6] Rad, F.M., Fung, A.S. and Rosen, M.A., 2017. An integrated model for designing a solar community heating system with borehole thermal storage. Energy for sustainable development, 36, pp.6-15.

[7] Rapantova, N., Pospisil, P., Koziorek, J., Vojcinak, P., Grycz, D. and Rozehnal, Z., 2016. Optimisation of experimental operation of borehole thermal energy storage. *Appl. Energy*, **181**, 464-476.
[8] Chapuis, S. and Bernier, M., 2009. Seasonal storage of solar energy in borehole heat exchangers.
[9] Jokiel, M., Rohde, D., Kauko, H. and Walnum, H.T., 2020. Integration of a high-temperature borehole thermal energy storage in a local heating grid for a neighborhood. In International Conference Organised by IBPSA-Nordic, 13th–14th October 2020, OsloMet. BuildSIM-Nordic 2020. Selected papers. SINTEF Academic Press.
[10] Midttomme, k., 2022. Borehole Thermal Energy Storage (BTES) – development of high temperature BTES. Norce Norwageian Research Centre.

IOP Conf. Series: Materials Science and Engineering 1294 (2023) 012060 doi:10.1088/1757-899X/1294/1/012060

Development of a surrogate model of a trans-critical CO_2 heat pump for use in operations optimization using an artificial neural network

T A S Sazon*, Q Zhang and H Nikpey

Faculty of Science and Technology, University of Stavanger, Norway

* Correspondence: thor.a.sazon@uis.no

Abstract. Conventional physics-based models can demand substantial computational resources when employed for operational optimization. To allow faster system simulations that can be employed for operational optimization, a surrogate model of the CO_2 heat pump has been developed using an artificial neural network (ANN). The ANN model takes in six (6) inputs: evaporator water-side mass flow, its temperature, gas cooler water-side mass flow, its temperature, set-point output temperature, and high-side heat pump pressure. The model's outputs comprise the electrical energy needed to run the heat pump, the heat from the gas coolers, the temperature of the heat pump-heated fluid, and the outlet temperature of the heat pump's evaporator. Data used for training, validating, and testing the ANN model were generated by running a calibrated Modelica model of the CO_2 heat pump for various combinations of input parameters obtained from Latin hypercube sampling. The ANN model developed includes an input layer with 6 inputs, 2 hidden dense layers, each with 30 neurons, and an output layer for 4 outputs (6-30-30-3). The ReLU activation function was implemented on each hidden layer and no regularizations were imposed. The Adam optimizer was used with a learning rate of 0.001 specified. Early stopping (patience = 2000) was implemented to ensure that the training data was not overfitted. A maximum of 30000 epochs was specified. The resulting Mean Square Error (MSE) obtained for the training, validation, and testing data sets were $1.38x10^{-5}$, $2.05x10^{-5}$, and $3.65x10^{-5}$, respectively. When tested against one-week operational runs generated by Modelica, the Root Mean Square Errors (RMSEs) for coefficient of performance (COP)s for spring, summer, autumn, and winter operations obtained were 0.232, 0.346, 0.089 and 0.076, respectively. The resulting surrogate ANN model can be integrated into the system model as a functional mock-up unit within Modelica to facilitate faster simulations for operational optimization.

1. Introduction

Powered by low-emissions electricity, heat pumps play a key role in the global transition to sustainable heating and cooling. It is estimated to have the potential to reduce global carbon dioxide (CO_2) emissions by at least 500 million tons in 2030, which is roughly the same amount of annual CO_2 emissions of all cars in Europe today [1]. With the pace of heat pump installation growing at record levels, it is necessary that this growth is accompanied using working fluids/refrigerants that have low global warming potential (GWP). Unfortunately, some of the commonly used working fluids nowadays have very high GWP. For instance, R134a, which is widely used in domestic hot water (DHW) production, has a GWP that is 1300 times higher than that of CO_2 [2]. These high-GWP refrigerants can leak during manufacturing, installation, maintenance, and disposal, offsetting the benefits that heat pumps can bring. However, using a low-GWP refrigerant for an incompatible system can result in an inefficient heat pump system, which can lead to using more energy and emitting more CO_2. Having good operational efficiency while using low-GWP refrigerants is the ideal set up.

Carbon dioxide is possibly one of the most promising low-GWP refrigerants. It has zero ozone depletion potential and a GWP of 1. Additionally, it is non-toxic, non-flammable, easily available, and affordable[3]–[5]. The modern use of CO_2 in a trans-critical cycle was first proposed by Lorentzen [6]. Its most distinct difference from the conventional cycle is its use of a gas cooler instead of a condenser, accommodating a relatively large temperature glide in the heat rejection process [7]. So far, it has been commercially applied in combined cooling, heating, ventilation, and air conditioning in supermarkets [8], [9]; water heating [10]; and car air conditioning [11].

A Modelica [12] model of a trans-critical CO_2 heat pump, based on the 6.5 kW test rig constructed and tested by Stene [13, 14], has been developed and calibrated and was then integrated into a system model that utilizes borehole heat exchangers (BHEs), solar thermal collectors (SC), and a thermal energy storage (TES) tank for space and water heating [15]. Typically, optimization has to be performed for every energy system installation due to vastly different weather conditions, occupancy profiles, energy tariffs, government tariffs, system components, and building types [16]. Sazon et al. [17] studied the optimization of the design of this CO_2 solar-assisted ground source heat pump (SAGSHP) system for West Norway climate and demand and showed that compared to a conventional SAGSHP system, it can perform at a slightly lower seasonal efficiency, but at a comparable levelized cost of heating (LCOH). To further improve the performance of this system, it has been recommended to explore the optimization of its operations as well.

Studies about optimizing the operation of a CO_2 heat pump system typically focus on (1) controlling the compressor operation to take advantage of the variability in electricity prices or (2) controlling the gas cooler pressure to maximize efficiency.

Heat pumps, when coupled with TES, can be operated flexibly to benefit from the variability of electricity tariffs. This is implemented by shifting electricity import from high-demand to low-demand periods, which helps in managing grid stress and averting peak plant operation [18]. By running the compressor during periods of low-electricity-price to charge the TES, stored heat can then be discharged later to meet demand when electricity prices are higher. Several studies have already investigated load shifting in conventional heat pumps [16, 18–21].

It is well known that CO_2 heat pumps exhibit an optimal gas cooler pressure due to the distinct working mode in the trans-critical region [22, 23]. Because of this, several studies have been implemented on developing empirical optimal pressure correlations or algorithms for maximizing the coefficient of performance (COP) or other performance metrics [24–32]. Typically, this is conducted offline using experiments, physics-based system modeling, and thermodynamic cycle simulations [24]. The experiment-based approach is straightforward but can be expensive and time-consuming. Correlations developed from this method can also be quite limited to the testing conditions during the conduct of the experiments. Thermodynamic cycle simulation is usually preferred over the physics-based approach because it requires fewer modeling efforts. However, simple optimal pressure correlations based on thermodynamic simulation usually ignore some parameters and cannot model complex non-linear thermal fluid behaviors [24].

The benefit of using Modelica to model the CO_2 SAGSHP system is that detailed simulation characteristics of the energy system can be obtained. This can be done by using component models of the system that are calibrated or validated. However, these tools can be computationally expensive, more importantly when used for optimization problems that involve predictive control [16, 21]. One of the ways to overcome this is to replace component models that require heavy computations with a surrogate model that uses machine learning methods such as an artificial neural network (ANN) [33]. ANN is a rapidly emerging technology that has been widely applied for system optimization and modeling in energy and process systems [34]. Conventional thermodynamic analysis involves many iterations that cause much time in calculation. In comparison, ANN can correlate complex non-linear relationships between input and output parameters more quickly, with no complicated iteration[34]. For CO_2 heat pumps, ANN has been utilized in some studies [7, 35].

In this study, a surrogate model of a trans-critical CO_2 heat pump in a SAGSHP system was developed using ANN. The model will eventually be utilized and integrated into the system model as a functional mock-up unit within Modelica to facilitate faster simulations for operations optimization studies.

2. Methods

A surrogate model of a trans-critical CO_2 heat pump, based on the 6.5 kW test rig constructed and tested by Stene [13, 14], has been developed using ANN. The surrogate model is meant to replace a calibrated model of a CO_2 heat pump in a SAGSHP system in the Modelica environment.

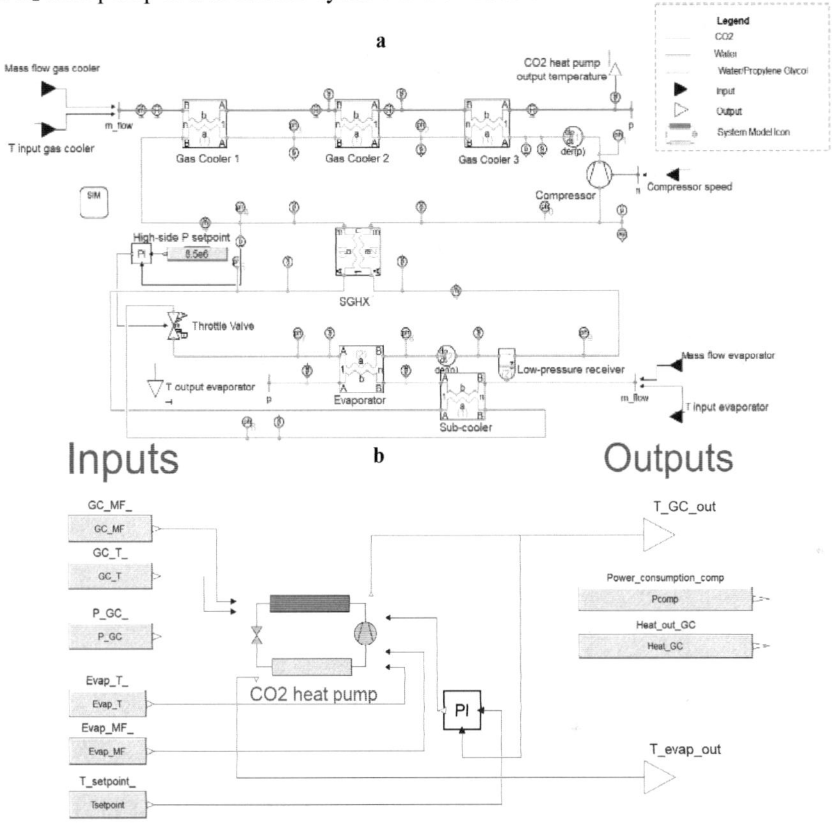

Figure 1. The Modelica Model of the trans-critical CO_2 heat pump: (a) showing the underlying components of the heat pump and (b) showing the inputs and outputs to the ANN model.

2.1. *The Modelica model of the trans-critical CO_2 heat pump*

The CO_2 heat pump model developed in this work was based on a 6.5 kW prototype unit. It consists of the counter-flow tripartite gas coolers, an evaporator, a compressor, a throttle valve, a suction gas heat exchanger (SGHX), a sub-cooler, and a low-pressure receiver (Figure 1a). Although the prototype can function in three modes (space heating only, domestic hot water (DHW) heating only, and simultaneous space and DHW heating), the model in this work was only calibrated for simultaneous space and DHW heating. It was developed using the Thermal Systems library [36] in the Dymola interface [37]. The Thermal Systems library, also synonymous with the TIL suite, is a commercial Modelica library for modeling thermo-fluid systems. It can be used to model various components, such as heat pump cycles, hydraulic networks, ventilation, and so on. It uses the TSMedia library to calculate the thermophysical properties of fluid and fluid mixtures. The model was calibrated against measured data at design conditions of the heat pump by adjusting the heat transfer coefficients of the heat exchangers (gas coolers, evaporator, internal heat exchanger) and the efficiency of the compressor. Details about the specifications of the CO_2 heat pump components and the calibration of the model are given in [15, 38]. In [17], this CO_2 heat pump model was integrated into a system that comprises of BHEs, SCs, and a TES and was used for space and DHW heating for climate and demand conditions in Western Norway.

2.2. Data preparation for training, validation, and testing of the ANN model

The Modelica model of the CO_2 heat pump was run for 10,000 combinations of input parameters to obtain the data needed to develop the ANN model. The inputs to the model were (1) the fluid mass flow to the water side of the evaporator (Evap_MF), (2) the temperature of this fluid (Evap_T), (3) the mass flow of the fluid to the water side of the gas coolers (GC_MF), (4) the temperature of this fluid (GC_T), (5) the set point output temperature (Tsetpoint) (controlled by adjusting the compressor speed), and (6) the gas cooler pressure set point (P_GC) (controlled by adjusting the throttle valve opening) (Figure 1b). Some of these parameters connect to the other components of the energy system while some can be controlled to optimize performance. The outputs chosen were (1) power consumption (Pcomp), (2) the total heat output (Heat_GC), (3) the output temperature of the heat pump (T_GC_out), and (4) the outlet temperature of the heat pump's evaporator (T_evap_out). The 10,000 combinations of input parameters that were simulated using the Modelica model were obtained using Latin hypercube sampling [39]. The value ranges (Table 1) for every input parameter were decided based on observed values from full-year simulations [17]. Each case was run for 15000 seconds to ensure that steady conditions of the outputs were obtained.

Table 1. The lower and upper bounds of the input parameters.

Input Parameters	Lower bound	Upper bound
Evaporator water side mass flow, kg/s	0.2	0.8
Evaporator water side temperature, °C	-5	25
Gas cooler water side mass flow, kg/s	0.0085	0.048
Gas cooler water side temperature, °C	15	55
Output temperature, °C	60	70
Gas cooler pressure, MPa	8.5	10

Some combinations of input parameters that are known to produce unfavorable heat pump performance, i.e., it both fails to reach the target output temperature and produces very low COPs, were filtered out from use in the development of the ANN model. This occurrence can happen when combinations of the following are met: the temperature of the fluid to the gas cooler is too high, the mass flow to the gas cooler is too low, or the evaporation temperature is too high. Only 6604 data points remained after doing this.

These data were then divided into training, testing, and validation sets. Twenty percent (20%) of the whole set was set for testing the ANN model. Afterward, 20% of the remainder was used for validation. All that remained were used for training the model. Given this, the training set, testing set, and validation set got 4226, 1321, and 1057 input/output combinations.

Since the different inputs and outputs have different magnitudes and units, it is necessary to subject all of them to normalization to remove the bias to parameters with high magnitudes. Here, min-max scaling was done. Min-max normalization is done by subtracting the minimum value from the data and then dividing this by the difference between and minimum and maximum values in the parameter set. This shifted and rescaled the values of every parameter, so they all ended up ranging from 0 to 1. Histograms of the normalized outputs in the training, testing, and validation sets for every input and output are given in Figures 2 and 3, respectively. It can be seen that the different values of the inputs and outputs seem to be equivalently represented in the data splits.

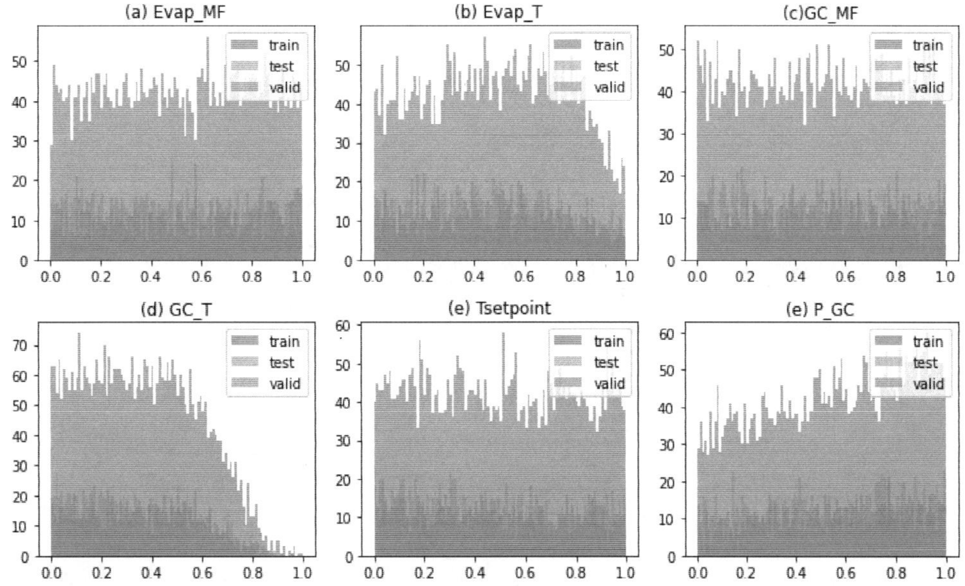

Figure 2. The histograms showing the splits of the inputs for training, testing, and validation: (a) mass flow of waterside evaporator fluid, (b) fluid temperature water side evaporator, (c) mass flow of waterside gas cooler fluid, (d) fluid temperature waterside gas cooler, (e) set point output temperature, and (f) set point gas cooler pressure.

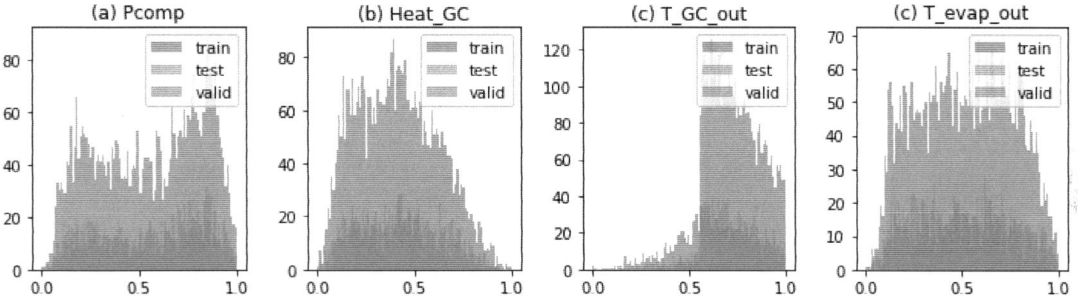

Figure 3. Histograms showing the splits of the labels for training, testing, and validation: (a) power consumption, (b) thermal energy generated, (c) output temperature of the heat pump, and (d) outlet temperature of the heat pump's evaporator.

2.3. *Hyperparameter tuning and ANN model evaluation*

The three (3) output parameters from the Modelica model were used as labels in the training of the ANN model. This was treated as a batch supervised regression problem and was implemented using Keras [40] in Python. The ANN was trained through backpropagation [41].

The hyperparameters tuned in this study include (1) the number of neurons, (2) the number of dense layers, (3) the optimizer, (4) the learning rate, (5) activation functions, and (6) the number of epochs. Early stopping was implemented as a callback to ensure that overfitting the training set was avoided.

The model was trained using mean square error (MSE) as the loss function.

$$MSE = \frac{1}{n}\sum_{i=1}^{n}\left(Y_i - \widehat{Y}_i\right)^2 \quad (1)$$

The metrics used to evaluate the model's performance after training and during model evaluation is the root mean square error (RMSE)

$$RMSE = \sqrt{\frac{1}{n}\sum_{i=1}^{n}\left(Y_i - \widehat{Y_i}\right)^2} \quad (2)$$

Where n is the number of data points, Y_i are the observed values, and $\widehat{Y_i}$ are the predicted values.

3. Results and discussions

The iterations involved in solving physics-based models make them computationally heavy. Comparably, ANN can correlate complex non-linear relationships between the inputs and outputs (labels) in a quicker way, without complicated iterations. When compared to statistical models, it has also been seen to perform at a higher accuracy [42]. After data generation and filtering, the ANN model was trained using backpropagation while considering six (6) input parameters (Evap_MF, Evap_T, GC_MF, GC_T, Tsetpoint, P_GC) and 3 outputs (Pcomp, Heat_GC, T_GC_out).

3.1. ANN model hyperparameter specifications

The training set, testing set, and validation set containing 4226, 1321, and 1057 input/output combinations, respectively were inputted to data frames in Python and used to develop the ANN model. The performance of an ANN significantly depends on its structure and hyper parameter settings. Here, the ANN performance was developed using trial and error approach. The loss function and metrics used were the MSE and the RMSE, respectively.

In developing an ANN model, there is no one-size-fits all approach. Often, it involves a combination of experimentation and best practices. Here, we started with the simplest model with 1 layer and 5 neurons while setting reasonable defaults for other hyper parameters. The number of neurons were increased until no significant improvement to the loss function was observed. By then, another hidden layer was added. The number of neurons investigated ranged from 5 to 50. The number of hidden layers from 1 to 3 were tested. Normalization of the inputs and outputs must be done to remove the biases to parameters with significant magnitudes. The min-max normalization was seen to give lower MSEs compared to standard scaling. The Adam optimizer outperformed other solvers, like SGD and RMSprop. Initially, SGD was giving better MSEs, but when the learning rate was reduced from 0.1 to 0.001, Adam performed better in terms of speed and accuracy. Among the activation functions tested (tanh, rectified linear unit (ReLU), sigmoid, scaled exponential linear unit (SELU)), ReLU gave the lowest MSEs. These observations underscore the iterative and dynamic nature of ANN model development, where a blend of experimentation and informed choices is essential to fine-tune the architecture and hyperparameters for optimal results.

The resulting model after tuning the hyperparameters contains an input layer with 6 inputs, 2 hidden dense layers, each with 30 neurons, and an output layer for 4 outputs (6-30-30-4 structure). The ReLU activation function [43] was implemented on each hidden layer and no regularizations were imposed. The optimizer used was Adam [44] and a learning rate of 0.001 was seen to produce the best results for training. Early stopping, with a patience equal to 2000, and a maximum epoch of 30000 were specified.

3.2. Training, testing, and validation results

The training of the ANN model terminated after 17627 epochs. Figure 4a below shows that the training and validation MSE decreased quickly at the start of the runs and remained almost constant until it satisfied the conditions for early stopping (patience = 2000). Figure 4b zooms into the last 500 epochs, showing the differences between the training and validation MSEs. Although a bit lower, the training losses obtained were comparable to the validation losses.

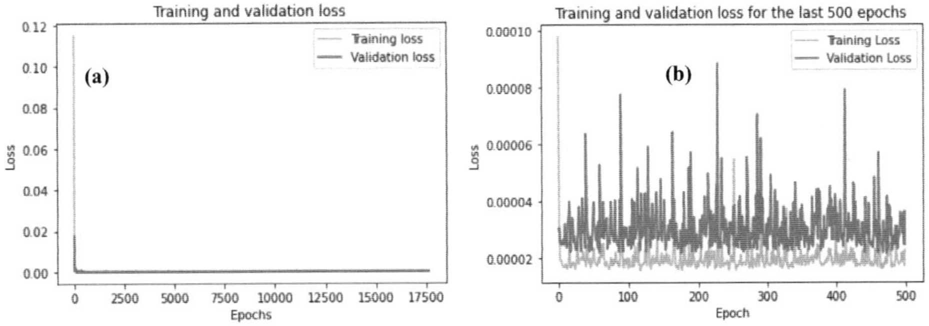

Figure 4. The MSE performance plot of the ANN during the training: (a) for the whole training period and (b) for the last 500 epochs.

The resulting values of the MSE and RMSE for training, testing, and validation at the end of the training process are given in Table 2. All of them seem to be sufficiently small, inferring a good match between the data and the simulation. Additionally, the MSE and RMSE for the testing data set are very close to that of the train set. This implies the capability of the developed ANN model for generalization. Figure 5 illustrates the ANN model-generated values of the outputs compared to those generated by the Modelica model. Most data points have been matched.

Table 2. MSE and RMSE for the training, validation, and testing sets.

	Training	Validation	Testing
MSE	$1.38\ x10^{-5}$	$2.05\ x10^{-5}$	$3.65\ x10^{-5}$
RMSE	0.0037	0.0045	0.0060

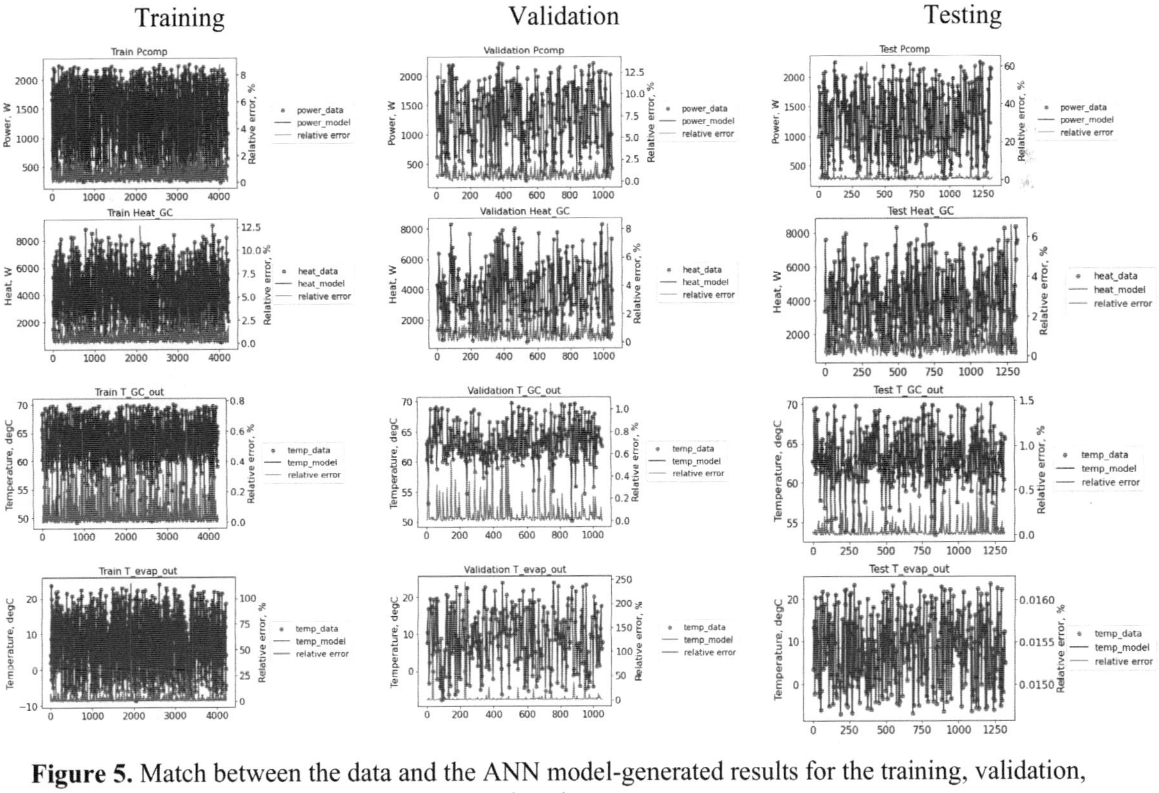

Figure 5. Match between the data and the ANN model-generated results for the training, validation, and testing sets.

3.3. Testing the ANN model for operational runs

To see if the training implemented is sufficient for the intended purpose, the ANN model was also tested against one-week operational runs generated in Modelica for different representative seasons (spring, summer, autumn, and winter). It can be seen that the model can follow the expected trends of output parameters of the CO_2 heat pump (Figure 6). This is slightly more evident for autumn and winter operations than for spring and summer. The RMSEs for each output for every season are summarized in Table 3. It shows that the average error the ANN model generates seems to be within reasonable levels. In particular, the COPs it generated differ from the Modelica-generated COPs only by 0.076 to 0.346.

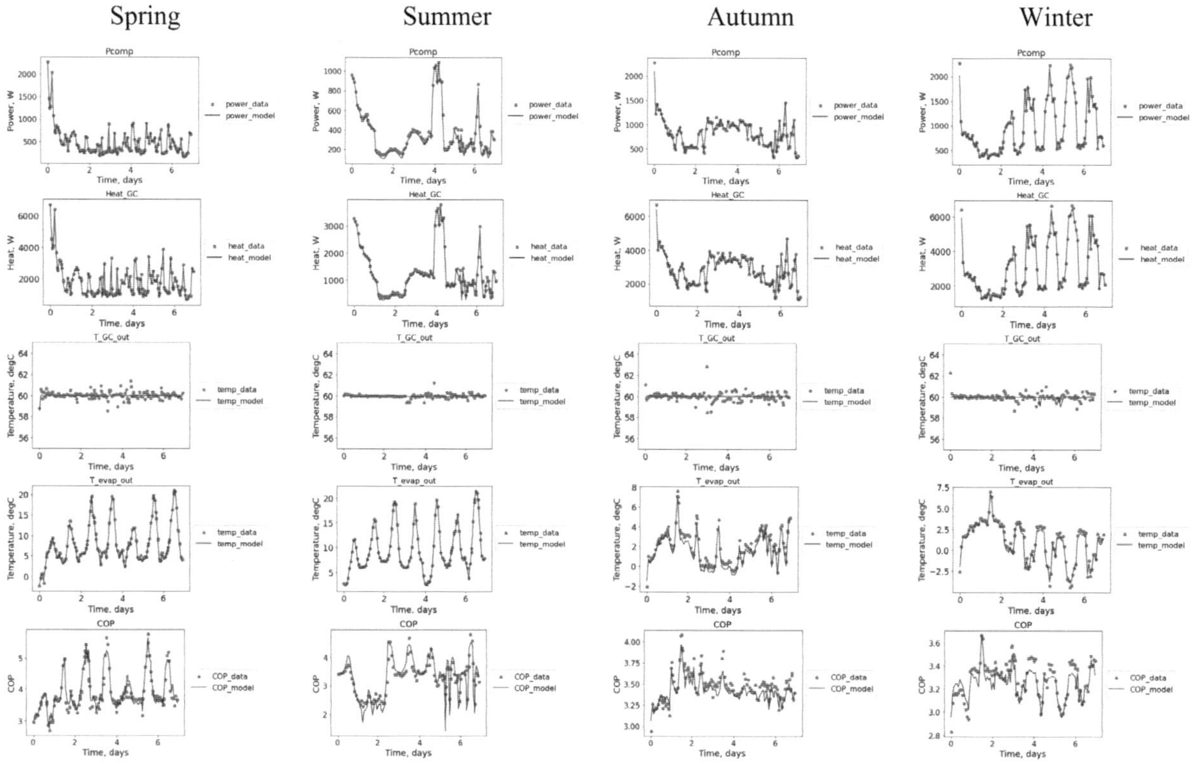

Figure 6. Match between the data and the ANN model-generated results for 1-week operational runs for different seasons

Table 3. RMSE for the model outputs for 1-week operation runs different seasons.

	RMSE				
	Pcomp, W	Heat_GC, W	T_GC_out, °C	T_evap_out, °C	COP
Spring	34.314	125.252	0.356	0.252	0.232
Summer	32.410	123.414	0.204	0.145	0.346
Autumn	31.990	95.921	0.489	0.346	0.089
Winter	31.859	79.933	0.413	0.293	0.076

4. Conclusion

Physics-based models of the trans-critical CO_2 heat pump can entail significant computational resources, particularly when used for operational optimization and predictive control. One way to bypass this is to replace them with a surrogate model developed using ANN. In this study, we successfully developed an

ANN-based surrogate model for a trans-critical CO_2 heat pump in a SAGSHP system. The key takeaways from this work can be summarized as follows:

- The selection of input parameters for the ANN model was based on three criteria: (1) controllability during operations, (2) strong correlation with the chosen outputs, and (3) dependence on inputs from other system components. In this work, the inputs considered include (1) the fluid mass flow to the water side of the evaporator (Evap_MF), (2) the temperature of this fluid (Evap_T), (3) the mass flow of the fluid to the water side of the gas coolers (GC_MF), (4) the temperature of this fluid (GC_T), (5) the set point output temperature (Tsetpoint), and (6) the gas cooler pressure set point (P_GC).
- The selected outputs should represent the CO_2 heat pump's performance and supply the necessary input data for interconnected system components. In this work, the outputs chosen were (1) power consumption (Pcomp), (2) the total heat output (Heat_GC), (3) the output temperature of the heat pump (T_GC_out), and (4) the outlet temperature of the heat pump's evaporator.
- Normalization is essential for both inputs and outputs to eliminate biases resulting from parameters with significant magnitudes. In this work, it was seen that min-max normalization generated lower MSEs and RMSEs.
- For the CO_2 heat pump considered in this work, an ANN architecture comprising an input layer with 6 inputs, 2 hidden dense layers, each with 30 neurons, and an output layer with 4 outputs (6-30-30-4) yielded satisfactory results with reasonable MSEs and RMSEs. The ReLU activation function was implemented on each hidden layer and no regularizations were imposed. The Adam optimizer was used with a learning rate of 0.001. Early stopping (patience = 2000) was implemented to ensure that the training data was not overfitted. A maximum of 30000 epochs was specified.
- The resulting MSE for the training, validation, and testing data sets were 1.38×10^{-5}, 2.05×10^{-5}, and 3.65×10^{-5}, respectively. Correspondingly, the RMSE for the training, validation, and testing data sets were 0.0037, 0.0045, and 0.0060, respectively. These results demonstrate a reasonable alignment between the model and the data.
- The ANN model was also tested against the Modelica model for one-week operational runs for different seasons. The results indicate favourable performance, with RMSEs for the COP to be 0.232, 0.346, 0.089, and 0.076, for spring, summer, autumn, and winter, respectively. This demonstrates its suitability for operational simulations.

References

[1] International Energy Agency, *The Future of Heat Pumps*. OECD, 2022. doi: 10.1787/2bd71107-en.

[2] K. Onno, "Report Annex 46 HPT-AN46-04: Refrigerants for Heat Pump Water Heaters," Heat Pump Centre c/o RISE – Research Institutes of Sweden, Borås, Dec. 2019. Accessed: Oct. 12, 2022. [Online]. Available: https://heatpumpingtechnologies.org/annex46/wp-content/uploads/sites/53/2020/10/hpt-an46-04-task-1-refrigerants-for-heat-pump-water-heaters-1.pdf

[3] J. Wang, M. Belusko, M. Evans, M. Liu, C. Zhao, and F. Bruno, "A comprehensive review and analysis on CO2 heat pump water heaters," *Energy Conversion and Management: X*, vol. 15, p. 100277, Aug. 2022, doi: 10.1016/j.ecmx.2022.100277.

[4] R. U. Rony, H. Yang, S. Krishnan, and J. Song, "Recent Advances in Transcritical CO2 (R744) Heat Pump System: A Review," *Energies*, vol. 12, no. 3, Art. no. 3, Jan. 2019, doi: 10.3390/en12030457.

[5] Navigant Consulting Inc., "Case Study: Transcritical Carbon Dioxide Supermarket Refrigeration Systems." Better Buildings Alliance - U.S. Department of Energy, Jan. 2015. Accessed: Oct. 12, 2022. [Online]. Available: https://betterbuildingssolutioncenter.energy.gov/sites/default/files/attachments/Transcritical_CO2_Supermarket_Refrigeration_Systems.pdf

[6] G. Lorentzen, "Trans-critical vapour compression cycle device," WO1990007683A1, Jul. 12, 1990 Accessed: Feb. 17, 2021. [Online]. Available: https://patents.google.com/patent/WO1990007683A1/en

[7] Y. Xu, C. Mao, Y. Huang, X. Shen, X. Xu, and G. Chen, "Performance evaluation and multi-objective optimization of a low-temperature CO2 heat pump water heater based on artificial neural network and new economic analysis," *Energy*, vol. 216, p. 119232, Feb. 2021, doi: 10.1016/j.energy.2020.119232.

[8] A. Hafner *et al.*, "Efficient and integrated energy systems for supermarkets," *11th IIR Gustav Lorentzen Conference on Natural Refrigerants: Natural Refrigerants and Environmental Protection, GL 2014*, pp. 311–319, Jan. 2014.

[9] Shecco, "World Guide to Transcritical CO2 Refrigeration," 2020. Accessed: Feb. 28, 2023. [Online]. Available: https://atmosphere.cool/fact_sheets/world-guide-to-transcritical-co2-refrigeration/

[10] K. Hashimoto, "Technology and Market Development of CO2 Heat Pump Water Heaters (ECO CUTE) in Japan," IEA Heat Pump Centre Newsletter, 2006. Accessed: Feb. 28, 2023. [Online]. Available: http://waterheatertimer.org/pdf/Market-develop-CO2.pdf

[11] AutomotiveTechinfo, "CO2 as a Refrigerant is Happening," Jun. 2021. Accessed: Feb. 28, 2023. [Online]. Available: https://automotivetechinfo.com/wp-content/uploads/2021/06/CO2-as-a-Refrigerant-is-Happening.pdf

[12] Modelica Association, "Modelica Language." Accessed: Sep. 02, 2021. [Online]. Available: https://modelica.org/modelicalanguage.html

[13] J. Stene, "Residential CO2 Heat Pump System for Combined Space Heating and Hot Water Heating," Fakultet for ingeniørvitenskap og teknologi, 2004. Accessed: May 23, 2022. [Online]. Available: https://ntnuopen.ntnu.no/ntnu-xmlui/handle/11250/233381

[14] J. Stene, "Residential CO2 heat pump system for combined space heating and hot water heating," *International Journal of Refrigeration*, vol. 28, no. 8, pp. 1259–1265, Dec. 2005, doi: 10.1016/j.ijrefrig.2005.07.006.

[15] T. A. Sazon and H. Nikpey, "Modeling and investigation of the performance of a solar-assisted ground-coupled CO2 heat pump for space and water heating," *Applied Thermal Engineering*, tbd.

[16] R. Renaldi, A. Kiprakis, and D. Friedrich, "An optimisation framework for thermal energy storage integration in a residential heat pump heating system," *Applied Energy*, vol. 186, pp. 520–529, Jan. 2017, doi: 10.1016/j.apenergy.2016.02.067.

[17] T. A. Sazon, Q. Zhang, and H. Nikpey, "Multi-Objective Optimization of a Solar-Assisted Ground-Source CO2 Heat Pump System for Space and Water Heating Using the Taguchi Method and Utility Concept," in *36th International Conference on Efficiency, Cost, Optimization, Simulation and Environmental Impact of Energy Systems (ECOS 2023)*, Las Palmas De Gran Canaria, Spain: ECOS 2023, 2023, pp. 769–780. doi: 10.52202/069564-0070.

[18] T. Xu, E. N. Humire, J. N. Chiu, and S. Sawalha, "Latent heat storage integration into heat pump based heating systems for energy-efficient load shifting," *Energy Conversion and Management*, vol. 236, p. 114042, May 2021, doi: 10.1016/j.enconman.2021.114042.

[19] A. V. Olympios, P. Sapin, J. Freeman, C. Olkis, and C. N. Markides, "Operational optimisation of an air-source heat pump system with thermal energy storage for domestic applications," *Energy Conversion and Management*, vol. 273, p. 116426, Dec. 2022, doi: 10.1016/j.enconman.2022.116426.

[20] D. Fischer, J. Bernhardt, H. Madani, and C. Wittwer, "Comparison of control approaches for variable speed air source heat pumps considering time variable electricity prices and PV," *Applied Energy*, vol. 204, pp. 93–105, 2017, doi: 10.1016/j.apenergy.2017.06.110.

[21] K. X. Le, M. J. Huang, C. Wilson, N. N. Shah, and N. J. Hewitt, "Tariff-based load shifting for domestic cascade heat pump with enhanced system energy efficiency and reduced wind

power curtailment," *Applied Energy*, vol. 257, p. 113976, Jan. 2020, doi: 10.1016/j.apenergy.2019.113976.

[22] Y. Song and F. Cao, "The evaluation of optimal discharge pressure in a water-precooler-based transcritical CO2 heat pump system," *Applied Thermal Engineering*, vol. 131, pp. 8–18, Feb. 2018, doi: 10.1016/j.applthermaleng.2017.11.092.

[23] Y. Song, C. Cui, X. Yin, and F. Cao, "Advanced development and application of transcritical CO2 refrigeration and heat pump technology—A review," *Energy Reports*, vol. 8, pp. 7840–7869, Nov. 2022, doi: 10.1016/j.egyr.2022.05.233.

[24] W.-J. Zhang and C.-L. Zhang, "A correlation-free on-line optimal control method of heat rejection pressures in CO2 transcritical systems," *International Journal of Refrigeration*, vol. 34, no. 4, pp. 844–850, Jun. 2011, doi: 10.1016/j.ijrefrig.2011.01.014.

[25] L. Cecchinato, M. Corradi, and S. Minetto, "A critical approach to the determination of optimal heat rejection pressure in transcritical systems," *Applied Thermal Engineering*, vol. 30, no. 13, pp. 1812–1823, Sep. 2010, doi: 10.1016/j.applthermaleng.2010.04.015.

[26] S. M. Liao, T. S. Zhao, and A. Jakobsen, "A correlation of optimal heat rejection pressures in transcritical carbon dioxide cycles," *Applied Thermal Engineering*, vol. 20, no. 9, pp. 831–841, Jun. 2000, doi: 10.1016/S1359-4311(99)00070-8.

[27] J. Sarkar, S. Bhattacharyya, and M. R. Gopal, "Optimization of a transcritical CO2 heat pump cycle for simultaneous cooling and heating applications," *International Journal of Refrigeration*, vol. 27, no. 8, pp. 830–838, Dec. 2004, doi: 10.1016/j.ijrefrig.2004.03.006.

[28] N. Agrawal, S. Bhattacharyya, and J. Sarkar, "Optimization of two-stage transcritical carbon dioxide heat pump cycles," *International Journal of Thermal Sciences*, vol. 46, no. 2, pp. 180–187, Feb. 2007, doi: 10.1016/j.ijthermalsci.2006.04.011.

[29] Y. Chen and J. Gu, "The optimum high pressure for CO2 transcritical refrigeration systems with internal heat exchangers," *International Journal of Refrigeration*, vol. 28, no. 8, pp. 1238–1249, Dec. 2005, doi: 10.1016/j.ijrefrig.2005.08.009.

[30] S. C. Kim, J. P. Won, and M. S. Kim, "Effects of operating parameters on the performance of a CO2 air conditioning system for vehicles," *Applied Thermal Engineering*, vol. 29, no. 11, pp. 2408–2416, Aug. 2009, doi: 10.1016/j.applthermaleng.2008.12.017.

[31] P.-C. Qi, Y.-L. He, X.-L. Wang, and X.-Z. Meng, "Experimental investigation of the optimal heat rejection pressure for a transcritical CO2 heat pump water heater," *Applied Thermal Engineering*, vol. 56, no. 1, pp. 120–125, Jul. 2013, doi: 10.1016/j.applthermaleng.2013.03.045.

[32] A. Okasha, N. Müller, and K. Deb, "Bi-objective optimization of transcritical CO2 heat pump systems," *Energy*, vol. 247, p. 123469, May 2022, doi: 10.1016/j.energy.2022.123469.

[33] W. S. McCulloch and W. Pitts, "A logical calculus of the ideas immanent in nervous activity," *Bulletin of Mathematical Biophysics*, vol. 5, no. 4, pp. 115–133, Dec. 1943, doi: 10.1007/BF02478259.

[34] M. Mohanraj, S. Jayaraj, and C. Muraleedharan, "Applications of artificial neural networks for refrigeration, air-conditioning and heat pump systems—A review," *Renewable and Sustainable Energy Reviews*, vol. 16, no. 2, pp. 1340–1358, Feb. 2012, doi: 10.1016/j.rser.2011.10.015.

[35] Z. G. Wu, J. Z. Zhang, Y. B. Tao, Y. L. He, and W. Q. Tao, "Application of artificial neural network method for performance prediction of a gas cooler in a CO2 heat pump," *International Journal of Heat and Mass Transfer*, vol. 51, no. 21, pp. 5459–5464, Oct. 2008, doi: 10.1016/j.ijheatmasstransfer.2008.03.009.

[36] Claytex Technia Company, "Thermal Systems Library & TIL Suite." 2022. Accessed: May 23, 2022. [Online]. Available: https://www.claytex.com/products/dymola/model-libraries/thermal-systems-library/

[37] Dassault Systèmes, "Dymola." Accessed: May 23, 2022. [Online]. Available: https://www.3ds.com/products-services/catia/products/dymola/

[38] T. A. Sazon and H. Nikpey, "Investigation of the performance of a ground-coupled CO2 heat pump for space and water heating," presented at the 14th IEA heat pump conference, Chicago, May 2023.

[39] M. D. McKay, R. J. Beckman, and W. J. Conover, "A Comparison of Three Methods for Selecting Values of Input Variables in the Analysis of Output from a Computer Code," *Technometrics*, vol. 21, no. 2, pp. 239–245, 1979, doi: 10.2307/1268522.

[40] F. Chollet and others, "Keras." GitHub, 2015. [Online]. Available: https://github.com/fchollet/keras

[41] D. E. Rumelhart, G. E. Hinton, and R. J. Williams, "Learning representations by back-propagating errors," *Nature*, vol. 323, no. 6088, Art. no. 6088, Oct. 1986, doi: 10.1038/323533a0.

[42] N. Gunasekar, M. Mohanraj, and V. Velmurugan, "Artificial neural network modeling of a photovoltaic-thermal evaporator of solar assisted heat pumps," *Energy*, vol. 93, pp. 908–922, Dec. 2015, doi: 10.1016/j.energy.2015.09.078.

[43] A. F. Agarap, "Deep Learning using Rectified Linear Units (ReLU)." arXiv, Feb. 07, 2019. doi: 10.48550/arXiv.1803.08375.

[44] D. P. Kingma and J. Ba, "Adam: A Method for Stochastic Optimization." arXiv, Jan. 29, 2017. doi: 10.48550/arXiv.1412.6980.

IOP Conf. Series: Materials Science and Engineering 1294 (2023) 012061 doi:10.1088/1757-899X/1294/1/012061

Review of recent advancement in performance, and thermal energy storage studies on indirect solar dryers for agricultural products.

G D Shekata[1,*], G S Tibba[2], and A T Baheta [1]

[1] Department Mechanical Engineering, Addis Ababa Science and Technology University, Addis Ababa, Ethiopia.
[2] Agro-Technical and Technology Collage (ATTC), Harar, Ethiopia.

[*] Correspondence: gadisa3224desa@gmail.com

Abstract. Efficient drying methods and post-harvest processes are crucial to reduce waste in fruits, vegetables, and agricultural products. Factors like moisture-related deterioration, climate change, mishandling, delayed shipping, improper storage, and sales delays contribute to post-harvest losses and quality degradation. Solar dryers, especially indirect-type solar dryers, provide a practical and environmentally friendly way to preserve these goods. In comparison to other types of solar dryers, indirect-type dryers have a number of advantages. This review focuses on evaluating the performance of different configurations of Indirect-type solar dryers in terms of drying time, maximum air temperature, drying efficiency for collectors, and overall dryer efficiency. The influence of various operating conditions on the thermal efficiency of Indirect-type solar dryers is also investigated. The study provides detailed information on the sensible and latent storage units and materials used in Indirect-type solar dryers, including those operating through natural or forced convection. The review also explores the utilization of advanced technologies, such as desiccant systems, recycling processes, the use of Nano fluids and nanoparticles, and thermal energy storage, to enhance the thermal performance of solar dryers. Additionally, the paper examines potential difficulties and suggestions for selecting, using, and testing thermal storage for indirect-type solar dryers. In summary, this review provides a comprehensive examination of indirect-type solar dryers, and their performance. It highlights the potential for improving drying efficiency through the integration of advanced technologies and thermal energy storage.

1. Introduction

The worldwide population, which was 7.7 billion in 2020, is expected to increase by approximately 2 billion in the next three decades. This growth poses significant challenges to global sustainability, with food supply and security emerging as major concerns alongside energy and water [1]. The rising energy consumption worldwide necessitates the prioritization of reliable, affordable, environmentally friendly, and renewable energy sources [2–4]. In both developed and developing countries, a significant portion of food production (around 30-40%) is lost during post-harvest and processing stages. In developed countries, this loss occurs mainly at retail and consumer levels [1, 5]. Implementing drying methods for raw agricultural products can substantially reduce post-harvest losses [6]. To preserve agricultural products while retaining their essential nutrients over an extended period, it is crucial to reduce their moisture content [7].

Solar drying, which involves exposing the materials to solar radiation directly or indirectly, is an effective process in enhancing the quality and shelf life of products. It facilitates the simultaneous transmission of heat and moisture [8-9]. Moisture is moved from a material's inside to its surface during the drying process, and then that surface moisture vaporizes into the environment. To change the temperature of the product being dried, heat transfer is required [10-11]. During the drying process, it is essential to remove moisture rapidly at a temperature that does not significantly affect the flavor, texture, and appearance of fruits, vegetables, and other food items [12-15]. Additionally, along with limiting losses, guaranteeing process control, and requiring less ground surface than open sun drying methods, the drying process also attempts to protect the items from damage, birds, insects, and unexpected rainfall [16, 17]. However, there is a substantial technology gap in the creation of low-cost, energy-efficient dryers that can handle items of excellent quality. Careful model and component selection is necessary for solar dryers to use solar energy as efficiently as possible [17, 18]. Due to the intermittent nature of solar radiation, manufacturing dried goods of sufficient quality via solar drying is extremely difficult. During times of low insolation or at night, when sun radiation is not accessible, dehydration is interrupted. There have been various attempts to solve this problem, but the use of thermal storage materials has proven to be the most successful strategy [19].

Phase-change materials (PCMs) have demonstrated significant potential among thermal storage materials for their ability to store thermal energy during periods of solar radiation and release it when sunlight is unavailable [20, 21]. Depending on the presence or absence of solar radiation, PCMs go through a phase change, absorbing and releasing thermal energy (latent heat) [22]. This property makes latent heat storage systems advantageous, as they can store a significant amount of heat with minimal temperature change and high storage density [23, 24]. Solar energy can be used more effectively as a dependable energy source by including thermal energy storage (TES), especially PCMs [25]. Concrete, boulders, sand, and bricks are examples of sensible materials that can be employed as solid-state thermal storage for heat storage in addition to PCMs [26]. The use of sensible heat storage devices accelerates the drying process and reduces product mass losses compared to conventional solar drying methods [27]. Additionally, using thermal energy storage allows for improved control over external drying factors including moisture content, temperature, and drying air flow rate while providing total protection for the goods against dust, rain, wind, insects, and animals [28].This leads to an improvement in product quality. A significant portion of current research on solar drying involves experimental investigations and performance evaluations.

Indirect-type solar dryers (ITSDs) are preferred over open sun drying (OSD) and direct-type solar dryers due to their ability to provide the necessary temperature, better drying control, color preservation, and crop protection. They are particularly recommended for photosensitive fruits, vegetables, and agricultural products [29, 30]. While there is some available literature on ITSDs, there is currently no specific study that comprehensively addresses their features, types, energy storage techniques, and performance. This review aims to fill this gap by thoroughly examining ITSD research and providing meaningful insights [19,31,32]. It explores various ITSD designs, operating characteristics, and energy storage techniques. The review's primary objective is to explore the fundamental concepts of solar drying methods, focusing on indirect solar drying mechanisms. The review analyzes the usefulness of various thermal energy storage materials used in ITSDs, covers several types of ITSDs for drying various agricultural goods, and assesses various design and performance factors. Ultimately, it aims to determine the most suitable dryer under different operating conditions for various product categories.

2. Advanced in solar dryer systems

In order to decrease post-harvest losses and boost product quality, many kinds of solar dryers have been designed recently. Only a small number of these dryer varieties are, however, commonly used or offered for sale [33]. The size, system design, and solar energy consumption mode of solar dryers for agricultural and marine products can all be categorized. The sun dryers shown in Figure 1 are passive, active, and hybrid, with natural or forced air circulation, as well as direct or indirect heat transfer [7]. Overviews of the benefits and drawbacks of each type of dryer are given in Table 1.

Open sun drying, which involves exposing products directly to sunlight, leads to significant losses due to factors such as birds, rodents, insects, microorganisms, and unpredictable weather conditions like storms or rain [7], [34]. Furthermore, it causes color fading from UV radiation, uneven drying, impurities from outside contaminants such as dirt, dust, insects, and bacteria, and uneven drying[18], [35], [36]. Open sun drying is a long procedure that can lead to significant losses, and the dried goods frequently fall short of quality standards [37].

Passive solar dryers have a simple construction and rely on natural air circulation driven by the buoyancy effect. Air that has been heated by the solar is forced through the drying chamber by air pressure, buoyancy forces, or a combination of both. Moist air is evacuated through a chimney or an exit at the top of the dryer [38], [39]. These dryers can be used to preserve fish, vegetables, and fruits at home by drying modest quantities of agricultural items at temperatures between 40°C and 50°C. Due to the fact that the goods being dried are not exposed to sun radiation directly, indirect passive solar dryers with forced convection have a drying chamber and a solar air collecting system, which solve the issue of cracking and safeguard vitamins and color [40].

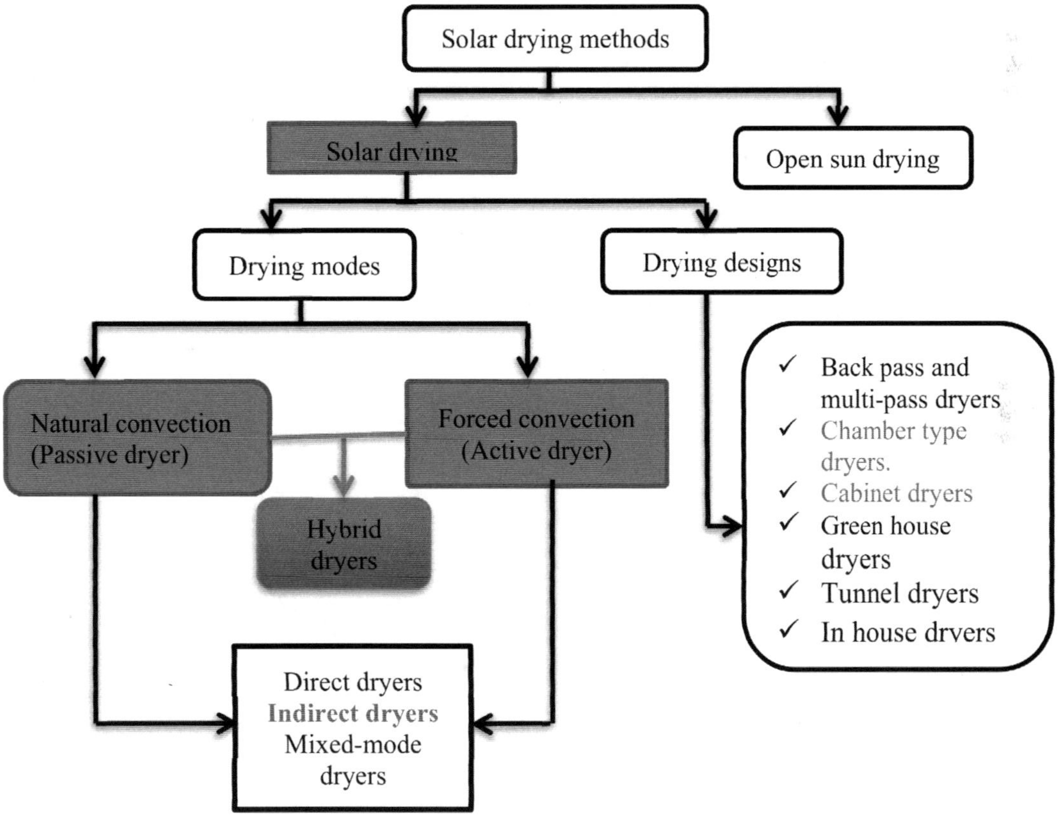

Figure 1: Solar drying methods [33, 41, 42].

In active dryers, such as the drying chamber or solar collector, hot air is circulated using a ventilation system. A fan maintains the required airflow rate in the dryer, resulting in more efficient moisture evaporation from the product [34]. Compared to passive dryers, active dryers require higher investment and are more complex to operate and maintain. In order to obtain high drying efficiency and product quality, temperature and airflow rate management must be done properly [18, 33].

A hybrid solar dryer is a form of solar dryer that also uses power from other sources, such as biomass or electricity, or combines energy storage technologies, such as thermal storage. As a result, drying can proceed even when there is no sunlight [40]. Simulated research on the drying of wood using a hybrid solar-electric dryer coupled with a thermal energy storage (TES) system was carried out by Lamrani and Draoui [43]. The TES keeps the drying chamber at a temperature that is between 4 and 20 degrees Celsius higher than the ambient air throughout the night to ensure that the wood is dried continuously.

Table 1: Here are the main advantages and limitations of various solar dryer types [5,19, 40, 42, 44, 45].

Type	Advantages	Limitations
Direct type	✓ Least expensive ✓ Simple design	- UV rays have the potential to damage the product; - Small capacity; - Lengthy drying times; and - Low efficiency. - Additionally, the product itself works as a solar energy absorber.
Indirect type	- Less damage to the product from high temperatures - Protection from UV radiation - High efficiency	- More complex and expensive compared to direct type
Mixed mode	- High efficiency - Less damage to the product from high temperatures	- UV rays might harm the product more difficult and expensive than direct sun drying
Hybrid type	- Ability to work without sunlight reduces dependence on solar energy availability - Minimizes product loss - Allows better control of drying process - High efficiency - Short drying time required - Good quality of the final product	- Expensive - May require fuel dependence

Indirect-type solar dryers (ITSD) are widely regarded as more advantageous compared to other solar dryer types, despite the benefits they offer. One of the key advantages is that they enable the production of clean final products that retain their color, flavor, and nutritional content. By incorporating energy storage systems to address the intermittent nature of solar energy, the drying period can be reduced, leading to further improvements in the quality of the end product.

2.1. Performance analysis of indirect-type solar dryer

2.1.1. Working principle of indirect-type solar dryer. The working principle of an indirect-type solar dryer (ITSD) involves two main components: a drying cabinet where materials are placed on trays, and a solar collector setup consisting of a glass plate and an absorber plate. The process begins by heating the food inside the drying cabinet and drawing out moisture by circulating hot air over it. The drying chamber contains wire mesh trays to hold the product being dried [46]. The solar collector setup, attached to the drying chamber at a specific inclination determined by the location's latitude, comprises a selectively coated absorber plate and a glass cover. The absorber plate is coated with a black hue to optimize sunlight absorption. When solar radiation reaches the glass cover, various heat transfer processes occur. Some energy is reflected back, some is transmitted through the glass, some is lost to the atmosphere, and some is absorbed by the absorber plate. Convection transfers heat to the air entering the drying chamber, causing the food item to absorb heat and undergo dehydration [47], [48].

2.1.2. Flat plate collector efficiency. The insufficient efficiency of solar dryers has been a point of contention from the start. Thermal performance, drying kinetics, environmental variables, economic evaluations, and dried product quality are just a few of the metrics that may be used to gauge how effective a solar drying system is [49]. The effectiveness of the solar flat plate collector depends on speed, ambient temperature, and humidity [50]. The glass's participation in energy [19], [51] is estimated by:

$$Q_{ic} = I_T A_c \tag{1}$$

$$Q_{ac} = \dot{m} C_{pa} (T_{co} - T_{ci}) \tag{2}$$

Q_{ic} is useful heat input, Q_{ac} is actual heat dispensed (W), I_T is solar intensity (W/m^2), $\dot{m}$ is air flow rate (kg/s), T_{ci} and T_{co} are the collector inlet and outlet air temperature (°C), and A_c = collector area (m^2). Collector efficiency (η_c) is estimated using the equation below [52]:

$$\eta_c = \frac{Q_{ac}}{Q_{ic}} = \frac{\dot{m} C_{pa} (T_{co} - T_{ci})}{I_T A_c} \tag{3}$$

2.1.3. Drying Efficiency. A certain degree of moisture content, which varies for different fruits and vegetables, must be reduced for long-term preservation. The quantity of heat, Q_d, required evaporating the product's water content from the surface and the core is calculated by:

$$Q_d = m_a C_{pa} (T_{co} - T_f) = m_w L_w \tag{4}$$

Where T_f is the temperature at the chamber's exit in degrees Celsius, L_w is latent heat at a mean temperature of $[(T_{co}+T_f)/2]$ in J/kg, and m_w is the mass of water removed in kilograms.
Quantity of water extracted (m_w) while drying is,

$$m_w = \frac{m_i (MC_a - MC_f)}{100 - MC_f} \tag{5}$$

Where m_i is the product's initial mass (drying load) in kilograms, while MC_i and MC_f are the product's initial and final moisture contents.
The product's moisture content can be estimated using:

$$\text{Moisture content, wet basis (wb), } \mathrm{MC_i} = \frac{(\mathrm{m_i} - \mathrm{m_d})}{\mathrm{m_i}} \tag{6}$$

$$\text{Moisture content, wet basis (db), } \mathrm{MC_i} = \frac{(\mathrm{m_i} - \mathrm{m_d})}{\mathrm{m_d}} \tag{7}$$

Where m_d is the product's dried mass and m_i is the product's initial mass before drying.
The power absorbed by the product (q) can be calculated using Equation (8) as:

$$q = \frac{m_a C_p (T_{co} - T_i)}{t} \tag{8}$$

IOP Conf. Series: Materials Science and Engineering 1294 (2023) 012061 doi:10.1088/1757-899X/1294/1/012061

The specific enthalpy of the air changes from h_i to h_{co} during heating in the collector, while the humidity ratio (ϕ) remains constant between ϕ_{co} and ϕ_f [11], [53].The mass of air required for moisture removal can be determined using Equation (9).

The specific enthalpy of the air rises from h_i to h_{co} during air heating in the collector, but the humidity ratio $\phi_i = \phi_{co}$ stays constant. Drying air in the chamber absorbs moisture from the product, changing the humidity ratio from ϕ_{co}, and ϕ_f [11], [53].

The mass of air needed for the moisture removal process can be calculated using:

$$m = \frac{m_w}{(\phi_f - \phi_{co})} \tag{9}$$

The thermal power needed to heat the air from the absorber plate is given by Equation (10):

$$\dot{q} = m(h_{co} - h_i) = \eta A_c I \tag{10}$$

The drying efficiency (%) can be estimated as follows [11]. For natural convection, drying efficiency (%)

$$, \eta_d = \frac{(m_w h_{fg})}{IA_c} \tag{11}$$

For forced convection[54], drying efficiency (%)

$$\eta_d = \frac{(m_w h_{fg})}{(IA_c + P_f)} \tag{12}$$

Where h_{fg} is the latent heat of vaporization (kJ/kg), P_f is the fan power. The power required to force air through the dryer can be determined by knowing the pressure drop (Δp) across the system and the efficiency of the blower/fan η_f [55].

$$P_f = \frac{(\dot{m}_a \Delta p)}{(\eta_f \rho_a)} \tag{13}$$

In this equation, $\dot{m}_a$ represents the mass flow rate of air, and ρ_a is the density of air. The moisture removal rate (MRR) represents the quantity of moisture eliminated from the product within a specific time period. It can be determined by measuring the change in mass of the product before and after the drying process, along with the duration of drying required to achieve the desired moisture level. The relationship for calculating the MRR is provided as follows [52]:

$$MRR = \frac{initial\ mass - final\ mass}{drying\ time} \tag{14}$$

The specific moisture extraction rate (SMER), which is the ratio of the mass of moisture removed in kilograms to the total energy supplied, can also be used to assess the drying performance.

$$SMER = \frac{mass\ of\ moisture\ removed(kg)}{total\ energy\ supplied} \tag{15}$$

The second law of thermodynamics, known as exergy, asserts that a flow of matter, heat, or work can generate the maximum possible work when it reaches equilibrium with a reference environment. Exergy analysis is a valuable tool for analysing, optimizing, and improving the energy efficiency of a process [46]. By conducting an exergy analysis, a more comprehensive understanding of the available useful energy at various system components, as well as the energy lost due to irreversibility, can be obtained. The equations (17) and (18) provide the exergy inflow and outflow of the drying chamber, respectively, as described in reference [52].

$$Ex_{in} = \dot{m}C_{pa}(T_o - T_a) - T_a \dot{m}C_{pa} ln \frac{T_o}{T_a} \quad (16)$$

$$Ex_{out} = \dot{m}C_{pa}(T - T_a) - T_a \dot{m}C_{pa} ln \frac{T}{T_a} \quad (17)$$

Where: Ta is ambient air temperature T is temperature of air leaving the drying chamber. The exergy efficiency is a measure of how effectively the available exergy is utilized for the drying of a product. This efficiency value provides insight into the system's ability to convert available exergy into useful work for the drying process.

$$\eta_{ex} = \frac{Ex_{in} - Ex_{out}}{Ex_{in}} \quad (18)$$

3. Thermal energy storage (TES)

Solar dryers utilize solar heat to dry various foods and agricultural products. However, their main limitation is the discontinuous drying process in the absence of sunlight. To address this issue, various technologies have been implemented, such as heat recovery mechanisms, integrating curtains and photovoltaic (PV) systems in solar tunnel dryers, modifying collector designs, incorporating sun-tracking mechanisms, and utilizing thermal energy storage (TES) systems [56]. These systems improve energy system efficiency by stabilizing energy supply and enhancing reliability, reducing the time gap or uncertainty between energy supply and demand. In many impoverished nations, inadequate thermal energy systems and TES lead to substantial food and grain losses, particularly during the harvest season. Additionally, the shrinkage process during drying reduces the volume of food and agricultural products, resulting in lower shipping costs [57].

Sensible heat, latent heat, and thermochemical processes are three ways that thermal energy storage (TES) materials can store energy. Heating or cooling a liquid or solid medium, such as water, sand, molten salts, or rocks, constitute sensible heat storage. Phase change materials (PCM) that switch between the solid and liquid states are used for latent heat storage. Chemical reactions are used in thermochemical storage (TCS) to store and release thermal energy [58]. Due to the complexity of chemical reactions involved and its limitations, such as chemical instability and being unable to stop the process, chemical energy storage is the most difficult of all technologies. The most common type of energy storage is sensible heat storage since it is straightforward and inexpensive. However, it suffers from inadequate storage capacity, requiring larger systems. Latent heat storage (LHS) is advantageous as it allows for increased heat storage without triggering chemical reactions. For LHS systems, a high power capacity is essential for both energy charging and discharging [59]. Comparing latent heat storage (LHS) to sensible heat storage (SHS), LHS has several advantages over SHS, including maintaining a nearly constant average temperature during energy absorption and diffusion, higher energy densities in phase change storages, 5–14 times more latent heat storage than sensible heat per unit volume, and a significant thermal storage capacity per unit mass and volume for small temperature differences [57]. Extensive research has been carried out on indirect-type solar dryers combined with thermal energy storage (TES) units, biogas units, fossil fuel heaters, and electric heaters. However, the use of electric heaters, fossil fuels, and biogas-based burners can lead to higher operational costs and environmental impact. Consequently, there is a growing interest in assessing previous studies on different thermal storage systems integrated with indirect solar dryers operated in natural convection (NC) and forced convection (FC) modes. The objective is to identify sustainable and efficient solutions for solar drying processes, with a focus on minimizing environmental harm and ensuring cost-effectiveness.

3.1. Sensible heat storage units (SHS)

The materials are heated in sensible heat storage units (SHS) in order to store additional solar energy. The specific heat capacity, mass, and temperature difference of the material are only a few variables that affect how much energy SHS can store. High density, great thermal conductivity, high specific heat, and long-term stability are all desired properties for SHS materials. Brick, NaCl, cast iron, cast steel, rock, sand, and concrete are materials that are frently utilized in SHS units [60]. The use of sensible heat storage in indirect sun dryers has been the subject of several studies that considered design, kinds, and qualities. Equation 19 can be used to figure out how much sensible heat is held in a substance.

$$Q = m_m C_p (T_f - T_i) = \rho C_p V (T_f - T_i) \quad (19)$$

Where Q is the heat stored, m_m is the mass of the material, C_p is the specific heat of the material, T_i is the initial temperature, ρC_p is the volumetric heat capacity of the solid material, V is the volume, and T_f is the final temperature.

Overall, sensible heat storage systems are essential for improving the effectiveness and performance of sun drying systems, enabling efficient energy storage, and shortening the drying time for a variety of agricultural products. Kareem et al. [61] tested the drying of Roselle (Hibiscus sabdariffa) in order to study the characteristics of multi-pass SAC. The mechanism was helped in storing SH by granite. Estimates showed that a number of efficiencies, including collector efficiency (64.08%), moisture pickup efficiency (67%), drying efficiency (36.22%), and system optical efficiency (70.53%), varied. The system consumed 21 hours less time than OSD. The anticipated payback period was 2.14 years. Cherry tomato dryers with integrated heat storage systems were created by Nabnean et al. [62]. 100 kg of tomatoes were dried over the course of four days at a temperature range of 30 to 65 °C. Insects, dust, and rain were all protected against by the system while the clothes were drying. Achieving efficiencies between 21% and 69%, the solar air collector (SAC) in this system. A triple-pass indirect-type forced convection solar dryer was constructed and tested in Vijayan's study [63]. With an average of 45%, the solar collector's thermal efficiency ranged from 12% to 66%. In the solar dryer, the pickup efficiency for drying potato slices ranged from 2.5% to 62.9%, with an average of 29.9%. The energy efficiency ranged from 2.8% to 87.02% throughout the day, with an average of 53.57%.

Dorouzi et al. [64] designed a solar dryer for drying tomato slices, which utilized a liquid desiccant system. This innovative dryer was self-sufficient in terms of electricity, eliminating the need for grid dependence. It operated at a drying temperature range of 60-65 °C and maintained a relative humidity (RH) of 28% during the desiccant regeneration process. Tekasakul et al. [65] investigated the use of sensible heat storage for drying rubber sheets. They conducted experiments with heat storage devices placed at heights of 50, 100, and 150 cm. When used at various heights, the thermal efficiency of the heat storage devices ranged from 33% to 28%. Rubber sheets dried over a 100 cm heat storage surface in 78 hours, with a 6.71% thermal efficiency for the drying system.

Dryers equipped with sensible heat storage systems offer advantages such as accelerated drying processes and improved product quality. The thermal energy storage (TES) system allows drying to continue even when there is no sunlight available. Without TES systems, uneven drying rates can occur in integrated thermal solar dryers (ITSDs) due to the lack of solar radiation after sunset. TES systems are crucial for crop drying applications to ensure consistent and efficient drying. Mathematical models can be used to determine the required amount of TES material based on the desired temperature inside the drying chamber and the mass of the material to be dried.

3.2. Latent heat storage units (LHS)

Through the phase change process of phase change materials (PCMs), latent heat storage units (LHS) are employed to store solar energy. Temperature control and energy storage are made possible by PCMs' ability to absorb and release heat throughout the melting and freezing stages. Based on the PCMs' thermo physical, kinetic, chemical, economic, and environmental properties. According to their composition, PCMs can be divided into three groups: organic, inorganic, and eutectic. For example, organic PCMs based on paraffin and inorganic PCMs containing metallic or salt-hydrate components are further subclasses [66]. Equation (20) can be used to determine how much heat is stored in an LHS material.

$$Q = m_m C_{ps}(T_m - T_i) + L + C_{pl}V(T_f - T_m) \tag{20}$$

where, Q represents the heat stored in the LHS material (J), m_m is the mass of the material (kg), C_{ps} is the specific heat capacity of the solid material (J kg^(-1) K^(-1)), T_i is the initial temperature of the material (°C), L is the latent heat of fusion (J kg^(-1)),T_m is the melting point temperature of the material (°C), T_f is the final temperature of the material (°C).

Gilago et al. [67] examined the efficiency of passive and active convection indirect solar drying systems for drying carrot slices utilizing paraffin wax as a heat storage medium in a current study on dryers with latent heat storage. In order to enhance the mass flow rate, they first developed passive indirect sun dryers (type I), which they later upgraded by adding solar-powered fans to create a type II active setup. Comparing the type-I system to the type-II system, the type-II system showed gains in real heat supply, activation energy, and specific energy consumption, with average values being 11.8%, 12.2%, and 20.7% better, respectively. The collector efficiency for type-II was also higher than type-I, with average values of 67.8% and 59.7%, respectively. Drying efficiency showed significant improvements, with average values of 14.2% and 11.1% for type-II and type-I, respectively, representing improvements of 27.93% and 13.6%, respectively. In a study by Gilago et al. [68], two setups of a forced convection indirect sun dryer for drying pineapples were compared. Setup-1 did not include thermal energy storage (TES) system, while Setup-2 incorporated a TES system. In comparison to Setup-2 (66.97% and 10.84%), Setup-1 had somewhat greater collection and drying efficiencies (67.04% and 7.23%, respectively). In terms of activation energy and targeted energy utilization, Setup-2 performed better. For Setup-1, the dryer and collector had average energy efficiencies of 2.33% and 34.66%, respectively, and 2.12% and 57.07% for Setup-2. Setup-2 also exhibited enhanced CO_2 emissions mitigation and credit. The efficiency of passive and active indirect type sundryers (PITSD and AITSD) with thermal energy storage units was assessed using drying ivy gourd slices in a different study by Gilago and Chandramohan [69]. Compared to AITSD, PITSD had an average collector efficiency of 69.87% and a drying efficiency of 15.2%. PITSD attained an average collector efficiency of 66.7% and a drying efficiency of 13.15 percent.

Ebrahimi et al. [70] investigated the impact of phase change materials (PCMs) positioned inside a flat plate collector (FPC) on its thermal performance and overall drying efficiency. The use of a collector with a PCM at the end section reduced the drying time of the slices by approximately 21.87%. The position of the PCM led to a thermal efficiency improvement of 5.02% to 10.13%. The overall efficiency varied from 21.92% to 25.72% depending on the PCM position. Mathew and Thangavel [71] designed and evaluated a novel evacuated tube heat pipe solar dryer with integrated thermal energy storage. The average energy and exergy efficiency of the solar collector ranged from 10% to 30% and 1.9% to 5.6%, respectively. The combined mass flow rate approach increased the collector output temperature to 67°C and shortened the drying time by 2 hours. Singh and Mall [72], used a paraffin wax layer as a thermal energy storage medium in an NC solar dryer for drying banana slices. The thermal efficiency of the dryer was 2.9%, and the banana slices were dried from 73.2% (w.b) to 20% (w.b) in 18 hours. Ebrahimi et al. [70] conducted a study to examine the impact of phase change materials (PCMs) placed inside a flat plate collector (FPC) on its thermal performance and overall drying efficiency. By incorporating a PCM at the end section of the collector, the drying time for the slices was reduced by approximately 21.87%. The position of the PCM also resulted in a thermal efficiency improvement ranging from 5.02% to 10.13%. The overall efficiency of the system varied between 21.92% and 25.72%, depending on the placement of the PCM. Mathew and Thangavel [71] developed and evaluated a novel evacuated tube heat pipe solar dryer with integrated thermal energy storage. The average energy and exergy efficiency of the solar collector ranged from 10% to 30% and 1.9% to 5.6%, respectively. By employing a combined mass flow rate approach, the collector output temperature was increased to 67°C, leading to a reduction of 2 hours in the drying time.

Singh and Mall [72] utilized a paraffin wax layer as a thermal energy storage medium in a natural convection (NC) solar dryer for drying banana slices. The thermal efficiency of the dryer was measured to be 2.9%, and the banana slices were dried from an initial moisture content of 73.2% (w.b) to a final moisture content of 20% (w.b) within a drying time of 18 hours. Vásquez, Reyes, and Pailahueque [73] conducted a thermodynamic analysis of solar dryers and highlighted the advantages of latent heat storage (LHS) dryers. LHS dryers maintain a nearly constant drying air temperature, even after sunset or during weak solar energy conditions. They are particularly suitable for drying medicinal plants with volatile oil content.

As a result of shortening the drying process' processing time, it significantly contributes to energy conservation and improves system functionality. Low melting temperatures should be present in the LH storage material of choice. Paraffin wax is frequently chosen as an LH storage medium due to its inexpensive cost, convenient availability, and limited melting temperature.

3.3. Both combination of latent and sensible heat thermal storage

Combining sensible and latent heat thermal energy storage systems allows for flexibility in using a range of energy sources, including solar heat, waste heat, heat pumps, and off-peak power. By combining sensible heat storage (SHS) and latent heat storage (LHS), this system takes advantage of both of their advantages while addressing some of their drawbacks. By using concrete as the sensible storage medium, passive thermal energy storage units with enhanced thermal conductivity and wider contact areas, such as tubular heat exchangers, are commonly employed. Hybrid materials that integrate latent and sensible heat storage materials can effectively combine the benefits of both techniques and mitigate concerns associated with SHS and LHS systems [74].

In an experimental investigation by Bhardwaj et al. [75], Valerian rhizomes were dried using an integrated air-based indirect sun dryer (AITSD) with thermal energy storage (TES). With an average drying rate of 0.051 kg/h, the solar air heater achieved energy and exergy efficiencies of 26.10% and 0.81%, respectively. The drying process's total efficiency (TE) and specific energy consumption (SEC) were recorded as 11.33 kWh/kg and 10.53%, respectively. Sensible heat storage (SHS), phase change material (PCM), and direct sunshine were used in an experimental examination of an indirect solar dryer for drying chilies in the Western Himalayas by Bhardwaj et al. [76]. When compared to solar drying without heat storage and direct sunshine, the drying time was lowered by 78.12% and 86.00%, respectively. The study also found that chilies dried in the solar dryer with SHS, PCM, and direct sunshine included increased quantities of capsaicin, carotene, thiamine, and vitamin C.

Atalay [77] evaluated two thermal energy storage technologies, packed bed thermal energy storage (PBTES) and phase change material (PCM), in terms of energy usage and economics. Pebble stones and paraffin wax with a melting point of 55-60 °C were used as the storage media. The maximum thermal energy storage during the charging phase was found to be 49.52 MJ for PCM and 52.59 MJ for PBTES. Lemon slices were dried using the stored energy, and the average drying time was 6.27 hours for PBTES and 6.23 hours for PCM. The energy efficiencies of PBTES and PCM were 68.2% and 68.55%, respectively, while PBTES had a lower initial investment cost than PCM by 10.47%.Combining sensible and latent heat thermal energy storage systems in indirect solar dryers helps compensate for daily variations in solar radiation intensity. Latent heat storage is generally more efficient than sensible heat storage for solar drying applications due to its ability to store and release large amounts of energy within a small temperature range. Simulation models should be used to assess the short- and long-term performance of drying systems with and without storage media, enabling estimation of drying curves and financial benefits before implementing solar drying systems on a larger scale. Atalay [77] compared two thermal energy storage technologies, packed bed thermal energy storage (PBTES) and phase change material (PCM), in terms of energy usage and economics. Pebble stones and paraffin wax with a melting point of 55-60 °C were employed as the storage media. Lemon slices were dried using the stored energy, and the average drying time was 6.27 hours for PBTES and 6.23 hours for PCM. The energy efficiencies of PBTES and PCM were 68.2% and 68.55%, respectively.

Integrating sensible and latent heat thermal energy storage systems in indirect solar dryers helps compensate for daily variations in solar radiation intensity. Latent heat storage is generally more efficient than sensible heat storage for solar drying applications due to its capacity to store and release large amounts of energy within a narrow temperature range. To assess the performance of drying systems with and without storage media, it is recommended to utilize simulation models. This approach enables the estimation of drying curves and financial benefits, facilitating the evaluation of solar drying systems before implementing them on a larger scale.

4. Challenges and future research

Several challenges and areas for future research have been identified in the field of solar drying. The efficiency of current solar dryers ranges from 2.5% to 54%, indicating moderate performance. One of the main limitations is their inability to function effectively during periods of insufficient or no solar energy, such as cloudy weather or at night. This makes solar dryers less suitable for extensive and industrial applications. To achieve higher performance with lower operating costs, there is a need for upgraded solar dryer designs. Here are some key areas for future research identified from the literature review:

- ✓ Optimization of indirect-type solar dryers to reduce equipment and operating costs while maintaining product quality.
- ✓ Investigation into the use of reflecting panels or lenses to focus sunlight into a smaller space, increasing the amount of solar energy absorbed by the absorber plate and increasing the temperature of the final product.
- ✓ Investigation of thermal drying using renewable energy sources other than solar energy to reduce pollution emissions.
- ✓ Development of methods to achieve regulated drying parameters, such as higher drying temperatures and lower relative humidity, while minimizing energy input consumption.
- ✓ Reduction of pressure drops in the drying chamber to enhance airflow over the products being dried, such as fruits, vegetables, and agricultural produce.
- ✓ Determination of the optimal value and position of latent heat thermal storage to improve the performance of solar cabinet dryers.
- ✓ Running computer simulations to determine the short- and long-term effectiveness of drying systems with and without storage media, calculating drying curves, and assessing the financial advantages of solar-drying agricultural goods.

By addressing these research areas, advancements can be made in enhancing the efficiency, effectiveness, and cost-effectiveness of solar drying technologies for various applications.

5. Conclusion

In this review, the focus was on indirect-type solar dryers (ITSDs), covering their designs, construction details, thermal energy storage, and performance evaluations. The following key points were summarized from the review:

- ➢ ITSDs are crucial equipment for mitigating post-harvest losses of agricultural produce, particularly in rural areas of developing countries.
- ➢ ITSDs can be categorized as passive solar dryers (natural convection) or active solar dryers (forced convection). Passive solar dryers are simpler and less expensive to construct but lack control over drying rate compared to active solar dryers.
- ➢ Forced convection ITSDs, incorporating double-pass or triple-pass collectors and reflectors, have demonstrated better performance, leading to improved drying quality and overall efficiency.

- Enhancements in ITSD performance efficiency can be achieved through improvements in solar absorber designs, configurations, and materials.
- Integration of thermal storage systems and the use of different thermos fluids show promise in enhancing the performance of ITSDs.
- The combination of solar dryers and thermal energy storage reduces the overall drying process time. Sensible heat storage materials can maintain the interior temperature of the dryer higher than the ambient temperature for an average of 2 to 6 hours, while latent heat storage materials can maintain it for 1.5 to 10 hours.
- The interior temperature of the solar dryer with sensible heat storage is typically 6 to 10°C higher than the ambient temperature, while with latent heat storage, it is 8 to 20°C higher. Latent heat storage systems are preferable as they provide a constant heat supply throughout the drying process.

References

[1] Ahmadi A et al. 2021 Energy , exergy , and techno-economic performance analyses of solar dryers for agro products : A comprehensive review *Sol. Energy*, **228**, 349–373, doi: 10.1016/j.solener.2021.09.060.

[2] Nurlaila W, Mat Y, Fudholi A and Yaakob Z 2020 Energy-economic-environmental analysis of solar drying system : a review, *Int. J. Power Electron. Drive Syst.* **11**(2), 1011–1018, doi: 10.11591/ijpeds.v11.i2.pp1011-1018.

[3] Missana W P, Park E and Kivevele T T 2020 Thermal Performance Analysis of Solar Dryer Integrated with Heat Energy Storage System and a Low-Cost Parabolic Solar Dish Concentrator for Food Preservation, *J. Energy*,

[4] Chauhan Y B and Rathod P P 2020 A comprehensive review of the solar dryer *International Journal of Ambient Energy*, **41**(3), 348–367, doi: 10.1080/01430750.2018.1456960.

[5] Getahun E, Delele M A, Gabbiye N, Fanta S W, Demissie P and Vanierschot M 2021 Importance of integrated CFD and product quality modeling of solar dryers for fruits and vegetables : A review, *Sol. Energy* **220**, 88–110, doi: 10.1016/j.solener.2021.03.049.

[6] Hussen A S and Zeru B A Design 2020 Optimization and CFD Simulation of Solar Air Heater with Jet Impingement on V-Corrugated Plate, *Int. Res. J. Eng. Technol.* 1805–1813.

[7] Gorjian S *et al.* 2021 Recent advancements in technical design and thermal performance enhancement of solar greenhouse dryers, *Sustain.* **13**(13), 1–32, doi: 10.3390/su13137025.

[8] Lakshmi D V N, Muthukumar P, Layek A, and Nayak P K 2018 Drying kinetics and quality analysis of black turmeric (Curcuma caesia) drying in a mixed mode forced convection solar dryer integrated with thermal energy storage, *Renew. Energy*, doi: 10.1016/j.renene.2017.12.053.

[9] Alonge O I and Obayopo S O 2019 Computational fluid dynamics and experimental analysis of direct solar dryer for fish, *Agric. Eng. Int. CIGR J.* **21**(2), 108–117.

[10] Norton B 2017 Characteristics of Different Systems for the Solar Drying of Crops, *Sol. Dry. Technol.* doi: 10.1007/978-981-10-3833-4.

[11] Gilago M C and Chandramohan V P 2022 Performance evaluation of natural and forced convection indirect type solar dryers during drying ivy gourd : An experimental study, *Renew. Energy*, **182**, doi: 10.1016/j.renene.2021.11.038.

[12] Kumar A *et al.* 2022 Design and Fabrication of Solar Dryer System for Food Preservation of Vegetables or Fruit, *J. Food Qual.*

[13] Lingayat A, Chandramohan V P and Raju V R K 2017 Design Development and Performance of Indirect Type Solar Dryer for Banana Drying, *Energy Procedia*, **109**, 409–416, doi: 10.1016/j.egypro.2017.03.041.

[14] Swami V M, Autee A T, and Anil T R 2018 Experimental analysis of solar fi sh dryer using phase change material, *J. Energy Storage* **20**(669), 310–315, doi: 10.1016/j.est.2018.09.016.

[15] Lamidi R O, Jiang L, Pathare P B, Wang Y D, and Roskilly A P 2019 Recent advances in sustainable drying of agricultural produce: A review, *Appl. Energy*, **233–234**, 367–385 doi:

10.1016/j.apenergy.2018.10.044.
[16] Dronachari M 2019 Application of Different Types Solar Dryers in Agriculture Crops- A Review, *Int. J. Pure Appl. Biosci.*, **7**(2), 303–326, doi: 10.18782/2320-7051.5586.
[17] Madhankumar S, Viswanathan K, and Wu W, 2021, Energy , exergy and environmental impact analysis on the novel indirect solar dryer with fi ns inserted phase change material, *Renew. Energy*, **176**, doi: 10.1016/j.renene.2021.05.085.
[18] Udomkun P *et al.* 2020 Review of solar dryers for agricultural products in Asia and Africa : An innovation landscape approach, *J. Environ. Manage.* **268**, 110730, doi: 10.1016/j.jenvman.2020.110730.
[19] Bhanudas A, Chandramohan V P, Raju V R K and Meda V 2020 A review on indirect type solar dryers for agricultural crops – Dryer setup , its performance , energy storage and important highlights, *Appl. Energy*, **258**, 114005, doi: 10.1016/j.apenergy.2019.114005.
[20] Amankwah E, Dzisi K A, Van Straten G, Van Willigenburg G, and. Van Boxtel A J B 2017 Distributed Mathematical Model Supporting Design and Construction of Solar Collectors for Drying, *Dry. Technol.* **3937**, doi: 10.1080/07373937.2016.1269806.
[21] Kumar P and Singh D 2020 Advanced technologies and performance investigations of solar dryers: A review, *Renew. Energy Focus*, **35**, 148–158, doi: 10.1016/j.ref.2020.10.003.
[22] Khosravi A, Malekan M, and Assad M. E. H. 2019 Numerical analysis of magnetic field effects on the heat transfer enhancement in ferrofluids for a parabolic trough solar collector, *Renew. Energy*, **134**, 54–63, doi: 10.1016/j.renene.2018.11.015.
[23] Li X, Cho S, and Han G G D 2023 Light-Responsive Solid-Solid Phase Change Materials for Photon and Thermal Energy Storage, *ACS Mater. Au*,. **3**(1), 37–42, doi: 10.1021/acsmaterialsau.2c00055.
[24] Yang G, Yim Y J, Lee J W, Heo Y J and Park S J 2019 Carbon-filled organic phase-change materials for thermal energy storage: A review, *Molecules*, **24**(11), doi: 10.3390/molecules24112055.
[25] Palomba V, Brancato V, and Frazzica A 2017 Experimental investigation of a latent heat storage for solar cooling applications, *Appl. Energy*, **199**, 347–358, doi: 10.1016/j.apenergy.2017.05.037.
[26] Atalay H 2019 Performance analysis of a solar dryer integrated with the packed bed thermal energy storage (TES) system, *Energy*, doi: 10.1016/j.energy.2019.02.023.
[27] Cetina-Quiñones A J, López López J, Ricalde-Cab L, El Mekaoui A, San-Pedro L and Bassam A 2021 Experimental evaluation of an indirect type solar dryer for agricultural use in rural communities: Relative humidity comparative study under winter season in tropical climate with sensible heat storage material, *Sol. Energy*, **224**, 58–75, 2021, doi: 10.1016/j.solener.2021.05.040.
[28] Elhage H, Herez A, Ramadan M, Bazzi H and Khaled M 2018 An investigation on solar drying : A review with economic and environmental assessment, *Energy*, doi: 10.1016/j.energy.2018.05.197.
[29] Chavan A, Vitankar V, Shinde N and Thorat B 2021 CFD simulation of solar grain dryer, *Dry. Technol.*, **0**(0), 1–13 doi: 10.1080/07373937.2020.1863422.
[30] Chavan A, Vitankar V, Mujumdar A and Thorat B 2021 Natural convection and direct type (NCDT) solar dryers: a review, *Dry. Technol.* **39**(13), 1969–1990, doi: 10.1080/07373937.2020.1753065.
[31] Mohamad F and Teh J 2018 Impacts of Energy Storage System on Power System Reliability : A Systematic Review, *Energies*, doi: 10.3390/en11071749.
[32] Selimefendigil F and Öztop H F 2023 Performance Analysis and Identification of an Indirect Photovoltaic Thermal Dryer with Aluminum Oxide Nano-Embedded Thermal Energy Storage Modification, *Sustainability*.
[33] Udomkun P *et al.* 2020 Review of solar dryers for agricultural products in Asia and Africa: An innovation landscape approach, *J. Environ. Manage.* **268**, doi: 10.1016/j.jenvman.2020.110730.
[34] Tiwari S, Tiwari G N and Al-helal I M 2016 Performance analysis of photovoltaic – thermal (PVT) mixed mode greenhouse solar dryer, *Sol. ENERGY*, **133**,421–428, doi: 10.1016/j.solener.2016.04.033.
[35] Gorjian S *et al.* 2021 Recent Advancements in Technical Design and Thermal Performance Enhancement of Solar Greenhouse Dryers, *sustainability*,.
[36] Joseph E 2021 Microorganisms in the Deterioration and Preservation of Cultural Heritage
[37] Udomkun *et al* P 2020 Review of solar dryers for agricultural products in Asia and Africa: An innovation landscape approach, *J. Environ. Manage.* **268**,110730, doi: 10.1016/j.jenvman.2020.110730.
[38] Gebremicheal G H *et al.* 2021 PERFORMANCE EVALUATION OF SOLAR AIR COLLECTOR BY CHIMNEY EFFECT FOR DRYING APPLICATIONS, *Acta Technol. Agric.* 159–165, doi: 10.2478/ata-2021-0027.

[39] Ahmad A *et al.* 2022 A Comprehensive State-of-the-Art Review on the Recent Developments in Greenhouse Drying, *Energies*, 1–42.
[40] Kamarulzaman A, Hasanuzzaman M, and Rahim N A 2021 Global advancement of solar drying technologies and its future prospects : A review, *Sol. Energy*, **221**, 559–582, doi: 10.1016/j.solener.2021.04.056.
[41] Taghinezhad E, Kaveh M and Szumny 2021 A Optimization and prediction of the drying and quality of turnip slices by convective-infrared dryer under various pretreatments by rsm and anfis methods, *Foods*, **10**(2), doi: 10.3390/foods10020284.
[42] Gorjian S *et al.* 2021 Recent Advancements in Technical Design and Thermal Performance Enhancement of Solar Greenhouse Dryers To cite this version : HAL Id : hal-03268850 Recent Advancements in Technical Design and Thermal Performance Enhancement of Solar Greenhouse Dryers," *Sustainability*.
[43] Lamrani B and Draoui A 2020 Thermal performance and economic analysis of an indirect solar dryer of wood integrated with packed-bed thermal energy storage system : A case study of solar thermal applications, *Dry. Technol* **0**(0) 1–18, doi: 10.1080/07373937.2020.1750025.
[44] Prakash O and Kumar A 2017 *Solar Drying Technology: Concept, Design, Testing, Modeling, Economics and Environment*, **4**(1).
[45] Visavale G 2021 Principles, Classification, and Selection of Dryers, *Handb. Ind. Dry.* **2012**, 29–58, , doi: 10.1201/9781420017618-11
[46] Vijayan S, Arjunan T V and Kumar A 2019 Exergo-Environmental Analysis of an Indirect Forced Convection Solar Dryer For Drying Bitter Gourd Slices, *Renew. Energy*, doi: 10.1016/j.renene.2019.08.066
[47] Ferahta F, Abid C, Ferahta F, Abid C, Ferahta F and Abid C 2020 Intensification of a flat solar collector efficiency, *Appl. Phys.*, **83**(3), 0–9,.
[48] ER Z 2017 Prevailing Parameter Evaluation with Heat Transfer Analysis of Absorber Plate in the Flat Solar Collector, *ACTA Phys. Pol. A*, **132**(3), doi: 10.12693/APhysPolA.132.1025.
[49] Mahesh S and Anil K 2023 Performance Assessment and Modeling Techniques for Domestic Solar Dryers, *Food Eng. Rev.*, 0123456789, doi: 10.1007/s12393-023-09335-5.
[50] Babu A K, Kumaresan G, Aroul V A and Velraj R 2018 Review of leaf drying : Mechanism and in fl uencing parameters, drying methods, nutrient preservation, and mathematical models, *Renew. Sustain. Energy Rev.* **90**, 536–556, doi: 10.1016/j.rser.2018.04.002.
[51] Reddy V, Gilago M C and Chandramohan V P 2022 Energy and exergy investigation of indirect solar dryer under natural and forced convection while drying muskmelon slices, *Energy Nexus*, **8**, 100153, doi: 10.1016/j.nexus.2022.100153
[52] Jangde P K, Singh A and Arjunan T V 2021 Efficient solar drying techniques: a review, *Environ. Sci. Pollut. Res.* **2012**.
[53] Wang H, Zhang S, Fan H, Zhang M, Hu N and Yang H 2023 Modeling and Experimental Study on Drying Characteristics of Corn Particles with Hot Air in Downward Moving Bed, *fluids*, 1–15.
[54] Suherman S, Widuri H, Patricia S and Susanto E E 2020 Energy Analysis of a Hybrid Solar Dryer for Drying Coffee Beans, *Int. J. Renew. Energy Dev.* **9**(1), 131–139.
[55] Jangde P K, Singh A and Arjunan T V 2022 Efficient solar drying techniques: a review, *Environ. Sci. Pollut. Res.* 50970–50983.
[56] Motahayyer M, Arabhosseini A andSamimi-Akhijahani H 2019 Numerical analysis of thermal performance of a solar dryer and validated with experimental and thermo-graphical data, *Sol. Energy*, **193**, 692–705, doi: 10.1016/j.solener.2019.10.001.
[57] Barghi Jahromi M S, Kalantar V, Samimi Akhijahani H and Kargarsharifabad H 2022 Recent progress on solar cabinet dryers for agricultural products equipped with energy storage using phase change materials, *J. Energy Storage*, **51**, 104434, doi: 10.1016/j.est.2022.104434.
[58] Ibrahim Dinc and Rosen M A 2018 Thermal Energy Storage Systems and Applications, Second Edition.
[59] Javadi F S, Metselaar H S C and Ganesan P 2020 Performance improvement of solar thermal systems integrated with phase change materials (PCM), a review, *Sol. Energy*, **206**, 330–352, doi: 10.1016/j.solener.2020.05.106.
[60] Gautam A and Saini R P 2020 A review on sensible heat based packed bed solar thermal energy storage system for low temperature applications, *Sol. Energy*, **207**,937–956, doi: 10.1016/j.solener.2020.07.027.
[61] Kareem M W, Habib K, Ruslan M H and Saha B B 2017 Thermal performance study of a multi-pass solar

air heating collector system for drying of Roselle (Hibiscus sabdariffa), *Renew. Energy*, **113**, 281–292, doi: 10.1016/j.renene.2016.12.099.

[62] Nabnean S, Janjai S, Thepa S, Sudaprasert K, Songprakorp R and Bala B K 2016 Experimental performance of a new design of solar dryer for drying osmotically dehydrated cherry tomatoes, *Renew. Energy*, **94**, 147–156, doi: 10.1016/j.renene.2016.03.013.

[63] Vijayan S K. 2018 Thermodynamic analysis of a triple-pass solar dryer for drying potato slices, *J. Therm. Anal. Calorim.* **0**, doi: 10.1007/s10973-018-7747-0.

[64] Dorouzi M, Mortezapour H, Akhavan H R and Moghaddam A G 2018 Tomato slices drying in a liquid desiccant-assisted solar dryer coupled with a photovoltaic-thermal regeneration system, *Sol. Energy*, **162**, 364–371, doi: 10.1016/j.solener.2018.01.025.

[65] Tekasakul P, Kumar A, Yuenyao C and Kirirat P 2017 Assessment of sensible heat storage and fuel utilization ef fi ciency enhancement in rubber sheet drying, *J. Energy Storage*, **10**, 67–74, doi: 10.1016/j.est.2017.01.002.

[66] Wang F, Lin W, Ling Z and Fang X 2019 A comprehensive review on phase change material emulsions: Fabrication, characteristics, and heat transfer performance," *Sol. Energy Mater. Sol. Cells*, **191**, 218–234, doi: 10.1016/j.solmat.2018.11.016.

[67] Gilago M C, Reddy V, Chandramohan M and Parvathy V 2023 Analysis and comparison of the performance parameters of passive and active indirect solar dryers with heat storage facility while drying carrot, *Environ. Sci. Pollut. Res.* 0123456789, doi: 10.1007/s11356-023-26400-y.

[68] Gilago M C, Reddy V and Chandramohan V P 2022 Investigation of exergy-energy and environ-economic performance parameters of active indirect solar dryer for pineapple drying without and with energy storage unit, *Sustain. Energy Technol. Assessments*, **53**, 102701, doi: 10.1016/j.seta.2022.102701.

[69] Gilago M Cand Chandramohan V P 2022 Performance parameters evaluation and comparison of passive and active indirect type solar dryers supported by phase change material during drying ivy gourd, *Energy*, **252**, 123998, doi: 10.1016/j.energy.2022.123998.

[70] Ebrahimi H, Akhijahani H S and Salami P 2021 Improving the thermal efficiency of a solar dryer using phase change materials at different position in the collector, *Sol. Energy*, **220**, 535–551, doi: 10.1016/j.solener.2021.03.054.

[71] Mathew A A and Thangavel V 2021 A novel thermal energy storage integrated evacuated tube heat pipe solar dryer for agricultural products: Performance and economic evaluation, *Renew. Energy*, **179**, 1674–1693, doi: 10.1016/j.renene.2021.07.029.

[72] Singh P and Mall D 2020 Experimental investigation of thermal performance of indirect mode solar dryer with phase change material for banana slices, *Energy Sources, Part A Recover. Util. Environ. Eff.*

[73] Vásquez J, Reyes A and Pailahueque N 2019 Modeling, simulation and experimental validation of a solar dryer for agro-products with thermal energy storage system, *Renew. Energy*, **139**, 1375–1390, doi: 10.1016/j.renene.2019.02.085.

[74] Elsayed K, Gambo A and Wang Q 2022 Thermal Performance Evaluation of Combined Sensible- Latent Heat Storage Tank with a Different Storage Medium, *Chem. Eng. Trans.* **94**, 1267–1272, doi: 10.3303/CET2294211.

[75] Bhardwaj A K, Kumar R, Kumar S, Goel B and Chauhan R 2021 Energy and exergy analyses of drying medicinal herb in a novel forced convection solar dryer integrated with SHSM and PCM, *Sustain. Energy Technol. Assessments*, **45**, 101119, doi: 10.1016/j.seta.2021.101119.

[76] Bhardwaj A K, Kumar R, Chauhan R and Kumar S 2020 Experimental investigation and performance evaluation of a novel solar dryer integrated with a combination of SHS and PCM for drying chilli in the Himalayan region, *Therm. Sci. Eng. Prog.* **20**, 100713, doi: 10.1016/j.tsep.2020.100713.

[77] Atalay H 2020 Assessment of energy and cost analysis of packed bed and phase change material thermal energy storage systems for the solar energy-assisted drying process, *Sol. Energy*, **198**,124–138, doi: 10.1016/j.solener.2020.01.051.

IOP Conf. Series: Materials Science and Engineering 1294 (2023) 012062 doi:10.1088/1757-899X/1294/1/012062

Numerical investigation on pillow plate heat exchangers: Effects of nanofluid and geometry

A Karimi

Department of Industrial Engineering, University of Naples Federico II, Italy

Correspondence: ali.karimi@unina.it

Abstract. The current research involves a comprehensive numerical simulation of nanofluid flow within a pillow-plate heat exchanger. Al_2O_3-water nanofluid and water are used as the working fluids, simulated using a two-phase mixture model. The study explores the influence of geometric properties on the heat exchanger's hydrodynamic and thermal performance. It also delves into the utilization of nanofluids as the working medium and its impact on dimensionless pressure drop, Nusselt numbers, and dimensionless temperature. An innovative aspect of this research lies in the integration of artificial intelligence (AI) techniques for heat exchanger design optimization. Specifically, a Random Forest Regressor (RFR) AI model is employed to predict crucial parameters, including heat transfer coefficient (HTC) and pressure drop, based on input design variables. Key findings reveal that non-linear hole layouts in heat exchangers significantly improve Nusselt numbers by up to 25 percent. Conversely, larger holes result in higher pressure drops. The use of nanofluids enhances thermal efficiency by up to 10% while increasing pressure drop by around 7%.

1. Introduction

Heat exchangers stand as fundamental components within various industries. The imperative need to enhance their efficiency holds the key to substantial reductions in fuel consumption [1], pollution levels, and overall industry efficiency [2]. Notably, achieving higher heat exchanger efficiency presents a tangible pathway to addressing global warming concerns and optimizing [3] energy utilization.

Numerous investigations have addressed the field of heat exchangers, spanning both numerical [4,5] and experimental [6–8] approaches. Several modifications in heat exchanger design, including the utilization of nanofluids [9], highly conductive materials[10], and geometry optimization[11], have been explored to enhance their thermal efficiency. For instance, Karimi et al. [12] conducted a numerical simulation study on a double-tube heat exchanger with a twisted tape, examining the impact of alumina/water nanofluid and pure water as working fluids. Their findings highlighted a 22% Nusselt number improvement with twisted tape and a 40% rise in pressure drop.

A recent passive heat transfer enhancement method, the "pillow plate heat exchanger," [13,14] has been applied to conventional plate heat exchangers. The pillow plate gets its name from its distinctive pillow-shaped surface, created by inflating two parallel plates at specific points through a welding process. This design results in continuous changes in the cross-sectional area and the presence of welding spots [15], both of which induce turbulent flow within the pillow channels, enhancing lateral mixing and heat transfer coefficients.

A limited number of literatures focused on evaluating the performance of pillow plate heat exchangers. Piper et al. [16] were pioneering researchers on this topic. Their study proposed a novel approach for determining key geometric parameters, such as hydraulic diameter, cross-sectional area, and heat transfer area, tailored explicitly for pillow-plate geometries. They have proposed a correlation for calculating these parameters to determine the Reynolds and Nusselt numbers. Furthermore, they have reported that pillow plate heat exchangers exhibit approximately 2-7% surface area enlargement compared to conventional plate heat exchangers, attributable to surface waviness.

Tran et al. [17] achieved similar outcomes in their research, focusing on pillow-plate condensers' performance and applicability in process industries. By using experimental setups at lab and pilot scales, their study incorporated non-conventional fiber-optic measurements to determine condensation channel axial temperature profiles, and their results were compared to calculated values for virtual plain tube bundles, presenting measured axial temperature profiles and heat transfer coefficients for comparison with conventional geometries.

In another study, Kumar et al. [18] conducted a numerical investigation on the flow and heat transfer characteristics of wavy pillow plate channels employed in pillow plate heat exchangers. Their findings revealed that in the thermally developing region, the recirculation zone significantly contributes to heat transfer compared to the thermally developed section, with high heat transfer occurring in the developing region due to lower average temperatures of the hot recirculating zones. Additionally, they conducted a parametric study involving variations in channel geometric parameters, focusing on transversal and equal-pitch pillow plate configurations. Furthermore, a sensitivity analysis was performed to quantify the influence of different geometric parameters and Reynolds numbers on the thermo-hydraulic performance of the pillow plate channel.

Based on existing literature, pillow plate heat exchangers have exhibited significant potential in enhancing heat transfer by promoting turbulent flow regimes. However, a crucial area that warrants exploration is the influence of spot layouts and their respective sizes and using different working fluids. Therefore, in this research, the study delves into assessing the effects of various spot layouts, sizes, and using nanofluids on the thermal and hydraulic performance of the pillow plate heat exchanger.

2. Mathematical modeling

2.1 Geometry of the problem

Figure 1 represents the geometry of the research problem. In this study, cold flow entered the pillow-plate heat exchanger through the inlet and exited after the heat transfer due to hot external walls of the heat exchanger. The hydraulic diameter is the same in all geometries and equal to 0.315 meters. In this paper, six different geometries with different hole sizes and layouts are investigated.

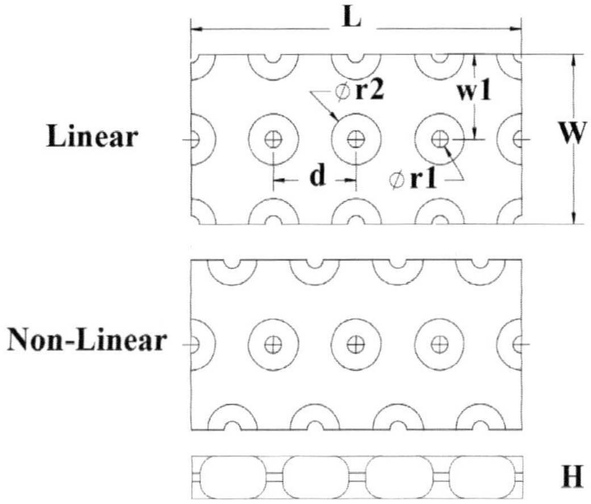

Figure 1. Geometry details of heat exchanger.

Table 1 presents the dimensionless dimensions of geometry. In this table, L = 1 m for all geometries, while the rest of dimensions can be calculated based on the values given in the table.

Table 1. Specifications of Linear (L) and Non-Linear (NL) pillow plate heat exchanger

Geometry type	W/L	H/L	w1/W	d	r_1/r_2	r_1/d
L1	0.5	0.25	0.5	0.25	0.2	0.05
L2	0.5	0.25	0.5	0.25	0.3	0.1
L3	0.5	0.25	0.5	0.25	0.5	0.2
NL1	0.5	0.25	0.5	0.25	0.2	0.05
NL2	0.5	0.25	0.5	0.25	0.3	0.1
NL3	0.5	0.25	0.5	0.25	0.5	0.2

2.2 Mixture model

Numerous methods are available for simulating nanofluids, with the single-phase model offering cost and speed advantages. However, it should be noted that this approach typically exhibits reduced accuracy compared to two-phase models. To enhance precision in this study, the mixture model is employed. Within the framework of the mixture model, the resolution of the conservation equations for mass, momentum, and energy necessitates the solution of the volumetric concentration equation for the second phase. Consequently, the relative velocity between the two phases is derived using algebraic methods. Subsequently, the conservation equations for mass, momentum, and energy in the context of the mixture model are provided [19]:

Mass:

$$\nabla.\left(\rho_{nf}\vec{V}\right) = 0 \tag{1}$$

Momentum:

$$\nabla.(\rho_{nf}\vec{V}\vec{V}) = -\nabla P + \nabla.(\mu_{nf}\nabla\vec{V}) \tag{2}$$

Energy:

$$\nabla.\left(\phi_K \vec{V}_k\left(\rho_k H_k + P\right)\right) = \nabla.\left(k\nabla T\right) \tag{3}$$

Volumetric concentration equation:

$$\nabla.\left(\rho_p\phi_p\vec{V}_m\right) = \nabla.\left(\rho_p\phi_p\vec{V}_{dr,p}\right) \tag{4}$$

Its parameters are calculated as follows [20]:

$$\mu_m = \sum_{k=1}^{n}\phi_k\mu_k \tag{5}$$

$$\rho_m = \sum_{k=1}^{n}\phi_k\rho_k \tag{6}$$

$$\vec{V}_m = \frac{\sum_{k=1}^{n}\phi_k\rho_k\vec{V}_k}{\rho_m} \tag{7}$$

$$H_k = h_k - \frac{P}{\rho_k} + \frac{v_k^2}{2} \tag{8}$$

In the momentum equation, drift velocity in the second phase (such as nanoparticles) is defined as follows:

$$\vec{V}_{dr,k} = \vec{V}_k - \vec{V}_m \tag{9}$$

Relative velocity (shear velocity) is defined as second-phase velocity (p) that is dependent on base-phase velocity:

$$\vec{V}_{p,f} = \vec{V}_p - \vec{V}_f \tag{10}$$

Relative drift velocity is defined as follows:

$$\vec{V}_{dr,p} = \vec{V}_{pf} - \sum_{k=1}^{n} \frac{\phi_k \rho_k}{\rho_m} V_{fk} \tag{11}$$

Relative velocity V_{pf} in the equation above is obtained from the following equation [21]:

$$\vec{V}_{pf} = \frac{\rho_p d_p^2}{18\mu_m f_{drag}} \frac{\left(\rho_{p-}\rho_m\right)}{\rho_p} \vec{a} \tag{12}$$

Where:

$$\vec{a} = \vec{g} - \left(\vec{V}_m . \nabla\right) \vec{V}_m \tag{13}$$

Table 2 shows the physical properties of the base fluid and nanoparticles that are used in this simulation. In this study, Al_2O_3- water is used as the working fluid of heat exchanger.

Table 2. Physical properties [27] of the base fluid and nanoparticles at T = 293 °K.

Properties	Water	Al_2O_3
C_p (J/kg K)	4182	880
K (W/m k)	0.6	36
ρ (kg/m³)	998.2	3890
μ_{water}= 0.001003 kg/m s		
size $_{nanoparticles}$ (nm)= 20		

In addition, nanofluid properties are calculated as follows:

Nanofluid density [22]:

$$\rho_{nf} = (1-\phi)\rho_{bf} + \rho_p \tag{14}$$

Nanofluid specific heat [23]:

$$c_p = (1-\varphi)cp_{bf} + \varphi.cp_p \tag{15}$$

Nanofluid dynamic viscosity [24]:

$$\mu_{nf} = (1 + 1.73\varphi + 123\varphi^2)\mu_{bf} \tag{16}$$

Nanofluid thermal conductivity [25]:

$$k_{nf} = (1 + 2.72\varphi + 4.97\varphi^2)k_{bf} \tag{17}$$

SST k-w transport equations [26]:

$$\frac{\partial}{\partial t}(\rho k)+\frac{\partial}{\partial x_i}(\rho k u_i)=\frac{\partial}{\partial x_j}\left(\Gamma_k \frac{\partial k}{\partial x_j}\right)+\tilde{G}_k - Y_k + S_k \tag{18}$$

$$\frac{\partial}{\partial t}(\rho \omega)+\frac{\partial}{\partial x_i}(\rho \omega u_i)=\frac{\partial}{\partial x_j}\left(\Gamma_\omega \frac{\partial \omega}{\partial x_j}\right)+\tilde{G}_\omega - Y\,\omega + D_\omega + S_\omega \tag{19}$$

2.3 Boundary conditions
In the inlets, constant velocity and dimensionless temperature are used. Also, non-slipping and constant temperature conditions are applied on external walls of heat exchangers. In the outlets, the pressure outlet condition is used. In this study, the outlet pressure assumed to be zero in order to facilitate the computation of absolute pressure differences. Symmetry conditions are used on sidewalls of heat exchangers too. Figure 2 presents the boundary conditions of geometry. The fluid enters to the heat exchanger at T=310.15 K and cools the walls of heat exchanger with T = 311.15 K.

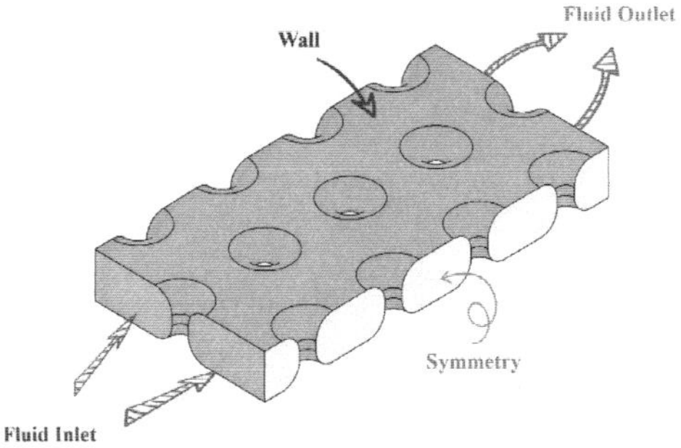

Figure 2. Geometry overview.

2.4 Numerical and evaluation methods
Various methods are available for simulating fluid flows and heat transfer. The Simple method, which has demonstrated high precision in numerous previous studies [28], is also employed in this research. Convergence criteria were set to a level of 10^{-6} to ensure accurate results.

Grid independence is a critical consideration in numerical simulations. The chosen mesh should not unduly influence the outcomes. Conversely, excessively fine grids can lead to increased computational costs and longer simulation times. To optimize the grid, a range of grid sizes was examined. Table 3 presents Nusselt numbers for different grid sizes. SP7.0 with an advanced meshing algorithm for structured grids is used in this research [29].

Table 3. The results of mesh study for pure water at Re= 3500

Geometry type	L1		L2		L3		NL1		NL2		NL3	
	Elements no.	Nu	Elements no.	Nu	Elements no.	Nu	Elements no.	Nu	Elements no.	Nu	Elements no.	Nu
Mesh 1	1254895	118.54	1236584	110.12	1235484	88.65	1245484	123.68	1234542	118.45	1245124	88.95
Mesh 2	1324548	119.48	1364484	115.45	1486531	91.48	1345984	128.94	1345844	124.05	1354512	90.18
Mesh 3	1456845	125.13	1436551	116.48	1564521	93.14	1453258	130.48	1453215	126.77	1456224	93.65
Mesh 4	**1578964**	**126.64**	**1598654**	**117.85**	**1598942**	**94.29**	**1568456**	**134.09**	**1568754**	**128.89**	**1565423**	**95.91**
Mesh 5	1687954	126.73	1699465	117.89	1689753	94.31	1687984	134.18	1687563	128.93	1648543	95.96

In Figure 3, structured mesh is presented for L3 geometry. To ensure correctness and precision, the numerical findings were compared to experimental data. Specifically, a pillow-plate heat exchanger with varying Reynolds numbers, containing water, was compared to the findings reported by Tran et al.

IOP Conf. Series: Materials Science and Engineering 1294 (2023) 012062 doi:10.1088/1757-899X/1294/1/012062

[30] findings. According to Figure 4, the given numerical results are in good agreement with the experimental results.

Figure 3. Mesh file of L3 geometry.

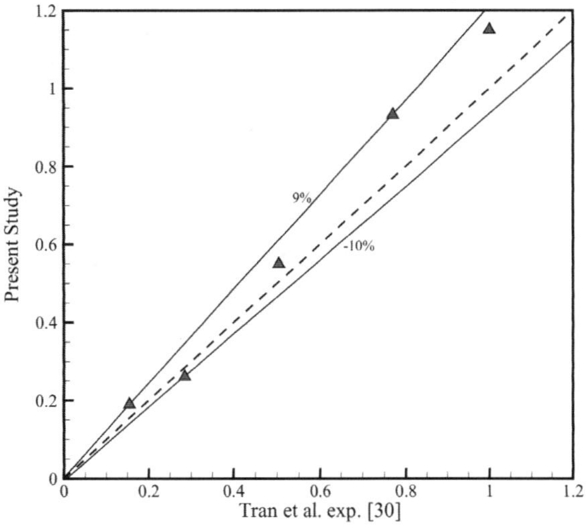

Figure 4. Comparison of Nu/No$_{mex}$ number obtained in the simulation with experimental data [30] for Re= 1500, 2000, 4000, 6000 and 8000.

3. Results and discussions

In this section, an examination of heat transfer analysis, and flow components for a pillow plate heat exchanger is conducted employing the two-phase mixture method. Numerical results have been acquired across a range of Reynolds numbers, spanning from 3000 to 5000, with varying volumetric concentrations from 0% to 3%.

3.1. Effect of hole size on thermal performance

In this paper, a pivotal aspect of the research revolves around the examination of how hole size impacts the thermal and hydrodynamic performance of heat exchangers. Within the scope of this study, three distinct hole sizes were thoroughly investigated to elucidate the relationship between hole dimensions and thermal performance.

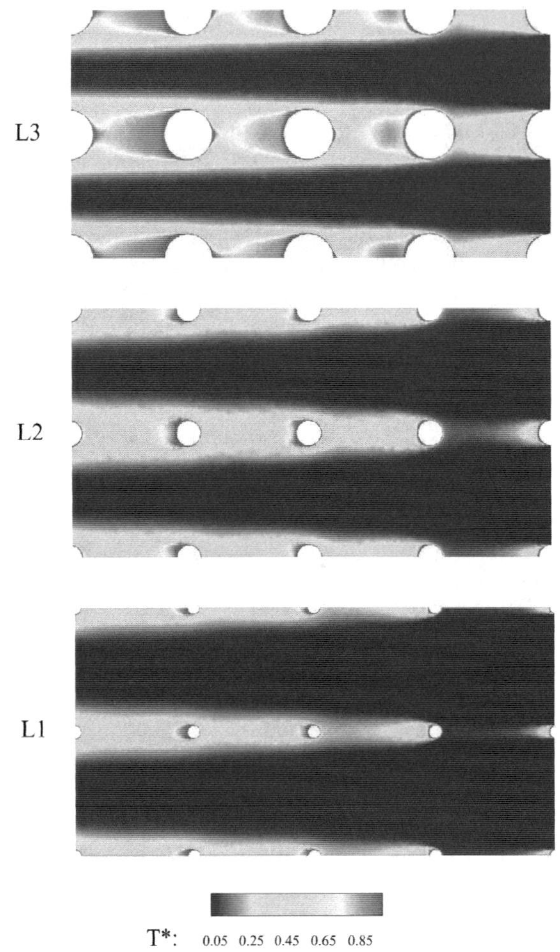

Figure 5. T* contours for pure water at Re= 3000 for different hole sizes.

Figure 5 illustrates the impact of varying hole sizes on the non-dimensional temperature, denoted as T^*. In order to evaluate the performance of the pillow-plate heat exchanger, a non-dimensional temperature metric, T*, has been introduced. Equation 20 provides the formulation for calculating T^*.

$$T^* = \frac{T_{Inlet} - T}{T_{Inlet} - T_{Wall}} \tag{20}$$

As per Equation (20), the optimal operating mode is achieved when the fluid temperature within the heat exchanger approaches the wall temperature, resulting in a T^* value nearing 1. Figure 6 displays the non-dimensional velocity for three distinct geometries. The non-dimensional velocity is defined by Equation (21).

$$V^* = \frac{V}{V_{Max}} \tag{21}$$

As shown in Figure 6, the size of the recirculation region exhibits a proportional increase with the enlargement of the hole diameters. Notably, the figure indicates that the size of the recirculation region closely approximates the dimensions of the hole sizes. Additionally, the streamlines visibly illustrate the presence of circulation zones situated behind the holes, which exert a substantial influence on both pressure drop and heat transfer. This topic will be further elaborated upon in the subsequent section.

IOP Conf. Series: Materials Science and Engineering 1294 (2023) 012062 doi:10.1088/1757-899X/1294/1/012062

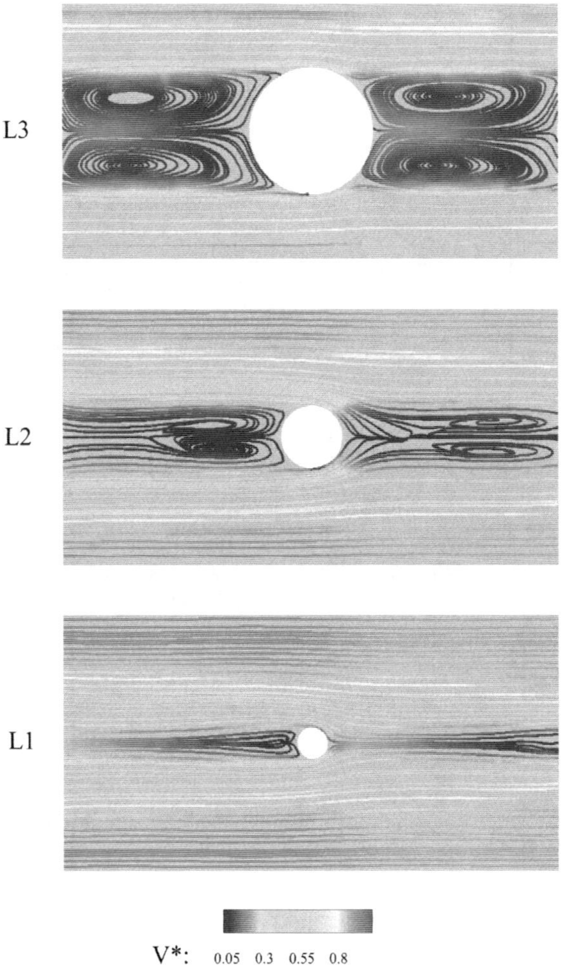

Figure 6. Streamlines for pure water at Re = 3000 for different hole sizes.

As illustrated in Figure 7, the utilization of larger holes resulted in an increment of up to 25 percent in the Nusselt number. This observation underscores the notable impact of hole sizes on the design and performance of heat exchangers. Furthermore, Figure 7 reveals a positive correlation between the Nusselt number and the Reynolds number. As the Reynolds number escalates, the boundary layer diminishes in size, leading to heightened velocity and temperature gradients on the pipe wall, subsequently enhancing heat transfer rates.

Figure 8 presents non-dimensional pressure for different geometries versus Reynolds number. As evident in Figure 8, in accordance with expectations, the L3 case exhibits the highest pressure drop. This outcome can be attributed to the presence of bigger holes in the L3 case, resulting in the partial occlusion of the fluid path within the pipe. Consequently, this restriction leads to an elevation in pressure drop, indicating an adverse impact on flow dynamics.

The graphical representation further highlights the similarity in pressure drop trends across various hole sizes concerning their response to changes in the Reynolds number. Specifically, the data indicates that with an increase in Reynolds number, the pressure drop consistently rises, underscoring the relationship between Reynolds number and increased pressure drop.

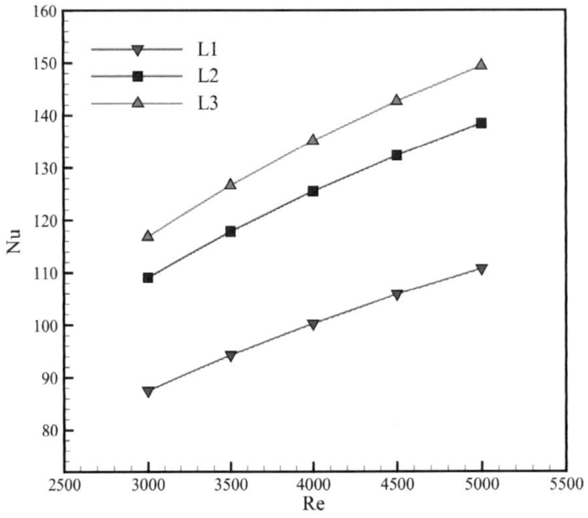

Figure 7. Nusselt number versus Reynolds number for different geometries.

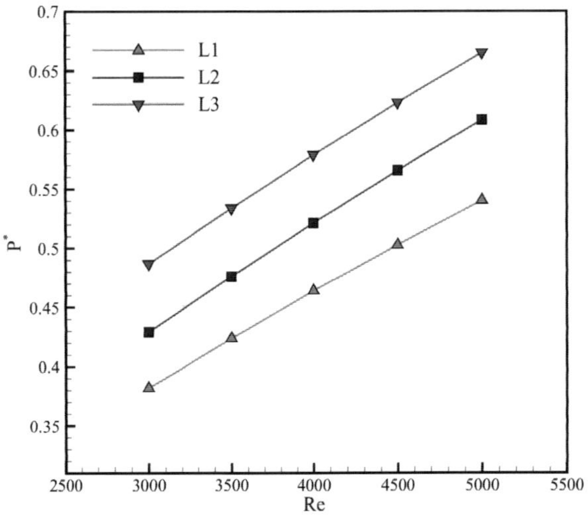

Figure 8. P* versus Reynolds number for different geometries.

3.2 Effect of hole layout on heat performance

Another influential parameter in the performance of pillow plate heat exchangers is the hole layout. This section delves into an examination of two distinct hole layouts. As Figure 9 illustrates, non-linear hole layouts tend to exhibit larger hot zones. However, as evident from Figure 10, these layouts correspond to higher flow turbulence, ultimately enhancing heat transfer performance. This phenomenon is further corroborated by the data presented in Figure 11.

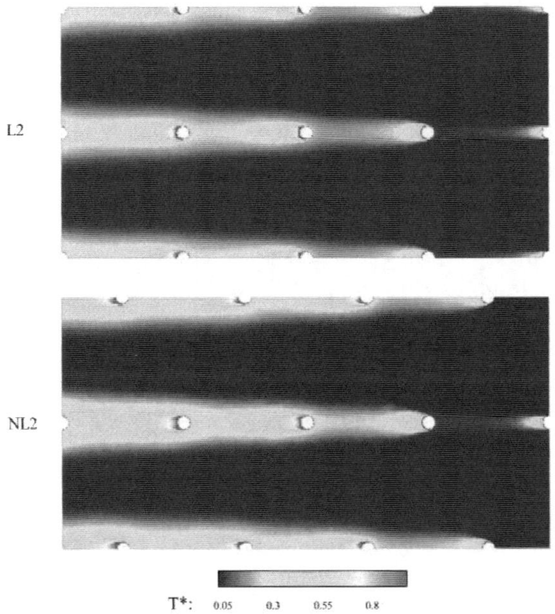

Figure 9. T* contours for pure water at Re= 3000 for different hole layout.

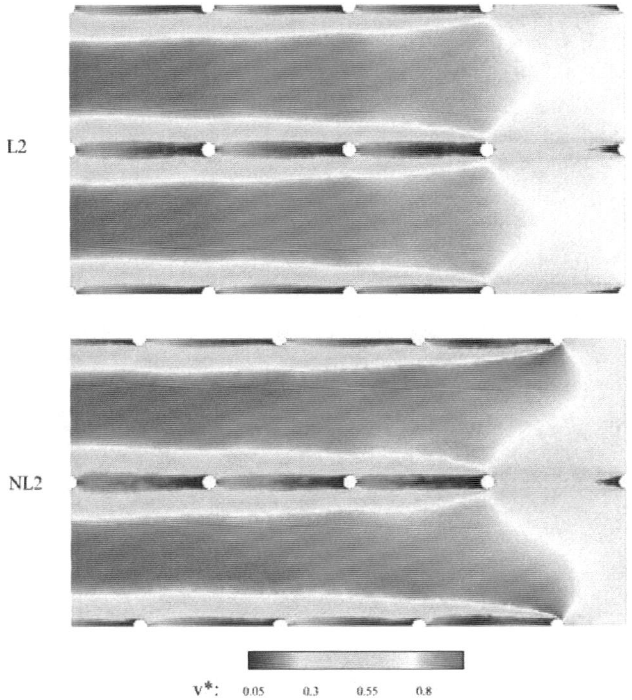

Figure 10. V* contours for pure water at Re= 3000 for different hole layout.

Figure 11 provides a visual representation of the Nusselt number versus Reynolds number for different hole layouts. As depicted in the figure, it becomes apparent that the choice of hole layout exerts a substantial influence on enhancing the Nusselt number. Notably, among the geometries with identical hole sizes, the non-linear hole layout geometry demonstrated a noteworthy increase in Nusselt number, reaching up to 25 percent, underscoring the significant impact of hole layout selection on heat transfer performance.

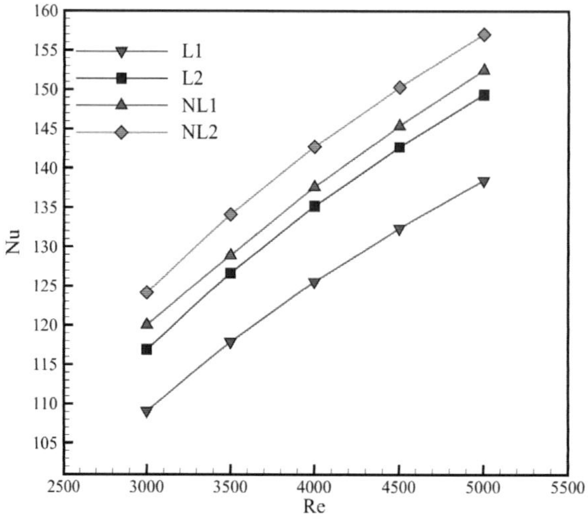

Figure 11. Nusselt number versus Reynolds number for different geometries.

This effect is attributed to the introduction of holes within the pillow-plate channels, which results in a pronounced deflection of the flow patterns and the emergence of secondary-flow phenomena, as evidenced in Figure 12. Near to these holes, the fluid pathway narrows significantly, introducing substantial flow resistance. Consequently, the fluid trajectory deviates radially away from the holes, giving rise to extensive recirculation regions in their wake.

In the case of the NL2 geometry, as depicted in Figure 12, the presence of holes causes a pronounced redirection of the primary flow, and the recirculation zones assume an approximately sinusoidal path. This intricate flow behavior contributes to an appreciable increase in pressure drop, a trend perceptible in Figure 13.

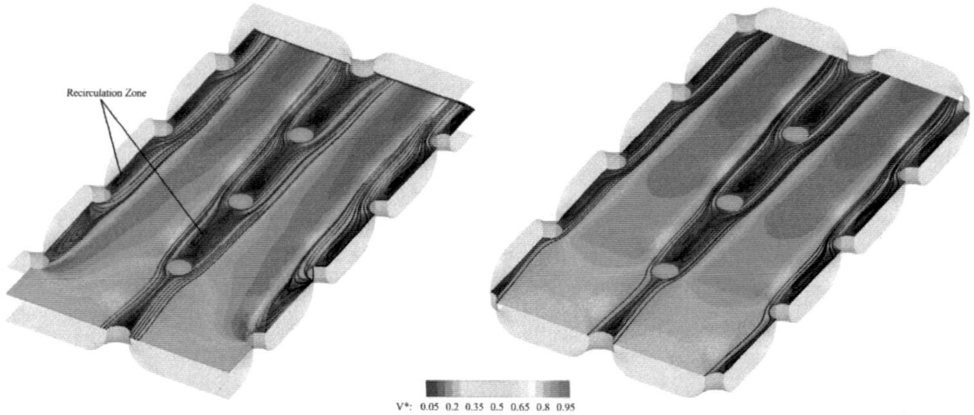

Figure 12. V* contours and streamlines for pure water at Re= 3000 for L2 and NL2.

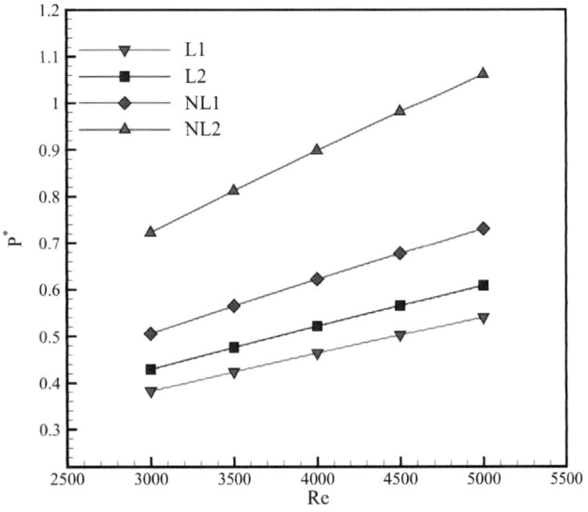

Figure 13. Nusselt number versus Reynolds number for different geometries.

Passive methods, while effective, are often more costly as they necessitate alterations to the specifications of the heat exchanger. There are scenarios where modifying the heat exchanger specifications is either impractical or unfeasible, such as when a heat exchanger has been procured. In such instances, active methods become a viable approach to enhance the heat exchanger's efficiency.

One of the active, and cost-effective strategies for optimizing the performance of a heat exchanger involves the adjustment of its operating conditions. Achieving an optimal flow rate, ensuring suitable ambient temperature, altering the inlet temperature, and selecting the appropriate working fluid can all significantly enhance the heat exchanger's performance. Although, in many cases, the operating conditions and inlet temperature of the exchanger cannot be controlled, changing the working fluid of the exchanger remains a potent avenue for effecting substantial improvements in the heat exchanger's overall efficiency.

3.3 The effect of nanofluid use on exchanger performance

The base fluids employed in heat exchangers typically possess low thermal conductivity. To enhance the heat transfer coefficient of these working fluids, numerous researchers have explored the introduction of nanoparticles into the base fluid. In this study, the influence of using alumina particles on the water-based fluid utilized in the heat exchanger has been rigorously investigated. Figure 14, Figure 15, and Figure 16 present the Nusselt numbers as a function of Reynolds number, elucidating the effects of varying nanofluid concentrations.

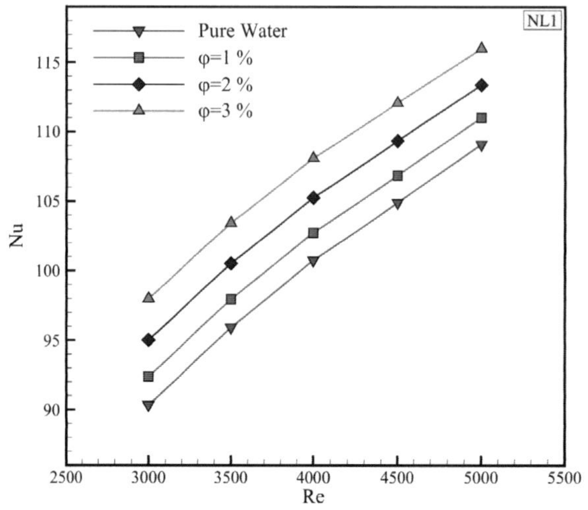

Figure 14. Nusselt number versus Reynolds number for different volume fractions in the NL1 geometry.

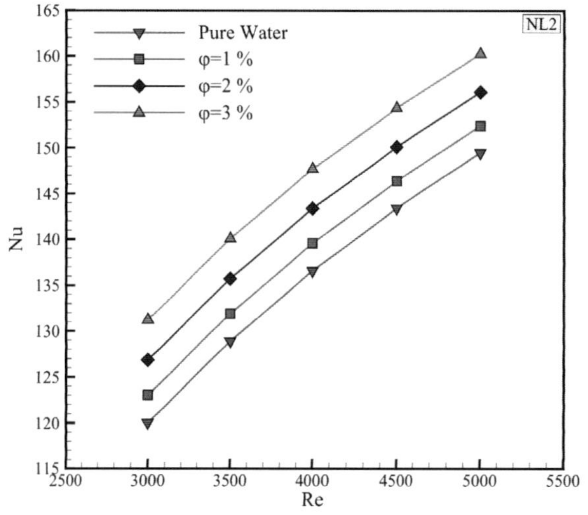

Figure 15. Nusselt number versus Reynolds number for different volume fractions in the NL2 geometry.

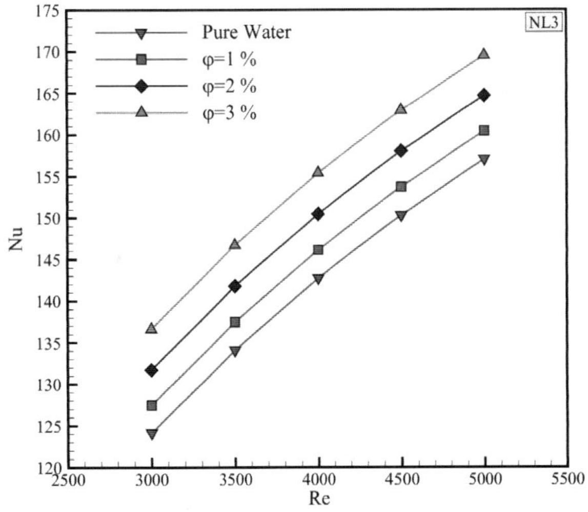

Figure 16. Nusselt number versus Reynolds number for different volume fractions in the NL3 geometry.

As indicated by Figure 14 to Figure 16, the performance of the heat exchanger exhibits enhancement as the nanofluid concentration increases. The incorporation of nanoparticles into the fluid leads to an elevation in the fluid's thermal conductivity, subsequently bolstering heat transfer, as reflected in the observed Nusselt numbers. It is noteworthy that even a slight 1% increase in nanofluid concentration yields a remarkable up to 10% increase in the Nusselt number.

However, it is imperative to acknowledge that while heat transfer is positively affected by higher nanofluid concentration, the introduction of nanoparticles into the fluid concurrently increases its viscosity. This heightened viscosity directly contributes to an increase in pressure drop, a negative factor within heat exchangers. The escalation in pressure drop necessitates greater pumping force, leading to associated cost increments. Hence, a trade-off emerges between improved heat transfer and the associated elevated operational costs incurred due to heightened viscosity and pressure drop.

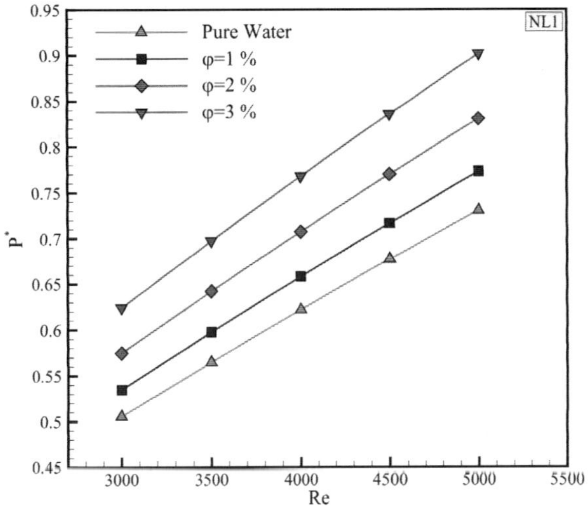

Figure 17. P^{*} versus Reynolds number for different volume fractions in the NL1 geometry.

Figure 17 provides a visualization of the dimensionless pressure drop versus Reynolds number at various nanofluid concentrations for the NL1 geometry. As evident from the figure, an increase in nanofluid concentration, and the resultant rise in fluid viscosity, leads to a notable increase in pressure drop. Furthermore, it is noteworthy that pressure drop also rises in direct correlation with increasing Reynolds number.

3.4. Data-driven optimization

In this study, the initial focus was on utilizing CFD simulation to create a comprehensive dataset outlining the performance of the heat exchanger. The CFD model was instrumental in simulating various design configurations and provided results for critical parameters. These simulations served as the foundation upon which the subsequent optimization efforts were built.

To enhance the optimization process and alleviate computational demands, an artificial intelligence (AI) model is used. The primary objective of this AI model was to predict two crucial parameters: HTC and pressure drop, based on the input design variables. The chosen AI model for this predictive task was the Random Forest Regressor (RFR) [31].

The RFR was selected for its proficiency in handling regression tasks and its ability to capture intricate, nonlinear relationships within the dataset. It was trained on the extensive dataset generated through CFD simulations, thus enabling it to provide precise predictions of HTC and pressure drop for various heat exchanger designs.

The development of the AI model for predicting HTC and pressure drop began with a meticulous selection and definition of the input variables. These variables were chosen to represent essential attributes of the heat exchanger design, including geometric features and operating conditions. Input variables encompassed hole geometry, nanofluid volume concentration, and Reynolds number, all carefully chosen due to their profound influence on heat exchanger performance. The model's input and output variables, along with the associated error, are detailed in Table 4.

Table 4. Model variables and associated errors.

Model	Random forest regressor	
Input variables	Hole layout, r1, r2, φ, Re	
Output variables	T_{outlet}, P^*, HTC, v_{outlet}	
	Model parameters	Value
	n_estimators	100
	Random_state	42
	Errors	value
	Mean absolute error	0.92
	Mean squared error	4.67
	Root mean squared error	2.16
	R-squared	0.93

RFR serves as an ensemble learning technique that amalgamates multiple decision trees to provide accurate predictions. Each decision tree in the ensemble is constructed through iterative data splits, ultimately leading to a leaf node where predictions are made. By aggregating the results from these individual trees, the RFR model leverages the independence of errors among them, culminating in robust and accurate predictions. Pertinent parameters in the model, such as 'n_estimators' (representing the number of decision trees) and 'random_state' (ensuring consistency), were meticulously configured to optimize predictive performance. Figure 18 shows the algorithm in this model.

By harnessing an AI model, the research conducted an optimization procedure to refine the geometric design parameters. This involved employing a systematic grid search method, fine-tuning the heat exchanger's configuration. The parameter space explored during this process included key variables such as r_1, r_2, nanofluid volume concentration, and Reynolds number. The aim was to identify the most favorable combination of these design parameters, maximizing HTC while minimizing pressure drop. To ensure the viability of the design, a well-balanced approach was maintained by assigning weights to HTC and P^*. Subsequently, the model's predictions were tested for their validity by conducting CFD simulations, a critical step to cross-verify the optimization results with actual CFD data. The entire optimization and validation procedure is depicted in Figure 19. This comprehensive approach not only significantly reduced computational resources and time requirements but also effectively realized the desired design objectives.

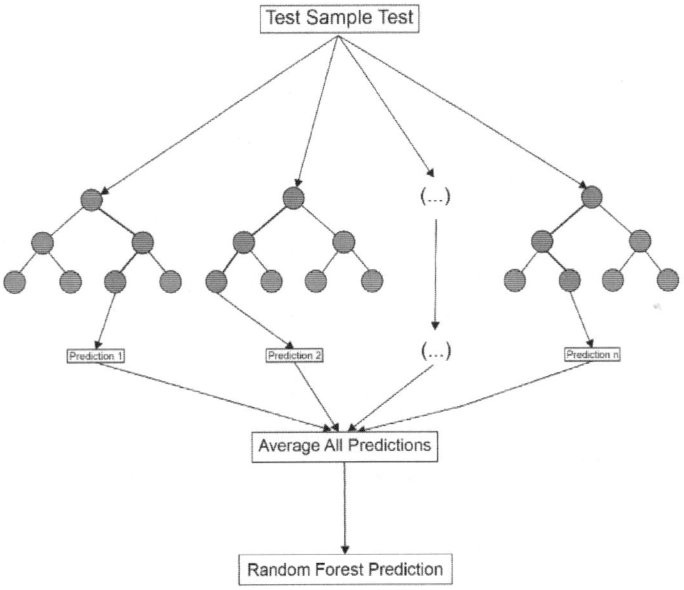

Figure 18. Random forest regressor algorithm.

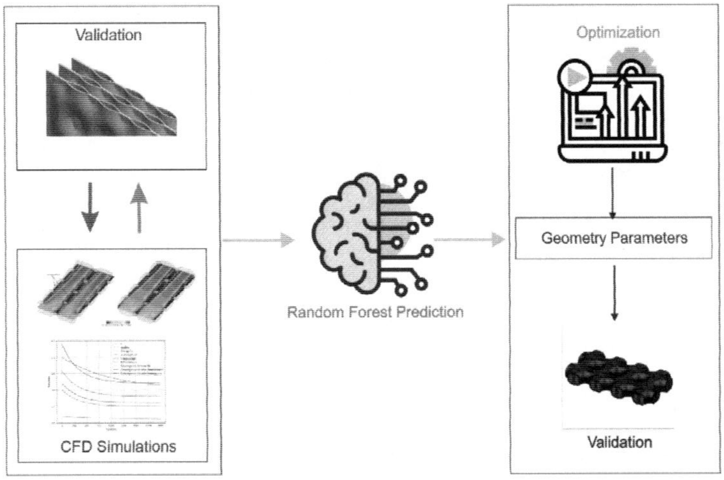

Figure 19. Procedure for AI-Driven optimization and CFD validation of heat exchanger design parameters.

The parameter correlation heatmap, as demonstrated in Figure 20, provides insights into the relationships between various design parameters in the heat exchanger optimization process. The heatmap showcases the strength and direction of correlations between parameters, aiding in the identification of influential factors. Notably, the heatmap reveals that HTC exhibits a positive correlation of 0.70 with P^* and 0.67 with Re. Furthermore, a correlation of 0.57 is observed between HTC and nanofluid volume concentration. In contrast, HTC demonstrates a negative correlation with the r_1 at -0.22 and r_2 at -0.25. For P*, it displays a robust positive correlation of 0.70 with HTC, 0.71

with Re, and 0.54 with φ, while maintaining relatively weaker positive correlations with r_1 and r_2, with values of 0.21 and 0.22, respectively.

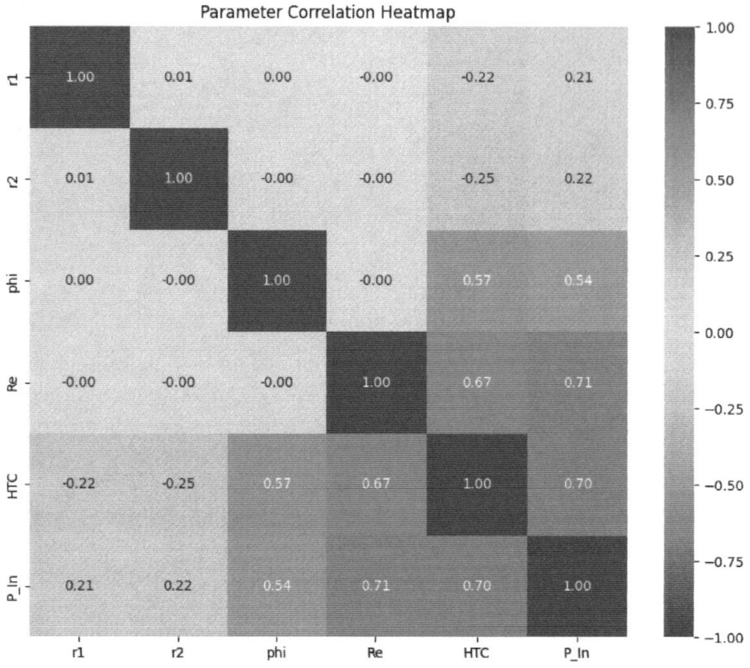

Figure 20. Parameter correlation heatmap.

Figure 21 depicts the results of the comprehensive optimization process, offers a clear visualization of the trade-offs in heat exchanger performance. This scatter plot highlights the relationship between HTC and P*, two key performance metrics. What's notably striking in the plot is the distinct presence of a Pareto line, which elegantly separates the optimized points from the non-optimal ones. The Pareto line signifies the boundary where further improvements in HTC can only be achieved at the expense of increased pressure drop, and vice versa. This graph serves as a valuable tool for design decision-making, enabling pinpointing the optimal design configurations by balancing the essential trade-offs between HTC and P^* for efficient and effective heat exchanger design.

After finishing the optimization process, a verification step was undertaken to assess the accuracy and reliability of the optimized Pareto Points. This was achieved through a CFD simulation conducted on Pareto Points to compare the predicted and actual results. The outcome was a thorough examination of the predictive capabilities of the AI model. The precision of these predictions was reaffirmed as the error, thoughtfully detailed in Table 4, indicated a good agreement between CFD and AI model: the error remained consistently less than 5%. This congruence between the predicted and actual results further strengthens the trustworthiness and applicability of the AI model in optimizing heat exchanger design, thereby offering a compelling avenue for future advancements in heat exchanger engineering.

Although the pressure drops increase with increasing nanofluid concentration, the choice of the optimal mode depends on the application of the exchanger. In cases such as reactors and sensitive systems where there is an urgent need to control the system temperature, thermal efficiency is more critical, and pumping cost is less important. The heat exchanger is very important in these applications.

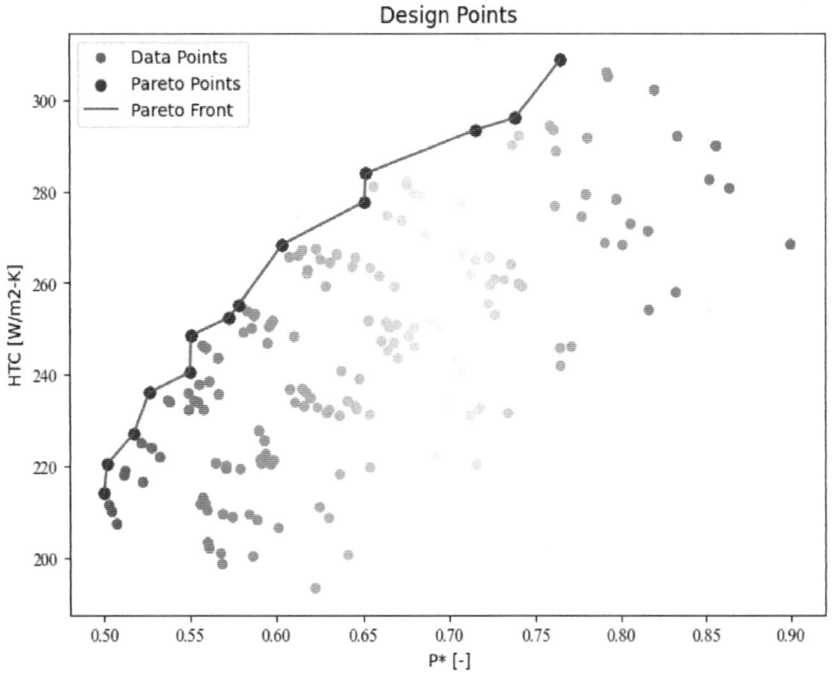

Figure 21. Scatter plot showing optimization results and pareto front.

Table 4. Optimization results vs CFD results.

Results from optimization						CFD results	
r_1 (mm)	r_2 (mm)	φ	Re	HTC [W/m^2.K]	P*	err. HTC [%]	err. P* [%]
25	125	0.02	3500	255.1079	0.5785	2.62	-1.98
25	125	0.01	4000	252.4526	0.572619	-0.17	5.15
25	100	0	3500	227.0102	0.517353	0.03	-4.15
25	125	0.01	4500	265.6777	0.607828	-0.56	-2.59
25	100	0.02	4000	266.0003	0.613052	0.28	0.64
25	125	0.01	3000	220.4693	0.501797	-3.73	-4.36
25	100	0	5000	265.6102	0.645832	3.89	-2.05
25	125	0.02	3000	235.938	0.549503	3.53	-0.09
25	100	0.02	3500	255.1079	0.5785	-1.40	0.66
25	100	0.04	5000	308.9095	0.764645	-0.01	3.23
25	125	0	3500	227.0102	0.517353	3.31	2.33
25	100	0	4000	240.5783	0.550246	-2.52	5.06
25	100	0.01	4000	252.4526	0.572619	3.15	4.17
25	125	0	4500	253.8293	0.583698	1.47	-3.07

4. Conclusion

The present study represents a comprehensive two-phase numerical study on the performance of a pillow-plate heat exchanger, with specific focus on heat exchangers featuring both linear and non-linear hole configurations and utilizing fluids with varying nanofluid concentrations. This research has sought to elucidate the impact of nanofluids, hole size, and hole arrangement on heat exchanger performance. The salient findings of this study are distilled as follows:

- The utilization of heat exchangers featuring larger holes leads to a notable improvement in Nusselt numbers and consequently enhances heat transfer efficiency. For instance, the transition from model L1 to L3 configurations yielded a substantial 25% increase in Nusselt numbers.
- It is imperative to recognize that augmenting the size of the holes also results in elevated pressure drop. The L3 geometry, when contrasted with the L1 geometry, exhibited a pressure drop increase of approximately 27%.
- Heat exchangers characterized by linear hole layout demonstrate lower pressure drops in comparison to their non-linear counterparts. This difference is about 20%
- Heat exchangers featuring non-linear hole configurations exhibit higher Nusselt numbers, denoting a superior heat transfer performance in comparison to heat exchangers with linear hole arrangements.
- The use of nanofluids as working fluids enhances heat transfer. A mere 1% increase in nanofluid concentration led to a significant up to 10% improvement in Nusselt numbers.
- However, the incorporation of nanoparticles also contributes to augmented pressure drops by elevating the fluid's viscosity. A 1% increase in nanofluid concentration leads to an approximate 7% increase in pressure drop.

These findings collectively underscore the intricate interplay of parameters affecting heat exchanger performance, thereby shedding light on the nuances of optimizing heat transfer efficiency while navigating the associated trade-offs, an essential endeavor in heat exchanger engineering.

References

[1] Cuan Z and Chen Y 2023 Evaluating two-phase fluid flow and heat transfer in pillow-plate heat exchangers with nanofluids for organic Rankine cycle in municipal solid waste power plant: A numerical simulation study *Eng Anal Bound Elem* **155** 814–25

[2] Trafczynski M, Markowski M and Urbaniec K 2023 Energy saving and pollution reduction through optimal scheduling of cleaning actions in a heat exchanger network *Renewable Sustainable Energy Rev.* **173** 113072

[3] Cuan Z and Chen Y 2023 Evaluating two-phase fluid flow and heat transfer in pillow-plate heat exchangers with nanofluids for organic Rankine cycle in municipal solid waste power plant: A numerical simulation study *Eng Anal Bound Elem* **155** 814–25

[4] Saedodin S, Zaboli M, Rostamian S H and Kharabati S 2023 Statistical analysis and shape optimization of a finned corrugated heat exchanger using RSM *Chem Eng Commun* **210** 716–39

[5] Arsenyeva O, Piper M, Zibart A, Olenberg A and Kenig E Y 2019 Investigation of heat transfer and hydraulic resistance in small-scale pillow-plate heat exchangers *Energy* **181** 1213–24

[6] Suresh S, Venkitaraj K P, Selvakumar P and Chandrasekar M 2012 A comparison of thermal characteristics of Al2O3/water and CuO/water nanofluids in transition flow through a straight circular duct fitted with helical screw tape inserts *Exp Therm Fluid Sci* **39** 37–44

[7] Suresh S, Venkitaraj K P and Selvakumar P 2011 Comparative study on thermal performance of helical screw tape inserts in laminar flow using Al2O3/water and CuO/water nanofluids *Superlattices Microstruct* **49** 608–22

[8] Amanowicz Ł 2018 Influence of geometrical parameters on the flow characteristics of multi-pipe earth-to-air heat exchangers – experimental and CFD investigations *Appl Energy* **226** 849–61

[9] Karimi A and Afrand M 2018 Numerical study on thermal performance of an air-cooled heat exchanger: Effects of hybrid nanofluid, pipe arrangement and cross section *Energy Convers Manag* **164** 615–28

[10] Jadhav P H, Gnanasekaran N and Mobedi M 2023 Analysis of functionally graded metal foams for the accomplishment of heat transfer enhancement under partially filled condition in a heat exchanger *Energy* **263** 125691

[11] Zeinali S and Neshat E 2023 Energy, exergy, economy analysis and geometry optimization of spiral coil heat exchangers *Case Studies in Thermal Engineering* **42** 102708

[12] Karimi A, Al-Rashed A A A A, Afrand M, Mahian O, Wongwises S and Shahsavar A 2019 The

IOP Conf. Series: Materials Science and Engineering 1294 (2023) 012062 doi:10.1088/1757-899X/1294/1/012062

effects of tape insert material on the flow and heat transfer in a nanofluid-based double tube heat exchanger: Two-phase mixture model *Int J Mech Sci* **156** 397–409
[13] Arsenyeva O, Tran J, Piper M and Kenig E 2019 An approach for pillow plate heat exchangers design for single-phase applications *Appl Therm Eng* **147** 579–91
[14] Yao Y, Ding J, Zhang Y, Wang W and Lu J 2022 Heat transfer performance of pillow plate heat exchanger with molten salt and supercritical carbon dioxide *Int J Heat Mass Transf* **183** 122211
[15] Mastani Joybari M, Selvnes H, Sevault A and Hafner A 2022 Potentials and challenges for pillow-plate heat exchangers: State-of-the-art review *Appl Therm Eng* **214** 118739
[16] Piper M, Olenberg A, Tran J M and Kenig E Y 2015 Determination of the geometric design parameters of pillow-plate heat exchangers *Appl Therm Eng* **91** 1168–75
[17] Tran J M, Sommerfeld S, Piper M and Kenig E Y 2015 Investigation of pillow-plate condensers for the application in distillation columns *Chemical Engineering Research and Design* **99** 67–74
[18] Kumar S, Premachandran B and Subbarao P M V 2019 Study on thermo–hydraulics in a Pillow Plate Channel *International Journal of Thermal Sciences* **145** 106020
[19] Nuim Labib M, Nine M J, Afrianto H, Chung H and Jeong H 2013 Numerical investigation on effect of base fluids and hybrid nanofluid in forced convective heat transfer *International Journal of Thermal Sciences* **71** 163–71
[20] Mojarrad M S, Keshavarz A and Shokouhi A 2013 Nanofluids thermal behavior analysis using a new dispersion model along with single-phase *Heat and Mass Transfer/Waerme- und Stoffuebertragung* **49** 1333–43
[21] Naumann Z and Schiller L 1935 A drag coefficient correlation *Z. Ver. Deutsch. Ing* **77** e323
[22] Moraveji M K and Ardehali R M 2013 CFD modeling (comparing single and two-phase approaches) on thermal performance of Al2o3/water nanofluid in mini-channel heat sink *International Communications in Heat and Mass Transfer* **44** 157–64
[23] Siginer D A and Wang H P 1995 Developments and applications of non-Newtonian flows 1995. Proceedings of a conference, San Francisco, November 1995 *[No source information available]*
[24] Kakaç S and Pramuanjaroenkij A 2009 Review of convective heat transfer enhancement with nanofluids *Int J Heat Mass Transf* **52** 3187–96
[25] El Bécaye Maïga S, Palm S J, Nguyen C T, Roy G and Galanis N 2005 Heat transfer enhancement by using nanofluids in forced convection flows *Int J Heat Fluid Flow* **26** 530–46
[26] Menter F R 2012 Two-equation eddy-viscosity turbulence models for engineering applications *AIAA J.* **32** 1598–605
[27] Asadi A, Asadi M, Rezaniakolaei A, Rosendahl L A, Afrand M and Wongwises S 2018 Heat transfer efficiency of Al2O3-MWCNT/thermal oil hybrid nanofluid as a cooling fluid in thermal and energy management applications: An experimental and theoretical investigation *Int J Heat Mass Transf* **117** 474–86
[28] Wang Z, Wang R, Li Z, Wang M and Wan L 2023 Numerical investigation on the effect of cylindrical turbulator on performance of corrugated plate-fin heat exchanger *Appl Therm Eng* **230** 120726
[29] GridPro v7.0SP1, available at: https://www.gridpro.com/products/ (accessed: 20.10.2023)
[30] Tran J M, H X and Kenig E Y 2015 Experimentelle Untersuchung des konvektiven Wärmeübergangs und Druckverlustes in einphasig durchströmten Thermoblechen. (Experimental investigation of convective heat transfer and pressure drop in pillow plates under single-phase through-flow conditions) *Chemie Ingenieur Technik* **87** 226–34
[31] Pedregosa F, Varoquaux G, Gramfort A, Michel V, Thirion B, Grisel O, Blondel M, Prettenhofer P, Weiss R, Dubourg V, Vanderplas J, Passos A, Cournapeau D, Brucher M, Perrot M and Duchesnay É 2011 Scikit-learn: Machine Learning in Python *Journal of Machine Learning Research* **12** 2825–30

AUTHOR INDEX